Advances in Plant Breeding Strategies

Volume 12

Series Editors

Jameel M. Al-Khayri, Biotechnology, King Faisal University, Al-Hassa, Saudi Arabia

Shri Mohan Jain, Department Agricultural Sciences, University of Helsinki, Helsinki, Finland

This book series is intended to provide a platform for plant scientists to present the latest advances in both innovative and traditional methodologies applied in plant breeding for the development of new crop varieties under different environmental conditions leading to sustainable food production, food and nutrition security, food supply chain, germplasm conservation and utilization. This series is an excellent reference source for plant breeders, geneticists, biotechnologists, and molecular biologists using molecular tools, bioinformatics, artificial intelligence, and other emerging technologies. The subject matter of all volumes will relate to postgraduate students specializing in agriculture, biotechnology, and molecular breeding as well as policymakers, seed companies, researchers, and professionals interested in plant genetic improvement.

Khaled F. M. Salem • Jameel M. Al-Khayri
Shri Mohan Jain
Editors

Breeding and Biotechnology of Grass and Bast Fiber Crops

Springer

Editors
Khaled F. M. Salem
Department of Biological Sciences
Shaqra University
Ad-Dawadimi, Saudi Arabia

Department of Plant Biotechnology,
Genetic Engineering and Biotechnology
Research Institute (GEBRI)
University of Sadat City
Sadat City, Egypt

Jameel M. Al-Khayri
Department of Agricultural Biotechnology
King Faisal University
Al-Ahsa, Saudi Arabia

Shri Mohan Jain
Department Agricultural Sciences
University of Helsinki
Helsinki, Finland

ISSN 3004-8737 ISSN 3004-8745 (electronic)
Advances in Plant Breeding Strategies
ISBN 978-3-032-00402-4 ISBN 978-3-032-00403-1 (eBook)
https://doi.org/10.1007/978-3-032-00403-1

This Springer imprint is published by the registered company Springer Nature Switzerland AG
The registered company address is: Gewerbestrasse 11, 6330 Cham, Switzerland

Preface of the Series

Contemporary plant breeders rely on conventional and emerging methodologies to provide sustainable food production worldwide. However, the human population is growing at an alarming rate, especially in developing countries, and food availability to feed new mouths could gradually become a serious problem. Agriculture production is negatively impacted by environmental pollution, rapid industrialization, water scarcity, soil erosion of fertile topsoil, limited areas of potentially arable land, lack of improvement of local plant types, erosion of genetic diversity, and dependence on just a few major crop species for the world's food supplies. Breeding programs of these crops have been very much dependent on the ready availability of genetic variation, whether spontaneous or induced. Plant breeders and geneticists are under constant pressure to sustain food production by using innovative breeding strategies and enhancing minor crops that are well adapted to the marginal land and source of nutrition as well as abiotic and biotic stresses. An inherent limitation of traditional breeding is that the introduction of one or few genes into a cultivar requires several generations of backcrossing, making it a time-consuming process.

With the development of new molecular tools, molecular marker-assisted breeding helped rapid introgression of transgene and reduction of linkage drag. The continued development and adaptation of plant biotechnology approaches including genetic engineering, genome editing, molecular markers, genomics, proteomics, bioinformatics, artificial intelligence, and nanotechnology has established new tools for the creation, analysis, and manipulation of genetic variation for the development of improved cultivars to achieve agricultural sustainability and food security. A multi-disciplinary approach combining traditional plant breeding, mutation breeding, plant biotechnology, and molecular biology would be strategically ideal for developing improved crop varieties to ensure global food security, provide natural pharmaceutical sources, and supply fiber and ornamental crops for industry, esthetic and environmental purposes. However, breeders continue to face challenged such as environmental stress, the limitation of arable land expansion, and erosion of genetic resources particularly under the impact of global climate change.

This book series is intended to provide a platform for plant scientists to present the latest advances in both innovative and traditional methodologies applied in plant breeding for the development of new crop varieties under different environmental conditions leading to sustainable food production, food and nutrition security, food supply chain, germplasm conservation, and plant sustainable utilization. This series is an excellent reference source for plant breeders, geneticists, biotechnologists, and molecular biologists using molecular tools, bioinformatics, artificial intelligence, and other emerging technologies. The subject matter relates to postgraduate students specializing in agriculture, biotechnology, and molecular breeding as well as policymakers, seed companies, researchers, and professionals interested in plant genetic improvement.

Al-Ahsa, Saudi Arabia Jameel M. Al-Khayri
Helsinki, Finland Shri Mohan Jain

Preface

Fiber crops serve various purposes in agriculture, industry, medical supplies, and the clothing and textile industries. Additionally, certain fiber crops' plant parts are essential for producing oil for human and animal consumption. In order to meet consumer demands for new, desired natural plant traits such as morphological traits, physiological traits, maturity traits, yield traits, and fiber quality traits like color, and improvements are necessary. It is projected that from 2025 to 2034, the global market for plant natural fibers will expand at a compound annual growth rate (CAGR) of over 7.2%, from an estimated USD 5.5 billion in 2024. One of the agricultural sectors experiencing rapid growth is the apparel, textile, and plant waste industries that are derived from them. Due to innovation driving this business, there is a growing need for plant breeders in both the public and private sectors to meet customer demands. Breeding mutations and creating novel fiber-specific features that benefit the apparel and textile industries are ideal tasks for biotechnological techniques such as genetic transformation, genomics, nanotechnology, gene editing, and induced mutagenesis. In addition, digital farming, artificial intelligence, and vertical farming are becoming increasingly important in this field.

The main idea of this book is to explore how traditional and novel plant breeding techniques can be applied to create new fiber crop varieties that are suitable for different environmental conditions in a changing global climate. The goal is to achieve sustainable production for use in the apparel and textile industries. In addition to traditional breeding methods, this book focuses on the latest advancements in molecular tools and biotechnology for a variety of grass and bast fiber crops, as well as their application in plant breeding.

The book contains 20 chapters supported by 153 color figures and 61 tables. The chapters are grouped into two parts. Part I titled Grass Fiber Crops covers the genera *Dendrocalamus*, *Hordeum*, *Zea*, *Avena*, *Sorghum*, *Saccharum*, and *Triticum*. Part II titled Bast Fiber Crops covers the genera *Musa*, *Linum*, *Cannabis*, *Apocynum*, *Corchorus*, *Hibiscus*, *Opuntia*, *Pongamia*, *Crotalaria*, *Cissus*, *Areca*, and *Acacia*. Each chapter discusses breeding techniques for specific species or a group of related species. Topics covered include current cultivation methods and challenges, biodiversity, germplasm conservation, breeding methods including conventional, molecular, tissue culture, genetic engineering, gene editing, hybridization, and mutation breeding. Each chapter concludes with future research directions and a comprehensive list of references for further reading. Seventy-nine globally recognized scientists from 11 different nations contributed to the chapters based on their expertise in the subject. The chapters underwent a thorough review process to ensure scientific accuracy and presentation quality.

The book is a valuable resource for professional plant breeders, geneticists, researchers, entrepreneurs, and advanced undergraduate and graduate students interested in biotechnology, molecular breeding, and agriculture.

The editors would like to thank the staff of Springer for the opportunity to publish this book and the contributing authors for their unwavering dedication and hard work.

Ad-Dawadimi, Saudi Arabia Khaled F. M. Salem
Al-Ahsa, Saudi Arabia Jameel M. Al-Khayri
Helsinki, Finland Shri Mohan Jain

Acknowledgment

The editors extend their appreciation to the support of the Deanship of Scientific Research, Vice Presidency for Graduate Studies and Scientific Research, King Faisal University, Saudi Arabia (Project No.KFU253809).

The original version of the frontmatter has been revised: An Acknowledgement has been added to the book's front matter

Contents

Editors and Contributors

About the Editors

Khaled F. M. Salem is affiliated with Biological Sciences Department, Faculty of Science and Humanities, Shaqra University, Ad-Dawadimi, Saudi Arabia, and Plant Biotechnology Department, Genetic Engineering and Biotechnology Research Institute, University of Sadat City, Egypt. He earned his BS in Crop Science with excellent honor degree in 1993 and MS with focused on rice breeding from Menoufiya University, Egypt, in 1998. He conducted his MS at Rice Research and Training Center, Field Crops Research Institute, Agriculture Research Center, Giza, Ministry of Agriculture, Egypt. In 2004, he completed his PhD in Plant Breeding and Genetics, Martin Luther University Halle-Wittenberg, Germany. He conducted his PhD project at IPK, Gatersleben, Germany. From 2001 to 2004 he worked as visiting scientist at Gene Bank Department and Gene and Genome Mapping Group, IPK, Germany. Additionally, he attended several training courses as visiting scientist in China, Italy, Spain, Turkey, and CIMMYT. He serves as reviewer for 50 journals and editor for several journals. He has published 78 articles, 30 book chapters, and 25 conference proceedings. In addition, he has edited plant gene bank reference book. His research has focused on cereal, fiber, and vegetable crops biotechnology for the past three decades. He has also mentored and supervised master's and PhD students.

Jameel M. Al-Khayri is affiliated with the Department of Agricultural Biotechnology, King Faisal University, Saudi Arabia. He received BS in Biology in 1984 from the University of Toledo, MS in Agronomy in 1988, and PhD in Plant Science in 1991 from the University of Arkansas. He is the National Correspondent of the International Association of Plant Tissue Culture and Biotechnology. He has authored 150 research articles and 86 chapters and has edited several journal special issues. In addition, he has edited 42 reference books on plant biotechnology, genetic resources, breeding, genomics, and nanotechnology. In addition to teaching, advising, and research, he held administrative responsibilities as the Assistant Director of Date Palm Research Center, Head of Department of Plant Biotechnology, and Vice Dean for Development and Quality Assurance. He served as a Member of Majlis Ash-Shura (Saudi Legislative Council) for the 2009–2012 term. He is interested in agricultural innovations and sustainability.

Shri Mohan Jain, (Adjunct) is currently associated with the Department of Agricultural Sciences, University of Helsinki, Helsinki, Finland. He received MPhil in 1973 and PhD in 1978 from Jawaharlal Nehru University, New Delhi, India. He served as a Technical Officer, Plant Breeding, and Genetics, International Atomic Energy Agency (IAEA), Vienna, Austria, from 1999 to 2005 when the agency was awarded Nobel Peace Prize, 2005. He is on the Editorial Board of *Euphytica*, *In Vitro*, and *Propagation of Ornamental Plants* and a reviewer in Plant Cell Tissue and Organ Culture, *Plant Cell Reports*, and *Plants*. He has published 225 peer-reviewed journals articles, book chapters, and conference proceedings. Additionally, he has edited 82 books; has been an invited speaker; and served as Chairperson in several international conferences worldwide. He is a former consultant to the IAEA, the European Union, the Government of Grenada, the Iranian Private Company, and the Egyptian government.

Contributors

Donia Abdallah Department of Biology, Faculty of Sciences of Tunis, University of Tunis El Manar, Tunis, Tunisia

El-Sayed S. Abdel Razik Livestock Research Department, City of Scientific Research and Technological Applications (SRTA-City), Arid Lands Cultivation Research Institute (ALCRI), New Borg El-Arab, Alexandria, Egypt

Sawsan Abd-Ellatif Bioprocess Development Department, Genetic Engineering and Biotechnology Research Institute (GEBRI), City of Scientific Research and Technological Applications (SRTA-City), New Borg El-Arab, Alexandria, Egypt

Ahmed M. Abdelmoghny Cotton Research Institute (CRI), Agricultural Research Center (ARC), Giza, Egypt

Farrag F. B. Abu-Ellail Department of Breeding and Genetics, Sugar Crops Research Institute, Agricultural Research Center (ARC), Giza, Egypt

Osama Alam Department of Biotechnology, University of Science & Technology Bannu, Khyber Pakhtunkhwa, Pakistan

Shahab Alam Department of Chemistry, University of Science & Technology Bannu, Khyber Pakhtunkhwa, Pakistan

Ahmad Ali College of Agronomy and Biotechnology, Southwest University, Chongqing, People's Republic of China

Salem Mesfir Al-Qahtani Biology Department, University College of Tayma, University of Tabuk, Tabuk, Saudi Arabia

P. Anantharaju Tamil Nadu Agricultural University, Coimbatore, Tamil Nadu, India

C. Appunu Division of Crop Improvement, ICAR-Sugarcane Breeding Institute, Coimbatore, Tamil Nadu, India

Jason Argyris Centre for Research in Agricultural Genomics (CRAG)–Institute of Agrifood Research and Technology (IRTA), Barcelona, Spain

Nyo Nyo Aung Sugarcane Research Center, Sugar Crops Division, Department of Agriculture, Pyinmana, NayPyiTaw, Myanmar

Aliya Baitelenova S.Seifullin Kazakh Agrotechnical University, Astana, Republic of Kazakhstan

Asmita Banerjee Department of Life Sciences, Presidency University, Kolkata, India

Ghada Baraket Department of Biology, Faculty of Sciences of Tunis, University of Tunis El Manar, Tunis, Tunisia

Sana Ben Mustapha Department of Biology, Faculty of Sciences of Tunis, University of Tunis El Manar, Tunis, Tunisia

Semantee Bhattacharyya Department of Life Sciences, Presidency University, Kolkata, India

Sukanya Chakraborty Department of Life Sciences, Presidency University, Kolkata, India

Department of Botany, Scottish Church College, Kolkata, India

Rashmi Chauhan G. B. Pant University of Agriculture and Technology, Pantnagar, Uttarakhand, India

Jiquan Chen Department of Germplasm and Breeding, Institute of Bast Fiber Crops, Chinese Academy of Agricultural Sciences (IBFC, CAAS), Changsha, China

Xiao Yan Chen Dali Bai Autonomous Prefecture Agricultural Science Extension Research Institute, Dali, Yunnan, China

Yanyun Chen Life Science School of Ningxia University, Yinchuan, China

Zhigang Dai Department of Germplasm and Breeding, Institute of Bast Fiber Crops, Chinese Academy of Agricultural Sciences (IBFC, CAAS), Changsha, China

Asmita Das Department of Life Sciences, Presidency University, Kolkata, India

Malay Das Department of Life Sciences, Presidency University, Kolkata, India

Nerhan A. Eid Department of Plant Protection Desert Research Center El-Matarya, Cairo, Egypt

Karima Mohamed El-Absy Biology Department, University College of Tayma, University of Tabuk, Tabuk, Saudi Arabia

Eco-physiology Unit, Plant Ecology and Ranges Department, Desert Research Center, Cairo, Egypt

Essam Fathy El-Hashash Department of Agronomy, Faculty of Agriculture, Al-Azhar University, Cairo, Egypt

N. Meenakshi Ganesan Tamil Nadu Agricultural University, Coimbatore, Tamil Nadu, India

Amani H. A. M. Gharib Vegetable, Medicinal and Aromatic Plant Breeding Research Department, Horticulture Research Institute (HRI), Agricultural Research Center (ARC), Giza, Egypt

Sara E. Gomaa Medicinal and Aromatic Plants Research Department, Horticulture Research Institute (HRI), Agricultural Research Center (ARC), Giza, Egypt

Hameed Gul College of Agronomy and Biotechnology, Southwest University, Chongqing, People's Republic of China

Shareef Gul Integrative Science Center of Germplasm Creation in Western China (CHONGQING) Science City and Southwest University, Chongqing, People's Republic of China

Hazaisi Hanipa Institute of Agricultural Science of Ili Prefecture, Ili, Xinjiang, China

Amel Salhi Hannachi Department of Biology, Faculty of Sciences of Tunis, University of Tunis El Manar, Tunis, Tunisia

Amira A. Ibrahim Botany and Microbiology Department, Faculty of Science, Arish University, Al-Arish, Egypt

Li Jiang State Key Laboratory of Desert and Oasis Ecology, Xinjiang Institute of Ecology and Geography, Chinese Academy of Sciences, Ürümqi, China

Abeer H. Jubily Department of Field Crops, Faculty of Agriculture, Tishreen University, Lattakia, Syria

Farah Kamal Chongqing Engineering Research Center for Rapeseed, College of Agronomy and Biotechnology, Southwest University, Chongqing, China

Rajamani Kandasamy Department of Medicinal and Aromatic Crops, Tamil Nadu Agricultural University, Coimbatore, Tamil Nadu, India

Qing Hua Kang Institute of Industrial Crops, Heilongjiang Academy of Agricultural Sciences, Harbin, Heilongjiang, China

Ei Ei Khaing Sugarcane Research Center, Sugar Crops Division, Department of Agriculture, Pyinmana, NayPyiTaw, Myanmar

Shahid Ullah Khan Integrative Science Center of Germplasm Creation in Western China (CHONGQING) Science City and Southwest University, College of Agronomy and Biotechnology, Southwest University, Chongqing, People's Republic of China

Department of Biochemistry, Women Medical College, Khyber Medical University, KPK, Pakistan

Evanylla Kharlyngdoh Institute of Bioresources and Sustainable Development, Meghalaya, Upper Shillong, Meghalaya, India

Sakhi Kundu Department of Life Sciences, Presidency University, Kolkata, India

Guoqi Li School of Ecology and Environment, Ningxia University, Yinchuan, China

Sarthak Mondal Department of Life Sciences, Presidency University, Kolkata, India

Amparo Monfort Centre for Research in Agricultural Genomics (CRAG)–Institute of Agrifood Research and Technology (IRTA), Barcelona, Spain

Girija Sangari Murugavelu Department of Botany, Avinashilingam Institute for Home Science and Higher Education for Women, Coimbatore, Tamil Nadu, India

Jajati Keshari Nayak G. B. Pant University of Agriculture and Technology, Pantnagar, Uttarakhand, India

Farkhanda Naz Biological Science Research Center, Academy for Advanced Interdisciplinary Studies, Southwest University, Chongqing, China

R. Nirmalaruban Indian Agricultural Research Institute, New Delhi, Delhi, India

Arif Nirsatmanto National Research and Innovation Agency (BRIN) the Republic of Indonesia, Bogor, Indonesia

Dinesh Pandey G. B. Pant University of Agriculture and Technology, Pantnagar, Uttarakhand, India

Jordi Petit Centre for Research in Agricultural Genomics (CRAG)—Institute of Agrifood Research and Technology (IRTA), Barcelona, Spain

Sharat Prabhakaran G. B. Pant University of Agriculture and Technology, Pantnagar, Uttarakhand, India

Cai Sheng Qiu Institute of Bast Fiber Crops, Chinese Academy of Agricultural Sciences, Changsha, Hunan, China

Hua Jiao Qiu Institute of Bast Fiber Crops, Chinese Academy of Agricultural Sciences, Changsha, Hunan, China

Ebtehag A. E. Sakr Botany Department, Faculty of Women for Arts, Science and Education, Ain Shams University, Cairo, Egypt

Maysoun M. Saleh Genetic Resources Department, General Commission for Scientific Agricultural Research, Damascus, Syria

Khaled F. M. Salem Department of Biological Sciences, Shaqra University, Ad-Dawadimi, Saudi Arabia

Department of Plant Biotechnology, Genetic Engineering and Biotechnology Research Institute (GEBRI), University of Sadat City, Sadat City, Egypt

H'ng Paik San Faculty of Forestry and Environment, Universiti Putra Malaysia (UPM), Serdang, Selangor Darul Ehsan, Malaysia

M. Sathyabhama Department of Biotechnology, PSG College of Arts & Science, Coimbatore, Tamil Nadu, India

Yadika Shouhereti Institute of Agricultural Science of Ili Prefecture, Ili, Xinjiang, China

A. Sivakumar Indian Agricultural Research Institute, New Delhi, Delhi, India

Xi Xia Song Institute of Industrial Crops, Heilongjiang Academy of Agricultural Sciences, Harbin, Heilongjiang, China

Jianguang Su Department of Germplasm and Breeding, Institute of Bast Fiber Crops, Chinese Academy of Agricultural Sciences (IBFC, CAAS), Changsha, China

Padmapriya Subramanian Department of Medicinal and Aromatic Crops, Tamil Nadu Agricultural University, Coimbatore, Tamil Nadu, India

S. Sugumar Indian Agricultural Research Institute, New Delhi, Delhi, India

Sri Sunarti National Research and Innovation Agency (BRIN) the Republic of Indonesia, Bogor, Indonesia

R. Suvitha Tamil Nadu Agricultural University, Coimbatore, Tamil Nadu, India

R. Viswanathan Division of Plant Pathology ICAR-Sugarcane Breeding Institute, Coimbatore, Tamil Nadu, India

Yu Fu Wang Institute of Bast Fiber Crops, Chinese Academy of Agricultural Sciences, Changsha, Hunan, China

Zemao Yang Department of Germplasm and Breeding, Institute of Bast Fiber Crops, Chinese Academy of Agricultural Sciences (IBFC, CAAS), Changsha, China

Dan Dan Yao Institute of Industrial Crops, Heilongjiang Academy of Agricultural Sciences, Harbin, Heilongjiang, China

Rakhiya Yelnazarky S.Seifullin Kazakh Agrotechnical University, Astana, Republic of Kazakhstan

Yin Yi School of Life Sciences, Guizhou Normal University, Guiyang, China

Xiaoyu Zhang Department of Germplasm and Breeding, Institute of Bast Fiber Crops, Chinese Academy of Agricultural Sciences (IBFC, CAAS), Changsha, China

Zheng Zhang Institute of Agricultural Science of Ili Prefecture, Ili, Xinjiang, China

Xuan Zhu Dali Bai Autonomous Prefecture Agricultural Science Extension Research Institute, Dali, Yunnan, China

Part I
Grass Fiber Crops

Chapter 1
Bamboo Breeding: Current Status and Future Prospects

Sakhi Kundu, Sarthak Mondal, Semantee Bhattacharyya, Asmita Banerjee, Asmita Das, Sukanya Chakraborty, Evanylla Kharlyngdoh, and Malay Das

Abstract The Majority of bamboo species render numerous applications and socio-economic relevance for people in rural areas of many Asian countries. It belongs to the Bambusoideae subfamily, which includes both woody and herbaceous bamboo. Bamboo is an ecologically sustainable option for achieving food and energy security. It can thrive in a variety of environments, including degraded lands, making it valuable for land restoration and preventing soil erosion. Bamboo can be propagated using vegetative organs like rhizomes, culms, and branches, in addition to seeds. Despite its versatility, there exists a dearth of knowledge about bamboo breeding, hindering its optimal and sustainable use as a bioresource. Breeding or genetically modified bamboo is challenging owing to issues such as the plant's perennial nature, rare flowering events and monocarpic behavior. Additionally, bamboo cultivators often face threats from various stresses, causing vegetation loss. Improving breeding technologies for bamboo would greatly benefit industrial, agronomical and ecological needs. This chapter reviews the recent status and advancements in bamboo breeding and discusses the potential for enhancing existing breeding practices.

Keywords Bamboo · Climate change · Breeding strategies · Gene alterations · Databases · Sustainable development · Artificial selection · Gene editing · Genetic transformation

---

S. Kundu · S. Mondal · S. Bhattacharyya · A. Banerjee · A. Das · M. Das (✉)
Department of Life Sciences, Presidency University, Kolkata, India
e-mail: malay.dbs@presiuniv.ac.in

S. Chakraborty
Department of Life Sciences, Presidency University, Kolkata, India

Department of Botany, Scottish Church College, Kolkata, India
e-mail: scbotn@scottishchurch.ac.in

E. Kharlyngdoh
Institute of Bioresources and Sustainable Development, Meghalaya, Upper Shillong, Meghalaya, India

K. F. M. Salem et al. (eds.), *Breeding and Biotechnology of Grass and Bast Fiber Crops*, Advances in Plant Breeding Strategies 12,
https://doi.org/10.1007/978-3-032-00403-1_1

1.1 Introduction

Bamboo is grouped in the family Poaceae, subfamily Bambusoideae. They are spread from a latitude of 51° N to 47° S and an altitude of 0–4000 m above sea level (Behari 2006). It can be found in every continent of the world except for Europe and Antarctica (Liese 2001; McClure 1993) (Fig. 1.1). Nevertheless, approximately 400 bamboo species have been introduced to Europe and are currently being cultivated (Gielis and Oprins 2000). Bamboo vegetation covers approximately 22 × 106 ha of forest areas worldwide (Zhou et al. 2011). Based on ploidy numbers, woody bamboos are distributed into three tribes. They are: Bambuseae (neotropical group: CC DD, paleotropical group: CC DD EE), Arundinarieae (temperate group: AA BB) and Olyreae (herbaceous bamboos: HH HH HH) (Triplett et al. 2014). Altogether 1670 species belonging to approximately 75 genera of bamboo are found globally (Das et al. 2008; Kelchner 2013; Liese and Kohl 2015). As the fastest growing grass, bamboo serves as a promising, sustainable bioresource for multiple purposes. Bamboo culms are useful for making houses, furniture, clothing, handicrafts, food, medicines, paper, plywood and biofuel (Biswas et al. 2022; Nirmala et al. 2018; Singhal et al. 2013; Sun et al. 2023). Similarly, their ecological utilities include carbon sequestration ability, replacing plastics with bamboo based fibers and biodiversity conservation (Embaye 2000; Hunter and Wu 2002; Liese 2001; Zhao et al. 2022). However, bamboo breeding is very challenging due to biological issues related to their perennial tree habit, long lasting clonal growth phase and ambiguous flowering time, followed by the sudden death of the population. Therefore, the application of conventional breeding programs is very limited on bamboo. The major focus here is to summarize our current knowledge of different aspects of bamboo breeding and also to discuss existing limitations that need to be overcome.

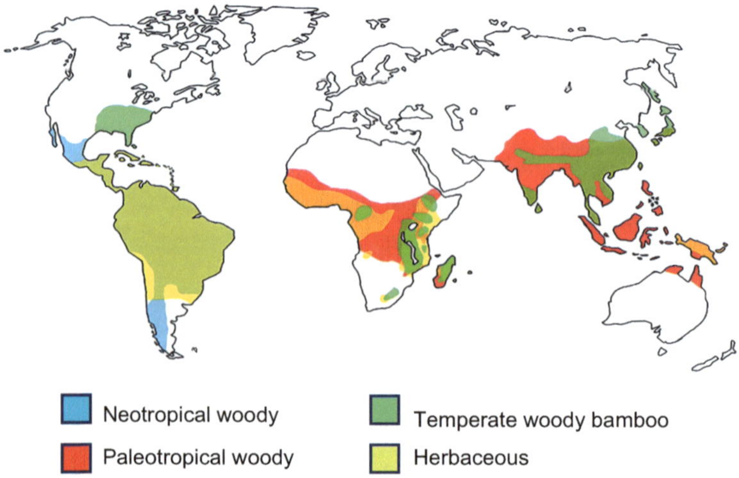

Fig. 1.1 Distribution of 4 bamboo tribes around the world. (The figure is redrawn by obtaining information from Basak et al. 2021)

1.1.1 Description, Taxonomy Classification and Distribution

Bamboos are placed in the family Bambusaceae within the tribe Graminae. They have either woody or herbaceous habit, and could be divided into three groups: Bambuseae, Arundinariae and Streptochaeteae. These taxonomic placements have been updated multiple times by including additional genera. For example, the tribe Bambusoideae was divided into four sub-tribes, which were Arundinarieae, Eubambuseae, Dendrocalameae and Melocanninae based on multiple morphological parameters (Bentham and Hooker 1883). This classification system proposed by Bentham was later revised by Gamble (1896), Rendle (1904), Camus (1913) and Bor (1938).

Kingdom Plantae—plantes, Planta, Vegetal, plants
  Subkingdom Viridiplantae—green plants
    Infrakingdom Streptophyta—land plants
      Superdivision Embryophyta
        Division Tracheophyta—vascular plants, tracheophytes
          Subdivision Spermatophytina—spermatophytes, seed plants, phanérogames
            Class Magnoliopsida
              Superorder Lilianae—monocots, monocotyledons, monocotylédones
                Order Poales
                  Family Poaceae—grasses, graminées

1.2 Bamboo Importance and Uses: Fiber and Its Textile Properties

Chemically, bamboo fibers are primarily composed of holocellulose (60–70%). In addition, they also contain pentosans (20–25%), lignin (20–30%), resins, waxes, tannins and inorganic salt (Tamolang et al. 1980). The presence of ample cellulose qualifies bamboo fiber for yarn and fabric production. Hydrolysis, alkalization and multiphase bleaching are required for the refinement of fibers from bamboo pulp (Prakash 2020). Fabric made of bamboo draws attention due to many good qualities such as high water absorbability, air permeability and softness (Erdumlu and Ozipek 2008). It also protects from UV radiation and allergies. Due to excellent thermal conductivity, textiles made out of bamboo fibers are comfortable to wear both during summer and winter. Bamboo fibers are antimicrobial due to a bioagent called "kun", thus preventing germs and odors (Afrin et al. 2012). Therefore, products of bamboo fiber cover a broad range of clothes such as shirts, sweaters, undergarments, towels, socks, blankets and mats. Due to their anti-microbial properties, they are also useful in making napkins, sanitary pads, masks and packaging bags

(Erdumlu and Ozipek 2008). Production of bamboo fibers is primarily restricted to the Asia-Pacific region, estimated at 40,000 tonnes per year (Prakash 2020).

1.3 Current Cultivation Practices and Challenges

Major bamboo vegetation around the world is currently dominated by three major hotspots: the Asia-Pacific Region (~accounting for approximately 45% of the world's bamboo forests), Latin America, and Africa. India boasts 125 indigenous species of bamboo and 11 exotic species from 23 different genera. Annual bamboo production worldwide is projected to be about 3.23 million tons, and China is leading the way with an average yield of 50 MT/Ha. China is home to 500 bamboo species from 39 different genera spread across seven million hectares of bamboo forest (https://nbm.nic.in/Documents/pdf/Concept_Note_Bamboo_25&Feb.pdf). Some case studies in Italy suggest that the Mediterranean climate is also suitable for cultivating bamboo species (Lombardo 2022). In India, bamboo vegetation is spread over land of 16 million hectares (Kumar et al. 2023). To produce high quality, and high yield bamboo, various factors must be taken into consideration. These factors include selecting high yielding species, utilizing appropriate propagation methods and ensuring quality propagules. Bamboos are primarily cultivated through asexual or vegetative propagation methods such as rhizome cutting, offset cutting and culm cutting, as well as sexual breeding through seed sowing (Banik 1995a, b).

1.3.1 Vegetative Propagation

Because of the unavailability of seeds or their low rate of germination, vegetative propagation remains the primary method of propagation for bamboo. Various vegetative parts, for example, rhizomes, offsets, and culm cuttings, are used as planting materials. The propagules are collected and planted on the nursery bed to generate new shoots. Commercial growers use procedures that include the planting of rhizome, offset, and culm part (Banik 1994, 1995a, b).

1.3.1.1 Rhizome Types and Cutting

A rhizome is a type of underground stem that aids in the growth of new culms. The nodes of rhizomes contain buds and root primordia. In bamboo, rhizomes are of three types (i) Pachymorph (ii) Leptomorph and (iii) Amphimorph rhizomes (Akinlabi et al. 2017a, b; Zamora 1994).

Pachymorph or sympodial rhizomes are short, and thick and have asymmetrical internodes (Fig. 1.2a). Because the rhizomes are short, emerging bamboo culms are clumped together, making them appear as clumpy bamboos. Bamboo genera with

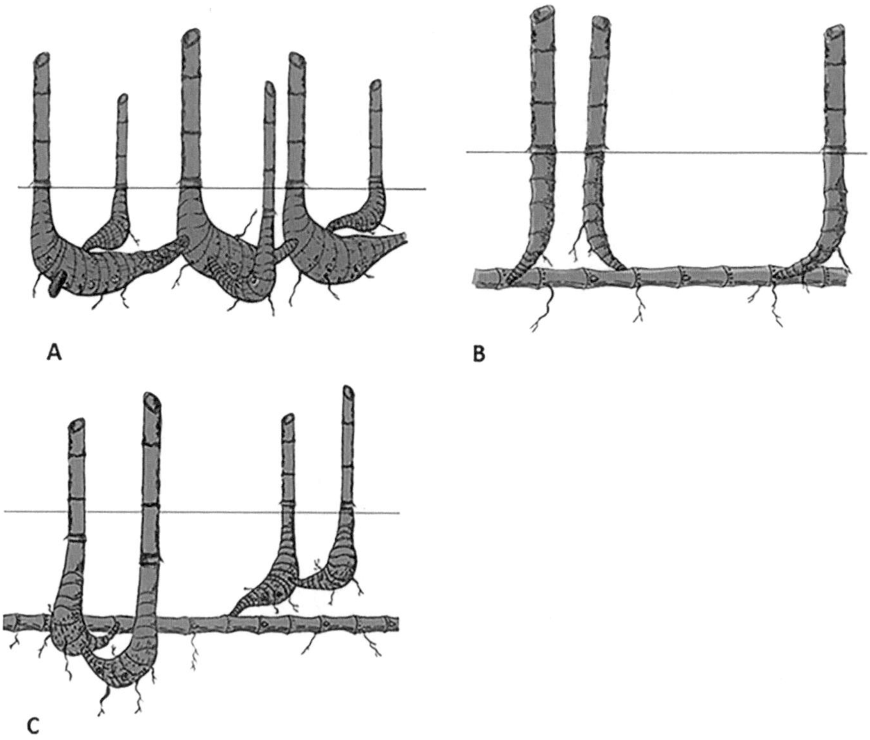

Fig. 1.2 (**a**) Pachymorph rhizome, (**b**) Leptomorph rhizome and (**c**) Amphimorph rhizome. (The figure is redrawn by using information obtained from https://www.guaduabamboo.com/blog/types-of-bamboo-rhizomes)

pachymorph rhizomes include *Bambusa, Cephalostachyum, Dendrocalamus, Thyrsostachys, Gigantochloa,* and *Schizostachyum* (Akinlabi et al. 2017a, b; Rao et al. 1989).

The leptomorph or monopodial rhizomes are narrow, long, hollow and have symmetrical internodes (Fig. 1.2b). Because the internodes are long, the culms that grow from them are distanced from the parent culm, resulting in non-clumpy bamboo species. They are observed in the species of *Phyllostachys* (Akinlabi et al. 2017a, b; Rao et al. 1989; Zamora 1994). The leptomorph type of rhizomes is primarily used for bamboo propagation.

Amphimorph rhizomes exhibit features of both Pachymorph and Leptomorph rhizome types (Fig. 1.2c). Bamboos like *Chusquea fendleri, Bashania, Indocalamus, Oligostachyum* and *Sasa* have this type of rhizome (McClure and Soderstrom 1973).

For propagation, rhizomes should be cut into pieces with a length of 50–60 cm containing nearby 10–15 nodes with roots attached (Fig. 1.3). A 3:1 mixture of soil and sand, proper watering and straw mulching are necessary to achieve optimal culm growth from rhizomes.

Fig. 1.3 Propagation of *Phyllostachys mannii* through rhizome cutting: Harvesting and washing of mother rhizomes having newly developed rhizome buds (**a** and **b**); Plantation of rhizome cuttings with newly developed rhizome buds (**c**). Source: Photographs obtained by Evanylla Kharlyngdoh

1.3.1.2 Offset Cutting

Young culms of 1–2 years of age are cut at a slant angle with approximately 3–5 nodes from the base, including rhizomes and roots (Rao et al. 1989). Offsets are usually planted before the rainy season and planted apart from other plants to reduce competition for nutrients (Midmore 2009). Due to weight, offsets can be difficult to move and are usually effective for species having thick walls (Fig. 1.4).

To enhance plantlet survival rate in the new environment, the following factors should be taken into consideration: (a) Bamboo offsets should be collected from a robust, healthy parent clump. (b) Care should be taken during the cutting process to avoid damaging the rhizomes and roots. (c) During offsetting, the rhizome should be separated from the clump, usually selecting culms from the outer ends and (d) the base of the uprooted offset culm should be covered with bags made of poly or jute and filled with wet soil or sawdust before transporting them to the cultivation site.

Fig. 1.4 Propagation of *Chimonocalamus griffithianus* through offset cutting: Harvesting and washing of mother rhizomes having newly developed rhizome buds (**a** and **b**); Development of new stems with leaves from the offset cuttings after 1 year of plantation (**c**). (Source: Photographs obtained by Evanylla Kharlyngdoh)

1.3.1.3 Culm Cutting

For culm cuttings, portions of the stem, typically containing one to three nodes along with buds or emerging branches, are used. Clumpy bamboo species can be grown vegetatively using their culms. Stems with branches and intact buds are cut and implanted in wet soil (Rao et al. 1989; Zamora 1994). The bed in the nursery is shielded by a protective covering and receives continuous watering until the cuttings begin to grow, and their roots reach a minimum level of growth. Subsequently, they are relocated to poly bags. Each node of the stem may sometimes give rise to multiple stems and roots, which are separated before being transferred to the field or the plantation site (Figs. 1.5 and 1.6).

To propagate bamboo culm cuttings effectively, beds have to be prepared that are one meter wide and 5–15 meters long. The beds should be formed on leveled ground, 20–30 cm deep, with a three-layer structure comprising loamy soil and 10–15 cm of sandy soil to facilitate rainwater drainage and prevent waterlogging during continuous rainfall (Banik 1994). Finally, the bamboo culm cuttings are horizontally placed on the prepared bed, positioning them properly for optimal growth.

Fig. 1.5 Propagation of *Dendrocalamus hookeri* through stem cutting: Cutting of a 1–2 years stem having a minimum of two nodes between the internodes (**a**); Planting of cut stems in the soil at 450 angles (**b**) and development of stems and branches after 1 year of planting (**c**). Photographs obtained by Evanylla Kharlyngdoh

Fig. 1.6 Propagation of *Phyllostachys mannii* through whole stem cutting: Harvesting of one year-old plant (**a**); Washing and soaking of roots and rhizomes in water after harvesting for propagation (**b**) and planting of one year-old *P. mannii* whole culms in the soil (**c**). Photographs obtained by Evanylla Kharlyngdoh

1.3.1.4 Bottlenecks of Vegetative Propagation

Offsets are not useful for mass propagation since the process requires extensive human effort, time and costs to transport offsets to the field. Rhizome cutting is also costly and the propagules are not easy to transport.

Vegetative propagation through culm, branch or rhizome cuttings is a common practice. However, it is not very effective for non-clumping bamboo species. Additionally, the failure rate is still quite high, especially when trying to establish large plantations. Despite severe attempts being carried out at various institutes, a unanimously applicable method for bamboo vegetative propagation has not so far been established (Liese 1985, 1991). Rao and Raghav (2002) also noted that the use of vegetative propagules, like planting of offsets or rooting of culms, is less useful for mass propagation of clump-forming sympodial types. Furthermore, their transportation to the planting sites renders another practical challenge.

Hence, the traditional methods of bamboo vegetative propagation seem inadequate to meet the demands for field-plantable saplings in large quantities for establishing extensive plantations.

1.3.1.5 Improvement Strategies

The macro proliferation technique to aid vegetative propagation was first developed by Banik (1997). The method can be applied universally and is effective for sympodial bamboo. The technique has the potential to solve existing difficulties of vegetative propagation of bamboo (Liese 1992).

1.3.2 Sexual Propagation

Growing bamboo from seeds is the primary method of sexual propagation. Scientific evidence supports the idea that clumps derived from seeds usually preserve their genetic authenticity. Additionally, a plantation of clump bamboo initially propagated from seeds produces high quality culms and contributes to the longevity of the clumps in the plantation. It is essential to keep the nursery bed in a shaded environment to protect both the seedbed and newly sprouting seedlings from direct sunlight, ensuring optimal germination rates. Regular watering is essential to maintain humidity in the seed beds and the soil around the seedlings. This task is best done in the early morning or evening hours, using finely calibrated rose-watering containers. The main advantage of sexual propagation is the inheritance of a plant's genetic diversity (Akinlabi et al. 2017a, b).

1.3.3 Digital Aid in Bamboo Farming

The Bamboost App is an android supported mobile application designed for farmers to help them sell their bamboo products. This app eliminates the middleman and helps farmers search for buyers in local and international markets directly (Diaz et al. 2021). Another initiative of digitalization was undertaken in Datang town, China. Locals led by Yang Changqin have set up an E-commerce model to showcase and sell the indigenous Chisui bamboo weaving products. It has been very successful in generating employment for 100 local villagers, creating an output value of 20 million Yuan for the company and most importantly popularizing the cultural heritage (Pan et al. 2023). Such application of digital platforms may help farmers reach their potential buyers and may be adopted by other regions as well.

1.4 Bamboo Germplasm Diversity and Conservation

To date, *ex situ* conservation of bamboo germplasm is carried out in botanical gardens, private gardens, institutions, seed banks, rhizome banks and cryogenic preservation. Distributed globally across every continent (except Antarctica), there are approximately 3026 botanical gardens present, among which 139 are set up as bambuseta. The majority of bambuseta are found in Asia (88), followed by Europe (23), North America (19), South America (4), Australia/Oceania (4), and Africa (1). Collectively, these facilities conserve around 560 species and 14 hybrids, with 28 genera and 228 species from Arundinarieae (out of a total of 31 genera and 581 species), 28 genera and 327 species from Bambuseae (out of a total of 73 genera and 966 species) and 4 genera and 5 species belonging to Olyreae (out of a total of 21 genera and 123 species) (Koshy 2023). The Bamboo Garden at the Royal Botanic Garden, Kew houses 130 bamboo species collected from China, the Himalayas, Japan and the Americas. (https://www.kew.org/kew-gardens/whats-in-the-gardens/bamboo-garden-and-minka-house). The largest known collection of bamboo can be found at the Fuji Bamboo Garden in Japan, which has over 450 taxa of bamboo. Notable bambuseta in China include the South China Botanical Garden (SCBG), Anji Bamboo Garden and Xishuangbanna Tropical Botanical Garden, initially conserved 222, 396 and 200 bamboo species respectively (https://www.bamboo-trip.com/scbg, https://www.chinahighlights.com/hangzhou/attraction/anji-bamboo-forests.htm, http://english.xtbg.cas.cn/gh/lc/201001/t20100122_50301.html). The Yunnan province of China boasts 108 nature reserves dedicated to *in situ* conservation of bamboo (Hui et al. 2014). In India, prominent bambuseta include the Indian Botanic Garden in West Bengal (36 species), the Jawaharlal Nehru Tropical Botanic Garden and Research Institute (JNTBGRI) in Thiruvananthapuram (82 species and 14 hybrids); the Forest Research Institute (FRI) in Dehradun, Uttarakhand (45 species); Van Vignan Kendra (VVK) in Chessa (40 species), and the Kerala Forest Research Institute (KFRI), in Peechi (63 species) (Alam 2011; Beniwal and Haridasan 1988; Biswas 1999, Christenhusz 2011; Koshy 2010; Wong 2011, https://www.kfri.res.in/).

1.5 Genetic Resources Conservation: Seed Preservation, Rhizome Bank and Cryopreservation

Conserving bamboo through seeds is not easy because of multiple practical issues, including the rare occurrence of flowering, the unusually extended flowering cycle (Janzen, 1976), and the very low viability of seeds. Even the pivotal Millennium Seed

Bank at the Royal Botanic Gardens at Kew lacks a comprehensive collection of bamboo seeds. Only the African bamboo *Oxytenanthera abyssinica* from Mali is hosted there (Williams 2007). In India, the ICAR-NBPGR Seed Genebank (URL: http://genebank.nbpgr.ernet.in/SeedBank/Default.aspx) maintains seeds of three bamboo species (*Arundinaria* sp., *Bambusa arundinacea, Dendrocalamus calostachys*). They also cryogenically preserve two species (*Dendrocalamus strictus, Chimonobambusa griffthiana*). Conservation of bamboo through rhizome banks has also been attempted. For example, rhizomes of eight bamboo species (*Bambusa bambos, Dendrocalamus asper, D. hamiltonii, D. membranaceus, D. strictus, Drepanostachyum falcatum, Melocanna baccifera,* and *Thyrsostachys siamensis)* have been conserved by developing a rhizome bank in Uttarakhand, India (Chandran 2008).

1.6 Traditional Breeding and Its Limitations in Bamboo

Traditional breeding in plants is mainly done by two methods, which are hybridization-based breeding and mutation breeding. Hybrid breeding requires crossing between parents and selection of the hybrid based on superior traits. An unusually long flowering period is the foremost barrier to bamboo hybridization. To overcome this problem, tissue culture can be exploited to initiate flowering (Nadgauda et al., 1990; Das and Rout and 1994) has become a promising tool.

Not all bamboo varieties are capable of seed production. Additionally, among those that do produce seeds, the occurrence is intermittent, as it requires approximately 30–120 years for certain species to bloom. The viability of seeds has also been proven to be a problem. Two types of flowering in bamboo are usually observed, which are gregarious and sporadic flowering. When bamboo undergoes gregarious flowering, entire groups of individuals within that species blossom, generate seeds and subsequently perish. Despite the abundant production of seeds, their viability is limited, often lasting only for a brief duration, ranging from a few days to a few months.

Storing seeds for an extended period is not ideal. Successful storage requires advanced drying methods, precise temperature control, and effective sealing. However, seeds stored in reduced humidity and temperature have a viability limit of approximately 2 years, and subsequently, they are no longer suitable for seedling production.

Consequently, the cultivation of bamboo from seed is rare. Growers primarily rely on vegetative propagation methods to successfully cultivate or propagate seedlings.

1.7 Hybridization

1.7.1 Artificial Selection, Conventional Hybridization to Improve Important Traits and Hybrid Cultivars

Selective breeding involves artificial selection, where a plant having superior traits is selected for breeding. This is how wild species were domesticated for cultivation (Sun et al. 2023). For example, twenty-one variants of Moso bamboo (*Phyllostachys edulis*), such as *P. edulis f. abbreviata*, *P. edulis f. anjiensis*, *P. edulis f. bicolor* and *P. edulis* 'Pachyloen', having stems with diverse colored stripes, have been observed (Zeng et al. 2018). They all offered different utilities. *P. edulis* 'Pachyloen', since the culms are strong, sturdy and very useful to make buildings and furniture (Shen et al. 2020). More genetic variants of *P. bambusoides* and *P. vivax*, having variations in stem colour and stripe patterns, have also been developed through artificial cultivation (Ma et al. 2014). More examples of elite individuals identified via selective breeding could be found in the International Cultivar Registration Center for Bamboos (64 cultivars), and the State Forestry and Grassland Administration of China (20 new varieties, Gao 2023).

Conventional hybridization has long been used as a method for crop improvement. Hybrid breeding is conducted by crossing parents and selecting the hybrid having superior inherited traits. In bamboo breeding, the prolonged vegetative phase and the uncertain flowering period have posed serious challenges to the success of hybridization. Induction of flowering by using tissue culture systems is a viable alternative to combat this.

Due to various factors, such as fragmented population size, stretched life cycle, low pollen viability (Fig. 1.7), reduced rate of pollination, high frequency of pollen sterility, mass culm death at post-flowering and insect damage to flower organs, artificial hybridization of bamboo has been challenging. However, a successful *in vivo* hand pollination experiment has been conducted on one of the dominant bamboos of

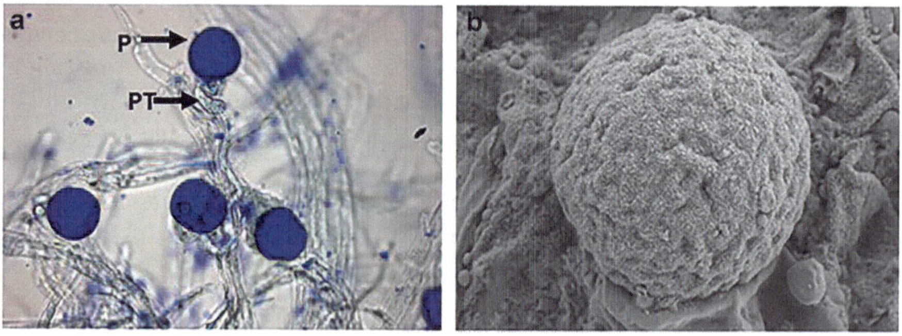

Fig. 1.7 Microscopic observations of pollen in *Bambusa tulda*. (**a**) Germinating pollen tube on stigma surface after cross-pollination and (**b**) Scanning electron microscopic (SEM) image of an individual pollen grain. Abbreviations used: PT Pollen tube, P Pollen. (Source: Photograph taken by Sukanya Chakraborty)

tropical India, *Bambusa tulda* (Chakraborty et al. 2021). This study identified the contribution of cross-pollination in increasing genetic variability in this species. In addition to this, the overall rate of seed set in *B. tulda* was found to be low, suggesting that distant populations and limited pollen availability are likely the two primary factors controlling the plant's breeding habit. Hybridization experiments have been tried on other bamboos to improve germplasm quality. For instance, crossing *Bambusa textilis* with *B. pervariabilis* has generated taller hybrids, which possessed better wood properties, and had long fibers ideal to make paper (Zhang and Chen 1980, 1986). Hybridization between *B. pervariabilis* and *Dendrocalamus daii* gave rise to hybrids appropriate for paper mills (Mo et al. 2005; Wang et al. 2005). Hybrids of *B. pervariabilis* and Ma bamboo yielded varieties having better qualities related to food consumption (Wang et al. 2020). Zhang and Chen (1986) crossed between *Bambusa pervariabilis* and *Dendrocalamus latiflorus* to obtain a new genetic variant ('Chengma 7') that offered better shoots to eat. By crossing *B. pervariabilis* × (*D. latiflorus* + *B. textilis*), a new variety called 'Chengmaqing 1' was obtained that demonstrated rapid growth, good timber, and also resistance to stress exposure (Yi 2008). Ning and Dai (1995) crossed between *B. pervariabilis* and *B. grandis* to develop the crossbreeds Chenglv 3, 6, 8, and 30. These varieties demonstrated many superior traits wanted by the paper industry. Over a long time, the 'Chenglv' hybrids have been grown in many areas of southern China. A list of the new hybrid cultivar varieties of woody bamboo developed through hybridization has been compiled (Table 1.1).

Lin et al. (2010a, b) crossed *P. kwangsiensis* (female parent) and *P. bambusoides* (male parent) for the first time, and obtained hybrid bamboo. They used the PCR based ISSR method and could authenticate two hybrids. The others were likely intra-species offspring of *P. kwangsiensis*. However, there was still a possibility that the remaining progenies were also hybrids because hybrids often demonstrate traits similar to either one of the parents (Duan et al. 2002). *Bambusa multiplex, B.*

Table 1.1 New hybrid varieties of bamboo are obtained by carrying out hybridization either between two bamboo species or between bamboo and rice

Inter-species crosses performed	Hybrid offsprings	Superior traits observed in offspring
♂ *D. latiflorus* × ♀ *B. perivariabilis*	Chengma 1, 7, 25	Grows faster and has enhanced tolerance to stress
♂*D. latiflorus, B.textilis* × ♀ *B. pervariabilis*	Chengmaqing 1	Speedy growth, good quality of material, tolerant to stress and better appearance
♂*B. grandis* × ♀ *B. pervariabilis*	Chenglv 3, 6, 8, 30	Large culm diameter (DBH = 10.1 cm), and good paper-making performance
♂*D. latiflorus* × ♀ *B. textilis*	Qingma 11	More attractive and good papermaking ability
♂ *D. latiflorus* × ♀ *D. hamiltonii*	Banma 1	Improved palatable characteristics of shoots
♂ *B. textilis* × ♀ *Oryza sativa*	Hybrid	The presence of lignified tissue confers strength, a highly proliferated root system, and resistance to stresses

Source: Published reports such as Xu (2019)

chungii, and *Dendrocalamus latiflorus* were used as parent stocks for distant and inbred hybridizations (Yuan et al. 2018). Two hybrid and three inbred populations were produced, in which growth morphology and fibers were examined. Two hybrid populations demonstrated superior qualities with regard to height, diameter at ground level and internode length.

Recently, patents have been filed on crossing between bamboo and rice in order to transfer important traits from bamboo to rice. The major traits include the presence of lignified tissue conferring strength, a highly proliferated root system, resistance to different abiotic stresses such as heat, drought, cold and resistance to biotic stresses such as disease and pest resistance (patents number: CN1973614A, CN103039352A). Such studies offer new directions to transfer genes controlling important traits from bamboo to crop plants.

1.8 Current Status of Mutation Breeding in Bamboos

The deliberate employment of mutations in plant breeding is known as "mutation breeding." It offers an advantage of improvement in an otherwise outstanding cultivar without compromising its agronomic and qualitative features. *In vitro* methods are useful since they can handle large mutagenized populations in a small area, have a faster progeny turnover rate in vegetatively propagated species, and can screen for a range of stress factors in the culture environment.

1.8.1 Conventional Seeds Mutagenesis

In orthodox mutagenesis, the unavailability of a large mutagenized population for screening and proper selection methods serves as a major hindrance. The utility of X-rays and other radiation altering the genetic material permanently prompted researchers to induce mutations employing chemical and physical agents (Ahloowalia et al. 2004). Since the occurrence of mutation is rare, and chances of success are increased by undertaking the screening of a larger mutagenized population. In addition to seeds, other plant propagules are also used successfully by utilizing the advantage of totipotency in single cells (Mba 2013). However, the formation of chimeras and the intricate nature of apical meristems also pose complications for the vegetative propagules to be used in conventional mutagenesis.

1.8.2 In Vitro Mutagenesis and Selection in Bamboo

Seeds of *D. strictus* were irradiated by gamma rays to induce precocious flowering (Kapoor and Sharma, 1992). On the other hand, Cai et al. (2007) optimized the radiation required to mutate seeds by probing the impacts of diverse doses on development-related parameters of Moso bamboo. Wang et al. (2008) investigated the effects of varying doses of gamma radiation on chromosomal structure, aberration, and mitosis in root tip cells of Moso bamboo. Transposable elements (TEs), also known as jumping genes, can change their positions in plant genomes (McClintock 1984) and are frequently used to study genome evolution (Feschotte et al. 2002; Jurka et al. 2007). Although their precise functions within the genome are unknown, TEs can cause natural genomic variations (Bourque et al. 2018). Recent efforts have utilized TEs as tools to study variations across the genome, gene tagging, and silent genes in a targeted manner, all of which can aid in engineering better crops (Cho 2018; Joly-Lopez et al. 2017; Makarevitch et al. 2015). TEs account for 63.3% of the Moso bamboo genome (Zhao et al. 2018), which is less than maize (85%) (Schnable et al. 2009) and analogous to sorghum (62%) (Paterson et al. 2009). TE insertions were often flanking around 26,366 genes, which is approximately 51.6% of all genes in Moso bamboo (Zhao et al. 2018). When transgenic Arabidopsis plants carrying an active LTR retroelement- *Phyllostachys heterocycla* retrotransposon 2 (PHRE2), were exposed to environmental stresses, gamma ray irradiation, and plant growth regulators, copy number enhancement for PHRE2 was recorded (Zhou et al. 2018). This transposition behavior in response to specific external stimuli enhances the application of a specific family of retrotransposons as a breeding resource, resulting in new agronomic alleles and restructuring of the gene regulatory network. Overall, TEs are very active in the Moso bamboo genome, altering both genetic and epigenetic architecture. Since most TEs can be influenced by environmental cues, they are potential breeding tools that may permit us to alter the architecture of the bamboo genome (Ramakrishnan et al. 2020).

1.8.3 Molecular Analysis to Screen Mutagenized Population

After inducing mutations, newer varieties are sequenced by whole genome re-sequencing to detect small and large sequence changes in order to reveal whole genome differences. While genome sequencing can accurately detect mutations due to its ability to pinpoint the precise site and type of mutation, its application to large populations is expensive and infrequently utilized in actual mutation breeding. Furthermore, heterozygous alleles can not be easily detected by direct sequencing. As a result, several techniques for searching sequence polymorphisms in amplicons in mutant populations have been developed.

TILLING is a reliable, high-throughput approach that works on a broad range of species. Using a reverse genetics method called TILLING, allelic differences within mutant populations are found by identifying sequence changes in target genes. Chemical mutagenesis across the complete genome is introduced to create TILLING populations, which are later spotted using an enzyme technique in conjunction with denaturing high-performance liquid chromatography or gel electrophoresis. It depends on the use of endonucleases to identify single-base mismatches in heteroduplexes. Based on a technique reported by Colbert et al. (2001), numerous TILLING populations have been created. This technique makes use of a mismatch-specific endonuclease present in celery.

Another widely used molecular analysis technique for mutation screening is PCR. PCR products are distinguished from one another using high-resolution melt analysis (HRM), which analyzes thermal denaturation characteristics. A PCR product's melting temperature (Tm) is determined by the duplex's length, sequence, and GC concentration. The sequence's variability can be seen in small variations in the melting curve.

Julio et al. (2008) suggested a unique method for high-throughput, low-cost mutant detection using the capillary electrophoresis-single strand conformation polymorphism (CE-SSCP) analysis. In order to create nicotine-free tobacco, they used it in a useful mutation breeding project that focused on two genes in the nornicotine biosynthesis pathway, ADC1 and NtabCYP82E4. In the M2 progeny of EMS-mutated tobacco, wild-type, heterozygous, and homozygous plants were distinguished using CE-SSCP for both genes.

These techniques have yet to be explored in bamboo mutation breeding, but their success in other species suggests the potential for their application in future studies.

1.9 In vitro Induction of Flowering as a Tool for Accelerated Breeding

Induction of flowering in bamboo through tissue culture methods can shorten the duration of clonal growth and quicken breeding opportunities. Plant growth regulators, for instance, cytokinins, are quite beneficial to induce flowering in regenerated bamboo seedlings (Yuan et al. 2017). This has been proven on Ma bamboo (*Dendrocalamus latiflorus*) (Lin et al. 2007; Zhang and Wang 2001), *D. giganteus* (Ramanayake et al. 2001) and *Bambusa edulis* (Lin et al. 2003a). Studies have demonstrated that the combined use of auxin, kinetin and GA is essential to cause flowering and seed development in a few species, such as *B. arundinacea.*

In vitro induction of flowers leading to the generation of viable seeds has been recorded in *B. arundinacea, Dendrocalamus brandisii, B. vulgaris, D. giganteus* and *D. strictus.* However, not all flowers obtained by *in vitro* methods could generate viable seeds (Yuan et al. 2017). For example, *in vitro* raised anthers could not yield fertile pollen or seeds after pollination in *B. edulis* (Lin and Chang 1998; Lin et al. 2003b, 2004), which could be due to species specific genetic factors (Yuan et al. 2017). Nevertheless, such flower-inducing procedure is faster compared to conventional flowering occurring *in vivo*.

To overcome the issue of uncertain flowering time, the initiation of flowering *in vitro* has been proposed as a viable alternative. Plant growth regulators, such as cytokinins, have been successfully used to induce *in vitro* flowering in many bamboo species (Yuan et al. 2017) such as Ma bamboo (*Dendrocalamus latiflorus*) (Zhang and Wang 2001; Lin et al. 2007), *D. giganteus* (Ramanayake et al. 2001), *Bambusa edulis* (Lin et al. 2003a) and *B. arundinacea* Willd, *D. brandisii* (Nadgauda et al. 1990).

1.10 Commercial Scale Micropropagation in Bamboo

For mass-scale propagation, conventional techniques are often incompetent, and tissue culture renders a viable alternative, in particular to attain the ever increasing demand (Das and Pal, 2005a, b). Therefore, the Oprins Plant, Belgium, has developed over 60 genotypes through tissue culture and many of them, such as *Arundinaria, Chimonobambusa, Fargesia, Phyllostachys, Pleioblastus, Sasa, Sasaella, Semiarundinaria, Shibataea* and *Yushania* (temperate bamboos) and *Bambusa, Dendrocalamus, Dinochloa*, and *Thyrsostachys* (tropical bamboos) are being produced at commercial scale (Gielis and Oprins 1998). Many of these bamboos have been selected due to their use in horticulture and forestry. A few other laboratories focusing on bamboo tissue culture include West Wind Technology, USA, and Bamboo World, Australia (Cusack 2000), Piccoplant, Germany, and Microflor, Belgium.

1.11 Molecular and Genomics Advancements and Resources to Facilitate Molecular Breeding in Bamboo

Recent progress in bamboo genomics may be utilized to implement molecular breeding programs in bamboo. Genomes of six bamboo species have been sequenced (Peng et al. 2013; Guo et al. 2019; Li et al. 2021; Basak et al. 2021 Zheng et al. 2022). Flow cytometric estimation identified that the tropical woody bamboo contains higher genomic DNA content than the temperate group (Gielis et al. 1997). The genome of *P. pubescens* is slightly over 2 Gb and similar in size of maize (Gui et al. 2007; Peng et al. 2013). It contains 31,987 protein coding genes, 43.9% GC and 59.0% transposable elements (Peng et al. 2013). A large number of partial genome sequences and expressed sequence tags (ESTs) are available in databases.

With the enhancement of genetic maps' resolution in major crops, it will be easier to link candidate genes discovered in model plants to those of corresponding loci in bamboo. This will also help breeders to pursue the application of moleculars in bamboo, like in other crops (Dekkers and Hospital 2002; Walsh 2001).

1.11.1 Bamboo Breeding Aided by Molecular Markers

Molecular markers are effective for the taxonomic identification of a specific bamboo species (Das et al. 2005, 2008), studying genetic diversity (Das et al. 2007; Loh et al. 2000) and testing clonal fidelity of *in vitro* regenerated plants (Das et al. 2007). The application of molecular markers in bamboo breeding programs is uncommon, but it can help to identify superior accessions for breeding drives (Nayak et al. 2003; Yeasmin and Ali 2021). Previous studies have identified molecular markers linked to important characteristic features of bamboo. For example, Pal et al. (2009) found a SCAR marker to identify elite *Bambusa balcooa* genotypes that contained high levels of cellulose-containing fibers. After performing whole-genome resequencing of 427 Moso bamboo individuals, Zhao et al. (2021) used genome-wide association analysis to identify 104 candidate genes linked to carbohydrate metabolism, cell wall, and adaptation to environmental perturbations.

1.11.2 Bioinformatics

Genomics and bioinformatics are quite instrumental in boosting the production rate of enhanced crop cultivars with cutting-edge glasshouse technology to quicken plant development (Watson et al. 2018). However, integrating diverse sets of data for breeding is challenging due to their enormous volume (Santos et al. 2017). This can, however, be resolved and breeding targets can be identified by combining phenotypes, genomes, and bioinformatics tools (Evans et al. 2013). Bioinformatics is deeply involved in the completion and evaluation of multiple genome sequences, including the bamboo genome, to identify molecular markers (Claverie and Notredame, 2003). Comparison of general genome features such as genome size, gene and chromosome number forms the foundation for comparative genomic analysis. For instance, MegaBlast uses a large sequence similarity search algorithm based on NCBI BLAST (Hesslop-Harrison, 2000). The Bamboo Genome Database (http://www.genobank.org/bamboo) has so far been the most useful database available for bamboo researchers. It provides draft genome, CDS and protein sequence data of four bamboo species, *Bonia amplexicaulis*, *Guadua angustifolia*, *Olyra latifolia* and *Raddia guianensis*. Another database called BambooNET (http://bioinformatics.cau.edu.cn/bamboo/) holds transcriptomic data for moso bamboo and has done co-expression analysis and functional module identification on it. For genetic diversity and phylogeny predictions, DNA markers are essential. Bamboo Microsatellite Database (http://14.139.223.245/bmdb/index.php) provides a platform that allows searching for simple and compound microsatellite repeat regions in the bamboo genome. Information can also be obtained to construct a genetic map and design primers for PCR. As the density of genetic maps in important crops continues to improve, it will become easier to identify syntenic genes and transfer information from reference plants to bamboo.

1.12 Advancement in Bamboo Transformation, Gene Editing Methodologies and Potential Application in Bamboo Breeding

Because of the erratic nature of flowering in bamboo, conventional breeding techniques used to make planned crosses between superior genotypes often meet severe challenges. Therefore, biotechnological tools such as genetic transformation have gained importance in recent years (Limera et al. 2017). Conventional transformation technologies such as *Agrobacterium*-mediated transformation, plant virus-mediated transformation, PEG-mediated and gene gun-mediated methods have been used to transform different bamboo species such as *Dendrocalamus farinosus*, and *D. latiflorus*, albeit with limited success (Jiang and Zhou 2014; Sood et al. 2014; Qiao et al. 2014). A summary of the different genetic transformation technologies developed, along with some successful applications in bamboo, is presented in Table 1.2.

Table 1.2 Summary of different bamboo transformation methods

Bamboo species	Transformation method	Targeted gene	Trait improvement	References
Dendrocalamus latiflorus (Ma bamboo)	*Agrobacterium*-mediated	Bacterial CodA encoding choline oxidase	Enhanced cold tolerance	Qiao et al. (2014)
Dendrocalamus farinosus	*Agrobacterium*-mediated	CoA ligase gene (4CL)	Better transformation efficiency of 9%	Li et al. (2012)
Ma bamboo	*Agrobacterium*-mediated	Glucuronidase gene (GUS) and hygromycin resistance genes	Hygromycin resistance, enhanced regeneration efficiency and transformation efficiency	Ye et al. (2017)
Ma bamboo	*Agrobacterium*-mediated	Maize leaf color gene (Lc)	Anthocyanin accumulation	Ye et al. (2017), Xiang et al. (2021)
Moso bamboo	Vacuum infiltration-assisted Agrobacterium-mediated	GUS gene	Transformation efficiency of 5%.	Huang et al. (2022)
Moso and Ma bamboos	Vacuum infiltration-assisted Agrobacterium-mediated	GUS gene and RUBY gene	Transformation efficiency of 20–30%	Chen et al. (2023)
Moso bamboo, *Phyllostachys aureosulcata*, f. *spectabilis* and *P. aureosulcata f. aureocaulis*	Vacuum infiltration and injection-assisted Agrobacterium-mediated	RUBY gene	Enhanced transformation efficiency, exogenous gene expression	Sun et al. (2023)
Moso and Ma bamboos	Bamboo Mosaic Virus (BaMV) mediated	Exogenous genes, EGFP and RUBY (in Moso and Ma bamboos) and endogenous genes, ACE1 and DEC1 (in Ma bamboo)	Large cargo capacity	Jin et al. (2023)

(continued)

Table 1.2 (continued)

Bamboo species	Transformation method	Targeted gene	Trait improvement	References
Moso bamboo	Gene gun-mediated	GUS gene	Direct delivery of DNA particles into cells, quick expression of exogenous genes, and shorter time for transformation	Hu et al. (2017a, b)
Moso bamboo	PEG-mediated	Green fluorescent protein (GFP) gene	Improved genetic transformation efficiency	Yuan et al. (2018)
Moso bamboo and Ma bamboo	PEG-mediated	Green fluorescent protein (GFP) gene	High transformation efficiency	Chen et al. (2023)
Moso bamboo	-Agrobacterium mediated	PLATZ transcription factor (PePLATZ8)	Increased tolerance to drought and salt stress	Zhang et al. (2024)

A few studies on genome editing have so far been conducted in bamboo (Huang et al. 2022; Ye et al. 2020). One of the initial studies observed that genetically silencing the *gibberellin-responsive gene 1* (*grg1*) in Ma bamboo (2n = 72, AABBCC) altered the height of the plant by enhancing internode elongation. In another study, plant regeneration was achieved from immature embryos of Moso bamboo, followed by the standardization of the transformation system, having an efficiency of 5% (Huang et al. 2022). Subsequently, the promoters from endogenous U3 small nuclear RNA were used to drive the sgRNA. Since the PeU3.1 promoter displayed elevated competence, it was used for genome editing. The CRISPR/Cas9 editing was utilized to obtain three lines of putative homozygous *pds1pds2* mutants of Moso bamboo displaying albino phenotypes. Furthermore, Sun et al. (2023) standardized an *in planta* transient expression system for editing the endogenous *violaxanthin de-epoxidase* gene (*PeVDE*) in Moso bamboo. Mutation of the PeVDE gene reduced the non-photochemical quenching of chlorophyll fluorescence under high light, making the PeVDE gene a native reporter for gene editing in bamboo. They also edited the *cinnamoyl-CoA reductase* gene (*PeCCR*), resulting in mutant plants with increased flavonoid content. Additionally, Jin et al. (2023) observed that BaMV-based vectors could express Cas9 to produce transgene-free, genome-edited bamboo plants for molecular breeding in the future. Findings from these studies can be used to conduct follow-up investigations to detect the molecular machinery behind rapid growth in bamboo. Other CRISPR/Cas-mediated genome editing tools, such as CRISPR interference (CRISPRi), CRISPR activation (CRISPRa), epigenetic engineering, and chromatin imaging, can be utilized to identify the role of each gene to advance bamboo growth. No genome editing studies have been conducted to date that are directly related to the breeding of bamboo. However, developments in transformation and genome editing methodologies may accelerate such efforts in the near future.

1.13 Genetics of Sexual Incompatibility in Grasses Providing Clues for Future Studies on Bamboos

Members of Poaceae usually exhibit GSI (Conner 1979). The multi-allelic *S-Z* loci system in grasses makes it quite challenging to study. Because of such a complex genetic background, the results of forward and reciprocal crosses may differ (Jiang et al. 2017). Also, there remain physiological factors that influence the SI response in grasses to a great extent. For instance, the short availability of grass pollens hinders the possibility of conducting crossing experiments (Ge et al. 2011). For example, in *B. tulda*, low viability of pollen (25%) and reduced germination (15%) are major obstacles to conducting pollination *in situ* (Chakraborty et al. 2021). More significantly, many factors such as dispersal of pollen, dry stigma, and quick SI response at the stigma surface are quite comparable to SSI responses (Yang et al. 2008). The precise timing and location of SI interaction may vary across grasses (Heslop-Harrison 1979). During strong SI interaction, pollen rejection takes place at the stigma surface. In contrast, if the interaction is weak, rejection happens at the intercellular spaces inside the stigma branch (Heslop-Harrison 1982). The physiological events leading to incompatible interaction may vary from one grass or bamboo to another and hence a universal SI response may not be feasible. For example, *Dendrocalamus membranaceus* and *D. sinicus* are principally cross-pollinated, although they can undertake self-pollination in special situations such as low availability of flowering culms or pollinators (Xie et al. 2019). Collectively, the SI/SC behavior of bamboo may depend on many external factors such as distance among flowering culms, pollinator frequency, and even whether it is gregarious or sporadic flowering (Basak et al. 2021; Chakraborty et al. 2021; Matsuo et al. 2014; Xie et al. 2019; Zhong et al. 2017).

1.14 Conclusion and Prospective

In comparison to major crop plants, fewer studies have been conducted on bamboo breeding. Only a limited traits have been investigated, and the nature of their genetic control is largely unknown. The unpredictable flowering time, low pollen abundance and viability pose serious challenges to carrying out breeding efforts. Despite these challenges, some limited efforts have been made to cross bamboo species in order to improve their genetic traits. Even crosses have been made between bamboo and rice to transfer important traits to rice. However, standardization of a more effective breeding program to improve bamboo quality depends on successful collaboration among breeders, researchers, and industry professionals. Advancements in transformation and genomic technologies will help in genome-assisted breeding in the future. Improved breeding to produce better qualities of bamboo would help it realize its potential, fostering the sustainable growth of the bamboo industry and acting as an ideal bio-economic candidate for achieving concerted ecological and industrial growth.

Acknowledgments The authors of this book chapter acknowledge research funding received from DBT-BIRAC (BT/EFO099/01/22). The host laboratory was also funded by the Council of Scientific and Industrial Research (CSIR), India [38(1386)/14/EMR-II], [38(1493)/19/EMR-II], Department of Biotechnology, India (BT/PR10778/PBD/16/1070/2014, BT/PR28859/FCB/125/3/2018), DBT_BUILDER program (BT/INF/22/SP45088/2022), Alexander von Humboldt Foundation, Germany and FRPDF grant of Presidency University.

Appendix 1.1 Prominent Research Institutes and Online Resources Available for Bamboo Breeding and Its Application

Institution	Area of specialization	Website
Kerala Forest Research Institute	Bamboo biology, utilization, and cultivation.	https://www.kfri.res.in/btsg.asp
Bamboo Research & Training Centre, Chichpalli, India	Training and research on bamboo-based product development	https://brtc.org.in/
Forest Research Centre for Bamboo & Rattan, Aizawl	Germplasm Bank of Bamboo and Canes, Bamboosetum and Canetum, Macro and Micropropagation, Genetic improvement, Technology for value addition, bamboo-based tools/machines for Bamboo working	https://rfri.icfre.gov.in/forest-research-centre-for%20 bamboo&rattan-aizawl
International Centre for Bamboo and Rattan (ICBR)	Bamboo and rattan germplasm resources, promotion and commercialization of the technologies.	http://eng.icbr.ac.cn/node/360
Guadua Bamboo	Producer and exporter of top-quality Guadua bamboo products	https://www.guaduabamboo.com/

Appendix 1.2 Genetic Resources for Bamboo Breeding

Bamboo cultivar type	Distribution	Cultivar Characteristics	Uses	References
Bambusa bambos (Tall form)	Southern India	Long (24–30 cm) and large-diameter (8–12 cm) culms.	Construction, ply bamboo and producing pulp. It is also used to make fences on farms	Banik (2020)
Bambusa bambos (Dwarf form)	Orissa, lower Bengal, the southeastern part of Bihar, Jharkhand, Chhattisgarh, and Uttar Pradesh (India), and Myanmar	Small culms with knots and bends. Thorns present	Used as a fence for farms	Banik (2016)
Dendrocalamus strictus (Large type)	North Bihar and Orissa	Long and big culms and internodes are straight and smooth	Construction, pulping and furniture, mats, baskets	Banik (2020)
Dendrocalamus strictus (Dwarf type)	Balaghat division of Madhya Pradesh	Small size, culms mostly absent	Building shelterbelts	Pande et al. (2012)
Dendrocalamus stocksii	Maharashtra and parts of the Western Ghats	Tall, solid, erect culms. Long internodes and fewer branches	Construction, furniture and House fabrication	Rane et al. (2019), Kumar (2011)
Bambusa balcooa (Shilbarak)	Upper Assam, Tripura (India); Sylhet (Bangladesh)	Thick-walled culms, with many short internodes, and swollen and elevated nodes.	One of the durable bamboos for construction	Banik (2020)
Bambusa balcooa (Nolbarak)	Tripura	Thin-walled culms and branches, with long internodes	Construction. Also used as incense sticks	Banik (2016, 2020), Karmakar et al. (2013)
Bambusa balcooa (Kanta Barak)	West Tripura, the eastern and northern parts of Bangladesh	Tall and strong culms, which make it hard to cut, are densely branched at the base	Construction of pillars, bridges and houses	Banik (2004)

Bamboo cultivar type	Distribution	Cultivar Characteristics	Uses	References
Bambusa vulgaris (Kenta Bizzya)	Orissa and West Bengal (India) Cox's Bazar and the south of Bangladesh	Short and bushy culms, branching present all over	Protection of coastal areas from cyclones	Banik (1995a, b)
Bambusa cacharensis (Moral bansh)	Tripura, Southern Assam, Meghalaya (India); Sylhet (Bangladesh)	Culms with long internodes and smooth nodes	Used to make cutlery items like chopsticks, spoons and tooth picks	Banik (2020)
Bambusa cacharensis (Jati Bom)	Tripura	Tall, straight and loosely growing culms	Furniture and flooring	Banik (2020)
Bambusa cacharensis (Pecheebom)	Tripura	Culms are short, compact, with small diameters and thick walls	Construction and fencing	Banik (2020)
Bambusa tulda	Tripura, Assam, West Bengal Chittagong and Sylhet (Bangladesh)	Tall, strong and big culms. Have thick walls and low branching. Durable in water	Making fishing nets, boats, roofs and furniture	Banik (2016), Bhattacharya et al. (2006)
Bambusa nutans	Tripura, Assam, Manipur, Nagaland, Meghalaya, West Bengal, Orissa, Himachal, Uttarakhand, Uttar Pradesh (India) Nepal, Bhutan	Straight, strong, moderate-sized culms	Construction, pulping, making furniture, baskets and mats	Adhikari and Shrestha (2008)
Bambusa polymorpha	Tripura, Assam, Manipur, Nagaland, Meghalaya, West Bengal, Orissa, Himachal, Uttarakhand, Uttar Pradesh	Tall, erect and shrubby culms.	Bamboo crafts and handicrafts. Also used to make incense sticks, chopsticks and straws Young shoots are edible	Banik (2016)
Melocanna baccifera	Tripura, Assam (India) Sylhet (Bangladesh)	Culms are tall and have small diameters. Some have curved rhizomes at the base	Construction of houses. Making incense sticks and umbrellas, paper pulp and edible shoots	Banik (1995a, b, 2010)
Thyrsostachys oliveri	Tripura	–	Furniture	Banik (2020)

Bamboo cultivar type	Distribution	Cultivar Characteristics	Uses	References
Dendrocalamus brandisii	Goa, southern Karnataka, northern Kerala and the Konkan region of Maharashtra	–	Furniture and construction industries	Kumar (2011)
Dendrocalamus hookeri	Arunachal Pradesh, Sikkim, Duars and the western Himalayas	–	Construction	Varmah and Bahadur (1980)
Dendrocalamus latiflorus	Manipur and Nagaland	–	Edible shoot and construction	Naithani (2011)
Himalayacalamus falconeri (Syn. *Thamnocalamus falconeri*)	Western Himalayas	–	For making baskets, mats and craft items.	Varmah and Bahadur (1980)
Ochlandra sp.	Southern India and Sri Lanka	Reed Bamboo	Housing, weaving, matting, and pulping	Kumar (1988)

References

Adhikari R, Shrestha K (2008) Infraspecific variation of *Bambusa nutans* subspecies nutans from six different sites of central Nepal. Sci World 6(6):81–84

Afrin T, Tsuzuki T, Kanwar RK, Wang X (2012) The origin of the antibacterial property of bamboo. J Text Inst 103(8):844–849

Ahloowalia BS, Maluszynski M, Nichterlein K (2004) Global impact of mutation-derived varieties. Euphytica 135:187–204

Akinlabi E, Anane-Fenin K, Akwada D (2017a) Regeneration, cultivation, and sustenance of bamboo. In: Bamboo. Springer, Cham. https://doi.org/10.1007/978-3-319-56808-9_2

Akinlabi ET, Anane-Fenin K, Akwada DR (2017b) Bamboo. The multipurpose plant, vol 268. Springer International Publishing

Alam MK (2011) 'Bamboos at TBGRI by KC Koshy. Bangladesh J Plant Taxon 18(2):209

Banik RL (1994) Studies on seed germination, seedling growth and nursery management of *Melocanna baccifera* (Roxb.) Kurz. In: Proceedings of the 4th international bamboo workshop on bamboo in Asia and the Pacific, 27–30 Nov 1991. FORSPA Publication No 6 IDRC FAOUNDP, Chiangmai, pp 113–119

Banik RL (1995a) A manual for vegetative propagation of bamboos. International Network for Bamboo and Rattan

Banik RL (1995b) Diversities, reproductive biology and strategies for germplasm conservation of bamboos. In: Bamboo and rattan: genetic resources and use. International Plant Genetic Resources Institute, Singapore, pp 1–22

Banik RL (1997) Domestication and improvement of bamboo. INBAR working paper/International Network for Bamboo and Rattan No. 10

Banik RL (2004) Bamboos of Tripura. Indian For 130(9):1081–1083
Banik RL (2010) Biology and silviculture of muli (*Melocanna baccifera*) bamboo. Department of Science & Technology, National Mission on Bamboo Applications (NMBA), TIFAC, New Delhi, pp 1–237
Banik RL (2016) Silviculture of South Asian priority bamboos. Springer Singapore, pp21–115
Basak M, Dutta S, Biswas S, Chakraborty S, Sarkar A, Rahaman T, Dey S, Biswas P & Das M (2021) Genomic insights into growth and development of bamboos: what have we learnt and what more to discover? Trees 35(6):1771–1791
Banik RL (2020) Distribution, diversity and prospects for propagation of industrially suitable bamboo species in India. J Bamboo Rattan (Kerala Forest Research Institute) 19(3):52–69
Behari B (2006) Status of bamboo in India. Compilation of papers for preparation of national status report on forests and forestry in India. Survey and Utilization Division, Ministry of Environment and Forest:109-120
Beniwal BS & Haridasan K (1988) Study of bamboos through establishment of Bambusetum in Arunachal Pradesh. Indian For [S.l.], p. 650-655
Bentham G, and Hooker JD (eds) (1883) Genera plantarum, III: London, pp 1094–1215
Bhattacharya S, Das M, Bar R, Pal A (2006) Morphological and molecular characterization of *Bambusa tulda* with a note on flowering. Ann Bot 98(3):529–535
Biswas S, Rahaman T, Gupta P, Mitra R, Dutta S, Kharlyngdoh E, Guha S, Ganguly J, Pal A, Das M (2022) Cellulose and lignin profiling in seven, economically important bamboo species of India by anatomical, biochemical, FTIR spectroscopy and thermogravimetric analysis. Biomass Bioenergy 158:106362
Bor NL (1938) A list of grasses of Assam. Indian Forest Rec New Ser (Botany) 1(3):47–102
Bourque G, Burns KH, Gehring M, Gorbunova V, Seluanov A, Hammell M, Imbeault M & Feschotte C (2018) Ten things you should know about transposable elements. Genome Biol 19(1):199
Cai C, Gao J, Mu S (2007) Effects of ^{60}Co γ rays radiation on seed vigor and young seedling growth of *Phyllostachys edulis*. Acta Agric Nucl Sin 21(5):436–440
Camus EG (1913) Les Bambusees: monographie, biologie, culture, principaux usages. Lechevalier, Paris, p 215
Chakraborty S, Biswas P, Dutta S et al (2021) Studies on reproductive development and breeding habit of the commercially important bamboo *Bambusa tulda* Roxb. Plants 10(11):2375
Chandran M (2008) Development of rhizome banks for bamboo multiplication. Indian Forester, 134(3):445-447
Cho J (2018) Transposon-derived non-coding RNAs and their function in plants. Front Plant Sci 9:600
Christenhusz MJ (2011) Book Reviewboj_1130 131. Bot J Linn Soc 167:131
Cusack V (2000) Technology for developing commercial bamboo plantations in a western culture. In: Bamboo 2000, Proceedings of the International Symposium by The Royal Project Foundation, Chiang Mai, Thailand, 2-4 August 2000, 23-34
Claverie JM, Notredame C (2003) Bioinformatics for dummies. Willey, New York, p 452
Colbert T, Till BJ, Tompa R et al (2001) High-throughput screening for induced point mutations. Plant Physiol 126(2):480–484
Conner HE (1979) Breeding systems in the grasses: a survey. N Z J Bot 17:547–574. https://doi.org/10.1080/0028825X.1979.10432571
Das P, Rout GR (1994) Analysis of current methods and approaches on the micropropagation of bamboo. Proc Natl Acad Sci India Sect B 64:235–235
Das M, Bhattacharya S, Pal A (2005) Generation and characterization of SCARs by cloning and sequencing of RAPD products: a strategy for species-specific marker development in bamboo. Ann Bot 95(5):835–841
Das M, Bhattacharya S, Basak J, Pal A (2007) Phylogenetic relationships among the bamboo species as revealed by morphological characters and polymorphism analyses. Biol Plant 51:667–672

Das M, Bhattacharya S, Singh P et al (2008) Bamboo taxonomy and diversity in the era of molecular markers. Adv Bot Res 47:225–268

Das M and Pal A (2005a) Clonal propagation and production of genetically uniform regenerants from axillary meristems of adult bamboo. Journal of Plant Biochemistry and Biotechnology 14: 185-188

Das M and Pal A (2005b) In vitro regeneration of Bambusa balcooa Roxb.: factors affecting changes of morphogenetic competence in the axillary buds. Plant Cell, Tissue and Organ Culture 81: 109-112

Dekkers JC, Hospital F (2002) The use of molecular genetics in the improvement of agricultural populations. Nat Rev Genet 3(1):22–32

Duan SH, Mao JN, Zhu YG (2002) Genetics analysis and identification of hybrid rice HL-type (Honglian-2) and their backbone parental with RAPD markers. J Wuhan Bot Res 20:171–178

Embaye K (2000) The indigenous bamboo forests of Ethiopia: An overview. Ambio, 29: 518–521

Erdumlu N, Ozipek B (2008) Investigation of regenerated bamboo fibre and yarn characteristics. Fibres Text East Eur 16(4):69

Evans K, Jung S, Lee T et al (2013) Addition of a breeding database in the genome database for Rosaceae. Database 2013. https://doi.org/10.1093/database/bat078

Feschotte C, Jiang N, Wessler SR (2002) Plant transposable elements: where genetics meets genomics. Nat Rev Genet 3(5):329–341

Gamble JS (1896) The Bambuseae of British India. Bengal Secretariat Press, Calcutta, p 133

Gao Z (2023) Research advances in bamboo genetics and breeding. World Bamboo Rattan (in Chinese) 23:1–9

Ge Y, Fu C, Bhandari H, Bouton J, Brummer EC, Wang ZY (2011) Pollen viability and longevity of Switchgrass (Panicum virgatum L.). Crop Sci 51:2698–2705. https://doi.org/10.2135/cropsci2011.01.0057

Gielis J, Everaert I, De Loose M (1997) Genetic variability and relationships in Phyllostachys using random amplified polymorphic DNA. In: Chapman GP (ed) The bamboos, Linnean Society Symposium, vol 19. Academic, London, pp 107–124

Gielis J and Oprins J (2000) Bamboo for Europe. EU-FAIR CT1747 Final Individual Report, 65 pp.

Gui Y, Wang S, Quan L et al (2007) Genome size and sequence composition of moso bamboo: a comparative study. Sci China Life Sci 50:700–705

Gielis J, Oprins J (1998) Biotechnological approaches to germplasm improvement in woody bamboos. Sustainable Agriculture for Food, Industry and Energy. James and James, London. improvement of bamboo, 195-203

Guo ZH, Ma PF, Yang GQ et al (2019) Genome sequences provide insights into the reticulate origin and unique traits of woody bamboos. Mol Plant 12(10):1353–1365

Heslop-Harrison J (1979) Pollen-stigma interaction in grasses: a brief review. New Zealand Journal of Botany, 17(4):537–546

Heslop-Harrison J (1982) Pollen-stigma interaction and cross-incompatibility in the grasses. Science, 215(4538):1358–1364

Heslop-Harrison JS (2000) Comparative genome organization in plants: from sequence and markers to chromatin and chromosomes. The Plant Cell, 12(5):617–635

Huang B, Zhuo R, Fan H et al (2022) An efficient genetic transformation and CRISPR/Cas9-based genome editing system for Moso Bamboo (*Phyllostachys edulis*). Front Plant Sci 13:822022. https://doi.org/10.3389/fpls2022822022

Hunter IR, Wu JQ (2002) Bamboo Biomass. International Network for Bamboo and Rattan Working Paper No 36

Hui C, Liang N, Yang XY, Chen F (2014) The characteristics of bamboo germplasm resources and its conservation from Yunnan. China Appl Mech Mater 522–524:1098–1101. https://doi.org/10.4028/www.scientific.net/AMM

Janzen DH (1976) Why bamboos wait so long to flower. Annu Rev Ecol Evol Syst 7:347–391

Jiang K, Zhou M (2014) Recent advances in bamboo molecular biology. J Trop Subtrop Bot 22:632–642

Jiang W, Bai T, Dai H, Wei Q, Zhang W, & Ding Y (2017) Microsatellite markers revealed moderate genetic diversity and population differentiation of moso bamboo (Phyllostachys edulis) -a primarily asexual reproduction species in China. Tree Genetics & Genomes, 13(6):130
Jin Y, Wang B, Bao M et al (2023) Development of an efficient expression system with large cargo capacity for interrogation of gene function in bamboo based on bamboo mosaic virus. J Integr Plant Biol 65:1369–1382
Joly-Lopez Z, Forczek E, Vello E et al (2017) Abiotic stress phenotypes are associated with conserved genes derived from transposable elements. Front Plant Sci 8:2027
Julio E, Laporte F, Reis S et al (2008) Reducing the content of nornicotine in tobacco via targeted mutation breeding. Mol Breed 21(3):369–381
Jurka J, Kapitonov VV, Kohany O, Jurka MV (2007) Repetitive sequences in complex genomes: structure and evolution. Annu Rev Genomics Hum Genet 8:241–259
Kapoor ML, Sharma VK (1992) Induced early flowering mutants in *Dendrocalamus strictus* Nees after gamma irradiation. Indian For 118(1):66
Koshy KC (2010) Bamboos at TBGRI. KC Koshy. Publisher: Tropical Botanic Garden and Research Institute, Palode, Thiruvananthapuram, Kerala IndiaKoshy K. C. (2023). Germplasm resources of bamboos. In Genetics, Genomics and Breeding of Bamboos (pp. 27-116). CRC Press.
Koshy KC (2023) Germplasm resources of bamboos. In Genetics, Genomics and Breeding of Bamboos (pp. 27-116). CRC Press
Karmakar K, Sinha S, Sinha RK (2013) Biochemical distinction in dimorphic culms of *Bambusa balcooa*–a potential Bamboo species of North-East India. NeBIO 4(3):6–9
Kelchner SA (2013) Higher level phylogenetic relationships within the bamboos (Poaceae: Bambusoideae) based on five plastid markers. Mol Phylogenet Evol 67:404–413
Kumar M (1988) Reed bamboos, Ochlandra in Kerala: Distribution and management. In: Bamboos: current research. Proceedings of the international bamboo workshop, November 14–18, 1988, Cochin, pp 39–43
Kumar S, Dev I, Kumar N (2023) Forestry and Agroforestry. In Trajectory of 75 years of Indian agriculture after independence (pp. 497-519). Singapore: Springer Nature Singapore
Kumar M (2011) Bamboos of peninsular India: all India coordinated project on taxonomy (AICOPTAX): Grasses and Bamboos Part-II. Art Options, New Delhi
Beniwal, B. S., & Haridasan, K. (1988). Study of bamboos through establishment of Bambusetum in Arunachal Pradesh.
Biswas RK (1999) Application of low cost non-conventional lining materials for seepage control in small irrigation channels. Agricultural Engineering Today, 23(3and4), 1–14
Li W, Shi C, Li K et al (2021) Draft genome of the herbaceous bamboo *Raddia distichophylla*. G3 11(2):jkaa049
Líese W (1985) Bamboo-biology, silvics, properties, utilization. Eschborn, pp 83–90
Liese W (2001) Advances in bamboo research. Journal of Nanjing Forestry University, 25:1–6
Liese W, Kohl, M (2015) Bamboo: The Plant and its Uses. Springer International Publishing, 362 pp
Liese W (1991) Progress in bamboo research. The Journal of the American Bamboo Society, 8(1)
Liese W (1992) The structure of bamboo in relation to its properties and utilization. In Zhu S, Li W, Zhang X, Wang Z ed., Bamboo and its use. Proceedings of the International symposium on Industrial Use of Bamboo, Beijing, China (pp. 7-11)
Limera C, Sabbadini S, Sweet JB, Mezzetti B (2017) New biotechnological tools for the genetic improvement of major woody fruit species. Front Plant Sci 8:1418. https://doi.org/10.3389/fpls201701418
Lin CS, Chang WC (1998) Micropropagation of *Bambusa edulis* through nodal explants of field-grown culms and flowering of regenerated plantlets. Plant Cell Rep 17:617–620
Lin CC, Lin CS, Chang WC (2003a) In vitro flowering of *Bambusa edulis* and subsequent plantlet survival. Plant Cell Tissue Organ Cult 72:71–78
Lin CS, Chen CT, Lin CC et al (2003b) A method for inflorescence proliferation. Plant Cell Rep 21:838–843

Lin CS, Lin CC, Chang WC (2004) Effect of thidiazuron on vegetative tissue-derived somatic embryogenesis and flowering of bamboo *Bambusa edulis*. Plant Cell Tissue Organ Cult 76:75–82

Lin CS, Liang CJ, Hsaio HW et al (2007) In vitro flowering of green and albino Dendrocalamus latiflorus. New For 34:177–186

Lin SY, Shi WW, Miu BB et al (2010a) Research advances in reproduction biology of bamboos. World Bamboo Rattan 8(2):1–6

Lin XC, Yuan XL, Lin R et al (2010b) Studies on floral biology of *Phyllostachys violascens*. J For Environ 30(4):333–337

Li X, Hu S, Cao Y, Lu X, Ren P, Wu, X, Zhou M (2012) Agrobacterium-mediated genetic transformation of 4CL gene from Neosinocalamus affinis into Dendrocalamus farinosus. Sci. Silvae Sin. 48:38–44

Loh JP, Kiew R, Set O et al (2000) A study of genetic variation and relationships within the bamboo subtribe Bambusinae using amplified fragment length polymorphism. Ann Bot 85(5):607–612

Lombardo E (2022) An overview of bamboo cultivation in Southern Italy. Advances in Bamboo Science, 1:100002

Ma NX, Lai GH, Zhang PX, Zhang HL (2014) The genus Fargesia in China. Zhejiang Science and Technology Press, China Hangzhou (in Chinese)

Makarevitch I, Waters AJ, West PT et al (2015) Transposable elements contribute to activation of maize genes in response to abiotic stress. PLoS Genet 11(1):e1004915

Matsuo A, Tomimatsu H, Suzuki JI, Saitoh T, Shibata S, Makita A, Suyama Y (2014) Female and male fitness consequences of clonal growth in a dwarf bamboo population with a high degree of clonal intermingling. Ann Bot 114:1035–1041. https://doi.org/10.1093/aob/mcu176

Mba C (2013) Induced mutations unleash the potentials of plant genetic resources for food and agriculture. Agronomy 3(1):200–231

McClure FA (1993) The Bamboos. Smithsonian Institution Press, Washington and London, 368 pp

McClure FA, Soderstrom TR (1973) Genera of bamboos native to the New World (Gramineae: Bambusoideae). Washington, Smithsonian Institution Press

McClintock B (1984) The significance of responses of the genome to challenge. Science, 226(4676):792-801

Midmore DJ (2009) Overview of the ACIAR bamboo project outcomes. Silvicultural management of bamboo in the Philippines and Australia for shoots and timber, 7

Mo JM, Wang SF, Lan RC, Huang YG, Zhou GL (2005) Pulping property analysis of crossbred bamboos *Bamdusa pervariadilis* x grandis. Paper Sci Technol (in Chinese) 24:28–31

Nadgauda RS, Parasharami VA, Mascarenhas AF (1990) Precocious flowering and seeding behaviour in tissue-cultured bamboos. Nature 344:335–336

Naithani HB (2011) Bamboos of Nagaland. Published by NEPED & NBDA, pp 1–206

Nayak S, Rout GR, Das P (2003) Evaluation of the genetic variability in bamboo using RAPD markers. Plant Soil Environ 49(1):24–28

Ning CQ, Dai QH (1995) Study on hybrid breeding of *Bambusa pervariabilis* × *Dendrocalamopsis daii*. Guangxi For Sci 24(4):167–168

Nirmala C, Bisht MS, Bajwa HK, Santosh O (2018) Bamboo: A rich source of natural antioxidants and its applications in the food and pharmaceutical industry. Trends in Food Science and Technology, 77: 91–99

Pal A, Ghosh JS, Das M, Bhattacharya S (2009) Identification of superior fiber-trait-yielding genetic resources of *Bambusa balcooa*: analysis of physicochemical properties of fibers. Volume 9 Material Properties of Bamboo 2

Pan C, Zhou G, Shrestha AK, Chen J, Kozak R, Li N, Wang, G (2023) Bamboo as a nature-based solution (NbS) for climate change mitigation: biomass, products, and carbon credits. Climate, 11(9):175

Pande VC, Kurothe RS, Rao BK et al (2012) Economic analysis of bamboo plantation in three major ravine systems of India. Agric Econ Res Rev 25(1):49–59

Paterson AH, Bowers JE, Bruggmann R et al (2009) The Sorghum bicolor genome and the diversification of grasses. Nature 457(7229):551–556

Peng Z, Lu Y, Li L et al (2013) The draft genome of the fast-growing 701 non-timber forest species moso bamboo (*Phyllostachys heterocycla*). Nat Genet 45(4):456–461
Prakash C (2020) Bamboo fibre. In: Handbook of natural fibres. Woodhead Publishing, pp 219–229
Qiao G, Yang H, Zhang L et al (2014) Enhanced cold stress tolerance of transgenic *Dendrocalamus latiflorus* Munro (Ma bamboo) plants expressing a bacterial CodA gene. In Vitro Cell Dev Biol Plant 50:385–391
Ramakrishnan M, Yrjälä K, Vinod KK et al (2020) Genetics and genomics of moso bamboo (*Phyllostachys edulis*): Current status, future challenges, and biotechnological opportunities toward a sustainable bamboo industry. Food Energy Secur 9(4):e229
Ramanayake SMSD, Wanniarachchi WAVR, Tennakoon TMA (2001) Axillary shoot proliferation and in vitro flowering in an adult giant bamboo, Dendrocalamus giganteus Wall. Ex Munro. In Vitro Cell Dev Biol Plant 37:667–671
Rane AD, Viswanath S, Sheshshayee MS, Sawardekar SV (2019) Population structure of *Dendrocalamus stocksii* along its geographical distribution. J Bamboo Rattan 18(3):44–54
Rao IU, Raghav R (2002) Control of in vitro flowering in *Bambusa bambos* from multiple shoots and somatic embryos. In: Bamboo for sustainable development. Brill, pp 367–372
Rao IV, Yusoff AM, Rao AN, Sastry C (1989) Propagation of bamboo and rattan through tissue culture. IDRC Bamboo and Rattan Research Network, Tanglin, SG. http://hdl.handle.net/10625/2356
Rendle AB (1904) Index florae sinensis: gramineae. J Linn Soc London 36:377–456
Santos R, Algar A, Field R, Mayes S (2017) Integrating GIScience and Crop Sci. datasets: a study involving genetic, geographic and environmental data (No. e2248v4). PeerJ Preprints
Schnable PS, Ware D, Fulton RS et al (2009) The B73 maize genome: complexity, diversity, and dynamics. Science 326(5956):1112–1115
Shen Z, Zhang YH, Zhang L et al (2020) Changes in the distribution of endogenous hormones in *Phyllostachys edulis* 'Pachyloen' during bamboo shooting. PLoS One 15:e0241806. https://doi.org/10.1371/journal.pone.0241806
Singhal, Poonam, Lalit M. Bal, Santosh Satya, Sudhakar, P, Naik SN (2013) Bamboo shoots: A novel source of nutrition and medicine. Critical Reviews in Food Science and Nutrition, 53: 517–534.
Tamolang FN, Lopez FR, Semana JA et al (1980) Properties and utilization of Philippine erect bamboos. In: Bamboo research in Asia: proceedings of a workshop held in Singapore, May 28–30, 1980. IDRC, Ottawa
Sood P, Bhattacharya A, Joshi R et al (2014) A method to overcome the waxy surface, cell wall thickening and polyphenol induced necrosis at wound sites the major deterrents to Agrobacterium-mediated transformation of bamboo, a woody monocot. J Plant Biochem Biotechnol 23:69–80
Sun H, Wang S, Zhu C et al (2023) A new biotechnology for in-planta gene editing and its application in promoting flavonoid biosynthesis in bamboo leaves. Plant Methods 19:20
Triplett JK, Clark LG, Fisher AE, Wen J (2014) Independent allopolyploidization events preceded speciation in the temperate and tropical woody bamboos. New Phytol, 204:66–73
Varmah JC, Bahadur KN (1980) Country report and status of research on bamboos in IndiaIndia. Indian Forest Rec (New Series) Bot 6:1–28
Walsh B (2001) Quantitative genetics in the age of genomics. Theor Popul Biol 59(3):175–184
Wang YX, Zhang GC, Li XW (2005) An evaluation on shoot quality of sympodial bamboo species and their hybrids. J Bamboo Res (in Chinese) 24:39–44
Wang X, Gao J, Yang P, Zhou L (2008) Cytological effects induced by $^{60}Co\gamma$ rays-irradiation on moso seed. J Nucl Agric Sci 22(6):770–775
Wang Y, Jiang H, Zheng RH et al (2020) Analysis on characteristics and pulping performance of Dendrocalamus mutates. J Sichuan For Sci Technol (in Chinese) 41:64–68
Williams N (2007) A billion seeds and counting. Current Biology, 17(11), R387-R388
Watson A, Ghosh S, Williams MJ et al (2018) Speed breeding is a powerful tool to accelerate crop research and breeding. Nat Plants 4(1):23–29
Wong KM, Low YW (2011) Hybrid zone characteristics of the intergeneric hybrid bamboo× Gigantocalamus malpenensis (Poaceae: Bambusoideae) in Peninsular Malaysia. Gardens' Bulletin Singapore, 63(1), 375-383

Xiang M, Ding W, Wu C, Wang W, Ye S, Cai C, Zhu, Q (2021) Production of purple Ma bamboo (Dendrocalamus latiflorus Munro) with enhanced drought and cold stress tolerance by engineering anthocyanin biosynthesis. Planta, 254(3):50

Xie N, Chen LN, Dong YR, Yang HQ (2019) Mixed mating system and variable mating patterns in tropical woody bamboos. BMC Plant Biol. 19:418

Xu PF (2019) Hybridization between *Phyllostachys edulis* and *P. violascens*, and identification and early growth characteristics of the hybrids. MD Dissertation, Zhejiang Agriculture and Forestry University, Hangzhou

Yang B, Thorogood D, Armstead I, Barth S (2008) How far are we from unravelling self-incompatibility in grasses? New Phytol 179:740–753. https://doi.org/10.1111/j.1469-8137.2008.02421.x

Ye S, Chen G, Kohnen MV et al (2020) Robust CRISPR/Cas9 mediated genome editing and its application in manipulating plant height in the first generation of hexaploid Ma bamboo (*Dendrocalamus latiflorus* Munro). Plant Biotechnol J 18(7):1501–1503

Yeasmin L, Ali N (2021) Molecular markers in bamboo genotyping: prospects for conservation and breeding. In: Biotechnological advances in Bamboo: the "Green Gold" on the Earth, pp 379–399

Yi TP (2008) Iconographia Bambusoidearum Sinicarum. Science Press, BeijingYe, S., Cai, C., Ren, H., Wang, W., Xiang, M., Tang, X., Zhu, C., Yin, T., Zhang, L. & Zhu, Q. (2017). An efficient plant regeneration and transformation system of ma bamboo (Dendrocalamus latiflorus Munro) started from young shoot as explant. Frontiers in Plant Science, 8, 1298.

Yuan JL, Yue JJ, Gu XP, Lin CS (2017) Flowering of woody bamboo in tissue culture systems. Front Plant Sci 8:1589

Yuan J, Yue J, Gu X (2018) A method for genetic transformation of bamboo. Chinese patent CN108660151A

Zeng Q, Yu L, Cheng P et al (2018) A study of culm color characteristic and type classification of *Phyllostachys edulis* cv. Tao Kiang. World Bamboo Rattan (in Chinese) 16:32–35

Zamora AB (1994) Review of micropropagation research on bamboos. Constrains to Production of Bamboo and Rattan, INBAR Technical Report, 5:45-100

Zhang GC, Chen FS (1980) Excellent sexual hybrids of bamboo: Chengmaqing no.1. Sci Sil Sin (in Chinese) 16:124–126

Zhang GC, Chen FS (1986) Study on bamboo hybrid breeding. Guangdong For Sci Technol (in Chinese) 3:1–5

Zhang GC, Wang YX (2001) Preliminary study on flowering of tube bamboo seedling. J Bamboo Res (in Chinese) 20:1–4

Zhang K, Lan Y, Zhang S, Wang L, Wu M, Xiang Y (2024) A PLATZ transcription factor PhePLATZ8 from Moso bamboo (Phyllostachys edulis) plays a positive role in regulating growth and abiotic stress tolerance. Industrial Crops and Products, 221:119334

Zhao H, Gao Z, Wang L et al (2018) Chromosome-level reference genome and alternative splicing atlas of moso bamboo (*Phyllostachys edulis*). Gigascience 7(10):giy115

Zhao H, Sun S, Ding Y (2021) Analysis of 427 genomes reveals Moso bamboo population structure and genetic basis of property traits. Nat Commun 12:5466

Zheng Y, Yang D, Rong J et al (2022) Allele-aware chromosome-scale assembly of the allopolyploid genome of hexaploid Ma Bamboo (*Dendrocalamus latiflorus* Munro). J Integr Plant Biol 64(3):649–670

Zhou M, Liang L, Hänninen H (2018) A transposition-active Phyllostachys edulis long terminal repeat (LTR) retrotransposon. J Plant Res 131:203–210.

Zhou G, Meng C, Jiang P, Xu Q (2011) Review of carbon fixation in bamboo forests in China. Bot Rev 77:262–270

Zhao J, Wan J, Liu C, Ren Z, Wang Z, Guan S, Zhang S, Zhao H (2022) The study on in situ asymmetric bending behavior of structural bamboo materials under combined tension-bending loads. J Mater Sci, 57(36):17304-17324

Zhong YB, Yue JJ, Lou C, Yuan JL, Gu XP (2017) Floral organ and breeding system of Dendrocalamus latiflorus. Sci Silvae Sin, 53:1–10

Chapter 2
Barley (*Hordeum vulgare* L.) Fiber Breeding

Essam Fathy El-Hashash, Salem Mesfir Al-Qahtani, and Karima Mohamed El-Absy

Abstract The challenges of world population rise include providing food, reducing the depletion of natural resources, developing the agricultural food industry, preserving biodiversity, and using alternative sources. Since lignocellulosic agro-waste is the most prevalent organic compound in the world, it has enormous potential for sustainable application. Because of its abundance, renewability, and biodegradability, it is an important and sought-after source for the creation of materials in many different industries. Barley (*Hordeum vulgare* L.) may be grown in many different climates and geographical locations, and it is currently farmed in all temperate climatic regions. Barley has the potential to be the perfect crop for making nutritious food, feed, and straw fiber. Barley straw falls into this group and is a valuable crop because of its availability and abundance. Barley straws can be used as a renewable supply of lignocellulosic fibers, which is a more sustainable way to reduce excessive levels of agro-residue. There is a high cellulose, lignin, and hemicellulose content in this waste. Barley cellulose fibers possess enormous potential as a useful raw material for a range of applications, including substitutes for wood fibers, the manufacture of textiles, paper industries (pulp, paper, wipes, fiber-based packages), bioplastics, and thermal-electric power, clay-fiber composites in construction, growth medium, insulation material, enzymatically decomposed intermediate products for targeted applications. Notably, there is a dearth of genetic research on barley fiber.

E. F. El-Hashash (✉)
Department of Agronomy, Faculty of Agriculture, Al-Azhar University, Cairo, Egypt
e-mail: dressamelhashash@azhar.edu.eg

S. M. Al-Qahtani
Biology Department, University College of Tayma, University of Tabuk, Tabuk, Saudi Arabia
e-mail: salghtani@ut.edu.sa

K. M. El-Absy
Biology Department, University College of Tayma, University of Tabuk, Tabuk, Saudi Arabia

Eco-physiology Unit, Plant Ecology and Ranges Department, Desert Research Center, Cairo, Egypt
e-mail: k.alabssi@ut.edu.sa

K. F. M. Salem et al. (eds.), *Breeding and Biotechnology of Grass and Bast Fiber Crops*, Advances in Plant Breeding Strategies 12,
https://doi.org/10.1007/978-3-032-00403-1_2

The application of biotechnology, like transgenics and molecular technologies, to various activities can benefit conventional barley fiber breeding. Therefore, this chapter delves into the world of barley fibers through its importance and uses, barley straws, barley fibers and industry, current cultivation practices and challenges, germplasm biodiversity and conservation, conventional breeding, molecular breeding, and biotechnology methods.

Keywords Barley (*Hordeum vulgare* L.) · Barley straw · Cellulose fibers · Importance · Traditional breeding · Biotechnology methods · Mutation · Hybridization

2.1 Introduction

Cereal crops are annual herbaceous plants that are cultivated for their edible seeds. They are members of the Triticeae tribe within the Poaceae (Gramineae) family of grasses. According to El-Hashash and El-Absy (2019), the ancient Roman goddess of harvest, Ceres, is the source of the word cereal. Straw has long been a valuable byproduct of cereal crops, mostly used in the past for animal bedding and feed. It is well known that cereal straws make up a sizeable portion of waste lignocellulosic biomass and that they contain roughly 30–50% cellulose, making their potential use as a source of cellulose encouraging (Borrega et al. 2022; Gunjan et al. 2023). Barley (*Hordeum vulgare* L.) is a highly significant grain on a global scale nowadays, ranking fourth globally in terms of production volume and acreage used for cereal crops, after maize, wheat, and rice (El-Hashash and El-Absy 2019). In November 2023/2024, the area, yield, and production of barley globally were 45.66 million hectares, 3.11 metric tons per hectare, and 141.80 million metric tons, respectively (USDA 2024). Because of this, a lot of agricultural wastes, like the husk that shields the grain and straw, are produced by barley and other cereal crops (Román-Gutiérrez et al. 2022). Comparing barley harvested area, yield, and production globally to other cereal crops and the leading countries in barley production in May 2024/2025 are presented in Tables 2.1 and 2.2.

Barley straw is an important raw material used in agriculture as feed for ruminants and in the production of cellulose as an energy source (Wisniewska et al. 2003). While the lignocellulose in straw can be improved for use as animal feed, a starting material for novel compounds, or bioconversion into bioethanol, barley grains offer a rich source of chemicals that can be investigated to produce various products to assist human health (Meng et al. 2023). Based on the variety of barley utilized and climate circumstances that affect plant growth, the amount of extracted barley fiber varies, with fiber yields varying from 14.82% to 19.59% (Kovačević et al. 2023).

Novel sources of cellulose fibers, which are currently extensively utilized by the forest industry, are necessary for the shift to sustainable, biodegradable, and recyclable materials. According to Hyvärinen and Kärki (2015), cellulose-based fibers,

Table 2.1 Barley harvested area, yield, and production worldwide in comparison to other cereal crops (November 2024/2025)

Crop	Area harvested (million ha)	Yield level (mt/ha)	Production (million mt)	Change in production from last year (%)
Wheat	222.22	3.58	794.73	0.55
Corn	203.07	6.00	1219.40	−0.79
Rice	168.79	4.72	533.81	2.23
Barley	45.66	3.11	141.80	−0.56
Sorghum	40.77	1.52	61.83	5.90
Oats	8.61	2.55	21.95	13.44
Rye	3.41	3.20	10.93	−6.66

Source: USDA (2024)

Table 2.2 Top countries for Barley production

Country	Area harvested (million ha)	Yield level (mt/ha)	Production (million mt)	Change in production from last year (%)
European Union	10.35	4.90	50.75	6.03
Russia	6.70	2.46	16.50	−19.51
Australia	4.45	2.74	12.20	12.96
Canada	2.32	3.23	7.50	−15.78
United Kingdom	1.20	6.00	7.20	3.40
Turkey	3.75	1.87	7.00	−12.50
Argentina	1.25	3.92	4.90	−3.92
Ukraine	1.50	3.60	5.40	−14.96
Iran	1.60	1.94	3.10	3.33
Kazakhstan	2.25	1.60	3.60	37.72

Source: USDA (2024)

which are increasingly viewed as low-cost viable substitutes, can also be found in agro-based materials like straw and stalks. Barley straw cellulose fibers have a variety of potential uses, including the paper and composites industries, wipes, pulp and bioplastics production, fiber-based packaging, anti-algae products, nonwoven mulches, growth medium, clay-fiber composites in construction, insulation material, textile fibers, enzymatically decomposed intermediate products for specific applications, and as a source of thermal-electric power (Borrega et al. 2022; Draga and Gąbka 2024; Fortunati et al. 2016; Koutous and Hilali 2021; Kovačević et al. 2023; Puglia et al. 2020; Oliver-Ortega et al. 2023). Since the fibers can be altered or mixed with other materials to successfully improve the end product's qualities, barley cellulose fibers that are separated from barley straw are frequently utilized in the composite and paper industries (Geng et al. 2022; Kovačević et al. 2023).

With the potential to boost output by roughly 50%, plant breeding has been a significant scientific strategy for enhancing crop productivity (Fehr 1984). The science of plant breeding has not diminished, as evidenced by the decades-long success in creating and introducing new cultivars to the world market (Gepts and Hancock 2006). Understanding the diversity, type, and size, heritability estimations, and predicted genetic advancement concerning available breeding materials is essential to starting an appropriate breeding program (Chand et al. 2008). A versatile crop, barley can thrive in a variety of environments, from the dry climate of the Middle East to the extreme cold of the high Andes. When compared to most other minor grain crops, like wheat, it is more productive and its yields fluctuate less. As a result, farmers with few resources and unfavorable soil and climate conditions frequently employ it (Newton et al. 2011). Barley is often resistant to adverse growing conditions, including drought, salt, alkalinity, and a variety of topographies, including plains, hills, and rainfed and irrigated areas. When growing wheat is not feasible, this is the farmers' preferred crop (El-Hashash and El-Absy 2019).

2.1.1 Botanical Classification and Distribution

In crop and agricultural sciences, it is crucial to have a standard crop classification system. The simplicity of plant breeding, research, gathering, and specialized development activities comes before plant classification. It is easier to communicate, disseminate, and retrieve scientific knowledge when plant names are standardized (El-Hashash and El-Absy 2019). More than 18,000 species of monocotyledons, blooming plants with a single seed leaf, are found in this order (Reid 1985). Important cereal crops belong to the Poaceae family's Triticeae tribe (Löve 1984). The world's main cereal crops are wheat, corn, rice, barley, rye, oats, sorghum, millets, and triticale. The barley crop (*Hordeum vulgare* L.) belongs to the Hordeum genus, which is a member of the Triticeae tribe within the Poaceae family, which is the most significant plant order in terms of economic importance. In Species Plantarum, Linnaeus published the first botanical description of barley in 1753 (von Bothmer and Jacobsen 1985).

The genus Hordeum has 32 species, or 45 taxa, according to Bothmer et al. (2003). According to Blattner et al. (2010), there are roughly 33 species in the genus Hordeum, with 23 of those found in North and South America and just 10 in other parts of the world. According to Jacobsen and von Bothmer (1992), there have been a variety of opinions expressed regarding the number and affinity of taxa within the genus Hordeum via taxonomic research conducted over the years. Barley was divided into four species by Harlan (1918), primarily based on the fertility of the lateral spikelets: (1) *H. vulgare* L. and *H. intermedium* Keke have all fertile spikelets (6-row barley), while (2) only the middle spikelets are fertile (2-row barley): *H. defidens* and *H. distichon* L. Steud. Nevski (1941) classified the 28 species of the genus Hordeum into 6 sections: Critesion, Stenostachys, Anisolepis, Hordeastrum, Bulbohordeum, and Crithe. Of these, 3 are now recognized as *H. vulgare* L. Dewey (1984) and Löve (1984) separated barley into two groups: all other species were classified in the genus Critesion, whereas *Hordeum s.str.* Consisted of either *H.*

vulgare or *H. vulgare* and *H. bulbosum*. The four parts of the genus Hordeum are Hordeum, Anisolepis, Stenostachys, and Critesion, according to von Bothmer and Jacobsen (1985). Petersen and Seberg (2003) used phylogenetic studies of single-copy nuclear and chloroplast DNA data to demarcate four distinct sections from von Bothmer et al. (1995). Blattner (2009) also used a variety of loci from the nuclear genomes, distribution regions, and chloroplasts to propose a tighter structure of infrageneric groupings.

There are various cultivars of barley, and there are numerous classification methods for them (El-Hashash and El-Absy 2019). The most well-known of these approaches is dividing barley based on how many rows of grains are visible while observing the stalk heads from above: (1) Two-row barley, as seen in Fig. 2.1, where two rows of barley kernels are present at the head of the stalk. Because the central floret is fertile and the two lateral florets are sterile, this variation yields one seed at each node, creating the appearance of a flat head (Kling 2004). From an overhead perspective, the spike appears to have two rows of kernels (Kumlehn and Stein 2014). There are 15–30 kernels per stalk. Two-row barley appears to perform best overall, with the greatest thousand kernel weight and starch content and slightly lower protein content (Bowman et al. 2011). Wild barley is a type of two-row barley. (2) Six-row barley, so named because there are six rows of barley kernels at the top of the stalk (Fig. 2.1). All florets are fertile. The lateral seeds typically have a minor asymmetry, whereas the core seeds are spherical and plump. Six rows of kernels occur when looking at the spike from the top (Kumlehn and Stein 2014). There are 25–60 kernels on each stalk. Up to 80 seeds can be produced by a single head of barley (Kling 2004). Many sources only differentiate between cultivars with two or six rows since four-row barley is a broad six-row barley. Six-row barley makes up the majority of barley cultivated (El-Hashash and El-Absy 2019).

How barley is used has not altered significantly; it is still utilized for food, feed, and the production of alcoholic drinks (Robinson 2007). Barley occupies a wider land area than practically any other grain crop. It is grown at the highest elevations in the Andes and Himalayas. Additionally, it is close to the Arctic Circle in northern Asia, Europe, and North America, as well as the deserts of Africa, the Middle East, and China (Blattner et al. 2010). In temperate regions of the world, the genus

Fig. 2.1 Barley (*Hordeum vulgare* L.) classification based on how many rows of grains are visible from above. (Source: El-Hashash and El-Absy 2019)

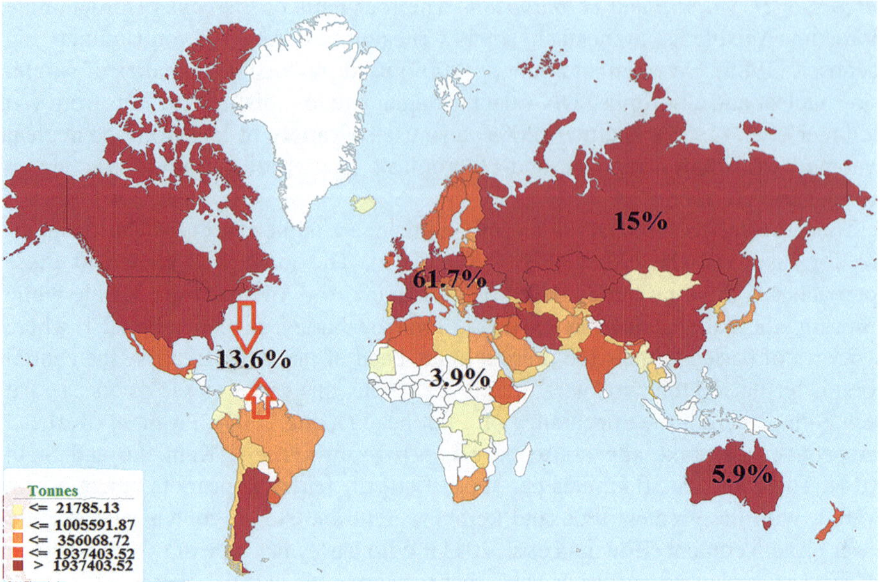

Fig. 2.2 Map of barley production and distribution worldwide: average metric tons, 1994–2022. (Source: https://www.fao.org/faostat/en/#data/QCL/visualize. Accessed February 15, 2024)

Hordeum is extensively spread in diversity centers, which are areas with the highest concentration of wild species. These regions include Central Asia, Southwest Asia, western North America, and southern South America (von Bothmer 1996). In Fig. 2.2, Asia (15%) is the second-most important continent in the world for barley growing after Europe (61.7%), followed by the Americas (13.6%), Oceania (5.9%) and Africa (3.9%).

2.1.2 *Importance and Uses*

One kind of edible grain that is used in bread, cereal, soups, and beers is barley. According to Yawson et al. (2017), barley makes up the majority of coarse grains utilized worldwide for animal feed. Beer, one of the most consumed alcoholic beverages worldwide, is frequently made from barley. A little over 70% of the barley crop has been utilized as fodder, 21% has gone toward the brewing, distilling, and malting businesses, and 6% has been eaten by humans (Tricase et al. 2018). It gives domestic animals energy and increases their intake of dietary fiber, making it a good choice as an element in feed. Compared to other crops like corn, barley grain has greater protein and crude fiber contents, and serves as a suitable supply of starch (Kumlehn and Stein 2014). Malt is the second-most significant application of barley. Thirty percent of the barley produced worldwide is utilized for malting (Akar et al. 2004). Cultivars with two or six rows are utilized for malting. Nonetheless, the majority of the globe prefers two-row barley (Kling 2004). Generally speaking, malting barley has less protein (<11%) than feed barley (>12%) (Kumlehn and Stein 2014). Approximately 90% of malted barley is used to make beer, with the remaining 10% being used as food ingredients (Akar et al. 2004).

Barley serves as the primary cereal grain crop utilized for human nourishment in many parts of the world, which brings us to our third significant use. One of barley's health benefits is that it contains β-glucan, which has been demonstrated to lower liver cholesterol levels (EFSA 2011). Because of its low glycemic index and high fiber content, it is a good choice for diabetics (Brennan and Cleary 2005). Additionally, it encourages the liver's production of fatty acids. It has also been observed that tocols (tocopherols, tocotrienols) reduce low-density lipoprotein cholesterol and total cholesterol (Wang et al. 1993). Additionally, barley is utilized in agriculture, such as animal bedding, and in industrial sectors, such as the production of paper, fiberboard, glucose, and maltose syrups (Akar et al. 2004; Zhang and Li 2009). Barley straw cellulose fibers can be used in many different ways (Borrega et al. 2022; Draga and Gąbka 2024; Fortunati et al. 2016; Koutous and Hilali 2021; Kovačević et al. 2023; Puglia et al. 2020; Oliver-Ortega et al. 2023). These uses include the paper and composites industries, wipes, pulp and bioplastics production, growth medium, fiber-based packaging, anti-algae products, insulation material, textile fibers, enzymatically decomposed intermediate products for specific applications, and as a source of thermal-electric power. To meet the anticipated demand for food, feed, and industrial uses by 2050, global barley production will need to rise by 54% from its level in 2000 (Yawson et al. 2017).

2.1.3 Barley Straws

Utilizing a byproduct from any industrial or agri-food process to produce goods with added value is one of the goals of the circular economy (Anastas and Warner 2000). Zohary and Hopf (1973) classify all parts of the barley plant as agro-residues, including the hull, grain derivatives, and straw. Straws from cereal crops are sometimes left in the fields to burn or decompose into fertilizer for the next harvest (Espinach et al. 2013). While there is no denying the advantages of these methods, they also emit carbon dioxide, are difficult to use with long cereal straws, and are only beneficial for stubble. Large volumes of barley straw are typically burned after harvest and left on the fields. This type of dumping harms the environment and could jeopardize ecosystems, particularly if treatment is not received (Awogbemi and Von Kallon 2022; Bala et al. 2023). The quantity of pollution in the globe today makes innovative solutions more important than ever, and greener alternatives are being used in a variety of applications (Giraldo et al. 2019). One such solution is biowaste made from barley straw, which is considered a second-generation biomass. Biowaste has emerged as a significant raw material in recent times. Barley straw is valuable because of its biowaste, which can help a crop increase even more

(Giraldo et al. 2019). This type of biowaste disposal requires a more environmentally friendly technique (Bala et al. 2023). Agro-based materials like straws and stalks can also produce cellulose-based fibers, despite having a lower cellulose content than wood. These fibers are becoming more and more popular as potential low-cost alternatives to wood fibers. Greater cellulose content, or a higher degree of cellulose polymerization and a lower microfibrillar angle, results in fibers with superior mechanical qualities. However, rather than the cellulosic content, the final usage of the fibers will depend on their quality. The amount of ash in wood fibers and straw differs significantly. Silica, which is included in ash, has several negative impacts, including blunting cutting equipment and complicating combustion (Hyvärinen and Kärki 2015).

Barley straw has historically been utilized for millennia in a variety of ways, including the creation of power and energy, building, bedding for animals, insulation, and soil inclusion (Borrega et al. 2022). In certain applications, such as the creation of composite materials and energy, the usage of alternative fiber sources is replacing wood feedstock due to increased awareness of deforestation, forest degradation, and its effects on biodiversity and global warming. Due to its availability and quantity, barley straw is a significant agricultural crop in this class (Serra-Parareda et al. 2020). Successful biofuel generation can be achieved from barley straw (Meyer et al. 2022). However, the utilization of barley straw as a source of biofuel can only occur if a treatment plant is sufficiently close, where transportation expenses (Serra-Parareda et al. 2020). Mulching barley straw has been shown to work well on soils impacted by crops and fires. Reduced soil erodibility and surface runoff are two benefits of straw mulch that were realized right away after the straw was applied (Prosdocimi et al. 2016).

Additionally, the relevance of barley straw biobased composites for semi-structural and engineering applications was reported by Serra-Parareda et al. (2020). Because barley straws have strong insulation qualities, they are utilized as a construction material in concrete and wall plaster (Hyvärinen and Kärki 2015). The analysis of sand reinforced with barley straw revealed that the fibers improved the stiffness, yield strain, and shear strength of the sand. Additionally, adding fiber raised the peak strength of the soil, where adding fiber at a weight ratio of 0.9% may raise the peak strength of the soil by as much as 60% (Vafaei et al. 2024). Consequently, the primary applications for barley straws are elastomers or concrete fillers (Masłowski et al. 2017). In composites, barley straw provides a substantial, affordable substitute for wood fibers in the form of polypropylene reinforcement. As a result, they work well as fillers in goods where high moisture content and mechanical attributes are not crucial (Hyvärinen and Kärki 2015). Barley straws have been suggested by other researchers as a way to manage algae (Draga and Gąbka 2024; Spencer and Lembi 2007). For many years, barley straw has been utilized throughout the world as an easy, affordable, and safe way to suppress the growth of phytoplankton (Draga and Gąbka 2024). In a decommissioned water supply reservoir, the presence of decaying barley straw at a rate of about 50 g m^{-3} seemed to dramatically lower the overall phytoplankton productivity as well as the cyanobacterial dominance (Everall and Lees 1996). Additionally, barley straw can

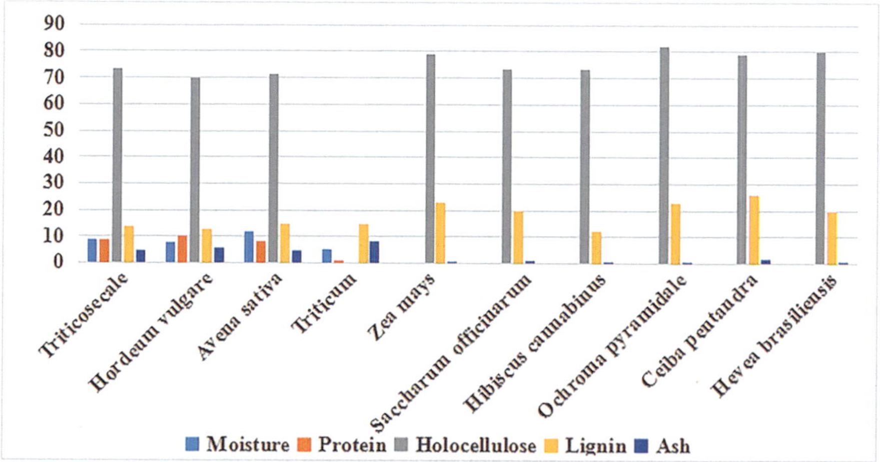

Fig. 2.3 Percentage of the chemical composition of barley straw compared with other crop straws based on dry weight. (Prepared by: E.F. El-Hashash)

reduce an agricultural residue that has no added value by serving as an additional raw material for hardwoods and softwoods in the manufacturing of cellulose, board, and paper (Vargas et al. 2015). Generally, barley straw is seen to have the potential as a raw material to make textile fibers, garden mulches, strawboards, waxes, and biochemicals (Borrega et al. 2022). Figure 2.3 displays the chemical composition percentage of barley straw compared with the other crop straws on a dry basis, according to Román-Gutiérrez et al. (2022).

2.1.4 *Barley Fibers and Industry*

Ecosystems are out of balance as a result of synthetic fibers, which account for 60% of world fiber production and are primarily made of polyester and polyamide (Kovačević et al. 2023). Agro-waste is another significant problem (Gunjan et al. 2023). Researchers are looking for new sources of textile fibers and materials that are sustainable, renewable, and biodegradable. Because of this, using natural fiber sources has become much more common (Kovačević et al. 2023). The benefits of natural fibers are their ease of handling, availability, and biodegradability. Although they differ based on the source, species, location, and many other factors, natural fibers exhibit exceptional mechanical and physical properties (Nathan 2023). Among these resources, natural fibers have distinguished themselves as a cost-effective, healthful, and ecologically sustainable choice (Kovačević et al. 2023). New sources of cellulose fibers, which are currently utilized in huge quantities by the forest industries, are necessary for the shift to recyclable, biodegradable, and sustainable materials (Borrega et al. 2022).

Cereal or barley straws can be used as a renewable supply of lignocellulosic fibers, which is a more sustainable way to reduce excessive levels of agro-residue. There is a high cellulose, lignin, and hemicellulose content in this waste. One of the main components present in plant cell walls is cellulose, a biopolymer that is biocompatible, renewable, and non-toxic. It has many hydroxyl functions. It makes use of technology for separation. The majority of plants have cellulose in a composite form with hemicellulose and lignin. Lignocellulose functions as an adhesive to hold hemicellulose and lignin together. Hemicellulose is a material that has cellulose and lignin in it. Cellulose is considered a crucial structural element of a plant and one of the most important natural resources (Gunjan et al. 2023).

The most common approach for extracting fibers from barley straw is chemical retting, which is followed by physical and mechanical procedures. Extreme temperature, pretreatment time, chemical, and other variables are used during these operations to isolate the relatively short fibers (Kovačević et al. 2023). Compared to other sources of natural fibers, barley straw is high in holocellulose and has a comparatively low lignin content (Serra-Parareda et al. 2020). Cellulose makes up the majority of barley fibers, with hemicelluloses, lignin, and pectins making up smaller amounts (Fortunati et al. 2016). The chemical properties of barley fibers are cellulose (30–31%), hemicellulose (9–15%), and lignin (4–8%) (Kovačević et al. 2023), compared with the chemical properties of cotton fibers, which were 82–96, 2–6.4, and 0–5% (Ansell and Mwaikambo 2009), respectively. Fortunati et al. (2016) report that the primary elements composed of barley straw fiber of raw materials are cellulose (56.2%), lignin (9.2%), hemicellulose (7.0%), and other compositions (lipids, pectins, proteins, organic acids, and sugars; 7.6%). Serra-Parareda et al. (2020) reported the physical and chemical makeup of barley thermomechanical fibers and barley straw, as shown in Fig. 2.4.

A horizontal incision in barley straw reveals fibers that tend to be uneven in the cuts. Fibers exhibit the opposite of a uniform cut as a result. However, because of its round or oval shape, it is evident that the straw contains an agglomeration of cellulose fibers. The cross sections of the various straw fibers show an external structure

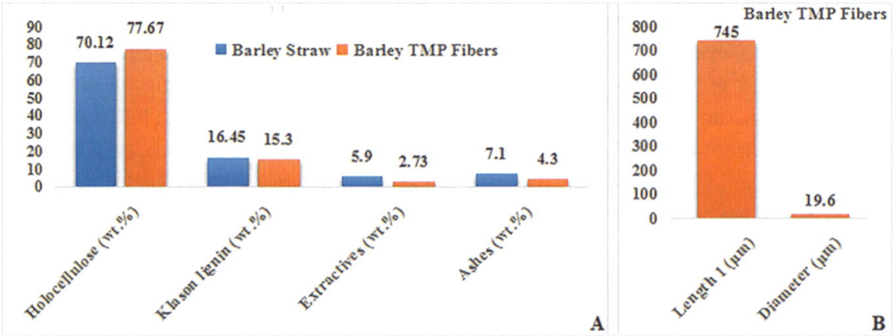

Fig. 2.4 Chemical (**a**) and morphological (**b**) composition of barley straw and barley thermomechanical (TMP) fibers, according to Serra-Parareda et al. (2020)

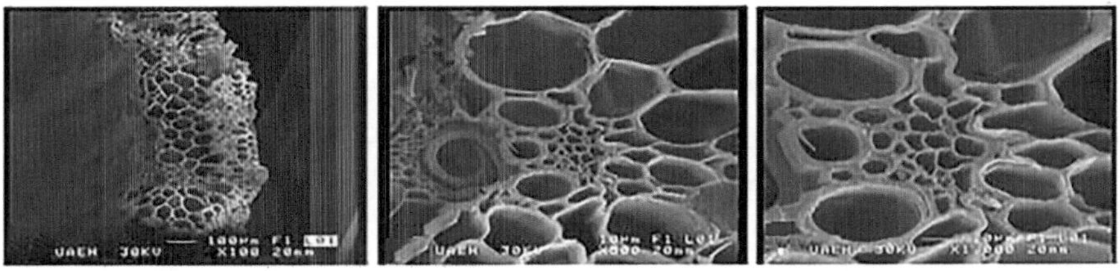

Fig. 2.5 Photomicrographs of straw barley. (Source: Román-Gutiérrez et al. 2022)

that is quite dense and varies in thickness, followed by a highly porous structure. Straw barley photomicrographs are shown in Fig. 2.5. According to barley fibers, moisture recapture ranged from 10.37% to 11.01%, which is comparable to traditional bast fibers. Furthermore, because of the improved adhesion between the fiber and the polymer matrix, low-moisture fibers perform better when utilized as reinforcement in composite materials (Kovačević et al. 2023). The density range of natural fibers is generally between 1.2 and 1.6 g/cm^3 (Thyavihalli Girijappa et al. 2019), which is the range in which the barley fibers are located (Kovačević et al. 2023).

In our pursuit of a sustainable future, lignocellulosic fibers, a type of biomass that is a product of the second generation, stand out as a potentially helpful approach. Furthermore, it possesses enormous potential as a useful raw material for a range of applications, including the manufacture of textiles, paper and pulp, thermal-electric power, and bioplastics (Fortunati et al. 2016; Kovačević et al. 2023). Due to their similar chemical composition, barley straws offer intriguing substitutes for wood fibers in lightweight material applications (Borrega et al. 2022). Barley cellulose fibers derived from barley straw are commonly used in the composite and paper industries because they may be modified or combined with other materials to successfully improve the attributes of the final product (Puglia et al. 2020; Oliver-Ortega et al. 2023). Barley straw can be used to create paper since, due to its reduced lignin concentration, it will lessen the bleaching process, which is a critical factor (Gallardo-Sánchez et al. 2019; Román-Gutiérrez et al. 2022). Also, compared to wood, barley straw has a less complex structure, which makes it possible for chemicals to effectively remove the components that are undesirable for the paper industry (Román-Gutiérrez et al. 2022). Fibers from barley straw can enhance rammed earth strengths similar to those of cement or lime, particularly in terms of tensile strength (Koutous and Hilali 2021). Wipes, fiber-based packaging, nonwoven mulches, clay-fiber composites in construction, growth media, insulating materials, enzymatically degraded intermediate compounds for particular uses, and textile fibers are among the potential uses for straw cellulose fibers (Borrega et al. 2022).

The paper business is the greatest place to use these short fibers, but longer fibers might be used more effectively as reinforcements in the composites sector or even in the apparel industry if they are long enough to be spun into yarn (Kovačević et al. 2023). The potential utilization of these fibers as reinforcement in plastic composites is inspired by their high holocellulose content and relatively high aspect ratio.

Therefore, completely biobased composites were made using barley fibers and a biobased polyethylene (BioPE) polymer matrix (Serra-Parareda et al. 2020). The findings supported low fiber density values (less than 1.5 g/cm^3), which is favorable for their application as lightweight reinforcements in a composite matrix meant for the automobile industry. The clothing industry may be another possible use for barley fibers from ancient cultivars since the obtained fiber tenacity of 20.31–23.08 cN/tex satisfies modest spinning conditions (Kovačević et al. 2023). One crucial aspect of textile fibers that influences how they are used in the future is their density. The product must be lightweight when utilized as reinforcement in composite materials, like those utilized in the automotive industry, since this lowers fuel consumption and costs (Shuvo 2020). All of the synthetic cellulose aerogel in barley straw demonstrated exceptional storage capacity and an oil adsorption capacity of up to 30 times its weight. According to Nguyen et al. (2022), this shows that cellulose aerogels can be utilized to clean up oil spills.

2.1.5 Domestication, Selection and Early Improvements

Barley, one of the earliest domesticated plants or grains by early humans in the Near East, is an ancient grain that served as a foundation crop for Old World agriculture (Robinson 2007). According to Zohary and Hopf's 1993 research, the barley grains discovered in various parts of the Fertile Crescent have archeological remnants that suggest the crop was domesticated in approximately 8000 BC. Ethiopia and the Middle East are where the crop first appeared in its natural state. The wild variety of barley, *Hordeum vulgare ssp. spontaneum*, is a useful genetic resource for enhancing the biotic and abiotic resilience of farmed barley (El-Hashash and El-Absy 2019). Numerous pieces of evidence suggest that populations of wild barley that are expanding in the Middle East have the majority of the species' genetic variability (Nevo 1992). Other genera of the tribe Triticeae, such as Taeniatherum, Psathyrostachys, Eremopyrum, Elymus, and Thinoypyrum, as well as the immediate ancestors of barley, many species strongly related to Triticum, Aegilops, and Hordeum, are still present in the Near East. Thus, for crop relatives, the Near East is among the world's most significant centers of genetic variation (von Bothmer 1996). In antiquity, it barely became fashionable. The Pharaohs, or ancient Egyptians, utilized it to create beer and bread, and it was their staple diet. Additionally, barley's usage as a high-energy diet and as an ingredient in bread preparation led to its adoption by the Greeks as their main cereal grain. Because barley is more tolerant of salty soils than wheat and because of historical irrigation techniques that raised soil salinity in the Tigris and Euphrates rivers, barley was the cereal crop of choice in many parts of Mesopotamia (El-Hashash and El-Absy 2019).

In Chinese tradition, the barley crop came to represent male masculinity. In the sixteenth and seventeenth centuries, barley was brought to the New World by European invasions. Because barley contains less gluten than wheat and rye and therefore cannot be converted into bread on its own, early farmers rapidly concluded that wheat was better for bread and barley was better for beer. Likewise, it lacks wheat's free nature, and husk removal is challenging (Robinson 2007). The earliest Egyptians and Mesopotamians likely brewed beer from barley before wine. The first recorded recipe for barley wine dates to 2800 BC and originates from Babylonia. Eventually, it gained significance mostly in regions where grapes cannot be produced, such as northern Europe (El-Hashash and El-Absy 2019).

The selection of suitable types of barley is complicated by the wide range of desired quality standards (Rani et al. 2024). The most significant economic characteristic, grain yield, is primarily influenced by environmental and genotypic variables and shows quantitative inheritance (Zali et al. 2023). Thus, indirect selection using other attributes that are highly correlated with yield is a useful substitute strategy (Rabiei et al. 2004). Selection using suitable selection indices is one technique of indirect selection. Selecting high-yielding varieties with unique features, like dwarfism, early maturity, etc., is sometimes the goal of selection (Zali et al. 2022; Zali et al. 2023), as well as barley straw fiber properties. Incorporating straw quality into breeding and selection criteria for barley has the potential to improve not only grain output intended for human consumption but also livestock productivity (Keno et al. 2021).

Since barley is a crop that self-pollinates, the vast majority of cultivars are based on pure line development (El-Hashash and El-Absy 2019). Breeding hybrids is also possible, and several hybrid cultivars have been introduced as a result of various hybridization processes (Longin et al. 2012). The study of novel barley genotypes with desirable morphological traits is unquestionably one of the appropriate methods for introducing commercial cultivars, breeding, and improving yields, all of which contribute to increased barley production, given the role that genetic diversity plays in the advancement of breeding programs (Zali and Barati 2020). While maintaining the increase in average yield and improving other traits like biotic and abiotic stress resistance as well as quality, barley breeding programs have significantly accelerated the time from first hybrid to cultivar release by combining modern technologies with traditional breeding methods (El-Hashash and El-Absy 2019). To ensure barley's success in the future, it will be crucial to maintain its production, tolerance to biotic and abiotic stressors, and qualitative attributes (Verstegen et al. 2014).

2.2 Current Cultivation Practices and Challenges

2.2.1 Current Cultivation Practices

Agronomic factors such as planting time, seed rate, and planting techniques are important in determining barley productivity. A good germination percentage (> 85%) and pure (free of weed seed, insect pests, and illnesses) seeds are essential before sowing. In particular, the seeds should not contain covered smut (U. horde), strip blotch (*Helimenthosporium graminum*), or loose smut (U. nigra, U. nuda). To prevent fungal diseases transmitted through seeds, fungicides are typically applied to seeds. Before planting, the seed needs to be carefully cleaned to get rid of salt. To ensure optimal and expedited germination when sowing in both alkaline and saline environments, the seeds ought to be left in water overnight at room temperature (El-Hashash and El-Absy 2019). Unlike wheat, barley farming is not sensitive. It is grown on more desolate terrain. Any soil type suitable for growing wheat can also be used to cultivate barley (Sarğın and Karaca 2023).

Barley is usually seeded between October 15 and November 15 in the Northern Hemisphere. However, the first to third weeks of November and the third to fourth weeks of October are the ideal times for rainfed and irrigated situations, respectively. Maho et al. (2022) have identified that the primary factor contributing to adaptation to climate change is the decision to grow barley in October rather than March. Barley can be planted either in rows or as a broadcast crop. The use of minimum cultivation techniques, or "min-till," has increased recently. This method involves quickly preparing a surface tilth that is only a few millimeters deep and frequently planting the crop in a single or small number of operations (Briggs 1978). However, planting in rows is recommended. The required planting rate is 90–100 kg ha^{-1} for rainfed circumstances and 80–100 kg ha^{-1} for irrigation conditions, depending on the quantity of seeds kg^{-1} and the expected establishment rate, respectively. To guarantee the desired plant population, 100–120 kg of seed ha^{-1} is utilized for saline and alkaline soils (El-Hashash and El-Absy 2019).

Higher weed infestations result in decreased barley grain production at lower sowing rates (Blokhin 2006). When barley is used as a cover crop for fodder, its seeding rate is reduced by one million seeds per hectare (Lopachev et al. 2001). In irrigated and rainfed environments, respectively, rows spaced 22–23 and 23–28 cm apart typically yield the maximum yields of barley. It is advised to space rows 20 cm apart in alkaline and saline soils. Barley has its maximum grain yields when its rows are placed no more than 25 centimeters apart. Row spacing of more than 25 cm is typically associated with low spike density. While the plant will make up for this by producing more kernels per spike and heavier kernels overall, grain output is decreased (Schillinger et al. 1999). Seeding depth under rainfed and irrigated conditions is 5.8 cm and 3–5 cm, respectively, depending on the initial moisture content of the soil. Shallow planting works well in clay soils or soils that tend to produce a crust. A crucial factor is the sowing technique, especially in rainfed environments. To guarantee even dispersion of seed at the ideal depth, seed should be dropped using a seed drill or a nai or pora, a wooden structure fastened to a country plow. The Kera method (spreading seeds and rotating the mold board plow), which involves manually dropping the seed into the furrows, can be used to plant seeds in irrigated areas. Sufficient soil moisture is necessary for healthy germination. Very little weeding or intercultural work is needed for barley. After seeding, two to three waterings are needed in arid regions (Duke 1983). Fertilizers and manures are important components in barley farming (Pandey 2022). It is challenging to choose the best flowering date because of the competing demands to complete grain filling before the late spring high temperatures and frequent dry spells, as well as to avoid frost damage around ear emergence and flowering (Shackley 2000). High-yielding barley cultivars generate roughly 30–35 quintals of grain and 40–45 quintals of straw per hectare when grown using advanced scientific techniques (Pandey 2022).

2.2.2 *Vertical Farming*

The rapid rise of the world's population, growing demand for food and feed, lack of consistent quality products throughout the year, limited natural resources (water), high fertilizer and chemical costs, and environmental issues induced by climate change have pushed global food security to the forefront. In recent decades, there has been a growing global focus on sustainable food production and resource efficiency (Yeşİl and Tatar 2020; Yeşİl et al. 2020). Therefore, research on food and forage production solutions is gaining attention. Advanced technical solutions, such as vertical farming, are emerging as a solution to this issue (Yeşİl and Tatar 2020), as shown in Fig. 2.6. Carrots, radishes, potatoes, tomatoes, peppers, peas, cabbage, spinach, lettuce, and strawberries are among the plants that are frequently cultivated in vertical farming systems (Yeşİl and Tatar 2020). Cultivating high-value fruits and vegetables through vertical farming is an established practice. Growing staple foods in vertical farms is still a very new and developing method, though (Zhu and Marcells 2023). While the production of cereal grains in vertical farming systems is currently not economically profitable, barley fodder production appears to have considerable promise given the notable increase in demand for forage crops (Yeşİl and Tatar 2020). In a vertical farming facility, an annual wheat crop yield of 11.7 kg per m^2 ground surface was recently displayed (Zhu and Marcells 2023).

Vertical farming, which offers various advantages, entails growing crops in tightly regulated indoor environments in layers that are stacked vertically (Zhu and Marcells 2023), where it is one of the most promising methods for maintaining continuous and sustainable plant production while safeguarding natural resources

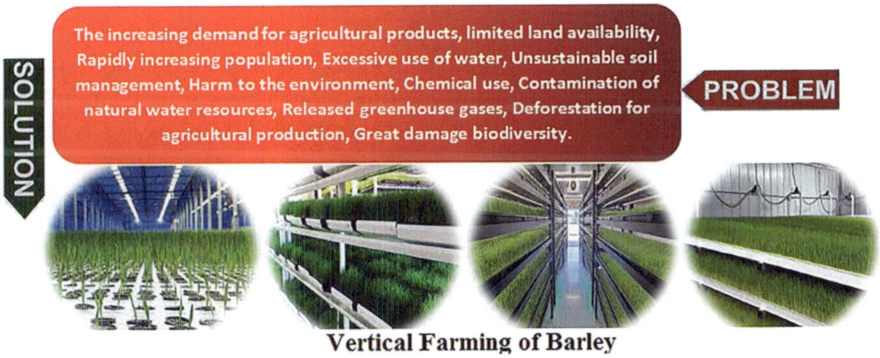

Fig. 2.6 Vertical farming as a solution to the problem of deteriorating global agricultural production. (Prepared by: E.F. El-Hashash)

(Yeşİl et al. 2020). Therefore, it has the potential to provide agricultural products that are both environmentally benign and sustainable (Despommier 2013). Compared to traditional agriculture, vertical farming achieves far more efficient use of area per unit of time, ignores the effects of seasonality, and eliminates the need for chemicals like pesticides and herbicides, as well as uses 70–80% less water (Yeşİl and Tatar 2020).

There are many advantages to vertical farming. Maximum yield per square meter and year-round output independent of weather are provided by vertical farming (Despommier 2013). In addition to reducing or eliminating the risk of other damaging pests and weeds, vertical farming also reduces or eliminates the utilize of pesticides and herbicides, increases nutrient and water efficiency, as well as reduces agricultural pollution (Yeşİl and Tatar 2020; Zhu and Marcells 2023). Vertical farming maximizes environmental factors, including light, temperature, humidity, and CO_2 levels, while shielding crop production from outside influences and meteorological events. In addition to increased fertilizer efficiency, less agricultural pollution, and greenhouse gas emissions, vertical farming claims water use efficiency through the reuse of drained water and the recapturing of transpired water (Van Delden et al. 2021). Vertical farming, which grows crops isolated from the outside world, can be carried out in cities, saving money on storage and shipping while bringing crops closer to consumers, an additional significant ecological benefit. It is possible to adapt technologies created for vertical farming for use in space exploration (Zhu and Marcells 2023).

Many nations, including Japan, Singapore, England, the USA, and the Netherlands, have already implemented vertical farming. According to Yeşİl and Tatar 2020, the vertical agricultural market is projected to grow by 25% by 2024, reaching 11.4 billion euros. One potential future alternative to supply the cattle industry with continuous and sustainably produced feed is to produce barley fodder using sophisticated technology in vertical farming systems (Yeşİl and Tatar 2020). Growing barley fodder in these systems will boost output and lessen agriculture's environmental effects while taking into account the benefits of vertical farming (Banerjee and Adenaeuer 2014). In a 0.1 ha vertical system with 10 layers and 52 growth periods, the production amount of barley fodder reaches 10.400–15.600 tons of fresh fodder annually; additionally, compared to conventional production, there is less water use (about 90%), no use of pesticides or herbicides, and no fertilizer application. These results are seen as quite alluring, though there are still unresolved issues regarding the quality, energy content, and water content. of the system's inputs and end products (Yeşİl and Tatar 2020).

The problem of economically cultivating staple crops, however, continues to be a major barrier. It is challenging to be financially successful in this field because of the high operating costs, especially the electricity costs related to the use of LED lighting (Benke and Tomkins 2017). According to Asseng et al. (2020), growing crops or vegetables by vertical farming is expected to be significantly more expensive than using traditional agronomic techniques. To overcome this obstacle and make it possible to use vertical farming to produce staple crops on a big scale, efforts must be made to concurrently reduce energy costs, create crops that are more suited for vertical farming, and maximize photosynthesis and growth dynamics for larger yields. Zhu and Marcells (2023) suggest that integrating vertical farming with photovoltaic-powered electricity generation could potentially improve the overall efficiency of light energy consumption.

Vertical farming can be used in barley production to obtain barley straw fibers and use them as raw materials for many industries based on barley straw fibers, which were mentioned above in this chapter.

2.2.3 Digital Farming and Artificial Intelligence

Recent centuries have seen tremendous advancements in agricultural resilience thanks to technological advancements, which have also greatly increased agricultural production to feed and support the world's fast-growing population (Ghosh and Kumpatla 2022). This goal is being assisted by several new and existing tools and techniques, including global positioning systems, artificial intelligence, big data analytics, remote sensing, the Internet of Things (IoT), and geographic information systems. When combined with other data, these technologies and tools offer data-driven insights for targeted or site-specific crop management that guarantees increased productivity (Delgado et al. 2019). The efficiency of farming operations has already started to rise thanks to digital technologies like big data, artificial intelligence, robotics, the Internet of Things, and virtual and augmented reality (Balaska et al. 2023).

Digital agriculture is sometimes referred to as smart (intelligent) or e-agricultural farming (Priyadarshan et al. 2024). Based on Shepherd et al. (2018), these agricultural techniques gather, store, process/analyze, and disseminate digital or electronic information in the agriculture sector. The application of data-driven technology in agriculture has led to the development of digital agriculture, which boosts productivity while reducing environmental effects (Liakos et al. 2018). Digitalizing agriculture is referred to as the "digital agricultural revolution" by the UN Food and Agriculture Organization (Klerkx et al. 2019). Agriculture may alter as a result of digital techniques. Revolutions in agriculture increase technology and productivity (El-Tabakh et al. 2024). Digital tools are finding their way into agriculture very quickly. Drones and self-driving tractors are just two of the solutions that governments, small and local businesses, and big technology firms are developing and funding to create the "smart" farmer (Pauschinger and Klauser 2022). Figure 2.7 displays the technologies utilized in digital agriculture, according to Priyadarshan et al. (2024). Global warming and the financial cost of extreme weather are examples of environmental challenges (Ebi et al. 2021). Digital agriculture shows potential in these kinds of situations (Lajoie-O'Malley et al. 2020).

If artificial intelligence applications are applied in the agricultural food sector, it has the potential to undergo a revolution and become more sustainable in many ways. According to Taneja et al. (2023), it might improve food security and sustainability, increase production, reduce waste, and assist farmers, food producers, and

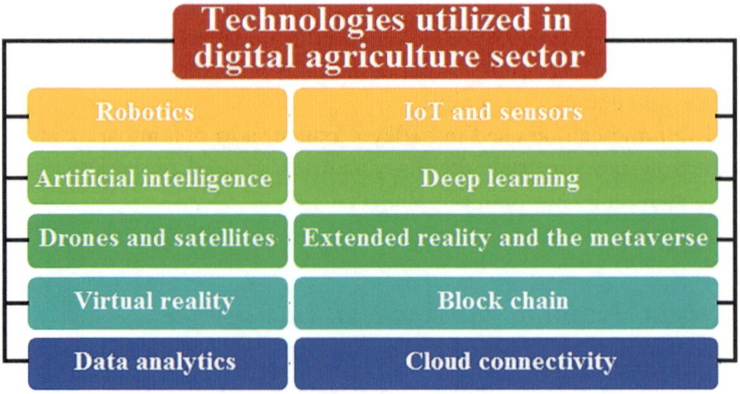

Fig. 2.7 The technologies utilized in digital agriculture. (Prepared by: E.F. El-Hashash)

distributors in making better decisions. Numerous additional agricultural processes, including managing water, classifying soil compatibility, crop selection, crop yield prediction, harvesting, crop demand forecasting, and commodity price prediction, also make use of artificial intelligence algorithms. Artificial intelligence and robotics-based systems could revolutionize crop protection techniques, which also have the potential to reduce the harmful impacts of chemical utilize on human health and the environment. Through the use of the system, farmers can maximize crop yields and advance sustainable agricultural practices by making informed decisions about crop protection strategies. Furthermore, a lot of location intelligence technology is used in conjunction with artificial intelligence-based sustainable agriculture techniques (Ben Ayed and Hanana 2021). This kind of smart farming with artificial intelligence provides a practical solution to the problems associated with agricultural sustainability that exist now.

Precision agricultural production and modern cultivation can produce quantifiable benefits, like an increase in the yield of harvested grain and straw, when paired with the application of the right amounts of chemical fertilizers for particular geographical and natural conditions, as well as the needs of the plants (Denisiuka 2020). Vertical farming employs modern technologies such as robotics, artificial intelligence, and sensors, and it provides viable solutions to several pressing issues confronting contemporary agriculture, including food security, climate management, water conservation, and land scarcity (Zhu and Marcells 2023). Utilizing Internet of Things technologies, precision farming in agriculture maximizes crop yields while reducing operating costs by optimizing the use of available resources. It reduces crop failure and lessens the chance of disastrous losses. Crop productivity is most affected by changes in soil moisture and temperature (Shamila et al. 2023).

Rapid breeding and artificial intelligence can be combined to improve barley straw fibers. Candidate genes for nitrogen usage efficiency have been found in barley collections recently; nevertheless, artificial intelligence will likely be needed for the genes' analysis and breeding application (Meng et al. 2023). To select the best-bet barley subsets for scald disease resistance from a large number of genebank accessions, according to Hiddar et al. (2021), machine learning was applied to refine the Focused Identification of Germplasm strategy method. This makes it possible for machine learning to explain how resistance and agroclimatic variables are related. The promise of computer vision and X-ray CT imaging in high-throughput barley phenotyping will speed up the development of new barley varieties with higher yields (Ling et al. 2023). By using precision farming technologies to support barley production, Denisiuka (2020) can increase not only the grain yield but also the straw yield, which is used as a raw material for energy. A lignocellulosic biomass, barley straw, can be used to make industrial goods with added value. Artificial neural networks are used to develop predictive models for the simulation and optimization of the barley straw delignification process (Serna-Diaz et al. 2016). Utilizing an artificial neural network technique, experimental data from the barley straw solubilization phase in the lignosulfonate production process were fitted well. Although the lignosulfonate production was adequately anticipated by both methods, the ANN method suited the data more closely (Serna-Diaz et al. 2016). Using a barley straw, the use of ozone technology for lignocellulosic biohydrogen production was investigated (Wu et al. 2013).

For all parties involved, grain handlers and traders-manually identifying barley cultivars via eye is difficult and time-consuming. Molecular protein weights and DNA-based technology are two expensive and time-consuming industrial approaches for barley variety identification that also require specialized laboratory equipment (Shah et al. 2022). According to Kozłowski et al. (2019), it was shown that computer vision algorithms may take the place of human experience when it comes to identifying faulty grains or different types of barley, where convolutional neural networks allow for barley classification with an accuracy rate of over 93%. To accurately classify barley seeds, data augmentation and transfer learning strategies were incorporated into deep traditional networks, where the results show that automatic barley seed classification is both feasible and effective, with high test and validation accuracies of 95.71% and 95.70%, respectively (Shi et al. 2022).

Experts claim that the use of artificial intelligence, big data analytics, and digital decision-support systems may boost agricultural output in several ways, including crop selection, yield prediction, biotic stress control, and climate forecasting (El-Tabakh et al. 2024). But for the foreseeable future, industrialized nations willing to incur significant upfront costs will probably be the ones taking advantage of big data technologies. Experts predict that by reducing crop losses and raising crop yields, big data and artificial intelligence can improve global food security (Lassoued et al. 2021).

2.2.4 Current Agricultural Challenges

Although there are many advantages to globally linked production processes, there are also issues with regulation and the need to direct them toward more sustainable and fair results (FAO 2017). Abiotic and biotic pressures have resulted in sluggish improvements in agricultural yields, which have made agriculture's various problems worse in recent years (Grassini et al. 2013). Global production is limited and prices rise as a result of climate shock, greenhouse gas emissions, rising carbon sequestration, and ensuing droughts and floods (FAO 2016). Among the most important environmental concerns of the twenty-first century is climate change (Vaghefi et al. 2017). Consequently, scientists, policymakers, and stakeholders now have difficulty in assessing and predicting future global and regional implications of climate change and mitigation (Latrech et al. 2024). Researchers studying barley and other cereal crops face challenges in creating varieties with high concentrations of these beneficial chemicals or in producing crops that can be used to make healthy foods. Food scientists are also faced with the task of creating barley food that is widely accepted in markets and has a high nutritional value (Geng et al. 2022). Barley is low sensitive to environmental factors than other plant crops. In light of this, attempts have been undertaken to enhance soil and water management as well as create new cultivars that are tolerant to drought and heat (El-Hashash and El-Absy 2019).

Millions of tons of agricultural waste are generated annually. Governments, stakeholders, and environmentalists continue to have valid concerns about the handling and disposal of these materials. These pollutants are a source of contamination for both terrestrial and aquatic environments, an eyesore to the environment, and a health risk when improperly disposed of and managed. Waste management that is economical, environmentally benign, and sustainable still involves turning agricultural wastes into valuable goods (Awogbemi and Von Kallon 2022). There has been a shift in emphasis toward discovering new uses for agricultural byproducts due to environmental policies and the growing demand for a wide range of biobased products. Agriculture-related wastes are abundant sources of bioactive chemicals and minerals. When discussing these materials' possible use in the creation of bioproducts, it is more accurate to refer to them as "feedstock" rather than "waste" due to the possibility of compositional change (Bala et al. 2023).

2.2.5 Breeding Objectives

Creating better barley varieties with high prospective yields (grain and straw yields), resilience to biotic and abiotic stresses, and excellent quality are among the goals of barley breeding. Another is creating production systems that are tailored to the specific needs of various barley production regions (Vivar and McNab 2001). Enhancing barley fiber further will lead to a multitude of industrial applications. Crop breeding offers many useful techniques to enhance barley traits, making the crop more appealing from an agronomic and financial standpoint. Breeders of barley must create cultivars that meet market demands for high-quality features, low moisture content, clean, bright grain, and improved resistance to disease (Knežević et al. 2004). The availability of suitable candidate genes and breeding methods to incorporate them into superior genotypes is a prerequisite for barley improvement

(Wiebe 1978). Transgenic barley breeding attempts to create a superior plant by expressing and passing on one or more desired genetic features from a donor plant or organism to a receptor plant (Newman and Newman 2008). Traditional breeding methods are still widely used by researchers who are breeding for yield to test elite lines and cultivars empirically. Biotechnological approaches, on the other hand, provide intriguing chances to get beyond many of the limitations of conventional breeding and to gain access to a wider variety of gene sources (El-Hashash and El-Absy 2019).

In recent years, a lot of contemporary technological advancements have been applied and have had an impact on barley breeding. Genetic engineering techniques, molecular markers, and doubled haploid lines have accelerated the breeding process, while other techniques have improved its precision and effectiveness (Verstegen et al. 2014). For barley genetics and special breeding, quantitative trait locus mapping is a helpful technique that calculates the economically significant functions linked to specific genomic locations (Knežević et al. 2004). Although it takes many years to create a barley cultivar through traditional breeding, methods that have made it possible for early generation and more focused selection are very important (Verstegen et al. 2014). Barley breeding hybridization methods can provide genetic variety in the progeny by utilizing wild species. This will require further crossing to rectify the selected genotype and the selection of a combination with desired traits (El-Hashash and El-Absy 2019). As demonstrated by the successful crossing of wild relatives in barley Hordeum chilense and durum wheat, as well as the creation of a unique hexaploid agricultural plant x Tritordeum Ascherson et Graebner, a broader hybridization effort may be an open trend to uncover new useful hybrids (Knežević et al. 2004). According to El-Hashash and El-Absy (2019), barley is unique in that cultivars can easily survive for 10 years or more. Many significant QTLs were found in barley, suggesting that this species may provide better perspectives in breeding for improved straw quality. Naz et al. (2017) showed the tremendous potential for molecular breeding of barley genotypes with improved straw quality for bioenergy applications. In particular, grain protein content and side streams of lignin polymer require further breeding to attain the necessary quality of grain and straw (Meng et al. 2023). The availability of whole reference genome sequences and cutting-edge technical methods like pQTL mapping facilitate data collection for barley proteomics investigations, which in turn helps designers create novel genotypes with better responses to dynamic environments (Kosová et al. 2023).

2.3 Germplasm Diversity and Conservation

The genetic material known as germplasm is what allows a species or population to be perpetuated. Although it has no reproductive value by itself, plant breeding can enhance germplasm to maximize crop performance (Acquaah 2007). Breeders‘selection process is based on their understanding of the degree of genetic variations both across and within barley accessions (Abebe et al. 2023). Since germplasm is essential to plant breeding, breeding cannot be done without it (El-Hashash and El-Absy 2019). Gathering and exploiting barley germplasm for breeding and research purposes involves the following steps: expedition, collection, characterisation, multiplication, regeneration, and field testing (Visioni et al. 2023). There are six major categories into which barley's genetic resources can be divided: (1) genetic and cytogenetic stocks; (2) neglected cultivars; (3) landraces; (4) wild cousins in the genus Hordeum; (5) recently adopted cultivars; and (6) breeding lines. The pedigrees of new cultivars often contain elite cultivars from the past (El-Hashash and El-Absy 2019). Table 2.5 lists all of the barley germplasm accessions that are available at the ten main global centers. Three criteria make biodiversity conservation urgently necessary: (1) the prevalence of natural disasters like wildfires and prolonged droughts; (2) human activities like selective breeding, the construction of new land settlements, careless land clearing, and the limited genetic base of developed cultivars; and (3) low genetic variation in both naturally occurring and artificially cultivated populations (Acquaah 2012). Barley has a broad variety of phenotypic and genetic variation because of its extensive agricultural history and ability to adapt to many settings worldwide. This diversity is maintained in international or governmental gene banks as well as germplasm collections in breeding companies, agricultural organizations, research facilities, and universities all over the world (Visioni et al. 2023). Major institutions engaged in barley research are presented in Table 2.6.

2.3.1 Germplasm Diversity

The term "germplasm diversity," originally coined as "genofond," was first used by Aleksandr Sergeevich Serebrovskii in the 1920s. Theodosius Dobzhansky introduced the idea to the United States and it was later interpreted as "*gene pooling*" (Graham 2013). The term "gene pool" refers to the entirety of all the genes that are present in a population of a particular species that can be passed down from parents to children (Gp). The variety increases with the size of the Gp population. Gp determines which phenotypes are present in the population at any given time. Plant breeders can benefit from the notion of genetic background (Gp), as it helps them choose which germplasm to employ in hybridization to improve plants. The Gp is a comprehensive summary of biosystematic data that is now accessible and mostly reflects the usage of crop germplasm in breeding (von Bothmer et al. 1992). The terms genetic resources, gene resources, and germplasm are interchangeable in the fields of crop breeding and genetics. They include any live organism that can be used in crop breeding, including both domestic and wild plants (Sun and Gong 2009).

In the larger framework of protecting and utilizing biodiversity, participatory plant breeding has long been a significant supplementary procedure in the barley industry (Ceccarelli and Grando 2009). By focusing on additional wild-related species and populations with climate change adaptations, more diversity can be gathered (Visioni et al. 2023). When collecting and using germplasm, each barley line among the accessions needs to be treated differently since it could be a source of genetic material (Abebe et al. 2023). To increase the genetic diversity and adaptive capacity of the barley germplasm overall, farmers are essential in locating, identifying, and protecting regional varieties, landraces, and wild relatives (Visioni et al. 2023). To diversify parental material for breeding and create a successful crop plants improvement program, a thorough evaluation of the prevalent variety within the genebank collection of barley plant germplasm is necessary. It has long been known that agro-morphological characterisation is crucial in determining genetic diversity, differentiating materials from various geographic locations, creating core collections, and ranking accessions for efficient breeding usage (Kaur et al. 2022). According to Sato et al. (2014), there are five categories of barley germplasm: cultivars, landraces, breeding lines, wild *Hordeum* species, and genetic stock. In characterizing the germplasm status of barley, genetic diversity can be subdivided into the different Gps and within the principal Gp in both cultivated and wild barley plants. Cultivars, landraces, and research material should all be separated even within cultivated barley (van Hintum and Menting 2003). Three GPS in the genus *Hordeum* can be used in breeding to provide novel, beneficial alleles that improve a variety of characteristics, including yield (El-Hashash and El-Absy 2019). Barley's significant genetic variety can be used to support sustainable agriculture and guarantee the crop's success for upcoming generations worldwide (Visioni et al. 2023).

2.3.2 Cultivars Characterization and Phylogeny

To create suitable selection criteria for breeding programs, it is crucial to have an understanding of genetic diversity (Elahe et al. 2017). There are various phenotypic differences between domesticated and wild barley, which are combined to form the domestication syndrome (Hammer 1984). Barley, both wild and cultivated, has been gathered throughout its range during the past century, and seed samples have been preserved and kept in ex-situ gene banks. According to Jakob et al. (2014), this stock of genotypes serves as the primary source of plant material used in investigations into many facets of genetic variation in barley. Barley is more resilient to many stresses than other cereals, such as heat and drought, probably because of its vast root system and early root formation (Herzig et al. 2018). Furthermore, in soil types like loams or deep sands, barley roots can spread down to depths of up to two meters (Gahoonia et al. 2001).

The complete sequencing of the barley genome (Morex genotype) offers a great chance to further barley research (Yazdani et al. 2020). ISSR can be utilized in molecular characterization procedures to distinguish between the various barley cultivars based on their morphological traits (Mohamed et al. 2021). 31,750 orthologous groups (OGs) across 8 genomes/sub-genomes were discovered by clustering analysis, and 1113 of these were determined to have significantly expanded in barley. Noteworthy is the increase of the gene family in five OGs, which comprise 20 barley DIR genes (HvDIRs) (Luo et al. 2022).

Thus, using genotypes of competitive barley is an economical way to gradually decrease the amount of weed seed banks and may also minimize the need for herbicides. Future cultivar selections of barley take into account the inclusion of weed-suppressive features, especially as farmers are very supportive of long-term, low-cost weed management techniques that don't call for new technological interventions on the farm (Mwendwa et al. 2022). Didon (2002) found that barley cultivars with strong to medium competitive power against weeds decreased the time it took for the plants to emerge after planting, in contrast to when the least competitive cultivar sprouted in the presence of annual broadleaf weeds. Spring barley varieties that were very active against weeds also exhibited early stem elongation in response to weed competition. Morphological traits like the length of the main shoot during the tillering stage, the length of the two initial internodes, and a modest leaf angle were also considered crucial in the competition for light (Didon 2002; Mwendwa et al. 2020).

2.3.3 *Genetic Resources Conservation Approaches*

The germplasm base of enhanced barley cultivars that are high-yielding and resistant to biotic and abiotic stressors can be expanded through the application of crop genetic resources (Singh 2006). Even though barley is very responsive to artificial selection due to its large degree of natural variation (Kling 2004), future attempts to improve barley production and quality may depend on the collection and maintenance of these genetic resources (El-Hashash and El-Absy 2019). Because barley genetic resources are so important, they must be conserved to meet the worldwide issues that human society is currently confronting, including food scarcity, climate change, natural disasters, and the emergence of novel pests and diseases. Furthermore, there are far fewer cultivated plants and their wild counterparts in terms of both quantity and variety (Dzyubenko 2018). Systemic molecular assessment utilizing molecular markers is essential for both the protection of genetic variety and the proper utilization of this accessible diversity in breeding. The preservation of genetic diversity and the best possible utilization of this available diversity in breeding both depend on systematic molecular assessment using molecular markers (Maanju et al. 2023). Ex-situ conservation, such as gene banks, involves keeping biodiversity outside of its native environments, whereas in-situ conservation involves keeping it there.

2.3.3.1 In Situ Conservation

Crops' wild relatives are valuable resources that can increase their ability to withstand biotic and abiotic stress. To maintain their intraspecific variety, planting wild relatives in situ is gaining more and more attention. Conserving species in their native habitat, instead of conserving seeds out of situ, encourages the emergence of novel adaptations to shifting environmental circumstances (Mewis et al. 2024). Preserving genetic resources in their natural environments is known as "in situ conservation"(Brush 2000). Landraces, wild cousins, and weed barley can all be kept in their native environments or in situ places where the plants have naturally evolved. It is a crucial method for preserving species for which ex-situ conservation is challenging. Targeting either the species or the ecosystem in which they reside, in situ conservation is frequently accomplished in protected areas or habitats (FAO 2010). The in situ strategy can preserve a larger percentage of the biological diversity than the ex-situ conservation techniques used in gene banks, which only cover a portion of the total diversity (Acquaah 2012).

In many places where it has evolved to certain environmental circumstances, barley is preserved on farms. These sites have either a long history of agriculture, usually in traditional agricultural systems, or are frequently represented by the natural habitats of their wild counterparts (Visioni et al. 2023). Priority locations for genetic reserves have been identified for barley (*Hordeum spp.*) in the Americas (Chile) in order to prepare for the establishment of a global network for the in situ conservation of barley wild relative species; the Near East has the highest priority locations for conserving the wild relatives of barley (*H. spontaneum, H. bulbosum*) (FAO 2010). The Near East (i.e., Fertile Crescent, which encompasses Asia Minor and the Caucasus), the European–Siberian region, the Ethiopian region, the East Asiatic region, and the New World centers (Americas and Oceania) are the main agro-ecological regions where barley has flourished and adapted. A wide range of agroecological groupings of varieties, including additional subdivisions, have emerged in each region as a result of a combination of notable climate changes and geographic isolation (Knüpffer et al. 2003; Visioni et al. 2023).

2.3.3.2 Ex Situ Conservation

To address national development planning for the agricultural, environmental, and socioeconomic sectors, a National Plan of Action for the Conservation and Management of Plant Genetic Resources for Food and Agriculture (PGRFA) is being established for the next 20 years (FAO 2018). According to FAO (2010), ex situ conservation is still the most important and widely utilized approach for protecting PGRFA. Ex-situ conservation refers to any conservation technique whereby a species or cultivar is removed from its natural habitat and maintained in an environment that was manufactured by humans (Acquaah 2007); examples of such environments include gene banks, research institute botanical gardens, and experimental stations. The genetic variety of cultivated plants and their wild cousins was permanently at risk of disappearing due to changing environmental circumstances and cultural activities. Ex situ plant gene bank collections were established at the start of the twentieth century as a result of this insight (Nagel et al. 2009). Gene banks run by governmental or private organizations can function independently or in conjunction with other organizations (FAO 2010). The vast expanse and diversity of germplasm collections kept in gene banks make it difficult to characterize, assess, make use of, and maintain these resources (El-Hashash and El-Absy 2019). Over the past few decades, there has been significant growth in the quantity and size of ex-situ germplasm collections, which has created challenges for the assessment, use, and preservation of genetic resources (Kaur et al. 2022). Access to germplasm for breeding and research is made easier by ex-situ conservation (Mewis et al. 2024). Table 2.7 lists the ex-situ germplasm collections of PGRC, NSGC, and ICARDA that contain barley genetic resources.

The Global Strategy for the Ex Situ Conservation and Use of Barley Germplasm was developed in order to conduct a global inventory of barley genetic resources in ex-situ collections (Visioni et al. 2023). Barley ranks third in terms of the overall number of ex-situ accessions, after wheat and rice (Fig. 2.8). All Hordeum taxa are included in the number of accessions: 1.7% of other wild species, 0.4% of *H. bulbosum*, 10% of *H. vulgare ssp. spontaneum*, and 88% of *H. vulgare* ssp. vulgare. Europe kept 41 barley accessions, followed by North America with 25, and the Middle East with 12%. The remaining accessions are spread across Asia, South America, Africa, and Australia, with 9, 7, 4, and 2% of each, according to van Hintum and Menting (2003). Cultivars, breeding lines, landraces, crop wild relatives, and genetic stocks in gene bank collections comprise 44, 17, 15, 15, and 9% of the overall genetic resources of barley (Bockelman et al. 2010). Therefore, 59% of barley accessions are accounted for by global barley holdings, which are made up of germplasm that has evolved via extended interaction with local environments and farming techniques, while 41% are accounted for by modern plant breeding and research (Bockelman et al. 2010).

According to data from FAO (2010), there are presently 18,469 fewer barley accessions registered in 1996 FAO (1996) than there are projected to be in the world's

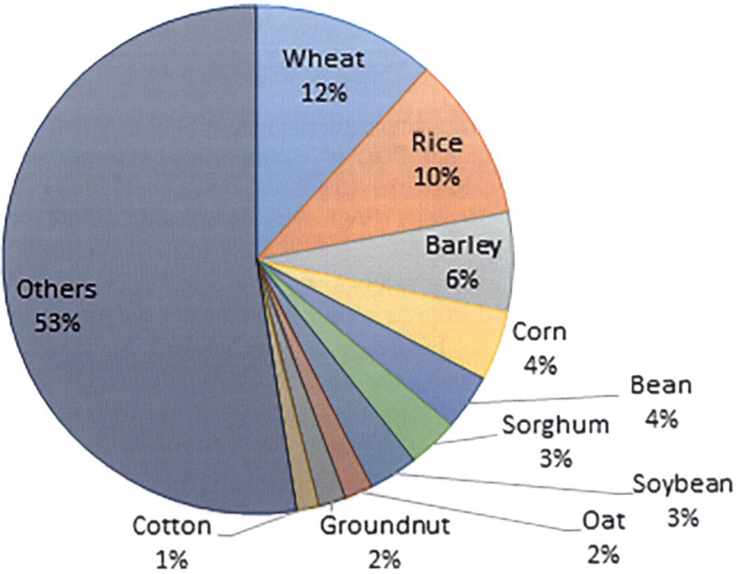

Fig. 2.8 Key crop groups' contributions to global ex situ conservation. (Source: FAO 2010)

ex-situ germplasm collection (466531). van Hintum and Menting (2003) revised the number of barley accessions to 371,000, and the Global Crop Diversity Trust (2008) revised it to 370,796. Landraces, breeding lines, cultivars, and selections are examples of accessions in a general collection. Worldwide, there are 47 barley collections with over 500 accessions and 25 smaller collections that house over 400,000 barley accessions (Alqudah et al. 2020; Marone et al. 2021). Finding new sources of genetic variety to add to barley breeding programs is currently the focus of an increasing number of breeding techniques (Czembor 2023). Collections with over 10,000 barley accessions are considered major. Table 2.8 lists the 10 largest collections of barley germplasm (El-Hashash and El-Absy 2019). Altogether, the ten collections preserve approximately 254,386 out of 466,531 accessions, or 54.5% of the total accessions worldwide. Plant Gene Resources of Canada (PGRC) is home to the world's biggest collection of barley germplasm. Its holdings make up 9% of the global total and 16.4% of 254,386. According to van Hintum and Menting (2003), the four major international collections—PGRC, NSGC, ICARDA, and CENARGEN—have a lot in common, especially when it comes to varieties (Global Crop Diversity Trust 2008).

Therefore, it is essential to increase the conservation of barley plant genetic resources by employing complementary ex-situ and in-situ measures to address the conflict between the encouragement of large-scale adoption of novel cultivars and the loss of barley biodiversity (Visioni et al. 2023). The importance of conservation activities won't change, but new tactics and the backing of regional farming systems are required to effectively meet evolving demands (Mascher et al. 2019). A move towards enhanced ex-situ conservation will need to represent a significant change in barley conservation strategies, given the threats that climate change, urbanization, and land-use change pose to natural habitats and traditional farming systems (Zair et al. 2021). A crucial component of crop improvement is the examination of genetic associations between crops in order to choose parental lines for hybridization, assess genotype genetic variability, and determine germplasm that must be conserved to protect the highest genetic variability (Maanju et al. 2023).

2.3.4 Cytogenetics

One of the crop plants that has been examined the most is barley, which is employed in cytogenetics and molecular cytogenetics research (Ramage 1985). In the genus *Hordeum*, four basic genomes, I, Xa, and Xu-have been found (Blattner 2009). The basic chromosomal number of x = 7 is shared by all 45 taxa in the genus *Hordeum*. Moreover, multiples of this fundamental number are obtained. Various researchers have claimed that the species *H. bulbosum*, *H. jubatum*, *H. murinum*, and *H. secalinum* possess seven or fourteen pairs (tetraploid, 2n = 4x = 28). It has been reported that *Hordium nodosum* is hexaploid (2n = 6x = 42), with either 7 or 21 pairs. Every economic species is diploid (2n = 2x = 14) with seven pairs of chromosomes (Hayes and Immer 1942). Compared to wheat (2n = 42), and maize (2n = 20), and rice (2n = 24), barley has the fewest chromosomes (2n = 14). According to Heneen (2011), the barley's genome size

(IC) is 4873–5096 Mbp, which is larger than that of rice (401–466 Mbp) and maize (2292–3313 Mbp), but smaller than wheat (15,966 Mbp). Barley's genome is diploid, with 5.1 billion base pairs and over 84% repetitive DNA, according to Doležel et al. (1998) (IBSC 2012). About 39,000 protein-coding genes, many of which are found in multiple copies, are found in the barley genome, according to Manuel Spannagl of the Helmholtz Zentrum München, who was in charge of the gene explanation. Mascher et al. (2017). Fifteen million nonredundant single-nucleotide variations (SNVs) were found after resequencing four distinct barley cultivars: Bowman, Barke, Igri, and Haruna Nijo. According to Rakshit and Ganapathy (2014), the frequency of these SNVs tends to drop toward the peri-centromeric regions of every chromosome.

In the early generations of wheat-barley crosses, barley chromosome 5H vanishes due to its extreme instability in the wheat genetic background. It contains important QTLs for abiotic stressor tolerance and plant adaptation. According to Ivanizs et al. (2018), it is crucial to promote pre-breeding and breeding studies on wheat and to accelerate functional genomic studies on barley chromosome 5H. Sixteen wheat × barley F_1 hybrid plants were analyzed, and the results showed that all of the wheat and barley chromosomes could be quickly and easily detected using Multiplex Polymerase Chain Reaction primer sets. Differential (sub)genome high-throughput analysis can also benefit from this approach. With accurate genome sequence data, the primer design approach shown here can be applied to a wide range of species (Ali et al. 2024). Due to its significance as a primary crop and its role as a model genome for other Triticeae genomes, such as those of wheat and rye, the International Barley Genome Sequencing Consortium (IBSC) was established to create a workable working agenda for the physical mapping and sequencing of the barley genome (Stein 2014). In addition to helping plant breeders create new, improved barley cultivars, growing sequencing data and the analysis that follows offer a valuable tool for comparative genomics (El-Hashash and El-Absy 2019).

2.4 Traditional Breeding

2.4.1 Traditional Breeding Methodologies and Improvement Strategies

Over the past 10,000 years, geneticists, farmers, and barley breeders have created progressively more sophisticated breeding techniques (Patrick and Alfonso 2013). There are several breeding strategies available, and while some techniques are widely acknowledged, other techniques are used in certain crops because they are more productive and successful, depending on the goals, resources, and crop type being mated. To put it another way, various breeding techniques are applied to enhance superiority relative to time and cost per unit. Additionally, based on the objectives of the breeding program, various approaches might be employed concurrently or at various program stages (Singh et al. 2023). El-Hashash and El-Absy

(2019) state that to enhance barley characteristics, including high grain yield, biotic resistance, abiotic resistance, and malting quality, breeders have employed a variety of conventional breeding techniques. The majority of contemporary elite barley cultivars were created by conventional breeding, which is still very successful. Male sterile-facilitated recurrent selection (MSFRS), diallel selective mating system (DSMS), hybridization, mutation, interspecific and intergeneric crosses, backcross, single seed descent, haploid production, doubled haploids, anther or microspore culture, Hordeum bulbosum, mass selection, pure line selection, pedigree selection, bulk selection, and single seed descent are examples of traditional breeding techniques used to improve barley yield and quality traits.

Barley breeding can make use of composite crossbreeding. According to Harlan (1957), it is the outcome of mixing several single crosses into a single, sizable mixture or composite. In contrast to other breeding techniques, the two approaches (MSFRS and DSMS) in barley promote genetic base expansion, disassemble preexisting linkage blocks, and yield a high degree of genetic variety (Anderson and Reinbergs 1985). Wiebe (1978) developed barley strains for agronomic and malt quality using the backcrossing method, while Hockett (1981) measured the impact of dietary fiber on barley's feed value. In barley breeding, adapting introduced germplasm to new production locations is regarded as a significant difficulty. This may be due to a variety of factors, such as the emergence of novel disease strains or shifting export market demands for malt quality. According to Horsley et al. (2009), these production locations frequently feature a unique combination of production restrictions or are marginal for arable crops. Since conventional breeding products might not be able to meet the standards, molecular breeding techniques must be actively used in contemporary breeding operations (Eglinton et al. 2006). According to Faheem Akhtar et al. (2023), hybridization is a technique used to acquire the required qualities from closely associated individuals and incorporate them into new cultivars. It also produces variation in traditional breeding. To address the interactions between genetics and the environment in barley breeding for climate change, automated techniques and predictors will be necessary (Cooper and Messina 2021). In order to integrate crop physiology and breeding and preserve yield stability, proximal and remote sensors, as well as computers, are being developed for crops. More breeding is required to produce the required grain and straw quality (Meng et al. 2023).

The outcome of breeding and selection may be populations that outperform earlier ones or new kinds or clones that, by some criterion or criteria, are superior to the standard commercially developed genotypes (checks) now in use (Singh et al. 2023). Although the method has certain drawbacks, conventional breeding has been widely utilized to create plant cultivars with desirable features (Numan et al. 2021). Early efforts in breeding concentrated on genetic improvement, which involved selecting pure lines from preexisting germplasm, as well as germplasm enhancement through collection, characterization, evaluation, and conservation (Assefa et al. 2011). By passing down the desired gene from generation to generation, these breeding techniques were used to create improved types (Begna 2021). DNA markers were added to conventional breeding methods at the close of the previous century. Moreover, molecular markers have been widely employed in QTL analysis and molecular genetic mapping (Rozanova and Khlestkina 2020).

2.4.2 Traditional Breeding Limitations

Because it is typically impossible to mix genes from unrelated species or make small alterations without disrupting the precise gene combination that makes a certain type distinctive, the fundamental constraints of conventional approaches originate from the restrictions of the sexual system. Furthermore, the quantity of genetic variety that is available within the crop (the gene pool) is limited. Because all characteristics that differ between parents are prone to segregation, finding exceptional people who possess the finest attributes from both parents may require examining vast populations and doing numerous generations of selection (Begna 2021). The problem facing barley breeders is that a high number of characters are influenced using separate loci spread throughout the genome, and for a high number of characters, the marker-trait relationships are frequently insufficiently strong to be used as a predictive tool (Stockinger 2021). The shortcomings of traditional breeding are most noticeable when populations that can be crossed sexually do not possess a necessary trait. Although mutations can be created, they are frequently harmful or associated with unfavorable outcomes (Begna 2021).

Improving every trait of interest at once is the challenge of conventional plant breeding. Genetic correlations between phenotypes, that pleiotropic genes can cause, physical connections between genes in chromosomes, or population genetic structure, make this process more difficult (Hartl et al. 1997). When one feature is chosen, linked attributes will alter as well-sometimes in a positive way and other times negatively (Falconer 1996). Selection may consequently lead to unanticipated changes that, according to Kok et al. (2008), are typically within the range that is typically observed in the plant and are therefore believed to present little danger to the environment or consumers. The small genetic background of the progenies produced by self-pollinated crop breeding is the largest bottleneck since breeders can only influence parental behavior on two individuals in a single cross, three and four in a triple cross, and four at most in a double cross. Thus far, efforts to enhance crop plants' resistance to abiotic stress using conventional breeding methods have yielded mixed results (Richards 1997).

Major drawbacks of conventional breeding include the requirement for numerous backcrosses to eliminate undesired characters, restriction to loci that result in observable phenotypes, and inadequacy if there is not enough diversity in the trait of interest in the gene pool. As a result, marker-assisted breeding is currently the main emphasis since it enables the "pyramiding" of desired features for faster crop development with less resource input (Begna 2021). According to Anjum et al. (2017), the traditional breeding methods employed to create novel crop varieties are frequently laborious and unable to meet the rapidly increasing need for food production. In order to handle the interactions between genetics and the environment, barley breeding for climate change will require automated methods and predictors (Cooper and Messina 2021). Additionally, proximal and remote sensors, along with computation, are being developed for crops in order to make it possible to integrate breeding and plant physiology to maintain plant yield stability (Meng et al. 2023).

In conventional breeding operations, hybridizations between high-yielding cultivars and wild barley may introduce certain features through back-crossing studies (Nevo 1992). Many excellent varieties of barley have been generated by traditional methods of breeding; however, future attempts would gain a great deal from combining genotypic data from whole genome scans using molecular markers with high-quality phenotypic data.

2.4.3 Role of Biotechnology

Traditional plant breeding and selection techniques are frequently imprecise and take a lot of time (Fig. 2.9) (Fehr 1987). Since the 1980s, when the barley genome was first molecularly mapped (Kleinhofs et al. 1988), biotechnology has been essential to the development of barley. This is due to its ability to outperform traditional breeding techniques for improving barley (El-Hashash and El-Absy 2019). Biotechnology approaches for genetic alteration work well at the organ, tissue, cell, protoplast, and molecular levels, in contrast to conventional plant breeding. For precise and successful plant breeding, these contemporary approaches are thought to be a supplement to more conventional approaches (Kang et al. 2007). Techniques for cultivating somatic and reproductive cells, tissues, and organs; Techniques for microbial bioprocessing; Plant biotechnology is made up of three interconnected technical components: molecular and cellular techniques for genome characterization and modification, as well as genetic material identification, recombination, cloning, transfer, and expression (El-Hashash and El-Absy 2019).

The completion of barley genome sequencing in 2012 opened up a wide range of new possibilities, including genomic selection and a more effective search for candidate genes governing economically significant traits (Rozanova and Khlestkina 2020). One of the primary barriers to the wider application of genetic transformation in barley plant breeding is genotype-limited plant regeneration. Therefore, the core of barley biotechnology continues to be the development of

Fig. 2.9 The distinction between traditional and genomics methods for breeding barley. (Source: El-Hashash and El-Absy 2019)

novel strategies that may enhance the responses of in vitro resistant genotypes (Nowak et al. 2024). Using sequencing techniques and quantitative trait locus mapping, among other genetic research techniques, genome-wide association scan (GWAS) investigations have identified novel genes, genomic areas, and molecular markers linked to barley's agronomic features. Furthermore, the most widely used techniques for assisting barley molecular breeding are haplotype-based GWAS and SNP marker technologies. The cv. Golden Promise is the greatest example of how genetic information is utilized for high-efficiency gene editing by CRISPR-Cas9 technology (Riaz et al. 2021). Expanding the application of epigenetic modifiers in *in vitro* plant regeneration research could potentially address issues impeding the widespread utilization of genetic transformation and genome editing in the breeding of significant crop plants, such as barley (Nowak et al. 2024). Bulk segregant ribonucleic acid sequencing is one sequencing-based genomic technique that has been applied to barley (Qi et al. 2019). A solid foundation for using genome editing technologies to provide material for the production of varieties with desired qualities is provided by the already available data on prospective target genes and the quality of the entire barley genome sequence (Rozanova and Khlestkina 2020).

Biotechnologies such as genetic engineering and molecular marker-aided technology have been used to increase barley's resistance to stem rust (Horvath et al. 2003), drought stress (Morran et al. 2011), salinity (Roy et al. 2013), thermostable β-glucanase (Horvath et al. 2000), thermostable β-amylase (Kihara et al. 2000), α-amylase activity (Tull et al. 2003), speed α-amylase, and pullulanase activity (Cho et al. 1999). Yi et al. (2021) state that in order to satisfy market demand, particularly from industries associated with bast fiber crops, new varieties of bast fiber crops must be developed. These crops must have high yield, high efficiency, resistance to disease and stress, suitability for light simplification and mechanization cultivation, high quality, and special purpose. We should also adapt to the production model of bast fiber plants. Existing natural variances of barley compounds can be combined with multiple mutant lines or utilized to develop improved cultivars. The technical specifications might not align with barley's intended purpose as feed or as a fiber and biofuel source. To enhance quality, barley breeders have applied new technologies, including genomic selection, marker-assisted selection, transgenics, site-directed mutagenesis, induced mutations, and machine learning (Meng et al. 2023).

2.5 Molecular Breeding

Plant breeders have focused on barley's crucial agronomic features and enhanced molecular breeding to find the ideal mix of traits to meet the needs of farmers and consumers (Riaz et al. 2021). Molecular markers comprise metabolite-based biomarkers, DNA markers (Fig. 2.10), and protein markers; however, only DNA markers are currently used in molecular marker applications (Singh and Singh 2015).

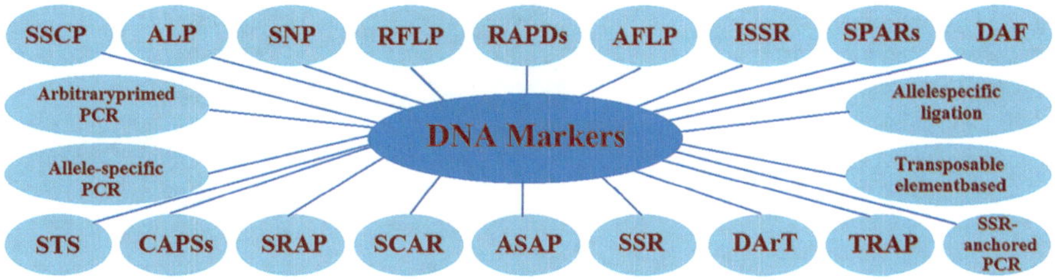

Fig. 2.10 Types of DNA markers. (Source: El-Hashash and El-Absy 2019)

Molecular breeding is the process of improving plants for particular uses by modifying their DNA using a range of instruments (Acquaah 2012). Based on genotyping technology, these identified markers can be used for molecular marker-assisted selection breeding or to ascertain the genetic links between various accessions/genotypes (Riaz et al. 2021). Molecular markers are thought to be the most suitable solution to meet the needs of plant breeders. These markers can be employed as genetic "flags," suggesting preferred genotypes for selection, using straightforward molecular procedures (Baidyussen et al. 2024).

2.5.1 Molecular Marker-Assisted Breeding

Crop breeders must figure out how to expedite the breeding process and enhance the efficiency of the selection process to meet the shifting market needs for plant varieties (Jiang 2015). The application of molecular marker-assisted breeding (MAB) has great promise for enhancing the precision and efficiency of traditional plant breeding techniques. In addition to traditional breeding, marker-assisted breeding is a helpful technique for choosing plants with desired characteristics (Jiang 2013). Additionally, it makes the process of plant breeding easier, helps identify genes of interest, and facilitates the study of genomic organization (Acquaah 2012). As a result, it's a tactic and an effective methodology for creating cultivars of barley grain that combine pertinent adaptation features, like resistance to abiotic and biotic stresses, with increased production and improved quality (El-Hashash and El-Absy 2019). Numerous breeding procedures are all referred to by the MAB, including genomic selection (GS) or genome-wide selection (GWS), marker-assisted backcrossing (MABC), marker-assisted selection (MAS), and marker-assisted recurrent selection (MARS), according to Ribaut et al. (2010). Based on the discovery of SNPs linked to the desired characteristic of barley and other crops, GWAS have surfaced (Afanasenko et al. 2022). Furthermore, MAS has been extensively employed in the breeding of barley and other cereal plants (Afanasenko et al. 2022), as it is an effective means of expediting the barley breeding process (Leišová-Svobodová et al. 2024).

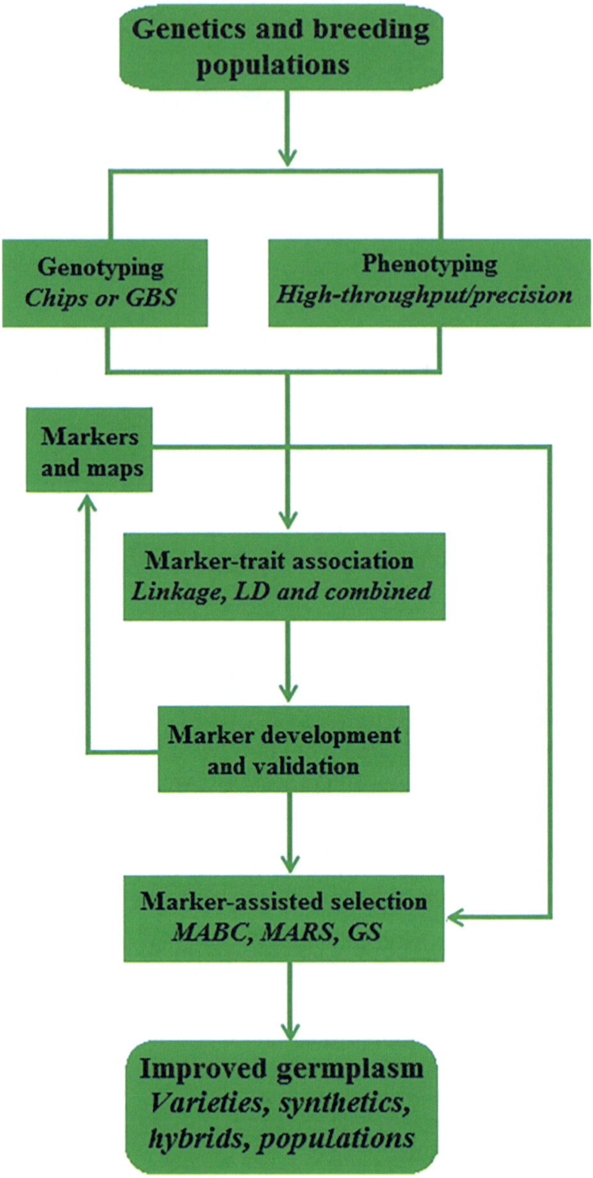

Fig. 2.11 Components of a marker-assisted breeding platform. (Source: Xu et al. 2013)

The advent of molecular marker techniques enabled high-throughput genotyping to map and tag agronomically valuable characteristics in barley, as shown in Fig. 2.11 (Verstegen et al. 2014). Several studies on the diversity of barley cultivars have employed these markers, including those by Bouhlal et al. (2022), Ghonaim et al. (2023), Guo et al. (2024), and Mohamed et al. (2021). According to Kleinhofs et al. (1988), the barley genome was first molecularly mapped in the 1980s. Since then, other researchers have produced a large amount of data, which has made it

Table 2.3 Molecular linkage map developed in barley

No.	Molecular map	Pedigree of the mapping population	Markers	Agronomic trait targeted	References
1	QTL map	Mv9kr1 (wheat) x Igri (barley)	EST and PCR	Morphological traits	Ivanizs et al. (2018)
2	Linkage map	3743 segregating progenies make up 11 biparental populations.	SNPs	Development of a genotyping tool	Abed et al. (2022)
3	High-density genetic map	134 recombinant inbred lines (RILs)	SLAF-seq and SNPs	Grain traits	Fang et al. (2020)
4	QTL mapping	Inbred lines from a hybrid between barley R90 and 1815D	SSR and InDel markers	Grain number and weight	Fan et al. (2024)

possible to create intricate barley chromosomal linkage maps (Table 2.3). These methods are currently being used extensively in barley breeding. Roughly half of the 1000 barley simple sequence repeats (SSRs) that have been published have undergone genetic mapping (Rao et al. 2016). Using molecular markers in MAS is an extremely appealing alternative. The ultimate goal of molecular markers is to be used practically in MAS, so that, depending on the molecular marker scores, desirable genotypes may be chosen and undesirable genotypes can be eliminated (Baidyussen et al. 2024).

Climate, nutritional availability, and exposure to biotic and abiotic pressures are only a few of the environmental variables that affect biomass composition (Tanger et al. 2015; Wang and Frei 2011). Lignin is sensitive to environmental cues since it aids in a plant's defense against a variety of abiotic and biotic stresses. The characteristics of straw composition were all sensitive to the drought. Straw composition is influenced by genetic variability within and between species, in addition to environmental influences. The necessary conditions for marker-assisted breeding aimed at bioenergy-related traits are provided by such intraspecies variation (Naz et al. 2017). Significant genotypic variation was seen in all straw compositional traits, which is necessary to take advantage of genetic diversity through genome-wide association studies (Huang and Han 2014).

Using biparental mapping populations, Grando et al. (2005) and Siahsar et al. (2009) discovered QTL linked to straw composition features in barley. This method only targets a small portion of genetic variation, and the number of chromosomal recombinations in a biparental cross limits the mapping precision, even though it has consistently shown remarkable agronomic promise (Morrell et al. 2012). These restrictions are circumvented in genomewide association studies, where high-density marker maps are used to screen a variety of populations of unrelated people and identify QTL based on linkage disequilibrium (Huang and Han 2014). Because of its strong QTL, high heritability, and strong association with energy efficiency in both biological and thermochemical conversion processes, thioglycolic acid lignin was shown to be an appropriate target characteristic for marker-assisted breeding. Thioglycolic acid lignin was discovered to have the highest number of QTL and to be essential in determining the bioenergy value. The robustness of these QTLs is indicated by their high heritability, high log-of-odds score, and high proportion of phenotypic variation explained (Naz et al. 2017).

2.5.2 Genomics

The botanist Hans Winkler is credited with coining the term genome in 1920. A self-replicating body's genome is its maximal unit (Saitou 2004). According to Griffiths et al. (2005), there are two main subfields within genomics: structural and functional genomics. Functional genomics is concerned with gene function, whereas structural genomics is concerned with genome sequencing (Acquaah 2007). Functional genomics focuses on the dynamic aspects of genomic information, such as gene transcription, translation, and protein-protein interactions, rather than the static components, such as DNA sequence or structures (Griffiths et al. 2005). In essence, the genome contains all of the instructions needed to make every type of protein. Since most genes are expressed as proteins, one popular method for deciphering the role of genes is to monitor the expression of proteins within cells, a process known as proteomics (Acquaah 2007). The functional genomics/systems biology platform is an extremely powerful and intricate method for figuring out how individual genes, networks, pathways, and genomes work in quality (El-Hashash and El-Absy 2019). Functional genomics'primary tenet is to examine and assess the whole cell or organism as a system in order to comprehend how different biological processes operate within it, how to regulate them, and how to put them into practice (Mittler and Shulaev 2013). Over the past years, insertional mutagenesis techniques have been employed to support functional gene analysis. These methods include the introduction of transposable elements into a gene of interest to produce loss-of-function mutations and the random genomic insertion of enhancer or promoter sequences (activation tagging) to produce dominant gain of function mutations (Ayliffe et al. 2007; Zhao et al. 2006).

The first barley reference genome assembly and ongoing genomics advancements have created new and intriguing opportunities as well as instruments for more accurate and productive barley breeding. Reference genome assemblies, for instance, have been used in conjunction with germplasm sequencing, QTL mapping, and GWAS research to find breeding. These studies have found novel genes, genomic areas, and molecular markers linked to agronomic characters. Consequently, more productive barley cultivars have been developed (Riaz et al. 2021). Transgenic plants containing the Ac and Ds elements are used to create insertion lines, and then they are crossed to cause Ds transposition. Ds elements had a high remobilization frequency and were primarily located in genic areas (Singh et al. 2006). These Ds launch pads were useful for future targeted gene tagging since they were represented by barley lines that each had a single-copy DS insertion at a specific location in the genome (El-Hashash and El-Absy 2019). In the huge barley genome, where loss-of-function mutations frequently do not produce symptoms due to gene redundancy, dominant overexpression phenotypes make it easier to research gene functions (Ayliffe et al. 2007). A more effective Agrobacterium-mediated barley genetic transformation technique based on immature embryos was developed for spring barley in order to functionally assess candidate genes found in functional genomic investigations (Hensel et al. 2008). Kumlehn et al. (2006) improved this approach by creating a transformation technique for winter barley that involves infecting androgenic pollen cultures with Agrobacterium. By doubling chromosomes, homozygous doubled haploid plants can be produced immediately and frequently using this method. Gene duplications specific to a certain lineage that are involved in the mobilization of carbohydrates in grains and the transport of nutrients to developing seeds are found in barley through gene family analysis. By examining the genomic partitioning of sequence variation in contemporary elite germplasm and identifying regions vulnerable to genetic degradation, they illustrate the significance of the barley reference sequence for breeding (Mascher et al. 2017).

The significance of creating a flexible and effective transformation protocol for functional genomics research is highlighted by the necessity of finishing the barley genome sequencing, the variety of genetic diversity, and the abundance of mutant populations (Rotasperti et al. 2020). Barley's special potential as an ideal system for functional genomics research can now be fully realized. Several molecular, bioinformatics, and genomics-based approaches that integrate genomic data with sequencing, re-sequencing, and genotyping datasets are being used to study key agricultural characteristics and the genes associated with them in barley (Riaz et al. 2021). Golden Promise is the most readily transformable barley genotype, with 95.2% of complete and single-copy genes found in the assembled genome of seven chromosomes totaling 4.13 Gb. This makes it particularly useful for functional genomics research (Rotasperti et al. 2020). The publishing of a barley pangenome is one of the most recent advancements in the fast-growing collection of genetic materials for barley (Schreiber et al. 2024). After the genome sequencing of bast fiber crops is finished, omics-based genetic improvement will be possible. This will make it easier to study how bast fiber forms, how bast fiber crop germplasms evolve, and how agronomic traits like fiber yield, fiber quality, stress tolerance, and disease resistance are formed at the molecular level. It will also make it easier to mine quality gene resources, produce novel germplasm, and build a high-throughput genotype-phenotype database. Furthermore, it will make it easier to create effective systems for rapid breeding technology by combining genome selection, transgenic technology, molecular marker-assisted selection, and other cutting-edge techniques (Yi et al. 2021).

2.5.3 *Bioinformatics*

Bioinformatics is the name of the knowledge-based theoretical field that uses information from DNA sequence analysis to forecast biological activity. Utilizing supercomputers and cutting-edge software to study and evaluate amassed datasets of

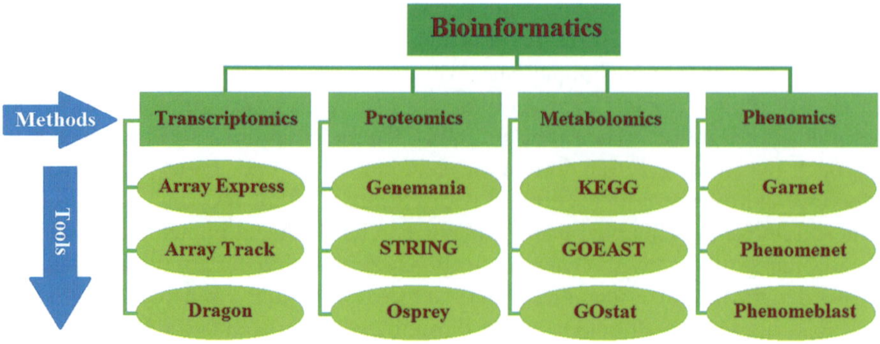

Fig. 2.12 Bioinformatics methods and tools of bioinformatics used in the systems biology of plants. (Source: El-Hashash and El-Absy 2019)

genome sequencing and other related project activities, bioinformatics is an information science application in biology (Acquaah 2007). Primary databases, which include original biological data such as raw DNA sequences and protein structure information from crystallography, and secondary databases, which contain the original data that has been processed to fit particular specialized applications, are the two categories into which the information used in bioinformatics research can be separated (Acquaah 2007). The application of many bioinformatics techniques, including transcriptomics, proteomics, metabolomics, and phenomics, has been made possible by recent developments in functional genomics. The bioinformatics techniques and instruments utilized in plant systems biology are depicted in Fig. 2.12 (El-Hashash and El-Absy 2019). Tanwar et al. (2022) employed bioinformatics methods to facilitate genome-wide identification, detailed gene characteristics, evolution, development, and stress impacts on the expression of gene families involved in the polyamine metabolic pathway.

Numerous studies have used RNA-sequencing (RNA-seq) to examine the dynamic expression patterns of transcribed genes, potentially yielding novel biological insights (Li et al. 2023). Compared with other main field crops, the functional annotation of the barley genome is quite inadequate. Furthermore, according to Lee et al. (2020), there are currently no bioinformatics tools available for system-wide analysis of barley omics data. Furthermore, wet labs and field researchers with limited computational resources and bioinformatics expertise still face significant challenges when it comes to processing and interpreting these enormous volumes of histology data (Li et al. 2023). Consequently, BarleyNet, a co-functional network of 26,145 barley genes, and a web server for network-based predictions were created by Lee et al. (2020). Furthermore, BarleyExpDB is expected to be a useful tool for the barley research community in utilizing the extensive publicly accessible RNA-seq datasets for crop molecular breeding and functional genomics studies (Li et al. 2023). Using resources from the ExPASy bioinformatics resource portal, Tanwar et al. (2022) were able to determine the isoelectric point, molecular weight, amino acid sequence length, and grand average of hydropathicity for every of the determined barley polyamine metabolic pathway proteins. Cao et al. (2024) separated 24 FADs from the barley genome using bioinformatics methods. According to Cao et al. (2024), phylogenetic analysis revealed that HvFADs are highly expressed in roots treated with alkali or salt, suggesting that they may help barley tolerate these conditions. For the first time, Zhou et al. (2024) used self-transcribing active regulatory region sequencing (STARR-seq) to find enhancers throughout the barley genome. This method illustrates the distinct characteristics of barley enhancers and makes them available for additional use.

2.6 Tissue Culture Applications

Plant tissues can be removed and grown on a nutritional medium as part of plant tissue culture. Numerous variations are employed rather widely, including anther or pollen culture to create haploid plants, cell suspension culture, tissue and organ culture, protoplast culture for mutagenic treatment, and meristem culture for virus-free plant breeding (Christos 2022; Morrison and Evans 1988). Researchers have access to special materials from plant tissue cultures that help speed up the creation of novel breeding genotypes and make research on off-type regenerates easier. The increasing variety of regenerants that come from plant tissue cultures could be beneficial or detrimental to the value of regenerated plants, and it could have both genetic and epigenetic roots. Therefore, it is imperative to ascertain the means of regulating the degree of variation in plant tissue cultures (Orłowska and Bednarek 2020). Only a few genotypes of barley can be routinely produced as transgenic lines in the culture of somatic tissue; the most popular genotype for agrotransformation is "Golden Promise" (Matres et al. 2021; Nowak et al. 2024). Applications of plant tissue culture are shown in Fig. 2.13.

2.6.1 Embryo Rescue

Among the most established and effective in vitro tissue culture methods for plant species are embryo rescue procedures. A group of techniques known as "embryo rescue"works to encourage the transformation of an immature or deadly embryo into a live plant (Rogo et al. 2023). The cultivar Golden Promise was employed in the first reports of agrobacterium-mediated transformation of immature barley embryos in 1997 (Tingay et al. 1997), and this cultivar has been used in most barley transformations subsequently (Orman-Ligeza et al. 2020). The application of biotechnological methods, such as embryo rescue, identification of possible compatible genotypes, and chemical alterations during seed development, can all help to overcome the sexual barriers that prevent cross-species hybridization (Pickering 1984). Furthermore, after 2–5 days of growth, the developing endosperm of domestic barley × bulbous barley hybrids aborts, meaning that the embryos must be saved

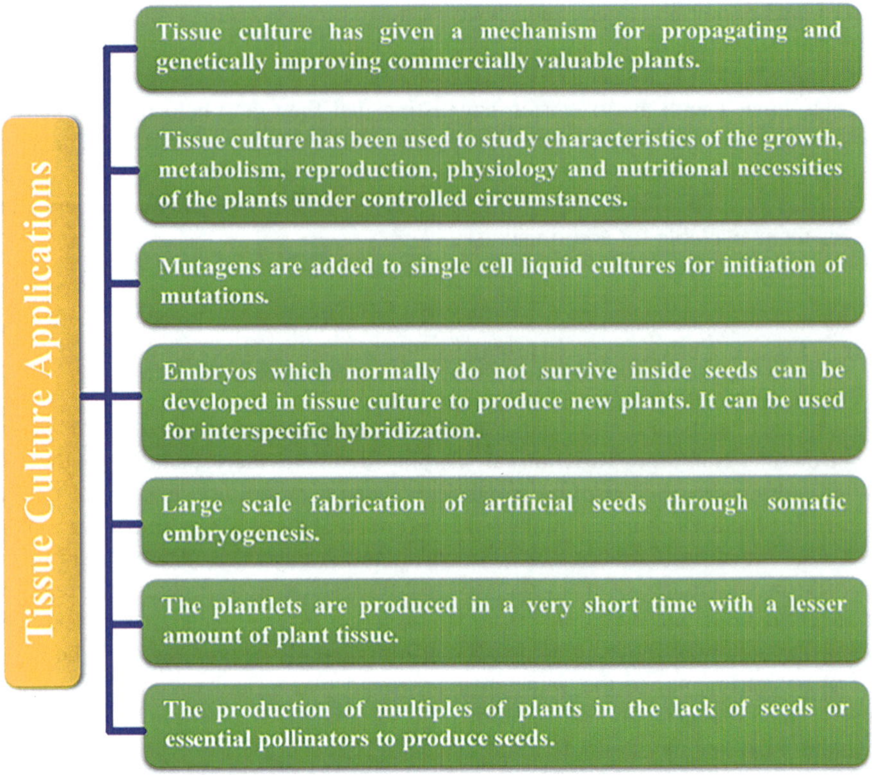

Fig. 2.13 Applications of plant tissue culture. (Prepared by: E.F. El-Hashash)

for them to finish developing. Therefore, to recover a crop from these sexually incompatible crossings, immature embryos must be cultivated on a nutritionally rich medium (Chahal and Gosal 2006; Pickering and Johnston 2005). In tissue culture for barley transformation and gene editing, embryogenic callus derived from microspores rather than scutellar tissues was used to overcome genotype dependence. This led to the successful production of transgenic and gene-edited plants from commercial types (Han et al. 2021). In the future, hybrid embryos should be assessed on culture media to maximize embryo rescue and seedling growth when new barley germplasm (wild and domestic) is obtained from seed banks for crossing and breeding studies (Mohammed et al. 2020).

2.6.2 *In Vitro Pollination*

For barley, an in vitro culture system has been created that produces high frequencies of good-quality embryos from microspores without requiring a callus phase. The shape and germination potential of the embryos are comparable to those of zygotic embryos (Datta and Potrykus 1989). After the flower bud is extracted from the plant and the leafy flower parts and anthers are dissected, it is placed on a tissue growth medium, where pollen is administered to the stigma of pistils grown in vitro. Fertilized ovules can be isolated and cultured in vitro after fertilization to avoid incompatibility with the maternal tissues around them. To fully overcome incompatibility barriers that are present in style and/or stigma, it is also feasible to remove the ovarian wall, as well as style and stigma, and perform placental pollination, which involves placing pollen on the placenta. Even when the ovarian wall has been removed or the ovules have been cultured in vitro, pollen can still come into direct contact with them (Van de Wiel et al. 2010). The most effective and practical in vitro approach for producing doubled haploid lines in barley is anther culture. When haploids and doubled haploidy are produced using these techniques, complete homozygous lines from heterozygous genotypes can be developed in a single step, which reduces the amount of time needed to develop homozygous genotypes when compared to traditional breeding (Patial et al. 2021). One of the primary barriers to the wider application of genetic transformation in barley breeding is genotype-limited plant regeneration. Therefore, the core of barley biotechnology continues to be the development of novel strategies that may enhance the responses of in vitro resistant genotypes (Nowak et al. 2024).

2.6.3 Synthetic Seeds

Murashige first proposed the idea of covering somatic embryos and utilizing them in the same way as a regular seed in 1977 (Siraree 2022). Artificially encapsulated somatic embryos, shoot tips, axillary buds, cell aggregates, or any other meristematic tissue which is utilized for seed sowing and can develop into entire plants in both in vitro and in vivo settings, as well as maintain their potential even after storage, are generally referred to as synthetic seeds (Capuano et al. 1998). Since an artificial seed is a somatic embryo wrapped in an artificial coat that resembles an immature zygotic embryo at most, possibly at the post-heart stage or early cotyledonary stage, it is sometimes referred to as a novel analog to a true seed (Bekheet 2006). In general, the production of synthetic seeds can be induced using somatic embryos, microshoots, protocorm-like bodies, and shoot buds as the encapsulating propagule (Awal et al. 2007). Synthetic seeds were somatic embryos used in field dispersal and plant culture. Following the introduction of artificial seeds, acceptable stand-ins for somatic embryos have also been found in the apical terminals of shoots, axillary buds, and nodal segments (Hussien et al. 2024). In general, artificial seed offers the following advantages:

1. Encourage the large-scale, economically viable reproduction of economically valuable plant species or superior plant genotypes (Reddy et al. 2012; Saiprasad 2001).

2. The mass manufacture of plant propagation systems in vitro has profited from synthetic seed technology (Awal et al. 2007). Better yet, clonal plants can be propagated similarly to seeds (Bekel 2021).
3. Breeding issues in seedless plants, such as male or female sterility, unsuccessful seed propagation in some crop species because of very small seed size, heterozygosity of seed, reduced endosperm, no germination in the absence of seed, and polyploid species multiplication, may be resolved by producing synthetic seeds in a tissue culture system (Awal et al. 2007).
4. The protection of germplasm by cryopreserving artificial seeds, especially in species that are susceptible to drought, and the preservation of uncommon plant species to increase biodiversity (Bekel 2021; Siraree 2022).
5. According to Awal et al. (2007), this type of encapsulation safeguards the microshoots and permits sensible handling, much like with genuine seeds.
6. A more regular and coordinated harvesting of agricultural crops (Bekel 2021).
7. Desiccation tolerance is a trait that somatic embryos might exhibit in response to environmental chemical or physical stressors. The stability of membranes in a dry state and the inhibition of biomolecules' oxidative destruction are two of the mechanisms behind desiccation tolerance. Encapsulation of embryos may regulate their intake of water, the release of nutrients, and the provision of the mechanical protection needed for field planting (Senaratna 1992).
8. The use of genetically identical somatic embryos to maintain the clonal nature of plants, the use of a suitable medium to transport novel plant lines obtained through biotechnological means directly to the greenhouse or field, the economically viable large-scale propagation of superior plant varieties, the absence of pathogens, ease of handling during storage, transport feasibility, long-term storage without loss of viability, etc. are additional benefits of synthetic seeds (Bekel 2021; Saiprasad 2001; Siraree 2022).

In vitro growing techniques are required for synthetic seed technologies to generate viable materials that can be widely transformed into plants (Abbas et al. 2022). Encapsulation, somatic embryogenesis, organogenesis, and improved auxiliary bud proliferation systems are efficient techniques for the rapid and widespread in vitro multiplication of elite and desired plant species. These methods enable the large-scale production of viable materials that can be converted into plants, which is necessary for the deployment of synthetic seed technology (Mudasir Magray et al. 2017). This technique can be modified for commercial production after it reduces the cost of producing elite plant genotypes (Abbas et al. 2022).

Beginning in 1989, artificial seed technology was applied to cereals (Bekel 2021). The development of artificial seed technology presents fresh opportunities for a viable plan to enhance a variety of economically important plant species, like cereal crops (Abbas et al. 2022). In autogamous species like wheat, barley, and oats, where commercial hybrid seed production is impractical due to high production costs, artificial seeds would be used to disperse hybrid seeds. In this instance, small-scale manual hybrid seed production would be followed by mass-scale artificial seed multiplication using technology (Kumar 1998). Barley artificial seeds germinated better than non-encapsulated embryos and had a more developed root system. On the other hand, non-encapsulated embryos did not survive in storage for more than 2 weeks, but artificial seeds retained their germination capacity for at least 6 months. Therefore, artificial seeds most likely offer a straightforward and universal method of delivering in vitro plantlets to a field or greenhouse (Datta and Potrykus 1989). High yield and quality are essential for fulfilling market demand and high output, and barley seeds that perform exceptionally well during grain growth, seed dormancy, and germination are essential for ensuring this (Xiong et al. 2024).

2.7 Genetic Engineering

Genetic engineering refers to manipulating plants using genetic modification, also known as transgenic (Fig. 2.14), which involves introducing genes from any organism, including bacteria, humans, other cereals, or even barley itself (Newman and Newman 2008). A transgenic plant expresses a *transgene*, which is a gene that has been inserted into an organism by recombinant DNA technology. *Agrobacterium* coculture or particle gun acceleration are two transformation techniques that are used to deliver the transgene into the plant cells once it has been incorporated into suitable plant expression vectors (Singh and Singh 2015). Since the 1980s, efforts to create transgenic plants by plant breeding with *Agrobacterium* have been underway (Bevan et al. 1983). The first transgenic plant to be allowed for commercial production was the tomato cultivar Flavr Savr, which was approved in 1994 (Singh

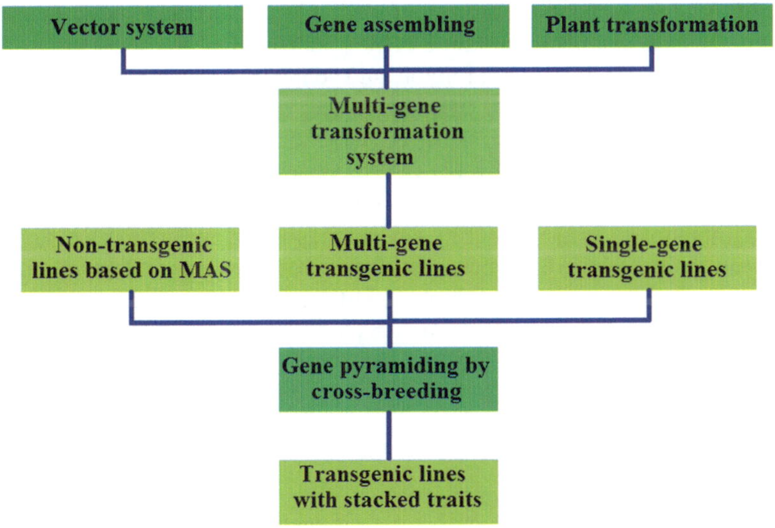

Fig. 2.14 Using transgenic pyramiding techniques to improve crops. (Source: Wan 2015)

Table 2.4 Transgenesis for specific characteristics and yield in barley

Gene	Function/stress tolerance	Reference
HvCKX1	Regulation of cytokinin content	Holubová et al. (2018)
CYTOKININ OXIDASE/ DEHYDROGENASE	Drought stress tolerance	Ramireddy et al. (2018)
HvAKT2	Drought tolerance	Feng et al. (2020)
HvGST4	Drought/heat/salt/cold/ hypoxia	Pan et al. (2022)
ALBOSTRIANS	Barley stripe mosaic virus	Tamilselvan-Nattar-Amutha et al. (2023)
GA2ox	Grain size and weight	Cheng et al. (2024)

and Singh 2015). Through the cocultivation of immature embryos with Agrobacterium tumefaciens, Tingay et al. (1997) achieved the first stable barley transformation. Numerous techniques, including direct gene transfer, plasmid DNA extraction, viral infection, and plant bacterial pathogens like *A. tumefaciens*, can be used to introduce recombinant DNA into barley cells (Kumlehn et al. 2014). Barley has been modified using a variety of methods, including particle bombardment, PEG-mediated, and electroporation-mediated (Joung et al. 2015). While the HVA1 gene conferred greater stress tolerance, the overexpression of the barley HvCBF4 gene in genetically engineered rice enhanced resistance to drought, low temperatures, and high salt (Oh et al. 2007; Rohila et al. 2002).

Stable genetic transformation in barley is necessary for the transfer of genetic reagents needed to trigger Cas9-mediated targeted genetic engineering (Lawrenson et al. 2015; Gerasimova et al. 2020). Over time, functional genes in barley under different stress conditions have been found, and with the advent of contemporary gene-editing tools, genetic advancement toward stress tolerance has taken a new direction. According to Karunarathne et al. (2023), CRISPR/CRISPR-associated protein 9 (Cas9) in particular is a reliable and adaptable tool for precisely creating mutations and improving traits. In addition to lowering the trade-off effects of Green Revolution semi-dwarfing genes on barley grain size and weight, the results provide a viable method of modifying GA2ox genes to fine-tune GA signaling and biofunctions in specific plant tissues. Encourage the development of barley cultivars that are more adaptable to future environmental changes (Cheng et al. 2024). Many transgenic barleys have been developed with desired traits like increased grain yield, tolerance to drought, high salinity, and low temperatures, as well as greater resistance to fungal diseases such as leaf rust and stem rust (Table 2.4).

2.8 Mutation Breeding

A mutation is a heritable alteration to genetic material; people who display altered traits as a result of these alterations are referred to as mutants (Mba 2013). In a broad sense, every alteration to a genome's nucleotide sequence can be referred to

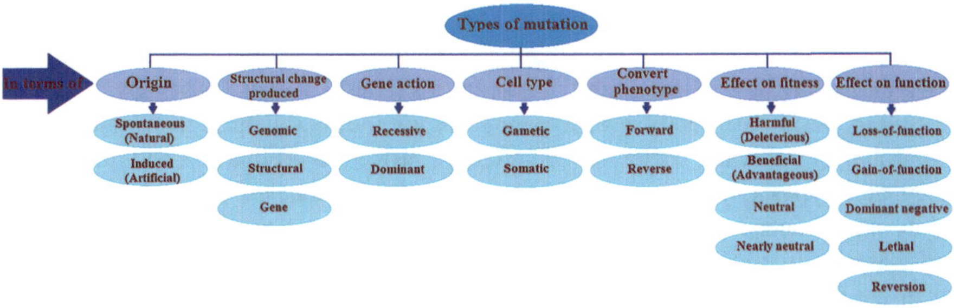

Fig. 2.15 Types of mutation. El-Hashash and El-Absy (2019)

as a mutation (Saitou 2013). Figure 2.15 displays the different types of mutations, as reported by El-Hashash and El-Absy (2019). Mutation breeding is the application of produced mutations for crop enhancement. Plant mutation breeding is a popular method for crop improvement because it is an effective way to identify biological processes, particular gene functions, and connections between mutations and phenotypes in the mutant plant (Lee et al. 2015). The ultimate cause of evolutionary change is a mutation; new alleles appear in all species, some of which develop naturally and others which are brought on by environmental exposure to chemicals and radiation (Griffiths et al. 2005). Prominent instances of mutation breeding include the creation of semi-dwarf mutant cultivars of barley (*H vulgare* L.) in the early 1970s, such as Golden Promise and Diamant (Spencer-Lopes et al. 2018). According to FAO/IAEA 2024, the Cargine cultivar is the first barley commercial mutant created in France by x-ray (100 Gy) seed irradiation. (FAO/IAEA 2024) states that the number of mutant cultivars of cereal crops was counted up until 2024, with rice (873) showing the largest number, followed by barley (307) and wheat (265). Maize (89) came in fourth. Although straw from grain cultivation is frequently seen as a straightforward byproduct of the crop that has been harvested, it is really utilized in the bioindustries to make biochemicals, textiles, paper, and biofuel (Guerriero et al. 2016). Enhancing the rigidity of barley culm to decrease lodging resulting from elevated harvest index, intense winds, and precipitation could capitalize on the numerous identified mutations (Dockter and Hansson 2015; Meng et al. 2023).

2.8.1 Conventional Mutagenesis

The breeding of barley dates back more than a century. Numerous natural and artificially produced mutants have been gathered and produced by extensive breeding efforts; these mutants are frequently stored in gene banks across the globe. Many historical mutants have come into sharper focus in recent years due to a growing interest in genetic variety. The collections are seen to be important resources for comprehending the genetic regulation of barley biology and breeding. Recent developments in genomic research have also contributed to the rising interest since they have made it possible to investigate and uncover the genetic variety of mutant collections using new methods and opportunities (Hansson et al. 2024). By employing mutagenesis to directly create genotypes or indirectly through crosses with other superior cultivars, barley breeders can improve the production and quality of barley genotypes (Anderson and Reinbergs 1985). Because seeds are the most widely used material for induced mutation, it is important to choose high-quality seeds. According to Spencer-Lopes et al. (2018), seeds must be of the greatest caliber, disease-free, uniform, representative of the cultivar, line, or genotype, and have a high germination rate of 90% or higher. According to El-Hashash and El-Absy (2019), physical and chemical mutagens are the two main categories into which artificial mutation can be separated. When pure barley line seed samples are exposed to a mutagen, they are classified as M0 seeds and become M1 plants upon germination. Following mutagenesis treatment of the seeds, the M2 or M3 generation is often when mutants are selected (Maluszynski et al. 1995). Although the M1 population contains some uncommon dominant mutants, selection is typically not used until later generations because of the chimeras and physiological abnormalities that M1 plants display (Szarejko and Forster 2007). Seldom is the segregation ratio for a recessive mutant (in the progeny of a single M1 plant) 3:1 when selection is carried out on a single plant basis in the M2 generation. Because of the chimeric nature of M plants and the somatic effects of mutagenesis therapy, recessive mutants will never be sufficient in M2 progeny. The homozygosity test must be performed (in the M3 or M4 generation) prior to employing the true-to-type homozygote mutant line for a cross-breeding program or direct agronomic evaluation (Szarejko et al. 1995). Before adding mutant lines into a crossing program, it can take up to four generations to find and evaluate mutations because many of the characters that attract plant breeders are quantitative, like yield and quality (Szarejko and Forster 2007).

Mutants that may be utilized in barley breeding initiatives were soon discovered. In terms of yield, seed size, straw stiffness, straw length, early maturity, tillering ability, variations in spikes, kernels, and awns, and colors, among other traits, a large number of highly attractive mutants were generated. Diamant, Golden Promise, and their hybrids are likely the most well-known mutant barley types (Ohnoutkova 2019). The Erectoides mutants were one of the most prevalent types; they were distinguished by an erect spike that was compact or dense and by a culm that was sometimes short and rigid. Culm strength characteristics include: easily lodged plants, fast, stiff straw, weak culm, and brittle culm at maturity (Hansson et al. 2024). Stiff straw was the Pallas mutant variety's primary improved trait. It has been extensively grown in several European nations. There have also been hybridization programs using the cultivar "Pallas." According to Ohnoutkova (2019), Golden Promise has short, stiff straws, a strong yield, and good malting quality. Hua30 microspores demonstrated some resistance to mutagenesis in Huang et al. (2023), ethyl methanesulfonate-induced mutagenesis of microspore cultures of landrace 'HTX' and cultivar 'Hua30' of barley. To increase the frequency of mutations in the barley genome, a dosage of more than 300 mg/L might be considered.

2.8.2 In Vitro Mutagenesis and Selection

A workable way to produce unique genetic variety is by mutagenesis employing in vitro plant cells and tissue cultures (Larkin and Scowcroft 1983). Since totipotent plant cells may be cultivated, multiplied in enormous quantities, and stimulated to regenerate entire plants, in vitro culture techniques are especially pertinent to mutagenesis. Radiation exposure or genetic alteration can be applied to rooted stems, cuttings, detached leaves, dormant buds/plants, shoot apices (apical buds), axillary buds, tubers, and other plant components (Penna et al. 2012). Any program involving mutation-breeding must include the selection and identification of suitable mutants (Penna et al. 2012). The availability of (1) a high variation of cells, (2) an easy way to apply the in vitro selection method, (3) a method for the regeneration of tolerant cells, and (4) the desired character to be inherited are all necessary for the in vitro selection technique to be successful in producing tolerant plants (Widoretno and Harran 2003; Yusnita 2005). The ability to treat and screen huge populations of cells prior to their regeneration into whole plants is one of the main advantages of utilizing in vitro cultures for mutagenesis (van Harten 1998).

In many areas of barley breeding and research, doubled haploids (DHs) for mutant induction and selection have been used among the different in vitro techniques (El-Hashash and El-Absy 2019). For direct screening, an efficient technique for producing several DH lines with homozygosity stability and mutant introduction using in vitro mutagenesis employing isolated microspore culture is offered (Gao et al. 2018). Given that induced mutations are primarily recessive and typically go undetected until the M_2 generation, the capacity of DH to repair mutations is an important factor. Targeting haploid cells, producing haploid cells for mutation treatment, and capturing the mutation in a homozygous, pure line are all made possible by the DH systems themselves (Szarejko and Forster 2007). Umba di-Umba et al. (1991), Szarejko et al. (1991, 1995), and Vagera et al. (2004) used M1 material of *H. vulgare* plants as examples of DH production. A rare mutation event is more likely if the mutant selection procedure can be carried out in vitro because the number of monocytes or embryos that can produce a significant number of mutations caused by classical mutations in barley. At the haploid or DH cell, embryo, callus, or plant level, selection variables can be applied (Szarejko et al. 1995). Tantau et al. (2004) and Chawla and Wenzel (1987) have documented the development of frost and disease resistance (*Fusarium* spp.) in barley through in vitro selection. Gao et al. (2018) state that boosting nitrogen uptake is necessary to improve nitrogen usage efficiency in barley and that high nitrogen uptake efficiency mutants can be produced by field screening and microspore mutagenesis.

2.8.3 Molecular Analysis

Site-directed mutagenesis, induced mutations, transgenics, genomic selection, marker-assisted selection, and machine learning are some of the new technologies that barley breeders have implemented to improve quality attributes (Meng et al.

2023). Any DNA segment of interest, including genes or regulatory regions, can have specific nucleotide changes targeted by the polymerase chain reaction (PCR) (Jost et al. 2019). Salvi et al. (2014) state that direct sequencing of PCR products can also be used to target induced local lesions in genomes (TILLING) if it is possible to differentiate between sequencing errors and real mutations. The formation of hetero-duplexes at mismatch positions between paired polymorphic DNA fragments during PCR is the basis for identifying induced single nucleotide polymorphisms (SNPs) in mutagenized plants (Jost et al. 2019). SNP genotyping and next-generation sequencing (NGS) techniques enable the cloning of target mutant genes in single-step experiments and the genetic and physical mapping of segregating populations (Salvi et al. 2014). The development of cv. Diamant from cv. Valticky was caused by a mutagen's effect on several genetic loci, which is larger than expected based on data from amplified fragment length polymorphisms (AFLPs) and simple sequence repeats (SSRs) (Mlčochová et al. 2004).

When it comes to causing point mutations in barley, the chemical mutagen N-methyl-N-nitrosourea (MNU) works better than the physical mutagen gamma rays. In the chosen sequences, the average density of mutations brought on by the MNU was 2.2 mutations per megabyte, or 1 mutation per 504 kb. The MNU-treated population had a level of AFLP polymorphism that was around three times higher than that of the gamma-irradiated group, according to the genomic barley survey's use of AFLP markers. According to Kurowska et al. (2012), the sequenced polymorphic AFLP products have undergone bioinformatic analysis, which suggests that they reflect the control of the barley genome. As a result, they can be utilized for genome-wide mutation screens in this species. Because radiation and bigger lesion introduction techniques like deletions and translocations are incompatible with PCR-based mutation detection, they should not be used to cultivate the TILLING population in barley (Jost et al. 2019).

The differences in the barley cytochrome P450 gene that exacerbate pathogen-induced programmed cell death and cutin layer instability were investigated by Ameen et al. in 2021. Furthermore, they asserted that our collection of mutant alleles, particularly the nec3.d mutant, will be a great tool for figuring out how the buildup of serotonin and maybe phenolic chemicals in necrotic lesions helps to prevent pathogen colonization or lesion expansion. A useful tool for researching the regulation of programmed cell death and hypersensitive reactions is the barley Nec3 gene. Genetic mapping and description of the gpa1.A mutation that causes a variegation phenotype in barley was done by Yang et al. (2021). On chromosome 2H, the Gpa1 gene was part of a gene cluster that was functionally associated with either chloroplast differentiation or photosynthesis. One potential Gpa1 candidate is the PTOX-encoding gene in the defined region. Consequently, the current work lays the groundwork for Gpa1's cloning, which will improve our comprehension of the molecular processes underpinning chloroplast biogenesis, especially in monocot plants (Yang et al. 2021). In barley, novel varieties with increased anthocyanin accumulation and various colors have been produced by a combination of conventional and molecular approaches (Meng et al. 2023). Using sequencing, 12,926 and 13,361 harmful SNPs were found in 48 wild barley samples from 10 groups. In response to

28 years of global warming, wild barley samples exhibited increased harmful mutations across seven chromosomes and enhanced selection sweeps. These findings reinforce the necessity to protect crop wild relatives and are important for understanding the genetic risks of losing wild barley due to global warming (Fu et al. 2024).

2.8.4 Enhanced Traits and Improved Cultivars

The 307 barley cultivars that have been released globally are listed in the FAO/IAEA (2024) Mutant Varieties Database. Both straight mutants and their hybrids are included in the database. The details contain the ID of the mutant variety, the name of the cultivar, the nation, the year of local registration, a brief description, information on character improvements, the kind of usage subcategory, and the type of mutant development. In addition to numerous other commercially significant characteristics, these cultivars have improved quality, earliness, good grain, straw stiffness, proanthocyanine-free content, drought and salt tolerance, disease and lodging resistance, cold and nematode resistance, and genetic male sterility (El-Hashash and El-Absy 2019). CVS for barley mutants. European brewing has greatly benefited from the excellent yields and small stature of Diamant and Golden Promise. The brewing and malting industries now have billions of USD more in value because of Golden Promise and Diamant (Ahloowalia et al. 2004). Golden Promise is thought to have contributed USD 417 million to Scotland's grain production alone (Ahloowalia et al. 2004). As a common cultivar that emerged from transformation studies, Golden Promise cv. is still widely grown and planted in Scotland. The enormous economic significance of the cv. Diamant, which is currently being crossed into other barley cultivars around the world, cannot be overstated (Ohnoutkova 2019). TILLING has been successfully applied in barley improvement, including the manipulation of spike morphology, starch content, and waterlogging tolerance by Gottwald et al. (2009), Sparla et al. (2014), and Mendiondo et al. (2016), respectively, because it leverages high-throughput mutation identification techniques and classical mutagenesis.

Natural variations of barley chemicals already in existence can be merged with numerous mutant lines or exploited to create superior cultivars. To find new mutations and get beyond the regulatory obstacles that are currently preventing agricultural advancements, it is advised that existing germplasms be screened utilizing TILLING (Meng et al. 2023). To identify the gene responsible for a trait in any species, it is crucial to connect the current probe set with the most recent whole-genome assembly. This allows for the inclusion of potentially unannotated genes in the study (Ameen et al. 2021). Site-directed mutagenesis in conjunction with natural variations can yield improved resources for customized health goods, as many of the genes governing barley's chemicals are known (Meng et al. 2023). A useful tool for determining the genetic variables underlying tiller number, leaf size, and orientation features in barley is the HorTILLUS population, which was created by mutagenesis of the spring two-row variety Sebastian. Mutant screens provide a potent method for achieving this goal. According to Mohammadi Aghdam et al. (2022), the chosen mutants offer good starting points for identifying the genetic variables governing tillering and leaf angle in barley.

To identify the barley mutant line with novel properties, Kihara et al. (2023) examined the traits of the barley mutant line “N68-411” with an enhanced balance of malt modification and seed dormancy. The N68-411 characteristic rapidly changed during germination, indicating that the germination period may have been shortened during the malting process. Analysis of progeny from 12WY375/N68-411 and Xena/N68-411 revealed that the “N68-411” phenotype was inherited and associated with a small number of genes. Kis et al. (2024) investigated how certain mutations in the GW2.1 gene could alter grain characteristics and reduce barley yield. They suggest that to optimize grain characteristics and yield, the barley GW2.1 gene is required. All things considered, our research demonstrates that the loss of GW2.1 gene activity in barley is linked to pleiotropic consequences that adversely impact the growth of generative organs and, in turn, grain yield.

2.9 Hybridization

To introduce novel varieties and start a breeding program, hybridization is the crossover or mating of two parental plants or pure lines that have interesting features. Hybrid breeding has also been used to produce new cultivars, as demonstrated by the development of barley and other crops (El-Hashash and El-Absy 2019).

2.9.1 Conventional Hybridization

You can make barley crosses in a field or a greenhouse. Every plant, with the exception of around 10–15, has its spike or panicle with florets that are meant to be used as a female parent removed from the head prior to anthesis. To prevent self-pollination, the remaining flowers’ heads are then put in a paper bag, and the stamens are removed with tiny forceps prior to the anthers dehiscing. The crossing process is finished when pollen grains from the male parent are applied to the stigma of the flowers of the female parent. According to Hayes and Immer (1942), the paper bag is placed over the pollinated head and remains there till harvest. A hybrid, or first generation (F1), is the term used to describe the seeds and offspring produced by hybridization. The F_1 offspring and subsequent generations, which are produced by selfing or merging F_1 and subsequent generations of plants, are the segregating generations. The hybridization results (F1) and the segregating generations (F_2, F_3, etc.) are frequently referred to as crosses. To create new cultivars, transgressive segregants in the strain that outperform their parents from the F2 generation and subsequent generations are selected. These strains are referred to as segregating generations and can be managed by bulk, backcross, or pedigree approaches. Some of the limitations of this strategy include the small number of recombinants generated in the F_1, F_2, and F_3 generations, as well as the low likelihood of genotype changes in subsequent generations because of the quick completion of homozygosity, as shown by the F_4 generation (87.5%) (Rauf et al. 2016). Numerous barley cultivars were created in various nations by utilizing genetically distinct individuals from the same species (intraspecific hybridization) or hybridizing individuals of different species (interspecific hybridization), according to von Bothmer et al. (1992).

2.9.2 *Somatic Cell Hybridization*

The goal of plant breeding is to introduce and manipulate genetic variants to create new genotypes. Combining genetic components by protoplast fusion has the potential to be a valuable strategy for producing somatic hybrid plants. When nuclear and/or cytoplasmic genomes recombine as a result of protoplast fusion, new genetic variations may occasionally be produced (Kisaka et al. 2001). By allowing for the gene transfer of the nuclear and cytoplasmic genomes, protoplast fusion offers a means of overcoming obstacles that impede sexual reproduction and enriches the gene pool of domesticated species. Moreover, protoplast fusion efficiently produces new germplasm for elite traditional crossbreeding and fosters agricultural enhancement in existing varieties (Wang et al. 2013). The isolation of protoplasts capable of dividing and yielding somatic embryos and plants resulted from the Gramineae's initial utilization of embryogenic cell suspension devices as a source of protoplasts (Vasil and Vasil 1979). *Hordeum vulgare* has been used to create numerous somatic hybrid cells and hybrid plants through protoplast fusion with other species, like rice (Kisaka et al. 1998), tobacco (Somers et al. 1986), carrot (Dudits et al. 1976; Kisaka et al. 1997), and soybean (Kao and Michayluk 1974). Mihók et al. (2024) created wheat × barley F_1 hybrids with various barley chromosome combinations that expressed genes encoding for either one (αCENH3) or both (α– and βCENH3) barley CENH3 protein variants. To determine the exact pattern of parental CENH3 deposition into the centromeric nucleosomes of wheat and barley, we developed particular antibodies that could differentiate between the wheat CENH3 proteins and the barley αCENH3. These antibodies were then used in conjunction with FISH probes. By examining the somatic and meiotic nuclei of the wheat × barley hybrids, it was demonstrated that the maternal CENH3 proteins, which are present in wheat, might fuse with the paternal centromeric nucleosomes, which are present in barley.

2.9.3 *Hybrid Cultivars*

Barley hybrid cultivars are defined as the offspring of an F1 generation cross between two or more unrelated pure lines; these progenies can be inbreds, clones, or other genetically distinct populations/lines; they are then released for commercial

	SY KINGSBARN Recommended List Yield	10.39 t/ha
Agronomic ratings	Yield (UK)	106%
	Yield (heavy soils)	105%
	Yield (light soils)	105%
	Specific weight	70.85 kg/hl
	Resistance to lodging with (+PGR)	7
	Maturity	0
Disease ratings	*Rhynchosporium*	6
	Net blotch	5
	Brown rust	5
	Mildew	7
	BaYMV	Resistant

Fig. 2.16 Overview of SY KINGSBARN hybrid. (Prepared by: E.F. El-Hashash)

cultivation and may possess one or more advantageous features. Abiotic and biotic tolerance, yield, grain quality, and other key agronomic qualities are all best achieved by hybrid barley cultivars (El-Hashash and El-Absy 2019). For instance, all of the hybrid barley varieties do well in the UK, where the SY KINGSBARN, SY NEPHIN, and SY BUZZARD hybrids hold the top three spots. UK growers are increasingly choosing hybrid barley because it produces large amounts of high-quality feed barley. In addition to having a high yield, hybrid barley is a beneficial crop choice. With good cause, the hybrid SY KINGSBARN is the most popular hybrid barley and the easiest to grow, ranking first in yield in the 2022 recommended list testing (Fig. 2.16). High and consistent yields with a track record of success on the farm; early maturity combined with excellent grain quality; good resistance to lodging and disease; and a special ability to reduce grass weeds due to hybrid vigor (https://www.syngenta.co.uk; accessed May 22, 2024).

In order to investigate the effects of plant hybridization in barley on the total thermomechanical performance of polymeric composites, including their by-products, Puglia et al. (2020) gathered and characterized straws from two F1 hybrid lines of barley (Tatoo and Zoo) and one self-pollinated line (Quench). They found that hybrid lines had somewhat greater Young's modulus and tensile strength than self-pollinated Quench. For polypropylene composites reinforced with hybrid lines, the stems' varied mechanical responses resulted in increased stiffness, while the barley straw line had no discernible effect on tensile strength.

2.10 Conclusion and Prospects

2.10.1 Current Research Initiatives to Combat Global Climate Change

Global food security is seriously threatened by climate change, which has an impact on all livable continents and every nation on Earth. Because there are so many variables, it is exceedingly challenging to forecast with precision how climate change may affect crop yields in the future. These consist of: (1) physical factors like temperature, precipitation patterns, and fertilization with carbon dioxide; (2) modifications to agricultural environmental systems like increasing occurrence of biotic stressors and pollinator loss; and (3) human systems' adaptive responses. We need to act today to address climate change since it is a pressing issue of our day. All facets of society will need to make an unparalleled effort to combat climate change, but there is still time. Ten developing nations had finished submitting their first national adaptation plans to address climate change, 168 parties had reported their first nationally identified contributions to the UN Framework Convention on Climate Change Secretariat, and 175 parties had ratified the 2016 Paris Agreement as of April 2018. There are 198 Parties to the Convention as of right now. Additionally, by 2020, industrialized nations will have pooled $100 billion a year for mitigation efforts. In November 2024, Baku, Azerbaijan, will host COP29, a climate change conference aimed at increasing motivation and speeding up action.

Due to the extensive and well-formed collections of landraces, wild genotypes (*Hordeum vulgare* ssp. spontaneum), and other Hordeum species that are significant sources of novel alleles, barley has emerged as a paradigm for establishing an excellent response to the effects of various climate change scenarios. Consequently, compared to wheat and the majority of other minor grains, barley yields indicate low volatility across climate change conditions. Some regions, such as the Andes (2.1%) and Finland (5.7%), will see lower barley harvests between 2010 and 2029 as a result of climate change. In the meantime, nations like the UK will profit from and see a boom in barley output due to climate change. Projected barley output varies from 4.6 million mt in 2030 to 9.0 million mt in 2050 under forecast land use and climate change. Landraces, wild species, and the larger *Hordeum* gene pool already have beneficial alleles to enhance barley; as a result, these resources are being used to counteract the decline in barley productivity brought on by climate change. Breeding efforts to enhance barley in drought-prone areas and with varying temperatures mustn't compromise the quality of the grain harvest. While planting barley that adapts to many climates and seasons is one way to prevent stress caused by climate change, traditional methods of reducing stress connected to climate change are not very efficient in tackling the many facets of climate change.

2.10.2 An Overview of the Current Status

Barley is the fourth most produced grain in the world, behind wheat, rice, and maize. It is used for both human consumption and animal feed. More so than other major grains, barley is grown all over the world and is resistant to cold, water deficit, salinity, and alkaline soils. Due to its availability and quantity, barley straw is currently a significant crop in agriculture. It is also used as a source of fiber to replace wood feedstock in some applications, like the creation of composite materials and energy. The world productivity of barley, which currently stands at around 149 million metric tons, shows that production increased by 5.49% in the 2023/2034 season as compared to the 2022/2023 season. Numerous experiments have been conducted worldwide to improve barley cultivars. To generate cultivars of barley fiber, future traditional and biotechnology breeding approaches should be more systematized and equipped with modern technical tools. Breeders are under pressure to expedite the breeding process due to worries regarding unfavorable conditions, anomalous weather patterns, and global warming. Crop straw, derived from barley, cornhusk, cornstalk, rice, soybeans, and wheat, is one of the most common plant fibers. Its lumen porosities are higher than 60%, which prevents the strong mechanical performance of the cell walls from transferring to the fiber scale. The type of plant, growth conditions, harvest maturity, precise fiber location within the plant, fiber extraction method, possible alkali treatment, storage conditions, and other aspects of the testing process all affect the physicochemical fiber properties.

2.10.3 Recommendations for Future Research

New sources of cellulose fibers, which are currently utilized in enormous quantities by the forest, textiles, and automobile industries, are necessary for the shift to recyclable, biodegradable, and sustainable products. Due to their similar chemical makeup, barley straws offer intriguing substitutes for wood fibers in lightweight material applications. Therefore, we hope to increase awareness of the potential and difficulties associated with using barley straw as a substitute source of cellulose fibers. Few details are available regarding the sustainability of barley fiber supplies in the future, even though a steady and sustainable supply of barley fiber is essential to the long-term viability of the barley fiber market worldwide. The fields of germplasm diversity, genetic resource conservation, cytology, cytogenetics, cell and tissue culture, conventional breeding techniques, conventional and somatic cell hybridization, molecular breeding, genetic engineering, and mutation breeding are notably understudied in the genetics of barley fiber. The application of biotechnology, like transgenics and molecular technologies, to various activities can benefit conventional barley fiber breeding. This allows programs to be implemented without introducing foreign genes into the produced cultivar. Barley straw fiber frequently necessitates considerable adjustments to the methods used for gathering, handling, processing, and conducting commerce. Stated differently, altering the sources of raw materials could necessitate creating new business models or modifying the ones that already exist. Generally speaking, creating a business model necessitates taking market demand into account in addition to other crucial components like raw material supply, processing, storage, and logistics. Moreover, close cooperation and modifications to the complete value chain are necessary to develop a viable business model that promotes a circular economy, like in the case of straw utilization. Therefore, all members of the barley research community must collaborate to develop cultivars that are resistant to both biotic and abiotic stressors and have high yield and fiber quality.

Appendix 2.1 Barley Genetic Resources

Table 2.5 Total barley germplasm accessions available at ten major world centers

Rank	Gene bank/ Institutes	Wild %	Landraces %	Breeding materials %	Genetic stocks %	Cultivars %	Unknown status %
1	PGRC	15	35	12	15	13	10
2	NSGC	7	44	13	11	15	12
3	ICARDA	7	59	17	2	9	6
4	CENARGEN	0	0	0	0	0	100
5	NIAS	4	75	0	20	0	0
6	IPK	6	56	10	2	23	2
7	CAAS	14	82	0	0	0	4
8	RDAGB-GRD	1	24	10	0	0	66
9	VIR	1	29	10	3	54	3
10	BCRI	0	100	0	0	0	0

Table 2.6 Major institutions engaged in barley research

Institution	Specialization and research activities	Contact information and website
International Barley Hub, James Hutton Institute	Barley research	Dundee, Scotland, UK, email: barleyhub@hutton.ac.uk, www.barleyhub.org
ICAR-Indian Institute of Wheat and Barley Research (IIWBR)	Increasing and stabilizing the barley production	Karnal-132,001, Haryana, India, Email: director.iiwbr@icar.gov.in, http://www.iiwbr.org/
Brewing and Malting Barley Research Institute	Malting barley research	P.O. Box 39,120 Lakewood PO, Saskatoon, Saskatchewan S7V 0A9, Canada, Email: gfeist@bmbri.ca., http://bmbri.ca/
Canadian Malting Barley Technical Centre	Providing technical support and market information to stakeholders of barley	1365–303 Main Street Winnipeg, Manitoba, Canada R3C 3G7, E-mail: cmbtc@cmbtc.com, http://cmbtc.com/

(continued)

Table 2.6 (continued)

Institution	Specialization and research activities	Contact information and website
Institute of Barley and Malt Sciences	Promote the informational and educational activities/resources on barley and malt at North Dakota State University	NDSU Dept. 7670, 166 Loftsgard Hall, N. Bolley Dr. Fargo, ND 58102. North Dakota State University, https://www.ag.ndsu.edu
Alberta's Field Crop Development Centre (FCDC)	The Field Crop Development Centre focuses on breeding barley for feed, malt and food	Toll-free in Alberta: 310-FARM (310–3276), Out of province 1–403–742-7901, Email: Ag Info Centre, https://www1.agric.gov.ab.ca
Global Crop Diversity Trust	International organization dedicated solely to conserving and making available barley and other crop diversity	Platz Der Vereinten Nationen 753,113 Bonn, Germany, info@croptrust.org, https://www.croptrust.org

Table 2.7 Barley genetic resources available in the ex situ germplasm collections of PGRC, NSGC and ICARDA

Species	Number of available accessions of barley		
	PGRC	NSGC	ICARDA
H. vulgare ssp. *spontaneum*	3875	1504	1696
H. vulgare ssp. *agriochriton*	–	1	9
H. vulgare ssp. *vulgare*	22,882	25,024	23,639
H. vulgare	9684	–	–
H. vulgare var. distichon	–	–	–
Hordeum mutant collection	–	–	–
Other species	5372	6647	6656
Total	41,813	33,176	32,000

Sources: PGRC, www.agr.gc.ca/pgrc-rpc; NSGC, www.ars-grin.gov/npgs; ICARDA, www.icarda.cgiar.org

Table 2.8 Total barley germplasm accessions available at ten major centers of the world

Rank	Country, city	Name/institution	Acronym	Code	Number of accessions
1	Canada, Saskatoon	Plant Gene Resources of Canada	PGRC	CAN004	41,813
2	USA, Aberdeen	USDA National Small Grain Collection	NSGC	USA029	33,176
3	Syria, Aleppo	International Centre for Agricultural Research in Dry Areas	ICARDA	SYR002	32,000
4	Brazil, Brasilia	Centro Nacional de Pesquisa de Recursos Genéticos e Biotec.	CENARGEN	BRA003	30,000
5	Japan, Ibaraki	National Institute of Agrobiological Sciences	NIAS	JPN003	23,471
6	Germany, Gatersleben	Institute of Plant Genetics and Crop Plant Research	IPK	DEU146	22,106

(continued)

Table 2.8 (continued)

Rank	Country, city	Name/institution	Acronym	Code	Number of accessions
7	China, Beijing	Institute of Crop Germplasm Resources	CAAS	CHN001	18,818
8	Republic Korea	Genetic Resources Division, National Institute of Agricultural Biotechnology, Rural Development Administration Suwon	RDAGB-GRD	KOR003	18,764
9	Russia, St. Petersburg	N.I. Vavilov Research Institute of Plant Industry	VIR	RUS001	17,850
10	Ethiopia, Addis Ababa	Biodiversity Conservation and Research Institute	BCRI	ETH001	16,388
11	Others countries				212,145
Total					466,531

References

Abbas MK, Mahood HE, Alhasan AS (2022) Production of synthetic seeds in vegetable crops: a review. IOP Conf Ser Earth Environ Sci 1060:012099. https://doi.org/10.1088/1755-1315/1060/1/012099

Abebe TD, Abate A, Leon J (2023) Genetic diversity within landraces of barley (Hordeum vulgare L.) and its implications on germplasm collection and utilization. Genet Resour Crop Evol 70:1985–1998. https://doi.org/10.1007/s10722-023-01549-0

Abed A, Badea A, Beattie A et al (2022) A high-resolution consensus linkage map for barley based on GBS-derived genotypes. Genome 65(2):83–94. https://doi.org/10.1139/gen-2021-0055

Acquaah G (2007) Principles of plant genetics and breeding, 1st edn. Blackwell, Malden

Acquaah G (2012) Principles of plant genetics and breeding, 2nd edn. Wiley-Blackwell, Oxford

Afanasenko O, Rozanova I, Gofman A et al (2022) Validation of molecular markers of barley net blotch resistance loci on Chromosome 3H for marker-assisted selection. Agriculture 12(4):439. https://doi.org/10.3390/agriculture12040439

Ahloowalia BS, Maluszynski M, Nichterlein K (2004) Global impact of mutation-derived varieties. Euphy 135:187–204

Akar T, Avci M, Dusunceli F (2004) Barley: post-harvest operations. http://www.fao.org/fileadmin/user_upload/inpho/docs/Post_Harvest_Compendium_-_BARLEY.pdf

Ali M, Polgári D, Sepsi A et al (2024) Rapid and cost-effective molecular karyotyping in wheat, barley, and their cross-progeny by chromosome-specific multiplex PCR. Plant Methods 20:37. https://doi.org/10.1186/s13007-024-01162-x

Alqudah AM, Sallam A, Stephen Baenziger P et al (2020) GWAS: fast-forwarding gene identification and characterization in temperate Cereals: lessons from Barley – a review. J Adv Res 22:119–135. https://doi.org/10.1016/j.jare.2019.10.013

Ameen G, Solanki S, Sager-Bittara L et al (2021) Mutations in a barley cytochrome P450 gene enhances pathogen induced programmed cell death and cutin layer instability. PLoS Gene 17(12):e1009473. https://doi.org/10.1371/journal.pgen.1009473

Anastas PT, Warner JC (2000) Green chemistry: theory and practice. Oxford University Press, New York

Anderson MK, Reinbergs E (1985) Barley breeding. In: Rasmusson D (ed) Barley, ASA monograph no 26, pp 231–268

Anjum SA, Ashraf U, Zohaib A et al (2017) Growth and developmental responses of crop plants under drought stress: a review. Zemdirbyste-Agriculture 104(3):267–276. https://doi.org/10.13080/z-a.2017.104.034
Ansell MP, Mwaikambo LY (2009) The structure of cotton and other plant fibres. In: Handbook of textile fibre structure. Woodhead Publishing, pp 62–94
Assefa K, Yu JK, Zeid M et al (2011) Breeding tef [Eragrostis tef (Zucc.) Trotter]: Conventional and molecular approaches. Plant Breed 130:1–9. https://doi.org/10.1111/j.1439-0523.2010.01782.x
Asseng S, Guarin JR, Raman M et al (2020) Wheat yield potential in controlled-environment vertical farms. Proc Natl Acad Sci USA 117(32):19131–19135. https://doi.org/10.1073/pnas.2002655117
Awal A, Taha RM, Hasbullah NA (2007) In vitro formation of synthetic seed from microshoots of Begonia x hiemalis Fotch. CATRINA 2(2):189–192
Awogbemi O, Von Kallon DV (2022) Pretreatment techniques for agricultural waste. Case Stud Chem Env Eng 6:100229. https://doi.org/10.1016/j.cscee.2022.100229
Ayliffe MA, Pallotta M, Langridge P, Pryor AJ (2007) A barley activation tagging system. Plant Mol Biol 64(3):329–347
Baidyussen A, Khassanova G, Utebayev M et al (2024) Assessment of molecular markers and marker-assisted selection for drought tolerance in barley (Hordeum vulgare L.). J Integr Agric 23(1):20–38. https://doi.org/10.1016/j.jia.2023.06.012
Bala S, Garg D, Sridhar K et al (2023) Transformation of agro-waste into value-added bioproducts and bioactive compounds: micro/nano formulations and application in the agri-food-pharma sector. Bioengineering 10(2):152. https://doi.org/10.3390/bioengineering10020152
Balaska V, Adamidou Z, Vryzas Z et al (2023) Sustainable crop protection via robotics and artificial intelligence solutions. Machines 11(8):774. https://doi.org/10.3390/machines11080774
Banerjee C, Adenaeuer L (2014) Up, up and away! The economics of vertical farming. J Agric Stud 2(1):40
Begna T (2021) Conventional breeding methods widely used to improve self-pollinated crops. Int J Res Stud Agric Sci 7(1):1–16. https://doi.org/10.20431/2454-6224.0701001
Bekel BG (2021) Review on production and application of synthetic seeds. GSJ 9(3):189–211
Bekheet SA (2006) A synthetic seed method through encapsulation of in vitro proliferated bulblets of garlic (*Allium sativum* L.). Arab J Biotech 9:415–426
Ben Ayed R, Hanana M (2021) Artificial intelligence to improve the food and agriculture sector. J Food Qual 2021:1–7. https://doi.org/10.1155/2021/5584754
Benke K, Tomkins B (2017) Future food-production systems: vertical farming and controlled-environment agriculture. Sustain Sci Pract Policy 13(1):13–26. https://doi.org/10.1080/15487733.2017.1394054
Bevan MW, Flavell RB, Chilton MD (1983) A chimeric antibiotic resistance gene as a selectable marker for plant cell transformation. Nature 304:184–187
Blattner FR (2009) Progress in phylogenetic analysis and a new infragenetic classification of barley genus *Hordeum* (Poaceae: Triticeae). Breed Sci 59:471–480
Blattner FR, Pleines T, Jakob SS (2010) Rapid radiation in the barley genus *Hordeum* (Poaceae) during the Pleistocene in the Americas. In: Glaubrecht M (ed) Evolution in action. Springer, Berlin, pp 17–33
Blokhin VI (2006) Peculiarities of cultural practices of barley in Tatarstan. J Zemledeliye 3:15–17
Bockelman HE, Valkoun J, Ullrich S (2010) Barley germplasm conservation and resources. Wiley-Blackwell, Oxford
Borrega M, Hinkka V, Hörhammer H et al (2022) Utilizing and valorizing oat and barley straw as an alternative source of lignocellulosic fibers. Materials 15(21):7826. https://doi.org/10.3390/ma15217826
Bouhlal O, Visioni A, Verma RPS et al (2022) CGIAR Barley Breeding Toolbox: A diversity panel to facilitate breeding and genomic research in the developing world. Front Plant Sci 13:1034322. https://doi.org/10.3389/fpls.2022.1034322

Bowman JGP, Blake TK, Surber LMM et al (2011) Feed-quality variation in the barley core collection of the USDA national small grains collection. Crop Sci 41:863–870
Bothmer R von, Sato K, Komatsuda T et al (2003) The domestication of cultivated barley. In: Bothmer von R, Hintum van TH, Knüpffer H, Sato K (eds) Diversity in barley (*Hordeum vulgare*). Elsevier Science B.V., Amsterdam, pp. 9–27
Brennan CS, Cleary LJ (2005) The potential use of Cerea l (1→3, 1→4)-β, 6-D-glucans as functional food ingredients. J Cereal Sci 42:1–13
Briggs DE (1978) Barley. Chapman and Hall, London. Community Plant Variety Office (CPVO), Angers. 2008. Available at http://www.cpvo.europa.eu/
Brush SB (2000) The issues of in situ conservation of crop genetic resources. In: Genes in the field. On-farm conservation of crop diversity. IPGRI, IDRC, Lewis Publishers, pp 3–26
Cao T, Du Q, Ge R et al (2024) Genome-wide identification and characterization of FAD family genes in barley. PeerJ 12:e16812. https://doi.org/10.7717/peerj.16812
Capuano G, Piccioni E, Standardi A (1998) Effect of different treatments on the conversion of M.26 apple rootstock synthetic seeds obtained from encapsulated apical and axillary micropropagated buds. J Hortic Sci Biotechnol 73:299–305
Ceccarelli S, Grando S (2009) Participatory plant breeding. Cereals, pp 395–414
Chahal GS, Gosal SS (2006) Principles and procedures of plant breeding. In: Biotechnological and conventional approaches. Alpha Science International Ltd, Harrow
Chand N, Vishwakarma SR, Verma OP, Kumar M (2008) Worth of genetic parameters to sort out new elite barley lines over heterogeneous environments. Barley Gene Newsl 38:10–13
Chawla HS, Wenzel G (1987) In vitro selection of barley and wheat for resistance against *Helminthosporium sativum*. Theor Appl Genet 74:841–845
Cheng J, Jia Y, Hill C et al (2024) Diversity of Gibberellin 2-oxidase genes in the barley genome offers opportunities for genetic improvement. J Adv Res 66:105–118. https://doi.org/10.1016/j.jare.2023.12.021
Cho MJ, Jiang W, Lemaux PG, Buchanan BB (1999) Over expression of thioredoxin leads to enhanced activity of starch branching enzyme (pullulanase) in barley grain. Proc Natl Acad Sci USA 25:14641–14646
Christos D (2022) Applications of plant tissue culture. Curr Synth Syst Biol 10:005. https://doi.org/10.35248/2332-0737.22.10.005
Cooper M, Messina CD (2021) Can we harness "Enviromics" to accelerate crop improvement by integrating breeding and agronomy? Front Plant Sci 12:735143. https://doi.org/10.3389/fpls.2021.735143
Czembor JH (2023) Barley genetic resources: advancing conservation and applications for breeding. Agronomy 13(12):2901. https://doi.org/10.3390/agronomy13122901
Datta SK, Potrykus I (1989) Artificial seeds in barley: encapsulation of microspore-derived embryos. Theor Appl Gene 77:820–824. https://doi.org/10.1007/BF00268333
Delgado JA, Short NM Jr, Roberts DP et al (2019) Big data analysis for sustainable agriculture on a geospatial cloud framework. Front Sustain Food Syst 3:54. https://doi.org/10.3389/fsufs.2019.00054
Denisiuka W (2020) Elements of precision agriculture in malting barley cultivation and the use of barley straw for energy purposes. Agric Eng 24(2):21–27. https://doi.org/10.1515/agriceng-2020-0013
Despommier D (2013) Farming up the city: the rise of urban vertical farms. Trends Biotechnol 31(7):388–389
Dewey RD (1984) The genomic system of classification as a guide to intergeneric hybridization with the perennial Triticeae. In: Gustafson JP (ed) Gene manipulation in plant improvement. Plenum, New York, pp 209–279
Didon UME (2002) Variation between barley cultivars in early response to weed competition. J Agron Crop Sci 188:176–184. https://doi.org/10.1046/j.1439-037X.2002.00566.x
Dockter C, Hansson M (2015) Improving barley culm robustness for secured crop yield in a changing climate. J Exp Bot 66(12):3499–3509. https://doi.org/10.1093/jxb/eru521

Doležel J, Greilhuber J, Lucretti S et al (1998) Plant genome size estimation by flow cytometry: inter-laboratory comparison. Ann Bot 82:17–26

Draga M, Gąbka M (2024) The beneficial effect of barley straw extract addition on the growth of two aquatic invasive alien species (Elodea nuttallii and Cabomba caroliniana) under laboratory conditions. Biologia 79:11–21. https://doi.org/10.1007/s11756-023-01550-z

Dudits D, Kao KN, Constabel F, Gamborg OL (1976) Fusion of carrot and barley protoplasts and division of heterokaryocytes. Can J Genet Cytol 19:263–269

Duke JA (1983) Handbook of energy crops. *Hordeum vulgare* L. Purdue University, Center for New Crops & Plants Products, Lafayette

Dzyubenko NI (2018) Vavilov's collection of worldwide crop genetic resources in the 21st century. Biopreserv Biobank 16(5):377–383

Ebi KL, Vanos J, Baldwin JW et al (2021) Extreme Weather and Climate Change: Population Health and Health System Implications. Annu Rev Public Health 42:293-315.https://doi.org/10.1146/annurev-publhealth-012420-105026

EFSA Journal (2011) Scientific opinion on the substantiation of a health claim related to barley beta-glucans and lowering of blood cholesterol and reduced risk of (coronary) heart disease pursuant to Article 14 of Regulation (EC), No 1924/2006. http://www.efsa.europa.eu/en/efsajournal/doc/2471.pdf, 9(12):2471

Eglinton J, Coventry S, Chalmers K (2006) Breeding outcomes from molecular genetics. In: Mercer CF (ed) Proceedings of the 13th Australasian Plant Breeding Conference, Christchurch, New Zealand 18–21 April 2006, pp 743–749

Elahe S, Kianoosh C, Leila Z (2017) Genetic diversity of characteristics in barley cultivars. Genetika 49(2):495–510. https://doi.org/10.2298/GENSR1702495S

El-Hashash EF, El-Absy KM (2019) Barley (*Hordeum vulgare* L.) breeding. In: Al-Khayri J, Jain S, Johnson D (eds) Advances in plant breeding strategies: cereals. Springer, Cham. https://doi.org/10.1007/978-3-030-23108-8_1

El-Tabakh MAM, Abd El-Samea MA, Roby YA (2024) Digital agriculture for enhancing yield, nutrition, and biological stress resistance. In: Priyadarshan PM, Jain SM, Penna S, Al-Khayri JM (eds) Digital agriculture. Springer, Cham. https://doi.org/10.1007/978-3-031-43548-5_15

Espinach FX, Julian F, Verdaguer N et al (2013) Analysis of tensile and flexural modulus in hemp strands/polypropylene composites. Compos Part B Eng 47:339–343. https://doi.org/10.1016/j.compositesb.2012.11.021

Everall NC, Lees DR (1996) The use of barley-straw to control general and blue-green algal growth in a Derbyshire reservoir. Water Res 30:269–276. https://doi.org/10.1016/0043-1354(95)00192-1

Faheem Akhtar S, Govardhana Rao E, Uike AR, Saatu M (2023) Chapter: 2: Plant breeding strategies: traditional and modern approaches. In: Genetic revolution in agriculture: unleashing the power of plant genetics, 1st edn. Elite Publishing House

Falconer DS (1996) Introduction to quantitative genetics. Pearson Education India

Fan C, Xu D, Wang C et al (2024) Natural variations of HvSRN1 modulate the spike rachis node number in barley. Plant Commun 5:100670. https://doi.org/10.1016/j.xplc.2023.100670

Fang Y, Zhang X, Zhang X et al (2020) A high-density genetic linkage map of SLAFs and QTL analysis of grain size and weight in barley (*Hordeum vulgare* L.). Front Plant Sci 11:620922. https://doi.org/10.3389/fpls.2020.620922

FAO (1996) FAO state of the world's plant genetic resources for food and agriculture. FAO, Rome

FAO (2010) The second report on the state of the world's plant genetic resources for food and agriculture. Commission on genetic resources for food and agriculture, food and agriculture organization of the United Nations, Rome. http://www.fao.org/docrep/013/i1500e/i1500e00.htm. Cited 28 Feb 2013

FAO (2016) The state of food and agriculture: climate change, agriculture and food security. Available from: http://www.fao.org/3/a-i6030e.pdf

FAO (2017) The future of food and agriculture – trends and challenges. FAO, Rome

FAO (2018) Commission on genetic resources for food and agriculture. FAO, Rome

FAO/IAEA (2024) Mutant variety database. https://nucleus.iaea.org/sites/mvd/SitePages/Home.aspx
Fehr WR (1984) Genetic contributions to yield gains of five major crop plants, Special Publication 7, ASA and CSSA, Madison
Fehr WR (1987) Principles of cultivar development: theory and technique, vol 1. McGraw-Hill, New York
Feng X, Liu WX, Qiu CW et al (2020) HvAKT2 and HvHAK1 confer drought tolerance in barley through enhanced leaf mesophyll H+ homoeostasis. Plant Biotechnol J 18(8):1683–1696. https://doi.org/10.1111/pbi.13332
Fortunati E, Benincasa P, Balestra GM et al (2016) Revalorization of barley straw and husk as precursors for cellulose nanocrystals extraction and their effect on PVA_CH nanocomposites. Ind Crop Prod 92:201–217. https://doi.org/10.1016/j.indcrop.2016.07.047
Fu Y-B, Peterson GW, Nevo E et al (2024) Intensified selection, elevated mutations, and reduced adaptation potential in wild barley in response to 28 years of global warming. Science 6(1):16. https://doi.org/10.3390/sci6010016
Gahoonia TS, Nielsen NE, Joshi PA et al (2001) A root hairless barley mutant for elucidating genetic of root hairs and phosphorus uptake. Plant Soil 235:211–219. https://doi.org/10.1023/A:1011993322286
Gallardo-Sánchez MA, Hernández JA, Casillas RR et al (2019) Obtaining Soluble-grade cellulose pulp from Agave tequilana Weber var. Azul Bagasse. BioResources 14(4):9867–9881
Gao R, Guo G, Fang C et al (2018) Rapid generation of barley mutant lines with high nitrogen uptake efficiency by microspore mutagenesis and field screening. Front Plant Sci 9:450. https://doi.org/10.3389/fpls.2018.00450
Geng L, Li M, Zhang G et al (2022) Barley: a potential cereal for producing healthy and functional foods. Food Qual Saf 6:fyac012. https://doi.org/10.1093/fqsafe/fyac012
Gepts P, Hancock J (2006) The future of plant breeding. Crop Sci 46:1630–1634
Gerasimova SV, Hertig C, Korotkova AM et al (2020) Conversion of hulled into naked barley by Cas endonuclease-mediated knockout of the NUD gene. BMC Plant Biol 20(Suppl 1):255. https://doi.org/10.1186/s12870-020-02454-9
Ghonaim MM, Attya AM, Aly HG et al (2023) Agro-morphological, biochemical, and molecular markers of barley genotypes grown under salinity stress conditions. BMC Plant Biol 23:526. https://doi.org/10.1186/s12870-023-04550-y
Ghosh P, Kumpatla SP (2022) GIS applications in agriculture. IntechOpen. https://doi.org/10.5772/intechopen.104786
Giraldo P, Benavente E, Manzano-Agugliaro F et al (2019) Worldwide research trends on wheat and barley: a bibliometric comparative analysis. Agronomy 9(7):352. https://doi.org/10.3390/agronomy9070352
Global Crop Diversity Trust (2008) Priority crops – barley. http://www.croptrust.org/main/priority.php?itemid¼148. Cited 28 Feb 2013
Gottwald S, Bauer P, Komatsuda T et al (2009) TILLING in the two-rowed barley cultivar 'Barke' reveals preferred sites of functional diversity in the gene HvHox1. BMC Res Notes 2:1–14
Graham L (2013) Lonely ideas: can Russia compete? MIT Press, Cambridge, MA
Grando S, Baum M, Ceccarelli S et al (2005) QTLs for straw quality characteristics identified in recombinant inbred lines of a *Hordeum vulgare* x *H. spontaneum* cross in a Mediterranean environment. Theor Appl Genet 110(4):688–695. https://doi.org/10.1007/s00122-004-1894-3
Grassini P, Eskridge KM, Cassman KG (2013) Distinguishing between yield advances and yield plateaus in historical crop production trends. Nat Commun 4:2918
Griffiths AJF, Wessler SR, Lewontin RC et al (2005) Introduction to genetic analysis, 8th edn. W H Freeman & Company, New York
Guerriero G, Hausman JF, Strauss J et al (2016) Lignocellulosic biomass: biosynthesis, degradation, and industrial utilization. Eng Life Sci 16:1–16. https://doi.org/10.1002/elsc.201400196
Gunjan, Chopra L, Manikanika (2023) Extraction of cellulose from agro waste – a short review. Mater Today Proc 92(Part 1):233–239. https://doi.org/10.1016/j.matpr.2023.04.378

Guo J, Zhao C, Gupta S et al (2024) Genome-wide association mapping for seedling and adult resistance to powdery mildew in barley. Theor Appl Genet 137(3):50. https://doi.org/10.1007/s00122-024-04550-y
Hammer K (1984) Das Domestikationssyndrom. Die Kult 32:11–34
Han Y, Broughton S, Liu L et al (2021) Highly efficient and genotype-independent barley gene editing based on anther culture. Plant Commun 2(2):100082. https://doi.org/10.1016/j.xplc.2020.100082
Hansson M, Youssef HM, Zakhrabekova S et al (2024) A guide to barley mutants. Hereditas 161(1):11. https://doi.org/10.1186/s41065-023-00304-w
Harlan HV (1918) The identification of varieties of barley. U.S. Dept. Agr. Bull. 622
Harlan JR (1957) One man's life with barley. Exposition Press, New York
Hartl DL, Clark AG, Clark AG (1997) Principles of population genetics, vol 116. Sinauer Associates, Sunderland
Hayes HK, Immer FR (1942) Methods of plant breeding. McGraw-Hill, New York
Heneen WK (2011) Cytogenetic and molecular cytogenetic of barley: a model cereal crop with a large genome. In: Ullrich SE (ed) Barley: production, improvement, and uses. Blackwell, London, pp 112–121
Hensel G, Valkov V, Williams JM, Kumlehn J (2008) Efficient generation of transgenic barley: the way forward to modulate plant-microbe interactions. J Plant Physiol 165(1):71–82
Herzig P, Maurer A, Draba V et al (2018) Contrasting genetic regulation of plant development in wild barley grown in two European environments revealed by nested association mapping. J Exp Bot 69(7):1517–1531. https://doi.org/10.1093/jxb/ery002
Hiddar H, Rehman S, Lakew B et al (2021) Assessment and modeling using machine learning of resistance to scald (*Rhynchosporium commune*) in two specific barley genetic resources subsets. Sci Rep 11:15967. https://doi.org/10.1038/s41598-021-94587-6
Hockett EA (1981) Registration of hulless and hulless short-awned spring barley germplasm. Crop Sci 21:146–147
Holubová K, Hensel G, Vojta P (2018) Modification of barley plant productivity through regulation of cytokinin content by reverse-genetics approaches. Front Plant Sci 9:1676. https://doi.org/10.3389/fpls.2018.01676
Horsley RD, Franckowiak JD, Schwarz PB (2009) Barley. In: Carena MJ (ed) Cereals. Springer Science + Business Media, pp 227–250
Horvath H, Huang J, Wong OT et al (2000) The production of recombinant proteins in transgenic barley grains. Proc Natl Acad Sci USA 97:1914–1919
Horvath H, Rostoks N, Brueggeman R et al (2003) Genetically engineered stem rust resistance in barley using the Rpg1 gene. Proc Natl Acad Sci USA 100:364–369
Huang X, Han B (2014) Natural variations and genome-wide association studies in crop plants. Ann Rev of Plant Biol 65:531–551. https://doi.org/10.1146/annurev-arplant-050213-035715
Huang L, Gao G, Jiang C et al (2023) Generating homozygous mutant populations of barley microspores by ethyl methanesulfonate treatment. aBIOTECH 4:202–212. https://doi.org/10.1007/s42994-023-00108-6
Hussien ET, Ahmed MF, Abdel-Gaber R et al (2024) Synthesis and characterization of artificial seeds plantlets of in vitro male Populus alba. Plant Cell Tissue Organ Cult 156:37. https://doi.org/10.1007/s11240-023-02664-7
Hyvärinen M, Kärki T (2015) The effects of the substitution of wood fiber with agro-based fiber (barley straw) on the properties of natural fiber/polypropylene composites. MATEC Web Conf 30:01014. https://doi.org/10.1051/matecconf/20153001014
ISBC (2012) International barley genome sequencing consortium. a physical, genetic and functional sequence assembly of the barley genome. Nature 491:711–716
Ivanizs L, Farkas A, Linc G et al (2018) Molecular cytogenetic and morphological characterization of two wheat-barley translocation lines. PLoS One 13(6):e0198758. https://doi.org/10.1371/journal.pone.0198758

Jacobsen N, von Bothmer R (1992) Supraspecific groups in the genus *Hordeum*. Hereditas 116:21–24

Jakob SS, Rödder D, Engler JO et al (2014) Evolutionary history of wild barley (*Hordeum vulgare* subsp. spontaneum) analyzed using multilocus sequence data and paleodistribution modeling. Genome Biol Evol 6(3):685–702. https://doi.org/10.1093/gbe/evu047

Jiang G-L (2013) Molecular markers and marker-assisted breeding in plants. InTech. https://doi.org/10.5772/52583

Jiang G-L (2015) Molecular marker-assisted breeding: a plant breeder's review. In: Al-Khayri JM, Jain SM, Johnson DV (eds) Advances in plant breeding strategies: breeding, biotechnology and molecular tools. Springer, pp 431–472

Jost M, Szurman-Zubrzycka M, Gajek K et al (2019) TILLING in barley. In: Harwood WA (ed) Barley: methods and protocols, methods in molecular biology. Springer Science+Business Media, pp 73–94

Joung YH, Choi PS, Kwon SY, Harn CH (2015) Chapter 9: Plant transformation methods and applications. In: Koh H-J, Kwon S-Y, Thomson M (eds) Current technologies in plant molecular breeding. Springer, Dordrecht, pp 297–343

Kang MS, Subudhi PK, Baisakh N, Priyadarshan PM (2007) Crop breeding methodologies: classic and modern. In: Kang MS, Priyadarshan PM (eds) Breeding major food staples. Blackwell, Hoboken, pp 5–40

Kao KN, Michayluk MR (1974) A method for high-frequency intergeneric fusion of plant protoplasts. Planta 115:355–367

Karunarathne S, Walker E, Sharma D et al (2023) Genetic resources and precise gene editing for targeted improvement of barley abiotic stress tolerance. J Zhejiang Univ Sci B 24(12):1069–1092. https://doi.org/10.1631/jzus.B2200552

Kaur V, Aravind J, Manju et al (2022) Phenotypic Characterization, Genetic Diversity Assessment in 6,778 Accessions of Barley (Hordeum vulgare L. ssp. vulgare) Germplasm Conserved in National Genebank of India and development of a core set. Front Plant Sci 13:771920. https://doi.org/10.3389/fpls.2022.771920

Keno MT, Tolemariam T, Demeke S et al (2021) Effect of barley variety on feed intake, digestibility, body weight gain and carcass characteristics in fattening lambs. Animals 11(6):1773. https://doi.org/10.3390/ani11061773

Kihara M, Okada Y, Kuroda H et al (2000) Improvement of β-amylase thermostability in transgenic barley seeds and transgene stability in progeny. Mol Breed 6:511–517

Kihara M, Hoki T, Hirota N et al (2023) Characteristics of the barley mutant line 'N68-411' with improved balance of malt modification and seed dormancy. J Cereal Sci 109:103618. https://doi.org/10.1016/j.jcs.2022.103618

Kis A, Polgári D, Dalmadi A (2024) Targeted mutations in the GW2.1 gene modulate grain traits and induce yield loss in barley. Plant Sci 340:111968. https://doi.org/10.1016/j.plantsci.2023.111968

Kisaka H, Kisaka M, Kanno A, Kameya T (1997) Production and analysis of plants that are somatic hybrids of barley (*Hordeum vulgare* L.) and carrot (*Daucus carota* L.). Theor Appl Genet 94:221–226

Kisaka H, Kisaka M, Kanno A, Kameya T (1998) Intergeneric somatic hybridization of rice (*Oryza sativa* L.) and barley (*Hordeum vulgare* L.) by protoplast fusion. Plant Cell Rep 17:362–367

Kisaka H, Kisaka M, Kanno A et al (2001) Somatic hybridization between *Hordeum vulgare* L. (Barley) and *Daucus carota* L. (Carrot). In: Nagata T, Bajaj YPS (eds) Somatic hybridization in crop improvement II, Biotechnology in agriculture and forestry, vol 49. Springer, Berlin/Heidelberg. https://doi.org/10.1007/978-3-642-56758-2_1

Kleinhofs A, Chao S, Sharp PJ (1988) Mapping of nitrate reductase genes in barley and wheat. In: Miller TE, Koebner RMD (eds) Proc 7th International Wheat Genetics Symposium, vol 1. Institute of Plant Science Research, Cambridge, pp 541–546

Klerkx L, Jakku E, Labarthe P (2019) A review of social science on digital agriculture, smart farming and agriculture 4.0: new contributions and a future research agenda. NJAS-Wageningen J Life Sci 90:100315

Kling J (2004) An introduction to barley- notes from CSS 330 world foods class. http://oregonstate.edu/instruct/css/330/five/BarleyOverview.htm. Accessed 18 Apr 2006
Knežević D, Pržulj N, Zečević V et al (2004) Breeding strategies for barley quality improvement and wide adaptation. Kragujevac J Sci 26:75–84
Knüpffer H, Terentyeva I, Karl Hammer K et al (2003) Ecogeographical diversity – a Vavilovian approach. In: Bothmer R, Hintum T, Knüpffer H, Sato K (eds) Developments in plant genetics and breeding, vol 7. Elsevier, pp 53–76. https://doi.org/10.1016/S0168-7972(03)80006-3
Kok EJ, Keijer J, Kleter GA et al (2008) Comparative safety assessment of plant-derived foods. Regul Toxicol Pharmacol 50(1):98–113
Kosová K, Prášil IT, Klíma M et al (2023) Proteomics of wheat and barley cereals in response to environmental stresses: Current state and future challenges. J Proteome 282:104923. https://doi.org/10.1016/j.jprot.2023.104923
Koutous A, Hilali E (2021) Reinforcing rammed earth with plant fibers: a case study. Case Stud Constr Mater 14:e00514. https://doi.org/10.1016/j.cscm.2021.e00514
Kovačević Z, Strgačić S, Bischof S (2023) Barley straw fiber extraction in the context of a circular economy. Fibers 11(12):108. https://doi.org/10.3390/fib11120108
Kozłowski M, Górecki P, Szczypiński PM (2019) Varietal classification of barley by convolutional neural networks. Biosyst Eng 184:155–165. https://doi.org/10.1016/j.biosystemseng.2019.06.012
Kumar U (1998) Synthetic seeds for commercial crop production. Vedams Books (P) Ltd, New Delhi, 160 pp
Kumlehn J, Stein N (2014) Biotechnological approaches to barley improvement. Springer, Berlin
Kumlehn J, Serazetdinova L, Hensel G et al (2006) Genetic transformation of barley (*Hordeum vulgare* L.) via infection of androgenetic pollen cultures with *Agrobacterium tumefaciens*. Plant Biotechnol J 4(2):251–261
Kumlehn J, Gurushidze M, Hensel G (2014) Genetic engineering. In: Kumlehn J, Stein N (eds) Biotechnological approaches to barley improvement. Springer, Berlin, pp 393–407
Kurowska M, Labocha-Pawłowska A, Gnizda D et al (2012) Molecular analysis of point mutations in a barley genome exposed to MNU and gamma rays. Mutat Res 738–739:52–70
Lajoie-O'Malley A, Bronson K, van der Burg S et al (2020) The future(s) of digital agriculture and sustainable food systems: an analysis of high-level policy documents. Ecosyst Serv45:101183.https://doi.org/10.1016/j.ecoser.2020.101183
Larkin PJ, Scowcroft WR (1983) Somaclonal variation and crop improvement. In: Kosuge T, Meredith C, Hollaender A (eds) Genetic engineering of plants. Plenum Press, New York, pp 289–314
Lassoued R, Macall DM, Smyth SJ, Phillips PWB, Hesseln H et al (2021) Expert insights on the impacts of, and potential for, agricultural big data. Sustainability 13(5):2521. https://doi.org/10.3390/su13052521
Latrech B, Hermassi T, Yacoubi S et al (2024) Comparative analysis of climate change impacts on climatic variables and reference evapotranspiration in Tunisian Semi-Arid Region. Agriculture 14(1):160. https://doi.org/10.3390/agriculture14010160
Lawrenson T, Shorinola O, Stacey N et al (2015) Induction of targeted, heritable mutations in barley and Brassica oleracea using RNA-guided Cas9 nuclease. Genome Biol 16:258. https://doi.org/10.1186/s13059-015-0826-7
Lee G, Kim D, Kwon S et al (2015) Identification of mutagenized plant populations. In: Koh H-J, Kwon S-Y, Thomson M (eds) Current technologies in plant molecular breeding, pp 205–239
Lee S, Lee T, Yang S et al (2020) BarleyNet: a network-based functional omics analysis server for cultivated barley, *Hordeum vulgare* L. Front Plant Sci 11:98. https://doi.org/10.3389/fpls.2020.00098
Leišová-Svobodová L, Psota V, Zavřelová M et al (2024) Malting quality molecular markers for barley breeding. Czech J Genet Plant Breed 60(2):70–78. https://doi.org/10.17221/72/2023-CJGPB
Li T, Li Y, Shangguan H et al (2023) BarleyExpDB: an integrative gene expression database for barley. BMC Plant Biol 23:170. https://doi.org/10.1186/s12870-023-04193-z

Liakos KG, Busato P, Moshou D et al (2018) Machine learning in agriculture: a review. Sensors (Basel, Switzerland) 18(8):2674. https://doi.org/10.3390/s18082674

Ling Y, Zhao Q, Liu W et al (2023) Detection and characterization of spike architecture based on deep learning and X-ray computed tomography in barley. Plant Methods 19:115. https://doi.org/10.1186/s13007-023-01096-w

Longin CFH, Muehleisen J, Maurer HP et al (2012) Hybrid breeding in autogamous cereals. Theor Appl Genet 125:1087–1096

Lopachev NA, Titova EM, Yatsina NV, Naumkin AV (2001) New cultural practices of barley. J Zemledeliye 4:10–11

Löve A (1984) Conspectus of the Triticeae. Feddes Repertorium 95:425–521

Luo R, Pan W, Liu W et al (2022) The barley DIR gene family: An expanded gene family that is involved in stress responses. Front Genet 13:1042772. https://doi.org/10.3389/fgene.2022.1042772

Maanju S, Jasrotia P, Yadav SS et al (2023) Deciphering the genetic diversity and population structure of wild barley germplasm against corn leaf aphid, Rhopalosiphum maidis (Fitch). Sci Rep 13:17313. https://doi.org/10.1038/s41598-023-42717-7

Maho A, Mero G, Maho F (2022) Changes in the agro-technique cultivation of barley as an adaptation to climate change. Ecol Quest 33(4):1–22. https://doi.org/10.12775/EQ.2022.038

Maluszynski M, Ahloowalia BS, Sigurbjôrnsson B (1995) Application of in vivo and in vitro mutation techniques for crop improvement. Euphytica 85:303–315

Marone D, Russo MA, Mores A et al (2021) Importance of landraces in cereal breeding for stress tolerance. Plants 10(7):1267. https://doi.org/10.3390/plants10071267

Mascher M, Gundlach H, Himmelbach A et al (2017) A chromosome conformation capture ordered sequence of the barley genome. Nature 544:427–433

Mascher M, Schreiber M, Scholz U et al (2019) Genebank genomics bridges the gap between the conservation of crop diversity and plant breeding. Nat Genet 51:1076–1081. https://doi.org/10.1038/s41588-019-0443-6

Masłowski M, Miedzianowska J, Strzelec K (2017) Natural rubber biocomposites containing corn, barley and wheat straw. Polym Test 63:84–91. https://doi.org/10.1016/j.polymertesting.2017.08.003

Matres JM, Hilscher J, Datta A et al (2021) Genome editing in cereal crops: an overview. Transgenic Res 30(4):461–498. https://doi.org/10.1007/s11248-021-00259-6

Mba C (2013) Induced mutations unleash the potentials of plant genetic resources for food and agriculture. Agronomie 3:200–231

Mendiondo GM, Gibbs DJ, Szurman-Zubrzycka M et al (2016) Enhanced waterlogging tolerance in barley by manipulation of expression of the N-end rule pathway E3 ligase PROTEOLYSIS6. Plant Biotechnol J 14:40–50

Meng G, Rasmussen SK, Christensen CSL et al (2023) Molecular breeding of barley for quality traits and resilience to climate change. Front Genet 13:1039996. https://doi.org/10.3389/fgene.2022.1039996

Mewis I, Bernhardt N, Bönisch M et al (2024) In situ conservation of *Helosciadium nodiflorum*: a crop wild relative of celery in Germany. Conserv Genet 25:227–243. https://doi.org/10.1007/s10592-023-01554-x

Meyer G, Okudoh V, van Rensburg E (2022) A rumen based anaerobic digestion approach for lignocellulosic biomass using barley straw as feedstock. S Afr J Chem Eng 41:98–104. https://doi.org/10.1016/j.sajce.2022.05.005

Mihók E, Polgári D, Lenykó-Thegze A et al (2024) Plasticity of parental CENH3 incorporation into the centromeres in wheat × barley F1 hybrids. Front Plant Sci 15:1324817. https://doi.org/10.3389/fpls.2024.1324817

Mittler R, Shulaev V (2013) Functional genomics, challenges and perspectives for the future. Physiol Plant 148:317–321

Mlčochová L, Chloupek O, Uptmoor R et al (2004) Molecular analysis of the barley cv. 'Valticky' and its X-ray-derived semidwarf-mutant 'Diamant'. Plant Breed 123(5):421–442

Mohamed AH, Omar AA, Attya AM et al (2021) Morphological and molecular characterization of some egyptian six-rowed barley (*Hordeum vulgare* L.). Plants 10(11):2527. https://doi.org/10.3390/plants10112527
Mohammadi Aghdam S, Abdollahi Mandoulakani B, Rossini L et al (2022) Identification of novel plant architecture mutants in barley. Cereal Res Commun 50:179–189. https://doi.org/10.1007/s42976-021-00178-6
Mohammed AH, Morrison JI, Baldwin BS (2020) Interspecific crosses between domestic and wild barley and embryo rescue to overcome sexual incompatibilities. Agrosyst Geosci Environ 3(1):e20130. https://doi.org/10.1002/agg2.20130
Morran S, Eini O, Pyvovarenko T et al (2011) Improvement of stress tolerance of wheat and barley by modulation of expression of DREB/CBF factors. Plant Biotechnol J 9:230–249
Morrell PL, Buckler ES, Ross-Ibarra J (2012) Crop genomics: advances and applications. Nat Rev Genet 13(2):85–96. https://doi.org/10.1038/nrg3097
Morrison R, Evans D (1988) Haploid plants from tissue culture: new plant varieties in a shortened time frame. Nat Biotechnol 6:684–690. https://doi.org/10.1038/nbt0688-684
Mudasir Magray M, Wani KP, Chatto MA et al (2017) Synthetic seed technology. Int J Curr Microbiol App Sci 6(11):662–674. https://doi.org/10.20546/ijcmas.2017.611.079
Mwendwa JM, Brown WB, Weidenhamer JD et al (2020) Evaluation of commercial wheat cultivars for canopy architecture, early vigour, weed suppression, and yield. Agronomy 10(7):983. https://doi.org/10.3390/agronomy10070983
Mwendwa JM, Brown WB, Weston PA et al (2022) Evaluation of barley cultivars for competitive traits in Southern New South Wales. Plants (Basel, Switz) 11(3):362. https://doi.org/10.3390/plants11030362
Nagel M, Vogel H, Landjeva S et al (2009) Seed conservation in ex situ genebanks-genetic studies on longevity in barley. Euphytica 170:5–14
Nathan VK (2023) Pulping and papermaking using pineapple leaves. In: Ainun ZMA, Sapuan SM, Ilyas RA (eds) Pulping and papermaking of nonwood plant fibers. Academic, pp 233–243. https://doi.org/10.1016/B978-0-323-91625-7.00016-3
Naz AA, Reinert S, Bostanci C et al (2017) Mining the global diversity for bioenergy traits of barley straw: Genomewide association study under varying plant water status. GCB Bioenergy 9(8):1356–1369. https://doi.org/10.1111/gcbb.12433
Nevo E (1992) Origin, evolution, population genetics and resources for breeding of wild barley, *Hordeum spontaneum* in the Fertile Crescent. In: Shewry PR (ed) Barley: genetics, biochemistry, molecular biology and biotechnology. CAB International, Wallingford, pp 19–43
Nevski SA (1941) Beiträge zur Kenntniss der wildwachsenden Gersten in Zusammenhang mit der Frage über den Ursprung von *Hordeum vulgare* L. and *Hordeum distichon* L. (Versuch einer Monographie der Gattung *Hordeum*). Trudy Bot Inst Akad Nauk SSSR, ser. 1,5: 64–255
Newman RK, Newman CW (2008) Barley for food and health: science, technology, and products. Wiley, Hoboken
Newton AC, Flavell AJ, George TS et al (2011) Crops that feed the world 4. Barley: a resilient crop? Strengths and weaknesses in the context of food security. Food Secur 3:141–178
Nguyen HSH, Phan HH, Huynh HKP et al (2022) Understanding the effects of cellulose fibers from various pre-treated barley straw on properties of aerogels. Fuel Process Technol 236:107425. https://doi.org/10.1016/j.fuproc.2022.107425
Nowak K, Wójcikowska B, Gajecka M et al (2024) The improvement of the in vitro plant regeneration in barley with the epigenetic modifier of histone acetylation, trichostatin A. J Appl Genet 65:13–30. https://doi.org/10.1007/s13353-023-00800-9
Numan M, Khan AL, Asaf S, Salehin M, Beyene G, Tadele Z, Ligaba-Osena A (2021) From Traditional Breeding to Genome Editing for Boosting Productivity of the Ancient Grain Tef [Eragrostis tef (Zucc.) Trotter]. Plants 10(4):628. https://doi.org/10.3390/plants10040628
Oh S, Kwon C, Choi D et al (2007) Expression of barley HvCBF4 enhances tolerance to abiotic stress in transgenic rice. Plant Biotechnol J 5:646–656

Ohnoutkova L (2019) Mutation breeding in barley: historical overview. In: Harwood W (ed) Barley, Methods in molecular biology, vol 1900. Humana Press, New York. https://doi.org/10.1007/978-1-4939-8944-7_2

Oliver-Ortega H, Julián F, Espinach FX et al (2023) Simulated environmental conditioning of PHB composites reinforced with barley fibres to determine the viability of their use as plastics for the agriculture sector. Polymers 15(3):579. https://doi.org/10.3390/polym15030579

Orłowska R, Bednarek PT (2020) Precise evaluation of tissue culture-induced variation during optimisation of in vitro regeneration regime in barley. Plant Mol Biol 103(1–2):33–50. https://doi.org/10.1007/s11103-020-00973-5

Orman-Ligeza B, Harwood W, Hedley PE et al (2020) TRA1: a locus responsible for controlling agrobacterium-mediated transformability in barley. Front Plant Sci 11:355. https://doi.org/10.3389/fpls.2020.00355

Pan R, Ding M, Feng Z et al (2022) HvGST4 enhances tolerance to multiple abiotic stresses in barley: evidence from integrated meta-analysis to functional verification. Plant Physiol Biochem 188:47–59. https://doi.org/10.1016/j.plaphy.2022.07.027

Pandey P (2022) Production technology of Barley under Indian condition. J Pharm Phytochem 11(1):14–17

Patial M, Chaudhary HK, Pramanic KK et al (2021) Doubled haploidy breeding in wheat and barley. Int J Trop Agric 39(4):403–405

Patrick H, Alfonso C-M (2013) New and renewed breeding methodology. In: Zhang G, Li G, Liu X (eds) Advance in barley sciences: proceedings of 11th international barley genetics symposium. Zhejiang University Press and Springer Science+Business Media, Dordrecht, pp 349–357

Pauschinger D, Klauser FR (2022) The introduction of digital technologies into agriculture: Space, materiality and the public–private interacting forms of authority and expertise. J Rural Studies 91:217-227.https://doi.org/10.1016/j.jrurstud.2021.06.015

Penna S, Vitthal SB, Yadav PV (2012) In vitro mutagenesis and selection in plant cultures and their prospects for crop improvement. Bioremed Biodiv Bioavail 6(1):6–14

Petersen G, Seberg O (2003) Phylogenetic analyses of the diploid species of *Hordeum* (Poaceae) and a revised classification of the genus. Syst Bot 28:293–306

Pickering RA (1984) The influence of genotype and environment on chromosome elimination in crosses between *Hordeum vulgare* L. × *Hordeum bulbosum* L. Plant Sci Lett 34:153–164. https://doi.org/10.1016/0304-4211(84)90138-X

Pickering R, Johnston PA (2005) Recent progress in barley improvement using wild species of Hordeum. Cyto Genome Res 109(1–3):344–349. https://doi.org/10.1159/000082418

Priyadarshan PM, Penna S, Jain SM et al (2024) Digital agriculture for the years to come. In: Priyadarshan PM, Jain SM, Penna S, Al-Khayri JM (eds) Digital agriculture. Springer, Cham. https://doi.org/10.1007/978-3-031-43548-5_1

Prosdocimi M, Jordán A, Tarolli P et al (2016) The immediate effectiveness of barley straw mulch in reducing soil erodibility and surface runoff generation in Mediterranean vineyards. Sci Total Environ 547:323–330. https://doi.org/10.1016/j.scitotenv.2015.12.076

Puglia D, Luzi F, Lilli M et al (2020) Straw fibres from barley hybrid lines and their reinforcement effect in polypropylene based composites. Ind Crop Prod 154:112736. https://doi.org/10.1016/j.indcrop.2020.112736

Qi J, Ni F, Wang X et al (2019) The anther-specific CYP704B is potentially responsible for MSG26 male sterility in barley. Theor Appl Genet 132:2413–2423. https://doi.org/10.1007/s00122-019-03363-8

Rabiei B, Valizdah M, Ghareyazie B et al (2004) Evaluation of selection indices for improving rice grain shape. Field Crop Res 89:359–367. https://doi.org/10.1016/j.fcr.2004.02.016

Rakshit S, Ganapathy KN (2014) Comparative genomics of cereal crops: status and future prospects. In: Kavi Kishor PB, Bandopadhyay R, Suravajhala P (eds) Agricultural bioinformatics. Springer India, pp 59–87

Ramage RT (1985) Cytogenetics in barley. In: Rasmusson DC (ed) ASA monograph No 26, pp 127–154

Ramireddy E, Hosseini SA, Eggert K et al (2018) Root engineering in barley: increasing cytokinin degradation produces a larger root system, mineral enrichment in the shoot and improved drought tolerance. Plant Physiol 177:1078–1095
Rani H, Bhardwaj RD, Sen R et al (2024) Deciphering the potential of diverse barley genotypes for improving the malt quality. J Stored Products Res 105:102247. https://doi.org/10.1016/j.jspr.2024.102247
Rao GJN, Reddy JN, Variar M, Mahender A (2016) Molecular breeding to improve plant resistance to abiotic stresses. In: Al-Khayri JM, Jain SM, Johnson DV (eds) Advances in plant breeding strategies: agronomic, abiotic and biotic stress traits. Springer, pp 283–326
Rauf S, Al-Khayri JM, Zaharieva M et al (2016) Breeding strategies to enhance drought tolerance in crops. In: Al-Khayri JM, Jain SM, Johnson DV (eds) Advances in plant breeding strategies: agronomic, abiotic and biotic stress traits. Springer, pp 397–445
Reddy MC, Murthy KSR, Pullaiah T (2012) Synthetic seeds: a review in agriculture and forestry. Afr J Biotechnol 11:14254–14275
Reid DA (1985) Morphology and anatomy of the barley plant. Barley 26:73–101. https://doi.org/10.2134/agronmonogr26.c4
Riaz A, Kanwal F, Börner A et al (2021) Advances in Genomics-Based Breeding of Barley: Molecular Tools and Genomic Databases. Agronomy 11(5):894. https://doi.org/10.3390/agronomy11050894
Ribaut J-M, de Vicente MC, Delannay X (2010) Molecular breeding in developing countries: challenges and perspectives. Curr Opin Plant Biol 13:1–6
Richards AJ (1997) Plant breeding systems. Garland Science
Robinson RA (2007) Crop histories, 2nd ed., Rev edn. Sharebook Publishing. www.sharebooks.ca
Rogo U, Fambrini M, Pugliesi C (2023) Embryo rescue in plant breeding. Plants 12(17):3106. https://doi.org/10.3390/plants12173106
Rohila JS, Rajinder K, Wua J, Wua R (2002) Genetic improvement of Basmati rice for salt and drought tolerance by regulated expression of a barley Hva1 cDNA. Plant Sci 163:525–532
Román-Gutiérrez AD, Duana-Ávila D, Hernández-Ávila J et al (2022) Reuse of barley straw for handmade paper production. Sustainability 14(19):12691. https://doi.org/10.3390/su141912691
Rotasperti L, Sansoni F, Mizzotti CT et al (2020) Barley's second spring as a model organism for chloroplast research. Plants 9(7):803. https://doi.org/10.3390/plants9070803
Roy SJ, Huang W, Wang XJ et al (2013) A novel protein kinase involved in Na(+) exclusion revealed from positional cloning. Plant Cell Environ 36:553–568
Rozanova IV, Khlestkina EK (2020) NGS sequencing in barley breeding and genetic studies. Vavilovskii zhurnal genetiki i selektsii 24(4):348–355. https://doi.org/10.18699/VJ20.627
Saiprasad GVS (2001) Artificial seeds and their applications. Resonance 6:39–47. https://doi.org/10.1007/BF02839082
Saitou N (2004) Genome and evolution. Shinyo-Sha, Tokyo (in Japanese)
Saitou N (2013) Introduction to evolutionary genomics, computational biology. Springer, London
Salvi S, Druka A, Milner SG, Gruszka D (2014) Induced genetic variation, TILLING and NGS-based cloning. In: Kumlehn J, Stein N (eds) Biotechnological approaches to barley improvement. Springer, Berlin, pp 287–310
Sarğın B, Karaca S (2023) Land suitability assessment for wheat barley cultivation in a semi-arid region of Eastern Anatolia in Turkey. PeerJ 11:e16396. https://doi.org/10.7717/peerj.16396
Sato K, Flavell A, Russell J et al (2014) Genetic diversity and germplasm management: wild barley, landraces, breeding materials. In: Kumlehn J, Stein N (eds) Biotechnological approaches to barley improvement. Springer, Berlin, pp 21–36
Schillinger WF, Cook RJ, Papendick RI (1999) Increased dryland cropping intensity with no-till barley. Agron J 91:744–752
Schreiber M, Wonneberger R, Haaning AM et al (2024) Genomic resources for a historical collection of cultivated two-row European spring barley genotypes. Sci Data 11:66. https://doi.org/10.1038/s41597-023-02850-4

Senaratna T (1992) Artificial seeds. Biotechnol Adv 10(3):379–392. https://doi.org/10.1016/0734-9750(92)90301-O

Serna-Diaz MG, Arana-Cuenca A, Medina-Marin J et al (2016) Modeling of sulfite concentration, particle size, and reaction time in lignosulfonate production from barley straw using response surface methodology and artificial neural network. BioResources 11(4):9219–9230. https://doi.org/10.15376/biores.11.4.9219-9230

Serra-Parareda F, Julián F, Espinosa E et al (2020) Feasibility of barley straw fibers as reinforcement in fully biobased polyethylene composites: macro and micro mechanics of the flexural strength. Molecules (Basel, Switzerland) 25(9):2242. https://doi.org/10.3390/molecules25092242

Shackley BJ (2000) Crop management. In: Anderson WK, Garlinge JR (eds) The wheat book: principles and practice, Bulletin 4443. Department of Agriculture and Food, Western Australia, Perth, pp 131–164

Shah SAA, Luo H, Putu Dita Pickupana PD et al (2022) Automatic and fast classification of barley grains from images: a deep learning approach. Smart Agric Technol 2:100036. https://doi.org/10.1016/j.atech.2022.100036

Shamila M, Kumari M, Pulikantham S (2023) Identification of soil features suitable for barley, maize and sugarcane cultivation using precision agriculture. E3S Web Conf 391:01158. https://doi.org/10.1051/e3sconf/202339101158

Shepherd M, Turner JA, Small B et al (2018) Priorities for science to overcome hurdles thwarting the full promise of the 'digital agriculture' revolution. J Sci Food Agric 100:5083-5092.https://doi.org/10.1002/jsfa.9346

Shi Y, Patel Y, Rostami B et al (2022) Barley Variety Identification by iPhone Images and Deep Learning. J Am Soc Brew Chem 80(3):215–224. https://doi.org/10.1080/03610470.2021.1958602

Shuvo II (2020) Fibre attributes and mapping the cultivar influence of different industrial cellulosic crops (cotton, hemp, flax, and canola) on textile properties. Bioresour Bioprocess 7:51. https://doi.org/10.1186/s40643-020-00339-1

Siahsar BA, Peighambari SA, Taleii AR et al (2009) QTL analysis of forage quality traits in Barley (Hordeum vulgare L.). Cereal Res Commun 37(4):479–488. https://doi.org/10.1556/CRC.37.2009.4.1

Singh RJ (2006) Utilization of genetic resources for barley improvement. In: Singh R, Prem J, Jauhar P (eds) Genetic resources, chromosome engineering, and crop improvement series volume 2 cereals. Taylor & Francis, pp 233–255

Singh BD, Singh AK (2015) Marker-assisted plant breeding: principles and practices. Springer India

Singh J, Zhang S, Chen C (2006) High-frequency Ds remobilization over multiple generations in barley facilitates gene tagging in large genome cereals. Plant Mol Biol 62(6):937–950

Singh AK, Singh A, Mahama AA (2023) Breeding methods. In: Suza WP, Lamkey KR (eds) Crop improvement. Iowa State University Digital Press. https://iastate.pressbooks.pub/cropimprovement/chapter/breeding-methods/

Siraree A (2022) Artificial Seed Technology. In: Misra V, Srivastava S, Mall AK (eds) Sugar beet cultivation, management and processing. Springer, Singapore. https://doi.org/10.1007/978-981-19-2730-0_8

Somers DA, Narayanan KR, Kleinhofs A et al (1986) Immunological evidence for transfer of the barley nitrate reductase structural gene to *Nicotiana tabacum* by protoplast fusion. Mol Gen Genet 204:296–301

Sparla F, Falini G, Botticella E et al (2014) New starch phenotypes produced by TILLING in barley. PLoS One 9(10):e107779

Spencer D, Lembi C (2007) Evaluation of barley straw as an alternative algal control method in Northern California rice fields. J Aquat Plant Manag 45:84–90

Spencer-Lopes MM, Forster BP, Jankuloski L (2018) Manual on mutation breeding, 3rd edn. FAO/IAEA, Vienna

Stein N (2014) Development of sequence resources. In: Kumlehn J, Stein N (eds) Biotechnological approaches to barley improvement. Springer, Berlin, pp 271–285

Stockinger EJ (2021) The breeding of winter-hardy malting barley. Plants 10(7):1415. https://doi.org/10.3390/plants10071415

Sun DF, Gong X (2009) Barley germplasm and utilization. In: Genetics and improvement of barley malt quality. Zhejiang University Press/Springer-Verlag Berlin, Hangzhou, pp 18–62

Szarejko I, Forster B (2007) Doubled haploidy and induced mutation. Euphy 158(3):359–370

Szarejko I, Maluszynski M, Polok K, Kilian A (1991) Doubled haploids in the mutation breeding of selected crops. In: Plant mutation breeding for crop improvement, vol 2. IAEA, Vienna, pp 355–378

Szarejko I, Guzy J, Davalos J et al (1995) Production of mutants using barley DH systems. In: Induced mutations and molecular techniques for crop improvement. IAEA, Vienna, pp 517–530

Tamilselvan-Nattar-Amutha S, Hiekel S, Hartmann F et al (2023) Barley stripe mosaic virus-mediated somatic and heritable gene editing in barley (Hordeum vulgare L.). Front Plant Sci 14:1201446. https://doi.org/10.3389/fpls.2023.1201446

Taneja A, Nair G, Joshi M et al (2023) Artificial intelligence: implications for the agri-food sector. Agronomy 13:1397. https://doi.org/10.3390/agronomy13051397

Tanger P, Vega-Sanchez ME, Fleming M et al (2015) Cell wall composition and bioenergy potential of rice straw tissues are influenced by environment, tissue type, and genotype. Bioenergy Res 8(3):1165–1182. https://doi.org/10.1007/s12155-014-9573-y

Tantau H, Balko C, Brettschneider B et al (2004) Improved frost tolerance and winter survival in winter barley (*Hordeum vulgare* L.) by in vitro selection of proline over accumulating lines. Euphytica 139:19–32

Tanwar UK, Stolarska E, Paluch-Lubawa E et al (2022) Unraveling the genetics of polyamine metabolism in barley for senescence-related crop improvement. Int J Biol Macromol 221:585–603. https://doi.org/10.1016/j.ijbiomac.2022.09.006

Thyavihalli Girijappa YG, Mavinkere Rangappa S, Parameswaranpillai J et al (2019) Natural fibers as sustainable and renewable resource for development of eco-friendly composites: a comprehensive review. Front Mater 6:226. https://doi.org/10.3389/fmats.2019.00226

Tingay S, McElroy D, Kalla R et al (1997) Agrobacterium tumefaciens-mediated barley transformation. Plant J 11:1369–1376. https://doi.org/10.1046/j.1365-313X.1997.11061369.x

Tricase C, Amicarelli V, Lamonaca E et al (2018) Economic analysis of the barley market and related uses. IntechOpen. https://doi.org/10.5772/intechopen.78967

Tull D, Phillipson BA, Kramhoft B et al (2003) Enhance amylolytic activity in germinating barley through synthesis of a bacterial alpha-amylase. J Cereal Sci 37:71–80

Umba di-Umba U, Maluszynski M, Szarejko I, Zbieszczyk J (1991) High frequency of barley DH-mutants from M1 after mutagenic treatment with MNH and sodium azide. MBNL 38:8–9

USDA (2024) World agricultural production. United States Department of Agriculture Foreign Agricultural Service, Office of Global Analysis

Vafaei A, Choobbasti AJ, Koutenaei RY et al (2024) Effect of barley straw fiber as a reinforcement on the mechanical behavior of babolsar sand. Transp Infrastruct Geotechnol 11:216–235. https://doi.org/10.1007/s40515-023-00281-7

Vagera J, Novotny J, Ohnoutkova L (2004) Induced and neurogenesis in vitro in mutated populations of barley, *Hordeum vulgare*. Plant Cell Tissue Organ Cult 77:55–61

Vaghefi SA, Abbaspour N, Kamali B et al (2017) A toolkit for climate change analysis and pattern recognition for extreme weather conditions – case study: California-Baja California Peninsula. Environ Model Softw 96:181–198. https://doi.org/10.1016/j.envsoft.2017.06.033

Van de Wiel C, Schaart J, Niks R (2010) Traditional plant breeding methods. Wageningen UR Plant Breeding, Wageningen. https://edepot.wur.nl/141713

Van Delden SH, Sharathkumar M, Butturini M et al (2021) Current status and future challenges in implementing and upscaling vertical farming systems. Nat Food 2(12):944–956. https://doi.org/10.1038/s43016-021-00402-w

Van Harten AM (1998) Mutation breeding: theory and practical applications. Cambridge Univ. Press, pp 163–251

Van Hintum T, Menting F (2003) Diversity in ex situ genebank collections of barley. In: von Bothmer R, van Hintum T, Knüpffer H, Sato K (eds) Diversity in barley (*Hordeum vulgare*). Elsevier Science, Amsterdam, pp 247–257

Vargas F, González Z, Rojas O et al (2015) Barley straw (Hordeum vulgare) as a supplementary raw material for Eucalyptus camaldulensis and Pinus sylvestris kraft pulp in the paper industry. BioResources 10(2):3682–3693. https://doi.org/10.15376/biores.10.2.3682-3693

Vasil V, Vasil IK (1979) Isolation and culture of cereal protoplasts. I. Callus formation from pearl millet (*Pennisetum americanum*) protoplasts. Z Pflanzenphysiol 92:379–383

Verstegen H, Köneke O, Korzun V, von Broock R (2014) The world importance of barley and challenges to further improvements. In: Kumlehn J, Stein N (eds) Biotechnological approaches to barley improvement. Springer, Berlin, pp 3–19

Visioni A, Basile B, Amri A et al (2023) Advancing the conservation and utilization of barley genetic resources: insights into germplasm management and breeding for sustainable agriculture. Plants 12(18):3186. https://doi.org/10.3390/plants12183186

Vivar HE, McNab A (2001) Breeding barley in the new millenium: proceedings of an international symposium. CIMMYT, Mexico

von Bothmer R (1996) Distribution and habitat preferences in the genus *Hordeum* in Iran and Turkey. Ann Natur Hist Mus Wien 98B(Suppl):107–116

von Bothmer R, Jacobsen N (1985) Origin, taxonomy, and related species. In: Rasmusson D (ed) Barley, ASA monograph no 26, pp 19–56

von Bothmer R, Seberg O, Jacobsen N (1992) Genetic resources in the Triticeae. Hereditas 116:141–150

von Bothmer R, Jacobsen N, Baden C et al (1995) An ecogeographical study of the genus *Hordeum*. In: Systematic and ecogeographic studies on crop gene pools. No. 7, 2nd edn. IBPGR, Rome

Bothmer von R, Sato K, Komatsuda T et al (2003) The domestication of cultivated barley. In: Bothmer von R, Hintum van TH, Knüpffer H, Sato K (eds) Diversity in barley (Hordeum vulgare). Elsevier Science BV, Amsterdam, Netherlands, pp 9–27

Wan B (2015) Transgenic pyramiding for crop improvement. In: Al-Khayri JM, Jain SM, Johnson DV (eds) Advances in plant breeding strategies: breeding, biotechnology and molecular tools. Springer, pp 369–396

Wang Y, Frei M (2011) Stressed food – the impact of abiotic environmental stresses on crop quality. Agric Ecosyst Environ 141:271–286. https://doi.org/10.1016/j.agee.2011.03.017

Wang L, Xue Q, Newman RK, Newman CW (1993) Enrichment of tocopherol, tocotrienol, and oil in barley by milling and pearling. Cereal Chem 70(5):499–501

Wang J, Jiang J, Wang Y (2013) Protoplast fusion for crop improvement and breeding in China. Plant Cell Tissue Organ Cult 112:131–142

Widoretno W, Harran S, Sudarsono (2003) Variation in qualitative and quantitative characters among somaclones of soybean derived from in vitro selected somatic embryos. Hayati 10:110–117

Wiebe GA (1978) Breeding. In: Barley: origin, botany, culture, winter hardiness, genetics, utilization, and pests, Agriculture handbook 338. USDA, Washington DC, pp 117–127

Wisniewska SK, Nalaskowski J, Witka-Jezewska E et al (2003) Surface properties of barley straw. Coll Surf B: Biointerfaces 29:131–142. https://doi.org/10.1016/S0927-7765(02)00178-9

Wu J, Ein-Mozaffari F, Simant Upreti S (2013) Effect of ozone pretreatment on hydrogen production from barley straw. Bioresour Technol 144:344–349. https://doi.org/10.1016/j.biortech.2013.07.001

Xiong M, Xu J, Zhou Z et al (2024) Salinity inhibits seed germination and embryo growth by reducing starch mobilization efficiency in barley. Plant Direct 8(2):e564. https://doi.org/10.1002/pld3.564

Xu Y, Xie C, Wan J et al (2013) Marker-assisted selection in cereals: platforms, strategies and examples. In: Gupta PK, Varshney RK (eds) Cereal genomics II. Springer Science, Dordrecht, pp 375–411

Yang S, Overlander M, Fiedler J (2021) Genetic analysis of the barley variegation mutant, grandpa1.a. BMC Plant Biol 21:134. https://doi.org/10.1186/s12870-021-02915-9

Yawson DO, Mulholland BJ, Ball T et al (2017) Effect of climate and agricultural land use changes on UK feed barley production and food security to the 2050s. Land 6(4):74. https://doi.org/10.3390/land6040074

Yazdani B, Sanjari S, Asghari-Zakaria R et al (2020) Revision of the barley WRKY gene family phylogeny and expression analysis of the candidate genes in response to drought. Biol Plant 64:9–19. https://doi.org/10.32615/bp.2019.109

Yeşİl V, Tatar Ö (2020) An innovative approach to produce forage crops: Barley fodder in vertical farming system. Sci Pap Ser A Agron, LXIII(1):723–728

Yeşİl V, Çakaloğulları U, Tatar Ö (2020) Effect of light intensity on dry matter accumulation of barley fodder in a vertical farming growth module. Sci Pap Ser A Agron LXIII(2):245–248

Yi X, Li-Lan Z, Jian-Min Q et al (2021) Genomics and genetic improvement in main bast fiber crops: advances and perspectives. Acta Agron Sinica 47(6):997–1019. https://doi.org/10.3724/SP.J.1006.2021.04121

Yusnita WS (2005) In vitro selection of peanut somatic embryos on medium containing culture filtrate of *Sclerotium rolfsii* and plantlet regeneration. Hayati 12(2):50–56

Zair W, Maxted N, Brehm JM et al (2021) Ex situ and in situ conservation gap analysis of crop wild relative diversity in the Fertile Crescent of the Middle East. Genet Resour Crop Evol 68:693–709. https://doi.org/10.1007/s10722-020-01017-z

Zali H, Barati A (2020) Evaluation of selection index of ideal genotype (SIIG) in other to selection of barley promising lines with high yield and desirable agronomy traits. JCB 12(34):93–104. https://doi.org/10.29252/jcb.12.34.93

Zali H, Barati A, Jabari M (2022) Using selection index of ideal genotype (SIIG) in selection of barley promising lines. J Sustain Agric Product Sci 32(1):293–307. https://sid.ir/paper/1050351/en

Zali H, Barati A, Pour-Aboughadareh A et al (2023) Identification of superior barley genotypes using selection index of ideal genotype (SIIG). Plants 12(9):1843. https://doi.org/10.3390/plants12091843

Zhang G, Li C (2009) Genetics and improvement of barley malt quality. Zhejiang University Press/Springer-Verlag, Hangzhou/Berlin

Zhao T, Palotta M, Langridge P (2006) Mapped Ds/T-DNA launch pads for functional genomics in barley. Plant J 47(5):811–826

Zhou W, Shi H, Wang Z et al (2024) Identification of highly repetitive barley enhancers with long-range regulation potential via STARR-seq. Genomics Proteomics Bioinformatics 22:qzae012. https://doi.org/10.1093/gpbjnl/qzae012

Zhu XG, Marcells L (2023) Vertical farming for crop production. Mod Agric 1(1):13–15. https://doi.org/10.1002/moda.4

Zohary D, Hopf M (1973) Domestication of Pulses in the Old World: Legumes were companions of wheat and barley when agriculture began in the Near East. Science (New York, NY) 182(4115):887–894. https://doi.org/10.1126/science.182.4115.887

Zohary D, Hopf M (1993) Domestication of plants in the old World. Clarendon Press, Oxford

Chapter 3
Breeding and Biotechnology Approaches for Genetic Enhancement in Maize (*Zea mays* L.)

Abeer H. Jubily, Essam Fathy El-Hashash, and Khaled F. M. Salem

Abstract Maize (*Zea mays* L.), commonly known as corn, is a crucial crop grown worldwide for human consumption, animal feed, and biofuel production. It is primarily cross-pollinated and has a wide genetic potential, with genetic resources from wild species preserved in gene banks. Maize is considered the "queen" of cereals due to its high genetic yield potential and ability to thrive in various agroecological conditions. C4 maize crops play a vital role in ensuring food security and sustainability in the face of climate change. Maintaining diverse genetic backgrounds through the study of genetic diversity is essential for developing climate-resilient maize varieties. Crossbreeding popular cultivars with indigenous ones enhances genetic diversity and introduces beneficial traits, leading to the emergence of new alleles. Genetic improvement can significantly reduce the impacts of biotic and abiotic stresses while increasing yield potential. The development of disease-resistant, stress-tolerant, and nutritionally enhanced cultivars involves both traditional and molecular techniques. This chapter provides a comprehensive overview of the origin and botanical classification of maize, its types, significance and uses, domestication, selection and early advancements, current cultivation practices, breeding goals, germplasm biodiversity and preservation, cultivar characterization and phylogeny, conservation of genetic resources, cytogenetics, traditional breeding

A. H. Jubily (✉)
Department of Field Crops, Faculty of Agriculture, Tishreen University, Lattakia, Syria
e-mail: abeerjubily.e@gmail.com

E. F. El-Hashash
Department of Agronomy, Faculty of Agriculture, Al-Azhar University, Cairo, Egypt
e-mail: dressamelhashash@azhar.edu.eg

K. F. M. Salem
Department of Biological Sciences, Shaqra University, Ad-Dawadimi, Saudi Arabia

Department of Plant Biotechnology, Genetic Engineering and Biotechnology Research Institute (GEBRI), University of Sadat City, Sadat City, Egypt
e-mail: khaled.salem@gebri.usc.edu.eg; khaledfathi@su.edu.sa

K. F. M. Salem et al. (eds.), *Breeding and Biotechnology of Grass and Bast Fiber Crops*, Advances in Plant Breeding Strategies 12,
https://doi.org/10.1007/978-3-032-00403-1_3

methods, improvement strategies, the role of biotechnology, molecular breeding, applications of tissue culture, genetic engineering and gene editing, mutation breeding, conventional mutagenesis, and maize hybridization.

Keywords Maize (*Zea mays* L.) · Traditional breeding · Biotechnology · Tissue culture · Conservation · Modern maize breeding

3.1 Introduction

Maize (*Zea mays* L.) is a crop of economic importance utilized for food, fuel, fiber, and animal feed. It is an essential component in numerous goods that influence global nutritional standards. Maize is among the three most significant cereal crops cultivated globally. In 2022, the worldwide area allocated for maize cultivation was roughly 200.53 million hectares, in contrast to 219.53 million hectares for wheat and 165.68 million hectares for paddy rice (USDA 2024a, b). In 2023, maize output exceeded 1.16 billion tonnes, surpassing wheat production at 789.17 million tonnes and rice production at 514.57 million tonnes (USDA 2024a, b).

Maize is a fundamental component of the diets of millions globally. In around 25 developing countries, maize constitutes 15–56% of the daily calorie consumption of individuals (Prasanna et al. 2001).

Maize breeding possesses distinct characteristics in contrast to other extensively farmed grain species, which are predominantly self-pollinated. Maize breeding integrates techniques from both self-pollinated and cross-pollinated species, facilitating the generation of substantial sample sizes and diverse breeding methodologies. The objective of these procedures is to produce enhanced populations, inbred lines, and hybrids for diverse markets. This chapter offers a comprehensive account of the origin and botanical classification of maize, its types, important applications, domestication, selection, early developments, current cultivation practices, breeding objectives, germplasm biodiversity and preservation, cultivar characterization and phylogeny, conservation of genetic resources, cytogenetics, traditional breeding methods, improvement strategies, the role of biotechnology, molecular breeding, tissue culture applications, genetic engineering and gene editing, mutation breeding, conventional mutagenesis, and maize hybridization.

3.1.1 Origin and Botanical Classification

Maize belongs to the Maydeae tribe within the Poaceae family of grasses. The term "*Zea*" originates from an ancient Greek word denoting a specific variety of grass. The genus *Zea* has four species, including *Zea mays* L., which exerts considerable economic influence. Another species of *Zea*, known as teosinte, is mostly a naturally occurring grass indigenous to Mexico and Central America. Zea mays has two chromosomes, or n = 20.

The tribe Maydeae comprises seven recognized genera, categorized into Old World and New World groupings. Zea and Tripsacum are classified within the New World group, whereas Coix (2n = 10/20), Chionachne (2n = 20), Sclerachne (2n = 20), Trilobachne (2n = 20), and Polytoca (2n = 20) belong to the Old World. There is widespread agreement that the American genera Zea and Tripsacum predominated in maize phylogeny, although it is also recognized that the genus Coix contributed to the evolutionary development of the species *Zea mays*.

Kingdom Plantae—plants, Planta, Vegetal
  Subkingdom Viridiplantae — green plants
    Infrakingdom Streptophyta — terrestrial plants
      Superdivision: Embryophyta
        Division Tracheophyta—vascular plants, tracheophytes
          Subdivision: Spermatophytina—spermatophytes, seed plants, phanerogams
            Class Magnoliopsida
              Superorder Lilianae—monocots, monocotyledons
                Order Poales
                  Family: Poaceae—grasses
                    Genus: Zea L.—maize
                      Zea mays L.—maize
                        Variety *Zea mays* var. Huehuetenangensis H.H. Iltis and Doebley—maize

A considerable body of research has been undertaken into the origins of maize; yet, the precise progenitors of contemporary farmed maize remain uncertain. Numerous specialists have examined this subject, although a conclusive answer remains elusive. Despite varying perspectives, the consensus is that cultivated maize originated in the Western Hemisphere (Weatherwax 1955; Goodman 1976; Wilkes 2004).

The variation in plant type, ear type, tassel type, and maturity among maize races, varieties, hybrids, and inbred lines is extensive, as demonstrated by the phenotypic distinctions. These variances, whether from adapted or introduced germplasm, consistently exist, particularly in breeding nurseries where germplasm is regularly integrated. The adaptability of maize to diverse latitudes and altitudes has resulted in the development of different cultivars over the 5000 years of human cultivation. The settlements of Native Americans and the contacts among various civilizations and migrations have enriched the extensive reservoir of maize germplasm. This genetic variety has been augmented by natural and artificial selection, leading to germplasm that markedly differs from its original parental material in genotype and phenotype.

3.1.2 Maize Types

Notwithstanding the great diversity of maize germplasm, no natural classification efforts were undertaken until the 1940s. Sturtevant (1899) was among the pioneers in classifying maize germplasm into six principal categories, predominantly according to endosperm composition. This approach was employed for more than 40 years without additional classification endeavors.

N. I. Vavilov and his associates determined that Central America was the progenitor of maize after gathering specimens from several global locations. Kuleshov (1933) further categorized maize according to endosperm types into specific classifications.

Corn varieties can be classified according to kernel type, comprising dent corn, flint corn, flour corn, sweet corn, and dried corn (Fig. 3.1). Excluding dent corn, these classifications are based on the quality, quantity, and arrangement of endosperm development in the kernel, and do not indicate natural relationships (Brown and Darrah 1985).

Dent corn, or toothed corn, is characterized by a hard endosperm on the sides and back of the kernel, but the central kernel contains a mushy, mealy endosperm that extends into the crown. It contracts to create a distinct depression upon drying, with the extent of the depression varying based on its genetic tendency. Dent corn is largely employed as animal feed, however, it also serves as an industrial raw material and a staple diet. It continues to be an important culinary and industrial commodity for humans, utilized in several specialized products within the dry or wet milling industry. White corn typically fetches a premium in the dry milling industry, where it is utilized in some human food products because of its lighter starch content. Dent corn is a common type of field corn; nevertheless, similar to flint corn, it is hard and necessitates extensive processing (i.e., drying and milling) prior to utilization. The term "dent corn" derives from the kernels' dented morphology, marked by a central depression around solid starch as they undergo drying. Dent corn, owing to its dense composition, is primarily utilized for livestock feed, although it can also be processed into varied corn flours or coarse grits.

Flint corn generally possesses a resilient, hard, glassy, or horny outer layer surrounding a diminutive, mushy, granular inside. The ratios of fine starch and horniness change across various varieties. The kernels are typically smooth and spherical, whilst the ears are elongated and slender, featuring a restricted number of rows or kernels. In temperate locations, flint corn generally matures earlier, has improved germination, demonstrates increased spring vigor, yields more stalks, and possesses fewer prop roots than other corn varieties. Flint corn is an ancient and diverse variety of maize, with native strains grown across North America. This maize variety is commonly processed due to its extremely low moisture content and the presence of a starch film encasing the kernel, which produces an extraordinarily hard texture. Subsequent to processing, it is frequently utilized for items including hominy, polenta, and masa. Consequently, flint corn is not the most suitable kind for cultivation in private gardens; yet, it yields numerous maize-derived items found in retail

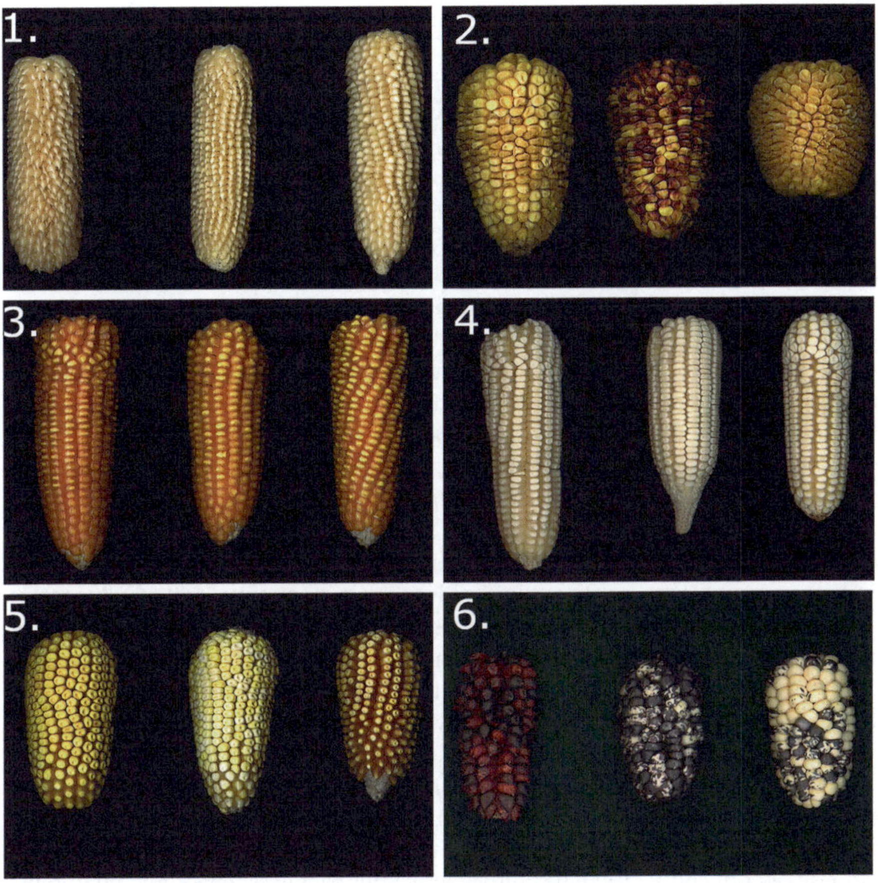

Fig. 3.1 Maize types. Various maize varieties exhibiting unique grain types: (1) Popcorn maize from the "Pisingallo" race, sourced from Argentina (accession number: CIMMYTMA 29717). Sweet maize of the "Chullpi" cultivar harvested in Peru (CIMMYTMA 18213). Flint maize from the "Cuban Flint" variety, obtained in Cuba (CIMMYTMA 2424). Dent maize from the "Tuxpeño" variety, sourced from Mexico (CIMMYTMA 679). Dent maize from the "Conico" variety, sourced from Mexico (CIMMYTMA 5645). Floury maize of the "Cuzco" variety was acquired in Peru (CIMMYTMA 32058). These accessions are conserved in the germplasm bank of the International Maize and Wheat Improvement Center (CIMMYT, Mexico). (Image attribution: CIMMYT Germplasm Bank. Source: Guzzon et al. 2021)

and is frequently utilized for decorative purposes. The colorful dried "Indian corn" typically showcased during the holidays is flint corn.

Flour corn is among the earliest types of maize, dating from the ancient Aztecs and Incas. Indigenous Americans and Indians have milled the little kernels into flour. Flour corn varieties possess a fine starch content and a negligible amount of hard, glassy endosperm, leading to an opaque kernel appearance. The kernels uniformly contract during drying, leading to negligible or no denting. It can be readily milled post-drying; nevertheless, it is prone to decay on the mature ear in humid conditions. In sweet corn, the sugar gene obstructs or postpones the usual conversion of sugar into starch during endosperm development, resulting in the accumulation of a water-soluble polymer called phytoglycogen in the kernel. Consequently, the dehydrated, delicious kernels exhibit a wrinkled and lustrous appearance. The elevated concentration of water-soluble polysaccharides contributes a qualitative textural dimension in addition to sweetness.

Sweet corn is the kind most frequently consumed in the summer. This warm-weather mainstay is collected at an immature stage, yielding very juicy and delicious kernels. The ideal harvesting period is referred to as the "milk stage," signifying that a milky white liquid would emerge from a pierced kernel when the corn attains its maximum sweetness. Sweet corn is distinguished by its elevated sugar content and tender kernels that remain intact when exposed to heat. It serves as a coating for fried ribs at a Mediterranean smokehouse, but for the more conventional home cook, it is superb whether consumed independently, incorporated into pasta and salads (such as succotash), or utilized in casseroles and chowders, where its sweet flavor and creamy texture are prominent.

Popcorn is undoubtedly the most fundamental among the several kinds. This kind of maize possesses a robust, rigid endosperm with minimal soft starch content. Popcorn is fundamentally a diminutive, hard kernel that can be either pointed, akin to rice, or round, resembling pearls. Recently produced popcorn cultivars possess a robust seed husk, although numerous semi-primitive kinds, such as Argentine popcorn, feature a delicate shell. This corn type is classified as flint; however, not all flint corns possess the specific characteristics required for optimal popping. Popcorn is a distinct variety of flint corn distinguished by its elevated natural internal moisture content, which facilitates ideal circumstances for expansion and conversion into popcorn upon cooking. This variety of maize is primarily utilized for popping. Various popcorn cultivars can be cultivated in your backyard, enabling you to relish fresh popcorn for cinematic evenings. Pod corn (Tunicate corn) is a more decorative variety. The principal gene implicated (Tu) generates lengthy husks surrounding each kernel, a characteristic also observed in numerous other grasses. The ear is likewise enveloped in husks, as observed in other maize species. Homozygous husk corn is generally predominantly self-sterile, whereas ordinary husk corn displays heterozygosity. Husk corn may have endosperm characteristics that are serrated, saccharine, waxy, tender, sleek, or mealy. They constitute an anomaly and are not cultivated for commercial objectives.

Waxy corn derives its name from the glossy appearance of the exposed endosperm in a precisely sliced cross-section. Standard cornstarch has roughly 73% amylopectin and 27% amylose, whereas waxy cornstarch consists solely of amylopectin, the branched molecular configuration of starch. Standard cornstarch exhibits a blue coloration when exposed to a 2% potassium iodide solution, whereas waxy cornstarch displays a reddish-brown hue. The waxy gene is expressed in pollen grains via this pigmentation response, facilitating reproduction. The food sector uses waxy maize derivatives as stabilizers and thickeners in pie fillings, puddings, sauces, gravies, tinned foods, salad dressings, and analogous products.

Supplementary waxy materials function as rewetting adhesives in the manufacture of tape and adhesives, as well as in the paper sector. Waxy grains are cultivated for consumption by cattle and dairy livestock.

High-amylose maize denotes maize with an amylose content of over 50%. The amylose extender (ae) mutation in the endosperm was initially found by R. P. Bear in 1950 and increases the amylose content of the endosperm to around 60% in numerous toothed backgrounds. Modifiers influence both the amylose concentration and the targeted agronomic traits of the grain. The expression of the amylose extender gene produces a pale, translucent, and occasionally almost intact kernel look. Starch obtained from high-amylose maize is utilized in the textile business, in the manufacture of chewing gum for its gelling attributes, and as an adhesive within the textile sector.

Kuleshov's classification, which can be affected by a singular genetic variation, is comparable to Sturtevant's classification. Kuleshov's classification effectively addresses kernel type but fails to adequately depict the morphological and polygenic variations of other germplasm factors. Kuleshov identifies eight systematic categories, distributed geographically as follows: (1) Yellow corn, prevalent across the Western Hemisphere, holds particular significance at the northern and southern limits of its production. Yellow corn was first identified in the Andes Valley in southern Colombia, Peru, and Bolivia, with the greatest diversity located in Peru. It is prevalent in areas south of the northern distribution of yellow corn in North America and the southwestern United States. Dent cultivars are prevalent in the American Corn Belt and certain areas of Mexico. The Dent group is not present in indigenous civilizations of South America, with the most diversified populations found in central and southern Mexico. Popcorn varieties have been sourced from diverse countries and areas, but have not been brought to North America, except via commercial processing. (2) Sweet corn is limited in southern and tropical climates, with harvesting predominantly taking place in the central and northeastern parts of the United States. (3) Bolivia and Peru account for almost all sugar-starch varieties, with Peru demonstrating the highest diversity. (4) Waxy variations are exclusively located in East Asia, and (5) no distinct geographic region has been recognized for the tunicate forms. Tunicate species can arise spontaneously in many locales, and the Tu allele is lacking in all individual species.

3.1.3 Importance and Uses

Maize grains comprise 72% carbohydrate, 10% protein, 4.8% oil, 8.5% fiber, 3.0% sugar, and 1.7% ash, indicating their substantial nutritional value (Chaudhary 1983). *Zea mays* is the primary grain crop and a source of cereal feed in semi-arid and arid tropical regions, utilized in both rainfed and irrigated agricultural systems (Alvi et al. 2003).

Maize (*Zea mays* L.) serves as an essential crop and a staple for millions, while increasingly becoming acknowledged as a high-energy silage crop. Silage maize, specifically cultivated for silage production, may produce substantial quantities of high-energy feed appropriate for extensive ruminant utilization. Silage maize is one of the optimal choices for supplementation, particularly in arid seasons, due to its capacity to generate dry matter per unit area, substantial green mass output per hectare, fermentability during storage, and favorable acceptance by livestock (Restle et al. 2006).

Maize is a vital crop owing to its substantial supply of oil, fiber, vitamins, minerals, and other nutrients, particularly concentrated in the embryo. The oil is utilized for culinary purposes and in soap manufacturing enterprises. Maize starch serves as a diluent in the pharmaceutical and cosmetic sectors. The stem is utilized for paper manufacturing, while the seeds are employed in alcohol manufacture. Small-scale farmers produce maize due to its affordability and high nutritional value, providing essential vitamins and minerals to rural populations.

Besides its therapeutic applications, maize possesses diuretic properties attributed to its principal component, potassium (Kumar and Jhariya, 2013). Maize silk has demonstrated kaliuretic and diuretic properties in rats experiencing water overload (Velazquez et al. 2005). In numerous locations, including Greece, France, and Spain, maize silk is utilized to address severe ailments such as nephrolithiasis and urinary tract infections. In China, it is commonly employed to address edema and jaundice. Kumar and Jhariya (2013) assert that corn silk reduces blood pressure, enhances liver function, and facilitates bile generation.

Corn oil, rich in linoleic acid, aids in the prevention of cardiovascular disorders and the regulation of blood pressure by managing blood cholesterol levels (Dupon et al. 1990; Sen et al. 2006). Corn oil is utilized for the treatment of dandruff. Maize is abundant in carotenoids, phenolic compounds, and phytosterols, generally referred to as phytochemicals (Jiang and Wang 2005; Lopez-Martinez et al. 2009). Bioactive compounds included in plants mitigate severe chronic diseases in humans (Liu 2004).

3.1.4 Domestication, Selection and Early Improvements

A distinct form of mutualism, termed "domestication," transpires when a domesticator influences the reproductive or propagative capacity (fitness) of a domesticated species to acquire resources or services. This establishes a distinctive coevolutionary dynamic (Purugganan 2022; Zeder 2015).

Approximately 7000–10,000 years ago, maize was domesticated from teosinte in Mexico. Maize production proliferated from Eastern Colorado to Nebraska in the 1950s, Indiana in the 1970s, the East Coast in the 1990s, and Europe in 1992. It is now regarded as one of the most important crops both regionally and internationally.

The genetic foundation of contemporary maize hybrids originates from two varieties: Northern Flint corn and Southern Dent corn. Substantial yield enhancements were attained through the creation of hybrids—initially double-cross and subsequently single-cross hybrids—alongside the shift to high-input agriculture. These improvements have been sustained as technology has progressed.

3.2 Current Cultivation

3.2.1 Current Cultivation Practices

Maize (*Zea mays* L.), commonly referred to as corn or maize, is a significant food crop that originated in Mexico and has subsequently proliferated throughout. Corn, a highly cultivated crop, is grown from the equator to approximately 50° north and south, and at altitudes ranging from sea level to 3000 meters (Morris 2002). Notwithstanding its global popularity, corn serves as the primary staple crop in numerous nations across Latin America, Eastern Africa, Central America, and Southeast Asia, including China. Corn is cultivated for consumption, livestock feed, and industrial applications in North America. More than 25% of cropland and over 40% of commercial fertilizers in the USA are allocated to maize production (Christensen 2002). Corn is distinct from other principal cereal crops, being a C4 grass that efficiently utilizes water, nutrients, and CO_2 to synthesize carbohydrates, which are stored in the leaves and stalks and subsequently harvested as grain starch. Oktem (2005) asserts that it ranks among the field crops yielding the highest dry matter (DM) per unit area of any crop.

Maize plants are vulnerable to heat stress over 30 °C, and prolonged exposure to temperatures above this threshold results in a significant reduction in grain output.

Variations in temperature between day and night, as well as during the growing season, are essential for optimal maize crop growth. The optimal temperature varies from 25 °C to 33 °C during the day and from 17 °C to 23 °C at night, with an average ideal temperature of 20–22 °C during the growing season (Neild and Newman 1987). The optimal temperature for maize plant germination is 25–28 °C (Farooq et al. 2008).

Maize can be effectively cultivated in several soil types, ranging from loamy sand to clay loam. Conversely, neutral pH soils characterized by substantial water retention and elevated organic matter content are considered advantageous for enhanced productivity. Prior studies indicate that maize necessitates a pH range of 5.8–7.0 for optimal growth (Albrecht et al. 2005; Crouse and Denny 2015; Deinum 1980; Mallarino 2011).

Plants are arranged 20–25 cm apart among rows of maize, which are spaced 60–75 cm apart. For optimal yield, a density of 60,000–75,000 plants per acre is recommended during harvest.

Weeds impede the growth of maize plants during the initial phases. Weeding within a row is performed manually; however, weeding between rows can be accomplished with tools towed by tractors or bullocks. Weeding the crop two to three times prior to earthing it up may be crucial for enhancing standability.

Post-flowering, intercultivation is unnecessary as it may damage the lateral roots. Chemical weed management can be effective if the fields are devoid of weeds. The dense population of maize plants effectively covers the soil surface, resulting in reduced weeding requirements for maize cultivation.

The maize crop designated for grain is harvested when the grain approaches dryness, exhibiting less than 20% moisture content. Maize stalks are incinerated or utilized as fodder for cattle. Every component of the maize plant is employed, including the cobs following the extraction of the grains.

For maize cultivation intended for fodder, it is optimal to harvest during the milk to early dough stages, as earlier harvests typically exhibit lower protein content and reduced yields. However, the late-stage dough is superior for silage.

Farmers cultivating hybrid maize should refrain from saving their seeds for future harvests, as the advanced hybrid seeds are likely to deliver a 25–30% reduction in output. Seeds from open-pollinated and composite varieties can be preserved when cultivated in isolation. Bulking must encompass a minimum of 500–1000 seeds from the highest-yielding, evenly spaced plants that exhibit resistance to prevalent pests and diseases. Ears must be dehusked, desiccated, and subjected to insecticidal treatment. Treated seeds are necessary as untreated seeds on ears are often badly infested by stored grain pests, which markedly diminishes germination rates.

A significant variation in grain production is evident. The kind and quantity of fertilizer utilized, rainfall patterns, and other factors all influence production levels under prescribed cultural practices and in an irrigated setting.

3.2.2 Breeding Objectives

Maize, the most versatile crop globally, is the leading cereal in production, yielding over 1.2 million metric tons annually and occupying more than 197 million hectares, utilized for food, feed, and industrial purposes (Erenstein et al. 2022). The demand for maize in the food and feed sectors is escalating significantly due to rising income levels and urbanization, especially in densely populated emerging countries where dietary preferences are evolving (Grote et al. 2021). The global demand for maize is expected to exceed that of other principal grains, largely due to the heightened need for livestock feed, principally influenced by dietary transitions towards animal products, particularly evident in Asia (Erenstein et al. 2022).

Climate change is expected to significantly affect the supply, price, and nutritional quality of maize due to rising temperatures, an increase in the frequency of extreme weather events (such as droughts and floods), alterations in agro-ecological conditions (such as crop seasons), and a decrease in the availability of suitable cultivated land (Grote et al. 2021; Thornton et al. 2014). Moreover, climate change exacerbates the incidence and severity of both existing and emerging diseases, along with insect pests, hence diminishing maize productivity (Cairns and Prasanna 2018; Deutsch et al. 2018). Notable examples of the potential impacts of climate change include the recent emergence of fall armyworm (De Groote et al. 2020) and maize lethal necrosis disease (Sileshi and Gebeyehu 2021) in African maize systems, along with the ensuing chaos and devastation they have caused in the region. More than 50% of the variability in maize yields across multiple regions of SubSaharan Africa and the Indo-Gangetic Plains is attributable to climate variability.

The overall plant yield is essential in silage maize cultivation. When developing silage hybrids, it is essential to examine specific quality traits, like elevated dry matter yield (DMY), suitably high dry matter content (DMC), superior feeding quality for ruminants, and substantial livestock consumption, alongside grain yield. Silage breeding differs from fodder hybrid breeding as the main target for silage is the grain, the maize plant's principal source of readily accessible carbohydrates.

Historically, it was postulated that genotypes yielding significant grain outputs would also generate substantial forage yields. Subsequent studies indicated that the physiological differences between silage maize hybrids and grain maize result in the highest silage yields not being correlated with high grain productivity. Silage maize breeders may need to prioritize the selection of high whole-plant biomass output above traits such as high grain yield, grain maturity, barrenness, and resistance to stalk lodging. Moreover, breeders may prioritize root lodging selection and overall plant health. Additional essential attributes for ensuring high-quality silage include dry matter yield, whole-plant digestibility, protein concentration, and the non-structural carbohydrate content of the stover. Dhillon et al. (1990) assert that breeders must simultaneously evaluate both the yield and quality attributes of stover and grain to enhance the selection process for feed performance. The yield and quality parameters of green fodder are equally significant in silage maize. Two areas necessitating focus are (1) elements of silage yield and (2) features of silage quality.

Grain yield is a vital characteristic in maize breeding. The principal objective of plant breeders is to supply farmers with enhanced varieties that can withstand adverse environmental circumstances, yield abundantly, and function effectively under stress (Rukuni et al. 2016).

Grain maize breeders prioritize safe growth, hence, they want an increased water supply to the grain prior to harvest. As climate change intensifies, yield stability is becoming an increasingly critical consideration alongside yield. Stability and resilience to drought stress are also essential. When selecting and growing the source population, it is essential to evaluate general adaptability, grain color, grain texture, maturity, disease resistance (Eid et al. 2025), abiotic stress tolerance (Saleh et al. 2021; Salem and Börner 2003; Salem et al. 2021a, b, d, e, f), heterotic pattern, heterotic response, combining ability, and additional value-added attributes (Eid et al. 2025). Methods for enhancing intrapopulation drought tolerance include individual plant selection, performance assessments, test crossings with individual plants, half-sib progenies, and parental testers (Bänziger et al. 2000).

In recent years, maize diseases have intensified due to causes including intensive fertility management, continuous cropping, straw reuse in fields, reduced genetic diversity of maize varieties, high-density planting, and extreme weather events (Duan et al. 2019; Sun et al. 2020). Excluding viral illnesses, annual yield losses from maize diseases are estimated to account for 4–14% of the global harvest. It is imperative to mitigate disease severity in this crop to sustain consistent maize yields and superior kernel quality, given that the extent of arable land allocated to this crop is unlikely to rise substantially. Breeders primarily concentrate on fungal illnesses to cultivate disease tolerance. Prominent instances encompass Fusarium, northern leaf blight, and maize ear rot. The incidence of various fungal diseases differs by region.

3.3 Germplasm Biodiversity and Conservation

3.3.1 Germplasm Diversity

Genetic diversity among genotypes utilized in crosses is universally recognized as essential (Salem and Ibrahim 2024). Historically, it has been frequently noted, though not consistently, that larger yields emerged from crosses between unrelated genotypes. Extensive evaluation trials were generally required to determine the appropriate combination of two paternal inbred lines. The superior hybrids comprised inbred lines derived from two or more genetic sources, notwithstanding the ambiguous genetic foundation of the importance of genetic variety (Hayes 1963). Wu (1939), Eckhardt and Bryan (1940), Johnson and Hayes (1940), Cowan (1943), and Griffing (1953) presented empirical evidence about the impact of genetic variety in parental inbred lines on hybrid performance. When parental inbred lines of related origin were mixed, the resultant crosses consistently yielded less than those with either a single or no common parent line. Notwithstanding a few anomalies, the predominant occurrence of high-yielding hybrids resulted from the crossbreeding of distinct varietal lines. Griffing (1953) showed that parental lines incorporating germplasm from beyond the US Corn Belt also contributed to crosses of diverse origins, despite the preponderance of cultivars being sourced from the US Corn Belt.

In 1939, the North Central Corn Breeding Research Committee (NCR-2, later renamed NCR-167 and currently NCCC167) established the A and B groupings of inbred lines, highlighting the significance of maintaining the genetic variety of the parent inbred lines. The classification of inbred lines into A and B groups was rather arbitrary, however, it sought to minimize gene exchange between the two gene pools. It was acknowledged that crossing and evaluating inbred lines between the two categories, instead of inside the groups, would enhance the likelihood of generating high-yielding hybrids; nevertheless, the rigorous implementation of this classification proved difficult. During that period, pedigree selection in F_2 populations of the most superior single-cross hybrids raised apprehensions regarding the genetic deterioration of the two groups, resulting in an increasing interest in integrating foreign germplasm into breeding initiatives. The crossing of Reid inbred lines with Lancaster inbred lines in the US Corn Belt yielded a hybrid with superior average yields. In Europe and other regions globally, crosses constructed from flint and dent variants were prevalent.

Conversely, the importance of genetic variety in crosses was broadly recognized, in contrast to correlation and visual selection. Enhanced heterosis was noted in crosses of more distantly related parents compared to those of more closely related parents (Moll et al. 1963; Melchinger 1999). The creation of new inbred lines as prospective seed sources for hybrids was notably affected by the establishment of heterotic patterns among types and lines. Crosses between genotypes that have significant heterosis are identified as heterotic patterns (Carena and Hallauer 2001).

These patterns were essential for classifying genotypes according to heterotic relationships and lineages. The application of molecular markers was anticipated to improve the accuracy of heterosis prediction, as the development of maize inbred lines is constrained by their membership in particular diversity groups characterized by a common pedigree or genetic link (e.g., heterotic groups). Although DNA marker data has restricted use in forecasting effective heterotic combinations, it has proven beneficial in augmenting pedigree information and categorizing various lines into heterotic groupings (Smith et al. 1997; Senior et al. 1998; Barata and Carena 2006; Sonnino et al. 2007). Evaluating the efficacy of crossings between genetically distinct parental groups is essential for identifying interesting heterotic patterns (Melchinger 1999). One method involves utilizing data from yield experiments (such as test crosses and mating designs) alongside molecular information (such as marker data) (Barata and Carena 2006), contingent upon the establishment of a robust mechanism to integrate both data types.

3.3.2 Cultivars Characterization and Phylogeny

Research on the comparative morphology of maize (*Zea mays* L.) is essential for the management of crop diversity. Farmers in traditional and subsistence agricultural cultures utilize morphology to inform their use of germplasm (Perales et al. 2005; van Etten 2006). Subsistence communities utilize advanced varietal management strategies that identify differences in the heritability of morphological traits, subsequently informing management practices. Louette et al. (1997) assert that farmer classifications of maize germplasm, based on ear features, can closely resemble those generated by intricate statistical analyses. The morphologies of inbred lines and hybrids of maize in the United States have developed due to the introduction of hybrids and modifications in crop management approaches (Duvick 1997, 2005; Lauer et al. 2011).

Morphological data established the basis for the initial taxonomic studies on maize (Sturtevant 1884) and remain a crucial element in the contemporary classification of maize races (Ortiz et al. 2008). Morphological data are essential for the management of preserved genetic resources in ex situ gene banks (Sanchez et al. 2000; Bioversity International 2007). The International Board for Plant Genetic Resources (IBPGR) and the International Center for the Improvement of Maize and Wheat (IBPGR 1991) have published numerous criteria for the morphological descriptors of maize (Bioversity International 2007). These guidelines are issued by

Bioversity International/CIMMYT (2009), the Council for Mutual Economic Aid (COMECON), and the N.I. Vavilov All-Russian Scientific Research Institute of Plant Industry (VIR), alongside several national initiatives such as the European Union Maize Landrace Database (EUMLDB 2010) and the USDA (2010), in addition to the International Union for the Protection of New Varieties of Plants (UPOV) (UPOV 1999, 2009).

Comparative analyses of morphological traits are essential for assessing eligibility for intellectual property protection (IPP) under UPOV, including Plant Variety Protection (PVP) and Plant Breeders' Rights (PBR). Proving that a candidate variety is morphologically distinct from previously recognized variations of that species is necessary for acquiring a PVP certificate. The "Guidelines for the conduct of tests for distinctness, uniformity, and stability for maize" (UPOV 1999, 2009) offer comprehensive descriptions of numerous attributes. UPOV additionally delineates subsets of features to facilitate the global harmonization of databases or to categorize related varieties in cultivation experiments. Furthermore, UPOV recognizes the possible impacts of both environmental factors and the examiner on the descriptions of identical varieties (UPOV 2003).

Morphological and agronomic criteria have served as the principal instruments for assessing genetic diversity across maize landraces (Ilarslan et al. 2002; Mansilla et al. 2021; Munaiz et al. 2021; Mayer et al. 2022). Molecular markers, including single-nucleotide polymorphism (SNP), simple sequence repeats (SSR), random amplified polymorphic DNA (RAPD), sequence-related amplified polymorphism (SRAP), amplified fragment length polymorphism (AFLP), inter-simple sequence repeat (ISSR), and inter-primer binding site (IPBS), serve as effective instruments for assessing the genetic diversity of plant germplasm resources (Filiz et al. 2015; Dumlupinar et al. 2016; Gungor et al. 2022; Hocaoglu-Ozyigit et al. 2022).

Plant relationships across all taxonomic levels have been elucidated through the extensive application of chloroplast DNA (cpDNA) and sequencing (Shaw et al. 2014). The trnL(UAA)-F (GAA) intergenic spacer (IGS) sequence, located between transfer RNA genes for leucine and phenylalanine, has been extensively employed for taxonomic and phylogenetic comparisons (Feng et al. 2022; Herman et al. 2023; Orton et al. 2017; Taberlet et al. 1991). The inter-simple sequence repeat (ISSR) method utilizes polymerase chain reaction (PCR) to produce multi-locus markers by employing microsatellite sequences as primers in PCR (Abouseada et al. 2023; Gemmill and Grierson 2021; Pradeep et al. 2002). The concerns of genetic diversity are critically significant for the future of agricultural products (Hocaoglu-Ozyigit et al. 2022; Singh and Kaundal 2023).

3.3.3 Genetic Resources Conservation

The genetic resources of maize are exceedingly useful to agriculture. In the later part of the twentieth century, the swift substitution of traditional landraces in Europe was partially attributable to the emergence of contemporary hybrids. Systematic collecting expeditions were commenced to avert the depletion of genetic variety in maize (*Zea mays* L.). Landraces must be meticulously saved, characterized, safeguarded from extinction, and preserved for breeding purposes, since they represent a vital reservoir of potentially advantageous features.

The ex-situ method is exceptionally advantageous for the preservation of seeds in gene banks (Abdullaev et al. 2025; Abulaila et al. 2025; Akparov et al. 2025; Al Lawati 2025; Aladele et al. 2025; Almarri et al. 2025; Arab et al. 2025; Békefi et al. 2025; Breidy 2025; Cañar et al. 2025; de Galarreta et al. 2025; Desheva et al. 2025; Díez et al. 2025; Galea-Gragera and Llera Cid 2025; Ghidan and Abdulmajid 2025; Gudaye et al. 2025; Harzli et al. 2025; Hidayatun and Kurniawan 2025; Ihsan and Prihatini 2025; Jatoi and Siddiqui 2025; Laghetti et al. 2025; Mallor and Luna 2025; Monteros-Altamirano et al. 2025; Osterc et al. 2025; Popović et al. 2025; Rocha et al. 2025; Wang et al. 2025). Conversely, in situ conservation aids in the preservation of biodiversity and the identification of types possessing beneficial and adaptive genes.

3.3.3.1 In-Situ Conservation

The genetic resources of maize are highly significant to agriculture. In the later part of the twentieth century, the swift substitution of traditional landraces in Europe was partially attributed to the emergence of modern hybrids. Systematic collecting expeditions were initiated to avert the loss of genetic diversity in maize (*Zea mays* L.). Landraces must be meticulously saved, characterized, safeguarded from extinction, and preserved for breeding purposes, as they represent a significant reservoir of potentially advantageous features.

The ex-situ method is particularly advantageous for the conservation of seed banks. Conversely, in situ conservation maintains biodiversity and recognizes types possessing beneficial and adaptable genes.

In situ conservation is a method of preserving agricultural genetic resources by protecting them within their natural environments (Brush 1991; Maxted et al. 1997). Preserving sustainable populations of Crop Wild Relatives (CWR) in their natural habitats facilitates the evolution of new, more adaptive characteristics. In situ conservation is essential because wild populations of CWR possess considerably greater genetic variation than a single accession in a genebank. Furthermore, sustaining viable populations in their natural habitats enables CWR to persist in their evolutionary processes, possibly resulting in the development of novel adaptive features.

CWRs are often the primary focus of in situ conservation efforts, although they may sometimes be unintentionally located within protected areas established for other purposes. The Sierra de Manantlán Biosphere Reserve in Jalisco, Mexico, exemplifies in situ conservation of crop wild relatives (CWR) by safeguarding the endemic relative of maize, *Zea diploperennis* (Hunter and Heywood 2011). The Apacunca Genetic Reserve was formed in Nicaragua in 1996 to safeguard the last remaining populations of teosinte, or *Zea nicaraguensis*, a wild relative of maize that exhibits resistance to salt, pests, diseases, and flooding. Member nations of the FAO Commission on Genetic Resources for Food and Agriculture (CGRFA) are presently deliberating the creation of a global network of in situ conservation sites for Crop Wild Relatives (CWR) (FAO 2009).

Recently, the national focal points of the CBD Programme of Work on Protected Areas (PoWPA) and the Secretariats of the Convention on Biological Diversity (CBD), the International Treaty on Plant Genetic Resources for Food and Agriculture (ITPGRFA), and Bioversity International informed their respective Contracting Parties of the necessity to enhance in situ conservation of plant genetic resources by integrating active conservation of crop wild relatives (CWR) within protected area networks and linking in situ conservation to sustainable utilization. To ensure our current and future food security, an immediate worldwide initiative is required to create a network of "genetic reserves" in CWR hotspot regions. This will guarantee their preservation and sustainable utilization.

3.3.3.2 Ex Situ Conservation

Ex-situ conservation entails the preservation of genetic resources threatened with depletion outside their native habitat. One approach to conservation is the maintenance of these resources largely in gene banks, botanical gardens, or research centers dedicated to agro-biodiversity (Plucknett et al. 1987). The ex-situ conservation of wild relatives of crops enhances their accessibility for breeders and researchers, facilitating their application in breeding and agricultural studies, while also permitting their preservation in various areas.

The propagation and use of crop wild relatives are facilitated by their ex-situ conservation. Utilizing their genetic material for breeding is challenging, if not unfeasible, without ex-situ conservation. Envision the necessity of traveling to a remote location each time a breeder requires seeds from a certain group. By collecting and conserving crop wild relatives in gene banks, breeders can more readily access and employ them for agricultural research and breeding purposes. Furthermore, it facilitates safety duplication, the process of storing samples in many locations for backup purposes. Crop wild relatives are preserved ex-situ in worldwide gene banks such as the Svalbard Global Seed Vault, the Millennium Seed Bank, and the CGIAR, in addition to national crop diversity collections. Approximately 700,000 crop wild-related accessions, constituting 10% of the total holdings, are presently maintained ex-situ in gene banks globally (FAO 2010).

Conserving the wild relatives of crops must not hinder the prospects for development for those tasked with creating and managing their natural habitats. Nonetheless, it enforces limitations on the introduction of external factors that may jeopardize the wild relative. One strategy is advocating for the utilization of an enhanced maize variety within the habitat of *Zea* diploperennis. The introduction of such a variety automatically displaces the typical cultigen, resulting in genetic erosion in the strict sense, and necessitates a weed-free field, fertilizers, and pesticides.

3.3.3.3 Cryopreservation

Cryopreservation, defined as preservation in liquid nitrogen at −196 °C, is one of the most effective techniques for the long-term, unaltered preservation of plant genetic resources. A variety of cryopreservation techniques are available, such as vitrification, rapid freezing, slow cooling, and encapsulation-dehydration methods. The encapsulation-dehydration method is intricate, labor-intensive, and yields minimal recovery growth; conversely, the preceding two protocols have demonstrated limited efficacy and seem to be genotype-dependent (Zhao et al. 2001).

The vitrification technique offers enhanced recovery and simplified protocols, regardless of encapsulation (Matsumoto 2002; Matsumoto et al. 1994). The significance of preserving crop germplasm has been reiterated repeatedly. Now, numerous countries are enhancing their germplasm banks (Table 3.1).

Plant material may generally be conserved for extended periods by seed-banking maize at low temperatures and reduced moisture levels (Cromarty et al. 1982). Nonetheless, depending on the species, prolonged storage may lead to a decline in viability. This suggests the necessity to explore alternative methods, such as cryopreservation, which could extend the shelf life of maize seeds and improve the conservation of the crop. Usman and Abdulmalik (2010) discovered that the vitrification cryopreservation technique may improve the preservation of maize germplasm. Their findings substantiate the notion that this technique can enhance the preservation of maize germplasm.

Table 3.1 Maize accessions in gene banks

Accessions	Term of storage	Institution'	Location
15,084	Medium, long	VIRb	Leningrad, USSR
15,000	Medium	Institute of Maize Research	Belgrade, Yugoslavia
14,000	Short, medium	CIMMYT	EI Batan, Mexico
11,000	Short, medium	INIA	Mexico City, Mexico
7619	Long	NSSLb	Fort Collins, United States
5006	Short	ICA	Colombia
3444	Medium	Universidad Nacional Agraria La Molina	Lima, Peru
3200	Short	Research Institute for Cereals and Technical Plants	Fundulea, Romania
3000	Medium, long	INTA	Pergamino, Argentina

Source: Plucknett et al. (1983)

CIMMYT Centro Intemacional de Mejoramiento de Maiz y Trigo, *ICA* Instituto Colombiano Agropecuario, *INIA* Instituto Nacional de Investigaciones Agricolas, *INTA* Instituto Nacional de Tecnologia Agropecuaria, *NSSL* National Seed Storage Laboratory, *VIR* All-Union Institute of Plant Industry. IBPGR-designated base collections

3.3.3.4 In Vitro Conservation

Tissue culture and transformation are critical elements in the effective production of transgenic crops that provide resistance or tolerance to biotic and abiotic stressors. Micropropagation using meristem and shoot culture generates substantial quantities of genetically identical specimens (Harding 2010; Feyisola et al. 2015; Olowe et al. 2014). Rapid clonal replication of crops is accomplished via the multiplication of auxiliary shoots, adventitious shoot formation, or somatic embryogenesis (Bouquet and Torregrosa 2003). A lower quantity of cytokinin combined with auxins, particularly 2,4-D, has been documented to stimulate embryogenic calli formation in maize and other cereal crops (Chang et al. 2003; Al-Abed et al. 2006). Regeneration of maize from immature embryos (Aguado-Santacruz et al. 1998; Green and Phillips 1975; Ombori et al. 2008), mature embryos (Huang and Wei 2004; Al-Abed et al. 2006), nodal regions (Vladmir et al. 2006), anthers (Barloy and Beckert 1993), and shoot meristems (Sairam et al. 2003) has been documented. The regeneration technique utilizing immature embryos in maize is arduous and protracted compared to direct regeneration from explants, which is more efficient and expedites the acquisition of the complete plant without the callus interphase that may induce somaclonal variation (Zapata et al. 1999). Somatic embryogenesis for plant regeneration was accomplished with leaves as explants (Steinmacher et al. 2007).

Akula et al. (1999) documented the utilization of mature maize embryos to initiate callus formation; nevertheless, plantlet regeneration did not occur. Huang and Wei (2004) also noted the regeneration of temperate maize lines from mature embryos. Direct shoot induction and regeneration have been documented for several cereals, including Sorghum (Baskaran and Jayabalan 2005), barley (Sharma et al. 2004), wheat (Sharma et al. 2005), and maize (Zhang et al. 1996). A reliable, straightforward, and efficient in vitro method utilizing direct organogenesis is required for the regeneration of multiple transgenic plants, facilitating both clonal propagation and effective genetic engineering, potentially resulting in the creation of new resistant varieties against abiotic and biotic stresses. This study examined the direct organogenesis approach for *in vitro* regeneration and multiple shoot proliferation of maize genotypes utilizing various amounts of auxin and cytokinin.

3.3.3.5 DNA and Seed Banks

Researchers have employed diverse conservation genetics techniques to mitigate the threat of biodiversity loss (Russello et al. 2020). One way is DNA conservation, supplemented by data on DNA architecture obtained from conserved genome sequencing (Yuan et al. 2018; Kolchanova et al. 2019). In recent decades, the utilization of molecular markers for assessing DNA polymorphism has proven particularly effective in investigations of genetic diversity (Ignjatovic-Micic et al. 2008; Drinic Mladenovic et al. 2012; Barcaccia et al. 2016). The molecular characterisation of accessions is vital for identifying and eliminating potential duplicates in collections (Reif et al. 2004), making it essential to comprehend the link between morphological and molecular diversity for conservation in gene banks.

During the later part of the twentieth century, hybrid maize supplanted local maize landraces in the territory of former Yugoslavia (western Balkan region) (Andjelkovic and Ignjatovic-Micic 2012). In subsequent decades, the collection, characterization, and natural classification initiatives led to the preservation of over 2000 local maize landraces at the Maize Research Institute Zemun Polje (MRIZP) gene bank, including 222 landraces sourced from the Republic of Macedonia (Andjelkovic and Ignjatovic-Micic 2012). Globally, around 135,000 maize seed samples are preserved in seed banks, encompassing the majority of cultivated varieties and a wild related, teosinte (Chang 1985; FAO/IPGRI 1994).

3.3.4 Cytogenetics

Maize possesses a substantial genome, rich in transposons, about equivalent in size to the human genome at 2.5 Gb (Schnable et al. 2009). Despite the maize genome functioning genetically as a straightforward diploid with 10 chromosomal pairs, its structure is notably intricate (Rhoades 1951; Wendel et al. 1986).

Contemporary microscopy has been utilized to enhance the application of meiotic chromosomes, incorporating immunolocalization (Fig. 3.2) (Shi and Dawe 2006) and advanced fluorescent in situ hybridization techniques (FISH; Fig. 3.3) (Kato et al. 2004) to efficiently identify both mitotic and meiotic chromosomes (Kato et al. 2011).

The maize (*Zea mays* L.) genome has been examined as a model for eukaryotic genetics, cereal crops, and monocot genome evolution (Chandler and Brendel 2002); however, its vastness and organizational complexity hinder the elucidation of its structure. The existence of extensive gene-deficient regions, segmental duplications, numerous retrotransposons, and microvariations among maize lines complicates the creation of a comprehensive physical map of the maize genome (Gaut et al. 2000; Swigonova et al. 2004; Paterson et al. 2005). Notwithstanding this complexity, various types of maps have been created to delineate its structure and function at the DNA and chromosome levels. A multitude of linkage maps have been created, encompassing those founded on mutant phenotypes (Emerson et al. 1935) and, more recently, those incorporating thousands of supplementary molecular markers, including restriction fragment length polymorphisms (RFLPs), simple sequence repeats (SSRs), single-nucleotide polymorphisms (SNPs), and insertion-deletion polymorphisms (indels) (Davis et al. 1999; Fu et al. 2006; Helentjaris et al. 1986; Taramino and Tingey 1996). Physical maps of overlapping clones have been generated and correlated with the linkage map utilizing molecular probes (Bowers et al. 2005; De Jong et al. 1999; Hass-Jacobus et al. 2006).

A cytological map, another form of physical map, is generated through direct microscopic examination of the chromosomes constituting the nuclear genome. Cytogenetic maps are significant as they accurately position genetic loci inside the

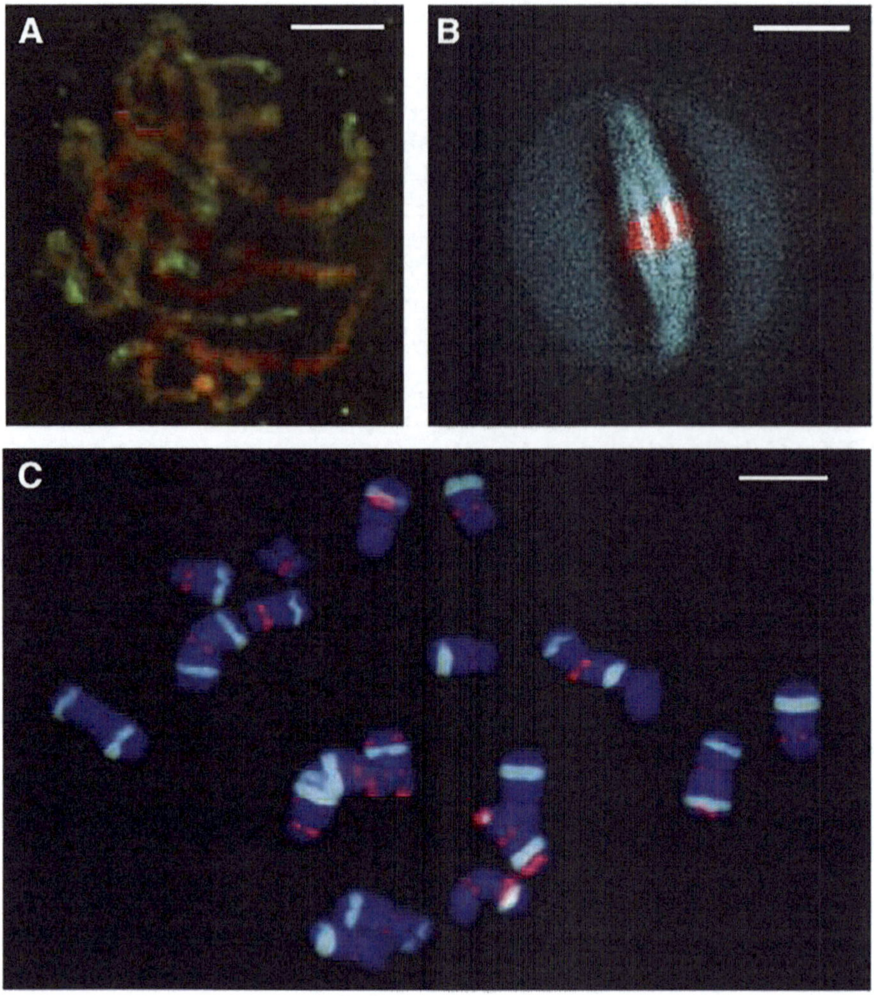

Fig. 3.2 Maize Cytogenetics. (**a**) Meiotic chromosomes can be extracted from immature tassels and observed using immunostaining. Pachytene chromosomes are marked with two antibodies: one that identifies heterochromatin (H3K27me2, red) and another that identifies euchromatin (H3K4me2, green). Scale bar: 5 μm. (**b**) Fluorescent lines are accessible for various tagged proteins; depicted here is a meiosis I cell exhibiting both tubulin-CFP (provided by Anne Sylvester) and histone H2B-mCherry (provided by Hank Bass). Scale bar: 15 μm. (**c**) Almost all of the 10 maize chromosomes may be distinguished from one another using simple two-color FISH. Chromosome spreads are generally conducted on mitotic (diploid) cells obtained from root tips, as illustrated above. Chromosomal dimensions, arm ratio, intensity of CentC staining (centromere repeat, green dots at the chromosomal center), and the position and kind of knobs (green and red patches near chromosome termini) are essential identifying characteristics. Scale bar: 10 μm. (Source: Nannas and Dawe 2015)

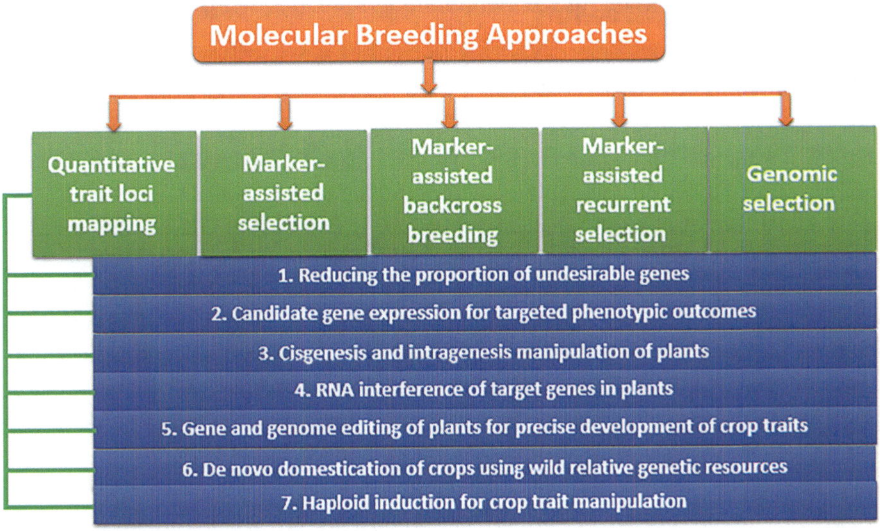

Fig. 3.3 Molecular breeding approaches and progression in plant breeding practices using these approaches. (Prepared by: E.F. El-Hashash)

complete chromosome, the definitive contig, offering insights on the location, sequence, and distribution of DNA sequences relative to other genetic markers along the chromosomes (Anderson et al. 2004; Sadder et al. 2000).

Previous research established the foundation for the meiotic chromosomal karyotype of maize, enabling the identification of the 10 distinct maize chromosomes by chromosome-spread preparations. Cytological chromosome stains indicate the existence of chromosomal features, including knobs and centromeres (Adawy et al. 2004; Carlson 1988; Dempsey 1994).

Mitotic and meiotic chromosomes have been effectively utilized to construct cytogenetic maps of maize. Mitotic chromosomes are simpler to prepare, but meiotic chromosomes possess longer axial fibers that enhance localization inside chromosomal arms (Pedersen and Linde-Laursen 1994; Cheng et al. 2001a, b; Kulikova et al. 2001). An alternative method for FISH mapping in maize involves utilizing substantial maize genomic DNA pieces alongside competitive in situ suppression hybridization (Sadder et al. 2000; Sadder and Weber 2002). Another technique involves utilizing genomic BAC clones derived from the small-genome cousin, sorghum (Koumbaris and Bass 2003). The cross-hybridization of DNA probes from one species to the target chromosomes of another species can mitigate detection limitations when the two species possess distinct or divergent classes of interspersed repetitive sequences, as exemplified by sorghum and maize. Transgenomic mapping and large-fragment FISH have been effectively employed in maize and many plant species for comparative genomics (Gomez et al. 1997; Hulbert et al. 1990; Jackson et al. 2000; Koumbaris and Bass 2003; Zwick et al. 1998).

3.4 Traditional Breeding

3.4.1 Improvement Strategies

The development of novel plant cultivars via conventional breeding excludes the application of advanced molecular breeding techniques (Abdel-Hamid and Salem 2021; Abu-Ellail et al. 2021; Aldahak et al. 2021; Elabd et al. 2020; Eltaher et al. 2017, 2018, 2021, 2022; El-Zanaty et al. 2013; Ghazy et al. 2020; Hassan et al. 2021; Ibrahim et al. 2023a, b; Ismail et al. 2013; Mourad et al. 2019). Conventional breeding does not contravene the fundamental principles of inheritance. Conventional breeding is the selective cultivation of plants based on their exceptional performance in specific qualities. Hybridization is employed to obtain essential characteristics from closely related species and incorporate them into new cultivars (Al-Khayri et al. 2016). Hybridization generates diversity in conventional breeding.

The paramount genotype is discerned by selection predicated on phenotype. The chosen cultivars undergo testing prior to their introduction as varieties. In traditional breeding, the lack of gamete segregation prevents the attainment of diversity with identical parental specimens (Acquaah 2009). Conventional breeding requires roughly a decade to create a new cultivar (Bharti and Chimata 2019). Conventional breeding predominantly depends on the phenotypic traits of crops to assess superiority (Aboshama et al. 2015; Alqahtani et al. 2024; Börner et al. 2009, 2010; Elabd and Salem 2009; El-Absawy et al. 2012, 2015; El-Zanaty et al. 2011; Emara et al. 2011, 2013; Esmail et al. 2009; Ibrahim et al. 2024; Mattar et al. 2009, 2016; Nawara et al. 2016, 2017; Saleh et al. 2020a, b; Salem 1997, 2004; Salem and Börner 2004; Salem et al. 2003, 2004a, b, 2015b, c, 2022). Thus, chosen cultivars are not without imperfections, as genotype by environment interactions considerably affect phenotypes (Lema 2018). Conventional breeding is an applied science that significantly relies on the experiences and observations of breeders (Allard 1961). The domestication of maize seems to have been a swift process rather than an extended one with numerous stages, coming from its wild progenitor, a subspecies of teosinte identified as *Zea mays* ssp. Parviglumis (Piperno et al. 2009; Ranere et al. 2009).

Since the domestication of plants, the predominant technique for genetic enhancement has been mass selection, wherein the finest ear from the healthiest plant is selected for planting in the subsequent year. The strategy progressed from basic mass selection to ear-to-row selection over time, although it proved ineffective for variables with low heritability, such as yield (Lee and Tracy 2009). Selection entails the identification of individuals for breeding based on their differences in preferred features, usually discernible or quantifiable traits. In one of the most extensive investigations worldwide, breeders at the Illinois Agricultural Experiment Station commenced the selection of maize kernels for elevated oil content in 1896. For nearly a century, maize breeders have continually selected to modify oil composition. After 100 generations, the oil content increased significantly from approximately 5% to around 20% in the high oil-producing line (Schlegel 2017).

The initial endeavors to transport pollen across plants commenced in 1867, with H.S. Bidwell implemented maize detasseling as a method of pollination control in his breeding program. Between 1878 and 1881, Professor W.J. Beal of Michigan State Agricultural College advocated for the hybridization of maize types to improve the yield of commercial cultivars. W.A. Kellerman and W.T. In 1891, Swingle performed the initial segregation counts on maize ears for the starchy gene. In 1893, at the World's Columbian Exposition in Chicago, R. and J. Reid's "Yellow Dent Maize" received the prestigious award as "the world's most beautiful corn." This maize became an essential progenitor of contemporary hybrid maize (Schlegel 2017). A significant advancement in maize breeding commenced with the identification of 'hybrid vigor' or heterosis in 1909, concurrently by E.M. East at Harvard University, in partnership with the Connecticut Agricultural Experiment Station (Byerlee 2020) and G.H. Schull at the Carnegie Institution Station at Cold Spring Harbor (Varshney and Tuberosa 2007; Byerlee 2020). This signified the commencement of scientific maize breeding. The inaugural hybrid maize, developed through the crossing of inbred strains, was introduced in 1933.

Contemporary corn breeding commenced in the 1970s and 1980s with the initial regeneration of a plant from cell culture and the sequencing of the first maize genes. In 1975, in vitro selection of maize revealed callus cultures resistant to the fungal disease Helminthosporium maydis T-toxin (Gengenbach and Green 1975). Molecular markers were initially employed in maize breeding during the 1980s and 1990s, with the first transgenic maize cultivar, *Bacillus thuringiensis* (Bt) hybrid, being sold in 1996 (Lundmark 2007; Troyer 2009). Subsequently, other transgenic hybrids exhibiting insect resistance, herbicide tolerance, elevated lysine content, and drought resistance were introduced (Lundmark 2007). Predictive breeding originated with the decreasing costs of genome sequencing from the late 1990s to early 2000s (Varshney and Tuberosa 2007), currently referred to as genomic prediction or genomic selection (Washburn et al. 2020; Fritsche-Neto et al. 2021). CRISPR technology for genome editing signifies the beginning of precision plant breeding (Kim and Kim 2019). The inaugural CRISPR-edited crop submitted to the US regulatory framework in 2016 was Pioneer high amylopectin maize from DuPont (Waltz 2016).

Innovative plant breeding methodologies utilizing gene editing, cisgenesis, and epigenetic strategies (Veillet et al. 2020) will be instrumental in the creation of novel plant cultivars, facilitating more accurate, foreseeable, and expedited breeding procedures.

3.4.2 *Traditional Breeding Methodologies and Limitations*

A recent molecular study indicates that maize domestication commenced approximately 8700 years ago in the central Balsas River Valley in southwestern Mexico. This domestication transpired swiftly, deriving from a wild lineage of *Zea mays* ssp.

Parviglumis, a subspecies of teosinte. This data corroborates the research conducted by Liu et al. (2017), Piperno et al. (2009), and Ranere et al. (2009). At the International Institute of Tropical Agriculture (IITA), maize has been enhanced in various quantitative features by conventional approaches. The enhancements encompassed characteristics such as Striga resistance, nitrogen use efficiency, drought tolerance, resilience to stem borers, reduction of aflatoxin buildup, yield potential, and improved nutritional quality (Alene et al. 2009; Gedil and Menkir 2019).

Conventional maize breeding entails the creation of new plant varieties based on the principles of natural genetics. This is the selection of plants based on their outstanding performance in particular qualities or attributes, as articulated by Lamichhane and Thapa (2022). Traditional breeding techniques for self-pollinating and cross-pollinating plants have been employed for a brief duration. An instance is the notion of pure-line selection, proposed by Peters in 1959, as recorded by. This technique entails generating pure lines via self-pollination of a singular, superior homozygous genotype. Following several years of multi-site trials, generally lasting about 6–7 years and incorporating comparisons with established test varieties, the superior genotypes are officially released as new maize varieties.

Pure-line selection is less efficacious due to diminished heritability resulting from environmental factors, as the genetic composition closely mirrors that of the parental genotype. Block selection, akin to purebred selection, utilizes highly heritable features for plant selection (Brown and Caligari 2020). Block selection can be executed in two manners: the first is single-parent selection, which involves the management of one type of gamete, and the second is biparent selection, which entails the control of a pair of gametes, one female and one male. The chosen individuals are subsequently cultivated in agricultural land and harvested upon reaching maturity. Post-harvest, the seeds are amalgamated and sown for the subsequent generation. In the subsequent year, the crop plants cultivated from the mixed seeds are hybridized with a test variety to assess heterosis. The chosen plants are introduced as new cultivars following multi-locus experiments.

Backcrossing incorporates advantageous characteristics from one plant to another while preserving other qualities through crossbreeding with a homozygous parent (Aleksoski 2018). In this procedure, the donor parents exhibit the desired trait, while the recipient parents acquire these selected genes. Following five to six generations of continuous hybridization with the recurrent parent, the hybrid progeny is expected to receive roughly 98% of the genome from the recurrent parent (Vogel 2009). In hybrid breeding, the newly developed variety predominantly acquires its genes from the recurrent parent, with a minimal contribution from the donor parent, as observed by Singh (1982).

Another strategy is repeated selection, a concept coined by Hull in 1945, initially utilized in maize breeding but subsequently expanded to other cereal crops, as noted by Ramya et al. (2022). This strategy entails the ongoing selection of advantageous traits across numerous generations to enhance their frequency via hybridization between high-performing individuals from heterozygous repeated parent and inbred individuals, as articulated by Bangarwa (2021). Hybridization is a technique for producing hybrids with advantageous characteristics by mating genetically divergent parents within the same species (intraspecific hybridization) or across distinct species (interspecific hybridization). Integrating traits from various progenitors to generate genetically enhanced progeny, whether via natural or artificial methods (Begna 2021).

Researchers at the Illinois Agricultural Experiment Station have successfully increased the oil content of corn through a compelling long-term study employing traditional breeding procedures. Initially, they commenced with a baseline of approximately 5% oil content, and throughout 100 generations, they cultivated maize lines with a remarkable oil content of 20% (Schlegel 2017).

3.4.3 Role of Biotechnology

Conventional plant breeding entails the repetitive selection of both progenitors and their descendants based on preferred traits. The importance of molecular breeding is considerable in today's world, particularly in developing nations where a limited percentage of the population participates in agriculture. This minority group holds the significant obligation of supplying sustenance for the majority of the nation's population (Breseghello and Coelho 2013). This accomplishment is attainable as plant breeding has successfully improved crop yields without necessitating an increase in arable acreage or agricultural labor force. This objective can be efficiently attained by crop improvement using molecular breeding approaches. Molecular breeding utilizes several methodologies, including the detection of basic qualities or quantitative trait loci (QTLs) in breeding lines or populations, the incorporation of genes from breeding lines or wild relatives, gene pyramiding, genetic diversity, marker-assisted recurrent selection (MARS), and marker-assisted backcrossing (MABC) (El Rabey et al. 2013; Salem 2004, 2009; Salem and El-Zanaty 2015; Salem and Mattar 2014a, b; Salem and Sallam 2016; Salem et al. 2007a, b, 2008a, b, 2010, 2015a, 2021c, 2023, 2024; Varshney et al. 2008).

Molecular MAS, commonly known as marker-aided selection, is an indirect method of selection that focuses on a specific feature with a marker (Salem 2015; Schlegel 2017). In the context of MAS, a marker is situated near a gene that regulates the characteristic, indicating the existence of a desired allele upon detection of the marker (Stevens 2008). Understanding the alleles at critical loci facilitates the formulation of optimal allele combinations to improve the agronomic value of the genotype. Marker-assisted selection is frequently employed for the pyramiding of resistance genes, which can confer comprehensive resistance across multiple plant generations until confronted by pathogen strains. Attaining gene pyramiding for resistance is highly difficult via traditional breeding techniques for specific traits, such as pathogen-induced disease resistance, when dominant resistance genes are involved (Breseghello and Coelho 2013). Genetic markers can be classified into three primary types. The first consists of visual or morphological indications, which are characteristics or phenotypic traits. The second group consists of biochemical indicators, specifically enzyme allelic variations known as isozymes. The third type consists of molecular or DNA markers that reveal variations in the DNA sequence (Winter and Kahl 1995; Jones et al. 1997).

For a nucleotide sequence to serve as an effective molecular marker in molecular breeding, it typically necessitates variation within its sequence. The variations in nucleotide sequences are revealed through molecular techniques such as restriction fragment length polymorphism (RFLP), amplified fragment length polymorphism (AFLP), random amplified polymorphic DNA (RAPD), single nucleotide polymorphism (SNP), microsatellite or simple sequence repeat polymorphism (SSRP), sequence-tagged site (STS), single-strand conformation polymorphism (SSCP), and cleavable amplified polymorphic sequences (CAPS), among others (Mohan et al. 1997). The application of this marker set depends on extensive prior research. This often necessitates multiple research stages for each trait, beginning with QTL mapping, advancing to fine mapping, and ultimately culminating in positional cloning (Bhullar et al. 2010). Key components for a successful MAS program encompass reliable markers, an effective DNA extraction technique, meticulously developed genetic maps, prompt and efficient data management, analysis of marker-trait associations, and expertise in utilizing tools for high-throughput marker detection (Kumawat et al. 2020).

MABC is a specific variant of MAS that employs genetic markers to streamline the backcrossing process, enhancing the selection of the recurrent parent and accelerating genome recovery. The MABC technique is widely utilized to eradicate undesirable traits, including insect and disease susceptibility, as well as anti-nutritional factors, from popular high-yielding varieties by incorporating QTLs or genes of interest from donor parents (Ribaut and Hoisington 1998; Hasan et al. 2015).

The utilization of DNA markers in a breeding program offers numerous advantages. DNA marker-based screening facilitates the early selection of traits by assessing plant genotypes at the seedling stage or from seeds, which may only become apparent in mature plants, including male sterility, fruit or grain quality, and photoperiod sensitivity. It accelerates the identification of alleles that are challenging to evaluate phenotypically, particularly for environmentally sensitive traits, hence streamlining and enhancing the breeding process. Individual plant selection, which may be unfeasible by phenotypic methods, becomes attainable when utilizing marker information. Marker-based selection eliminates the issue of low heritability. Moreover, for traits exhibiting intricate inheritance patterns, one can selectively identify the particular genes that contribute to the phenotype. Numerous features that often exhibit epistatic interactions can be stabilized and maintained through the utilization of molecular markers. Furthermore, Lema (2018) elucidates that homozygous and heterozygous plants can be differentiated using codominant markers, hence eliminating the necessity to test progeny in each generation to maintain recessive genes. To enhance nutrient utilization (Mehta et al. 2020; Prasanna et al. 2020; Baveja et al. 2021; Das et al. 2021; Yathish et al. 2022), address the intricate inheritance patterns associated with drought tolerance (Ribaut and Ragot 2007; Wang et al. 2016; Prasanna et al. 2020), and augment resistance to diseases, parasites, and insect pests in maize (Gedil and Menkir 2019; Yang et al. 2017; Xu et al. 2020).

3.5 Molecular Breeding

At the turn of the twentieth century, the scientific foundation for crop breeding and genetics was formed by the seminal discoveries of Darwin and Mendel (Moose and Mumm 2008). Traditional breeding techniques, such as population improvement and hybrid development, made significant contributions to the improvement of corn, primarily in terms of productivity and oil content (Choudhary et al. 2023). According to Singh et al. (2023), traditional methods of breeding maize that involve assessing progenies and then selecting for intricate characteristics like yield with low heritability are sluggish. The complexity of abiotic stress intensity, frequency, duration, and timing, the linkage drag of undesirable traits/genes with desirable traits, and the transfer of favorable genes/alleles from diverse plant genetic resources limited by gene pool barriers make molecular breeding a viable option for producing plant genotypes that can withstand stress environments, even though conventional breeding has made significant progress in producing tolerant genotypes (Gazal et al. 2018).

Rapid developments in plant biotechnology, including marker-assisted selection, genomic selection, transgenic technology/gene modification, haploid induction, gene editing, synthetic biology, molecular design, and intelligent breeding, have ushered in a new and more effective era of molecular breeding in plants (Jez et al. 2016; Wan et al. 2024) (Fig. 3.3). Speed breeding, genome editing, and cisgenesis are modern biotechnological techniques that have changed breeding practices. These advancements have increased breeding efficiency by combining genotypic and phenotypic characteristics (Sanchez et al. 2023). Breeders eventually began to switch to molecular breeding due to the enormous time and resource commitment involved in conventional breeding and the availability of molecular markers. Furthermore, the intricate characteristics of both biotic and abiotic stressors forced breeders to combine genomics and traditional breeding methods, leading to the development of genomics-assisted breeding (Choudhary et al. 2023). Conventional breeding approaches lack precision in manipulating and selecting certain genes, which might lead to unforeseen outcomes. Nonetheless, plant breeders have started using a range of molecular and genetic markers as molecular tools have improved, enabling a thorough and rapid evaluation of genetic variation among crop offspring. The genetic potential of common landraces, crop variants, and wild relatives has been unleashed through the combination of molecular and biotechnological treatments. This would make it possible to choose desirable traits more precisely and effectively, encouraging the development of improved plant varieties that more effectively meet societal and agricultural demands (Abdul Aziz and Masmoudi 2024).

There are now many opportunities to more efficiently develop nutritionally enriched cultivars of both grain and specialty corn because of a number of factors, including the availability of a large genetic diversity for the target traits, improvements in our understanding of key biochemical pathways for metabolite biosynthesis, analytical tools for screening germplasm for quality traits, and the potential to use molecular markers and genome editing techniques to speed up product development (Reynolds et al. 2019). Plant breeding has continuously evolved over the past century, moving from conventional to contemporary techniques, motivated by the need to more efficiently meet human needs (Abdul Aziz and Masmoudi 2024).

Molecular breeding is common in many crops. The application of molecular markers and, more recently, high-throughput genome sequencing projects has significantly improved the capacity to evaluate genetic variation in the germplasm pool for practically any agricultural species (Moose and Mumm 2008). The interdisciplinary field of molecular plant breeding is transforming twenty-first-century agricultural improvement. It was made possible by the recent combination of conventional plant breeding methods with developments in genomic research, biotechnology, and molecular marker applications (Moose and Mumm 2008). According to Singh et al. (2023), integrating science-led novel molecular breeding with a speed breeding platform appears to be a viable way to accelerate genetic gain in response to time, resources, and demand. For purely hereditary features, molecular breeding has shown considerable promise. On the other hand, possible marker-based techniques, like genomic and marker-assisted recurrent selection, have been created and combined to accommodate complicated features. However, the sympathetic orientation of breeders toward such techniques with obvious benefits can hasten their acceptance and inclusion in traditional breeding programs to expedite the development of new products (Singh et al. 2023). According to Meng et al. (2018), molecular markers are extensively utilized in marker-assisted selection, genetic polymorphism assessment, germplasm characterization, and genetic diversity evaluation of plant germplasm. For a trustworthy identification of a plant's genetic variety, the number of markers applied to the given genome in genetic analyses and the employment of an appropriate molecular approach is crucial (Chňapek et al. 2024).

Molecular breeding methods concentrate on finding the optimal combinations of alleles or haplotypes, gene networks, and certain genomic regions in order to quickly generate crops with desired traits (Abdul Aziz and Masmoudi 2024). The development of superior cultivars of crops with a wide range of commercially beneficial traits has been made possible by advancements in molecular methods. Plant breeders can create high-yielding, disease-resistant corn varieties more efficiently by utilizing molecular breeding techniques like genomic selection (GS) or genome-wide selection (GWS), marker-assisted recurrent selection (MARS), marker-assisted backcrossing breeding, and genomic selection (GS), as outlined by Abdul Aziz and Masmoudi (2024). Innovative techniques like MARS and GS are used to enhance precision, reduce the cost of phenotyping, and shorten the time required for developing disease-resistant traits, particularly for abiotic stressors where multiple quantitative trait loci (QTLs) influence expression (Gazal et al. 2018). Various molecular breeding techniques have been employed to enhance maize's resistance to drought. Thanks to advancements in molecular research, numerous molecular factors associated with maize's ability to withstand water deficit stress have been identified, leading to the creation of hundreds of drought-tolerant corn genotypes (Rasheed et al. 2023). The initial step in the molecular breeding process is to identify QTLs that confer resistance to abiotic stress. Once markers linked to desired traits have been identified, marker-assisted backcrossing can be utilized to introduce candidate QTLs or genes into elite lines (Gazal et al. 2018).

Through QTL mapping and genome-wide association studies, genomics-assisted breeding has identified significant main QTLs for yield, quality, and resistance to biotic and abiotic stressors. Thus, genomics-assisted breeding has the potential to accelerate global maize enhancement initiatives (Choudhary et al. 2023). Further research using GWAS analysis can be conducted to explore the metabolic basis of water deficit tolerance in maize. The latest molecular breeding methods must be employed to modify candidate genes identified in GWAS investigations (Rasheed et al. 2023). Genes and QTLs associated with drought tolerance in maize have been pinpointed through QTL mapping and GWAS analysis. These genes can be leveraged for potential breeding applications in other drought breeding programs. Drought breeding initiatives have accelerated due to the identification of potential genes and protein groups associated with drought resistance made possible by transcriptome and transcription factor research. By utilizing genetic engineering, transgenic corn genotypes tolerant to drought stress have been developed. Clustered Regularly Interspaced Short Palindromic Repeats (CRISPR/Cas9) offer a new avenue for editing desired genes to enhance drought tolerance in maize, reducing the time and cost associated with genotype development (Rasheed et al. 2023). Genomic selection and doubled haploids have the potential to significantly improve genetic gains, enhance efficiency, and shorten breeding cycles for a variety of crops (Peixoto et al. 2024).

GWAS has identified several genes, QTLs, and single nucleotide polymorphisms associated with drought resistance in corn, making this molecular technique valuable for developing drought-tolerant maize cultivars. Single-sequence repeats were utilized for molecular screening of drought tolerance. Transcriptome analysis is considered the most effective approach as it sheds light on the molecular pathways underlying maize's drought resistance. CRISPR/Cas9 can facilitate rapid molecular breeding for crop tolerance to abiotic stress (Rasheed et al. 2023). Molecular breeding, coupled with advancements in phenotyping, has made it easier to achieve breeding goals for various nutrients in maize. By incorporating breeding for nutrient enrichment into maize breeding initiatives, the goal going forward should be to give farmers and consumers high-performing, climate-resilient maize cultivars that offer increased nutritional quality. In order to alleviate frequent nutrient deficits, especially in sub-Saharan Africa, Latin America, and Asia, efforts should be made to develop exceptional maize cultivars that are stacked with numerous nutrients. Furthermore, dietary diversity and nutrient enhancement by agronomic and/or food processing treatments are complementary techniques that should be used in conjunction with biofortification (Prasanna et al. 2020).

3.6 Tissue Culture Applications

An effective in vitro plant regeneration system is crucial for the success of all biotechnological methods aimed at improving characteristics, including genetic engineering, haploid induction, and somaclonal variation (Hussain et al. 2012). Tissue culture was initially proposed a century ago to regenerate entire plants in vitro using somatic cells (Haberlandt 1902). The concepts of Gottlieb Haberlandt at the beginning of the twentieth century and Schwann and Schleiden's cell theory from 1838 serve as the foundation for plant cell and tissue culture, commonly known as in vitro culture (Thorpe 2007). Tissue culture involves the in vitro aseptic cultivation of cells, tissues, organs, or entire plants under carefully regulated environmental and nutritional conditions. This process is often used to create plant clones that match the selected genotype exactly. The culture is provided with a favorable environment for growth and multiplication through controlled settings, including appropriate pH balance, a suitable gaseous and liquid environment, and an adequate supply of nutrients (Hussain et al. 2012).

Plant tissue cultures are increasingly utilized in basic and applied research across various biological species. Globally, the commercial growth of plant species and extraction of bioactive components through in vitro technology has become a profitable industry (Yancheva and Kondakova 2018). Tissue culture techniques have been well-developed for various target cells and tissues, such as protoplasts, calli, cell suspension cultures, and immature embryos (Ishida et al. 2020). Plants can regenerate or repair damaged structures through a process known as plant regeneration, with tissue culture being one of the most popular regeneration technologies. Recent advancements have been made in plant regeneration research (Long et al. 2022).

Ovarian and ovule culture involves cultivating unpollinated female gametophytes (ovary, ovule, and flower buds) in vitro on suitable solid or liquid nutrient media. Monoclonal variants can result from tissue culture methods (Wijerathna-Yapa et al. 2022). In several crucial crops for agriculture, including maize, a second haploid production technique called in vitro gynogenesis is employed (Watts et al. 2018).

Particle bombardment and Agrobacterium tumefaciens techniques are well-established and commonly used in tissue culture regimens. Numerous components, including culture conditions, techniques for protecting cells during gene delivery, and handling before gene delivery, have been meticulously adjusted for various genotypes. Plant regeneration from altered cells in elite germplasm and tissue culture are the rate-limiting steps in both gene editing and transformation. Despite prolonged efforts, large varietal variability in tissue culture responses in maize remains a significant challenge. Protocols utilizing morphogenic regulator genes, including Wus2 and Bbm, have shown promise for effectively transforming resistant cultivars. These techniques are beneficial for both transferring gene editing tools and optimizing tissue culture for gene editing (Ishida et al. 2020).

Effectively screening plants for biotic and abiotic stressors can be done using plant tissue culture, which is efficient and cost-effective (Thorpe 2007). Tissue culture is an effective and economical method for developing elite and superior genotypes through stress tolerance breeding, a practice that has been widespread for many years. Tissue culture is an invaluable tool for building stress tolerance, screening for stress tolerance, and understanding physiological and biochemical changes under stress. Selection carried out in vitro in controlled climatic conditions is cost-effective and sustainable (Wijerathna-Yapa and Hiti-Bandaralage 2023).

The generation of somatic embryos and embryogenic callus is crucial for genetically modifying crops through gene editing and genetic transformation. Micropropagation techniques in many crop species heavily rely on regenerable embryogenic cultures (McFarland and Kaeppler 2024).

Maize, with its significant genetic potential, output, and productivity, is highly versatile. In the past few decades, plant tissue culture and transformation techniques have played a key role in enhancing maize through the introduction of beneficial transgenes or modification of endogenous gene expression (Kumar et al. 2021). Mature embryos are the most regenerative explants in maize, capable of generating high frequencies of plantlet regeneration in vitro. In addition to medium, explants, and hormones, in vitro traits are greatly influenced by genotypic variation (Muppala et al. 2020). Shahzad et al. (2017) culture has successfully led to the development of haploid somatic cells in maize, providing a way to generate genetic diversity. Using the embryogenic callus of immature maize embryos, the entire genome profile of DNA methylation has been investigated. In the callus, primary regenerated plants, and their selfed progenies, DNA methylation losses occur more frequently than gains, leading to an epigenetic footprint and somatically heritable variation in selfed progenies throughout the regeneration process (Wijerathna-Yapa et al. 2022).

As maize tissue culture evolved, the ontogeny of somatic embryos and embryogenic callus was discovered and described by scientists. The ability to create embryogenic cultures is heavily dependent on genotypes in several species, including maize (McFarland and Kaeppler 2024).

Creating pure lines of maize using tissue culture is considered a more efficient way to enhance the production of high-yielding hybrids. Somaclonal variants in wide hybrids may be advantageous for the exchange of genetic material. Tissue culture techniques, applicable to all known infectious agents but particularly useful in removing viruses and viroids from vegetatively propagated crops, can produce disease-free superior clones. In vitro methods for producing clean planting material can significantly increase crop production and quality. A synthetic medium is superior to natural components in the medium for inducing and growing calluses (Ahsan et al. 2000). Maize tissue culture, a fundamental aspect of maize genetic transformation, has gained increasing attention, particularly with significant advancements made using mature embryos as explants (Wang et al. 2012). Genotypic variations in tissue culture responses across different maize genotypes are significant, with elite genotypes proving challenging to culture and transform. Identifying genes that support tissue culture may be possible by searching for quantitative trait loci (QTLs) regulating tissue culture response (Ishida et al. 2020). Tissue culture and somatic embryogenesis are crucial and pivotal processes in maize transformation biology (Albert et al. 2021). Functional analysis and genetic improvement of plant genes are significantly influenced by genetic transformation (Liu et al. 2023).

Advanced genomics, genome editing, and maize transformation work together to create a potent, interdependent trifecta for better molecular breeding and functional genomics research. The use of cutting-edge technologies to edit the genome and modify maize will have a significant impact on global agriculture, ushering in a new era of molecular breeding and varietal development (Albert et al. 2021). In order to promote embryonic cell growth and increase the effectiveness of genetic transformation, the maize gene Wuschel-like homeobox protein 2a, also known as Wox2a, is essential and may hasten maize genetic improvement (Liu et al. 2023). A variety of factors, including genotypes, explant types, and media compositions, greatly influence maize's ability to regenerate in vitro. While some genotypes are resistant to tissue culture, others are more receptive to creating embryogenic calli through tissue culture. Genotypic variations in morphogenesis and organogenesis are well-known and may result from differences in endogenous hormone levels. Immature zygotic embryos are typically used to generate embryogenic calluses as maize's in vitro regeneration ability typically decreases throughout tissue development (Kumar et al. 2021).

One obstacle to the widespread use of plant genome editing is plant transformation. A breakthrough in somatic embryogenesis induction in maize was achieved through the co-transformation of the morphogenic transcription factors WUSCHEL and BABY BOOM. This method significantly boosts transformation efficiency when combined with modified tissue culture media. Despite the clear need for greater transformation capability in academic settings, this method has not received much attention (Aesaert et al. 2022).

Significant advancements in maize biotechnology are expected to come from emerging technologies that use tissue culture-independent methods and either stable or transient expression. These technologies include haploid-inducer-mediated genome editing, pollen transformation, simultaneous double haploid manufacturing and editing, and direct genome editing with an RNA-guided endonuclease system acting as an in vivo desired-target mutator (Yassitepe et al. 2021). Advances in DNA and protein delivery mechanisms and tissue culture-free techniques for corn modification will also help advance gene-editing technologies (Li et al. 2017; Vejlupkova et al. 2020). Novel approaches, such as double haploid induction linked to gene editing techniques, will accelerate precision breeding (Kelliher et al. 2019; Wang et al. 2019).

3.7 Genetic Engineering and Gene Editing

Due to the decreasing amount of arable land, population growth, and climate change, sustainable agriculture has become more challenging. It is essential to develop modified crops with enhanced quality, productivity, and resistance to various biotic and abiotic stressors. Genetic modification of plant crops has enabled the creation of efficient production systems that offer significant benefits to farmers and the community, aligning with the principles of sustainable agriculture—environmental protection, improved human health, and economic growth (Abdul Aziz et al. 2022). Key techniques used for maize improvement include transgenic breeding, cross-breeding, and mutation breeding (Varotto 2024). Since the early 1990s, genetic engineering has been utilized in agriculture to suppress undesirable genes and introduce desirable traits in crops. However, the actual production impacts of this technology are limited by laws prohibiting genetically modified materials in many countries and financial barriers hindering its adoption in authorized regions (Scheitrum et al. 2020).

Genetic engineering can significantly enhance grain yield and disease resistance, driven by mechanistic understanding. Despite this, pinpointing the exact causative gene for a desired trait remains challenging due to the complexity of identifying candidate genes within a large interval (Yang and Yan 2021).

Genetically modified corn production experiments have been conducted globally since the 1990s (Farooq et al. 2023). Over 150 distinct transgenic corn events have been approved for commercial cultivation, with the first transgenic hybrids entering the market over two decades ago. Advancements in genome sequencing and genomic tools have provided scientists with the knowledge needed to attempt genome alteration at specific genomic loci (Varotto 2024). Following the commercial release of genetically engineered seeds for key field crops in the US in 1996, adoption rates gradually increased. By 2024, 90% of maize acres in the US were planted with herbicide-resistant seeds, and the cultivation of Bt corn increased significantly over the years (USDA 2024a, b). GE maize farming has led to reduced human exposure to mycotoxins and improved grain quality (Pellegrino et al. 2018).

Genetically engineered maize MON 94804 was developed to reduce plant height by incorporating the GA20ox_SUP suppression cassette. Molecular and bioinformatic analyses have not identified any safety concerns related to food/feed. Ear height, plant height, and forage carbohydrate levels are the only variations between maize MON 94804 and its conventional equivalent that need more research; none of these variances pose any threats to human health or nutrition (EFSA et al. 2024). The genetically modified organisms panel has concluded that corn MON 94804 is as safe as its conventional and non-genetically modified counterparts, with no need for post-market monitoring of food or feed (EFSA et al. 2024). However, concerns remain regarding the development of genetically modified plant crops (Fig. 3.4), as noted by Abdul Aziz et al. (2022).

The development of biotechnology crops for more sustainable food production has been greatly aided by recent advancements in genome editing (Hernandes-Lopes et al. 2023). Robust genome editing techniques are now a component of the plant breeding toolkit (Valentine et al. 2024). According to Manzoor et al. (2024), there are several methods for editing genomes, including meganucleases (MNs), base editing, zinc finger nucleases (ZFNs), transcription activator-like effector nucleases (TALENs), and clustered regularly interspaced short palindromic repeats (CRISPR), and CRISPR-associated protein (Cas9), CRISPR/Cas13. Nucleases are used by targeted genome editing tools to cause double-strand breaks (DSBs) in DNA. There are two primary methods for repairing double-strand breaks (DSBs) in plant cells: homology-directed repair (HDR) and nonhomologous end-joining

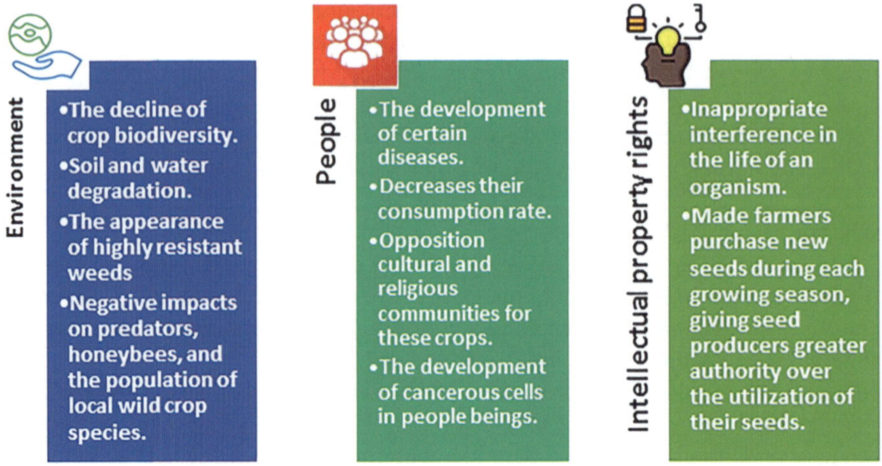

Fig. 3.4 Concerns and associated problems with the development of genetically modified plant crops. (Prepared by: E.F. El-Hashash)

(NHEJ) (Varotto 2024). At the moment, genome editing is often carried out at one locus or several loci simultaneously. Massively parallel genomics platforms, flexible genome editing tools, and flexible transformation systems enable targeted variation at nearly any location, throughout the spectrum of genotypes within a species (Valentine et al. 2024). By more accurately and effectively targeting various features, genome editing has become one of the promising techniques for improving crop attributes through targeted mutagenesis, gene knockouts, knockdowns, changes, and activation of target genes. This approach has proven successful in improving disease resistance in plant crops (Manzoor et al. 2024). Plant breeding may now be done with greater accuracy and efficiency thanks to genome-editing techniques like the CRISPR/Cas9 system (Yamada et al. 2024). Crop breeding techniques are currently using CRISPR/Cas9 technique to enhance features like disease resistance, nutrition, and drought tolerance (Ndudzo et al. 2024). The most flexible genome-editing technique, CRISPR/Cas, has demonstrated the ability to modify the genome in a variety of ways, from allele-specific alterations to gene knockouts and modulations of gene expression patterns, all to design superior genotypes with numerous enhanced agronomic qualities. Delivering CRISPR/Cas to plant crops that are less receptive to transformation and regeneration, however, is a common bottleneck. To overcome transformation recalcitrance, several methods have lately been developed, such as ectopic/transient expression of genes encoding morphogenic regulators and HI-Edit/IMGE (Hernandes-Lopes et al. 2023). According to Hernandes-Lopes et al. (2023), these technologies enable the breakdown of the barriers that prevent crops from being used for genome editing.

It will be extremely beneficial to agriculture and humanity to use genome-editing technologies on maize, one of the world's most important cereal crops (Yamada et al. 2024). Svitashev et al. (2015) provided the first comprehensive report in a scientific journal on the use of the Cas9-gRNA genome editing technique in corn, with direct delivery of sgRNA in the form of RNA molecules. These genome editing techniques have been used in maize to improve its agronomic qualities and increase yield. These protein-based nucleotides are believed to be ideal instruments for gene editing (Farooq et al. 2023). Genome editing is currently a potent tool for precise maize modification thanks to the considerable advancements brought about by CRISPR/Cas systems from their initial uses to the present (Varotto 2024). Agrobacterium tumefaciens-based genetic transformation protocols or biolistic delivery techniques in a few temperate varieties appropriate for Agrobacterium infection and regeneration have been used in the majority of maize genome editing investigations (Kausch et al. 2021; Yassitepe et al. 2021). However, the widespread use of genome editing in maize is hampered by the lack of a transformation methodology that can be utilized for a larger range of genotypes and the generally low transformation efficiency. Recently, methods such as trans genome editing and the use of morphogenic regulators (MRs) have been proposed to get around these limitations. Globally, programs based on tropical germplasm should benefit from the widespread application of such techniques in maize breeding (Hernandes-Lopes et al. 2023). A Cas9-guide RNA expression cassette and a selectable marker are inserted into the genome DNA of maize utilizing conventional genome-editing techniques; however, to allay legislative concerns about gene-modified organisms, the inserted DNA cassettes must be removed. Recalcitrant genotypes/cultivars represent a significant obstacle to the development of an effective and widely applicable DNA-free genome-editing system for corn, as the production of transgenic and/or genome-edited maize depends on cell/tissue culture and its subsequent regeneration into plantlets (Yamada et al. 2024).

Because they play a critical role in controlling cell-to-cell communication, the CLAVATA3/EMBRYO SURROUNDING REGION-RELATED (CLE) genes directly affect yield. CRISPR-Cas9 was utilized to target the CLE genes, producing null alleles of one partially redundant compensatory CLE gene and weaker promoter alleles of the CLE genes, which led to progeny with higher grain yields (Liu et al. 2021). In an unrelated study, Wang et al. (2021) applied CRISPR-Cas9 to enhance the scent of corn grain by altering the BETAINE ALDEHYDE DEHYDROGENASE 2 (BADH2) genes. A promising strategy to increase crop improvement and lessen the effect of abiotic stresses on crop yields in Africa and other regions is the development of drought-tolerant corn genotypes through CRISPR-mediated genome editing (Sami et al. 2021). This will also support sustainable agriculture. The ability of CRISPR-Cas9 technology to generate novel allelic variations and lines with improved resistance to drought stress was demonstrated by the researchers. The endogenous mRNA expression of the targeted ARGOS8 gene is low and varies geographically. The CRISPR-Cas9-mediated production of two variants of the ARGOS8 gene, ARGOS8-v1 and ARGOS8-v2, exhibited the potential to modify endogenous genes to create novel variations that enhance maize's drought tolerance under water-limiting conditions in the field (Ndudzo et al. 2024; Liu et al. 2021; Shi et al. 2017; Wang et al. 2021).

Multiplex gene editing was accomplished in corn by co-expressing Cas9 and many gRNAs, each of which targeted a different location and was delivered conventionally. Theoretically, a single construct with many guide RNAs can expedite and enhance the transformation processes. Two methods were created: the first relied on a multi-guide RNA that was activated by a single promoter and split into multiple sgRNAs through tRNA motif-mediated self-cleavage (Qi et al. 2016); the second method used tandem repeats of various U3 and U6 promoters, every of which controlled a single guide RNA (Char et al. 2017). Targeted mutagenesis produced changes in both cases that were primarily deletions or insertions of a few nucleotides, most likely as a result of NHEJ (Varotto 2024). In order to create a genome-editing system that does not require DNA for resistant genotypes or cultivars of maize, Cas9-gRNA ribonucleoproteins were injected directly into zygotes that were separated from the pollinated flowers of the B73 cultivar. The zygotes underwent effective development and were subsequently co-cultured with phytosulfokine, a peptide phytohormone, to regenerate into genome-edited plantlets. With the help of the technique presented here, it was possible to efficiently obtain genome-edited recalcitrant maize genotypes/cultivars that were devoid of DNA and selectable markers. Regardless of their stubborn traits, this approach can improve the molecular breeding of corn and other significant crops (Yamada et al. 2024).

In the next 30 years, the viability of genetically modified crops should be assessed in light of their contribution to human development and sustainable agriculture. Not only do genetically modified crops carry some hazards and concerns, but there are disadvantages to any approach used in food production. However, genetically modified crops are as safe as traditionally bred crops thanks to the application of genome editing technologies and regulation (Abdul Aziz et al. 2022). These crops can also be the main drivers of sustainable food security. Furthermore, international adoption of current and future regulatory rules will determine how CRISPR-Cas technology is utilized in the breeding of new varieties of maize (Varotto 2024).

3.8 Mutation Breeding

3.8.1 Conventional Mutagenesis

Mutagenesis is a technique that induces mutations, resulting in abrupt, heritable alterations in an organism's genetic composition using chemical, physical, or biological agents, as opposed to gene segregation or recombination (Roychowdhury and Tah 2013). Mutation breeding employs three types of mutations: site-directed mutagenesis, which entails creating a mutation at a precise location within a DNA molecule; induced mutations, generated through radiation (gamma rays, X-rays, and ion beams) or exposure to chemical mutagens; and insertional mutations, resulting from DNA insertion via genetic transformation, T-DNA insertion, or the activation of transposable elements (Kharkwal and Shu 2009; Forster and Shu 2012).

Table 3.2 Country-specific status of crop mutants categorized by crop type worldwide

S. No.	Top 10 countries who has developed the highest number of crop mutants		Top 10 crops having highest number of crop mutant	
1	China	810	Rice	833
2	Japan	479	Barley	305
3	India	341	Chrysanthemum	285
4	Russian	216	Wheat	264
5	The Netherlands	176	Soybean	175
6	Germany	171	Maize	89
7	United States	139	Groundnut	78
8	Bulgaria	76	Rose	67
9	Bangladesh	75	Common bean	57
10	Viet Nam	58	Cotton	48

Source: FAO/IAEA Mutant Variety Database (FAO/IAEA 2020)

Enhancing agriculture necessitates genetic variety in plant breeding for beneficial features (Novak and Brunner 1992). Currently, plant breeders systematically employ mutation techniques to discover novel genetic resources for traits associated with nutritional quality, yield potential, herbicide resistance, biotic and abiotic stress tolerance, and climatic resilience (Choudhary et al. 2023). Jankowicz-Cieslak et al. (2017) demonstrate that induced mutations effectively identify and map significant genes and quantitative trait loci (QTLs), investigate gene function, and analyze the molecular, biochemical, and metabolic pathways associated with the development of the mutant phenotype. Mutations are the fundamental cause of all genetic alterations observed in every species, including plants. Natural selection and evolution rely on stable and resultant genetic differences (Forster and Shu 2011). Numerous countries have employed induced mutations in recent decades to develop enhanced crop varieties, significantly influencing food supply and consumption (Kharkwal and Shu 2009; Suprasanna and Jain 2017). The FAO/IAEA Mutant Variety Database (http://mvd.iaea.org/) (FAO/IAEA 2020) comprises data on roughly 3332 mutant varieties across 190 crop species, generated and uploaded by 2020. Germany is ranked sixth globally in the development of mutant varieties, having generated 89 mutant maize varieties. The most significant crop mutations globally were created for rice, succeeded by barley (Table 3.2).

3.8.2 *In Vitro Mutagenesis and Selection*

Due to the ability of totipotent plant cells to be cultured, extensively multiplied, and induced to regenerate complete plants, in vitro culture techniques are particularly pertinent to mutagenesis. A range of plant components, including rooted stems, cuttings, detached leaves, dormant buds or plants, shoot apices (apical buds), axillary buds, and tubers, are vulnerable to radiation or mutation. One of the primary advantages of employing in vitro cultures for mutagenesis is the capacity to treat and screen extensive populations of cells prior to their regeneration into complete plants (van Harten 1998). Callus and cell suspension cultures exhibit significant regeneration capacity, making them an attractive source. Somatic embryogenesis is the most efficient in vitro method for mutagenesis, as it generally initiates from a single cell. Moreover, numerous subcultures can be efficiently conducted for chimera separation and to augment the population of mutagenized individuals for selection. Consequently, non-chimeric mutants can be isolated from the irradiated explants by callus proliferation. The induction of secondary or recurrent embryogenesis in cultures substantially enhances the likelihood of generating desirable mutants.

M1 (irradiated explants) are frequently utilized to facilitate in vitro subcultures for three, and occasionally four, cycles. The primary variables influencing the regeneration process include the passage of in vitro culture, the specific gene or trait selected and expressed during the selection phase, and the mutagen therapy employed. Regeneration durations can be reduced with the prompt regeneration of certain variants alongside ideal mutation-selection conditions.

Haploid callus cultures derived from microspores or ovules are the optimal targets for mutagenesis. In vitro research on mutant selection benefits from haploid cell and protoplast cultures, as predominantly recessive mutations can be readily identified in later generations. Moreover, the capacity to rectify abnormalities via doubled haploidy (DH) is essential, particularly as induced mutations are predominantly recessive and generally remain undetectable until the M2 generation at the earliest (Szarejko and Forster 2007). Mutagen treatment can be administered at different phases: either during the development of the parent cultivar to generate M1 plants from which haploid explants are cultured to yield doubled haploid mutant lines or during the maturation of the M1 plants to produce M1 plants for culturing microspores or anthers for subsequent selection and doubled haploid mutants. Successful selection for herbicide tolerance (chlorsulfuron, imidazolinone), blackleg resistance, elevated oleic acid, reduced linoleic acid, and diminished saturated fatty acid levels have been accomplished via haploid embryos, subsequently leading to haploid plants and doubled haploids (Szarejko et al. 1991; Maluszynski et al. 1995). This was achieved through the application of microspore culture and mutagenesis.

Callus induction is a critical stage in the transformation of maize. Type I callus is a less dense and less regenerative kind of callus generated by most maize varieties. Conversely, Type II calluses are malleable, fragile, and capable of embryogenesis (Armstrong et al. 1991; Hansen and Wright 1999). The F1 hybrid line is designated as Hi-II. The F2 population resulting from the A188 × B73 cross, after four generations of tissue culture, selfing, and sib-pollination, yielded its progenitors, "Hi-II Parent A" and "Hi-II Parent B." Immature embryo explants from Hi-II have been documented to elicit a 100% Type II response (Armstrong and Green 1985). Vega et al. (2008) reported that the Hi-II genotype had the highest transformation efficiency, varying between 12% and 18%. A188 has been intermittently shown to provide a 30% transformation efficiency via the Agrobacterium-mediated method (Ishida et al. 1996). Line Gz 643, an Egyptian genotype, exhibited a regeneration frequency of 42.2% (El-Itriby et al. 2003). Rasha et al. (2013) indicated that IL3 was the most transformable tropical maize genotype, surpassing the A188 temperate genotype, with transformation frequencies of 31.7% and 5.82%, respectively. The primary targets for maize transformation are embryogenic callus obtained from immature embryos and embryogenic callus cell suspension cultures (Armstrong 1999; Hansen and Wright 1999; Torney et al. 2007).

A crucial element of every mutation-breeding program is the selection and identification of advantageous mutations. In the context of controlled selection following mutagenesis therapy, in vitro cultured explants provide greater alternatives than methods necessitating the treatment of in vivo specimens. In vitro screening can accommodate large populations, hence resolving the limitations associated with utilizing a limited number of individuals in vivo plant material. Protoplast cultures and mutagenized cell suspension cultures are advantageous due to their greater genetic homogeneity compared to calli, embryos, and other explants.

The following elements are essential for the in vitro selection technique to yield tolerant plants: (i) significant cellular variation; (ii) a straightforward in vitro selection method; (iii) an effective method for regenerating tolerant cells (Widoretno et al. 2003); and (iv) the trait intended for inheritance (Yusnita et al. 2005).

The trait of interest should be selectable at the cellular level and expressed in the regenerated plants by any selection strategy. Certain traits, including yield, seed color, and plant height, are predominantly regulated by several genes and are not exclusively selected at the cellular level. In plant cell cultures, selection agents can be employed to isolate certain traits of fundamental interest and those of agronomic importance. Selected in vitro qualities include disease resistance, stress tolerance—particularly to salinity and drought—enhanced nutritional quality, and herbicide tolerance. Mutants have been generated and isolated in multiple plant species (Predieri 2001).

In vitro selections can be categorized into two types: single-step selection and multi-step selection. An inhibitor or antimetabolite is often included in the culture media at a quantity sufficient to induce cell death or impede growth in mutagenized cells. In the single-step selection procedure, cultures are maintained through many subculture regimes following the introduction of the inhibitor to the culture medium at a concentration of at least two to three times the maximum inhibitory concentration (MIC).

Examples include selection for herbicide resistance (Chaleff and Parsons 1978), resistance to lysine and threonine (Hibberd and Green 1982), salt tolerance (Bressan et al. 1985), and disease resistance (Gengenbach and Green 1975). In a multi-step selection method, a sub-lethal concentration (below the MIC) is added into the medium to facilitate the proliferation of in vitro cultures, with subsequent subcultures maintaining a gradual increase in the inhibitor concentration. It is expected that the mutant trait selected by this strategy will often exhibit greater stability and expression because of the variant cells' continuous exposure to increasing doses of the inhibitor (Miao et al. 1983; McCoy 1987; Patade et al. 2006).

3.8.3 Molecular Analysis

The renewed interest in induced mutagenesis stems from molecular biology research, which emphasizes approaches for developing and utilizing molecular markers, as well as identifying, mapping, and discovering novel genes and alleles. Seed companies have transitioned from aneuploid and open-pollinated varieties to hybrids, offering breeders criteria to discern original kinds based on genetic and phenotypic characteristics (Panella 2008). Focusing on locally produced genomic lesions represents a novel stage in the advancement of experimental mutagenesis (TILLING). Ethyl methanesulfonate (EMS), a chemical mutagen, has been employed to produce point mutations in many crops. TILLING integrates molecular screening methods for mutations with induced mutagenesis to generate a population subjected to mutagens that yield single nucleotide polymorphisms (SNPs). This approach facilitates the acquisition of diverse allele variants in genes essential for selection (Kornienko and Butorina 2013). Functional genomics can be developed via TILLING (Till et al. 2003). Initiatives to build functional genomics for several crops, such as barley, wheat, rice, maize, sorghum, beans, rapeseed, soybean, and other species, are presently underway in multiple countries (Cooper et al. 2008; Porch et al. 2009; Tadege et al. 2009).

The majority of reverse genetic resources in maize have employed the native transposon families Mutator (Mu) and Activator (Ac). Mutator transposable elements are frequently present in high copy numbers and typically insert into or adjacent to genes (Lisch 2002), often resulting in gene knockout or loss-of-function alleles due to insertions in regulatory regions. Resources such as the Trait Utility System for Corn (Bensen et al. 1995), the Maize Targeted Mutagenesis project (May et al. 2003), and the Photosynthetic Mutant Hunt (http://chloroplast.uoregon.edu/) are accessible for identifying Mu insertions in specific genes. A modified Mutator element has been created to facilitate the insertion of flanking genomic DNA into E. Coli and rescue via a plasmid (Raizada et al. 2001), enabling the establishment of an insertional database like the T-DNA database for Arabidopsis and permitting the sequencing of tagged genes. Low-copy number elements known as activator transposons preferentially integrate into or adjacent to genes, akin to Mu (Cowperthwaite et al. 2002). The large-scale isolation and sequencing of Ac elements for a reverse genetics database is advantageous since it enables the acquisition of single to low-copy number insertions.

Strategies derived from mutations generated by reverse genetics may possess extensive uses. Two methods have been delineated for plants. Fast neutron bombardment deletional mutagenesis is effective for eliminating tandemly duplicated genes (Li et al. 1991). An alternative method is TILLING (Targeting Induced Local Lesions IN Genomes), which involves chemical mutagenesis that produces point mutations, later identified in target genes by a sensitive technique for detecting single-nucleotide changes (McCallum et al. 2000). TILLING can generate an allelic series comprising missense and knockout mutations, which are valuable in traditional forward genetic studies.

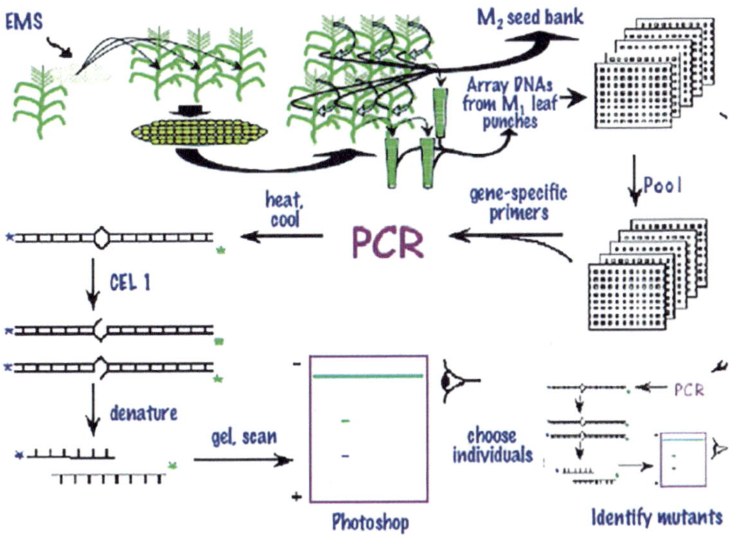

Fig. 3.5 Depicts the schematic representation of maize TILLING. Post-harvest, fresh pollen undergoes treatment with ethylmethanesulfonate (EMS) to induce mutations. Subsequently, silks from wild-type plants possessing identical genetic backgrounds are fertilized with the treated pollen. The M1 generation plants are cultivated from the seeds of the resultant ears, exhibiting heterozygosity for all introduced mutations. Tissue is extracted from almost every M1 plant, or each plant. Ten M2 progeny are generated from the self-pollination of the M1 generation. Ten to twelve M2 siblings are randomly crossbred to produce M3 seeds, which serve as the principal source for subsequent research. Tissue samples are taken, DNA is extracted and subsequently mixed to optimize screening. Sequence-specific primers are employed in PCR to amplify the target region for the identification of mutations. Following amplification, samples are subjected to thermal denaturation and subsequently reannealed to create heteroduplexes between mutant and wild-type amplicons. Denaturing polyacrylamide gel electrophoresis is employed to visualize heteroduplexes after cleavage by the CEL I endonuclease. (Source: Till et al. 2003)

TILLING is a sequence of processes that begins with the chemical mutation of reference people and concludes with DNA sequencing to identify mutant base pairs (McCallum et al. 2000). High-throughput TILLING employs the CEL I mismatch-cleavage enzyme on heteroduplexes, with end-labeled cleavage products identified on electrophoretic gels (Colbert et al. 2001; Till et al. 2004). Electromagnetic stimulation (EMS) was employed to address the pollen (Fig. 3.5).

The sole method capable of refining the search for missense mutations to a particular portion of a protein, such as an individual domain within a multidomain protein, is TILLING, which offers an allelic series of mutations. Tilling lines can be produced in a uniform wild-type genetic background, circumventing complications associated with heterogeneity, which are frequently required for insertional mutagenesis, particularly in maize. There is probable agricultural interest in developing phenotypic variants without incorporating foreign DNA into a plant's genome, due to the substantial expenses of intellectual property and regulations linked to transgenics, as well as prevailing concerns around genetically modified crops (Tax and Vernon 2001).

3.9 Hybridization

By combining genetically distant parents within the same species (intraspecific hybridization) or between separate species (interspecific hybridization), hybridization is another technique for producing hybrids with desired features. Whether through natural or artificial means, it entails merging traits from other parents to produce genetically superior offspring (Begna 2021). To ensure that the world's food need is met, plant breeding must produce high-yield hybrids (Dong et al. 2024). In southern Mexico, a single event approximately 9000 years ago led to the domestication of corn, which was derived from teosinte (*Zea mays* ssp. parviglumis) found in the Balsas River (Yang et al. 2019). Physiology, developmental biology, genetics, and, more recently, genomic studies have all historically considered maize to be a fundamental topic (Singh et al. 2024). The first instance of natural hybridization in corn was noted by Cotton Mather in 1716 (Begna 2021). Edward East's groundbreaking work on inbreeding depression and hybrid vigor in 1908 marked the beginning of the genetic study of maize (*Zea mays* L.) (Singh et al. 2024). Since the 1930s, breeding initiatives have led to modifications in the composition of hybrid maize plants. (Duvick 2005). According to reports, genetic gain contributed 79% of the improvements in maize productivity from 1930 to 2011 in the target production environment (Smith et al. 2014). The widespread planting of maize in single crosses since the 1960s is an example of how heterosis is used in maize (Dong et al. 2024).

Corn stigmas may be initially pruned before hybridization occurs. Male inflorescences in maize will be removed as soon as they appear to prevent self-pollination, and the cobs will be covered with insulation made of paper that is resistant to moisture to prevent unintentional pollination. Afterward, pollen from other lines of maize was used to fertilize the parental lines. Single-cross, three-way, double-cross, top-cross and varietal cross are common types of maize hybrids. In general, single-cross hybrids yield more than other varieties of hybrids. Because every plant has the same genetic makeup, they all have fairly consistent appearances. However, because the female in a single-cross hybrid is descended from an inbred line, the seed output is lower than in other hybrids. As a result, single-cross hybrid seed is the most costly; nevertheless, this is typically justified by the seed's high production potential. A single-cross hybrid is used as the female in three-way and double-cross hybrids, which results in a high seed yield. Double-crosses can produce the best seed yield of all hybrid varieties, due to the vigor and copious pollen production of the single-cross male; a high female: male ratio can be employed. Therefore, double-cross hybrid seed is the least expensive; however, the grain yield is typically lower than that of a three-way or single-cross hybrid, and the resulting hybrid crop is more variable (MacRobert et al. 2014).

Wu et al. (2016) created a unique hybridization platform in maize that makes use of nuclear male sterility. One of the most important parts of this platform is the Seed Production Technology (SPT) procedure. This method incorporates a transgenic SPT maintainer line that can generate nuclear male sterile, nontransgenic lines for use as female parents in hybrid development. The maize SPT maintainer line is the homozygous recessive male sterility modified with an SPT construct including three genes. (1) a gene that indicates the color of seeds, (2) a gene that inhibits pollination by α-amylase, and (3) a gene that restores fertility in wild-type males. Only half of the pollen grains have the SPT transgenes; the sporophytic wild-type allele and the recessive mutant work in tandem to produce the grains. All of the pollen grains carry the recessive allele. Because α-amylase is expressed, pollen grains containing the SPT transgenes show starch deficiency and cannot germinate. To produce nontransgenic male-sterile progeny that can be utilized as female parents, pollen grains without the SPT transgenes are nontransgenic and capable of fertilizing homozygous mutant plants. Using mechanical color sorting, transgenic SPT maintainer seedlings can be easily identified and distinguished from non-SPT transgene seeds due to the expression of red fluorescent protein. In the production of commercial maize hybrid seeds, the SPT technique has the potential to displace existing approaches for pollen control. It can also unleash the potential for increased hybrid yield by increasing the parental germplasm pool, which is significant for other cross-pollinating crops (Wu et al. 2016).

Researchers have uncovered the origins of maize and determined that its ancestors had ten chromosomes in total (Wei et al. 2007). Maize originated as a result of hybridization rather than self-duplication, according to a detailed study of duplicated genes. The approximate size of the corn genome is 2.4 Gb (Haberer et al. 2005; Yang et al. 2019). To produce high-magnitude heterosis and facilitate the creation of high-yield hybrid varieties by crossing, the majority of maize breeders establish complementing heterotic groups. Based on F_1 hybrids, scientists have created models to explain the genetic mechanism, which are useful tools for studying heterosis and implementing research findings (Dong et al. 2024). The genetic potential of maize and the utilization of heterosis F_1 hybrids in production have resulted in an annual 3–5% increase in global production. However, the requirement for the yearly production of first-generation seeds poses serious issues and is an expensive technique in terms of the utilization of planting areas and the level of scientific validation it receives (Sokolov et al. 2024). Although planting densities have increased due to changes in management approaches, maize cultivars have been able to withstand higher levels of stress thanks to genetic variations. A combination of advancements in crop breeding and changes in management strategies has resulted in a century-long enhancement in corn grain yields (Tranel 2024).

Selection against hybrids and hybrid derivatives might be countered by a high enough frequency of spontaneous hybridization between the parental taxa. On the other hand, hybridization may occur extremely seldom, although individuals with hybrid ancestry may persist mostly due to selective benefit. Suppose natural hybridization rates are far higher than those suggested by hand pollination. In that case, gene flow pressure alone may account for the high frequency of hybrid-derived plants in sympatry areas. However, if hand-crosses are a good indicator of a very low spontaneous hybridization rate, then additional mechanisms need to be looked into to explain the longevity of hybrid swarms (Ellstrand et al. 2007). The use of the F_1 population to analyze the genetic makeup of complex phenotypic traits has advanced significantly. Nonetheless, it has rarely been emphasized that the normal breeder population has complementary heterotic groups, especially in maize (Dong et al. 2024). Using multiplexed in situ hybridization, molecular cartography is a technique for analyzing the spatiotemporal localization of a selection of genes related to auxin and potential vein patterning regulators in maize leaves (Perico et al. 2024). Many examples of permanent, fertile hybrids resulting from sexual hybridization may be found in crop species; oat and maize are two particularly intriguing examples, as they are closely related plant species capable of sexual reproduction. Examining the maize genome has become easier after it was discovered that oat plants with retained maize chromosomes exist after pollinating with corn and producing an oat line with retained maize chromosomes called OMA (oat × maize addition) (Ananiev et al. 1998; Rines et al. 2009). Breeders may create an effective breeding program to produce novel, drought-tolerant varieties with high grain-yield potential for food and nutritional security with the aid of this information, which would enable them to choose more desirable parents (Warzecha et al. 2023).

Wang et al. (2017) created a multiple-hybrid population of maize F1 hybrids to examine the feasibility of genome-wide association studies (GWAS). They concluded that modern genetics, genomics, and breeding may make extensive use of the mega population. Using a GWAS of maize's quantitative characteristics, the effectiveness of genetic analysis in a typical breeder hybrid population was demonstrated by Dong et al. (2024). Hybrid populations provide more information about the genetic architecture of agronomy features, including particular effects like dominance and epistatic effects than do traditional populations of inbred maize. Moreover, they discovered additional candidate genes for features connected to grain yield, including Zm00001d044211. The findings offer fresh theoretical and molecular direction for research on maize genetics and agronomy trait breeding. F_1 hybrids have a variety of genetic effects, like additive, dominant, and several epistatic effects, in contrast to inbred lines. Additionally, they suggested using a GWAS model to analyze the genetic impacts and identify potential genes for additional biological research and subsequent molecular breeding.

3.10 Conclusions and Prospects

Hunger and poverty have worsened due to disruptions in the global food supply chain caused by biotic and abiotic stressors on crops, as well as the rising cost of food, leading to impending economic repercussions. This situation highlights the importance of using cutting-edge molecular breeding technologies in plant breeding to achieve long-term food security. It is now essential to combine traditional breeding methods, such as hybridization, with genome editing and genetic engineering.

Reducing the time, space, and resources needed to genetically select and improve superior maize genotypes, accurately introducing genomic variations and chromosomal alterations into plants, combining new alleles by overcoming breeding barriers, efficiently selecting elite genotypes and lines with novel traits, and speeding up the development of superior high-performance maize genotypes with desirable traits are just a few advantages that can result from the application of effective maize breeding strategies and the use of contemporary molecular technology tools.

The development of novel maize cultivars with intricate quantitative traits altered by biotechnological means may be imminent given the current developments in genome editing. While biotech maize's resistance to pests and herbicides has significantly impacted global maize production, meeting the growing demand for corn-derived products will require further enhancements in complex quantitative traits like yield, nutrient acquisition and use efficiency, and heat and drought tolerance. A variety of resources are now available to maize breeding projects to facilitate the precise and rapid development of new cultivars, including high-throughput phenotyping, advanced statistics, computational techniques, crop models, novel genome sequencing and prediction methods, and emerging genome editing technologies. By combining these tools with speed breeding, maize breeding can be expedited to produce superior varieties that exhibit desired traits in various environmental conditions.

Appendix 3.1 List of Major Institutes engaged in Research on Maize (*Zea mays* L.)

Country	Institute name	Website
Mexico	International Maize and Wheat Improvement Center (CIMMYT), Mexico—CIMMYT for Centro Internacional de Mejoramiento de Maíz y Trigo	https://www.cimmyt.org/
Vietnam	National Maize Research Institute, Vietnam	https://www.gfar.net/organizations/national-maize-research-institute
India	ICAR—Indian Institute of Maize Research, Ludhiana, Punjab	https://iimr.icar.gov.in/
India	ICAR—Indian Institute of Maize Research, Winter Nursery, Rajendra Nagar, Hyderabad	https://iimr.icar.gov.in/
India	CIMMYT, Hyderabad	https://www.cimmyt.org/funder_partner/indias-ministry-of-agriculture-and-farmers-welfare/icar/indian-institute-of-maize-research-icar-iimr/
Bangladesh	Bangladesh Agricultural Research Institute (BARI)	

(continued)

Country	Institute name	Website
Egypt	Horticulture Research Institute (HRI), Agricultural Research Center (ARC)	Nadi Elsaid St., Field Crops Research Institute (HRI), Agricultural Research Center (ARC), Dokki, Giza, Egypt. http://www.arc.sci.eg/InstsLabs/Default.a
India	Central Research Institute for Jute and Allied Fibres (ICAR)	Bus Stop, Barrackpore—Barasat Rd, near Saheb Bagan Lane, Nilganj, Barrackpur Cantonment, West Bengal 700120, India https://crijaf.icar.gov.in/
Malaysia	Malaysian Agricultural Research and Development Institute (MARDI)	Ibu Pejabat MARDI, Persiaran MARDI-UPM, 43400 Serdang, Selangor, Malaysia
Malaysia	Malaya University and University of Kebangsaan Malaysia (UKM)	43600 UKM Bangi, Selangor Malaysia https://www.ukm.my/portalukm/contact-us/
Malaysia	Malaysian Nuclear Agency	59, Jalan Kemudi 2b, Taman Bernam Jaya, 44000 Kerling, Selangor, Malaysia https://www.nuclearmalaysia.gov.my/
Sudan	National Centre for Research (NCR)	El Mek Nimir Avenu, Khartoum, Sudan https://www.ncr.gov.sd

Appendix 3.2 World List of Recommended Maize (*Zea mays* L.) Varieties

Country	Recommended varieties	Type
India	PMH 10 (Hybrid)	Normal
India	Pusa HM-9 Improved (AQH-9)	Quality Protein Maize (QPM)
India	HQPM-5	Quality Protein Maize (QPM)
India	20 A 55 HX	Transgenic, Triple hybrid, Early; Orange
India	BRS 2020	Conventional, Early; Orange, Double hybrid
European	INDEM668	Very late maturity, high grain yield, better values, the high potential fields secure an agronomic profile, kernel type: dent
CIMMYT	Mex 17 Early	Earliness, resistance to lodging
CIMMYT	lodgingComp 4	High yield, lodging resistance
CIMMYT	Mamaba (hybrid)	High yield, drought tolerant (hybrid), lodges heavily in certain conditions.
CIMMYT	consumptionHonampa	Open-pollinated variety, yellow, source of provitamin A
Zambia	GV662A	Hybrid
Zambia	GV664A	Hybrid
Zambia	GV665A	Hybrid
Nigeria	Ife maizehyb-3	Hybrid

Country	Recommended varieties	Type
Nigeria	Ife maizehyb-4	Hybrid
Nigeria	Sammaz 38	OPV
Nigeria	Sammaz 39	OPV
China	CML161	QPM Inbred
China	CML171	QPM Inbred
Egypt	SC125 variety	Hybrid
Egypt	SC123 variety	Hybrid
Egypt	SC162 variety	Hybrid
Egypt	SC166 variety	Hybrid
Egypt	SK93Y variety	Hybrid
Egypt	SC10 variety	Hybrid

References

Abdel-Hamid AME, Salem KFM (2021) Breeding strategies of garden pea (*Pisum sativum* L.). In: Al-Khayri JM, Jain SM, Johnson DV (eds) Advances in plant breeding strategies: vegetable crops. Springer, Cham. https://doi.org/10.1007/978-3-030-66969-0_9

Abdul Aziz M, Masmoudi K (2024) Molecular breakthroughs in modern plant breeding techniques. Horticult Plant J 11:15–41. https://doi.org/10.1016/j.hpj.2024.01.004

Abdul Aziz M, Brini F, Rouached H et al (2022) Genetically engineered crops for sustainably enhanced food production systems. Front Plant Sci 13:1027828. https://doi.org/10.3389/fpls.2022.1027828

Abdullaev FK, Rustamov NSU, Yakubov MD (2025) Uzbekistan Genebank. In: Al-Khayri JM, Salem KFM, Jain SM (eds) Plant Gene Banks. Springer, Singapore. https://doi.org/10.1007/978-981-99-4236-7_57-1

Aboshama HM, Salem KFM, Hassan HMF, Tahoun MH (2015) Estimating heterosis and combining ability for *in vivo* and *in vitro* traits using diallel cross on *tomato* (*Solanum lycopersicum* L.). Minufiya. J Agric Res 40(3):757–770

Abouseada HH, Mohamed AS, Teleb SS et al (2023) Genetic diversity analysis in wheat cultivars using SCoT and ISSR markers, chloroplast DNA barcoding and grain SEM. BMC Plant Biol 23:1–5. https://doi.org/10.1186/s12870-023-04196-w

Abu-Ellail FFB, Salem KFM, Saleh MM et al (2021) Molecular breeding strategies of beetroot (*Beta vulgaris* ssp. vulgaris var. conditiva Alefeld). In: Al-Khayri JM, Jain SM, Johnson DV (eds) Advances in plant breeding strategies: vegetable crops. Springer, Cham. https://doi.org/10.1007/978-3-030-66965-2_4

Abulaila K, Al Sane KO, Tahabsom Z (2025) Jordan seed bank. In: Al-Khayri JM, Salem KFM, Jain SM (eds) Plant Gene Banks. Springer, Singapore. https://doi.org/10.1007/978-981-99-4236-7_45-1

Acquaah G (2009) Principles of plant genetics and breeding. Wiley, Hoboken

Adawy SS, Stupar RM, Jiang J (2004) Fluorescence in situ hybridization analysis reveals multiple loci of knob-associated DNA elements in one-knob and knobless maize lines. J Histochem Cytochem 52:1113–1116

Aesaert S, Impens L, Coussens G et al (2022) Optimized transformation and gene editing of the B104 Public maize inbred by improved tissue culture and use of morphogenic regulators. Front Plant Sci 13:883847. https://doi.org/10.3389/fpls.2022.883847

Aguado-Santacruz GA, Garcia-Moya E, Aguilar-Acuna JL (1998) In-vitro techniques and mutagenesis for the improvement of vegetatively propagated plants. In: Jain SM, Brar DS, Ahloowalia BS (eds) Somaclonal variation and induced mutations in crop improvement. Kluwer Academic Publishers, Dordrecht, pp 293–309

Ahsan M, Mehdi SS, Khaliq I (2000) Tissue culture and breeding of maize (*Zee mays* L.) "a review". Pak J Biol Sci 3:1985–1988. https://doi.org/10.3923/pjbs.2000.1985.1988

Akparov ZI, Mammadova SA, Kalantarova NS, Mirzaliyeva IA, Mammadov AT (2025) National Genebank of Azerbaijan. In: Al-Khayri JM, Salem KFM, Jain SM (eds) Plant Gene Banks. Springer, Singapore. https://doi.org/10.1007/978-981-99-4236-7_31-1

Akula C, Akula A, Henry R (1999) Improved regeneration efficiency from mature embryos of barley cultivars. Biol Plant 42:505–513

Al Lawati AH (2025) Plant genetic resource conservation of Oman. In: Al-Khayri JM, Salem KFM, Jain SM (eds) Plant Gene Banks. Springer, Singapore. https://doi.org/10.1007/978-981-99-4236-7_50-1

Al-Abed D, Rudrabhatla S, Talla R, Goldman S (2006) Slit seed: a new tool for maize researchers. Planta 223(223):1355–1360

Aladele SE et al (2025) Plant genetic resources conservation and utilization in Nigeria. In: Al-Khayri JM, Salem KFM, Jain SM (eds) Plant Gene Banks. Springer, Singapore. https://doi.org/10.1007/978-981-99-4236-7_17-1

Albert PK, Kimberly N-V, Michael T et al (2021) Maize tissue culture, transformation, and genome editing. In Vitro Cell Dev Biol Plant 57:653–671. https://doi.org/10.1007/s11627-021-10196-y

Albrecht G, Ketterings QM, Beckman J (2005) Soil pH for field crops: agronomy fact sheet series. New York. Retrieved from http://nmsp.css.cornell.edu

Aldahak L, Salem KFM, Al-Salim SHF, Al-Khayri JM (2021) Advances in chicory (*Cichorium intybus* L.) breeding strategies. In: Al-Khayri JM, Jain SM, Johnson DV (eds) Advances in plant breeding strategies: vegetable crops. Springer, Cham. https://doi.org/10.1007/978-3-030-66969-0_1

Aleksoski J (2018) The effect of backcross method in tobacco breeding. J Agric Plant Sci 16:9–19

Alene AD, Menkir A, Ajala SO, Badu-Apraku B, Olanrewaju AS, Manyong VM et al (2009) The economic and poverty impacts of maize research in West and Central Africa. Agric Econ 40:535–550

Al-Khayri JM et al (2016) Advances in plant breeding strategies: breeding, biotechnology and molecular tools. Springer International Publishing Switzerland

Allard R (1961) Principles of plant breeding. Soil Sci 91(6):414

Almarri NB, Ahmad S, Alsubaie GTG, Elshal MH (2025) Saudi Arabia Gene Bank. In: Al-Khayri JM, Salem KFM, Jain SM (eds) Plant Gene Banks. Springer, Singapore. https://doi.org/10.1007/978-981-99-4236-7_54-1

Alqahtani MM, Saleh MM, Alwutayd KM et al (2024) Performance and genotypic variability in diverse date palm (*Phoenix dactylifera* L.) cultivars for fruit characteristics. Genet Resour Crop Evol 71:1759–1772. https://doi.org/10.1007/s10722-023-01719-0

Alvi MB, Rafique M, Tariq MS, Hussain A, Mahmood T, et al (2003) Hybrid vigour of some quantitative characters in maize (Zea mays L). Pakistan J Biol Sci 6:139-141

Ananiev EV, Phillips RL, Rines HW (1998) Chromosome-specific molecular organization of maize (*Zea mays* L.) centromeric regions. Proc Natl Acad Sci USA 95(22):13073–13078. https://doi.org/10.1073/pnas.95.22.13073

Anderson LK, Salameh N, Bass HW et al (2004) Integrating genetic linkage maps with pachytene chromosome structure in maize. Genetics 166:1923–1933

Andjelkovic V, Ignjatovic-Micic D (2012) Maize genetic resources-science and benefits. Serbian Genetic Society and Maize Research Institute Zemun Polje, Belgrade

Arab S et al (2025) National Gene Bank of Egypt. In: Al-Khayri JM, Salem KFM, Jain SM (eds) Plant Gene Banks. Springer, Singapore. https://doi.org/10.1007/978-981-99-4236-7_5-1

Armstrong CL (1999) The first decade of maize transformation: a review and future perspective. Maydica 44:101–109

Armstrong CL, Green CE (1985) Establishment and maintenance of friable, embryogenic maize callus and the involvement of L-proline. Planta 164:207–214. https://doi.org/10.1007/BF00396083

Armstrong CL, Green CE, Sand Phillips RL (1991) Development and availability of germplasm with high Type II culture formation response. Maize Genet Cooperat Newsl 65:92–93

Bangarwa S (2021) Recurrent selection- definition and types. North Carolina, USA. Available from: Biotecharticles.com

Bänziger M, Edmeades GO, Beck D, Bellon M (2000) Breeding for drought and nitrogen stress tolerance in maize: from theory to practice. CIMMYT, Mexico

Barata C, Carena MJ (2006) Classification of North Dakota maize inbred lines into heterotic groups based on molecular and testcross data. Euphytica 151:339–349

Barcaccia G, Volpato M, Gentili R et al (2016) Genetic identity of common buckwheat (*Fagopyrum esculentum* Moench) landraces locally cultivated in the Alps. Genet Resour Crop Evol 63:639–651

Barloy D, Beckert M (1993) Improvement of regeneration ability of androgenic embryos by early anther transfer in maize plant. Plant Cell Tissue Organ Cult 33:45–50

Baskaran P, Jayabalan N (2005) *In vitro* plant regeneration and mass propagation system for *Sorghum bicolor*-a valuable major cereal crop. J Agric Technol 1:345–363

Baveja A, Muthusamy V, Panda KK et al (2021) Development of multinutrient-rich biofortified sweet corn hybrids through genomics-assisted selection of shrunken2, opaque2, lcyE and crtRB1 genes. J Appl Genet 62:419–429

Begna T (2021) Conventional breeding methods widely used to improve self-pollinated crops. Int J Res 7(1):1–16. https://doi.org/10.20431/2454-6224.0701001

Békefi Z, Szilágyi S, Desiderio F, Demku T, Varga J (2025) Prunus Gene Bank in Hungarian university of agriculture and life sciences. In: Al-Khayri JM, Salem KFM, Jain SM (eds) Plant Gene Banks. Springer, Singapore. https://doi.org/10.1007/978-981-99-4236-7_71-1

Bensen RJ, Johal GS, Crane VC et al (1995) Cloning and characterization of the maize An1 gene. Plant Cell 7:75–84. https://doi.org/10.1105/tpc.7.1.75

Bharti G, Chimata MK (2019) Review on new plant breeding techniques. Int J Sci Res 8(4):723–730

Bhullar NK, Zhang Z, Wicker T, Keller B (2010) Wheat gene bank accessions as a source of new alleles of the powdery mildew resistance gene Pm3: a large-scale allele mining project. BMC Plant Biol 10:1–3

Bioversity International (2007) Guidelines for the development of crop descriptor lists, Bioversity tech bull series. Bioversity International, Rome

Bioversity International, International Maize and Wheat Improvement Center (CIMMYT) (2009) Key access and utilization descriptors for maize genetic resources. International Maize and Wheat Improvement Center (CIMMYT), p 6

Börner A, Landjeva S, Salem KFM, Lohwasser U (2009) Pflanzengenetische Ressourcen als Grundlage für die Züchtung klimatoleranter Sorten. Proceedings of a German Conference 'Weizenforum = Wheat Forum' Norddeutsches Weizenforum Schriftenreihe des Inst. Für Pflanzenbau und –züchtung der CAU Kiel, Germany, 63, 23–27.

Börner A, Landjeva S, Salem KFM, Lohwasser U (2010) Plant genetic resources- a prerequisite for drought tolerance breeding in cereals. In: Proceedings of the Jahrestagung der Vereinigung der Pflanzenzüchter und Saatgutkaufleute Österreichs, 24–26 November 2009, Raumberg-Gumpenstein, pp 11–13

Bouquet A, Torregrosa L (2003) Micropropagation of the grapevine (*Vitis* spp.). In: Micropropagation of woody trees and fruits. Springer Netherlands, Dordrecht, pp 319–352

Bowers JE, Arias MA, Asher R et al (2005) Comparative physical mapping links conservation of microsynteny to chromosome structure and recombination in grasses. Proc Natl Acad Sci USA 102:13206–13211

Breidy J (2025) Lebanon Gene Bank. In: Al-Khayri JM, Salem KFM, Jain SM (eds) Plant Gene Banks. Springer, Singapore. https://doi.org/10.1007/978-981-99-4236-7_48-1

Breseghello F, Coelho AS (2013) Traditional and modern plant breeding methods with examples in rice (*Oryza sativa* L.). J Agric Food Chem 61:8277–8286

Bressan RA, Singh NK, Handa AK et al (1985) Stable and unstable tolerance to NaCl in cultured tobacco cells. In: Freeling M (ed) Plant genetics, proceedings of the third annual ARCO plant cell research institute-UCLA symposium on plant biology, symposia on molecular and cellular biology, April 13–19, 1985, UCLA, New series. A.R. Liss, New York, pp 755–769

Brown J, Caligari P (2020) An introduction to plant breeding. Wiley, Hoboken

Brown WL, Darrah LL (1985) Origin, adaptation, and types of corn. In: National corn handbook, pp 1–6

Brush S (1991) A farmer-based approach to conserving crop germplasm. Econ Bot 45:153–165

Byerlee D (2020) The globalization of hybrid maize, 1921–1970. J Glob Hist 15:101–122

Cairns JE, Prasanna BM (2018) Developing and deploying climate-resilient maize varieties in the developing world. Curr Opin Plant Biol 45:226–230. https://doi.org/10.1016/j.pbi.2018.05.004

Cañar DY et al (2025) Colombia Gene Bank. In: Al-Khayri JM, Salem KFM, Jain SM (eds) Plant Gene Banks. Springer, Singapore. https://doi.org/10.1007/978-981-99-4236-7_95-1

Carena MJ, Hallauer AR (2001) Response to inbred progeny recurrent selection in leaming and midland yellow dent populations. Maydica 46:1–10

Carlson WR (1988) The cytogenetics of corn. In: Sprague GF, Dudley JW (eds) Corn and corn improvement. American Society of Agronomy, Madison, pp 259–344

Chaleff RS, Parsons RE (1978) Direct selection in vitro for herbicide-resistant mutants of *Nicotiana tobaccum*. Proc Natl Acad Sci USA 75:5104–5107

Chandler VL, Brendel V (2002) The Maize Genome Sequencing Project. Plant Physiol. 130:1594–1597

Chang TT (1985) Preservation of crop germplasm. Iowa State J Res 59:365–378

Chang H, Chakrabarty SD, Hahn EJ, Pack KY (2003) Micropropagation of Calla Lily (*Zantedeschia albomaculata*) via in vitro shoot tip proliferation. In Vitro Cell Dev Biol Plant 39:129–134

Char SN, Neelakandan AK, Nahampun H (2017) An *Agrobacterium*-delivered CRISPR/Cas9 system for high-frequency targeted mutagenesis in maize. Plant Biotechnol J 15(2):257–268. https://doi.org/10.1111/pbi.12611

Chaudhary A (1983) Maize in Pakistan, Punjab Agri. Research Coordination Board Univ of Agri Faisalabad

Cheng Z, Buell CR, Wing RA et al (2001a) Toward a cytological characterization of the rice genome. Genome Res 11:2133–2141

Cheng Z, Presting GG, Buell CR et al (2001b) High-resolution pachytene chromosome mapping of bacterial artificial chromosomes anchored by genetic markers reveals the centromere location and the distribution of genetic recombination along chromosome 10 of rice. Genetics 157:1749–1757

Chňapek M, Balážová Ž, Špalcková A et al (2024) Genetic diversity of maize resources revealed by different molecular markers. Genet Resour Crop Evol 2024:1–19. https://doi.org/10.1007/s10722-024-01908-5

Choudhary M, Choudhary J, Kumar P et al (2023) Conventional and molecular breeding for genetic improvement of maize (*Zea mays* L.). In: Raina A, Wani MR, Laskar RA et al (eds) Advanced crop improvement, vol 2. Springer, Cham. https://doi.org/10.1007/978-3-031-26669-0_12

Christensen LA (2002) Soil, nutrient and water management systems used in US corn production, Agriculture Information Bulletin No. 774. United States Department of Agriculture (USAID)

Colbert T, Till BJ, Tompa R et al (2001) High-throughput screening for induced point mutations. Plant Physiol 126:480–484. https://doi.org/10.1104/pp.126.2.480

Cooper JL, Till BJ, Laport RG et al (2008) TILLING to detect induced mutations in soybean. BMC Plant Biol 8:9. https://doi.org/10.1186/1471-2229-8-9

Cowan JR (1943) The value of double cross hybrids involving inbreds of similar and diverse genetic origin. Sci Agric 23:287–296

Cowperthwaite M, Park W, Xu Z et al (2002) Use of the transposon Ac as a gene-searching engine in the maize genome. Plant Cell 14:713–726. https://doi.org/10.1105/tpc.010468

Cromarty AS, Ellis RH, Roberts EH (1982) The design of seed storage facilities for genetic conservation. Int Board Plant Gene Res, Rome, p 96

Crouse K, Denny G (2015) Soil pH and fertilizers. Plant Soil Sci 372:7–8

Das AK, Gowda MM, Muthusamy V et al (2021) Development of maize hybrids with enhanced vitamin-E, vitamin-A, lysine, and tryptophan through molecular breeding. Front Plant Sci 12:659381
Davis GL, McMullen MD, Baysdorfer C et al (1999) A maize map standard with sequenced core markers, grass genome reference points and 932 expressed sequence tagged sites (ESTs) in a 1736-locus map. Genetics 152:1137–1172
de Galarreta JIR, Barandalla L, Ritter E, Velázquez ME, Rios D (2025) Spanish potato germplasm Bank. In: Al-Khayri JM, Salem KFM, Jain SM (eds) Plant Gene Banks. Springer, Singapore. https://doi.org/10.1007/978-981-99-4236-7_114-1
De Groote H, Kimenju SC, Munyua B et al (2020) Spread and impact of fall armyworm (*Spodoptera frugiperda* J.E. Smith) in maize production areas of Kenya. Agric Ecosyst Environ 292:106804. https://doi.org/10.1016/j.agee.2019.106804
De Jong JH, Fransz P, Zabel P (1999) High-resolution FISH in plants-techniques and applications. Trends Plant Sci 4:258–263
Deinum B (1980) Forage maize: fertiliser requirements. In: Field crops research. Potash Development Association, York. https://doi.org/10.1016/0378-4290(80)90013-1
Dempsey E (1994) Traditional analysis of maize pachytene chromosomes. In: Freeling M, Walbot V (eds) The Maize Handbook. Springer-Verlag, New York
Desheva G, Valchinova E, Deshev M (2025) Bulgarian gene bank. In: Al-Khayri JM, Salem KFM, Jain SM (eds) Plant Gene Banks. Springer, Singapore. https://doi.org/10.1007/978-981-99-4236-7_66-1
Deutsch CA, Tewksbury JJ, Tigchelaar M et al (2018) Increase in crop losses to insect pests in a warming climate. Science 361(6405):916–919. https://doi.org/10.1126/science.aat3466
Dhillon BS, Paul C, Zimmer E et al (1990) Variation and covariation in stover digestibility traits in diallel crosses of maize. Crop Sci 30:931–936. https://doi.org/10.2135/cropsci1990.0011183x003000040035x
Díez MJ, Valcárcel JV, Solbes E, Muñoz E (2025) Forty years conserving Spanish biodiversity in the Gene Bank of the Universitat Politècnica de València, Spain. In: Al-Khayri JM, Salem KFM, Jain SM (eds) Plant Gene Banks. Springer, Singapore. https://doi.org/10.1007/978-981-99-4236-7_86-1
Dong Y, Li G, Zhang X et al (2024) Genome-wide association study for maize hybrid performance in a typical breeder population. Int J Mol Sci 25(2):1190. https://doi.org/10.3390/ijms25021190
Drinic Mladenovic S, Andjelkovic V, Ignjatovic Micic D (2012) Genetic diversity of maize landraces as sources of favorable traits p. In: Caliskann M (ed) The molecular basis of plant genetic diversity. InTech, Rijeka, pp 89–112
Duan C, Song F, Sun S et al (2019) Characterization and molecular mapping of two novel genes resistant to pythium stalk rot in maize. Phytopathology 109(5):804–809. https://doi.org/10.1094/phyto-09-18-0329-R
Dumlupinar Z, Brown R, Campbell R et al (2016) The art of attrition: development of robust oat microsatellites. Plant Breed 135:323–334. https://doi.org/10.1111/pbr.12362
Dupon J, White P, Carpenter M et al (1990) Food uses and health effects of corn oil. J Am Coll Nutr 9:438–470
Duvick DN (1997) What is yield? In: Edmeades GO, Bänziger B, Mickelson HR, Pena-Valdivia CB (eds) Developing drought and low N-tolerant maize. CIMMYT, El Batan, pp 332–335
Duvick DN (2005) The contribution of breeding to yield advances in maize (*Zea mays* L.). Adv Agron 86:83–145. https://doi.org/10.1016/s0065-2113(05)86002-x
Eckhardt RC, Bryan AA (1940) Effect of method of combining the four inbreds of a double cross of maize upon the yield and variability of the resulting hybrid. J Am Soc Agron 32:347–353
EFSA Panel on Genetically Modified Organisms (GMO), Mullins E, Bresson J-L (2024) Assessment of genetically modified maize MON 94804 (application GMFF-2022-10651). EFSA J 22(4):e8714. https://doi.org/10.2903/j.efsa.2024.871410.1111/pbi.12611

Eid NA, Ibrahim AMH, Elsharawy AA, Salem KFM (2025) Nanofertilizers for plant viral disease management. In: Kumar P, Dubey RC (eds) Nanofertilizers for sustainable agriculture. Springer, Cham. https://doi.org/10.1007/978-3-031-78649-5_7

El Rabey H, Salem KFM, Mattar MZ (2013) The genetic diversity and relatedness of eight rice (*Oryza sativa* L.) cultivars as revealed by AFLP and SSRs markers. Life Sci J 10(1):1471–1479

Elabd AB, Salem KFM (2009) Detection of genetic diversity in rice (*Oryza sativa* L.) varieties using microsatellite DNA markers and Quantitative Traits. In: 6th Egyptian plant society of breeding conference, 4–7 May 2009, Ismailia, Egypt

Elabd AB, Saleh MM, Aldahak L, Salem KFM (2020) Identification of superior rice (*Oryza sativa* L.) genotypes under irrigated conditions based on multivariate analysis. In: 7th International Scientific Conference of Genetic and Environment, 8–9th December 2020, Baghdad, Iraq

El-Absawy EA, Salem KFM, Abdellatif KF, ElShamy H (2012) Estimation of genetic diversity among some Egyptian barley (*Hordeum vulgare* L.) cultivars using RAPD markers and agronomic traits. In: 1st international conference on biotechnology applications in agriculture, Moshtohor-Hurghada, 18–22 February 2012, Faculty of Agriculture, Benha University, Egypt

El-Absawy EA, Nowar AA, Salem KFM et al (2015) Estimation of genetic diversity among some bread wheat (*Triticum aestivum* L.) genotypes using molecular markers and agronomic traits. Minufiya J Agric Res 40(3):223–233

El-Itriby HA, Assem SK, Hussein EHA et al (2003) Regeneration and transformation of Egyptian maize inbred lines via immature embryo culture and a biolistic particle delivery system. In Vitro Cell Dev Biol Plant 39:524–531. https://doi.org/10.1079/IVP2003439

Ellstrand NC, Garner LC, Hegde S et al (2007) Spontaneous hybridization between maize and teosinte. J Hered 98(2):183–187. https://doi.org/10.1093/jhered/esm002

Eltaher S, Sallam A, Belamkar V et al (2017) Genome-wide association mapping for drought tolerance traits at seedling stage using Nebraska winter wheat. In: International conference on productive crop design (genome to phenome), April 2017, USA

Eltaher S, Sallam A, Belamkar V et al (2018) Genetic diversity and population structure of $F_{3:6}$ Nebraska winter wheat genotypes using genotyping-by-sequencing. Front Genet 9:76. https://doi.org/10.3389/fgene.2018.00076

Eltaher S, Baenziger S, Belamkar V et al (2021) GWAS revealed effect of genotype x environment interactions for grain yield of Nebraska winter wheat. BMC Genomics 22:1–14

Eltaher S, Sallam A, Emara HA et al (2022) Genome-wide association mapping revealed SNP alleles associated with spike traits in wheat. Agronomy 12:1469. https://doi.org/10.3390/agronomy12061469

El-Zanaty AM, Salem KFM, Esmail RM (2011) Detection of genetic diversity in Egyptian cotton (*Gossypium barbadense* L.) varieties using RAPD markers and morphological traits. J Am Sci 7(12):1107–1115

El-Zanaty AM, Salem KFM, Ismail M, Esmail RM (2013) Impact of plant breeding on genetic diversity of the Egyptian cotton (*Gossypium barbadense* L.) varieties as revealed by RAPD markers. J Agric Chem Biotechn Mansoura Univ 4:217–227

Emara HA, Nower AA, Salem KFM, Eltaher S (2011) Estimating heterosis and combining ability for *in vivo* and *in vitro* traits using diallel cross in Maize (*Zea mays* L). J Product Dev 16:101–120

Emara HA, Salem KFM, Abdel-Halim AA, Abd El-Kowy RNF (2013) An *in vivo* and *in vitro* traits analysis of a diallel cross in rice (*Oryza sativa L.*). Egypt J Plant Breed 17(2):591–610

Emerson R, Beadle G, Fraser A (1935) A summary of linkage studies in maize. Cornell Univ Agric Exp Stn Memoir 180:1–83

Erenstein O, Jaleta M, Sonder K et al (2022) Global maize production, consumption and trade: trends and R&D implications. Food Secur 14(5):1295–1319. https://doi.org/10.1007/s12571-022-01288-7

Esmail RM, Salem KFM, El-Zanaty AM (2009) Genetic diversity studies of Egyptian cotton (*Gossypium barbadense* L.) varieties using RAPD markers and agronomic traits. In: 6th Egyptian Plant Society of Breeding Conference, 4–7 May 2009, Ismalia, Egypt

EUMLDB 2010 http://www.ensam.inra.fr/gap/resgen88/results.htm

FAO (2009) Establishment of a global network for in-situ conservation of crop wild relatives: status and needs, by N. Maxted & S. Kell. Commission on Genetic Resources for Food and Agriculture, Rome, p 211

FAO/IAEA (2020) Mutant variety database. Available online: https://mvd.iaea.org/. Accessed 25 June 2020

FAO/IPGRI (1994) Genebank standards. Food and Agriculture Organization of the United Nations, Rome and International Plant Genetic Resources Institute, Rome

Farooq M, Aziz T, Basra S et al (2008) Chilling tolerance in hybrid maize induced by seed priming with salicylic acid. J Agron Crop Sci 194:161–168

Farooq S, Shahzadi A, Razzaq A et al (2023) Advances in genome editing for maize improvement. In: Wani SH, Dar ZA, Singh GP (eds) Maize improvement. Springer, Cham. https://doi.org/10.1007/978-3-031-21640-4_9

Feng J, Liao F, Kong D et al (2022) Genetic diversity of the cultivated *Salvia miltiorrhiza* populations revealed by four intergenic spacers. PLoS One 17:e0266536. https://doi.org/10.1371/journal.pone.0266536

Feyisola RT, Odutayo OI, Godonu KG et al (2015) In vitro proliferation of plantain using different concentration of auxin and Cytokinin. J Biol Agric Health 5:77–82

Filiz E, Birbilener S, Ozyigit II et al (2015) Assessment of genetic variations of silver lime (*Tilia tomentosa* Moench.) by RAPD markers in urban and forest ecosystems. Biotech Biotechnol Equip 29:631–636. https://doi.org/10.1080/13102818.2015.1042049

Food and Agriculture Organization of the United Nations (2010) The second report on the state of the world's plant genetic resources for food and agriculture. Food & Agriculture Organization of the UN (FAO)

Forster BP, Shu QY (2011) Plant mutagenesis in crop improvement: basic terms and applications. In: Shu QY, Forster BP, Nakagawa H (eds) Plant mutation breeding and biotechnology. Joint FAO/IAEA Division of Nuclear Techniques in Food and Agriculture, International Atomic Energy Agency, Vienna, pp 9–71

Forster BP, Shu QY (2012) Plant mutagenesis in crop improvement: basic terms and applications. In: Shu QY, Forster BP, Nakagawa H (eds) Plant mutation breeding and biotechnology. CABI, Wallingford, pp 9–20

Fritsche-Neto R, Galli G, Borges KLR et al (2021) Optimizing genomic-enabled prediction in small-scale maize hybrid breeding programs: a roadmap review. Front Plant Sci 12:658267

Fu Y, Wen TJ, Ronin YI et al (2006) Genetic dissection of intermated recombinant inbred lines using a new genetic map of maize. Genetics 174:1671–1683

Galea-Gragera FA, Llera Cid F (2025) Spain germplasm Bank of Annual Forage Legumes Seeds at the scientific and technological research Center of Extremadura (CICYTEX). In: Al-Khayri JM, Salem KFM, Jain SM (eds) Plant Gene Banks. Springer, Singapore. https://doi.org/10.1007/978-981-99-4236-7_102-1

Gaut BS, Le Thierry d'Ennequin M, Peek AS, Sawkins MC (2000) Maize as a model for the evolution of plant nuclear genomes. Proc Natl Acad Sci USA 97(13):7008–7015

Gazal A, Dar ZA, Lone AA (2018) Molecular breeding for abiotic stresses in maize (*Zea mays* L.). InTech. https://doi.org/10.5772/intechopen.71081

Gedil M, Menkir A (2019) An integrated molecular and conventional breeding scheme for enhancing genetic gain in maize in Africa. Front Plant Sci 10:1430

Gemmill CCE, Grierson ERP (2021) Inter-simple sequence repeats (ISSR), microsatellite-primed genomic profiling using universal primers. In: Besse P (ed) Molecular plant taxonomy: methods in molecular biology, vol 2222. Humana, New York, pp 249–262

Gengenbach BG, Green CE (1975) Selection of T-cytoplasm maize callus cultures resistant to Helminthosporium maydis race T pathotoxin. Crop Sci 15:645–649

Ghazy MI, Salem KFM, Sallam A (2020) Utilization of genetic diversity and marker-trait to improve drought tolerance in rice (*Oryza sativa* L.). Mol Biol Rep 48:157–170. https://doi.org/10.1007/s11033-020-06029-7

Ghidan WF, Abdulmajid DA (2025) Egyptian Rice Gene Bank. In: Al-Khayri JM, Salem KFM, Jain SM (eds) Plant Gene Banks. Springer, Singapore. https://doi.org/10.1007/978-981-99-4236-7_106-1

Gomez M, Nurul Islam-Faridi M, Woo SS et al (1997) FISH of a maize sh2-selected sorghum BAC to chromosomes of Sorghum bicolor. Genome 40:475–478

Goodman MM (1976) Maize. In: Simmonds NW (ed) Evolution of crop plants, pp 128–136

Green CE, Phillips RL (1975) Plant regeneration from tissue culture of maize. Crop Sci 15:417–421

Griffing JB (1953) An analysis of tomato yield components in terms of genotypic and environmental effects. The Iowa Agricultural Experiment Station (IAES) Research Bulletin, 397

Grote U, Fasse A, Nguyen TT, Erenstein O (2021) Food security and the dynamics of wheat and maize value chains in Africa and Asia. Front Sustain Food Syst 4:17009. https://doi.org/10.3389/fsufs.2020.617009

Gudaye TY, Lakew BF, Selamo MM, Tessema ES (2025) Ethiopian crop Gene Bank. In: Al-Khayri JM, Salem KFM, Jain SM (eds) Plant Gene Banks. Springer, Singapore. https://doi.org/10.1007/978-981-99-4236-7_7-1

Gungor H, İlhan E, Kasapoğlu AG et al (2022) Genetic diversity and population structure of barley cultivars released in Turkey and Bulgaria using iPBS-retrotransposon and SCoT markers. J Agric Sci:14–14. https://doi.org/10.15832/ankutbd.886221

Guzzon F, Arandia Rios LW, Caviedes Cepeda GM et al (2021) Conservation and use of Latin American maize diversity: pillar of nutrition security and cultural heritage of humanity. Agronomy 11(1):172. https://doi.org/10.3390/agronomy11010172

Haberer G, Young S, Bharti AK et al (2005) Structure and architecture of the maize genome. Plant Physiol 139(4):1612–1624. https://doi.org/10.1104/pp.105.068718

Haberlandt G (1902) Culturversuche mit isolierten Pflanzenzellen. Sitzungsber Akad Wiss Wien Math Nat 111:69–91

Hansen G, Wright MS (1999) Recent advances in transformation of plants. Trends Plant Sci 4:226–231. https://doi.org/10.1016/S1360-1385(99)01412-0

Harding K (2010) Plant and algal cryopreservation: issues in genetic integrity, concepts in cryobionomics and current applications in cryobiology. Aspac J Mol Biol Biotechnol 18(1):151–154

Harzli I, Mostafa K, Abdulla MF, Mehmood M (2025) Sustaining Turkish genetic resources. In: Al-Khayri JM, Salem KFM, Jain SM (eds) Plant Gene Banks. Springer, Singapore. https://doi.org/10.1007/978-981-99-4236-7_88-1

Hasan MM, Rafii MY, Ismail MR et al (2015) Marker-assisted backcrossing: a useful method for rice improvement. Biotechnol Biotechnol Equip 29:237–254

Hassan MN, Mekkawy SA, Mahdy M et al (2021) Recent molecular and breeding strategies in lettuce (Lactuca spp.). Genet Resour Crop Evol 68:3055–3079. https://doi.org/10.1007/s10722-021-01246-w

Hass-Jacobus BL, Futrell-Griggs M, Abernathy B, Westerman R, Goicoechea JL et al (2006) Integration of hybridization-based markers (overgos) into physical maps for comparative and evolutionary explorations in the genus Oryza and in sorghum. BMC Genomics 7:199

Hayes HK (1963) A professor's story of hybrid corn. Burgess publishing company, Minneapolis, MN. Wu, S. K. 1939. The relationship between the origin of selfed lines of corn and their value in hybrid combinations. J Am Soc Agron 31:131–140

Helentjaris T, Weber DF, Wright S (1986) Use of monosomics to map cloned DNA fragments in maize. Proc Natl Acad Sci USA 83:6035–6039

Herman AK, Wansyah AR, Asmania UR, Anzaelina DN, Oktaviano Z, Lestari W, Adiwirman RD (2023) The ability of matK and trnL-trnL-trnF intergenic spacer to discern certain species accessions of the families Solanaceae and Fabaceae. SABRAO J Breed Genet 55:97–106. https://doi.org/10.54910/sabrao2023.55.1.9

Hernandes-Lopes J, Yassitepe JECT, Koltun A et al (2023) Genome editing in maize: toward improving complex traits in a global crop. Genet Mol Biol 46(Suppl 1):e20220217. https://doi.org/10.1590/1678-4685-GMB-2022-0217

Hibberd KA, Green CE (1982) Inheritance and expression of lysine plus threonine resistance selected in maize tissue cultures. Proc Natl Acad of Sci USA 79:559–563

Hidayatun N, Kurniawan H (2025) Conservation of agricultural genetic resources in Indonesia. In: Al-Khayri JM, Salem KFM, Jain SM (eds) Plant Gene Banks. Springer, Singapore. https://doi.org/10.1007/978-981-99-4236-7_40-1

Hocaoglu-Ozyigit A, Ucar B, Altay V, Ozyigit II (2022) Genetic diversity and phylogenetic analyses of Turkish cotton (*Gossypium hirsutum* L.) lines using ISSR markers and chloroplast trnL-F regions. J Nat Fibers 19:1837–1850. https://doi.org/10.1080/15440478.2020.1788493

Huang X, Wei Z (2004) High frequency plant regeneration through callus initiation from mature embryos of maize (*Zea Mays* L.). Plant Cell Rep 22:793–800

Hulbert SH, Richter TE, Axtell JD, Bennetzen JL (1990) Genetic mapping and characterization of sorghum and related crops by means of maize DNA probes. Proc Natl Acad Sci USA 87:4251–4255

Hull F (1945) Recurrent selection for specific combining ability in corn 1. Agron J 37:134–145

Hunter D, Heywood V (2011) Crop wild relatives: a manual of in situ conservation, 1st edn. Earthscan, London

Hussain A, Ahmed I, Nazir H et al (2012) Plant tissue culture: current status and opportunities. InTech. https://doi.org/10.5772/50568

IBPGR (1991) Descriptors for maize. International Wheat and Maize Improvement Center/International Board for Plant Genetic Resources, Mexico City/Rome

Ibrahim AA, Salem KFM, Abdein MA, Ramadan SM (2023a) Advances in summer squash (*Cucurbita pepo* L.) molecular breeding strategies. In: Singh S, Sharma D, Sharma SK, Singh R (eds) Smart plant breeding for vegetable crops in post-genomics era. Springer, Singapore. https://doi.org/10.1007/978-981-19-5367-5_8

Ibrahim AA, Salem KFM, Ramadan S (2023b) *Allium sativum* and stressful conditions. In: Abdel Latef AAH (ed) Book medicinal plant responses to stressful conditions. CRC Press

Ibrahim AA, Abd-Ellatif S, Abdel Razik ESS, Salem KFM (2024) Biodiversity of cereal crops and utilization in food and nutritional security. In: Al-Khayri JM, Jain SM, Penna S (eds) Sustainable utilization and conservation of plant genetic diversity, Sustainable development and biodiversity, vol 35. Springer, Singapore. https://doi.org/10.1007/978-981-99-5245-8_2

Ignjatovic-Micic D, Mladenovic Drinic S, Nikolic A, Lazic-Jancic V (2008) SSR analysis for genetic structure and diversity determination of maize local populations from former Yugoslavia territories. Russ J Genet 44(11):1317–1324

Ihsan F, Prihatini R (2025) Indonesian tropical fruit Gene Bank. In: Al-Khayri JM, Salem KFM, Jain SM (eds) Plant Gene Banks. Springer, Singapore. https://doi.org/10.1007/978-981-99-4236-7_41-1

Ilarslan R, Kaya ZE, Kandemir I, Bretting PK (2002) Genetic variability among Turkish pop, flint and dent corn (*Zea mays* L. spp. mays) races: morphological and agronomic traits. Euphytica 128:173–182. https://doi.org/10.1023/A:1020896805265

Ishida Y, Saito H, Ohta SH et al (1996) High efficiency transformation of maize (*Zea mays* L.) mediated by Agrobacterium tumefaciens. Nat Biotechnol 14:745–750. https://doi.org/10.1038/nbt0696-745

Ishida Y, Hiei Y, Komari T (2020) Tissue culture protocols for gene transfer and editing in maize (Zea mays L.). Plant Biotechnol (Tokyo) 37(2):121–128. https://doi.org/10.5511/plantbiotechnology.20.0113a

Ismail M, Salem KFM, El-Zanaty AM, Esmail RM (2013) Association mapping of fiber and agronomic characters in Egyptian cotton (*Gossypium barbadense* L.) genotypes. J Agric Chem Biotech Mansoura Univ 4:205–216

Jackson SA, Cheng Z, Wang ML, Goodman HM, Jiang J (2000) Comparative fluorescence in situ hybridization mapping of a 431-kb Arabidopsis thaliana bacterial artificial chromosome contig reveals the role of chromosomal duplications in the expansion of the Brassica rapa genome. Genetics 156:833–838

Jankowicz-Cieslak J, Mba C, Till BJ (2017) Mutagenesis for crop breeding and functional genomics. In: Jankowicz-Cieslak J, Tai T, Kumlehn J, Till B (eds) Biotechnologies for plant muta-

tion breeding. Joint FAO/IAEA Division of Nuclear Techniques in Food and Agriculture International Atomic Energy Agency, Vienna, pp 3–18
Jatoi SA, Siddiqui SU (2025) National Gene Bank of Pakistan. In: Al-Khayri JM, Salem KFM, Jain SM (eds) Plant Gene Banks. Springer, Singapore. https://doi.org/10.1007/978-981-99-4236-7_51-1
Jez JM, Lee SG, Sherp AM (2016) The next green movement: plant biology for the environment and sustainability. Science 353:1241–1244. https://doi.org/10.1126/science.aag1698
Jiang Y, Wang T (2005) Phytosterols in cereal by-products. J Am Oil Chem Soc 82:439–444
Johnson IJ, Hayes HK (1940) The value of hybrid combinations of inbred lines of corn selected from single crosses by the pedigree method of breeding. J Am Soc Agron 32:479–485
Jones CJ, Edwards KJ, Castaglione S et al (1997) Reproducibility testing of RAPD, AFLP and SSR markers in plants by a network of European laboratories. Mol Breed 3:381–390
Kato A, Lamb JC, Birchler JA (2004) Chromosome painting using repetitive DNA sequences as probes for somatic chromosome identification in maize. Proc Natl Acad Sci USA 101:13554–13559
Kato A, Lamb JC, Albert PS et al (2011) Chromosome painting for plant biotechnology. In: Birchler J (ed) Plant chromosome engineering. Methods in molecular biology, vol 701. Humana Press, Totowa. https://doi.org/10.1007/978-1-61737-957-4_4
Kausch AP, Wang K, Kaeppler HF et al (2021) Maize transformation: history, progress, and perspectives. Mol Breed 41(6):38. https://doi.org/10.1007/s11032-021-01225-0
Kelliher T, Starr D, Su X et al (2019) One-step genome editing of elite crop germplasm during haploid induction. Nat Biotechnol 37(3):287–292. https://doi.org/10.1038/s41587-019-0038-x
Kharkwal MC, Shu QY (2009) The role of induced mutations in world food security. In: Shu QY (ed) Induced plant mutations in the genomics era. Food and Agriculture Organization of the United Nations, Rome, pp 33–38
Kim JI, Kim JY (2019) New era of precision plant breeding using genome editing. Plant Biotechnol Rep 13:419–421
Kolchanova S, Kliver S, Komissarov A et al (2019) Genomes of three closely related Caribbean amazons provide insight for species history and conservation. Genes 10:54. https://doi.org/10.3390/genes10010054
Kornienko AV, Butorina AK (2013) Induced mutagenesis in sugar beet (*Beta vulgaris* L.): obtained results and prospects for use in development of TILLING project. Biol Bull Rev 3(2):152–160
Koumbaris GL, Bass HW (2003) A new single-locus cytogenetic mapping system for maize (Zea mays L.): overcoming FISH detection limits with marker-selected sorghum (S. propinquum L.) BAC clones. Plant J 35:647–659
Kuleshov NN (1933) World's diversity of phenotypes of maize. J Am Soc Agron 25:688–700
Kulikova O, Gualtieri G, Geurts R, Kim DJ, Cook D et al (2001) Integration of the FISH pachytene and genetic maps of Medicago truncatula. Plant J 27:49–58
Kumar D, Jhariya AN (2013) Nutritional, medicinal and economical importance of corn: A mini review. Res J Pharm Sci
Kumar K, Abhishek A, Jha AK et al (2021) In vitro regeneration in maize (Zea mays L.). Maize J 10(1):6–12
Kumawat G, Kumawat CK, Chandra K, Pandey S, Chand S, Mishra UN et al (2020) Insights into marker assisted selection and its applications in plant breeding. In: Plant breeding-current and future views. IntechOpen, London
Laghetti G, Piergiovanni AR, De Paola D, Direnzo P, Bucci G, Sonnante G (2025) Italy IBBR gene bank. In: Al-Khayri JM, Salem KFM, Jain SM (eds) Plant Gene Banks. Springer, Singapore. https://doi.org/10.1007/978-981-99-4236-7_73-1
Lamichhane S, Thapa S (2022) Advances from conventional to modern plant breeding methodologies. Plant Breed Biotechnol 10:1–4
Lauer S, Hall B, Mulaosmanovic E, Anderson SR, Nelson B, Smith S (2011) Morphological changes in parental inbred lines of maize hybrids in the Central Corn Belt of the United States. Crop Sci 52:1033–1043

Lee EA, Tracy WF (2009) Modern maize breeding. In: Bennetzen JL, Hake S (eds) Handbook of maize: genetics and genomics. Springer, New York, pp 141–160; ISBN 978-0-387-77863-1.
Lema M (2018) Marker assisted selection in comparison to conventional plant breeding: review article. Agric Res Technol 14:555914
Li X, Song Y, Century K, Straight S, Ronald P, Dong X, Lassner M, Zhang YA (1991) Fast neutron deletion mutagenesis-based reverse genetics system for plants. Plant J 27:235–242. https://doi.org/10.1046/j.1365-313x.2001.01084.x
Li C, Liu C, Qi X et al (2017) RNA-guided Cas9 as an in vivo desired-target mutator in maize. Plant Biotechnol J 15(12):1566–1576. https://doi.org/10.1111/pbi.12739
Lisch D (2002) Mutator transposons. Trends Plant Sci 7:498–504. https://doi.org/10.1016/S1360-1385(02)02347-6
Liu RH (2004) Potential synergy of phytochemicals in cancer prevention: mechanism of action. J Nutr 134:3479S–3485S
Liu C, Li X, Meng D, Zhong Y, Chen C, Dong X et al (2017) A 4-bp insertion at ZmPLA1 encoding a putative phospholipase A generates haploid induction in maize. Mol Plant 10:520–522
Liu L, Gallagher J, Arevalo ED et al (2021) Enhancing grain-yield-related traits by CRISPR–Cas9 promoter editing of maize CLE genes. Nat Plants 7:287–294. https://doi.org/10.1038/s41477-021-00858-5
Liu Q, Zhang XS, Su YH (2023) Application of Wox2a in transformation of recalcitrant maize genotypes. aBIOTECH 4:386–388. https://doi.org/10.1007/s42994-023-00116-6
Long Y, Yang Y, Pan G et al (2022) New insights into tissue culture plant-regeneration mechanisms. Front Plant Sci 13:926752. https://doi.org/10.3389/fpls.2022.926752
Lopez-Martinez LX, Oliart-Ros RM, Valerio-Alfaro G, Lee CH, Parkin KL et al (2009) Antioxidant activity, phenolic compounds and anthocyanins content of eighteen strains of Mexican maize. LWT-Food Sci 42:1187–1192
Louette D, Charrier A, Berthaud (1997) In situ conser-vation of maize in Mexico: genetic diversity and maize seed management in a traditional commu-nity. Econ Bot 51:20–38
Lundmark C (2007) Genetically modified maize. Bioscience 57:996
MacRobert JF, Setimela P, Gethi J et al (2014) Maize hybrid seed production manual. CIMMYT, Mexico
Mallarino AP (2011) Corn and soybean response to soil pH level and liming. In: Integrated crop management conference, pp 93–102
Mallor C, Luna E (2025) The genetic diversity of the Spanish vegetable Genebank BGHZ-CITA: conservation and utilization. In: Al-Khayri JM, Salem KFM, Jain SM (eds) Plant Gene Banks. Springer, Singapore. https://doi.org/10.1007/978-981-99-4236-7_109-1
Maluszynski M, Ahloowalia BS, Sigurbjoornsson B (1995) Application of in vitro mutation techniques for crop improvement. Euphytica 85:303–315
Mansilla PS, Bongianino NF, Nazar MC, Perez GT (2021) Agronomic and chemical description of open-pollinated varieties of opaque-2 and purple maize (*Zea mays* L.) adapted to semi-arid region of Argentina. Genet Resour Crop Evol 68:2351–2366. https://doi.org/10.1007/s10722-021-01133-4
Manzoor S, Nabi SU, Rather TR et al (2024) Advancing crop disease resistance through genome editing: a promising approach for enhancing agricultural production. Front Genome Ed 6:1399051. https://doi.org/10.3389/fgeed.2024.1399051
Matsumoto T (2002) Cryopreservation of in vitro grown apical shoot tips of wasabi (Wasabia japonica) by different procedures. In: Towill LE, Bajaji YPS (eds) Cryopreservation of plant germplasm ll, Biotechnology in Agriculture and Forestry 50. Springer, Berlin, pp 180–195
Matsumoto T, Sakai A, Yamada K (1994) Cryopreservation of in vitro grown apical meristems of wasabi (Wasabia japonica) by vitrification and subsequent high regeneration. Plant Cell Rep 13:442–446
Mattar MZ, Eissa OA, Salem KFM (2009) Genotypic variability in some wheat genotypes (*Triticum aestivum* L.) exposed to salinity at germination stage. Egypt J Exp Biol (Bot) 5:143–148

Mattar MZ, Salem KFM, El-Abed AB (2016) Identification of candidate microsatellite markers associated with agronomic traits in rice (*Oryza sativa* L.). J Bot 15(5):213–232

Maxted N, Ford-Lloyd B, Hawkes JG (1997) Plant genetic conservation: the in-situ approach. Chapman & Hall, London

May BP, Liu H, Vollbrecht E, Senior L, Rabinowicz PD, Roh D, Pan X, Stein L, Freeling M, Alexander D, Martienssen R (2003) Maize-targeted mutagenesis: a knockout resource for maize. Proc Natl Acad Sci USA 100:11541–11546. https://doi.org/10.1073/pnas.1831119100. https://chloroplast.uoregon.edu/

Mayer M, Holker AC, Presterl T, Ouzunova M, Melchinger AE, Schon CC (2022) Genetic diversity of European maize landraces: dataset on the molecular and phenotypic variation of derived doubled-haploid populations. Data Brief 42(108164):1–9. https://doi.org/10.1016/j.dib.2022.108164

McCallum CM, Comai L, Greene EA, Henikoff S (2000) Targeted screening for induced mutations. Nat Biotechnol 18:455–457. https://doi.org/10.1038/74542

McCoy TJ (1987) Characterization of alfalfa (*Medicago sativa* L.) plants regenerated from selected NaCl tolerant cell lines. Plant Cell Rep 6:417–422

McFarland FL, Kaeppler HF (2024) History and current status of embryogenic culture-based tissue culture, transformation and gene editing of maize (Zea mays L.). Plant Genome:e20451. https://doi.org/10.1002/tpg2.20451

Mehta BK, Muthusamy V, Zunjare RU, Baveja A, Chauhan HS, Chhabra R et al (2020) Biofortification of sweet corn hybrids for provitamin-A, lysine and tryptophan using molecular breeding. J Cereal Sci 96:103093

Melchinger AE (1999) Genetic diversity and heterosis. In: Coors JG, Pandey S (eds) The genetics and exploitation of Heterosis in crops. ASA, CSSA, SSSA, Madison, pp 99–118

Meng Z, Song F, Liu T (2018) Genetic diversity and genetic structure analysis of maize (*Zea mays*) landraces in Tibet. Int J Agric Biol 20:791–798

Miao S, Duncan DR, Widholm J (1983) Selection of regenerable maize callus cultures resistant to 5-methyl tryptophan, S-2-aminoethyl-L-cysteine and high levels of L-lysine plus L-threonine. Plant Cell Tissue Organ Cult 14:3–14

Mohan M, Nair S, Bhagwat A, Krishna TG, Yano M, Bhatia CR et al (1997) Genome mapping, molecular markers and marker-assisted selection in crop plants. Mol Breed 3:87–103

Moll RH, Thompson DL, Harvey PH (1963) A quantitative genetic study of the inheritance of resistance to brown spot (*Physoderma maydis*) of corn. Crop Sci 3:389–391

Monteros-Altamirano Á et al (2025) Plant genetic resources in Ecuador. In: Al-Khayri JM, Salem KFM, Jain SM (eds) Plant Gene Banks. Springer, Singapore. https://doi.org/10.1007/978-981-99-4236-7_99-1

Moose SP, Mumm RH (2008) Molecular plant breeding as the foundation for 21st century crop improvement. Plant Physiol 147(3):969–977. https://doi.org/10.1104/pp.108.118232

Morris ML (2002) Impacts of international maize breeding research in developing countries 1966–1998. CIMMYT, Mexico

Mourad AMI, Alomari DZ, Alqudah AM et al (2019) Recent advances in wheat (*Triticum* spp.) breeding. In: Al-Khayri J, Jain S, Johnson D (eds) Advances in plant breeding strategies: cereals. Springer, Cham. https://doi.org/10.1007/978-3-030-23108-8_15

Munaiz ED, Albrecht KA, Ordas B (2021) Genetic diversity for dual use maize: grain and second-generation biofuel. Agronomy 11:230.https://doi.org/10.3390/agronomy11020230

Muppala S, Gudlavalleti PK, Pagidoju S et al (2020) Distinctive response of maize (*Zea mays* L.) genotypes in vitro with the acceleration of phytohormones. J Plant Biotechnol 47:26–39. https://doi.org/10.5010/JPB.2020.47.1.026

Nannas NJ, Dawe RK (2015) Genetic and genomic toolbox of Zea mays. Genetics 199(3):655–669

Nawara HM, Mattar MZ, Salem KFM, Eissa OA (2016) Estimating heterosis and combining ability for agronomic and tissue culture traits using diallel cross on bread wheat (*Triticum aestivum* L.). Egypt J Bot 2016:177–200

Nawara HM, Mattar MZ, Salem KFM, Eissa OA (2017) Diallel study on some in vitro traits of bread wheat (*Triticum aestivum* L.) under salt stress. Int J Agric Environ Res 3(1):1988–2006

Ndudzo A, Makuvise AS, Moyo S et al (2024) CRISPR-Cas9 genome editing in crop breeding for climate change resilience: implications for smallholder farmers in Africa. J Agric Food Res 16:101132. https://doi.org/10.1016/j.jafr.2024.101132

Neild RE, Newman JE (1987) Growing season characteristics and requirements in the Corn Belt; Cooperative Extension Service. Iowa State University, Ames

Novak FJ, Brunner H (1992) Plant breeding: induced mutation technology for crop improvement. IAEA Bull 4:25–33

Oktem A (2005) Response of sweet corn (*Zea mays saccharata* Sturt) to nitrogen and intra row spaces in semi-arid region. Pak J Biol Sci 8:160–163

Olowe OM, Adesoye AI, Ojobo OA et al (2014) Effects of sterilization and phytohormones on Shoot tip culture of *Telferia occidentalis*. J Nat Sci Res 4:53–58

Ombori O, Gitonga NM, Machuka S (2008) Somatic embryogenesis and plant regeneration from immature embryos of tropical maize (*Zea mays* L.) inbred lines. Biotechnology 7(2):224–232

Ortiz R, Crossa J, Franco J, Sevilla R, Burgueno J (2008) Classification of Peruvian highland maize races using plant traits. Genet Resour Crop Evol 55:151–162

Orton LM, Burke SV, Wysocki WP, Duvall MR (2017) Plastid phylogenomic study of species within the genus Zea: rates and patterns of three classes of microstructural changes. Curr Genet 63:311–323. https://doi.org/10.1007/s00294-016-0637-8

Osterç G, Gostinčar K, Hudina M, Usenik V, Luthar Z (2025) Slovene fruit Gene Bank. In: Al-Khayri JM, Salem KFM, Jain SM (eds) Plant Gene Banks. Springer, Singapore. https://doi.org/10.1007/978-981-99-4236-7_84-1

Panella L (2008) Utilizing genetic resources for prebreeding of stress-resistant sugar beet germplasm: using molecular tools. In: Conventional and molecular breeding of field and vegetable crops. Novi Sad, November 24–27, pp 107–111

Patade VY, Suprasanna P, Kulkarni UG, Bapat VA (2006) Selection for abiotic (salinity and drought) stress tolerance and molecular characterization of tolerant lines in sugarcane. BARC Newsl 273:244–257

Paterson AH, Freeling M, Sasaki T (2005) Grains of knowledge: genomics of model cereals. Genome Res 15:1643–1650

Pedersen C, Linde-Laursen I (1994) Chromosomal locations of four minor rDNA loci and a marker microsatellite sequence in barley. Chromosom Res 2:65–71

Peixoto MA, Coelho IF, Leach KA, Lübberstedt T, Bhering LL, Resende MFR (2024) Use of simulation to optimize a sweet corn breeding program: implementing genomic selection and doubled haploid technology. G3 Gene Geno Genet 14(8):jkae128. https://doi.org/10.1093/g3journal/jkae128

Pellegrino E, Bedini S, Nuti M et al (2018) Impact of genetically engineered maize on agronomic, environmental and toxicological traits: a meta-analysis of 21 years of field data. Sci Rep 8:3113. https://doi.org/10.1038/s41598-018-21284-2

Perales HR, Benz BF, Brush SB (2005) Maize diver-sity and ethnolinguistic diversity in Chiapas, Mex-ico. PNAS 102:949–954

Perico C, Zaidem M, Sedelnikova O et al (2024) Multiplexed in situ hybridization reveals distinct lineage identities for major and minor vein initiation during maize leaf development. Proc Natl Acad Sci USA 121(28):e2402514121. https://doi.org/10.1073/pnas.2402514121

Piperno DR, Ranere AJ, Holst I, Iriarte J, Dickau R (2009) Starch grain and phytolith evidence for early ninth millennium BP maize from the Central Balsas River Valley. Mexico. Proc Natl Acad Sci 106:5019–5024

Plucknett DL, Smith NJH, Williams JT, Murthi Anishetty N (1983) Crop germplasm conservation and developing countries. Science 220:163–169

Plucknett DL, Smith NJH, Williams JT et al (1987) Gene banks and the world's food. Princeton University Press, Princeton

Popović V et al (2025) Economic seed production and storage of fiber crops in gene banks in Serbia. In: Al-Khayri JM, Salem KFM, Jain SM (eds) Plant Gene Banks. Springer, Singapore. https://doi.org/10.1007/978-981-99-4236-7_108-1

Porch TG, Blair MW, Lariguet P et al (2009) Generation of a mutant population for TILLING common bean genotype BAT93. Am Hortic Soc 134:348–355

Pradeep RM, Sarla N, Siddiq EA (2002) Inter simple sequence repeat (ISSR) polymorphism and its application in plant breeding. Euphytica 128:9–17. https://doi.org/10.1023/A:1020691618797

Prasanna BM, Vasal SK, Kassahun B, Singh NN (2001) Quality protein maize. Curr Sci 81:1308–1319

Prasanna BM, Palacios-Rojas N, Hossain F et al (2020) Molecular breeding for nutritionally enriched maize: status and prospects. Front Genet 10:1392. https://doi.org/10.3389/fgene.2019.01392

Predieri S (2001) Induced mutation and tissue culture in fruits. Plant Cell Tissue Organ Cult 64:185–210

Purugganan MD (2022) What is domestication? Trends Ecol Evol 37(8):663–671. https://doi.org/10.1016/j.tree.2022.04.006

Qi W, Zhu T, Tian Z et al (2016) High-efficiency CRISPR/Cas9 multiplex gene editing using the glycine tRNA-processing system-based strategy in maize. BMC Biotechnol 16(1):58. https://doi.org/10.1186/s12896-016-0289-2

Raizada MN, Nan GL, Walbot V (2001) Somatic and germinal mobility of the RescueMu transposon in transgenic maize. Plant Cell 13:1587–1608. https://doi.org/10.1105/tpc.13.7.1587

Ramya P, Singh GP, Jain N, Singh PK, Pandey MK, Sharma K et al (2022) Effect of recurrent selection on drought tolerance and related morpho-physiological traits in bread wheat. PLoS One 11:e0156869. Euphytica Sep;218(9):127

Ranere AJ, Piperno DR, Holst I, Dickau R, Iriarte J (2009) The cultural and chronological context of early Holocene maize and squash domestication in the central Balsas River Valley, Mexico. Proc Natl Acad Sci USA 106:5014–5018

Rasha AO, Matheka JM, Ali AM, Machuka J (2013) Transformation of tropical maize with the NPK1 gene for drought tolerance. Int J Genet Eng 3:7–14

Rasheed A, Jie H, Ali B et al (2023) Breeding drought-tolerant maize (*Zea mays*) using molecular breeding tools: recent advancements and future prospective. Agronomy 13(6):1459. https://doi.org/10.3390/agronomy13061459

Reif JC, Xia XC, Melchinger AE, Warburton ML, Hoisington DA, Beck D et al (2004) Genetic diversity determined within and among CIMMYT maize populations of tropical, subtropical, and temperate germplasm by SSR markers. Crop Sci 44:326–334

Restle J, Pacheco PS, Alves Filho DC, Freitas AKD, Neumann M, Brondani IL et al (2006) Silagem de diferentes híbridos de milho para produção de novilhos superjovens. Rev Bras Zootec 35:2066–2076. https://doi.org/10.1590/s1516-35982006000700026

Reynolds M, Borrell A, Braun H et al (2019) Translational research for climate resilient, higher yielding crops. Crop Breed Genet Genomics 1:e190016. https://doi.org/10.20900/cbgg20190016

Rhoades M (1951) Duplicate genes in maize. Am Nat 85:105–110

Ribaut JM, Hoisington D (1998) Marker-assisted selection: new tools and strategies. Trends Plant Sci 3:236–239

Ribaut JM, Ragot M (2007) Marker-assisted selection to improve drought adaptation in maize: the backcross approach, perspectives, limitations, and alternatives. J Exp Bot 58:351–360

Rines HW, Phillips RL, Kynast RG et al (2009) Addition of individual chromosomes of maize inbreds B73 and Mo17 to oat cultivars starter and Sun II: maize chromosome retention, transmission, and plant phenotype. Theor Appl Genet 119(7):1255–1264. https://doi.org/10.1007/s00122-009-1130-2

Rocha F et al (2025) Portugal Gene Bank. In: Al-Khayri JM, Salem KFM, Jain SM (eds) Plant Gene Banks. Springer, Singapore. https://doi.org/10.1007/978-981-99-4236-7_79-1

Roychowdhury R, Tah J (2013) Mutagenesis–a potential approach for crop improvement. In: Hakeem KR, Ahmad P, Ozturk M (eds) Crop improvement: new approaches and modern techniques. Springer, New York, pp 149–187

Rukuni M, Tawonezvi P, Munyuki-hungwe M, Dube E, Kilian W, Mwadzingeni L, Sosibo NZ, Barnard A, Tsilo TJ, Rice E et al (2016) Agricultura revolution. Crop Sci 51:185–192

Russello M, Amato G, DeSalle R, Knapp M (2020) Conservation genetics and genomics. Genes 11(3):318. https://doi.org/10.3390/genes11030318

Sadder MT, Weber G (2002) Comparison between genetic and physical maps in Zea mays L. of molecular markers linked to resistance against Diatraea spp. Theor Appl Genet 104:908–915

Sadder MT, Ponelies N, Born U, Weber G (2000) Physical localization of single-copy sequences on pachytene chromosomes in maize (*Zea mays* L.) by chromosome in situ suppression hybridization. Genome 43:1081–1083

Sairam RV, Paran M, Franklin G, Lifeng Z, Smiyh B, MacDougall J, Wilber C, Sheikhi H, Kashikar N, Meekar, Sairam RV, Parani M, Franklin G et al (2003) Shoot meristem: an ideal explant for *Zea mays* L. transformation. Genome 46(2):323–329

Saleh MM, Ali WA, Alwan MH et al (2020a) Multi-environmental evaluation of wheat tetraploid genotypes for agronomic traits under rainfed conditions in Syria. J Aridland Agric 6:1–5

Saleh MM, Salem KFM, Elabd AB (2020b) Definition of selection criterion using correlation and path coefficient analysis in rice (*Oryza sativa* L.) genotypes. Bull Natl Res Cent 44(1):1–6

Saleh MM, Alnaddaf LM, Almuhammady AK et al (2021) Applications of plant-derived nanomaterials in mitigation of crop abiotic stress. In: Al-Khayri JM, Ansari MI, Singh AK (eds) Nanobiotechnology. Springer, Cham. https://doi.org/10.1007/978-3-030-73606-4_9

Salem KFM (1997) Breeding studies on rice (*Oryza sativa* L.). M Sc Thesis, Faculty of Agriculture, Minufiya University, Shibin El-Kom

Salem KFM (2004) The inheritance and molecular mapping of genes for post-anthesis drought tolerance (PADT) in wheat. Martin-Luther-University, Halle-Wittenberg, p 124

Salem KFM (2009) Relationship between genetic diversity based on SSRs markers with heterosis and combining ability in diallel cross of bread wheat (*Triticum aestivum* L.). Minufiya J Agric Res 6(34):2159–2178

Salem KFM (2015) Allelic state at the microsatellite locus *Xgwm261* marking the dwarfing gene *Rht8* in Egyptian bread wheat (*Triticum aestivum* L.) genotypes released from 1947 to 2004. Genetika 47:741–750

Salem KFM, Börner A (2003) Evaluation of wheat accessions from Gatersleben Genebank for post-anthesis drought tolerance in the field. In: Gesellschaft für Pflanzenzüchtung, GPZ-Tagung, Vortragsveranstaltung zum Thema: "Abiotischer Streß – Herausforderung für die Pflanzenzüchtung", 11-12 Juni 2003, Groß Lüsewitz, Germany

Salem KFM, Börner A (2004) Post-anthesis drought tolerance in genus *Triticum*. In: Botanikertagung, 5–10 September 2004, Braunschweig, Germany

Salem KFM, El-Zanaty AM (2015) Assessing genetic diversity of old and modern Egyptian rice (Oryza sativa L.) using microsatellite markers. Minufiya J Agric Res 40(6):1507–1518

Salem KFM, Ibrahim AA (2024) Plant biodiversity in the context of food security under climate change. In: Al-Khayri JM, Jain SM, Penna S (eds) Sustainable utilization and conservation of plant genetic diversity, Sustainable development and biodiversity, vol 35. Springer, Singapore. https://doi.org/10.1007/978-981-99-5245-8_1

Salem KFM, Mattar MZ (2014a) Genetic diversity in old and modern Egyptian bread wheat (*Triticum aestivum* L.) varieties revealed by simple sequence repeats. Egypt J Genet Cytol 43:143–156

Salem KFM, Mattar MZ (2014b) Identification of microsatellite alleles for salt tolerance at seedling stage in wheat (*Triticum aestivum* L.). Life Sci J 11(12s):1064–1073

Salem KFM, Sallam A (2016) Analysis of population structure and genetic diversity of Egyptian and exotic rice (*Oryza sativa* L.) genotypes using SSR markers. Comptes Rendus Biol 339:1–9. https://doi.org/10.1016/j.crvi.2015.11.003

Salem KFM, Röder MS, Börner A (2003) Mapping of quantitative trait loci (QTLs) for post-anthesis drought tolerance in bread wheat (Triticum aestivum L.). In: Gesellschaft für Pflanzenzüchtung, GPZ-Tagung, 11. Vortragsveranstaltung "Harnessing Genetic Diversity: Genomics and Allele Mining"., 16-17.09.2003, Institut für Pflanzengenetik und Kulturpflanzenforschung (IPK), Gatersleben, Germany

Salem KFM, Röder MS, Börner A (2004a) Drought tolerance in wheat germplasm- evaluation and molecular mapping. In: Institut für Pflanzengenetik und Kulturpflanzenforschung, IPK Tagung, 22–24/08/2004, Ag Ressourcengenetik und Reproduktion, Abteilung Genbank, Gatersleben, Germany

Salem KFM, Röder MS, Börner A (2004b) Molecular mapping of quantitative trait loci (QTLs) determining post-anthesis drought tolerance in hexaploid wheat (*Triticum aestivum* L.). In: 7 Gesellschaft für Pflanzenzüchtung, GPZ-Tagung, Vortragsveranstaltung zum Thema: " Klimatische und edaphische Sortenanpassung und Züchtung für Nachwachsende Rohstoffe". 3-5 March 2004, Halle/Saale, Germany. Vort Pflanzenzüchtung 64:21–24

Salem KFM, Röder MS, Börner A (2007a) Evaluation of some barley varieties for the presence of thermostable alleles of *ßamylase*. In: 8th African crop science society conference, El-Minia, Egypt, 27–31 October 2007, pp 643–648

Salem KFM, Röder MS, Börner A (2007b) Identification and mapping quantitative trait loci for stem reserve mobilisation in wheat (*Triticum aestivum* L.). Cereal Res Common 35:1367–1374

Salem KFM, Varshney RK, Röder MS, Börner A (2008a) Detection of genetic diversity in Egyptian barley genotypes using barley expressed sequence tag-SSR markers. In: 10th International Barley Genetics Symposium (IBGS), Alexandria, Egypt, April 5–10, 2008. The International Center for Agricultural Research in the Dry Areas (ICARDA) and the Bibliotheca Alexandrina, Alexandria

Salem KFM, El-Zanaty AM, Esmail RM (2008b) Assessing wheat (*Triticum aestivum* L.) genetic diversity using morphological characters and microsatellite markers. World J Agric Sci 4:538–544

Salem KFM, Varshney RK, Röder MS, Börner A (2010) EST-SSR based estimates on functional genetic variation in a barley (*Hordeum vulgare* L.) collection from Egypt. Genet Resour Crop Evol 57(4):515–521

Salem KFM, Röder MS, Börner A (2015a) Assessing genetic diversity of Egyptian hexaploid wheat (*Triticum aestivum* L.) using microsatellite markers. Genet Resour Crop Evol 62:377–385

Salem KFM, Yaseen YM, El-Abd AB, Ayoub A (2015b) Inheritance of some yield components and grain quality traits in rice *(Oryza sativa* L.). J Prod Dev 20:161–177

Salem KFM, Yaseen YM, El-Abd AB, Ayoub A (2015c) Breeding studies on some agronomic and quality traits in rice (Oryza sativa L.). J Prod Dev 20(2):211–226

Salem KFM, Alloosh MT, Saleh MM et al (2021a) Utilization of nanofertilizers in crop tolerance to abiotic stress. In: Al-Khayri JM, Ansari MI, Singh AK (eds) Nanobiotechnology. Springer, Cham. https://doi.org/10.1007/978-3-030-73606-4_11

Salem KFM, Elabsawy A, Aldahak L et al (2021b) Evaluation of stem reserve mobilization in Egyptian bread wheat (*Triticum aestivum* L.) genotypes and F_1 hybrids under post-anthesis chemical desiccation stress. J Genet Environ Resour Conserv 9(1):176–182

Salem KFM, El-Absawy EA, Khidr YA, El-Shamy H (2021c) Assessment genetic diversity among some Egyptian bread wheat (*Triticum aestivum* L.) cultivars using RAPD markers. J Genet Environ Resour Conserv 9(1):168–175

Salem KFM, Saleh MM, Abu-Ellail FFB et al (2021d) Role of quantum dots, polymeric NPS and dendrimers in emphasizing crops tolerate biotic and abiotic stresses. In: Faizan M, Hayat S, Yu F (eds) Sustainable agriculture reviews, vol 53. Springer, Cham. https://doi.org/10.1007/978-3-030-86876-5_1

Salem KFM, Saleh MM, Abu-Ellail FFB et al (2021e) The role of salicylic acid in crops to tolerate abiotic stresses. In: Hayat S, Siddiqui H, Damalas CA (eds) Salicylic acid- a versatile plant growth regulator. Springer, Cham. https://doi.org/10.1007/978-3-030-79229-9_7

Salem KFM, Saleh MM, Aldahak L, Elabd AB (2021f) Assessment phenotypic diversity of rice (*Oryza sativa* L.) genotypes by multivariate analysis. J Arid Agric 7:52–59. https://doi.org/10.25081/jaa.2021.v7.6516
Salem KFM, Alghuthaymi MA, Elabd AB et al (2022) Prediction of heterosis for agronomic traits in half-diallel cross of rice (*Oryza sativa* L.) under drought stress using microsatellite markers. Plants 11:1532. https://doi.org/10.3390/plants11121532
Salem KFM, Tawfik E, Mahdy M et al (2023) Advances in lettuce (*Lactuca* spp.) molecular breeding strategies. In: Singh S, Sharma D, Sharma SK, Singh R (eds) Smart plant breeding for vegetable crops in post-genomics era. Springer, Singapore. https://doi.org/10.1007/978-981-19-5367-5_11
Salem KFM, Safhi FA, Alwutayd KM et al (2024) Analysis of genetic diversity, population structure and phylogenetic relationships of rice (*Oryza sativa* L.) cultivars using simple sequence repeat (SSR) markers. Genet Resour Crop Evol. https://doi.org/10.1007/s10722-023-01789-0
Sami A, Xue Z, Tazein S et al (2021) CRISPR–Cas9-based genetic engineering for crop improvement under drought stress. Bioengineered 12(1):5814–5829. https://doi.org/10.1080/21655979.2021.1969831
Sanchez JJ, Goodman MM, Stuber CW (2000) Isozy-matic and morphological diversity in the races of maize in Mexico. Econ Bot 54:43–59
Sanchez D, Sadoun SB, Mary-Huard T, Allier A, Moreau L, Charcosset A (2023) Improving the use of plant genetic resources to sustain breeding programs' efficiency. Proc Natl Acad Sci USA. Apr 4, 120(14):e2205780119. https://doi.org/10.1073/pnas.2205780119
Scheitrum D, Schaefer KA, Kjersti Nes K (2020) Realized and potential global production effects from genetic engineering. Food Policy 93:101882. https://doi.org/10.1016/j.foodpol.2020.101882
Schlegel RHJ (2017) History of plant breeding. CRC Press, Boca Raton
Schnable PS, Ware D, Fulton RS et al (2009) The B73 maize genome: complexity, diversity, and dynamics. Science 326:1112–1115
Sen CK, Khanna S, Roy S (2006) Tocotrienols: vitamin E beyond tocopherols. Life Sci 78:2088–2098
Senior ML, Murphy JP, Goodman MM, Stuber CW (1998) Utility of SSRs for determining genetic similarities and relationship in maize using agarose gel system. Crop Sci 38:1088–1098
Shahzad A, Parveen S, Sharma S et al (2017) Plant Tissue Culture: Applications in Plant Improvement and Conservation. In: Abdin M, Kiran U, Kamaluddin Ali A (eds) Plant Biotechnology: Principles and Applications. Springer, Singapore.https://doi.org/10.1007/978-981-10-2961-5_2
Sharma VK, Hansch R, Mendel RR, Schulze J (2004) A highly efficient plant regeneration system through multiple shoot differentiation from commercial cultivars of barley using meristematic shoot segments excised from germinated mature embryos. Plant Cell Rep 23:9–16
Sharma VK, Hansch R, Mendel RR, Schulze J (2005) Mature embryo axis-based high frequency somatic embryogenesis and plant regeneration from multiple cultivars of barley. J Exp Bot 56:1913–1922
Shaw J, Shafer HL, Leonard OR, Kovach MJ, Schorr M, Morris AB (2014) Chloroplast DNA sequence utility for the lowest phylogenetic and phylogeographic inferences in angiosperms: the tortoise and the hare IV. Am J Bot 101:1987–2004. https://doi.org/10.3732/ajb.1400398
Shi J, Dawe RK (2006) Partitioning of the maize epigenome by the number of methyl groups on histone H3 lysines 9 and 27. Genetics 173:1571–1583
Shi J, Gao H, Wang H et al (2017) ARGOS 8 variants generated by CRISPR-Cas9 improve maize grain yield under field drought stress conditions. Plant Biotechnol J 15(2):207–216. https://doi.org/10.1111/pbi.12603
Sileshi GW, Gebeyehu S (2021) Emerging infectious diseases threatening food security and economies in Africa. Glob Food Sec 28(June 2020):100479. https://doi.org/10.1016/j.gfs.2020.100479
Singh SP (1982) Alternative methods to backcross breeding. Ann Rep Bean Improv Coop 25:11–12

Singh V, Kaundal A (2023) Genomic selection in maize breeding. In: Maize improvement: current advances in yield, quality, and stress tolerance under changing climatic scenarios. Springer International Publishing, Cham, pp 99–108

Singh NK, Joshi A, Smrutishree Sahoo S et al (2023) Molecular breeding for improving yield in maize: recent advances and future perspectives. In: Wani SH, Wang D, Singh GP (eds) QTL mapping in crop improvement. Academic, pp 75–99. https://doi.org/10.1016/B978-0-323-85243-2.00010-6

Singh KV, Pundir S, Chaturvedi D et al (2024) Enhancing maize (Zea mays L.) crop through advanced techniques: a comprehensive approach. IntechOpen. https://doi.org/10.5772/intechopen.114029

Smith JS, Chin ECL, Shu H et al (1997) An evaluation of the utility of SSR loci as molecular markers in maize (*Zea mays* L.): comparisons with data from RFLPs and pedigree. Theor Appl Genet 95:163–173

Smith S, Cooper M, Gogerty J et al (2014) Maize. In: Smith S, Diers B, Specht J et al (eds) Yield gains in major us field crops, CSSA special publications, vol 33. American Society of Agronomy and Soil Science Society of America, Madison, pp 125–171

Sokolov VA, Panikhin PA, Plotnikov KO et al (2024) Development of apomictic 56-Chromosomal maize–tripsacum hybrids: a potential breakthrough in heterosis fixation. Plants 13(15):2138. https://doi.org/10.3390/plants13152138

Sonnino A, Carena MJ, Guimaraes EP et al (2007) An assessment of the use of molecular markers in developing countries. In: Guimaraes EP, Ruane J, Scherf BD et al (eds) Markers-assisted selection: current status and future perspectives in crops, livestock, forestry, and fish. Food and Agriculture Organization of the United Nations, Rome, pp 15–26

Steinmacher DA, Cangahuala-Inocente GC, Clement CR, Guerra MP (2007) Somatic embryogenesis from peach palm zygotic embryos. In Vitro Cell Develop Biol Plant 43:124–132

Stevens R (2008) Prospects for using marker-assisted breeding to improve maize production in Africa. J Sci Food Agric 88:745–755

Sturtevant EL (1884) Maize: an attempt at classification. Democrat and Chronicle Print, Rochester

Sturtevant EL (1899) Varieties of corn. USDA Off Exp Stn Bull 57:1–108

Sun X, Qi X, Wang W et al (2020) Etiology and symptoms of maize leaf spot caused by *Bipolaris* spp. in Sichuan, China. Pathogens 9(3). https://doi.org/10.3390/pathogens9030229

Suprasanna P, Jain SM (2017) Mutant resources and Mutagenomics in crop plants. Emir J Food Agric 29:651–657

Svitashev S, Young JK, Schwartz C et al (2015) Targeted mutagenesis, precise gene editing, and site-specific gene insertion in maize using Cas9 and guide RNA. Plant Physiol 169(2):931–945. https://doi.org/10.1104/pp.15.00793

Swigonova Z, Lai J, Ma J, Ramakrishna W, Llaca V et al (2004) Close split of sorghum and maize genome progenitors. Genome Res 14:1916–1923

Szarejko I, Forster BP (2007) Doubled haploidy and induced mutation. Euphytica 158:359–370. https://doi.org/10.1007/s10681-006-9241-1

Szarejko I, Maluszynski M, Polok K, Kilian A (1991) Doubled haploids in the mutation breeding of selected crops. In: Plant mutation breeding for crop improvement, vol 2. IAEA, Vienna, pp 355–378

Taberlet P, Gielly L, Pautou G, Bouvet J (1991) Universal primers for amplification of three non-coding regions of chloroplast DNA. Plant Mol Biol 17:1105–1109. https://doi.org/10.1007/BF00037152

Tadege M, Wang TL, Wen J et al (2009) Mutagenesis and beyond! Tools for understanding legume biology. Plant Physiol 151:978–984

Taramino G, Tingey S (1996) Simple sequence repeats for germplasm analysis and mapping in maize. Genome 39:277–287

Tax FE, Vernon DM (2001) T-DNA-associated duplication/translocations in Arabidopsis. Implications for mutant analysis and functional genomics. Plant Physiol 126:1527–1538. https://doi.org/10.1104/pp.126.4.1527

Thornton PK, Ericksen PJ, Herrero M, Challinor AJ (2014) Climate variability and vulnerability to climate change: a review. Glob Chang Biol:3313–3328. https://doi.org/10.1111/gcb.12581
Thorpe TA (2007) History of plant tissue culture. Mol Biotechnol 37(2):169–180. https://doi.org/10.1007/s12033-007-0031-3
Till BJ, Colbert T, Tompa R et al (2003) High-throughput TILLING for functional genomics. In: Grotewald E (ed) Plant functional genomics: methods and protocols. Humana Press, Clifton, pp 205–220
Till BJ, Burtner C, Comai L, Henikoff S (2004) Mismatch cleavage by single-strand specific nucleases. Nucleic Acids Res 32:2632–2641. https://doi.org/10.1093/nar/gkh599
Torney F, Moeller L, Scarpa A, Wang K (2007) Genetic engineering approaches to improve bioethanol production from maize. Curr Opin Biotechnol 18:193–199. https://doi.org/10.1016/j.copbio.2007.03.006
Tranel E (2024) Origin of yield gains in maize hybrids. Nat Sci Educ 53:e20138. https://doi.org/10.1002/nse2.20138
Troyer AF (2009) Development of hybrid corn and the seed corn industry. In: Bennetzen JL, Hake S (eds) Handbook of maize: genetics and genomics. Springer, New York, pp 87–114; ISBN 978-0-387-77863-1
UPOV (1999) Guidelines for the conduct of tests for distinctness, uniformity and stability; maize. TG/2/6 + Corr, Revision of TG/2/4. UPOV, Geneva
UPOV (2003) Progress report on the work of the technical committee, the technical working parties and the working group on biochemical and molecular techniques, and DNA-profiling in particular. C/37/10. UPOV, Geneva
UPOV (2009) Guidelines for the conduct of tests for distinctness, uniformity and stability; maize. TG/2/7. UPOV, Geneva
USDA 2010 http://www.ars-grin.gov/cgi-bin/npgs/html/desclist.pl?89
USDA (2024a) Recent Trends in GE Adoption. https://www.ers.usda.gov/data-products/adoption-of-genetically-engineered-crops-in-the-u-s/recent-trends-in-ge-adoption/. Accessed 11 Sept 2024
USDA (2024b) World agricultural production. United States, Department of Agriculture, Foreign Agricultural Service, Circular Series WAP 3-24. https://ipad.fas.usda.gov/
Usman IS, Abdulmalik MM (2010) Cryopreservation of embryonic axes of maize (Zea mays L.) by vitrification protocol. Afr J Biotechnol 9(52):8955–8957
Valentine M, Butruille D, Achard F et al (2024) Simultaneous genetic transformation and genome editing of mixed lines in soybean (Glycine max) and maize (Zea mays). aBIOTECH 5:169–183. https://doi.org/10.1007/s42994-024-00173-5
Van Etten J (2006) Changes in farmers' knowledge of maize diversity in highland Guatemala, 1927/37-2004. J Ethnobiol Ethnomed 2:1–17
Van Harten AM (1998) Mutation breeding: theory and practical applications. Cambridge University Press, pp 163–251
Varotto S (2024) Current status and future prospective of genome editing application in maize. In: Ricroch A, Eriksson D, Miladinović D, Sweet J, Van Laere K, Woźniak-Gientka E (eds) A roadmap for plant genome editing. Springer, Cham. https://doi.org/10.1007/978-3-031-46150-7_11
Varshney RK, Tuberosa R (2007) Genomics-assisted crop improvement: an overview. In: Varshney RK, Tuberosa R (eds) Genomics-assisted crop improvement: Vol. 1: genomics approaches and platforms. Springer Netherlands, Dordrecht, pp 1–12; ISBN 978-1-4020-6295-7
Varshney RK, Salem KFM, Röder MS et al (2008) Assessment and comparison of genetic diversity in barley germplasm collection using gSSR, eSSR and SNP markers. Plant Genet Res 6:167–174
Vega JM, Yu W, Kennon A, Chen X, Zhang Z (2008) Improvement of Agrobacterium-mediated transformation in Hi-II maize (*Zea mays*) using standard binary vectors. Plant Cell Rep 27:297–305. https://doi.org/10.1007/s00299-007-0463-z
Veillet F, Durand M, Kroj T, Cesari S, Gallois JL (2020) Precision breeding made real with CRISPR: illustration through genetic resistance to pathogens. Plant Commun 1:100102

Vejlupkova Z, Warman C, Sharma R et al (2020) No evidence for transient transformation via pollen magnetofection in several monocot species. Nat Plants 6(11):1323–1324. https://doi.org/10.1038/s41477-020-00798-6

Velazquez D, Xavier H, Batista J, Castro-Chaves C (2005) *Zea mays* L. extracts modify glomerular function and potassium urinary excretion in conscious rats. Phytomedicine 12:363–369

Vladmir S, Gilbertson L, Adae P, Duncan D (2006) Agrobacterium-mediated transformation of seedling-derived maize callus. Plant Cell Rep 25:320–328

Vogel KE (2009) Backcross breeding. In: Transgenic maize: methods and protocols, pp 161–169

Waltz E (2016) CRISPR-edited crops free to enter market, skip regulation. Nat Biotechnol 34:582–583

Wan X, Wu S, Wei X et al (2024) Plant biotechnology and molecular breeding for global food security. ACS Agric Sci Technol 4(5):521–523. https://doi.org/10.1021/acsagscitech.4c00187

Wang H, Cheng J, Cheng Y (2012) Study Progress on tissue culture of maize mature embryo. Phys Procedia 25:2225–2227. https://doi.org/10.1016/j.phpro.2012.03.374

Wang X, Wang H, Liu S, Ferjani A, Li J, Yan J et al (2016) Genetic variation in ZmVPP1 contributes to drought tolerance in maize seedlings. Nat Genet 48:1233–1241

Wang H, Xu C, Liu X et al (2017) Development of a multiple-hybrid population for genome-wide association studies: theoretical consideration and genetic mapping of flowering traits in maize. Sci Rep 10(7):40239. https://doi.org/10.1038/srep40239

Wang B, Zhu L, Zhao B et al (2019) Development of a haploid-inducer mediated genome editing system for accelerating maize breeding. Mol Plant 12(4):597–602. https://doi.org/10.1016/j.molp.2019.03.006

Wang Y, Liu X, Zheng X et al (2021) Creation of aromatic maize by CRISPR/Cas. J Integr Plant Biol 63(9):1664–1670. https://doi.org/10.1111/jipb.13105

Wang J et al (2025) China National Gene Bank (CNGB). In: Al-Khayri JM, Salem KFM, Jain SM (eds) Plant Gene Banks. Springer, Singapore. https://doi.org/10.1007/978-981-99-4236-7_103-1

Warzecha T, Bathelt R, Skrzypek E et al (2023) Studies of oat-maize hybrids tolerance to soil drought stress. Agriculture 13(2):243. https://doi.org/10.3390/agriculture13020243

Washburn JD, Burch MB, Franco JAV (2020) Predictive breeding for maize: making use of molecular phenotypes, machine learning, and physiological crop models. Crop Sci 60:622–638

Watts A, Kumar V, Raipuria RK et al (2018) In vivo haploid production in crop plants: methods and challenges. Plant Mol Biol Report 36:685–694. https://doi.org/10.1007/s11105-018-1132-9

Weatherwax P (1955) History and origin of corn. I. Early history of corn and theories as to its origin. In: Sprague GF (ed) Corn and corn improvement. Academic, New York, pp 1–16

Wei F, Coe E, Nelson W et al (2007) Physical and genetic structure of the maize genome reflects its complex evolutionary history. PLOS Gene 3(7):e123. https://doi.org/10.1371/journal.pgen.0030123

Wendel J, Stuber C, Edwards M, Goodman M (1986) Duplicated chromosome segments in *Zea mays L.:* further evidence from hexokinase enzymes. Theor Appl Gen 72:178–185

Widoretno W, Harran S, Sudarsono (2003) Variation in qualitative and quantitative characters among somaclones of soybean derived from in vitro selected somatic embryos. Hayati 10:110–117

Wijerathna-Yapa A, Hiti-Bandaralage J (2023) Tissue culture-a sustainable approach to explore plant stresses. Life 13(3):780. https://doi.org/10.3390/life13030780

Wijerathna-Yapa A, Ramtekey V, Ranawaka B et al (2022) Applications of in vitro tissue culture Technologies in Breeding and Genetic Improvement of wheat. Plants 11(17):2273. https://doi.org/10.3390/plants11172273

Wilkes G (2004) Corn, strange, and marvelous: but is a definitive origin known? In: Smith CW, Betran J, Runge ECA (eds) Corn: origin, history, technology, and production. Wiley, Hoboken, pp 3–63

Winter P, Kahl G (1995) Molecular marker technologies for plant improvement. World J Microbiol Biotechnol 11:438–448

Wu SK (1939) The relationship between the origin of selfed lines of corn and their value in hybrid combinations. J Am Soc Agron. 31:131–40

Wu Y, Fox TW, Trimnell MR et al (2016) Development of a novel recessive genetic male sterility system for hybrid seed production in maize and other cross-pollinating crops. Plant Biotechnol J 14(3):1046–1054. https://doi.org/10.1111/pbi.12477

Xu Z, Hua J, Wang F, Cheng Z, Meng Q, Chen Y et al (2020) Marker-assisted selection of qMrdd8 to improve maize resistance to rough dwarf disease. Breed Sci 70:183–192

Yamada H, Kato N, Ichikawa M et al (2024) DNA- and selectable-marker-free genome-editing system using zygotes from recalcitrant maize inbred B73. Plant and Cell Physiol 65(5):729–736. https://doi.org/10.1093/pcp/pcae010

Yancheva S, Kondakova V (2018) Plant tissue culture technology: present and future development. In: Pavlov A, Bley T (eds) Bioprocessing of plant in vitro systems, Reference series in phytochemistry. Springer, Cham. https://doi.org/10.1007/978-3-319-54600-1_16

Yang N, Yan J (2021) New genomic approaches for enhancing maize genetic improvement. Curr Opin Plant Biol 60:101977. https://doi.org/10.1016/j.pbi.2020.11.002

Yang Q, He Y, Kabahuma M, Chaya T, Kelly A, Borrego E et al (2017) A gene encoding maize caffeoyl-CoA O-methyltransferase confers quantitative resistance to multiple pathogens. Nat Genet 9:1364–1372

Yang CJ, Samayoa LF, Bradbury PJ et al (2019) The genetic architecture of teosinte catalyzed and constrained maize domestication. Proc Natl Acad Sci USA 116(12):5643–5652. https://doi.org/10.1073/pnas.1820997116

Yassitepe JECT, da Silva VCH, Hernandes-Lopes J et al (2021) Maize transformation: from plant material to the release of genetically modified and edited varieties. Front Plant Sci 12:766702. https://doi.org/10.3389/fpls.2021.766702

Yathish KR, Karjagi CG, Gangoliya SS, Kumar A, Preeti J, Yadav HK et al (2022) Introgression of the low phytic acid locus (lpa2) into elite maize (*Zea mays* L.) inbreds through marker-assisted backcross breeding (MABB). Euphytica 218(9):127

Yuan Y, Zhang P, Wang K et al (2018) Genome sequence of the freshwater Yangtze finless porpoise. Genes 9:213. https://doi.org/10.3390/genes9040213

Yusnita, Widodo, Sudarsono (2005) In vitro selection of peanut somatic embryos on medium containing culture filtrate of Sclerotium rolfsii and plantlet regeneration. Hayati 12(2):50–56

Zapata C, Srivatanakul M, Park SH, Lee BM, Salas MG, Smith RH (1999) Improvements in shoot apex regeneration of two fiber crops: cotton and kenaf. Plant Cell Tissue Organ Cult 56:185–191

Zeder M (2015) Core questions in domestication research. Proc Natl Acad Sci USA 112(11):3191–3198. https://doi.org/10.1073/pnas.1501711112

Zhang S, Zhang H, Zhang MB (1996) Production of multiple shoots from shoot apical meristems of oat (*Avena sativa* L.). J Plant Physiol 148:667–671

Zhao Y, Wu Y, Engelmann F, Zhou M (2001) Cryopreservation of axillary buds of grape Vitis vinifera in vitro plants. Cryo Letters 22:321–328

Zwick MS, Islam-Faridi MN, Czeschin DGJ et al (1998) Physical mapping of the liguleless linkage group in Sorghum bicolor using rice RFLP-selected sorghum BACs. Genetics 148:1983–1992

Chapter 4
Oat (*Avena sativa* L.): Genetic Diversity, Breeding Innovations, Technological Progress, and Conservation Approaches

Sana Ben Mustapha, Donia Abdallah, Amel Salhi Hannachi, and Ghada Baraket

Abstract Oat (*Avena sativa* L.), commonly known as wild oats, belongs to the family *Poaceae.* This *genus is* characterized by variable ploidy levels with a main chromosome number of x = 7. Oat is one of the most cultivated cereal crops in the world. Historically, oat production mainly focused on the animal food market. Due to its composition rich in vitamins, minerals, and antioxidants, oat is considered a crucial part of the human diet and has gained a good reputation as a healthy and nutritious food. This crop represents significant genetic resources with ecological and agricultural importance. In recent years, the genus *Avena*, particularly *Avena sativa*, has been suffering from a loss of diversity and severe genetic erosion. Threatening the remaining germplasm. This chapter highlights the cultural value and contributions of the genus *Avena* to biodiversity. It also analyzes existing conservation strategies employed to preserve *A. sativa* genetic diversity, including "ex-situ" and "in situ" conservation methods, germplasm banks, and the use of molecular tools for precision breeding. The challenges of plant breeding lie in improving all traits of interest in order to produce varieties with no risk to consumers or the environment. However, numerous challenges hinder effective conservation efforts, including habitat loss, invasive species, climate change, and socio-economic factors. Oat cultivation practices have evolved to accommodate changing agricultural landscapes and demands. Innovative approaches are being adopted in crop research. Addressing these challenges requires interdisciplinary collaboration and innovative conservation approaches to ensure the long-term survival and sustainable use of *A. sativa* resources.

Keywords *Avena sativa* · Biodiversity · Biotechnology · Challenges · Conservation · Oat

---

S. B. Mustapha (✉) · D. Abdallah · A. S. Hannachi · G. Baraket
Department of Biology, Faculty of Sciences of Tunis, University of Tunis, Tunis, Tunisia
e-mail: amel.hannachi@fst.utm.tn; ghada.baraket@fst.utm.tn

K. F. M. Salem et al. (eds.), *Breeding and Biotechnology of Grass and Bast Fiber Crops*, Advances in Plant Breeding Strategies 12,
https://doi.org/10.1007/978-3-032-00403-1_4

4.1 Introduction

The world's nutritional status is influenced by many factors, including food availability, economic access to adequate food, and agricultural practices. Cereals play a crucial role in the world's diet and are of major importance for human nutrition. The Food and Agriculture Organization (FAO) plays an important role in addressing global food security challenges, including the cultivation of cereals, which are essential crops for sustaining human life. Cereals are essential food for billions of people around the world, providing a major source of calories and nutrients. The FAO also supports farmers by conducting research and technology transfer to reduce the effects of climate change on cereal crops and improve their resilience. It promotes agroecological approaches, conservation agriculture, and efficient resource management techniques to minimize the environmental footprint of cereal production. These efforts contribute to conserving biodiversity, reducing greenhouse gas emissions, and promoting ecosystem health. Cereals, including oats, have significant importance in agriculture. Oats, also known as *Avena sativa*, have a long history as one of humanity's oldest crops. Evidence of oats dating back to 10,500 BCE was discovered in the Franchthi caves in Greece (Youngs et al. 1982). Archives mentioning oats date back to the Roman occupation of Europe, while some sources suggest that oats were known to the ancient Chinese civilizations. Because they contain a wealth of nutrients (Vitamins, dietary fiber, proteins, and carbohydrates) that are valuable for human nutrition, animal feed, healthcare, and cosmetics, oats stand out from other cereals. This last year, the genus *Avena*, in particular *Avena sativa*, has been suffering the loss of diversity and severe genetic erosion, which also threatens the remaining germplasm due to various factors like monoculture farming, use of pesticides and herbicides, and land use changes. In this critical situation, it is important to establish conservation and management strategies to ensure food security, biodiversity, and sustainable agriculture. This chapter highlights the cultural significance and contributions of the genus Avena to biodiversity. It also explores current conservation tactics used to preserve genetic diversity in *A. sativa*. This includes both "ex-situ" and "in-situ"conservation methods, germplasm banks, and the use of molecular technologies for precision breeding.

4.2 Botanical Classification and Taxonomy of Genus *Avena*

Oats belong to the subfamily *Avenaea* of the family *Poaceae* and genus *Avena* L. It is a self-pollinated cereal crop that includes many species, both cultivated and wild, characterized by variable ploidy levels with a main chromosome number of x = 7 (Li et al.2000). *A. sterilis*, *A. byzantina*, and *A. sativa* are hexaploid

(2n = 6x = 42), while *A. strigosa* is diploid with 2n = 4 = 14. As described by recent literature, the genus *Avena* L. can be divided into two subgenera: *Avenastrum* C. and *Avena*. The subgenus *Avenastrum* includes the perennial, tetraploid species *A. macrostachya* Balan (2n = 4x = 28), whereas the subgenus *Avena* includes other wild and domesticated species (Loskutov 2016). Based on karyotype analysis and chromosomal pairs of the hybrid, four genomes (A, B, C, and D) have been identified in *Avena*. Genomes B and D have been derived from the A genome. Many researchers have shown that the C genome is considered one of the basic genomes in oats and is unchanged across all ploidy levels with only minor variation (Cp and Cv). The A genome developed independently from the C genome and produced many genome variants (Al, Ad, Ac, As), resulting in the cultivated diploid species (AA, CC) *A. strigosa.* The genesis of tetraploid species occurs either by doubling the chromosome number of one of the diploid species or through spontaneous hybridization of two closely related (AB = AA) diploid species, as A. *Abyssinica*, containing the AB genome. Cultivated Oat (*A. sativa*) is an allohexaploid with A, C, and D genomes. It evolves in three steps: first, two diploid species (AA and CC) are hybridized to create a tetraploid species; next, the number of chromosomes doubles, and a third diploid species is hybridized (Rines et al. 2006). This species is distinguished by its large and complex genome, which contains approximately 121 thousand protein-coding genes (Amosova Alexandra et al. 2024).

Different species within the genus *Avena* L. exhibit significant morphological and ecological diversity across various geographic regions. The description of the species in this genus began in the sixteenth century and continues to this day. Authors have described various species since the establishment of the nomenclature proposed by Linneaus (1753). In modern taxonomy, a detailed analysis of *Avena* species shows that those with the C and AC genomes which feature a bidentate lemma tip typical of hexaploid species in the *Avenae* section, are considered transitional ancestral forms (such as *A. ventricosa, A. canariensis, and A. magna*)Diploid species like *A. ventricosa* Balan, *A. bruhnsiana* Grun, *A. canariensis* Baum, and *A. agadiriana* Baum ET Fed, along with tetraploid species like *A. magna* Murph ET Terr., *A. murphyi* Ladiz, and *A. insularis* Ladiz, only have disarticulated spikelets and do not directly correspond to cultivated varieties. In contrast, polyploid plants are more advanced than diploids, exhibiting larger seeds, fruits, and storage organs (Ridley 1997).

The cultivated species *A. sativa L.* (common oat), characterized by white glumes, serves as the type species of the genus. Interestingly, Linnaeus originally described *A. sativa* with black glumes, regarding those with white glumes as a variant of the former, which he designated as "β" (Loskutov et al. 2020). Kole et al. (2011) identified four additional species related to *A. diffusa*: *A. brevis, A. hispanica, A. nuda, and A. lusitanica.*

4.3 Origins and Geographic Distribution of Genus *Avena*

4.3.1 Origins

Oats are a cereal crop that originated from Mediterranean regions, including the Iberian Peninsula, Northwest Africa, and Middle East countries such as Iran, Iraq, and Turkey (Stevens et al. 2004). The origin of the oats species has caused, until now, some confusion due to the presence of diploid and hexaploid cultivated of this species. The domestication of oat species and their forms from wild species is essentially shown by morphogeographic data, physiology, biochemistry, cytogenetics, and molecular biology.

According to research conducted by Maltsev in 1930, the western part of the Mediterranean regions, like the Atlas Mountains and the Pyrenees, is the primary center of origin for all diploid oat species. This region has the highest concentration of diversity of these forms. Igor et al. (2008) suggest that the axes of modern areas of distribution of all 14-chromosome species radiate towards the East. Two independent branches can be observed in the evolution of tetraploid oat species: those with the AB genome and those with the AC genome. Diploids with a variant of the A genome are the origin of the genomic structure of the species with the AB genome. *Avena barbata* Pott is an autotetraploid of one of the diploid species produced by chromosome duplication (Ladizinsky and Johnson 1972). According to Thomas (1995), the weed species *A. barbata* was introduced into Ethiopia through oat seeds and used to cultivate *A. abyssinica Hochst*. Since then, *A. barbata* has been invading oat fields in the region. The species *A. vaviloviana (Malz.) Mordv.* and *A. abyssinica Hochst*., according to several morphological characters, might be the relics of the ancient African floras (Baum 1971).

Scientific research indicates that A. magna Murph. & Fedis likely has a hybrid origin, resulting from the interaction between the distribution areas of two diploid species: A. canariensis Baum and A. ventricosa Balan. This hybridization led to the formation of an allotetraploid species. Coffman (1977) demonstrated that the progenitor of the hexaploid forms was A. sterilis L., which originated in Asia. This species is believed to have contributed to the creation of the cultivated species A. byzantina C. Koch and A. fatua L., the latter of which is considered a weed in agricultural crops.

Furthermore, a close genetic relationship was identified between A. byzantina and A. sterilis accessions from Northern Mesopotamia, as well as between A. sativa L. and A. sterilis from Eastern Anatolia, according to research on chromosome translocations in oats and the geographical distribution of hexaploid forms (Loskutov I et al. 2008). Additionally, studies have shown that two independently domesticated species, A. sativa and A. byzantina, exist (Jellen and Beard 2000). Thomas (1995) proposed that the hexaploid originated in South-Western Asia and later spread to China. Loskutova et al. (2021) reviewed multiple studies and published findings based on the identification of oat species using various genetic markers in molecular genetics investigations.

4.3.2 Geographic Distribution

The geographical distribution of *Avena L.* was studied through field investigation and review of specimens and literature. All 29 species of *Avena* are found across Europe, the Mediterranean, Northern Africa, Western Asia, and America. Europe is the homeland of oats, where they grow the most; North America ranks second, and Africa is the last.

In China, four species of *Avena* are distributed in high-altitude areas in the Northern, Northwestern, and southwestern provinces.

4.4 Production

In the past, oat production has primarily focused on the animal feed market. Around 75% of oat grain is still used for animal feed worldwide (Webster and Wood 2011). In recent decades, researchers have confirmed that oats are a healthy food with high nutritional value for humans. Oats rank sixth or seventh in the world in terms of area and importance, following wheat, maize, rice, barley, and sorghum (Kumar et al. 2023). In countries like Finland, oats are the second most important crop, while in Bulgaria, they rank fifth. While oats are primarily grown for their grains, they can also be used as a whole crop for green forage or silage. Oats are native to Europe, where they are cultivated across large areas. North America comes second in oat production, with Africa ranking last. In Europe, the largest producers of oats are Russia, Poland, Ukraine, Finland, and Spain. In 2022, global oat cultivation covered 9,508,645 hectares, yielding a total output of 26,385,330.04 tons (FAO 2023) (Fig. 4.1).

4.5 Oat Development

Oat cultivation follows a similar pattern to wheat and barley, with both spring and winter varieties grown in regions like the UK, Portugal, and the Mediterranean with mild winters. However, unlike wheat and barley, oats are less tolerant to frost and typically cannot survive temperatures below minus 14 °C. Studies have shown that only 64% of overwintering oat plants (*Avena sativa*) are suitable for the winter pea variety Pleven 10. During flowering, high temperatures can have a negative effect on oats. For instance, high temperatures at 35–38 °C reduce oats‘hardiness compared to wheat.

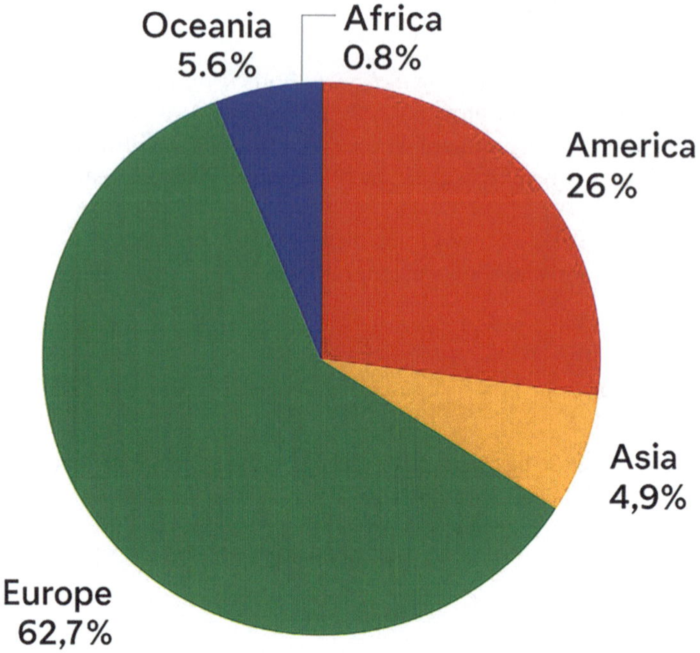

Fig. 4.1 Production share of Oats by region (1994–2022). (Source: FAO Stat 2024)

4.6 Oat Composition and Nutritive Values

Oat is one of the most nutritious cereal crops with a highly nutritious plant cultivated globally. The eighteenth-century writings described it as a principal food of horses and cattle in England and other parts of Europe. From 1500 to 1700. It was considered a crucial part of the human diet and gained as well, a good reputation as a healthy and nutritious food.

The components of the oat grain are the pericarp, testa and aleurone (9%), starchy endosperm (63%), hull (25%), and embryo (3%). Oats are high in protein and could vary from 13% to 20%. In addition, 30% occur in the embryo (Paudel et al. 2021). The dominant protein is a venalin, and cannot be found easily in any other type of grain.

Oats also contain phenolic compounds, sterols, and other antioxidant compounds that have beneficial health effects. Numerous vitamins, minerals, and special plant compounds can be found in it (Table 4.1).

Table 4.1 Health benefits and nutritional composition of oats

Vitamins and minerals contained in Oats	Effects	References
Manganese	Development, growth and metabolism	Aschner and Dorman (2006)
Phosphorus	Bone health and tissue maintenance	Takeda et al. (2012)
Copper	Antioxidant mineral Heart health	Allen and Klevay Copper (1994)
Vitamin E	Antioxidant Preventing premature aging, chronic diseases, cancer and strokes	Skoglund et al. (2008)
Vitamin B1		
Selenium	Antioxidant Reduced selenium levels are linked to a higher chance of dying young and impaired mental function.	Rayman (2012)
Magnesium	Important for numerous processes in the body (cellular processes, DNA replication)	Blaszczyk and Duda Chodak (2013)
Zinc	Participate in many chemical reactions in the body Important for overall health	Prasad (2009)
Avenanthramides (AVA,AVC)(exclusive in oats)	Power Antioxidant AVC derived from germinated oats has been proposed as a potential therapy for mast cell-mediated allergic inflammation by inhibiting the release of inflammatory mediators	Dhakal et al. (2019)
Ferulic Acid	High antioxidantactivate	Soycan et al. (2019)
Phytic Acid	Inhibit the generation of metal-induced free radicals	Soycan et al. (2019)

4.7 Oats Utilities

In addition to being high in vitamins, minerals, and antioxidants, oats are a source of fiber, especially beta-glucan (β-glucan). In fact, oat husks have lignocellulose fiber in their composition, which is formed by cellulose, hemicelluloses, and lignin. De Oliveira et al. (2017) extracted cellulose fiber from oats to produce hydrogels. These are a three-dimensional network of hydrophilic polymers that can absorb some of the water. According to Chang et al. (2010), several fields like the pharmaceutical, medical, agricultural, and food industries use hydrogels because of their properties.

Another study by Borrega et al. (2022) discusses the business opportunities and system-level requirements of using oat traw as an alternative source of lignocellulosic fibers that support a circular economy. On the other hand, β-glucan in humans is involved in natural defense against infections caused by viruses, bacteria, and fungi, in addition to reducing the glucose and cholesterol levels in the blood (Knezevic 2014). Besides, the β-glucan in oats differs from that of barley or wheat in many physicochemical properties, which makes it more efficient in the gastrointestinal tract and reduces blood glucose and insulin response.

Also, oats are recognized for their potential in preventive and therapeutic health care, like coronary heart disease (Nwachukwu et al. 2015). According to Gorash et al. (2017), the industry uses isolates like protein and starch for cosmetics, β-glucan and antioxidants as food additives, and lipids for food and pharmaceutical supplements. Additionally, as the number of vegetarians rises, oat milk has emerged as a vegan substitute for dairy milk. People who are lactose intolerant or allergic to dairy products can choose oat milk because it is a sustainable option with lower greenhouse gas emissions during production. Oats are popular for their high protein content (around 12–20%), high digestibility (about 90–94%), and high amounts of fiber, such as about 4–8% beta-glucan. For this reason, it is applied to food industries, including flakes, porridge, cereal bars, cookies, and noodles. In agriculture, oats serve multiple purposes, such as silage, pasture, or hay, contributing significantly to sustainable farming practices. These grass species offer forage during winter months when warm-season grass production is limited by freezing temperatures (Kim et al. 2017).

4.8 Germplasm Biodiversity and Conservation

4.8.1 Germplasm Diversity

The genus *Avena* comprises colossal and diverse species characterized by the coexistence of both cultivated and wild varieties. There are several species of oats, including domesticated species *Avena sativa* (white oat), *Avena byzantina* (red oat), *Avenastrigosa, Avena abyssinica,* and *Avena nuda* L., all of which are annuals. *Avena macrostachya* is a perennial species. Winterfeld et al. (2009) considered this last as an 'intermediate taxon between *Avena* and *Helictotrichon*. This suggests that *A. macrostachya* may exhibit characteristics or genetic similarities that place it between the two genera. The naked oats are the most widely cultivated of all the species. Nowadays, some of the wild oat species are rare in their natural habitats and threatened by genetic erosion or even extinction (Boczkowska et al. 2016).

According to the FAO, the South and East Mediterranean sub-region, East Asia, and Europe are considered regions of diversity for Avena spp. In contrast, Loskutov (2001) identified three primary centers of diversity for wild diploid and tetraploid Avena species: the western Mediterranean (including Spain, Morocco, and Algeria), the Middle East, and the Caspian Sea region. For cultivated varieties, Loskutov (2005) outlined seven centers of diversity: Spain and Portugal (A. strigosa), Great Britain (A. nuda L. = A. strigosa subsp. nudibrevis), Ethiopia (A. abyssinica), Algeria and Morocco (A. byzantina), Iran, Georgia, and Tatarstan (hulled forms of A. sativa), and finally, Mongolia and China (hull-less forms of A. sativa).

4.8.2 Cultivars Characterization and Phylogeny

Cultivar identification is essential to maintaining the quality of oat production and the interests of producers. Morphological, physiological, biochemical, and genetic characteristics are the main pillars of traditional cultivar discrimination techniques. Morphological characteristics like seed shape, size, color, and field seedling growth characteristics are typically used to identify cultivars. The limitations of phenotypic characterization are reflected in the high similarity and abundance of new oat cultivars. Thus, molecular detection has been regarded as a highly accurate and successful technique for classification and identification (Saccomanno et al. 2020).

Kumar et al. (2023) studied twelve agro-morphological traits of thirty-eight oat germplasms using twenty-two inter-simple sequence repeats (ISSR) markers. The ISSR markers are dominant, so phenotypes can be differentiated at each locus in a Mendelian pattern. One hundred fifty-eight discriminative alleles were found and high-yielding genotypes were identified using DNA fingerprint-based principal component analysis. On the other hand, eight cultivars were identified as superior under high grain yield.

However, the study of genetic diversity in the *Avena sativa* species has been subject to many other concerns for a long time. Many oat accessions have been classified according to simple sequence repeat (SSR) (Pipan et al. 2021) and expressed sequence tag (EST) (Montilla-Bascón et al. 2013) techniques based on genetic distance. Additionally, molecular markers are also used for the identification and characterization of oats, like restriction fragment length polymorphism (RFLP) and amplified fragment length polymorphism (AFLP) (Paczos-Grzeda 2004). Also, the selection of cross-compatible oat germplasm by molecular markers can be used to incorporate new traits into elite oat backgrounds. Due to the continuous development and application of molecular marker technology in cultivar identification, new techniques have a growing interest, like imaging techniques, NIR spectroscopy or precise remote sensors to address the constraints associated with traditional methods. The image analysis technology can provide an alternative to manual identification based on the measurement of the geometric shape of a single seed (Sumathi and Balamurugan 2013). Near-infrared spectroscopy (NIRS) is utilized for analyzing the chemical composition of the seed (both surface and inside the seed) (Boelt et al. 2018). Fu et al. (2023) categorize 16 oats cultivars according to their morphological characteristics and spectral attributes combined with multivariate analysis. Their results illustrate a clear difference among oats accessions and clear differences can easily be visualized through multispectral imaging.

4.8.3 Genetic Resources Conservation

The preservation of biological diversity and exploiting its components for human benefit is one of the major challenges facing the world today. The "ex-situ" conservation of biological materials of plants is seen to be a more successful alternative than on-farm or "in-situ" conservation, or conservation in the natural areas in which they have evolved. For several reasons (changing lifestyles, food habits, and an increase in demand for oats), oats should take a prominent place in future crop improvement programs. However, the loss of genetic diversity observed within cultivated and semi-cultivated species, worldwide collections of *Avena* species consist of approximately 131,000 accessions preserved by 63 countries.

Currently, there are 4 cultivated species of oats at the N.I. Vavilov Institute of Plant Industry and 22 wild species. There are more than 11,700 cultivated and 2000 wild oat accessions there. This collection presents the global genetic diversity of *Avena* genus with a large variability of morphological, agronomical, and biochemical characters for oat breeding. In addition, *Avena* is one of four model crops represented in Europe for the European Genebank Integration Strategy (AEGIS), as well as representing the European regional strategy. Spain is a European country that holds a diversity of oats. It currently monopolizes the third largest land area behind barley and wheat in Spanish agriculture, with an increasing annual production since 1996.

The Regional Strategy for the Conservation of Plant Genetic Resources in West Asia and North Africa highlights that oats rank third in importance among cereals, behind barley and wheat. The region's ex-situ collection consists of four major holdings, with a total of 4.089 oat accessions.

According to Germeier (2008), the regional strategy reports the following oat accessions: 2133 at INRA Morocco, 814 at NGB-AARI Turkey, 540 at NARC Pakistan, 500 at NGB Iran, 30 at MOA Tunisia, 20 at NGB Egypt, 20 at GCSAR Syria, 18 at ARC Libya, nine at INRA Algeria, three at AREA-NGRC Yemen, one at NCARTT Jordan, and one at MAF Oman. Despite the Maghreb's rich oat genetic resources, the region heavily depends on introduced varieties for the sustainability and viability of its livestock production. Financial constraints in Algeria, Tunisia, and Morocco have led to a shortage of staff to manage these resources.

The government does not prioritize oats as a crop and is reducing financial support for *Avena* collections. Extreme genetic erosion is expected due to climate change.

Genetic diversity studies, conducted using various tools such as DNA markers, offer valuable information for genetic conservation and the effective development of new commercial varieties.

4.9 Current Cultivation Practices and Challenges

In the past, the oat crop has often been neglected and cultivated in areas less suitable for crops. In recent years, there has been a growing interest in oat cultivation because of its dietary benefits and potential therapeutic applications for human health. Despite the widely acknowledged benefits, the oat crop's status remains precarious for several reasons:

- The cultivation area is the smallest compared with other cereals and commercial efforts in oat breeding are less.
- Some areas distinguish seasonal changes and offer both opportunities and challenges for oat production. For example, in the Southern United States or Australia, oat production faces challenges related to hydric stress, which can affect oat yields and qualities. In contrast, regions with high altitudes and cool temperatures (such as South America or the Himalayas) create favorable conditions for oat cultivation.
- Oats are susceptible to various pathogens (crown rust, for example), which can cause damage to the root system.
- Limited land due to increasing urbanization and environmental pollution.

A key factor in maintaining a stable level of oat production and increasing grower interest is increased productivity with reduced risk. This can be achieved through appropriate oat cropping systems utilizing good agronomic practices and cultivation techniques. Alternatively, oats can be considered a viable break/specialty crop with clear market incentives.

Oat cultivation practices have evolved to accommodate changing agricultural landscapes and demands. Innovative approaches are adopted in crop research. Implementing and adopting technologies in agriculture can have many advantages and can facilitate many tasks. An intelligent agriculture system can collect and process big data while determining and initiating the best course of action.

Cutting-edge technology in agriculture plays an important role in improving, in a non-invasive manner, the efficiency of detecting environmental parameters and accurately assessing complex traits in plants (Tripodi et al. 2022). Utilization of digital technologies, artificial intelligence (AI) and nanotechnology in agricultural practices, particularly managing water, energy, and food resources, can significantly contribute to sustainable development efforts (Dayioğlu and Türkera 2021). Here are some use cases for AI in agriculture.

- Optimisionsautomate irrigation system.
- Detecting leaks or damage to the irrigation system.
- Monitoring livestock health.

Also, the agriculture sector is adopting genome editing, smart breeding technology and digital AI with microbial soil mapping to enhance crop quality and develop seeds with pest resistance (Javaid et al. 2022). All these technologies hold the promise of revolutionizing agriculture towards Agriculture 4.0. Farmers use Agriculture 4.0 to adopt a more organized approach to cultivating with a high level of precision (Ferrag et al. 2021).

4.10 Traditional Breeding Methods and Biotechnology

4.10.1 Traditional Breeding Methods

Developing new cereal cultivars that can resist biotic and abiotic stresses is a constant challenge in crop breeding. The selection of variants in the wild and cultivated farms was the most primitive method of breeding selection. With the discovery of the laws of heredity, the importance of hybridization in plant breeding was recognized. The traditional approach to breeding oats centered around three categories: Introduction, Selection, and Hybridization. The main goal of this process is to obtain high quality seeds with good yields, which resist lodging and diseases. Some diploid and tetraploid species seem to have better agronomic traits that would be valuable if incorporated into domestic hexaploid oats (Brown and Petterson 1992). Following the initial hybridization of the parents, it was common to use backcrossing, pedigree selection, single-seed descent, bulk method, multilink breeding, and other current selection techniques. Genotype evaluation takes place in both field and laboratory settings. Biochemical, physiological, and biometric analyses and evaluation tests were conducted. Increasing β-glucan levels in oats has been a part of a variety of breeding programs. The economic and nutritional value of oats can be enhanced by developing varieties with higher β-glucan content. In Tunisia, over a period of 10 years, 11 oat genotypes were selected for further evaluation based on their resistance to foliar diseases, height, maturity, and potential for forage and grain yield. These selected lines include: Av.3, Av.14, Av.72, Av.77, Av.81, Av.86, Av.88, Av.89, Av.90, Av.93, and Av.95 (Hammami et al. 2008).

4.10.2 Traditional Breeding Limitations

Numerous crops, including peanuts, common beans, safflower, chickpeas, wheat, tall fescue, soybeans, wheatgrass, barley, and maize, have historically been produced by traditional breeding methods. In this conventional approach, DNA recombination occurs through separate chromosomal assortment and crossing-over during the sexual cycle, utilizing existing genetic diversity. Continuous application of traditional breeding methods within specific species may result in the reduction of genetic diversity available for developing cultivars. This could make crops more susceptible to both biotic and abiotic stresses, and impact future advancements.

Population breeding involves successive rounds of selection and recombination. However, it relies on an efficient crossing system, which can pose challenges for self-pollinating species. This method, especially when applied repeatedly in small populations, can deplete genetic variation and reduce genetic progress. To prevent this, it is essential to maintain a large effective population size to prevent genetic drift from outweighing other forces driving genetic improvement.

In order to create cultivars, plant breeders cross regionally adapted germplasm with desirable agronomic traits to create populations. In some areas, breeders frequently exchange lines and cultivars, but this is less common across continents. Therefore, a relatively small gene pool may be used by European spring oat breeding programs to produce both agronomic and quality characteristics (Koroluk et al. 2022). However, through years of selection, valuable alleles have been lost, and modern breeding explores new germplasm in wild ancestors and landraces (Boczkowska et al. 2004). However, the oat genome presents challenges due to its large and complicated nature, and despite advancements in DNA sequencing, molecular genetics research in oats still significantly lags behind that of other cereal crops (Hilli and Kapoor 2023).

4.10.3 Role of Biotechnology

Biotechnology provides an alternative approach to oat breeding, utilizing in vitro cultivation of mature embryos, the selection of somaclonal lines, and the identification of regeneration with enhanced cold and drought resistance. Table 4.2 displays experimental data on advanced breeding techniques aimed at improving cold resistance, conducted at the Institute of Agriculture in Karnobat, Bulgaria.

Cereal has recently been biotechnologically improved through the application of in vitro regeneration techniques. *Avena sativa*, like other *Poaceae* species, has the characteristic of tissue regeneration. Many researchers are currently studying it in cell culture and somatic embryogenesis. Chen et al. (1995) analyzed regeneration from leaf tissues of six different oat cultivars and presented them as Coolabah, Cooba, Blackbutt, Mortlock, Victorgrain, and HVR. Zhang et al. (1996) utilized shoot apical meristems of *A. sativa* to create an efficient micro-propagation method.

4.10.3.1 Double Haploid Technology

The discovery of theories of totipotency led to the introduction of haploid and double-haploid (DH) plants in breeding programs. For oat species, the generation of DH has become an essential tool and opens novel possibilities for developing new techniques

Table 4.2 Advanced breeding lines with high cold resistance

AdvancedBreedingLines	Cold Resistance	%
CR 619-4	−11.6	110
CR 645-3 3Y	−11.6	108
CR 619-4	−11.6	106
3У 757-4-3	−10.9	104
Z 501-1-8	−11.6	103

Source: Savovatodorka et al. (2014)

to minimize the time needed for cultivar production (Warcol et al. 2023). Multiple studies have focused on employing DH technology in oats through the utilization of anther and microspore culture techniques. The resulting homozygosity in DH plants offers a clear manifestation of genotypes, making them valuable for genetic analyses and trait evaluation (Pathi and Sprink 2023). The first successful anther culture was described by Kiviharju et al. (1997) using the wild, hexaploid, highly self-pollinating oat species *A. sterilis*. The researchers confirmed that this oat species is capable of producing embryogenic structures that can give rise to both green and albino plantlets. Additional studies were conducted to examine the effects of Zinc, Copper, and Silver ions, in combination with cold treatment, on androgenesis in oats. The findings indicated that modifying the concentrations of copper, zinc, and silver in the culture medium can enhance androgenesis in oats. The highest number of embryo-like structures was observed in anthers treated with CuSO4 during pretreatment or when AgNO3 was added to the induction medium (Warchoł et al. 2021).

4.10.3.2 Tissue Culture

The in vitro cultivation of plant cells presents numerous applications, further enhanced by the totipotent nature of plant cells, which greatly expands their potential in plant biotechnology. Originating in the early 1960s, plant cell culture has evolved into a fundamental aspect of modern biotechnology. In oat, many efforts have been made to establish the effective utilization of plant tissue cultures and genetic transformation.

For a long time, researchers conducted studies on the impact of Indole-3-Acetic Acid (IAA) on cell division and cell enlargement of *Avena* internode (Kaufman. 1965) and the impact of high auxin medium for callus formation from *A. sativa* hypocotyle (Brenneman and Galston 1975). Later, Shekhawat et al. (1984) described a method of totipotent culture from young secondary tillers of plants aged 60–75 days, employing small explants. Several studies have been carried out successfully, allowing to induce of callus and plant regeneration from six genotypes of oat leaves (Chen et al. 1995).

Kelley et al. (2004) explored the regenerative potential of seven oat genotypes by employing three culture techniques: callus, multiple meristems, and multiple meristem-to-callus combinations. In the same year, Jung-Hung et al. regenerated Korean oat accessions from mature embryos and leaf base segments. Al Mamun et al. (2020) reported, in vitro, oat callus induction under NaCl salt stress as well as the subsequent regeneration of plants.

4.10.3.3 Wide Hybridization

Wide hybridization and inbreeding involve crossing two parents from different species or genera. This process can lead to the elimination of one parental genome during early embryonic divisions, potentially causing developmental failure. It serves as a valuable technique for generating double haploid plants in plant breeding programs. Haploid individuals of cultivated oats (A. sativa L.) have been successfully produced through wide hybridizations with species from the Panicoidae family, particularly maize (Zea mays) (Pathi and Sprink 2023). The cross between oat and maize results in the retention of maize chromosomes in the recovered oat plant. Additionally, these hybridizations can promote partial self-fertility in haploid oat plants. However, the low germination success of oat haploid embryos may be attributed to the excessive production of reactive oxygen species, reduced levels of low-molecular-weight antioxidants, and stress hormones (Pathi and Sprink 2023).

4.10.3.4 Genetic Transformation

In order to prove genetic transformation in oats, numerous studies were conducted. For instance, Somers et al. (1992) used a biolistic gene transfer technique to create a transgenic oat plant. As selection markers, the BAR (phosphonothricin) and Escherichia coliuidA (GUS) genes were used; the former gave the plants resistance to herbicides. Later, Zhang et al. (1999) used shoot meristematic culture to transform oat plants, resulting in seven distinct transgenic lines, five of which were self-fertile. An Agrobacterium-mediated transformation system was described by Dattgonde et al. (2019), who also looked at various co-cultivation techniques with various incubation times.

4.11 Molecular Breeding

Traditional oat breeding methods rely on phenotypic traits such as yield, disease resistance, and quality characteristics. However, molecular breeding integrates molecular markers, genomics, and other advanced biotechnology tools to improve the efficiency and precision of oat breeding programs. Molecular techniques such as marker-assisted selection (MAS), genomic selection, and molecular trait loci (QTLs) mapping offer a more precise and efficient way to select desired characteristics. Breeders can develop varieties rapidly by identifying specific regions of the oat genome associated with traits like disease resistance and nutritional quality.

The first focus of the resistance-breeding program was to supply genes for disease(s) resistance, insect pests, and nematodes. Previous studies have shown that the presence of minor gene additive APR to PCA (resistance genes of crown rust in oat) can be combined within the hexaploid oat to give a high level of resistance (Park et al. 2022). The region of interest was first identified on a consensus map in 2018, either directly or through comparative mapping using the CMap tool in the Grain Genes database. The analysis of candidate genes was performed for the CsIF gene (3–12) for β-glucan, the ACC gene for oil and the heading-related genes VRN1, VRN3, and CO1 (Tinker et al. 2022).

Increasing yield is a goal of the crop-breeding program. Herrmann et al. (2014) used AFLP markers in a backcross population for genetic mapping of yield-related traits, particularly ßglucan content in grain. They discovered several genetically mapped agronomic traits, including six thousand grain weights. Similarly, in 2015, Gaoyuan et al. (2015) investigated two hexaploid naked oat hybrids using SSR markers to analyze grain length and width in an F_2 population. They conducted QTLs mapping on thousands of grain weights and identified a total of 17 genes associated with grain-related traits. Genome-wide association Analysis (GWAS) was performed, resulting in the identification of genes related to lodging. In general, Marker-assisted breeding (MAB) is a valuable tool for breeders, offering efficient and rapid selection based on DNA markers. The availability of diverse DNA markers and genetic maps enables MAB to be applied to traits influenced by both qualitative and quantitative genes. This approach streamlines breeding efforts and provides more precise and targeted breeding strategies. Technological advancements in DNA sequencing have significantly transformed genomic research in oats, addressing previous limitations such as the high polymorphic DNA markers and the complexity of the oat genome (Rines et al. 2006). This progress has considerably enhanced the application of molecular markers, enabling the discovery of genes influencing quality and adaptation in various crop species. The DArT genotyping by sequencing (DArTseq™) is a more informative and widely used technique based on next-generation Illumina short-read sequencing (Koroluk et al. 2023). DArTseq-derived SNPs were used to genotype breeding lines of Polish oats, from the three largest Polish breeding companies, to assess the genetic relatedness and population structure of Polish breeding materials (Koroluk et al. 2023). Indeed, Szerszunowicz et al. (2017) used Computer analysis of protein maps obtained from the separation of proteins with two-dimensional polyacrylamide gel electrophoresis (2-DE) combined with mass spectrometry (MS) and bioinformatics analysis for the identification of oat proteins.

Morcia et al. (2024) applied an innovative and sensitive chip-based digital PCR, a third-generation PCR technology, to detect some fungal species in oat grains, and compared its performance with a qPCR method.

In Oher hand, Crop-specific databases are very useful in explicating the biology of a particular crop using data from several bioinformatics platforms. Given the agronomic importance of oats, more emphasis should be placed on developing crop-specific databases that may offer additional options in breeding programs.

4.12 Mutagenesis and TILLING

Throughout the twentieth century, mutagenesis played an important role in creating vast genetic variability and continues to play a significant role in plant breeding programs worldwide. Mutagenesis is defined as the introduction and isolation of a mutation in the organism. The objective of mutagenesis in breeding is to induce maximum genomic variation while minimizing the impact on plant viability.

Chemical mutagens have become widely used due to their ease of application, lack of need for specialized equipment, and ability to produce a high mutation frequency. Unlike radiological methods, chemical mutagens typically cause single base-pair changes or single-nucleotide polymorphisms (SNPs) rather than large deletions or chromosomal translocations. Other mutagens, such as Azide (NaN3) and MNU (Methyl-NitrosoUrea), are also used in both diploid and hexaploid oats (Rines 1985), often in combination as an Az-MNU solution.

With the development of TILLING (Targeting Induced Local Lesions in Genomes) technology, an allelic series of genes for a particular trait of interest can be generated. This technique is a reverse genetic approach that utilizes traditional mutagenesis to discover spontaneous mutation. In fact, in the last few years, the development of the TILLING population within hexaploid oats, exhibiting a moderate mutation frequency, would be a successful approach for the breeding tool (Chawade et al. 2010). Chawade et al. (2010) developed 2550 different mutagenized lines in the Swedish oat *AsPAL1* and *AsCslF6* genes—key genes in the lignin and β-glucan biosynthetic pathways, respectively. The sequences for AsPAL1 and As CslF6 were submitted to the gene bank (accession GQ373155 and GQ379900, respectively) and the mutagenized population will be submitted to NordGen when enough seeds from propagation are available. Next-generation sequencing (NGS) technology was used to detect mutations at the whole-genome level.

4.13 Advances in Oat Genomics

The genome size of *Avena Sativa* is 1C = 12.8 Gb, comprising 21 chromosome pairs originating from three distinct ancestral genomes. Oliver et al. (2013) were the first to develop the initial consensus map of oats. The construction of this map was based on 985 SNPs assessed across 390 recombinant inbred lines (RILs) originating from six hexaploid oat biparental populations. Yan et al. (2016) performed high-density marker profiling and genome analysis across 27 oat species. Their study successfully determined the ancestral origins of the 21 mapped chromosomes within hexaploid oats. Genotype by Sequencing (GBS) marker analysis by Latta et al. (2019) provides additional information in the maps of the diploid A genome and the tetraploid AB genome. For hexaploid oats, breeding is hard to breed due to their large genome size, which is highly repetitive, and because of numerous significant intra- and intergenic rearrangements. For open-source applications, *A. sativa* had its first genome published in 2020 (https://wheat.pw.usda.gov/GG3/graingenes_downloads/oat-ot3098-pepsico). First, *A. saiva* line OT3098 was published (Graingene), followed by the cultivar Sang (Kamal et al. 2022). In May 2021, scientists from China and the UK collaborated to create the diploid draft genome sequence of *A. strigosa*, a diploid oat species. These findings represent a considerable advancement in oat genomics research and will open the way for a new opportunity in genetic improvement of traits of interest in oats, including high-throughput mapping and cloning of crown rust resistance genes.

4.14 Conclusion and Prospects

Oat production and productivity have fluctuated over the years. The impacts of climate change on oat production are a direct consequence of his change. In the short term, some countries in northern Europe may even witness an increase in oat yields. Although oats can be grown under a wide range of adverse environmental conditions, they make a good break crop in crop rotation cycles. Elevated CO_2 levels are detrimental to both yield and quality.

Strategies for sustainable agriculture are essential for the welfare of the environment and the economic viability of agriculture in the future. Numerous studies on oat germplasm, particularly wild species, could provide valuable sources of new practical genes for breeding. The genetic pool of A. *sativa* cultivars is narrow, and expanding it is crucial. However, molecular biology and genomics support modern breeding techniques. In the case of oats, this area is significantly less developed than in other cereals. Sequencing the oat genome would create new opportunities for breeders to generate masses of sequence-based markers such as SNPs. These markers will help identify genes for a particular trait of interest by exploring linkage disequilibrium mapping and genomic selection techniques.

Acknowledgments The authors would like to thank all those who have contributed to this work.

References

Al Mamun SM, Islam M, Alam NEK, Noushin S (2020) In vitro Callus Induction under NaCl salt stress and subsequent plant regeneration in Oats (*Avena sativa* L.). Plant Tiss Cult Biotechnol 30:253–266

Allen KG, Klevay Copper LM (1994) An antioxidant nutrient for cardiovascular health. Curr Opin Lipidol 5(1):22–28. https://doi.org/10.1097/00041433-199402000-00005

Amosova Alexandra V, Gnutikov AA, Rodionov AV et al (2024) Genome variability in artificial allopolyploid hybrids of *Avena sativa* L. and Avena macrostachya Balansa ex Coss. et durieu based on marker sequences of satellite DNA and the ITS1–5.8Sr DNA Region. J Mol Sci 25:5534. https://doi.org/10.3390/ijms25105534

Aschner M, Dorman DC (2006) Manganese: pharmacokinetics and molecular mechanisms of brain uptake. Toxicol Rev 25:147–154. https://doi.org/10.2165/00139709-200625030-00002

Baum BR (1971) Taxonomic studies in Avena abyssinica and A. vaviloviana, and related species. Canadian J of Botany.49:2227–2232. https://doi.org/10.1139/b71-311

Blaszczyk U, Duda-Chodak A (2013) Magnesium: its role in nutrition and carcinogenesis RoczPanstwZakl Hig.64(3):165–71

Boczkowska M, Łapinski B, Kordulasinska I et al (2004) Promoting the use of common Oat genetic resources through diversity analysis and core collection construction. PLoS One 11:e0167855

Boelt B, Shrestha S, Salimi Z, Jørgensen J, Nicolaisen M, Carstensen J (2018) Multispectral imaging—a new tool in seed quality assessment? Seed Sci Res 28(3):222–228. https://doi.org/10.1017/S0960258518000235

Borrega M, Hinkka V, Hörhammer H, Kataja K, Kenttä E, Ketoja JA (2022). Utilizing and valorizing oat and barley straw as an alternative source of lignocellulosic fibers materials. 15:7826. https://doi.org/10.3390/ma15217826

Brenneman FN, Galston AW (1975) Experiments on the Cultivation of Protoplasts and Calli of Agriculturally Important Plants I. Oat (*Avena sativa* L.). Biochem Physiol Pflanz 168:453–471

Brown JA, Petterson DS (1992) Oats for animal feed. In RW Welch (Ed), The oat crop: Production and utilization. 289–310. London: Chapman and Hall

Chang C, Zhang L, Zhou J, Zhang L, Kennedy JF (2010) Structure and properties of hydrogel prepared from cellulose in NaOH/urea aqueous solutions. Carbohydr Polym 82:122–127

Chawade A, Sikora P, Bräutigam M, Larsson M, Vivekanand V, Nakash MA, Chen T, Olsson O (2010) Development and characterization of an Oat TILLING population and identification of mutations in lignin and β-glucan biosynthesis genes. BMC Plant Biol 10:86. http://www.biomedcentral.com/1471-2229/10/86

Chen H, Xu G, Loschke DC et al (1995) Efficient callus formation and plant regeneration from leaves of Oats (*Avena sativa* L.). Plant Cell Rep 14:393–397

Coffman FA (1977) Oat history, identification and classification. USDA-ARS, Washington

Dattgonde N, Tiwari S, Sapre S, Gontia-Mishra I (2019) Genetic Transformation of Oat Mediated by Agrobacterium is enhanced with Sonication and Vacuum Infiltration. Iran J Biotechnol 17:563–573

Dayıoğlu, Türker (2021) Digital Transformation for Sustainable Future - Agriculture 4.0: A review. J of Agricultural Sciences (Tarim Bilimleri Dergisi). 27(4):373–399

De Oliveira JP, Bruni GP, Lima KO, El Halal SLM, Silveira G, da Rosa ARGD, da Rosa Zavareze E (2017) Cellulose fibers extracted from rice and Oat husks and their application in hydrogel. Food Chem 221:153–160

Dhakal H, Yang EJ, Lee S et al (2019) Avenanthramide C from germinated oats exhibits anti-allergic inflammatory effects in mast cells. Sci Rep. 9:1–11

FAOSTAT (2024) FAOSTAT Database. Rome: Food and Agriculture Organization of the United Nations. https://www.fao.org/faostat/en/#data

Ferrag MA, Shu L, Djallel H, Choo KKR (2021) Deep learning-based intrusion detection for distributed denial of service attack in Agriculture 4.0. Electronics 10(11):1257

Food and Agriculture Organization of the United Nations (FAO) (2023) Statistical Yearbook 2023: World Food and Agriculture. Rome, Italy: FAO

Fu X, Bai M, Xu Y, Wang T, Hui Z, Hu X (2023) Cultivars identification of Oat (*Avena sativa* L.) seed via multispectral imaging analysis. Front Plant Sci 14:1113535. https://doi.org/10.3389/fpls.2023.1113535

Gaoyuan Song, Pengjie Huo, Bin Wu, Zongwen Zhang (2015). A genetic linkage map of hexaploid naked oat constructed with SSR markers. The Crop Journal 3:353–357. https://doi.org/10.1016/j.cj.2015.01.005

Germeier CU (2008) Global strategy for the ex situ conservation of Oats (*Avena* spp.). Federal Centre for Breeding Research on Cultivated Plants (BAZ), Quedlinburg

Hammami I, Allagui MB, Chakroun M, El-Gazzeh M (2008) Agronomic characterization of tunisian spontaneous Oat accessions resistant to Oat crown rust and potential in plant breeding. Tunis J Plant Prot 3:1–9

Herrmann MH, Yu JZ, Beuch S, Weber WE (2014) Quantitative trait loci for quality and agronomic traits in two advanced backcross populations in Oat (*Avena sativa* L.). Plant Breed 133:588–601

Hilli HJ, Kapoor R (2023) An overview of breeding objectives to improve the economically important traits in Oat. Curr Agric Res J 11(1):18–27. ISSN: 2347-4688

Javaid M, Haleem A, Singh RP, Suman R (2022) Enhancing smart farming through the applications of Agriculture 4.0 technologies. Int J Intell Networks 3:150–164

Jellen EN, Beard JL (2000) Geographical distribution of a chromosome 7C and 1 intergenomic translocation in cultivated Oat. Crop Sci 40:256–263

Kamal N, Tsardakas Renhuldt N, Bentzer J, Gundlach H, Haberer G, Juhász A, Lux T, Bose U, Tye-Din JA, Lang D et al (2022) The mosaic Oat genome gives insights into a uniquely healthy cereal crop. Nature 606:113–119

Kaufman PB (1965) The Effects of Growth Substances on Intercalary Growth and Cellular Differentiation in Developing Internodes of Avena sativa I. The Effects of Indole-3-acetic Acid. Physiol. Plant.18:424–443

Kelley RY, Zipf AE, Wesenberg DE, Sharma GC (2004) Comparative evaluation of three tissue culture methods for the improvement of plant regeneration from diverse Oat (Avena spp. L.) genotypes. Cereal Res Commun 32:113–118

Kim K, Webb S, Newell M, Anderson J, Butler T (2017) Variation of winter forage production in four small grain species: Oat, rye, triticale and wheat. Pak J Bot 49(2):553–559

Kiviharju E, Puolimatka M, Pehu E (1997) Regeneration of anther-derived plants of *Avena sterilis*. Plant Cell Tissue Organ Cult 48:147–152

Knezevic D, Dragovic AY, Przulj N, Siskina A, Menkovska M (2014) Variability of β-glucans contents in grain of barley cutivars fifth international scientific agricultural symposium: Agrosym 2014. Original scientific paper. https://doi.org/10.7251/agsy1404416k

Kole C (Ed.) (2011) Wild Crop Relatives: Genomic and Breeding Resources. Cereals. Springer Berlin Heidelberg. https://doi.org/10.1007/978-3-642-14228-4

Koroluk A, Paczos-Grzęda E, Sowa S et al (2022) Diversity of Polish Oat cultivars with a glance at breeding history and perspectives. Agronomy 12:2423. https://doi.org/10.3390/agronomy12102423

Boczkowska M, Onysk A, Waga J (2016) Unused genetic resources: a case study of Polish common oat germplasm. Annals of Applied Biology, 169(1):155–165. https://doi.org/10.1111/aab.12345

Koroluk A, Sylwia S, Boczkowska M, Edyta Paczos-Grz E (2023) Utilizing genomics to characterize the common Oat gene pool—the story of more than a century of polish breeding. Int J Mol Sci 24:6547. https://doi.org/10.3390/ijms24076547

Kumar SV, Jayaswal D, Jayaswal K et al (2023) Agro-morphological and genetic variability analysis in Oat germplasms with special emphasis on food and feed. PLoS One 18:e0280450. https://doi.org/10.1371/journal.pone.0280450

Ladizinsky G, Johnson BL (1972) Seed protein homologies and the evolution of polyploidy in Avena. Canadian J of Genetics and Cytology.14:875–888. https://doi.org/10.1139/g72-110

Latta RG, Bekele WA, Wight CP, Tinker NA (2019) Comparative linkage mapping of diploid, tetraploid, and hexaploidy Avena species suggests extensive chromosome rearrangement inancestral diploids. Scientific Reports.9:12298. https://doi.org/10.1038/s41598-019-48639-7

Linneaus C (1753) Species plantarum, vol 1. A facsimile of the first edition. London, pp 1957–1959

Loskutov I (2001) Interspecific crosses in the genus Avena. Russ J Genet 37:581–590

Loskutov IG (2005) Contribution of N.I. Vavilov to the problem of the origin and geography of cultivated plants. Rome: FAO. https://www.fao.org/3/y5609e/y5609e00.htm

Loskutov Igor (2008) On evolutionary pathways of Avena species. Genetic Resources and Crop Evolution.55:211–220. https://doi.org/10.1007/s10722-007-9229-2

Loskutov IG, Blinova EV, Gavrilova OP, Gagkaeva T, Yu (2016) Diversity of cultivated oat (Avena L.) genotypes according to economically valuable traits and their relationship with resistance to Fusarium. VavilovJ of Genetics and Breeding. 20(3):286–294. https://doi.org/10.18699/VJ16.151

Loskutov IG, Shelenga TV, Konarev AV, Vargach YI, Porokhovinova EA, Blinova E V, Gnutikov AA, Rodionov AV (2020) Modern approach of structuring the variety diversity of the naked and covered forms ofcultural oats (Avena sativa L.). Ecological Genetics. 18(1):27–41. https://doi.org/10.17816/ecogen12977

Loskutova IG, Gnutikova AA, Blinovaa EV, Rodionov AV (2021) The origin and resource potential of wild and cultivated species of the genus of Oats (*Avena* L.). Russ J Genet 57:642–661

Montilla-Bascón G, Sánchez-Martín J, Rispail N, Rubiales D, Mur L, Langdon T et al (2013) Genetic diversity and population structure among Oat cultivars and landraces. Plant Mol Biol Report 31(6):1305–1314

Morcia C, Terzi V, Ghizzoni R, Carrara I, Gazzetti K (2024) Looking for *Fusarium* resistance in Oats: an update. Agronomy 14:505. https://doi.org/10.3390/agronomy14030505

Nwachukwu ID, Devassy JG, Aluko RE, Jones PJH (2015) Cholesterol-lowering properties of Oat β-glucan and the promotion of cardiovascular health: did Health Canada make the right call? Appl Physiol Nutr Metab 40:535–542

Oliver RE, Herrmann DD, Beckmann MW, Shapter FM (2013) A consensus genetic linkage map of the hexaploid oat genome. Theoretical and Applied Genetics. 126(5):1331–1344. https://doi.org/10.1007/s00122-013-2073-9

Paczos-Grzeda E (2004) Pedigree, RAPD and simplified AFLP-based assessment of genetic relationships among *Avena sativa* L. cultivars. Euphytica 138(1):13–22. https://doi.org/10.1023/B:EUPH.0000047055.99322.7a

Park RF, Boshof WHP, Cabral AL et al (2022) Breeding Oat for resistance to the crown rust pathogen Puccinia coronata f. sp. avenae: achievements and prospects. Theor Appl Genet 135:3709–3734

Pathi KM, Sprink T (2023) From petri dish to field: plant tissue culture and genetic engineering of Oats for improved agricultural outcomes. Plants 12:3782. https://doi.org/10.3390/plants12213782

Paudel D, Dhungana B, Caffe M, Krishnan P (2021) A review of health-beneficial properties of Oats. Foods 10:2591. https://doi.org/10.3390/foods10112591

Pipan B, Sinkovič L, Rutar R, Meglic V (2021) The use of DNA markers for genetic differentiation of common (*Avena sativa* L.) and naked Oat (*Avena nuda* L.). J Cent Eur Agric 22:329–340. https://doi.org/10.5513/JCEA01/22.2.3114

Prasad AS (2009) Impact of the discovery of human zinc deficiency on health. J Am Coll Nutr 28(3):257–265. https://doi.org/10.1080/07315724.2009.10719780

Rayman MP (2012) Selenium and human health. Lancet 379(9822):1256–1268. https://doi.org/10.1016/S0140-6736(11)61452-9

Ridley AM, White RE, Simpson RJ, Callinan L (1997) Water use and drainage under phalaris, cocksfoot, and annual ryegrass pastures. Australian J of Agricultural Research. 48(7):1011–1024. https://doi.org/10.1071/A96157

Rines HW (1985) Sodium azide mutagenesis in diploid and hexaploid Oats and comparison with ethylmethanesulfonate treatments. Environ Exp Bot 25:7–16

Rines HW, Molnar SJ, Tinker NA, Phillips RL (2006) Oat. In: Kole C (ed) Genome mapping and molecular breeding in plants, Cereals and Millets, vol 1. Springer, Berlin/Heidelberg, pp 211–242

Saccomanno B, Wallace M, O'Sullivan DM, Cockram J (2020) Use of genetic markers for the detection of off-types for DUS phenotypic traits in the inbreeding crop, barley. Mol Breed 40(1):13. https://doi.org/10.1007/s11032-019-1088-y

Savovatodorka VD, Toshka P, Djjugerovborjana DV (2014) Main trends in Oat (*Avena sativa* L.) Breeding at the institute of agriculture in karnobat, bulgaria. I. Winter Oats. Trakya Univ J Nat Sci 15(2):75–81. ISSN 2147–0294

Shekhawat NS, Gordon PN, Galston AW (1984) Highly regenerative tissue cultures from axillary tiller buds of sexually mature Oat plants (*Avena sativa* L.). In Vitro 20:707–711

Skoglund M, Peterson DM, Andersson R, Nilsson J, Dimberg LH (2008) Avenanthramide content and related enzyme activities in Oats as affected by steeping and germination. J Cereal Sci 48:294–303

Somers DA, Rines HW, Gu W, Kaeppler HF, Bushnell WR (1992) Fertile, transgenic Oat plants. Bio/Technology 10:1589–1594

Song G, Huo P, Wu B, Zhang Z (2015) A genetic linkage map of hexaploid naked oat constructed with SSR markers. The Crop J, 3:353–357. https://doi.org/10.1016/j.cj.2015.01.005

Soycan G, Schär MY, Kristek A, Boberska J, Alsharif SN, Corona G, Shewry PR, Spencer JP (2019) Composition and content of phenolic acids and avenanthramides in commercial Oat products: are Oats an important polyphenol source for consumers? Food Chem X 3:100047

Stevens EJ, Armstrong KW, Bezar HJ, Griffin WB, Hampton JG (2004) Fodder Oats: an overview. In: Suttie JM, Reynolds SG (eds) Fodder Oats: a world overview. Food and Agriculture Organization of the United Nations, Rome, pp 1–9

Sumathi S, Balamurugan P (2013) Characterization of Oats (*Avena sativa* L.) cultivars using machine vision. Pak J Biol Sci 16(20):1179–1183. https://doi.org/10.3923/pjbs.2013.1179.1183

Szerszunowicz I, Nałęcz D, Dziuba M (2017) Selected Bioinformatic Tools and MS (MALDI-TOF, PMF) Techniques Used in the Strategy for the Identification of Oat Proteins After 2-DE. Methods Mol Biol.1536:253-270. https://doi.org/10.1007/978-1-4939-6682-0_18

Takeda E, Yamamoto H, Yamanaka-Okumura H, Taketani Y (2012) Dietary phosphorus in bone health and quality of life. Nutr Rev 70(6):311–321. https://doi.org/10.1111/j.1753-4887.2012.00473

Thomas H (1995) 29 Oats. Avena spp. (Gramineae–Aveneae). In: Smartt J, Simmonds NW (eds) Evolution of crop plants, 2nd edn. Longman, New York

Tinker N, Wight CP, Bekele WA, Yan W, Jellen EN, Nikos TR, Sirijovski N, Lux T, Spannagl M, Mascher M (2022) Genome analysis in *Avena sativa* reveals hidden breeding barriers and opportunities for Oat improvement. Commun Biol 5:474. https://doi.org/10.1038/s42003-022-03256-5

Tripodi P, Nicastro N, Pane C (2022) Digital applications and artificial intelligence in agriculture toward next-generation plant phenotyping. Crop Pasture Sci. https://doi.org/10.1071/CP21387

Warchoł M, Juzoń K, Dziurka K, Czyczyło-Mysza I, Kapłonia K, Marcińska I, Skrzypek E (2021) The effect of zinc, copper, and silver ions on Oat (*Avena sativa* L.) androgenesis. Plants 10:248

Warchoł M, Skrzypek E, Juzo ́n-Sikora K, Jakovljevi ́c D (2023) Oat (Avena sativa L.) In Vitro Cultures:Prospects and Challenges for Breeding. Agronomy.13:2604. https://doi.org/10.3390/agronomy13102604

Webster FH, Wood PJ (Eds.) (2011) Oats: Chemistry and Technology (2nd ed.). AACC International Press

Winterfeld G, Doring E, Roser M (2009) Chromosome evolution in wild Oat grasses (Aveneae) revealed by molecular phylogeny. Genome 52:361–380. https://doi.org/10.1139/g09-012

Yan H, Bekele WA, Wight CP, Peng Y, Langdon T, Robert G. Latta, Yong-Bi Fu, Axel Diederichsen Catherine J, Howarth, Eric N, Jellen, Brian Boyle, Yuming Wei, Nicholas A, Tinker (2016) High-density marker profiling confirms ancestral genomes of Avena species and identifies D-genome chromosomes of hexaploid oat Theor Appl Genet 129:2133–2149. https://doi.org/10.1007/s00122-016-2762-7

Youngs VL, Peterson DM, Brown CM (1982) Oat. In: Pomerons Y (ed) Advances in cereal science and technology, vol 5. AACC International, NC, St. Paul

Zhang S, Zhang H, Zhang MB (1996) Production of multiple shoots from shoot apical meristems of Oat (*Avena sativa* L.). J Plant Physiol 148:667–671

Zhang S, Cho MJ, Koprek T, Yun R, Bregitzer P, Lemaux PG (1999) Genetic transformation of commercial cultivars of Oat (*Avena sativa* L.) and barley (Hordeum vulgare L.) using in vitro shoot meristematic cultures derived from germinated seedlings. Plant Cell Rep 18:959–966

Chapter 5
Genetic Improvement of Sorghum (*Sorghum bicolor* L. Moench) Through Biotechnology

Khaled F. M. Salem, Farrag F. B. Abu-Ellail, Amira A. Ibrahim, Maysoun M. Saleh, El-Sayed S. Abdel Razik, Sawsan Abd-Ellatif, and Ebtehag A. E. Sakr

Abstract Conventional breeding methods have helped sorghum to be considerably better recently. Still, there is potential for rising *Sorghum* output overall using modern genetic engineering approaches to identify and introduce new genes improving yield and tolerance against biotic and abiotic challenges. Since sorghum is an enemy, developing genetically modified plants in a lab becomes a difficult chore. The creation of a useful and economically feasible result depends on stable and

K. F. M. Salem (✉)
Department of Biological Sciences, Shaqra University, Ad-Dawadimi, Saudi Arabia

Department of Plant Biotechnology, Genetic Engineering and Biotechnology Research Institute (GEBRI), University of Sadat City, Sadat City, Egypt
e-mail: khaledfathi@su.edu.sa; khaled.salem@gebri.usc.edu.eg

F. F. B. Abu-Ellail
Department of Breeding and Genetics, Sugar Crops Research Institute, Agricultural Research Center (ARC), Giza, Egypt
e-mail: farrag_abuellail@arc.sci.eg

A. A. Ibrahim
Department of Botany and Microbiology, Faculty of Science, Arish University, Al-Arish, Egypt
e-mail: amiranasreldeen@sci.aru.edu.eg

M. M. Saleh
Genetic Resources Department, General Commission for Scientific Agricultural Research, Damascus, Syria
e-mail: mzainsamasaleh@gcsar.gov.sy

E.-S. S. Abdel Razik
Livestock Research Department, City of Scientific Research and Technological Applications (SRTA-City), Arid Lands Cultivation Research Institute (ALCRI), New Borg El-Arab, Alexandria, Egypt
e-mail: eshabaan@srtacity.sci.eg

K. F. M. Salem et al. (eds.), *Breeding and Biotechnology of Grass and Bast Fiber Crops*, Advances in Plant Breeding Strategies 12,
https://doi.org/10.1007/978-3-032-00403-1_5

effective genomic transformation to generate transgenic plants. Plant genetic transformation driven by Agrobacterium is the most widely used method. Advances in Agrobacterium-mediated sorghum transformation have led to several genetically modified technologies like recombinant protein overexpression, RNAi-mediated target gene knockdown, and target genome editing for genetic enhancement. New genome-editing technologies have, nevertheless, made some progress in the forward direction. This chapter addresses the benefits of contemporary genetic technologies for strengthening desired features as well as the method of genetically upgrading sorghum using tissue culture-dependent genetic transformation. Additionally, this chapter provides an overview of the botanical classification and distribution, diversity and conservation of sorghum germplasm, importance and uses, domestication, selection and early improvements, breeding objectives, traditional breeding methodologies, the role of biotechnology, mutation breeding, tissue culture applications, genetic engineering and enhanced traits.

Keywords Biotechnology · Genetic improvement · Modern breeding · Traditional breeding · Sorghum (*Sorghum bicolor* L. Moench)

5.1 Introduction

5.1.1 Background

Grown mostly in the summer, feed sorghum, technically known as Sorghum (*Sorghum bicolor L. Moench*). The blossoming of this short-day shrub accelerates with shorter days and longer nights. Sorghum is a basic staple food for many people, especially those living in dry or rural locations. Grown for its several uses—food, feed, fodder, and fuel—sorghum is grown throughout the dry and semi-arid tropical areas of the Americas, Asia, and Africa (Khalilian et al. 2022; Morris et al. 2013). Effective crop-breeding strategies that guarantee constant crop production, including cereal and strower yield, nutritional and plant protection features (both biotic and abiotic), and so on depend on a thorough knowledge of crop botany, variation, and heredity (Abdel-Hamid and Salem 2021; Aboshama et al. 2015; Abu-Ellail et al. 2021; Aldahak et al. 2021; Alqahtani et al. 2024; Elabd et al. 2020; Emara

S. Abd-Ellatif
Bioprocess Development Department, Genetic Engineering and Biotechnology Research Institute (GEBRI), City of Scientific Research and Technological Applications (SRTA-City), New Borg El-Arab, Alexandria, Egypt
e-mail: sabdellatif@srtacity.sci.eg

E. A. E. Sakr
Botany Department, Faculty of Women for Arts, Science and Education, Ain Shams University, Cairo, Egypt
e-mail: ebtehag.abdelfattah@women.asu.edu.eg

et al. 2011, 2013; Hassan et al. 2021; Ibrahim et al. 2023a, b; Mourad et al. 2019; Nawara et al. 2016, 2017; Saleh et al. 2020b; Salem 1997, 2004, 2009; Salem et al. 2007, 2015a, b, 2021b, 2022, 2023).

Sorghum is among the most significant food crops worldwide. Mostly in the developing world, about 500 million people consume it daily in more than 30 nations. Furthermore, spread over 40 million hectares in 90 countries in the Americas, Oceania, Asia, and Africa is its reach. Its adaptability makes sorghum unique among crops (Khalilian et al. 2022; Reddy et al. 2006). Benevolent behind wheat, rice, maize, barley, and sorghum, the fifth most important cereal crop worldwide remains unknown. Ranked fifth in acreage, it makes a 3.5% contribution to total cereal grain output. Sorghum has grown to be a major crop for bioenergy generation of sugar, biomass, and biofuels (de Siqueira et al. 2013; Rooney et al. 2007; Tavazoh et al. 2024; Zohry and Ouda 2022). As a C4 plant that can resist high temperatures, drought, and strong heat while still growing in poor soils. With an average yield of 1459.2 kg per hectare (FAO 2021), sorghum is grown on 44.79 million hectares worldwide, producing 70.83 million metric tonnes of grain. Growing sorghum is best in dry areas, especially those with shallow, compact clay soils. A lot of factors prevent Sorghum from growing. Among the several studies on sorghum published in the fields of genetic diversity, ecology, and botany are Abreha et al. (2022), Dwivedi et al. (2013), and Mickelbart et al. (2015).

Grain sorghum, sweet sorghum, Sudan grass and broom corn are the four cultivated varieties of sorghum shown in Fig. 5.1. Food and feed come from grain sorghum; sweet sorghum is utilized for animal grazing; Blake honey comes from broom corn; broom making uses broom corn. Asia and Africa account for around 80% of the arable land available worldwide. Apart from their several functions as fruit, feed, fiber, and building materials, grown sorghums show a great variety of morphological traits, including height and inflorescence characteristics (Dillon et al. 2007b; Frankowski et al. 2022; Kumari et al. 2024).

Ideal for burning, roofing, fencing, sugar, alcohol, syrup, and jaggery, grain and stems from sweet sorghum (*Sorghum bicolour* L.) Moench). No else grown produces these goods. For almost 150 years, it has been used to make concentrated syrup, which is excellent. Sweet sorghums are also extensively used in the growing of silage and fodder crops for animal feed. During the 1973 and 1976 oil crises, commercial sweet sorghum farming for biological conversion into ethyl alcohol for fuel or as an additive to petrol attracted a rebirth in interest. A one-hectare sweet sorghum plot can yield, according to Fig. 5.1, a range of products in set quantities, including grain, leaves, bagasse, jaggery, syrup, or alcohol. As efforts are being made to raise the general production and the quality of these commodities, farmers raising this crop are finding more income. Our hybrid "Madhura" has become a main crop in India for sweet sorghum ethanol generation. Many distilleries all around have started running on it.

Sweet sorghum is under considerable conjecture as a possible energy crop. Unlike sugarcane, a tropical plant, sweet sorghum is suited for almost every temperate or tropical climate. NARI developed a technique for sweet sorghum that allows one to make alcohol. Rural India uses this fuel for cooking and lighting. The fermenting studies

Fig. 5.1 The images of sorghum (*Sorghum bicolor* L.) grain. (Source: Ropelewska and Nazari 2021)

carried out at the Institute show that the NCIM 3319 strain of Saccharomyces cerevisiae performed well in the batch fermentation of unsterilized juice to generate ethanol. Among sixteen strains under evaluation, this one performed the best. After 48–72 h of fermentation in juice with 10–11% total fermentable sugars (w/w), roughly 6% ethanol (by volume) was generated. The 200-litre batch of juice was not vitamin fortified.

5.1.2 *Biology of Sorghum*

5.1.2.1 Origin and Domestication

Sorghum was initially cultivated in the northeastern region of Africa, predominantly in what is now Sudan and Ethiopia, between 4000 and 3000 BC, according to Cowan et al. (2022). Dillon et al. (2007b), Harlan and de Wet (1971), Liu et al. (2017), and Price et al. (2005). Modern cultivated sorghums derive from the wild *Sorghum bicolour* subspecies Arundinaceum (Doggett 1988). Harlan and de Wet (1971) and Harlan and de Wet (1972) list five primary varieties of sorghum: bicolour, guinea, caudate, kafir, and durra.

Early phases of domestication took advantage of disruptive selection. Originally, it was believed that domesticated lines would replace wild kinds with smaller seeds and more open panicles with bigger, non-crushing seeds and more dense panicles (Doggett 1965). These developments raised yields, which in turn drove sorghum to change into a great range of forms when compared to the initial landrace varieties (Dillon et al. 2007b). *S. bicolour* was thus split into three categories: those cultivated and improved, those wild, and those intermediate (Kimber 2000).

Although it is difficult to say exactly when sorghum first arrived in different regions of the globe, we can deduce its introduction by recognized trade routes and economic ties (de Wet et al. 1970). According to Murdock (1959), Mande people living close to the Niger River's headwaters raised sorghum. Archaeological evidence points to wheat farming being transported to Ethiopia from Egypt approximately 3000 BC, per Doggett (1965). Sorghum was originally domesticated most likely during that period.

From northern Africa to other areas of the continent, enhanced sorghum cultivars most certainly traveled by trade routes and human migration between 1500 and 1000 BC. Before making its way to the Far East, sorghum most certainly was traded and carried from Africa to the Middle East and India (900–700 BC). Originally brought to the United States in the latter half of the nineteenth century, sorghum first came as a result of the slave trade (Doggett 1988; FAO 1995; Venkateswaran et al. 2019).

5.1.2.2 Geographic Distribution

Table 5.1 demonstrates that growing sorghum, a tropical crop, and semi-arid sites are most often used. From the southernmost point of Africa to Sudan, Ethiopia, and Somalia, Ethiopia is a vast and always-growing continent. Among the nations growing it are Burundi, Zambia, Egypt, Uganda, Kenya, and Tanzania. Major producers are the arid areas of Brazil, Venezuela, Argentina, France, Italy, and Spain, as well as India, Thailand, Pakistan, central and northern China, Australia, and the United States. With 42–43 million hectares planted, the world's sorghum harvest is

Table 5.1 Geographic distribution of sorghum races in the world

No	Snowdenian species	Distribution	No.	*Snowdenian species*	Distribution
	S.bicolor ssp. halepense			*var. bicolor race guinea*	
18	*S. controversum*	Eastern and southern India	17	*S. conspicuum*	Tanzania to Mosambique
			28	*S. exertum*	West Tropical Africa
31	*S. halepense*	The Mediterranean to northern India	29	*S. gambicum*	West Tropical Africa
			30	*S. guineense*	West Tropical Africa to Uganda
45	*S. miliaceum*	North West Pakistan to Uganda India	39	*S. margaretiferum*	Sierra Leone to northern Nigeria
55	*S. propinquum*	South East Asia	43	*S. mellitum*	West Africa to South Africa
	S. bicolor ssp. bicolor var. aethiopicum		60	*S. roxburghii*	East Africa to India and Burma
1	*S. aethiopicum*	Northern Nigeria to Ethiopia		*Race hafir*	
			13	*S. caffrorum*	Widely distributed in Africa
34	*S. lanceolatum*	Senegal to Sudan			
73	*S. virgatum*	Egypt and Sudan			
			15	*S. caudatum*	Equatorial Africa
			19	*S. coriaceum*	Tanzania to Congo
	var. arundinaceum		24	*S. dulcicaule*	Congo
5	*S. arundinaceum*	West Africa	48	*S. nigricans*	Tropical Africa
74	*S. vogelianum*	Tropical West Africa		*race durra*	
			16	*S. cernuum*	Asia Minor to India
	var. verticilliflorum		25	*S. durra*	East Africa to Arabia and India
11	*S. brevicarinatum*	Kenya and Tanzania	59	*S. rigidum*	Sudan
14	*S. castaneum*	Northeastern Congo	67	*S. subglabrescens*	East Africa to Arabia and India
37	*S. macrochaeta*	Congo to Sudan		*race bicolor*	
53	*S. panicoides*	Ethiopia	4	*S. ankolib*	North-East Africa
56	*S. pugionifolium*	Punjab of India	8	*S. basutorum*	South Africa
63	*S. somaliense*	Somaliland	9	*S. bicolor*	Arabia to Burma
70	*S. usambarense*	Tanzania	22	*S. dochna*	India to Burma
72	*S. verticilliflorum*	Southward from Ethiopia	26	*S. elegans*	East Africa to West Africa
	Weed-crop hybrids	Sudan and Ethiopia			
32	*S. hewisonii*	Sudan and Egypt	42	*S. melaleucum*	Noth East Africa

(continued)

Table 5.1 (continued)

No	Snowdenian species	Distribution	No.	*Snowdenian species*	Distribution
68	*S. sudanense*	West Africa and the upper Nile	44	*S. membranaceum*	East Africa, India, and China
6	*S. aterrimum*	West Tropical Africa			
23	*S. drummondii*	Tanzania along the river banks	46	*S. miliiforme*	East Africa and North East India
50	*S. nitens*	Kenya to Congo and Sudan	47	*S. nervosum*	Eastern Asia
49	*S. niloticum*	Uganda	52	*S. notabile*	Northern Nigeria to Sudan
27	*S. elliottii*		62	*S. simulans*	Malawi
			64	*S. splendidum*	South East Asia

Source: de Wet and Huckabay (1967)

59–60 million tonnes. According to Hari et al. (2017), over the past 18 years (1994–2011), worldwide sorghum output averaged 59–60 million tonnes.

5.1.2.3 Botanical Classification

Members of the complex genus Sorghum (*Sorghum bicolour* L. Moench), all varieties of farmed sorghums as well as other semi-wild species usually linked with them as weeds. Clayton and Renvoize (1986) claim that the genus Sorghum belongs to the Poaceae family of grasses, more especially the Panicoideae subfamily, the Andropogoneae tribe, and the Sorghinae subtribe. Two of Andropogonea's most important crops are maize (Zea mays L.), and sugarcane (Saccharum spp.). Great morphological variation in Sorghum (*S. bicolour*) led Snowden (1955) to first classify 28 domesticated and 24 nearly comparable wild species. However, as de Wet (1968) noted, these taxa are all members of the same genus and lack genetic barriers separating them. The great variety of sorghum species makes it challenging to distinguish domesticated versions from their wild counterparts (Wiersema and Dahlberg 2007). Five morphologically based subgenera—chaeto sorghum, Stiposorghum, Heterosorghum, Parasorghum, and Eusorghum—define the classification (Celarier 1958; Price et al. 2005; USDA 2015).

Kingdom Plantae—Plants
- Subkingdom Tracheobionta—Vascular plants
  - Superdivision Spermatophyta—Seed plants
    - Division Magnoliophyta—Flowering plants
      - Class Liliopsida—Monocotyledons
        - Subclass Commelinidae
          - Order Cyperales
            - Family Poaceae—Grass family
              - Genus *Sorghum* Moench—sorghum P
                - Species *Sorghum bicolor* (L.) Moench—sorghum P

There are three accepted taxa and three subspecies; i) one lower-ranking taxon. Scientifically named *Sorghum bicolour*, grain sorghum is a subspecies of Sorghum (L.) Moench ssp. arundinaceum (Desv.). de Wet & Harlan. Whereas *Sorghum bicolour* is a subspecies of *Sorghum* (L.), Moench ssp. drummondii (Nees ex Steud.) de Wet & Harlan is a subspecies of Sorghum (L.). As demonstrated by the great number of species among numerous subgenera, Sorghum is a quite varied genus.

Whereas most polyploid species have 2n = 10, 20, 30, or 40, where n is the haploid chromosomal number, the lowest known haploid chromosome count in Parasorghum and Stiposorghum is five. With a minimum of 10 haploid chromosomes and three allopolyploids—that is, 2n = 20, 40—Eu-sorghum has integral multiples of twenty in its chromosomal count. Celarier (1958) claims that since their 2n = 40 makes them both allopolyploids, Chaetosorghum and Heterosorghum.

Their closest wild relatives, as well as all domesticated sorghum cultivars, are categorized as Eusorghum. Genes from wild sorghum can be transferred to cultivated sorghum since several species and subspecies within the subgenus of sorghum are sexually compatible. Sorghum halepense is one of the most common weeds, along with the other two species of the subgenus Eusorghum, *Sorghum propinquum*, and *Sorghum bicolor* (de Wet 1978). There are three subspecies of S. bicolor as noted by Wiesema and Dahlberg (2007). The general nomenclature for all cultivated sorghums is subspecies bicolor of Sorghum. Table 5.2 shows ten intermediate races produced from combinations of the five fundamental races in *S. bicolor* subsp. bicolor.

5.1.2.4 Importance

In terms of grown area and yield, sorghum ranks sixth among the major cereal crops worldwide. Since it adds to main food dishes and beverages worldwide, especially in African countries, this is rather important in reaching food security. Apart from

Table 5.2 Cultivated races of *Sorghum* (*Sorghum bicolor* subsp. bicolor) with genotype abbreviated in brackets

Basic races	Intermediate races
Race 1: Bicolor	Race 6: Guinea-bicolor
Race 2:Guinea	Race 7: Kafir-bicolor
Race 3: Caudatum	Race 8: Caudatum-bicolor
Race 4: Kafir	Race 9: Durra-bicolor
Race 5: Durra	Race 10: Guinea-caudatum
	Race 11: Guinea-kafir
	Race 12: Guinea-durra
	Race 13: Kafir-caudatum
	Race 14: Durra-caudatum
	Race 15: Kafir-durra

Source: Adapted from Harlan and de Wet (1972)

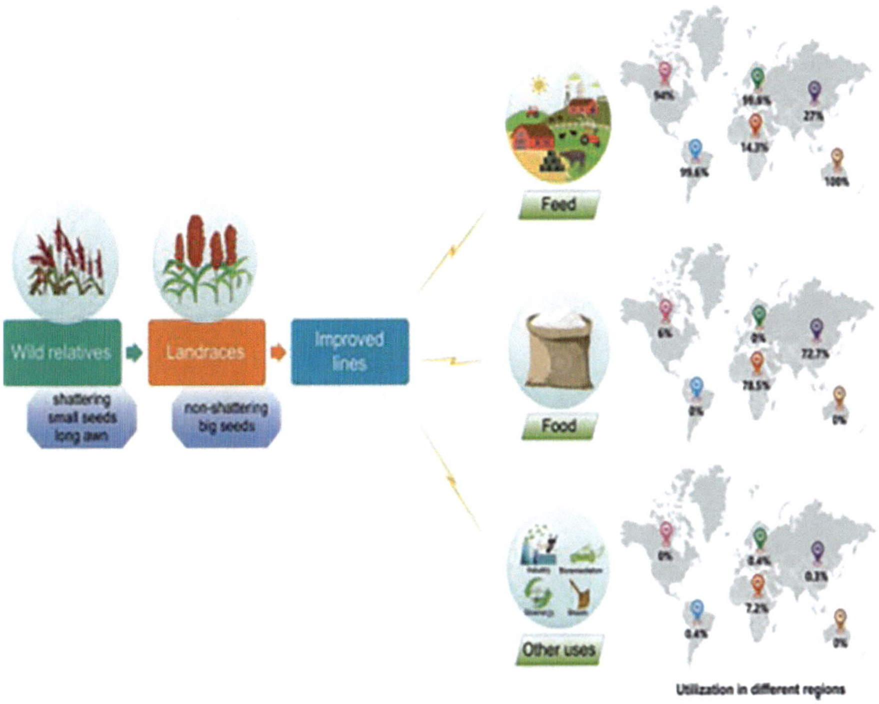

Fig. 5.2 An illustration presenting the different uses of sorghum in the world. (Source: Hao et al. 2020)

the advantages of sorghum as a fuel and its usage in broom manufacture throughout many nations (Figs. 5.2 and 5.3), it is also fed.

Liu et al. (2020) underlined sorghum as a phytoremediation technique for contaminated soils. Different countries and regions employ different sorghum genotypes, claims Batey (2017). Stefoskaneedham et al. (2015) validated the part sorghum plays as a source of energy molecules that increase physical fitness and help prevent diseases, particularly chronic ones. Scientists respect sorghum because of its several applications in business and agriculture, as well as because of its potential as a food crop.

Although sorghum is widely used as animal feed in developed countries such as the United States, Mella (2011) claims that people eat it in many African and Asian countries. As Parthasarathy et al. (2010) note, sorghum grain flour is used to make beverages and snacks like porridge and popped grains. Studies by Heuzé et al. (2013) find that silage produced from sorghum grains helps sheep and cattle. House (2005) claims that the dried sorghum stems are utilized as cooking fuel; the fibers are used to create fences, package biodegradable goods, and dye leather clothes.

According to Tokach et al. (2010) that ethanol could be produced from sorghum grains. The amounts of carbohydrates, protein, fat, minerals, vitamins, and phenolic components determine the nutritional value of sorghum grain, per Hao et al. (2020).

Fig. 5.3 Broom made of Serbian sorghum fields 2010. (Sources: Dahlberg 2019)

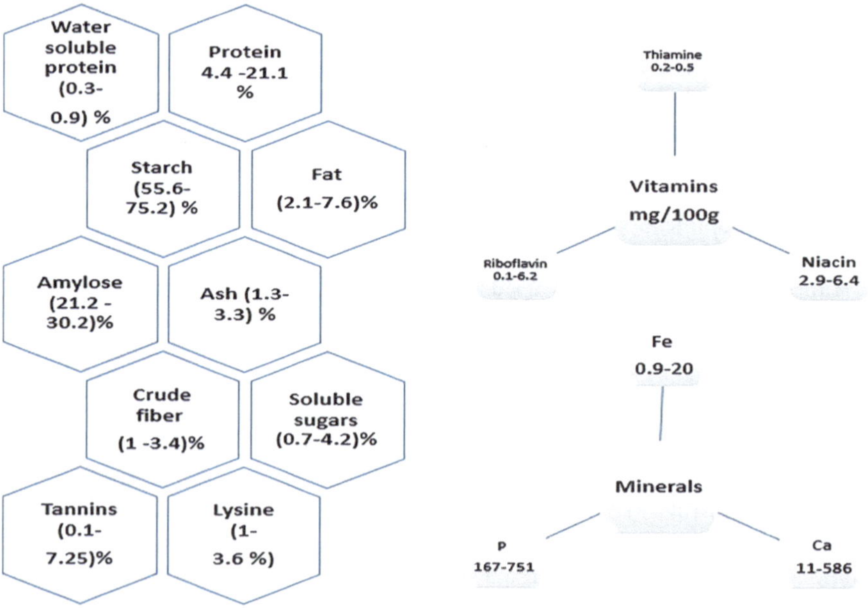

Fig. 5.4 Sorghum grain composition. (Constructed by M.M. Saleh based on Ratnavathi and Patil 2013)

Ratnavathi and Patil (2013) claim that the grain composition of sorghum exhibits genotype-dependent variation (Fig. 5.4).

Hay, green plants, pellets, and silage are regarded as essential feed for animal fattening; Matei (2016) looked at the idea of substituting sorghum for maize given its several applications, including pet, poultry, and cattle. Made from sorghum are paper, fences, brushes, and wallboard. From this crop, one can extract the energy for heating, electricity generation, gas, and fuel (Khalilian et al. 2022; Popescu and Condei 2014).

Because it includes more than 9% sucrose, 6% glucose and 4% fructose, and 23% additional sugars, the sugary syrup derived from sorghum stems is often used in desserts and beverages (Babeanu et al. 2017). Leder (2004) underlined the nutritional value of maize and sorghum grains and their similar composition. He claims sorghum grains have calcium, iron, and trace amounts of copper, magnesium, and nickel in addition to the B vitamins riboflavin, thiamine, and niacin. Billa et al. (1997) claim that sorghum leaves and stems are rather high in hemicelluloses and cellulose.

5.2 Conserving Sorghum Genetic Resources

Conservation of genetic resources in plant gene banks through various approaches is highly beneficial for seed preservation (Abdullaev et al. 2025; Abulaila et al. 2025; Akparov et al. 2025; Al Lawati 2025; Aladele et al. 2025; Al-Khayri et al. 2025; Almarri et al. 2025; Arab et al. 2025; Békefi et al. 2025; Breidy 2025; Cañar et al. 2025; de Galarreta et al. 2025; Desheva et al. 2025; Díez et al. 2025; Galea-Gragera and Llera Cid 2025; Ghidan and Abdulmajid 2025; Gudaye et al. 2025; Harzli et al. 2025; Hidayatun and Kurniawan 2025; Ihsan and Prihatini 2025; Jatoi and Siddiqui 2025; Laghetti et al. 2025; Mallor and Luna 2025; Monteros-Altamirano et al. 2025; Osterc et al. 2025; Popović et al. 2025; Rocha et al. 2025; Wang et al. 2025). On the other hand, conservation helps preserve biodiversity and identify varieties with beneficial and adaptive genes.

Ex situ collections stored in gene banks are valuable genetic resources that must be safely maintained, economically viable, and readily available to a wide variety of users. Numerous ex-situ seed gene banks have been developed globally to preserve plant genetic resources with natural genetic variety.

5.3 Conventional Breeding Techniques in Sorghum

Whether they reproduce by self-pollination or by crossing their genes with those of other plants, crop plants can be either homozygosity or heterozygosity. The breeding method applied to any crop is mostly determined by the propagation method and the objectives of genetic enhancement. Since a regularly cross-pollinated crop has a cross-pollination rate ranging from 5 to 15%, ideal cross-pollinated crops can profit from breeding techniques that work for both types of pollination (Poehlman 1987). A benefit for sorghum growers is the possibility to produce a variety from either a homozygous or heterozygous line.

5.3.1 Sorghum Breeding Objectives

Before beginning a breeding program, it is imperative to establish objectives so that everything is carried out as expected and precisely. The breeder has to monitor the work schedule, the areas needing development, and above all, the anticipated outcomes of the investigation. Along with the degrees of biotic and abiotic stresses in the target areas, consumer tastes, the processing sector, and farming techniques, sorghum breeders must also know what farmers, millers, and bakers need. According to Rakshit and Gomashe (2013), sorghum breeding seeks to maximize output while also guaranteeing that yield stays constant.

5.3.2 Yield Traits

Sorghum grain production is a complicated trait affected by several elements. Any breeder worth their salt will tell you that their first concern is rising production. Yield is not a suitable indicator of utility for selection due to its poor heritability and great sensitivity to external elements. Thus, it is imperative to find traits that directly affect yield, including kernel size, harvest index, 1000-kernel weight, kernel number, various morphophysiological factors, and so on. By enhancing these qualities, breeders may get even more benefits.

Plant breeders mostly aim to increase economic productivity using genetic improvement, as farmers aim for better economic yields from every genotype or F_1 hybrid. To boost grain output, use transgressive segregants or create crosses using enough heterosis. Raskhit et al. (2014) claim that the efficiency of sorghum beats what farmers are now producing.

5.3.3 Biotic Stresses

Nearly all crops farmed for human consumption are affected by diseases and insect pests, which dramatically reduce crop yields and quality (Eid et al. 2025; Salem et al. 2021c). Farmers in dry areas lack chemical and agronomic means of controlling biotic stress. Consequently, creating resistant genotypes using several breeding methods is the best and most affordable approach to lowering these losses. Using an interdisciplinary and multifarious strategy combining phenotypic and genotypic approaches helps one to better understand and utilize wild relatives (Elabd and Salem 2009; El-Absawy et al. 2012, 2015; Eltaher et al. 2017, Eltaher et al. 2018, 2021, 2022; El-Zanaty et al. 2011, 2013; Esmail et al. 2009; Ghazy et al. 2020; Kamala et al. 2016; Ismail et al. 2013; Mattar et al. 2009, 2016; Saleh et al. 2020a; Salem and Börner 2004; Salem and Matter 2014; Salem et al. 2003, 2004a, 2004b, 2008). The most successful sources of resistance, according to Raskhit et al. (2016) and Singh (2004), are wild variations and their ancestral forms. The tertiary sorghum gene pool consists of four kinds of sorghum: chieto, hetero, stipo, and para. Based on Kamala et al. (2002) estimated that resistance genes could arise at 36 sites. Among the illnesses that could strike sorghum during the wet season are grain mold, downy mildew, ergot, and anthracnose. Conversely, in the post-wet season, root and stalk rot as well as viral infections are rather widespread (Das 2013). Worldwide, sorghum crops are attacked by more than 150 distinct types of insects. Still, few species endanger economies in a significant manner. The most often occurring pests in the wet season include head bugs, midges, stem bores, and shoot flies. In the period after rain, sugarcane aphids, shot bugs, and shoot flies are the most often occurring pests (Bhagwat et al. 2013). A continual process, resistance breeding exploits genotypes discovered in germplasm banks as donor genotypes in subsequent breeding projects.

5.3.4 Abiotic Stresses

Among the natural stresses plants deal with are heat, cold, salinity, mineral scarcity, toxicity, and oxidative stress (Börner et al. 2010; Ibrahim et al. 2023a, b; Saleh et al. 2021; Salem and Börner 2003; Salem and Ibrahim 2024; Salem et al. 2021a, d). Two main dangers to modern agriculture worldwide are heat stress and drought, since their detrimental consequences affect phenotypic development (Patil and Sanjana 2013). Cold stress in sorghum is manifested in excessive tillering during the vegetative phase and hampered seed development during the flowering phase (Patil and Sanjana 2013). Its avoidance of the crop from reaching its maximum yield, as indicated by its genetic composition, results in a significant limitation on agricultural output. The detrimental effects of environmental stress are said to be the reason behind crops only reaching 25% of their optimum potential yield (Boyer 1982). Cold weather inhibits seed germination in seed production areas. An abiotic stressor's frequency, strength, and length of duration depend on where it occurs. The fact that stressors can strike at any stage of the sorghum life cycle shows the flexibility of agricultural plants and their yields. Especially in tillering, flowering, and grain-filling, breeding for stress tolerance is vital. Though much research has been done over the years, conventional approaches have only demonstrated little success in raising drought and heat stress tolerance (Deshpande et al. 2016).

5.3.5 Quality and Days to Maturity Traits

Improving the quality of the final product is another important objective of plant breeding since it impacts its suitability and customer inclination. Production of sorghum depends much on grain and feed quality. Food quality is influenced by the physical and biological properties of grain sorghum (Bello et al. 1995; Rhodes et al. 2017; Sleper and Poehlman 2006). Among the most important traits affecting food quality in sorghum is endosperm texture (Rakshit and Gomashe 2013; Rooney et al. 1986). Superior molecular weight polymers like amylose and amylopectin, high-level starch solubility and outstanding dehulling qualities define good-grade cultivars (Cagampang et al. 1985).

Many agricultural plants have a limited lifetime to satisfy the season duration given and produce continuously under low rainfall incidence following the crop season. Along with improved cropping intensity, it offers new crop rotations and fewer management techniques as well. Four chromosomal sites in sorghum pinpoint the maturation characteristics Ma1, Ma2, Ma3, and Ma4. While certain genes are dominant, the recessive genes ma1 and ma3 produce early maturity; but late maturity results when all four genes (Ma1, Ma2, Ma3, Ma4) are dominant (Quinby 1967).

5.4 Procedures of Breeding Techniques

5.4.1 Breeding Techniques

A blooming plant, sorghum has one floret, including both sexual organs. Effective crossover of hybrid seeds depends on the exact emasculation of female parents. Two male fertile lines can be crossed using hand emasculation (Fig. 5.5), hot water or the plastic bag/bulk emasculation technique (Rakshit and Bellundagi 2019) (Fig. 5.6).

5.4.2 Breeding Procedures

The type of genotype (varieties or hybrids) to be generated influences the breeding techniques as well as the crop fertilization behavior. Since sorghum is essentially a self-pollinated crop, methods of breeding for such crops are applied. It allows some cross-pollination and the availability of male sterility genotypes makes using heterosis in hybrids easy for breeders.

To produce sorghum, producers of the crop have used numerous conventional selections and breeding techniques. Among these techniques are bulk selection, pure-line selection, pedigree approach, bulk method, backcross method, single seed

Fig. 5.5 Hand emasculation steps: choosing the fluorescence (**a**); removing the florets (**b**); manually removing the anthers (**c** and **d**); and covering the emasculated fluorescence with a brown cover bag and labeling (**e** and **f**). (Source: Rakshit and Bellundagi 2019)

descent (SSD) technique and others. Although in segregating populations, different approaches including pedigree, bulk, backcross, and SSD systems can be used (Rakshit and Bellundagi 2019), in naturally varied and homozygous populations, bulk and pure-line selection methods can be applied.

Fig. 5.6 Plastic bag emasculation steps: fluorescence selection and floret removal (**a**) and (**b**); fluorescence covering with a plastic bag (**a** and **b**); moisture accumulation after 1 day (**c**); vigorous tapping of the inflorescence to remove anthers (**d**); accumulation of removed anthers in a plastic bag (**e**); emasculated fluorescence (**f**); remaining steps (**e**) and (**f**). (Source: Rakshit and Bellundagi 2019)

5.4.2.1 Selection Procedures

A great number of phenotypically identical plants are selected in bulk and their kernels are mixed to produce a superior mass variability (Allard 1960). Mass selection is based on the external appearance of plants for particular traits; progeny testing is not applied. Generally, homogeneity in agronomic characteristics was desired, as well as strong resistance to common pests and diseases. It is usually used to purify pure-line genotypes and new arrivals when their genetic variability becomes evident. Mass-selected cultivars have great genetic variation for quantitative traits even if they have an agronomically similar appearance.

The pure-line selection process chooses from a single, homozygous, and self-fertilized crop a sizable number of genetically identical plants. It differs from bulk selection in that advanced progeny-generating individuals are distributed and collected separately without mixing the seeds, depending on the performance of the offspring and single select plants are subjected to offspring testing. Selection of single plants is predicated on a naturally varied, homogeneous basis population. The progenies of certain plants allow one to evaluate yield and related traits as well as consistency for agronomic variables. Selected are the best offspring lines matching the maximum yield and necessary features. After that, the best line for the new introduction is found by comparing the performance of several variants with that of the check genotypes. Though there will be a genetic difference among the offspring of selected plants, there will not be any genetic diversity in the individual offspring of particular plants. Mechanical mixings, natural cross-pollination, and spontaneous mutations produce genetic variability in pure-line cultivars. Many sorghum varieties based on the pure-line selection strategy have been found and marketed for commercial farming in India, including M-35-1, Sel-3, Yashoda and Maulee (RSLG-262) (Rakshit and Bellundagi 2019).

5.4.2.2 Hybridization Methods

Hybridizing two types that are naturally different from one another helps to generate a great range of offspring with unique traits. Before Mendelian genetics, hybridizing genotypes was a rather frequent procedure. Hybrids create children that mix the best traits of both of their parents or that surpass either one of them (Meena et al. 2024). wide hybridization in sorghum using a tertiary gene pool for parental line diversification. Agricultural Science. Transgressive segregation is the phenomenon whereby improved recombinants show up in generations that have been segregating.

The types of hybrids produced by a hybridization process depend on the number of essentially broad parents utilized; these could be single-cross (A x B), double-cross ([A x B] x [C x D]) or three-way cross([A x B] x C). Multiparent advanced generation intercross was used historically to create sorghum populations that combined desired features from several parents into one genetic framework (Rakshit et al. 2012). These populations increase the success rate of recombination and simplify the integration of genes from several donors. Usually, one uses a pedigree, bulk, or backcross technique to manage the generational difference when crossing two essentially distinct genotypes.

The pedigree approach chooses individual specimens from the F_2 generation and follows segregating generations to achieve effective homogeneity. Every generation, separating the plants is assessed in terms of the performance of the progeny of those plants. Depending on how well their offspring do, plants that are outstanding in creating new offspring can self-fertilize and harvest themselves. Since the F_2 generation has the highest genetic variety for the traits, it calls for the biggest population size. Though an exact count is not known, a realistic estimate falls between 1500 and 2000 (Singh 2004). Selection takes place both inside and between families during the first three generations (F_3 and F_4); but, during generations six and seven, it predominantly occurs inside families as the offspring of the lineage achieve higher degrees of homozygosity and homogeneity. The panicle-row approach evaluates sorghum holistically and effectively progresses each generation by selection until the number of families is dropped to a reasonable level (Audilakshmi and Aruna 2008). Using a pedigree technique is a typical method to establish sorghum varieties; this has produced several current varieties, including CSV 15 (SPV 475 x SPV 462), CSV 17 (SPV 946 x SPV 462), and CSV 20 (SPV 946 x Hh89-246) (Rakshit and Bellundagi 2019).

Originally applied in plant breeding, the bulk method—often known as the population or mass approach was devised by Nilsson-Ehle (1908). In F_2 and following generations that are separated, the selected people are gathered jointly rather than separately. Individual plants are selected depending on how well their offspring perform and are harvested separately during the bulking period, which spans many generations. Individual plants are assessed by the pedigree breeding technique. Artificial selection is not applied, hence natural selection takes the front stage during the bulking period.

This approach allows homozygous superior lines to be extracted following processing, therefore utilizing the inclination of natural selection for highly adaptable and prolific genotypes. Very little work and supplies are needed when separating generations and looking for homozygous lines with this technique. The population is maintained at the F_6 generation using the pedigree method.

Drawing on the variety of hundreds to thousands of people to improve upon F_2, F_2-F_4 lets natural selection flourish. Natural selection favors versatile and productive genotypes. This method helps to extract homozygous superior lines more easily after processing. To identify the homozygous lines and monitor the generational separation. This method requires minimal materials and effort. Following the pedigree strategy helps one to achieve the F_6 generation in the population. Natural selection selects for those with greater degrees of produced variability, hence improving F_2 plants across the F_2–F_4 generations. The F_5 or F_6 generation is the beginning of the family selection process, whereby the nurturing of individual plant descendants carries on to the next generations. A small sample of notable families is thoroughly evaluated (Rakshit and Gomashe 2013).

Widely applied in many crops, the single-seed descent (SSD) approach is a successful and flexible version of the mass strategy. This approach mostly aims to employ off-season nurseries to accelerate the generations of crossings, control the great variation generated by the hybrids, and advance the generations of crossings. Other techniques, such as mass and pedigree, clearly outperform this one. SSD is a refined method whereby one seed from the F_2-$F_{6:7}$ generation advances each panicle or plant. The most often used products of this technique are recombinant inbred lines (RILs) (Takele et al. 2022). Several mapping populations in sorghum were produced with this method. This method mostly helps since all F_2 plants are included in the ultimate population. This lets breeders employ all the crosses produced in terms of variation. With little area, labor, or resources needed, it makes the most of the off-season and greenhouse assets. Many studies have started using single panicles instead of individual seeds from every plant to preserve a distinct plant and maintain F_2 generated variation. This approach can be used to designate populations for various highly efficient features, including pest tolerance (like shoot fly) or disease resistance (like charcoal rot), using either recombinant inbred lines (RILs) or near-isogenic lines (NILs). Though they remain the same between the lines, RILs derived from SSDs show a lot of diversity, which is a reflection of the variability in the F_2 generation (Ibrahim 2015; Kong et al. 2013; Phuong et al. 2013). Because of more homozygosity, which results from higher variety among lines generated by additive gene activation (Mangoendidjojo 2003), the following generation of SSD lines gains more from selection.

5.4.2.3 Population Improvement

Population development consists of selecting advanced inbreds from a basic population to act as parents in crossings and synthetics, hence increasing the frequency of desired alleles and genotypes while reducing inbreeding (Nath 1972; Singh 2004). Recurrent selection is a technique for enhancing populations whereby plants are selected either for their general performance (bulk selection) or their offspring's performance. Hybrids are therefore produced, which guarantees a continuous supply of modern genotypes in the breeding nursery (House 1985). A random crossover among selected individuals is ensured during population increase. Since sorghum is usually self-fertilizing (ms3, ms7), the recessively inherited ms system makes random mating in sorghum possible. The method starts with the hybridizing of donors into the population using MS lines and proceeds with more hybridizing of the fertile F_1 generation using MS lines. The families are tied down the road. In homes of the next generation, male fertility (mf) plants will be 1:1 in female fertility (ms) plants. Anthesis is the period during which male plants are found and suitable hybridizing plants are chosen. The families so engage in close coordination. Combining and fertilizing equal numbers of seeds from each parent plant produces the following generation of hybrid plants. Finding the Ms. plants and compiling equal numbers of kernels from every plant marks the first stage in designing a mix for the next random mating cycle. Plants are either chosen in bulk or, depending on the quality of their progeny, at the end of each life cycle. Using updated bulk selection techniques in sorghum evolution (Goyal et al. 2020; Jan-Orn et al. 1976; Doggett 1972; Obilana and El-Rouby 1980) yields plant height and days to flowering. At the International Crops Research Institute for the Semi-Arid Tropics (ICRISAT), yield stability, quality, agronomic desirability, and resistance to pests and diseases were improved using a recurrent selection approach. Among these illnesses and pests were midge, stem borers, and shoot flies, as well as grain mold, charcoal rot, and several leaf diseases. Population breeding has produced a great number of improved cultivars fit for use. While A36001, A3872, A3895, A6072, and 42A are sorghum genotypes approved for commercial use in China, M36248 is a sorghum genotype created from a population in Myanmar. Using a diallel population development breeding MS line, several hybrids were produced and dispersed over China, India, and Brazil (Premalatha et al. 2006; Ribeiro et al. 2021). Reddy and Ashok Kumar (2008) claim that ICRISAT created and disseminated HT pop-S5-98, a better tillering fodder sorghum population, for agricultural use.

5.4.2.4 Mutation Breeding

Mutation techniques aim to enhance crop plants using either natural or synthetic mutations. Using induced mutations in sorghum has attracted great interest. Hydrazine effectively caused alterations in sorghum chlorophyll. Reddy and Rao (1981) and Wei et al. (2022) reported that nitroso-methyl urea and ethyl methanesulfonate were strong mutagens capable of causing favorable mutations at a concentration of 0.015 M. Reddy and Rao (1981) found that gamma radiation in the 20e35 kr range can effectively generate several physical mutagens. Induced mutagenesis, which can be brought on by either physical or chemical mutagens, is another source of variety. Both Wei et al. (2022) and Hanna (1982) noted efforts in sorghum mutation induction. Most studies of induced mutagenesis have produced mutations in chlorophyll. Mohan and Axtell successfully produced opaque endosperm mutations with high lysine in 1975 using a vitreous endosperm genotype and a 0.1% diethyl sulphate solution. P-721 was a mutant having more lysine and protein and an enhanced plant form. The Indonesian National Nuclear Agency has found the following mutant genotypes: GH-ZB-41-07, YT30–39-07, B-76, and B-92 for better grain yield; ZH30–29-07, ZH30–30-07, and ZH30–35-07 for generation of biofuels; BTx623dw5 for dwarf features; and ZH30–35-07, ZH30–30-07, ZH30–29-07 for mutations in sweet sorghum. These lines exceeded the unique genotype (Chen et al. 2019; Soeranto et al. 2010), hence their commercial distribution permission was approved in these lines. The most recent multi-cut fodder sorghum genotype in India is an EMS mutant derived from the fodder genotype CoFS 29, CSV 33MF (Rakshit and Bellundagi 2019).

5.5 Transgenesis-Transformation Methods

Transgenic plants are those that have been genetically altered to incorporate extra, permanently connected genes, typically from another species. These are genetically modified organisms, or LMOs, sometimes known as GMOs. Genetic transformation, sometimes known as transgenesis, is the process by which genes are passed from one organism to another. Transgenesis is a fresh approach to creating transgenic, or one-gene, crops. Unlike the conventional strategy, the technique introduces just the desired cloned agronomic gene(s) instead of the undesired genes from the donor. The recipient genotype is unchanged, hence, fewer backcrosses are needed. The transformation method opens a large gene pool since genes can be obtained from a wide range of sources, including viruses, bacteria, fungi, insects, animals, people, unrelated plants, or even laboratory-produced DNA. Agrobacterium, physicochemical DNA absorption, and liposomes help to visually show gene transfer mechanisms. Among the developed technologies are encapsulation, protoplast electroporation, microinjection, DNA injection into healthy plants, pollen tube pathway, microbeam laser application, tissue/embryo electroporation, particle bombardment, and in-plant transformation. Mostly using Agrobacterium and particle gun technologies, plant genetic modification is achieved (Ahmed et al. 2018).

5.5.1 Making Transgenic Plants

Transgenes allow one to put into plants safety procedures, an adjustable tag/reporter gene, promoter and terminator sequence data, and an attention gene. After that, different biological techniques and morphological laboratory tests help to identify the produced plants (Zhu et al. 2010).

5.5.1.1 Advances in Genetic Transformation

Successful genetic engineering of better crops depends on fast gene transfer, strong integration, and predictable transgenic expression. By introducing genes into plant genomes that provide resistance to several biotic and abiotic stressors, recombinant DNA-based gene transformation technologies have opened the path to increasing production. Using gene transfer techniques and repeatable tissue culture methods, agronomically beneficial genes from other genera can be introduced into the genome of Sorghum plants, therefore augmenting their yields and nutritional quality. Although gene transfer for cultivable crops has advanced a lot, sorghum has gotten quite little attention. Because of tissue culture limits, including a poor regeneration rate and the accumulation of phenolic pigments, sorghum has lagged behind other cereals in genetic transformation. Sorghum has lagged in the race to change crops by tissue culture, regeneration, and genetic engineering since it is not as receptive to transgenic technologies as other cereal crops. Transgenic sorghum is not being farmed commercially at this point. Starting now are agronomically valuable genes and selective marker modifications into sorghum. Table presents a synthesis of the research on genetic engineering in sorghum. Although sorghum tissue culture has advanced, genetic transformation is still not widespread or easily comprehensible.

5.5.1.2 Genetic Transformation

Protoplast-mediated transformation, Agrobacterium-mediated transformation and particle bombardment have all been used to genetically change sorghum.

5.5.1.3 Protoplast-Based

Transformation

Although the first study on sorghum genetic transformation involved the electroporation of DNA into protoplasts and the selection of modified cells, no indication of plant regeneration was observed (Battraw and Hall 1991). We investigated under what conditions the electroporation of chimeric neomycin phosphotransferase II (nptII) genes into sorghum protoplasts is affected. Sorghum cell suspensions can first be produced; yet, over prolonged times, they show no evidence of regeneration. Reduced regenerative capacity and elongation tendency of cells impede isolating and regeneration protoplasts. Among the several disadvantages of the protoplast technique are restricted recovery potential, time-consuming nature, genotype-specific character, and the need for specialized ability. Separating protoplasts in dicotyledons using mesophyll tissues taken from leaves proved one way to address these challenges. In 1999, Sairam et al. isolated protoplasts from Sorghum seedling leaf tissues and raised plants. Still, no additional research focused on either mesophyll protoplasts or genetic manipulation.

Particle Bombardment

One of the monocot species that has profited much from the use of gene transfer instead of the Agrobacterium-mediated procedure in the biolistic approach is Sorghum. Hagio et al. (1991) offered proof of the first known continuous expression of the *hph* and *nptII* genes, which confer tolerance to some dosages of hygromycin and kanamycin. Casas et al. (1993, 1997) pioneered the first genetically modified sorghum plants by use of microprojectile bombardment. This approach did, however, have some shortcomings, including a low conversion rate of 0.2%, a 7-month long culture period, and the adoption of the less than ideal yielding PI898012 genotype. Using genes from maize that code for anthocyanin signalling molecules R and C1, Casas et al. (1993) enhanced DNA transport factors in developing sorghum embryos. As evidenced by the transient expression of gus, which generated less than 20 blue foci per embryo, genotyping effects caused transgenic expression in sorghum to be lower than in maize. Transgenic expression levels in sorghum were probably lower than in maize because of genotyping effects, genetic-acceleration pressure interactions, or other natural characteristics of sorghum scutellar tissue. Using explants of immature embryos and inflorescences, Kononowicz et al. (1995) investigated the effects of particle bombardment on plant regeneration. Early phases of embryonic development saw varying responses to regeneration techniques among different sorghum genotypes; the most promising were genotypes PS98012 (immature embryos) and SRN39 (immature inflorescence). Through selection on geneticin, Genotype 214,856 exhibits efficacy basis producing high-quality callus and reduced sensitivity to phenolic compounds as well as optimal transformation needs, microprojectile bombardment parameters, and promoter efficacy (Tadesse et al. 2003). Using innovative blasting techniques, robust monocot gene promoters, and an antibiotic selection mechanism, scientists produced transgenic sorghum. We investigated the biological and physical elements causing the transitory expression of the *gus* reporter gene without lowering the frequency of somatic embryogenesis or regeneration capacity. Several types of explants and several stressors were used to probe the extent of transient gene expression. The main plant tissues obtained for their possible progenitor cells in embryogenesis were young embryos and shoot tips. To evaluate the transient expression of the *gus* reporter gene, Tadesse et al. (2003) employed four distinct promoters under various conditions; they discovered that the three monomeric promoters (ubi1, act1, and adh1) were more active in sorghum than the CaMV35S promoter. For the continuous expression of foreign genes in immature sorghum embryos and shoot-tip explants, the ubi1 and rice act1 gene promoters proved to be favorable. Transformation rates for shoot tips were 0.5% and for immature embryos were 1.3%. Shoots regenerated with minimum growth or no root formation were seen under selection with phosphinothricin or geneticin. When bombarded with callus explants, genotypes P898012 (Casas et al. 1993; Rathus et al. 2004), SRN 39 (Casas et al. 1997), and Tx430 (Zhu et al. 1998) showed minimal incidence of transgenic sorghum. This implies that methods of improving sorghum transformation depend on a genotype-independent plant regeneration mechanism.

Several investigations have indicated that the P898012 sorghum genotype generates more high-quality embryogenic tissue than immature embryos and is rather responsive to coconut water (Casas et al. 1993; Carvalho et al. 2004; Kaeppler and Pederson 1997). Table 5.3 reveals P898012, Tx430, and C401 as the most often occurring sorghum genotypes for successful transformation. An early embryo is one of the finest targets for sorghum among tissues. Because of their photoperiod sensitivity, some genotypes with great agricultural value are not always accessible at specific times of year. Improved methods for rapidly isolating and developing embryos could make expanded research on transformation possible (Raju et al. 2007). An interesting target for genetic modification, buds developed at the apex provide year-round advantages. Starting at the tip of the shoot, repeated shoot budding presents a possible organ for genetic engineering that can tackle regeneration and supply challenges all year long. Although callus developed from immature inflorescences has a high rate of reproduction after 30–50 days, it deteriorates rapidly and so transgenic plant repair becomes more challenging. Liu and Godwin (2012) created a quite successful microprojectile transformation method for sorghum using Tx430 genotype embryos. Twenty-25 distinct transgenic events were produced from one hundred and twelve disrupted immature cells following optimal tissue culture conditions and microprojectile transformation parameters. Obtained

Table 5.3 Genetic transformation studies in sorghum

Target gene/s	Transformation method	Explant	Genotype	Transformation efficiency (%)	References
gus, *nptII*	E	P	NK300	–	Battraw and Hall (1991)
gus, *hph*, *nptII*	PB	IE[a]	–	–	Hagio et al. (1991)
bar, *gus*	PB	IE	P898012	0.08	Casas et al. (1993)
bar, *gus*	PB	IF	P898012	0.33	Casas et al. (1997)
bar, *chiII*	PB	IE	Tx430, SRN 39	0.09	Zhu et al. (1998)
bar, *gus*	AT	IE	P898012, PHI391	2.1	Zhao et al. (2000)
bar, *gfp*	PB	IE	P898012	1.0	Able et al. (2001)
bar, *chiII*	PB	IE	Tx430, SRN39	–	Krishnaveni et al. (2004)
bar, *gus*	PB	IE	Tx430	0.18	Emani et al. (2002)
bar, *gus*	PB	IE	214856, 213108	1.3	Tadesse et al. (2003)
gfp, *G11*, *tlp*	AT	IE	Tx430, C401, Wetland	–	Jeoung et al. (2002)
gfp, *tlp*	AT	IE	Tx430, C401	2.5	Gao et al. (2005a)
*man*A	AT	IE	P8505, C401	2.88, 3.3	Gao et al. (2005b)
Bt cry1Ac	PB	SA	BTx623	1.5	Girijashankar (2005)
nptII, gusplus	AT	IE	Tx430, C2-97	0.3–4.5	Howe et al. (2006)
gus, *hpt*	AT	IE	Sensako 85/1191	5.0	Nguyen et al. (2007)
Sgfp, *man*A	AT	IE	P898012	7.7	Gurel et al. (2009)
bar, *gus*	AT	IE	P898012	0.4, 0.7	Lu et al. (2009)
mtlD	PB	SM	SPV462	4.0–7.0	Maheswari et al. (2010)
hpt	PB	IE	Ramda	0.09	Raghuwanshi and Birch (2010)
bar, *pmi*	PB	IE	P898012	0.77	Grootboom et al. (2010)
gus, *hpt*	AT	IE	P898012, RTx430	1.1–7.2	Kumar et al. (2011)
nptII, *gfp*	PB	IE	Tx430	20.7	Liu and Godwin (2012)
Bt cry1Aa, *cry1B*	PB and AT	IE	CS3541, 296B, SSV84, RSSV9	0.1–0.3	Visarada et al. (2014)

(continued)

Table 5.3 (continued)

Target gene/s	Transformation method	Explant	Genotype	Transformation efficiency (%)	References
moPAT, pmi, *dsRED*	AT	IE	Tx430	8–13.4	Wu et al. (2014)
gus	AT	IE	P898012	–	Urriola and Rathore (2014)
CYP79A1	PB	SM	CSV 15	–	Pandey et al. (2019)

[a] Derived from the suspension culture of Sorghum. Source: adapted from Balakrishna et al. (2019)
bar phosphinothricin acetyltransferase, *chiII* rice chitinase, *gfp* green fluorescent protein, *gus* β-glucuronidase, *hph* hygromycin phosphotransferase, *mtlD* mannitol-1-phosphate dehydrogenase, *manA/pmi* phosphomannose isomerase, *moPAT* codon-modified phosphinothricin acetyltransferase gene, *nptII* neomycin phosphotransferase II, *tlp* thaumatin-like protein. Electroporation, particle bombardment (PB), and Agrobacterium-mediated transformation (AT). *P* protoplast, *IE* immature embryos, *IF* inflorescences, *SA* shoot apices, *SM* shoot meristem

in three different trials, the highest notable rate observed thus far was 20.7%. The authors of the study claim that silence based on methylation is prevalent in sorghum and most likely explains why the crop has transgene deactivation in the past.

Emani et al. (2002) employed a cytidine 5-azacytidine (azaC) derivative in sorghum to fight methylation-induced transgene silence. Calli produced from T1 and T2 embryos, as well as T1 seedlings treated with 5-azacytidine, effectively induced gus gene expression (azaC.). Their results indicate that silence by methylation is ubiquitous in sorghum and most likely explains why transgenes have been deactivated historically.

Agrobacterium-Mediated Transformation

Agrobacterium-mediated transformation, which has become somewhat popular among scientists, is one of the often-used techniques for producing transgenic crops with low- or single-copy insertions. Zhao et al. (2000) demonstrated the first successful sorghum transformation using this approach concerning the line P898012. Four elements influencing sorghum transformation were found: (1) the sensitivity of the embryos to Agrobacterium infection; (2) the growth needs of the donor plant; (3) the features of the plantlets; and (4) the co-cultivation circumstances. The necrotic response in micropropagation is a key element in raising product yield in co-cultivation. One of the most crucial elements in improving yield was the degree of embryo death during co-cultivation; sorghum embryos were demonstrated to be quite sensitive to invasion by Agrobacterium. Thus, one should take into account elements such as the quantity of Agrobacterium cells in the inoculant, the selected sorghum genotypes, and the usage of less likely to be infected explants. Adding coconut water simplified callus expansion, lowered pigment synthesis, and raised the number of embryos that developed calluses (Carvalho et al. 2004). While some genotypes like P898012 respond favorably to coconut water, there may be other uses for it with other genotypes, as they do not all react the same way (Kaeppler and Pederson 1996). Visarada et al. (2003) primarily looked at the transient overexpression of GUS and investigated the impact of Agrobacterium-mediated transformation on sorghum genes using several plantlets in their work. Young embryos and the calli derived from immature inflorescences proved to be the best tissues to induce GUS expression in, according to our results. Wu et al. (2014) were able to attain enhanced transformation rates (10–33%) by adding copper sulphate and 6-benzyl amino purine to the resting and selecting media. This was carried out utilizing the Tx430 genotype while Agrobacterium infected immature embryos. Zhao et al. (2000) found the ideal conditions for the consistent Agrobacterium-mediated transformation of sorghum, thereby preparing the foundation for the next studies. The N6 medium improved both the callus responses in embryos (76% with MS and 20% with N6) and the synthesis of phenolic pigments. Two percent or less changed overall. Following careful medium and another factor adjustment for Agrobacterium-mediated sorghum transformation, a transformation incidence of 2.1% was attained. When the vir operon was activated with a certain concentration (100 M), however, the acetosyringone concentration was unaffected. L-cysteine was added to plant growth media according to Sai Kishore et al. (2004), to help with recovering potential transformants, increasing temporary GUS expression, and making transformed explants more stable. By customizing in vitro techniques among particular genotypes, one can produce genetically modified sorghum plants. Using small treatments to sterilize Agrobacterium improved the lifetime of transgenic plants in resistant genotypes. Natural selection and sterilizing agents made the sorghum-resistant calli non-regenerative (Sai Kishore et al. 2011). Furthermore, repeated sub-culturing assisted lower regeneration in the early phases of post-transformation since sweet sorghums produce polyphenols in the growth medium. By use of mild treatments for Agrobacterium disinfection, transgenic plants in resistant genotypes survived. Although the choice of sorghum yielded resistant calluses, the detrimental effects of cleaning agents and selection pressure meant that they regenerated somewhat slowly (Sai Kishore et al. 2011). Moreover, early on in post-transformation, repeated sub-culturing served to mitigate the decreased regeneration brought on by polyphenols expelled into the growing media by sweet sorghums. Plant byproducts did, however, show somewhat low Bt protein expression. The in-planta technique offers numerous choices to increase agro-infiltration liquid for transformation, like L-cysteine and Tween-20, so addressing the regeneration challenge. It's reasonable, quick, and simple. Prasad Sant refined the standards for Agrobacterium-mediated transformation and particle bombardment in 2011 using the sorghum cultivars SA281, 296B, and Tx430. Heat the plants to 43 °C for 3 min, then inoculate them with media (pH 5.2) that contained the surfactant Pluronic F-68 (0.02 percent w/v) and Agrobacterium strain LBA4404 at an inoculant intensity of 0.5 OD 600 nm was found to be the best way to achieve elevated transient GFP expression in immature embryos and callus development from immature inflorescences. The plants were co-cultivated for 3 days in darkness at 22 °C. Using 3–7 day old embryos, calli that were 4 weeks old, osmoticum treatments spanning 4 h before and after the bombardment, 0.6 m gold microparticles, helium pressure of 1500 kPa, and a 15 cm target distance defined the ideal conditions for particle bombardment. This method resulted in transient GFP expression for 14, 30, and 50 days in SA281 and 296B, respectively; particle bombardment induced fewer ulcerations than Agrobacterium-mediated transformation.

Though they perform effectively on sorghum, both the Agrobacterium and particle bombardment transformation technologies offer advantages and disadvantages. Agrobacterium-mediated transformation (Gao et al. 2005b; Howe et al. 2006; Zhao et al. 2000) has produced, thus far, a range of 2.1% to 4.5% yield. Microprojectile bombardment was demonstrated to be the most effective method thus far, with a transformation efficiency of 20.7%; Liu and Godwin 2012. In sorghum, Wang et al. (2007) found a pollen-mediated method. The ultrasonicated plasmid DNA and pollen soaked in a 0.3 mol/l sucrose solution were then utilized to pollinate the stigmas of the male sterile line A2V4A. The Southern blotting technique produced data consistent with gene integration.

5.5.2 Gene Transfer Methods

Fromm et al. 1985; Grimsley et al. 1987; Gunther and Spangenberg 1990; Herrera-Estrella 1983; Bechtold et al. 1993; Christou 1994; Feldmann and Marks 1987; Fraley et al. 1980; Frame et al. 1994; The range of genetic modification has been greatly expanded by the CRISpen/Cas9 system (Arora and Narula 2017; Wang and Russa 2016) and RNA interference (RNAi), which operates via RNA molecules to suppress gene expression or translation by targeting specific mRNA sequences (Gupta et al. 2013; Kim and Rossi 2008; Younis et al. 2014).

Many screenable biomarkers are used to assess the transgenics generated from transgene insertion (Rakosy-Tican et al. 2007; Shimada et al. 2010, 2011). Evaluating transformation concerning scorable markers or reporter genes can be done directly with GUS transcription. Maintaining an X-gluc solution at 37 °C (in the absence of light), transformed cells are kept for 1–2 weeks. Blue patches or sectors indicate their transgenic nature. Cultivating transgenic cells for at least 2 weeks in a medium including selectable markers, such as antibiotics or herbicides, at suitable dosages helps to separate them. To help the whole plant regenerate, selected explants are grown on suitable media under the control of a selective agent. Analyses of physical and genetic traits follow the regeneration of plants (Deom et al. 1990; Guttikonda et al. 2016).

5.6 Conclusion and Prospects

Developing crop varieties has benefited from historical methods of plant breeding. Bulk or pure-line selections provide high-yielding homozygous lines marked by great regularity in output by self-fertilized crops. Genetic changes produced by hybridization methods are then used to choose better plants depending on their progeny performance. Individual plant selection and crossing methods provide variations in sorghum, both naturally occurring and hybrids. Most crosses use A1 cytoplasm; nonetheless, other methods of male sterility are available to improve heterosis in sorghum hybrids. Although modern molecular crop development tools, including marker-assisted selection and phenomics approaches, have emerged, traditional breeding procedures remain indispensable for the improvement of sorghum cultivars. Along with quality attributes, basic genetic features, including disease and pest resistance, are passed along via the backcross approach.

Highly efficient transformation systems have high-throughput processing capability and are independent of genotype. Despite the complete availability of the sorghum genome sequence, site-directed integration of transgenes in sorghum poses the greatest challenges to obtaining successful sorghum genetic transformation. The deployment of transgenic sorghum in agricultural areas depends critically on the choice of successful events. Effective sorghum transgenesis calls for the combining of tissue culture, genetic modification, and, above all, plant breeding.

Successful production of transgenic events depends on the standardization of methods for sorghum tissue development. Apart from Bt genes connected to abiotic stress tolerance, insect pest resistance, and disease resistance used to raise sorghum yield and quality, there were few well-characterized gene discoveries. Research on genetic transformation presently lags behind that of other cereal crops due to the aggressive features of sorghum, which emphasizes the need for more efforts to create stable genetic transformers for this crop. Improving agricultural genetics depends on advanced mutant approaches like genome editing more and more.

Appendix 5.1 List of Major Institutes Engaged in Research on Sorghum (*Sorghum bicolor*)

Country	Institute name	Website
Egypt	Field Crops Research Institute (FCRI)	http://www.arc.sci.eg/InstsLabs/Default.aspx?OrgID=3&lang=en
Ethiopia	Ethiopian Institute of Agricultural Research.	http://www.eiar.gov.et/about-eiar/
India	The International Crops Research Institute for the Semi-Arid Tropics (ICRISAT)	https://www.icrisat.org/
India	ICAR-Central Plantation Crops Research Institute	https://cpcri.icar.gov.in/
South Africa	South African Sugarcane Research Institute	https://sasri.org.za/
South Africa	African Centre for Crop Improvement	https://acci.org.za/contact-us/

Country	Institute name	Website
Tanzania	Tanzania Agricultural Research Institute (TARI)	https://www.tari.go.tz/
Mauritius	**Mauritius Sugarcane Industry Research Institute**	http://www.msiri.mu/index.php?langue=eng&rub=50
Nigeria	The Institute for Agricultural Research (IAR) Samaru	https://iar.gov.ng/about
USA	Sugarcane Research: Houma, LA	https://www.ars.usda.gov/southeast-area/houma-la/sugarcane-research/
Australia	Sugar Research Australia	https://sugarresearch.com.au/
Brazil	CTC—Centro de Tecnologia Canavieira	https://ctc.com.br/en/
Sri Lanka	Sugarcane Research Institute	https://sugarres.lk/
Pakistan	Sugarcane Research Institute, Faisalabad	https://aari.punjab.gov.pk/
USA	Sugar Research Station, FL	https://www.lsuagcenter.com/portals/our_offices/research_stations/sugar

Appendix 5.2 List of World Sorghum (*Sorghum bicolor* L.) Varieties

Country	Recommended sugarcane varieties
China	Z6, Z27, and HC4
Egypt	Shandaweel-1, Shandaweel-305, Shandaweel-306, SOHAG-1, SOHAG-3 and SOHAG-4 and Dorado
Ethiopia	Jiru, Adelle, Fendisha-1, Dibaba
India	CSV-216R,CSV-18, CSV-22, CSH-15R, CSH-19R, CSV-26
India, ICRISAT	S 9830, ICSB 587, ICSB 570, ICSB 576
Nigeria	SAMSORG 47, SAMSORG 48 and SAMSORG 49

References

Abdel-Hamid AME, Salem KFM (2021) Breeding strategies of garden pea (*Pisum sativum* L.). In: Al-Khayri JM, Jain SM, Johnson DV (eds) Advances in plant breeding strategies: vegetable crops. Springer, Cham. https://doi.org/10.1007/978-3-030-66969-0_9

Abdullaev FK, Rustamov NSU, Yakubov MD (2025) Uzbekistan Genebank. In: Al-Khayri JM, Salem KFM, Jain SM (eds) Plant gene banks. Springer, Singapore. https://doi.org/10.1007/978-981-99-4236-7_57-1

Able JA, Rathus C, Godwin ID (2001) The investigation of optimal bombardment parameters for transient and stable transgene expression in sorghum. In Vitro Cell Dev Biol Plant 37:341–348

Aboshama HM, Salem KFM, Hassan HMF, Tahoun MH (2015) Estimating heterosis and combining ability for in vivo and in vitro traits using diallel cross on tomato (*Solanum lycopersicum* L.). Minufiya. J Agric Res 40(3):757–770

Abreha KB, Enyew M, Carlsson AS et al (2022) Sorghum in dryland: morphological, physiological, and molecular responses of sorghum under drought stress. Planta 255:20. https://doi.org/10.1007/s00425-021-03799-7

Abu-Ellail FFB, Salem KFM, Saleh MM et al (2021) Molecular breeding strategies of beetroot (*Beta vulgaris* ssp. vulgaris var. conditiva Alefeld). In: Al-Khayri JM, Jain SM, Johnson DV (eds) Advances in plant breeding strategies: vegetable crops. Springer, Cham. https://doi.org/10.1007/978-3-030-66965-2_4

Abulaila K, Al Sane KO, Tahabsom Z (2025) Jordan Seed Bank. In: Al-Khayri JM, Salem KFM, Jain SM (eds) Plant gene banks. Springer, Singapore. https://doi.org/10.1007/978-981-99-4236-7_45-1

Ahmed RI, Ding A, Xie M, Kong Y (2018) Progress in optimization of agrobacterium-mediated transformation in sorghum (*Sorghum bicolor*). Int J Mol Sci 19:2983. https://doi.org/10.3390/ijms19102983

Akparov ZI, Mammadova SA, Kalantarova NS, Mirzaliyeva IA, Mammadov AT (2025) National Genebank of Azerbaijan. In: Al-Khayri JM, Salem KFM, Jain SM (eds) Plant Gene Banks. Springer, Singapore. https://doi.org/10.1007/978-981-99-4236-7_31-1

Al Lawati AH (2025) Plant genetic resource conservation of Oman. In: Al-Khayri JM, Salem KFM, Jain SM (eds) Plant Gene Banks. Springer, Singapore. https://doi.org/10.1007/978-981-99-4236-7_50-1

Aladele SE et al (2025) Plant genetic resources conservation and utilization in Nigeria. In: Al-Khayri JM, Salem KFM, Jain SM (eds) Plant Gene Banks. Springer, Singapore. https://doi.org/10.1007/978-981-99-4236-7_17-1

Aldahak L, Salem KFM, Al-Salim SHF, Al-Khayri JM (2021) Advances in chicory (*Cichorium intybus* L.) breeding strategies. In: Al-Khayri JM, Jain SM, Johnson DV (eds) Advances in plant breeding strategies: vegetable crops. Springer, Cham. https://doi.org/10.1007/978-3-030-66969-0_1

Al-Khayri JM, Salem KFM, Jain SM (2025) Plant gene banks. Springer Singapore. https://doi.org/10.1007/978-981-99-4236-7

Allard RW (1960) Principles of plant breeding. Wiley, New York

Almarri NB, Ahmad S, Alsubaie GTG, Elshal MH (2025) Saudi Arabia Gene Bank. In: Al-Khayri JM, Salem KFM, Jain SM (eds) Plant Gene Banks. Springer, Singapore. https://doi.org/10.1007/978-981-99-4236-7_54-1

Alqahtani MM, Saleh MM, Alwutayd KM et al (2024) Performance and genotypic variability in diverse date palm (*Phoenix dactylifera* L.) cultivars for fruit characteristics. Genet Resour Crop Evol 71:1759–1772. https://doi.org/10.1007/s10722-023-01719-0

Arab S et al (2025) National Gene Bank of Egypt. In: Al-Khayri JM, Salem KFM, Jain SM (eds) Plant Gene Banks. Springer, Singapore. https://doi.org/10.1007/978-981-99-4236-7_5-1

Arora L, Narula A (2017) Gene editing and crop improvement using CRISPR-Cas9 system. Front Plant Sci 2017(8). https://doi.org/10.3389/fpls.2017.01932

Audilakshmi S, Aruna C (2008) Breeding methods in sorghum. In: Reddy BVS, Ramesh S, Ashok Kumar A, Gowda CLL (eds) Sorghum improvement in the new millennium. ICRISAT, pp 28–30

Babeanu C, Matei G, Dodocioiu AM (2017) Evaluation of sugar content in stem juice from sweet sorghum hybrids grown in Southwest Romania. Ann Univ Craiova Agric Montanol Cadastre Ser 47:25–29

Balakrishna D, Vinodh R, Madhu P et al (2019) Tissue culture and genetic transformation in *Sorghum bicolor*. In: Breeding Sorghum for diverse end uses. Woodhead Publishing, pp 115–130

Batey I (2017) The diversity of uses for cereal grains. In: Wrigley C, Batey I, Miskelly D (eds) Cereal grains. Elsevier, Amsterdam, pp 41–45

Battraw M, Hall TC (1991) Stable transformation of *Sorghum bicolor* protoplast with chimeric neomycin phototransferase II and β-glucuronidase genes. Theor Appl Genet 82:161–168

Bechtold N, Ellis J, Pelletier G (1993) In planta agrobacterium-mediated gene transfer by infiltration of adult Arabidopsis thaliana plants. Acad Sci Paris Life Sci 316:1194–1199

Békefi Z, Szilágyi S, Desiderio F, Demku T, Varga J (2025) Prunus Gene Bank in Hungarian university of agriculture and life sciences. In: Al-Khayri JM, Salem KFM, Jain SM (eds) Plant Gene Banks. Springer, Singapore. https://doi.org/10.1007/978-981-99-4236-7_71-1

Bello AB, Waniska RD, Gomez MH, Rooney LW (1995) Starch solubilization and retrogradation during preparation of to (a food gel) from different sorghum cultivars. Cereal Chem 72:80–84

Bhagwat VR, Shyam Prasad G, Kalaisekhar A et al (2013) Major insect pests of sorghum and their scoring procedure. In: Rakshit S, Patil JV (eds) Basics of sorghum breeding and AICSIP data management. Directorate of Sorghum Research, Hyderabad, pp 51–61

Billa E, Koullas DP, Monties B et al (1997) Structure and composition of sweet sorghum stalk components. Ind Crops Products 6:297–302

Börner A, Landjeva S, Salem KFM, Lohwasser U (2010) Plant genetic resources- a prerequisite for drought tolerance breeding in cereals. In: Proceedings of the Jahrestagung der Vereinigung der Pflanzenzüchter und Saatgutkaufleute Österreichs, 24-26 November 2009, Raumberg-Gumpenstein, pp 11–13

Boyer JS (1982) Plant productivity and environment. Science 218:443–448

Breidy J (2025) Lebanon Gene Bank. In: Al-Khayri JM, Salem KFM, Jain SM (eds) Plant Gene Banks. Springer, Singapore. https://doi.org/10.1007/978-981-99-4236-7_48-1

Cagampang GB, Creek B, Kirleis AW (1985) Properties of starches isolated from sorghum floury and corneous endosperm. Starch 37:253–257

Cañar DY et al (2025) Colombia Gene Bank. In: Al-Khayri JM, Salem KFM, Jain SM (eds) Plant Gene Banks. Springer, Singapore. https://doi.org/10.1007/978-981-99-4236-7_95-1

Carvalho CHS, Zehr UB, Gunaratna N et al (2004) Agrobacterium-mediated transformation of sorghum: factors that affect transformation efficiency. Genet Mol Biol 27:259–269

Casas AM, Kononowicz AK, Zehr UB et al (1993) Transgenic sorghum plants via microprojectile bombardment. Proc Natl Acad Sci USA 90:11212–11216

Casas AM, Kononowicz AK, Haan TG et al (1997) Transgenic sorghum plants obtained after microprojectile bombardment of immature inflorescences. In Vitro Cell Dev Biol Plant 33:92–100

Celarier RP (1958) Cytotaxonomy of the Andropogoneae. III. Sub-tribe Sorgheae, genus *sorghum*. Cytologia 23:395–418

Chen JP, Xin ZG, Laza H (2019) Registration of BTx623dw5, a new sorghum dwarf mutant. J Plant Reg 13(2):254–257

Christou P (1994) Applications to plants. In: Yang NS, Christou P (eds) Particle bombardment technology for gene transfer. Oxford University Press, New York, pp 71–99

Clayton WD, Renvoize SA (1986) Genera graminum. Grasses of the world, Kew bulletin series XIII. Her Majesty's Stationery Office, London, pp 338–345

Cowan M, Møller BL, Norton S et al (2022) Cyanogenesis in the sorghum genus: from genotype to phenotype. Genes 13(1):140

Dahlberg J (2019) The role of sorghum in renewables and biofuels. In: Zhao ZY, Dahlberg J (eds) Sorghum: methods and protocols, methods in molecular biology, vol 1931, pp 269–277. https://doi.org/10.1007/978-1-4939-9039-9_1

Das IK (2013) Major sorghum diseases and scoring techniques. In: Rakshit S, Patil JV (eds) Basics of sorghum breeding and AICSIP data management. Directorate of Sorghum Research, Hyderabad, pp 62–69

de Galarreta JIR, Barandalla L, Ritter E, Velázquez ME, Rios D (2025) Spanish potato germplasm Bank. In: Al-Khayri JM, Salem KFM, Jain SM (eds) Plant Gene Banks. Springer, Singapore. https://doi.org/10.1007/978-981-99-4236-7_114-1

de Siqueira FS, Nishiyama MY Jr, Paterson AH, Souza GM (2013) Biofuel and energy crops: high yield Saccharinae take center stage in the post-genomics era. Genome Biol 14(6):210. https://doi.org/10.1186/gb-2013-14-6-210

de Wet JMJ (1968) The origin of weediness in plants. Proc Okla Acad Sci 47:14–17 de Wet JMJ (1978) Systematics and evolution of *sorghum* sect. *Sorghum* (Gramineae). Am J Bot 65:477–484

de Wet JMJ, Huckabay JP (1967) The Origin of *Sorghum bicolor*. II. Distribution and domestication. Evolution 21(4):787–802. https://www.jstor.org/stable/2406774

De Wet JMJ, Harlan JR, Price EG (1970) Origin of variability in the Spontanea complex of *Sorghum bicolor*. Am J Bot 57:704–707. https://doi.org/10.2307/2441294

Deom CM, Schubert KR, Wolfs S et al (1990) Molecular characterization and biological function of the movement protein of tobacco mosaic virus in transgenic plants. Proc Natl Acad Sci USA 87:3284–3288

Desheva G, Valchinova E, Deshev M (2025) Bulgarian gene bank. In: Al-Khayri JM, Salem KFM, Jain SM (eds) Plant Gene Banks. Springer, Singapore. https://doi.org/10.1007/978-981-99-4236-7_66-1

Díez MJ, Valcárcel JV, Solbes E, Muñoz E (2025) Forty years conserving Spanish biodiversity in the Gene Bank of the Universitat Politècnica de València, Spain. In: Al-Khayri JM, Salem KFM, Jain SM (eds) Plant Gene Banks. Springer, Singapore. https://doi.org/10.1007/978-981-99-4236-7_86-1

Dillon SL, Shapter FM, Henry RJ et al (2007b) Domestication to crop improvement: genetic resources for *Sorghum* and *Saccharum* (Andropogoneae). Ann Bot 100:975–989

Doggett H (1965) The development of cultivated sorghums. In: Hutchinson JB (ed) Essays in crop plant evolution. Cambridge University Press, Cambridge, pp 50–69

Doggett H (1972) The improvement of sorghum in East Africa. In: Rao NGP, House LR (eds) Sorghum in seventies. Oxford and IBH Publishing Co, New Delhi

Doggett H (1988) Sorghum, 2nd edn. Longman Scientific and Technical, Essex

Dwivedi S, Sahrawat K, Upadhyaya HD, Ortiz R (2013) Food, nutrition and agrobiodiversity under global climate change. In: Sparks DL (ed) Advances in agronomy. Academic/Elsevier Inc, pp 1–128

Eid NA, Ibrahim AMH, Elsharawy AA, Salem KFM (2025) Nanofertilizers for plant viral disease management. In: Kumar P, Dubey RC (eds) Nanofertilizers for sustainable agriculture. Springer, Cham. https://doi.org/10.1007/978-3-031-78649-5_7

Elabd AB, Salem KFM (2009) Detection of genetic diversity in rice (*Oryza sativa* L.) Varieties using microsatellite DNA markers and Quantitative Traits. In: 6th Egyptian Plant Society of Breeding Conference, 4–7 May 2009, Ismailia, Egypt

Elabd AB, Saleh MM, Aldahak L, Salem KFM (2020) Identification of superior rice (*Oryza sativa* L.) genotypes under irrigated conditions based on multivariate analysis. In: 7th international scientific conference of genetic and environment from 8th –9th December 2020, Baghdad, Iraq

El-Absawy EA, Salem KFM, Abdellatif KF, ElShamy H (2012) Estimation of genetic diversity among some Egyptian barley (*Hordeum vulgare* L.) cultivars using RAPD markers and agronomic traits. In: 1st international conference on biotechnology applications in agriculture, Moshtohor-Hurghada, 18–22 February 2012. Faculty of Agriculture, Benha University, Egypt

El-Absawy EA, Nowar AA, Salem KFM et al (2015) Estimation of genetic diversity among some bread wheat (*Triticum aestivum* L.) genotypes using molecular markers and agronomic traits. Minufiya. J Agric Res 40(3):223–233

Eltaher S, Sallam A, Belamkar V et al (2018) Genetic diversity and population structure of F3:6 Nebraska winter wheat genotypes using genotyping-by-sequencing. Front Genet 9:76. https://doi.org/10.3389/fgene.2018.00076

Eltaher S, Baenziger S, Belamkar V et al (2021) GWAS revealed effect of genotype x environment interactions for grain yield of Nebraska winter wheat. BMC Genomics 22:1–14

Eltaher S, Sallam A, Emara HA et al (2022) Genome-wide association mapping revealed SNP alleles associated with spike traits in wheat. Agronomy 12:1469. https://doi.org/10.3390/agronomy12061469

El-Zanaty AM, Salem KFM, Esmail RM (2011) Detection of genetic diversity in Egyptian cotton (*Gossypium barbadense* L.) varieties using RAPD markers and morphological traits. J Am Sci 7(12):1107–1115

El-Zanaty AM, Salem KFM, Ismail M, Esmail RM (2013) Impact of plant breeding on genetic diversity of the Egyptian cotton (*Gossypium barbadense* L.) varieties as revealed by RAPD markers. J Agric Chem Biotechnol Mansoura Univ 4:217–227

Emani C, Sunilkumar G, Rathore KS (2002) Transgene silencing and reactivation in sorghum. Plant Sci 162:181–192

Emara HA, Nower AA, Salem KFM, Eltaher S (2011) Estimating heterosis and combining ability for in vivo and in vitro traits using diallel cross in maize (*Zea mays* L). J Product Dev 16:101–120

Emara HA, Salem KFM, Abdel-Halim AA, Abd El-Kowy RNF (2013) An in vivo and in vitro traits analysis of a diallel cross in rice (*Oryza sativa* L.). Egypt J Plant Breed 17(2):591–610

Esmail RM, Salem KFM, El-Zanaty AM (2009) Genetic diversity studies of Egyptian cotton (*Gossypium barbadense* L.) varieties using RAPD markers and agronomic traits. In: 6th Egyptian Plant Society of Breeding Conference, 4–7 May 2009, Ismalia, Egypt

FAO (2021) FAOSTAT statistical database. https://www.fao.org/giews/countrybrief/country.jsp?code=BOL&lang=ar

Feldmann KA, Marks MD (1987) Agrobacterium-mediated transformation of germinating seeds of Arabidopsis thaliana: a non-tissue culture approach. Mol Gen Genet 208(1–2):1–9

Fraley R, Wilschut J, Düzgüneş N et al (1980) Studies on the mechanism of membrane fusion: role of phosphate in promoting calcium ion-induced fusion of phospholipid vesicles. Biochemistry 19(26):6021–6029

Frame BR, Drayton PR, Bagnall V et al (1994) Production of fertile transgenic maize plants by silicon carbide whisker-mediated transformation. Plant J 6(6):941–948

Frankowski J, Przybylska-Balcerek A, Stuper-Szablewska K (2022) Concentration of pro-health compound of sorghum grain-based foods. Foods 11(2):216. https://doi.org/10.3390/foods11020216

Fromm M, Taylor LP, Walbot V (1985) Expression of genes transferred to monocot and dicot plant cells by electroporation. Proc Natl Acad Sci USA 82:5824–5828

Galea-Gragera FA, Llera Cid F (2025) Spain germplasm Bank of Annual Forage Legumes Seeds at the scientific and technological research Center of Extremadura (CICYTEX). In: Al-Khayri JM, Salem KFM, Jain SM (eds) Plant Gene Banks. Springer, Singapore. https://doi.org/10.1007/978-981-99-4236-7_102-1

Gao Z, Jayaraj J, Muthukrishnan S et al (2005a) Efficient genetic transformation of sorghum using a visual screening marker. Genome 48(2):321–333

Gao Z, Xie X, Ling Y et al (2005b) Agrobacterium tumefaciens-mediated sorghum transformation using a mannose selection system. Plant Biotechnol J 3(6):591–599

Ghazy MI, Salem KFM, Sallam A (2020) Utilization of genetic diversity and marker-trait to improve drought tolerance in rice (*Oryza sativa* L.). Mol Biol Rep 48:157–170. https://doi.org/10.1007/s11033-020-06029-7

Ghidan WF, Abdulmajid DA (2025) Egyptian Rice Gene Bank. In: Al-Khayri JM, Salem KFM, Jain SM (eds) Plant Gene Banks. Springer, Singapore. https://doi.org/10.1007/978-981-99-4236-7_106-1

Girijashankar V (2005) Effect of promoters on Bt transgene expression in sorghum (*Sorghum bicolor* Moench). PhD Biotechnology thesis, Dept of Biotechnology, Jawaharlal Nehru Technological University, Hyderabad, AP, India

Goyal M, Sohu RS, Bhardwaj R et al (2020) Genetic variance and predicted response for three types of recurrent selection procedures in forage sorghum [*Sorghum bicolor* (L.) Moench]. Range Manag Agrof 41(1):43–51

Grimsley N, Hohn T, Daview JW, Hohn B (1987) Agrobacterium-mediated delivery of infectious maize streak virus into maize plants. Nature 325:177–179

Grootboom AW, Mkhonza NL, O'Kennedy MM et al (2010) Biolistic mediated sorghum (*Sorghum bicolor* L. Moench) transformation via mannose and Bialaphos based selection systems. Int J Bot 6(2):89–94

Gudaye TY, Lakew BF, Selamo MM, Tessema ES (2025) Ethiopian crop Gene Bank. In: Al-Khayri JM, Salem KFM, Jain SM (eds) Plant Gene Banks. Springer, Singapore. https://doi.org/10.1007/978-981-99-4236-7_7-1

Gunther N, Spangenberg G (1990) Plant transformation by microinjection techniques. Physiol Plant 79:213–217

Gupta B, Saha J, Sengupta A, Gupta K (2013) Recent advances on virus-induced gene silencing (VIGS): plant functional genomics. J Plant Biochem Physiol 1:e116. https://doi.org/10.4172/2329-9029.1000e116

Gurel S, Gurel E, Kaur R et al (2009) Efficient, reproducible Agrobacterium-mediated transformation of sorghum using heat treatment of immature embryos. Plant Cell Rep 28(3):429–444

Guttikonda SK, Marri P, Mammadov J et al (2016) Molecular characterization of transgenic events using next-generation sequencing approach. PLoS One 11(2):e0149515

Hagio T, Blowers AD, Earle ED (1991) Stable transformation of sorghum cell cultures after bombardment with DNA-coated microprojectiles. Plant Cell Rep 10:260–264

Hanna WW (1982) Mutation breeding of pearl millet and sorghum. Mutat Breed Rev 1:1–13

Hao H, Li Z, Leng C et al (2020) Sorghum breeding in the genomic era: opportunities and challenges. Theor Appl Genet 134:1899–1924. https://doi.org/10.1007/s00122-021-03789-z

Harlan JR, de Wet JMJ (1971) Toward a rational classification of cultivated plants. Taxon 20:509–517

Harlan JR, de Wet JMJ (1972) A simplified classification of cultivated sorghum. Crop Sci 12:172–176

Harzli I, Mostafa K, Abdulla MF, Mehmood M (2025) Sustaining Turkish genetic resources. In: Al-Khayri JM, Salem KFM, Jain SM (eds) Plant Gene Banks. Springer, Singapore. https://doi.org/10.1007/978-981-99-4236-7_88-1

Hassan MN, Mekkawy SA, Mahdy M et al (2021) Recent molecular and breeding strategies in lettuce (*Lactuca* spp.). Genet Resour Crop Evol 68:3055–3079. https://doi.org/10.1007/s10722-021-01246-w

Herrera-Estrella L (1983) Transfer and expression of foreign genes in plants. PhD thesis. Laboratory of Genetics, Gent University, Belgium

Heuzé V, Tran G, Bastianelli et al (2013) Sorghum by-products. www.feedipedia.org/node/752] site visited 28/10/2013

Hidayatun N, Kurniawan H (2025) Conservation of agricultural genetic resources in Indonesia. In: Al-Khayri JM, Salem KFM, Jain SM (eds) Plant Gene Banks. Springer, Singapore. https://doi.org/10.1007/978-981-99-4236-7_40-1

House LR (1985) A guide to sorghum breeding. ICRISAT, India

House LR (2005) Sorghum. One of the world's great cereals. Afr Crop Sci J 3(2):130–135

Howe A, Sato S, Dweikat I et al (2006) Rapid and reproducible agrobacterium-mediated transformation of sorghum. Plant Cell Rep 25(8):784–791

Ibrahim HM (2015) Effectiveness of breeding methods for production of superior genotypes and maintenance of genetic variance in Faba Bean (*Vicia faba* L.). Am J Life Sci 3(1):11–16

Ibrahim AA, Salem KFM, Abdein MA, Ramadan SM (2023a) Advances in summer squash (*Cucurbita pepo* L.) molecular breeding strategies. In: Singh S, Sharma D, Sharma SK, Singh R (eds) Smart plant breeding for vegetable crops in post-genomics era. Springer, Singapore. https://doi.org/10.1007/978-981-19-5367-5_8

Ibrahim AA, Salem KFM, Ramadan S (2023b) Allium sativum and stressful conditions. In: Abdel Latef AAH (ed) Book medicinal plant responses to stressful conditions. CRC Press

Ihsan F, Prihatini R (2025) Indonesian tropical fruit Gene Bank. In: Al-Khayri JM, Salem KFM, Jain SM (eds) Plant Gene Banks. Springer, Singapore. https://doi.org/10.1007/978-981-99-4236-7_41-1

Ismail M, Salem KFM, El-Zanaty AM, Esmail RM (2013) Association mapping of fiber and agronomic characters in Egyptian cotton (*Gossypium barbadense* L.) genotypes. J Agric Chem Biotechnol Mansoura Univ 4:205–216

Jan-Orn J, Gardner CO, Ross WM (1976) Quantitative genetic studies of the NP3R random mating grain sorghum population. Crop Sci 16:489–496

Jatoi SA, Siddiqui SU (2025) National Gene Bank of Pakistan. In: Al-Khayri JM, Salem KFM, Jain SM (eds) Plant Gene Banks. Springer, Singapore. https://doi.org/10.1007/978-981-99-4236-7_51-1

Jeoung MJ, Krishnaveni S, Muthukrishnan S et al (2002) Optimization of sorghum transformation parameters using genes for green fluorescent protein and â-glucuronidase as visual markers. Hereditas 137:20–28

Kaeppler HF, Pederson JF (1996) Media effects on phenotype of callus cultures initiated from photoperiod insensitive, elite inbred sorghum lines. Maydica 41(2):83–89

Kaeppler HF, Pederson JF (1997) Evaluation of 41 elite and exotic inbred sorghum genotypes for high-quality callus production. Plant Cell Tissue Organ Cult 48:71–75

Kamala V, Singh SD, Bramel PJ, Rao DM (2002) Sources of resistance to downy mildew in wild and weedy sorghums. Crop Sci 42:1357–1360

Kamala V, Sivaraj N, Pandravada SN et al (2016) Potential of wild relatives in sorghum improvement through molecular approaches. In: Conference on National Priorities in plant health management, February 4–5, Tirupati, Telangana

Khalilian ME, Habibi D, Golzardi F et al (2022) Effect of maturity stage on yield, morphological characteristics, and feed value of sorghum [*Sorghum bicolor* (L.) Moench] cultivars. Cereal Res Commun 50:1095–1104. https://doi.org/10.1007/s42976-022-00244-7

Kim DH, Rossi JJ (2008) RNAi mechanisms and applications. BioTechniques 44:613–616

Kimber CT (2000) Chapter 1.1: Origin of domesticated sorghum and its early diffusion to India and China. In: Smith CW, Frederiksen RA (eds) Sorghum: origin, history, technology and production. Wiley, New York, pp 3–98

Kong W, Jin H, Franks CD et al (2013) Genetic analysis of recombinant inbred lines for *Sorghum bicolor* × *Sorghum propinquum*. G3 3(1):101–108. https://doi.org/10.1534/g3.112.004499

Kononowicz AK, Casas AM, Tomes DT et al (1995) New vistas are opened for sorghum improvement by genetic transformation. Afr Crop Sci J 3:171–180

Krishnaveni S, Jeoung JM, Mutukrishnan S, Liang GH (2004) Biolistic transformation of sorghum and influence of a transgenic chitinase gene. In: Seetharama N, Godwin ID (eds) Sorghum tissue culture and transformation. Oxford Publishers, New Delhi, pp 65–74

Kumar V, Campbell LM, Rathore KS (2011) Rapid recovery and characterization of transformants following Agrobacterium-mediated T-DNA transfer to sorghum. Plant Cell Tissue Organ Cult 104:137–146

Kumari P, Kharor N, Pahuja SK et al (2024) Morphological and biochemical traits attribution for fodder yield and quality in forage sorghum using principal component analysis. Vegetos 37:2059–2068. https://doi.org/10.1007/s42535-024-01017-9

Laghetti G, Piergiovanni AR, De Paola D, Direnzo P, Bucci G, Sonnante G (2025) Italy IBBR gene bank. In: Al-Khayri JM, Salem KFM, Jain SM (eds) Plant Gene Banks. Springer, Singapore. https://doi.org/10.1007/978-981-99-4236-7_73-1

Leder I (2004) Sorghum and millets, cultivated plants as food sources. v.1. www.eolsd.net/samplechapters/c10/E5-02-01-04.pdf

Liu G, Godwin ID (2012) Highly efficient sorghum transformation. Plant Cell Rep 31(6):999–1007

Liu Q, Liu H, Wen J (2017) Relationships between *Sorghum bicolor* (Poaceae) and its close relatives based on genomic in situ hybridization evidence. Turk J Bot 41:11–24

Liu ZQ, Li HL, Zeng XJ et al (2020) Coupling phytoremediation of cadmium-contaminated soil with safe crop production based on a sorghum farming system. J Clean Prod 275:123002

Lu L, Wu X, Yin X et al (2009) Development of marker-free transgenic sorghum [*Sorghum bicolor* (L.) Moench] using standard binary vectors with bar as a selectable marker. Plant Cell Tissue Organ Cult. https://doi.org/10.1007/s11240-009-9580-4

Maheswari M, Varalaxmi Y, Vijayalakshmi A et al (2010) Metabolic engineering using mtlD gene enhances tolerance to water defi cit and salinity in sorghum. Biol Plant 54(4):647–652

Mallor C, Luna E (2025) The genetic diversity of the Spanish vegetable Genebank BGHZ-CITA: conservation and utilization. In: Al-Khayri JM, Salem KFM, Jain SM (eds) Plant Gene Banks. Springer, Singapore. https://doi.org/10.1007/978-981-99-4236-7_109-1

Matei G (2016) Study on yield features of sweet sorghum hybrids grown in south west of Romania. In: SGEM2016, Conference proceedings, vol 1, pp 783–790

Mattar MZ, Eissa OA, Salem KFM (2009) Genotypic variability in some wheat genotypes (*Triticum aestivum* L.) exposed to salinity at germination stage. Egypt J Exp Biol (Bot) 5:143–148

Mattar MZ, Salem KFM, El-Abed AB (2016) Identification of candidate microsatellite markers associated with agronomic traits in rice (*Oryza sativa* L.). J Bot 15(5):213–232

Meena K, Jacob J, Bhavyasri M et al (2024) Wide hybridization in sorghum with tertiary gene pool for diversification of parental lines. Crop Sci 64:2589–2259

Mella ON (2011) Effects of malting and fermentation on the composition and functionality of sorghum flour. MSc thesis, University of Nebraska, Lincoln, Nebraska, p 90

Monteros-Altamirano Á et al (2025) Plant genetic resources in Ecuador. In: Al-Khayri JM, Salem KFM, Jain SM (eds) Plant Gene Banks. Springer, Singapore. https://doi.org/10.1007/978-981-99-4236-7_99-1

Morris GP, Ramu P, Deshpande SP et al (2013) Population genomic and genome-wide association studies of agroclimatic traits in sorghum. Proc Natl Acad Sci USA 110(2):453–458

Mourad AMI, Alomari DZ, Alqudah AM et al (2019) Recent advances in wheat (*Triticum* spp.) breeding. In: Al-Khayri J, Jain S, Johnson D (eds) Advances in plant breeding strategies: cereals. Springer, Cham. https://doi.org/10.1007/978-3-030-23108-8_15

Murdock GP (1959) Africa its people and their culture history. McGraw-Hill, New York. 1960. Staple subsistence crops of Africa. Geogr Rev 50:521–540

Nath B (1972) Population breeding techniques in sorghum: sorghum in the eighties. In: Proceedings of the international symposium on sorghum, vol 1. ICRISAT Center, Patancheru, Hyderabad

Nawara HM, Mattar MZ, Salem KFM, Eissa OA (2016) Estimating heterosis and combining ability for agronomic and tissue culture traits using diallel cross on bread wheat (*Triticum aestivum* L.). Egypt J Bot 2016:177–200

Nawara HM, Mattar MZ, Salem KFM, Eissa OA (2017) Diallel study on some in vitro traits of bread wheat (*Triticum aestivum* L.) under salt stress. Int J Agric Environ Res 3(1):1988–2006

Nguyen TV, Thu TT, Claeys M, Angenon G (2007) Agrobacterium-mediated transformation of sorghum using an improved in vitro regeneration system. Plant Cell Tissue Organ Cult 91:155–164

Nilsson-Ehle H (1908) Einige Ergebnisse von Kreuzungen bei Hafer und Weizen. Botan Notizen:257–294

Obilana AT, El-Rouby MM (1980) Recurrent mass selection for yield in two random mating populations of sorghum [*Sorghum bicolor* (L.) Moench]. Maydica 25:127–133

Osterc G, Gostinčar K, Hudina M, Usenik V, Luthar Z (2025) Slovene Fruit Gene Bank. In: Al-Khayri JM, Salem KFM, Jain SM (eds) Plant Gene Banks. Springer, Singapore. https://doi.org/10.1007/978-981-99-4236-7_84-1

Pandey AK, Madhu P, Bhat BV (2019) Down-regulation of *CYP79A1* gene through antisense approach reduced the cyanogenic glycoside Dhurrin in [*Sorghum bicolor* (L.) Moench] to improve fodder quality. Front Nutr 6:122. https://doi.org/10.3389/fnut.2019.00122

Parthasarathy RP, Basavaraj G, Ahmad W et al (2010) An analysis of availability and utilization of sorghum grain in India. J SAT Agric Res 8:1–8

Patil JV, Sanjana RP (2013) Practical considerations in rabi sorghum breeding. In: Rakshit S, Patil JV (eds) Basics of sorghum breeding and AICSIP data management. Directorate of Sorghum Research, Hyderabad, pp 95–102

Phuong N, Stützel H, Uptmoor R (2013) Quantitative trait loci associated to agronomic traits and yield components in a *Sorghum bicolor* L. Moench RIL population cultivated under pre-flowering drought and well-watered conditions. Agric Sci 4:781–791. https://doi.org/10.4236/as.2013.412107

Poehlman JM (1987) In: Reinhold v N (ed) Breeding field crops, 3rd edn. University of Missouri, Columbia. https://doi.org/10.1007/978-94-015-7271-2

Popescu A, Condei R (2014) Some considerations on the prospects of sorghum crop. Sci Pap Ser Manag Econ Eng Agric Rural Dev 14(3):295–304

Popović V et al (2025) Economic seed production and storage of fiber crops in gene banks in Serbia. In: Al-Khayri JM, Salem KFM, Jain SM (eds) Plant Gene Banks. Springer, Singapore. https://doi.org/10.1007/978-981-99-4236-7_108-1

Premalatha N, Kumaravadivel N, Veerabadhiran P (2006) Heterosis and combining ability for grain yield and its components in sorghum [*Sorghum bicolor* (L.) Moench]. Indian J Genet 66:123–126

Price HJ, Dillon SL, Hodnett G et al (2005) Genome evolution in the genus *Sorghum* (Poaceae). Ann Bot 95:219–227

Quinby JR (1967) The maturity genes of sorghum. Adv Agron 19:267–305

Raghuwanshi A, Birch RG (2010) Genetic transformation of sweet sorghum. Plant Cell Rep 29(9):997–1005

Raju RK, Nagaraju K, Annapurna A et al (2007) Simple methods to generate ample explant sources for genetic transformation of sorghum. In: Poster presented at 8th Agricultural Science Congress held during 15–17 Feb 2007. Tamil Nadu Agricultural University (TNAU), Coimbatore

Rakosy-Tican E, Aurori CM, Dijkstra C et al (2007) The usefulness of the *gfp* reporter gene for monitoring agrobacterium-mediated transformation of potato dihaploid and tetraploid genotypes. Plant Cell Rep 26(5):661–671

Rakshit S, Bellundagi A (2019) Conventional breeding techniques in sorghum. In: Aruna C, Visarada KBRS, Bhat BV, Tonapi VA (eds) Breeding sorghum for diverse end uses. Woodhead Publishing, Duxford, pp 77–91. https://doi.org/10.1016/B978-0-08-101879-8.00005-X

Rakshit S, Gomashe S (2013) Basics of plant breeding with reference to sorghum. In: Rakshit S, Patil JV (eds) Basics of sorghum breeding and AICSIP data management. Directorate of Sorghum Research, Hyderabad, pp 9–16

Rakshit S, Rakshit A, Patil JV (2012) Multi-parent intercross populations in analysis of quantitative traits. J Genet 91:111–117

Rathus C, Nguyen T, Able JA et al (2004) Optimizing sorghum transformation technology via somatic embryogenesis. In: Seetharama N, Godwin ID (eds) Sorghum tissue culture, transformation and genetic engineering. ICRISAT and Oxford Publishers, New Delhi

Ratnavathi CV, Patil JV (2013) Sorghum utilization as food. Nutr Food Sci 4(1):1–8

Reddy BVS, Ashok Kumar A (2008) Population improvement in sorghum. In: Reddy BVS, Ramesh S, Ashok Kumar A, Gowda CLL (eds) Sorghum improvement in the new millennium. International Crops Research Institute for the Semi-Arid Tropics, Patancheru, pp 93–104

Reddy CS, Rao NGP (1981) Induced mutations in sorghum improvement. Proc Indian Natl Sci Acad B 47(3):427–446

Reddy BVS, Ramesh S, Reddy PS (2006) Sorghum genetic resources, cytogenetics and improvement. In: Singh RJ, Jauhar PP (eds) Genetic resources, genetic engineering and crop improvement, vol 2. Cereals, CRC Taylor and Francis, pp 309–363

Rhodes DH, Hoffmann L, Rooney WL et al (2017) Genetic architecture of kernel composition in global sorghum germplasm. BMC Genomics 18(1):15

Ribeiro LP, Tardin FD, de Menezes CB et al (2021) Combining yield, earliness and plant height in a single genotype: a proposal for breeding in grain sorghum (*Sorghum bicolor* L.). Revista de la Facultad de Ciencias Agrarias UNCuyo 53(1):11–21

Rocha F et al (2025) Portugal Gene Bank. In: Al-Khayri JM, Salem KFM, Jain SM (eds) Plant Gene Banks. Springer, Singapore. https://doi.org/10.1007/978-981-99-4236-7_79-1

Rooney LW, Kirleis AW, Murty DS (1986) Traditional foods from sorghum: their production evaluation and nutritional value. Adv Cereal Sci Technol 8:317–353

Rooney WL, Blumenthal J, Bean B, Mullet JE (2007) Designing sorghum as a dedicated bioenergy feedstock. Biofuels Bioprod Biorefin 1(2):147–157

Ropelewska E, Nazari L (2021) The effect of drought stress of sorghum grains on the textural features evaluated using machine learning. Eur Food Res Technol 247:2787–2798. https://doi.org/10.1007/s00217-021-03832-9

Sai Kishore N, Visarada KBRS, Rao SV, Seetharama N (2011) Progress and prospects for agrobacterium-mediated genetic transformation in sorghum in comparison to other cereals. Transgenic Plant J 5(1):27–34

Saleh MM, Ali WA, Alwan MH et al (2020a) Multi-environmental evaluation of wheat tetraploid genotypes for agronomic traits under rainfed conditions in Syria. J Aridland Agric 6:1–5

Saleh MM, Salem KFM, Elabd AB (2020b) Definition of selection criterion using correlation and path coefficient analysis in rice (*Oryza sativa* L.) genotypes. Bull Natl Res Cent 44(1):1–6

Saleh MM, Alnaddaf LM, Almuhammady AK et al (2021) Applications of plant-derived nanomaterials in mitigation of crop abiotic stress. In: Al-Khayri JM, Ansari MI, Singh AK (eds) Nanobiotechnology. Springer, Cham. https://doi.org/10.1007/978-3-030-73606-4_9

Salem KFM (1997) Breeding studies on rice (Oryza sativa L.). MSc thesis, Faculty of Agriculture, Minufiya University, Shibin El-Kom, Egypt

Salem KFM (2004) The inheritance and molecular mapping of genes for post-anthesis drought tolerance (PADT) in wheat. Martin-Luther-University, Halle-Wittenberg, 124 p

Salem KFM (2009) Relationship between genetic diversity based on SSRs markers with heterosis and combining ability in diallel cross of bread wheat (*Triticum aestivum* L.). Minufiya. J Agric Res 6(34):2159–2178

Salem KFM, Börner A (2003) Evaluation of wheat accessions from Gatersleben Genebank for post-anthesis drought tolerance in the field. In: Gesellschaft für Pflanzenzüchtung, GPZ-Tagung, Vortragsveranstaltung zum Thema: "Abiotischer Streß – Herausforderung für die Pflanzenzüchtung", 11-12 Juni 2003, Groß Lüsewitz, Germany

Salem KFM, Börner A (2004) Post-anthesis drought tolerance in genus Triticum. In: Botanikertagung, 5–10 September 2004, Braunschweig, Germany

Salem KFM, Ibrahim AA (2024) Plant biodiversity in the context of food security under climate change. In: Al-Khayri JM, Jain SM, Penna S (eds) Sustainable utilization and conservation of plant genetic diversity, Sustainable development and biodiversity, vol 35. Springer, Singapore. https://doi.org/10.1007/978-981-99-5245-8_1

Salem KFM, Matter MZ (2014) Identification of microsatellite alleles for salt tolerance at seedling stage in wheat (*Triticum aestivum* L.). Life Sci J 11(12s):1064–1073

Salem KFM, Röder MS, Börner A (2003) Mapping of quantitative trait loci (QTLs) for post-anthesis drought tolerance in bread wheat (*Triticum aestivum* L.). In: Gesellschaft für Pflanzenzüchtung, GPZ-Tagung, 11. Vortragsveranstaltung "Harnessing Genetic Diversity: Genomics and Allele Mining"., 16-17.09.2003. Institut für Pflanzengenetik und Kulturpflanzenforschung (IPK), Gatersleben

Salem KFM, Röder MS, Börner A (2004a) Drought tolerance in wheat germplasm- evaluation and molecular mapping. In: Institut für Pflanzengenetik und Kulturpflanzenforschung, IPK Tagung, 22–24/08/2004. Ag Ressourcengenetik und Reproduktion, Abteilung Genbank, Gatersleben

Salem KFM, Röder MS, Börner A (2004b) Molecular mapping of quantitative trait loci (QTLs) determining post-anthesis drought tolerance in hexaploid wheat (*Triticum aestivum* L.). In: 7 Gesellschaft für Pflanzenzüchtung, GPZ-Tagung, Vortragsveranstaltung zum Thema: " Klimatische und edaphische Sortenanpassung und Züchtung für Nachwachsende Rohstoffe". 3–5 March 2004, Halle/Saale, Germany. Vort Pflanzenzüchtung 64:21–24

Salem KFM, Röder MS, Börner A (2007) Identification and mapping quantitative trait loci for stem reserve mobilisation in wheat (*Triticum aestivum* L.). Cereal Res Common 35:1367–1374

Salem KFM, El-Zanaty AM, Esmail RM (2008) Assessing wheat (*Triticum aestivum* L.) genetic diversity using morphological characters and microsatellite markers. World J Agric Sci 4:538–544

Salem KFM, Yaseen YM, El-Abd AB, Ayoub A (2015a) Inheritance of some yield components and grain quality traits in rice *(Oryza sativa* L.). J Product Dev 20:161–177

Salem KFM, Yaseen YM, El-Abd AB, Ayoub A (2015b) Breeding studies on some agronomic and quality traits in rice (*Oryza sativa* L.). J Product Dev 20(2):211–226

Salem KFM, Alloosh MT, Saleh MM et al (2021a) Utilization of nanofertilizers in crop tolerance to abiotic stress. In: Al-Khayri JM, Ansari MI, Singh AK (eds) Nanobiotechnology. Springer, Cham. https://doi.org/10.1007/978-3-030-73606-4_11

Salem KFM, Elabsawy A, Aldahak L et al (2021b) Evaluation of stem reserve mobilization in Egyptian bread wheat (*Triticum aestivum* L.) genotypes and F1 hybrids under post-anthesis chemical desiccation stress. J Genet Environ Resour Conserv 9(1):176–182

Salem KFM, Saleh MM, Abu-Ellail FFB et al (2021c) Role of quantum dots, polymeric NPS and dendrimers in emphasizing crops tolerate biotic and abiotic stresses. In: Faizan M, Hayat S, Yu F (eds) Sustainable agriculture reviews, vol 53. Springer, Cham. https://doi.org/10.1007/978-3-030-86876-5_1

Salem KFM, Saleh MM, Abu-Ellail FFB et al (2021d) The role of salicylic acid in crops to tolerate abiotic stresses. In: Hayat S, Siddiqui H, Damalas CA (eds) Salicylic acid- a versatile plant growth regulator. Springer, Cham. https://doi.org/10.1007/978-3-030-79229-9_7

Salem KFM, Alghuthaymi MA, Elabd AB et al (2022) Prediction of heterosis for agronomic traits in half-diallel cross of rice (*Oryza sativa* L.) under drought stress using microsatellite markers. Plants 11:1532. https://doi.org/10.3390/plants11121532

Salem KFM, Tawfik E, Mahdy M et al (2023) Advances in lettuce (*Lactuca* spp.) molecular breeding strategies. In: Singh S, Sharma D, Sharma SK, Singh R (eds) Smart plant breeding for vegetable crops in post-genomics era. Springer, Singapore. https://doi.org/10.1007/978-981-19-5367-5_11

Sanford JC, Skubik KA, Reisch BI (1985) Attempted pollen-mediated plant transformation employing genomic donor DNA. Theor Appl Genet 69:571–574

Shimada TL, Shimada T, Hara-Nishimura I (2010) A rapid and nondestructive screenable marker, FAST, for identifying transformed seeds of Arabidopsis thaliana. Plant J 61(3):519–528

Shimada T, Ogawa Y, Shimada T, Hara-Nishimura I (2011) A non-destructive screenable marker, OsFAST, for identifying transgenic rice seeds. Plant Signal Behav 6(10):1454–1456

Singh BD (2004) Plant breeding, principles and methods, 7th edn. Kalyani Publishers, New Delhi

Sleper DA, Poehlman JM (2006) Breeding field crops, 5th edn. Blackwell Publishing, Iowa

Snowden JD (1955) The wild fodder sorghums of the section *Eu-Sorghum*. Bot J Linnean Soc London Bot 55(358):191–260. https://doi.org/10.1111/j.1095-8339.1955.tb00011.x

Soeranto H, Trikoesoemaningtyas T, Sihono S, Sungkono S (2010) Development of sorghum tolerant to acid soil using induced mutation with gamma irradiation. Atom Indones 36(10):11–15

Stefoskaneedham A, Beck EJ, Johnson SK et al (2015) Sorghum: an underutilized cereal whole grain with the potential to assist in the prevention of chronic disease. Food Rev Int 31:401–437

Tadesse Y, Sagi L, Swennen R, Jacobs M (2003) Optimization of transformation conditions and production of transgenic sorghum (*Sorghum bicolor*) via microparticle bombardment. Plant Cell Tissue Organ Cult 75:1–18

Takele A, Feyissa T, Disasa T (2022) Quantitative trait loci mapping of stem sugar content and stem diameter in sorghum recombinant inbred lines using genotyping-by-sequencing. Mol Biol Rep 49:3045–3054. https://doi.org/10.1007/s11033-022-07131-8

Tavazoh M, Habibi D, Golzardi F et al (2024) Effect of drought stress on morpho-physiological characteristics, nutritive value, and water-use efficiency of sorghum [*Sorghum bicolor* (L.) Moench] varieties under various irrigation systems. Brazil. J Biol 84:e286121

Tokach MD, Goodband B, De Rouchey J (2010) Sorghum in swine production. Feeding guide sorghum check off. Kansas State University, Manhattan, p 172

Urriola J, Rathore KS (2014) Temporal and spatial activities of a rice glutelin promoter in transgenic sorghum. Plant Cell Tissue Organ Cult 116:227–234

USDA (2015) ARS, United States Department of Agriculture, ARS (ed) National Genetic Resources Program: Genetic Resources Information Network (GRIN)

Venkateswaran K, Elangovan M, Sivaraj N (2019) Origin, domestication and diffusion of *Sorghum bicolor*. In: Breeding sorghum for diverse end uses. Woodhead Publishing, pp 15–31

Visarada KBRS, Saikishore N, Balakrishna SD, Rao SV (2003) Transient *gus* expression studies in sorghum to develop a simple protocol for Agrobacterium-mediated genetic transformation. J Genet Breed 57:147–154

Visarada KBRS, Padmaja PG, Sai Kishore N et al (2014) Production and evaluation of transgenic sorghum for resistance to stem borer. In Vitro Cell Dev Biol Plant 50:176–189

Wang H, Russa ML, Qi LS (2016) CRISPR/Cas9in genome editing and beyond. Annu Rev Biochem 85:227–264

Wang W, Wang J, Yang C et al (2007) Pollen-mediated transformation of *Sorghum bicolor* plants. Biotechnol Appl Biochem 48(2):79–83

Wang J et al (2025) China National Gene Bank (CNGB). In: Al-Khayri JM, Salem KFM, Jain SM (eds) Plant Gene Banks. Springer, Singapore. https://doi.org/10.1007/978-981-99-4236-7_103-1

Wei WA, Chen CH, Yubin GE et al (2022) EMS mutagenesis of sweet sorghum to create new germplasm. J Nucl Agric Sci 36(1):7. https://doi.org/10.11869/j.issn.100-8551.2022.01.0007

Wiersema JH, Dahlberg J (2007) The nomenclature of *Sorghum bicolor* (L.) Moench (Gramineae). Taxon 56:941–946

Wu E, Lenderts B, Glassman K et al (2014) Optimized agrobacterium-mediated sorghum transformation protocol and molecular data of transgenic sorghum plants. In Vitro Cell Dev Biol Plant 50:9–18

Younis A, Siddique MI, Kim CK, Lim KB (2014) RNA interference (RNAi) induced gene silencing: a promising approach of hi-tech plant breeding. Int J Biol Sci 10(10):1150–1158

Zhao Z, Cai T, Taglini L, Miller M, Wang N, Pang H, Rudert M, Schroeder S, Hondred D, Seltzer J, Pierce D (2000) Agrobacterium-mediated sorghum transformation. Plant Mol Biol 44:78–798

Zhu H, Muthukrishnan S, Krishnaveni S et al (1998) Biolistic transformation of sorghum using a rice chitinase gene. J Genet Breed 52:243–252

Zohry AEH, Ouda S (2022) Climate-resilient crops. In: Climate-smart agriculture. Springer, Cham. https://doi.org/10.1007/978-3-030-93111-7_6

Chapter 6
Genetic Improvement of Sugarcane (*Saccharum* spp.) for Sugar, Fibre and Biomass Energy Through Breeding and Biotechnology

Farrag F. B. Abu-Ellail, Ebtehag A. E. Sakr, Amira A. Ibrahim, Nyo Nyo Aung, Ei Ei Khaing, M. Sathyabhama, R. Viswanathan, C. Appunu, Girija Sangari Murugavelu, and Khaled F. M. Salem

Abstract The capacity of Sugarcane (*Saccharum* spp.) to synthesize and deposit an extraordinarily elevated amount of sucrose is widely recognized and appreciated. The four primary byproducts of the sugarcane business are cane tops, bagasse fiber, filter mud, and the molasses that remain after the plant is crushed. Noble high--yielding canes have been developed through breeding techniques and intergeneric hybridization. Natural fiber is frequently utilized in construction. Bagasse and straw, which are primarily produced as waste during sugarcane processing for ethanol and sugar production must be properly disposed of and recycled using the waste fiber from sugarcane Natural materials are extremely valuable in construction because

F. F. B. Abu-Ellail (✉)
Department of Breeding and Genetics, Sugar Crops Research Institute, Agricultural Research Center (ARC), Giza, Egypt
e-mail: farrag_abuellail@arc.sci.eg

E. A. E. Sakr
Botany Department, Faculty of Women for Arts, Science and Education, Ain Shams University, Cairo, Egypt
e-mail: ebtehag.abdelfattah@women.asu.edu.eg

A. A. Ibrahim
Department of Botany and Microbiology, Faculty of Science, Arish University, Al-Arish, Egypt
e-mail: amiranasreldeen@sci.aru.edu.eg

N. N. Aung
Sugarcane Research Center, Sugar Crops Division, Department of Agriculture, Pyinmana, Naypyitaw, Myanmar

E. E. Khaing
Sugarcane Research Center, Sugar Crops Division, Department of Agriculture, Pyinmana, Naypyitaw, Myanmar

K. F. M. Salem et al. (eds.), *Breeding and Biotechnology of Grass and Bast Fiber Crops*, Advances in Plant Breeding Strategies 12,
https://doi.org/10.1007/978-3-032-00403-1_6

they can improve the ductility and crack control of brittle concrete. Moreover, construction projects using sugarcane can lessen environmental pollutants. Using molecular markers of the RAPD and RFLP types, the first sugarcane genetic maps were created. Currently, the creation of genetic and molecular maps mostly uses gene or genomic microsatellite molecular markers (SSRs). The recommended sugarcane fiber can serve as a starting point to generate high-value products like biofuels, biopolymers, and other useful chemicals through fermentation, biocatalysis, and chemo-catalysis. This chapter gives a broad perspective on the botanical classification and distribution, origin, domestication, and early improvements of sugarcane germplasm. It also discusses the importance and uses of sugarcane, fiber and textile properties, domestication, selection, and early improvements. Additionally, it covers current cultivation practices and challenges, bioconversion of sugarcane residues into sustainable products, and bioproducts from sugarcane bagasse (SCB). The chapter also delves into topics such as artificial intelligence, breeding objectives, germplasm biodiversity and conservation, cytogenetics, traditional breeding methodologies, the role of biotechnology, molecular breeding, molecular marker-assisted breeding, genomics, bioinformatics, tissue culture applications, micropropagation, embryo rescue, synthetic seeds, genetic engineering, and gene editing.

Keywords Bioinformatics · Biotechnology · Molecular breeding · Genetic improvement · Modern breeding · Fiber · Sugarcane (*Saccharum* spp.) · Tissue culture

---

M. Sathyabhama
Department of Biotechnology, PSG College of Arts & Science, Coimbatore, Tamil Nadu, India
e-mail: sathyabhama@psgcas.ac.in

R. Viswanathan
Division of Plant Pathology, ICAR-Sugarcane Breeding Institute, Coimbatore, Tamil Nadu, India
e-mail: r.viswanathan@icar.gov.in

C. Appunu
Division of Crop Improvement, ICAR-Sugarcane Breeding Institute, Coimbatore, Tamil Nadu, India
e-mail: C.Appunu@icar.gov.in

G. S. Murugavelu
Department of Botany, Avinashilingam Institute for Home Science and Higher Education for Women, Coimbatore, Tamil Nadu, India

K. F. M. Salem (✉)
Department of Biological Sciences, Shaqra University, Ad-Dawadimi, Saudi Arabia

Department of Plant Biotechnology, Genetic Engineering and Biotechnology Research Institute (GEBRI), University of Sadat City, Sadat City, Egypt
e-mail: khaled.salem@gebri.usc.edu.eg; khaledfathi@su.edu.sa

6.1 Introduction

Sugarcane (*Saccharum* spp. hybrids) is a perennial grass that uniquely stores its carbohydrate reserves as sucrose. Its commercial value is derived from its stalks and the sugar (sucrose) they produce. The four main byproducts of the sugarcane industry are cane tops, fiber (bagasse), filter mud, and molasses, which are obtained after crushing. Sugar is derived from sugarcane in tropical and subtropical areas, while sugar beet is cultivated in temperate regions globally. The cultivation of sugarcane occurs in regions positioned between 37°N and 32°S of the equator. Alternatively, sugar beet is predominantly grown beyond the 35° latitude mark, reaching up to nearly 60° latitude. As of now, about 76% of the world's sugar is manufactured from sugarcane, whereas roughly 24% is produced from sugar beet (Babar et al. 2011). Sugarcane is highly polyploid, heterozygous, and genetically complex (Khan et al. 2012). Modern sugarcane cultivars for sugar extraction are derived from interspecific hybrids of *Saccharum spontaneum* and *Saccharum officinarum*, followed by repeated backcrossing to *S. officinarum*. Today's commercial sugarcane varieties are usually developed by crossing commercial or pre-commercial hybrids. Sugarcane's cultivation as a garden crop dates back to around 2500 BC (Singh and Tomar 2005). This chapter gives a broad perspective on the botanical classification and distribution, origin, domestication, and early improvements of sugarcane germplasm, its importance and uses, fiber and textile properties, domestication, selection and early improvements, current cultivation practices and challenges, bioconversion of sugarcane residues into sustainable products, bioproducts from sugarcane bagasse (SCB), artificial intelligence, breeding objectives, germplasm biodiversity and conservation, cytogenetics, traditional breeding methodologies, the role of biotechnology, molecular breeding, molecular marker-assisted breeding, genomics, bioinformatics, tissue culture applications, micropropagation, embryo rescue, synthetic seeds, genetic engineering and gene editing.

6.1.1 Botanical Classification

Saccharum L., a genus within the grass family (Poaceae), has historically been placed in the tribe Andropogoneae. The grain genera Zea and Sorghum, along with tropical and subtropical grasses, belong to this group. The tribe is further subdivided into the Saccharinae Benth, which includes sugarcane. After that, it can be further subdivided into two groups, with sugarcane belonging to the Saccharastra, also known as Saccharininae. However, Daniels and Roach (1987) note that this level of grouping does not have an official classification in the International Code for Botanical Nomenclature (ICBN). Since plants from five genera exhibit similar traits and form the "Saccharum complex," an interbreeding group with close genetic ties, the taxonomy and phylogeny of sugarcane are notably complex. Saccharum, Erianthus section Ripidium, Miscanthus section Diandra, Narenga, and

Sclerostachya are the members of the Saccharum complex. These genera exhibit high polyploidy and often have unbalanced chromosome numbers, complicating taxonomy and leading to frequent revisions of their relationships (Sreenivasan et al. 1987). The genus Saccharum has traditionally been made up of six species: *S. spontaneum*, *S. officinarum*, *S. robustum*, *S. edule*, *S. barberi*, and *S. sinense* (D'Hont et al. 1998). Sugarcane is a perennial grass, and modern cultivars are hybrids of various species from the *Saccharum* genus. The ancestral lineages of the *Saccharum* genus include the following families (https://www.itis.gov/servlet/SingleRpt/SingleRpt?search_topic=TSN&search_value=42054#null):

Kingdom Plantae plantes, Planta, Vegetal, plants
Subkingdom Viridiplantae – green plants
Infrakingdom Streptophyta – land plants
Superdivision Embryophyta
Division Tracheophyta – vascular plants, tracheophytes
Subdivision Spermatophytina – spermatophytes, seed plants, phanérogames Class Magnoliopsida
Superorder Lilianae – monocots, monocotyledons, monocotylédones
Order Poales
Family Poaceae – grasses, graminées
Genus *Saccharum* L. – sugarcane, sugar cane

Taxonomists and cytogeneticists have put forward various hypotheses about the provenance of Saccharum. Besides Saccharum, which belongs to the Andropogoneae tribe, nine related genera include Miscanthus, Erianthus, Narenga, Sclerostachya, Miscanthidium, Eccoilopus, Erichrysis, and Imperata. Among these, Saccharum, Erianthus, Narenga, Miscanthus, and Sclerostachya are especially valuable for cane breeders due to their morphological similarities, breeding compatibility, and evolutionary relationships. Understanding the provenance of sugarcane relies on these closely linked interbreeding groups, which are categorized as the Saccharum complex (Mukherjee 1957).

6.1.2 Importance and Uses

Sugarcane yields approximately six times more sugar than sugar beet. It is cultivated for its sucrose content and is primarily used for producing refined sugar and other products. The extraction process at sugarcane mills involves shredding the cane, crushing it to obtain juice, and clarifying the juice with heated lime. In 2022, global sugarcane production was around 1.8 billion tonnes, harvested from 25.8 million hectares, with Brazil being the top producer at 590 million tonnes (FAO 2022). India's sugar output was around 16.5 million tonnes, while traditional sweeteners totaled ten million tonnes (Kansal 1998). Other countries that ranked in the top 20 producers that year included Argentina, Vietnam, South Africa, the USA,

Cuba, Egypt, Peru, Myanmar, and Ecuador (FAO 2022). Sugarcane is an enduring crop with many years of safe and trusted agricultural use. In sugarcane-growing regions, juice is extracted by squeezing or chewing the cane, often sold fresh or bottled. Artisanal methods produce unrefined sugar cakes like jaggery, rapadura, and panela. Sugarcane also yields various products such as refined sugar, molasses, alcohol, and bagasse. From 100 tonnes of sugarcane, about 14.3 tonnes of raw sugar, 27.2 tonnes of bagasse, and other byproducts are produced (Allen et al. 1997). Some sucrose is broken down into ethanol via fermentation (Schubert 2006). In 2006, 47% of Brazil's sugarcane was used to produce 17.8 billion liters of ethanol, supplying about 40% of car fuel (Goldemberg and Guardabassi 2009; Orellana and Neto 2006). Molasses, a viscous byproduct left after extracting sucrose from clarified sugar juice, is commonly used in alcoholic fermentation, as a livestock feed additive, and as a fertilizer for sugarcane cultivation (Mackintosh 2000). Trash refers to the plant remaining after the sugarcane stalks have been harvested. About 10 tons of trash per hectare is roughly produced by sugarcane fields (Karve et al. 2001). The wax on the sugarcane stalk, especially at the nodes, and the lipids in the cells together create sugarcane wax (Allen et al. 1997). Sugarcane wax serves as an ingredient in cosmetics and cholesterol-lowering pharmaceuticals, while sugarcane ash from burning bagasse and filter cake from cane juice filtration are frequently used as fertilizers on cane farms (Cheavegatti-Gianotto et al. 2011).

6.1.3 Bioconversion of Sugarcane Residues into Sustainable Products

Bagasse and straw, which are primarily produced as waste during the production of ethanol and sugar from sugarcane, must be properly disposed of and recycled (Fig. 6.1) (Leal et al. 2013). As sugarcane reaches maturity, secondary cell walls are made of cellulose, hemicellulose (mainly glucuronoarabinoxylan), and lignin form between the primary cell wall and the plasma membrane (Nasution et al. 2022). Sugarcane residues must undergo several processes before microorganisms can use their sugars to bioconvert them into desired products. Pretreatment and saccharification or hydrolysis are two of these procedures. The effective conversion of (hemi) cellulose carbohydrates into bioproducts is significantly hampered by lignin (Sukumaran et al. 2021). To make crystalline cellulose and hemicellulose more accessible for subsequent hydrolysis, lignin must first be removed from lignocellulose (Antoniêto et al. 2022). Pretreatment can be chemical, physical, physicochemical, biological, or a combination of the four. Chemical pretreatment uses acids, alkalis, oxidizing agents, and organic solvents. Physical pretreatment uses mechanical methods, ultrasounds, microwaving, milling, and grinding (Mankar et al. 2021; Patel and Shah 2021; Usmani et al. 2020). The procedures used to produce bioproducts from sugarcane wastes and their conversion are illustrated in the following sections.

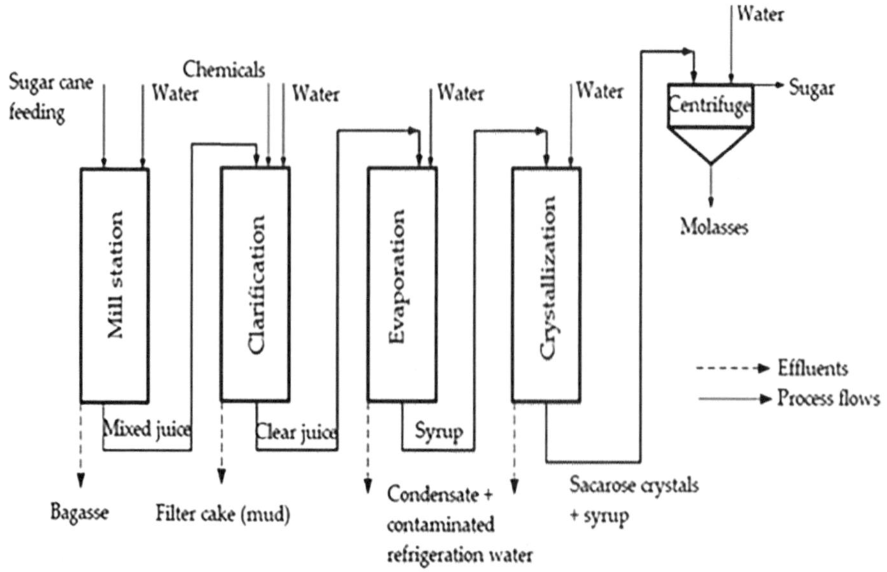

Fig. 6.1 Operations in sugarcane processing using fresh water and generation of wastewater. (Source: Ingaramo et al. (2009))

6.1.3.1 Fiber and Textile Properties

Bagasse is the fibrous residue remaining after extracting juice from sugarcane. It is anticipated that each tonne of sugarcane processed produces 240–280 kg of bagasse (Cheavegatti-Gianotto et al. 2011). The composition of bagasse includes long fibers in the rind and finer, softer fibers in the pith. Its cellulose fibers (1–1.5 mm) are ideal for papermaking, being longer than hardwood fibers but shorter than softwood fibers. Bagasse is widely used in paper production across various nations (Allen et al. 1997). The pith portion of bagasse, accounting for 25–35% of its dry weight, must be eliminated to ensure premium-grade paper (Dunlap and Callihan 1969). Particleboard, made from bagasse, is commonly used in cabinets, laminate flooring, and fiberboard (Almazan 1994). More recently, bagasse has been used as the primary material for the fibers and the resin in panel construction (Hoareau et al. 2006). Bagasse serves as fodder, though its low digestibility, even for ruminants, can be improved with steam treatment, making it suitable for fattening calves (Allen et al. 1997). Bagasse has also been used to feed prawns (Freeman et al. 1992). The combustion of bagasse generates steam for sugar refineries, with any excess energy produced power for grid distribution in several locations, including Australia (Mackintosh 2000), Brazil (Cheavegatti-Gianotto et al. 2011) and Mauritius (Deepchand 2005).

Fig. 6.2 Sustainable valorization of sugarcane residues and potential strategies for fuels and chemicals production. (Source: Jugwanth et al. (2020))

6.1.3.2 Bioproducts from Sugarcane Bagasse (SCB)

The main byproduct of the sugarcane industry is bagasse, formed during the crushing of sugarcane stalks for juice extraction (Clauser et al. 2016). An estimated 100 million metric tonnes of bagasse are produced globally (Reena et al. 2022). The crystalline cellulose with amorphous areas is gathered into microfibril clusters and encapsulated within a loosely organized hemicellulose-lignin network that makes up SCB, which is made of holo-cellulose. Furthermore, the biomass composition also contains trace amounts of ashes, extractives, and pectin. Lignin acts as a physical barrier, limiting the biodegradability of the polysaccharide matrix (Nwajiaku et al. 2018). SCB, being a plentiful, renewable and affordable feedstock (Fig. 6.2), may serve as a foundation for producing value-added products like biofuels, biopolymers, and other useful chemicals through fermentation, biocatalysis, and chemo-catalysis (Singh et al. 2022).

6.1.4 Distribution, Origin, Domestication, and Early Improvements

Across more than 100 countries and on every continent (FAO 2021, Fig. 6.3) sugarcane is grown between the latitudes of 30°N and 30°S (Bull and Glasziou 1979). Intensive selective breeding of *Saccharum* species has led to the development of commercial sugarcane hybrid cultivars, primarily by crossing *S. officinarum and S. spontaneum. S. officinarum* has high levels of sucrose in the stem but is susceptible to diseases (Lakshmanan et al. 2005), while *S. spontaneum*, has low sucrose levels, slender stalks, and increased fiber content but exhibits high polymorphism and is resistant or tolerant to many pests and diseases (Jackson 2005). *S. officinarum*, a cultivated species not found in the wild, is believed to have originated in the Indonesia/New Guinea region, where it has been grown since 8000 BC (Fauconnier 1993; Sreenivasan et al. 1987). *S. officinarum* is thought to have evolved from

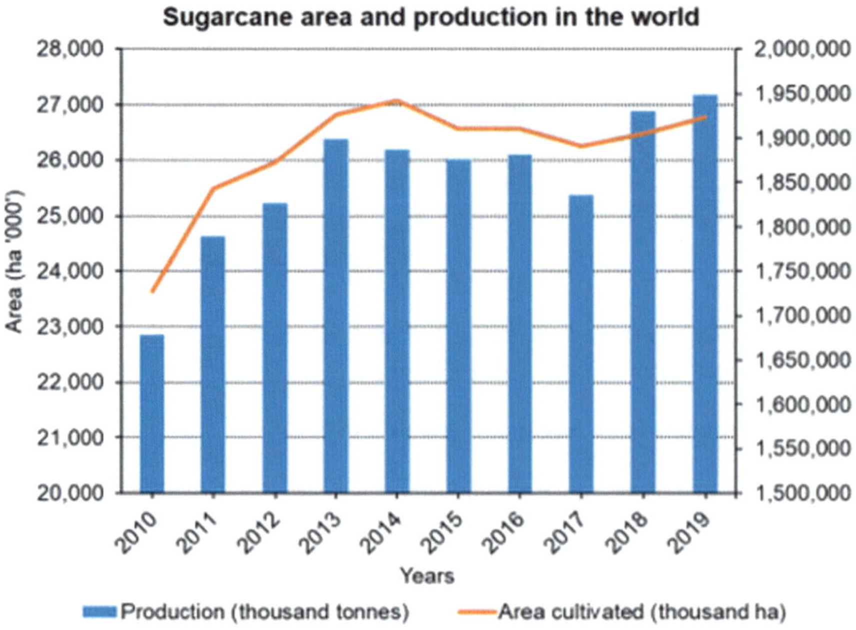

Fig. 6.3 Sugarcane distribution area and production throughout the world from 2010 to 2019. (Source: FAOSTAT (2021))

selected sweet forms of *S. robustum*. Its canes were used for construction, fencing, and archery, with animals like pigs or rats possibly influencing the selection of sweeter varieties (Daniels and Roach 1987). Sugarcane cultivation spread through human migration to Southeast Asia, India, and the Pacific, where it hybridized with wild varieties. By 500 B.C., it reached the Mediterranean and expanded to Morocco, Egypt, Crete, Greece, and Sicily, dominating production until the fifteenth century. It was later introduced to West Africa, Central and South America, and the West Indies.

6.2 Current Cultivation Practices and Challenges

6.2.1 Land Preparation and Planting Tim

As a long-duration crop with high nutrient needs, sugarcane thrives in heavy, fertile soil. This type of soil delivers the essential nutrient content and creates a favorable environment in terms of pH, climatic factors, humidity and organic substances and supports the crop throughout its entire growth period. Sugarcane is adaptable to various types of soil, ranging from sandy loam to clay. Land preparation is required every 3–5 years. Proper land preparation is crucial for achieving successful plant and ratoon crops, as inadequate preparation will burden the land. The key purpose of field conditioning is to make a well-tilled soil, which facilitates rapid absorption of moisture, easy sprouting of buds, deep root growth, and the release of nutrients, ultimately establishing a robust crop stand (Abu-Ellail et al. 2021a, b; El-Shafai and Ismail 2006). The timing of planting is determined by the temperature needed for bud germination. Optimal germination occurs at temperatures between 27 and 32 °C. Therefore, any season that coincides with these favorable temperatures is appropriate for planting. In equatorial regions, where the temperature range remains between 25 and 30 °C throughout the year, sugarcane crops can be effectively grown in any month. However, in tropical regions and areas beyond the subtropics, planting is mostly limited to the spring (March) and autumn (October) months (Abu--Ellail et al. 2020).

6.2.2 Seed Rate Depending on Planting Methods

The plant population in cane fields depends on the seed quantity, germination efficiency, and tillering capacity of the varieties. The seed quantity is a fundamental factor in establishing a basic crop stand. To ensure a good crop stand, it is crucial to consider the right quantity of high-quality seed. Varieties, planting periods, types of soil, seedbed setup, and planting techniques all determine the seed rate. This rate is equivalent to 50,000 two-eye-bud setts or 30,000 three-eye-bud setts per hectare in India (Hunsigi 1989). For wide-row cultivation (1.5 m apart rows) in the USA, 3.5–4 tons per hectare is usually used when planting is done by machine. Developing nations typically rely on manual planting with bullocks or tractors for field tasks and furrow preparation, whereas developed countries use fully mechanized systems.

6.2.3 Irrigation and Fertilization

The first irrigation for a plant crop must be well-timed. To maintain a shallow root system during early growth, frequent irrigation is essential. However, to encourage deeper root development, it is important to allow slight moisture stress in the early plant growth stage. Neglecting proper irrigation in the stage of fully grown crops, particularly in July and later months is a major cause of lodging. Consequently, care must be taken when watering the crop. Excessive irrigation can weaken the root system, increasing the risk of lodging, particularly in high winds. It is better to opt for light, frequent irrigation instead. To provide the root system more support, the soil needs to be periodically kept both moist and dry rather than being constantly wet. This can be achieved through judicious irrigation practices (Abu-Ellail et al. 2023).

6.2.4 Fertilizer Application

Very fertile soil produces vigorous crops but these heavy crops are more prone to lodging compared to weaker crops. Heavy soils require careful and timely fertilizer application. Nitrogen must be applied by the end of May. Applying it later causes the crop to become more succulent and more vulnerable to lodging. Applying nitrogen fertilizer in a ratoon crop is less efficient than in a plant crop, mainly due to poor stubble root development, delayed root growth, and an imbalanced shoot-root ratio early in development. Yadav (1986) reported that fertilizer absorption efficiency was 1030 kg cane per kg N for plant crops, in relation to 850 kg and 902 kg cane per kg N for ratoon crops. Clements (1980) found that phosphorus fertilization for ratoon crops requires twice the amount of P needed for plant crops. Since P is often fixed in the soil, its uptake can be improved by combining it with farmyard manure (FYM) or filter press cake (FPC). Prasad (1976) showed that applying 20 t/ha of FPC boosts cane production by 60% in plant crops and 30% in ratoon crops. Phosphorus use efficiency is similar for both crop types (Hunsigi 1993) and adding P can increase juice Pol and enhance purity at optimal levels (Hunsigi 1993). The recommended NPK ratio is 2:1:2 for plant crops and 2:1.4:3.0 for ratoon crops (Humbert 1958).

6.2.5 Maturity and Harvesting Technique

Well-managed ratoon crops increase the economic benefits of cane production. In comparison to plant cane age, early and fast shoot growth enlarges cane stalks, which causes quicker ripening and larger sugar content at harvest. As harvesting progresses through January, this difference lessens depending on the variety. Ratoon crops cut production costs by reducing land preparation, seed investment and planting expenses, saving farmers 30–35% of the overall cultivation cost. Ratoon crops mature 3–5 weeks earlier than plant cane, producing higher sugar yields. To foster appropriate sprouting and root growth, cane should be harvested 2–3 cm below the soil surface to keep shoots from drying out. Stubble shaving after harvest encourages sprouting from underground buds; it is commonly done with a tractor-operated stubble shaver but should be avoided in shallow-rooted kinds or loose soil. Winter harvesting of plant crops results in gaps of 18–26%, compared to 11–13% for February harvesting (Mathur 1986; Abu-Ellail et al. 2020).

6.2.6 Vertical Farming

In vertical farming, crops are grown in layered stacks under controlled environmental techniques like aquaponics, hydroponics, and aeroponics, which do not rely on soil. With limited farmland and increasing food demands, this method, introduced by Columbia University professor Dickson Despommier in 1999, offers a potential solution. There are two types of vertical farms. The first type involves repurposing abandoned buildings for vertical farming, although such buildings aren't always readily available. New buildings can also be constructed for vertical farming based on specific requirements. The second category is underground vertical farms, referred to as 'Deep Farms,' located in tunnels, old mine shafts, or other below-ground environments. With stable temperature and humidity, less energy is needed for heating, and underground water sources provide irrigation. These farms can yield 7–9 times more food than traditional farms. Studies of sugarcane root water relations reveal significant variation in root hydraulic conductivity, suggesting vertical farms may not be economical (Bonnett 2013).

Two primary approaches in vertical farming are hydroponics and aeroponics. Hydroponics is a soil-free growing method where plant roots are immersed in a nutrient-rich solution containing elements like magnesium, nitrogen, potassium, and calcium. This method has been shown to increase yield by 11 times compared to conventional farms while using 13 times less water. This makes hydroponics the preferred technique in vertical farming. Aeroponics, on the other hand, uses air to grow plants instead of mediums like solids or liquids. The plants are exposed to a sprayed liquid solution in the air, which allows them to take in nutrients. This method is ideal as it requires no water or soil, only a growing medium. Unpredictable weather patterns and natural disasters, including floods, droughts, and wildfires, often disrupt traditional farming. In contrast, vertical farming in a controlled environment eliminates these factors, making the supply chain process less susceptible to disruption. Vertical farming also helps in environmental conservation by reducing deforestation, ultimately saving resources in the long run. However, vertical farming is not suitable for crops like sugarcane, which have long stalks and require land for support, making them susceptible to root infection. In a study by Hari et al. (2017), sugarcane root growth, development, and traits were investigated under hydroponic conditions (Fig. 6.4). Twenty-five commercial hybrids were evaluated for root traits, lateral root development, and response to injury, showing variations among varieties in root disease response with an increase in water compared to soil.

6.2.7 Digital Farming

Precision agriculture is regarded as a significant advancement and enhancement of traditional farming practices. The initial phase of precision agriculture focused on machines equipped with Global Positioning Systems (GPS), and productivity maps. Precision agriculture aims to address spatial variability, optimize economic gains, and reduce environmental harm. As a strategy, precision agriculture leverages information technologies to gather diverse data and guide crop production choices. Precision agriculture relies on a three-step approach: reading, interpretation, and action, forming a continuous cycle. Reading entails mapping the parameter of interest to a specific location, including details such as latitude, longitude, and time

Fig. 6.4 Five-month-old sugarcane in hydroculture. (Source: Hari et al. (2017))

stamps. Data with geographic indexing allows for interpretation via mapping and analysis with geostatistical methods. Studies on the geographical dependency of information and its relationships with other characteristics necessitate the use of thematic maps and significant amounts of data. Therefore, Geographic information systems (GIS) and tools are integrated into Information Technology. Progress in precision agriculture for sugarcane cultivation has been slower than for corn, despite sugarcane's technological edge in ethanol production. Although harvest monitors for combines have only recently entered the market (Magalhaes and Cerri 2007; Molin and Menegatti 2004), commercial adoption is still limited due to few machines in operation. The lack of tools and support for assessing spatial differences in crop performance has hindered the application of precision agriculture in sugarcane, particularly for soil correction. This involves recommending input application rates based on samples from two or more hectares, resulting in significant fertilizer savings (Barbieri et al. 2008; Menegatti et al. 2006; Sanchez et al. 2008). Pezeto et al. (2008). Automation aims to replace human labor to increase efficiency and yield. While new technologies are being explored to maintain sugarcane cultivation's competitiveness, there is still potential to enhance conventional mechanization in planting, cultivation, and harvesting. The use of on-the-go sensors during farming activities is a major focus of industry and research However, monitoring variables like pests, diseases and nitrogen levels in plants often require ground--based instruments, making efficient sampling challenging. For instance, creating a soil property map with a 20-meter resolution across 200 hectares requires 5000

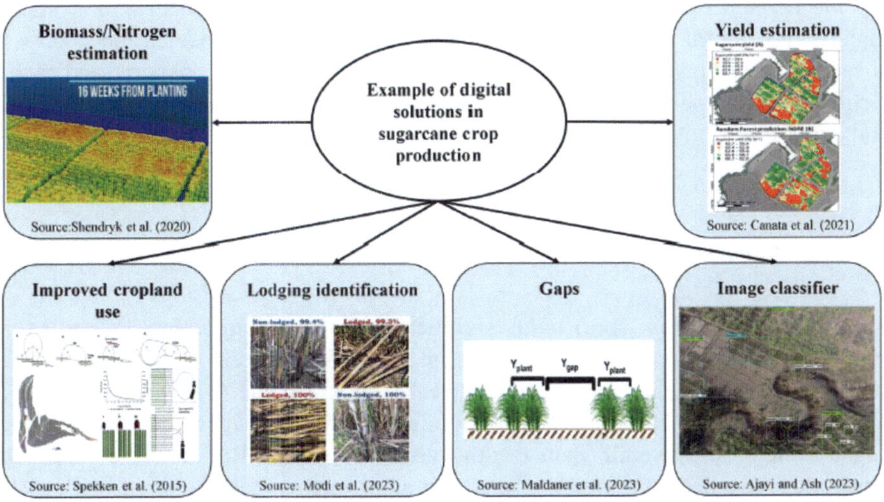

Fig. 6.5 An example of digital solutions used in the production of sugarcane. Cropland utilization, yield estimation, gap identification, image classification, and lodging identification were all improved by biomass and nitrogen estimation. (Source: Molin et al. (2024))

samples, making manual collection impractical (Fig. 6.5). Sugarcane is a significant crop in India, accounting for 7% of the total agricultural production in the country (Aruna 2016).

6.2.8 *Current Agricultural Challenges*

6.2.8.1 Low Productivity and Lack of Good Agricultural Practices

Manual labor drives up sugarcane production costs, accounting for half of the total expenses. To improve sustainability, mechanization is required. Low production, delayed payments, and unscientific price fixing are major issues that prevent farmers from growing sugarcane farming. A recovery-rate-based pricing scheme may motivate mills to enhance recovery rates, while farmers focus on better farm management to increase yield. Policies supporting mechanization, timely input supply, scientific crop management, and motivating the younger generation will benefit both farmers and sugar mills by increasing crop production and profitability (Pokharel et al. 2019).

6.2.8.2 Low Price and Unscientific Pricing Mechanism

The difficulties of contract sugarcane farming are apparent when reviewing farmers' income statements, where net incomes amounted to 32%, 31%, and 34% of their gross incomes, respectively. With companies keeping over 60% of the gross income per ton of sugarcane, significant disparities in income distribution arise, highlighting moral, administrative, and governmental concerns. Farmers across sites blamed company deductions for their low earnings, citing transport, seed cane, fertilizer, and harvesting costs as key issues, which vary by management practices and community culture (Waswa et al. 2012).

6.2.9 Insects and Diseases

Sugarcane (*Saccharum* spp.) holds significant economic importance, mainly for producing sugar and ethanol. However, insects and pathogens including viruses, bacteria, and fungi can cause various diseases in sugarcane (Viswanathan 2016). These diseases have the potential to significantly reduce sugarcane productivity, sugar content, and overall plant health (Viswanathan and Rao 2011). Numerous epidemics have been reported in the past century, such as red rot, smut, wilt, rust, leaf scald, and yellow leaf (YLD). Many sugarcane varieties have been replaced due to their susceptibility to new diseases or pathogenic strains. Red rot is the dominant sugarcane disease in India, with wilt and smut also posing significant threats to cane yield (Viswanathan 2021). Maintaining the productivity and financial success of sugarcane production depends on the management of sugarcane insects and diseases all over the globe. Using insect and disease-resistant cultivars, agricultural practices, crop rotation, sanitation, chemical treatments, and, occasionally, biological control techniques are examples of disease management measures (Wongkaew 2012).

6.2.9.1 Sugarcane Diseases

Can be categorized into several groups as follows:

Viral Diseases

Sugarcane Mosaic (Sugarcane mosaic pot virus)

Viruses had a significant impact on sugarcane worldwide, causing notable epidemics and significant losses (Viswanathan and Rao 2011). In Brazil, the two primary viruses affecting sugarcane are the sugarcane mosaic virus (SCMV) and the sugarcane yellow leaf virus (SCYLV) (Xie et al. 2009). Mosaic is a viral disease that affects sugarcane, maize, sorghum, and other Poaceae plants worldwide (Shukla et al. 1989). These diseases are caused by viruses that infect sugarcane plants, often resulting in mosaic patterns, yellowing of leaves, and reduced growth (Fig. 6.6). Aphids and leafhoppers are typical vectors for viral transmission (Ramawas 1983).

The disease is most prominent on the lower portion of younger foliage causing chlorosis. Heavy infections cause yellow streaks on the leaf sheaths and stalks, long necrotic patches on the stalks, and stem splitting. The internodes also suffer from necrotic lesions, leading to a stunted and chlorotic plant. The virus is predominantly spread through the use of infected canes as seed.

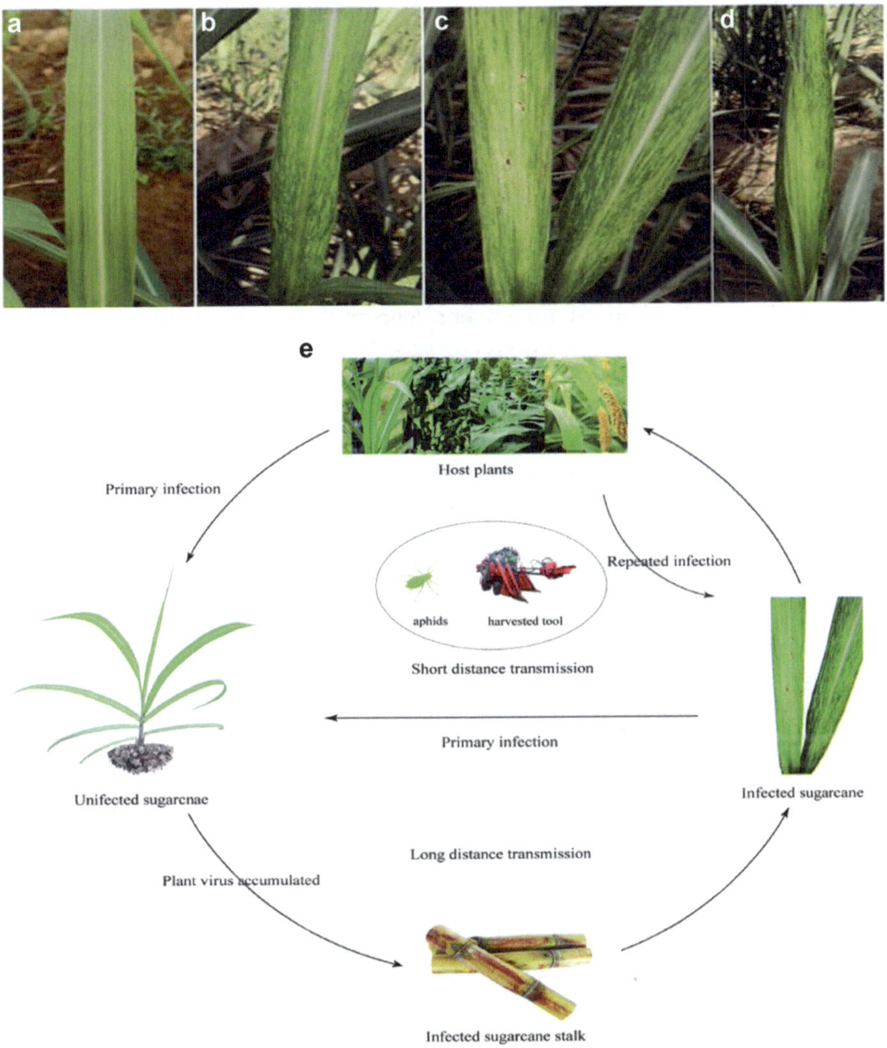

Fig. 6.6 Symptoms of mosaic infection in sugarcane leaves. (**a**) healthy leaves, (**b**) infected leaves, (**c**) severely infected leaves, (**d**) abnormally twisted top leaves of infected plant and (**e**) The transmission pathway of sugarcane mosaic disease. (Source: Lu et al. (2021))

Sugarcane Streak Mosaic Virus

Sugarcane streak mosaic virus (SCSMV) is a major concern in sugarcane agriculture as it can lead to reduced yields and lower sugar quality. Aphids are the primary carriers of SCSMV, as they feed on infected sugarcane plants and spread the virus to healthy ones when they consume them (Shah et al. 2015). Additionally, SCSMV can also spread through infected planting material, such as cane stalks used for propagation (Putra et al. 2014). Typically, SCSMV-infected sugarcane plants display a variety of symptoms. These symptoms may include yellow, green, or both, colored leaves with streaking or mosaic patterns (Fig. 6.7). Leaf curling slowed development, and decreased cane production are further typical indicators. Variables such as viral strain, weather conditions, and sugarcane type can all affect the severity of the symptoms (Carr and Knox 2011).

Bacterial Diseases

Sugarcane bacterial infections are a major concern for sugarcane growers and the global sugar industry. These infections are caused by various species of bacteria that can infect sugarcane plants, leading to reduced yields, lower sugar content, and economic losses. Symptoms of bacterial infections in sugarcane include leaf scorch, red stripes on leaves, and other visible signs (Huang et al. 2018). These infections ten affect the plant's vascular system, disrupting nutrient and water transport (Christopher 1930).

Fig. 6.7 Sugarcane streak mosaic virus symptoms. (Source: Putra et al. (2014))

Red Stripe (Pseudomonas rubrilineans)

The disease begins at the base of young leaves with narrow, water-soaked, chlorotic streaks that turn reddish-brown. These 0.5–1 mm wide, 5–100 mm long stripes run parallel to the midrib and remain on the lower half of the leaf. Whitish flakes spread to the shoot's growing points, forming yellowish stripes that eventually turn reddish-- brown (Fig. 6.8). Rotting may start at the shoot tip and progress downwards. The core becomes discolored, shriveled and forms a hole in the center. In severely infected fields, a strong and repulsive stench may be present. The pathogen can survive in the soil and infected plant residues. The bacterium spreads mainly via infected canes, with secondary transmission through rain, irrigation, and insects (Rangaswami and Rajagopalan 1973).

Gummosis (Xanthomonas axonopodis)

The disease manifests in two different phases namely the foliar stage and the systemic stage. During the foliar stage, 3–7 mm wide streaks emerge on mature leaves near the damaged veins at the tip beginning at the margin and continuing down vascular bundles, ranging from yellow or orange with red flecks to necrotic grey. Infected canes have shorter internodes, giving them a bushy appearance. In extremely susceptible cultivars during the systemic stage, the bacterium infects the stalk, resulting in reddish discoloration of the vascular bundles, bacterial slime, and gum pockets. Stalk deformation, leaf chlorosis, and growth point death can all occur, especially in poor growing conditions (Vauterin et al. 1992).

Fig. 6.8 Red stripe disease symptoms. (Source: Huang et al. (2018))

Fungal Diseases

Fungal infections pose a major threat in sugarcane (*Saccharum* spp.) agriculture as they can reduce yield and sugar content in harvested cane. Pathogenic fungi that infect various parts of the plant result in sugarcane fungal infections (Karamchandani et al. 2022). Given the potential impact on sugarcane quality and productivity, sugarcane growers and the sugar industry to prioritize the management of these diseases. Fungi can infect multiple areas of the sugarcane plant, leading to symptoms such as smut on the stems and leaves, red rot in the internodes, and rust on the leaf surfaces., all of which negatively affect sugarcane quality and production (Sivanesan and Waller 1986).

Smut (Ustilago scitaminea)

Calcareous smut is a fungal pathogen in rice that causes stunted growth and transforms the central grow into a whip-like form, known as a smut whip. Initially enclosed by a white membrane, the whip eventually ruptures, releasing black smut spores. Infected plants are slender, rigid, and grow at a sharp angle. Each shoot or tiller may develop a whip-like structure, and smutted clumps can produce mummified arrows with normal flowers at the base and a smut whip at the top. Rarely, smut sori forms on leaves and stems. The fungal hyphae are intercellular, producing tiny black spores between the host's vascular bundles. These light brown, spherical spores measure 6.5–8.5 micrometers in diameter. Smut spores can remain active in the soil for a maximum of 10 years. The disease primarily spreads through infected seed pieces (setts), and spores in the soil can be dispersed by rain and irrigation water, leading to further infection. Fungi that infect sugarcane leaves are known as foliar diseases, which can impair photosynthesis and limit the plant's ability to produce sugars. Smut infection is a common example of a foliar disease (Alexander and Viswanathan 1996) (Fig. 6.9).

Red Rot (Colletotrichum falcatum)

Colletotrichum falcatum, a fungus that causes red rot, which affects sugarcane at all growth stages. Symptoms appear after 4–5 months and include yellowing and drying leaves, withering canes (Fig. 6.10) and purplish or brown discoloration of the pith, which becomes pithy with boat-shaped cavities. The fungus produces a cottony white mycelium, and its conidia are oval to elliptical microconidia, with straight macroconidia containing 3–5 septa, and blastoconidia that are straight, sickle-shaped, or pike-shaped. The disease is transmitted through infected setts, irrigation water, tools, rain splash, and air currents, which spread conidia from diseased to healthy plants. Red rot weakens the plant, increasing its susceptibility to lodging and reducing its capacity to deliver water and nutrients. It is a common disease of sugarcane stalks (Viswanathan 2020).

Fig. 6.9 Sugarcane smut symptoms. Different forms of whip morphology in smut-infected sugarcane. (**a**) Long whip. (**b**) Closed whip. (**c**) Twisted whip. (**d**) Short whip. (**e**) Multiple whips. (Source: Sundar et al. (2012))

6.2.9.2 Insects

Sugarcane Aphid (*Melanaphis sacchari*)

One of the most noticeable symptoms is the yellowing or chlorosis of sugarcane leaves. Infested sugarcane plants often exhibit stunted growth, with shorter and thinner stalks (Fig. 6.11). Aphid feeding can cause leaves to curl or roll upward.

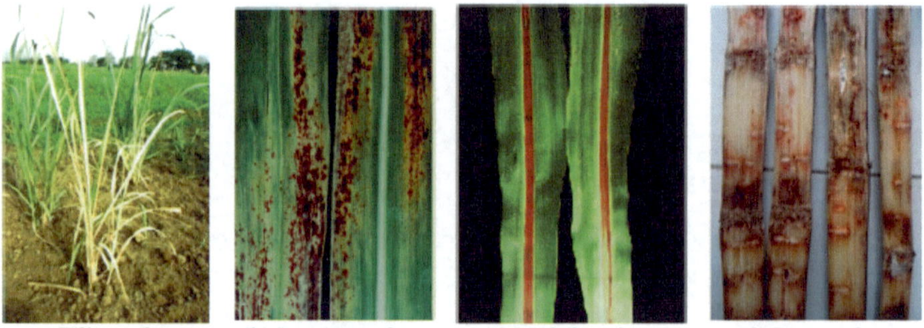

Fig. 6.10 Shows typical red rot disease symptoms on various sugarcane plant parts and stages in the Observations of red rot in the tillering stage (**a**), lamina red rot (**b**), midrib red (**c**), and stem (**d**) Later on in the sugarcane's life cycle, red rot was noticed. (Sourse: Hossain et al. (2020))

Fig. 6.11 Melanaphis sacchari adults, nymphs, and exuvia (top); and colony on the underside of a Sugarcane leaf (bottom). (Source: Wilson (2019))

This curling can protect the aphids against natural enemies and environmental factors. Aphids feed on infected plants and acquire sugarcane viruses. The viruses replicate within aphids and can be transmitted for an extended period. Aphids transmit the viruses to healthy sugarcane plants as they continue to feed. Infected plants may show symptoms over time, including leaf yellowing and reduced growth. Infected plants serve as sources for more aphids, leading to disease spread. According to Yang (1986), sugarcane and corn leaf aphids efficiently spread SMV across sorghum, corn, and sugarcane.

Sugarcane Borer (*Diatraea saccharalis*)

The sugarcane borer is classified as a pest, not a disease, and poses a significant threat to sugarcane crops. Several types of borers affect sugarcane, with the most common being the sugarcane stalk borer (Fig. 6.12). These pests bore into the sugarcane stalks, causing damage that weakens the plant and reduces yield (Costa et al. 2013). Adult borers lay eggs on the surface of the sugarcane stalks. Entry holes, often small and round, can be visible on the stalks where the larvae bore into the plant. Larvae tunnel through the stalks, creating tunnels or galleries within the plant. As they feed on the inner tissues, these tunnels disrupt the transport of water and nutrients, weakening the stalks. Infested sugarcane plants may exhibit wilting, yellowing, or drooping leaves due to the disruption in nutrient flow caused by the borers. Stunted growth and reduced vigor are common symptoms as well. Severely infested stalks may show signs of damage such as broken or collapsed stalks, hollowed-out sections, or even the presence of frass (excrement) near entry holes (Tomaz et al. 2018).

Fig. 6.12 The major sugarcane borers. (Source: Li et al. (2024))

6.2.10 Breeding Objectives

The key objectives of cane breeding are to refine sugarcane varieties and enhance sugar production, maximizing the tonnage of sugar per hectare (Fig. 6.13). Over the past 70–80 years, significant progress has been made in developing sugarcane varieties. Breeding advancements have resulted in a 25–50% increase in sugar yields across various countries. These varieties have been refined to address the needs of both growers and the sugar industry. Requirements for cane varieties differ based on factors such as weather, soil conditions, cultivation practices, and outbreaks of diseases and pests. Additionally, advancements in mechanization for cultivation and harvesting have created specific demands for certain varieties. Sugarcane varieties must also fulfill the criteria set by sugar mills for effective milling and processing. To solve this issue, genetic alterations must be implemented by either traditional breeding or biotechnology. Sugarcane breeders' major goal is to develop novel mutations that interact efficiently with a variety of environmental conditions, including soil types and farming practices. Breeding is necessary to improve genetic production potential and reduce yield-limiting variables since sugarcane is grown in a variety of climatic, environmental, and cultural settings and confronts risks from diseases and pests. Evans (1980) outlines four key focus areas for cane breeders: adaptability to evolving agronomic and management practices, disease and pest resistance through incorporating desired genetic traits, exploration of genetic

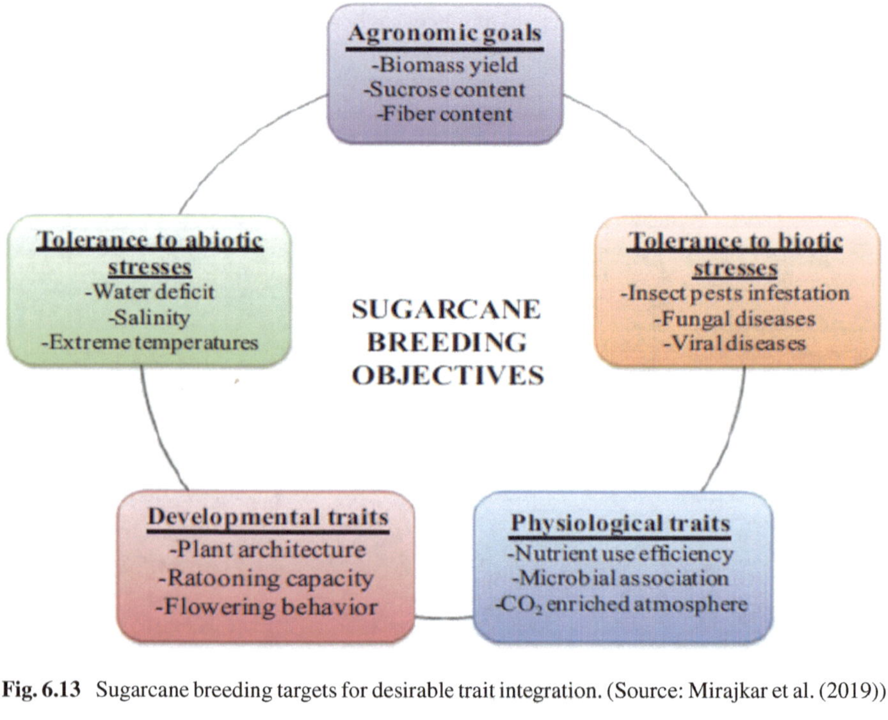

Fig. 6.13 Sugarcane breeding targets for desirable trait integration. (Source: Mirajkar et al. (2019))

engineering if conventional methods fall short, and an organized breeding strategy that includes selecting parental lines, studying hybridization compatibility, applying successful crossbreeding strategies, and evaluating for various environments.

6.3 Germplasm Biodiversity and Conservation

6.3.1 Germplasm Diversity

Sugarcane breeding programs around the world have utilized collections for variety development. The reported number of accessions or cultivars are as follows: Australia (4220), Brazil (3736), the United States (5020), Barbados (2567), Cuba (3386), India (3979), and Fiji (6000) accessions (INICA 2003). Along with genetic resources, germplasm collections are equipped with quarantine protocols to avoid quarantine pest invasion.

The diversity center for *S. officinarum* is believed to be in Papua New Guinea, as initially suggested by Daniels and Roach (1987) and later supported by AFLP marker analysis (Aitken et al. 2006). *S. spontaneum* likely evolved in Southern Asia and is characterized by lower sucrose content, with slender stalks and more fiber than *S. officinarum* (Jackson 2005). This adaptable species thrives in a diverse range of habitats and altitudes, from tropical to temperate regions, spanning latitudes 8°S to 40°N across three geographical zones.

In sugarcane breeding, germplasm collections serve as the genetic base for breeding new cultivars with enhanced sugar content, optimal agronomic characteristics and resistance to microbial and insect attacks comprising germplasm (Saccharum species and related genera) and hybrids. The global sugarcane germplasm collection, with replicas in India and the USA, is centered in Miami, Florida, and primarily focuses on basic germplasm. It comprises 1394 accessions from sugarcane species and related grasses, including *Saccharum officinarum* (397), *S. barberi* (58), *S. sinense* (42), *S. robustum* (85), *S. spontaneum* (348), *Saccharum* spp. (229), commercial hybrids (193), as well as Erianthus (23), Narenga (1), and Miscanthus (18) (Ming et al. 2006).

6.3.1.1 *Saccharum officinarum* L. ($2n = 80$)

Saccharum officinarum seems to have originated from the selection and domestication of *S. robustum*, with early clones chosen for their multicolored stalks, high sucrose content in the sap, thick soft and juicy stems with low fiber, large stalk diameter, and reduced tillering and flowering. It is primarily important for sugar and bioenergy production. This species is often recognized as a noble cane. Under ideal conditions, it produces a high cane and sugar yield. Because of its softness and sweetness, certain types are popular as chewing cane. Some strains have resistance to Fiji, downy mildew, and smut, but they are normally susceptible to many significant cane diseases. This species has long been used as a primary breeding stock for sucrose genes, and it normally has $2n = 80$ chromosomes (Vijayalakshmi 1967). Numerous researchers have recognized 10 as the fundamental chromosomal number of this species and characterized it as an octaploid (Roche and Daniels 1987).

6.3.1.2 *Saccharum spontaneum* L. ($2n = 40$–128)

Originating from northern India (Daniels and Daniels 1975), this perennial herbaceous plant is closely related to sugarcane and is a key parent in interspecific hybridization. It is commonly bred with *S. officinarum* to produce vigorous and adaptable germplasm for diverse environments. It consists of small stalks, with lower sugar content. It is an abundant tillering plant with compact, dense kinds ranging from slender-leaved, thin-stemmed types to tall, upright varieties with wide leaves, thick stems, and a strong root system. The stalks are usually firm and pithy, although they can also be hollow in the center. While it generally has low sugar content and poor juice quality, some varieties have shown superior juice quality. Highly adaptable, this species is resistant to diseases, cold, drought, and other challenging growing conditions (Vijayalakshmi 1967). Clones are resistant to severe mosaic disease as well as resilient to red rot. They can withstand drought, cold, and salinity well. This species has improved commercial varieties by increasing vigor, hardness, tillering capacity, and resilience to environmental stress, diseases, and pests (Mohan Naidu and Srinivasan 1982).

6.3.1.3 *Saccharum barberi* Jesw ($2n = 111$–120)

Hailing from northern India, this species is generally referred to as Indian canes and is named after the prominent breeder C. A. Barber. Barber (1913) proposed that this species originated from *S. spontaneum* through processes of mutation and selection. Parthasarathy (1948) reported that this species emerged via the natural hybridization of *S. officinarum* and *S. spontaneum* in Bengal, Bihar, and Orissa. The canes are typically tiny, rigid, and fibrous, with short, thin internodes and a robust root structure. While not suitable for industrial sugar production, they were historically used to manufacture ‘gur,’ with the juice from the ‘Katha’ variety used as a beverage. These canes can withstand harsh temperatures, drought, and waterlogging, and they are resistant to serious illnesses.

6.3.1.4 *Saccharum sinense* ($2n = 116$–118)

The species *S. sinese* is significant in cane breeding due to its deep root system and drought resistance. Despite its advantages, its application is limited due to inadequate blooming and fertility, which are most likely caused by aneuploidy and irregular meiosis of hybrid origin. This species comprises Chinese and Japanese canes, featuring hard, slender, and fibrous stalks with broad leaves, making them a common choice for fodder. The Uba cultivar, which has been farmed in China since prehistoric times, is thick, soft, and sweet, with a satisfactory sucrose content but poor yield. In China, it is largely utilized for chewing and, to a less extensive amount, for generating raw sugar. Grassl (1964) suggested that the species likely evolved in the China-Japan region through introgression between *S. officinarum* and tetraploid Miscanthus.

6.3.1.5 *Saccharum robustum* (2*n* = 60–80)

The stalks are robust, tall, and straight, including a few clones exceeding ten meters in height. Since some varieties have a high Brix, they are hard, high in fiber, and frequently pithy at the center, with little juice and sugar. They are more dense near the nodes with puffy growth rings. This species is distributed in the wild, predominantly in New Guinea and Indonesia, with a small population in the Philippines. Geographically, it ranks as the most immediate wild relative of *S. officinarum*. According to Daniels and Roach (1987), the species exhibits population diversity, probably due to introgression between *S. spontaneum* and other genera (Eriathus, Arrundinaceum, and Miscanthus) in New Guinea.

6.3.1.6 *Saccharum edule* Hassk (2*n* = 60–80)

This plant, which is traditionally grown in Melanesian settlements in New Guinea and Fiji owing to its edible inflorescence, has hairy leaves, a bloated and abortive inflorescence, and is sterile. Despite having minimal value in breeding efforts. Grassl (1967) postulated that it could be a mutant variety of *S. robustum*, most likely the product of *S. robustum*-related introgression. According to Roche and Daniels (1987), this species evolved as a result of introgression between *S. officinarum* or *S. robustum* and other related genera. This species' wild germplasm has been examined for its ability to transmit characteristics critical for biomass improvement, tolerance to harsh environments such as severe weather, flooding, and lodging, as well as resistance to diseases and pests.

6.3.2 *Cultivars Characterization and Phylogeny*

Before the end of the nineteenth century, *S. officinarum, S. sinense*, and *S. barberi* were the primary sources for most cultivated varieties (D'Hont et al. 1996). The development of the majority of commercial sugarcane hybrids today stems from hybridization between *S. officinarum* and *S. spontaneum* (Bull and Glasziou 1979). The fundamental breeding technique focused on merging the robust growth, ratooning ability, and resilience to abiotic stressors of *S. spontaneum* with the superior sucrose levels of *S. officinarum* and disease resistance (Berding et al. 2004). Other Saccharum species have also contributed to the parental lines. According to an examination conducted in the 1980s, commercial cultivar breeding projects comprised two clones of *S. sinense*, S. *barberi*, and *S. robustum*, as well as 19 clones of *S. officinarum* and multiple clones of *S. spontaneum* (Roach 1989). Other researchers have proposed that modern cultivars are primarily based on just 20 derivatives of *S. officinarum* and fewer than 10 derivatives of *S. spontaneum* (Patade and Suprasanna 2008). This interspecific hybridization has broadened the geographical extent of commercially sustainable sugarcane cultivation (Berville et al. 2005).

Estimates using GISH and AFLP markers indicate that chromosomes from *S. officinarum* make up 80% of the genome in commercial hybrid cultivars, containing 10% *S. spontaneum* chromosomes, while the rest are recombined from these genetic sources (D'Hont et al. 1996; Hoarau et al. 2001). A subsequent investigation using GISH and other approaches determined that the genetic makeup of various cultivars mainly comprises *S. officinarum*, with *S. spontaneum* contributing 15–20% chromosomes and under 5% translocated or recombinant chromosomes (Cuadrado et al. 2004). The hybridization of *S. officinarum* with *S. spontaneum* resulted in the emergence of the cultivar POJ2878, also known as the "Java Wondercane," in 1921 in Java, Indonesia. This cultivar increased sugar output by 35% compared to the best preceding varieties (Cox et al. 2000; Jeswiet 1929). This cultivar made a substantial contribution to the genetic basis of many modern sugarcane types. The International Society of Sugar Cane Technologists (ISSCT) manages an international cultivar naming system that includes letters and numbers (e.g., POJ2878). The letters represent the country and breeding station, while the numbers represent the year of selection and a sequence number (for example, POJ for "Proefstation Oost Java" in Indonesia). Important international collections of these cultivars are kept in India, Brazil, South Africa, and the United States for breeding purposes (Fauconnier 1993).

6.3.3 Genetic Resources Conservation Approaches

Conservation of genetic resources in plant gene banks through various approaches is highly beneficial for seed preservation (Abulaila et al. 2025; Abdullaev et al. 2025; Akparov et al. 2025; Aladele et al. 2025; Al Lawati 2025; Almarri et al. 2025; Arab et al. 2025; Békefi et al. 2025; Breidy 2025; Cañar et al. 2025; de Galarreta et al. 2025; Desheva et al. 2025; Díez et al. 2025; Galea-Gragera and Llera Cid 2025; Ghidan and Abdulmajid 2025; Gudaye et al. 2025; Harzli et al. 2025; Hidayatun and Kurniawan 2025; Ihsan and Prihatini 2025; Jatoi and Siddiqui 2025; Laghetti et al. 2025; Mallor and Luna 2025; Monteros-Altamirano et al. 2025; Osterc et al. 2025; Popović et al. 2025; Rocha et al. 2025; Wang et al. 2025). On the other hand, in situ conservation helps preserve biodiversity and identify varieties with beneficial and adaptive genes.

6.3.3.1 Genetic Resources

The primary storage of sugarcane genetic resources is in two global repositories: the Sugarcane Breeding Institute (SBI) in India and the World Collection of Sugarcane and Related Grasses (WCSRG) in Miami, Florida, USA. The WCSRG houses approximately 1002 accessions, which include significant alleles influencing biomass productivity, vital agronomic traits, and resilience to stress (Nayak et al. 2014). The biggest in situ germplasm collection center in the world, the SBI gathers a variety of *Saccharum* species and related taxa either on its own or in collaboration with other national organizations.

6.3.3.2 Field Gene Bank

The SBI Research Centre in Kannur, Kerala is home to the majority of Saccharum species (*Saccharum officinarum, S. barberi, S. sinense, S. robustum* and *S. edulle*) while Coimbatore, Tamil Nadu, is home to wild species such as *Saccharum spontaneum* and *Erianthus* spp. and related genera. *S. spontaneum* from Arunachal Pradesh and other high-elevation clones from Meghalaya are preserved at the IARI Regional Station in Wellington, Tamil Nadu. Additionally, there are large sugarcane germplasm fields in Miami, Florida and Home, Louisiana, USA (Mirajkar et al. 2019).

6.3.3.3 Seed Bank

At the Sugarcane Breeding Institute (SBI), the National Active Germplasm Site (NAGS) is responsible for the registration and release of genetic stocks. Commercial varieties are distributed to authorized researchers and kept as active collections for reference. In the field, 226 documented varieties and registered germplasm are being preserved as of April 2016. Annually, true polycross seeds of *S. spontaneum* and NAG variants are gathered, germination evaluated, and viable seeds are kept at the National Bureau of Plant Genetic Resources (NBPGR), New Delhi, in the National Gene Bank (Mirajkar et al. 2019).

6.3.3.4 In Vitro Conservation

To preserve germplasm, SBI's Kannur Research Centre has developed an in vitro breeding facility. The Miami World Collection Centre and Kannur, India, organize and preserve the world's sugarcane germplasm collections. The Miami collection is thought to be less stress-free than the Kannur Centre (Berding et al. 2004). The Indian World Collection is maintained at the field gene bank of the Sugarcane Breeding Institute (SBI) Research Centre. This center has created over 2600 Co (Coimbatore) selections, which dominate sugarcane production areas throughout India (Amalraj and Balasundaram 2006). Field-based conservation strategies are more vulnerable to biotic and abiotic stressors (Börner et al. 2009, 2010; Eid et al. 2025; Elabd et al. 2020; Ibrahim et al. 2023a, b; Saleh et al. 2020, 2021a, b; Salem and Börner 2003, 2004; Salem et al. 2021a, b, d, e, 2023a, b, c), hence in vitro conservation is the recommended way for protecting sugarcane germplasm (Glaszmann et al. 1996). Various long- and short-term strategies, including the use of entire plantlets and cryopreservation of in vitro shoot apices, have been effectively employed for in vitro germplasm storage (Martínez-Montero et al. 2008). Plantlets kept in tissue culture at 18 and 24 °C, with medium supplemented with various concentrations of sucrose and abscisic acid for more than 4 years, successfully demonstrated shoot retrieval and proliferation (Banasiak and Snyman 2017). The use of cryopreservation for preserving in vitro shoot meristems led to survival rates ranging from 41.7% to 69.4%.

6.3.4 Molecular Cytogenetics

Sugarcane species possess a large number of small chromosomes, typically ranging from 35 to over 200 (Ming et al. 2006). Several molecular cytogenetic studies have estimated the genome size of *S. spontaneum* to be between 3.05 and 5.31 pg, while *S. officinarum* ranges from 6.32 to 6.66 pg. The genome sizes of certain Saccharum hybrids from Canal Point vary between 6.30 and 7.5 pg (Grivet et al. 1994). In advanced sugarcane cultivars, 70% to 80% of chromosomes are from *S. officinarum* and 10% to 20% from *S. spontaneum*, with a small proportion arising from recombination. Clones with a higher chromosome count are classified as hybrids or atypical. For *S. officinarum*, with primary cytotypes of $2n = 60–80$, the basic chromosome number is likely $x = 10$ (Butterfield et al. 2001; Ming et al. 2006). *S. spontaneum* exhibits a broad range of chromosome numbers, from $2n = 36$ to $2n = 128$, with five primary cytotypes: $2n = 64$, 80, 96, 112, and 128. Using immunofluorescence, D'Hont et al. (1998) mapped the physical locations of 18 s–25 s rDNA and 5S rDNA genes on the chromosomes of various *S. spontaneum* cytotypes. The study revealed that the total chromosome number correlates with the number of rDNA sites mapped, leading to the conclusion that the basic chromosome number for *S. spontaneum* is $x = 8$.

Modern sugarcane cultivars (Saccharum spp. hybrids) are produced by crossing *S. officinarum* ($2n = 8x = 80$), known for high sugar yield, with *S. spontaneum*, a wild species with chromosome numbers ranging from $2n = 5x = 40$ to $n = 16x = 128$ (Butterfield et al. 2001; Ming et al. 2006; Le Cunff et al. 2008). Interspecific hybrids, typically with *S. officinarum* as the female parent and *S. spontaneum* as the male, often exhibit a triploid (AAB) chromosome number. For example, a cross between *S. officinarum* ($2n = 10x = 80$) and *S. spontaneum* ($2n = 8x = 112$) results in hybrids with $2n = 136$ chromosomes, showing low sugar content, slender stalks, high fiber, strong ratooning, and resistance to biotic and abiotic stresses (Sreenivasan et al. 1987).

6.4 Traditional Breeding

In traditional sugarcane breeding, *S. officinarum* was crossed with the wild, robust *S. spontaneum*, and the resulting hybrids were backcrossed to *S. officinarum* (Berding and Roach 1987) to increase disease resistance and production. Current cultivars have anywhere from 100 to 130 chromosomes, with 80% coming from *S. officinarum* 10% to 15% from *S. spontaneum*, and 5% to 10% being recombinant chromosomes (D'Hont et al. 1996; Piperidis et al. 2020). The expansion of sugarcane breeding programs throughout the twentieth century was ultimately aimed at maximizing sugarcane production. Given the current state of sugar management systems and the carbon partition between sucrose accumulation and plant development (Jackson 2005; Moore 2005), surpassing current sugar yields will be challenging. Over the past 50 years, particularly in the last decade (FAO 2019), yield instability in the US, Brazil, and globally highlights the early success of traditional breeding, followed by a plateau in genetic improvements for sugarcane.

6.4.1 *Improvement Strategies*

Traditionally, the sugar industry has relied on imported sugarcane varieties. Based on research records, POJ 2878, a sugarcane variety imported from Java, was the first. The varieties underwent evaluation and were released through methods of varietal introduction and selection. Unfortunately, many imported varieties were not well suited to local conditions and often fell victim to pests and diseases. The failure of these imported varieties prompted the establishment of sugarcane breeding programs in various breeding stations. Sugarcane hybridization programs have been implemented for two main reasons: (1) it is challenging to find new varieties from introduced ones due to on limitation on importing exotic varieties because of Intellectual Property Rights, and (2) Locally bred varieties tend to be more adaptable than introduced ones (Aung and Soe 2005).

6.4.2 *Traditional Breeding Methodologies*

Sugarcane breeding programs are designed to address the specific needs of different regions, with a primary goal of increasing sugar yield per unit area. This is achieved by improving factors such as stalk height, diameter, number, and sugar accumulation. Enhancing sugar yield has always been a central focus of these breeding efforts. Sugarcane breeding programs aim to improve cane yield, sugar content, pest resilience, and ratooning potential. Growers, especially under cane weight-based procurement systems, often favor thick-stalked varieties like Thailand's K-95/84. Yield differences between plant and ratoon crops are common, with most farmers cultivating one plant cane and one ratoon crop. The low ratooning ability of current varieties is a significant issue, making its enhancement a key breeding goal. Additionally, breeding efforts may focus on improving drought and waterlogging tolerance for specific regions (Aung et al. 2021).

The Sugarcane Breeding Program consists of three components: (i) the collection, characterization, and maintenance of sugarcane germplasm; (ii) hybridization (cross-pollination of selected parents) and (iii) the selection process based on phenotype and agro-chemical characteristics. (i) Collection, characterization, and maintenance of sugarcane germplasm are crucial due to the limited genetic diversity in current sugarcane varieties. The program focuses on collecting germplasm and identifying elite genetic stocks with desirable economic traits. Germplasm collection is carried out through international cooperation in Egypt, as Myanmar and India are centers of diversity for *Saccharum* spp. Wild canes like *Saccharum spontaneum* offer traits like robust growth, abundant tillering, and resilience to extreme climates making them valuable for breeding programs. While Myanmar researchers collected wild canes, their characterization and utilization in the breeding program were lacking. A cooperative effort between Myanmar researchers and Indian sugarcane breeders began in 2009, resulting in the collection of 76 genotypes, including *Saccharum officinarum*, *Saccharum spontanum*, and *Erianthus* spp. from various regions of Myanmar (Table 6.1) (Aung et al. 2022). The collected germplasm is maintained in the research center, and research is conducted to aid in selecting parental lines for the breeding program. This includes (a) germplasm characterization based on morphological and agro-chemical data, (b) investigation into the synchronization flowering among parental lines, (c) studying the effect of altitude and latitude on flowering abilities, (d) assessing pollen fertility levels in different varieties used for crossing and (a) characterizing germplasm based on morphological and agrochemical data: In sugarcane breeding, genetic variability is essential for selecting desirable traits. 121 sugarcane germplasms were grouped into eight clusters based on 11 qualitative traits (Fig. 6.14)

6.4.3 Traditional Breeding Limitation

The main challenges in sugarcane hybridization are a limited genetic base and low F_1 seed germination rates, with fuzz germination often below 5%. Seed counts per gram vary, and some fuzz fails to germinate. The crossing period, particularly

Table 6.1 Collection of wild canes in Myanmar (2021)

Sr	Wild cane	Number
1	*Saccharum officinarum*	18
2	*Saccharum spontanum*	6
3	*Erianthus* spp.	52
Total		76

Fig. 6.14 Wildcane collection in Myanmar. (Source: Photo by Nyo Nyo Aung and Ei Ei Khaing)

during cold nights in January, reduces viable fuzz formation while synchronizing male and female flowering remains difficult. These issues hinder effective parent selection and crossing of early and late-flowering cultivars (Abu-Ellail and Mohamed 2019). For the breeding program to succeed, knowledge of modern technologies and approaches as well as experienced and professionally trained breeders are required. Other constraints include (c) Aligning the blooming periods of male and female flowers in some sugarcane varieties (d) lack of photoperiodic facilities and limited experience, and (e) inadequate qualified staff in this field. To overcome the narrow genetic base in sugarcane breeding, many stations are turning to sugarcane-related genera like Erianthus to broaden the genetic pool and develop varieties with improved commercial traits. Study visits should be arranged for researchers to international sugarcane breeding institutes to enable them to gain more technical knowledge and research experience (Aung et al. 2021; Abu-Ellail and McCord 2019).

6.4.4 Role of Biotechnology

Sugarcane production struggles with environmental challenges and a limited variety of options. Most varieties are developed through inter-specific hybridization and back-crossing, a lengthy process to release elite varieties. The conventional breeding methods are now being supplemented by biotechnology-based breeding. Because sugarcane is a polyploid, the selection cycle can be shortened and the necessary gene can be identified for marker-assisted selection. The sugar industry is working to incorporate these methods to advance sugarcane research using modern biotechnology. This will require collaboration among various sugarcane research stations (Edwardson and Graham 1994).

6.5 Molecular Breeding

As molecular breeding advances, identifying quantitative trait loci (QTLs) in successful parents and tracing them within lineages could help compare them with less successful cultivars (Abdel-Hamid and Salem 2021; Abu-Ellail et al. 2021a, b; Aldahak et al. 2021; Hassan et al. 2021; Ibrahim et al. 2023a, b; Ismail et al. 2013; Mattar et al. 2016; Mourad et al. 2019; Salem 2004; Salem and Mattar 2014b; Salem et al. 2003, 2004a, b, 2007b, 2015b, c, 2023a, b, c). A combined pedigree and QTL mapping approach was used to understand Mandalay's genetic contribution to Australian varieties and elite parental material. The biotechnology effort is primarily supported by government funds. The biotech laboratory was constructed in the sugarcane research center, with the government providing basic infrastructure and equipment for biotechnology research. Our division addresses the shortage of skilled biotechnology manpower in Myanmar, which is due to limited facilities, equipment, technical expertise, research and development, and training resources. Biotech research was conducted to investigate improving disease tolerance such as smut disease and white leaf disease on sugarcane, using marker-assisted selection. This year, the study will concentrate on analyzing genomic variability and exploring genotypic relationships within sugarcane varieties using SSR markers. Accurately estimating genetic diversity among germplasm sources could enhance plant breeding efficiency (Alqahtani et al. 2024; El Rabey et al. 2013; Elabd and Salem 2009; El-Absawy et al. 2012, 2015; Eltaher et al. 2018; El-Zanaty et al. 2011, 2013; Esmail et al. 2009; Ghazy et al. 2020; Hassan et al. 2021; Ibrahim et al. 2024; Mattar et al. 2009; Salem and El-Zanaty 2015; Salem 2009; Salem and Ibrahim 2024; Salem and Mattar 2014a; Salem and Sallam 2016; Salem et al. 2008a, b, 2010, 2015a, 2021c, f, 2022, 2024; Varshney et al. 2008). Understanding genetic diversity, relationships among breeding genomes, and their polymorphic nature and codominance is crucial for advancing crop improvement (Burner and Legendre 1993; Burner 1997; Tai 1988).

6.5.1 Molecular Marker Assisted Breeding

Marker-assisted selection (MAS) is where molecular markers truly shine in improving crop breeding efficiency (Salem 2015; Salem et al. 2007a). While sugarcane molecular tools have been developed, they have not been particularly effective in addressing the complex breeding challenges presented by sugarcane genetics. The sugarcane genetic map has an entire extent of 2600 cM and a marker density of 7.3 cM, consisting of over 1100 molecular markers (including various types) (Garcia et al. 2006), which has similarities with different crop species (Casu et al. 2005). Target genes in Sugarcane breeding lines have been identified and mapped using molecular markers derived from both DNA regions (Tabasum et al. 2010) and expressed RNA sequences (Wei et al. 2010). Genetic maps containing sugarcane disease and pest resistance genes have been successfully developed using bulked segregant analysis (BSA) and molecular marker analysis of quantitative trait loci (Asnaghi et al. 2004; Dussle et al. 2003; Salem et al. 2023a, b, c). A study found no notable associations between stalk features and sugar output in the tested population. However, molecular markers associated with yield were found in 27 areas of the sugarcane genome in a hybrid of the Australian variety Q165 and *S. officinarum* (Aitken et al. 2008; Essa et al. 2023). Each of the 30 random amplification of polymorphic DNA (RAPD) markers used in a study of 40 sugarcane genotypes, including *S. officinarum* and *S. barberi*, identified more than one unique allele (Ibrahim et al. 2023a, b; Safhi et al. 2022; Tabasum et al. 2010). A potentially valuable large--scale approach is diversity array technology (DArT), which utilizes a high throughput microarray platform as a tool for revealing a vast array of polymorphic markers across an entire genome in a single analysis (Wei et al. 2010; Abd El-Moneim et al. 2021; El-Mansy et al. 2021). Gaining insights into genetic diversity and relationships within breeding genomes, focusing on their polymorphism, codominance, and relevant materials, significantly impacts crop improvement. DNA markers applicable in sugarcane breeding include Polymerase Chain Reaction (PCR) (Burner 1997; Burner and Legendre 1993; Tai 1988). Using Nei's (1978) approach, genetic distance was measured for SSR data from 15 sugarcane varieties, and the resulting relationships between accessions were depicted in a dendrogram and Table 6.2 and Fig. 6.15, showing genetic similarity values ranging from 77% to 100% among the 15 sugarcane varieties.

6.5.2 *Bioinformatics*

Advances in technology have led to an explosion of biological data that must be organized and analyzed to deepen investigations and offer benefits. Bioinformatics aims to categorize this data and present it in a user-friendly manner, as well as develop tools to uncover previously unseen connections within the data. For many plant and animal species, extensive resources such as genomes, transcriptomes, proteomes, and metabolomes are readily available The complexity and immense size of the sugarcane genome, estimated at 10,000 megabases for current varieties, make sequencing a significant challenge (D'Hont and Glaszmann 2001). Genetic anomalies hamper the sugarcane genome sequencing process, making comparative genome

Table 6.2 Name of the primers used for detection of polymorphism in sugarcane genotypes

Primer no	Primer sequence (F/R)	Annealing temperature
SSR21	GCA CAC CAC TCG AAT AAC GGA T AGC TGA ATC GTG GTG AAC AA	55
SSR31	TCA TAC AAG CAG CAA GGA TAG GAG CCG CAA GGA AGC GAC	58
SSR 9	AGC TCT ATC AGT TGA AAC CGA GCC AAA GCA AGG GTC ACT AGA	51

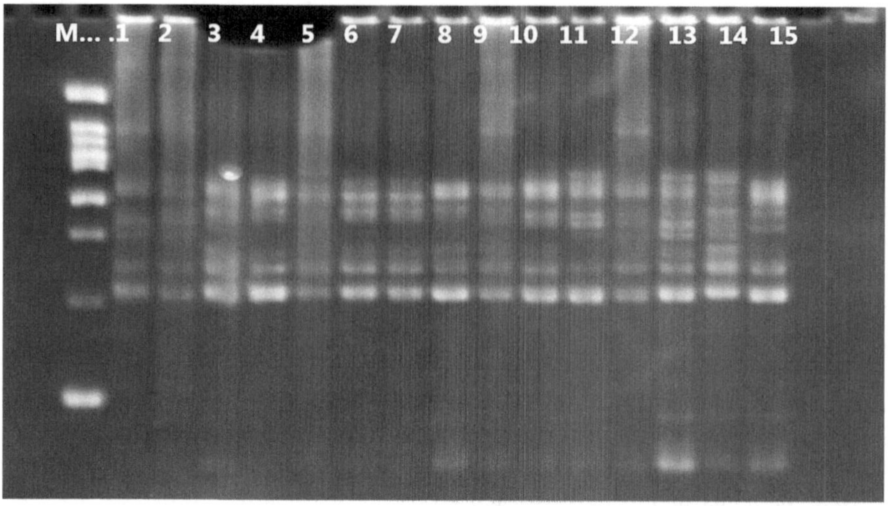

Fig. 6.15 SSR-PCR products of 14 sugarcane varieties on agarose gel (DNA size markers are shown in the left lane (Research on Sugarcane Research Center, Pyinmana, Myanmar, unpublished paper). (Photo by Nyo Nyo Aung and Ei Ei Khaing)

analysis crucial for gene alignment, particularly on homologous chromosomes. Genetic maps and resources are vital for identifying markers and moving mapping forward. However, there are few genetic maps for sugarcane; only two maps of *Saccharum spontaneum* from 1993 to 1995 are published on Grain Genes, and neither is still in use (Ruiz et al. 2004). The only publicly available mapping resource for sugarcane to date has been developed by CIRAD as part of TropGENE-DB (http://tropgenedb.cirad.fr/).

6.6 Tissue Culture Application

Plant tissue culture (PTC) is essential to plant biotechnologies and represents a dynamic field in both basic and applied sciences, offering substantial opportunities for further research (Aboshama et al. 2015; Emara et al. 2011, 2013; Nawara et al. 2016, 2017). It serves as a crucial tool for speeding up genetic transformation and enhancing breeding efforts (Alves et al. 2011). Sugarcane (*Saccharum* spp.) is commercially important due to its primary products and byproducts. However, seed multiplication of new varieties is a major challenge, as it can take 6–7 years to generate enough improved seed material from a desired clone, creating a bottleneck in breeding programs. Micropropagation offers an efficient solution for large-scale cloning, addressing the limitations of traditional methods in sugarcane breeding. Plant tissue culture and genetic engineering are now essential for sugarcane biotechnology, with established protocols for shoot tip culture, callus culture, embryo culture, virus-free plants, somatic embryogenesis, and genetic transformation. Meristem culture is particularly effective for producing healthy, pathogen-free sugarcane seed material. The three-tier nursery program now successfully uses tissue culture-derived planting materials to produce foundation and certified seeds from breeder seeds obtained through tissue culture laboratories (Anjali Singh 2021).

6.6.1 Micropropagation

Tissue culture, or micropropagation, enables the swift and extensive propagation of plants under regulated conditions. This technique offers a thousand-fold rate of multiplication and is the quickest method available for sugarcane propagation. Micropropagation in sugarcane provides a rapid way to produce healthy seeds of new varieties and rejuvenate old run-down varieties. Micropropagation involves artificially propagating plants vegetatively using tissue culture or cell culture techniques. Tissue culture is a process in which plant tissues are cultivated *in vitro* using artificial nutrient media containing a carbon source, minerals, vitamins, and plant growth regulators under aseptic conditions (Aboshama et al. 2015; Emara et al. 2011, 2013; Nawara et al. 2016, 2017). This technique, also known as the micropropagation technique, is used for large-scale production of true-to-type and disease-free sugarcane plantlets using the apical meristem culture technique. Sugarcane is a tropical and subtropical crop cultivated as a cash crop and a key commercial crop used for food and biofuels. It contributes significantly to obtaining vigorous seed cane with excellent yield and quality. The outbreak of sugarcane white leaf disease in many areas and the poor quality of current sugarcane varieties are the main reasons for producing healthy seed cane. There are three methods used for clean seed cane production. The tissue culture technique is more effective for producing clean seed cane than chemical therapy and heat therapy. Therefore, the objectives of tissue culture in sugarcane are to rapidly multiply and mass produce in a short period obtain healthy seed cane true to to the mother plant, and regenerate old varieties.

6.6.1.1 Tissue Culture Steps

Tissue culture steps are usually divided into the following stages (Fig. 6.16). Stage 1: Culture establishment: The initial planting materials for micropropagation are the shoot tips containing the meristem with four to five youngest leaves. The top portion of the apical meristem tissue of a 2.5-month-old plant bearing the spindle tissue was used as the source of the plant to initiate the culture. First plant pieces are cleaned with pure water and trimmed to a size of about 3–5 cm. Then the prepared plant pieces are soaked in 20% Clorox and swirled for 20 minutes. They are then washed with sterile steamed distilled water three times in an aseptic chamber and only apical meristems or shoot tips are kept on MS-C media (callus formation media) These

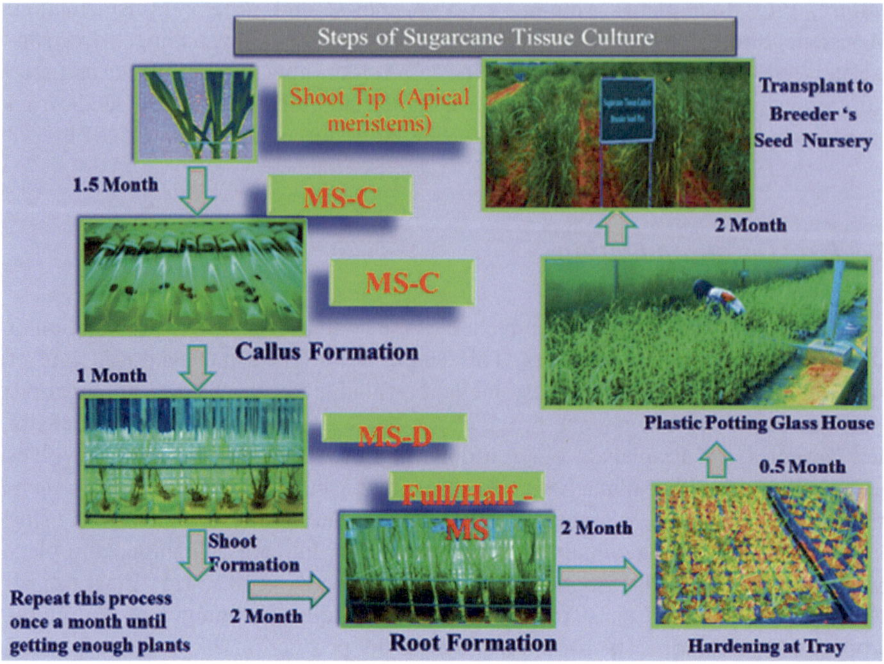

Fig. 6.16 Steps of sugarcane tissue culture. (Source: 5th Annual paper conference, Department of Agriculture in Myanmar, 2023) (unpublished paper): Editing by Nyo Nyo Aung and Ei Ei Khaing)

are excised aseptically and then planted in a liquid culture medium (Anjali Singh 2021). Stage II: Shoot multiplications. Shoots that emerge are sterilized, detached from the parent shoot, and transferred to a new medium to promote multiplication. New shoots emerge from the original shoot tip, while the separated shoots undergo multiplication within 10–15 days. This process can be reiterated by transferring the newly formed shoots to a fresh multiplication medium. Multiplication of shoots persists until the required number is reached. Stage III: Rooting. Shoots are transferred to the rooting medium. This is the root development and hardening stage in preparation for nursery growing. Sufficient rooting ready for transfer to the soil can be obtained in 10–15 days.

Stage IV: Nursery planting. After sufficient roots have formed, the plantlets are separated, sorted out, and transplanted in plant trays containing sterilized soil medium with compost. The plantlets are grown in shaded areas with humid conditions or a greenhouse or under plant shades for 1–2 weeks. Then they are exposed to direct sunlight for another 3–5 weeks. This period allows the plantlets to further develop and harden for final field transplanting. Sufficient moisture and foliar fertilizer are critical inputs in this stage (Fig. 6.17). Stage V: Transplanting in the field. Grown plantlets are transferred to the field with moist soil conditions. They must be irrigated immediately after transplanting to have a high survival percentage. Plant population could be from 2.0 to 2.5 lacs per hectare or at a planting distance of 0.4–0.5 m apart, respectively, with a 1 m furrow distance. This wider planting

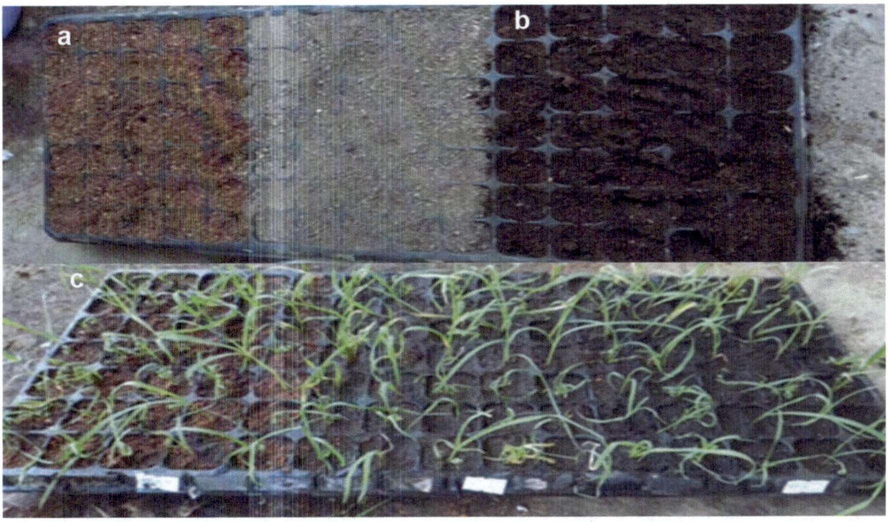

Fig. 6.17 Media test for the hardening stage. (**a**) Substrate Media Coconut Husk, (**b**) Organic Fertilizer; (**c**) Substrate Media. (Source: 5th Annual paper conference, Department of Agriculture in Myanmar, 2023. Photo by Nyo Nyo Aung and Ei Ei Khaing)

Table 6.3 Cane yield comparison of mother plant and tissue Culture plant

Sr	Tested variety (K-95/84)	Cane yield (ton/ha) 2022–2023 Cane season (Plantcane)	2023–2024 Cane season (Ratoon)
1	Mother plant	32.01	62.98
2	Tissue plant (breeder)	57.00	75.58
	Increased productivity (%)	78	20

distance is recommended because micropropagated plants produce more tillers than ordinary cane points. After 4–5 months, plants are ready for cutback. Immature leaves are preferred target tissues for developing embryogenic calli due to their high proliferation rates among various explants (Ali et al. 2010; Snyman et al. 2006), and this was a significant breakthrough in sugarcane biotechnology research.

Sugarcane multiplication via tissue culture is encouraged to generate planting materials that are free from viral and phytoplasma infections. By using tissue culture plants for Breeders' seed nursery in the three-tier system of sugarcane seed multiplication, several advantages can be achieved: (i) a higher seed multiplication rate compared to the conventional method, (ii) true to type plantlets, (iii) good germination and quicker germination ability, (iv) obtaining disease free seed material from tissue plantlets and (v) increased sugarcane yield I due to continuous tillering leading to the production of the increased cane yield per unit area (Anjali Singh 2021). The productivity of sugarcane in tissue culture fields was higher than in the mother sugarcane field (explant) both in plant and ratoon (Table 6.3). Sugarcane micropropagation can increase field productivity by up to 25%, while Pérez et al. (2000) suggest that integrating in vitro sanitation with rejuvenation could boost sugar yields by 10–15% per unit area.

6.6.2 Synthetic Seeds

Using artificial seeds developed from somatic rather than zygotic embryos presents new opportunities in agriculture. Artificial seeds provide an effective method for propagating transgenic plants, non-seed-producing species, polyploids with desirable traits, and plant lines that face challenges with seed propagation. This clonal technique simplifies the time-consuming process of traditional recombination breeding and offers a cost-effective way for farmers to benefit from biotechnological advancements. The artificial seed plays the role of a seed and can germinate into complete plantlets (Gantait and Mitra 2019). An artificial seed contains various plant structures, such as somatic embryos, buds, shoots, or other meristematic tissues enclosed in a coating, which can be sown like a regular seed to grow a new plant. Germplasm exchange is made easier through synthetic seed technology, allowing for short-term preservation. Synthetic seed technology, an emerging field in plant tissue culture, holds the potential for efficient crop replication but confronts practical hurdles in widespread adoption among farmers and producers. Synthetic seeds, which originate from plant tissue culture, often use somatic embryos or tissue-derived components to provide environmental protection. This method allows for short- and long-term storage at low temperatures, as well as an effective delivery system, which opens up new paths for plant biotechnology research (Roy and Tulsiram 2013).

6.6.2.1 Types of Synthetic Seeds

There exist two categories of synthetic seeds: (i) Desiccated Synthetic Seeds, which involve the encapsulation of multiple somatic embryos followed by desiccation and (ii) Hydrated Synthetic Seeds, which involve the encapsulation of a single somatic embryo in hydrogel capsules.

6.6.2.2 Applications of Synthetic Seeds

Synthetic seeds are valuable for multiplying plants that do not produce seeds, including ornamental hybrids and polyploids, and are also useful for propagating endangered species. Cryopreserved synthetic seeds aid in germplasm preservation, with synthetic seed technology enabling short-term germplasm exchange. Cryopreservation offers an effective solution for the prolonged conservation of plant specimens at extremely low temperatures.

6.6.2.3 Limitations of Synthetic Seed Production Technology

The production of synthetic seed is hindered by several major issues, which are impeding their commercialization. The advancement of synthetic seed technology is facilitating mass production of high-quality micropropagules, yet it continues to be a constraint. Another obstacle is the low survival and germination rates of encapsulated somatic embryos. Poor germination in synthetic seeds can stem from inadequate (Roy and Tulsiram 2013).

6.7 Genetic Engineering and Gene Editing

6.7.1 Genetic Engineering

The application of genetic engineering and gene editing techniques in sugarcane holds significant promise for transforming the crop's productivity, disease resistance, and overall sustainability. These technologies allow scientists to manipulate the genetic composition of sugarcane to enhance desirable qualities and eliminate undesirable characteristics. One example of genetic engineering being used in sugarcane is the enhancement of sucrose production. Scientists are working on developing genetically modified sugarcane varieties that can produce higher amounts of sucrose, ultimately leading to increased sugar production. The goal is to raise the total sugar content in crops by incorporating genes associated with sucrose production or regulation (Al-Khayri et al. 2025;Aitken et al. 2005; Barreto et al. 2019; Bilal et al. 2015; Hoarau et al. 2002; Racedo et al. 2016; Yang et al. 2019). Another area of research focuses on improving disease resistance in sugarcane through genetic engineering techniques. Sugarcane is susceptible to various diseases, including viruses, bacteria, and fungi, which can significantly reduce crop productivity. Researchers are exploring the incorporation of genes that confer resistance to these diseases, to reduce the need for chemical treatments and improve overall plant health. Gene editing technologies, such as CRISPR-Cas9, provide precise methods for making targeted alterations to the genetic makeup of sugarcane. This advancement enables scientists to modify the genetic material of plants, with accuracy, enhancing desirable traits like drought and salinity tolerance, as well as improving nutritional composition. For example, researchers can use gene editing techniques to manipulate genes related to water and nutrients in sugarcane plants enabling them to succeed in challenging environments (Carrasco-Valenzuela et al. 2019).

The implementation of genetic engineering and gene editing techniques shows promise in improving sustainability and reducing the environmental impact of sugarcane cultivation. Researchers are exploring genetic modification related to nitrogen fixation, which could help reduce the need for chemical fertilizers. Additionally, gene editing techniques could help develop sugarcane cultivars that require less water and energy, without sacrificing productivity levels (Koonin and Makarova

2019). Despite these potential benefits, the use of genetic engineering and gene editing in sugarcane farming faces regulatory and public acceptance challenges. It is imperative to address concerns about the safety of genetically modified organisms (GMOs) and their potential ecological effects transparently to ensure the responsible and ethical use of these technologies in sugarcane cultivation (Hoang et al. 2017).

Saccharum barberi, Saccharum edule, Saccharum officinarum, Saccharum robustum, Saccharum sinense, and Saccharum spontaneum are all species of sugarcane. They share a common ancestor with sorghum and belong to the Poaceae family, which also includes genera like *Erianthus* and *Miscanthus* (Amalraj and Balasundaram 2006; Souza et al. 2019). Today's sugarcane varieties (*Saccharum* spp.) exhibit high levels of polyploidy and aneuploidy resulting from interspecific hybridization. Souza et al. (2011) estimate the sugarcane genome to be approximately 10 gigabase pairs (Gbp), within multiple copies of several genes. Depending on the ploidy level, the monoploid genome size ranges from 800 to 900 Mb (Garsmeur et al. 2018). The emergence of gene-editing (GE) tools has revolutionized biology. By enabling selective modification of specific genes. Plant breeders now have more opportunities to enhance yields through genetic improvements using gene editing. Through gene editing, specific alleles can be eliminated, or genes can be introduced to enhance functional traits. This method, as discussed previously (Boettcher and McManus 2015; Smith et al. 2017), can generate knock-out phenotypes, that are highly desirable in many applications. Both Agrobacterium-based and biolistic transformation techniques for sugarcane have been thoroughly established and are commonly used in global research settings. A recent summary of advancements in sugarcane transformation methods was published by Krishnan and Mohan (2017).

6.7.2 Gene Editing

Scientists specializing in molecular biology have greatly benefited from gene--editing technology, as it allows for precise manipulation and targeting of individual genes. Plant breeders have utilized gene editing methods to create improved cultivars with preferred characteristics. Gene editing can silence or introduce any gene to produce a desired effect (Smith et al. 2017). Clustered regularly interspaced short palindromic repeats (CRISPR), transcription activator-like effector nucleases (TALENs) (Li et al. 2018a, b; Li et al. 2020), and zinc finger nucleases (ZFNs) (Carroll 2011; Cong et al. 2013; Jaganathan et al. 2018) are just a few examples of transgenic systems used to manipulate eukaryotic genomes. The CRISPR/Cas9 system is a user-friendly, affordable and direct method for editing. By using CRISPR/Cas9, site-specific double-strand breaks (DSBs) can be induced in targeted genes leading to modifications. This overcomes a major constraint of ZFNs and TALENs which require lengthy and laborious technical design processes to achieve desired outcomes. By generating double-strand breaks, the CRISPR/Cas9 system enables repair through cellular mechanisms, including non-homologous end joining (NHEJ) and homologous-directed repair (HDR) pathways (Watterworth et al. 2011). Various gene editing techniques, both simple and complex, are being utilized to create new crop varieties with preferred agro-botanical and physiological traits, as depicted in Fig. 6.18.

Many different genetic engineering (GE) methods for manipulating plant genomes and other eukaryotic genomes have been created in the past. Meganucleases (Puchta et al. 1993), zinc finger nucleases (ZFNs) (Carroll 2011), transcription activator-like effector nucleases (TALENs) (Li et al. 2012), and the popular Clustered Regularly Interspaced Palindromic Repeat (CRISPR)-associated protein 9 system (CRISPR-Cas9) (Cong et al. 2013) are all examples of such systems. ZFN and TALEN are two older methods that use intricate technical designs to accomplish precise targeting, but they are time-consuming and arduous due to their

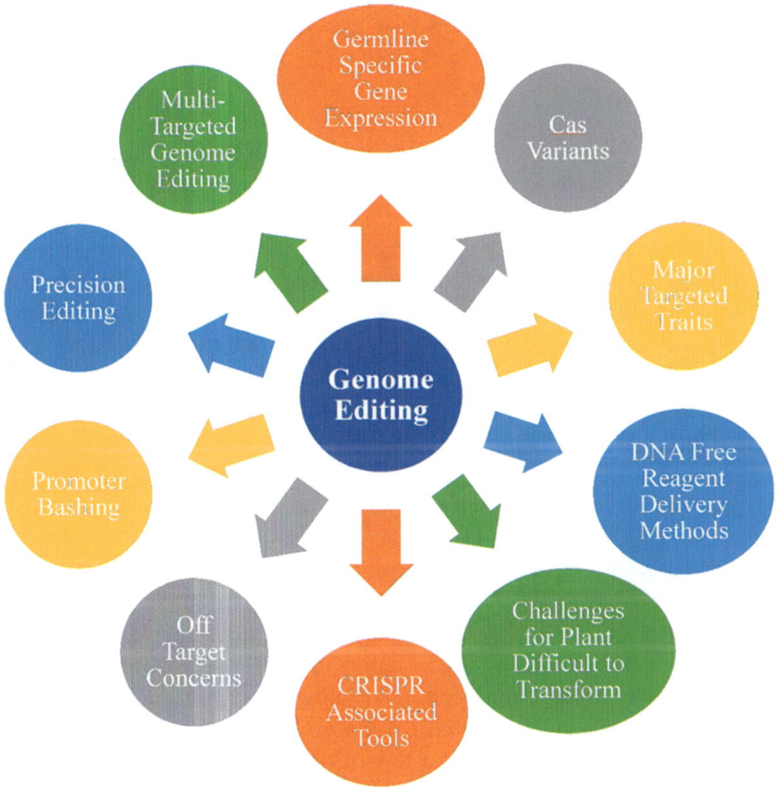

Fig. 6.18 Methods of modifying DNA sequences to create new plant varieties with specific characteristics. (Source: Hussin et al. (2022))

complexity. In contrast, CRISPR-Cas-based genetic engineering is an efficient, low-cost alternative (Marone et al. 2023).

Sugarcane has been the subject of numerous reports of investigations using genetic modification. Both Agrobacterium-based and biolistic transformation techniques for sugarcane have been thoroughly established and are commonly used in global research settings (Budeguer et al. 2021). Despite exhibiting insect and glyphosate resistance in field conditions, a transgenic sugarcane line has shown reduced agronomic and industrial performance when compared to non-transformed control plants in several studies (Wang et al. 2017). Disease-resistant transgenic sugarcane has been studied in the field; for example, two cultivars altered for resistance to sugarcane mosaic virus (SCMV) strain E showed wide variance in yield characteristics and disease resistance in the field (Gilbert et al. 2005). Transgenic lines altered to be resistant to sugarcane yellow leaf virus (SCYLV) performed worse than the original genotype during three-year research assessing agronomic variables of interest and virus resistance (Gilbert et al. 2009). In contrast, transgenic sugarcane lines expressing the coated protein gene CP of SCMV have been reported to have a positive outcome, with higher yields and reduced SCMV disease incidence over two growing seasons and four experimental locations in China (Yao et al. 2017).

6.7.3 Application of Gene Editing

For vegetatively propagated crops like sugarcane, the ability to create transgene--free mutants in a single generation thanks to advances in genome editing technology is of great benefit. These methods eliminate the need for selecting transgene-free plants through meiotic recombination in future generations. Figure 6.19 depicts a few of the methods that could be used to create transgene-free genetically engineered lines in sugarcane.

6.7.3.1 Editing of Genomes Caused by Viruses

To create genetically engineered (GE) plants without using transgenes, virus--induced genome editing (VIGE) is a potential method. The guide RNA (gRNA) and Cas endonucleases can be introduced into plant systems via either DNA or RNA viruses. Both the gRNA and Cas9 can be delivered into the plant genome simultaneously, or the gRNA can be introduced into plants that have Cas9 pre-inserted into their genomic DNA (Zhang et al. 2022). Tobacco, Arabidopsis, wheat, rice, maize, soybean, potato, and tomato are just a few of the model plants and economically significant crops where VIGE has been successfully utilized (Zhang et al. 2022). Various types of viruses Including DNA viruses like bean yellow dwarf virus (BeYDV), wheat dwarfing virus (WDV), cotton leaf crumple virus (CLCrV), and

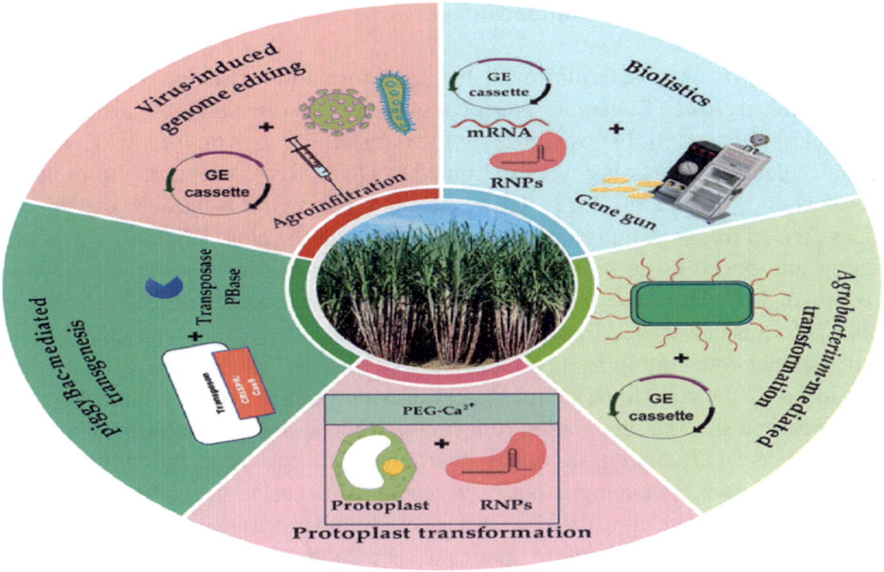

Fig. 6.19 Methods for generating genetically engineered (GE) sugarcane lines devoid of transgenes. (Source: Surya Krishna et al. (2023))

cabbage leaf curl virus (CaLCuV), as well as RNA viruses like tobacco rattle virus (TRV), potato virus X (PVX), pea early browning virus (PEBV), and tobacco mosaic virus (TMV) have been employed for gRNA delivery with great success. Given that sugarcane is commonly propagated using setts, the following technique may be valuable for implementing the VIGE strategy in this crop. The first step involves agroinfiltration of viral vectors into developing canes to create transgenic, virus-free GE plants through meristem culture of the resulting setts grown in a container. Canes can also be agroinfiltrated, and the resulting offspring can be used as a source of propagating material for transgene-free GE plants. Despite its advantages over tissue culture, the latter method has limited applicability due to the regional and seasonal specificity of sugarcane seed production and the higher frequency of recombinants produced by sugarcane's heterozygous nature (Moore and Nuss 1987; Pereira et al. 1983; Surya Krishna et al. 2023). Key benefits of VIGE are high gRNA expression from rapid virus proliferation, bypassing tissue culture in seed-derived plants, and minimizing off-target mutations and non-GMO plants (Zhang et al. 2022; Uranga and Daròs 2022). Since viruses cannot replicate in the apical meristem or reproductive organs, producing heritable mutations is challenging. To overcome this obstacle and generate heritable mutations without residual transgenes, it is possible to fuse mobile RNA sequences like Flowering Locus T (FT) with gRNA and transmit the virus to the shoot apical meristem.

6.7.3.2 Biolistic Particle Bombardment

In addition to Agrobacterium-mediated transformation (Ellison et al. 2020; Budeguer et al. 2021), Sugarcane genetic transformation often utilizes particle bombardment/biolistics. Biolistics allows for the intracellular delivery of a wide variety of molecules, encompassing, yet not limited to DNA, RNA, plasmid, and proteins (Laforest and Nadakuduti 2022). GE plants are termed GMOs because exogenous genes derived from plasmids have been integrated into the plant DNA, with genome editing introduced using biolistics. Particle bombardment for transferring ribonucleoproteins (RNPs) made of in vitro synthesized gRNA and Cas9 offers a viable solution to this challenge (Tsanova et al. 2021). Cas endonuclease and gRNA molecules, as genome editing agents, remain active for only a short time before being eliminated by the cell's internal machinery. Cas9 protein and gRNA molecules are combined in a specific preparation to form the RNP, along with additional components such as a buffer (Svitashev et al. 2016). Genome editing in agricultural crops including rice, wheat, and maize has been accomplished using the biolistics delivery of RNPs (Banakar et al. 2019, 2020). Particle bombardment employing embryogenic callus is commonplace in sugarcane, and biolistics has been employed to create numerous successful transgenic events (Mohan et al. 2020). The genome editing cassette using CRISPR/Cas9 was successfully transferred into sugarcane callus via biolistics by Oz et al. (2021) and Eid et al. (2021). Herbicide resistance was achieved through editing acetolactate synthase (ALS) gene alleles by Oz et al. (2021). Herbicide nicosulfuron was used on both wild and GE varieties, killing the former while sparing the latter. Chlorophyll production depends on the magnesium chelatase gene and when Eid et al. (2021) altered the many alleles of this gene, the resulting plants were yellow instead of green. These investigations underscored the application of biolistics in CRISPR/Cas-driven genome modification to generate specific mutations in sugarcane.

6.8 Mutation Breeding

6.8.1 Somaclonal Variation

Various plant cells (callus, protoplasts) and tissues (stems, roots, embryos) can be maintained and grown in aseptic in vitro conditions using a variety of techniques that fall under the umbrella term of tissue culture (Hartmann et al. 2001). In vitro tissue culture has its origins in the cell theory of totipotency (Vasil 2008). According to research by Kumar and Loh (2012), plants with totipotency can develop into mature plants from a single cell or tissue. The generation of a whole sugarcane plantlet from callus tissue first suggested the potential of an in vitro culture technique for use with sugarcane (Heinz and Mee 1969).

Plant cells cultured *in vitro* can experience both dedifferentiation and redifferentiation, initiating a novel biological pathway that results in somatic embryo formation. Through the use of established methods like organogenesis and somatic embryogenesis (Snyman et al. 2011), sugarcane plants can be grown *in vitro*. Regeneration of plants from explants can occur either directly, with little or no callus, or indirectly, with an initial callus stage. The creation and establishment of efficient plant regeneration protocols are crucial initial steps for their biotechnological use (Loyola-Vargas and Ochoa-Alejo 2016), researchers have studied a wide range of parameters that affect somatic embryogenesis. Figure 6.20 illustrates the agronomic trait diversity induced by somaclonal variation in CoJ 64 sugarcane plants regenerated from callus cultures of young leaf segments.

Fig. 6.20 Development of Somaclonal variation in sugarcane cultivar CoJ 64 (**a**) Cultured young leaf segments showing callus development at their cut edges (**b**) Callus development (**c**) Shoot formation from callus cultures maintained long-term (**d**) Extension of shoots (**e**) Root formation in shoots (**f**) Field-grown plants regenerated from callus (**g**) Variability in tiller production in plants grown in the field (**h**) Cane girth variability (**i**) Internode length variability. (Source: Manchanda et al. (2018))

6.8.2 Chemical Mutagens

Aspergillus was the first organism to have chemically induced mutations documented (Auerbach et al. 1947) using mustard gas (1, 5-dichloro-3-thiopentone) as a mutagen. Various compounds, including alkylating agents, base analogs, nitroso compounds, azides and acridine dyes, facilitate positive genetic changes in crops. Alkylating agents are the most frequently used chemical mutagens, including EMS, ethyl-2-chloroethyl sulfide, 2-chloroethyl-dimethyl, ethyleneimine, diazomethane, and N-ethyl-N-nitrosourea (Acquaah 2012). Based on DNA sequence analysis, chemical mutagens are more efficient, safer, and less expensive to use than other methods (Alsharari et al. 2023; Koetle et al. 2018; Lodish et al. 2000). Deletions, inversions, and translocations are some of the chromosomal abnormalities that can be caused by chemical mutagens (Robert and Pelletier 2018; Savkovic-Stevanovic 2018). Some chemical mutagens cause chromosome duplication rather than changes in DNA sequence (Ntuli and Zobolo 2008; Alshamrani et al. 2022). Although chemical mutagens lead to more damaging alterations than physical ones, they are widely applied in sugarcane due to their ease of use (Oladosu et al. 2016). Siddiqui and Javed (1982) observed callus growth stimulation at low gamma doses. Proliferation was greatest in the control, then at 10 Gy for all three genotypes, and lowest at 40 Gy.

6.9 Hybridization

To improve sugar, ethanol, or biomass yields in sugarcane, hybridization (crossing) has been the primary method of creating new genetic recombination events. Several closely related genera, including Erianthus, Miscanthus, Narenga, and Sclerostachya, as well as the six species that make up the genus *Saccharum*, together, enhance the genetic diversity available for breeding sugarcane and are together recognized as the Saccharum Complex (Daniels and Daniels 1975). Various sugarcane breeding projects around the world have developed their own techniques. Briefly, conventional breeding can be broken down into three stages: (i) selecting the parents, (ii) creating hybrids, and (iii) choosing the best offspring. Parental value, defined as the ability to produce healthy offspring, is employed as a selection criterion. Genetically distinct populations can be created through either biparental or poly-cross breeding. Polycrosses make it difficult to determine the precise male parent of offspring since multiple pollen sources are combined and used to fertilize a single female, biparental crosses benefit from having both male and female parents identified. In this situation, the male parent can be determined by molecular markers in the event of a superior genotype selection. Biparental crosses are more informative because they allow the prediction of both the General Combining Ability (GCA) and the Specific Combining Ability (SCA) between the parents, while polocrosse is only useful for predicting the GCA. Flowering stalks of the selected male parent are removed,

Fig. 6.21 Sugarcane pollination facilitated by lanterns. (Source: Centro de Cana—IAC; Katia et al. (2012))

labeled, and moved to the breeding station, where they are positioned just above the female plant to allow gravity to pollinate it. To avoid cross-pollination, the arrangement is covered with a 'lantern' (Fig. 6.21).

6.10 Conclusions and Prospects

Sugarcane is renowned for having a high fiber content. Biomass, resistance to waterlogging and drought, and pests along with the capacity for multi-ratoon and disease resistance. Analysis of these genes' functional genomics would allow for the complex's dissection in abiotic stress qualities into the constituent characters and would provide sources for different metrics in the abiotic breeding process's ability to handle stress. Reports indicating close connections among the potential genes and QTLs granting abiotic.

Sugarcane has developed from a fibrous grass to a sweet, juice-filled crop through traditional breeding procedures and novel trait improvement strategies. Current study endeavors have refined sugarcane into a hardy, productive crop. Researchers are aiming to have a more precise molecular understanding of the sugarcane crop. Several factors contribute to tolerance to water deficit stress.

Sugarcane is significant because it is not only a significant sugar source but also a renewable source of power. Through fermentation, biocatalysts, and chemo--catalysis, the recommended sugarcane fiber can be used as a starting point to produce economically significant products like biofuels, biopolymers, and other helpful chemicals.

Appendix 6.1 List of major institutes engaged in research on sugarcane (*Saccharum* spp.)

Country	Institute name	Website
India	**ICAR-Sugarcane Breeding Institute**	https://sugarcane.icar.gov.in/
India	ICAR - Indian Institute of Sugarcane Research, Lucknow	https://iisr.icar.gov.in/iisr/
Egypt	Sugar Crops Research Institute	http://www.arc.sci.eg/InstsLabs/Default.aspx?OrgID=11&lang=en
South Africa	South African Sugarcane Research Institute	https://sasri.org.za/
Mauritius	**Mauritius Sugarcane Industry Research Institute**	http://www.msiri.mu/index.php?langue=eng&rub=50
USA	Sugarcane Research: Houma, LA	https://www.ars.usda.gov/southeast-area/houma-la/sugarcane-research/
Australia	Sugar Research Australia	https://sugarresearch.com.au/
Brazil	CTC - Centro de Tecnologia Canavieira	https://ctc.com.br/en/
Sri Lanka	Sugarcane Research Institute	https://sugarres.lk/ /
Pakistan	Sugarcane Research Institute, Faisalabad	https://aari.punjab.gov.pk/
USA	Sugar Research Station, FL	https://www.lsuagcenter.com/portals/our_offices/research_stations/sugar

Appendix 6.2 World list of recommended varieties of sugarcane (*Saccharum* spp.)

Country	Recommended sugarcane varieties
Egypt	GT54-9, G. 99-103 G84.47 ,G2003-47,G2004-27
India	Co 205, Co 213, Co 214, Co 281, Co 285, Co 290, Co 312, Co 419, Co 453, Co 740, Co 997, Co 1148, Co 62,175, Co 6304 and Co 6806, Co 0118 and Co 0238
USA	CP L 01-299, HoCP 14-885, HoCP 96-540, HoCP 09-804, L 11-183, L 12-201, Ho 13-739, L 14-267 and HoCP 14-885. L 15-306 and HoL 15-508, CP 88-1762, CP 89-2143, CP 80-1743, and CP 78 1628, CP 89-2376 and CP 00-1101
South Africa	NCo310,NCo376, NCo382, N551805, N19, N21, N22,N23, N24, N25, N26, N27 N28 N29 N30 N31 N32 N33 N35 N36 N37 N39 N40 N41 N42 N43 N44 N45 N46 N47 N48 N49 N50 N51 N52 N53 N54 N55 N56 N5
Brazil	RB867515, RB966928, RB92579, CTC4, RB855156, RB855453, SP81-3250, SP83-2847, CTC15, RB855536
Australia	Q99 Q133 Q146 Q149 Q151 Q171A Q177A Q182A Q200A Q212A KQ228A Q232, QC75-326
China	GT11, YT57-423, YT63–237, GT42, GT32
Pakistan	SPF 234, BF 129,

References

Abd El-Moneim D, ELsarag EIS, Aloufi S et al (2021) Quinoa (*Chenopodium quinoa* Willd.): genetic diversity according to ISSR and SCoT markers, relative gene expression, and morpho-physiological variation under salinity stress. Plants 10:2802. https://doi.org/10.3390/plants10122802

Abdel-Hamid AME, Salem KFM (2021) Breeding strategies of garden pea (*Pisum sativum* L.). In: Al-Khayri JM, Jain SM, Johnson DV (eds) Advances in plant breeding strategies: vegetable crops. Springer, Cham. https://doi.org/10.1007/978-3-030-66969-0_9

Abdullaev FK, Rustamov NSU, Yakubov MD (2025) Uzbekistan Genebank. In: Al-Khayri JM, Salem KFM, Jain SM (eds) Plant gene banks. Springer, Singapore. https://doi.org/10.1007/978-981-99-4236-7_57-1

Aboshama HM, Salem KFM, Hassan HMF, Tahoun MH (2015) Estimating heterosis and combining ability for in vivo and in vitro traits using diallel cross on tomato (*Solanum lycopersicum* L.). Minufiya. J Agric Res 40(3):757–770

Abu-Ellail FFB, Attafy TM, Attia AE, El-Bakry A (2023) Variability, heritability, and performance of some sugarcane varieties under three surface irrigation regimes. J Sohag Agrisci (JSAS) 8(2):220–239. https://doi.org/10.21608/JSASJ.2023.334537

Abu-Ellail FFB, Gadallah AFI, El-Gamal ISH (2020) Genetic variance and performance of five sugarcane varieties for physiological, yield and quality traits influenced by various harvest age. J Plant Prod Mansoura Univ 11(5):429–438. https://doi.org/10.21608/jpp.2020.102763

Abu-Ellail FFB, Makhlouf BSI, Attia AE (2021a) Genetic variability and response of some sugarcane varieties to application methods of N-fixation organism (*Azospirillum brasilense*) under different inorganic nitrogen Levels. J Sugarcane Res 11:37–55. https://doi.org/10.37580/JSR.2021.1.11.37-55

Abu-Ellail FFB, McCord PH (2019) Temperature and relative humidity effects on sugarcane flowering ability and pollen viability under natural and seminatural conditions. Sugar Tech 21:83–92. https://doi.org/10.1007/s12355-018-0644-9

Abu-Ellail FFB, Mohamed BD (2019) Effect of photo initiation treatments on flowering, pollen viability and seed germinatibilty treatments on of sugarcane clones. J Sugarcane Rese. https://doi.org/10.37580/JSR.2019.2.9.138-149

Abu-Ellail FFB, Salem KFM, Saleh MM et al (2021b) Molecular breeding strategies of beetroot (*Beta vulgaris* ssp. vulgaris var. conditiva Alefeld). In: Al-Khayri JM, Jain SM, Johnson DV (eds) Advances in plant breeding strategies: vegetable crops. Springer, Cham. https://doi.org/10.1007/978-3-030-66965-2_4

Abulaila K, Al Sane KO, Tahabsom Z (2025) Jordan Seed Bank. In: Al-Khayri JM, Salem KFM, Jain SM (eds) Plant gene banks. Springer, Singapore. https://doi.org/10.1007/978-981-99-4236-7_45-1

Acquaah G (2012) Principles of plant genetics and breeding. Wiley-Blackwell, Oxford, UK, pp 261–300

Aitken K et al (2006) AFLP analysis of genetic diversity within *Saccharum officinarum* and comparison with sugarcane cultivars. Australian J Agric Res 57(11):167–184. https://doi.org/10.1071/AR05391

Aitken KS, Hermann S, Karno K et al (2008) Genetic control of yield related stalk traits in sugarcane. Theor Appl Genet 117(7):1191–1203

Aitken KS, Jackson PA, McIntyre CL (2005) A combination of AFLP and SSR markers provides extensive map coverage and identification of homo(eo)logous linkage groups in a sugarcane cultivar. Theor Appl Genet 110:789–801

Akparov ZI, Mammadova SA, Kalantarova NS, Mirzaliyeva IA, Mammadov AT (2025) National Genebank of Azerbaijan. In: Al-Khayri JM, Salem KFM, Jain SM (eds) Plant gene banks. Springer, Singapore. https://doi.org/10.1007/978-981-99-4236-7_31-1

Al Lawati AH (2025) Plant genetic resource conservation of Oman. In: Al-Khayri JM, Salem KFM, Jain SM (eds) Plant gene banks. Springer, Singapore. https://doi.org/10.1007/978-981-99-4236-7_50-1

Aladele SE et al (2025) Plant genetic resources conservation and utilization in Nigeria. In: Al-Khayri JM, Salem KFM, Jain SM (eds) Plant gene banks. Springer, Singapore. https://doi.org/10.1007/978-981-99-4236-7_17-1

Aldahak L, Salem KFM, Al-Salim SHF, Al-Khayri JM (2021) Advances in chicory (*Cichorium intybus* L.) breeding strategies. In: Al-Khayri JM, Jain SM, Johnson DV (eds) Advances plant breeding strategies: vegetable crops. Springer, Cham. https://doi.org/10.1007/978-3-030-66969-0_1

Alexander KC, Viswanathan R (1996) Major diseases affecting sugarcane production in India and recent experiences in quarantine. In: Croft BJ, Piggin CM, Wallis ES, Hogarth DM (eds) Sugarcane germplasm conservation and exchange report of an international workshop held in Brisbane, Queensland, Australia, 28–30 June, pp 46–48

Ali S, Iqbal J, Khan MS (2010) Genotype independent in vitro regeneration system in elite varieties of sugarcane. Pak J Bot 42:3783–3790

Allen CJ, Mackay MJ, Aylward JH, Campbell JA (1997) New technologies for sugar milling and by-product modification. In: Keating BA, Wilson JR (eds) Intensive sugarcane production: meeting the challenges beyond 2000, Proceedings of the Sugar 2000 Symposium. CABI, Wallingford, UK, pp 267–285

Almarri NB, Ahmad S, Alsubaie GTG, Elshal MH (2025) Saudi Arabia gene bank. In: Al-Khayri JM, Salem KFM, Jain SM (eds) Plant gene banks. Springer, Singapore. https://doi.org/10.1007/978-981-99-4236-7_54-1

Almazan O (1994) Past, present and future of sugarcane by-products in Cuba. Cooperative Sugar 26(4):253–258

Alqahtani MM, Saleh MM, Alwutayd KM et al (2024) Performance and genotypic variability in diverse date palm (*Phoenix dactylifera* L.) cultivars for fruit characteristics. Genet Resour Crop Evol 71:1759–1772. https://doi.org/10.1007/s10722-023-01719-0

Alshamrani M, Safhi S, Alshaya FA et al (2022) Genetic diversity using biochemical, physiological, karyological and molecular markers of *Sesamum indicum* L. Front Genet 13:1035977

Alsharari SF, Ibrahim AA, Okasha SA (2023) Combining ability for yield, oil content, and physio-biochemical characters of canola (*Brassica napus* L.) Under salt stress conditions. SABRAO J Breed Genet 55(4):1003–1024. https://doi.org/10.54910/sabrao2023.55.4.1

Alves SAO, de Lemos OF, Santos Filho BG, da Silva ALL (2011) In vitro embryo rescue of interspecific hybrids of oil palm (Elaeis oleifera x Elaeis guineensis). J Biotechnol Biodiv 2:1–6

Amalraj VA, Balasundaram N (2006) On the taxonomy of the members of *Saccharum* complex. Genet Resour Crop Evol 53(1):35–41. https://doi.org/10.1007/s10722-004-0581-1

Anjali Singh A (2021) Application and limitation of micropropagation. https://www.plantcelltechnology.com/blog/application-and-limitation-of-micropropagation/

Antoniêto ACC, Nogueira KMV, Mendes V et al (2022) Use of carbohydrate-directed enzymes for the potential exploitation of sugarcane bagasse to obtain value-added biotechnological products. Inter J Biol Macromol 221:456–471

Arab S et al (2025) National gene bank of Egypt. In: Al-Khayri JM, Salem KFM, Jain SM (eds) Plant gene banks. Springer, Singapore. https://doi.org/10.1007/978-981-99-4236-7_5-1

Aruna DD (2016) A survey on different disease and image processing techniques in sugarcane crops. IJSRD 3(11):323–325

Asnaghi C, Roques D, Ruffel S et al (2004) Targeted mapping of a sugarcane rust resistance gene (Bru1) using bulked segregant analysis and AFLP markers. Theor Appl Genet 108(4):759–764

Auerbach C, Robson JM, Carr J (1947) Chemical production of mutations. Science 105:243–247

Aung NN, Khaing EE, Mon YY (2022) History of sugarcane breeding, germplasm development and related research in Myanmar. Sugar Tech 24(1):243–253. https://doi.org/10.1007/s12355-021-01079-y

Aung NN, Soe OM (2005) Study on selection efficiency of cross of sugarcane breeding program at Pyinmana Research and Seed Farm. In: Proceeding of research papers, sugarcane research

seminar held in 2005, commemorating the 80th anniversary of Pyinmana sugarcane research center, pp 8–29
Barber, H. S. (1913). The remarkable life-history of a new family (Micromalthidae) of beetles
Babar LK, Iftikhar T, Khan HN, Hameed MA (2011) Agronomic trials on sugarcane crop under Faisalabad conditions. Pakistan Pak J Bot 43(2):929–935
Banakar R, Eggenberger AL, Lee K et al (2019) High-frequency random DNA insertions upon co-delivery of CRISPR-Cas9 ribonucleoprotein and selectable marker plasmid in rice. Sci Rep 9:19902
Banakar R, Schubert M, Collingwood M et al (2020) Comparison of CRISPR-Cas9/Cas12a ribonucleoprotein complexes for genome editing efficiency in the rice phytoene desaturase (OsPDS) gene. Rice 13:4
Banasiak M, Snyman SJ (2017) Exploring in vitro germplasm conservation options for sugarcane (Saccharum spp. hybrids) in South Africa. In Vitro Cell Dev Biol Plant 53:402–409
Barbieri DM, Marques Junior J, Pereira GT et al (2008) Dependência espacial dos custos de fertilizantes para aplicação em taxa variada em cana-de-açúcar. In: Congresso Brasileiro de Agricultura de Precisão, 2008, Piracicaba-SP. Anais Piracicaba. Esalq, 1 CD-ROM
Barreto Z, Rosa JR, Balsalobre TW et al (2019) Genome-wide association study identified loci for yield component traits in sugarcane (*Saccharum* spp.). PLoS One 14:1–22
Békefi Z, Szilágyi S, Desiderio F, Demku T, Varga J (2025) Prunus gene bank in Hungarian University of Agriculture and Life Sciences. In: Al-Khayri JM, Salem KFM, Jain SM (eds) Plant gene banks. Springer, Singapore. https://doi.org/10.1007/978-981-99-4236-7_71-1
Berding N, Hogarth M, Cox M (2004) Plant improvement of sugarcane. In: Glyn J (ed) Sugarcane. Blackwell Science Publication, Oxford, pp 20–53
Berding N, Roach BT (1987) Germplasm collection, maintenance, and use. In: Heinz GD (ed) Sugarcane improvement through breeding, 1st edn. Elsevier, Amsterdam, pp 143–210
Berville A, Breton C, Cunliffe K, Darmency H et al (2005) Issues of ferality or potential for ferality in oats, olives, the Vigna group, ryegrass species, safflower, and sugarcane. CRC Press, pp 231–255. https://doi.org/10.1201/9781420037999.ch15
Bilal M, Saeed M, Nasir IA et al (2015) Association mapping of cane weight and tillers per plant in sugarcane. Biotechnol Biotechnol Equip 29:617–623
Boettcher M, McManus MT (2015) Choosing the right tool for the job: RNAi, TALEN or CRISPR. Mol Cell 58(4):575–585
Bonnett GD (2013) Developmental stages (phenology). In: Moore PH, Botha FC (eds) Sugarcane: physiology, biochemistry, and functional biology. Wiley, Chichester, UK
Börner A, Landjeva S, Salem KFM, Lohwasser U (2009) Pflanzengenetische Ressourcen als Grundlage für die Züchtung klimatoleranter Sorten. In: Proceedings of a German Conference 'Weizenforum = Wheat Forum' Norddeutsches Weizenforum Schriftenreihe des Inst. Für Pflanzenbau und –züchtung der CAU Kiel, Germany, 63, pp 23–27
Börner A, Landjeva S, Salem KFM, Lohwasser U (2010) Plant genetic resources- a prerequisite for drought tolerance breeding in cereals. In: Proc Jahrestagung der Vereinigung der Pflanzenzüchter und Saatgutkaufleute Österreichs, 24–26 November 2009, Raumberg--Gumpenstein, pp 11–13
Breidy J (2025) Lebanon gene bank. In: Al-Khayri JM, Salem KFM, Jain SM (eds) Plant gene banks. Springer, Singapore. https://doi.org/10.1007/978-981-99-4236-7_48-1
Budeguer F, Enrique R, Perera MF et al (2021) Genetic transformation of sugarcane, current status and future prospects. Front Plant Sci 12:2467
Bull TA, Glasziou KT (1979) Sugarcane. In: Lovett JV, Lazenby A (eds) Australian field crops, Tropical cereals, oilseeds, grain legumes and other crops, vol 2. Angus and Robertson Publishers, Sydney, pp 95–113
Burner DM (1997) Chromosome transmission and meiotic stability of sugarcane (*Saccharum* spp.) hybrid derivaties. Crop Sci 33(3):600–606
Burner DM, Legendre BL (1993) Sugarcane genome amplification for the subtropic: a twenty-year effort. Sugarcane 3:5–10

Butterfield M, D'hont A, Berding N (2001) The sugarcane genome: a synthesis of current understanding, and lessons for breeding and biotechnology. Proc South Afr Sugar Technol Assoc 75:1–5

Cañar DY et al (2025) Colombia gene bank. In: Al-Khayri JM, Salem KFM, Jain SM (eds) Plant gene banks. Springer, Singapore. https://doi.org/10.1007/978-981-99-4236-7_95-1

Carr MKV, Knox JW (2011) The water relations and irrigation requirements of sugar cane (*Saccharum officinarum*): a review. Exp Agric 47(1):1–25

Carrasco-Valenzuela T, Munñoz-Espinoza C, Riveros A et al (2019) Expression QTL (eQTLs) analyses reveal candidate genes associated with fruit flesh softening rate in Peach [Prunus persica (L.) Batsch]. Front Plant Sci 10:1581

Carroll D (2011) Genome engineering with zinc-finger nucleases. Genetics 188:773–782

Casu R, Manners J, Bonnett G et al (2005) Genomics approaches for the identification of genes determining important traits in sugarcane. Field Crop Res 92(2–3):137–147

Cheavegatti-Gianotto A, De Abreu HMC, Arruda P et al (2011) Sugarcane (Saccharum X officinarum): a reference study for the regulation of genetically modified cultivars in Brazil. Trop Plant Biol 4:62–89. https://doi.org/10.1007/s12042-011-9068-3

Christopher WN (1930) Bacterial stripe diseases of sugarcane in Louisiana. J Agric Res 41:259

Clauser NM, Gutiérrez S, Area MC et al (2016) Small-sized biorefineries as strategy to add value to sugarcane bagasse. Chem Eng Res Des 107:137–146

Clements HF (1980) Sugarcane crop logging and crop control: principles and practice. Univ. Hawaii Press, Honolulu

Costa, S. M., Mazzola, P. G., Silva, J. C., Pahl, R., Pessoa Jr, A., & Costa, S. A. (2013). Use of sugar cane straw as a source of cellulose for textile fiber production. Industrial Crops and Products, 42, 189–194

Cong L, Ran FA, Cox D et al (2013) Multiplex genome engineering using CRISPR/Cas systems. Science 339:819–823

Cox M, Hogarth M, Smith G (2000) Cane breeding and improvement. In: Hogarth M, Allsopp P (eds) Manual of cane growing, bureau of sugar experimental stations. Indooroopilly, Australia, pp 91–108

Cuadrado A, Acevedo R, Díaz M, de la Espina S et al (2004) Genome remodelling in three modern S. officinarum x S. spontaneum sugarcane cultivars. J Exp Bot 55:847–854. https://doi.org/10.1093/jxb/erh093

D'Hont A, Grivet L, Feldmann P et al (1996) Characterisation of the double structure of modern sugarcane cultivars (*Saccharum* spp.) by molecular cytogenetics. Mol Gen Genet 250(4):405–413

D'Hont A, Ison D, Alix K, Roux C, Glaszmann JC (1998) Determination of basic chromosome numbers in the genus Saccharum by physical mapping of ribosomal RNA genes. Genome 41(2):221–225

Daniels J, Daniels CA (1975) Geographical, historical and cultural aspects of the origin of the Indian and Chinese sugarcanes *S. barberi* and *S. sinense*. Sugarcane Breed Newsl 36:4–23

Daniels J, Roach BT (1987) Taxonomy and evolution. In: Heinz DJ (ed) Sugarcane improvement through breeding, vol 11. Elsevier, Amsterdam, pp 7–84 ISBN: 978-0444427694

de Galarreta JIR, Barandalla L, Ritter E, Velázquez ME, Rios D (2025) Spanish potato germplasm bank. In: Al-Khayri JM, Salem KFM, Jain SM (eds) Plant gene banks. Springer, Singapore. https://doi.org/10.1007/978-981-99-4236-7_114-1

Deepchand K (2005) Sugar cane bagasse energy cogeneration—Lessons from Mauritius. Parliamentarian Forum on Energy Legislation and Sustainable Development, Cape Town, South Africa

Desheva G, Valchinova E, Deshev M (2025) Bulgarian gene bank. In: Al-Khayri JM, Salem KFM, Jain SM (eds) Plant gene banks. Springer, Singapore. https://doi.org/10.1007/978-981-99-4236-7_66-1

D'Hont A, Glaszmann JC (2001) Sugarcane genome analysis with molecular markers: a first decade of research. In: Hogarth DM (ed) 24th Congress of the International Society of Sugar Cane Technologists. Australian Soc Sugar Cane Technologists, Brisbane, Australia, pp 556–559

Díez MJ, Valcárcel JV, Solbes E, Muñoz E (2025) Forty years conserving Spanish biodiversity in the gene bank of the Universitat Politècnica de València, Spain. In: Al-Khayri JM, Salem KFM, Jain SM (eds) Plant gene banks. Springer, Singapore. https://doi.org/10.1007/978-981-99-4236-7_86-1

Dunlap CE, Callihan CD (1969) Fermentative utilization of sugar cane bagasse. Am Soc Sugar Cane Technol 16:82–90

Dussle CM, Quint M, Melchinger AE et al (2003) Saturation of two chromosome regions conferring resistance to SCMV with SSR and AFLP markers by targeted BSA. Theor Appl Genet 106:485–493

Edwardson W, Graham M (1994) Biotechnology in developing countries. International Development Research Centre Environment and Natural Resources Division, Ottawa, OM

Eid A, Mohan C, Sanchez S, Wang D et al (2021) Multiallelic, targeted mutagenesis of magnesium chelatase with CRISPR/Cas9 provides a rapidly scorable phenotype in highly polyploid sugarcane. Front Genome Ed 3:654996

Eid NA, Ibrahim AMH, Elsharawy AA, Salem KFM (2025) Nanofertilizers for plant viral disease management. In: Kumar P, Dubey RC (eds) Nanofertilizers for sustainable agriculture. Springer, Cham. https://doi.org/10.1007/978-3-031-78649-5_7

El Rabey H, Salem KFM, Mattar MZ (2013) The genetic diversity and relatedness of eight rice (*Oryza sativa* L.) cultivars as revealed by AFLP and SSRs markers. Life Sci J 10(1):1471–1479

Elabd AB, Saleh MM, Aldahak L, Salem KFM (2020) Identification of superior rice (*Oryza sativa* L.) genotypes under irrigated conditions based on multivariate analysis. In: 7th International Scientific Conference of Genetic and Environment from 8th –9th December 2020, Baghdad, Iraq

Elabd AB, Salem KFM (2009) Detection of genetic diversity in rice (*Oryza sativa* L.) Varieties using microsatellite DNA-Markers and Quantitative Traits. In: 6th Egyptian Plant Society of Breeding Conference, 4–7 May 2009, Ismailia, Egypt

El-Absawy EA, Nowar AA, Salem KFM et al (2015) Estimation of genetic diversity among some bread wheat (*Triticum aestivum* L.) genotypes using molecular markers and agronomic traits. Minufiya J Agric Res 40(3):223–233

El-Absawy EA, Salem KFM, Abdellatif KF, ElShamy H (2012) Estimation of genetic diversity among some Egyptian barley (*Hordeum vulgare* L.) cultivars using RAPD markers and agronomic traits. In: 1st International Conference On Biotechnology Applications In Agriculture, Moshtohor-Hurghada, 18–22 February 2012, Faculty of Agriculture, Benha University, Egypt

Ellison EE, Nagalakshmi U, Gamo ME et al (2020) Multiplexed heritable gene editing using RNA viruses and mobile single guide RNAs. Nat Plants 6:620–624

El-Mansy AB, Abd El-Moneim D, SM AL et al (2021) Genetic diversity analysis of tomato (*Solanum lycopersicum* L.) with morphological, cytological, and molecular markers under heat stress. Horticulturae 7:65

El-Shafai AM, Ismail AMA (2006) Effect of row spacing on yield and quality of some promising sugarcane varieties. Egypt J Appl Sci 21(11):32–46

Eltaher S, Sallam A, Belamkar V et al (2018) Genetic diversity and population structure of F3:6 Nebraska winter wheat genotypes using genotyping-by-sequencing. Front Genet 9:76. https://doi.org/10.3389/fgene.2018.00076

El-Zanaty AM, Salem KFM, Esmail RM (2011) Detection of genetic diversity in Egyptian cotton (Gossypium barbadense L.) varieties using RAPD markers and morphological traits. J Am Sci 7(12):1107–1115

El-Zanaty AM, Salem KFM, Ismail M, Esmail RM (2013) Impact of plant breeding on genetic diversity of the Egyptian cotton (*Gossypium barbadense* L.) varieties as revealed by RAPD markers. J Agric Chem Biotechn Mansoura Univ 4:217–227

Emara HA, Nower AA, Salem KFM, Eltaher S (2011) Estimating heterosis and combining ability for in vivo and in vitro traits using diallel cross in Maize (*Zea mays* L). J Prod Dev 16:101–120

Emara HA, Salem KFM, Abdel-Halim AA, Abd El-Kowy RNF (2013) An in vivo and in vitro traits analysis of a diallel cross in rice (*Oryza sativa* L.). Egyptian J Plant Breed 17(2):591–610
Esmail RM, Salem KFM, El-Zanaty AM (2009) Genetic diversity studies of Egyptian cotton (*Gossypium barbadense* L.) varieties using RAPD markers and agronomic traits. In: 6th Egyptian Plant Society of Breeding Conference, 4–7 May 2009, Ismalia, Egypt
Essa NM, Ibrahim AA, Soliman MI (2023) Association study between some cultivated species and their wild relatives from Apiaceae, Asteraceae and Brassicaceae families based on molecular and DNA barcoding in Egypt. Genet Resour Crop Evol 71:1125–1143. https://doi.org/10.1007/s10722-023-01681-x
Evans LT (1980) The natural history of crop yield. Amer Scientist 68:388–397
FAOSTAT (2021) Crops: sugarcane. Available online: http://www.fao.org/faostat/en/#data/QC. Accessed on 28 July 2021
Fauconnier R (1993) Sugar cane. Macmillan, London, pp 1–140
Freeman DW, Duerr EO, Leber KM (1992) Use of bagasse as a feed input to semi-intensive shrimp growout ponds. J World Aquac Soc 23(1):23–30. https://doi.org/10.1111/j.1749-7345.1992.tb00747.x
Galea-Gragera FA, Llera Cid F (2025) Spain Germplasm Bank of Annual Forage Legumes Seeds at the Scientific and Technological Research Center of Extremadura (CICYTEX). In: Al-Khayri JM, Salem KFM, Jain SM (eds) Plant gene banks. Springer, Singapore. https://doi.org/10.1007/978-981-99-4236-7_102-1
Gantait S, Mitra M (2019) Applications of synthetic seed technology for propagation, storage, and conservation of orchid germplasms. In: Faisal M, Alatar A (eds) Synthetic Seeds. Springer, Cham. https://doi.org/10.1007/978-3-030-24631-0_13
Garcia AA, Kido EA, Meza AN et al (2006) Development of an integrated genetic map of a sugarcane (*Saccharum* spp.) commercial cross, based on a maximum-likelihood approach for estimation of linkage and linkage phases. Theor Appl Genet 112:298–314
Garsmeur O, Droc G, Antonise R et al (2018) A mosaic monoploid reference sequence for the highly complex genome of sugarcane. Nat Commun 9:2638
Ghazy MI, Salem KFM, Sallam A (2020) Utilization of genetic diversity and marker-trait to improve drought tolerance in rice (*Oryza sativa* L.). Mol Biol Rep. https://doi.org/10.1007/s11033-020-06029-7
Ghidan WF, Abdulmajid DA (2025) Egyptian rice gene bank. In: Al-Khayri JM, Salem KFM, Jain SM (eds) Plant gene banks. Springer, Singapore. https://doi.org/10.1007/978-981-99-4236-7_106-1
Gilbert RA, Gallo-Meagher M, Comstock JC et al (2005) Agronomic evaluation of sugarcane lines transformed for resistance to strain E. Crop Sci 45:2060–2067
Gilbert RA, Glynn NC, Comstock JC, Davis MJ (2009) Agronomic performance and genetic characterization of sugarcane transformed for resistance to sugarcane yellow leaf virus. Field Crop Res 111:39–46
Glaszmann JC, Rott P, Engelmann F et al (1996) Role of in vitro maintenance of sugarcane for germplasm conservation and exchange. In: Croft BJ, Piggin CM, Wallis ES, Hogarth OM (eds) Sugarcane germplasm conservation and exchange. ACIAR Proceedings No. 67, p 134
Goldemberg J, Guardabassi P (2009) Are biofuels a feasible option? Energy Policy 37(1):10–14. https://doi.org/10.1016/j.enpol.2008.08.031
Grassl, C. O. (1964). Problems relating to the origin and evolution of wild and cultivated Saccharam. Le Cunff, L., Garsmeur, O., Raboin, L. M., Pauquet, J., Telismart, H., Selvi, A., ... & d'Hont, A. (2008). Diploid/Polyploid syntenic shuttle mapping and haplotype-specific chromosome walking toward a rust resistance gene (Bru1) in highly polyploid sugarcane (2 n~ 12 x~ 115). Genetics, 180(1), 649–660.
Grivet L, D'Hont A, Dufour P et al (1994) Comparative genome mapping of sugar cane with other species within the Andropogoneae tribe. Heredity 73:500–508

Gudaye TY, Lakew BF, Selamo MM, Tessema ES (2025) Ethiopian crop gene bank. In: Al-Khayri JM, Salem KFM, Jain SM (eds) Plant gene banks. Springer, Singapore. https://doi.org/10.1007/978-981-99-4236-7_7-1
Hari K, Vasantha S, Anna Durai A et al (2017) Sugarcane root growth and development in hydroponics system. J Sugarcane Res 7(2):71–82
Hartmann H, Kester D, Davies F, Geneve R (2001) Principles of tissue culture and micropropagation H. In: Hartmann DK (ed) Plant propagation, principles and practices, 7th edn. Pearson Education LTD, pp 639–714
Harzli I, Mostafa K, Abdulla MF, Mehmood M (2025) Sustaining Turkish genetic resources. In: Al-Khayri JM, Salem KFM, Jain SM (eds) Plant gene banks. Springer, Singapore. https://doi.org/10.1007/978-981-99-4236-7_88-1
Hassan MN, Mekkawy SA, Mahdy M, Salem KFM et al (2021) Recent molecular and breeding strategies in lettuce (Lactuca spp.). Genet Resour Crop Evol 68:3055–3079. https://doi.org/10.1007/s10722-021-01246-w
Heinz DJ, Mee GW (1969) Plant differentiation from callus tissue of Saccharum species. Crop Sci 9:346–348
Hidayatun N, Kurniawan H (2025) Conservation of agricultural genetic resources in Indonesia. In: Al-Khayri JM, Salem KFM, Jain SM (eds) Plant gene banks. Springer, Singapore. https://doi.org/10.1007/978-981-99-4236-7_40-1
Hoang NV, Furtado A, Mason PJ et al (2017) A survey of the complex transcriptome from the highly polyploid sugarcane genome using full-length isoform sequencing and de novo assembly from short read sequencing. BMC Genomics 18:395
Hoarau JY, Grivet L, Offmann B et al (2002) Genetic dissection of a modern sugarcane cultivar (Saccharum spp.) II. Detection of QTLs for yield components. Theor Appl Genet 105:1027–1037
Hoarau J-Y, Offmann B, D'Hont A et al (2001) Genetic dissection of a modern sugarcane cultivar (*Saccharum* spp.). I. Genome mapping with AFLP markers. Theor Appl Genet 103:84–97
Hoareau W, Oliveira FB, Grelier S et al (2006) Fiberboards based on sugarcane bagasse lignin and fibers. Macromol Mat Eng 291(7):829–839. https://doi.org/10.1002/mame.200600004
Hossain MI, Ahmad K, Siddiqui Y, Saad N, Rahman Z, Haruna AO, Bejo SK (2020) Current and prospective strategies on detecting and managing *Colletotrichumfalcatum* Causing Red Rot of Sugarcane. Agronomy 10:1253. https://doi.org/10.3390/agronomy10091253
Huang YK, Li WF, Zhang RY, Wang XY (2018) Diagnosis and control of sugarcane important diseases. In: Color illustration of diagnosis and control for modern sugarcane diseases, pests, and weeds. Springer, Singapore. https://doi.org/10.1007/978-981-13-1319-6_1
Humbert RP (1958) Potash Fertilization in Hawaiian Sugar Industry. Proc. Cong. Int. Potash Instt, Madrid, Spain
Hunsigi G (1989) Ratooning in sugarcane. Outlook on Agriculture, 18(4):175–184
Hunsigi G (1993) Production of sugarcane: theory and practices. Springer, Berlin, p 245
Hussin SH, Liu X, Li C et al (2022) An uUpdated overview on insights into sugarcane genome editing via CRISPR/Cas9 for sustainable production. Sustainability 14:12285. https://doi.org/10.3390/su141912285
Ibrahim AA, Abd-Ellatif S, Abdel Razik ESS, Salem KFM (2024) Biodiversity of cereal crops and utilization in food and nutritional security. In: Al-Khayri JM, Jain SM, Penna S (eds) Sustainable utilization and conservation of plant genetic diversity, Sustainable development and biodiversity, vol 35. Springer, Singapore. https://doi.org/10.1007/978-981-99-5245-8_2
Ibrahim AA, Salem KFM, Abdein MA, Ramadan SM (2023a) Advances in summer squash (Cucurbita pepo L.) molecular breeding strategies. In: Singh S, Sharma D, Sharma SK, Singh R (eds) Smart plant breeding for vegetable crops in post-genomics era. Springer, Singapore. https://doi.org/10.1007/978-981-19-5367-5_8
Ibrahim AA, Salem KFM, Ramadan S (2023b) Allium sativum and stressful conditions. In: Abdel Latef AAH (ed) Book medicinal plant responses to stressful conditions. CRC Press

Ihsan F, Prihatini R (2025) Indonesian tropical fruit gene bank. In: Al-Khayri JM, Salem KFM, Jain SM (eds) Plant gene banks. Springer, Singapore. https://doi.org/10.1007/978-981-99-4236-7_41-1

Ingaramo A, Heluane H, Colombo M, Mario Cesca M (2009) Water and wastewater eco-efficiency indicators for the sugar cane industry. J Clean Prod 17:487–495

Ismail M, Salem KFM, El-Zanaty AM, Esmail RM (2013) Association mapping of fiber and agronomic characters in Egyptian cotton (Gossypium barbadense L.) genotypes. J Agric Chem Biotech Mansoura Univ 4:205–216

Jackson PA (2005) Breeding for improved sugar content in sugarcane. Field Crop Res 92:277–290. https://doi.org/10.1016/j.fcr.2005.01.024

Jaganathan D, Ramasamy K, Sellamuthu et al (2018) CRISPR for crop improvement: an update review. Front Plant Sci 9:985

Jatoi SA, Siddiqui SU (2025) National gene bank of Pakistan. In: Al-Khayri JM, Salem KFM, Jain SM (eds) Plant gene banks. Springer, Singapore. https://doi.org/10.1007/978-981-99-4236-7_51-1

Jeswiet J (1929) The development of selection and breeding of the sugarcane in Java. In: Proceedings of the Third Congress of the International Society of Sugar Cane Technologists, vol 3, pp 44–57

Jugwanth Y, Sewsynker-Sukai Y, Kana EG (2020) Valorization of sugarcane bagasse for bioethanol production through simultaneous saccharification and fermentation: optimization and kinetic studies. Fuel 262:116552

Kansal S (1998) Factors determining Indian sugar production and its comparative advantage. In: Proceedings of the Fiji/FAO 1997 Asia Pacific Sugar Conference, Food and Agriculture Organization, Rome, pp 78–92. www.fao.org/docrep/005/x0513e/x0513e16.htm

Karamchandani BM, Chakraborty S, Dalvi SG, Satpute SK (2022) Chitosan and its derivatives: promising biomaterial in averting fungal diseases of sugarcane and other crops. J Basic Microbiol 62(5):533–554

Karve P et al (2001) A chain of technologies for using sugarcane trash as a household fuel. Boiling Point 47:16–18

Katia, L., & Lebraty, J. F. (2012). L'open Data: nouvelle pratique managériale risquée?. Management & Prospective, 29(4), 103–116

Khan IA, Bibi S, Yasmin S, Khatri A, Seema N, Abro SA (2012) Correlation studies of agronomic traits for higher sugar yield in sugarcane. Pak J Bot 44(3):969–971

Koetle MJ, Evans DL, Singh V et al (2018) Watt Agronomic evaluation and molecular characterisation of the acetolactate synthase gene in imazapyr tolerant sugarcane (*Saccharum* hybrid) genotypes. Plant Cell Rep 37:1201–1213

Koonin EV, Makarova KS (2019) Origins and evolution of CRISPR-Cas systems. Philos Trans R Soc B 374:20180087

Krishnan SR, Mohan C (2017) Methods of sugarcane transformation. In: Mohan C (ed) Sugarcane biotechnology: challenges and prospects. Springer, Cham, pp 51–60

Kumar PP, Loh CS (2012) Plant tissue culture for biotechnology. In: Altman A, Hasegawa PM (eds) Plant, biotechnology and agriculture: prospects for the 21st Century. Academic, London, pp 131–138

Laforest LC, Nadakuduti SS (2022) Advances in delivery mechanisms of CRISPR gene-editing reagents in plants. Front Genome Ed 4:830178

Laghetti G, Piergiovanni AR, De Paola D, Direnzo P, Bucci G, Sonnante G (2025) Italy IBBR gene bank. In: Al-Khayri JM, Salem KFM, Jain SM (eds) Plant gene banks. Springer, Singapore. https://doi.org/10.1007/978-981-99-4236-7_73-1

Lakshmanan P et al (2005) Sugarcane biotechnology: the challenges and opportunities. In Vitro Cell Dev Biol 41:345–363

Leal MRLV, Galdos MV, Scarpare FV et al (2013) Oliveira Sugarcane straw availability, quality, recovery and energy use: a literature review. Biomass Bioenergy 53:11–19. https://doi.org/10.1016/j.biombioe.2013.03.007

Li AM, Chen ZL, Liao F et al (2024) Sugarcane borers: species, distribution, damage and management options. J Pest Sci 97:1171–1201. https://doi.org/10.1007/s10340-024-01750-9

Li C, Yue Y, Chen H et al (2018a) The ZmbZIP22 transcription factor regulates 27-kD γ-zein gene transcription during maize endosperm development. Plant Cell 30:2402–2424

Li J, Yang J, Li Y, Ma L (2020) Current strategies and advances in wheat biology. Crop J 8:879–891

Li J, Zhang X, Sun Y et al (2018b) Efficient allelic replacement in rice by gene editing: a case study of the NRT1.1B gene. J Integr Plant Biol 60:536–540. [CrossRef]

Li T, Liu B, Spalding MH et al (2012) High-efficiency TALEN-based gene editing produces disease-resistant rice. Nat Biotechnol 30(5):390–392

Lodish H, Berk A, Zipursky SL et al (2000) DNA damage and repair and their role in carcinogenesis Molecular Cell Biology, 4th edn. B. H. Freeman and Company USA, Bethesda, MD

Loyola-Vargas VM, Ochoa-Alejo N (2016) Somatic embryogenesis. An overview. In: Loyola--Vargas VM, Ochoa-Alejo N (eds) Somatic embryogenesis: fundamental aspects and application. Springer, Cham, pp 1–8

Lu G, Wang Z, Xu F, Pan Y-B, Grisham MP, Xu L (2021) Sugarcane mosaic disease: characteristics, identification and control. Microorganisms 9(9):1984. https://doi.org/10.3390/microorganisms9091984

Mackintosh D (2000) "Sugar milling", Chapter 17. In: Hogarth M, Allsopp P (eds) Manual of cane growing. Bureau of Sugar Experiment Stations, Indooroopilly, pp 369–377. www.sugarresearch.com.au/icms_docs/166965_Chapter_17_Sugar_Milling.pdf

Magalhaes PSG, Cerri DGP (2007) Yield monitoring of sugarcane. Biosyst Eng 96(1):1–6 3

Mallor C, Luna E (2025) The genetic diversity of the Spanish vegetable genebank BGHZ-CITA: conservation and utilization. In: Al-Khayri JM, Salem KFM, Jain SM (eds) Plant gene banks. Springer, Singapore. https://doi.org/10.1007/978-981-99-4236-7_109-1

Manchanda P, Kaur A, Gosal SS (2018) Somaclonal variation for sugarcane improvement. In: Gosal S, Wani S (eds) Biotechnologies of crop improvement, vol 1. Springer, Cham. https://doi.org/10.1007/978-3-319-78283-6_9

Mankar AR, Pandey A, Modak A, K.K. (2021) Pant, Pretreatment of lignocellulosic biomass: a review on recent advances. Bioresour Technol 334:125235. https://doi.org/10.1016/J.BIORTECH.2021.125235

Marone D, Mastrangelo AM, Borrelli GM (2023) From transgenesis to genome editing in crop improvement: applications, marketing, and legal issues. Int J Mol Sci 24(8):7122. https://doi.org/10.3390/ijms24087122

Martínez-Montero ME, Martínez J, Engelmann F (2008) Cryopreservation of sugarcane somatic embryos. CryoLetters 29(3):229–242

Mathur PS (1986) Practical approach to improving sugarcane ratoon management under cultivation conditions. Coop Sugarcane 17(9):695–699

Mattar MZ, Eissa OA, Salem KFM (2009) Genotypic variability in some wheat genotypes (Triticum aestivum L.) exposed to salinity at germination stage. Egyptian J Exp Biol (Bot) 5:143–148

Mattar MZ, Salem KFM, El-Abed AB (2016) Identification of candidate microsatellite markers associated with agronomic traits in rice (*Oryza sativa* L.). J Bot 15(5):213–232

Menegatti LAA, Molin JP, Goes SL et al (2006) Benefícios econômicos e agronômicos da adoção de agricultura de precisão em usinas de açúcar. In: Con¬gresso Brasileiro de Agricultura de Precisão, São Pedro-SP. Anais... Piracicaba: Esalq, 2006. 1 CD-ROM

Ming R et al (2006) Sugarcane improvement through breeding and biotechnology, Chapter 2. In: Janick J (ed) Plant breeding reviews, vol 27. Wiley, pp 15–117. https://doi.org/10.1002/9780470650349.ch2

Mirajkar SJ, Devarumath RM, Nikam et al (2019) Sugarcane (Saccharum spp.): breeding and genomics. In: Al-Khayri J, Jain S, Johnson D (eds) Advances in plant breeding strategies: industrial and food crops. Springer, Cham. https://doi.org/10.1007/978-3-030-23265-8_11

Mohan C, Ashwin Narayan J, Esterling M, Yau Y-Y (2020) Current transformation methods for genome–editing applications in energy crop sugarcane. In: Kumar A, Yau Y, Ogita S, Scheibe

R (eds) Climate change, photosynthesis and advanced biofuels: the role of biotechnology in the production of value-added plant bio-products. Singapore, Springer, pp 369–388

Mohan Naidu K, Srinivasan TR (1982) Effect of Potassium on sugarcane crop subjected to moisture stress. In: P R II Res., Review Series, vol 2, pp 75–80

Molin JP, Menegatti LAA (2004) Field-testing of a sugarcane yield monitor in Brazil 2004. In: ASAE Annual International Meeting, pp 733–744

Molin JP, Wei MCF, da Silva ERO (2024) Challenges of digital solutions in sugarcane crop production: a review. AgriEngineering 6:925–946. https://doi.org/10.3390/agriengineering6020053

Monteros-Altamirano Á et al (2025) Plant genetic resources in Ecuador. In: Al-Khayri JM, Salem KFM, Jain SM (eds) Plant gene banks. Springer, Singapore. https://doi.org/10.1007/978-981-99-4236-7_99-1

Moore PH (2005) Integration of sucrose accumulation processes across hierarchical scales: towards developing an understanding of the gene-to-crop continuum. In: Field crops research. Elsevier, pp 119–135

Moore PH, Nuss K (1987) Flowering and flower synchronization. In: Heinz DJ (ed) Sugarcane improvement through breeding, vol 11. Elsevier, Amsterdam, pp 273–311

Mourad AMI, Alomari DZ, Alqudah AM et al (2019) Recent advances in wheat (*Triticum* spp.) breeding. In: Al-Khayri J, Jain S, Johnson D (eds) Advances in plant breeding strategies: cereals. Springer, Cham. https://doi.org/10.1007/978-3-030-23108-8_15

Mukherjee SK (1957) Origin and distribution of Saccharum. Bot Gaz 119:55–61

Nasution MH, Lelinasari S, Kelana MGS (2022) A review of sugarcane bagasse pretreatment for bioethanol production. IOP Conf Ser Earth Environ Sci 963(1):012014

Nawara HM, Mattar MZ, Salem KFM, Eissa OA (2016) Estimating heterosis and combining ability for agronomic and tissue culture traits using diallel cross on bread wheat (Triticum aestivum L.). Egypt J Bot 2016:177–200

Nawara HM, Mattar MZ, Salem KFM, Eissa OA (2017) Diallel study on some in vitro traits of bread wheat (Triticum aestivum L.) under salt stress. Int J Agric Environ Res 3(1):1988–2006

Nayak SN, Song J, Villa A et al (2014) Promoting utilization of Saccharum spp. genetic resources through genetic diversity analysis and core collection construction. PLoS One 9(10):e110856. https://doi.org/10.1371/journal.pone.0110856

Nei M (1978) Estimation of average heterozygosity and genetic distance from a small number of individuals. Genetics 89:583–590

Ntuli NR, Zobolo AM (2008) Effect of water stress on growth of colchicine induced polyploidy in *Coccinia palmata* and *Lagenaria sphaerica* plants. Afr J Biotechnol 7:3548–3652

Nwajiaku IM, Olanrewaju JS, Sato et al (2018) Change in nutrient composition of biochar from rice husk and sugarcane bagasse at varying pyrolytic temperatures. Int J Recycl Org Waste Agric 7:269–276. https://doi.org/10.1007/s40093-018-0213-y

Oladosu Y, Rafii MY, Abdullah N et al (2016) Principle and application of plant mutagenesis in crop improvement: a review. Biotechnol Biotechnol Equip 30:1–16

Orellana C, Neto RB (2006) Brazil and Japan give fuel to ethanol market. Nat Biotechnol 24:232. https://doi.org/10.1038/nbt0306-232

Osterc G, Gostinčar K, Hudina M, Usenik V, Luthar Z (2025) Slovene fruit gene vank. In: Al-Khayri JM, Salem KFM, Jain SM (eds) Plant gene banks. Springer, Singapore. https://doi.org/10.1007/978-981-99-4236-7_84-1

Oz MT, Altpeter A, Karan R et al (2021) CRISPR/Cas9-mediated multi-allelic gene targeting in sugarcane confers herbicide tolerance. Front Genome Ed 3:673566

Parthasarathy N (1948) Origin of noble sugarcanes (*Saccharum offcinarum* L.). Nature 161:608

Patade V, Suprasanna P (2008) Radiation induced in vitro mutagenesis for sugarcane improvement. Sugar Tech 10(1):14–19

Patel A, Shah AR (2021) Integrated lignocellulosic biorefinery: gateway for production of second generation ethanol and value-added products. J Bioresour Bioprod 6:108–128. https://doi.org/10.1016/J.JOBAB.2021.02.001

Pereira AR, Barbieri V, Nova NAV (1983) Climatic conditioning of flowering induction in sugarcane. Agric Meteorol 29:103–110
Pérez JN, Suárez M, Orellana P (2000) Posibilidades y potencial de la propagación masiva de plantas en Cuba. Biot Veg 1:3–12
Pezeto RJ, Marques Junior J, Pereira et al (2008) Identificação de limites de zonas específicas de manejo utilizando autocor¬relograma, em áreas sob cultivo de cana-de-açúcar. In: Congresso Brasileiro de Agricultura de Precisão, 2008, Piracicaba-SP. Anais... Piracicaba: Esalq,. 1 CD-ROM
Piperidis G, Piperidis N, D'Hont A (2020) Molecular cytogenetic investigation of chromosome composition and transmission in sugarcane. Mol Gen Genomics 284(1):65–73
Pokharel D, Uprety R, Mehata S et al (2019) Dynamics and economic analysis of sugarcane production in eastern plains of Nepal. Curr Agric Rese J 7(2):201–212
Popović V et al (2025) Economic seed production and storage of fiber crops in gene banks in Serbia. In: Al-Khayri JM, Salem KFM, Jain SM (eds) Plant gene banks. Springer, Singapore. https://doi.org/10.1007/978-981-99-4236-7_108-1
Prasad M (1976) Response of sugarcane to filter press mud and NPK fertilizers: effect on sugarcane yield sucrose content. Agron J 68:539–543
Putra LK, Kristini A, Achadian EM et al (2014) Sugarcane streak mosaic virus in Indonesia: distribution, characterisation, yield losses and management approaches. Sugar Tech 16:392–399. https://doi.org/10.1007/s12355-013-0279-9
Puchta, H., Dujon, B., & Hohn, B. (1993). Homologous recombination in plant cells is enhanced by in vivo induction of double strand breaks into DNA by a site-specific endonuclease. Nucleic acids research, 21(22), 5034–5040
Racedo J, Gutiérrez L, Perera MF et al (2016) Genome-wide association mapping of quantitative traits in a breeding population of sugarcane. BMC Plant Biol 16(1):142
Ramawas SZ (1983) Effect of selfing on wild sugarcane (*Saccharum Spontaneum*) Clones. University of Hawai'i at Manoa
Rangaswami G, Rajagopalan S (1973) Bacterial plant pathology. Tamil Nadu Agriculture University, Coimbatore
Reena R, Alphy MP, Reshmy et al (2022) Sustainable valorization of sugarcane residues: efficient deconstruction strategies for fuels and chemicals production. Bioresour Technol 361:127759
Roach BT (1989) Origin and improvement of the genetic base of sugarcane. In: Proceedings of the Australian Society of Sugar Cane Technologists, vol 11, pp 34–47. www.assct.com.au/media/pdfs/1989_pa_ag7.pdf
Robert F, Pelletier J (2018) Exploring the impact of single-nucleotide polymorphisms on translation. Front Genet 9:1–11
Rocha F et al (2025) Portugal gene bank. In: Al-Khayri JM, Salem KFM, Jain SM (eds) Plant gene banks. Springer, Singapore. https://doi.org/10.1007/978-981-99-4236-7_79-1
Roy, Tulsiram (2013) Synthetic seed of rice: an emerging avenue of applied biotechnology. Rice Genom Genet 4:14–27
Ruiz M, Rouard M, Raboin LM et al (2004) TropGENE-DB, a multi-tropical crop information system. Nucleic Acids Res 32:D364–D367
Safhi FA, ALshamrani SM, Jalal AS et al (2022) Genetic characterization of some Saudi Arabia's accessions from *Commiphora gileadensis* using physio-biochemical parameters, molecular markers, DNA barcoding analysis and relative gene expression. Genes 13:2099
Saleh MM, Ali WA, Alwan MH et al (2020) Multi-environmental evaluation of wheat tetraploid genotypes for agronomic traits under rainfed conditions in Syria. J Aridland Agric 6:1–5
Saleh MM, Alnaddaf LM, Almuhammady AK et al (2021a) Applications of plant-derived nanomaterials in mitigation of crop abiotic stress. In: Al-Khayri JM, Ansari MI, Singh AK (eds) Nanobiotechnology. Springer, Cham. https://doi.org/10.1007/978-3-030-73606-4_9
Saleh MM, Mahmoud AS, Abbas HS et al (2021b) Nanotechnological approaches for efficient delivery of plant ingredients. In: Faizan M, Hayat S, Yu F (eds) Sustainable agriculture reviews, vol 53. Springer, Cham. https://doi.org/10.1007/978-3-030-86876-5_11

Salem KFM (2004) The inheritance and molecular mapping of genes for post-anthesis drought tolerance (PADT) in wheat. Martin-Luther-University, Halle-Wittenberg, p 124

Salem KFM (2009) Relationship between genetic diversity based on SSRs markers with heterosis and combining ability in diallel cross of bread wheat (Triticum aestivum L.). Minufiya J Agric Res 6(34):2159–2178

Salem KFM (2015) Allelic state at the microsatellite locus Xgwm261 marking the dwarfing gene Rht8 in Egyptian bread wheat (*Triticum aestivum* L.) genotypes released from 1947 to 2004. Genetika 47:741–750

Salem KFM, Abd-Ellatif S, Abdel Razik ES et al (2023a) Effect of nanomaterials on water and solutes translocation in plants. In: Al-Khayri JM, Alnaddaf LM, Jain SM (eds) Nanomaterial interactions with plant cellular mechanisms and macromolecules. Springer, Cham

Salem KFM, Alghuthaymi MA, Elabd AB et al (2022) Prediction of heterosis for agronomic traits in half-diallel cross of rice (Oryza sativa L.) under drought stress using microsatellite markers. Plants 11:1532. https://doi.org/10.3390/plants11121532

Salem KFM, Alloosh MT, Saleh MM et al (2021a) Utilization of nanofertilizers in crop tolerance to abiotic stress. In: Al-Khayri JM, Ansari MI, Singh AK (eds) Nanobiotechnology. Springer, Cham. https://doi.org/10.1007/978-3-030-73606-4_11

Salem KFM, Börner A (2003) Evaluation of wheat accessions from Gatersleben Genebank for post-anthesis drought tolerance in the field. In: Gesellschaft für Pflanzenzüchtung, GPZ--Tagung, Vortragsveranstaltung zum Thema: "Abiotischer Streß – Herausforderung für die Pflanzenzüchtung", 11–12 Juni 2003, Groß Lüsewitz

Salem KFM, Börner A (2004) Post-anthesis drought tolerance in genus Triticum. In: Botanikertagung. Braunschweig

Salem KFM, Elabsawy A, Aldahak L et al (2021b) Evaluation of stem reserve mobilization in Egyptian bread wheat (Triticum aestivum L.) genotypes and F1 hybrids under post-anthesis chemical desiccation stress. J Genet Environ Resour Conserv 9(1):176–182

Salem KFM, El-Absawy EA, Khidr YA, El-Shamy H (2021c) Assessment genetic diversity among some Egyptian bread wheat (Triticum aestivum L.) cultivars using RAPD markers. J Genet Environ Resour Conserv 9(1):168–175

Salem KFM, El-Zanaty AM (2015) Assessing genetic diversity of old and modern Egyptian rice (*Oryza sativa* L.) using microsatellite markers. Minufiya J Agric Res 40(6):1507–1518

Salem KFM, El-Zanaty AM, Esmail RM (2008b) Assessing wheat (Triticum aestivum L.) genetic diversity using morphological characters and microsatellite markers. World J Agric Sci 4:538–544

Salem KFM, Ibrahim AA (2024) Plant biodiversity in the context of food security under climate change. In: Al-Khayri JM, Jain SM, Penna S (eds) Sustainable utilization and conservation of plant genetic diversity, Sustainable development and biodiversity, vol 35. Springer, Singapore. https://doi.org/10.1007/978-981-99-5245-8_1

Salem KFM, Mattar MZ (2014a) Genetic diversity in old and modern Egyptian bread wheat (*Triticum aestivum* L.) varieties revealed by simple sequence repeats. Egypt J Genet Cytol 43:143–156

Salem KFM, Mattar MZ (2014b) Identification of microsatellite alleles for salt tolerance at seedling stage in wheat (*Triticum aestivum* L.). Life Sci J 11(12s):1064–1073

Salem KFM, Röder MS, Börner A (2003) Mapping of quantitative trait loci (QTLs) for post-anthesis drought tolerance in bread wheat (*Triticum aestivum* L.). In: Gesellschaft für Pflanzenzüchtung, GPZ-Tagung, 11. Vortragsveranstaltung "Harnessing Genetic Diversity: Genomics and Allele Mining", 16–17.09.2003, Institut für Pflanzengenetik und Kulturpflanzenforschung (IPK), Gatersleben

Salem KFM, Röder MS, Börner A (2004a) Drought tolerance in wheat germplasm- Evaluation and molecular mapping. In: Institut für Pflanzengenetik und Kulturpflanzenforschung, IPK Tagung, 22–24/08/2004, Ag Ressourcengenetik und Reproduktion, Abteilung Genbank, Gatersleben

Salem KFM, Röder MS, Börner A (2004b) Molecular mapping of quantitative trait loci (QTLs) determining post-anthesis drought tolerance in hexaploid wheat (*Triticum aestivum* L.). In:

7 Gesellschaft für Pflanzenzüchtung, GPZ-Tagung, Vortragsveranstaltung zum Thema: " Klimatische und edaphische Sortenanpassung und Züchtung für Nachwachsende Rohstoffe". 3–5 March 2004, Halle/Saale, Germany. Vort Pflanzenzüchtung 64:21–24

Salem KFM, Röder MS, Börner A (2007a) Evaluation of some barley varieties for the presence of thermostable alleles of ßamylase. In: 8th African Crop Science Society Conference, El-Minia, Egypt, 27–31 October 2007, pp 643–648

Salem KFM, Röder MS, Börner A (2007b) Identification and mapping quantitative trait loci for stem reserve mobilisation in wheat (*Triticum aestivum* L.). Cereal Res Common 35:1367–1374

Salem KFM, Röder MS, Börner A (2015a) Assessing genetic diversity of Egyptian hexaploid wheat (Triticum aestivum L.) using microsatellite markers. Genet Resour Crop Evol 62:377–385

Salem KFM, Safhi FA, Alwutayd KM et al (2024) Analysis of genetic diversity, population structure and phylogenetic relationships of rice (*Oryza sativa* L.) cultivars using simple sequence repeat (SSR) markers. Genet Resour Crop Evol. https://doi.org/10.1007/s10722-023-01789-0

Salem KFM, Saleh MM, Abu-Ellail FFB et al (2021d) Role of quantum dots, polymeric NPS and dendrimers in emphasizing crops tolerate biotic and abiotic stresses. In: Faizan M, Hayat S, Yu F (eds) Sustainable agriculture reviews, vol 53. Springer, Cham. https://doi.org/10.1007/978-3-030-86876-5_1

Salem KFM, Saleh MM, Abu-Ellail FFB et al (2021e) The Role of salicylic acid in crops to tolerate abiotic stresses. In: Hayat S, Siddiqui H, Damalas CA (eds) Salicylic acid- A versatile plant growth regulator. Springer, Cham. https://doi.org/10.1007/978-3-030-79229-9_7

Salem KFM, Saleh MM, Aldahak L, Elabd AB (2021f) Assessment phenotypic diversity of rice (*Oryza sativa* L.) genotypes by multivariate analysis. J Arid Agric 7:52–59. https://doi.org/10.25081/jaa.2021.v7.6516

Salem KFM, Sallam A (2016) Analysis of population structure and genetic diversity of Egyptian and exotic rice (Oryza sativa L.) genotypes using SSR markers. Comptes Rendus Biol 339:1–9. https://doi.org/10.1016/j.crvi.2015.11.003

Salem KFM, Tawfik E, Mahdy M et al (2023b) Advances in lettuce (*Lactuca* spp.) molecular breeding strategies. In: Singh S, Sharma D, Sharma SK, Singh R (eds) Smart plant breeding for vegetable crops in post-genomics era. Springer, Singapore. https://doi.org/10.1007/978-981-19-5367-5_11

Salem KFM, Varshney RK, Röder MS, Börner A (2008a) Detection of genetic diversity in Egyptian barley genotypes using barley expressed sequence Tag-SSR markers. In: 10th International Barley Genetics Symposium (IBGS), Alexandria, Egypt 5–10 April 2008. The International Center for Agricultural Research in the Dry Areas (ICARDA) and the Bibliotheca Alexandrina, Alexandria

Salem KFM, Varshney RK, Röder MS, Börner A (2010) EST-SSR based estimates on functional genetic variation in a barley (*Hordeum vulgare* L.) collection from Egypt. Genet Resour Crop Evol 57(4):515–521

Salem KFM, Yaseen YM, El-Abd AB, Ayoub A (2015b) Inheritance of some yield components and grain quality traits in rice (*Oryza sativa* L.). J Product Dev 20:161–177

Salem KFM, Yaseen YM, El-Abd AB, Ayoub A (2015c) Breeding studies on some agronomic and quality traits in rice (*Oryza sativa* L.). J Product Dev 20(2):211–226

Salem KFM et al (2023c) Advances in lettuce (Lactuca spp.) molecular breeding strategies. In: Singh S, Sharma D, Sharma SK, Singh R (eds) Smart plant breeding for vegetable crops in post-genomics era. Springer, Singapore. https://doi.org/10.1007/978-981-19-5367-5_11

Savkovic-Stevanovic J (2018) Toxic effects and DNA transformation Biochem. Biotechnol Res 6:15–19

Schubert C (2006) Can biofuels take center stage? Nat Biotechnol 24:777–784

Shah MA, Khan AA, Junaid JM et al (2015) Aphid vectored viral diseases and their management. In: Insect pests management of fruit crops, pp 511–554

Shukla DD, Tosic M, Jilka J et al (1989) Taxonomy of potyviruses infecting maize, sorghum and sugarcane in Australia and the United States as determined by reactivities of polyclonal antibodies directed towards virus-specific N-termini of coat proteins. Phytopathology 79(2):223–229

Siddiqui SH, Javed M (1982) Mutation breeding in sugarcane (Saccharum sp. Hybrid) by gamma irradiation of cuttings and tissue culture. In: Induced mutations in vegetatively propagated plants II. Proc. IAEA, Vienna, pp 155–166

Singh D, Tomar PK (2005) Studies on critical period of weed removal in sugarcane ratoon. Coop Sugar 36:911–914

Singh RV, Sharma P, Sambyal K (2022) Application of sugarcane bagasse in chemicals and food packaging industry: potential and challenges. Circ Econ Sustain. https://doi.org/10.1007/s43615-022-00167-9

Sivanesan A, Waller JM (1986) Sugarcane diseases

Smith I, Greenside PG, Natoli T et al (2017) Evaluation of RNAi and CRISPR technologies by large-scale gene expression profiling in the connectivity map. PLoS Biol 15:e2003213

Snyman SJ, Meyer GM, Koch AC et al (2011) Applications of in vitro culture systems for commercial sugarcane production and improvement. In Vitro Cell Dev Biol Plant 47:234–249

Snyman SJ, Meyer GM, Richards JM et al (2006) Refining the application of direct embryogenesis in sugarcane: effect of the developmental phase of leaf disc explants and the timing of DNA transfer on transformation efficiency. Plant Cell Rep 25:1016–1023. https://doi.org/10.5376/rgg.2013.04.0004

Souza GM, Berges H, Bocs S et al (2011) The sugarcane genome challenge: strategies for sequencing a highly complex genome. Trop Plant Biol 4:145–156

Souza GM, Sluys MAV, Lembke CG et al (2019) Assembly of the 373k gene space of the polyploidy sugarcane genome reveals reservoirs of functional diversity in the world's leading biomass crop. Giga Sci 8:1–18

Sreenivasan T, Ahloowalia B, Heinz D (1987) Cytogenetics. In: Heinz D (ed) Sugarcane improvement through breeding. Elsevier Press, Amsterdam, pp 211–253

Sukumaran RK, Christopher M, Kooloth-Valappil P et al (2021) Addressing challenges in the production of cellulases for biomass hydrolysis: targeted interventions into the genetics of cellulase producing fungi. Bioresour Technol 329:124746. https://doi.org/10.1016/j.biortech.2021.124746

Sundar AR, Barnabas EL, Malathi P et al (2012) A mini-review on smut disease of sugarcane caused by *Sporisorium scitamineum*. In: Botany. InTech, Shanghai, p 226

Surya Krishna S, Harish Chandar SR, Ravi M et al (2023) Transgene-free genome editing for biotic and abiotic stress resistance in sugarcane: prospects and challenges. Agronomy 13:1000. https://doi.org/10.3390/agronomy13041000

Svitashev S, Schwartz C, Lenderts B et al (2016) Genome editing in maize directed by CRISPR--Cas9 ribonucleoprotein complexes. Nat Commun 7:13274

Tabasum S, Khan FA, Nawaz S, Iqbal MZ et al (2010) DNA profiling of sugarcane genotypes using randomly amplified polymorphic DNA. Genet Mol Res 9(1):471–483. ISSN 1676-5680

Tai PYP (1988) Long-term storage of *Saccharum spontaneum* L. pollen at low temperature. Sugar Cane (Spring supppl):12–16. https://byjus.com/biology/micro-propagation/

Tomaz AC, Coutinho AE, Soares et al (2018) Assessing resistance of sugarcane varieties to sugarcane borer *Diatraea saccharalis* Fab. (Lepidoptera: Crambidae). Bull Entomol Res 108(4):547–555

Tsanova T, Stefanova L, Topalova L et al (2021) I. DNA-free gene editing in plants: a brief overview. Biotechnol Biotechnol Equip 35:131–138

Uranga M, Daròs JA (2022) Tools and targets: the dual role of plant viruses in CRISPR–Cas genome editing. Plant Genome e20220

Usmani Z, Sharma M, Gupta P, Y. et al (2020) Ionic liquid based pretreatment of lignocellulosic biomass for enhanced bioconversion. Bioresour Technol 304:123003. https://doi.org/10.1016/J.BIORTECH.2020.123003

Varshney RK, Salem KFM, Röder MS et al (2008) Assessment and comparison of genetic diversity in barley germplasm collection using gSSR, eSSR and SNP markers. Plant Genet Res 6:167–174

Vasil IK (2008) A history of plant biotechnology: from the cell theory of Schleiden and Schwann to biotech crops. In: Plant Cell Rep, vol 27, p 1423

Vauterin L, Yang P, Hoste B et al (1992) Taxonomy of xanthomonads from cereals and grasses based on SDS-PAGE of proteins, fatty acid analysis and DNA hybridization. Microbiology 138(7):1467–1477

Vijayalakshmi U (1967) Taxonomy of sugarcane. ISSCT Sugarcane Breed Newsle 20:52–64

Viswanathan R (2016) Varietal degeneration in sugarcane and its management in India. Sugar Tech 18:1–7

Viswanathan R (2020) Fusarium diseases affecting sugarcane production in India. Indian Phytopathol 73(3):415–424

Viswanathan R (2021) Sustainable sugarcane cultivation in India through threats of red rot by varietal management. Sugar Tech 23(2):239–253

Viswanathan R, Rao GP (2011) Disease scenario and management of major sugarcane diseases in India. Sugar Tech 13:336–353

Wang J et al (2025) China National Gene Bank (CNGB). In: Al-Khayri JM, Salem KFM, Jain SM (eds) Plant gene banks. Springer, Singapore. https://doi.org/10.1007/978-981-99-4236-7_103-1

Wang WZ, Yang BP, Feng XY et al (2017) Development and characterization of transgenic sugarcane with insect resistance and herbicide tolerance. Front Plant Sci 8:1535

Waswa F, Gweyi-Onyango JP, Mcharo M (2012) Contract sugarcane farming and farmers' incomes in Lake Victoria basin Kenya. J Appl Biosci 52:3685–3695

Watterworth WM, Drury GE, Bray CM, West CE (2011) Repairing breaks in the plant genome: the importance of keeping it together. New Phytol 192:805–822

Wei X, Jackson PA, Hermann S et al (2010) Simultaneously accounting for population structure, genotype by environment interaction, and spatial variation in marker-trait associations in sugarcane. Genome 53(11):973–981. ISSN 0831-2796

Wilson BE (2019) Hemipteran pests of sugarcane in North America. Insects 10:107. https://doi.org/10.3390/insects10040107

Wongkaew P (2012) Sugarcane white leaf disease characterization, diagnosis development, and control strategies. Funct Plant Sci Biotechnol 6:73–84

Xie Y, Wang M, Xu D et al (2009) Simultaneous detection and identification of four sugarcane viruses by one-step RT-PCR. J Virol Methods 162(1–2):64–68

Yadav RL (1986) Sustaining ratoon cane yields with efficient 'N' management practices. Fert News 32(6):17–22

Yang CR (1986) Transmission of sugarcane mosaic virus by three kinds of aphids. Chin J Entomol 6:43–49

Yang L, Zambrano Y, Hu et al (2010) Sugarcane metabolites produced in CO 2-rich temporary immersion bioreactors (TIBs) induce tomato (Solanum lycopersicum) resistance against bacterial wilt (*Ralstonia solanacearum*). In Vitro Cell Dev Biol Plant 46:558–568

Yao W, Ruan M, Qin L et al (2017) Field performance of transgenic sugarcane lines resistant to sugarcane mosaic virus. Front Plant Sci 8:104

Zhang C, Liu S, Li X et al (2022) Virus-induced gene editing and its applications in plants. Int J Mol Sci 23:10202

Chapter 7
Molecular Breeding and Biotechnology for Wild Sugarcane (*Saccharum spontaneum* L.)

Rashmi Chauhan, Jajati Keshari Nayak, Dinesh Pandey, and Sharat Prabhakaran

Abstract Wild sugarcane (*Saccharum spontaneum* L.) is a perennial wild grass, which is a member of the Saccharum genus, well known to possess properties like high fiber content, low sucrose levels, slender stalks, excellent regrowth capabilities, rhizomatous tillering and adaptability to various environmental stresses. Most often used as donor clones for the development of modern sugarcane cultivars. It originated in India and is distributed across a range of geographical zones. The present chapter describes the role of this grass as an alternative to synthetic fiber. Despite its well-known use as a donor parent, studies at the genomic level are limited. The present chapter aims to provide a clear picture of various aspects including origin, habitat, and genetics. Further, the chapter sheds light on the various modern and conventional crop improvement strategies for disease, pest tolerance, herbicide and abiotic stress resistance. Plant tissue culture and other propagation approaches are also discussed. This chapter also includes its various secondary uses like cellulase production and bioremediation.

Keywords Hybridization · Nobilization · Ratooning · Molecular marker · Traditional breeding · Textile

7.1 Introduction

Wild sugarcane (*S. spontaneum* L.) is a perennial grass with tall, slender culms and robust rhizomatous tillering. Better varieties act as excellent options for increasing quality and crop production. In the era before the twentieth century, the worldwide sugarcane industry heavily depended on the esteemed canes from India and China.

R. Chauhan · J. K. Nayak · D. Pandey · S. Prabhakaran (✉)
G. B. Pant University of Agriculture and Technology, Pantnagar, Uttarakhand, India
e-mail: 55493@gbpuat.ac.in; dinesh.pandey@gbpuat-cbsh.ac.in

K. F. M. Salem et al. (eds.), *Breeding and Biotechnology of Grass and Bast Fiber Crops*, Advances in Plant Breeding Strategies 12,
https://doi.org/10.1007/978-3-032-00403-1_7

However, these canes had limitations in terms of their quantity, potential for pests, yield, and susceptibility to diseases. Nonetheless, a noteworthy advancement in sugarcane breeding occurred when successful hybrids were created by crossing *S. officinarum* (esteemed cane) with the wild clone of sugarcane (*S. spontaneum* L.). These hybrids exhibited both resistance to disease along with high sugar yield (Heinz 1987). The recent advancements in crop genetics have increased the interest of breeders and researchers. The expanding understanding of crop genetic diversity, coupled with the innovative applications of genetic tools, breeding techniques, and modern biotechnology, has significantly enriched the field of crop improvement. *S. spontaneum* L. is used differently (a) for producing a better variety of sugarcane (b) for bioenergy production (c) for removing heavy metal and dye (d) in phytoremediation. Breeders have utilized this wild sugarcane (*S. spontaneum* L.) for producing sugarcane variety which is also resistant to abiotic stress as well as for ratooning ability by utilizing hybridization techniques, and molecular marker techniques. This chapter provides a comprehensive overview of the distribution, taxonomic classification, and conservation. It also provides an understanding of molecular markers and their use for crop improvement combined with tissue culture, mutation as well as traditional breeding methods.

7.1.1 Botanical Classification and Distribution

Wild sugarcane (*Saccharum spontaneum* L.) is a grass within the class Monocotledonae, order Poales and family Poaceae. Many plants are coming under this class like Wheat, corn, and oat (Hassan et al. 2024).

Kingdom: Plantae, Planta
- Subkingdom Viridiplantae – green plants
  - Infrakingdom Streptophyta – land plants
    - Superdivison Embryophyta
      - Division Tracheophyta – vascular plants
        - Subdivision Spermophytina – spermatophytes, seed plants
        - Class Magnoliopsida
          - Superorder Lilianae – monocots
            - Order Poales
              - Family Poaceae – grasses
                - Genus Saccharum L. – sugarcane, sugar cane
                  - Species *Saccharum spontaneum* L. – wild sugarcane, ahlek, loa, wild cane

The genus Saccharum, consisting of six species, includes wild sugarcane (*Saccharum spontaneum* L.). The term "Saccharum," arises from the Greek word "sakcharon," meaning sugar, and was first mentioned by Linnaeus in his book "Species Plantarum" in 1753. A wealth of literature provides insights into the origin and diversity of wild sugarcane (*S. spontaneum* L.). Literature indicates, its origination is mainly in the temperate regions of subtropical India and is distributed across a wide range, from 8°S to 40°N. This distribution encompasses three distinct geographical zones: the eastern regions (South Pacific Islands, Japan, Taiwan, Philippines, China, Thailand, Myanmar, Vietnam, and Malaysia), the central areas (India, Nepal, Bangladesh, Sri Lanka, Pakistan, Afghanistan, Iran, and the Middle East), and the western regions (Uganda, Sudan, Kenya, Egypt, Tanzania, as well as other Mediterranean countries). These zones form natural cytogeographical clusters, characterized by variations in chromosome numbers at each location (Daniels and Roach 1987).

In India, it is noted that different types of sugarcane plants are found in different regions. The ones with lower chromosome numbers are mostly found in the north, while those with higher ploidy levels are more common in the east and south, particularly in Indonesia. The most diverse among these types are those with $2n = 60–70$, which have both rhizomatous and weedy characteristics (Panje 1970). This species is typically classified into two subspecies: subsp. *spontaneum*, which can be accessed in tropical as well as warm temperate regions of Asia, whereas subsp. *aegyptium*, which is found in tropical Africa and extends from Malawi to Syria (Clayton and Renvoize 1982).

7.1.2 Domestication, Selection and Early Improvement

Wild sugarcane (*S. spontaneum* L.) is characterized by white, green, ivory green, or grey culms and is hard in texture which is pithy and many times void at the center. The grass bears long, silky white panicles and has narrow-lanceolate leaves with silicaceous structures. Panicles are typically 50–60 cm in length and spikelets remain covered with hairs and 3–4 mm long. Glumes are long and pointed with ciliate margins, and the lemma is very small or absent (CABI Compendium). Wild sugarcane (*Saccharum spontaneum* L.) thrives in a wide variety of environments, from rich soils that allow it to stretch up to 5 meters in height, to less fertile soils where it generally reaches about 3 meters tall (Roxburgh 1971). It can be found along riverbanks, roadsides, abandoned lands, as well as the edges of ponds and lakes (Holm et al. 1997).

New cultivars are produced through different types of breeding programs that include, the preservation of germplasm stocks. Lines with desirable genetic characteristics are employed for hybridizations in order to create fresh lines. The process of enhancing sugarcane cultivars through breeding is lengthy (Cox et al. 2020). Sugarcane breeding initiatives heavily depend on the interbreeding of superior cultivars, typically through the process of cross-pollination. When it comes to self-pollination, inflorescence is either isolated from other clones or shielded with bags (Sleper and Poehlman 2006). Viable pollen is recognized usually through staining of iodine or aceto-carmine (McIntyre and Jackson 2001; Melloni et al. 2013).

In scientific terms, cultivars that possess a pollen viability of 20% (or 25% in the case of Barbados) are identified as male. On the other hand, cultivars with an intermediate range of pollen viability, ranging from 10% to 20%, are categorized as bisexual and can act as either female or male parents. This classification was outlined by McIntyre and Jackson (2001). In the states of Florida and Louisiana, the evaluation of pollen production involves a visual assessment in which closed yellow anthers and stigma devoid of pollen are observed in female plants (McIntyre and Jackson 2001).

The method of emasculation, achieved through the application of heated water or the reduction of pollen viability via the cultivation of plants under chilly conditions, is used in cultivating male sterile plants for use as female progenitors in breeding initiatives (Heinz 1987; McIntyre and Jackson 2001; Song et al. 2024).

Sugarcane breeding programs face significant constraints due to the unique flowering patterns exhibited by each sugarcane cultivar, particularly when it comes to reduced flowering and pollen viability in regions with higher latitudes (Moore and Nuss 1987). The ability to perform crosses is limited to cultivars that have overlapping flowering periods. Numerous strategies have been devised to promote flowering, including photoperiod manipulation to ensure the availability of flowers for crossbreeding purposes (Das Bhowmik et al. 2016). However, it should be noted that altering the timing of flowering may also have an impact on fertility (Midmore 1980; Song et al. 2024).

In commercial breeding programs, assisted crosses are carried out between hybrids of *Saccharum spp.* under highly climatically suitable conditions. One specific approach involves cutting off the flowering stalks and placing them in buckets filled with a crossing solution. This solution is a diluted mixture of acids that aids in the preservation of the stalks and provides certain nutrients (Cox et al. 2020). Female and male flowers are enclosed within pollen-resistant canvas bags called lanterns, with the male flowers positioned above the female flowers to allow the downward shedding of pollen onto the female flowers (Cox et al. 2020). Once pollination occurs, the stalks are submerged in the crossing solution for a maturation period of approximately 12–14 days (Buzacott 1965). In order to facilitate specific crosses, sugarcane stalks can also be preserved through the process of marcotting or air layering (Bischoff and Gravois 2004). The field of breeding improved for various characteristics, such as the generation of biomass, the ability to withstand stress, adaptability to drought, resilience in low temperatures, and disease defense (Ming et al. 2006; Shearman et al. 2022). An examination of plant germplasm utilized in sugarcane breeding initiatives during the 1980s revealed that the development of commercial cultivars at that time involved two *S. barberi*, *S. sinense*, and *S. robustum*, along with 19 *S. officinarum* and a restricted number of wild sugarcane (*S. spontaneum* L.) clones (Roach 1989). Different studies have proposed that the genetic makeup of current cultivars is restricted to 20 *S. officinarum* and fewer than 10 wild sugarcane (*S. spontaneum* L.) lineages. This crossbreeding between different species has expanded the geographical scope of sugarcane production for economic purposes (Patade and Suprasanna 2008; Dalvi et al. 2021). Hybridization between *S. officinarum* (female parent) and *S. spontaneum* L. (male parent) produces triploid progeny exhibiting a chromosome number ranging from 100 to 130 ($2n + n$) (Sreenivasan et al. 1987). This phenomenon arises from the $2n$ chromosome contribution of the maternal parent and the n chromosome contribution of the paternal *S. spontaneum* L. parent. This uneven chromosome transmission was also observed in the initial backcross generation to *S. officinarum*, attributed to either endoreduplication or nuclear fusion during meiosis (Lu et al. 1994).

The development of high sucrose cultivars can now be achieved with fewer rounds of backcrossing, according to a study by Sreenivasan et al. (1987). The Dutch breeders referred to this process as "nobilisation". Using special techniques (GISH, AFLP markers), scientists have found that commercial sugarcane varieties get about 80% of their chromosomes from *S. officinarum*, 10% from wild sugarcane (*S. spontaneum* L.), and the rest are recombinant chromosomes formed during cell division (meiosis) (Garsmeur et al. 2018; Zhang et al. 2018).

7.2 Current Cultivation Practices

7.2.1 Current Agricultural Problems and Challenges

Wild sugarcane (*S. spontaneum* L.), a remarkable perennial species, employs both vegetative and seed propagation methods to proliferate. Rhizomes and stem fragments are the primary modes of vegetative propagation (Artschwager 1942). The seed production of this species varies considerably depending on the location. Pancho (1964) reported an average of 12,800 seeds per plant in the Philippines, while Datta and Banerjee (1973) observed a mean of 3042 seeds per plant in India. Wind dispersal is facilitated by callus hairs forming a parachute-like structure; seed aggregation into woolly masses can also occur, aiding long-distance dispersal (Sharma and Tiagi 1979). These authors also identified three distinct biotypes: xerophilous, hygrophilous, and an intermediate type thriving in loamy soils.

7.2.2 Genetic Improvement Objectives

Isolating key genes responsible for adaptation to low-input environments could contribute to reducing the expenses of cultivating advantageous crops. Nevertheless, traditional breeding techniques are costly and time-consuming as they may inadvertently introduce undesirable gene variants (Paterson et al. 1996). To mitigate this problem, researchers have created molecular markers linked to sugar content in wild sugarcane (*S. spontaneum* L.). These markers aid in the more efficient incorporation of beneficial gene variants into sugarcane (Ming et al. 2002).

The successful integration of wild sugarcane (*Saccharum spontaneum* L.) in the enhancement of sugarcane characteristics has been a significant milestone in the field of plant breeding. Among all other cultivated plants, no other wild species has been as effectively harnessed to fulfill human requirements. Wild sugarcane (*S. spontaneum* L.), a wild species, displays remarkable resilience against a broad array of biotic and abiotic constraints (Lu et al. 2024). Roach's (1989) study, outlined three distinct stages in the progress of sugarcane improvement since approximately 1890. The initial stage centered on the process of crossing and selecting among different clones of the *S. officinarum* species. Many early sugar industries worldwide were established using multiple clones of this species. In the second stage of breeding, *S. officinarum* was hybridized with other species, primarily wild sugarcane (*S. spontaneum* L.). The resulting hybrids were then utilized in the third stage of sugarcane improvement strategies. The raw sugar market is forecast to grow at a compound annual growth rate (CAGR) of 3.26% between 2022 and 2027. In addition, the bioethanol market, valued at USD 63.13 billion in 2022, is expected to reach USD 116.06 billion by 2032, representing a 7% CAGR in revenue over the forecast period. (Karp et al. 2022).

Research has shed light on the significant role that various genes play in responding to cold environments, thereby improving tolerance to abiotic factors. Additionally, efforts are underway to breed new energy cane cultivars, with a specific focus on enhancing yield and fortifying resistance against pests and environmental adversities through the utilization of wild sugarcane (*S. spontaneum* L.) (Lu et al. 2024).

7.3 Textile (Yarn Production)

The production of synthetic fiber has been identified as a significant contributor to greenhouse gas emissions. Consequently, due to environmental concerns, the exploration of alternative cellulosic fibers has emerged as a promising strategy to mitigate the ecological impact associated with synthetic fibers (Laitala et al. 2018). Currently, the market offers few cellulosic fibers as well as natural yarn options, with Sisal, Pineapple, and commercially available cotton being among the limited potential substitutes for synthetic fibers (Ari et al. 2023).

Saccharum sp. is also an option as a feasible alternative for synthetic textiles, providing environmental benefits as well as economic opportunities for local communities through cultivation.

A combination of fibers by making yarn blends consisting of 30% wild sugarcane (*S. spontaneum* L.) and 70% pure cotton by following a series of steps that involved cutting, drying, blending, and spinning fibers extracted from leaves and stalks onto bobbins (Santos and Torres 2019). Treating the fibers with alkali creates a significant positive effect on their mechanical, morphological properties, and crystallinity which is measured by mechanical property testing, XRD, TGA/DTG, FTIR spectroscopy, XRD, AFM, and SEM analysis (Aravindh et al. 2022). To enhance wild sugarcane (*S. spontaneum* L.) properties like fibers, 3 wt%, 5 wt%, and 7 wt% varying concentrations of alkali were used and 5% alkali exhibited the most notable elevated average activation energy, ranging from 137 ± 3.1 kJ/mol to 194 ± 2.4 kJ/mol compared to the untreated fibers also better thermal and mechanical properties than previously studied fibers (Devnani and Sinha 2019).

The copolymerization technique is used to enhance the characteristics of natural fibers while maintaining their original qualities. It leads to resistance to acids, bases, and heat, as well as reduces moisture absorption which is good for fiber characteristics. This technique has diverse applications like in packaging. Different combinations of vinyl monomers, like MMA + AA, MMA + AAm, and MMA + AN are incorporated into wild sugarcane (*S. spontaneum* L.) fibers also ferrous ammonium sulfate/potassium persulfate is used as a redox initiator influenced by the physical and chemical properties of the fibers Kaith et al. (2009). Similarly, in (2010) Kaith et al, (2010) colleagues developed a material of wild sugarcane (*S. spontaneum* L.) fiber and corn starch with different substances like acrylic acid (AA), acrylamide (AAm), methyl methacrylates (MMA), and acrylonitrile (AN). Tyagi and Dutta (2007) used wild sugarcane (*S. spontaneum* L.) for producing high-quality writing as well as printing papers due to having favorable characteristics for chemical-grade pulp production, such as low lignin content and high holocellulose content. Its unique structure, including a thin cell wall, wide lumen, good flexibility, and open fiber structure, allows for efficient collapse during cooking, resulting in a ribbon-like structure that maximizes surface area for bonding. Additionally, wild sugarcane (*S. spontaneum* L.) responds well to oxygen pre-bleaching followed by multi-stage bleaching sequences, reduced amount of bleach chemicals required compared to wood pulps.

7.4 Germplasm Conservation

7.4.1 Genetic Resources Conservation

To effectively address the current and upcoming global challenges, it is essential to thoroughly investigate, gather safeguard and responsibly utilize plant genetic resources (PGR). Furthermore, it is the collective duty of all nations to systematically survey, explore, collect, preserve and sustainably utilize PGRs. Policymakers, Scientists, and strategists have initiated efforts to carefully plan for the sustainable use and conservation of PGRs, aiming to benefit humankind (Pandotra and Gupta 2015).

7.4.2 In Situ Conservation

Modern conservation strategies often *propose* a synergistic approach that combines in-situ and ex-situ conservation. In-situ conservation is generally prioritized for preserving genetic resources, whereas ex-situ conservation serves as a supplementary measure. As our understanding of taxonomic evolutionary histories deepens, the implementation conservation approach for specific taxa becomes easier (Luo et al. 2019).

7.4.3 Ex Situ Conservation

Within the scientific community, researchers undertake the task of assembling and preserving a wide array of plant genetic variations in order to mitigate the potential decline in biodiversity and genetic erosion associated with conventional landraces and indigenous cultivars, especially as they are supplanted by modern agriculture breeds (Khoury et al. 2015). Gauchan and Smale (2003) emphasized the requirement of acquiring also conserving additional genetic types across different plant species of various plant species to ensure sustainable conservation. Institutions like Sugarcane Breeding Institute in Coimbatore (India), and USDA-ARS National Germplasm Repository in Miami (Florida) are dedicated to storing vegetative clones of sugarcane and grasses (Schnell et al. 1997; Orton and Ayeni 2022). Maintaining a worldwide huge collection of wild sugarcane (*Saccharum spontaneum* L.) clones demands considerable labor as well as expense. Careful selection of core subsets can effectively capture the diversity within the collection, thereby improving conservation strategies and unlocking valuable breeding potential. Through the evaluation of 342 accessions from the USDA-ARS National Germplasm Repository in Miami (Florida), various sampling strategies were assessed to establish a core collection of 75 clones based on geographic origin (Tai and Miller 2001, 2002). The germplasm collection housed at the SBI Research Centre in Kannur encompasses 3368 accessions. This assortment comprises specimens of various sugarcane species, including wild sugarcane (*Saccharum spontaneum* L.) (Chandran 2010). In a separate study, a core collection of wild sugarcane (*S. spontaneum* L.) has been proposed for the accessions from the collection in Coimbatore (India) using different sampling methods (Amalraj et al. 2006). Various international repositories are in India, Brazil, the United States House, and South Africa, significant cultivars for breeding purposes (Cursi et al. 2022).

7.4.4 In Vitro Conservation

In vitro culture conservation is a reliable method used to preserve plant parts for several years under controlled conditions. It is especially useful when preserving genetic resources is not possible through other means. The method of in vitro conservation is widely acknowledged by international organizations like the IBPGR for its ability to safely transport plants while maintaining regulated phytosanitary control. This technique offers several benefits, including the production of insect and disease-free materials, the ability to mass multiply plants, prevention of genetic erosion, reduced space and labor needs, and a shorter timeframe for obtaining new plant varieties (Chandran et al. 2023). The World Collection of Sugarcane and Related Grasses currently maintains clones of wild sugarcane (*Saccharum spontaneum* L.) through vegetative methods in Miami, Florida, USA, and Coimbatore, India. In order to safeguard the genes, present in this collection, true seeds of select wild sugarcane clones were produced at Canal Point, Florida. These seeds were then stored as a foundational collection at the National Seed Storage Laboratory in Fort Collins, Colorado (Tai and Miller 2001). Sreenivasan and Sreenivasan (1984) conducted research on the effectiveness of in vitro techniques in micropropagation and crop enhancement. They also reported on the use of the meristem culture technique for germplasm storage in sugarcane.

7.4.5 Cryopreservation

Storing sugarcane germplasm as defuzzed true seed at low temperatures can effectively preserve its germination ability. It has been observed that seed stored at −20 °C for 5 years maintained its germination potential without significant reduction. This method offers practical advantages for maintaining and utilizing sugarcane germplasm in breeding programs (Doule and Balasundaram 2006).

7.4.6 DNA and Seed Bank

The SBI Research Centre in Kannur, India, is responsible for curating the gene bank containing sugarcane germplasm, whereas wild species like wild sugarcane (*S. spontaneum* L.), *Erianthus* spp., and related genera are curated in Coimbatore, India. Certain clones sourced from high-altitude regions such as Arunachal Pradesh and Meghalaya failed to grow and flower in Coimbatore, due to this reason they are preserved at the IARI Regional Station in Wellington (Chandran et al. 2023). All germplasm-related information is listed in Tables 7.1, 7.2, 7.3, 7.4 and 7.5. Furthermore, viable seeds resulting from poly crossbreeding of wild sugarcane (*S. spontaneum* L.) are forwarded for long-term storage at the National Gene Bank in NBPGR, New Delhi (India) (Mirajkar et al. 2019).

Table 7.1 Germplasm holdings in India

Institute	Species	ISSCT accessions	Indian accessions
SBI, Coimbatore	*S. spontaneum* L.	Nil	878
SBI, Kannur	*S. spontaneum* L.	79	305
IARI, Wellington	*S. spontaneum* L.	Nil	47

Source: ICAR-SBI

Table 7.2 Registration of sugarcane hybrids as germplasm by ICAR

Clone No.	Registration No.	Characters
CYM 08-903	IC0594482; INGR13019	Sugarcane hybrid with cytoplasm from wild *Erianthus arundinaceus*. Derivative of (*Erianthus arundinaceus* × *Saccharum spontaneum*) × sugarcane hybrid. Red rot resistant.
CYM 04-420	IC556971;INGR08039	*Saccharum spontaneum* × *Erianthus arundinaceus* hybrid with Erianthus cytoplasm

Source: ICAR-SBI

Table 7.3 Sugarcane Germplasm registered with NBPGR (National Bureau of Plant Genetic Resources, New Delhi, India)

Donor identity	Year	INGR_No	National identity	Pedigree	Novel unique features	Developer
CYM 08-922	2020	INGR20071	IC635054	CYM 07-971 X CoC 671	Drought tolerance.	K Mohanraj, M N Premachandran, S Vasantha, C Appunu, VP Sobhakumari, Adhini S Pazhani, G Hemaprabha

Source: ICAR-SBI

Table 7.4 Sugarcane germplasm registered with USDA

Germplasm	Characteristics
L 79-1002	Released on 26 Apr. 2007 by the Louisiana State University Agricultural Center in cooperation with the USDA-ARS and the American Sugarcane League, Inc. The cross for L 79-1002, an F1 hybrid, was made in 1974 using 'CP 52-68' as the female parent and Tainan, a *S. spontaneum* clone, as the male parent.
Co 205	First commercially successful inter-specific hybrid in the world.

Source: USDA-ARS

7.5 Genetic Information and Importance

Modern sugarcane cultivars (*Saccharum* spp.), characterized by 100–120 chromosomes, have been extensively studied to understand their origins. These cultivars predominantly arise from the hybridization of two distinct species: *Saccharum officinarum* and wild sugarcane (*S. spontaneum*) (D'Hont et al. 1996). *Saccharum officinarum*, with 80 chromosomes, is renowned for its thick stalks, low fiber content, and high sucrose content but exhibits limited regrowth potential after harvest (Daniels and Roach 1987). In contrast, wild sugarcane (*S. spontaneum*), possessing 40–128 chromosomes, is characterized by high fiber, slender stalks, low sucrose content and remarkable resilience to various environmental stresses, including cold and drought (Daniels and Roach 1987).

Table 7.5 List of wild sugarcane (*Saccharum spontaneum* L.) lines, commercial cultivars

Accession line	Chromosome number	Origin	References
IK76-41	–	Indonesia	–
MOL1032	64	Unknown	Aitken et al. (2018)
NG28-101	114,80	Papua new guinea	Aitken et al. (2018)
US56-15-2	–	Thailand	Aitken et al. (2018)
IS76-173	–	Indonesia	Aitken et al. (2018)
SES516	96	Thailand	Aitken et al. (2018)
IMP1944	–	Indonesia	Aitken et al. (2018)
IK76-86	–	Indonesia	Aitken et al. (2018)
SES343	64	India	Aitken et al. (2018)
US71-4-1	–	Unknown	Aitken et al. (2018)
IK76-49	–	Indonesia	Aitken et al. (2018)
OKINAWA	112	Japan	Aitken et al. (2018)
SES208A	64	India	Aitken et al. (2018)
DJANTOER_2	112	Indonesia	Aitken et al. (2018)
SES182	52	India	Aitken et al. (2018)
KLOET	112–114	Indonesia	Aitken et al. (2018)
COIMBATORE	64	India	Aitken et al. (2018)
GLAGAH-1286	58	Indonesia	Aitken et al. (2018)
MANDALAY	96	Indonesia	Aitken et al. (2018)

Dashes indicate no data available. Source Bhuiyan et al. (2019)

7.6 Cytogenetics

Saccharum species are polyploids with varying genome structures, which has made it challenging for molecular breeding. The utilization of classical and contemporary molecular cytogenetics has been contribute significantly to understanding genome structure, and the evolution of Saccharum. One of the pioneers in this field was Barber, who conducted a cytological study on sugarcane in the 1910s to aid in the establishment of a classification system (Barber 1916). By examining the number of chromosomes, Bremer determined that when *S. officinarum* and wild sugarcane (*S. spontaneum* L.) are bred together, the offspring have a chromosome count of $2n + n$. Bremer proposed that canes that have undergone this process have higher chromosome numbers (ranging from 100 to 130) compared to their parents. This phenomenon is believed to be the primary factor behind the enhanced productivity and adaptability seen in modern sugar cultivars. Classical cytogenetics has greatly contributed to elucidating the process of nobilization in contemporary cultivar breeding (Heinz 1987). Utilizing FISH techniques, D'Hont conducted a chromosomal analysis of *S. officinarum* and wild sugarcane (*S. spontaneum* L.). The results revealed a basic chromosome number (x) of 8 for *S. spontaneum* L. (D'Hont et al. 1996). Further analysis using genomic DNA revealed that 10–25% of chromosomes in modern cultivars descended from wild sugarcane (*S. spontaneum* L.) (Cuadrado et al. 2004; Piperidis et al. 2010). 31 cytotypes of wild sugarcane (*S. spontaneum* L.) have been identified, maximizes diversity among the Saccharum species. Panje and Babu (1960)conducted chromosome counts in wild sugarcane (*S. spontaneum* L.) and identified 400 clones with an array of 2*n* counts between 40 and 128, covering all known cytotypes (Sreenivasan 1975). Different studies revealed that diverse chromosome numbers are found in sugarcane. It mentions that the basic chromosome numbers suggested are $x = 5, 6, 7, 8, 10$, and 12. However, through cytotypes and phylogenetic analysis, it is confirmed that the basic chromosome number for most cytotypes is $x = 8$ 12 (D'Hont et al. 1996; Meng et al. 2018). The reduction from $x = 10$ to $x = 8$ in wild sugarcane (*S. spontaneum* L.) has been shown to involve two chromosome rearrangement events (Garsmeur, 2018). The researcher mentioned the discovery of wild sugarcane (*S. spontaneum* L.) clones with basic chromosome numbers of $x = 10$ and $x = 9$, indicating a two-step process in the evolution of wild sugarcane (*S. spontaneum* L.) from $x = 10$ to $x = 9$–8 (Panje and Babu 1960; Piperidis and D'Hont 2020).

Determining the ploidy of wild sugarcane (*S. spontaneum* L.) clones has been challenging due to their large array of chromosome numbers and the diverse basic chromosome numbers recoomended. The use of rDNA probes for fluorescence in situ hybridization (FISH) has been an unreliable approch in indicating ploidy levels due to variations in rDNA locus numbers (D'Hont et al. 1996). However, FISH studies using chromosome-specific oligo probes have determined seven different ploidy types among wild sugarcane (*S. spontaneum* L.) clones, including the highest-ploidy clone ever recorded. In total, eight ploidy types have been revealed for wild sugarcane (*S. spontaneum* L.) clones, discovery indicated that clone Glagah was a unique tetra-decaploid (14*x*) (Piperidis and D'Hont 2020). Sugarcane cultivars are primarily created through crosses between noble canes with the wild species *S. spontaneum* L., subsequent backcrossing procedures are then employed to introduce desirable traits from the wild species into the cultivated background. The breeding program benefits from female restitution, a chromosome transmission phenomenon. This transmission can result in increased chromosome numbers in the nobilized canes. However, this phenomenon is not consistent in all interspecific crosses, so cytogenetic analysis is crucial to confirm successful nobilized hybridization in sugarcane breeding programs (Heinz 1987; Wang et al. 2023). Using a technique called FISH, researchers discovered that the sugarcane variety R570 has chromosomes that are derived from both *S. officinarum* as well as wild sugarcane (*S. spontaneum* L.), with approximately 80% coming from *S. officinarum* while 10% from wild sugarcane (*S. spontaneum* L.) D'Hont et al. (1996). They also found evidence of interspecific recombination in modern cultivars, contradicting earlier beliefs. Another study introduced a new FISH strategy to distinguish wild sugarcane (*S. spontaneum* L.) chromosomes in sugarcane cultivars, identifying specific retrotransposons. This new method showed a higher percentage of interspecific recombination than previous methods, but further confirmation is needed due to differences in the cultivars used (Huang et al. 2020; Piperidis and D'Hont 2020). An analysis of ten modern sugarcane cultivars revealed an intriguing pattern of chromosome rearrangements. While exchanges of genetic material between chromosomes of the wild species *S. spontaneum* were uncommon, a surprisingly high frequency of translocations was observed within the cultivated species *S. officinarum*, averaging 2.15 translocations per chromosome. This finding suggests frequent recombination between non-homologous chromosomes during the breeding of *S. officinarum*. Furthermore, the study documented instances of genetic exchange between the two species, as exemplified by cultivar YT55, which possesses a chromosome containing segments from both *S. spontaneum* Chromosome 1 and an unidentified fragment of an *S. officinarum* chromosome (Wang et al. 2023).

7.7 Traditional Breeding

7.7.1 Improvement Strategies

Sugarcane breeding is unique among other crops because of its high polyploidy, heterozygosity, and cytogenetic features such as unreduced gametes, parthenogenesis, and chromosome elimination. During the initial phases of the sugarcane breeding program, the focus was on finding suitable varieties for the sub-tropical regions of India. The goal was to replace the underperforming Indian Sugarcane with higher-yielding alternatives. The breeding process involved selecting and prioritizing characteristics like quality improvement, adaptability, yield, as well as resistance to disease resistance. This was achieved by utilizing different species of sugarcane, namely wild sugarcane (*S. spontaneum* L.), *S. officinarum*, and *S. barberi*. As a result, exceptional varieties were developed that not only had higher yields but also demonstrated resistance to major diseases like red rot. The Sugarcane Breeding Institute (SBI) in Coimbatore conducted the first breeding experiments for better the quality of subtropical sugarcane. The selected parents for crossbreeding possessed a wider genetic diversity compared to other nations (Verma 2001).

In the field of sugarcane breeding, the primary focuses remain on germplasm collection with characterization, selecting appropriate parental lines, synchronizing flower growth, carrying out crossing or hybridization, and finally, selecting and testing the resulting progeny. These aspects continue to be of utmost importance across all sugarcane breeding programs (Pan 2006). Before the early 1990s, breeders possessed limited knowledge regarding the genetic makeup of the sugarcane genome and its molecular composition. Consequently, their selections were solely reliant on evaluations of observable traits (Burner and Legendrc 1993). Conventional selection methods are not effective for economic traits, particularly in F1 hybrids. Selection choices often rely on subjective assessments and short-term observations. Objective verification is needed for hybrids with wild sugarcane (*S. spontaneum* L.) cytoplasm due to their self-pollination and vigorous growth (Burner 1997; Tai et al. 1994). The availability of diverse germplasm from Saccharum species and related genera is crucial for targeted breeding programs to improve sugarcane (Shrivastava and Srivastava 2016). Despite the ongoing release of productive cultivars for commercial use, the nuclear genome of these varieties predominantly consists of a limited number of clones from *S. sinensis*, *Saccharum officinarum*, *S. barberi* *S. robustum*, and wild sugarcane (*S. spontaneum* L.) (Pompidor et al. 2021). Breeding for red rot resistance is the only solution to controlling this disease, as there are no chemical controls available. To successfully breed for resistance, breeders need to understand host-parasite interactions and resistance testing methods, sources of resistance, inheritance patterns, and the types of pathotypes present (Verma 2001). Wild sugarcane (*Saccharum spontaneum* L.) clones were assessed for their resistance to red rot disease and some were found to be resistant (Viswanathan 2018).

In the second stage, following the successful adaptation to various environments, developing resistance to diseases, insects, pests, and abiotic stress, as well as the ability to regrow, a select few indigenous canes (*S. robustum, S. barberi*, and *S. spontaneum* L.) were strategically selected for hybridization and subsequent nobilization to improve existing varieties (Shrivastava and Srivastava 2016).

Exploring previously untouched genetic material has shown great potential. By utilizing Mandalay, a type of wild sugarcane (*S. spontaneum* L.) from Burma, the Australian 'Early CCS Canes Programme' achieved success and contains low sugar content. Additionally, the use of another wild sugarcane (*S. spontaneum* L.) variant, US56-15-8, resulted in the creation of LCP 85-384. This Louisiana variety is known for its high yield, high sugar content, early ripening, reduced nitrogen requirements, and ability to withstand cold temperatures. As a result, LCP 85-384 has been widely adopted across a large geographical area (Shrivastava and Srivastava 2016). Since 1980, the Sugarcane Breeding Institute (SBI) in Coimbatore has incorporated clones of *S. robustum, S. barberi, S. officinarum* and *S. spontaneum* into breeding programs to develop interspecific hybrid clones that exhibit both high stalk yields and superior juice quality (Shrivastava and Srivastava 2016).

The success of breeding canes in North India was uncertain due to the region's challenging climate. However, by using wild sugarcane (*S. spontaneum* L.) in their breeding process, Barber and Venkatraman were able to achieve success with the Co. 205 seedling in Punjab. The institute is now organizing an exploration to collect wild forms of Saccharum and allied genera from all over India (Dutt 1953). Wild sugarcane (*Saccharum spontaneum* L.) has been a major source of important traits for modern sugarcane cultivars, like vigor as well as ratooning potential. Identifying wild sugarcane (*S. spontaneum* L.) clones with high early vigor and tillering ability is crucial for hybridization programs. A study evaluated 43 wild sugarcane (*S. spontaneum* L.) accessions and found high heritability and significant genetic improvement for leaf width and stalk length, suggesting the importance of additive gene action for these traits. Six superior clones were identified as potential donors for bioenergy canes (Pathy et al. 2022).

Since the 1960s, various breeding programs have incorporated wild sugarcane (*S. spontaneum* L.) into their activities (Berding and Roach 1987; Bremer 1961; Giamalva et al. 1984; Pange 1972; Sreenivasan and Sreenivasan 1984; Wang et al. 2008). Out of all these programs, the most extended and consistent utilization of wild sugarcane (*S. spontaneum* L.) germplasm has taken place in Australia,

Barbados, and Louisiana (Ming et al. 2006). In Houma, USA, a breeding initiative was launched in 1972 to enhance the genetic diversity of sugarcane and develop better sugarcane varieties. This was achieved by crossing wild sugarcane (*S. spontaneum* L.) with commercially cultivated sugarcane material. The methodology implemented in this program encompassed an exponential augmentation in the average chromosome count with successive generations, culminating in the achievement of $2n = 108$ by the BC3 stage (Bremer 1961). Wild sugarcane (*S. spontaneum L.*) has been extensively utilized in sugarcane breeding programs, both to enhance genetic diversity and, more recently, to develop bioenergy crops. In the USA, for example, the prominent energy cane cultivar, Ho02-113, was derived from a cross between wild sugarcane (*S. spontaneum)* and a cultivated sugarcane variety (Hale et al. 2013). Wild sugarcane (*S. spontaneum* L.) exhibits several unique characteristics within the *Saccharinae* species, including non-preferential chromosome pairing, the ability to reproduce both sexually and asexually, and its perennial nature.

7.7.2 Role of Biotechnology

The application of biotechnological methods is necessary for the examination of plant genetic resources, their preservation, and their integration into breeding programs. It is vital to prioritize the exploration of alleles and genes associated with valuable traits in uncultivated species and related wild plants. By effectively utilizing plant genetic resources, we can address the challenges that hinder crop productivity. To ensure future food security, it is crucial to employ high-capacity genotypic and phenotypic methods to utilize genetic resources sustainably (Chaudhary et al. 2023).

The molecular makeup can enhance the visual description and more accurately determine optimal pairings between accessions for breeding objectives (Guden and Qureshi 2024). Molecular markers are valuable resources for identifying genetic variations within the genome and have been employed to examine the genetic diversity of Saccharum spp. accessions (Barreto et al. 2019). Aitken et al. (2018), investigated genetic variation in *S. spontaneum* L. by analyzing 724 markers in 430 accessions from 21 countries. Additionally, they examined 255 cultivars from two breeding programs. The results of this study suggest that diversity analysis can aid in the selection of appropriate wild *S. spontaneum* L. accessions, which can subsequently be integrated into breeding programs to enhance genetic diversity. These findings can be combined with phenotypic data for a more comprehensive approach.

7.8 Molecular Breeding

Modern molecular techniques are suitable to enhance breeding efficiency as well as cultivar identification. These molecular markers facilitate a deeper understanding of genetic diversity within parental lines used in breeding programs and also act as genetic tags, enabling the identification of genes linked to desirable traits (Hotta et al. 2010; Manners and Casu 2011). Molecular markers have been developed in sugarcane breeding to accelerate breeding efforts along with the study of genetic diversity (Lakshmanan et al. 2005). Expressed sequence tags (ESTs) have been identified from various sugarcane tissues, including meristematic tissue, stem tissue, and red-rot infected sugarcane (Srivastava 2020). A Brazilian consortium, SUCEST, has generated 238,000 ESTs from 26 cDNA libraries that represent diverse developmental stages, organs, and tissues (Arruda 2001). These EST projects have helped to understand sugarcane biology and genomics at the molecular level by providing a wealth of sequence data and potential functions have been predicted for numerous genes associated with economically significant traits. QTLs linked to stalk number, suckering, sugar content, and yield-related traits have been identified (Aitken et al. 2006, 2008; Hoarau et al. 2002; Jordan et al. 2004; Ming et al. 2002).

7.8.1 Molecular Marker Assisted Breeding

Before the advent of molecular markers, the selection of F1 progeny in Louisiana was based solely on phenotypic evaluations. This conventional method of selection was not always efficient, because many important traits for farmers (like yield and disease resistance) are influenced by many genes and are hard to measure precisely. This is especially tricky with F1 hybrids, where the offspring are a mix of traits and decisions are often made based on how the plants look, which can be unreliable. Researchers have noted that this type of selection is not always reliable, as it can be difficult to distinguish between different genotypes based on morphology alone (Burner and Legendrc 1993; Burner 1997). Another challenge with conventional selection methods is that hybrid selection can be compromised when F_1 and selfed progeny share similar morphology. This makes it difficult to select for desirable traits, as the traits of the selfed progeny may be masked by the traits of the F_1 progeny. To overcome these challenges, researchers have developed molecular markers that are suitable for identifying and selecting specific traits in F_1 progeny. These markers allow researchers to make more informed decisions about which progeny to select, based on their genetic makeup rather than their morphology alone.

Scientific advancements since the mid-1990s have given rise to various types of molecular markers for sugarcane breeding. These markers, which are specific to both species and traits, include PCR markers derived from the 5S-rDNA spacer region (Piperidis et al. 2000; Pan et al. 2001), RAID markers (McIntyre and Jackson 1995; Mudge et al. 1996; Burner et al. 1997), RFLP markers (Ming et al. 2002), microsatellite markers (Cordeiro et al. 2003; da Silva 2001; Pan et al. 2003), and AFLP markers (Besse et al. 1998; Hoarau et al. 2001).

Microsatellites or RAPD markers were found to be effective in discriminating between inter-clonal or inter-specific hybrids (Pan et al. 2004). Molecular techniques indicated a clear distinction between wild sugarcane (*Saccharum spontaneum* L.) and other species based on cytoplasmic DNA, while cluster analyses of nuclear DNA corroborate this differentiation (Shanthi et al. 2023). Pan et al. (2004) were the first to develop Saccharum hybrids with the cytoplasm of wild sugarcane (*S. spontaneum* L.), aiming to augment the genetic diversity of Saccharum germplasm and foster the creation of novel sugarcane cultivars. Researchers performed breeding experiments by crossing the wild sugarcane species *Saccharum spontaneum* (clone Djatiroto) with two elite sugarcane clones, CP 62-258 and LCP 85-384. The researchers employed sugarcane microsatellite markers and RAPD markers to confirm the hybrid nature of the progeny. Further Pan (2006) used an SSR marker to identify the progeny with the cytoplasm of wild sugarcane (*S. spontaneum* L.) produced by the cross of commercial-type sugarcane and wild sugarcane (*S. spontaneum* L.).

Molecular markers were developed from the gene expression patterns of sucrose synthase in wild sugarcane. These markers were then utilized to identify genetic variants within the wild species that are linked to higher sugar accumulation. The goal was to identify and incorporate these beneficial genes into commercially grown sugarcane varieties (Medeiros et al. 2020). Yang et al. (2019) employed a genome-wide association study to identify genetic regions associated with resistance to yellow leaf virus and orange rust diseases in sugarcane. A recent study by Zhang et al. (2023a) investigated the genetic basis of yield in sugarcane by analyzing a diverse set of 159 genotypes, including those derived from wild relatives like *Saccharum spontaneum* L., *Tripidium arundinaceum*, and modern sugarcane genotypes. This genome-wide association study yielded significant results, identifying 12 stable QTLs associated with yield-related traits. Importantly, these QTLs harbor 35 candidate genes that may play crucial roles in determining sugarcane yield potential.

Molecular breeding, also referred to as Marker-Assisted Selection (MAS), is a plant breeding approach that leverages DNA markers linked to specific desirable traits to improve the efficiency of the selection process. Molecular breeding relies heavily on identifying and analyzing suitable DNA markers, such as Single Nucleotide Polymorphisms (SNPs), Simple Sequence Repeats (SSRs) and Amplified Fragment Length Polymorphisms (AFLPs) (Cooper et al. 2014). Molecular breeding approaches have been employed to enhance various aspects of sugarcane varieties, including yield, sugar content, maturity, and resistance to both diseases and environmental stresses (Dhansu et al. 2018).

7.8.2 Functional Genomics

Modern sugarcane cultivars are the result of interspecific hybridization between two species, S. *officinarum* ($2n = 8x = 80$; $x = 10$) and wild sugarcane (*S. spontaneum* L.) ($2n = 5x–16x = 40–128$; $x = 8$). Approximately 70–80% of the sugarcane genome originates from *S. officinarum*, with 10–20% coming from wild sugarcane (*S. spontaneum* L.) (Piperidis and D'Hont 2020). The remaining 10% of the genome results from interspecific recombination. This genome is exceptionally large, about ten gigabytes, making it the largest among the Poaceae family members. Additionally, it is highly complex, being aneuploid, polyploid, and heterozygous (Aitken et al. 2016). Most sugarcane cultivars have between 110 and 130 chromosomes. For comparative genomics, sorghum is often used due to its close synteny with sugarcane (Aono et al. 2021). As of February 7, 2024, the National Center for Biotechnology Information (NCBI) database contains a total of 28,217 entries for wild sugarcane (*Saccharum spontaneum* L.). Along with diploid plants, polyploid species, such as sugarcane benefitted from DNA marker applications like the construction of gene tagging and linkage maps (Aitken 2022).

The creation of high-resolution genetic maps facilitates the discovery of potential genes responsible for specific crop traits). In the early 1990s, molecular markers were introduced to sugarcane for linkage mapping and gene tagging utilizing expressed sequence tags (Da Silva 2001; Ming et al. 2002). Multiple molecular maps of sugarcane, with wild sugarcane (*S. spontaneum* L.) as a parent, have been constructed to identify genetic regions (QTLs) associated with sugar yield (Zhang et al. 2023a,b). The genotype SES208, which was mapped, demonstrated polysomic segregation commonly found in autopolyploids. The estimated genome size was 330 cM (Da Silva et al. 1993; Zhang et al. 2023a,b). Through a thorough examination of this genetic map, valuable information regarding chromosome assortment and pairing, as well as chromosome number, was inferred from an evolutionary standpoint. Sharma et al. (2018) generated bacterial artificial chromosome (BAC) libraries from *S. officinarum* variety LA Purple ($2n = 8x = 80$) and wild sugarcane (*S. spontaneum* L.) haploid clone AP85-441 ($2n = 4x = 32$), further 97 BAC clones sequenced. The assembled sequences comprised 5,847,280 base pairs from *S. officinarum* and 5,011,570 base pairs from wild sugarcane (*S. spontaneum* L.). 749 gene models were applied for annotation purposes. Their findings suggest that both *S. officinarum* and wild sugarcane (*S. spontaneum* L.) underwent at least two rounds of independent polyploidization after they diverged from a common ancestor. The researchers, da Silva et al. (1993), successfully created a genetic linkage map for the wild sugarcane species wild sugarcane (*Saccharum spontaneum* L.) ($2n = 8x = 40–128$) using RFLP markers. These markers were obtained from a total of seven DNA libraries, including three sugarcane cDNA libraries, one oat cDNA library, one rice cDNA library, one barley cDNA library, and one sugarcane genomic library. The genetic linkage map consisted of 216 loci distributed across 44 linkage groups, covering a total of 2107 cM. This map accounted for approximately 86% of the sugarcane genome, with an average distance of 25 cM between adjacent markers. To understand the process of polyploidy nature and genome evolution some techniques are applied by researchers.

Shan Yang et al. (2018) have developed a novel approach to distinguish wild sugarcane (*Saccharum spontaneum* L.) and its offspring in the Saccharum genus by analyzing nuclear ribosomal DNA internal transcribed spacer (nrDNA-ITS) sequences and primer for tetra-primer amplification refractory mutation system (ARMS) PCR. Vilela et al. (2017) sought to investigate the process of polyploidization in greater detail within both wild sugarcane (*Saccharum spontaneum*) and cultivated sugarcane (*Saccharum officinarum*). To achieve this, they utilized 27 BACs by utilizing the sugarcane R570 cultivar, which contained the potential single-copy genes of PHYC, LFY, and TOR. Through the analysis of SNPs, amino acid substitutions, and indels, SNPs determination of the origin of the TOR haplotypes became possible, whether they came from *S. officinarum* or *S. spontaneum*. This valuable information allowed the researchers to estimate that these two ancestral sugarcane species diverged approximately 2.5–3.5 million years ago.

7.8.3 Bioinformatics

Advancements in genomics and data analysis have significantly impacted our understanding of how genomes function. Next-generation sequencing technologies have revolutionized genome analysis, enabling detailed investigations into genome structure, gene function, and how organisms respond to environmental challenges. Furthermore, the development of sophisticated bioinformatics tools allows researchers to effectively analyze large datasets, extract meaningful insights, and identify promising genes for crop improvement (López de Maturana et al. 2019). Publicly available genomic datasets have allowed researchers to identify and characterize important genes related to specific traits in wild sugarcane (*Saccharum spontaneum L.*). Li et al. (2021) conducted an *in silico* genomic study of wild sugarcane (*Saccharum spontaneum* L.) to elucidate the characterization of DREBs genes implicated in abiotic stress tolerance and identified Twelve DREB1 genes as well as six DREB2 genes, all exhibited a conserved AP2/ERF domain in their proteins. Expression profiling revealed up-regulation of SsDREB1L, SsDREB1F, SsDREB2F and SsDREB2D under drought and cold conditions, indicating their involvement in both dehydration and cold responses in sugarcane. Through *in silico* characterization of the WRKY transcription factor family in wild sugarcane (*S. spontaneum* L.), 154 WRKY members were identified, and classified into four groups based on phylogenetic relationships and WRKY domain structure. Phylogenetic and collinearity analyses revealed key aspects of the evolutionary history of SsWRKY genes. Notably, these analyses strongly support the hypothesis that gene duplication events have significantly contributed to the expansion of this gene family. Expression analysis indicated their potential roles in development, sugar metabolism, and photosynthesis (Sun et al. 2023).

Yang et al. (2020) investigated the genetic mechanisms that contribute to increased sugar levels and improved cold tolerance in a unique type of wild sugarcane (haploid). In a subsequent study (Yang et al. 2023), they focused on a specific family of genes called hexokinases (SsHXKs) in wild sugarcane. They identified 20 SsHXK genes distributed across seven chromosomes and classified them into three distinct subfamilies. Importantly, they observed that the expression of several SsHXK genes increased significantly when the plants were subjected to cold or drought stress. These findings provide crucial insights into the biological functions of SsHXK genes in wild sugarcane. To further understand how the genetic material inherited from a particular ancestral species contributes to sucrose accumulation in sugarcane, Zhao et al. (2023) conducted a comprehensive analysis of gene expression across the entire transcriptome to understand.

7.9 Tissue Culture

7.9.1 Micropropagation

Tissue culture is utilized to introduce somaclonal variation, enables the modification of genetic traits and the creation of improved sugarcane genotypes compared to their predecessors. The narrow range of genetic material found in commercially grown Indian sugarcane underscores the urgent need to increase its diversity. This can be accomplished by combining traditional breeding techniques with modern approaches. These approaches include utilizing wild relatives of sugarcane, employing tissue culture methods, and genetic engineering technologies employed" (Larkin and Scowcroft 1981).

Tissue culture variants have proven to be more effective than donor clones in increasing disease resistance, sugar yield, and cane yield, sugar yield (Krishnamurthy and Tlaskal 1974; Liu and Chen 1978). In the realm of research, microspores and anther culture techniques are utilized to produce haploid plants. Hinchee and Fitch (1984) explored microspore culture, focusing on the development of calli and procalli through sporophytic development Through the optimization of factors such as sucrose concentration, nitrogen sources, and microspore density, further they discovered specific conditions that favored the development of procalli. Their findings emphasized the significance of manipulating different stages for the successful development of microspores into calli and procalli.

Moderate concentrations of 2,4-D, cool temperatures, high microspore density, and moderately high sucrose concentrations were found to promote microspore viability. While, high concentrations of 2,4-D, low sucrose concentrations, low microspore density, and cool temperatures were beneficial for multicellular microspore development. The best conditions for microspore pro-callus development were low sucrose and low microspore density. Fitch and Moore (1983), conducted an experiment using anther culture to produce haploid plants. They exposed panicle branches to a prolonged cold treatment at 10 °C for 21 days. Their results supported the hypothesis proposed by other scientists that subjecting microspores to stressful conditions, such as cold temperatures and abscisic acid, could trigger a shift from reproductive cell formation to plant structure formation. Interestingly, Fitch and Moore also investigated the outcome of different sucrose levels in the culture medium. They found that high sucrose levels (ranging from 90 to 150 g/l) were necessary for inducing plant structure development without cold treatment. However, excessively high sucrose concentrations did not result in callus formation. The division process of nuclei can be accelerated by cold pretreatment, leading to the formation of structures containing 30 cells in just 7–10 days. The researchers highlighted the significant impact of the developmental stage of microspores on the production of haploid structures, which is influenced by the interaction with cold pretreatment. Despite not knowing the exact reasons for this interaction, it is clear that it plays a crucial role. A study conducted by Fitch and Moore (1984) examined methods to enhance the production of haploid plants from anther cultures of wild sugarcane (*Saccharum spontaneum* L.) plants by subjecting the branches to extended periods of cold treatment, utilizing low salt media during the cold treatment, substituting picloram for 2,4-D in the cultivation process, and maintaining a specific nitrate/ammonium ratio in the liquid culture. The experiments also revealed that anthers cultured in liquid media or placed on liquid media for 1–8 weeks before further cultivation formed visible calli within a 30-day timeframe. By reducing the concentration of picloram, the calli transitioned to a green coloration and eventually developed into plants. Additionally, the researchers determined the ploidy level of the regenerated plants by examining stomata guard cell lengths.

7.9.2 Embryo Rescue

In 2013, Takahashi and Takamizo (2013) successfully utilized a wild sugarcane clone called *Saccharum spontaneum* L. 'Glagah Kloet' as a valuable genetic resource. They aimed to develop high-yielding sugarcane cultivars that could be used for various purposes, such as forage, bioethanol production, and sugar production. Through their observations, they discovered that apical meristems from shoots cultured in a medium containing 0.01 mg/L benzyladenine and 5 mg/ L 2,4-D could generate embryogenic callus. Additionally, they found that the use of 0.3 mg/L gibberellic acid or 3 g/L activated charcoal induced further shoot generation. In (2009), Rajeswari conducted a study aiming to enhance the quality of sugarcane genotypes by utilizing tissue culture methods for inter-species and inter-generic hybridization. The study involved the creation of hybrids across different species and genera, resulting in the generation of 700 subclones through tissue culture methods. These subclones exhibited variations in internode length, fiber content, stalk length, single clump weight, and the amount of pure cane sugar that could be obtained. The research demonstrated that tissue culture techniques can introduce diversity in these traits, ultimately enhancing the profitability of sugarcane cultivation (Godwin et al. 1997; Taylor et al. 1995).

7.9.3 Synthetic Seeds

Novel methods for lowering the expense of storing synthetic sugarcane seeds have been discovered, involving the use of sugarcane bagasse and various additives within the gel matrix. A study was conducted to develop a concise and efficient method for mass propagation and preservation of synthetic seeds. The study found that encapsulation with Na-alginate was effective for synthetic seed conservation.

The highest shoot proliferation was observed with specific concentrations of BA and NAA in the gel matrix. Optimal synthetic seed germination was achieved with a specific concentration of sucrose. ABA supplementation in the synthetic endosperm hindered the regrowth of synthetic seeds, while sugarcane bagasse supplementation led to maximized regrowth while minimizing storage costs. Overall, the developed protocol offers a cost-effective solution for the mass propagation and storage of sugarcane synthetic seeds (Badrelden 2018).

7.10 Genetic Engineering and Gene Editing

Genetic engineering offers a promising approach for enhancing the agricultural potential of sugarcane by introducing desirable features like increased yield, pest resistance, and disease tolerance. This could substantially enhance the efficiency and sustainability of sugarcane cultivation. Furthermore, genetic modification can be utilized to optimize sugarcane for biofuel production. By overexpressing cellulolytic enzymes, which break down cellulose into fermentable sugars, genetically modified sugarcane varieties could yield a higher quantity of second-generation bioethanol. This would potentially reduce production costs and increase the economic viability of biofuel derived from sugarcane (Kumar et al. 2024). Agrobacterium-mediated transformation was used to transform embryogenic calli (Lee et al. 2011; Cui et al. 2020). Transformation cassettes designed for sugarcane typically contain genes that confer resistance to antibiotics (Wang et al. 2015), herbicide resistance genes (Wang et al. 2017), or phosphomannose isomerase (PMI) (Jain et al. 2007). Sugarcane plants genetically modified through particle bombardment exhibit significant genetic instability. Transgenic plants frequently exhibit intricate integration patterns, characterized by the insertion of multiple copies of the transgene into the host genome. This can include tandem insertions or insertions at various locations anywhere in the genome.

While these engineered plants often demonstrate elevated expression levels of transgene in young seedlings, this expression typically diminishes and is eventually silenced in mature plants within the first few ratoon cycles (Birch et al. 2010).

The first successful reports of generating transgenic sugarcane plants emerged in the 1990s (Arencibia et al. 1998; Gallo-Meagher and Irvine 1996). The pioneering achievement of obtaining the first transgenic sugarcane plant can be attributed to the work of Bower and Birch (1992). They successfully optimized the process and managed to generate transgenic sugarcane plants featuring a commercially valuable trait—herbicide resistance (Gallo-Meagher and Irvine 1996). Arencibia et al. (1998), reported the first successful application of this *A. tumefaciens* method in sugarcane. They fine-tuned and adapted this method for sugarcane and introduced the reporter gene gusA, marking a milestone in the field. Shortly thereafter, Enríquez-Obregón et al. (1998) documented their achievement in creating glufosinate-resistant sugarcane through A. *tumefaciens*-mediated transformation. They accomplished this by introducing the bar gene, which encodes a bacterial PPT acetyltransferase capable of conferring resistance to the broad-spectrum herbicide glufosinate-ammonium. These significant advancements in sugarcane transformation techniques paved the way for the development of transgenic sugarcane plants with enhanced traits, holding great promise for agricultural innovation.

7.11 Mutation Breeding

Scientists have used mutation breeding to increase the genetic variation found in sugarcane (Patade and Suprasanna 2008; Snyman et al. 2011). This approach involves the use of tissue culture techniques to induce somaclonal variation, which encompasses both genetic and epigenetic changes. Through the screening of plants following tissue culture, somaclonal variants resistant to eyespot disease (*Helminthosporium sacchari*) have been successfully generated (Larkin and Scowcroft 1983). In certain cases, selection for specific traits, such as resistance to eyespot toxin or smut, has been applied (Patade and Suprasanna 2008; Rodrigues et al. 2001). Furthermore, Mutagenesis in tissue cultures has produced sugarcane plants with traits such as delayed flowering, red-rot resistance, waterlogging tolerance, altered maturity (Patade and Suprasanna 2008), downy mildew resistance, as well as improved yield (Larkin and Scowcroft 1981).

7.11.1 Conventional Mutagenesis

Mutation breeding is a methodology used for enhancing crop characteristics by exposing seeding materials like a seed, vegetative propagules or gametes to physical or chemical mutagens. This approach proves particularly valuable for improving crops that are propagated asexually since mutated traits can be retained through non-sexual reproduction. Within this group of crops, many cultivars possess unique and well-suited combinations of genes for adaptation. However, if these cultivars were subjected to a sexual breeding cycle to enhance specific traits, their distinctive genetic makeup would likely be permanently altered. Sugarcane is highly heterozygous and polyploidy in nature, created through hybridization involving multiple *Saccharum* species. Consequently, conventional breeding programs for sugarcane aimed at enhancing particular desirable traits can be time-consuming due to the intricate genetic makeup of these cultivars. In this area, sugarcane breeders have explored the potential of mutation breeding as an alternative. This method allows for the correction of specific deficiencies or the enhancement of desired traits in commercial cultivars, all without engaging in a sexual reproduction cycle. Furthermore, the capacity to perpetuate the mutant genotype through vegetative propagation offers an additional advantage in the context of sugarcane improvement (Xu et al. 2019).

7.11.2 In Vitro Mutagenesis and Selection

Heinz et al. (1977) observed variations in the susceptibility to eye spot disease caused by *Helminthosporium sacchari* in sugarcane. Many resistant clones were identified, and interestingly, these resistant clones showed no visual differences from the parent plants. To broaden the scope, frequency, and efficiency of selection, different types of chemical and physical mutagens along with in vitro selections were employed. However, over the period, somaclones developed from mutagenesis and showed resistance to eye-spot disease. Balasundaram (1981) conducted a study to enhance the sucrose content and yield in commercial cultivar Co 419 (PoJ 2878 x Co 290). He found the occurrence of higher sucrose content mutant is high compared to cane yield. Two mutants namely 368 and 419/1 consistently have higher sucrose content compared to parent Co 419. In these mutants, researchers observed that there is chromosomal loss from $2n = 110–115$ in the parent to $2n = 80–85$ in the offspring. This suggests that extensive chromosome losses acted as a factor in the origin of these mutants. In sugarcane, smut disease also causes a massive loss in the production. Smut disease resistance can also be produced through mutation. Radhakrishnan (1990) used the CoC 671 variety for gamma-ray treatment for the production of disease-resistant mutants. Irradiated setts were planted up to sixth generations with continuous selection and he developed three mutants that were morphologically superior to the parent concerning juice content, resistant to smut, and, sucrose content. Mohanraj (1996) developed 120 superior mutants resistant to red rot disease from the lineage of CoC 671 and CoJ 64 known for high sucrose content and early maturity.

7.12 Hybridization

In the field of sugarcane breeding, it has been found that hybridization plays a significant role (Wu et al. 2019). In scientific terms, a novel cellular system obtains two sets of chromosomes by merging artificial cells or through the crossing of different species. This procedure undeniably triggers alterations in the genome. The field of somatic cell genetics relies on the deliberate elimination of specific chromosomes. This elimination takes place gradually as cells divide following the fusion of cells from different species (Schulz-Schaeffer 2012). Hybridization is quite common, especially among plants. The emergence of a new species through hybridization (either through diploidy or allopolyploidy) necessitates swift genomic changes to coordinate the genomes (Peterson and Arick 2019). In cases of vegetative propagation, such as in sugarcane, this remodeling process can persist for extended periods during non-sexual cellular divisions.

7.12.1 Conventional Hybridization

The contemporary sugar cane genome is an intricate combination of aneuploidy and polyploidy. It originates from the interbreeding of various Saccharum species, specifically *S. officinarum* and wild sugarcane (*S. Spontaneum* L.). The significance of wild sugarcane (*S. spontaneum* L.) species ($x = 8$, $2n = 40–128$) was recognized following its prosperous crossbreeding with the cultivated sugar-producing species. It is distinguished by its low sugar concentration, slender stalks, abundant fiber, remarkable ability to regenerate, and strong resistance to biotic and abiotic stresses (Singh et al. 2010). The BC1-1 generation was produced by crossing wild sugarcane (*S. spontaneum* L.) and *S. officinarum* F_1 hybrids, with the *S. officinarum* parent exhibiting favorable heterosis in terms of percent CT, percent SCD and percent VBC. Within the BC1-1 population, certain clones displayed promising attributes such as leaf anatomy and stalk weight, making them highly suitable for future clone selections (Wiangwiset et al. 2023). Zhu et al. (2011) crossed wild sugarcane (*Saccharum spontaneum* L.) Yunge F_1, F_2, Yacheng 71-374 (F_2) with commonly used parents and identified maximum heritability contribution was in cross combination (89.6%) also identified as potentially suitable for female parents in the production of high biomass and high cane yield.

7.12.2 Hybrid Cultivars

Hybridization is a widely utilized technique in crop breeding to create new hybrid cultivars by crossing two inbred lines. The resulting hybrid cultivars exhibit superior traits in comparison to their parent lines (Abu-Ellail et al. 2021). Until the late 1800s, the main sources for commercial hybrid cultivars of sugarcane were *S. sinense*, *S. officinarum*, and *S. barberi* (D'Hont et al. 1996). However, today, the majority of commercial hybrid varieties of sugarcane are the outcome of crossbreeding between wild sugarcane (*S. spontaneum* L.) and *S. officinarum* species (Bull and Glasziou 1979). In 2014, Sobhakumari and Dutta created a hybrid by *S. officinarum* clones and wild sugarcane (*S. spontaneum* L.) clones, which were evaluated based on plant height, internode length, number of internodes, stalk diameter, number of millable canes, Brix %, and sucrose %. Hybrids were also resistant to Red rot, IJ 76-501 x IND 81-151 exhibiting the highest percentage of resistant progenies at 63%. In another study, Terajima et al. (2007) explored the potential of wild sugarcane (*S. spontaneum* L.) in breeding programs. They crossed this wild species with other sugarcane varieties to create new hybrid plants. The goal was to develop high-yielding sugarcane clones that could be used for various purposes, such as biofuel production. As a result, the first backcross generation (BC_1) showed a remarkable ability to regrow after harvest (ratooning). While initial yields during the first planting (plant cane) were similar to the standard sugarcane variety (NCo310), the new variety significantly outperformed NCo310 in subsequent regrowths (first and second ratoons). This improvement was substantial, with the dry matter, yields of cane, sugar content, fiber content, and sugar concentration (Brix) in the BC1 variety reaching double that of the reference variety.

7.13 Economic Importance

7.13.1 Cellulase Production

Cellulases are among the most sought-after industrial enzymes worldwide, ranking third in demand. This high demand stems from their remarkable ability to break down cellulose. They have diverse applications across various industries, including pulp and paper production, food processing (starch hydrolysis, animal feed), textile manufacturing, brewing, pharmaceuticals, and biofuel production (bioethanol, alcoholic fermentation) (Cunha et al. 2012; Gusakov 2011). Researchers are constantly striving to improve cellulase production by using more cost-effective materials and optimizing the fermentation process. One promising candidate for cellulase production is wild sugarcane (*S. spontaneum* L.), due to its low lignin content. Cellulases play a crucial role in bioethanol production, but their high production costs pose a challenge. By using affordable lignocellulosic substrates, it is possible to improve the economic viability of the fermentation process (Gottschalk et al. 2010). Cellulases are a diverse group of enzymes, categorized by their unique amino acid sequences and three-dimensional structures (Henrissat 1991). To effectively break down cellulose into glucose, a combination of three key enzymes, exo-1,4-β-glucanase, endo-1,4-β-glucanase, and β-glucosidase—work together synergistically. This glucose is a vital starting material for the production of bioethanol (Bayer et al. 1998; Henrissat 1994). Endo-1,4-b-glucanase (EG), also referred to as endoglucanase, works by breaking the intermolecular b-1,4-glucosidic linkages within cellulose sequences randomly. The reduction in viscosity of carboxymethyl cellulose (CMC) solutions is typically used to measure the activity of endoglucanases. These enzymes decrease viscosity by hydrolyzing the cellulose through both intramolecular and limited cleavages (Teeri et al. 1998). In the field of bioethanol production, it is crucial to industrialize the conversion of lignocellulose using cellulase enzymes. Research is focused on the production of cellulase from pretreated wild sugarcane (*S. spontaneum* L.) using different bacteria under conditions of submerged fermentation (Ahmed et al. 2009). Various parameters, such as pretreatment for delignification, substrate concentration, nutritional content in the growth media, pH, optimum temperature, inoculum concentration, the stimulator (surfactant), and the effect of metal ions, play critical roles in evaluating the activity of cellulase production. In their 2015 study, Iqtedar et al. (2015) explored the use of cellulase produced by *Trichoderma viride* HN1 to break down lignocellulosic biomass derived from wild sugarcane (*Saccharum spontaneum* L.).

They used NaOH for substrate pretreatment, which resulted in a reduction of lignin content through delignification. They found that the maximum cellulase yield occurred with a 6% substrate concentration, reaching 2555 ± 28.39 U/mL/min of carboxymethyl cellulase (CMCase), indicating the optimal conditions for *T. viride* HN1. Initially, the microbes efficiently utilized the abundant nutritional content, leading to a high enzyme yield. However, over time, the enzyme yield decreased due to nutrient depletion (Singh et al. 2009). The production of cellulase was also affected by fluctuations in temperature and variations in the concentration of inoculum. Ghazanfar et al. (2018) aimed to find the best conditions for producing cellulase using Bacillus subtilis K-18. They used wild sugarcane as the starting material and pretreated it using different combinations of sulfuric acid concentration, substrate amount, and treatment time, including steam treatment. To determine the optimal conditions, they employed the Box-Behnken design (BBD), a statistical method, to maximize sugar production during the enzymatic breakdown of the pretreated sugarcane. In another study researcher aimed to maximize cellulase production by using Bacillus species, specifically focusing on the utilization of wild sugarcane (*S. spontaneum* L.). Through physical and chemical mutations induced in Cellulase-producing Bacillus PC-BC6, Bacillus N3 was selected as it exhibited the highest activities of carboxymethyl cellulase (CMCase) and filter paperase (FPase). Among the various media tested for cellulase production, the M_2 medium, which incorporated pretreated wild sugarcane (*S. spontaneum* L.), proved to be the most conducive for the mutant Bacillus N3 to achieve maximum enzyme production. In addition, they examined how different amounts of pretreated wild sugarcane (*S. spontaneum* L.) influenced cellulase production when used as the carbon source for both Bacillus PC-BC6 and its mutant derivative, Bacillus N3. In a separate experiment, *A. terreus* was utilized for cellulase production using wild sugarcane (*S. spontaneum* L.) as the carbon source and solid support in solid-state fermentation. This study demonstrated that wild sugarcane (*S. spontaneum* L.) was effective in supporting cellulase production by *A. terreus* and also reduces the costs of cellulase production at the commercial level (Ilyas et al. 2012). An alternative study focused on optimizing three physical parameters for thermophilic fungi to enhance cellulase production. Through the use of response surface methodology (RSM), *Talaromyces thermophilus* was found to be the most effective cellulase-producing fungus among the thermophilic strains tested. This highlights the significant impact of pH, temperature, and incubation period on cellulase production (Abdullah et al. 2021).

7.13.2 Bioethanol Production

The potential of wild sugarcane (*S. spontaneum* L.) as a sustainable bioresource for second-generation ethanol and environmentally friendly polymer composites is increasingly being acknowledged. This recognition is attributed to its abundance of cellulosic and hemicellulosic components. Microcrystalline cellulose (MCC), a partially depolymerized form of cellulose derived from α-cellulosic sources through the use of mineral acids, exhibits novel physicochemical and mechanical properties. These properties include renewability, low density, non-toxicity, large surface area, biodegradability, high strength, notable moisture absorption, and biocompatibility. MCC has a wide range of applications in various industries, serving as binders, fillers, thickeners, stabilizers, emulsifiers, gelling agents, water retainers, viscosity regulators, and reinforcing agents. Its versatility makes MCC an essential component in pharmaceuticals, food products, ointments, lotions, and polymer composites, as highlighted by Scordia et al. (2010). Baruah et al. (2020) conducted a research study, that employed a highly effective approach to extract Microcrystalline Cellulose (MCC) from wild sugarcane (*S. spontaneum* L.). The process prioritized eco-friendly practices by incorporating acid hydrolysis treatments, alkaline delignification, and chlorine-free bleaching. To optimize the hydrolysis conditions, the Taguchi orthogonal array design was utilized, considering important factors like H2SO4 concentration, pulp-solution ratio, temperature and reaction time. By determining the ideal operational parameters as 5% H2SO4 concentration, 50 °C temperature, and the pulp-to-solution ratio of 1:15 were maintained for 120 minutes, the highest MCC crystallinity, yield, and degradation peak temperature were achieved, with values of 83%, 74.06%, and 338 °C, respectively. This extraction process, which is environmentally friendly due to its lower acid concentration and chlorine-free bleaching, reduces the use of hazardous chemicals. Moreover, the versatility of this method extends to other lignocellulosic sources, providing both economic feasibility and environmental sustainability.

7.13.3 In Removal of Dye and Heavy Metal

The clothing and textile sector has become notorious for its detrimental impact on the environment, characterized by a significant ecological footprint resulting from excessive energy, water, and chemical usage. Additionally, the problem is compounded by textile waste and the release of microfibers during the laundry process, exacerbating environmental challenges (Niinimäki et al. 2020).

Among the various stages, the dyeing and finishing processes play a significant role and account for up to 20% of pollution in industrial wastewater (Morlet et al. 2017). In light of the urgent need for solutions, there is a growing demand for sustainable alternatives that not only mitigate the environmental impact but also reduce overall pollution. Wild sugarcane (*S. spontaneum* L.), with its unique properties, offers a promising solution that has the potential to revolutionize traditional textile practices. By carefully examining different parameters, wild sugarcane (*S. spontaneum* L.) has made significant contributions to the textile industry, particularly in the area of dye removal through parameter fixing.

Wild sugarcane (*S. spontaneum* L.) is frequently utilized as an absorbent material for dye removal or absorption purposes. Several studies have recognized the critical parameters for the efficient elimination of various dyes, including the adjustment of mesh particle size, pH, contact time, adsorbent dosage, and temperature.

Narayanan et al. (2015) investigate the removal of titan yellow dye using wild sugarcane (*Saccharum spontaneum* L.) as an absorbent in batch mode. To achieve optimal removal, the researchers focused on the size of the adsorbent particles, ensuring they were below 63 μm to maximize the surface area for effective adsorption. The suitable range of adsorbent quantity was found to be between 50 and 200 mg, resulting in absorption rates ranging from 55.89% to 93.10% respectively. The researchers identified the critical factor of optimal pH level to be 1.0, with a temperature of 30 °C proving to be the most beneficial. They observed that as the pH levels increased, the dye uptake decreased due to the higher mobility of H+ ions, which favored preferential adsorption of the titan yellow dye at lower pH levels. The maximum contact time for adsorption was determined to be 3 hours, resulting in a 98.45% adsorption rate. Furthermore, the researchers noted that as the concentration of the titan yellow dye increased from 10 to 60 mg/L, the adsorption rate decreased from 94.21% to 58.84%. It is worth mentioning that even in small quantities of dye traces, wild sugarcane (*S. spontaneum* L.) was effectively removed.

Wild sugarcane (*Saccharum spontaneum* L.) has also the potential to eliminate heavy metals and dyes, specifically malachite green dye (MG) and arsenite (As(III)), from water solutions. MG, which is used in aquaculture to combat fungi and bacteria, needs to be removed due to its harmful effects. To minimize environmental impact, regulatory standards limit MG residues in aquaculture products to a maximum of 2 μg/kg, while water should have MG concentrations within the range of 0.5–100 μg/L. Arsenic in irrigation water also poses risks, with the World Health Organization (WHO) recommending its concentration in drinking water to be below 10 μg/L. absorption processes have gained recognition for their cost-effectiveness, efficiency, and simplicity when it comes to MG removal (Wang et al. 2014). To remove MG, wild sugarcane (*Saccharum spontaneum* L.) is utilized to produce Biochar through a gasification process. This process resulted in 250 g of biochar derived from wild sugarcane (BWS), with a reported yield of 15.74%. BWS displayed an alkaline nature with a pH of 10.6 and a specific surface area of 16.03 $m^2\,g^{-1}$. The researchers determined that a contact time of 120 minutes was necessary to achieve sorption equilibrium. BWS exhibited impressive sorption capacities, eliminating 97.46% of MG and 82.36% of arsenite using the pseudo-second-order model in both cases. An examination of functional groups like – C = C, OH, and C = O underscored their significance in the absorption of dye or heavy metals. The research emphasized that BWS exhibits the greatest capacity for absorption among biochars, which can be attributed to a range of characteristics such as zeta potential, cation exchange capacity, functional groups, and, elemental composition (Barría et al. 2023).

7.13.4 Phytoremediation

Phytoremediation is a technique that utilizes hyperaccumulator plants to effectively remove harmful environmental pollutants by harnessing the impressive ability of plants to absorb contaminants while still growing. According to research by Rini and

Hidayati (2021), Saccharum sp. was identified as an excellent choice for remediation due to its resilience in polluted surroundings and its potential for heavy metal accumulation. The plant's tolerance to heavy metals involves several processes, including uptake, extraction, detoxification, chelation, root accumulation, as well as translocation to shoots. An essential enzyme *called γ-glutamylcysteine dipeptidyl transpeptidase*, or *Phytochelatin synthase* (PCS) is vital for heavy metal tolerance. It catalyzes the synthesis of phytochelatins from glutathione, which then bind and sequester heavy metals within the vacuole, thereby minimizing cellular damage (Su et al. 2020).

In wild sugarcane (*Saccharum spontaneum* L.), the accumulation of lead (Pb) can be observed in both the roots and shoots. A study conducted by Rini and Hidayati (2021) revealed that exposing wild sugarcane (*Saccharum spontaneum* L.) plants to Pb treatments ranging from 0 to 400 ppm for 24 hours resulted in higher Pb sequestrate in the roots compared to the shoots. The expression of the *Phytochelatin synthase* (PCS) gene in response to Pb exposure showed a threefold increase.

A recent study conducted by George and Tembhurkar (2022) has delved into the capabilities of wild sugarcane (*S. spontaneum* L.) as a sorbent for biological systems for removing fluoride from contaminated water. They utilized Taguchi methodology to identify the optimal conditions for this process. The results revealed that the most influential parameter was pH, accounting for 56.21% of the effect, followed by initial, temperature (1.01%), biosorbent dose (5.49%), contact time (3.35%), and adsorbate concentration (33.94%).

Furthermore, the study examined the kinetics of the sorption process and found that the Pseudo-second-order kinetic model was the most favorable. The thermodynamic assessment indicated that the process was spontaneous and endothermic, with an ΔH value of 16.268 kJ/mol. Notably, they demonstrated that the biosorbent could be regenerated and reused for up to three cycles, highlighting its potential for repetitive applications in fluoride removal. In summary, these findings highlight the remarkable potential of wild sugarcane (*S. spontaneum* L.) as a natural remediation tool for heavy metal and fluoride pollutants. This research offers promising prospects for a more environmentally friendly and sustainable future.

7.14 Conclusion and Prospects

Notwithstanding the availability of the draft genome sequence of *Saccharum spontaneum* L., the genomics of this species remains insufficiently explored in a comprehensive manner. This limitation restricts its potential utilization in marker-assisted breeding programs and the incorporation of beneficial genetic traits.

However, the genome of *Saccharum spontaneum* L. can serve as a foundation for comparative genomics and the identification of economically significant genes. Core collection not only optimizes resources and enhances conservation research but also safeguards significant components of the World Collection while facilitating efficient utilization of *S. spontaneum* L. Additionally, there is a need for research policies that prioritize the development and utilization of *Saccharum spontaneum*

L., as well as increased awareness of its nutritional importance. The combination of various genomics, genetics, and biotechnology techniques will lead to improving sugarcane yield, quality, and ability to withstand stress. This will help meet the growing demands of the food market as well as bioethanol production. Accurately determining the chromosome number in many sugarcane varieties, especially wild sugarcane (*Saccharum spontaneum* L.) and modern cultivars, remains a significant hurdle. A promising technique to overcome this challenge is fluorescence in situ hybridization (FISH) using centromere-specific probes. Oligo-based FISH probes have provided valuable insights into the complex genomes of modern sugarcane cultivars. These studies have revealed frequent instances of genetic material exchange (recombination) between different species and even within individual chromosomes. Surprisingly, rearrangements involving chromosomes from the *S. officinarum* species are common, suggesting that non-homologous chromosomes often pair up during breeding. Furthermore, recombination events between non-homologous chromosomes from different species have also been observed. However, the true extent of these events might be underestimated due to limited investigations. While accurately counting chromosomes in sugarcane remains difficult, the centromere repeat-based FISH method holds significant potential to address this challenge.

Appendix 7.1 Research institutes and online resources available for *Sacaccharum spontaneum* breeding

Institution	Research activities
Khon Kaen University, Thailand	The objective is to develop germplasm suitable for dual purposes, encompassing both sugar production and bioenergy (biomass) and sugar production production.
Sugarcane Breeding Institute (SBI), Coimbatore, (India)	It is a leading global center for sugarcane research, specializing in breeding sugarcane varieties with desirable traits like high yield, disease resistance, and sugar content.
Woodford Pathology Farm, Sugar Research Australia, Woodford, Queensland, (Australia)	Efforts on developing cultivars resistant to root-knot nematode (*Meloidogyne javanica*), and root-lesion nematode (*Pratylenchus zeae*).
ICAR-Sugarcane Breeding Institute (India)	It offers sugarcane growers a sustainable means to enhance their revenue, contributing to the goal of doubling their income.
Sugarcane Research Institute of Yunnan Academy of Agricultural Sciences	It developed a technique to overcome the challenges of distant hybridization between sugarcane genera.

Appendix 7.2 Genetic resources for wild sugarcane (*Saccharum spontaneum* L.) breeding

Species organization	Important traits function	Cultivation location
China's National Germplasm Repository of Sugarcane (Kaiyuan)	The world's largest repository of sugarcane germplasm resources	China
USDA Sugarcane Research Institute (Miami)	One of the world's centers of sugarcane germplasm resources	United States
Indian Institute of Sugarcane Breeding (Kannur)	One of the world's centers of sugarcane germplasm resources	India

Appendix 7.3 Wild sugarcane (*Saccharum spontaneum* L.) functional and genomic resources available in the public domain

Use for	Database	Website
MicroRNA regulatory networks of homeobox family Genes and expression profiling in Sugarcane *Saccharum spontaneum* L.	Saccharum genome database	sugarcane.zhangjisenlab.cn/sgd/html/download.html
Plant comparative genomics resource	Phytozome v13	phytozome.jgi.doe.gov/pz/portal.html
Annotation	Browser, BLAST Ensembl	plants.ensembl.org
It is dedicated to advancing the technical development of the cane sugar industry and its associated co-products.	International Society of Sugarcane Technologists (ISSCT) Database	issct.org

References

Abdullah R, Tahseen M, Nisar K et al (2021) Statistical optimization of cellulases by *Talaromyces thermophilus* utilizing *Saccharum spontaneum*, a novel substrate. Electron J Biotechnol 51:79–87

Ahmed S, Bashir A, Saleem H et al (2009) Production and purification of cellulose degrading enzymes from filamentous fungus *Trichoderma harzianum*. Pak J Bot 41(3):1411–1419

Aitken K, Farmer A, Berkman P et al (2016) Generation of a 345K sugarcane SNP chip. In Proc Aust Soc Sugar Cane Technol 29:1165–1172

Aitken K, Li J, Piperidis G et al (2018) Worldwide genetic diversity of the wild species *Saccharum spontaneum* and level of diversity captured within sugarcane breeding programs. Crop Sci 58(1):218–229

Aitken KS (2022) History and development of molecular markers for sugarcane breeding. Sugar Tech 24(1):341–353

Aitken KS, Herman S, Karno K et al (2008) Genetic control of yield-related stalk traits in sugarcane. Theor Appl Genet 117:1191–1203

Aitken KS, Li JC, Jackson P et al (2006) AFLP analysis of genetic diversity within *Saccharum officinarum* and comparison with sugarcane cultivars. Aust J Agric Res 57(11):1167–1184

Amalraj VA, Balakrishnan R, Jebadhas AW et al (2006) Constituting a core collection of *Saccharum spontaneum* L. and comparison of three stratified random sampling procedures. Genet Resour Crop Evol 53:1563–1572

Andra S, Muthalagu M, Jeevanandam J, Sekar DD, Ramamoorthy R (2019) Evaluation and development of antibacterial fabrics using Pongamia pinnata extracts. Res J Text Appar 23(3):257-268

Aono AH, Pimenta RJG, Garcia ALB et al (2021) The wild sugarcane and sorghum kinomes: insights into expansion, diversification, and expression patterns. Front Plant Sci 12:668623

Aravindh M, Sathish S, Ranga Raj R et al (2022) A review on the effect of various chemical treatments on the mechanical properties of renewable fiber-reinforced composites. Adv Mater Sci Eng 2022(1):1–24

Arencibia AD, Carmona ER, Tellez P et al (1998) An efficient protocol for sugarcane (*Saccharum* spp. L.) transformation mediated by *Agrobacterium tumefaciens*. Transgenic Res 7:213–222

Ari A, Karahan M, Ahmed HAM et al (2023) A review of cellulosic natural fibers' properties and their suitability as reinforcing materials for composite panels and applications. AJOR 10(3):163–183

Arruda P (2001) Sugarcane transcriptome: a landmark in plant genomics in the tropics. Genet Mol Biol 24:1–4

Artschwager E (1942) A comparative analysis of vegetative characteristics of 'some variants of *Saccharum spontaneum*. In: Asian-Pacific Weed Science Society Conference, Asian Pacific Weed Science Society, vol 1, pp 87–91

Badrelden A (2018) New approaches for reducing the cost of the synthetic seeds storage using sugarcane bagasse and different additives to the gel matrix for sugarcane plant: in vitro. Egypt J Bot 58(1):1–10

Balasundaram N (1981) Yield and quality of induced mutants in sugarcane. Indian J Agric Sci 51:1–4

Barber CA (1916) The classification of indigenous Indian canes. Agric J India 11(4):371–376

Barreto FZ, Rosa JRBF, Balsalobre TWA et al (2019) A genome-wide association study identified loci for yield component traits in sugarcane (Saccharum spp.). PLoS One 14(7):e0219843

Barría Y, Burbano A, James A et al (2023) Sorption capacity of biochars obtained by gasification of rice husks and wild sugarcane: removal of malachite green and arsenic from water solutions. Biomass Convers Biorefinery:1–13. https://doi.org/10.1007/s13399-023-04325-3

Baruah J, Deka RC, Kalita E (2020) Greener production of microcrystalline cellulose (MCC) from *Saccharum spontaneum* (Kans grass): statistical optimization. Int J Biol Macromol 154:672–682

Basumatary SF, Das B, Brahma S, Basumatary S (2025) Musa ABB (Kachkal) banana waste derived heterogeneous nanocatalyst for transesterification of binary oil mixture of Jatropha curcas and Pongamia pinnata tobiodiesel. Bioresour Technol Rep 29:102018

Bayer EA, Chanzy H, Lamed R et al (1998) Cellulose, cellulases and cellulosomes. Curr Opin Struct Biol 8(5):548–557

Belide S, Sajjalaguddam RR, Paladugu A (2010) Cytokinin preconditioning enhances multiple shoot regeneration in Pongamia pinnata (L.) Pierre-a potential, non-edible tree seed oil source for biodiesel. Electron J Biotechnol 13(6):3–4

Berding N, Roach BT (1987) Germplasm collection, maintenance, and use. In: Heinz DJ (ed) Sugarcane improvement through breeding. Elsevier, New York, pp p143–p210

Besse P, Taylor G, Carrol B et al (1998) Assessing genetic diversity in a sugarcane germplasm collection using an automated AFLP analysis. Genetica 104:143–153

Bhuiyan SA, Garlick K, Piperidis G (2019) *Saccharum spontaneum*, a novel source of resistance to root-lesion and root-knot nematodes in sugarcane. Plant Dis 103(9):2288–2294

Birch RG, Bower RS, Elliott AR (2010) Highly efficient, 5′-sequence-specific transgene silencing in a complex polyploid. Trop Plant Biol 3:88–97

Bischoff KP, Gravois KA (2004) The development of new sugarcane varieties at the LSU AgCenter. J Am Soc Sugar Cane Technol 24:142–164

Bower R, Birch RG (1992) Transgenic sugarcane plants via microprojectile bombardment. Plant J 2(3):409–416

Bremer G (1961) Problems in breeding and cytology of sugar cane. Euphytica 10:59–78

Bull TA, Glasziou KT (1979) Sugarcane. Chapter 4. In: Lovett JV, Lazenby A (eds) Australian field crops. Angus and Robertson Publishers, pp 95–113

Burner DM (1997) Chromosome transmission and meiotic behavior in various sugarcane crosses. J Am Soc Sugar Cane Tech 17:38–50

Burner DM, Legendrc BL (1993) Sugarcane genome amplification for the subtropics: a twenty-year effort. Sugar Cane 3:5–10

Burner DM, Pan YB, Webster RD (1997) Genetic diversity of North American and Old World Sacchari.4m assessed by RAPD analysis. Genet Resour Crop Evol 44:235–240

Buzacott JH (1965) Cane varieties and breeding. In: Mungomery RW, Huges C (eds) Kim NJ. Manual of cane growing, Sydney, pp 220–253

CABI Compendium. https://doi.org/10.1079/cabicompendium.48162

Chandran K (2010) In vitro multiplication and conservation of Saccharum germplasm. IJPGR 23(01):65–68

Chandran S, Raghu AV, Mohanan KV (2023) In vitro conservation of rare, endangered, and threatened plants. In: Conservation and sustainable utilization of bioresources. Springer, Singapore, pp 391–408

Chaudhary N, Salgotra RK, Chauhan BS (2023) Genetic enhancement of cereals using genomic resources for nutritional food security. Genes 14(9):1770

Clayton WD, Renvoize SA (1982) Flora of tropical East Africa. Graminea (Part 3). A.A. Balkema, Rotterdam, The Netherlands, p 448

Cooper M, Messina CD, Podlich D et al (2014) Predicting the future of plant breeding: complementing empirical evaluation with genetic prediction. Crop Pasture Sci 65(4):311–336

Cordeiro GM, Pan Y-B, Henry RJ (2003) Sugarcane microsatellites for the assessment of genetic diversity in sugarcane germplasm. Plant Sci 165:181–189

Cordeiro GM, Taylor GO, Henry RJ (2000) Characterisation of microsatellite markers from sugarcane (*Saccharum* sp.), a highly polyploid species. Plant Sci 155:161–168

Cox MC, Hogarth M, Smith G (2020) Cane breeding and improvement. In: Hogarth, Allsopp (eds) Manual of cane growing. Bureau of Sugar Experiment Stations, pp 91–108

Cuadrado A, Acevedo R, Díaz M, de la Espina S (2004) Genome remodelling in three modern *S. officinarum*× *S. spontaneum* sugarcane cultivars. J Exp Bot 55(398):847–854

Cui ML, Liu C, Piao CL et al (2020) A stable agrobacterium rhizogenes-mediated transformation of cotton (Gossypium hirsutum L.) and plant regeneration from transformed hairy root via embryogenesis. Front Plant Sci 11:604255

Cunha FM, Esperança MN, Zangirolami TC et al (2012) Sequential solid-state and submerged cultivation of *Aspergillus Niger* on sugarcane bagasse for the production of cellulase. Bioresour Technol 112:270–274

Cursi DE, Hoffmann HP, Barbosa GVS et al (2022) History and current status of sugarcane breeding, germplasm development and molecular genetics in Brazil. Sugar Tech 24(1):112–133

da Silva JA (2017) The importance of the wild cane *Saccharum spontaneum* for bioenergy genetic breeding. Sugar Tech 19(3):229–240 da Silva JAG (2001) Preliminary analysis of microsatellite markers derived from sugarcane expressed sequence tags (ESTs). Genet Mol Biol 24(1–4):155–159

Da Silva JAG, Sorrells ME, Burnquist WL et al (1993) RFLP linkage map and genome analysis of *Saccharum spontaneum*. Genome 36:782–791

Dalvi SG, Tawar PN, Suprasanna P et al (2021) MS-based in vitro mutagenesis and mutant screening for smut resistance with agronomic traits in sugarcane. Sugar Tech 23:854–864

Daniels J, Roach BT (1987) Taxonomy and evolution. In: Heinz DJ (ed) Sugarcane improvement through breeding. Elsevier, New York, pp 7–84

Das Bhowmik SS, Brinin AK, Williams B et al (2016) Sugarcane biotechnology: tapping unlimited potential. In: Sugarcane-based biofuels and bioproducts. Wiley, pp 23–52

Datta S, Banerjee A (1973) Weight and number of weed seeds. In: Proceedings of the 4th

Devnani GL, Sinha S (2019) Effect of nanofillers on the properties of natural fiber reinforced polymer composites. Mater Today Proc 18(3):647–654

Dhansu P, Kumar A, Mann A et al (2018) Insights into biotechnological interventions for sugarcane improvement. In: Biotechnology to enhance sugarcane productivity and stress tolerance. CRC Press, pp 115–136

D'Hont A, Grivet L, Feldmann P et al (1996) Characterisation of the double genome structure of modern sugarcane cultivars (Saccharum spp.) by molecular cytogenetics. MGG 250:405–413

Doule RB, Balasundaram N (2006) Relative efficiency of cane characters in selection for cane yield in sugarcane (Saccharum spp. hybrid). Indian J Genet Plant Breed 66(04):347–348

Dutt NL (1953) The breeding of sugarcane. In: Proceedings/Indian Academy of Sciences, vol 38(2). Springer, New Delhi, pp 81–88

Enríquez-Obregón GA, Vázquez-Padrón RI, Prieto-Samsonov DL et al (1998) Herbicide-resistant sugarcane (*Saccharum officinarum* L.) plants by Agrobacterium-mediated transformation. Planta 206:20–27

Fitch MM, Moore PH (1983) Haploid production from anther culture of *Saccharum spontaneum* L. J Plant Physiol 109(3):197–206

Fitch MM, Moore PH (1984) Production of haploid *Saccharum spontaneum* L.-comparison of media for cold incubation of panicle branches and for float culture of anthers. J Plant Physiol 117(2):169–178

Gallo-Meagher M, Irvine JE (1996) Herbicide resistant transgenic sugarcane plants containing the bar gene. Crop Sci 36(5):1367–1374

Garsmeur O, Droc G, Antonise R et al (2018) A mosaic monoploid reference sequence for the highly complex genome of sugarcane. Nat Commun 9(1):2638. https://doi.org/10.1038/s41467-018-05051-5

Gauchan D, Smale M (2003) Choosing the 'right' tools to assess the economic costs and benefits of growing landraces: an example from Bara District, Central Terai, Nepal. Plant Genet Res Newsl 134:18–25

George AM, Tembhurkar AR (2022) Taguchi L16 orthogonal array approach for optimization of fluoride removal from aqueous solution using *Saccharum spontaneum* weed grass novel biosorbent. Sustain Chem Pharm 30:100833

Ghazanfar M, Irfan M, Tabssum F et al (2018) Effect of different pretreatment conditions on *Saccharum spontaneum* for cellulase production by *B. subtilis* K-18 Through Box–Bhenken design. Iran J Sci Technol Trans A: Sci 42:313–320

Godwin ID, Sangduen N, Kunanuvatchaidach R et al (1997) RAPD polymorphisms among variant and phenotypically normal rice *Oryza sativa* var indica somaclonal progenies. Plant Cell Rep 16:320–324

Gottschalk LMF, Oliveira RA, Da SilvaBon EP (2010) Cellulases, xylanases, β-glucosidase and ferulic acid esterase produced by Trichoderma and Aspergillus act synergistically in the hydrolysis of sugarcane bagasse. Biochem Eng J 51(1–2):72–78

Guden B, Qureshi M, Wang et al (2024) Molecular breeding to design ideal sesame crop. In: Allele mining for genomic designing of oilseed crops. CRC Press, pp 217–238

Gusakov AV (2011) Alternatives to *Trichoderma reesei* in biofuel production. Trends Biotechnol 29(9):419–425

Hale AL, Dufrene EO, Tew TL et al (2013) Registration of 'Ho02-113'sugarcane. J Plant Regist 7(1):51–57

Hassan A, Malik K, Naqvi SAM et al (2024) A comprehensive review of *Saccharum spontaneum*, its traditional uses, phytochemistry and pharmacology. Ethnobot Res Appl 29:1–13

Heinz DJ, Krishnamurthi M, Nickell LG et al (1977) Cell, tissue and organ culture in sugarcane improvement. In: Applied and fundamental aspects of plant cell, tissue and organ culture. Springer, pp 1–17

Henrissat B (1991) A classification of glycosyl hydrolases based on amino acid sequence similarities. Biochem J 280(2):309–316

Henrissat B (1994) Cellulases and their interaction with cellulose. Cellulose 1:169–196

Hinchee MA, Fitch MM (1984) Culture of isolated microspores of *Saccharum spontaneum*. Z Pflanzenphysiol 113(4):305–314

Hoarau JY, Grivet L, Offman B et al (2002) Genetic dissection of a modern sugarcane cultivar (*Saccharum* spp.). II. Detection of QTLs for yield components. Theor Appl Genet 105:1027–1037

Hoarau JY, Offmann B, D'Hont A et al (2001) Genetic dissection of a modern sugarcane cultivar (Saccharum spp.). I. Genome mapping with AFLP markers. Theor Appl Genet 103:84–97

Holm L, Doll J, Holm E et al (1997) World weeds: natural histories and distribution. Wiley

Hotta CT, Lembke CG, Domingues DS et al (2010) The biotechnology roadmap for sugarcane improvement. Trop Plant Biol 3:75–87

Huang J, Khan MT, Perecin D et al (2020) Sugarcane for bioethanol production: potential of bagasse in Chinese perspective. Renew Sust Energ Rev 133:110296

Ilyas U, Ahmed S, Majeed A et al (2012) Biohydrolysis of *Saccharum spontaneum* for cellulase production by *Aspergillus terreus*. Afr J Biotechnol 11(21):4914–4920

Iqtedar M, Nadeem M, Naeem H et al (2015) Bioconversion potential of Trichoderma viride HN1 cellulase for a lignocellulosic biomass Saccharum spontaneum. Nat Product Res 29(11):1012–1019

Jain M, Chengalrayan K, Abouzid A et al (2007) Prospecting the utility of a PMI/mannose selection system for the recovery of transgenic sugarcane (Saccharum spp. hybrid) plants. Plant Cell Rep 26:581–590

Jordan DR, Casu RE, Besse P et al (2004) Markers associated with stalk number and suckering in sugarcane with tillering and rhizomatousness QTLs in sorghum. Genome 47:988–993

Kaith BS, Jindal R, Jana AK et al (2010) Development of corn starch based green composites reinforced with Saccharum spontaneum L fiber and graft copolymers–evaluation of thermal, physico-chemical and mechanical properties. Bioresour Technol 101(17):6843–6851

Kaith BS, Jindal R, Jana AK et al (2009) Ferrous-persulphate induced graft copolymerization of monomer mixtures onto *Saccharum spontaneum*-L natural fibre. Iranian Polymer J 18:789–800

Karmee SK, Chadha A (2005) Preparation of biodiesel from crude oil of Pongamia pinnata. Bioresour technol 96(13):1425–1429

Karp SG, Burgos WJ, Vandenberghe LP et al (2022) Sugarcane: a promising source of green carbon in the circular bioeconomy. Sugar Tech 24(4):1230–1245

Khoury CK, Heider B, Castañeda-Álvarez NP et al (2015) Distributions, ex situ conservation priorities, and genetic resource potential of crop wild relatives of sweetpotato (*Ipomoea batatas* (L.) Lam., I. series Batatas). Front Plant Sci 6:251

Krishnamurthy M, Tlaskal J (1974) Figi disease resistant *Saccharum officinarum* var. Pindar subclones from tissue cultures. Proc Congr Int Soc Sugar Cane Technol South Africa 15:130–137

Kumar T, Wang JG, Xu CH et al (2024) Genetic engineering for enhancing sugarcane tolerance to biotic and abiotic stresses. Plants 13(13):1739

Laitala K, Klepp I, Henry BK (2018) Does use matter? Comparison of environmental impacts of clothing based on fiber type. Sustainability 10:2524. https://doi.org/10.3390/su10072524

Lakshmanan P, Geijskes RJ, Aitken KS et al (2005) Sugarcane biotechnology: the challenges and opportunities. In Vitro Cell Dev Biol Plant 41:345–363
Larkin PJ, Scowcroft WR (1983) Somaclonal variation and crop improvement. In: Genetic engineering of plants: an agricultural perspective. Boston, MA, Springer, pp 289–314
Larkin PJ, Scowcroft WR (1981) Somaclonal variation- a novel source of variability from cell cultures for plant improvement. Theor Appl Genet 60:197–214
Lee KW, Kim KY, Kim KH et al (2011) Development of antibiotic marker-free creeping bentgrass resistance against herbicides. Acta Biochim Biophys Sin 43(1):13–18
Li Z, Wang G, Liu X et al (2021) Genome-wide identification and expression profiling of *DREB* genes in *Saccharum spontaneum*. BMC Genomics 22:456
Liu MC, Chen WH (1978) Tissue and cell culture as aids to sugarcane breeding. II. Performance and yield potential of callus derived lines. Euphytica 27(1):273–282
López de Maturana E, Alonso L, Alarcón P et al (2019) Challenges in the integration of omics and non-omics data. Genes 10(3):238
Lu G, Liu P, Wu Q et al (2024) Sugarcane breeding: a fantastic past and promising future driven by technology and methods. Front Plant Sci 15:1375934
Lu YH, D'Hont A, Walker DIT et al (1994) Relationships among ancestral species of sugarcane revealed with RFLP using single copy maize nuclear probes. Euphytica 78:7–18
Luo D, Li Y, Zhao Q et al (2019) Highly resolved phylogenetic relationships within order acipenseriformes according to novel nuclear markers. Genes 10:38
Manners JM, Casu RE (2011) Transcriptome analysis and functional genomics of sugarcane. Trop Plant Biol 4(1):9–21
McIntyre CL, Jackson PA (2001) Low level of selfing found in a sample of crosses in Australian sugarcane breeding programs. Euphytica 117(3):245–249
McIntyre L, Jackson P (1995) Does selfing occur in sugarcane? In: Plant Animal Genome, V Conference, vol P165
Medeiros C, Balsalobre TWA, Carneiro MS (2020) Molecular diversity and genetic structure of Saccharum complex accessions. PLoS One 15(5):e0233211
Melloni MLG, Scarpari MS, de Mendonça JR et al (2013) Comparison of two staining methods for pollen viability studies in sugarcane. Sugar Tech 15:103–107
Meng Z, Zhang Z, Yan T et al (2018) Comprehensively characterizing the cytological features of *Saccharum spontaneum* by the development of a complete set of chromosome-specific oligo probes. Front Plant Sci 9:1624
Midmore DJ (1980) Effects of photoperiod on flowering and fertility of sugarcane (Saccharum spp.). Field Crop Res 3:65–81
Ming R, Moore PH, Wu KK et al (2006) Sugarcane improvement through breeding and biotechnology. Plant Breed Rev 27:15–118
Ming R, Wang Y, Draye X et al (2002) Molecular dissection of complex traits in autopolyploids: mapping QTLs affecting sugar yield and related traits in sugarcane. Theor Appl Genet 105:332–345
Mirajkar SJ, Devarumath RM, Nikam AA et al (2019) Sugarcane (*Saccharum* spp.)*:* Breeding and genomics. In: Al-Khayri JM, Jain SM, Johnson DV (eds) Advances in plant breeding strategies: Industrial and food crops. Springer, pp 363–406
Mohanraj D (1996) Evaluation of red rot toxin resistant somaclones for disease reaction. Sugarcane Breeding Institute Annual Report 1995–96:32
Moore PH, Nuss KJ (1987) Flowering and flower synchronization. In: Developments in crop science, vol 11. Elsevier, pp 273–311
Morlet A, Opsomer R, Herrmann S et al (2017) A new textiles economy: redesigning fashion's future. Ellen MacArthur Foundation
Mudge J, Andersen WR, Kehrer RL et al (1996) A RAPD genetic map of *Saccharum officinarum*. Crop Sci 36(5):1362–1366

Narayanan AL, Dhamodaran M et al (2015) Thermodynamics and kinetics of adsorption of azo dye titan yellow from aqueous solutions on natural plant material *Saccharum spontaneum*. Der Pharma Chemica 7(8):36–45

Niinimäki K, Peters G, Dahlbo H et al (2020) The environmental price of fast fashion. Nat Rev Earth Environ 1(4):189–200

Orton T, Ayeni A (2022) Specialty crop germplasm and public breeding efforts in the United States. Agronomy 12(2):239

Pan YB (2006) Highly polymorphic microsatellite DNA markers for sugarcane germplasm evaluation and variety identity testing. Sugar Tech 8(4):246–256

Pan YB, Burner DM, Wei Q (2001) Developing species-specific DNA markers to assist in sugarcane breeding. Proc Int Soc Sugar Cane Technol 24:337–342

Pan YB, Cordeiro GM, Richard EP Jr et al (2003) Molecular genotyping of sugarcane clones with microsatellite DNA markers. Maydica 48:319–329

Pan YB, Wei Q, Cordeiro GM et al (2004) New Saccharum hybrids in *S. spontaneum* cytoplasm developed through a combination of conventional and molecular breeding approaches. Genet Res 2(2):131–139

Pancho JV (1964) Seed sizes and production capacities in common weed species of the rice fields of The Philippines. Philipp Agric 48:307–316

Pandotra P, Gupta S (2015) Biotechnological approaches for conservation of plant genetic resources and traditional knowledge. In: Plant genetic resources and traditional knowledge for food security. Springer, pp 121–135

Panje RR (1970) The evolution of a weed. PANS Pest Artic News Sum 16(4):590–595

Panje RR, Babu CN (1960) Studies on *Saccharum spontaneum*. Distribution and geographical association of chromosome numbers. Cytologia 25:152–172

Patade VY, Suprasanna P (2008) Radiation induced in vitro mutagenesis for sugarcane improvement. Sugar Tech 10:14–19

Paterson AH, Lan T-H, Reischmann KP et al (1996) Toward a unified map of higher plant chromosomes, transcending the monocot-dicot divergence. Nat Genet 14:380–382

Pathy TL, Mohanraj K, Alarmelu S et al (2022) Population structure of Coimbatore canes developed in a century of sugarcane breeding in India. Sugar Tech 24(5):1449–1460

Peterson DG, Arick M (2019) Sequencing plant genomes. In: Progress in botany, vol 80, pp 109–193

Piperidis G, Christopher MJ, Carroll BJ et al (2000) Molecular contribution to selection of intergeneric hybrids between sugarcane and the wild species *Erianthus arundinaceu*.s. Genome 43:1033–1037

Piperidis G, Piperidis N, D'Hont A (2010) Molecular cytogenetic investigation of chromosome composition and transmission in sugarcane. Mol Genet Genomics 284(1):65–73

Piperidis N, D'Hont A (2020) Sugarcane genome architecture decrypted with chromosome-specific oligo probes. Plant J 103(6):2039–2051

Pompidor N, Charron C, Hervouet C et al (2021) Three founding ancestral genomes involved in the origin of sugarcane. Ann Bot 127(6):827–840

Radhakrishnan CM (1990) Sugarcane smut disease resistant mutants of CoC 671. Sugarcane Breeding Institute Annual Report 1989-90:31

Rajeswari S, Thirugnanakumar S, Anandan A et al (2009) Somaclonal variation in sugarcane through tissue culture and evaluation for quantitative and quality traits. Euphytica 168:71–80

Rini DS, Hidayati N (2021) Analysis of metal accumulation and Phytochelatin Synthase (PCS) gene expression in *Saccharum spontaneum* L. as regards its function for remediation of lead (Pb) contamination. IOP Conf Ser Earth Environ Sci 905(1):012117

Roach BT (1989) Origin and improvement of the genetic base of sugarcane. Proc Aust Soc Sugar Cane Technol 11:34–47

Rodrigues RDCLB, Felipe MDGDA, Silva JB et al (2001) The influence of pH, temperature and hydrolyzate concentration on the removal of volatile and nonvolatile compounds from sugarcane bagasse hemicellulosic hydrolyzate treated with activated charcoal before or after vacuum evaporation. Braz J Chem Eng 18:299–311

Roxburgh W (1971) Flora indica on description of Indian plants. Today & Tomorrow's Printers & Publishers, New Delhi

Santos RMA, Torres MDD (2019) Talahib (*Saccharum spontaneum* Linn.) as alternative to commercial cotton yarn. In: IEEE Integrated STEM Education Conference (ISEC). IEEE, pp 23–25

Sarma AK, Konwer D, Bordoloi PK (2005) A comprehensive analysis of fuelproperties of biodiesel from Koroch seed oil. Ener Fuel 19:656657

Schnell RJ, Tai PYP, Miller JD (1997) History and current status of the World Collection of sugarcane and related grasses maintained at National Germplasm Repository, Miami, Florida. Sugar Cane 1:15–17

Schulz-Schaeffer J (2012) Cytogenetics: plants, animals, humans. Springer

Scordia D, Cosentino SL, Jeffries TW (2010) Second generation bioethanol production from *Saccharum spontaneum* L. ssp. aegyptiacum (Willd.) Hack. Bioresour Technol 101(14):5358–5365

Shanthi RM, Vigneshwari R, Hemaprabha G et al (2023) Sucrose-specific genic microsatellites to analyse the genetic structure among the commercial hybrids and clones of interspecific and intergeneric origin in Saccharum complex. Sugar Tech 25(1):185–197

Sharma A, Song J, Lin Q et al (2018) Comparative analysis of homologous sequences of *Saccharum officinarum* and *Saccharum spontaneum* reveals independent polyploidization events. Front Plant Sci 9:1414

Sharma S, Tiagi B (1979) Flora of North East Rajasthan. Kalyani Publishers, New Delhi

Shearman JR, Pootakham W, Sonthirod C et al (2022) A draft chromosome-scale genome assembly of a commercial sugarcane. Sci Rep 12(1):20474

Shrivastava AK, Srivastava S (2016) Diversity of the germplasm of Saccharum species and related genera available for use in directed breeding programmes for sugarcane improvement. Curr Sci 111:475–482

Singh A, Singh N, Bishnoi R (2009) Production of Cellulase by *Aspergillus heteromorphus* from wheat straw under submerged fermentation. Int J Env Sci Technol 1:23–26

Sleper DA, Poehlman JM (2006) Breeding field crops, 5th edn. Wiley, p 424

Snyman SJ, Meyer GM, Koch AC et al (2011) Applications of in vitro culture systems for commercial sugarcane production and improvement. In Vitro Cell Dev Biol Plant 47:234–249

Song J, Zhang X, Jones T et al (2024) Identification of male sterility-related genes in *Saccharum officinarum* and *Saccharum spontaneum*. Plant Reprod 37:1–18

Sreenivasan J, Sreenivasan TV (1984) *In vitro* propagation of *Saccharum officinarum* (L.) and *Sclerostachya fusca* (Roxb.) A. Camus hybrid Theor Appl Genet 67:171–174

Sreenivasan TV (1975) Cytogenetical studies in *Saccharum spontaneum*. In: Proceedings/Indian Academy of Sciences, vol 81(3). Springer, New Delhi, pp 131–144

Sreenivasan TV, Ahloowalia BS, Heinz DJ (1987) Cytogenetics. In: Heinz DJ (ed) Sugarcane improvement through breeding. Elsevier, Holland

Srivastava PKS (2020) Omics approaches for sugarcane crop improvement. IJST 35(02):1–4

Su H, Zou T, Lin R et al (2020) Characterization of a phytochelatin synthase gene from Ipomoea pes-caprae involved in cadmium tolerance and accumulation in yeast and plants. Plant Physiol Biochem 155:743–755

Sun T, Chen Y, Feng A et al (2023) The allene oxide synthase gene family in sugarcane and its involvement in disease resistance. Ind Crop Prod 192:116136

Tai PYP, Miller JD (2001) A core collection for *Saccharum spontaneum* L. from the world collection of sugarcane. Crop Sci 41(3):879–885

Tai PYP, Miller JD (2002) Germplasm diversity among four sugarcane species for sugar composition. Crop Sci 42(3):958–964

Tai PYP, Miller JD, Legendre BL (1994) Preservation of *Saccharum spontaneum* germplasm through storage of true seeds. Sugar Cane 6:3–8

Takahashi W, Takamizo T (2013) Plant regeneration from embryogenic calli of the wild sugarcane (*Saccharum spontaneum* L.) Clone Glagah Kloet. Bull NARO Inst Livest Grassl Sci 13:23–32

Taylor PWJ, Geijskes JR, Ko HL et al (1995) Sensitivity of random amplified polymorphic DNA analysis to detect genetic change in sugarcane during tissue culture. Theor Appl Genet 90:1169–1173. https://doi.org/10.1007/BF00222939

Teeri TT, Koivula A, Linder M et al (1998) *Trichoderma reesei* cellobiohydrolases: why so efficient on crystalline cellulose? Biochem Soc Trans 26:173–178

Terajima Y, Matsuoka M, Irei S et al (2007) Breeding for high biomass sugarcane and its utilization. Japan Proc Int Cong Sugarcane Technol 26:759–763

Tyagi CH, Dutta D (2007) Studies on pulp and paper making characteristics of *Saccharum spontaneum*. IPPTA 19(2):109

Verma DK (2001) Resistance breeding in sugarcane (Saccharum officinarum L.). Indian J Genet Plant Breed 61(03):256–262

Vilela MDM, Del Bem LE, Van Sluys MA et al (2017) Analysis of three sugarcane homo/homeologous regions suggests independent polyploidization events of *Saccharum officinarum* and *Saccharum spontaneum*. Genome Biol Evol 9(2):266–278

Wang H, Yuan X, Zeng G et al (2014) Removal of malachite green dye from wastewater by different organic acid-modified natural adsorbent: kinetics, equilibriums, mechanisms, practical application, and disposal of dye-loaded adsorbent. Environ Sci Pollut Res 21:11552–11564

Wang K, Zhang H, Khurshid H et al (2023) Past and recent advances in sugarcane cytogenetics. Crop J 11(1):1–8

Wang W, Yang B, Cai W et al (2015) Establishment of mannose selection system in sugarcane transformation. Biotechnol Bull 31(1):92–97

Wang WZ, Yang BP, Feng XY et al (2017) Development and characterization of transgenic sugarcane with insect resistance and herbicide tolerance. Front Plant Sci 8:1535

Wu J, Wang Q, Xie J et al (2019) SSR marker-assisted management of parental germplasm in sugarcane (Saccharum spp. hybrids) breeding programs. Agronomy 9(8):449. https://doi.org/10.3390/agronomy9080449

Xu F, He L, Gao S et al (2019) Comparative analysis of two sugarcane ancestors *Saccharum officinarum* and *S. spontaneum* based on complete chloroplast genome sequences and photosynthetic ability in cold stress. Int J Mol Sci 20(15):3828

Yang N, Zhou S, Wang Y et al (2023) Downscaled satellite solar-induced chlorophyll fluorescence detects the early response of sugarcane to drought stress in a major sugarcane-planting region of China. Remote Sens Environ 15(16):3937. https://doi.org/10.3390/rs15163937

Yang S, Li X, Huang F et al (2018) A new method based on SNP of nrDNA-ITS to identify *Saccharum spontaneum* and its progeny in the genus Saccharum. PLoS One 13(5):e0197458

Yang X, Luo Z, Todd J et al (2020) Genome-wide association study of multiple yield traits in a diversity panel of polyploid sugarcane (Saccharum spp.). TPG 13(1):e20006

Yang X, Sood S, Luo Z et al (2019) Genome-wide association studies identified resistance loci to orange rust and yellow leaf virus diseases in sugarcane (Saccharum spp.). Phytopathology 109(4):623–631

Zhang B, Huang Y, Zhang L et al (2023a) Genome-wide association study unravels quantitative trait loci and genes associated with yield-related traits in sugarcane. J Agric Food Chem 71(44):16815–16826

Zhang J, Zhang X, Tang H et al (2018) Allele-defined genome of the autopolyploid sugarcane Saccharum spontaneum L. Nat Genet 50:1565–1573. https://doi.org/10.1038/s41588-018-0237-2

Zhang X, Meng Z, Han J et al (2023b) Characterization of meiotic chromosome behavior in the autopolyploid *Saccharum spontaneum* reveals preferential chromosome pairing without distinct DNA sequence variation. Crop J 11(5):1550–1558

Zhao J, Li S, Xu Y et al (2023) The subgenome *Saccharum spontaneum* contributes to sugar accumulation in sugarcane as revealed by full-length transcriptomic analysis. J Adv Res 54:1–13

Zhu YJ, Mc Cafferty H, Osterman G et al (2011) Genetic transformation with untranslatable coat protein gene of sugarcane yellow leaf virus reduces virus titers in sugarcane. Transgenic Res 20:503–512. https://doi.org/10.1007/s11248-010-9432-3

Chapter 8
Modern Breeding Strategies and Tools in Bread Wheat (*Triticum aestivum* L.)

Khaled F. M. Salem and Nerhan A. Eid

Abstract One of the most significant cereal crops is self-pollinating bread wheat (*Triticum aestivum* L; Gramineae). Wheat is a valuable crop used for food, feed, seeds, and industry worldwide. A vast genetic resource collection of wheat germplasm, including wild Triticum species, has been preserved in gene banks. Triticum has many species, each with a unique genome and number of chromosomes. Geographical regions, biological status, morpho-agronomic features, and ploidy level all contribute to the great diversity found in Triticum. By mixing different wheat genetic resources, such as current varieties with regionally adapted types, novel alleles can be introduced, increasing genetic diversity and pre-selection for desired traits, necessary to guarantee significant natural variation at the phenotypic level. The primary goals of wheat breeders and geneticists are to improve wheat for quality features, yield attributes, and biotic and abiotic stress resistance qualities. The use of contemporary genomics techniques to support conventional breeding operations can help achieve these goals. In addition to modern plant breeding tools like marker-assisted breeding, genetic engineering, and genome editing, this chapter provides an overview of the diversification and conservation of wheat germplasm, as well as the goals and phases of wheat breeding programs, farming, and traditional breeding techniques. This chapter focuses on the origin and distribution of wheat, its sources and dissemination, description, taxonomy classification and distribution, importance and uses, fiber and textile properties, domestication, selection, and early advancements, current cultivation and traditional breeding, present-day farming methods such as vertical farming, digital farming, and artificial

K. F. M. Salem (✉)
Department of Biological Sciences, Shaqra University, Ad-Dawadimi, Saudi Arabia

Department of Plant Biotechnology, Genetic Engineering and Biotechnology Research Institute (GEBRI), University of Sadat City, Sadat City, Egypt
e-mail: khaled.salem@gebri.usc.edu.eg; khaledfathi@su.edu.sa

N. A. Eid
Department of Plant Protection, Desert Research Center El-Matarya, Cairo, Egypt

K. F. M. Salem et al. (eds.), *Breeding and Biotechnology of Grass and Bast Fiber Crops*, Advances in Plant Breeding Strategies 12,
https://doi.org/10.1007/978-3-032-00403-1_8

intelligence, bread wheat breeding objectives, germplasm biodiversity and conservation, diversity of germplasm, phylogeny and characterization of wheat cultivars, methods for conserving wheat genetic resources, cytogenetics, traditional breeding, wheat breeding program goals and phases, traditional breeding methodologies, traditional breeding limitations, molecular breeding, DNA markers, genetic engineering, genome editing, and breeding by mutation.

Keywords Bread wheat (*Triticum aestivum* L.) · Biotechnology · Breeding · Genetic map · Conservation · Genetic diversity

8.1 Introduction

The earliest cereal crop, wheat (*Triticum* spp. L; Gramineae), can be grown in a variety of soil types and climates. It is well-suited to temperate climates with annual rainfall between 30 and 90 cm. Wheat originates primarily from South Asia. There are numerous Triticum species, each with a unique genome and number of chromosomes. Taxonomically, wheat is a self-pollinating crop belonging to the *Triticum* genus (Linnaeus 1753), Triticeae tribe, Poaceae family (Gramineae), and Cyperales order (Briggle and Reitz 1963). Inflorescences are sessile, alternating on both sides of the rachis, and consist of one to numerous flowering spikelets.

Wheat is economically valuable for industrial, food, feed, and seed purposes (Nachit 1992). Both rainfed and irrigated environments are used for wheat cultivation from seeds. Traditional plant breeding plays a significant role in introducing novel alleles by crossing genotypes from various plant genetic resources, including modern varieties and locally adapted strains (Abdel-Hamid and Salem 2021; Abu-Ellail et al. 2021; Aldahak et al. 2021; Ibrahim et al. 2023; Hassan et al. 2021; Salem 1997; Salem et al. 2015a, b, 2023a, b). This enhances genetic diversity and allows for the selection of desirable traits such as high grain yield, early maturity, improved grain quality, and resistance to lodging, biotic, and abiotic stresses (El Rabey et al. 2013; El-Absawy et al. 2015; El-Zanaty et al. 2011, 2013; Ibrahim et al. 2024; Mattar et al. 2016; Saleh et al. 2020b; Salem 2007a, b, 2009, 2015; Salem and El-Zanaty 2015; Salem and Ibrahim 2024; Salem and Mattar 2014a, 2014b; Salem et al. 2007a, b, 2008, 2010, 2015a, b, c, 2021b, d, e, 2022, 2024; Varshney et al. 2008). Traditional plant breeding methods remain crucial initial steps in developing new wheat cultivars with desired characteristics. Advancements in wheat biotechnology, utilizing DNA sequencing and molecular techniques, have garnered interest from plant breeders and geneticists (Hassan et al. 2021; Nawara et al. 2016, 2017; Salem 2004; Salem and Mattar 2009). This chapter covered cytogenetics, traditional breeding, wheat breeding program goals and phases, constraints, molecular breeding, DNA markers, genetic engineering, genome editing, breeding by mutation, and the origin and distribution of wheat. The chapter also discusses the sources and dissemination of wheat, its description, taxonomy classification, distribution, importance, and uses. It covers fiber and textile properties, domestication, selection, and early advancements in wheat cultivation. Additionally, it touches on current cultivation and traditional breeding, and modern farming practices like digital farming, artificial intelligence, and vertical farming. Specifically, the chapter delves into bread wheat breeding objectives, germplasm biodiversity, and conservation. It explores the diversity of germplasm, phylogeny, and characterization of wheat cultivars, as well as the conservation of wheat genetic resources.

8.1.1 Origin and Distribution

Understanding the origins of wheat is a critical stage in improving it through breeding efforts. Wheat primarily originated in South Asia. The Fertile Crescent and its neighboring countries have a high level of genetic variety. Many species of Triticum have varied genomes and chromosomal counts as shown in Table 8.1.

Of all the crops, including other cereals, Triticum aestivum is the most widely grown and common species of wheat. It is allohexaploid wheat with 42 chromosomes and three distinct genomes (A, B and D). It was created by crossing an unknown wheat (2n = 14, BB) with Triticum monococum (2n = 14, AA). After spontaneously doubling, the F_1 (AB, 2n = 14) formed tetraploid wheat (2n = 28, AABB). The hexaploid wheat (2n = 42, AABBDD) was created by spontaneously doubling the F_1 (2n = 21, ABD) after the latter was crossed to *Aegilops squrrosa* or *T. tauschii* (2n = 14, DD). Breeding wheat is greatly impacted by this evolutionary

Table 8.1 List of different wheat species

Common name	Specie name	Number of chromosomes	Ploidy level	Type of genome
Einkorn wheat	*Triticum monococum*	14	Diploid	AA
Wild diploid wheat	*Triticum urartu*	14	Diploid	AA
Wild diploid wheat	*Triticum boeoticum*	14	Diploid	AA
Wild diploid wheat	*Aegilops speltoids*	14	Diploid	BB
Wild diploid wheat	*Aegilops caudate*	14	Diploid	CC
Wild diploid wheat	*Aegilops speltoids*	14	Diploid	DD
Wild diploid wheat	*Aegilops Tauschii*	14	Diploid	DD
Wild tetraploid wheat	*Triticum timopheevii*	28	Tetraploid	A^tA^tGG
Wild tetraploid wheat	*Triticum turgidum*	28	Tetraploid	AABB
Emmer wheat	*Triticum dicoccum*	28	Tetraploid	AABB
Durum wheat	*Triticum durum*	28	Tetraploid	AABB
Common wheat	*Triticum aestivum*	42	Hexaploid	AABBDD
Club wheat	*Triticum compactum*	42	Hexaploid	AABBDD
Spelt wheat	*Triticum spelta*	42	Hexaploid	AABBDD

process because it: (i) increases genetic diversity among wheat and its relatives; and (ii) increases genetic redundancy, which is the state of having multiple genes that code for similar proteins.

8.1.2 Source and Distribution

One of the most crucial aspects of improving wheat through breeding programs is knowing where it comes from. Wheat is primarily from South Asia. The Fertile Crescent and its neighboring countries have high levels of genetic variety. Of all the crops, including other cereals, *Triticum aestivum* is the most widely grown and common species of wheat. It is allohexaploid wheat with 42 chromosomes and three distinct genomes (A, B, and D). It was created by crossing Unknown wheat (2n = 14, BB) with *Triticum monococum* (2n = 14, AA). After spontaneously doubling, the F_1 (AB, 2n = 14) formed tetraploid wheat (2n = 28, AABB). The hexaploid wheat (2n = 42, AABBDD) was created by the spontaneous doubling of the F_1 (2n = 21, ABD) and the *A. tauschii* (2n = 14, DD). In terms of wheat breeding, this evolutionary process has two major effects: (i) it has expanded genetic diversity among wheat and its relatives, and (ii) it has raised genetic redundancy, which is the state of having multiple genes that code for similar proteins. While the second impact was detrimental to diploid species since it made chromosomal manipulations and breeding tactics more difficult, the first impact was critical for wheat plant breeding to increase desired qualities, resulting in wheat becoming the world's most important cereal crop.

8.1.3 Description, Taxonomy Classification and Distribution

Numerous varieties of wheat, many of which are hybrids, have evolved over 10,000 years of agriculture as a result of both natural and artificial selection. The name of the wheat has become very confusing as a result of this variation. Wheat's classification is influenced by its genetic and physical traits; numerous common and botanical names are now in use. Wheat belongs to the genus Triticeae, tribe Triticeae, subtribe Triticinae, family Poaceae (Gramineae), order Poales (Glumiflorae), class Monocotyledonopsida, and phylum Angiospermatophyta. This subtribe also contains the genera *Triticum, Secale, Aegilops, Eremopyron, Amblyopyrum, Agropyron* and *Haynalidia*. The first major turning point in the history of taxonomy was reached in 1753 with the publication of Carl Linnaeus' book Species Plantarum. Bread wheat (*T. aestivum* L.) is the common wheat identified by Linnaeus. Common wheat's scientific name has undergone several changes in the last 250 years. The taxonomy and nomenclature of wheat species changed when new information became available. First, morphological differences were used to classify the species of Aegilops and wheat. The first contemporary classification of the *Triticum* genus was developed by Schulz in 1913, and it divided all species into three groups: einkorn, emmer, and dinkel (spelled that way). According to Sakamura (1918) and Kihara (1924), these clusters make up a polyploid series with distinct somatic chromosome counts. Subsequent analyses showed that seven basic chromosomes were present in all *Triticum* species (Sax and Sax 1924; Kihara 1937; McFadden and Sears 1946). Because the majority of the above-listed systems are still in use, some species' names and classifications remain unclear. Karyological analyses of the Aegilops genus, which contributed to the genesis of polyploid wheat, showed that the species in this genus had both diploid and polyploid forms and that the genus's fundamental chromosomal number was likewise seven (Kihara 1937). This new information also led to changes in the taxonomy of the genus Aegilops, Zhukovsky (1928) divided the genus into nine segments, but Eig (1929), Kihara (1954), and Witcombe (1983) only divided it into six. There isn't much agreement on how to classify the genus at the moment. Since polyploid wheat species arose through allopolyploidization due to spontaneous crossings between *Triticum* and Aegilops species, Bowden (1959) placed the entire Aegilops genus under the *Triticum* genus. Although MacKey (1966, 1988) disagreed, several authors adopted and used this classification (Morris and Sears 1967; Kimber and Sears 1987; Kimber and Feldman 1987). This was because fertile amphiploids were also obtained. Linnaeus conducted a systematic taxonomic study of wheat in 1953 (https://www.itis.gov/, Table 8.2).

8.1.4 Importance and Uses

For the majority of people on the planet, wheat is the first crucial and strategic cereal crop. Additionally, it is the primary food source of carbs for over two billion people globally or one-third of the world's population (36%). Approximately 55% of the

Table 8.2 A systematic taxonomic study of wheat

Kingdom	Plantae—Planta, plants
Subkingdom	Viridiplantae—green plants
Infrakingdom	Streptophyta—land plants
Superdivision	Embryophyta
Division	Tracheophyta—vascular plants, tracheophytes
Subdivision	Spermatophytina—spermatophytes, seed plants, phanérogames
Class	Magnoliopsida
Superorder	Lilianae—monocots, monocotyledons, monocotylédones
Order	Poales
Family	Poaceae—grasses, graminées
Genus	Triticum L.—wheat
Species	*Triticum aestivum* L.—common wheat, wheat
Subspecies	*Triticum aestivum* ssp. aestivum L.

world's carbs and 20% of its calories come from wheat (Breiman and Graur 1995). After maize and rice, which are regarded as the most significant grain crops in the world, wheat comes in third place in terms of production and cultivated area (FAO 2017). Wheat may be grown in a variety of climates. The Poaceae (Gramineae) family, which contains several important crops like rice (*Oryza sativa* L.), barley (*Hordeum vulgare* L.), rye (*Secale cereale* L.), maize (*Zea mays* L.), and bread wheat (*Triticum aestivum* L.), comes after wheat. One of the tribes with around 15 genera and 300 species, including wheat, is Triticeae. The self-pollinating crop wheat is a member of the Poaceae (Gramineae) family (Briggle and Reitz 1963). Aegilops are made up of wheat (*Triticum*) and ryes (Secale), whereas the subtribe Triticineae is made up of *Haynalidia*, *Agropyron*, and *Eremopyron* (Simmonds 1976). In 1753, Linnaeus is credited with classifying wheat for the first time (Linnaeus 1753). Each species' chromosomal count was reported by Sakamura in 1918. The *Triticum* classification underwent a sea change at this time. Three groups of wheat were created. Wheat is used to make wheat bread (*T. aestivum*). The other two (*T. durum* and *T. compactum*) primary forms are compactum. These three are the results of natural hybridization between ancestors that are no longer being grown for profit (Briggle 1967).

8.1.5 Fiber and Textile Properties

A major food crop in the world, wheat (*Triticum aestivum*) is rich in carbs as well as other essential elements such as proteins, trace levels of fats, vitamins, minerals, and phytochemicals (Shewry and Hey 2015). Dietary fiber consists of carbohydrate oligomers and polymers that cause partial or total fermentation in the human large intestine because they are difficult for the small intestine to digest and absorb (Andersson et al. 2013). According to Murphy et al. (2008). Furthermore, the resistant starch included in wheat grain's dietary fiber is not broken down in the small intestine and can pass through to the large intestine and colon mostly undigested (Hemery et al. 2007). Wheat grains are made up of an embryo (germ), an exterior layer (bran) that is high in fiber, and an endosperm that is rich in starch. Whole wheat or white flour, which is formed from wheat grains that have had the bran and germ removed, is used to make wheat-derived goods (Lu et al. 2004). While it has been recommended that people regularly consume whole grain items as part of their diet, white flour-based goods which are made by milling white flour to separate the bran and germ—remain the most popular in the majority of countries. Fiber consumption may be increased by raising the fiber content of white flour preparations and improving the overall palatability of whole wheat preparations. Eating meals made from wheat that are high in β-glucan, arabinoxylan, and resistant starch can lessen the impact of cardiovascular, gastrointestinal, and other disorders, including type 2 diabetes (Sima et al. 2018; Park et al. 2009). Furthermore, nutritional interventions are being promoted more and more as a means of reducing diabetes and obesity and as having positive impacts on human health. Clinical studies have shown that dietary fiber positively affects blood pressure and serum cholesterol levels and protects against the development of some malignancies, especially breast and colorectal cancers (Hazard et al. 2020). In this paper, we provide an overview of the three main components of wheat's dietary fiber: resistant starch, β-glucan, and arabinoxylan. We outline their main enzymes, biosynthesis processes, and compositions, as well as how these elements affect the health advantages of wheat fiber. The difficulties and viewpoints about future research targeted at enhancing dietary fiber in wheat are examined, along with a summary of the most recent developments in QTL/gene mapping and genetic enhancement of wheat fiber.

8.1.6 Fiber Sector

When it comes to market segmentation for wheat fiber, the end-use industry category is crucial. The market is dominated by the "Food and Beverage" sector, which is fueled by consumers' growing desire for foods and drinks that are higher in fiber. The Pet Food business also makes a significant contribution since wheat fiber helps pets' digestive systems and provides nutritional advantages. In addition, the Healthcare sector is expanding as more people become aware of the health advantages of wheat fiber, such as its capacity to control blood sugar and encourage normal digestion. Furthermore, the 'Paper and Packaging' sector uses wheat fiber as a biodegradable and sustainable substitute for conventional materials. The end-use industry segment is anticipated to rise significantly in the upcoming years due to the increased emphasis on sustainability and the growing demand for plant-based goods, which will help the wheat fiber market as a whole.

8.1.7 Domestication, Selection and Early Advancements

Since the beginning of human civilization, wheat has been domesticated. Since then, early farmers have enhanced a variety of features, such as tolerance to salinity and drought stresses, insect pests, diseases, and grain yield and quality, by selection, followed by deliberate hybridization following the discovery of Mendel's laws (Mattar et al. 2009; Saleh et al. 2020a; Salem et al. 2021a). Mutagens were used to create genetic variability in the 1950s, and superior mutants were then selected. To increase grain production, tolerance to drought, salt, terminal heat, and herbicides, as well as nutritional quality, research has been concentrated on creating superior hybrids, starting marker-assisted selection and targeted breeding, and creating genetically modified wheat during the past three decades. The end-users acceptance of genetically modified wheat continues to be a significant barrier to its release into the environment will be crucial (Gohar et al. 2022).

Triticum urartu (AA genome), *Aegilops speltoides* (controversial, BB genome), and *Aegilops tauschii* (DD genome) are the three distinct genomes that make up the hexaploid wheat (AABBDD) (Feldman et al. 1995; Nesbitt and Samuel 1996). Southeast Turkey is where wheat originated, based on archeological data. The tetraploid viable wheat was first produced by hybridizing the progenitor species with AA and BB subgenomes (Aaronsohn 1910) and then tripling the number of chromosomes. *Triticum turgidum* (AABB). The Fertile Crescent is where wild emmer, or *T. turgidum*, was domesticated. Then *T. turgidum* and a diploid species *Aegilops tauschii* hybridized, leading to the development of hexaploid wheat (AABBDD). The Fertile Crescent is where the hexaploid bread wheat emerged. Notably, tetraploid progenitors expanded within the diploid species' native range. Due to its widespread acceptance as the ultimate calorie source, it expanded through various paths around the world. Today's cultivated bread wheat is the product of centuries of hexaploid wheat cultivation and selection in many geographic locations following domestication. Einkorn wheat, or *Triticum monococcum*, is regarded as the first domesticated hulled wheat among diploids. It was domesticated 12,000 years ago, according to historical records. Pre-Pottery Neolithic times, 8500 years ago (Zaharieva and Monneveux 2014). Nonetheless, the domesticated tetraploids *Triticum durum* (tetraploid durum) and *Triticum dicoccum* (wheat emmer) both descended from wild progenitors.

Before spreading to other nations worldwide, the wheat first reached Greece, Cyprus, India, and Egypt. Compared to its wild counterpart, domesticated wheat had larger hulled seeds that were affixed to the ear. In species that were first farmed, the wheat grain was small, thin, and long. Non-brittle rachis and naked grain, which led to the domestication of wheat, were the first spontaneously altered features in the wild germplasm (Pourkheirandish et al. 2018). The modern farmed wheat was greatly influenced by the process of natural selection. Early farmers selected their choices based on phenotypic characteristics such as grain size, color, and non-shattering type. While lower-quality seeds were thrown away, the seeds from the best plants were used to produce the following generation.

8.2 Current Cultivation Practices and Challenges

8.2.1 Current Cultivation and Use of the Wheat Crop

Wheat is one of the first cereal crops. It may be grown in a variety of soil types and temperatures, but it is best suited to temperate regions with rainfall ranging from 30 to 90 cm. There are two primary wheat varieties: winter wheat and spring wheat. Spring wheat is sown in the spring, while winter wheat is always planted in the fall. China, the European Union, India, Russia, the United States, Canada, Pakistan, Australia, Ukraine, and Turkey were the top ten wheat producers globally in 2024. Hexaploid wheat, also called bread wheat (*Triticum aestivum* L.), is the genus to which the bread wheat varieties belong. Strong gluten and 11–15% protein are characteristics of hardened wheat genotypes that are typically produced in the dry zone.

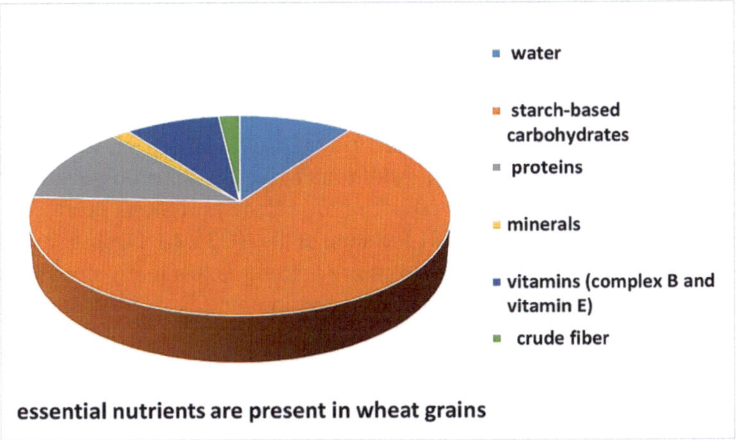

Fig. 8.1 Essential nutrients are present in wheat grains. (Prepared by Dr. Nerhan A. Eid)

Bread wheat's robust gluten captures carbon dioxide (CO_2) produced during the dough's fermentation process, allowing the fermented dough to rise. Humidity-grown wheat cultivars have poor gluten, 8–10% protein, and are mushy. Cakes, biscuits, pastries, and flour are all made from soft wheat flour. One of the best sources of semolina is durum wheat (*Triticum turgidum* L.), which can be used to make pasta, oxalic, borage, and other goods (Nachit 1992). Conversely, diploid wheat is not grown since it is not a crop with any commercial value anywhere in the globe. About 10% of the wheat produced is used in industry (to make starch, paste, dextrose, and gluten), whereas the majority of the grain is grown for human consumption. All essential nutrients are present in wheat grains according to chemical analysis: 12% water, 60–80% starch-based carbohydrates, 8–15% proteins Fig. 8.1.

8.2.2 *Present-Day Farming Methods Vertical Farming*

The situation of vertical farming has emerged as a result of the loss of viable agricultural land. In the vertical orientation, this kind of farming is primarily regarded as soilless farming. Hydroponic, aeroponic, and aquaponic are three popular soilless vertical farming techniques Fig. 8.2. Despite its lack of popularity in poorer nations, many European nations have invested in vertical farming and attempts to use it as a commercial commodity are progressing. This sort of farming can also solve challenges related to food security (Naqvi et al. 2022).

Due to several problems, including unchecked population increase, unequal resource distribution, the depletion of natural resources, overcrowded urbanization, and the loss of rich soils, feeding the globe in the twenty-first century is not an easy undertaking. One way to reduce the threats to food security is through vertical farming. One acre of vertical farming may produce as much food as thirty acres of conventional farming, making it a considerably more efficient method of producing crops continuously. Diseases and insect/pest assaults are less likely to occur because of the controlled environmental conditions, which can also remove the possibility of using chemicals during farming operations. Controlled environmental conditions also remove several environmental problems that lead to crop loss, such as drought, flood, and hail (Saleh et al. 2021a, b). Additionally, vertical farming contributes to a 70% reduction in water loss and a reduction in carbon emissions produced during agronomic activities (Kulak et al. 2013; Salem et al. 2023a, b). Despite its ability to address food security concerns, this cropping technique has many drawbacks. The cost-benefit analysis is the primary obstacle to overcome Fig. 8.3. Although the cost

Fig. 8.2 Soilless vertical farming techniques. (Prepared by Dr. Nerhan A. Eid)

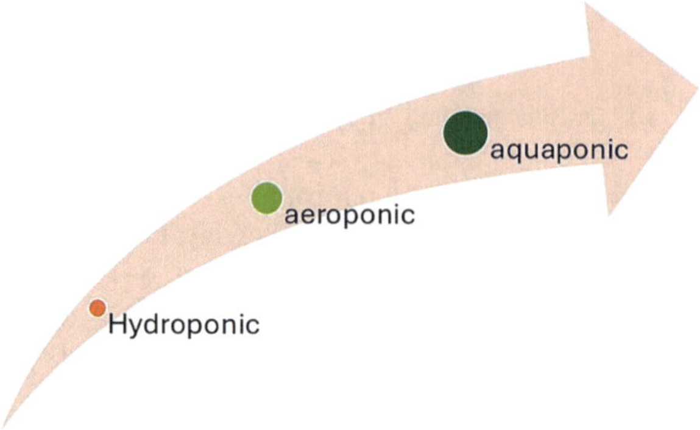

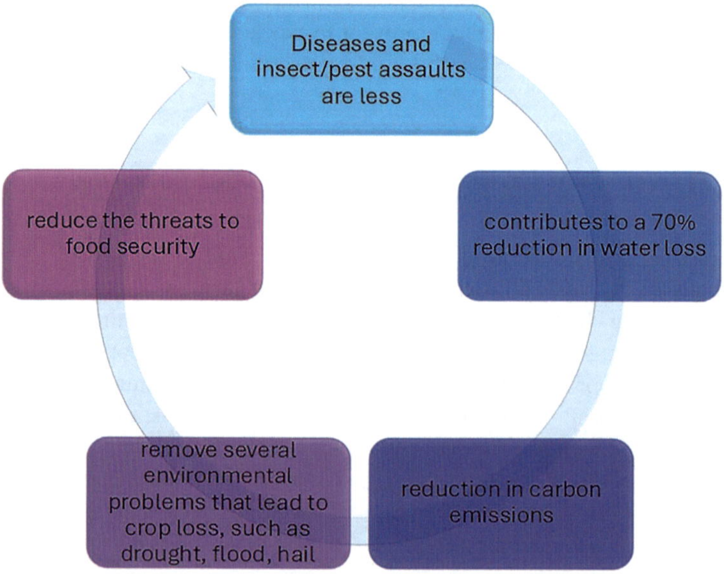

Fig. 8.3 Benefits of vertical farming. (Prepared by Nerhan A. Eid)

of land and construction may differ from one location to another, vertical farming is not feasible in large cities due to expense. Cost-effectiveness may be hampered by the energy and operating expenses required to maintain and regulate the interior environmental conditions of a farming setup. Only a few crops can be grown vertically, and they also need to be manually pollinated while growing. Although there are certain economic inefficiencies in the system, vertical farming will become popular in the future to prevent food shortages, and the system's cost will be decreased by reusing necessary construction structures.

8.2.3 Digital Farming

Farmers employ technology to link field-level and financial records for full farm activity management, a practice known as "digital farming." Each plot's data may be examined to provide information on soil, weather, and crop growth patterns, providing actionable, timely insights that are geographically relevant and assist reduce losses while maximizing the production of each farm plot. Through applications on mobile phones, farmers may even oversee the supply chain directly and get their questions answered. Digital farming seeks to take over every facet of farming from farm to fork by managing farms both before and after harvest.

Farmers may implement best practices and manage farms more effectively by using digital farming, which minimizes losses and maximizes revenues by providing them with timely and important insights. There are numerous ways to adjust to modern farming through technology. The Internet of Things in agriculture consists of sensors, drones, and computer imagery, as well as analytical tools that generate helpful insights. Physical equipment is installed on farms to monitor and collect data, which is then used to gain insights. Satellite photography, machine learning, and cloud data storage have all made significant contributions to predictive analytics software. Since it is user-friendly and extremely scalable.

Precision and smart farming are combined in digital farming, which is accomplished by using hardware and software with intelligence. A common definition of precision farming is a technology-enabled method of managing farms that monitors, gauges, and assesses the requirements of specific areas and crops. Utilizing data from several sources (historical, geographic, and instrumental) to manage farm operations is the main focus of smart farming.

8.2.4 Artificial Intelligence

The pressing challenges of climate change and global food security in agriculture, particularly in crop breeding, call for creative solutions. In wheat breeding, artificial intelligence (AI) is a game-changing technology that offers fresh methods for enhancing crops' resilience and yields in response to shifting climatic conditions.

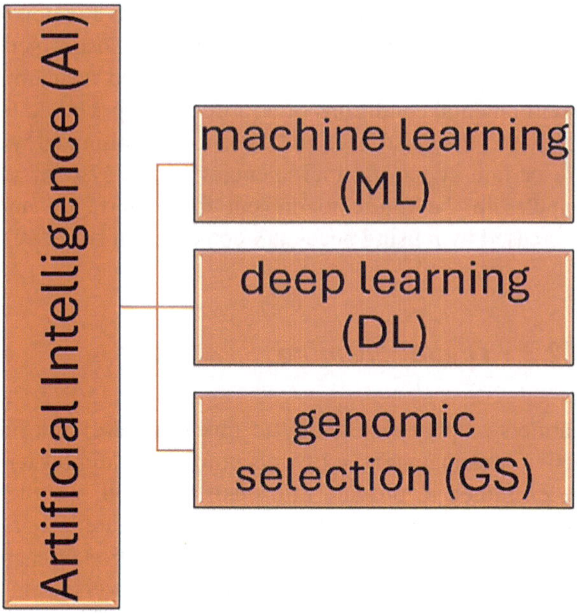

Fig. 8.4 Artificial intelligence (AI) technology. (Prepared by Nerhan A. Eid)

The integration of artificial intelligence (AI) into traditional wheat breeding techniques is examined in this review paper, with an emphasis on how AI can help address the complexities of modern agriculture. The purpose of this review is to evaluate how AI technologies can improve wheat breeding programs' sustainability, accuracy, and efficiency. By accelerating the rapid analysis and interpretation of large datasets, these technologies enhance the process of choosing and breeding wheat varieties that are suitable for a variety of environmental conditions Fig. 8.4. The results of the studies under review show that artificial intelligence has significantly advanced wheat breeding. While genomic selection has shortened the length of breeding cycles, machine learning methods have improved the accuracy of phenotypic trait prediction. High-throughput phenotyping, which makes use of artificial intelligence, enables careful analysis of plant traits in various stress conditions, making it easier to identify resilient cultivars. Additionally, when compared to traditional techniques, AI-driven models have demonstrated higher anticipated accuracies for crop productivity and disease resistance. AI technologies significantly improve crop performance and adaptability, making them essential to the modernization of wheat breeding. In light of climate change, this integration enhances global food security by promoting the development of wheat cultivars that can tolerate harsh environments and provide high yields (Mushtaq et al. 2024).

The state in which people have access to all the resources required to eat a healthy diet and stay well is not what is meant by food security. Eating enough food is essential for nutrition and overall wellness. Malnutrition may result from food scarcity (Mbow et al. 2020). When people don't have enough food to consume, food insecurity frequently results. Natural catastrophes including earthquakes, floods, droughts, and wars can occasionally be the reason. But more often than not, it is due to long-term poverty rather than only natural calamities. A person's income, abilities, and privileges all affect their ability to get food. To investigate this, scientists employ two primary methodologies (Seaman et al. 2014). The first method looks at the big picture and connects family reactions to climate change. The second method looks into how households and communities adjust to climate change. The more comprehensive strategy makes use of intricate models that combine economic, crop, and climate data to predict how climate change would affect crop trade. People's adaptations to climate change are not fully taken into account by this method. Household surveys are used in smaller-scale research to learn how individuals and communities adapt to climate change. The International Institute links the outcomes of the food supply, market, and prices with scenarios of climate change. It also looks at the financial effects of variables like food availability and consumption that influence food accessibility. These ramifications include children's nutritional status and food energy intake. Changes in local weather patterns are referred to as climate change. It has major effects on both civilization and the environment and is brought on by natural variations in the climate system. Applications of artificial intelligence (AI) are revolutionizing genetics and plant breeding to address the problems caused by climate change. Breeding climate-resilient crops is accelerated by AI techniques like machine learning (ML) and big data analytics, which allow for the study of enormous volumes of genetic and phenotypic data. For instance, by connecting intricate biological processes with plant reactions to environmental changes, researchers have been using next-generation AI to combine multi-omics data and improve the breeding process (Harfouche et al. 2019). Furthermore, high-throughput omics technologies are used in ML-assisted plant breeding programs to create novel crop varieties that thrive in a variety of climatic settings (Yoosefzad Najafabadi et al. 2023). To secure food production in the face of global climate change, these technologies are essential for boosting yield performance, strengthening resistance to pests and diseases, and boosting overall resilience to climate change.

Food systems may be impacted by climate change in several ways, including modifications to market dynamics or crop output. Each region is affected differently in terms of food security. Food systems are also changing as a result of urbanization and globalization (Gregory et al. 2005). The impact of climate change on food security varies by nation. The effect is less severe and is anticipated to improve in areas with lower rates of hunger. There is a chance for changes, therefore ignoring climate change does not immediately provide serious hazards for malnutrition and undernutrition. However, changes brought about by climate change are expected to widen gaps in access to food. To counterbalance rapid adjustment for a few decades, vulnerable communities must take quick adaptation measures due to their increased susceptibility to catastrophic weather occurrences. Previous greenhouse gas emissions are linked to this urgency. Risks will rise if catastrophic weather events occur more frequently in the future (Wheeler and Von Braun 2013). Impacts must be formally identified by contrasting observed changes with a predetermined reference point. It is challenging to create a reference point in the food production example due to non-climate elements. These characteristics are difficult to measure and require thorough characterization (Porter et al. 2014). The yields of maize improve by 1.5% annually, whereas those of wheat, rice, and soybeans increase by 1.0% annually. A minimum tolerable rate of yield progress is 1.1% annually to avoid a 30% increase in grain prices by 2050 (Fischer et al. 2014). Careful nutrition management and enhanced farming practices are needed to meet the growing demand for food. Systems of production that are both sustainable and productive are necessary for food security. To maximize nutrient use, combined nutrient management is advised. Research findings must be put into practice by following practical principles. To promote sustainable food security, more research is required (Roy et al. 2006). By utilizing natural agrobiodiversity and farmers' traditional knowledge, traditional breeding techniques have been crucial in creating wheat types that are climate-robust. But by using genomics, marker-assisted selection, and high-throughput phenotyping platforms to find novel QTLs and genes for crop improvement as well as stress-tolerant genotypes, genomic breeding techniques have greatly improved the process (Gesesse et al. 2023). These techniques have proven especially successful in creating wheat cultivars that are resistant to heat and drought, which is essential for fending off the dangers posed by climate change. Breeding programs can capture environmental adaptations and genetic differences by combining genomics and participatory varietal selection. This allows for the creation of drought-adapted, high-performing wheat varieties that are customized to meet the demands of local farmers (Devate et al. 2023). Food security for the expanding population is ensured by the development of high-yielding, climate-resilient wheat varieties that can flourish in harsh climatic conditions thanks to the combination of traditional and genomic breeding techniques. In light of climate change, artificial intelligence (AI) is transforming wheat breeding programs by increasing crop output and resilience. To enhance genomics-assisted breeding (GAB), recent developments in AI have made it possible to use non-linear modeling techniques that capture intricate relationships in genomic selection (GS) and genome-wide association studies (GWAS) (Robles-Zazueta et al. 2024). AI methods have also made it easier to incorporate speed breeding, which has greatly decreased the amount of time needed for traditional breeding and ensured increased precision and effectiveness in creating wheat cultivars that can withstand stress (Bhat et al. 2023).

8.2.5 Bread Wheat Breeding Objectives

Genetically enhancing resistance to diseases (such as rust, smut, bunt, and *Fusarium*), and tolerance to abiotic stressors (such as drought, and heat) (Salem et al. 2021a, d) are the primary challenges facing wheat breeders and geneticists (Salam et al. 2018). According to Salem (2004), wheat can grow in a range of climates, including warm, humid, dry, high rainfall, and irrigated temperate zones. Therefore, resolving the problem and setting objectives are essential elements of any wheat breeding program's development. For breeding programs to improve wheat crops for desired qualities, genetic variety is essential. According to Baenziger et al. (2016), there are five essential steps for a successful breeding program. Every step must be finished for the breeding program to be effective.

Even though wheat may be cultivated in a variety of conditions, wheat breeders put a lot of effort into genetically enhancing the crop to address significant issues that impede wheat output and productivity. Each climate has its own set of problems, including low input conditions, insects, viruses, heat stress, salt stress, and drought stress. Therefore, describing the problem is necessary before deciding on the best breeding program to acquire the desired features. Additionally, the breeding program may change based on the developmental stage. For instance, the growth stage of the wheat that is impacted by drought determines how to improve drought resilience in that crop (Salem et al. 2007a, b).

8.3 Germplasm Biodiversity and Conservation

The fundamental concept behind plant breeding is natural diversity, which comprises both genotypic and phenotypic variability (Alqahtani et al. 2024; Börner et al. 2010). The idea is to select a positive variety for future generations. The variation may exist in nature as a result of previous allele recombination, crossover, or mutation. Wheat diversity comes from wild, landrace, and modern cultivars, as well as breeding resources and gene bank accessions (Abulaila et al. 2025; Abdullaev et al. 2025; Akparov et al. 2025; Aladele et al. 2025; Al Lawati 2025; Almarri et al. 2025; Arab et al. 2025; Békefi et al. 2025; Breidy 2025; Cañar et al. 2025; de Galarreta et al. 2025; Desheva et al. 2025; Díez et al. 2025; Galea-Gragera and Llera Cid 2025; Ghidan and Abdulmajid 2025; Gudaye et al. 2025; Harzli et al. 2025; Hidayatun and Kurniawan 2025; Ihsan and Prihatini 2025; Jatoi and Siddiqui 2025; Laghetti et al. 2025; Mallor and Luna 2025; Monteros-Altamirano et al. 2025; Osterc et al. 2025; Popović et al. 2025; Rocha et al. 2025; Wang et al. 2025).

8.3.1 Diversity of Germplasm

Without a doubt, wheat's allopolyploid origin and nature are responsible for its diversity, allowing it to grow and adapt to various circumstances. For example, wheat has different genome levels as an allohexaploid crop.

The development of wheat crops relies on genetic variety and natural variation for selection through landraces and wild relatives, accelerating genetic gain in breeding operations. Old wheat germplasm, like landraces and wild varieties, is a crucial genetic resource for enhancing contemporary wheat with novel alleles. The wheat diversity bottleneck, which reduced genetic variation, was influenced by various biological events, starting with domestication as the initial bottleneck. As wheat spread from the domesticated fertile Crescent region to Europe and Asia, it gradually adapted to local environments, resulting in a loss of genetic variation. The loss of desired alleles from the gene pool due to diploid and tetraploid crossings for hexaploid wheat creation, along with early farmer selection, further reduced wheat's genetic diversity. Bread wheat diversity declined due to breeding methods that separated environmental and genetic effects, replacing native landraces with improved varieties (Novoselović et al. 2016). Plant breeders are concerned about decreasing genetic variation in the grain, despite a noticeable decrease in wheat genetic diversity over the past century (Novoselović et al. 2016).

To enhance the robustness of results while considering the set of variables for the current task, analyzing factors allows for investigating qualities and uncovering potential connections. It is crucial to use indicators that enhance genetic gain and selection boldness. According to Olivoto and Lúcio (2020), the multitrait genotype-ideotype distance index (MGIDI) enables efficient selection and recommendation based on multivariate data, distinguishing it from conventional selection indices. This index helps determine the ideal genotype and adjusts traits of interest. Additionally, MGIDI has the potential to improve genotype selection and breeding program strategies (Olivoto et al. 2021).

8.3.2 Phylogeny and Characterization of Wheat Cultivars

A potent tool for characterizing both new and ancient materials and comprehending natural variation and its use in breeding is the examination and characterization of wheat Gene Bank collections. High-throughput genotyping can be used to characterize large-scale collections using the latest technological advancements. The genetic diversity in wheat germplasms, particularly ancient collections, is now very appealing due to the introduction of DNA markers (Börner et al. 2000a; Salem et al. 2015a; Sehgal et al. 2015). Understanding and tracking genetic variation can be aided by population structure utilizing molecular markers in a large, diversified collection that includes landrace, contemporary, and wild species. By using the approach that reliably determines the individuals' relatedness and groups them according to their genetic data, which is frequently employed in association mapping research (Ismail et al. 2013). Understanding the intricate history of wheat distribution and genetic variation is aided by phylogenetic analysis. Numerous phylogenetic analyses have been carried out that have successfully divided wheat collections according to biological status, ploidy level, and geographic location (Golovnina et al. 2007; Goncharov et al. 2009; Demir et al. 2015).

Temperature fluctuations, altered rainfall patterns, decreased soil fertility, elevated salinity, biological stressors, elevated pollution, and dwindling biodiversity are all consequences of climate change that affect the Earth's environment (Cramer et al. 2018). Given that multiple factors influence plant growth and development, the harsh effects of climate change on crop production could be amplified (Sinha et al. 2021). It is crucial to evaluate and preserve wheat genetic resources for potential future use (Guzzon et al. 2022). To increase the productivity of wheat cultivars, genomics-assisted breeding techniques take advantage of pre-breeding material and cultivars (Rasheed et al. 2018a). To evaluate genetic diversity in crop plants, DNA markers have been created and used (Bhandari et al. 2017).

Pour-Aboughadareh et al. (2017) evaluated the genetic diversity of 70 accessions of Iranian *Triticum* species using 15 SCoT primers. Mohamed et al. (2017) also used the SCoT markers to genetically describe 14 North African wheat varieties. Additionally, eight Asian wheat varieties were distinguished using SCoT. CBDP and SCoT markers were used to evaluate molecular variability and relationships among the samples (Ghobadi et al. 2021). In the same setting, CBDP and ISSR markers were used to evaluate the genotypic and phenotypic diversity of 96 durum wheat genotypes. Hebert et al. (2021) first used DNA barcoding to identify plant species, and Hollingsworth et al. (2009) claimed that the technique had been refined. The maturase kinase (matK) and ribulose-b phosphate carboxylase (rbcL) major subunits of the chloroplast genes are essential barcodes for plant species. Furthermore, the ribosomal internal transcripted spacer (ITS), psbA, and chloroplast spacer sequences are frequently used barcodes at the species level. Additionally, DNA barcoding offers comparative analyses and insights into genetic diversity in populations under study. DNA barcoding has recently emerged as a useful method for evaluating biodiversity in plant evolution and phylogenetic reconstruction. The main forage plants were barcoded using a combination of nuclear and chloroplast DNA sequences. Little is known about the genetic diversity of the 12 bread wheat cultivars, both within Northern African types and among cultivars in other nations. The significance of choosing the aforementioned wheat cultivars stems from their capacity to adapt to the harsh climates of the nations in which they were created. Several pilot and preliminary measurements were conducted on previously published wheat cultivars in the same context. These measurements included morphological characteristics (grain length, width, color), physical analyses (moisture percentage, falling number, ash content, wet gluten, dry gluten), and grain weight (the weight of one to twenty grains) (data not shown) and disclosed concentrating on the twelve chosen cultivars under study. In addition to other genes that encode transcription factors that are essential for plant growth and development, certain genotypes also carry abiotic-resistant genes. Cereals contain diagnostic qualities that aid in cultivar identification and separation, as well as their impact on production and quality for growers, consumers, and plant breeders. SEM screening of the grain surface revealed a significant degree of similarity across the four genotypes of cereal bowls (Abou- Taleb and Abd – El Maksoud 2013).

8.3.3 Methods for Conserving Wheat Genetic Resources

Conservation of genetic resources in plant gene banks through various approaches is highly beneficial for seed preservation (Abulaila et al. 2025; Abdullaev et al. 2025; Akparov et al. 2025; Aladele et al. 2025; Al Lawati 2025; Almarri et al. 2025; Arab et al. 2025; Békefi et al. 2025; Breidy 2025; Cañar et al. 2025; de Galarreta et al. 2025; Desheva et al. 2025; Díez et al. 2025; Galea-Gragera and Llera Cid

2025; Ghidan and Abdulmajid 2025; Gudaye et al. 2025; Harzli et al. 2025; Hidayatun and Kurniawan 2025; Ihsan and Prihatini 2025; Jatoi and Siddiqui 2025; Laghetti et al. 2025; Mallor and Luna 2025; Monteros-Altamirano et al. 2025; Osterc et al. 2025; Popović et al. 2025; Rocha et al. 2025; Wang et al. 2025). On the other hand, conservation helps preserve biodiversity and identify varieties with beneficial and adaptive genes.

Ex situ wheat collections stored in gene banks are valuable genetic resources that must be safely maintained, economically viable, and readily available to a wide variety of users. Numerous ex-situ seed gene banks have been developed globally to preserve plant genetic resources with natural genetic variety. A list of important gene banks for the conservation of plant genetic resources around the world is provided in Table 8.3.

Plant ex-situ gene banks are used for ex-situ conservation, which involves growing the accession in a garden or storing it in a seed bank. Preventing a species' local, regional, or worldwide extinction and preserving as much of its genetic variety as feasible are the general objectives of ex-situ conservation. One of the main obstacles to ex-situ conservation is preserving genetic integrity because of contamination from foreign pollen or improper treatment during multiplication (Börner 2006). Because most wheat collections are developed and supplied by farmers or breeders,

Table 8.3 A list of important wheat gene banks for the conservation of plant genetic resources around the world

Gene Bank	Country	Website
Czech Republic Gene Bank	Czech Republic	http://genbank.vurv.cz/wheat/pedigree/
Genetics Resources for Wheat Sciences Gene Bank	Japan	https://shigen.nig.ac.jp/wheat/komugi/
ICAR-National Bureau of Plant Genetic Resources (NBPGR)	India	http://www.nbpgr.ernet.in/
Instituto del Germoplasma	Italy	https://biodiversitapuglia.it/la-banca-del-germoplasma-dellibbr-cnr-bari/
International Center for Agricultural Research in the Dry Areas (ICARDA), ICARDA Gene Bank	Egypt	http://www.icarda.org/
Leibniz Institute of Plant Genetics and Crop Plant Research (IPK) Gatersleben Gene Bank	Germany	https://www.ipk-gatersleben.de/en/gbisipk-gaterslebendegbis-i/
National Gene Bank	China	http://www.cgris.net/cgris_english.html
National Small Grains Collection Gene Bank	USA	https://www.ars.usda.gov/paci
The International Maize and Wheat Improvement Center (CIMMYT) Gene Bank	Mexico	https://www.cimmyt.org/seed-request/
Winter Cereal Collection Gene Bank	Australia	www.dpi.nsw.gov.au/about-us/research-development/centres/

they are primarily composed of local or regional resources, which are probably duplicates of accessions. As a result, the plant genetic resource centers preserve the utilization of wheat germplasms, which is crucial for future studies.

8.3.4 Cytogenetics

After identifying chromosomal segment translocations in wheat cultivars through cytogenetic analysis, breeders employed marker-assisted breeding to select the desired genotypes. This technique is useful in breeding elements to determine greater productivity and phytopathogen resistance.

Grain quality was high and fungal disease resistance was high in Russian wheat types with intact wheatgrass chromosomes (Salina et al. 2015). Cytogenetic techniques are still essential for describing genetic variation and their use in breeding research, despite the widespread advent of high-throughput molecular markers.

8.4 Traditional Breeding

Traditional plant breeding techniques remain the most important and initial steps in creating new wheat cultivars with desired features, even if modern plant breeding that makes use of DNA technology has drawn the attention of geneticists and wheat breeders. The main goal of conventional wheat breeding is to produce a wheat plant that can readily grow and flourish in a variety of environments. As global challenges emerge, it is important to take into account changes in farmable areas, harsh cropping systems, and food security (Davis et al. 2004). By using plant breeding techniques, which allow for the selection of genotypes with desired genes or QTLs influencing significant traits, these worldwide concerns can be addressed in wheat.

The amount of dietary fiber in wheat is directly linked to the physicochemical characteristics of resistant starch, β-glucan, and arabinoxylan. To improve the nutritional content of grain, wheat breeding should focus on further optimizing the manufacture of resistant starch, β-glucan, and arabinoxylan in wheat endosperm. However, there are still a lot of obstacles to get beyond.

8.4.1 Wheat Breeding Program Goals and Phases

The key to any successful wheat breeding effort is addressing the issue and establishing goals. To improve the wheat crop for specific attributes in breeding programs, genetic diversity is essential. Five key phases for a successful breeding program were identified by Baenziger (2016). For a breeding program to be successful, each step is crucial. According to Salem et al. (2004), wheat can grow in a range of conditions, including warm, humid, and dry, with considerable rainfall, and irrigated temperate zones. Therefore, resolving the problem and setting objectives are essential elements of any wheat breeding program's development. For breeding programs to improve wheat crops for desired qualities, genetic variety is essential. According to Baenziger et al. (2016), there are five essential steps for a successful breeding program. Every step must be finished for the breeding program to be effective. Addressing the Issue and Establishing the Goals Although wheat may be cultivated in several conditions, wheat breeders strive to genetically enhance the wheat crop to address significant issues that impede wheat output and productivity. Each climate has its own set of problems, including low input conditions, insects, viruses, heat stress, salt stress, and drought stress. Therefore, describing the problem is necessary before deciding on the best breeding program to acquire the desired features. Additionally, the breeding program may change based on the developmental stage. For instance, the growth stage of the wheat that is impacted by drought determines how to improve drought resilience in that crop (Salem et al. 2007a, b).

The final phase in a breeding effort is deciding whether to release a cultivar if it outperforms at least one key target attribute. Wheat grain yield is typically one of the breeding program's goals. Regional differences exist in cultivar release protocols.

8.4.2 Traditional Breeding Methodologies

Plant breeders still employ several traditional breeding techniques for crop development (Baenziger 2016). Two key factors determine the variations in breeding techniques: (i) population inbreeding and (ii) the selection procedure.

8.4.2.1 Pedigree Methods

Using this approach, wheat breeders chose the best plants from this generation by evaluating F_2 plants for target attributes. The chosen seeds are cultivated in a progeny row for the next generation. Out of all the plants that are grown, the best ones are chosen. Until any segregation is noticed inside the progeny row, the section is carried out every generation. Considering the number of segregating features in the population under investigation, the number of inbred generations, and the phenotypic data are among the many criteria that often determine segregation within a row.

8.4.2.2 Pure Lines Methods

Several carefully chosen individual plants from a cultivar are cultivated in rows using this breeding technique, and the best rows ought to be chosen. To determine which selection is better, the best rows are then grown in duplicate (Baenziger 2016).

8.4.2.3 Bulk Methods

The majority of the offspring produced by a particular cross are grown by wheat breeders. Next harvest, a portion of the bulk seed is replanted for the next year. Following numerous plantings and mass harvests, the population is deemed to have a variety of primarily homozygous lineages (Tee and Qualset 1975).

8.4.2.4 Mass Selection Methods

The most basic technique for crop enhancement is mass selection. Many plants with similar phenotypes are chosen by breeders. The new variety is created by combining the seeds of a few chosen plants. There would be a lot of genetic variation in the variety created this way. Therefore, in such a variety, mass selection could be carried out once more (Marais and Botes 2009).

8.4.2.5 Backcross Methods

Incorporating easily target-acquired features from unadapted donor genotypes into recipient genotypes is the primary goal of backcross techniques. Backcross techniques involve selecting the desired characteristic to be transferred after multiple cycles of crossing to the recurrent parent (Kenaschuk 1975). To introduce a single variety with good character, breeders repeatedly cross the F_1 to that variety. Many beneficial traits, including disease resistance, were transferred via this technique. For instance, common wheat inherited numerous genes from Triticum species that are resistant to stem, leaf, and stripe rust (Marais et al. 2003). The backcross technique increased winter wheat's resistance to fusarium head blight (Clark et al. 2016).

8.4.3 Traditional Breeding Limitations

Even though wheat may be produced in a wide variety of conditions, wheat breeders strive to genetically enhance the crop in order to address significant issues that restrict wheat productivity and production. Diseases, low input settings, insects, heat stress, salt stress, drought stress, and other issues are unique to each location. In order to choose the best breeding program to enhance the desired qualities, it is crucial to define the issue. Furthermore, depending on the growth stage, the breeding program may also vary. For instance, the growth stage of wheat that is subjected to drought determines how to improve drought tolerance in that crop (Salem et al. 2004, 2007a, b). Both the seedling and grain-filling stages are susceptible to drought (Salem et al. 2004, 2007a, b). Because genotypes may react differently to drought depending on the growth stage, some studies found no link between drought

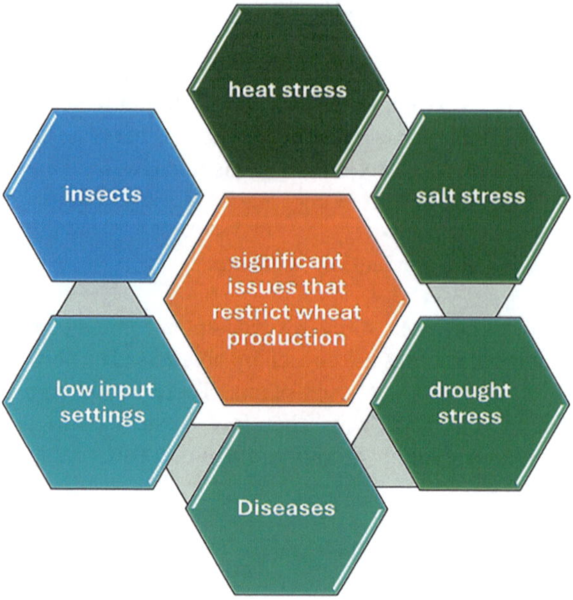

Fig. 8.5 Significant issues that restrict wheat productivity and production. (Prepared by Dr. Nerhan A. Eid)

tolerance in the grain-filling stage and drought tolerance in the seedling stage (Salem et al. 2007a, b) (Fig. 8.5).

8.4.4 Utilizing a Key Source of Genetic Diversity and Variability

The crossing of two or more different genes is the main way that genetic diversity is introduced in wheat. Most crosses between extremely homozygous (pure) lineages are done to create F_1. Several studies have examined the inheritance of important traits in the F_1 generation using a variety of diallele analytic approaches. Some try to create children with a genetic identity for one or more traits (such as early flowering) that are closer to the parents by crossing the F_1 with the target parent. The F_2 population can be a highly interesting target to investigate the segregation of target traits. Most recent studies have focused on the bi-parental population, where parent-to-parent crossing can result in highly homozygous lineages through single-seed descent (SSD). These bi-parental lines are usually used to identify important quantitative trait loci (QTLs) influencing target attributes. More interestingly, more and more genome-wide association studies (GWAS) are being carried out on different populations. Numerous genotypes from all over the world make up this diverse population. For germplasm to achieve the goals of any breeding effort, there must be a significant degree of genetic variability among plant materials (Salem et al. 2015a).

8.4.5 Phenotypic Selection and Inbreeding in the Outcome Variants

Selecting which selection and inbreeding techniques to employ is the next step for wheat breeders. Heterozygosity gives way to homozygous lines as a result of inbreeding. A small number of the best homozygous lines are then chosen via selection to be included in the following phases of the breeding cycle. The breeding program will eliminate the remaining unselected lines.

One of the fundamental techniques of conventional plant breeding is selection. Selection can occur naturally (by the force of nature) or arterially (by humans). Wheat breeders should use extreme caution when testing plant material because selection in breeding programs varies depending on the environment (Baenziger 2016). For instance, selection should take place in settings that enable breeders to have a range of desirable qualities among the tested genotypes if the goal is to choose wheat genotypes for winter hardiness.

8.4.6 Selected Elite Lines for Particular Goals

The most critical step before releasing the cultivar is screening and assessing the germplasm for the particular goal, even if all five phases are critical to a successful breeding effort. Accurate phenotypic data collection is essential to the selection process. Over years, replications, places, or conditions, plant material should be assessed and screened. This step is solely dependent on the goals that the breeding program must accomplish. To successfully select and improve the wheat crop genetically. The estimated heritability of a chosen characteristic should be high. Any growth stage or a variety of growth phases could be used for the evaluation (Salem 2004). To select genotypes with high tolerance at both growth stages, a bi-parental population was tested for drought tolerance at the seedling and grain filling stages (Salem et al. 2004, 2007a, b). The selection process should give careful consideration to the relationship between genes and the environment.

For instance, a biparental population's drought tolerance was assessed at the seedling and grain-filling stages. Before making a choice, the relationship between genotype and environment should be considered.

8.4.7 Release of Cultivars

The choice to release a cultivar if it excels in at least one key desired attribute is the final step in a breeding effort. Generally speaking, one of the desired characteristics for every breeding program should be wheat grain yield. Each country has different cultivar release protocols.

8.5 Molecular Breeding

Before releasing a new cultivar, a lot of genotypes must be evaluated to determine the causes of genetic diversity. Superior genotypes could be employed as parents in breeding programs if the genetic variation that is accessible has been identified. Hundreds of thousands of genotypes may be examined in a crop as diverse as wheat. It takes a lot of money and effort to evaluate so many genotypes in the field. One potential method that could assist reduce the number of genotypes that are examined and save time and money is the use of a molecular marker. For instance, in contrast to traditional backcrossing programs, molecular markers are typically employed to track a limited number of loci during backcrossing programs. This helps the breeder identify the germplasm near the recurrent parent and requires less work (Langridge 2003).

8.5.1 DNA Markers

Because of its size and complexity, the use of molecular markers in the wheat genome is more difficult than in other crop genomes. Nonetheless, numerous attempts were made to use various molecular markers to comprehend the wheat genome. The first wheat genome map employing SSR markers was published in 1998 as a result of these efforts (Röder et al. 1998). Numerous studies that used various types of molecular markers to map numerous quantitative loci linked to significant features followed this one (Börner et al. 2000b; Börner et al. 2002; Salem et al. 2007a, b; Echeverry-Solarte et al. 2015; Bhusal et al. 2017). Molecular markers and molecular maps are the two main categories into which the various kinds of molecular breeding tools can be divided.

8.5.2 DNA Markers

Over the past 50 years, wheat breeding has made extensive use of molecular markers. New kinds of molecular markers are typically emerging more quickly than the previously created kinds as a result of the ongoing advancements in biotechnology. Molecular markers employed in wheat can be broadly divided into three categories: PCR-based markers, hybridization-based DNA markers, and DNA chip and sequencing-based DNA markers.

8.5.2.1 Bread Wheat Molecular Maps

It is now possible to quickly determine the sequence of many genotypes thanks to advancements in genome sequencing techniques and the introduction of next, third, and fourth sequencing technologies. One widely utilized example of the revolution in sequencing techniques is genotyping-by-sequencing (GBS). Numerous SNP markers are produced by GBS and are dispersed throughout the whole genome. According to He et al. (2014), these SNPs may find application in molecular marker identification, genetic diversity, genome-wide association studies (GWAS), and genetic linkage research. Thanks to this advancement, wheat breeders may now use several techniques, including genome-wide association studies and quantitative trait loci, to pinpoint the position of quantitative loci linked to significant genes.

The 21-chromosome sequence is now accessible and may be used to find gene models, create physical maps, annotate these models, and assign gene sequences to particular chromosomes by using these novel sequence approaches on various wheat populations.

GWAS stands for Genome-Wide Association Study. Various genotypes with known genetic information are used in this kind of investigation. Although GWAS analysis can be performed with any number of genotypes, a minimum of 100 genotypes are needed (Kumar et al. 2011). GWAS finds quantitative loci that are responsible for the phenotypes under study using the genomic and morphological data that are available (Chang et al. 2018). Several potential genes that are responsible for the examined characteristic could be found by examining the linkage disequilibrium (LD) between the loci that have been discovered. Genes governing biotic stress resistance were found in wheat through numerous GWAS investigations. Candidate genes linked to wheat resistance to leaf rust, stripe rust, and tan spots were discovered by Juliana et al. (2018). Using GWAS, Pariyar et al. (2016) found potential genes governing nematode resistance in wheat. Ando et al. (2018) identified the potential genes governing spring wheat resistance to Hessian fly, Septoria blotch, and stripe rust. Wheat breeders will be able to create a map that shows the chromosomal location of the resistance genes by combining the findings of several GWAS research. For instance, https://maswheat.ucdavis.edu/ currently offers maps of genes that resist stripe rust and wheat stem rust.

Finding potential genes that regulate abiotic stressors was made possible in large part by GWAS. Sukumaran et al. (2018), for instance, discovered potential genes governing durum wheat's resistance to heat and drought. Five potential genes that regulate root length under water stress circumstances were found by Ayalew et al. (2018) on four distinct chromosomes. Using SSR markers, Liu et al. (2018a) identified 24 candidate genes governing salt tolerance on 17 chromosomes.

Furthermore, numerous studies have identified candidate genes that regulate significant agronomic variables, including plant height, 1000-kernel weight, grain yield and related traits (Wang et al. 2017; Garcia et al. 2019), spike-related traits (Liu et al. 2018a), and more (Daba et al. 2018). Finding the candidate genes that govern significant wheat features will enhance wheat breeding, particularly if GWAS is followed by a thorough examination of the identified loci utilizing haplotype-block analysis. However, the accuracy of the phenotypic data that is currently available influences the GWAS's accuracy. High phenotyping platforms and skilled researchers are needed in this situation. GWAS has a bright future in improving wheat thanks to developments in bioinformatics, statistics, and sequencing techniques.

Numerous QTL mappings have been conducted on wheat using various marker types. Understanding the variety of the examined germplasm and pyramiding numerous target genes in a single genotype will be made easier by all of these tasks (Khan 2015).

8.5.2.2 Molecular Marker Assisted Breeding

Genomic selection (GS) is widely employed in breeding programs for both plants and animals. GS's primary goal is to save breeding time by predicting genotype performance using genomic data, analyzing a small number of genotypes, and forecasting the performance of the remaining population. GBS generates a large number of SNP markers that are dispersed throughout the genome. Even for complicated genomes like wheat, these SNPs could be utilized to create precise genomic prediction (GP) techniques. However, several factors that impact the GP's accuracy should be considered before implementing it. These parameters include marker density, population size, and population structure. According to a study by Norman et al. (2018), the accuracy of GP increased when the tested set was quite diverse, the markers utilized were high-density, and no more than 2000 genotypes were employed.

To determine the optimal amount of genomic selection in wheat, a lot of work was done. According to Belamkar et al. (2018), initial yield trials in winter wheat could use 50% genomic prediction. According to Poland et al. (2012), genomic prediction accuracy for grain yield ranged from 0.28 to 0.45. Using various prediction models, GP accuracy varied from 50% to 80% for autograph measurements, protein content, gluten index, and grain yield (Haile et al. 2018).

According to the GWAS analysis, a small number of loci with negligible effects were revealed to regulate Fusarium head blight (FHB). According to Dong et al. (2018), the GS accuracy for this feature was 60%. Utilizing GS in wheat breeding programs appears to boost breeding progress and result in discoveries in functional genomics and breeding.

8.6 Genetic Engineering and Genome Editing

Direct manipulation of wheat genes is accomplished by genetic engineering or genetic modification. By improving wheat production efficiency and preventing losses from biotic and abiotic stresses by creating high tolerance lines under a variety of situations, it is anticipated to enhance conventional breeding for future wheat production.

8.6.1 Genetic Transformation Methodologies

By utilizing the developments in wheat genome sequencing. Transformation systems, for example. Agronomically significant genes have been used to successfully apply *Agrobacterium tumefaciens* to wheat. Ever since wheat transformation with A was first reported. Cheng et al. (1997) developed the *A. tumefaciens* strategy, which has since spread throughout the world and created transgenic wheat plants with the required genes or genes of agronomic features (Habib et al. 2014).

8.6.2 Gene Editing Methodologies

Agronomically significant genes have been successfully used to apply *Agrobacterium tumefaciens* to wheat. The *A. tumefaciens* strategy has been widely adopted worldwide, leading to the creation of transgenic wheat plants with necessary agronomic genes (Habib et al. 2014). Recent advancements in engineered nucleases, such as ZFNs, TALENs, and CRISPR/Cas9, have made this approach more prevalent. CRISPR/Cas9, in particular, has shown precision and effectiveness in developing wheat lines with desired traits (Gil-Humanes et al. 2017; Liang et al. 2017).

Genetic engineering in wheat must evolve beyond conventional transformation methods to hold promise for enhancing performance and yield. The objective of genetic engineering is to produce plants with multiple desired traits that exhibit consistent inheritance and expression in subsequent generations. The application of CRISPR/Cas9 to key agronomic traits in wheat is promising for achieving significant yield gains. Reviews by Ishida et al. (2015) and Shrawat and Armstrong (2018) have highlighted the use of transformation and gene editing techniques to enhance various agronomic characteristics in stable wheat transgenics.

The use of Cas9-based multiplexed gene editing (MGE) to knockout *TaGW2*, *TaLpx-1*, and *TaMLO* genes has resulted in notable improvements in thousand-grain weight, grain area, width, length, and resistance to *Powdery mildew* and *Fusarium graminearum* (Wang et al. 2018). Although CRISPR technology, as demonstrated by Jiang et al. (2013), offers an economical and time-efficient approach to genome editing, it can be complex when dealing with polyploid germplasm like the wheat genome, requiring editing of each copy of the targeted gene. Nevertheless, several studies have shown the effectiveness of CRISPR in modifying specific wheat genes. For example, Kim et al. (2018) documented successful modifications using CRISPR/Cas9 technology on protoplasts.

8.6.3 Improved Traits

The goal of genetic engineering is to create plants that exhibit several desired features with consistent inheritance and expression in subsequent generations. With novel gene editing, significant yield gains would be anticipated. CRISPR/Cas9 was applied to key agronomic parameters of wheat. According to reviews by Ishida et al. (2015) and Shrawat and Armstrong (2018), transformation and gene editing approaches have been used to improve a variety of agronomic properties to create stable wheat transgenics. Additionally, it has been demonstrated that CRISPR/Cas9 enhanced bread wheat's resistance to *Blumeria graminis* f. Sp. mutant TaMLO genes that cause infection (powdery mildew) (Wang et al. 2014). TaVIT2 has been edited using CRISPR/Cas9 to increase the nutritional value of wheat, particularly the Fe content (Connorton et al. 2017). Wheat dehydration response element binding protein 2 (TaDREB2). Ploidy levels in wheat, such as hexaploidy wheat, present a barrier to gene editing techniques. The results cited above demonstrated that CRISPR/Cas9 might be used to improve wheat.

8.7 Tissue Culture Applications

Numerous useful tools are provided by biotechnology, which advances breeding procedures to enhance genetic traits, including salt tolerance, in economically viable crops. Plant breeding programs can benefit from the diverse genetic variety that tissue culture produces in plant species. Mutants having beneficial agronomic features, like as disease resistance or tolerance to salt or drought, can be quickly discovered using in vitro selection. However, the genetic stability of somaclonal variation in the following generations is crucial to its effective utilization (Mercado et al. 2000; Jain 2001; El-Aref 2002). When it helps introduce genetic material from wild relatives to cultivate or manufacture DHLs, embryo rescue becomes even more significant. The significance of genes from wild relatives in wheat breeding efforts was documented by Friebe and Gill (1996).

In wheat, numerous investigations into the genetic regulation of in vitro characteristics utilizing immature embryos have also been documented (Nawara et al. 2017). In genetic research, mating designs like the diallel and line x tester are frequently used to examine how certain genotypes acquire phenotypes in vitro (Nawara et al. 2017). The methods of Griffing (1956) are typically used to analyze diallel data. According to Nawara et al. (2017), these approaches divide the total variance of diallel data into GCA of the parents and SCA of the crosses. Numerous useful tools are provided by biotechnology, including cell, tissue, and organ culture, which advances breeding procedures to enhance genetic traits, including salt tolerance, in economically viable crops. Plant breeding programs can benefit from the diverse genetic variety that tissue culture produces in plant species. Mutants having beneficial agronomic features, like as disease resistance or tolerance to salt or drought, can be quickly discovered using in vitro selection. However, the genetic stability of somaclonal variation in the following generations is crucial to its effective application. When it helps introduce genetic material from wild relatives to cultivate or manufacture DHLs, embryo rescue becomes even more significant. The significance of genes from wild relatives in wheat breeding efforts was documented by Friebe and Gill (1996).

8.8 Wheat Mutation Breeding

The mutation's alteration is unintended and could result in the creation of new materials. All genes differ from one another due to mutations. Through mutation breeding, desired mutations are created and used to improve crops. It is frequently employed to develop characteristics in crops like larger seeds and self-pollinated crops like wheat. The process of creating variations is called mutation. A mutation has limited use in enhancing the variety of seed-bearing crops like wheat if it cannot be passed down through the germ line and is therefore not inherited. Mutations that cannot be passed down through a pure line to the following generation cannot be inherited and passed down from one generation to the next. As a result, they are of little use in the development of wheat varieties that reproduce by seed.

8.8.1 Mutation Breeding Program

The M0 generation, whose wheat seed would undergo mutagenesis treatment, initiated the mutation breeding program. Additionally, the first year of treated seeds is when the M1 generation of wheat grains is developed, and each plant is harvested independently. M2 generation (second year): Individual plants are developed in a line, and the plants with the mutant allele, phenotypic alterations, or mutations are chosen and collected separately. This generation's M3 (third year) saw the growth of individual plants and the selection of stable lines from every row. M4 (fourth year) Comparative yield trial using a suitable check and superior lines chosen for the preliminary yield testing. These generations' M5-7 (fifth through seventh years) reproduced yield trials at multiple places, and the exceptional line was introduced as a new variety. M8 seed multiplication for farmer distribution (eight-year) Fig 8.6.

8.8.2 Multiplication of Seeds and Introduction of New Varieties

The mutation's alteration is unintended and could result in the creation of new materials. All genes differ from one another due to mutations. Through mutation breeding, desired mutations are created and used to improve crops. It is frequently employed to develop characteristics in crops like larger seeds and self-pollinated crops like wheat. The process of creating variations is called mutation. A mutation has limited use in enhancing the variety of seed-bearing crops like wheat if it cannot be passed down through the germ line and is therefore not inherited. Mutations that cannot be passed down through a pure line to the following generation cannot be inherited and passed down from one generation to the next. As a result, they are of little use in the development of wheat varieties that reproduce by seed.

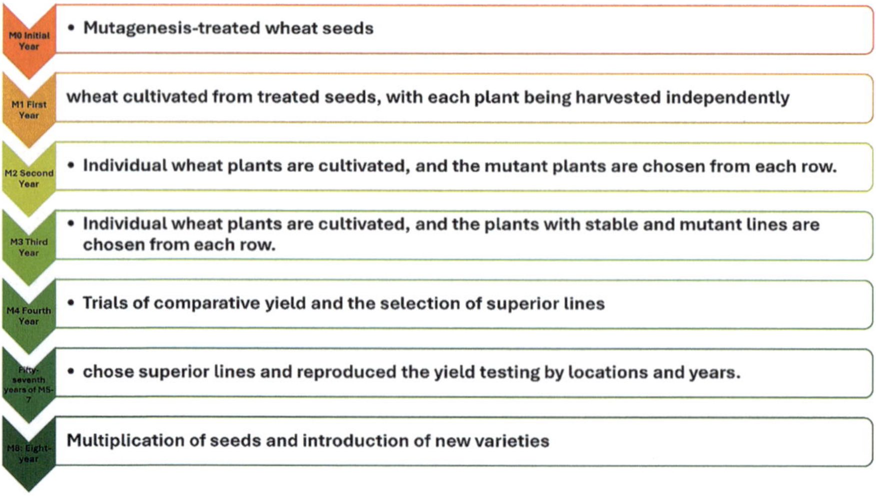

Fig. 8.6 Program for Mutation Breeding. (Prepared by Dr. Nerhan A. Eid)

The M0 generation, whose wheat seed would undergo mutagenesis treatment, initiated the mutation breeding program. Additionally, the first year of treated seeds is when the M1 generation of wheat grains is developed, and each plant is harvested independently. M2 generation (second year): Individual plants are developed in a line, and the plants with the mutant allele, phenotypic alterations, or mutations are chosen and collected separately. This generation's M3 (third year) saw the growth of individual plants and the selection of stable lines from every row. M4 (fourth year) Comparative yield trial using a suitable check and superior lines chosen for the preliminary yield testing. These generations' M5-7 (fifth through seventh years) reproduced yield trials at multiple places, and the exceptional line was introduced as a new variety. M8 seed multiplication for farmer distribution (eight years).

First-generation markers are the name given to this type of marker. These molecular identifiers were created using differences in the length of DNA fragments generated by a particular restriction enzyme. Usually, it is employed for fingerprinting or to differentiate between two or more people. This kind of marker includes restriction fragment length polymorphism. To raise the amount of β-glucan in wheat, genes from distantly related wild and cultivated relatives can be introduced using the techniques of interspecific hybridization and chromosomal engineering. Because barley grain had the highest β-glucan content (Burton et al. 2008).

Finding QTLs or genes and creating molecular indicators for wheat's arabinoxylan content.

Using various wheat populations, several QTLs for arabinoxylan and WE-AX content have been discovered in recent years (Yang et al. 2016; Quraishi et al. 2011). Genes for these properties have also been discovered. According to reports, the arabinoxylan production pathway involves four genes: a methyltransferase, a glucosyltransferase, a peroxidase, and a-glucosidase (Marcotuli et al. 2015). Nine of the 19 QTLs that Marcotuli et al. discovered to arabinoxylan content (Mitchell et al.

2007) coincide with annotated genes that encode enzymes involved in arabinoxylan biosynthesis (Anders et al. 2012; Charmet et al. 2009; Lee et al. 2023).

8.8.3 QTLs and Genes Linked to Wheat B-Glucan Concentration

In recent years, studies have been conducted utilizing several wheat populations to examine genetic variability related to wheat β-glucan concentration. QTLs were linked to the β-glucan content in *Aegilops biuncialis*. A different QTL that was co-located with the glutathione S-transferase-encoding GTS3L was involved in several metabolic processes. However, the majority of these QTLs are not environment-invariant (Lombard et al. 2014; Ivanizs et al. 2022), Three homologs of TaCslF6 have been localized close to the centromeres on wheat chromosomes 7AL, 7BL, and 7DL (Jääskeläinen et al. 2013), making it a key candidate for wheat β-glucan biosynthesis (Marcotuli et al. 2018). When RNA interference (RNAi) was used to precisely reduce TaCslF6 in hexaploid wheat grain, the amount of β-glucan decreased by 30–52% (Marcotuli et al. 2018). Marcotuli et al. found that the amount of β-glucan in durum wheat is positively correlated with CslF6 expression (Danilova et al. 2019).

8.8.4 Molecular Markers and QTLs/Genes for Wheat-Resistant Starch Content

The offspring were also descended from Ae. When peregrine and KU37 were crossed, a connection between resistant starch content and BGRANULE CONTENT 1 (BGC1) was discovered (Chia et al. 2020). In early grain development, BGC1 prevents the creation of A-granules, while in intermediate grain development, it encourages the synthesis of B-granules. Generally speaking, resistant starch is higher in starches with a higher amylose content. Therefore, genes and QTLs controlling the amount of amylose may also influence the amount of resistant starch. Qhams7A.1, a stable QTL linked to amylose content, was discovered by analysis of an F_2 population created by crossing the high-yielding variety ‘WH 1105’ with the high-amylose donor ‘TAC 75’. The GBSSI gene is part of this QTL. High-amylose starch is linked to the mutant protein GBSSI_L539P, which is produced by a high-amylose line with the missense mutation g.35767184 T > C in GBSSI. By altering its starch-binding capacity, the L539P mutation increases the activity of GBSSI, indicating that this mutant allele may be a molecular target for increasing the amount of amylose and resistant starch in wheat grain (Sharma et al. 2022). Numerous other molecular indicators of resistant starch concentration have also been created. For instance, the SSR marker Xbarc59, which is found on spring wheat chromosomes 2D and 5B, was discovered using bulk segregation analysis (Mishra et al. 2016).

The Xbarc59 marker was effectively used to screen for genotypes with high, medium, and low resistant starch levels in the F_2 generation that resulted from a cross between Wuchun 4 and M344. Wheat lines with high amylose and high resistant starch contents have been effectively created using ethyl methanesulfonate (EMS)-targeted mutagenesis. The amount of amylose in regular wheat grains ranges from 20% to 30%. Using 101 EMS-induced mutant lines, Mishra et al. created a population of bread wheat with resistant starch content varying from 1% to 41% and amylose content ranging from ~3% to 76% (Irshad et al. 2022). A higher percentage of resistant starch can result from altering starch biosynthesis to raise the amylose content by either increasing the expression of the GBSSI gene in the amyloplast or decreasing the expression of the SBE or SS genes. GBSSI is encoded by the waxy (Wx) gene. By altering its capacity to bind starch, a mutation in Wx-B1 decreased GBSSI activity, suggesting that altering GBSSI activity can alter amylose content.

One of the main components of wheat grain's cell wall is arabinoxylan. According to Freeman et al. (2017), arabinoxylan makes up 70% of the cell wall polysaccharides in wheat starchy endosperm, with a quantity ranging from 5% to 27% based on whole wheat bran. One of the main sources of dietary fiber in the human diet, arabinoxylan also plays a critical role in defining the quality of wheat near the end of its useful life, including its use in distillation and animal feed (Lu et al. 2000). In healthy adult subjects, postprandial glycemic reactions are negatively connected with the consumption of bread high in arabinoxylan (Saulnier et al. 2007). Arabinoxylan gets entangled with starch and other carbs in the gastrointestinal tract, which delays the absorption of carbohydrates and lowers blood sugar levels after meals.

Wheat's arabinoxylan composition and biosynthesis.

Two fractions are commonly used to classify arabinoxylan: water-extractable (WE-AX) and water-unextractable (WUAX). About 25–30% of the wheat grain's arabinoxylans are WE-AXs (Zannini et al. 2022). A linear (1,4) connected backbone of β-D-xylopyranosyl (xylose) residues makes up the arabinoxylan found in white flour (Pellny et al. 2012). Due to the presence of several isoforms of enzymes involved in biosynthesis, xylose and arabinose, the monomers that make up the arabinoxylan polysaccharide, are biosynthesized in the cytoplasm and Golgi. Uridine diphosphate-sugar transporters are among the membrane channels that control the intricate interaction between metabolites and the cytosol during the biosynthesis of the arabinoxylan backbone in the Golgi. Glycosyltransferases (GTs) and glycoside hydrolases (GHs), two key categories of enzymes, participate in the biosynthesis of the final arabinoxylan molecule. By delivering the nucleoside diphosphate sugars—which function as a sugar donor with a nucleoside phosphate—to a specific receptor, GTs accelerate the formation of glycosidic linkages.

Enhancement of wheat's arabinoxylan content through genetic means RNA interference (RNAi) downregulating the expression of TaGT47_2 or TaGT43_2 led to a 50% reduction in overall arabinoxylan levels and a 25–30% increase in arabinoxylan levels (Charmet et al. 2009). According to immunolabeling, the starchy endosperm cell walls in these lines had 50% less cell wall thickness and fewer xylan and arabinoxylan epitopes than the wild-type control. Additionally, Freeman et al.

observed that downregulating TaGT43_2 and TaGT47_2 in wheat decreased WE-AX levels (Pellny et al. 2020). About 65% less arabinoxylan was present in a mutant with TaGT43_2 three homoeologous were knocked out (Cavallero et al. 2002). TaXAT1 gene in the GT61, which is most highly expressed in starchy endosperm, is silenced, α-(1,3)-linked arabinofuranosyl (Araf) residues in arabinoxylan are reduced by 70–80%, and β-(1,4)-D-xylopyranosyl (Xylp) residues are substituted (Lovegrove et al. 2013). All things considered, our findings support the importance of particular genes in regulating the amount of arabinoxylan in wheat. Nevertheless, arabinoxylan content has not yet been enhanced by RNA interference.

8.8.5 Wheat Grain B-Glucan

β-glucan makes up around 20% of all cell wall polysaccharides (Freeman et al. 2017). According to Vetvicka and Vetvickova (2009), β-glucan effectively lowers postprandial blood glucose levels. To reduce the adsorption of biliary salts in the bowel (Izydorczyk and Dexter 2008). Compared to barley or oats, wheat grains have a significantly lower β-glucan content. In particular, barley and oats have βglucan contents of 2.5–11.5% and 2–7.8%, respectively, while wheat only has 0.4–1.4% (Danilova et al. 2018). Wheat is a prime candidate for increasing its β-glucan concentration due to its extensive consumption.

Long or short-chain polymers of glucose subunits joined by (1,3) and (1,4) connections are known as β-glucans. Lichenase, a particular enzyme, breaks down single (1,3) links into two or three (1,4) connections, producing trisaccharide and tetrasaccharide molecules (Jääskeläinen et al. 2013). The subaleurone layer of the wheat seed has a higher concentration of β-glucans, while the endosperm has a lower concentration (Bulone et al. 2019). Enzymes from the cellulose synthase-like superfamily, which is a subfamily of the large glycosyltransferase GT2 family, biosynthesize β-glucans in cereals (Marcotuli et al. 2016).

Furthermore, the β-glucan concentration was much higher in a wheat addition line that carried the HvCslF9 gene (Marcotuli et al. 2019). All things considered, introducing barley HvCslF6 to wheat helps increase the amount of β- Among the two *Aegilops* species they discovered were Ae, UU genome, and umbellulata. The possible options for improving the βglucan content of wheat were *Aegilops markgrafii*, CC genomes, which showed a higher β-glucan content than *Triticum* species (Sestili et al. 2015). In particular, cultivated hexaploid bread wheat (approximately 0.83%) appeared to have a significantly lower β-glucan level than species with the U genome (around 5.3%). An additional potential gene is ISA1, which codes for Isoamylase 1 (ISA1), a debranching enzyme of the isoamylase type that is crucial to the production of amylopectin. When Sestili et al. used RNA interference to knock down Isa1 in durum wheat, the transgenic lines that resulted had lower starch contents and somewhat higher levels of phytoglycogen and β-glucan (Regina et al. 2015). These results show that the amount of β-glucan in wheat grains can be controlled by using a transgenic approach.

The primary polysaccharide in wheat is starch, which is divided into amylose and amylopectin according to the form of glucose polymerization (Englyst et al. 1992). In the wheat grain endosperm, amylose, a linear structure joined by α1, 4 bonds, is closely packed inside the starch granule (Englyst et al. 1992). The complex structure that takes up the majority of the space inside the starch granule of wheat grain endosperm is created by amylopectin, a branching molecule that has both α1, 4 and α1, 6 links. Amylopectin usually makes up more than 70% of the overall composition of wheat grain, whereas amylose makes up less than 30%. Depending on its post-consumption behavior, starch is typically classified as resistant starch, rapidly digested starch, or slowly digestible starch (Sajilata et al. 2006). Dietary starch that ferments only in the colon and cannot be broken down in the small intestine is referred to as resistant starch (Harris 2019). Resistant starch consumption has been demonstrated to help prevent it (Park et al. 2009). According to Zeeman et al. (2010), increasing the amount of amylose can slow down the pace at which starch is digested and increase the amount of resistant starch type 2, one of the five forms of resistant starch (type 1–5). The production of high-amylose grains with greater quantities of resistant starch type 2 and, as a result, more dietary fiber in refined flour is made possible by concentrating on metabolic pathways (Zeeman et al. 2010).

Biosynthesis of resistant starch involves the biosynthesis of sucrose in leaves, which is subsequently carried into grains and broken down by SUS into fructose and UDP-glucose. After that, AGPase transforms the products into ADP-glucose. The starch production is then started when soluble ADP-glucose is delivered into the amyloplasts as a direct substrate (Park et al. 2009). While SBE, DBE, and SS are in charge of amylopectin production, GBSSI regulates amylose biosynthesis on its own (MacNeill et al. 2017). According to Kumar et al. (2016), the percentage of amylose is favorably correlated with the amount of resistant starch. Thus, the most popular way to enhance the amount of resistant starch in wheat endosperm is to either increase GBSSI or decrease the expression of SS and SBE genes.

Effective techniques for increasing resistant starch content include marker-assisted selection and plant breeding with crop types that have high levels of resistant starch. Three mutant lines were also created by the same authors using mutagenesis and traditional marker-assisted breeding (Botticella et al. 2016). Transgenic wheat lines with higher amylose content have also been produced via RNA interference. While simultaneously downregulating SBEIIa and SBEIIb produced starch that contained more than 70% amylose, downregulating SBEIIb alone did not influence the amount of amylose. Several measures of large-bowel function improved in rats fed whole meal flour from the RNAi double downregulated lines. Similarly, in durum wheat, RNAi-mediated downregulation of SBEIIa increased amylose content by up to 75.05% (Wang et al. 2021). Lastly, amylose content (31.4%) was significantly increased by RNAi-induced SSI suppression (Wang et al. 2021). Transgene-free lines with different amylose contents have been successfully produced using this technique in crops like maize and rice (Wang et al. 2021).

The SBEII, SSII, and SSIII mutants' elevated amylose and resistant starch levels are linked to modifications in grain properties that might restrict their applicability in breeding initiatives. Yield penalties result from these mutants' decreased starch levels and grain weight (Fahy et al. 2022). For instance, Hazard et al. found that mutations in the durum wheat genes SBEIIa/bAB were linked to a 15% decrease in grain output and an average 5.2% decrease in kernel weight (Fahy et al. 2022). According to Schönhofen et al. (2016), SBEII mutants showed 2.8% and 5.8% decreases in kernel weight and grain yield, respectively (Fahy et al. 2022).

By altering the SBEII, SSII, or SSIII genes, high-amylose wheat varieties have been effectively created in recent decades. These wheat varieties have a lower glycemic index and are better for human health (Fahy et al. 2022). Unfortunately, this had a detrimental effect on the quality of grain or flour that could be processed. SBEII mutations significantly affect the quality of bread. End-use quality characteristics include increased protein content, water absorption, starch degradation, and grain hardness (Botticella et al. 2016). According to Li et al. (2022), bread manufactured with high-amylose wheat flour had a higher hardness, a more dense crumb structure, and a lower loaf volume than wheat bread made with wild-type wheat flour. They also discovered that raising the amylose concentration significantly decreased cooking quality. According to Schönhofen et al. (2017), durum wheat SBEIIa/b-AB mutant lines performed well in terms of pasta stiffness but poorly in terms of pasta color and semolina extraction. Comparing these mutants to the wild-type lines, the average difference in firmness was 12.4%. In comparison to the wild-type line, pasta prepared with the SSIIa null mutant's semolina showed a much tougher texture even when overdone, higher cooking loss, shorter cooking times, and less water absorption. When compared to the wild type, the SSIIa null mutant's cooked and uncooked pasta had less brightness. As a result, initiatives to increase the amount of dietary fiber must keep an eye on how processing quality is affected and choose characteristics that have the least negative effects on quality.

Increasing the thickness of the bran or aleurone layer in wheat can improve its dietary fiber content. Unlike barley, wheat has a lower and less uniform distribution of β-glucans in the endosperm, with a high concentration in the aleurone layer, making up about 29% of its dry weight (Li et al. 2022). Adding wheat bran to food as a dietary fiber supplement is effective in supporting gastrointestinal and cardiovascular health.

Therefore, one effective way to enhance the dietary fiber in wheat is by increasing the thickness of the bran or aleurone layer. Recent studies have shown that increasing the thickness of the aleurone layer in rice can improve its nutritional qualities (Li et al. 2022). Additionally, Chen et al. identified a low-gluten protein lgp1 mutant in wheat that had significantly higher bran and dietary fiber content compared to the wild type (Song et al. 2022). This mutant could be utilized to enhance the dietary fiber content of wheat. Furthermore, mild alterations in the expression of genes linked to starch production, which are abundantly expressed in the endosperm, may increase the amount of dietary fiber without negatively affecting rice yield and quality characteristics. In particular, Xu et al. used cytidine base editors (CBEs) to create a series of Wxb mutants that increased the amylose content by 1.4–11.9% without compromising the appearance or quality of milled rice. Since mutations in these genes typically result in mild alterations in starch production, limiting undesirable effects, another strategy would be to target genes related to starch biosynthesis that are expressed at relatively low levels in the endosperm. Targeting gene regulatory regions with CRISPR/Cas9 allowed several other researchers to effectively fine-tune gene expression (Song et al. 2022). The effectiveness of these methods in raising wheat's dietary fiber content without compromising yield or end-use quality has yet to be determined. One of the main issues that needs to be resolved in the future is this one.

8.9 Conclusion and Prospects

Due to the significant economic importance of wheat and its various uses in food, feed, seed, and industry, a large amount of scientific research has been conducted globally. In order to ensure meaningful natural variation at the phenotypic level, plants from different genetic resources can be crossed to introduce novel alleles. For instance, crossing modern varieties with locally adapted ones can increase genetic diversity and pre-select for desired traits. This is necessary due to the time-consuming and challenging nature of traditional breeding techniques. Although new biotechnology approaches for wheat utilizing DNA sequences and molecular techniques have garnered attention from plant breeders and geneticists, traditional breeding methods remain crucial as the first steps in developing new wheat cultivars with desired traits. Nonetheless, several promising cultivars have been developed that are adapted to biotic and abiotic stresses and climate change, necessitating new breeding methods for high grain production and the creation of wheat varieties resilient to stressors.

The main challenges in wheat breeding programs include biotic and abiotic stresses, as well as food demands. Wheat growth is particularly affected by stressors occurring before and after flowering, with frost and high temperatures being key climate variables of interest to breeders worldwide. Consequently, breeding programs and the education of new plant pathologists and breeders require increased funding. Efforts must be intensified to develop highly adaptable wheat varieties that can thrive in the face of abiotic stresses like drought and salt to mitigate the impacts of global warming. Recent biotechnology methods have shown promise in creating new wheat cultivars with desirable traits. Additionally, the sequencing of the wheat genome has caught the attention of geneticists and plant breeders interested in developing new wheat cultivars with resistance to biotic and abiotic stresses, as well as advantageous agronomic traits.

Wheat is globally utilized for both industrial and food purposes, holding substantial economic value. Introducing novel alleles through the crossing of plants from different genetic resources, such as modern varieties with locally adapted ones, enhances genetic diversity and pre-selection for traits necessary to ensure significant natural phenotypic variation. This is due to the labor-intensive and time-consuming nature of traditional breeding methods. Traditional breeding techniques remain essential for producing new wheat cultivars with desired traits, despite the appeal of new biotechnology approaches utilizing DNA sequences and molecular techniques. Nevertheless, several promising cultivars have been developed that can withstand biotic and abiotic stressors, as well as climate change, necessitating innovative wheat breeding methods to create resilient new varieties with higher yields.

The major obstacles in wheat breeding include biotic and abiotic stresses, as well as meeting food demands. Worldwide environmental changes, particularly frosts and warmer temperatures, are of great interest to wheat breeders. Diseases like Cercospora leaf spot, Aphanomyces root rot, pests, and worms are expected to spread rapidly and impact global food security. More work is needed to breed new wheat varieties capable of thriving under abiotic stresses such as salinity and drought. Recent biotechnology techniques have been successful in creating promising new cultivars with desirable traits. The sequencing of the wheat genome and advancements in molecular techniques are encouraging breeders and geneticists to develop new cultivars with desired agronomic traits and tolerance to biotic and abiotic stresses.

There are numerous reasons to further study wheat development to enhance both quantitative and qualitative qualities. Essential biotic and abiotic stresses, physiology, crop yield, and their constituent features should all be linked directly to the creation of DNA markers. Genes or QTLs related to both qualitative and quantitative traits should be identified to improve these characteristics. Additionally, upgrading biotechnology and germplasm is necessary to accelerate the development of promising new genotypes with high yields and quality. Further research is needed on how wheat interacts with microorganisms like rhizobia and mycorrhiza to create crops resistant to biotic stresses.

Haploids and doubled haploid lines (DHLs) are crucial for producing pure lines that can be used in breeding programs, as most commercially important wheat crop traits lack genetically distinct pure lines. It is also essential to establish DNA markers closely linked to important physiological, vegetable, grain, biotic and abiotic stress properties. Genes or QTLs should be identified to enhance both qualitative and quantitative attributes. Additionally, upgrading biotechnology and germplasm is necessary to expedite the development of new promising lines with high yields and grain quality. Producing haploids and doubled haploid lines (DHLs) is essential to generate pure lines for the majority of commercially significant crop traits in wheat that lack genetically distinct pure lines. Developing DNA markers strongly associated with key biotic and abiotic stress, physiological, vegetable, grain production, and constituent features is crucial. To improve both qualitative and quantitative traits, genes or QTLs need to be identified. Upgrading biotechnology and germplasm is crucial to accelerate and facilitate the development of new promising lines with high yields and grain quality.

Appendix 8.1 List of Major Institutes Engaged in Research on Wheat

Institution	Specialization and research activities	Contact information	Website
International Center for Agricultural Research in the Dry Areas (ICARDA) ICARDA	Breeding and molecular breeding program	Department of Wheat Breeding/ Genetics, International Center for Agricultural Research in the Dry Areas (ICARDA), Rabat, Morocco	http://www.icarda.org/
Leibniz Institute of Plant Genetics and Crop Plant Research (IPK), Gatersleben	Breeding and molecular breeding research	Genbank Department, Leibniz Institute of Plant Genetics and Crop Plant Research (IPK), Gatersleben, Germany	https://www.ipk-gatersleben.de/
The International Maize and Wheat Improvement Center (CIMMYT)	Breeding and molecular breeding program	Wheat Program, The International Maize and Wheat Improvement Center (CIMMYT), Mexico	https://www.cimmyt.org
Wheat Department, Agriculture Research Center (ARC)	Breeding program	Department of Wheat, Agriculture Research Center (ARC), Giza, Egypt	http://www.arc.sci.eg/

Appendix 8.2 World List of Recommended Wheat Varieties

Cultivars	Year of release	Pedigree	Cultivation location / Breeding sites	Important traits
Bima1	1951	Mazha/Biyu	Shaanxi Province, China	None dwarf
Changhan58	2004	Changwu112/PH 82-2	Shaanxi Province, China	Rht-B1b
Fengchan3	1964	Danmai 1/Xinong 6028×Bima1	Shaanxi Province, China	None dwarf
Gemmiza 10	2004	Maya74"s"/On//1160147/3/Bb/4/Chat"s"/5/Ctow	Egypt	High yield, resistance to rust and smut
Giza 168	1999	Mil/Buc//Seri	Egypt	High yield, resistance to rust and smut
Mazha	1940	Landrace	Shaanxi Province, China	None Dwarf genes

Cultivars	Year of release	Pedigree	Cultivation location / Breeding sites	Important traits
Sakha 94	2004	Opata/Rayon//Kauz	Egypt	High yield, resistance to rust and smut
Seds 9	1994	Maya"s"/Mon"S"/4// CMH72.428/MRC//jip/3/ CMH74A582/5/ Giza157*2SD10003	Egypt	High yield, susceptible to rust, resistance to smut, long spike
Xiaoyan6	1981	(ST2422×464)/Xiaoyan96	Shaanxi Province, China	Rht-B1b+Rht8

References

Aaronsohn A (1910) Agricultural and botanical explorations in Palestine. US Government Printing Office, Washington, DC

Abdel-Hamid AME, Salem KFM (2021) Breeding strategies of garden pea (*Pisum sativum* L.). In: Al-Khayri JM, Jain SM, Johnson DV (eds) Advances in plant breeding strategies: vegetable crops. Springer, Cham. https://doi.org/10.1007/978-3-030-66969-0_9

Abdullaev FK, Rustamov NSU, Yakubov MD (2025) Uzbekistan genebank. In: Al-Khayri JM, Salem KFM, Jain SM (eds) Plant gene banks. Springer, Singapore. https://doi.org/10.1007/978-981-99-4236-7_57-1

Aboshama HM, Salem KFM, Hassan HMF, Tahoun MH (2015) Estimating heterosis and combining ability for *in vivo* and *in vitro* traits using diallel cross on *tomato* (*Solanum lycopersicum* L.). Minufiya J Agric Res 40(3):757–770

Abou- Taleb SM, Abd – El Maksoud HS (2013) Comparative study on four cereal genotypes. Micromorphological charachteristics of leaf and grain by using S.E.M. New York Sci J 6(12):185–192

Abu-Ellail FFB, Salem KFM, Saleh MM et al (2021) Molecular breeding strategies of beetroot (*Beta vulgaris* ssp. vulgaris var. conditiva Alefeld). In: Al-Khayri JM, Jain SM, Johnson DV (eds) Advances in plant breeding strategies: vegetable crops. Springer, Cham. https://doi.org/10.1007/978-3-030-66965-2_4

Abulaila K, Al Sane KO, Tahabsom Z (2025) Jordan seed bank. In: Al-Khayri JM, Salem KFM, Jain SM (eds) Plant gene banks. Springer, Singapore. https://doi.org/10.1007/978-981-99-4236-7_45-1

Akparov ZI, Mammadova SA, Kalantarova NS, Mirzaliyeva IA, Mammadov AT (2025) National Genebank of Azerbaijan. In: Al-Khayri JM, Salem KFM, Jain SM (eds) Plant Gene Banks. Springer, Singapore. https://doi.org/10.1007/978-981-99-4236-7_31-1

Al Lawati AH (2025) Plant genetic resource conservation of Oman. In: Al-Khayri JM, Salem KFM, Jain SM (eds) Plant gene banks. Springer, Singapore. https://doi.org/10.1007/978-981-99-4236-7_50-1

Aladele SE et al (2025) Plant genetic resources conservation and utilization in Nigeria. In: Al-Khayri JM, Salem KFM, Jain SM (eds) Plant gene banks. Springer, Singapore. https://doi.org/10.1007/978-981-99-4236-7_17-1

Aldahak L, Salem KFM, Al-Salim SHF, Al-Khayri JM (2021) Advances in chicory (*Cichorium intybus* L.) breeding strategies. In: Al-Khayri JM, Jain SM, Johnson DV (eds) Advances plant breeding strategies: vegetable crops. Springer, Cham. https://doi.org/10.1007/978-3-030-66969-0_1

Al-Khayri JM, Salem KFM, Jain SM (2025) Plant gene banks. Springer, Singapore. https://doi.org/10.1007/978-981-99-4236-7

Almarri NB, Ahmad S, Alsubaie GTG, Elshal MH (2025) Saudi Arabia Gene Bank. In: Al-Khayri JM, Salem KFM, Jain SM (eds) Plant gene banks. Springer, Singapore. https://doi.org/10.1007/978-981-99-4236-7_54-1

Alqahtani MM, Saleh MM, Alwutayd KM et al (2024) Performance and genotypic variability in diverse date palm (*Phoenix dactylifera* L.) cultivars for fruit characteristics. Genet Resour Crop Evol. https://doi.org/10.1007/s10722-023-01719-0

Anders N, Wilkinson MD, Lovegrove A, Freeman J, Tryfona T et al (2012) Glycosyl transferases in family 61 mediate arabinofuranosyl transfer onto xylan in grasses. Proc Natl Acad Sci USA 109:989–993

Andersson AAM, Andersson R, Piironen V, Lampi A-M, Nyström L et al (2013) Contents of dietary fibre components and their relation to associated bioactive components in whole grain wheat samples from the HEALTHGRAIN diversity screen. Food Chem 136:1243–1248

Ando K, Rynearson S, Muleta KT, Gedamu J et al (2018) Genome-wide associations for multiple pest resistances in a Northwestern United States elite spring wheat panel. PLoS One 13(2):e0191305. https://doi.org/10.1371/journal.pone.0191305

Arab S et al (2025) National Gene Bank of Egypt. In: Al-Khayri JM, Salem KFM, Jain SM (eds) Plant gene banks. Springer, Singapore. https://doi.org/10.1007/978-981-99-4236-7_5-1

Ayalew H, Liu H, Börner A et al (2018) Genome-wide association mapping of major root length QTLs under PEG induced water stress in wheat. Front Plant Sci 9:1–9. https://doi.org/10.3389/fpls.2018.01759

Baenziger PS (2016) Wheat breeding and genetics. Reference Module in Food Sciences. https://doi.org/10.1016/B978-0-08-100596-5.03001-8

Békefi Z, Szilágyi S, Desiderio F, Demku T, Varga J (2025) Prunus gene bank in Hungarian University of Agriculture and Life Sciences. In: Al-Khayri JM, Salem KFM, Jain SM (eds) Plant gene banks. Springer, Singapore. https://doi.org/10.1007/978-981-99-4236-7_71-1

Belamkar V, Guttieri MJ, Hussain W et al (2018) Genomic selection in preliminary yield trials in a winter wheat breeding program. G3 (Bethesda) 8(8):2735–2747. https://doi.org/10.1534/g3.118.200415

Bhandari HR, Bhanu AN, Srivastava K et al (2017) Assessment of genetic diversity in crop plants – an overview. Adv Plants Agric Res 7(3):279–286

Bhat JA, Feng X, Mir ZA, Raina A, Siddique KH (2023) Recent advances in artificial intelligence, mechanistic models, and speed breeding offer exciting opportunities for precise and accelerated genomics-assisted breeding. Physiol Plant 175:e13969

Bhusal N, Sarial AK, Sharma P, Sareen S (2017) Mapping QTLs for grain yield components in wheat under heat stress. PLoS One 12(12):e0189594. https://doi.org/10.1371/journal.pone.0189594

Börner A (2006) Preservation of plant genetic resources in the biotechnology era. Biotechnol J 1:1393–1404. https://doi.org/10.1002/biot.200600131

Börner A, Chebotar S, Korzun V (2000a) Molecular characterization of the genetic integrity of wheat (Triticum aestivum L.) germplasm after long-term maintenance. Theor Appl Genet 100:494–497. https://doi.org/10.1007/s001220050064

Börner A, Röder MS, Unger O, Meinel A (2000b) The detection and molecular mapping of a major gene for non-specific adult plant disease resistance against stripe rust (*Puccinia striiformis*) in wheat. Theor Appl Genet 100:1095–1099

Börner A, Schumann E, Fürste A et al (2002) Mapping of quantitative trait loci determining agronomic important characters in hexaploidy wheat (*Triticum aestivum* L.). Theor Appl Genet 105:921–936

Borner A, Landjeva S, Salem KFM, Lohwasser U (2009) Pflanzengenetische Ressourcen als Grundlage für die Züchtung klimatoleranter Sorten. Proceedings of a German Conference 'Weizenforum = Wheat Forum' Norddeutsches Weizenforum Schriftenreihe des Inst. Für Pflanzenbau und –züchtung der CAU Kiel, Germany, 63, 23–27

Börner A, Landjeva S, Salem KFM, Lohwasser U (2010) Plant genetic resources- a prerequisite for drought tolerance breeding in cereals. Proc Jahrestagung der Vereinigung der Pflanzenzüchter und Saatgutkaufleute Österreichs, 24–26 November 2009, Raumberg-Gumpenstein, pp 11–13

Botticella E, Sestili F, Ferrazzano G, Mantovani P, Cammerata A et al (2016) The impact of the SSIIa null mutations on grain traits and composition in durum wheat. Breed Sci 66:572–579

Bowden WM (1959) The taxonomy and nomenclature of the wheats, barleys, and ryes and their wild relatives. Can J Bot 37:647–687

Breidy J (2025) Lebanon gene bank. In: Al-Khayri JM, Salem KFM, Jain SM (eds) Plant gene banks. Springer, Singapore. https://doi.org/10.1007/978-981-99-4236-7_48-1

Breiman A, Graur D (1995) Wheat evolution. Israel J Pl Sci 43:85–98

Briggle LW (1967) Morphology of the wheat plant. In: Quisenberry KS, Reitz LP (eds) Wheat and wheat improvement. American Society of Agronomy, Madison, pp 89–116

Briggle LW, Reitz LP (1963) Classification of *Triticum* species and of wheat varieties grown in the United States. Tech Bull 1278. US Department of Agriculture, Washington, DC

Bulone V, Schwerdt JG, Fincher GB (2019) Co-evolution of enzymes involved in plant cell wall metabolism in the grasses. Front Plant Sci 10:1009

Burton RA, Jobling SA, Harvey AJ, Shirley NJ, Mather DE et al (2008) The genetics and transcriptional profiles of the cellulose synthaselike HvCslF gene family in barley. Plant Physiol 146:1821–1833

Cañar DY et al (2025) Colombia gene bank. In: Al-Khayri JM, Salem KFM, Jain SM (eds) Plant gene banks. Springer, Singapore. https://doi.org/10.1007/978-981-99-4236-7_95-1

Cavallero A, Empilli S, Brighenti F, Stanca AM (2002) High (1→3, 1→4)-β-glucan barley fractions in bread making and their effects on human glycemic response. J Cereal Sci 36:59–66

Chang M, He L, Cai L (2018) An overview of genome-wide association studies. In: Huang T (ed) Computational systems biology, Methods Mol Biol, vol 1754. Humana Press, New York

Charmet G, Masood-Quraishi U, Ravel C, Romeuf I, Balfourier F et al (2009) Genetics of dietary fibre in bread wheat. Euphytica 170:155–168

Cheng M, Fry JE, Pang S et al (1997) Genetic transformation of wheat mediated by *Agrobacterium tumefaciens*. Plant Physiol 115:971–980. https://doi.org/10.1104/pp.115.3.971

Chia T, Chirico M, King R, Ramirez-Gonzalez R, Saccomanno B et al (2020) A carbohydrate-binding protein, B-GRANULE CONTENT 1, influences starch granule size distribution in a dose-dependent manner in polyploid wheat. J Exp Bot 71:105–115

Clark AJ, Sarti-Dvorjak D, Brown-Guedira G et al (2016) Identifying rare FHB-resistant segregants in intransigent backcross and F_2 winter wheat populations. Front Microbiol 7:277. https://doi.org/10.3389/fmicb.2016.00277

Connorton JM, Jones ER, Rodríguez-Ramiro I et al (2017) Wheat vacuolar iron transporter TaVIT2 transports Fe and Mn and is effective for biofortification. Plant Physiol 174:2434–2444. https://doi.org/10.1104/pp.17.00672

Cramer W, Guiot J, Fader M, Garrabou J, Gattuso JP, Iglesias A et al (2018) Climate change and interconnected risks to sustainable development in the Mediterranean. Nat Clim Chang 8(11):972–980. https://doi.org/10.1038/s41558-018-0299-2

Daba SD, Tyagi P, Brown-guedira G, Mohammadi M (2018) Genome-wide association studies to identify loci and candidate genes controlling kernel weight and length in a historical United States wheat population. Front Plant Sci 9:1–14. https://doi.org/10.3389/fpls.2018.01045

Danilova TV, Friebe B, Gill BS, Poland J, Jackson E (2018) Development of a complete set of wheat-barley group-7 Robertsonian translocation chromosomes conferring an increased content of βglucan. Theor Appl Genet 131:377–388

Danilova TV, Poland J, Friebe B (2019) Production of a complete set of wheat-barley group-7 chromosome recombinants with increased grain β-glucan content. Theor Appl Genet 132:3129–3141

Davis DR, Epp MD, Riordan HD (2004) Changes in USDA food composition data for 43 garden crops, 1950 to 1999. J Am Coll Nutr 23:669–682 de Galarreta JIR, Barandalla L, Ritter E, Velázquez ME, Rios D (2025) Spanish Potato Germplasm Bank. In: Al-Khayri JM, Salem KFM, Jain SM (eds) Plant gene banks. Springer, Singapore. https://doi.org/10.1007/978-981-99-4236-7_114-1

Demir P, Onde S, Severcan F (2015) Phylogeny of cultivated and wild wheat species using ATR–FTIR spectroscopy. Spectrochimica Acta Part A Mol Biomol Spectrosc 135:757–763. https://doi.org/10.1016/j.saa.2014.07.025

Desheva G, Valchinova E, Deshev M (2025) Bulgarian gene bank. In: Al-Khayri JM, Salem KFM, Jain SM (eds) Plant gene banks. Springer, Singapore. https://doi.org/10.1007/978-981-99-4236-7_66-1

Devate NB, Manjunath KK, Ghajghate R, Shashikumara P, Reddy UG, Kumar M, Krishna H, Jain N, Singh P, Pratap Singh G (2023) Strategies to Develop Heat and Drought–Tolerant Wheat Varieties Following Physiological Breeding. In: Translating Physiological Tools to Augment Crop Breeding. Springer, Berlin/Heidelberg, pp 19–52

Díez MJ, Valcárcel JV, Solbes E, Muñoz E (2025) Forty years conserving spanish biodiversity in the gene bank of the Universitat Politècnica de València, Spain. In: Al-Khayri JM, Salem KFM, Jain SM (eds) Plant gene banks. Springer, Singapore. https://doi.org/10.1007/978-981-99-4236-7_86-1

Dong H, Wang R, Yuan Y et al (2018) Evaluation of the potential for genomic selection to improve spring wheat resistance to fusarium head blight in the Pacific Northwest. Front Plant Sci 9:1–15. https://doi.org/10.3389/fpls.2018.00911

Echeverry-Solarte M, Kumar A, Kianian S et al (2015) New QTL alleles for quality-related traits in spring wheat revealed by RIL population derived from supernumerary × non-supernumerary spikelet genotypes. Theor Appl Genet 128(5):893–912. https://doi.org/10.1007/s00122-015-2478-0

Eig A (1929) Monographisch-kritische Übersicht der Gattung Aegilops. Feddes Repert 55:1–228

El Rabey H, Salem KFM, Mattar MZ (2013) The genetic diversity and relatedness of eight rice (*Oryza sativa* L.) cultivars as revealed by AFLP and SSRs markers. Life Sci J 10(1):1471–1479

El-Absawy EA, Nowar AA, Salem KFM et al (2015) Estimation of genetic diversity among some bread wheat (*Triticum aestivum* L.) genotypes using molecular markers and agronomic traits. Minufiya J Agric Res 40(3):223–233

Eltaher S, Sallam A, Belamkar V et al (2018) Genetic diversity and population structure of $F_{3:6}$ Nebraska winter wheat genotypes using genotyping-by-sequencing. Front Genet 9:76. https://doi.org/10.3389/fgene.2018.00076

Eltaher S, Baenziger S, Belamkar V et al (2021) GWAS revealed effect of genotype x environment interactions for grain yield of Nebraska winter wheat. BMC Genomics 22:1–14

Eltaher S, Sallam A, Emara HA et al (2022) Genome-wide association mapping revealed SNP alleles associated with spike traits in wheat. Agronomy 12:1469. doi.org/10.3390/agronomy12061469

El-Zanaty AM, Salem KFM, Esmail RM (2011) Detection of genetic diversity in Egyptian cotton (*Gossypium barbadense* L.) varieties using RAPD markers and morphological traits. J Am Sci 7(12):1107–1115

El-Zanaty AM, Salem KFM, Ismail M, Esmail RM (2013) Impact of plant breeding on genetic diversity of the Egyptian cotton (*Gossypium barbadense* L.) varieties as revealed by RAPD markers. J Agric Chem Biotechnol Mansoura Univ 4:217–227

Emara HA, Nower AA, Salem KFM, Eltaher S (2011) Estimating heterosis and combining ability for *in vivo* and *in vitro* traits using diallel cross in Maize (*Zea mays* L*)*. J Prod Dev 16:101–120

Emara HA, Salem KFM, Abdel-Halim AA, Abd El-Kowy RNF (2013) An *in vivo* and *in vitro* traits analysis of a diallel cross in rice (*Oryza sativa L.*). Egypt J Plant Breed 17(2):591–610

Englyst HN, Kingman SM, Cummings JH (1992) Classification and measurement of nutritionally important starch fractions. Eur J Clin Nutr 46(Suppl 2):S33–S50

Fahy B, Gonzalez O, Savva GM, Ahn-Jarvis JH, Warren FJ et al (2022) Loss of starch synthase IIIa changes starch molecular structure and granule morphology in grains of hexaploid bread wheat. Sci Rep 12:10806

FAO (2017) Production Year Book. FAO, Rome. http://www.fao.org/faostat/en/#rankings/countries_by_commodity

Feldman M, Lupton F, Miller T (1995) Wheats. In: Smartt J, Simmonds NW (eds) Evolution of crop plants, 2nd edn. Longman Scientific, London, pp 184–192

Fischer R, Byerlee D, Edmeades G (2014) Crop Yields and Global Food Security. ACIAR/Bruce, Canberra, pp 8–11

Freeman J, Ward JL, Kosik O, Lovegrove A, Wilkinson MD et al (2017) Feruloylation and structure of arabinoxylan in wheat endosperm cell walls from RNAi lines with suppression of genes responsible for backbone synthesis and decoration. Plant Biotechnol J 15:1429–1438

Friebe B, Gill BS (1996) Chromosome banding and genome analysis in diploid and cultivated polyploid wheats. In: Jauhar PP (ed) Methods of genome analysis in plants. CRC Press, Boca Raton/New York/London/Tokyo, pp 39–60

Galea-Gragera FA, Llera Cid F (2025) Spain Germplasm Bank of Annual Forage Legumes Seeds at the Scientific and Technological Research Center of Extremadura (CICYTEX). In: Al-Khayri JM, Salem KFM, Jain SM (eds) Plant Gene Banks. Springer, Singapore. https://doi.org/10.1007/978-981-99-4236-7_102-1

Garcia M, Eckermann P, Haefele S et al (2019) Genome-wide association mapping of grain yield in a diverse collection of spring wheat (*Triticum aestivum* L.) evaluated in southern Australia. PLoS One:1–19. https://doi.org/10.25909/5becfa45c176f

Gesesse CA, Nigir B, de Sousa K, Gianfranceschi L, Gallo GR, Poland J, Kidane YG, Abate Desta E, Fadda C, Pè ME (2023) Genomics-driven breeding for local adaptation of durum wheat is enhanced by farmers' traditional knowledge. Proc Natl Acad Sci USA 120:e2205774119

Ghazy MI, Salem KFM, Sallam A (2020) Utilization of genetic diversity and marker-trait to improve drought tolerance in rice (Oryza sativa L.). Mol Biol Rep. doi.org/10.1007/s11033-020-06029-7

Ghidan WF, Abdulmajid DA (2025) Egyptian rice gene bank. In: Al-Khayri JM, Salem KFM, Jain SM (eds) Plant gene banks. Springer, Singapore. https://doi.org/10.1007/978-981-99-4236-7_106-1

Ghobadi G, Etminan A, Mehrabi AM, Shooshtari L (2021) Molecular diversity analysis in hexaploid wheat (*Triticum aestivum* L.) and two Aegilops species (*Aegilops crassa* and *Aegilops cylindrica*) using CBDP and SCoT markers. J Genet Eng Biotechnol 19(1). https://doi.org/10.1186/s43141-021-00157-8

Gil-Humanes J, Wang Y, Liang Z et al (2017) High-efficiency gene targeting in hexaploid wheat using DNA replicons and CRISPR/Cas9. Plant J 89:1251–1262

Gohar S, Sajjad M, Zulfiqar S, Liu J, Wu J, Mehboob-ur- Rahman (2022) Domestication of newly evolved hexaploid wheat—a journey of wild grass to cultivated wheat. Front Genet 13:1022931

Golovnina KA, Glushkov SA, Blinov AG et al (2007) Molecular phylogeny of the genus Triticum L. Pl Syst Evol 264:195–216

Goncharov NP, Golovnina KA, Kondratenko EY (2009) Taxonomy and molecular phylogeny of natural and artificial wheat species. Breed Sci 59:492–498

Gregory PJ, Ingram JS, Brklacich M (2005) Climate change and food security. Philos Trans R Soc B Biol Sci 360:2139–2148

Gudaye TY, Lakew BF, Selamo MM, Tessema ES (2025) Ethiopian crop gene bank. In: Al-Khayri JM, Salem KFM, Jain SM (eds) Plant gene banks. Springer, Singapore. https://doi.org/10.1007/978-981-99-4236-7_7-1

Guzzon F, Gianella M, Giovannini P, Payne TS (2022) In: Reynolds MP, Braun H-J (eds) Conserving wheat genetic resources BT – wheat improvement: food security in a changing climate. Springer International Publishing, Cham, pp 299–318

Habib I, Rauf M, Qureshi J et al (2014) Optimization of somatic embryogenesis and *Agrobacterium*-mediated transformation of elite wheat (*Triticum aestivum*) Cultivars of Pakistan. Int J Agric Biol 16:1098–1104

Haile JK, Diaye AN, Clarke F et al (2018) Genomic selection for grain yield and quality traits in durum wheat. Mol Breed 38:1–18

Harfouche AL, Jacobson DA, Kainer D, Romero JC, Harfouche AH, Mugnozza GS, Moshelion M, Tuskan GA, Keurentjes JJ, Altman A (2019) Accelerating climate resilient plant breeding by applying next-generation artificial intelligence. Trends Biotechnol 37:1217–1235

Harris KF (2019) An introductory review of resistant starch type 2 from high-amylose cereal grains and its effect on glucose and insulin homeostasis. Nutr Rev 77:748–764

Harzli I, Mostafa K, Abdulla MF, Mehmood M (2025) Sustaining Turkish genetic resources. In: Al-Khayri JM, Salem KFM, Jain SM (eds) Plant gene banks. Springer, Singapore. https://doi.org/10.1007/978-981-99-4236-7_88-1

Hassan MN, Mekkawy SA, Mahdy M et al (2021) Recent molecular and breeding strategies in lettuce (Lactuca spp.). Genet Resour Crop Evol (2021). https://doi.org/10.1007/s10722-021-01246-w

Hazard B, Trafford K, Lovegrove A, Griffiths S, Uauy C et al (2020) Strategies to improve wheat for human health. Nat Food 1:475–480

He J, Zhao X, Laroche A et al (2014) Genotyping-by-sequencing (GBS), an ultimate marker-assisted selection (MAS) tool to accelerate plant breeding. Front Plant Sci 5:1–8. https://doi.org/10.3389/fpls.2014.00484

Hemery Y, Rouau X, Lullien-Pellerin V, Barron C, Abecassis J (2007) Dry processes to develop wheat fractions and products with enhanced nutritional quality. J Cereal Sci 46:327–347

Hidayatun N, Kurniawan H (2025) Conservation of agricultural genetic resources in Indonesia. In: Al-Khayri JM, Salem KFM, Jain SM (eds) Plant gene banks. Springer, Singapore. https://doi.org/10.1007/978-981-99-4236-7_40-1

Hollingsworth ML, Andra Clark A, Forrest LL, Richardson J, Pennington RT, Long DG, Cowan R, Chase MW, Gaudeul M, Hollingsworth PM (2009) Selecting barcoding loci for plants: evaluation of seven candidate loci with specieslevel sampling in three divergent groups of land plants. Mol Ecol Resour 9:439–457

Ibrahim AA, Salem KFM, Abdein MA, Ramadan SM (2023) Advances in summer squash (*Cucurbita pepo* L.) molecular breeding strategies. In: Singh S, Sharma D, Sharma SK, Singh R (eds) Smart plant breeding for vegetable crops in post-genomics era. Springer, Singapore. https://doi.org/10.1007/978-981-19-5367-5_8

Ibrahim AA, Abd-Ellatif S, Abdel Razik ESS, Salem KFM (2024) Biodiversity of cereal crops and utilization in food and nutritional security. In: Al-Khayri JM, Jain SM, Penna S (eds) Sustainable utilization and conservation of plant genetic diversity. Sustainable development and biodiversity, vol 35. Springer, Singapore. https://doi.org/10.1007/978-981-99-5245-8_2

Ihsan F, Prihatini R (2025) Indonesian tropical fruit gene bank. In: Al-Khayri JM, Salem KFM, Jain SM (eds) Plant gene banks. Springer, Singapore. https://doi.org/10.1007/978-981-99-4236-7_41-1

Irshad A, Guo H, Ur Rehman S, Gu J, Wang C et al (2022) Screening of induced mutants led to the identification of starch biosynthetic genes associated with improved resistant starch in wheat. Int J Mol Sci 23:10741

Ishida Y, Hiei Y, Komari T (2015) High efficiency wheat transformation mediated by *Agrobacterium tumefaciens*. In: Ogihara Y, Takumi S, Handa H (eds) Advances in wheat genetics: from genome to field. Springer, Tokyo, pp 167–173

Ismail M, Salem KFM, El-Zanaty AM, Esmail RM (2013) Association mapping of fiber and agronomic characters in Egyptian cotton (*Gossypium barbadense* L.) genotypes. J Agric Chem Biotech Mansoura Univ 4:205–216

Ivanizs L, Marcotuli I, Rakszegi M, Kalapos B, Szőke-Pázsi K et al (2022) Identification of new QTLs for dietary fiber content in Aegilops biuncialis. Int J Mol Sci 23:3821

Izydorczyk MS, Dexter JE (2008) Barley β-glucans and arabinoxylans: molecular structure, physicochemical properties, and uses in food products–a Review. Food Res Int 41:850–868

Jääskeläinen AS, Holopainen-Mantila U, Tamminen T, Vuorinen T (2013) Endosperm and aleurone cell structure in barley and wheat as studied by optical and Raman microscopy. J Cereal Sci 57:543–550

Jatoi SA, Siddiqui SU (2025) National Gene Bank of Pakistan. In: Al-Khayri JM, Salem KFM, Jain SM (eds) Plant Gene Banks. Springer, Singapore. https://doi.org/10.1007/978-981-99-4236-7_51-1

Jiang W, Zhou H, Bi H et al (2013) Demonstration of CRISPR/Cas9/sgRNA-mediated targeted gene modification in Arabidopsis, tobacco, sorghum and rice. Nucleic Acids Res 41:1–12. https://doi.org/10.1093/nar/gkt780

Juliana P, Singh RP, Singh PK et al (2018) Genome wide association mapping for resistance to leaf rust, stripe rust and tan spot in wheat reveals potential candidate genes. Theor Appl Genet 131:1405–1422. https://doi.org/10.1007/s00122-018-3086-6

Kenaschuk EO (1975) Flax breeding and genetics. In: Harapiak JT (ed) Oilseed and pulse crops in western Canada-A symposium. Western co-operative fertilizers Ltd, Calgary, pp 203–221

Khan S (2015) QTL mapping: a tool for improvement in crop plants. Res J Recent Sci 4:7–12

Kihara H (1924) Cytologische und genetische Studien bei wichtigen Getreidearten mit besonderer Rücksicht auf das Verhalten der Chromosomen und die Sterilität in den Bastarden. Mem Coll Sci Kyoto Imp Univ 1:1–200

Kihara H (1937) Genomanalyse bei Triticum und Aegilops VII. Kurze Übersicht über die Ergebnisse der Jahre 1934–36. Mem Coll Sci Kyoto Imp Univ 41:1–61

Kihara H (1954) Considerations on the evolution and distribution of Aegilops species based on the analyser-method. Cytologia 19:336–357

Kim D, Alptekin B, Budak H (2018) CRISPR/Cas9 genome editing in wheat. Funct Integr Genomics 18:31–41. https://doi.org/10.1007/s10142-017-0572-x

Kimber G, Feldman M (1987) Wild wheat, an introduction, Special Report 353. College of Agriculture, University of Missouri, Columbia, 146 p

Kimber G, Sears ER (1987) Evolution in the genus Triticum and the origin of cultivated wheat. In: Heyne EG (ed) Wheat and wheat improvement, 2nd edn. American Society of Agronomy, Madison, pp 154–164

Kulak M, Graves A, Chatterton J (2013) Reducing greenhouse gas emissions with urban agriculture: a life cycle assessment perspective. Landsc Urban Plan 111:68–78

Kumar J, Pretep A, Solanki R et al (2011) Advances in genomics resources for improving food legume crops. J Agric Sci 150:289–318

Kumar R, Kumar A, Sharma NK, Kaur N, Chunduri V et al (2016) Soft and hard textured wheat differ in starch properties as indicated by Trimodal distribution, morphology, thermal and crystalline properties. PLoS One 11:e0147622

Laghetti G, Piergiovanni AR, De Paola D, Direnzo P, Bucci G, Sonnante G (2025) Italy IBBR gene bank. In: Al-Khayri JM, Salem KFM, Jain SM (eds) Plant gene banks. Springer, Singapore. https://doi.org/10.1007/978-981-99-4236-7_73-1

Langridge P (2003) Molecular breeding of wheat and barley. In: Tuberosa R, Philips RL, Gale M (eds) Proceedings of the International Congress, In the wake of the double helix: from the green revolution to the gene revolution, Bologna, Italy, pp 279–286

Lee MH, Park J, Kim KH, Kim KM, Kang CS et al (2023) Genomewide association study of arabinoxylan content from a 562 hexaploid wheat collection. Plants 12:184

Li C, Dhital S, Gidley MJ (2022) High-amylose wheat bread with reduced in vitro digestion rate and enhanced resistant starch content. Food Hydrocoll 123:107181

Liang Z, Chen K, Li T et al (2017) Efficient DNA-free genome editing of bread wheat using CRISPR/Cas9 ribonucleoprotein complexes. Nat Commun 8:14261. https://doi.org/10.1038/ncomms14261

Linnaeus C (1753) Species plantarum, exhibentes plantas rite cognitas, ad genera relatas, cum differentiis specificus, nominibus trivialibus, synonymis selectis, locis natalibus, secundum systema sexuale digestas. T1 Impensis Laurentii Salvii, Holmiae

Liu J, Wu X, Yao X, Yu R, Larkin PJ et al (2018a) Mutations in the DNA demethylase OsROS1 result in a thickened aleurone and improved nutritional value in rice grains. Proc Natl Acad Sci 115:11327–11332

Lombard V, Golaconda Ramulu H, Drula E, Coutinho PM, Henrissat B (2014) The carbohydrate-active enzymes database (CAZy) in 2013. Nucleic Acids Res 42:D490–D495

Lovegrove A, Wilkinson MD, Freeman J, Pellny TK, Tosi P et al (2013) RNA interference suppression of genes in glycosyl transferase families 43 and 47 in wheat starchy endosperm causes large decreases in arabinoxylan content. Plant Physiol 163:95–107

Lu ZX, Walker KZ, Muir JG, Mascara T, O'Dea K (2000) Arabinoxylan fiber, a byproduct of wheat flour processing, reduces the postprandial glucose response in normoglycemic subjects. Am J Clin Nutr 71:1123–1128

Lu ZX, Walker KZ, Muir JG, O'Dea K (2004) Arabinoxylan fibre improves metabolic control in people with Type II diabetes. Eur J Clin Nutr 58:621–628

MacKey J (1966) Species relationships in Triticum. Proc 2nd Int Wheat Genet Symp Hereditas (Suppl) 2:237–276

MacKey J (1988) A plant breeder's perspective on taxonomy of cultivated plants. Biologisches Zentralblatt 107:369–379

MacNeill GJ, Mehrpouyan S, Minow MAA, Patterson JA, Tetlow IJ et al (2017) Starch as a source, starch as a sink: the bifunctional role of starch in carbon allocation. J Exp Bot 68:4433–4453

Mallor C, Luna E (2025) The genetic diversity of the Spanish Vegetable Genebank BGHZ-CITA: conservation and utilization. In: Al-Khayri JM, Salem KFM, Jain SM (eds) Plant gene banks. Springer, Singapore. https://doi.org/10.1007/978-981-99-4236-7_109-1

Marais G, Botes W (2009) Recurrent mass selection for routine improvement of common wheat: a review. In: Lichtfouse E (ed) Organic farming, pest control and remediation of soil pollutants, Sustainable Agric Rev, vol 1. Springer, Dordrecht, pp 85–105

Marais GF, Pretorius ZA, Marais AS, Wellings CR (2003) Transfer of rust resistance genes from Triticum species to common wheat. S Afr J Plant and Soil 20(4):193–198. https://doi.org/10.1080/02571862.2003.10634934

Marcotuli I, Houston K, Waugh R, Fincher GB, Burton RA et al (2015) Genome wide association mapping for arabinoxylan content in a collection of tetraploid wheats. PLoS One 10:e0132787

Marcotuli I, Houston K, Schwerdt JG, Waugh R, Fincher GB et al (2016) Genetic diversity and genome wide association study of βglucan content in tetraploid wheat grains. PLoS One 11:e0152590

Marcotuli I, Colasuonno P, Blanco A, Gadaleta A (2018) Expression analysis of cellulose synthase-like genes in durum wheat. Sci Rep 8:15675

Marcotuli I, Colasuonno P, Cutillo S, Simeone R, Blanco A et al (2019) β-glucan content in a panel of Triticum and Aegilops genotypes. Genet Resour Crop Evol 66:897–907

Mattar MZ, Eissa OA, Salem KFM (2009) Genotypic variability in some wheat genotypes (*Triticum aestivum* L.) exposed to salinity at germination stage. Egypt J Exp Biol (Bot) 5:143–148

Mattar MZ, Salem KFM, El-Abed AB (2016) Identification of candidate microsatellite markers associated with agronomic traits in rice (*Oryza sativa* L.). Egypt J Bot 15(5):213–232

Mbow C, Rosenzweig CE, Barioni LG, Benton TG, Herrero M, Krishnapillai M, Ruane AC, Liwenga E, Pradhan P, Rivera-Ferre MG (2020) Food security. Geneva, IPCC

McFadden ES, Sears ER (1946) The origin of Triticum spelta and its free-threshing hexaploid relatives. J Hered 37(81–89):107–116

Mishra A, Singh A, Sharma M, Kumar P, Roy J (2016) Development of EMS-induced mutation population for amylose and resistant starch variation in bread wheat (Triticum aestivum) and identification of candidate genes responsible for amylose variation. BMC Plant Biol 16:217

Mitchell RA, Dupree P, Shewry PR (2007) A novel bioinformatics approach identifies candidate genes for the synthesis and feruloylation of arabinoxylan. Plant Physiol 144:43–53

Mohamed ASH, Ibrahim M, Teleb SS, Tantawy ME (2017) SEM and SCoT Markers unveil new taxonomic and genetic insights about some Northern African Triticum aestivum L. Cultivars. Vegetos 30(1). Available from: https://www.scitechnol.com/peer-review/sem-and-scot-markers-unveil-new-taxonomic-and-genetic-insights-about-some-northern-african-triticum-aestivum-l-cultivars-Txqb.php?article_id=5884

Monteros-Altamirano Á et al (2025) Plant genetic resources in Ecuador. In: Al-Khayri JM, Salem KFM, Jain SM (eds) Plant gene banks. Springer, Singapore. https://doi.org/10.1007/978-981-99-4236-7_99-1

Morris R, Sears ER (1967) The cytogenetics of wheat and its relatives. In: Quinsberry KS, Reitz LP (eds) Wheat and wheat improvement. American Society of Agronomy, Madison, pp 19–87

Murphy MM, Douglass JS, Birkett A (2008) Resistant starch intakes in the United States. J Am Diet Assoc 108:67–78

Nachit MM (1992) Durum wheat breeding for Mediterranean dryland of North Africa and West Asia. In: Rajram S, Saari EE, Hetel GP (eds) Durum wheats: challenges and opportunities. CIMMYT, Ciudad Obregon, pp 14–27

Naqvi SMZA, Saleem SR, Tahir MN, Hussain S, Haq SIU, Awais M, Qamar S (2022) Vertical farming—current practices and its future. Environ Sci Proc 23(1):4

Nawara HM, Mattar MZ, Salem KFM, Eissa OA (2016) Estimating heterosis and combining ability for agronomic and tissue culture traits using diallel cross on bread wheat (*Triticum aestivum* L.). Egypt J Bot 2016:177–200

Nawara HM, Mattar MZ, Salem KFM, Eissa OA (2017) Diallel study on some in vitro traits of bread wheat (*Triticum aestivum* L.) under salt stress. Int J Agric Environ Res 3(1):1988–2006

Nesbitt M, Samuel D (1996) From staple crop to extinction? The archaeology and history of the hulled wheat. In: Padulosi S, Hammer K, Heller J (eds) Hulled Wheats, Promoting the conservation and used of underutilised and neglected crops. IPGRI, Rome, pp 40–99

Norman A, Taylor J, Edwards J, Kuchel H (2018) Optimising genomic selection in wheat: effect of marker density, population size and population structure on prediction accuracy. G3 (Bethesda) 8:2889–2899. https://doi.org/10.1534/g3.118.200311

Novoselović D, Bentley AR, Šimek R et al (2016) Characterizing Croatian wheat germplasm diversity and structure in a European context by DArT markers. Front Plant Sci 7:184. https://doi.org/10.3389/fpls.2016.00184

Olivoto T, Lúcio AD (2020) metan: an R package for multi-environment trial analysis. Methods Ecol Evol 11:783–789

Olivoto T, Diel MI, Schmidt D, Lúcio ADC (2021) Multivariate analysis of strawberry experiments: where are we now andwhere can we go? bioRxiv, 2020–12.

Osterc G, Gostinčar K, Hudina M, Usenik V, Luthar Z (2025) Slovene fruit gene bank. In: Al-Khayri JM, Salem KFM, Jain SM (eds) Plant Gene Banks. Springer, Singapore. https://doi.org/10.1007/978-981-99-4236-7_84-1

Pariyar SR, Dababat AA, Sannemann W et al (2016) Genome-wide association study in wheat identifies resistance to the cereal cyst nematode Heterodera filipjevi. Phytopathology 10c6:1128–1138

Park Y, Brinton LA, Subar AF, Hollenbeck A, Schatzkin A (2009) Dietary fiber intake and risk of breast cancer in postmenopausal women: the National Institutes of Health-AARP Diet and Health Study. Am J Clin Nutr 90:644–651

Pellny TK, Lovegrove A, Freeman J, Tosi P, Love CG et al (2012) Cell walls of developing wheat starchy endosperm: comparison of composition and RNA-Seq transcriptome. Plant Physiol 158:612–627

Pellny TK, Patil A, Wood AJ, Freeman J, Halsey K et al (2020) Loss of TaIRX9b gene function in wheat decreases chain length and amount of arabinoxylan in grain but increases cross-linking. Plant Biotechnol J 18:2316–2327

Poland J, Endelman J, Dawson J et al (2012) Genomic selection in wheat breeding using genotyping-by-sequencing. Plant Genome 5:103–113. https://doi.org/10.3835/plantgenome2012.06.0006

Popović V et al (2025) Economic seed production and storage of fiber crops in gene banks in Serbia. In: Al-Khayri JM, Salem KFM, Jain SM (eds) Plant gene banks. Springer, Singapore. https://doi.org/10.1007/978-981-99-4236-7_108-1

Porter JR, Xie L, Challinor AJ, Cochrane K, Howden SM, Iqbal MM, Lobell DB, Travasso MI (2014) Food security and food production systems. In: Climate Change 2014: impacts, adaptation, and vulnerability. Cambridge University Press, Cambridge

Pour-Aboughadareh A, Ahmadi J, Mehrabi AA, Etminan A, Moghaddam M (2017) Assessment of genetic diversity among Iranian Triticum germplasm using agro-morphological traits and start codon targeted (SCoT) markers. Cereal Res Commun 45(4):574–586

Pourkheirandish M, Dai F, Sakuma S, Kanamori H, Distelfeld A, Willcox G et al (2018) On the origin of the non-brittle rachis trait of domesticated einkorn wheat. Front Plant Sci 8:2031

Quraishi UM, Murat F, Abrouk M, Pont C, Confolent C et al (2011) Combined meta-genomics analyses unravel candidate genes for the grain dietary fiber content in bread wheat (Triticum aestivum L.). Funct Integr Genomics 11:71–83

Rasheed A, Mujeeb-Kazi A, Ogbonnaya FC, He Z, Rajaram S (2018a) Wheat genetic resources in the post-genomics era: promise and challenges. Ann Bot 121(4):603–616

Regina A, Berbezy P, Kosar-Hashemi B, Li S, Cmiel M et al (2015) A genetic strategy generating wheat with very high amylose content. Plant Biotechnol J 13:1276–1286

Robles-Zazueta CA, Crespo-Herrera LA, Piñera-Chavez FJ, Rivera-Amado C, Aradoir GI (2024) Climate change impacts on crop breeding: targeting interacting biotic and abiotic stresses for wheat improvement. Plant Genome 17:e20365

Rocha F et al (2025) Portugal Gene Bank. In: Al-Khayri JM, Salem KFM, Jain SM (eds) Plant gene banks. Springer, Singapore. https://doi.org/10.1007/978-981-99-4236-7_79-1

Röder MS, Korzun V, Wendehake K et al (1998) A microsatellite map of wheat. Genetics 149:2007–2023

Roy RN, Finck A, Blair G, Tandon H (2006) Plant nutrition for food security. A guide for integrated nutrient management. FAO Fertilizer and Plant Nutrition Bulletin, Rome, p 16

Sajilata MG, Singhal RS, Kulkarni PR (2006) Resistant starch – A review. Compr Rev Food Sci Food Saf 5:1–17

Sakamura T (1918) Kurze mitteilung über die chromosomenzahlen und die verwandtschaftsverhältnisse der Triticum arten. Bot Mag Tokyo 32:150–153

Saleh MM, Ali WA, Alwan MH et al (2020a) Multi-environmental evaluation of wheat tetraploid genotypes for agronomic traits under rainfed conditions in Syria. J Aridland Agric 6:1–5

Saleh MM, Salem KFM, Elabd AB (2020b) Definition of selection criterion using correlation and path coefficient analysis in rice (*Oryza sativa* L.) genotypes. Bull Natl Res Cent 44(1):1–6

Saleh MM, Alnaddaf LM, Almuhammady AK et al (2021a) In: Al-Khayri JM, Ansari MI, Singh AK (eds) Applications of plant-derived nanomaterials in mitigation of crop abiotic stress. Nanobiotechnology, Springer, Cham. https://doi.org/10.1007/978-3-030-73606-4_9

Saleh MM, Mahmoud AS, Abbas HS et al (2021b) Nanotechnological approaches for efficient delivery of plant ingredients. In: Faizan M, Hayat S, Yu F (eds) Sustainable agriculture reviews, vol 53. Springer, Cham. https://doi.org/10.1007/978-3-030-86876-5_11

Salem KFM (1997) Breeding studies on rice (*Oryza sativa* L.). M Sc Thesis, Faculty of Agriculture, Minufiya University, Shibin El-Kom, Egypt

Salem KFM (2004) The inheritance and molecular mapping of genes for post-anthesis drought tolerance (PADT) in wheat. Martin-Luther-University, Halle-Wittenberg, 124 p

Salem KFM (2009) Relationship between genetic diversity based on SSRs markers with heterosis and combining ability in diallel cross of bread wheat (*Triticum aestivum* L.). Minufiya. J Agric Res 6(34):2159–2178

Salem KFM (2015) Allelic state at the microsatellite locus Xgwm261 marking the dwarfing gene Rht8 in Egyptian bread wheat (Triticum aestivum L.) genotypes released from 1947 to 2004. Genetika 47:741–750

Salem KFM, El-Zanaty AM (2015) Assessing genetic diversity of old and modern Egyptian rice (Oryza sativa L.) using microsatellite markers. Minufiya J Agric Res 40(6):1507–1518

Salem KFM, Ibrahim AA (2024) Plant biodiversity in the context of food security under climate change. In: Al-Khayri JM, Jain SM, Penna S (eds) Sustainable utilization and conservation of plant genetic diversity. Sustainable development and biodiversity, vol 35. Springer, Singapore. https://doi.org/10.1007/978-981-99-5245-8_1

Salem KFM, Mattar MZ (2009) Allelic detection at the microsatellite XGwm261 locus linked to the Rht8 dwarfing gene in wheat. Egypt J Biotechnol 32:226–235

Salem KFM, Mattar MZ (2014a) Genetic diversity in old and modern Egyptian bread wheat (*Triticum aestivum* L.) varieties revealed by simple sequence repeats. Egypt J Genet Cytol 43:143–156

Salem KFM, Mattar MZ (2014b) Identification of microsatellite alleles for salt tolerance at seedling stage in wheat (*Triticum aestivum* L.). Life Sci J 11(12s):1064–1073

Salem KFM, Sallam A (2016) Analysis of population structure and genetic diversity of Egyptian and exotic rice (*Oryza sativa* L.) genotypes. Comptes Rendus Biologies 339:1–9. https://doi.org/10.1016/j.crvi.2015.11.003

Salem KFM, Röder MS, Börner A (2004) Molecular mapping of quantitative trait loci (QTLs) determining post-anthesis drought tolerance in hexaploid wheat (Triticum aestivum L.). In:7 Gesellschaft für Pflanzenzüchtung, GPZ-Tagung, Vortragsveranstaltung zum Thema: Klimatische und edaphische Sortenanpassung und Züchtung für Nachwachsende Rohstoffe. 3–5 March 2004, Halle/Saale. Germany Vort Pflanzenzüchtung 64:21–24

Salem KFM, Röder MS, Börner A (2007a) Evaluation of some barley varieties for the presence of thermostable alleles of *βamylase*. In: 8th African Crop Science Society Conference, El-Minia, Egypt, 27–31 October 2007, pp 643–648

Salem KFM, Röder MS, Börner A (2007b) Identification and mapping quantitative trait loci for stem reserve mobilisation in wheat (*Triticum aestivum* L.). Cereal Res Common 35:1367–1374

Salem KFM, El-Zanaty AM, Esmail RM (2008) Assessing wheat (*Triticum aestivum* L.) genetic diversity using morphological characters and microsatellite markers. World J Agric Sci 4:538–544

Salem KFM, Varshney RK, Röder MS, Börner A (2010) EST-SSR based estimates on functional genetic variation in a barley (*Hordeum vulgare* L.) collection from Egypt. Genet Resour Crop Evol 57(4):515–521

Salem KFM, Röder MS, Börner A (2015a) Assessing genetic diversity of Egyptian hexaploid wheat (Triticum aestivum L.) using microsatellite markers. Genet Resour Crop Evol 62:377–385

Salem KFM, Yaseen YM, El-Abd AB, Ayoub A (2015b) Inheritance of some yield components and grain quality traits in rice (*Oryza sativa* L.). J Product Dev 20:161–177

Salem KFM, Yaseen YM, El-Abd AB, Ayoub A (2015c) Breeding studies on some agronomic and quality traits in rice (Oryza sativa L.). J Product Dev 20(2):211–226

Salem KFM, Elabsawy A, Aldahak L et al (2021a) Evaluation of stem reserve mobilization in Egyptian bread wheat (*Triticum aestivum* L.) genotypes and F_1 hybrids under post-anthesis chemical desiccation stress. J Genet Environ Resour Conserv 9(1):176–182

Salem KFM, El-Absawy EA, Khidr YA, El-Shamy H (2021b) Assessment genetic diversity among some Egyptian bread wheat (*Triticum aestivum* L.) cultivars using RAPD markers. J Genet Environ Resour Conserv 9(1):168–175

Salem KFM, Saleh MM, Abu-Ellail FFB et al (2021c) Role of quantum dots, polymeric NPS and dendrimers in emphasizing crops tolerate biotic and abiotic stresses. In: Faizan M, Hayat S, Yu F (eds) Sustainable agriculture reviews, vol 53. Springer, Cham. https://doi.org/10.1007/978-3-030-86876-5_1

Salem KFM, Saleh MM, Abu-Ellail FFB et al (2021d) The role of salicylic acid in crops to tolerate abiotic stresses. In: Hayat S, Siddiqui H, Damalas CA (eds) Salicylic acid- a versatile plant growth regulator. Springer, Cham. https://doi.org/10.1007/978-3-030-79229-9_7

Salem KFM, Saleh MM, Aldahak L, Elabd AB (2021e) Assessment phenotypic diversity of rice (*Oryza sativa* L.) genotypes by multivariate analysis. J Arid Agric 7:52–59. https://doi.org/10.25081/jaa.2021.v7.6516

Salem KFM, Alghuthaymi MA, Elabd AB et al (2022) Prediction of heterosis for agronomic traits in half-diallel cross of rice (*Oryza sativa* L.) under drought stress using microsatellite markers. Plants 11:1532. https://doi.org/10.3390/plants11121532

Salem KFM, Abd-Ellatif S, Abdel Razik ES et al (2023a) Effect of nanomaterials on water and solutes translocation in plants. In: Al-Khayri JM, Alnaddaf LM, Jain SM (eds) Nanomaterial interactions with plant cellular mechanisms and macromolecules. Springer, Cham

Salem KFM, Tawfik E, Mahdy M et al (2023b) Advances in lettuce (*Lactuca* spp.) molecular breeding strategies. In: Singh S, Sharma D, Sharma SK, Singh R (eds) Smart plant breeding for vegetable crops in post-genomics era. Springer, Singapore. https://doi.org/10.1007/978-981-19-5367-5_11

Salem KFM, Safhi FA, Alwutayd KM et al (2024) Analysis of genetic diversity, population structure and phylogenetic relationships of rice (*Oryza sativa* L.) cultivars using simple sequence repeat (SSR) markers. Genet Resour Crop Evol. https://doi.org/10.1007/s10722-023-01789-0

Salina EA, Adonina IG, Badaeva ED et al (2015) A thinopyrum intermedium chromosome in bread wheat cultivars as a source of genes conferring resistance to fungal diseases. Euphytica 204:91–101

Saulnier L, Sado PE, Branlard G, Charmet G, Guillon F (2007) Wheat arabinoxylans: exploiting variation in amount and composition to develop enhanced varieties. J Cereal Sci 46:261–281

Sax K, Sax J (1924) Chromosome behavior in a genus cross. Genetics 9:454–464

Schönhofen A, Hazard B, Zhang X, Dubcovsky J (2016) Registration of common wheat germplasm with mutations in SBEII genes conferring increased grain amylose and resistant starch content. J Plant Regist 10:200–205

Schönhofen A, Zhang X-q, Dubcovsky J (2017) Combined mutations in five wheat STARCH BRANCHING ENZYME II genes improve resistant starch but affect grain yield and breadmaking quality. J Cereal Sci 75:165–174

Schulz A (1913) Die Geschichte der kultivierten Getreide, vol 2. L. Neberts Verlag, Halle a.d.S

Seaman JA, Sawdon GE, Acidri J, Pey C (2014) The Household Economy Approach. Managing the impact of climate change on poverty and food security in developing countries. Clim Risk Manag 4:59–68

Sehgal D, Vikram P, Sansaloni CP et al (2015) Exploring and mobilizing the GeneBank biodiversity for wheat improvement. PLoS One 10:e0132112. https://doi.org/10.1371/journal.pone.0132112

Sestili F, Palombieri S, Botticella E, Mantovani P, Bovina R et al (2015) TILLING mutants of durum wheat result in a high amylose phenotype and provide information on alternative splicing mechanisms. Plant Sci 233:127–133

Sharma V, Jahan K, Kumar P, Puri A, Sharma VK et al (2022) Mechanistic insights into granule-bound starch synthase I (GBSSI. L539P) allele in high amylose starch biosynthesis in wheat (Triticum aestivum L.). Funct Integr Genomics 23:20

Shewry PR, Hey SJ (2015) The contribution of wheat to human diet and health. Food Energy Secur 4:178–202

Shrawat AK, Armstrong CL (2018) Development and application of genetic engineering for wheat improvement. Crit Rev Plant Sci 37:335–421

Sima P, Vannucci L, Vetvicka V (2018) β-glucans and cholesterol (Review). Int J Mol Med 41:1799–1808

Sinha R, Fritschi FB, Zandalinas SI, Mittler R (2021) The impact of stress combination on reproductive processes in crops. Plant Sci 311:111007

Song X, Meng X, Guo H, Cheng Q, Jing Y et al (2022) Targeting a gene regulatory element enhances rice grain yield by decoupling panicle number and size. Nat Biotechnol 40:1403–1411

Sukumaran S, Reynolds MP, Sansaloni C (2018) Genome-wide association genome-wide association analyses identify QTL hotspots for yield and component traits in durum wheat grown under yield potential, drought and heat stress environments. Front Plant Sci 9:1–16. https://doi.org/10.3389/fpls.2018.00081

Tee TS, Qualset CO (1975) Bulk populations in wheat breeding: comparison of single-seed descent and random bulk methods. Euphytica 24:393–405. https://doi.org/10.1007/BF00028206

Varshney RK, Salem KFM, Röder MS et al (2008) Assessment and comparison of genetic diversity in barley germplasm collection using gSSR, eSSR and SNP markers. Plant Genet Res 6:167–174

Vetvicka V, Vetvickova J (2009) Effects of yeast-derived β-glucans on blood cholesterol and macrophage functionality. J Immunotoxicol 6:30–35

Wang Y, Cheng X, Shan Q et al (2014) Simultaneous editing of three homoeoalleles in hexaploid bread wheat confers heritable resistance to powdery mildew. Nat Biotechnol 32:947. https://doi.org/10.1038/nbt.2969

Wang S, Zhu Y, Zhang D et al (2017) Genome-wide association study for grain yield and related traits in elite wheat varieties and advanced lines using SNP markers. PLoS One 12:1–14

Wang W, Pan Q, He F et al (2018) Transgenerational CRISPR-Cas9 activity facilitates multiplex gene editing in allopolyploid wheat. The CRISPR J 1:65–74. https://doi.org/10.1089/crispr.2017.0010

Wang Y, Liu X, Zheng X, Wang W, Yin X et al (2021) Creation of aromatic maize by CRISPR/Cas. J Integr Plant Biol 63:1664–1670

Wang J et al (2025) China National Gene Bank (CNGB). In: Al-Khayri JM, Salem KFM, Jain SM (eds) Plant gene banks. Springer, Singapore. https://doi.org/10.1007/978-981-99-4236-7_103-1

Wheeler T, Von Braun J (2013) Climate change impacts on global food security. Science 341:508–513

Witcombe JR (1983) A guide to the species of *Aegilops* L.: their taxonomy, morphology, and distribution. International Board for Plant Genetic Resources (IPGRI), Rome, 74 p

Yang L, Zhao D, Yan J, Zhang Y, Xia X et al (2016) QTL mapping of grain arabinoxylan contents in common wheat using a recombinant inbred line population. Euphytica 208:205–214

Yoosefzad Najafabadi M, Hesami M, Eskandari M (2023) Machine learning-assisted approaches in modernized plant breeding programs. Gene 14:777

Zaharieva M, Monneveux P (2014) Cultivated einkorn wheat (Triticum monococcum L. subsp. monococcum): the long life of a founder crop of agriculture. Genet Resour Crop Evol 61:677–706. https://doi.org/10.1007/s10722-014-0084-7

Zannini E, Bravo Núñez Á, Sahin AW, Arendt EK (2022) Arabinoxylans as functional food ingredients: a review. Foods 11:1026

Zeeman SC, Kossmann J, Smith AM (2010) Starch: its metabolism, evolution, and biotechnological modification in plants. Annu Rev Plant Biol 61:209–234

Zhukovsky PM (1928) A critical-systematical survey of the species of the genus Aegilops. Bull Appl Bot Genet Plant Breed 18:417–609

Part II
Bast Fiber Crops

Chapter 9
Breeding and Biotechnology Approaches for Genetic Enhancement in Bananas (*Musa* spp.)

Amira A. Ibrahim and Khaled F. M. Salem

Abstract Bananas are a crucial commodity for food security and agricultural economies, as they are one of the most consumed fruits worldwide. Thanks to advancements in breeding methods, new banana cultivars with improved traits such as disease resistance, yield, and flavor have been developed. Despite several botanical limitations, conventional breeding has successfully produced a variety of disease- and pest-resistant hybrids. The primary approach involves creating new tetraploid cultivars by crossing triploids with wild or improved diploid clones showing resistance, or by using secondary triploids produced from the hybridization of evolved tetraploids with diploid clones. The limitation of the 3x/2x method is the reduced fertility of the triploid cultivar's gametophyte targeted for improvement. This limitation has led to the development of an alternative approach that focuses on producing triploid hybrids directly from crossings between diploid cultivars. This strategy leverages the reproductive sterility of male and female diploid hybrids at the diploid level. The 3x with 2x approach has enabled the adoption of high-yielding, biotic stress-resistant hybrids resulting from intensive breeding programs in various countries, supported by the International Banana Testing Program. Recent insights into the genetic variability of bananas obtained through molecular techniques have significantly contributed to the selection of parental populations. Advances in molecular markers associated with key traits are expected to expedite improvement efforts. Mutation breeding, offering a distinct alternative method for banana enhancement, has been utilized to create new cultivars. This chapter explores the botanical

A. A. Ibrahim (✉)
Department of Botany and Microbiology, Faculty of Science, Arish University, Al-Arish, Egypt
e-mail: amiranasreldeen@sci.aru.edu.eg

K. F. M. Salem
Department of Biological Sciences, Shaqra University, Ad-Dawadimi, Saudi Arabia

Department of Plant Biotechnology, Genetic Engineering and Biotechnology Research Institute (GEBRI), University of Sadat City, Sadat City, Egypt
e-mail: khaled.salem@gebri.usc.edu.eg; khaledfathi@su.edu.sa

K. F. M. Salem et al. (eds.), *Breeding and Biotechnology of Grass and Bast Fiber Crops*, Advances in Plant Breeding Strategies 12,
https://doi.org/10.1007/978-3-032-00403-1_9

taxonomy of bananas, their distribution, origin, diversity, importance, and uses. It also covers banana fiber, its textural characteristics, extraction, and use, current cultivation techniques, agronomic challenges, breeding objectives, germplasm diversity, and conservation. Additionally, it delves into conventional, mutation, and molecular breeding, genetic engineering, molecular marker-assisted breeding, genomics, hybridization markers, and polymerase chain reaction (PCR)-based markers. The chapter also discusses the use of genomic technologies to identify critical genes associated with key traits, develop hybridization procedures for producing new banana cultivars, and molecular markers for marker-assisted selection as recent advances in banana breeding. Banana breeders are employing these advanced technologies to address issues such as black sigatoka, Panama disease, impacts of climate change, and changing consumer preferences. The integration of advanced methods and technologies in banana breeding holds great promise in enhancing the resilience and productivity of this vital crop amidst increasing global challenges.

Keywords Biodiversity · Biotechnology · Breeding · Bananas (*Musa* spp.) · Molecular Breeding

9.1 Introduction

Bananas are now grown as a crop in more than 130 tropical and subtropical nations worldwide. They were originally from Southeast Asia. Worldwide, bananas are grown on 5.14 million hectares, yielding about 103 million metric tons per year (FAO 2014). Many people choose bananas and plantains because they can withstand a wide range of temperatures, they bear fruit all year round, and tHhey have health benefits and medicinal uses. Bananas are the fourth most important crop in the developing world, behind rice, wheat, and maize, and they provide food for almost 400 million people in tropical regions. While carbohydrates make up the bulk of tropical meals, protein and energy are also provided by plantains, bananas, and root vegetables. In rural areas, where 75% of the population resides, food security and poverty can be alleviated by the cultivation of bananas and plantains. Among starchy staples in emerging countries, banana output has grown at the fastest rate, increasing by 70% in the last 40 years.

Farmers can't make a living off of fruits like plantains and bananas. Most of the world's bananas, grown by small-scale farmers, end up in local marketplaces. The survival of these farmers depends on a diverse range of cultivars that have been suited to the specific conditions of their region. Cultivars that farmers have chosen for themselves account for most of the current diversity. Due to the ever-changing landscape and the constant presence of biotic threats, the extensive gene pool is of great material value.

The competing demands of a growing human population and dwindling arable land have put traditional plant breeding under closer scrutiny. Conventional or classical plant breeding helped many domesticated plant species advance in the 1900s (Aboshama et al. 2015; Emara et al. 2011, 2013; Johnson 2000). Classical breeding relies on a thorough evaluation of the material under different conditions in comparison to known cultivars. The most cost-effective and environmentally friendly way to control the many pests and diseases that harm banana crops is to use host plant resistance (Jones 2000; Lorenzen et al. 2010). Lorenzen et al. (2010) note substantial progress in developing hybrids that are resistant to various biotic stressors.

Here we will go over some of the bioinformatics tools and resources that are accessible for studying *Musa*, including gene annotation, transcriptomics, proteomics, and data integration. These advancements have the potential to improve our understanding of the genetic basis of important agronomic traits, which could make breeding efforts easier.

9.1.1 Botanical Classification and Distribution

Ensete and *Musa* L. are the two genera that make up the Musaceae family (https://www.itis.gov/servlet/SingleRpt/SingleRpt?search_topic=TSN&search_value=42388#null). The four subgenera that make up the genus Musa are as follows: two of them, Callimusa and Australimusa, have ten chromosomes apiece, while the other two, Eumusa and Rhodochlamys, have eleven. The upright inflorescences of many wild Callimusa species, which have bracts of varying colors and a propensity for quick reproduction, make them quite attractive plants. The fruit-bearing Musa fei and the milk-producing Musa textilis are the two most prominent of the five or six species found in Australia. The Indo-Burman border and other regions of Southeast Asia are home to *M. textiles*, whereas the Pacific Islands are the main commercial centers for Fe′i banana cultivation. Rhodochlamys has a reputation for being a beautiful plant due to its lean stature and erect inflorescences that are ornamented with bracts of various colors. Because they are viable, the females of the Eumusa species can breed with those of other species, producing a new generation that is mostly Rhodochlamys as shown in Fig. 9.1.

Kingdom: Plantae—plantes, Planta, Vegetal, plants
- Subkingdom: Viridiplantae– green plants
  - Infrakingdom: Streptophyta– land plants
    - Super division: Embryophyta
      - Division: Tracheophyta– vascular plants, tracheophytes
        - Subdivision: Spermatophytina– spermatophytes, seed plants, phanérogames
          - Class: Magnoliopsida
            - Superorder: Lilianae– monocots, monocotyledons, monocotylédone
              - Order: Zingiberales

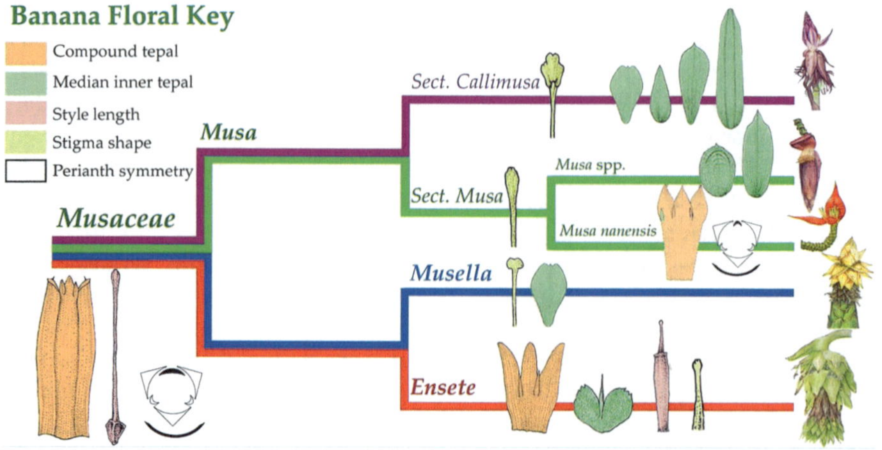

Fig. 9.1 Musaceae taxonomy based on floral structure in male plants. (Source: Inta et al. (2023))

Family: Musaceae—banana
Genus: *Musa L.—banana*
Species: *Musa acuminata* Colla—edible banana Common names: Banana

9.1.2 Taxonomy

Ensete and Musa L. are the two genera that make up the Musaceae family. There are four subgenera within the genus *Musa*. Ten chromosomes are found in Callimusa and Australimusa, and eleven in Eumusa and Rhodochlamys. There are a lot of ornamental Callimusa species in nature. Rhizomatous stems facilitate rapid plant multiplication, and bracts of varying hues embellish the upright inflorescences of these plants. Musa textilis, which produces fiber, and Musa fe'i, which bears fruit, are two of the most prominent species in the genus Australimusa. Although *M. textilis* is most commonly found in Southeast Asia, there have been reports of sightings as far north as the Indo-Burman border; in contrast, Fe′i bananas are native to the Pacific Islands. Members of the genus Rhodochlamys are solely decorative; they are easily recognizable by their distinctive, tall, erect inflorescences and colorful bracts. The females of these self-propagating plants are reproductive, and the offspring often display characteristics of the Rhodochlamys species.

If you're looking for edible bananas, your only options are Fe′i or Eumusa. A strong pseudostem, inflorescences that can be horizontal, angular, or pendulous, and parthenocarpic fruits are common characteristics of these bananas. Musa acuminata is the progenitor of all edible bananas, while *Musa balbisiana* is the progenitor of the B genome. According to Stover and Simmonds (1987), several combinations

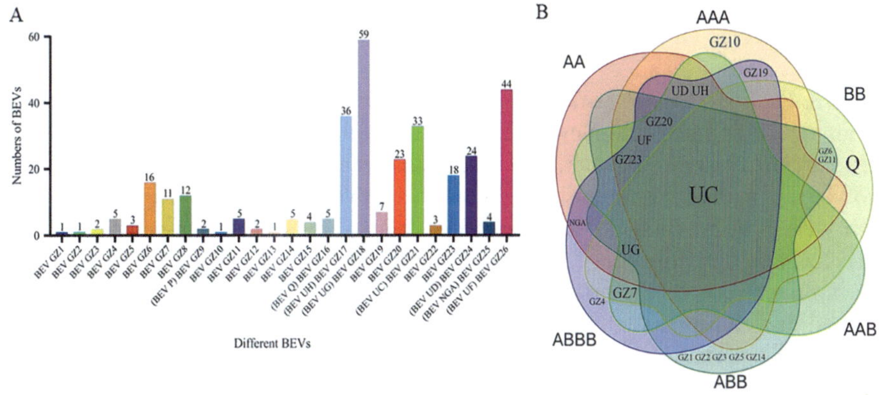

Fig. 9.2 Identifying banana endogenous badnavirus sequences (BEVs) in various banana genotypes. (**a**) Various BCEVs. (**b**) Banana BEVs across several A and B genotypes are shown as a six-way Venn diagram. (Source: Rao et al. (2023))

have resulted from the contributions of each genome. These combinations include AA, AAA, AB, AAB, ABB, BB, BBB, and ABBB (Fig. 9.2).

9.1.3 Origin, Distribution and Diversity of Bananas

Most of today's banana varieties are hybrids of the two wild species *Musa acuminata* Colla and *Musa balbisiana* Colla, which are both diploid (2n = 2x = 22). The seeded fruit of diploid Musa species is unmarketable because of its low starch and pulp content. A crucial part of the development of edible bananas in the Eumusa family was the wild Musa acuminata subspecies, which is cultivated in Southeast Asia. Musa acuminata, a seedless diploid species, evolved from its native acuminate due to vegetative parthenocarpy and female sterility. The hybridization with the hardier *M. balbisiana* was a major event. AAw and BBw stand for diploid *M. acuminata* and *M. balbisiana*, respectively. Triploids were the outcome of fertilization between haploid pollen and viable diploid cells, which were generated during the second stage of meiosis (Jones 2000).

While there are just a handful of acuminata subtypes found in India, there are numerous wild variations of *M. balbisiana*. Evolutionary theory has been illuminated by the discovery that cultivated and wild banana cultivars contain the S and T genomes of two diploid wild species, *M. textilis* and *M. schizocarpa*, which led to the banana's natural introgression (Jones 2000). By analyzing DNA sequences, D'Hont et al. (2000) were able to distinguish between S and T genomes. Six out of the eleven species of wild Musas on Earth call India home. The southern part of India, the Andaman and Nicobar Islands, and the northeastern region are home to the ancestors of cultivated bananas, M. acuminata and *M. balbisiana*. Some other species found in nature, mostly in the northeast, include *M. itinerens*, *M.*

nagensium, *M. sikkimensis* and *M. cheesmani*. India is home to just three of the eight subspecies of Musa acuminata found in Asia: banksii, burmanica, and burmannicoides.

M. balbisiana does not originate in Central Malaysia but rather in dry parts of India, the Philippines, and New Guinea. There are no known subspecies, however, it is hardier, drought-resistant, and disease-tolerant than *M. acuminata*. Bananas evolved from their non-pulpy, seedy wild relatives to parthenocarpy, palatable, high-yield breeds. Bananas have benefited from vegetative propagation, sterility, polyploidy, and parthenocarpy. Genomes in plastids, mitochondria, and the nucleus may influence crucial agronomic traits. In Musa acuminata, Faure et al. (1994) discovered that chloroplast DNA is transmitted maternally and mitochondrial DNA is transferred paternally. After analyzing 300 Musa genotypes, Carreel et al. (2002) found that the mitochondrial genomes of the majority of cultivars are related to the banksii and errans subspecies of Musa acuminata.

Clonal multiplication, survival, and geographical diffusion are all made possible by human intervention because most bananas and plantains are sexually sterile. According to Purseglove (1975), somatic mutations caused by foreign materials are responsible for secondary diversification in areas where wild Musa plants do not exist.

The edible bananas from the Eumusa series have genomes that start with AA and go up to AAAA, ABB, AB, and ABBB. There are more than fifty different types of bananas. Notable banana varieties include AAA dessert bananas, AAA highland cooking bananas, AAB plantains, and ABB cooking bananas, as well as East African beer bananas.

Genetic composition and ploidy can be inferred from the taxonomic approach to banana cultivar genomes by analyzing features such as petiolar canal morphology, bract attributes, and pseudostem color (Simmonds and Shepherd 1955). Sortiz et al. (1998) found that numerical taxonomy made this strategy even better. Reliable estimations can be obtained without plants growing using flow cytometry (Dolezel and Bartos 2005), and results from molecular approaches are correlated. Within each genetic group, clones and cultivars can be further identified by visual traits; for example, numerous clones exhibit varying fruit and bunch morphologies, pigmentations, and heights. According to Menon (2000) and Uma et al. (2005a), taxonomic complexity is heightened when a single clone or cultivar has multiple names and synonyms. When it comes to identifying and classifying bananas, it's important to have a fundamental grasp of cultivar and landrace evolution, taxonomic status, and traits. One useful tool for this is the banana descriptor, which was developed by IPGRI, INIBAP, and CIRAD in 1996.

There are 'groups' and subgroups of landraces and cultivars in a genome, and 'types' of wild accessions. Taller leaves and thinner pseudostems are characteristics of diploids AA or AB. Triploid cultivars, which are classified by the genetic codes AAA, AAB, and ABB, are more robust and produce bigger fruits compared to diploids. Tetraploid cultivars with a strong pseudostem and drooping leaves are rare; AAAA, AABB, AAAB, and ABBB are the available ones. When haploid pollen fertilizes triploid egg cells, the outcome is a tetraploid (Jones 2000).

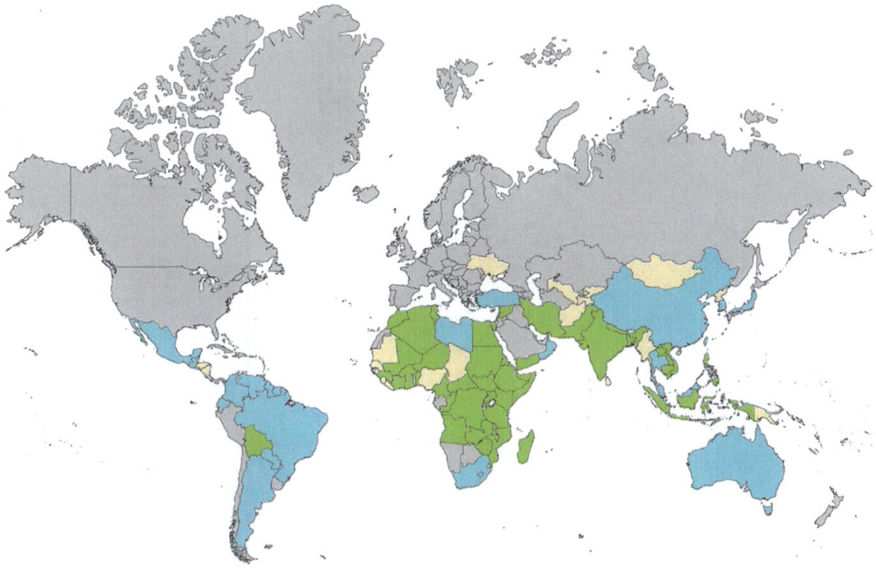

Fig. 9.3 Countries that both meet and exceed the World Population Review's standards for development (yellow), as well as those in Asia, Africa, Australia, and Latin America and the Caribbean (blue) that produce a significant number of bananas (green). (Source: Waraczewski et al. (2024))

Vegetative parthenocarpy allows bananas to produce pulp independently of pollination. To avoid shriveling, pollinate wild-seeded bananas before fruit development. The walls and septa of the ovary encase the hard black seeds in a thin, sweet pulp when the fruit is ripe. Without pollination, seeded banana ovaries cannot form. Bananas that are grown vegetatively produce pulp that contains ovules that shrivel up early, without the need for pollination. Banana parthenocarpy development is regulated by auxin, which is independently produced by the mature ovary (Purseglove 1975; Stover and Simmonds 1987). The size, shape, and color of the seeds can vary greatly; they can be anywhere from 4 to 20 mm in diameter, and their shapes can be round, triangular, or oval. The seed coats can be warty or smooth, brown, or even pitch black. The origin of bananas in the world is characterized by maps of developing countries (Fig. 9.3).

9.1.4 Banana Importance and Uses

Especially outside of tropical countries, most people's mental image of a banana is the sweet dessert variety. The fact that it's tasty, creamy, and peels off easily could be to blame (Robinson 1996). Having said that, there are many kinds of bananas. Included in the umbrella word "banana" is starchy plantain, a versatile ingredient in many dishes. The *Musa* spp. the family includes two distinct types of bananas: those used for cooking and those used for dessert (Jones 2000).

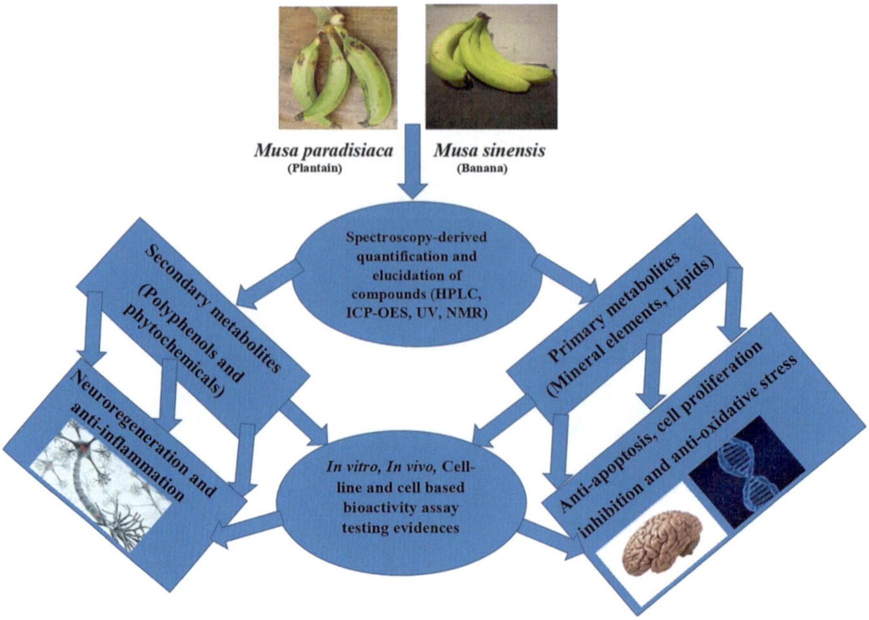

Fig. 9.4 Appropriateness of plantain and banana fruits for reducing neurodegenerative disorders. (Source: Waraczewski et al. (2024))

Dessert bananas, which account for 43% of the world's banana harvest, are renowned for their exceptionally sweet flavor. Enjoy them fresh, while they're ripe and golden, for the finest flavor. In tropical regions, when other dessert options are few, bananas are sometimes served as a standalone dessert. Because bananas contain a lot of carbohydrates, even when they are ripe, 57% of the world's bananas are used for cooking. To bring out their full flavor, boil, fry, or bake these bananas. Of all bananas grown, 23% are plantains, a kind that is quite popular in the kitchen (Fig. 9.4).

Steaming or boiling unripe bananas is another method of preparation. A starchy staple food with equal nutritional value to potatoes can be made using this technology (Simmonds 1966). The highlands and central East African regions are also home to a special kind of banana called an East African Highland banana. The perfect circumstances for growing bananas and plantains have led to their cultivation in more than 130 tropical and subtropical nations.

Among the most important tropical food sources, bananas rank second in terms of gross output value, behind cereal mainstays like rice, wheat, and maize. When it comes to exports, they are second only to citrus fruits in terms of both volume and value. Worldwide banana production in 2005 was over 100 million metric tons, with exports making up less than 10% of that total (FAO, 2014). In 2004, bananas were exported in more than 15.9 million metric tons around the world. In 2004, just eleven countries produced nearly three-quarters of the world's bananas. The vast majority of the world's bananas come from developing countries. Global banana production is dominated by four countries: India, Uganda, Brazil, and Ecuador. Costa Rica, Guatemala, Ecuador, and the Philippines are the top five countries that export bananas. A small amount of bananas and plantains is exported by Brazil, Uganda, and India, even though they are the top three banana growers. The majority of bananas produced by countries like China, India, and Uganda find their way into local dishes, with each country's own special way of preparing and enjoying this fruit (Sharrock and Frison 1998). With annual per capita banana consumption ranging from 200 to 250 kilos, East African states surrounding the Great Lakes region and New Guinea have relatively high banana consumption rates. Bananas and plantains are not only vital dietary staples, but they also facilitate local trade, which provides millions of people worldwide with substantial income. In terms of employment, income, and export earnings, the banana industry is vital to developing nations in Latin America, the Caribbean, Asia, and Africa, as the global value of banana exports exceeds US$4.7 billion per year (UNCTAD Secretariat from FAO statistics, www.unctad.org accessed 28/1/11). For several of these countries, banana exports are the lifeblood. The FAO states that there is a strong correlation between banana revenue and food security for households on a local and regional scale (FAO 2014). The uses of bananas are illustrated in Fig. 9.5.

9.1.5 Banana Fiber and Textile Properties

Bahama fiber, a sort of vernacular cellulosic fiber, originates in the banana plant's pseudostem. Banana fiber is known for its exceptional mechanical properties when used as a bast fiber. Comparable to glass fiber, banana fiber is a common material with exceptional tensile strength. The thickness of the material is reduced when

Fig. 9.5 Uses of bananas in different fields. (Source: Oyeyinka et al. (2022))

glass filaments are present. Arranged in a barrel shape, a cluster of leaf-tail bases constitutes a pseudo-stem. The fiber from bananas is a waste product that isn't always put to good use. By finding practical applications for these threads, we can standardize the interest and lower the expenses. You can't beat the biodegradability, strength, lightness, extensibility, fireproofness and solid moisture absorption of banana strands. Plus, they're quite adaptable and have incredible potential.

People are thinking about making clothes and cleaning supplies out of banana fiber. The remarkable qualities of banana fiber for papermaking add to the already captivating quality of handcrafted paper. Banana fiber has a wide variety of uses, including but not limited to channel paper, paper bags, welcome cards, decoration papers, rope, mats, composite material, light stands, pen stands, and ornamental papers. Banana fiber is already in circulation in German currency, and India is considering doing the same. Mercedes and other high-end vehicles have underfloor security boards made of polypropylene reinforced with banana fiber. Banana fiber is commonly used in DIY and craft projects. Fireproofing and structural sheets are two applications for banana fiber composite. Research has shown that paper made from this fiber has a long shelf life of over a century, as it is the strongest of the long filaments and can be compressed three thousand times more than ordinary strands. Reports like composition archive preserver, which can generate revenue and value, often employ it.

9.1.6 Extraction of Banana Fiber

Based on their research, Das et al. (2019) determined that plant bast, fruit, seed, leaf, and sheath can be used to produce vegetable fibers. The materials in question include cotton, which is composed of separate units; lignocellulosic meshy materials like jute and mesta; ramie, flax, sisal, jute, and pineapple leaf fiber (PALF); and short-staple materials like areca nut and kapok. The strength, fineness, and high length-to-breadth aspect ratio of materials like ramie and cotton make them perfect for spinning into yarn. Clothing, luggage, ropes, linens, and money paper blanks are just a few of the many practical uses for natural fibers such as cotton, jute, mesta, ramie, sisal, flax, and sun hemp. Ramie has promising specialized applications due to its status as the strongest vegetable fiber. Grain, sugar, potato, and onion packaging have traditionally made use of certain vegetable fibers. Crops like jute, mesta, sisal, and PALF can be transformed into a wide variety of goods through the use of various technologies. The growing demand for fine and mixed yarn among fabric and ready-to-wear manufacturers is driving the exploration of new spinning technologies to keep up with supply.

Biotechnology has the potential to revolutionize agriculture in developing countries by making crops more resilient to extreme weather. Biotechnology has made agriculture more sustainable and competitive by opening up new non-food markets for crops. The present investigation seeks to maximize the potential of biotechnology in non-food plant products by improving the fiber extraction method from

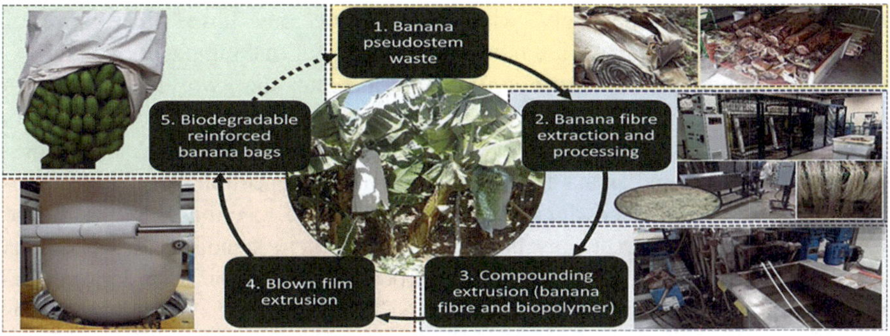

Fig. 9.6 Production of Fiber from banana. (Source: Bordón et al. (2021))

banana plants using a pectinase enzyme. The Raspador machine has enhanced fiber yield and quality by enzymatically treating the green stem and trunk of banana plants, which improves the pseudo-stem fiber from banana plants. This fiber has several uses, including making high-quality handmade paper. This has the potential to boost farmers' profits while also supplying artisans with high-quality materials for making paper.

In an effort to create banana fiber of superior quality, Vadivel et al. (2014) describe the design and building of a machine to extract fiber from banana pseudo stems. Although hand extraction results in fiber of superior quality, it is an expensive and time-consuming process. Although India does have access to mass-production machinery, the quality of its output is inadequate. Problems, including fiber breakage, knot formation, and varying fiber lengths, are taken care of by the updated machine model. Figure 9.6 shows the production of fibers from bananas.

9.1.7 Utilization of Banana Fiber

9.1.7.1 Banana Fiber Textile Sector

Research conducted by Arafat et al. (2018) indicates that banana fiber is a naturally occurring, very strong fiber that forms composites effectively when combined with cotton or synthetic fibers. One of the major environmental issues caused by fiber extraction is the massive amount of lignocellulosic waste that is generated. The purpose of this study is to describe the chemical and physical characteristics of EBF and WBF in order to lay out a process for making handmade paper. The longer fiber length, lower α-cellulose content, and greater lignin levels were the characteristics used to identify WBF. To pulp them, EBF and WBF were boiled at a boiling temperature for varied durations with different active alkalis. Pulp content increased to 35.9% after 120 minutes of cooking WBF with an 8% alkali charge. Drainage resistance degrees (SR values) for EBF were 65 in their initial form and for WBF, they were 71. The tensile, burst, and tear indices of WBF were 23.7 N.m/g, 2.2 kPa.m2/g, and 5.0 mN.m2/g, respectively, which were noticeably lower than those of EBF. However, these numbers are more than enough for making paper by hand.

Research conducted by Shroff et al. (2015) indicates that, worldwide, bananas are most produced in India. Banana fiber has limited textile applications due to its high stiffness and low cohesion. The goal of the research was to develop needle-punch nonwovens from banana fiber by eliminating the noncellulosic sticky components. Various dosages and durations of hemicellulase, pectinase, and cellulase enzymes were used to treat grey banana fibers. Prior to standardization, various enzyme combinations were evaluated after process optimization. In order to determine the effect of the enzymes on the fibers, SEM investigations, subjective touch-and-feel evaluations, fiber whiteness index, strength loss, and weight loss were employed. Nonwovens made from the highest quality softened fibers were created at NIRJAFT Kolkata using the needle punch technique.

9.1.7.2 Characteristics of Banana Fiber

According to research by Ramesh et al. (2014), contemporary engineering companies strive to supply eco-friendly products. Biodegradability, low processing costs, low weight, and excellent strength are some of the distinctive properties offered by natural fibers as compared to manmade fibers such as glass and carbon. Aside from jute, coir, bamboo, and hemp, banana fibers are among the many annual natural fibers that are both strong and easily accessible. This article describes in full the process of banana fiber extraction as well as investigations on the mechanical, chemical, and physical properties of the extracted fibers. Hybridization characterization, surface morphology, processing methods, chemical treatments, and processing techniques for banana fiber are all covered in the article. Dilleswara et al. (2017) found that researchers and scientists are highly interested in natural fiber-based composites due to their unique features and eco-friendliness. Natural fibers are convenient since they are both accessible and easy to work with. The procedure does not correlate with the result of ordinary filament manufacture. Distinct fiber composites' biodegradability is one of the attractive aspects that has piqued the curiosity of numerous studies seeking to mitigate the nursery effect. You can get natural filaments in a stringy form from plants like leaves and stems, and they're cheap and easy to harvest. There are a few unique fibers that have better mechanical properties than banana fiber. This research delves into the mechanical properties of banana fiber. The article delves into the mechanical properties of various banana-reinforced composites and covers the most popular methods for assembling these types of materials.

Through their experimental effort, mechanical properties affected the blend of treated banana-hemp fibers and glass fibers. By utilizing enzymatic techniques, the lignin was successfully extracted from the fibers of hemp and bananas. An alkaline solution was used to treat the fibers in order to boost their mechanical strength. The hand layup technique was used to create hybrid composites that were then reinforced with an epoxy matrix utilizing glass, hemp, and banana fibers. The laminates were made stronger by using several fiber orientations during manufacture. In accordance with ASTM guidelines, test specimens were constructed from laminates containing successive layers of glass/banana fiber, glass/hemp fiber, and glass/banana/hemp fiber. Impact, tensile, and flexural tests were used to ascertain the mechanical characteristics of the produced samples. An interfacial analysis was carried out using the scanning electron microscope to estimate voids, fractures, and fiber pull. According to the data, the combination of hemp fibers, glass, and bananas had the strongest tensile strength compared to the other two combinations. The composites with the greatest flexural strength were those including hemp and glass fibers, rather than those containing glass, hemp, and banana fibers. The impact energy value of composites composed of banana fibers and glass ranged from 7.33 to 9.33 Joules, and their impact resistance was limited. These composites showed excellent performance under all mechanical loading conditions, which could lead to their use in a wide range of fields.

Surface treatments performed on banana fibers (BFs) can cause substantial changes in their superficial structure, which can be used to create composites with higher mechanical strength, as mentioned in a recent study by Cecci etal. (2020). The objective of their research was to evaluate the physical and mechanical properties of banana fibers from the Prata species, which were taken from the inner and outer leaf sheaths and subjected to UV light ($\lambda max = 400$ nm) for 7 (UV7) and 15 (UV15) days, respectively. Microstructural and structural analyses utilizing FT-IR spectroscopy and Scanning Electron Microscopy (SEM) on both irradiated and non-irradiated fibers proved that UV irradiation altered the surface and chemical composition of BFs. The EA responsible for the thermal degradation of In Natura fiber, which measured 188.2 kJ. mol^{-1}, was determined using Differential Thermal Analysis (DTA/TG). Results from mechanical characterization show that compared to In Natura fibers, UV7 fibers have much higher tensile strength (89.77 MPa) and elastic modulus (238.94 MPa). The two-way analysis of variance results showed that the mechanical properties of BFs in the outer and inner leaf sheaths were similar. Utilizing ultraviolet radiation, a non-toxic method of treating the surfaces of BF fibers can enhance the mechanical properties of composites manufactured with these fibers.

Bhuvneshwar et al. (2011) reviewed the literature on polyester banana fiber composites that used the coagulum of euphorbia as a binder. Euphorbia coagulum, the sticky mass of dried latex, contains 60–80% terpenes and polyisoprene. Composite physico-mechanical and fire properties of polyester banana fibers have been investigated in relation to varying coagulum percentages. Because banana fiber contains polar groups that make it water-sensitive, banana composites delaminate and degrade in wet conditions. The addition of alkali-treated banana fiber improves the mechanical properties and water absorption of the polyester banana fiber composite. The ideal physico-mechanical properties were obtained by substituting coagulum for polyester resin for 40% of the total. Improvements in impact strength (45%), bending modulus (30%), and bending strength (50%) were seen, along with a 68% decrease in water absorption. With the addition of coagulum to a combination of polyester and banana fibers, the limiting oxygen index increased from 18% to 21%

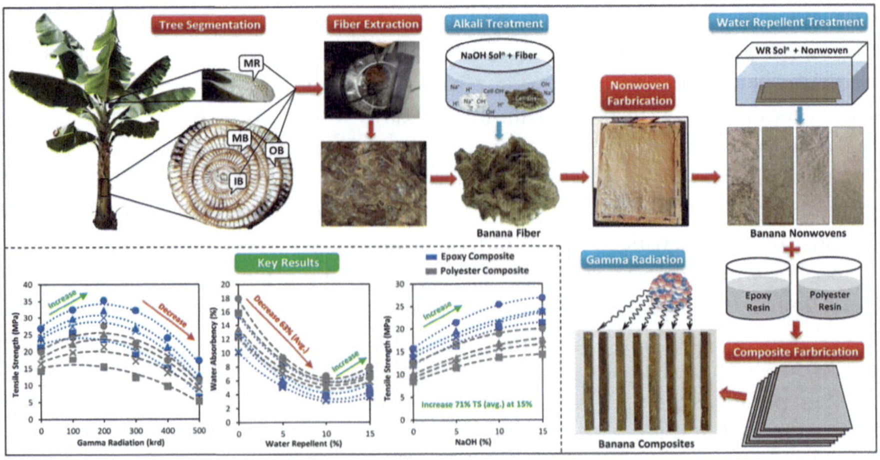

Fig. 9.7 Banana fiber characteristics. (Source: Motaleb et al. (2021))

and the smoke density decreased. Among the many possible uses for the manufactured composite material are particle boards, partition boards, and others. In order to produce composites in an environmentally conscious manner, this study investigates the possibility of using renewable resources as feedstock instead of petroleum products. Figures 9.7 and 9.8 show a schematic diagram of the characteristics of banana fibers.

9.2 Current Techniques and Challenges in Cultivation

9.2.1 Current Cultivation Practices

Vegetative propagation is the sole option for bananas because almost all cultivated cultivars are triploid, seedless, or seed sterile. Plantains are not an exception. Cronauer and Krikorian (1984) state that corms, large and tiny suckers, and sword suckers are the materials employed in traditional propagation. The seeds are exclusively used by breeding programs. Various-sized suckers can be used, however small or sickly suckers should be avoided. Planting whole suckers harvested from earlier plants is possible. Suckers are best transplanted when they have attained the size of their first true leaves. When propagating new plants from corms, it is common practice to disinfect them to prevent the spread of soil-borne diseases. It is not uncommon to start new plantations with tissue-grown plantlets because they are pest-free. Placing them in containers, typically black polyethylene bags, is the first step in getting them ready for field planting.

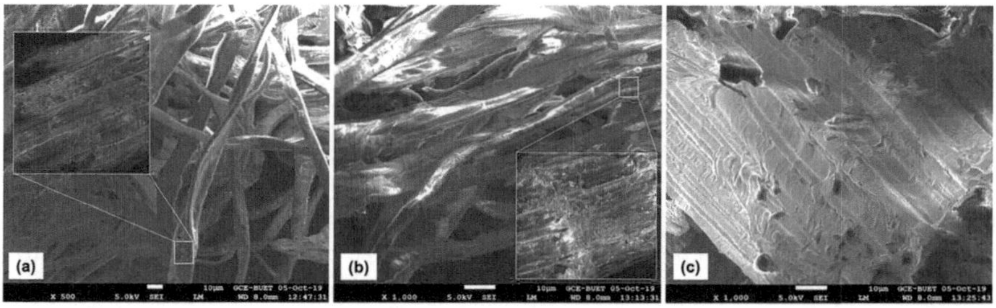

Fig. 9.8 Scanning electron microscope (SEM) micrograph of (**a**) raw banana fibers, (**b**) alkali-treated banana fibers, and (**c**) water-repellent-treated banana fiber nonwovens. (Source: Motaleb et al. (2021))

The large pest population, sluggish growth rate, bulkiness, and lack of phytosanitary quality in conventional planting materials make them an unsuitable choice for propagation. In vitro banana propagation has many important advantages over more conventional methods. Faster growth in the early stages compared to conventional materials, a shorter harvest interval, a steady supply of disease-free material, physiological uniformity, a high rate of multiplication, and the ability to rapidly disseminate new plant materials globally are all advantages (Daniells and Smith 1991). Reducing off-types induced by tissue culture variability is the primary obstacle to successfully micropropagating banana plants. Variants emerge frequently, necessitating careful control to reduce their occurrence.

9.2.2 Spacing Between Banana Plants

Factors such as cultivar type, soil fertility, pruning techniques, irrigation systems, weed management, bunch size requirements, fruit quality, and herbicide usage are among those that might affect the spacing between plants in banana farms.

There is a large global variation in plant spacing. The standard spacing for plantings is 360–680 plants per acre, with spacing ranging from 2.4–2.4 m to 3.4–3.4 m. The benefits of closer spacing include fewer weeds and higher yields, but there are some drawbacks, such as a possible delay in blooming, lower bunch weight, and fewer hands per bunch. When producers pick their spacing, they need to think about a lot of things, like how much land and plants will cost, how they will be watered, what fertilizer programs they will use, and what tools they will use for harvesting and spraying. Plantations often use hexagonal or bedded layouts because this shape provides the ideal spacing for plants per square foot. Plant in two or three rows for the best results while using machinery and harvesting.

9.2.3 Deforestation of Banana Trees

Pruning your banana plant is essential if you desire abundant fruit and healthy leaves. Cutting out non-essential suckers is the most undervalued aspect of caring for banana plants. In a perfect banana mat, there should only be one mother plant and two or three suckers. One sucker should be allowed to grow every four to six months. Disease development and poor fruit quality can result from letting too many pseudostems grow from a single mat. There is just one fruiting of a banana pseudostem. Gather the stalk by chopping it at the base, then scatter the small pieces on the ground; this will form mulch. The next harvest is generated when the rhizome (or mat) is allowed to create additional pseudostems (follow-up stalks).

It is important to provide plants with plenty of room to grow without stifling their growth or making them compete for water, light, or nutrients. To control the amount and replacement of pseudostems, it is helpful to remove newly generated suckers as soon as they emerge. In a perfect world, each mat would have one little sucker, one pseudostem that is halfway grown, and one pseudostem that is both flowering and fruiting.

First, eradicate any suckers at ground level to eliminate the underground bud. After that, get rid of all the leftover buds by using a metal digging rod. Make sure the internal bud is dead to stop suckers from growing because regrowth happens fast. Even though it's a pain, you should remove any dead leaves or leaves that touch the bunch. You can speed up the fruiting process by snipping the barren blooming stalk that hangs below the last banana hand.

9.2.4 Fundamental Requirements According to Weather Conditions and Topography

Bananas are grown worldwide in tropical and subtropical regions that get 200 cm of equally distributed rainfall per year and average temperatures of up to 20 °C, according to Wardlaw (1972). Because altitude affects meteorological factors like temperature, precipitation, humidity, and light intensity, Green (1963) says that producing bananas at elevations above 1000 meters could be difficult. Cold weather and dry spells are also not good for bananas. During the summer, banana plants produce three or four leaves per month, but during the winter, they only produce one leaf every month. If cold weather hits just as the flower is about to open, the bud may not be able to separate from the stem. It is possible to delay the ripening of fruits that have already begun to form by a few months or even prevent them from reaching maturity. When left exposed to the wind, banana plants do not do well. To protect the banana fruit from the wind, windbreaks are commonly planted around banana orchards. Banana farms in the West Indies were forcibly relocated to Central America, Ecuador, and Colombia due to the extensive damage caused by storms and cyclones (Morton 1987). Planting bananas in rich, well-drained soil brings out their best results, while they can tolerate a wide range of soil types. The perfect location would be a defunct compost heap. Bananas do best in slightly acidic soils (5.5 to 6.5 pH). However, excessively saline soils will cause their plants to die (www.crfg.org).

Soils that drain well or can be improved are ideal for growing bananas due to their high water requirement and susceptibility to waterlogging (Morton 1987). Because bananas require a lot of water, careful planning and execution of irrigation systems are necessary to ensure maximum yields. Before planting, the system's design should take into consideration and account for a wide range of factors that could affect productivity, such as soil type, landform, water quality, water availability, manpower, and budget. When precipitation is less than 2 inches weekly, it's best to water plants. The banana plant requires frequent fertilizer applications due to its fast growth and large fruit production. During the first three or four months of growth, the plant's developmental type dictates the bunch weight and the number of hands. Therefore, it is critical to provide first-rate care right now. Two pounds of complete fertilizer per mat every two to three months is a decent rule of thumb to follow in the absence of soil or tissue test findings. Due to their high potassium requirements, banana plants thrive when fertilized with potassium-rich soil. When soil micronutrients are low, it's best to use manures or foliar sprays that contain these nutrients. The optimal timing for applying fertilizer is two to three months after planting and should be maintained throughout the growing season, including throughout flowering and any further harvests from the same mat. For optimal banana yields, low-fertility soils (such as sandy and calcareous soils) require four or six applications of fertilizer.

9.2.5 Weed Management

The spread of disease and the competition for fertilizers and irrigation water are two ways in which weeds diminish agricultural production. Weeds can be effectively controlled by utilizing a mix of herbicides, mulching, mowing, and hand hoeing. To protect the plantation's delicate roots from mechanical cultivation, it is important to work only in weed-free zones surrounding the planting. In the early phases of plant growth, before the plants can produce enough shade for themselves, herbicides are an easy and effective technique to manage weeds. It is crucial to do a spot test before spraying herbicides on a larger area since their effects on different plants and soils can vary. The best time to apply a herbicide is either before or just after the weeds begin to sprout. Applying a pre-emergent pesticide to bare soil will stop weed seeds from growing. Once weeds have sprouted, it's time to apply a post-emergent treatment. If you want to avoid spraying newly planted suckers, which can cause leaf burn and stunt plant growth, it's best to apply herbicides first thing in the morning when the wind is calm.

9.2.6 Crop Protection

Pests and diseases can affect any crop, even bananas. Diseases caused by bacteria, fungi, and viruses can damage different areas of the plant. While some diseases affect only the roots, others might spread to the fruit or even the foliage. A terrible and pervasive disease, Panama disease is another name for Fusarium vascular wilt disease. This illness is caused by parasitic *Fusarium oxysporum* f. cubense strains. Three interrelated aspects dictate the disease's complexity: the host plant's genetic makeup, the pathogenic features of the fungal strains, and the impact of environmental conditions on the host and the pathogen. Although it has been recorded in nearly every banana-growing nation, Stover and Simmonds (1987) propose that the India-Malaysia region is its likely birthplace. Infection can easily spread from young suckers to immature plants and even older plants that have developed a cluster at the top. The clusters rarely mature, though, because the plants eventually die (Wardlaw 1972). Growing resistant types or using quarantine and exclusion procedures are the sole viable ways to control fusarium wilt. Other Cavendish cultivars are, nevertheless, vulnerable to certain fungal infections, in contrast to the economically significant and very resistant Cavendish Nana variety (Stover 1987). After soil contamination develops, the only method to diminish the population is to have extended fallow periods with crops that are not hosts.

Black Sigatoka leaf spots are caused by the fungal disease *Mycosphaerella fijiensis*, which is a major concern for plant health. The plant's capacity to photosynthesize is compromised as the fungal infection makes its way into its leaves. Because it grows best on wet leaves, this fungus is a known problem that reduces the yields of bananas and plantains around the world. This disease takes several forms before it kills the plant, and it might stop fruits from ripening or make the pulp seem different. Consequently, fungicides are used by many farmers.

Large plantations often employ aircraft because the irregular spacing of banana trees makes it difficult to maneuver ground vehicles through the foliage (Wardlaw 1972). Numerous insects find bananas to be an attractive food source. Bananas are susceptible to damage from almost 400 different species of bugs and mites. In this group, 250 are leaf feeders, 10 are pseudostem eaters, 70 are root and rhizome eaters, 130 are fruit and flower eaters, and 10 or more are disease vectors. The 58 different kinds of ants found in Central American banana farms are quite remarkable. All of the main banana-growing regions are infested by the banana borer, scientifically called *Cosmopolites sordidus* (Ostmark 1974). Potential crop losses can reach 100%, which is especially dangerous for small-scale farmers. Borers are accused of causing damage by rhizome rot and nematodes, among other things (Ostmark 1974). Before replanting bananas in fields infested with nematodes, it is essential to implement crop rotation with non-host species. Additionally, subsistence farmers must be provided with alternative control strategies, such as the identification of natural predators, and protocols for effectively sanitizing planting materials should be established.

Most Central American farms avoided employing pesticides to control borer infestations for a long time since the weevil is resistant to them (Stover 1987). If you want to buy high-quality bananas of high quality, make sure they are green and not damaged by bugs. Even though most flaws are external rather than inside, protection from fruit-eating insects is still important. In order to maximize bunch weight and decrease leaf scarring, it is common practice on farms to place fruit bunches in perforated polyethylene bags at about two weeks of age. To prevent fruit peel damage, place a pesticide inside the bag before sealing it to create a fumigation chamber. Then pests like aphids and thrips won't be able to get to the bag's contents (Ostmark 1974).

Banana leaf-eating pests can hinder photosynthate delivery to the rhizome and, eventually, the fruit (Turner 1970). The two most dangerous defoliators of bananas in central and northern South American regions are *Antichloris eriphia* and *Anthochloris eriphia*. In order to avoid losing fruit due to pests, surveys are carried out every two weeks during periods of heavy infestation (Stover 1987). When pesticides aren't an option, natural control mechanisms like predators and parasites can step in. If methods of controlling pests using nature fail, aerial pesticide spraying becomes an attractive alternative (Ostmark 1974).

Cavendish types are more susceptible to nematode-caused root and rhizome rot, which can be worsened by specific fungi and bacteria. Following Black Sigatoka in terms of severity, this condition ranks second. For the majority of the Americas (Caribbean, Central, and South), the Philippines, and Africa, nematicides are an absolute must for crop production optimization (Stover 1987).

9.2.7 Harvesting and Producing

Annual yields can range from 4 to 10 tons per acre, subject to several variables. Factors such as weather, pests, diseases, and farming techniques affect how long banana plants live. Bananas are picked in bunches when they're ripe but not quite yellow. It could not be cost-effective to harvest and ripen bananas on a small scale. The optimal time for homeowners to harvest fruit is seven to fourteen days before it ripens on the plant. Fruits ripened off the vine, in a cool, shaded spot, have a more concentrated flavor. For optimal quality when used for commercial purposes, bananas that have been picked recently should be refrigerated or kept in a cool place with a relative humidity of 90–95% and temperatures between 13–14 °C. When the temperature falls below 13 °C, cooling damage happens. In commercial settings, bananas ripen best at temperatures of 18–20 degrees Celsius. A process that is started and accelerated by ethylene gas is ripening. Specialized storage rooms in business settings utilize ethylene gas generators to control ripening. By placing a polyethylene bag over the bunch, homeowners can hasten the ripening process.

9.2.8 Current Agriculture Challenges

9.2.8.1 Biotic Stresses Regarding Bananas

New diseases like *Xanthomonas* wilt have emerged, expanding the potential areas of research on biotic stresses affecting bananas. Efforts to create transgenic plants resistant to Xanthomonas wilt, nematodes, and black Sigatoka have been made, but these plants have not yet proven successful in field trials.

9.2.8.2 Ecological Considerations

Fungal diseases, nematodes, viruses, bacteria, and environmental variables like temperature and air pollution have garnered the majority of the attention in banana research, while abiotic challenges like drought, flooding, salinity, metal toxicity, mineral deficiency, unfavorable pH, and insect infestations have been given less consideration.

9.2.8.3 Ecological Transformations

Various regions are being affected in different ways by the changes in temperature and precipitation levels caused by global warming. Since bananas are most productive in hot and humid regions, they stand to lose out in the face of climate change. Soil moisture, temperature, and the health of both helpful and pest species are all impacted by weather, which in turn impacts banana development and production. Drought poses a significant risk to banana crops, which are highly dependent on rainfall levels. The combined effects of drought and salt might cause over half of the world's arable land to become severely salinized by the year 2050. Because of the impact of climate change, studies on Musa's ability to withstand heat and drought are crucial.

9.2.8.4 Drought Tolerance

Because drought is a serious environmental problem that impacts farming all over the world, plants that can withstand dry conditions are highly prized for their economic value (Ibrahim et al. 2023b; Börner et al. 2009, 2010; Elabd et al. 2020; Saleh et al. 2020a, b; 2021a,b; Salem and Börner 2004; Salem et al. 2021a, b, d, e; Salem et al. 2004a). Bananas have not been improved by conventional breeding methods to make them more resistant to drought. In order to develop Musa for consistent yields in areas with limited water, researchers in Uganda sought out banana varieties that could endure dry spells. In order to create novel methods for growing bananas, it is essential to understand the molecular mechanisms that control their drought tolerance. We now know more about how plants react to drought thanks to genetic engineering, which has helped us identify the genes that play a role in these reactions. Transgenic plants and gene expression studies can shed light on Musa's mechanisms for drought resistance. In order to confirm the genetic resilience of Musa cultivars, researchers studying drought tolerance should employ a variety of methods to identify and isolate those cultivars. To better understand the complex processes of drought tolerance in Musa, it is necessary to integrate genomes, physiology, proteomics, and transcriptomics. This integration has the potential to lead to the development of banana cultivars that are more resistant to drought.

9.2.8.5 Cold Tolerance in Banana

Research on banana varieties that can withstand freezing temperatures has been documented in the literature (Chen et al. 2008). According to earlier research, a number of banana types have shown field resilience to frost. Of the dessert accessions tested, four cultivars showed very high tolerance levels and six showed moderate tolerance. Twelve of the cooking banana accessions were determined to be fairly tolerant, while seven were determined to be highly tolerant. A higher proportion of the A genome was associated with more frost injury in cultivars, while a higher proportion of the B genome was associated with greater frost injury tolerance in accessions, according to other study findings.

9.2.8.6 Impact of Salinity on Banana

Much of the prior research on Musa's susceptibility to salt stress has concentrated on banana varieties specifically, "Pacovan," as well as on arbuscular mycorrhizal fungal infections (Yano-Melo et al. 2003) and induced mutations (Miri et al. 2009). While some banana cultivars appear to be able to withstand different degrees of salt stress, more investigation into this topic is required to draw any firm conclusions.

9.2.9 Breeding Objectives

Bananas and plantains (*Musa* spp. L.) are significant export crops for several nations in Latin America, the Caribbean, and portions of Western and Central Africa. Nevertheless, only over 10% of the world's output goes to export. Bananas and plantains are major cash crops and staple foods for rural populations in producing countries. Plantains account for over 25% of the caloric intake for over 70 million people in West and Central Africa alone, which is particularly significant in the equatorial region of Africa (Robinson 1996). Although plantain exports to the US and EU are on the rise, the majority of the world's supply—73 percent—is grown and consumed in Western and Central Africa (WCA) (Marin et al. 2003).

There has been considerable morphological variety in this region even though bananas and plantains were introduced to Africa approximately three thousand years ago. The main cultivars also go by a plethora of regional names, which adds to the already rich linguistic, genetic, and ecological diversity (Santayana et al. 2014). Similar interactions between language and genealogy accompanied the diversification of the Musaceae family in Asia, where it was first found. Many native species can be impacted by a range of pests and diseases. In the highlands, plantain crops are primarily at risk from the yellow Sigatoka (*Mycosphaerella musicola*) and in the lowlands, from the black Sigatoka (*M. fijiensis*).

Management costs associated with black sigatoka, the most expensive and severe disease affecting plantains and bananas, are 27 percent of overall production expenses in commercial plantations (Stover 1987). Banana weevils, Fusarium wilt, *Helicotylenchus multicinctus*, *Radopholus similis*, and a complex of plant-parasitic nematodes have all been known to negatively impact banana and plantain production for quite some time. In recent times, bacterial wilt (*Xanthomonas campestris* pv. musacearum) has become the most devastating banana disease in East Africa's highlands and Great Lakes region. Meanwhile, in Central Africa and, to a lesser degree, West Africa, the banana bract mosaic virus has caused concern. Black Sigatoka became an epidemic in Africa after an unintentional import in the late 1970s (Wilson and Buddenhagen 1986). Deterioration of the plant's leaves, caused by the uncontrolled spread of the illness, decreases the area that plants may use for photosynthesis. According to Mobambo et al. (1993), this can lead to drastically reduced or nonexistent crop production. The native variety' yields have been cut in half due to the introduction and spread of black Sigatoka, which does not possess any intrinsic resistance.

There are two main reasons why people in areas hit by Black Sigatoka don't have enough food: first, food scarcity was already there before the disease hit, and second, bananas and plantains were more resilient than staple grains to production variables like droughts, floods, diseases, pests, and storage losses.

Plantains became a luxury good in metropolitan Nigeria and other African countries after the Black Sigatoka illness drastically reduced their supply, driving up prices. Since chemical protection is both costly and environmentally damaging, host-plant resistance is the most practical and long-term solution for disease control for resource-limited farmers in the developing world, where the majority of production takes place. Since the fungicides used by commercial banana exporters and producers are out of reach and too expensive for the numerous small-scale plantain growers in Africa, the only long-term option is to create hybrids that are resistant to black Sigatoka.

To combat pests and diseases without sacrificing the exquisite flavor of Musa, breeders are creating new cultivars. Developing hybrids with high yields per unit area and time while being resistant to biotic stressors is one goal of genetically altering Musa. Thus, ideal hybrids would exhibit low winter harvest delay, high photosynthetic efficiency, and rapid maturity in the first growth cycle (Ortiz and Vuylsteke 1994a). In addition to being short and having a robust root system, this plant also has the desirable qualities of improved nutrient absorption and resistance to wind damage.

Growing bananas, like most crops, is primarily about I getting more fruit out of them, (ii) making them more resistant to pests and diseases, (iii) changing their shape in ways like dwarfism, (iv) making them mature faster, and (v) making them more resistant to things like drought, salt, high temperatures, and metal ions.

So far, efforts in banana breeding have primarily concentrated on creating disease- and pest-resistant cultivars that can keep yields stable. This happened because banana farming was almost eradicated as a commercial enterprise due to devastating fungal illnesses. Although improving commercial banana varieties was the original aim of banana breeding programs, it soon became apparent that smallholder farmers' crops were just as vulnerable to pests and illnesses. Since bananas and plantains are the primary sources of nutrition for millions of people in tropical locations across the globe, a decline in banana output could endanger their way of life. There are a lot of hybrid bananas thanks to the main banana breeding programs. The fact that people aren't buying the new hybrids, though, is a major problem. This is beginning to change, though, since new hybrids are being rapidly adopted by farmers in many nations. Rising climate change alarmism and the present emphasis on the role of diet in human health highlight the urgent need to include new banana breeding goals to address future demands.

9.3 Germplasm Biodiversity and Conservation

9.3.1 Types of Germplasm

Genetic diversity of both wild and cultivated banana accessions has been studied by researchers using molecular markers since the 1980s. Multiple studies have found markers that indicate genetic diversity within the Musa species. These studies include Carreel et al. (2002), Ude et al. (2002), Horry and Jay (1988), Jarret et al. (1992), Lebot et al. (1993), Hippolyte et al. (2010), Pillay et al. (2000) and Wong et al. (2001). A banana-specific module is included in TropGENE-DB, a tropical crop genetic database created by Ruiz et al. (2004). Thanks to its intuitive web interfaces, this database facilitates thorough and quick research, with findings presented in a way that's easy to comprehend.

Around 1100 banana entries are present in the TropGENE-DB module at the moment. The following details are included in these entries: the place of origin, details about the collection, agro-morphological data, resistance to diseases and pests, reactions to environmental conditions, family tree, history of breeding, and extensive phenotypic data. In addition to sequencing data, genetic maps, and information on markers (RFLP, AFLP, microsatellites, isozymes), the database also contains information about markers and their details. Users can also see pictures of fruits, gel plates, and how plants react to diseases.

9.3.2 Conservation and Biodiversity of Germplasm

Conservation of genetic resources in plant gene banks through various approaches is highly beneficial for seed preservation (Abulaila et al. 2025; Abdullaev et al. 2025; Akparov et al. 2025; Aladele et al. 2025; Al Lawati 2025; Almarri et al. 2025; Arab et al. 2025; Békefi et al. 2025; Breidy 2025; Cañar et al. 2025; de Galarreta et al. 2025; Desheva et al. 2025; Díez et al. 2025; Galea-Gragera and Llera Cid 2025; Ghidan and Abdulmajid 2025; Gudaye et al. 2025; Harzli et al. 2025; Hidayatun and Kurniawan 2025; Ihsan and Prihatini 2025; Jatoi and Siddiqui 2025; Laghetti et al. 2025; Mallor and Luna 2025; Monteros-Altamirano et al. 2025; Osterc et al. 2025; Popović et al. 2025; Rocha et al. 2025; Wang et al. 2025). On the other hand, conservation helps preserve biodiversity and identify varieties with beneficial and adaptive genes.

Because banana improvement projects depend on the complete spectrum of variability, it is crucial to preserve and document current cultivars and their wild relatives. Several techniques are used to preserve the biodiversity of bananas, including cryopreservation, in vitro plantlets, and field gene banks that hold full-size plants (Amalraj et al. 1993; Menon 2000; Uma et al. 2005a). A total of sixty national collections hold the genetic material from various species of Musa. An important component of Bioversity International's worldwide banana collection is the in vitro collection of 1200 Musa accessions housed in the International Transit Centre (ITC) at Belgium's KU Leuven University. The Musa Germplasm Information System (MGIS) includes evaluation and characterization data from twenty-two separate collections, as mentioned by Channelie're et al. (2011). Banana bunchy top virus, banana bract mosaic virus, weevil borers, Sigatoka leaf spot, Fusarium wilt, and other diseases are major dangers to germplasm, small-scale production systems, and field repositories. Some rare varieties of backyard clones and homesteads are in danger of going extinct. Deforestation, feral animal introductions, and agricultural land conversion pose threats to several Musa species in their natural habitats, especially in the forests of the North Eastern area and the Western Ghats. That is why it is critical to protect genetic resources, including endangered species, for the sake of their future usage. Agrawal et al. (2007) state that the NBPGR in New Delhi regularly shares banana germplasm by using in vitro conservation procedures. Additionally, NRCB, Trichy, and other state agricultural universities keep field collections. Funding for germplasm conservation operations is provided by the Indian Council of Agricultural Research through its All India Coordinated Research Project on Tropical Fruits.

9.3.3 In Vitro, In Situ, Cryopreservation and Gene Bank

Because banana improvement efforts depend on the full spectrum of genetic variety, it is vital to effectively preserve and document present cultivars and their wild relatives. Preservation of banana biodiversity is achieved through cryopreservation, in vitro plantlets, and field gene banks. Amalraj et al. (1993), Menon(2000), and Uma et al. (2005a) all mention full-sized plants as a possible method. The genetic material of *the Musa* species is preserved in about sixty national collections. As part of Bioversity International's global banana collection, the International Transit Centre (ITC) at Belgium's KU Leuven University houses an in vitro collection of 1200 Musa accessions. The Musa Germplasm Information System (MGIS) incorporates characterization and assessment data from 22 different collections, according to Channelie're et al. (2011). Germplasm, small-scale production methods, and field repositories are all in danger from a variety of illnesses. Blossom top banana virus, banana bract mosaic virus, weevil borer, Sigatoka leaf spot, and Fusarium wilt are all examples of such diseases. There are a number of these unusual backyard/homestead clones that are going extinct. The change of natural habitats into agricultural land, the invasion of feral animals, and deforestation pose threats to many Musa species in the Western Ghats and the Northeastern region. For this reason, preserving genetic resources, especially those of endangered species, is essential. In order to preserve and disseminate banana germplasm, the National Banana Plant Genetic Resources Reserve (NBPGR) in New Delhi routinely uses in vitro conservation methods (Agrawal et al. 2007). At NRCB, Trichy, and other state agricultural universities, field collections are carefully kept. Funding for germplasm conservation operations is provided by the Indian Council of Agricultural Research through its All India Coordinated Research Project on Tropical Fruits.

9.3.4 Phylogenetic Relationships

Genomic variety identification, in situ and ex-situ conservation management, and future banana cultivar breeding all depend on accurately identifying banana accession data.

According to Valmayor et al. (2000), plantain bananas (*Musa paradisiaca* Linn.) and dessert bananas (*Musa sapientum* Linn.) were classified by Linnaeus in 1953 and 1959, respectively. That is why you will hear both names used most frequently in Southeast Asia.

Banana gin caused disorientation, according to studies by Valmayor et al. (2000) and Singh et al. (2001). One of the biggest problems that Southeast Asian horticulturists and taxonomists have is the proliferation of banana cultivar names and synonymous cultivars across many languages, dialects, and areas (Valmayor et al. 2000; Wahyudi et al. 2020). Thus, the innovative taxonomy put forward by Simmonds and Shepherd (1955) that was unanimously approved in 1999 (Valmayor et al. 2000) could be the answer to this dilemma. A new technique, the three-tier system, has been developed for classifying bananas. The process begins with the species designation, then includes the ploidy and genomic groupings from their ancestors, and finally ends with the local cultivar names (Simmonds and Shepherd 1955; Valmayor et al. 2000). The three-tier system was put in place because the B genome ($x = 11$)

was produced via hybridization of *Musa acuminata* Colla and *M. balbisiana*, which led to the development of banana cultivars in Southeast Asia.

Numerous researchers have explored this subject, including Singh et al. (2001), De Langhe et al. (2009), Li et al. (2010), and Häkkinen (2013), to name a few... Hapsari et al. (2015) found that bananas can have a wide variety of ploidy levels and genomic configurations, including AAA (Ambon, Berlin, Mas), AAB (Pisang Raja), and ABB (Kapok, Ebung). Morphological traits were used to evaluate the genetic classifications of banana cultivars (Hapsari et al. 2015). However, morphological methods are subjective, take a long time to process, and are susceptible to environmental influences; thus, they aren't always reliable (Probojati et al. 2019).

Modern methods for determining banana cultivar classifications and analyzing genetic makeup have advanced to the level of molecular markers. A large body of research (Poerba and Ahmad 2010; de Jesus et al. 2013; Lamare and Rao 2015; Sundari et al. 2017; Probojati et al. 2019; Wahyudi et al. 2020) shows that DNA sequence-based molecular methods are more accurate and efficient than morphological techniques and PCR-based analyses.

Although PCR-based methods can be used to categorize banana cultivar genomes, there is still no clear answer on the parents of these banana varieties (Li et al. 2013). Based on DNA sequences derived from the chloroplast genome, particularly rbcL, intron trnK, trnL-F, and matK (maturase K), the genome of banana cultivars can be delineated, according to multiple studies (Bieniek et al. 2015; Janssen et al. 2016; Nikmah et al. 2016; Liu et al. 2010; Wahyudi et al. 2013; Udensi et al. 2017).

The DNA of chloroplasts is structurally stable and does not undergo recombination, according to multiple studies (Costion et al. 2011; Shekhar et al. 2019; Yuan et al. 2015).

The matK gene is the go-to choice for plant identification because of its excellent specificity and low mutation rate. Further, the matK gene was suggested by the CBOL Plant Working Group et al. (2009) for use in generic plant identification. Reconstructing phylogenetic trees and determining genetic divergence time have both been accomplished using the matK gene (Liu et al. 2009; Udensi et al. 2017). Nevertheless, there has been a lack of research on the matK gene sequence and the genetic links of banana cultivars in Java, making it difficult to predict when these varieties diverged. Figure 9.9 shows different types of conservation of germplasm.

9.4 Mutation Breeding

Banana chromosomes, which are usually between 1 and 2 micrometers in size, cause hyperpigmented lesions because they don't attach well to traditional DNA staining methods. Because of their condensation, the little chromosomes are difficult to describe; even with sophisticated staining methods, measuring the long and short arms remains an enormous challenge (Adeleke et al. 2002; Pillay and Adeleke 2001). Banana mitotic metaphase chromosomes look like tiny dots without any cytological markers when the cell is fully contracted. The chromosomes are quite

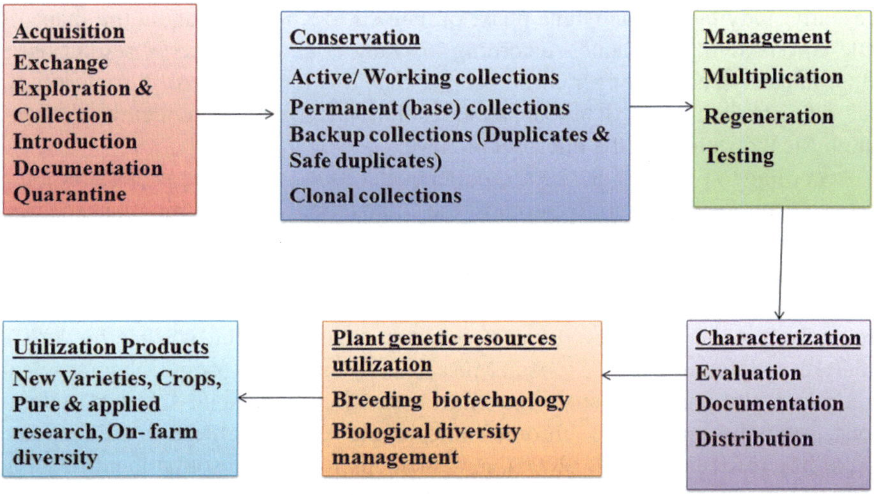

Fig. 9.9 General conservation of germplasm. (Source: Priyanka et al. (2021))

similar to one another, which makes it difficult to differentiate them and prevents them from forming karyotypes.

In their study of the Musa karyotype, Dantas et al. (1993) found that the two longest chromosomes, which are also known as satellite chromosomes, are the only ones of note. Another study by Wang et al. (1993) found one set of satellite chromosomes in AA bananas found in nature. It appears that the remaining chromosomes display orientations ranging from metacentric to submetacentric. Using cytological markers collected through chromosomal banding methods, homologous pairs and individual karyotype chromosomes have been easily detectable since the 1960s. These methods make use of chromatin condensation in euchromatic and heterochromatic regions to achieve varied staining intensity across chromosomes (Danilova and Birchler 2008).

Different banding processes have different biochemical implications, which in turn define the resulting banding pattern (Stace 1980). More often than not, plant researchers will use C, G, Q, R, or Hy-banding. According to Dolezelova et al. (1998), banana chromosomes could not be correctly banded by their lab or any other top cytogenetics lab. It may be required to improve current approaches and do additional research to determine the possible uses of banding techniques for Musa and their breeding.

Mitotic metaphase chromosomes have been the primary focus of Musa cytological investigations. According to Adeleke et al. (2002), using mitotic prometaphase or pachytene chromosomes might improve the process of producing Musa karyotypes. According to unpublished research, chromosomes in mitotic prometaphase tend to be three to five times longer than chromosomes in completely condensed metaphase. Chromosomes at this stage of metaphase have much less DNA condensation, making the primary and secondary constrictions easier to see. Scientists are presently studying the pachytene phase of meiosis to learn more about the shape of little chromosomes in plants. According to Wang et al. (2009), axial resolution is better in pachytene for longer chromosomes than it is for metaphase chromosomes. According to De Jong et al. (1999), axial resolution is the ability to identify the tiniest physical distances at the microscopic level.

According to Fransz et al. (2000), Zoller et al. (2004), and Wang et al. (2006), the degree of mitotic and meiotic chromosome compaction varies within plant species. For example, in Arabidopsis, it is 25-fold, in rye it is 6.2-fold, and in maize it is 15.9-fold. The ability to discriminate between eu and heterochromatin, visualize centromeres, chromomeres, telomeres, knobs, and nucleoli, and identify individual chromosomes can be achieved using pachytene analysis (Schulz-Schaeffer 1980). Adeleke et al. found a way to look at Musa's pachytene chromosome. Natural selection and evolutionary pressures may have caused chromosomal structural alterations; studying pachytene chromosomes could disclose these modifications. According to Hribova et al. (2008), new methods for chromosomal identification, such as fluorescent in situ hybridization (FISH), could be useful for bananas.

9.4.1 Conventional Mutagenesis

According to Kumar and Soorianathasundaram (2007), scientists at TNAU in Coimbatore are using gamma-ray and electromagnetic spectrum (EMS)-induced in vitro mutations to cultivate a variety of banana types, including Robusta, Rasthali, Nendran, Poovan, Karpooravalli, red banana, and Monthan. Four Robusta mutants have reached the V4M1 generation and produced bunches over 28.0 kg in field settings, but no viable mutants were obtained from Rasthali. Researchers have examined the effects of antimitotic agents on theoretical diploid populations, such as Sannachenkadali (AA), Anaikomban (AA), Kunnan (AB), and Thattillakunnan (AB), to induce tetraploidy experimentally. For this reason, chemists utilize colchicine and oryzalin. We are now evaluating and incorporating 24 tetraploids into our breeding program. Induced mutations at NRCB and Trichy are improving Rasthali (AAB Silk group). The genetic fidelity of 54 separate mutants was assessed using molecular markers, which revealed polymorphism ranging from 30.2% to 50%. Also, in comparison to Rasthali plants that were not modified, mutants showed a variance of 19–21%.

9.4.2 *In Vitro* Mutagenesis and Selection

Musa's genetic improvement has not made full use of tissue culture's ability to produce genetic variants. The ability to generate large amounts of genetic variation is an advantage of tissue culture in plant breeding (Jain 2001). In order to create genetic variety, somaclonal variation, which encompasses somatic embryogenesis and callus cultivation, has long been acknowledged (Larkin and Scowcroft 1981). Through organogenesis and somatic embryogenesis, musa has been demonstrated to generate somaclones that are advantageous to agriculture (Jain et al. 2011). Plant breeders have made use of in vitro mutagenesis methods like ethyl-methane-sulfonate (EMS) treatment, fast neutron radiation, and insertional mutagenesis to successfully breed for crop stress, whether it be biotic (Bhagwat and Duncan 1998; Kowalski and Cassells 1999) or abiotic (Khan et al. 2001; Fuller and Eed 2003). The requirement for a large number of individuals makes these procedures problematic when it comes to identifying the target feature. However, in vitro selection methods make this problem much easier to solve. It is possible to assess the viability of stress-resistant lines generated by both traditional breeding techniques and transgenic procedures using in vitro selection. Abiotic stressors, for which screening mechanisms may be inadequate, are an area where this shines. Despite the increasing popularity of genetic transformation and high-throughput technology, an alternate breeding technique is provided by somaclonal variation, in vitro mutagenesis, and in vitro selection (Dita et al. 2006).

9.5 Molecular Breeding

Molecular cytogenetics has made great strides in the last 20 years in elucidating the genome structure and evolution of plants (Valarik et al. 2004). Using FISH and GISH, researchers have been able to physically map genes on chromosomes (Dolezelova et al. 1998; Fuchs et al. 1998), integrate genetic and physical maps with marker-tagged BAC clones (Fahrenkrug et al. 2001; Yuan et al. 2000), analyze the genomic distribution of mobile genetic elements (Balint-Kurti et al. 2000), and more (Gortner et al. 1998; Heslop-Harrison 1996).

According to studies conducted by Dolezelova et al. (1998), FISH was beneficial in identifying banana genomes and chromosomes. Some banana chromosomes have visible hybridization signals in their centromeric regions, which could be used to identify specific Musa chromosomes. Different hybridization signals along chromosomes may be explained by the arrangement of different types of repetitive DNA. Our unpublished research indicates that metaphase and prometaphase chromosomes in Musa display distinct patterns of contraction. It seems that the centromeric region, which has the maximum density compared to the distal regions, is where chromosome condensation in Musa begins. Because of the delay in this process, the distal segments can look like slightly stained "tails" from time to time; this gives another explanation for the stronger hybridization signals seen in the central sections. Prometaphase chromosomes, according to Fukui (1996), have unusual patterns of staining, particularly in plants that have relatively tiny chromosomes. Differential condensation on the chromatin fiber causes condensation patterns to form. In the proximal sections, primary condensations are more common, whereas in the interstitial or terminal regions, you can find faint, unstable, or tiny condensations. Also known for their noticeable staining are centromeric areas in other short-chromosome species like Arabidopsis (Maluszynska and Heslop-Harrison 1993).

FISH studies on Musa revealed that its 18S–25S rDNA was found on one set of sites (Dolezelova et al. 1998; Osuji et al. 1997). However, the 5S rDNA genes were found on two to four separate chromosomes. All known repetitive DNA sequences were useless for chromosomal identification in M. banana because they are present on every chromosome. There was a high density of monkey transposons in the nucleolar organizer regions of the genome of the small plant *M. Cucuminata* (Balint-Kurti et al. 2000). Radka14, a repetitive DNA clone, and one segment of the retrotransposon monkey were found to be localized to the secondary constriction (Balint-Kurti et al. 2000; Valarik et al. 2002). It is not necessary to use FISH to detect secondary chromosomal constrictions. Cytogenetic markers can be derived from DNA sequences found in large insert libraries like BAC, according to research by Hribova et al. (2008). Finding BACs and other DNA sequence sites on chromosomes requires molecular cytogenetic mapping. Supporting the development of physical maps and map-based cloning initiatives, cytogenetic maps greatly aid in the placement of genes within large heterochromatic regions, where linkage distances are not precisely defined due to reduced meiotic recombination (Zhong et al. 1999).

Among the many banana-related BAC libraries currently available are those from *M. acuminata* cv. Tuu Gia, M. balbisiana cv. Pisang Klutuk Wulung, and Acuminata cv. Calcutta 4 (Vilarinhos et al. 2003, 2006). Another example is the binary BAC (BIBAC) library of M. (Ortiz-Vanquez et al. 2005). The site in M. infected with acuminata could not be determined using BAC-FISH, even though the technique was successful in identifying BAC clones on several plant chromosomes (e.g., rice, cotton, and potato) (Hribova et al. 2008). Out of eight BAC clones tested, only one produced a single locus signal on the chromosomes of the little plant *M. Cucuminata*.

The clone signal was in agreement with the 5S rDNA gene positions. When it comes to developing cytogenetic markers in bananas, BAC libraries don't work, according to Hribova et al. (2008). Similar issues with cotton were pointed out by Wang et al. (2009). According to Wang et al. (2009), cotton has many organic components, such as polysaccharides and phenolics, which could have prevented the hybridization probe from reaching the target DNA sequences in the BAC-FISH investigation. Similar reasons may contribute to the low BAC-FISH resolution in banana chromosomes. Overly compressed mitotic metaphase chromosomes hinder the optical resolution of nearby FISH targets, rendering them useless for FISH (Kulikova et al. 2001). Cömert et al. (2025) states that the optical axial resolution of metaphase chromosomal FISH targets is 1–3 Mbp. Because their complements are 10–40 times longer and display a clear pattern of euchromatic and heterochromatic areas, pachytene chromosomes are more suited for higher resolution, according to Zhong et al. (1996).

According to De Jong et al. (1999), pachytene spreads are more technically demanding than mitotic metaphase spreads, but they may be necessary for mapping

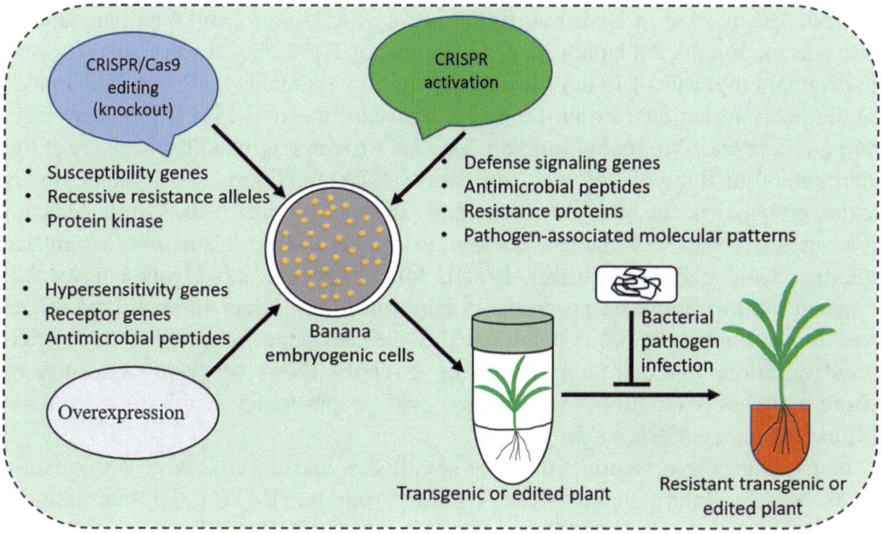

Fig. 9.10 Molecular breeding in banana using. (Source: Tripathi et al. (2022))

BACs on banana chromosomes. Cotton (Wang et al. 2009) and rice are two examples of crops whose BACs have been precisely and thoroughly mapped at good resolution. To bring the banana's physical and cytogenetic maps into harmony, FISH-based mapping is going to be crucial. The molecular breeding of bananas is illustrated in Fig. 9.10.

9.6 Selection

There was a formal start to banana improvement initiatives following the catastrophic effects of Panama disease, caused by *Fusarium oxysporum* f. sp. cubense, on huge commercial dessert banana plantations. Since there were no effective cultural or chemical methods of control, this fungal disease spread rapidly from Surinam to Honduras, the world's largest producer of commercial bananas, in the early 1920s. In the 1950s, a disease nearly eradicated the main 'Gros Michel' cultivar, threatening the future of commercial dessert banana production. Finding a suitable replacement for 'Gros Michel' necessitated looking into existing accessions as a first step. For commercial candies, the Cavendish subgroup of AAA bananas quickly became and remains the go-to variety. Genetic improvement efforts were further pushed forward in the early 1970s by the environmental and economic costs of chemical treatment, as well as the fact that no cultivar at the time exhibited a natural resistance to the newly acquired Sigatoka fungal infections.

Banana breeding attempts, unfortunately, started at the now-defunct Imperial College of Tropical Agriculture in Trinidad. The oldest ongoing breeding program was founded in 1958 in La Lima, Honduras, by the United Fruit Company and is known as the Fundación Honduría de Investigación Agrícola. Dessert bananas were the original emphasis of FHIA's initiative, but now plantains and cooking bananas are the main attractions. From the mid-1970s until the mid-1990s, larger projects were put in place. Two main programs for East African Highland bananas are at the International Institute of Tropical Agriculture (IITA) in Nigeria and Uganda; one is for dessert bananas, the other is for plantains, and both are run by the Centre Africain de Recherches sur les Bananiers et Plantains (CARBAP) in Cameroon. Countries including Côte d'Ivoire, Austria, Brazil, India, Taiwan, and Uganda have also launched modest domestic programs in addition to the larger international initiatives. Induced mutagenesis is being used by the Taiwanese program to find somaclonal variations that can increase genetic diversity. Every program makes use of different diversity-creation mechanisms, with a particular emphasis on cross-pollination, except CNRA's program.

To facilitate the dissemination of new cultivars and improve access to genetic stocks via the International Musa Testing Program (IMTP), the International Network for Improvement of Banana and Plantains (INIBAP, France) was established in 1985. INIBAP oversees networks of interdisciplinary banana researchers throughout Asia, Africa, and Latin America who are committed to improving banana genetics. Belgian university Katholieke Universiteit Leuven (KULeuven) runs the INIBAP International Transit Center, the world's principal banana ex-situ in vitro gene bank. One of the most important databases for banana research, INIBAP, is now part of Bioversity International (Italy). Little breeding initiatives exist for one of the world's most important crops, bananas, despite the fruit's centrality to international trade and marketing.

9.7 Traditional Breeding

9.7.1 Banana Ploidy Breeding

Diploids, which include plants taken from wild, semi-wild, or parthenocarpic populations, have a lot of different genes, a lot of offspring, and a lot of tolerance to diseases and pests. Tomekpe et al. (2004) noted that these parents are exceptional role models and that they also make excellent cytogenetic and genetic study subjects. According to Rowe and Rosales (1994), creating diploids with better agronomic traits and increased disease resistance is a major goal in banana and plantain breeding. They detailed the efforts of FHIA in the field of diploid breeding. The creation of SH-2095, the first diploid hybrid chosen during the first decade of the program, was a watershed moment in this respect. It laid the groundwork for future diploids that were chosen from populations that were separating. With extensive pollination with the naturally resistant AA diploid species "Pisang Jaribuaya," the diploid hybrid SH-3142 has developed resistance to burrowing worms. Contrary to the sterile 'Pisang Jaribuaya,' this hybrid is good for breeding because it has both sexes. The ability to withstand Fusarium wilt allowed for the creation of the FHIA-01 tetraploid hybrid. The goal of the following series of crosses, which used SH-3723 as a progenitor line, was to produce an improved diploid with a wide range of resistance characteristics. In the pursuit of genetic improvement, several enhanced diploids have been employed.

Researchers at the Institute for Tropical Agriculture (IITA) found that among many wild acuminata subspecies (microcarpa, burmannica, malaccensis, and banksii), a model of partial disease resistance was more persistent than a gene-for-gene hypersensitive reaction (Sofan et al. 2020). Two enhanced diploid parents with resistance were produced through crossbreeding with resistant sources that showed resistance to R: TMB2X51051 and TMB2X91283 (AA). related to Tenkouano et al. (2003). The authors Tenkochuano and Swennen (2004) used a combination of recurrent diploid breeding and 4x and 2x crosses to select seven diploid hybrids that were resistant to worms and black Sigatoka.

Vuylsteke et al. (1993) observed that plantain cultivars impact the fraction of diploids in the F_1 population obtained from plantain. According to Vuylsteke and Ortiz (1993), the diploids known as TMP2x, which come from plantain, are important for improving the 2× level of germplasm and for providing plantain alleles. Because plantain genes make up half of a diploid plantain plant's genome, it's more likely to produce plants with plantain-like traits than plants pollinated with Calcutta 4 pollen. Out of plantains have come a plethora of diploids that share many of its desirable characteristics, including resistance to leaf spots and high-quality fruit. For enhanced resistance to Fusarium wilt and Mycosphaerella leaf spot, a diploid cultivar called M53 was developed. A wide variety of diploid hybrids, including both interspecific AB and monospecific AA varieties, have been developed by the National Research Centre for Banana and Tamil Nadu Agricultural University in India (Tomekpe et al. 2004).

9.7.2 Banana 3x and 2x Procedure

By pairing delicate triploids with male-viable, disease-resistant diploids, the 3x x 2x method has been extensively used to produce tetraploid hybrids that exhibit considerable agronomic utility and disease resistance. Because triploid cultivars have lower fertility, more crossings are needed to increase their population. Hybrid tetraploid quality is affected by pollination, seed production ability and male and female fertility, which are all inherited from the fertile diploid. The initial tetraploids produced by this process were crossed with diploids to produce triploids that resembled the original triploid cultivars.

Recombination during the second stage of tetraploid meiosis reduces the genetic advantage gained through nuclear restitution in the first step (creation of tetraploids), as noted by Tomekpe et al. (2004). What happens throughout the F_1 and F_2 phases of a secondary triploid hybrid's regeneration depends on the diploid that was used.

When these cultivars' B genomes activate viral sequences, AAAB hybrids, which are offspring of AAB and AA, show signs of BSV (Tenkouano and Swennen 2004).

This approach can be used by FHIA, IITA, and EMBRAPA to identify secondary triploid hybrids of high quality. By using this method, CARBAP was able to produce a large number of secondary triploids from the initial tetraploid dwarf plantains (Tomekpe et al. 2004).

The fourth generation of Fusarium wilt, black Sigatoka, and burrowing nematodes are the targets of the longest-running FHIA program, which seeks to breed dwarf dessert, plantain, and cooking banana hybrids. 'Dwarf Prata' (AAB Pome) has proven to be an advantageous triploid line for dessert banana breeding, even though the procedure started with 'Highgate' (a dwarf mutant of 'Gros Michel') as a viable triploid seed to produce tetraploid hybrids (Rowe 1990). Opportunities for breeding disease-resistant plantains have arisen because French plantain (AAB plantain) can produce tetraploid progeny with better bunch characteristics (Rowe 1987). Growing 'Gaddatu' (ABB Gubao) cooking bananas was a success (Rowe and Rosales 1994). The breeding efforts produced three exceptional hybrids.

This tetraploid hybrid, FHIA-01, was produced by crossing the diploid (AA) burrowing nematode-resistant SH-3142 with 'Dwarf Prata.' Everyone agrees that this is a huge win for the FHIA program (Rowe and Rosales 1994). Because it is resistant to race 4 of Fusarium wilt, the hybrid can be grown without black Sigatoka management. The purported benefits include outperforming Cavendish in terms of finger size, shelf life, and resistance to restricted growth circumstances. According to Daniells et al. (1995), FHIA-01 was formally introduced as Goldfinger in Australia in 1995. The FHIA program has so far chosen FHIA-21, a hybrid of the French plantain and SH-3142, as its most promising plantain hybrid. According to Rowe and Rosales (1994), it has the same form, color, and pulp flavor as a plantain. FHIA-21 outperformed native plantain in bunch size and showed better resilience and resistance to black Sigatoka. Semidwarf hybrid (ABBB) FHIA-03 is a descendant of the Gaddatu cultivar. Being more resilient and productive than the commonly grown cooking banana 'Bluggoe,' it is the superior choice. Not only is it drought and poor soil tolerant, but it also resists black sigatoka.

Although studies conducted by Simmonds (1962), Stover and Buddenhagen (1987), and Stover and Simmonds (1987) showed that Cavendish cultivars are female sterile, there was less focus on improving this type of dessert banana using traditional breeding methods until 2002. Uncontrolled *Musa acuminata* ssp. Eventually set seed, though. When pollinated with 'Valery' pollen, malaccensis is fertilized. Following this, other dwarf diploids were successfully crossed. Another significant development was the tetraploid plant FHIA-02, which was created by a hybridization process involving the 'Williams cultivar and the enhanced diploid SH-3142 variety. With these factors in mind, conventional breeding was able to produce a Cavendish replacement that was resistant to BLS and/or Foc TR4, which allowed for the selection of many resistant triploid hybrids (Morán 2013).

Low reproductive fertility, triploidy, and delayed propagation were factors that impeded the advancement of plantain breeding. IITA evaluated germplasm for resistant sources and selected viable female plantain cultivars as part of their breeding attempts to fight black Sigatoka. Several French and False Horn plantain cultivars were found to be partially female fertile after being pollinated by the wild diploid banana 'Calcutta 4'; this was proven by the production of viable seeds. 'Obino l'Ewai' and 'Bobby Tannap', two plantains, were descendants of these. Longer growing seasons and different cultivars used as mother plants affected the seed set rate. When subjected to high temperatures, intense sun radiation, and low relative humidity, Ortiz and Vuylsteke (1995b) discovered that seed output increased. The hybridization process involved adding additional resistant diploids to produce tetraploid offspring from Plantain × Calcutta 4, which exhibited some resistance to black Sigatoka. Ortiz and Vuylsteke (1994a) noted that crossing resistant diploids with diploids produced from plantains was another strategy for diversifying the resistant population.

The pure subspecies of Musa acuminata. Despite poor bunch characteristics, Swennen and Vuylsteke (1990) found that resistance to black Sigatoka was significantly heritable in the burmannicoides clone Calcutta 4. The successful mating of tetraploid hybrids with diploid Musa accessions resulted in secondary triploids, which in turn prevented seed set in hybrids and rendered the hybrids female and male sterile. The "PITA-14" (AAAB) secondary triploid hybrid was selected as the best plantain variety as part of the improvement effort.

By crossing 3x with 2x, diploids, triploids, and tetraploids were generated. Phylogenetic analysis, stomatal dimensions/density, and morphological features were employed for ploidy determination. One major benefit of flow cytometry over traditional cytology methods was its speed. Ortiz and Vuylsteke (1994a) used second division restitution (SDR) to find that the 3x plantain genome segregated and recombinated during the production of 2n eggs, which helped recover diversity from plantain crosses. Given the stability of the triploid female genome and the fact that recombination only occurs in diploid male progenitors, the earlier idea that plantain improvement could only be achieved through diploid breeding was disproved by this (Rowe 1984; Simmonds 1986).

With just 80 known triploid clones, bananas cultivated in the East African highlands (AAA) are highly specialized and have very little genetic variation; these bananas are popular in the kitchen and in beer making. Half of the landraces exhibited female reproductive characteristics compared to the wild Calcutta 4, which might be improved by conventional breeding techniques. Breeding could increase genetic diversity by increasing the population size and promoting new forms of allelic recombination (Pillay et al. 2004).

Breeding at CARBAP in Cameroon resulted in the CRPB-39 tetraploid plantain hybrid. It was developed by a hybridization process involving the native plantain variety 'French Clair' and the synthetic diploid M53. Cohan et al. (2003) found that the hybrid outperforms "French Clair" in agronomic metrics and shows resistance to BSLD. The breeding program at EMBRAPA in Brazil has produced numerous clones that are resistant to Fusarium wilt (Shepherd et al. 1974). In order of designation, 'Prata AnÍ', 'Pacovan', and 'Prata' are PA 03–22 (AAAB), PV 03–44 (AAAB), and PC12 05 (AAAB), respectively. Because they are resistant to race 1, these clones are being studied in several nations.

9.7.3 Banana 4x and 2x Procedure

Creating triploid hybrids by crossing a diploid and a tetraploid parent is known as the 4x x 2x technique. An ancestral diploid or an enhanced diploid treated with colchicine can have its chromosomes doubled to produce a tetraploid parent. Triploid banana cultivars that occur naturally may have descended from diploid parents through hybridization, which resulted in the unintended creation of gametes from one diploid parent, according to Simmonds (1962). Here, the idea of evolution serves as the foundation. Research by Bakry et al. (1997) has shown that chromosome duplication from a single progenitor can fix a meiotic mistake that causes an unreduced gamete. This strategy resulted in sterile triploid hybrids, which made using traditional methods more difficult (Tomekpe et al. 2004).

The goal of the 4x x 2x method is to develop novel types by addressing the drawbacks of triploid x diploid crossings. These crosses frequently fail to provide the desired results because triploid progenitors have poor gametic fertility. Finding effective specialized combining abilities between diploids and doubled diploids as sources of diplogametes is the goal of this technique, which aims to maximize heterosis in triploid progenies. Some agronomic considerations to keep in mind while deciding between natural and modified diploid forms of bananas include tolerance to diseases and insects, fertility in males and females at diploid and tetraploid levels, and the sort of banana you prefer—dessert or cooking. Jenny et al. (2013) examined extensive data on genetic resources and the linkages between cultivated and ancestral varieties to help with the selection of parental lines according to specific demands.

9.7.4 Triploid Hybrids Banana

Using this method, CIRAD scientists were able to create AAA dessert bananas and send them to banana farmers in the Caribbean and the French West Indies to be tested extensively (Horry 2011). Jenny et al. (2013) found that the success of this new triploid approach was proven by the subsequent creation of AAB x ABB banana hybrids. Key findings of the investigation are summarized below: i) Diploids'gametic fecundity is different from the original organisms. ii) Some AAA hybrids were viable at the diploid and tetraploid stages, while others were sterile at the tetraploid level; still, some AA diploids were utilized to produce AAA. iii) All interspecific AB clones had male and female reproductive traits at the tetraploid level, even though they were infertile at the diploid level. The dramatic rise in population was thought to be caused by AB's birth rate, which was doubled. Despite the hybrids' wild origins, the study also revealed the positive heterosis effect and the production of parthenocarpy (edible) fruits in the examined progenies. Dolezel and Bartos (2005) used flow cytometry to confirm that hybrids resulting from crossings involving Kunnan 4x were triploid. One intriguing hybrid variety, AAB '2006–22/III9'

(*M. acuminata* ssp. malaccensis, × Kunnan 4x), resembles the Mysore subgroup in terms of appearance, major bunch weight, and flavor. The '2005/25-L9' sweet-acid AAB banana hybrid, which is a cross between 'IDN110/AAcvRose' and 'Kunnan 4X,' can produce abundant harvests. In ABB, you can find a hybrid variety that looks like Pisang Awak ('2008/12-I6'), which is a cross between Kunnan 4x and balbisiana 'CMR', and another hybrid variety that is a cross between Kunnan 4X and 'Pisang Klutuk'. The resistance to Sigatoka leaf spot appears to be mostly passed down from the diploid progenitor, based on the hybrids' capacity to resist the disease in real-world conditions. In conventional agricultural methods, where Fusarium wilt is an issue, the researchers found that ABB hybrids might be utilized in place of Pisang Awak clones or other naturally occurring ABB clones (Jenny et al. 2013).

9.7.5 Utilizing Molecular Markers for Banana Breeding

Researchers have used a wide range of molecular marker techniques to study Musa species. Horry and Jay (1988) and Jarret and Litz (1986a, b) previously examined the genetic diversity of Musa germplasm using isozyme and anthocyanin analysis. According to Heslop-Harrison and Schwarzacher (2007), non-DNA biochemical markers yield more precise results with less sophisticated laboratory equipment and in less time. Musa germplasm in both wild and cultivated forms has been characterized using various DNA markers such as RFLP, AFLP, PCR-RFLP, microsatellites, and RAPD (Tenkouano et al. 1999; Ude et al. 2002; Teo et al. 2005; Ravishankar et al. 2013; Saraswathi et al. 2011; Jones et al. 1997). Finding polymorphisms and investigating interactions among Musa accessions are both facilitated by PCR-based approaches. Heslop-Harrison and Schwarzacher (2007), Buhariwalla et al. (2005), and Ravishankar et al. (2012) are only a few of the many studies that have looked at simple sequence repeats (SSRs) as markers for codominant chromosomes and to evaluate genetic links among accessions. Markers such as Inter-retroelement amplified polymorphisms (IRAP) and Diversity Arrays Technology (DArT) are frequently employed.

In plant breeding, the use of molecular markers linked to critical features is paramount. Finding these markers can help with gene introgression and marker-assisted selection. Direct screening in controlled conditions is inefficient compared to molecular markers, which enable improved cultivars to be selected indirectly (Ortiz 1995). Our understanding of disease-related indicators is poor. *Mycosphaerella fijiensis* toxin (black sigatoka) resistance has been linked to methylation-sensitive amplification polymorphism (MSAP) markers (Giménezet al. 2006). An extremely sensitive molecular technique for recognizing Cubense (Foc) race *Fusarium oxysporum* f. sp. isolates in Taiwan was developed by Lin et al. (2010).

The National Research Centre for Bananas (NRCB) in India has found what may be a resistance indicator for Sigatoka in the RAPD gene. We are currently validating this marker by studying the progeny of Sigatoka (*M. musicola*) fish lines with varying degrees of disease resistance. It is now a sequence-characterized amplified region (SCAR) marker after being cloned and sequenced. In a related study, a RAPD marker for nematode resistance was also found. An AFLP marker that distinguishes between fruit parthenocarpy (low or high pulp content) and seedlessness was found by Tenkouano and Swennen in 2004. The "Dwarf Cavendish" cv. clones that were chosen through genetic engineering had an RAPD marker that indicates salt tolerance (Miri et al. 2009).

9.7.6 Evaluation of Improved Banana Hybrids

After rigorous selection for improved agronomic and fruit processing traits, IITA sent seven secondary triploid hybrids to farmers in Nigeria (PITA PITA 21–26 and BITA 7). Farmers in other West African countries have also used a wide variety of hybrids in their crops. Tekonouano and Swennen (2004) found that when it comes to Sigatoka, the enhanced hybrids perform two to five times better than the native plantains that do not receive any pest management. In 1989, Bioversity International (then known as INIBAP) began a worldwide program called the International Musa Testing Programme (IMTP) to assess potential germplasm accessions and superior Musa hybrids. A total of 25 hybrids, including reference cultivars and local controls, were evaluated throughout three stages in different parts of the world (Molina et al. 2007).

The INIBAP Transit Centre (ITC) in Leuven, Belgium, houses the world-renowned Musa collection, which India has had access to since 1994. There are wild species, natural cultivars, and improved hybrids in this collection. In vitro gene banks at NBPGR (New Delhi) and other sites of the All India Coordinated Research Project on Tropical Fruits preserve the accessions, which are used for need-based evaluations at NRCB and elsewhere (Agrawal et al. 2007). Now these unusual and rare accessions are protected in field gene banks across India. According to Menon (2000) and Menon et al. (2005), field research in Kerala has led to the selection of numerous natural cultivars for implementation.

At AICRP facilities, the National Research Center for Banana coordinated the International Musa Testing Programme's stages II and III, which evaluated the performance of exotic hybrids and reference clones. Menon et al.(2002) found that out of four hybrids tested in phase II, SH-3436-9(AAAA) and FHIA-23(AAAA) had the best bunch yield and demonstrated tolerance to Sigatoka leaf spot. The final stage involved testing nine different hybrids. With bunch weights ranging from 16 kg for FHIA-18 to 36.5 kg for FHIA-25, the hybrids outperformed the standard cultivars, and their heights ranged from medium to tall. A range of resistance levels to Sigatoka leaf spot was observed in the hybrids. These compounds were subjected to additional testing: SH-3640, TMBX 5295-1, FHAI-03, FHIA-17, FHIA-25, FHIA-21, and CRPB-39. At BRS, Kannara, three of these hybrids were found. Several agricultural plots have advanced in the on-farm assessments of SH-3640, TMBX 5295-1, FHIA-03, and FHIA-21 (Menon et al. 2005). Based on research done at the NRCB, three hybrids (FHIA-01, FHIA-03, and FHIA-23), sourced from Trichy, have been sent to farmers for evaluation. Processors decided that FHIA-01 was the better choice because of its high sugar-acid ratio and minimal polyphenol oxidation, the latter of which causes pulp browning. Uma et al. 2005b found that FHIA-03 may be used in banana recipes with no problems.

9.7.7 Traditional Breeding Limitations

The shortcomings of conventional Musa breeding have been thoroughly discussed by Pillay and Tripathi (2007) and Pillay et al. (2000). Essentially, the following are the primary limitations: 1) A large majority of banana varieties used for human consumption are triploid ($2n = 3x = 33$), meaning they either aren't fertile or have low fertility rates in both sexes. 2) Pest and disease-resistance genes are absent in most edible cultivars. Transferring desirable features, such as pest and disease resistance, from wild and cultured diploid species ($2n = 2x = 22$) is essential. 4) To choose offspring that closely resemble the cultivated variety, lengthy and laborious backcrossing techniques are necessary due to the unfavorable features of wild species. Many banana clones are inherently sterile, which restricts breeding possibilities. 6) Germination rates of these seeds in the soil can decrease below 1% in cultivars with female fertility, leading to very poor seed production per bunch (Ortiz and Vuylsteke 1995b). When compared to annual crops, bananas' long generation time, up to 18 months at times, extends breeding efforts. Aneuploids, undesirable hyperpolyploids (>5x), 2x, 3x, and 4x euploids, and multiploid and autopolyploid chromosomes' inherent characteristics cause their random occurrence (Simmonds 1966). Molecular breeding appears to be an appropriate strategy for Musa based on the examination.

9.7.8 Role of Biotechnology

The sustained profitability of Musa production could be greatly enhanced by plant biotechnology, according to Hansen and Wright (1999). There aren't any GM bananas or plantains on the shelves right now. The genetic engineering of Musa species to increase their resistance to pests and diseases might be greatly improved with the use of existing transformation techniques, according to Tripathi (2007). It is possible to drastically shorten the time it takes to breed plants that are resistant to nematodes, fungi, bacteria, and viruses by employing the right approaches. When it's possible to introduce numerous features at once, this becomes much more helpful. An important quality to look for might be drought resilience; this would open up new areas for growing bananas and plantains, which would have a significant influence on reducing poverty and ensuring safe food supplies.

9.8 Genetic Engineering

The majority of banana and plantain cultivars utilized in conventional breeding for genetic improvement are sterile triploids, which means they yield fruits without seeds. Mutation breeding is a novel method for improving vegetatively propagated crops by altering desired cultivar features while preserving their genetic composition. Various approaches to mutant breeding were investigated by Krikorian and Cronauer (1984). According to Suprasanna and Rao (1997), in vitro selection and mutagenesis work hand in hand to enhance mutant production, selection, and recovery. Using tissue culture differences as selection criteria, mutant breeding is a common method for improving bananas in Taiwan. According to Tang and Hwang (1994), this method has been effective in producing plants with features like resistance to Fusarium wilt, shorter plant height, faster flowering, and larger bunch weight. Because of the recessive nature of induced mutations and the smaller genome size of diploid plants, they are preferred over triploids in mutation breeding. Nevertheless, due to the decreased likelihood of acquiring homozygous recessive traits, features may manifest less frequently in heterozygous triploids (Roux et al. 1994). Using two commercially available lines, Roux et al. (2004) demonstrated the mechanism of gamma-ray-induced mutations. Novaria, a subgroup of the Grand Naine of Cavendish, was formally introduced in 1995. Compared to the original clone, its flowering duration was 10 weeks shorter. The huge, cylindrical bunches of larger fruits are a hallmark of the Thai cultivar Klue Hom Thong. A successful in vitro mutagenesis program requires the development of an effective in vitro regeneration technique, the improvement of mutagenic treatments of cultures, and the evaluation of populations for specific modifications. Six elite banana cultivars were subjected to in vitro mutation induction with gamma irradiation to isolate mutants; subsequent field studies confirmed the results (Bapat et al. 2007). Since mutations impact just one cell, identifying induced mutations can be difficult since chimeras can form after mutagenic treatment of multicellular organisms. We provide approaches for multicellular structures that generate large mutant sectors, as well as therapies for plant structures derived from single cells, to tackle this problem.

9.9 Molecular Breeding

9.9.1 Molecular Markers-assisted Breeding

In order to speed up the selection process in crop breeding, molecular markers do away with the necessity for labor-intensive direct screening approaches in controlled environments like greenhouses and fields (Alqahtani et al. 2024; El Rabey et al. 2013; Elabd and Salem 2009; El-Absawy et al. 2012, 2015; El-Zanaty et al. 2011, 2013; Esmail et al. 2009; Ibrahim et al. 2024; 2023a, Ismail et al. 2013;

Mattar et al. 2009, 2016; Salem 1997, 2009, 2015; Salem and El-Zanaty 2015; Salem and Ibrahim 2024; Salem and Mattar 2014 a, b; Salem and Sallam 2016; Salem et al. 2007a,b; 2008a,b; 2010; 2015a,b; 2021c, f; 2022; 2024; Varshney et al. 2008). Finding characteristics regulated by a combination of genes is where molecular markers shine. With the right molecular markers found and the relevant QTLs linked to an agronomic trait mapped, it is feasible to produce an improved cultivar by transferring multiple QTLs at once or by combining QTLs for different agronomic traits (Abdel-Hamid and Salem 2021; Abu-Ellail et al. 2021; Aldahak et al. 2021; Dita et al. 2006; Hassan et al. 2021; Salem 2004; Eltaher et al. 2017, 2018, 2021, 2022; Ghazy et al. 2020; Mourad et al. 2019; Salem and Börner 2003; Salem et al. 2003, 2004b, 2007b, 2023). Although there are potential benefits to using marker-assisted selection to improve Musa, the lack of understanding of several agronomic criteria and insufficient financing are preventing its widespread deployment.

9.9.2 Genomics

The development of long-term pest and disease resistance in host plants is a major obstacle for plant breeders. Combining different resistance genes into one genotype is one method. Several studies, including those by Richardson et al. (2006), offer numerous instances of advantageous allele and QTL pyramiding and introgression in species used for agriculture. Nevertheless, Musa has not yet provided proof of gene pyramiding. Cultivars that are resistant to different stresses are created using biotechnological approaches like as marker-assisted selection, mutagenesis, and transgenesis. A single cultivar with numerous resistance genes can be produced by combining these procedures.

Crop development stands to benefit greatly from innovative technology derived from molecular biology developments. These breakthroughs are demonstrated by molecular markers that can be used to identify target genes. According to Lefebvre et al. (1995), genetic markers help in genome analysis by serving as chromosomal landmarks. Markers are "genetic tags" that are passed down through generations in plants according to inheritance principles; they designate specific loci within the DNA sequence (Drew 1997). For heritable features like pest and disease resistance in host plants, researchers have utilized DNA-based markers to identify specific genes or chromosomal regions (Drew 1997). Once discovered, genetic markers can be used in breeding programs to predict features simply by looking for them. Another approach is to clone genes linked to a certain trait after locating them using genetic marker techniques. The analysis of complicated multigenic characteristics is made easier by these methods. Genomic activity is the only criterion for sorting genes using mapping approaches. When genes are located, sequenced, and cloned, they can be used to transfer them to other species (Drew 1997).

Molecular markers are based on natural DNA polymorphisms, which include base pair deletions, substitutions, and additions. There are a variety of markers available, some based on DNA and others on morphology or proteins. While DNA markers show changes in DNA, biochemical markers show variations in proteins. Proteins that can be isolated by electrophoresis for allele identification are biochemical indicators that are created by gene expression. Protein markers often include isozymes, which are distinct enzyme variations. Variegated genomic sequences are observed in codominant protein markers. Their usage is limited, nonetheless, due to their vulnerability to post-translational changes and the fact that they are few across all crop species (Staub et al. 1996). There are two main categories of DNA polymorphisms, identified by different methods of detection: hybridization and polymerase chain reaction (PCR). Polymorphisms that are dependent on hybridization can be found at loci like restriction fragment length polymorphisms (RFLP) and variable number tandem repeats (VNTR) (Rogstad 2001). Hybridization of random genomic clones, microsatellite sequences (SSR), and mini-satellite sequences with restriction enzyme-digested DNA at these loci is done through the use of screens. Point mutations, inversions, deletions, and translocations are the causes of RFLPs, whereas variations at VNTR loci arise from changes in the total number of repetitions.

9.9.3 Molecular Markers

Molecular markers have greatly enhanced breeding procedures. They are not impacted by environmental influences and immediately indicate genetic variety through DNA analysis (Staub et al. 1996). According to Patterson et al. (1991), marker-assisted selection (MAS) was made possible by the advent of numerous molecular markers for plant genomic research. This allowed for genotype selection instead of phenotypic selection. Barone (2004) lists RFLP, AFLP, RAPD, and SSR as common markers for molecular analysis. There may be some variation among the best MAS markers concerning dominance/codominance, cost, identification effort, and locus representation. Musa has made extensive use of molecular markers for a variety of applications, including but not limited to analyzing irradiated banana genotypes, detecting genetic variation and relationships in banana germplasm, identifying trait markers for breeding programs, and identifying duplicate accessions in field and tissue culture germplasm banks. Tissue culture specimens' genetic stability has been evaluated with their assistance as well (Nawara et al. 2016, 2017). Notable Musa markers and new variants with potential future usefulness are discussed in this chapter.

9.9.3.1 Hybridization-Based Markers

Restriction Fragment Length Polymorphisms (RFLPs)

Alterations in the banding patterns of DNA fragment lengths, known as restriction fragment length polymorphisms (RFLPs), can be detected by electrophoresis of DNA samples using restriction digests. A restriction enzyme cleavage site's existence can differ from person to person, leading to frequent changes. The codominant nature of RFLP allows it to differentiate between homozygotes and heterozygotes. Because RFLP does not require any prior knowledge of an organism's sequence, it is reliable, portable, and easy to transfer across labs. This makes it a priceless resource for building Musa-specific genomic linkage maps.

The fact that synteny conservation is locus-specific makes it possible to obtain estimates. The use of single locus probes allows for discrimination at the population level. Some of the disadvantages include high research costs, large amounts of DNA needed, a time-consuming and laborious process, the need for an appropriate probe library, the possibility of radioactive labeling, and a lack of automation when compared to other methods like AFLP and diversity array technology (DArT) or multiplexing of SSRs. The use of RFLPs dates back to the beginning of Musa mapping operations. Hippolyte et al. (2010) report that they have lately been associated with resistance gene analog polymorphisms.

High specificity and reproducibility are characteristics of RFLPs, which are made possible by using a probe with a known sequence. Researchers have used RFLPs from many genotypes and variations in the chloroplast genome to study the taxonomy and phylogeny of Musa species (Gawel and Jarret 1991a, b; Jarret et al. 1992). The sole publication that addresses their use in differentiating between closely related data is Bhat et al. (1995). The high cost and technical complexity of this technology make it unsuitable for use in traditional breeding programs.

Variable Number of Tandem Repeats (VNTR)

Minisatellites (ten to forty-five base pairs) and microsatellites (two to six base pairs) are two terms for the same type of short DNA sequences called tandem repeats or variable number tandem repeats (VNTRs). The development of PCR technology has made the use of radioactive probes to identify VNTRs a more viable option than in the past. Tourmente et al. (1994) found that eukaryotic genomes include an abundance of VNTRs. These sequences can hybridize with various loci and have a high variability, making them valuable for genetic studies. Minisatellites can have repeat units ranging from five to one hundred base pairs, whereas microsatellites have units of five base pairs or less (Buard and Vergnaud 1994).

9.9.3.2 PCR-Based Markers

There is a great deal of variation among Musa breeding populations, and this variation can be identified using PCR-based genetic markers. High-throughput screening of these populations is best accomplished using PCR-based techniques. Consequently, scientists are interested in exploring the Musa genome for possible applications of the polymerase chain reaction. Several molecular markers have been used to assess genetic diversity and phylogenetic relationships among farmed and wild banana accessions. According to several studies (Grapin et al. 1998; Loh et al. 2000; Wong et al. 2001; Carreel et al. 2002; Ude et al. 2002, 2003; Creste et al. 2003a,b; Nwakanma et al. 2003), these markers encompass AFLP, RAPD, microsatellites, and PCR-based markers of varying types. A great deal of PCR amplification technology has been created since its discovery. Weising et al. (1995) and Gupta et al. (1994) both cite the SPAR assay as an example of such tests. A strategy for researching minisatellites is the SPAR method, which is sometimes called direct amplification of minisatellite-region DNA (DAMD). This method targets regions with high minisatellite repeats by using the core sequence of minisatellites as a single primer (Heath et al. 1993; Zhou et al. 1997). A quick way to find small changes in DNA sequence that are made through polymerase chain reaction is single-strand conformation polymorphism (SSCP) analysis (Orita et al. 1989). By integrating microsatellite hybridization with randomly or semi-specifically primed polymerase chain reaction (PCR), RAMPO can supposedly produce many distinct and polymorphic genetic fingerprints on a single electrophoretic gel (Richardson et al. 1995). The genetic diversity and connections among forty Musa accessions were studied by Muhammad et al. (2010) using the sequence-related amplified polymorphism (SRAP) approach. In order to improve open reading frames (ORFs), this method is being used.

Random Amplified Polymorphic DNA (RAPD)

Although RAPD's simplicity and low cost make it a popular choice, the method's repeatability and reliability are highly dependent on the skill of the operator. Results from RAPD analysis have varied between laboratories, even when using the same PCR settings and circumstances. No prior knowledge of an organism's genome is necessary for the RAPD technique because it relies on short, random primers. Among the many benefits of the RAPD method are its speed, ease of use, and low DNA template required. Finding variations in germplasm is facilitated by employing this method. RAPDs have been extensively utilized to investigate genetic relationships in banana germplasm (Uma et al. 2006; Jain et al. 2007; Nsabimana and van Staden 2007) for the identification of duplications among accessions in field and tissue culture germplasm banks, as well as for monitoring the genetic stability of tissue culture material (i.e., somaclonal variation) (Bairu et al. 2006, 2008; Ray et al. 2006; Lakshmanan et al. 2007), analyzing and distinguishing irradiated banana genotypes (Toruan-Mathius and Haris 1999; Finalet et al. 2000; Imelda et al. 2001;

Hautea et al. 2004), and identifying trait markers for application in cross- and mutation-breeding programs. Genomic analysis would be a huge help in managing plant patents and securing legal safeguards for newly developed cultivars of asexual crops like Musa (Kaemmer et al. 1997). Primers with specific uses can be created from very informative RAPD markers using a technique called sequence-characterized amplified regions (SCARs). For backcross breeding and marker-assisted introgression, RAPD tests are strong and effective. Researchers have found and used specific RAPD markers for the Musa A and B genomes in studies by Pillay et al. (2000) and Oselebe et al. (2006).

Two major drawbacks of RAPD analysis that might prevent its use in marker-assisted selection are the marker system's predominant features and difficulties with reproducibility. Several studies have used the RAPD methodology to successfully differentiate different Musa germplasm. These include Well et al. (1994), Bhat and Jarret (1995), Pillay et al. (2001), Uma et al. (2006), Jain et al. (2007), and Nsabimana and van Staden (2007). When building a molecular linkage map, RAPD is one of the marker approaches used (Faure et al. 1993). Musa genomes have been classified into particular groups and closely related genotypes have been identified using RAPD analysis (Bhat and Jarret 1995). The ability to micropropagate bananas from true-to-type clones is essential for the successful propagation of banana species. One major issue with in vitro culture is the somaclonal variability that can occur among subclones that originate from the same parental line. In order to assess clonal fitness in banana plantlets, Bennici et al. (2004), Carvalho et al. (2004), and Ray et al. (2006) used RAPD and inter-simple sequence repeat (ISSR) methods. Somaclones evaluated by PCR-based testing showed stability and seldom reverted to normal phenotypes, as demonstrated by Ramage et al. (2004). High annealing temperature (HAT)-RAPD was a rigorous RAPD method that Sitthipron et al. (2005) used to assess several varieties of Thai bananas. Despite concerns about the technology, RAPD is still being used in banana genome research, according to recent publications (Miri et al. 2009; Khatri et al. 2009).

Simple Sequence Repeats (SSRs)

Creste et al. (2003a, b), Jarret et al. (1995), and Grapin et al. (1998) all utilized microsatellite primers for Musa since RAPD markers had limitations. Microsatellites, also known as SSR loci, are marked by several markers that are distributed and distinct to each locus. The DNA motifs that make them up are repetitive and can be anywhere from four to six base pairs long, with a total length of several tens of base pairs. According to Crouch et al. (2000), a large number of plant species have SSRs, or microsatellites, scattered throughout their genomes. It is thought that DNA replication slippage is the mechanism by which mutations in the motif occur. The consistency, availability, and polymorphism of SSRs make them dependable markers across different systems, according to a prior study by Campbell et al. (2003).

Microsatellites are a typical tool for assessing plant genetic diversity. In plant genetics, mapping, and fingerprinting, these markers have seen extensive use.

According to research (Crouch et al. 2000), SSRLP analysis can uncover a great deal of individual variation in breeding populations of Musa. Nevertheless, microsatellite isolation is an expensive and labor-intensive process. However, automated DNA sequencing facilities, better methods for choosing suitable clones, and genomic libraries enhanced for SSR are making SSR isolation more widespread. Consequently, several microsatellites from the Musa B genome were swiftly isolated by researchers not long ago (Buhariwalla et al. 2005). According to many studies (Jarret et al. 1995; Crouch et al. 2000; Creste et al. 2003a, b; Wang et al. 2009), there are now several hundred SSR markers in *M. acuminata*, and the number is projected to increase as genomic technologies progress.

Among the numerous possible uses for these markers, two of the most important are Marker-Assisted Selection (following co-localization with particular traits) and their incorporation into genetic maps. Cheung and Town (2007) discovered 352 SSR loci in M. 4 BAC end-sequences from *Bacillus acuminata* Calcutta. There aren't as many distinct SSR loci in Musa as there are in other agricultural species, which limits genetic research. This is because different cultivars of Musa may have identical or absent alleles. Five complete M genomes have yielded more than 40 microsatellite sites. Miller et al. (2010) used four BAC consensus datasets (AC186748, AC186749, AC186954, AC186747, and AC186750) from the plant acuminata Calcutta to validate polymorphisms on genotypes demonstrating variability in host plant tolerance to Sigatoka infections.

Irradiation-Reverse-Amplified Polymorphism (IRAP)

As a supplementary dominant marker system, the IRAP method can detect polymorphisms in retrotransposon insertions. Retrotransposons are ubiquitous and common in plant genomes; in fact, they make up over 50% of nuclear DNA in many plant species (Kumar 1999). Their mutagenic properties, according to Heslop-Harrison (2000), make them possible biodiversity sources. Retroviruses, retroelements that are similar to copia or gypsy, retroviruses, long interspersed elements (LINES), and short interspersed elements (SINES) are the main categories. In order to classify the Musa genome constitution and to identify and characterize banana cultivars, Teo et al. (2005) employed IRAP markers. (Pearce et al. 2000; Yu and Wise 2000) Several plant species have made substantial use of Ty1-copia-like retrotransposons as molecular markers. Musa acuminata diverged before Ty3-gypsy-like retrotransposons were introduced to the genus Musa, according to Balint-Kurti et al. (2000). To differentiate between the A and B genomes, Salvia and *M. vellutina* made use of retrotransposons similar to Ty3. Diploid M. pisang manis and M. 4 of the acuminata Calcutta cultivars were found to contain sequences that resembled the "monkey" sequence, as pointed out by Aert et al. (2004). This highlights the interconnection of their A genomes. Using IRAP markers, Nair et al. (2005) found out which genes are in Indian bananas. The B genome was also identified using a PCR-based molecular marker. The B-specific band was not seen in any AA or AAA cultivars, except one AAB cultivar. On the other hand, this band was present in all AB,

AAB, and ABB cultivars. While four cultivars having ABB were classified as AAB, one AAA cultivar showed a pattern similar to ABB-genome cultivars. The writers seemed to be implying that the plants were misidentified.

Affinity-Frequency-Length Polymorphisms (AFLP)

DNA restriction fragments are subjected to severe conditions for selective PCR amplification in order to obtain DNA fingerprinting using AFLP (Vos et al. 1995). It applies to DNA of any complexity and origin, and it is trustworthy and consistent (Vos et al. 1995). The strength of PCR and the dependability of RFLP are brought together in AFLP. The selective amplification of restriction fragments from digested entire genomic DNA is an essential part of AFLP. Numerous studies have shown that AFLP markers have become more popular in different plant species due to their high reproducibility, comprehensive genome coverage, and capacity to detect multiple polymorphisms in a single assay (Cervera et al. 1998; Shim and Jørgensen 2000; Vos et al. 1995; Vuylsteke et al. 1999).

Research has used AFLP to evaluate the genetic diversity of Musa, both in domesticated and wild types (Wong et al. 2001; Ude et al. 2002, 2003; Opara et al. 2010). While these banding patterns were originally thought to be dominant identifiers, their information richness makes them more likely to be codominant markers in populations that are segregating (Masiga and Turner 2004). A person's band intensity can be used by software to identify if they are homozygous or heterozygous.

Research using AFLP is technically difficult and expensive because of the complicated DNA alterations and necessary visualization methods. A large quantity of high-quality DNA is required since using low-quality DNA can cause erroneous polymorphisms and insufficient digestion. Banana streak badnavirus infection was determined to be present or absent in Musa hybrids that were co-segregated with ten AFLP markers by Lheureux et al. (2003). Now, AFLPs and SSRs are being used to identify banana and plantain indicators of fruit parthenocarpy, dwarfism, and apical dominance. The best methods for marker-assisted breeding in Musa, according to Crouch et al. (2000) and Hautea et al. (2004), are microsatellite markers and AFLP analysis.

Methods based on AFLP have been developed for the analysis of cytosine methylation at CCGG loci by Reyna-Lopez et al. (1997) and Fraga and Esteller (2002). In addition to its importance in gene transcription regulation and gene silencing, DNA methylation is involved in a number of other biological processes that are essential for controlling plant development (Finnegan et al. 1996). Molecular markers based on methylation are distinct from other markers concerning their genetic features, transmission, and evolutionary/mutational processes, according to research by Xiong et al. (1999) and Cervera et al. (2002). The phenotypic variety of plantain landraces can be better understood with the help of these markers, which can zero in on certain genomic regions and shed light on epigenetic mechanisms.

The lack of a definitive correlation between the methylation diversity in plantains and notable phenotypic and genotypic qualities has left the predictability of traits in plants lacking a clear answer (Crouch et al. 2000). Unlike random mutation, the mechanisms controlling the evolution, transmission, and regulation of methylation are still a mystery, even though methylation is associated with phenotype (Cervera et al. 2002). Xu et al. (2001) and Bradeen and Simon (1998) found that AFLP bands may be sequenced and used as SCAR markers in PCR analysis by using agarose gel electrophoresis. It is also possible to screen pooled BAC libraries using these markers. One useful way to generate Musa linkage maps is the AFLP technique, which was used to construct the M53 (selfed) and AFCAM (an F2 population from a "Calcutta 4" x "Madang" cross) mapping populations.

Technology for Diversity Arrays (DArT)

DArT is a DNA hybridization-based genotyping method that has recently made possible the cost-effective whole-genome profiling of crops in the absence of previous sequence data. To simplify things, DArT finds constant fragments, those that are in every genomic "representation" of a specific cultivar or species and variable (polymorphic) fragments, also called molecular markers, that are in some but not all "representations" (www.diversityarray.com). A DArT marker can be either dominant (present) or codominant (two dosages instead of one) depending on the situation. With this technique, we may quickly phenotype-separate populations utilizing a battery of polymorphism markers for quantitative trait locus (QTL) research.

Creating accurate physical genome maps is an essential part of any sequencing effort. One efficient and economical way to accomplish this is to link sequenced DArT markers to BAC contigs. Under the auspices of the Generation Challenge Program, CIRAD is building a Musa framework map with 1500 DArT markers culled from a diversified "metagenome" (Kilian 2007) of Musa accessions. Additionally, the BORLI map that was recently developed at CIRAD utilized 380 of these markers (Hippolyte et al. 2008, 2010).

According to Wenzl et al. (2007), DArT has been shown to be useful for bulked segregant analysis in a Hordeum array. The array, a microarray printer/scanner, and computers for data processing/storage/management are all necessary but not sufficient infrastructure components for this technology, which is a drawback. However, the markers are sequence-ready, so they can be used in PCR investigations using regular electrophoresis equipment.

Single-Nucleotide Polymorphisms (SNPs)

More than 1% of the time, an organism's DNA contains single-nucleotide polymorphisms (SNPs). The bulk of single-nucleotide polymorphisms (SNPs) are found outside key gene domains, while approximately 3–5% of a cell's DNA encodes proteins (Sherry et al. 2001). Because of their clear correlation with desired features, researchers are drawn to single-nucleotide polymorphisms (SNPs) found in a gene of interest. Thanks to technological developments, SNPs are being employed more and more for genetic trait analysis and selection. Because of their potential influence on protein function and, by extension, the organism's phenotype, single-nucleotide polymorphisms (SNPs) have become an integral part of useful markers, particularly when found in coding areas, namely expressed SNPs (eSNPs or cSNPs in cDNA) (Brookes 1999). So far, the banana genome has not been shown to contain any single-nucleotide polymorphisms (SNPs).

EcoTILLING

When searching the Musa genome for polymorphisms, Till et al. (2010) employed EcoTilling. Ecotilling is a high-throughput method that they describe in order to aid in the detection and characterization of SNPs and indels (Till et al. 2010). Developed for the Targeting Induced Local Lesions in Genome (TILLING) reverse-genetic technique, it is a variant of the enzymatic mismatch cleavage and fluorescence detection methods originally described by Colbert et al. (2001) and Comai et al. (2004). The approach was originally designed for use with Arabidopsis ecotypes, which is why it was named ecotilling. Many creatures have benefited from this precise, low-cost, high-throughput method of identifying and assessing nucleotide diversity. In order to amplify gene target areas between 700 and 1600 bp, enzyme mismatch cleavage is employed to fluorescently identify gene-specific primers for ecotilling. Heteroduplexed molecules are formed when polymorphic amplicons undergo hybridization; this process begins with denatured and annealed samples following polymorphic amplicon PCR. Afterward, mismatched areas in ordinarily double-stranded duplexes are cleaved using a crude celery juice extract that contains the single strand specific nuclease CEL I. According to Till et al. (2006), the cleavage products are resolved by denaturing polyacrylamide gel electrophoresis (PAGE) and subsequently seen using fluorescence detection.

Accessions can be categorized based on shared banding patterns that demonstrate haplotype clustering using denaturing polyacrylamide gel electrophoresis (PAGE) with base pair resolution (Comai et al. 2004). Sequence validation on a single or a small number of samples can produce group-wide base polymorphism data with less strain on informatics and more efficiency than sequencing methods. Banding patterns can also be used independently to measure genetic diversity and similarity among genes. Independently screened samples can be added to a database of heterozygous nucleotide variations using ecotilling. In order to detect homozygous polymorphisms, reference DNA might be introduced to each sample prior to screening. Additionally, because of the assay's great sensitivity, many samples can be combined to specifically identify rare SNPs.

9.10 Conclusion and Prospects

Persistent droughts, caused by the fruit's lengthy growth cycle, threaten the world's banana harvest. Research into the physiological changes that bananas experience during drought stress has, so far, predominated over efforts to develop drought-resistant banana cultivars. Few studies have looked at this problem from a biochemical and molecular perspective. We have stopped screening cultivars for drought tolerance and sensitivity, therefore, we can't evaluate or document them in controlled or field environments. Consequently, it is crucial to develop cultivars that are resistant to drought. This can be achieved by conventional breeding methods or by identifying potential genes and incorporating them into vulnerable kinds.

While innovative DNA technologies are mainly used in human genetic studies, they are finding some uses in botanical sciences as well. Because of their abundance, polymorphism, ease of measurement, and potential to uncover hidden variants that other methods may overlook, high-throughput procedures using SNPs or small-scale indels are more efficient than classical markers (RFLP, RAPD, or AFLP). Alleles and haplotypes can be identified with great precision using single-nucleotide polymorphisms (SNPs).

A dosage-sensitive marker system that can differentiate between heterozygous genotypes with several haplotypes is essential for polyploid crop efficiency. Reduced analysis time and no need for post-PCR processing or separation are only two of the many benefits of high-resolution DNA melting (HRM) analysis over other genotyping methods. We take a look at the three approaches to HRM genotyping and variant screening. HRM has been used on various crops and has great promise for cultivar identification, especially in polyploids, mapping, polymorphism detection, gene candidate identification, and quantitative trait locus (QTL) or association study location of genomic regions linked to important traits. Researchers are eagerly awaiting the implementation of these technologies on Musa. In addition, phenotypic tools and procedures should be improved for use in controlled and field environments.

To hasten the development of bananas that can withstand drought, recent technical advances should be put to use, such as marker-assisted selection, high-throughput precision phenotyping, and genetic engineering. These technologies allow for the accurate and comprehensive selection of candidates that are resistant to drought. It is also possible to integrate drought tolerance genes into banana genomes and silence or remove genes associated with drought susceptibility using genome editing with CRISPR/Cas9 technology, a type of genetic engineering.

References

Abdel-Hamid AME, Salem KFM (2021) Breeding strategies of garden pea (Pisum sativum L.). In: Al-Khayri JM, Jain SM, Johnson DV (eds) Advances in plant breeding strategies: vegetable crops. Springer, Cham. https://doi.org/10.1007/978-3-030-66969-0_9

Abdullaev FK, Rustamov NSU, Yakubov MD (2025) Uzbekistan genebank. In: Al-Khayri JM, Salem KFM, Jain SM (eds) Plant gene banks. Springer, Singapore. https://doi.org/10.1007/978-981-99-4236-7_5

Aboshama HM, Salem KFM, Hassan HMF, Tahoun MH (2015) Estimating heterosis and combining ability for in vivo and in vitro traits using diallel cross on tomato (Solanum lycopersicum L.). Minufiya. J Agric Res 40(3):757–770

Abu-Ellail FFB, Salem KFM, Saleh MM et al (2021) Molecular breeding strategies of beetroot (Beta vulgaris ssp. vulgaris var. conditiva Alefeld). In: Al-Khayri JM, Jain SM, Johnson DV (eds) Advances in plant breeding strategies: vegetable crops. Springer, Cham. https://doi.org/10.1007/978-3-030-66965-2_4

Abulaila K, Al Sane KO, Tahabsom Z (2025) Jordan Seed Bank. In: Al-Khayri JM, Salem KFM, Jain SM (eds) Plant Gene Banks. Springer, Singapore. https://doi.org/10.1007/978-981-99-4236-7_45-1

Adeleke MT, Pillay M, Okoli BE (2002) An improved method for examining meiotic chromosomes in Musa L. Hort Sci 37(6):959–961

Aert R, Volckaert G, Sagi L (2004) Gene content and density in banana (*Musa acuminata*) as revealed by genomic sequencing of BAC clones. Theor Appl Genet 109:129–139

Agrawal A, Sharma H, Dhillon BS (2007) In vitro conservation of Musa germplasm in India. In: Singh HP, Uma S (eds) Banana-technological advancements. Association for the Improvement in Production and Utilization of Banana (AIPUB), National Research Centre for Banana, Trichy, pp 87–108

Akparov ZI, Mammadova SA, Kalantarova NS, Mirzaliyeva IA, Mammadov AT (2025) National genebank of Azerbaijan. In: Al-Khayri JM, Salem KFM, Jain SM (eds) Plant gene banks. Springer, Singapore. https://doi.org/10.1007/978-981-99-4236-7_31-1

Al Lawati AH (2025) Plant genetic resource conservation of Oman. In: Al-Khayri JM, Salem KFM, Jain SM (eds) Plant gene banks. Springer, Singapore. https://doi.org/10.1007/978-981-99-4236-7_50-1

Aladele SE et al (2025) Plant genetic resources conservation and Utilization in Nigeria. In: Al-Khayri JM, Salem KFM, Jain SM (eds) Plant gene banks. Springer, Singapore. https://doi.org/10.1007/978-981-99-4236-7_17-1

Aldahak L, Salem KFM, Al-Salim SHF, Al-Khayri JM (2021) Advances in chicory (Cichorium intybus L.) breeding strategies. In: Al-Khayri JM, Jain SM, Johnson DV (eds) Advances in plant breeding strategies: vegetable crops. Springer, Cham. https://doi.org/10.1007/978-3-030-66969-0_1

Al-Khayri JM, Salem KFM, Jain SM (2025) Plant gene banks. Springer, Singapore. https://doi.org/10.1007/978-981-99-4236-7

Almarri NB, Ahmad S, Alsubaie GTG, Elshal MH (2025) Saudi Arabia Gene Bank. In: Al-Khayri JM, Salem KFM, Jain SM (eds) Plant Gene Banks. Springer, Singapore. https://doi.org/10.1007/978-981-99-4236-7_54-1

Alqahtani MM, Saleh MM, Alwutayd KM et al (2024) Performance and genotypic variability in diverse date palm (Phoenix dactylifera L.) cultivars for fruit characteristics. Genet Resour Crop Evol 71:1759–1772. https://doi.org/10.1007/s10722-023-01719-0

Amalraj VA, Velayudhan KC, Agarwal RC, Rana RS (1993) Banana genetic resources. National Bureau of Plant Genetic Resources, New Delhi

Arab S et al (2025) National gene bank of Egypt. In: Al-Khayri JM, Salem KFM, Jain SM (eds) Plant Gene Banks. Springer, Singapore. https://doi.org/10.1007/978-981-99-4236-7_5-1

Arafat KMY, Nayeem J, Quadery AH et al (2018) Handmade paper from waste banana fiber. Banglad J Sci Ind Res 53(2):83–88

Bairu MW, Fennell CW, van Staden J (2006) The effect of plant growth regulators on somaclonal variation in Cavendish banana (Musa AAA cv. 'Zelig'). Sci Hort 108:347–351

Bairu MW, Stirk WA, Dolezal K, Staden J (2008) The role of topolins in micropropagation and somaclonal variation of banana cultivars 'Williams' and Grand Naine (*Musa* spp. AAA). Plant Cell Tissue Organ Cult 95:373–379

Bakry F, Carreel F, Caruana ML et al (1997) Les bananiers. In: Jacquot M, Hamon S, Nicolas D (eds) Charrier A. L'amelioration des plantes tropicales. CIRAD and ORSTOM, pp 109–140

Balint-Kurti PJ, Clendennen SK, Dolezelova M et al (2000) Identification and chromosomal localization of the monkey retrotransposon in Musa sp. Mol Gen Genet 263:908–915

Bapat VA, Ganapathi TR, Kulkarni VM et al (2007) Cellular and molecular approaches for improvement of banana. In: Singh HP, Uma S (eds) Banana-technological advancements. Association for the Improvement in Production and Utilization of Banana (AIPUB), National Research Centre for Banana, Trichy, pp 160–173

Barone A (2004) Molecular marker-assisted selection for potato breeding. Amer J Potato Res 81:111–117

Békefi Z, Szilágyi S, Desiderio F, Demku T, Varga J (2025) Prunus Gene Bank in Hungarian University of Agriculture and Life Sciences. In: Al-Khayri JM, Salem KFM, Jain SM (eds) Plant Gene Banks. Springer, Singapore. https://doi.org/10.1007/978-981-99-4236-7_71-1

Bennici A, Anzidei M, Vendramin GG (2004) Genetic stability and uniformity of *Foeniculum vulgare* Mill. Regenerated plants through organogenesis and somatic embryogenesis. Plant Sci 166:221–227

Bhagwat B, Duncan EJ (1998) Mutation breeding of banana cv Highgate (Musa spp., AAA Group) for tolerance to Fusarium oxysporum f. sp. cubense using chemical mutagens. Sci Hort 73:11–22

Bhat KV, Jarret RL (1995) Random amplified polymorphic DNA and genetic diversity in Indian Musa germplasm. Genet Resour Crop Evol 42:107–118

Bhuvneshwar R, Kumar G, Diwan RK, Khandal RK (2011) Study on effect of euphorbia coagulum on physicomechanical and fire-retardant properties of polyester-banana fiber composite. Indian J Sci Tech 4(4):443–446

Bieniek W, Mizianty M, Szklarczyk M (2015) Sequence variation at the three chloroplast loci (matK, rbcL, trnH-psbA) in the Triticeae tribe (Poaceae): comments on the relationships and utility in DNA barcoding of selected species. Plant Syst Evol 301:1275–1286

Bordón P, Paz R, Peñalva C, Vega G, Monzón M, García L (2021) Biodegradable polymer compounds reinforced with banana fiber for the production of protective bags for banana fruits in the context of circular economy. Agronomy 11, no. 2:242. https://doi.org/10.3390/agronomy11020242

Borner A, Landjeva S, Salem KFM, Lohwasser U (2009) Pflanzengenetische Ressourcen als Grundlage für die Züchtung klimatoleranter Sorten. Proceedings of a German Conference 'Weizenforum = Wheat Forum' Norddeutsches Weizenforum Schriftenreihe des Inst, vol 63. Für Pflanzenbau und –züchtung der CAU Kiel, Germany, pp 23–27

Börner A, Landjeva S, Salem KFM, Lohwasser U (2010) Plant genetic resources—a prerequisite for drought tolerance breeding in cereals. Proc Jahrestagung der Vereinigung der Pflanzenzüchter und Saatgutkaufleute Österreichs, 24–26 Nov 2009, Raumberg-Gumpenstein, pp 11–13

Bradeen JM, Simon PW (1998) Conversion of an AFLP fragment linked to the carrot Y2 locus to a simple, codominant, PCR-based marker form. Theor Appl Genet 97:960–967

Breidy J (2025) Lebanon gene bank. In: Al-Khayri JM, Salem KFM, Jain SM (eds) Plant gene banks. Springer, Singapore. https://doi.org/10.1007/978-981-99-4236-7_48-1

Brookes AJ (1999) The essence of SNPs. Gene 234:177–186

Buard J, Vergnaud G (1994) Complex recombination events at the hypermutable minisatellite CEB1 (D2S90). EMBO J 13:3203–3210

Buhariwalla HK, Jarret RL, Jayashree B et al (2005) Isolation and characterization of microsatellite markers from *Musa balbisiana*. Mol Ecol Notes 5:327–330

Campbell D, Duchesne P, Bernatchez L (2003) AFLP utility for population assignment studies: analytical investigation and empirical comparison with microsatellites. Mol Ecol 12:1979–1991

Cañar DY et al (2025) Colombia gene bank. In: Al-Khayri JM, Salem KFM, Jain SM (eds) Plant Gene Banks. Springer, Singapore. https://doi.org/10.1007/978-981-99-4236-7_95-1

Carreel F, Gonzalez de León D et al (2002) Ascertaining maternal and paternal lineage within Musa by chloroplast and mitochondrial DNA RFLP analysis. Genome 45:679–692

Carvalho LC, Goulao L, Oliveira C et al (2004) RAPD assessment for identification of clonal identity and genetic stability of in vitro propagated chestnut hybrids. Plant Cell Tissue Organ Cult 77:23–27

CBOL Plant Working Group, Hollingsworth PM, Forrest LL et al (2009) A DNA barcode for land plants. PNAS 106(31):12794–12797

Cecci RRR, Passos AA, de Aguiar Neto TC et al (2020) Banana pseudostem fibers characterization and comparison with reported data on jute and sisal fibers. SN Appl Sci 2:20. https://doi.org/10.1007/s42452-019-1790-8

Cervera MT, Cabezas JA, Sancha JC et al (1998) Application of AFLPs to the characterization of grapevine Vitis vinifera L. genetic resources. A case study with accessions from Rioja (Spain). Theor Appl Genet 97:51–59

Cervera MT, Ruiz-Garcia L, Martanez-Zapater JM (2002) Analysis of DNA methylation in Arabidopsis thaliana based on methylation-sensitive AFLP markers. Mol Genet Genomics 268:543–552

Channelie're S, Van den Houwe I, Arnaud E et al (2011) Standardized procedure for Musa Germplasm characterization. Acta Hortic 897:113–121

Chen JY et al. (2008) Role of phenylalanine ammonia- lyase in heat pretreatment-induced chilling tolerance in banana fruit. Physiologia Plantarum, Malden, 132(3), p. 318-328

Cheung F, Town CD (2007) A BAC end view of the musa acuminata genome. BMC Plant Biol 7:29

Cohan JP, Abadie C, Tomekpe K et al (2003) Agronomic performance and resistance to black leaf streak of the hybrid 'CRBP-39'. Info 12(1):29–32

Colbert T, Till BJ, Tompa R et al (2001) High-throughput screening for induced point mutations. Plant Physiol 126:480–484

Comai L, Young K, Till BJ et al (2004) Efficient discovery of DNA polymorphisms in natural populations by Ecotilling. Plant J 37:778–786

Cömert N, Deniz T, Göktürk D (2025) Sustainable Fisheries and Non-Target Species Management: A Seasonal and Depth-Based Study in the Deep-Sea Fisheries of Antalya Bay. Sustainability, 17:5040. https://doi.org/10.3390/su17115040

Costion C, Ford A, Cross H, Crayn D, Harrington M, Lowe A (2011) Plant DNA barcodes can accurately estimate species richness in poorly known floras. PLoS One 6(11):e26841

Creste C, Neto TA, Sebastino OS, Figuira A (2003a) Genetic characterization of banana cultivars (Musa spp.) from Brazil using microsatellite markers. Euphytica 132:259–268

Creste S, Neto TA, Silva SO, Figueira A (2003b) Genetic characterization of banana cultivars (Musa spp.) from Brazil using microsatellite markers. Euphytica 132:259–268

Crouch HK, Crouch JH, Madsen S, Vuylsteke DR, Ortiz R (2000) Comparative analysis of phenotypic and genotypic diversity among plantain landraces (Musa spp, AAB group). Theoretical and Applied Genetics,101:1056–1065. https://doi.org/10.1007/s001220051580

d'Hont A, Paget-Goy A, Escoute J, Carreel F (2000) The interspecific genome structure of cultivated banana. *Musa* spp. revealed by genomic DNA in situ hybridization. Theor Appl Genet 100:177–183

Daniells J, Smith M (1991) Post-flask management of tissue-cultured bananas. ACIAR Technical Reports 8

Daniells J, Pegg K, Searle C et al (1995) Goldfinger in Australia: a banana variety with potential. Info 4(1):5–6

Danilova TV, Birchler JA (2008) Integrated cytogenetic map of mitotic metaphase chromosome 9 of maize: Resolution, sensitivity, and banding paint development. Chromosoma 117:345–356

Dantas JLL, Shephard K, W. dos S. Soares Filho et al (1993) Citogenetica e melhoramento genetico da bananeira (Musa spp). Documentos EMBRAPA-CNMPF:48

Das PK, Nag D, Debnath S, Nayak LK (2019) Machinery for extraction and traditional spinning of plant fibers. Ind J Tradit Knowl 65

de Galarreta JIR, Barandalla L, Ritter E, Velázquez ME, Rios D (2025) Spanish potato germplasm bank. In: Al-Khayri JM, Salem KFM, Jain SM (eds) Plant Gene Banks. Springer, Singapore. https://doi.org/10.1007/978-981-99-4236-7_114-1

de Jesus ON, Silva SO, Amorim EP et al (2013) Genetic diversity and population structure of Musa accessions in ex situ conservation. BMC Plant Biol 13:41

De Jong JH, Fransz P, Zabel P (1999) High resolution FISH in plants: techniques and applications. Trends Plant Sci 4:258–263

De Langhe E, Vrydaghs L, de Maret P, Perrier X, Denham T (2009) Why bananas matter: an introduction to the history of banana domestication. Ethnobot Res Appl 7:165

Desheva G, Valchinova E, Deshev M (2025) Bulgarian gene bank. In: Al-Khayri JM, Salem KFM, Jain SM (eds) Plant Gene Banks. Springer, Singapore. https://doi.org/10.1007/978-981-99-4236-7_66-1

Díez MJ, Valcárcel JV, Solbes E, Muñoz E (2025) Forty years conserving Spanish biodiversity in the Gene Bank of the Universitat Politècnica de València, Spain. In: Al-Khayri JM, Salem KFM, Jain SM (eds) Plant Gene Banks. Springer, Singapore. https://doi.org/10.1007/978-981-99-4236-7_86-1

Dilleswara Rao P, Venkata Rao D, Lakshumu Naidu A, Raju Bahubalendruni MVA (2017) Mechanical properties of banana fiber reinforced composites and manufacturing techniques: a review. Inter J Res Develop Tech 8(5)

Dita MA, Rispail N, Prats E, Rubiales D, Singh AB (2006) Biotechnology approaches to overcome biotic and abiotic stress in legumes. Euphytica 147:1–24

Doležel J, Doleželová M, Roux N, Van den houwe I (1998) A novel method to prepare slides for high resolution chromosome studies in Musa spp. Infomusa, 7:3–4

Dolezel J, Bartos J (2005) Plant DNA flow cytometry and estimation of nuclear genome size. Ann Bot 95:99–110

Dolezelova M, Valárik M, Swennen R et al (1998) Physical mapping of the 18S-25S and 5S ribosomal RNA genes in diploid bananas. Biol Plant 41:497–505. https://doi.org/10.1023/A:1001880030275

Douglas H, Marín Ronald A, Romero, Mauricio Guzmán, Turner B (2003) Sutton Plant Disease, 87(3):208-222

Drew RA (1997) The application of biotechnology to the conservation and improvement of tropical and subtropical fruit species. Food and Agriculture Organization of the United Nations, Rome

El Rabey H, Salem KFM, Mattar MZ (2013) The genetic diversity and relatedness of eight rice (Oryza sativa L.) cultivars as revealed by AFLP and SSRs markers. Life Sci J 10(1):1471–1479

Elabd AB, Salem KFM (2009) Detection of genetic diversity in rice (*Oryza sativa* L.) Varieties using microsatellite DNA markers and Quantitative Traits. 6th Egyptian Plant Society of Breeding Conference, 4–7 May 2009, Ismailia

Elabd AB, Saleh MM, Aldahak L, Salem KFM (2020) Identification of superior rice (Oryza sativa L.) genotypes under irrigated conditions based on multivariate analysis. In: 7th International Scientific Conference of Genetic and Environment from 8th –9th December 2020, Baghdad

El-Absawy EA, Salem KFM, Abdellatif KF, ElShamy H (2012) Estimation of genetic diversity among some Egyptian barley (Hordeum vulgare L.) cultivars using RAPD markers and agronomic traits. 1st International Conference on Biotechnology Applications In Agriculture, Moshtohor-Hurghada, 18–22 February 2012, Faculty of Agriculture, Benha University, Egypt

El-Absawy EA, Nowar AA, Salem KFM et al (2015) Estimation of genetic diversity among some bread wheat (Triticum aestivum L.) genotypes using molecular markers and agronomic traits. Minufiya. J Agric Res 40(3):223–233

Eltaher A, Sallam V, Belamkar HA et al (2017) Genome-wide association mapping for drought tolerance traits at seedling stage using nebraska winter wheat. In: International Conference on Productive crop design (genome to phenome). April 2017, USA

Eltaher S, Sallam A, Belamkar V et al (2018) Genetic diversity and population structure of F3:6 Nebraska winter wheat genotypes using genotyping-by-sequencing. Front Genet 9:76. https://doi.org/10.3389/fgene.2018.00076

Eltaher S, Baenziger S, Belamkar V et al (2021) GWAS revealed effect of genotype x environment interactions for grain yield of Nebraska winter wheat. BMC Genomics 22:1–14

Eltaher S, Sallam A, Emara HA et al (2022) Genome-wide association mapping revealed SNP alleles associated with spike traits in wheat. Agronomy 12:1469. https://doi.org/10.3390/agronomy12061469

El-Zanaty AM, Salem KFM, Esmail RM (2011) Detection of genetic diversity in Egyptian cotton (Gossypium barbadense L.) varieties using RAPD markers and morphological traits. J American Sci 7(12):1107–1115

El-Zanaty AM, Salem KFM, Ismail M, Esmail RM (2013) Impact of plant breeding on genetic diversity of the Egyptian cotton (*Gossypium barbadense* L.) varieties as revealed by RAPD markers. J Agric Chem Biotechn Mansoura Univ 4:217–227

Emara HA, Nower AA, Salem KFM, Eltaher S (2011) Estimating heterosis and combining ability for in vivo and in vitro traits using diallel cross in Maize (Zea mays L). J Prod Dev 16:101–120

Emara HA, Salem KFM, Abdel-Halim AA, Abd El-Kowy RNF (2013) An in vivo and in vitro traits analysis of a diallel cross in rice (Oryza sativa L.). Egyptian. J Plant Breed 17:(2):591–610

Esmail RM, Salem KFM, El-Zanaty AM (2009) Genetic diversity studies of egyptian cotton (Gossypium barbadense L.) Varieties Using RAPD Markers and Agronomic Traits. 6th Egyptian Plant Society of Breeding Conference, 4–7 May 2009, Ismalia

Fahrenkrug SC, Rohrer GA, Freking BA et al (2001) A porcine BAC library with tenfold genome coverage: a resource for physical and genetic map integration. Mamm Genome 12:472–474

FAO (2014) Banana market review and banana statistics 2012–2013, p 39. http://www.Fao.org

Faure S, Noyer JL, Horry JP et al (1993) A molecular marker-based linkage map of diploid bananas (Musa acuminata). Theor Appl Genet 87:517–526

Faure' S, Noyer JL, Carreel F et al (1994) Maternal inheritance of chloroplast genome and paternal inheritance of mitochondrial genome in bananas (Musa). Curr Genet 25:265–269

Finalet J, Peteira B, Leon O et al (2000) Application of RAPD markers to the differentiation of *Musa* spp. mutants. Revista de Proteccion Vegetal 15:60–63

Finnegan EJ, Peacock WJ, Dennis ES (1996) Reduced DNA methylation in Arabidopsis thaliana results in abnormal plant development. Proc Natl Acad Sci USA 93:8449–8554

Fraga MF, Esteller M (2002) DNA methylation: a profile of methods and applications. BioTechniques 33:632–634

Fransz PF, Armstrong S, de Jong JH et al (2000) Integrated cytogenetic map of chromosome arm 4S of A. thaliana: Structural organization of heterochromatic knob and centromere region. Cell 100:367–376

Fuchs J, Kuhne M, Schubert I (1998) Assignment of linkage groups to pea chromosomes after karyotyping and gene mapping by fluorescent in situ hybridization. Chromosoma 107:272–276

Fukui K (1996) Plant chromosomes at Mitosis. In: Fukui K, Nakayama S (eds) Plant chromosomes: laboratory methods. CRC Press, Boca Raton, pp 1–18

Fuller MP, Eed MHI (2003) The development of multiple stress-resistant cauliflower using mutagenesis in conjunction with a microshoot tissue culture technique. Acta Hort 618:71–76

Galea-Gragera FA, Llera Cid F (2025) Spain germplasm bank of annual forage legumes seeds at the scientific and technological research center of extremadura (CICYTEX). In: Al-Khayri JM, Salem KFM, Jain SM (eds) Plant Gene Banks. Springer, Singapore. https://doi.org/10.1007/978-981-99-4236-7_102-1

Gawel N, Jarret RL (1991a) Cytoplasmic genetic diversity in bananas and plantains. Euphytica 52:19–23

Gawel NJ, Jarret RL (1991b) Chloroplast DNA restriction fragment length polymorphisms (RFLPs) in Musa species. Theor Appl Genet 81:783–786

Ghazy MI, Salem KFM, Sallam A (2020) Utilization of genetic diversity and marker-trait to improve drought tolerance in rice (Oryza sativa L.). Mol Biol Rep. https://doi.org/10.1007/s11033-020-06029-7

Ghidan WF, Abdulmajid DA (2025) Egyptian rice gene bank. In: Al-Khayri JM, Salem KFM, Jain SM (eds) Plant Gene Banks. Springer, Singapore. https://doi.org/10.1007/978-981-99-4236-7_106-1

Giménez C, Palacios G, Colmenares M (2006) Musa methylated DNA sequences associated with tolerance to Mycosphaerella fijiensis toxins. Plant Mol Biol Report 24:33–43

Gortner G, Nenno M, Weising K et al (1998) Chromosomal localization and distribution of simple sequence repeats and the Arabidopsis-type telomere sequence in the genome of *Cicer arietinum* L. Chrom Res 6:97–104

Grapin A, Noyer JL, Carreel F et al (1998) Diploid Musa acuminata genetic diversity assayed with sequence-tagged microsatellite sites. Electrophoresis 19:1374–1380

Green GC (1963) The effect of weather and climate upon the keeping quality of fruit. Part 2. The banana plant. Technical Note No. 53. World Meteorological Organisation, Geneva

Gudaye TY, Lakew BF, Selamo MM, Tessema ES (2025) Ethiopian crop gene bank. In: Al-Khayri JM, Salem KFM, Jain SM (eds) Plant Gene Banks. Springer, Singapore. https://doi.org/10.1007/978-981-99-4236-7_7-1

Gupta M, Chyi YS, Romero-Severson J, Owen JL (1994) Amplification of DNA markers from evolutionarily diverse genomes using single primers of simple-sequence repeats. Theor Appl Genet 89:998–1006

Häkkinen M (2013) Reappraisal of sectional taxonomy in Musa (Musaceae). Taxon 62(4):809–813

Hansen G, Wright MS (1999) Recent advances in transformation of plants. Trends Plant Sci, 4:226-231

Hapsari L, Wahyudi D, Azrianingsih R, Arumingtyas EL (2015) Genome identification of bananas (Musa L.) from East Java Indonesia, assessed with PCR-RFLP of the internal transcribed spacers nuclear ribosomal DNA. Intl. J Biosci 7(3):42–52

Harzli I, Mostafa K, Abdulla MF, Mehmood M (2025) Sustaining Turkish genetic resources. In: Al-Khayri JM, Salem KFM, Jain SM (eds) Plant Gene Banks. Springer, Singapore. https://doi.org/10.1007/978-981-99-4236-7_88-1

Hassan MN, Mekkawy SA, Mahdy M et al (2021) Recent molecular and breeding strategies in lettuce (Lactuca spp.). Genet Resour Crop Evol (2021). https://doi.org/10.1007/s10722-021-01246-w

Hautea DM, Molina GC, Balatero CH et al (2004) Analysis of induced mutants of Philippine with molecular markers. In: Jain S, Swennen R (eds) Banana Improvement: cellular, molecular biology and induced mutations. Kluwer Academic Publishers, Dordrecht, pp 45–57

Heath DD, Lwama GK, Devlin RH (1993) PCR primed with VNTR core sequences yields species-specific patterns and hypervariable probes. Nucl Acid Res 21:5782–5785

Heslop-Harrison JS (1996) Electron microscopy in plant chromosomes. In: Fukui K, Nakayama S (eds) Plant chromosomes: laboratory methods. CRC Press, Boca Raton, pp 225–245

Heslop-Harrison JS (2000) Comparative genome organization in plants: from sequence and markers to chromatin and chromosomes. Plant Cell 12:617–636

Heslop-Harrison JS, Schwarzacher T (2007) Domestication, genomics and the future of banana. Ann Bot 100(5):1073–1084

Hidayatun N, Kurniawan H (2025) Conservation of agricultural genetic resources in Indonesia. In: Al-Khayri JM, Salem KFM, Jain SM (eds) Plant gene banks. Springer, Singapore. https://doi.org/10.1007/978-981-99-4236-7_40-1

Hippolyte I, Seguin M, Bakry F et al (2008) Musa genetic mapping. In: XVI International Plant and Animal Genome Conference (PAG), San Diego

Hippolyte I, Bakry F, Seguin M et al (2010) A saturated SSR/DArT linkage map of Musa acuminata addressing genome rearrangements among bananas. BMC Plant Biol 10:65

Horry JP (2011) The use of molecular markers in the CIRAD banana breeding programme. Acta Hortic 897:235–241

Horry J, Jay M (1988) Distribution of anthocyanins in wild and cultivated banana varieties. Phytochemistry 27:2667–2672

Hribova M, Dolezelova M, Dolezel J (2008) Localization of BAC clones on mitotic chromosomes of *Musa acuminate* using fluorescence in situ hybridization. Biol Plant 52(3):445–452

Ibrahim AA, Salem KFM, Abdein MA, Ramadan SM (2023a) Advances in summer squash (*Cucurbita pepo* L.) molecular breeding strategies. In: Singh S, Sharma D, Sharma SK, Singh

R (eds) Smart plant breeding for vegetable crops in post-genomics era. Springer, Singapore. https://doi.org/10.1007/978-981-19-5367-5_8

Ibrahim AA, Salem KFM, Ramadan S (2023b) Allium sativum and stressful conditions. In: Abdel Latef AAH (ed) Book medicinal plant responses to stressful conditions. CRC Press

Ibrahim AA, Abd-Ellatif S, Abdel Razik ESS, Salem KFM (2024) Biodiversity of cereal crops and utilization in food and nutritional security. In: Al-Khayri JM, Jain SM, Penna S (eds) Sustainable utilization and conservation of plant genetic diversity. Sustainable development and biodiversity, vol 35. Springer, Singapore. https://doi.org/10.1007/978-981-99-5245-8_2

Ihsan F, Prihatini R (2025) Indonesian tropical fruit gene bank. In: Al-Khayri JM, Salem KFM, Jain SM (eds) Plant Gene Banks. Springer, Singapore. https://doi.org/10.1007/978-981-99-4236-7_41-1

Imelda M, Estiati A, Hartati NS (2001) Induction of mutation through gamma radiation in three cultivars of banana. Ann Bogorien 7:75–82

Inta W, Traiperm P, Ruchisansakun S et al (2023) Evolution and classification of Musaceae based on male floral morphology. Plants 12(8):1602. https://doi.org/10.3390/plants12081602

IPGRI INIBAP/CIRAD (1996) Descriptors for Banana (Musa Spp) International Plant Genetic Resources Institute, Rome, Italy/International Network for the Improvement of Banana and Plantain, Montpellier. France. Centre de Coop'eration internationale en recherché agronomoque pourle d'eveloppement, Montpellier

Ismail M, Salem KFM, El-Zanaty AM, Esmail RM (2013) Association mapping of fiber and agronomic characters in Egyptian cotton (Gossypium barbadense L.) genotypes. J Agric Chem Biotech Mansoura Univ 4:205–216

Jain SM (2001) Tissue culture-derived variation in crop improvement. Euphytica 118:153–166

Jain PK, Saini MK, Pathak H, Gupta VK (2007) Analysis of genetic variation in different banana (*Musa* species) variety using random amplified polymorphic DNAs (RAPDs). Afri J Biotechnol 6:1987–1989

Jain SM, Till J, Suprasanna P, Roux N (2011) Mutations and cultivar development of banana. In: Pillay M, Tenkouano A (eds) Banana breeding: progress and challenges. CRC Press, Boca Raton, pp 203–217

Janssen SB, Vandelook F, Langhe ED et al (2016) Evolutionary dynamics and biogeography of Musaceae reveal a correlation between the diversification of the banana family and the geological and climatic history of Southeast Asia. New Phytol 210(4):1453–1465

Jarret RL, Litz RE (1986a) Enzyme polymorphism in *Musa acuminata* Colla. J Hered 77:183–188

Jarret RL, Litz RE (1986b) Isozymes as genetic markers in bananas and plantains. Euphytica 35:539–549

Jarret R, Gawel N, Whittemore A, Sharrock S (1992) RFLP-based phylogeny of *Musa* species in Papua New Guinea. Theor Appl Genet 84:579–584

Jatoi SA, Siddiqui SU (2025) National gene bank of Pakistan. In: Al-Khayri JM, Salem KFM, Jain SM (eds) Plant gene banks. Springer, Singapore. https://doi.org/10.1007/978-981-99-4236-7_51-1

Jenny C, Holtz Y, Horry JP, Bakry F (2013) Synthesis of new interspecific hybrids from AB germplasm in banana (*Musa* sp). Acta Hortic 986:209–218

Johnson R (2000) Classical plant breeding for durable resistance to diseases. J Plant Pathol 82(1):3–7

Jones DR (2000) Diseases of banana, abaca and enset. CAB International, Wallingford, p 544

Jones CJ, Edwards KJ, Castiglione S et al (1997) Reproducibility testing of RAPD, RFLP and SSR markers in plants by a network of European laboratories. Mol Breed 3:381–390

Kaemmer D, Fischer D, Jarret RL et al (1997) Molecular breeding in the genus Musa: a strong case for STMS marker technology. Euphytica 96:49–63

Khan AJ, Hassan S, Tariq M, Khan T (2001) Haploidy breeding and mutagenesis for drought tolerance in wheat. Euphytica 120:409–414

Khatri A, Dahot MU, Khan I et al (2009) Use of RAPD for the assessment of genetic diversity among exotic and commercial banana clones. Pak J Bot 41:2995–2999

Kilian A (2007) Towards Effective Deployment of Diversity Arrays Technology (DArT). In: Banana Genomics and Sequencing. Plant and Animal Genomes XV Conference, January 13–17, San Diego

Kowalski B, Cassells AC (1999) Mutation breeding for yield and Phytophthora infestans (Mont.) de Bary foliar resistance in potato (Solanum tuberosum L-cv. Golden Wonder) using computerized image analysis in selection. Potato Res 42:121–130

Krikorian AD, Cronauer SS (1984) Handbook of plant cell culture. In: Sharp WR, Evans DA, Ammirato P, Yamada Y (eds) Crop species, vol 2. Macmillan, New York, pp 327–348

Kulikova O, Gualtieri G, Geurts R, Kim DJ, Cook D, Huguet T, de Jong JH, Fransz PF, Bisseling T (2001) Integration of the FISH pachytene and genetic maps of Medicago truncatula. Plant J 27:49–58. https://doi.org/10.1046/j.1365-313x.2001.01057.x

Kumar LS (1999) DNA markers in plant improvement: an overview. Biotechnol Adv 17:143–182

Kumar N, Soorianathasundaram K (2007) Progress and current strategies in banana improvement in Tamil Nadu. In: Singh HP, Uma S (eds) Banana-technological advancements. Association for the Improvement in Production and Utilization of Banana (AIPUB), National Research Centre for Banana, Trichy, pp 128–137

Laghetti G, Piergiovanni AR, De Paola D, Direnzo P, Bucci G, Sonnante G (2025) Italy IBBR Gene Bank. In: Al-Khayri JM, Salem KFM, Jain SM (eds) Plant gene banks. Springer, Singapore. https://doi.org/10.1007/978-981-99-4236-7_73-1

Lakshmanan V, Venkataramareddy SR, Neelwarne B (2007) Molecular analysis of genetic stability in long-term micropropagated shoots of banana using RAPD and ISSR markers. Electron J Biotechnol 10:106–113

Lamare A, Rao SR (2015) Efficacy of RAPD, ISSR and DMD markers in assessment of genetic variability and population structure of wild Musa acuminata Colla. Physiol Mol Biol Plants 21(3):349–358

Larkin PJ, Scowcroft WR (1981) Somaclonal variation—a novel source of variability from cell cultures for plant improvement. Theor Appl Genet 60:197–214

Lebot V, Aradhya KM, Manshardt R, Meilleur B (1993) Genetic relationships among cultivated bananas and plantains from Asia and the Pacific. Euphytica 67:163–175

Lefebvre V, Palloix A, Caranta C, Pochard E (1995) Construction of an intraspecific integrated linkage map of pepper using molecular markers and doubled-haploid progenies. Genome 38:112–121

Lheureux F, Carreel F, Jenny C et al (2003) Identification of genetic markers linked to banana streak disease expression in interspecific Musa hybrids. Theor Appl Genet 106:594–598

Li LF, Häkkinen M, Yuan YM et al (2010) Molecular phylogeny and systematics of the banana family (Musaceae) inferred from multiple nuclear and chloroplast DNA fragments, with a special reference to the genus Musa. Mol Phylo Evol 57(1):1–10

Li LF, Wang HY, Zhang C et al (2013) Origins and domestication of cultivated banana inferred from chloroplast and nuclear genes. PLoS One 8(11):e80502

Lin YH, Chang JY, Liu ET et al (2010) Development of a molecular marker for specific detection of *Fusarium oxysporum* f.sp cubense race 4. Eur J Plant Pathol 123:353–365

Liu N, Zhu Y, Wei Z et al (2009) Phylogenetic relationships and divergence times of the family Araucariaceae based on the DNA sequences of eight genes. Sci Bull 54(15):2648–2655

Liu AZ, Kress WJ, Li DZ (2010) Phylogenetic analyses of the banana family (Musaceae) based on nuclear ribosomal (ITS) and chloroplast (trnL-F) evidence. Taxon 59(1):20–28

Loh JP, Kiew R, Set O et al (2000) Amplified Fragment Length Polymorphism Fingerprinting of 16 banana cultivars (*Musa* cvs.). Mol Phylogenet Evol 17:360–366

Lorenzen J, Tenkouano A, Bandyopadhyay R et al (2010) Overview of banana and plantain (Musa spp.) improvement in Africa: past and future, proc. IC on banana & plantain in Africa. Acta Hort 879:595–603

Mallor C, Luna E (2025) The genetic diversity of the spanish vegetable genebank BGHZ-CITA: conservation and utilization. In: Al-Khayri JM, Salem KFM, Jain SM (eds) Plant gene banks. Springer, Singapore. https://doi.org/10.1007/978-981-99-4236-7_109-1

Maluszynska J, Heslop-Harrison JS (1993) Molecular cytogenetics of the genus Arabidopsis, in situ hybridization of rDNA sites, chromosome numbers and diversity in centromeric heterochromatin. Ann Bot 71:479–484

Marin DH, Romero RA, Guzmán M, Sutton TB (2003) Plant Disease, 87(3):208-222

Masiga DK, Turner CMR (2004) Amplified (restriction) fragment length polymorphism (AFLP) analysis. In: Melville SE (ed) Methods in Molecular Biology Humana Press, pp 173–186

Mattar MZ, Eissa OA, Salem KFM (2009) Genotypic variability in some wheat genotypes (Triticum aestivum L.) exposed to salinity at germination stage. Egyptian J Exp Biol (Bot) 5:143–148

Mattar MZ, Salem KFM, El-Abed AB (2016) Identification of candidate microsatellite markers associated with agronomic traits in rice (*Oryza sativa* L.). J Bot 15(5):213–232

Menon R (2000) Ex situ conservation of Musa genetic resources in Kerala. In: Abstracts of international conference on science and technology for managing plant genetic diversity in the 21st century, 12–16 June 2000, Kuala Lumpur

Menon R, Suma A, Cherian AK, Nair S, Sudheesh MV (2002) Conservation and utilization of Musa AA diploids in breeding. In: Global conference on banana and plantain, Bangalore, October 28–31, 2002. Abstract, p 52

Menon R, Agrawal A, Reddy BMC (2005) Status of introduced Musa germplasm at Banana Research Station, Kannara. Ind J Plant Genet Resour 18(1):81–83

Miller RNG, Passos MAN, Menezes NNP et al (2010) Characterization of novel microsatellite markers in *Musa acuminata* subsp. burmannicoides, var. Calcutta 4. BMC Res Notes 3:149

Miri SM, Mousavi A, Naghavi MR et al (2009) Analysis of induced mutants of salinity resistant banana (Musa acuminata cv. Dwarf Cavendish) using morphological and molecular markers. Iran J Biotechnol 7:86–92

Mobambo KN, Gauhl F, Vuylsteke D, Ortiz R, Pasberg-Gauhl C, Swennen R (1993) Yield loss in plantain from black sigatoka leaf spot and field performance of resistant hybrids. Field Crops Research, 35(1):35–42

Molina AB, Escalant VJ, Van Den Bergh I (2007) International Musa Testing Programme (IMTP): a worldwide programme to evaluate Musa varieties. In: Singh HP, Uma S (eds) Banana. Technological advancements. Association for Improvement in Production and Utilization of Banana (AIPUB), NRCB, Trichy, pp 54–61

Monteros-Altamirano Á et al (2025) Plant genetic resources in Ecuador. In: Al-Khayri JM, Salem KFM, Jain SM (eds) Plant gene banks. Springer, Singapore. https://doi.org/10.1007/978-981-99-4236-7_99-1

Morán JFA (2013) Improvement of Cavendish banana cultivars through conventional breeding, Fundación Hondureña de Investigación Agrícola (FHIA) LaLima, Honduras. In: Van den Bergh I et al (eds) Proceedings of the International ISHS-ProMusa Symposium on bananas and plantains: towards sustainable global production and improved uses. Acta Hort 986:205–208

Morton J (1987) Fruits of warm climates. Julia F. Morton Publ, Miami, FL, pp 29–46

Motaleb KZM, Abdul, Abdul Ahad, Laureckiene G, Milasius R (2021) Innovative banana fiber nonwoven reinforced polymer composites: pre and post-treatment effects on physical and mechanical properties. Polymers 13, no. 21:3744. https://doi.org/10.3390/polym13213744

Mourad AMI, Alomari DZ, Alqudah AM et al (2019) Recent advances in wheat (*Triticum* spp.) breeding. In: Al-Khayri J, Jain S, Johnson D (eds) Advances in plant breeding strategies: Cereals. Springer, Cham. https://doi.org/10.1007/978-3-030-23108-8_15

Muhammad Y, James AC, Rivera-Madrid R et al (2010) Musa genetic diversity revealed by SRAP and AFLP. Mol Biotechnol 47:189–199. https://doi.org/10.1007/s12033-010-9328-8

Nair AS, Teo CH, Schwarzacher T, Harrison PH (2005) Genome classification of banana cultivars from South India using IRAP markers. Euphytica 144:285–290

Nawara HM, Mattar MZ, Salem KFM, Eissa OA (2016) Estimating heterosis and combining ability for agronomic and tissue culture traits using diallel cross on bread wheat (*Triticum aestivum* L.). Egypt J Bot 2016:177–200

Nawara HM, Mattar MZ, Salem KFM, Eissa OA (2017) Diallel study on some in vitro traits of bread wheat (Triticum aestivum L.) under salt stress. Inter J Agric. Environ Res 3(1):1988–2006
Nikmah IA, Azrianingsih R, Wahyudi D (2016) Genetic diversity of porang populations (Amorphophallus muelleri Blume) in Central Java and West Java based on LEAFY second intron marker. J Trop Life Sci 6(1):23–27
Nsabimana A, Van Staden J (2007) Assessment of genetic diversity of highland bananas from the National Banana Germplasm Collection at Rubona, Rwanda using RAPD markers. Sci Hort 113:293–299
Nwakanma DC, Pillay M, Okoli BE, Tenkouano A (2003) PCR-RFLP of the ribosomal DNA internal transcribed spacers (ITS) provides markers for the A and B genomes in Musa L. Theor Appl Genet 108:154–159
Opara UL, Jacobson D, Al-Saady NA (2010) Analysis of genetic diversity in banana cultivars (Musa cvs.) from the South of Oman using AFLP markers and classification by phylogenetic, hierarchical clustering and principal component analyses. J Zhejiang Univ Sci 11:332–341
Orita M, Suzuki Y, Sekiya T, Hayashi K (1989) Rapid and sensitive detection of point mutations and DNA polymorphisms using the polymerase chain reaction. Genomics 5:874–879
Ortiz R (1995) Musa genetics. In: Gowen S (ed) Bananas and plantains. Chapman and Hall, London, pp 84–109
Ortiz R, Vuylsteke D (1994a) Plantain breeding at IITA. In: Jones DR (ed) The improvement and testing of Musa: a global partnership. Proc. Conf. FHIA, Honduras, 27–30, April, 1994. INIBAP, Montpellier
Ortiz R, Vuylsteke D (1995b) Factors influencing seed set in Triploid Musa spp. L and production of diploid hybrids. Ann Bot 75:151–155
Ortiz R,Vuylsteke D (1998) Quantitative variation and phenotypic correlations in banana and plantain, Scientia Horticulturae, 72(3–4):239–253
Ortiz-Vázquez E, Kaemmer D, Zhang HB et al (2005) Construction and characterization of a plant transformationcompetent BIBAC library of the black Sigatoka-resistant banana Musa acuminata cv. Tuu Gia (AA). Theor Appl Genet 110:706–713. https://doi.org/10.1007/s00122-004-1896-1
Oselebe HO, Tenkouano A, Pillay M et al (2006) Ploidy and genome segregation in Musa breeding populations assessed by flow cytometry and randomly amplified polymorphic DNA markers. Hort Sci 131:780–786
Osterc G, Gostinčar K, Hudina M, Usenik V, Luthar Z (2025) Slovene fruit gene bank. In: Al-Khayri JM, Salem KFM, Jain SM (eds) Plant Gene Banks. Springer, Singapore. https://doi.org/10.1007/978-981-99-4236-7_84-1
Ostmark HE (1974) Economic insect pests of bananas. Annu Rev Entomol 19:161–176
Oyeyinka BO, Afolayan AJ (2022) Suitability of banana and plantain fruits in modulating neurodegenerative diseases: implicating the in vitro and in vivo evidence from neuroactive narratives of constituent biomolecules. Foods 11(15):2263. https://doi.org/10.3390/foods11152263
Patterson AH, Damon S, Hewitt JD, Zamir D, Rabinowitch HD, Lincoln SE, Lander ES, Tanksley SD (1991) Mendelian factors underlying quantitative traits in tomato: comparison across species, generations and environments. Genetics, 127:181—197
Pearce SR, Knox M, Ellis TNM et al (2000) Pea Tyl-copia group retrotransposon: transitional activity and use as makers to study genetic diversity in Pisum. Mol Gen Genomics 263:898–907
Pillay M, Adeleke MTV (2001) Silver staining of *Musa* L. chromosomes. Cytologia 66:33–37
Pillay M, Tripathi L (2007) Banana breeding. In: Kang MS, Priyadarshan PM (eds) Breeding major food staples. Blackwell Science, Boston, pp 393–428
Pillay M, Nwakanma DC, Tenkouano A (2000) Identification of RAPD markers linked to A and B genome sequences in Musa L. Genome 43:763–767
Pillay M, Ogundiwin E, Nwakanma DC, Ude G, Tenkouano A (2001) Analysis of genetic diversity and relationships in East African banana germplasm. Theor Appl Genet 102:965–970
Pillay M, Ssebuliba R, Hartman J et al (2004) Conventional breeding strategies to enhance the sustainability of biodiversity conservation for endemic cultivar. Afr Crop Sci J 12(1):59–65

Poerba YS, Ahmad F (2010) Genetic variability among 18 cultivars of cooking bananas and plantains by RAPD and ISSR markers. Biodiversitas 11(3):118–123

Popović V et al (2025) Economic seed production and storage of fiber crops in gene banks in Serbia. In: Al-Khayri JM, Salem KFM, Jain SM (eds) Plant Gene Banks. Springer, Singapore. https://doi.org/10.1007/978-981-99-4236-7_108-1

Priyanka V, Kumar R, Dhaliwal I, Kaushik P (2021) Germplasm conservation: instrumental in agricultural biodiversity—a review. Sustainability 13, no. 12:6743. https://doi.org/10.3390/su13126743

Probojati RT, Wahyudi D, Hapsari L (2019) Clustering analysis and genome inference of pisang raja local cultivars (*Musa* Spp.) from Java Island by random amplified polymorphic DNA (RAPD) marker. J Trop Biodiv Biotechnol 4(2):42–53

Purseglove JW (1975) Tropical crops monocotyledons, vol 1. Wiley, Halsted Press Division, p 519

Ramage CM, Borda AM, Hamill SD, Smith MK (2004) A simplified PCR test for early detection of dwarf off-types in micropropagated Cavendish bananas (Musa spp. AAA), Scientia Horticulturae, 103(1):145-151

Ramesh M, Sri T, Ananda Atreya, Aswin US, Eashwar H, Deepa C (2014) Processing and Mechanical property evaluation of banana fiber reinforced polymer composites, Procedia Eng, 97(5):563–572

Rao X, Chen H, Lu Y et al (2023) Distribution and Location of bevs in different genotypes of bananas reveal the coevolution of BSVs and bananas. Inter J Mol Sci 24(23):17064. https://doi.org/10.3390/ijms242317064

Ravishankar KV, Vidhya L, Cyriac A et al (2012) Development of SSR markers based on a survey of genomic sequences and their molecular analysis in banana (Musa spp.). J Hortic Sci Biotechnol 87:84–88

Ravishankar KV, Raghavendra KP, Athani V et al (2013) Development and characterisation of microsatellite markers for wild banana (*Musa balbisiana*). J Hortic Sci Biotechnol 88(5):605–609

Ray T, Saha P, Roy SC (2006) Commercial production of Cordyline terminalis (L) Kunth from shoot apex meristem and assessment for genetic stability of somaclones by isozyme markers. Scientia Hortic,108:289–294

Rekha A, Ravishankar KV, Ambika DS (2011b) Evaluation of segregating populations from Musa acuminata ssp. burmannicoides (Calcutta-4). ISHS/ProMusa symposium, Bananas and plantains: toward sustainable global production and improved uses. Bahia Othon Palace Hotel, Salvador, Bahia, Brazil, 10–14 Oct 2011. Abstract. p 150

Reyna-López GE, Simpson J, Ruiz-Herrera J 1997 Differences in DNA methylation patterns are detectable during the dimorphic transition of fungi by amplification of restriction polymorphisms. Molecular and General Genetics 253:703–710

Richardson DR, Tran EH, Ponka P (1995) The potential of iron chelators of the pyridoxal isonicotinoyl hydrazone class as effective antiproliferative agents. Blood, 86(11):4295–306

Richardson M, Pimenta AMC, Bemquerer MP, Santoro MM, Beirao PSL, Lima ME, Figueiredo SG, Bloch C, Vasconcelos EAR, Campos FAP, Gomes PC, Cordeiro MN (2006) Comparison of the partial proteomes of the venoms of Brazilian spiders of the genus Phoneutria, Comparative Biochemistry and Physiology Part C: Toxicology and Pharmacology, 142(3–4):173-187

Robinson JC (1996) Bananas and plantains. Crop production Science in horticulture 5. CAB International, Wallingford, p 238

Rocha F et al (2025) Portugal Gene Bank. In: Al-Khayri JM, Salem KFM, Jain SM (eds) Plant Gene Banks. Springer, Singapore. https://doi.org/10.1007/978-981-99-4236-7_79-1

Rogstad SH, Keane B, Keiffer CH et al (2001) DNA extraction from plants: The use of pectinase. Plant Mol Biol Rep, 19:353–359. https://doi.org/10.1007/BF02772833

Roux N, Afza R, Brunner H, Morpurgo R, van Duren M (1994) Complementary approaches to cross breeding and mutation breeding for Musa improvement. In: Jones DR (ed) The improvement and testing of Musa: a global partnership-proceedings of the first global conference of the International Musa Testing Program at FHIA, Honduras, 27–30 April 1994. INIBAP, Montpellier, pp 213–218

Roux NS, Jain SM, Swennen R (2004) Mutation induction in Musa review. In: Banana improvement: cellular, molecular biology, and induced mutations. Proceedings of a meeting held in Leuven, Belgium, 24–28 September 2001. Science Publishers, Inc., pp 23–32

Rowe PR (1984) Breeding bananas and plantain. Plant Breed Rev 2:135–155

Rowe PR (1987) Breeding plantains and cooking bananas. In: International cooperation for effective plantain and banana research: Proceedings of the 3rd meeting of the International Association for Research on Plantain and Bananas. Abidjan, 23–31 May 1985, pp 21–23

Rowe PR (1990) New genetic combination in breeding bananas and plantains resistant to diseases. In: Identification of genetic diversity in the genus Musa: proceedings of an international workshop held at Los Banos, 5–10 Sept 1988, pp 114–123

Rowe P, Rosales F (1994) Musa breeding at FHIA. In: Jones DR (ed) The improvement and testing of Musa, a global partnership, Proceedings of the global conference of the international testing programme, FHIA, Honduras, 27–30 Apr 1994, pp 117–129

Ruiz M, Rouard M, Raboin LM et al (2004) TropGENE-DB, a multi-tropical crop information system. Nucleic Acids Res 32:D364–D367

Saleh MM, Ali WA, Alwan MH et al (2020a) Multi-environmental evaluation of wheat tetraploid genotypes for agronomic traits under rainfed conditions in Syria. J Aridland Agric 6:1–5

Saleh MM, Salem KFM, Elabd AB (2020b) Definition of selection criterion using correlation and path coefficient analysis in rice (Oryza sativa L.) genotypes. Bull Natl Res Cent 44(1):1–6

Saleh MM, Alnaddaf LM, Almuhammady AK et al (2021a) In: Al-Khayri JM, Ansari MI, Singh AK (eds) Applications of plant-derived nanomaterials in mitigation of crop abiotic stress. Nanobiotechnology, Springer, Cham. https://doi.org/10.1007/978-3-030-73606-4_9

Saleh MM, Mahmoud AS, Abbas HS et al (2021b) Nanotechnological approaches for efficient delivery of plant ingredients. In: Faizan M, Hayat S, Yu F (eds) Sustainable agriculture reviews, vol 53. Springer, Cham. https://doi.org/10.1007/978-3-030-86876-5_11

Salem KFM (1997) Breeding studies on rice (Oryza sativa L.). M Sc Thesis, Faculty of Agriculture, Minufiya University, Shibin El-Kom, Egypt

Salem KFM (2004) The inheritance and molecular mapping of genes for post-anthesis drought tolerance (PADT) in wheat. Martin-Luther-University, Halle-Wittenberg, p 124

Salem KFM (2009) Relationship between genetic diversity based on SSRs markers with heterosis and combining ability in diallel cross of bread wheat (Triticum aestivum L.). Minufiya. J Agric Res 6(34):2159–2178

Salem KFM (2015) Allelic state at the microsatellite locus Xgwm261 marking the dwarfing gene Rht8 in Egyptian bread wheat (Triticum aestivum L.) genotypes released from 1947 to 2004. Genetika 47:741–750

Salem KFM, Börner A (2003) Evaluation of wheat accessions from Gatersleben Genebank for post-anthesis drought tolerance in the field. In: Gesellschaft für Pflanzenzüchtung, GPZ-Tagung, Vortragsveranstaltung zum Thema: "Abiotischer Streß—Herausforderung für die Pflanzenzüchtung", 11–12 Juni 2003, Groß Lüsewitz, Germany

Salem KFM, Börner A (2004) Post-anthesis drought tolerance in genus Triticum. In: Botanikertagung, 5–10 September 2004, Braunschweig, Germany

Salem KFM, El-Zanaty AM (2015) Assessing genetic diversity of old and modern Egyptian rice (Oryza sativa L.) using microsatellite markers. Minufiya J Agric Res 40(6):1507–1518

Salem KFM, Ibrahim AA (2024) Plant biodiversity in the context of food security under climate change. In: Al-Khayri JM, Jain, SM, Penna S (eds) Sustainable utilization and conservation of plant genetic diversity. Sustainable development and biodiversity, vol 35, Springer, Singapore. https://doi.org/10.1007/978-981-99-5245-8_1

Salem KFM, Mattar MZ (2014) Genetic diversity in old and modern Egyptian bread wheat (Triticum aestivum L.) varieties revealed by simple sequence repeats. Egypt J Genet Cytol 43:143–156

Salem KFM, Matter MZ (2014) Identification of microsatellite alleles for salt tolerance at seedling stage in wheat (Triticum aestivum L.). Life Sci J 11(12s):1064–1073

Salem KFM, Sallam A (2016) Analysis of population structure and genetic diversity of Egyptian and exotic rice (*Oryza sativa* L.) genotypes using SSR markers. Comptes Rendus Biol 339:1–9. https://doi.org/10.1016/j.crvi.2015.11.003

Salem KFM, Röder MS, Börner A (2003) Mapping of quantitative trait loci (QTLs) for post-anthesis drought tolerance in bread wheat (Triticum aestivum L.). In: Gesellschaft für Pflanzenzüchtung, GPZ-Tagung, 11. Vortragsveranstaltung "Harnessing Genetic Diversity: Genomics and Allele Mining", 16–17.09.2003. Institut für Pflanzengenetik und Kulturpflanzenforschung (IPK), Gatersleben

Salem KFM, Röder MS, Börner A (2004a) Drought tolerance in wheat germplasm- Evaluation and molecular mapping. In: Institut für Pflanzengenetik und Kulturpflanzenforschung, IPK Tagung, 22–24/08/2004. Ag Ressourcengenetik und Reproduktion, Abteilung Genbank, Gatersleben

Salem KFM, Röder MS, Börner A (2004b) Molecular mapping of quantitative trait loci (QTLs) determining post-anthesis drought tolerance in hexaploid wheat (*Triticum aestivum* L.). In: 7 Gesellschaft für Pflanzenzüchtung, GPZ-Tagung, Vortragsveranstaltung zum Thema: Klimatische und edaphische Sortenanpassung und Züchtung für Nachwachsende Rohstoffe, vol 64. 3–5 March 2004, Halle/Saale, Germany. Vort Pflanzenzüchtung, pp 21–24

Salem KFM, Röder MS, Börner A (2007a) Evaluation of some barley varieties for the presence of thermostable alleles of ßamylase. 8th African Crop Science Society Conference, El-Minia, Egypt, 27–31 October 2007, pp 643–648

Salem KFM, Röder MS, Börner A (2007b) Identification and mapping quantitative trait loci for stem reserve mobilisation in wheat (*Triticum aestivum* L.). Cereal Res Common 35:1367–1374

Salem KFM, Varshney RK, Salem KFM, Röder MS, Börner A (2008a) Detection of genetic diversity in Egyptian barley genotypes using barley expressed sequence Tag-SSR markers. In: 10th International Barley Genetics Symposium (IBGS), Alexandria, Egypt 5–10 April 2008. In: The International Center for Agricultural Research in the Dry Areas (ICARDA) and the Bibliotheca Alexandrina. Alexandria, Egypt

Salem KFM, El-Zanaty AM, Esmail RM (2008b) Assessing wheat (*Triticum aestivum* L.) genetic diversity using morphological characters and microsatellite markers. World J Agric Sci 4:538–544

Salem KFM, Varshney RK, Röder MS, Börner A (2010) EST-SSR based estimates on functional genetic variation in a barley (*Hordeum vulgare* L.) collection from Egypt. Genet Resour Crop Evol 57(4):515–521

Salem KFM, Yaseen YM, El-Abd AB, Ayoub A (2015a) Inheritance of some yield components and grain quality traits in rice (*Oryza sativa* L.). J Product Dev 20:161–177

Salem KFM, Yaseen YM, El-Abd AB, Ayoub A (2015b) Breeding studies on some agronomic and quality traits in rice (Oryza sativa L.). J Product Dev 20(2):211–226

Salem KFM, Alloosh MT, Saleh MM et al (2021a) In: Al-Khayri JM, Ansari MI, Singh AK (eds) Utilization of nanofertilizers in crop tolerance to abiotic stress. Nanobiotechnology, Springer, Cham. https://doi.org/10.1007/978-3-030-73606-4_11

Salem KFM, Elabsawy A, Aldahak L et al (2021b) Evaluation of stem reserve mobilization in Egyptian bread wheat (Triticum aestivum L.) genotypes and F1 hybrids under post-anthesis chemical desiccation stress. J Genet Environ Resour Conserv 9(1):176–182

Salem KFM, El-Absawy EA, Khidr YA, El-Shamy H (2021c) Assessment genetic diversity among some Egyptian bread wheat (Triticum aestivum L.) cultivars using RAPD markers. J Genet Environ Resour Conserv 9(1):168–175

Salem KFM, Saleh MM, Abu-Ellail FFB et al (2021d) Role of quantum dots, polymeric NPS and dendrimers in emphasizing crops tolerate biotic and abiotic stresses. In: Faizan M, Hayat S, Yu F (eds) Sustainable agriculture reviews, vol 53. Springer, Cham. https://doi.org/10.1007/978-3-030-86876-5_1

Salem KFM, Saleh MM, Abu-Ellail FFB et al (2021e) The Role of salicylic acid in crops to tolerate abiotic stresses. In: Hayat S, Siddiqui H, Damalas CA (eds) Salicylic acid—a versatile plant growth regulator. Springer, Cham. https://doi.org/10.1007/978-3-030-79229-9_7

Salem KFM, Saleh MM, Aldahak L, Elabd AB (2021f) Assessment phenotypic diversity of rice (*Oryza sativa* L.) genotypes by multivariate analysis. J Arid Agric 7:52–59. https://doi.org/10.25081/jaa.2021.v7.6516

Salem KFM, Alghuthaymi MA, Elabd AB et al (2022) Prediction of heterosis for agronomic traits in half-diallel cross of rice (Oryza sativa L.) under drought stress using microsatellite markers. Plants 11:1532. https://doi.org/10.3390/plants11121532

Salem KFM, Tawfik E, Mahdy M et al (2023) Advances in lettuce (Lactuca spp.) molecular breeding strategies. In: Singh S, Sharma D, Sharma SK, Singh R (eds) Smart plant breeding for vegetable crops in post-genomics era. Springer, Singapore. https://doi.org/10.1007/978-981-19-5367-5_11

Salem KFM, Safhi FA, Alwutayd KM et al (2024) Analysis of genetic diversity, population structure and phylogenetic relationships of rice (Oryza sativa L.) cultivars using simple sequence repeat (SSR) markers. Genet Resour Crop Evol 71:2213–2227. https://doi.org/10.1007/s10722-023-01789-0

Santayana M, Rossel G, Núñez J et al (2014) Molecular characterization of cultivated species of the genus pachyrhizus rich ex DC by AFLP Markers: Calling for More Data. Tropical Plant Biol, 7:121–132. https://doi.org/10.1007/s12042-014-9143-7

Saraswathi MS, Uma S, Vadivel E et al (2011) Diversity analysis in Indian cooking bananas (Musa ABB) through morphotaxonomic and molecular characterization. Acta Hortic 897:123–129

Schulz-Schaeffer J (1980) Cytogenetics, plants and humans. Springer, Berlin

Sharrock S, Frison E (1998) Musa Production around the World: Trends, Varieties and Regional Importance; INIBAP Annual Report: Montpellier, France, pp. 42–47

Shekhar R, Bhavya G, Prakash HS, Geetha N (2019) DNA barcoding-based authentication of Musa acuminata var. nanjangud rasabale. Jetir 6(3):5

Shepherd RL (1974) Breeding root-knot-resistant Gossypium hirsutum L. using a resistant wild G. barbadense L. Crop Sci, 14:687-691

Sherry ST, Ward MH, Kholodov M, Baker J, Phan L, Smigielski EM, Sirotkin K (2001) dbSNP: the NCBI database of genetic variation. Nucleic Acids Res 29:308–311

Shim SI, Jørgensen RB (2000) Genetic structure in cultivated and wild carrots (Daucus carota L.) revealed by AFLP analysis. Theor Appl Genet, 101:227–233

Shroff A, Karolia A, Shah J (2015) Bio-softening of banana fiber for nonwoven application. Inter J Sci Res 2015:32–43

Simmonds NW (1962) The evolution of the bananas. Green, Longmans, p 170

Simmonds NW (1966) Bananas, 2nd edn. Longmans, London

Simmonds NW (1986) Classification and breeding of bananas. In: Persley GJ, De Langhe EA (eds) Banana and plantain breeding strategies. Proc. of International Workshop, Cairns, pp 69–73

Simmonds NW, Shepherd K (1955) The taxonomy and origins of the cultivated bananas. J Linn Soc Lond 55:302–312

Singh HP, Uma S, Sathiamoorthy S (2001) A tentative key for identification and classification of bananas and plantains in India. National Research Center for Banana, Trichy, p 61

Sitthipron S, Anutalabhochai S, Dumampai N et al (2005) Investigation of genetic relationships and hybrid detection in Longan by high annealing temperature RAPD. Acta Hort 665:161–169

Sofan P, Bruce D, Jones E, Khomarudin MR, Roswintiarti O (2020) Applying the Tropical Peatland Combustion Algorithm to Landsat-8 Operational Land Imager (OLI) and Sentinel-2 Multi Spectral Instrument (MSI) Imagery. Remote Sens, 12:3958. https://doi.org/10.3390/rs12233958

Stace C (1980) Plant taxonomy and biosystematics. Edward Arnold, London

Staub JE, Sequen FC, Gupta M (1996) Genetic markers, map construction, and their application in plant breeding. HortScience 31:742–762

Stover RH (1987) In: Persley GI, De Langhe EA (eds) Somaclonal variation in Grande Naine and Saba bananas in the nursery and in the field. Banana and Plantain

Stover RH, Simmonds NW (1987) Bananas, 3rd edn. Longman Scientific and Technical, Harlow. 468pp

Sundari, Arumingtyas EL, Hakim L et al (2017) Genetic variability of local durian (Durio zibethinus Murr.) in Ternate Island based on RAPD markers. Plant Cell Biotechnol Mol Biol 18:68–75
Suprasanna P, Rao PS (1997) Selection of mutants using plant cell and tissue culture techniques. In: Hemantarajan A (ed) Advances in plant physiology, vol 1. Scientific Publishers, Jodhpur, pp 103–122
Swennen R, Vuylsteke D (1990) Aspects of plantain breeding at IITA. In: Fullerton RA, Stover RH (eds) Sigatoka leaf spot diseases of bananas. INIBAP, Montpellier, pp 252–266
Tang CY, Hwang SC (1994) Musa mutation breeding in Taiwan. In: Jones DR (ed) The improvement and testing of Musa: a global partnership-proceedings of the first global conference of the International Musa Testing Program at FHIA, Honduras, 27–30 Apr 1994, pp 219–227
Tenkouano A, Swennen RL (2004) Progress in breeding and delivering improved plantain and banana to African farmers, horticultural science focus. Chronica Hortic 44(1):9–15
Tenkouano AJH, Crouch HK, Crouch J et al (1999) Comparison of DNA marker and pedigree-based methods of genetic analysis of plantain and banana (Musa spp.) clones. I. Estimation of genetic relationships. Theor Appl Genet 98:62–68
Tenkouano A, Vuylsteke D, Okoro J et al (2003) Diploid banana hybrids TMB2x5105-1 and TMB2x9128-3 with good combining ability, resistance to black Sigatoka and nematodes. HortSci 38:468–472
Tenkouano A, Swennen R (2004) Plantains and banana: progress in breeding and delivering improved plantain and banana to African farmers. Chronica Horticulturae, 44:9–15
Teo CH, Tan SH, Ho CL et al (2005) Genome constitution and classification using retrotransposon-based markers in the orphan crop banana. J Plant Biol 48:96–105
Tomekpe K, Jenny C, Escalant JV (2004) A review of conventional improvement strategies for Musa. Info 13(2):2–6
Toruan-Mathius N, Haris N (1999) Induction of genetic variation of banana cv. Nangka by gamma Co-60 irradiation and fusaric acids. Menara Perkebunan 67:13–22
Tourmente S, Deragon JM, Lafl euriel J et al (1994) Characterization of minisatellites in Arabidopsis thaliana with sequence similarity to the human minisatellite core sequence. Nucl Acid Res 22: 3317–3321
Tripathi L, Ntui VO, Tripathi JN (2022) Control of bacterial diseases of banana using CRISPR/Cas-based gene editing. Int J Mol Sci 23, no. 7:3619. https://doi.org/10.3390/ijms23073619
Turner DW (1970) Leaf tearing in bananas. Agric Gaz NSW 81:372–373
Ude G, Pillay M, Nwakanma D, Tenkouano A (2002) Analysis of genetic diversity and sectional relationships in Musa using AFLP markers. Theor Appl Genet 104:1239–1245
Ude G, Pillay M, Ogundiwin E, Tenkouano A (2003) Genetic diversity in an African plantain core collection using AFLP and RAPD markers. Theor Appl Genet 107:248–255
Udensi OU, Ita EE, Ikpme EV et al (2017) Sequence analysis of maturase K (matk): a chloroplast-encoding gene in some selected pulses. Glob JPAS 23:213–230
Uma S, Sathiamoorthy S (2007) Status of genetic diversity in banana in India. In: Singh HP, Uma S (eds) Banana-technological advancements. Association for the Improvement in Production and Utilization of Banana (AIPUB), National Research Centre for Banana, Trichy, pp 71–84
Uma S, Sathiamoorthy S, Dorai P (2005a) Banana-Indian genetic resources and catalogue. National Research Centre for Banana, Trichy, p 268
Uma S, Sathiamoorthy S, Saraswathi MS et al (2005b) Evaluation and utilization of introduced Musa Germplasm in India. Indian J Plant Genet Resour 18(1):79–81
Vadivel K, Vijayakumar A, Solomon S, Santhoshkumar R (2014) A review paper on design and fabrication of banana fiber extraction machine and evaluation of banana fiber properties. Int J Adv Res Electr Electron Instrum Eng:63–63
Valarik M, Bartos J, Kovarova P et al (2004) High-resolution FISH on super-stretched flow-sorted plant chromosomes. Plant J 37:940–950
Valmayor RV, Jamaluddin SH, Silayoi B et al (2000) Banana cultivar names and synonyms in Southeast Asia. International Network for the Improvement of Banana and Plantain—Asia and the Pacific Office, Los Ban˜os, Laguna, p 24

Varshney RK, Salem KFM, Röder MS et al (2008) Assessment and comparison of genetic diversity in barley germplasm collection using gSSR, eSSR and SNP markers. Plant Genet Res 6:167–174
Vos P, Hogers R, Bleeker M et al (1995) AFLP: a new technique for DNA fingerprinting. Nucl Acid Res 23:4407–4414
Vuylsteke D, Ortiz R (1993) Diploid plantains with black Sigatoka resistance. Mus Africa 2:1–2
Vuylsteke D, Ortiz R, Swennen R (1993) Genetic improvement of plantain and banana at IITA. Info 2(1):10–12
Vuylsteke M, Mank R, Antonise R et al (1999) Two high-density AFLP linkage maps of *Zea mays* L.: analysis of distribution of AFLP markers. Theor Appl Genet 99:921–935
Wang Z, Lin Z, Pan K (1993) Cytogenetical studies in Musa (Eumusa). In current banana research and development in China (SCAU). South China Agricultural. Univ, pp 29–43
Wang C-JR, Harper L, Zacheus Cande W (2006) High-resolution single-copy gene fluorescence in situ hybridization and its use in the construction of a cytogenetic map of maize chromosome 9. Plant Cell 18:529–544
Wang JY, Zheng LS, Huang BZ et al (2009a) Development, characterization, and variability analysis of microsatellites from a commercial cultivar of Musa acuminata. Genet Resour Crop Evol 57:553–563
Wang K, Zaijie Y, Shu C et al (2009b) Higher axial-resolution and sensitivity pachytene fluorescence in situ hybridization protocol in tetraploid cotton. Chromosom Res 17:1041–1050
Wang J et al (2025) China National Gene Bank (CNGB). In: Al-Khayri JM, Salem KFM, Jain SM (eds) Plant gene banks. Springer, Singapore. https://doi.org/10.1007/978-981-99-4236-7_103-1
Waraczewski, Robert, Sołowiej BG (2024) Potential valorization of banana production waste in developing countries: bio-engineering aspects. Fibers 12, no. 9:72. https://doi.org/10.3390/fib12090072
Wardlaw CW (1972) Banana diseases including plantains and abaca. Longman, London
Weising K, Atkinson RG, Gardner RC (1995) Genomic fingerprinting by microsatellite-primed PCR: a critical evaluation. Genome Res 4:249–255
Wenzl P, Raman H, Wang J et al (2007) A DArT platform for quantitative bulked segregant analysis. BMC Genomics 8:196–196
Wong C, Kiew R, Loh JP et al (2001) Genetic diversity of the wild banana Musa acuminata Colla in Malaysia as evidenced by AFLP. Ann Bot 88:1017–1025
Xiong LZ, Xu CG, Saghai Maroof MA, Zhang Q (1999) Patterns of cytosine methylation in an elite rice hybrid and its parental lines, detected by a methylation-sensitive amplification polymorphism technique. Mol Gen Genet 261:439–446
Yano-Melo AM, Saggin OJ, Maia LC (2003) Tolerance of mycorrhized banana (Musa sp. cv. Pacovan) plantlets to saline stress. Agric Ecosyst Environ 95:343–348
Yu GX, Wise RP (2000) An anchored AFLP- and retrotransposon-based map of diploid Avena. Genome 43:736–749
Yuan Q, Liang F, Hsiao J et al (2000) Anchoring of rice BAC clones to the rice genetic map in silico. Nucl Acids Res 28:3636–3641
Yuan Q-J, Bin Z, Dan J et al (2015) Identification of species and Materia medica within Angelica L. (Umbelliferae) based on phylogeny inferred from DNA barcodes. Mol. Ecol Res 15(2):358–371
Zhou Z, Bebeli PJ, Somers DJ, Gustafson JP (1997) Direct amplification of minisatellite-region DNA with VNTR core sequences in the genus Oryza. Theor Appl Genet 95:942–949
Zoller JF, Herrmann RG, Wanner G (2004) Chromosome condensation in mitosis and meiosis of rye (*Secale cereale* L.). Cytogenet Genome Res 105:134–144

Chapter 10
Advances in Flax (*Linum usitatissimum* L.) Breeding and Biotechnology Strategies

Yu Fu Wang, Qing Hua Kang, Xuan Zhu, Zheng Zhang, Hua Jiao Qiu, Dan Dan Yao, Xiao Yan Chen, Yadika Shouhereti, Cai Sheng Qiu, Xi Xia Song, Rakhiya Yelnazarky, Aliya Baitelenova, and Hazaisi Hanipa

Abstract Flax (*Linum usitatissimum* L.) is an annual herbaceous plant that self-pollination. It is an important crop for both fiber and oil production, with a history of cultivation spanning 6000–8000 years. Flax is grown in various regions around the world, including Europe, Asia, North America, and Africa. In China, flax has been grown for over 5000 years, mainly for seeds or oil. Flax has only a history of nearly 90 years as a fiber for large-scale cultivation. Flax dress is not only green, and natural, but also with a millennium of cultural atmosphere, so the application of flax fiber is ageless and everlastingly new, but the use of flax fiber is not only for clothing. Due to the good mechanical properties of flax fiber composites, which can replace glass fiber, carbon fiber and other fiber composites, the study and application of flax fiber composites have developed vigorously in recent years. Additionally, flax seed oil is recognized as a functional food with high nutritional value, boasting antioxidant, anti-inflammatory, anti-cancer, antihypertensive and cardiovascular disease prevention properties. With the increasing use of flax and the challenges

Yu Fu Wang and Qing Hua Kang contributed equally with all other contributors.

Y. F. Wang · H. J. Qiu · C. S. Qiu (✉)
Institute of Bast Fiber Crops, Chinese Academy of Agricultural Sciences, Changsha, Hunan, China
e-mail: qiuhuajiao@caas.cn; qiucaisheng@caas.cn

Q. H. Kang · D. D. Yao · X. X. Song
Institute of Industrial Crops, Heilongjiang Academy of Agricultural Sciences, Harbin, Heilongjiang, China

X. Zhu · X. Y. Chen
Dali Bai Autonomous Prefecture Agricultural Science Extension Research Institute, Dali, Yunnan, China

Z. Zhang · Y. Shouhereti · H. Hanipa
Institute of Agricultural Science of Ili Prefecture, Ili, Xinjiang, China

R. Yelnazarky · A. Baitelenova
S.Seifullin Kazakh Agrotechnical University, Astana, Republic of Kazakhstan

K. F. M. Salem et al. (eds.), *Breeding and Biotechnology of Grass and Bast Fiber Crops*, Advances in Plant Breeding Strategies 12,
https://doi.org/10.1007/978-3-032-00403-1_10

posed by climate change, it is necessary to adjust breeding strategies and goals accordingly. This chapter provides a summary of the current production status, market trends, price fluctuations, breeding technologies, and advancements in new variety selection for flax. It also discusses breeding objectives and strategies, serving as a valuable reference for future flax breeding efforts.

Keywords Anther culture · Breeding · Flax (*Linum usitatissimum* L.) · Germplasm · Haploid · Male-sterility · Tissue culture

10.1 Introduction

10.1.1 Botanical Classification and Distribution

Cultivated flax (*Linum usitatissimum* L.), abbreviated as flax, is an annual or autumn-sown perennial herbaceous plant of the flax family. It has a chromosome count of 2n = 30. The flax stem is upright, with a lignified base that is hairless, and the fibers in the phloem are strong, tough, and elastic. The leaves are alternate, with a linear lanceolate blade. The flowers are solitary and can be found on the top or upper leaf axils of branches, in cyme inflorescences. They are typically blue or white, with a small portion of pink or purple. The capsule is spherical, turning brownish-yellow when dried. The seed coat can be brown or milky yellow, and the flowering period is from June to August, with the fruiting period lasting from July to October.

Flax can be categorized into fiber flax, fiber and seed dual-purpose flax, and oil flax based on their different uses (Fig. 10.1). Oil flax is commonly known as linseed. In the past decade (2013–2022), fiber flax has been mainly distributed in France, Belgium, Belarus, Russia, China, the UK, the Netherlands, Egypt, Chile, and Argentina (Fig. 10.2). Therespective raw or retted flax yields for these countries are provided in Table 10.1.

10.1.2 Fiber and Textile Properties

The natural spindle-shaped structure of flax fiber (Fig. 10.3) and its unique pectin oblique structure gives it excellent characteristics such as moisture absorption, heat dissipation, breathability and functions like being anti-static, anti-ultraviolet light, flame retardant and anti-allergy properties. Wearing linen fabric in the summer naturally makes people cool and comfortable, not sticky, or stuffy, earning flax fiber the nickname "breathing fiber", and "fiber queen". Linen clothing also has the reputation of being a" natural air conditioner. Due to its green, natural qualities and

Fig. 10.1 Flax (*Linum usitatissimum* L.). (1) Fiber flax, (2) Fiber seed and flax and (3) Oil flax. (Source: Photo taken by Wang YF)

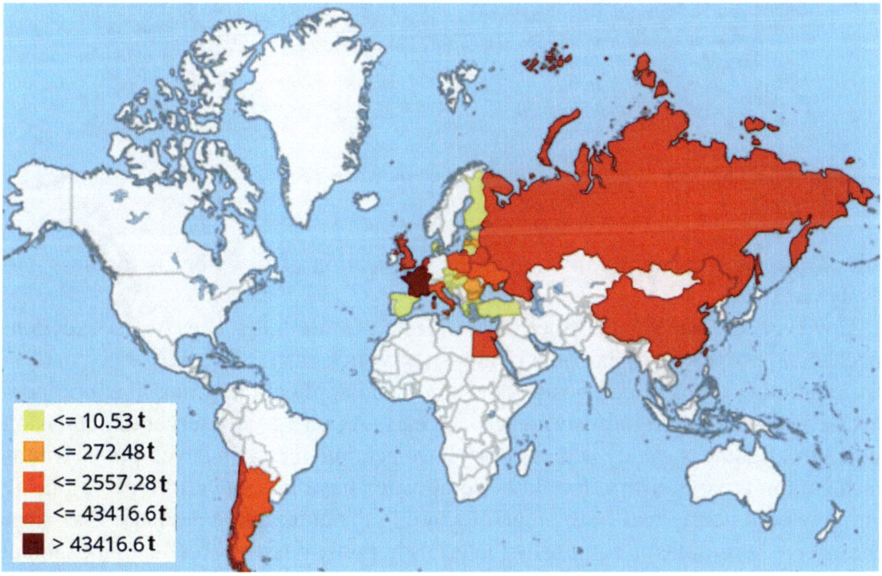

Fig. 10.2 Production quantities of Flax, raw or retted by country (Average 2013–2022). (Source: FAOSTAT)

Table 10.1 The raw or retted flax yield top 10 countries (Average 2013–2022)

Number	Country	Yield(t)
1	France	629,139.67
2	Belgium	79,349.30
3	Belarus	43,416.60
4	Russia	36,596.85
5	China	21,688.51
6	UK	14,558.85
7	The Netherlands	11,064.60
8	Egypt	7846.72
9	Chile	3068.95
10	Argentina	2557.28

Source: The data comes from FAOSTAT

Fig. 10.3 Flax fiber and yarn. (Source: Photo taken by Wang YF)

cultural heritage spanning millennia. The application of flax fiber is timeless and always fresh.

However, the use of flax fiber is not limited to clothing. Due to the excellent mechanical properties of flax fiber composites, which can replace glass fiber, carbon fiber and other types of composites, the study and application of flax fiber composites have developed vigorously in recent years. As energy conservation, emission reduction, lightness, safety and comfort, have become the main development trends in the automotive industry, flax fiber composites have been able to replace plastic and now hold nearly half of the market share in the automotive industry. Flax fiber reinforced thermoplastic composite materials not only possess excellent mechanical properties and are cost-effective, but they are also renewable, biodegradable, and environmentally friendly. Theproduction process emits less CO_2, aligning with the principle of energy conservation and emission reduction.

Moreover, flax fiber reinforced composite materials have a low density, high specific stiffness and strength, good molding process performance, customizable material properties, good fatigue resistance, excellent vibration reduction capabilities, and strong thermal stability, leading to rapid development in the field. Flax fiber composite materials produced through various processes have found widespread application in automobile, aircraft, construction, geoengineering, transportation and other industries.

10.1.3 Seeds and Its Utilization Value

In addition to its fiber use, flax seed also has a wide range of other uses. The oil content of flax seed is about 30–45%, with the content of α-linolenic acid in flax seed oil reaching up to 62%. Thisα-linolenic acid is a necessary fatty acid that cannot be synthesized by the human body, playing an important role in improving the composition of people's dietary fatty acids and maintaining human health. Flax seed oil is also rich in vitamin E, flavonoids and other active substances, making it a high-end edible oil in China and abroad. Flax seed is also rich in flax seed pectin, protein, dietary fiber, antioxidant peptides, lignan and other functional active ingredients. Therefore, flax seed is considered a functional food with high nutritional value, offering antioxidant, anti-inflammatory, anti-cancer, antihypertensive and cardiovascular disease prevention benefits. Due to its many excellent characteristics, flax fiber is in short supply worldwide, and its price the price of flax fiber has been continuously rising in recent years. In order to meet market demand, there is an urgent need to breed more high-yield and high-quality flax varieties.

Flax is a high-quality multi-purpose industrial crop used not only for fiber and edible oil but also for health care food, medicine and composites. In recent years, new uses of flax have gained widespread attention from the global industry. For example: flax lignan has great potential in the pharmaceutical industry. Many studies have proven its effectiveness as an antioxidant, anti-tumor, anti-cardiovascular disease, and in the prevention of osteoporosis and diabetes (Li et al. 2024b). Therefore, flax is an important multi-purpose industrial crop and breeding efforts should focus on these new uses.

10.2 Development of Flax Production

Fiber flax requires moderate temperatures (18–20 °C) and moderate insulation (cloud cover) to ensure optimum yields (Agosti et al. 2005). Linseed as a crop of continental climate is more resistant to drought, sunny and warm weather conditions (Casa et al. 1999). Fiber flax is mainly distributed in more than 20 countries,

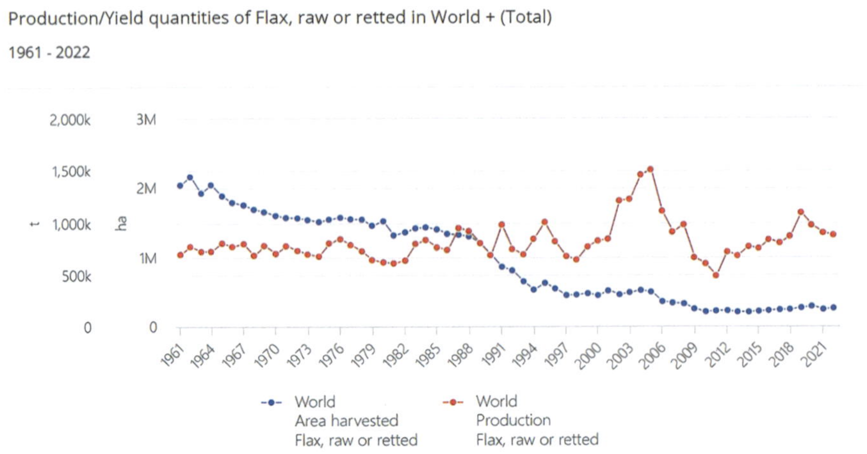

Fig. 10.4 Global fiber flax planting area and yield from 1961 to 2022. (Source: FAOSTAT)

including the former Soviet Union, France, China, Belgium, Belarus, Russia, Poland, the Netherlands, the Czech Republic, the United Kingdom, Ukraine, Bulgaria, Egypt, Estonia, Finland, Germany, Latvia, Portugal, Sweden and Ireland. According to FAO data, from 1961 to 2022, the average annual planting area of fiber flax worldwide was 973,434 ha. The planting area has decreased from 2,041,125 ha in 1961 to 256,540.5 ha in 2022 (Fig. 10.4). The planting area has significantly decreased, but due to the development of new varieties, improvement of planting techniques, and progress in production levels, the yield of flax, whether raw or retted in the world has not significantly decreased, with a yield of 875,994.92 t in 2022, an increase of 179,415.92 t compared to 1961. Although the planting area in 2022 is only 12.6% of what it was in 1961, the yield is 125.8% higher. The yield per hectare has also increased significantly, from 341.3 kg in 1961 to 3414.6 kg in 2022. The progress in breeding technology and the adoption of new varieties have been key factors in the increased yield of flax. The rise in total flax production indicates a growing demand for natural, environmentally friendly products and the expansion of flax multiple uses in recent years.

In the new century (2001–2022), the top ten countries with an average annual planting area of flax are as follows: France with 83,390.2 ha and 536,788.22 t, Russia with 58,715.8 ha, and 42,897.11 t, Belarus with 57,915.6 ha, and 43,621.32 t, China with 48,246.6 ha, and 185,643.55 t, Belgium with 14,715.3 ha, and 50,691.05 t, the United Kingdom with 11,686.4 ha, and 17,017.38 t, Egypt with 9245.1 ha, and 8421.93 t, Ukraine with 7908.9 ha, and 3941.73 t, the Netherlands with 2823.3 ha, and 15,329.5 t, and the Czech Republic with 2452.8 ha, and 6744.46 t. Fiber flax cultivation in these countries occupies an important position in the world (FAOSTAT 2024). France's flax fiber production accounts for over 50% of the world's total, representing 58.9%. This aligns with its high scientific research and production levels. The production and research level of flax in countries such as

France, the Netherlands, and Belgium in Western Europe are at the forefront of the world, with their original stem yield reaching 6000–7500 kg/ha and fiber yield reaching 1200–1800 kg/ha. The production of flax in Europe is mechanized throughout the entire process from sowing to fiber processing. Western Europe uses self-propelled flax pulling machines for harvesting, with a working efficiency of about 1.5 ha/hour. Due to large-scale and mechanized cultivation, coupled with government subsidies, the flax planting area in Western Europe remains relatively stable. For example, the fiber flax planting area in France has been steadily increasing, benefiting from government support, which has kept the industry in a sustainable development state. The fiber flax planting area in Russia is 35,483 ha, which has decreased by more than 80% compared to its peak in 1992. Some Eastern European countries, like Poland, the Czech Republic, and Lithuania, have essentially ceased production.

Oil flax is more widely planted in the world, but the planting area and yield have been declining from 1961 to 2007. In 1964, the maximum planting area reached 8,049,735 ha, with a flax seed yield of 3,276,909 t. By 2007, it had decreased to 1,977,659 ha, with a flax seed yield of 1,658,238 t (Fig. 10.5). The main reason for the decrease in the planting area is that during this period, flax seed was primarily used for edible oil, but its yield is relatively low, only 600 kg/ha, so the planting enthusiasm is not high. However, in the past decade or so, people's awareness of health has gradually increased, and the health functions of flax seeds have been recognized. The consumption of flax seed oil and related products has gradually increased. By 2022, the planting area of oil flax had reached 4,533,187 ha, with flax seed production reaching 397,3931.78 t, surpassing the production in 1964 and more than double the production. The yield has also significantly increased, from 401.6 kg/ha in 1961 to 876.6 kg/ha in 2022 (FAOSTAT 2024). This increase can be attributed to new varieties, planting techniques, and advancements in productivity.

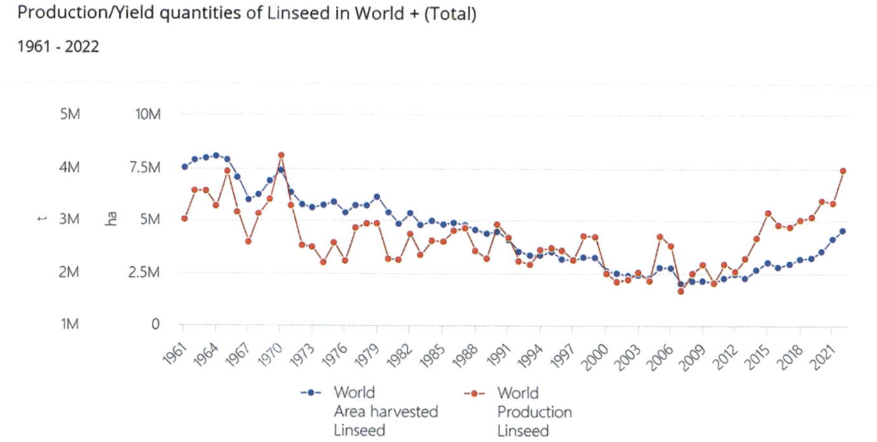

Fig. 10.5 Global oil flax planting area and yield from 1961 to 2021. (Source: FAOSTAT)

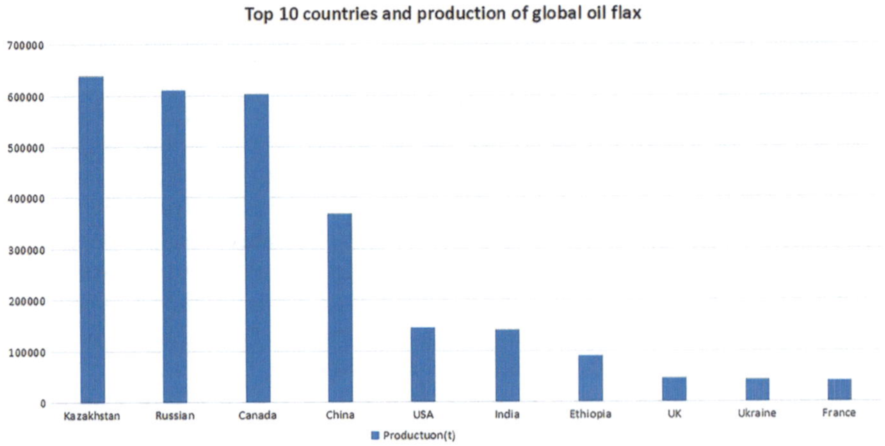

Fig. 10.6 Top 10 countries and production of global oil flax from 2012 to 2021. (Source: The data comes from FAOSTAT)

In the past 10 years, countries with a relatively large cultivation area of oil flax include Kazakhstan, Russia, Canada, China, the United States, India, Ethiopia, the United Kingdom, Ukraine, and France. Their flax seed yields are as follows: 638,394.62; 611,920.29; 604,740.8; 368,347.53; 145,392.0; 140,947; 90,571.07; 46,411.4; 43,197 and 42,062.92 tons, respectively. The yields per hectare are 745, 820, 1425, 1293, 1200, 536, 1059, 1847, 1056and 1848 kg/ha (Fig. 10.6).

Due to the recent increase in demand for flax fiber and the decrease in fiber production, the price of flax fiber has skyrocketed. Long fiber flax prices have risen from 3.94 US$/kg in January 2022 to 10.07 USD/kg in March 2024. The current price is 2.56 times that of January 2022. Short fiber prices have also increased from 1.61 US$/kg in January 2022 to 3.27 US$/kg in March 2024. The current price is 2.03 times higher than that of January 2022. This means that the price of flax fiber has more than doubled in just over two years (Fig. 10.7).

10.3 The impact of Market and Environmental Changes on Flax Breeding Strategies

10.3.1 Changes in the Global Flax Fiber Market

Clothing: In recent years, with the improvement of people's living standards and the awakening of health awareness, the pursuit of fabrics has shifted towards pure nature. People now require fabrics to have characteristics such as breathability, moisture absorption, nonstick, wrinkle resistance, antibacterial, anti-mold, anti-UV, anti-radiation, and anti-static properties. They also seek fabrics that are easy to maintain and offer a comfortable skin feel. The unity of external beauty and internal

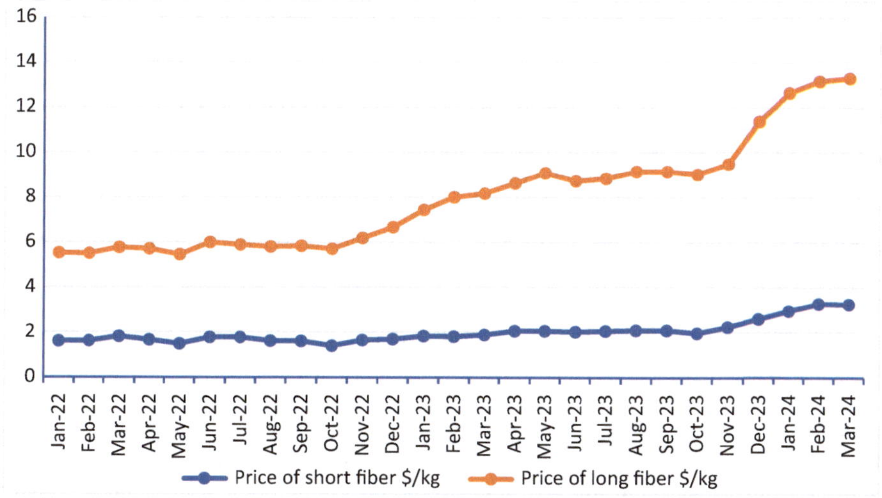

Fig. 10.7 The prices of long and short flax fiber from January 2022 to March 2024 (US$/kg). (Source: The data comes from the China Bast and Leaf Fiber Textile Association Website)

health functions has become a new trend in clothing consumption. Linen clothing, with its natural texture and light, transparent fabric caters to people's needs with a strong artistic atmosphere and an unrestrained artistic conception. With the increasing preference for natural fiber fabrics and growing environmental awareness, the demand for flax fiber in the fashion clothing industry is gradually increasing.

Home decoration: Flax fiber is commonly used for home decoration, such as curtains, tablecloths, carpets, bedding, cool mats and sofa covers. The natural texture and minimalist style of flax fiber align with modern home design and are favored by many consumers.

Industrial use: Flax fiber has high wear resistance and strength, making it suitable for situations requiring durability and breathability. It can also be used in industrial fields, such as making rope, tarpaulin, work clothes, and protective clothing. Creative handicrafts: The texture and color of flax fiber are suitable for making handicrafts, such as oil painting canvas, fashion bags, fabric dolls, and embroidered works. These handicrafts have an original and natural style and are deeply loved by consumers. Overall, flax fiber has a wide demand in the fields of fashion, home decoration, industrial use, and handicrafts. In the future, as consumers pursue natural materials and environmentally friendly products, the demand for flax fiber may further increase. This increased demand requires the development of new flax varieties with higher yields.

The above requirements are all extensions of traditional applications, and currently, the biggest emerging use of flax fiber is in the application of composite materials. The flax fiber composite is more eco-friendly than corresponding petroleum-based fiber composites (More 2022).

Flax fiber composite materials have gradually replaced glass and other fiber composite materials due to their excellent mechanical properties, showing huge market potential in the construction and automotive industries in recent years (Ahmad-Rasyid et al. 2016). The application of low-carbon and safe woven composite of flax in automobiles, aviation, and sports goods is on the rise.

With increasing environmental concerns, there is an urgent need for the development of light-weight, cost-effective and sustainable materials as alternatives to conventional materials. The use of bio-fibers to replace glass fibers as reinforcement materials in polymer composites has gained popularity for various engineering applications.

At present, the application and development of flax fiber composite materials in the construction industry are rapidly growing. In addition to industries such as automotive parts, composite pipes made of flax fiber are lighter in concrete, which can reduce labor and construction time. In the fields of railways and bridges, flax fiber composite materials have gradually become a new type of building material that replaces wood. Flax fiber composite materials also have great consumption potential in other emerging markets, such as tiles, flowerpots, pallets, disposable daily necessities, waste bins, buckets, and ocean docks. As a new type of biodegradable material, it has better corrosion resistance than wood and less pollution than plastic, making it highly competitive in the market (Jiang et al. 2019). The application of flax in composite materials will greatly promote the market demand for flax fibers. Therefore, new flax varieties with higher yields urgently need to be bred.

Li et al. (2022) studied the lightweight and environmentally friendly properties of automotive interior materials. They selected oil flax fiber as the reinforcement and epoxy resin as the matrix and used vacuum assisted resin transfer molding to prepare oil flax fiber reinforced epoxy resin composite materials for automotive interiors. The results showed that when using 30% oil flax fiber and 2.5% silane coupling agent, the relevant mechanical indicators of the oil flax fiber reinforced epoxy resin composite reached the best, with tensile strength, bending strength, and compressive strength of 54.32, 127.19, and 211.95 MPa, respectively. The optimal process parameters were determined through experiments to obtain high-performance boards (Li et al. 2022). However, the fiber yield of oil flax is very low, and there may be efficiency issues in processing oil flax fiber. Oil flax fiber can be well applied in composite materials. The fiber and seed dual-purpose flax fibers can better meet the requirements of composite materials. We believe that developing fiber and seed dual-purpose flax is a good direction. Planting fiber and seed dual-purpose flax can not only harvest a large amount of fiber but also increase planting income by harvesting the seeds at the same time. So in the future, planting fiber and seed dual purpose flax will be an important development direction.

In addition, in recent years, due to the specialized demand for product processing, the demand for special professional varieties has gradually increased. Flax seed pectin, flax seed lignin, α-linolenic acid, and other substances have various biological activities and health benefits and are of great significance in the fields of food, health products, medicine, and more. At present, there are professional processing factories for products such as flax seed pectin, lignans, flax seed protein powder, and α-linolenic acid. Therefore, the demand for professional varieties rich in these effective ingredients is also rapidly increasing.

10.3.2 The Impact of Climate Change on Flax

Agriculture is sensitive to climate change, and once it occurs, it will have a significant impact on agricultural production, with multiple implications. On one hand, it affects the stability of agricultural development while on the other hand, it impacts the yield and quality of crops.

From the perspective of agricultural production, climate change, through changes in precipitation and temperature, can lead to changes in factors such as air, soil, water, and light. This can shorten the crop growth period in agricultural ecosystems, offsetting the effect of extending the annual growth period of crops, ultimately affecting the potential for agricultural production. Climate change causes large-scale abnormal atmospheric circulation, increasing the survival rate of crop pests, advancing the occurrence and migration period of pests, prolonging the growth cycle of pests, and increasing the risk of crop pests and diseases. Prolonged dry and rainy weather can cause soil drought in farmland, reduce soil fertility, cause surface vegetation degradation, exacerbate soil salinization and desertification, and ultimately lead to a decrease in the resilience of agricultural ecosystems. This will increase the instability of agricultural production systems, increasing the risk of crop yield reduction and ultimately affecting food security (Dai MH and Yu FW 2024).

To cope with climate change, it is possible to strengthen the cultivation of stress resistant varieties. Relevant personnel can cultivate stress resistant crop varieties that can adapt to climate change, maintaining high yields and stable quality under extreme weather conditions. Stress resistant varieties have good adaptability to extreme weather events such as drought, high temperatures, and floods, helping to reduce the negative impact of climate change on crops (Zhang 2023).

Flax is a plant that is cultivated in spring in an oceanic climate. Therefore, Normandy is a major global producer of flax fiber, and its flax fiber is mainly exported to China for textile manufacturing. It requires regular watering and excessive heat should be avoided. The temperature rise will lead to a reduction in the length of the vegetative cycle of spring-sown flax fiber. Flax is also at risk of lodging during heavy rainfall (Beauvais et al. 2022). Adopting a comprehensive approach involving various bioinformatics, breeding, and genetic engineering technologies will greatly facilitate the introduction of climate-change-adaptive traits into flax varieties, thus maintaining their productivity in the context of global climate change (Shivaraj et al. 2019).

10.3.3 Flax Breeding Strategies to Cope with Changes in the Market and Natural Environment

Flax breeding has observed lots of achievements in the world. The new breeding methods were developed and improved, specifically the studies on molecular biology related breeding techniques have received increasing attention from scientific and technical workers. Lots of fiber flax and oil flax variations were bred. As the new varieties yield, seeds yield, fiber content and oil content of flax were increased, the fiber quality and resistance to disease and lodging were improved too, which has promoted the development of the flax industry.

However, the market for flax products, the areas where flax is grown, and the ecological environment are all changing. In order to cope with changes in the market and natural environment, corresponding strategies should be adopted in flax breeding.

10.3.3.1 Flax Breeding Objectives

The breeding goal of flax is to cultivate improved varieties and adjust them according to the needs of farmers, processors, and consumers. Different users have diverse requirements for the quality of flax. Integrating all the desired characteristics into one genotype is a challenging task for breeders. The ideal fiber flax variety should possess such characteristics as high fiber yield, easy separation from the stem and bast, quick processing, low impurity content, being colorless, and having stable properties across different locations and seasons.

Flax varieties should exhibit resistance to both biotic and abiotic stresses while achieving high yield and good quality. The fiber flax breeding program aims to cultivate varieties that can resist biotic stresses in the environment and enhance their resistance to abiotic stresses, including extreme temperatures, humidity, and saline-alkali soils (Obert et al. 2009). The period of insufficient water will damage the quality and quantity of the yield. Excessive moisture, especially when the sowing density and soil nitrogen content are high, may lead to lodging. In terms of resistance to biotic stresses, priority should be given to cultivating flax varieties that are resistant to fungal strains such as *Fusarium oxysporum*, *Melampsora lini*, and *Erysiphe polygoni* (Jankauskien and Gruzdevien 2008; Kozłowski et al. 2000).

The important breeding goals for fiber flax include yield, disease resistance, lodging resistance, fiber content, resistance to Fusarium, and good adaptation to local soil and climate conditions. The breeding goals for oil flax focus on stabilizing yield in different environments, enhancing oil production and quality, improving omega-3 fatty acid content, obtaining durable resistance to wilt and rust diseases (Rahman and Hoque 2023), improving lodging resistance and other agronomic traits, and adjusting crop phenology according to regional climatic limitations.

In conclusion, several main breeding goals can be identified: breeding for quantitative traits (increasing fiber or seed yield, including improving fiber content of the stem and oil content of the seeds), breeding for quality traits (of seeds or fiber, including improving omega-3 fatty acid content), and breeding for resistance (lodging resistance, drought resistance, saline soil resistance, disease and pest resistance).

10.3.3.2 Breeding and Application of Fiber and Seed Dual Purpose Varieties

Due to the excellent mechanical properties of flax fiber composite materials, they can replace glass fiber, carbon fiber, and other fiber composite materials. In recent years, the study and application of flax fiber composite materials have flourished, with energy conservation, emission reduction, light weighting, safety and comfort becoming the main development trends in the automotive industry. Flax fiber composite materials can replace plastics and have occupied nearly half of the market share in the automotive industry. Flax fiber reinforced thermoplastic composites not only have excellent mechanical properties and low cost, but also have renewable, biodegradable, and environmentally friendly properties, resulting in low carbon dioxide emissions during the production process. The requirements for fiber quality in composite materials are different from those of fine spinning, the fiber of the oil flax can be used for composite. This provides development space for the cultivation of fiber and seed dual-purpose flax. In addition, it will have a positive impact on the traditional processing of flax seed oil, flax seed pectin, flax seed powder, protein powder, flax seed oil powder, food and lignin. It is worth noting that there is no significant difference in the total phenolic and flavonoid content as well as other activities between fiber flax and oil flax. Even the bioactive substances of fiber flax seed are better than those of oil flax seed. Fiber flax seeds would be valuable candidates as functional products and dietary supplement production due to their higher biological values as well as linseeds. This trend of production development has created unprecedented opportunities for the application of flax varieties with the dual purpose of fiber and seeds (Wang et al. 2017). Therefore, in terms of variety types, while strengthening the original goal of high-quality and high-yield breeding, in order to cope with changes in market demand, it is necessary to accelerate the breeding and application of fiber and seed dual-purpose flax varieties.

10.3.3.3 Breeding of Specialized Flax Varieties with Special Functions

Flax seed contains many beneficial active ingredients for the human body, such as lignans, α-linoleic acid, pectin, flavonoids, and cyclic peptides. These bioactive substances are being produced and sold in the market. With the increasing awareness of people's health, the demand for these products will continue to rise. Therefore, it is necessary to breed specialized flax varieties with unique functions to make significant contributions to human health.

10.3.3.4 Breeding of Flax Varieties Adaptable to Abiotic Stresses

Due to climate change, land salinization, desertification and drought are becoming increasingly serious. In order to adapt to drought, salinization, rainy weather, wind disasters and other natural environments and disasters, as well as to ensure food security, flax planting in saline-alkali land, mudflat land and arid and barren land should be developed to expand planting space and promote the improvement and utilization of saline-alkali land, mud beach land and winter fallow land. By breeding new varieties, their adaptability and salt alkali tolerance can be improved. Due to the influence of rain and extreme weather, flax is prone to lodging, diseases, etc. Therefore, it is necessary to strengthen the cultivation of flax varieties for lodging resistance and disease resistance.

10.3.3.5 Cultivating Winter Flax Varieties

Europe is actively developing winter flax, experiencing strong growth in this area. The winter flax variety, planted in late autumn, matured about one month earlier than traditional spring flax. Its growth cycle is longer, making it more resistant to intense weather events such as droughts, heavy rain and storms, which have been observed during spring in recent years. China also cultivated 70,000–80,000 ha of winter flax with high yields at the end of the last century. In the future, as the demand for flax fiber continues to increase, better winter flax varieties should be selected and promoted to meet market demand.

10.3.3.6 Strengthening Research on Practical New Technologies in Molecular Biology

In recent years, molecular biology research on flax has received unprecedented attention, leading to a large number of studies being conducted. However, there is a lack of technologies that specifically focus on application, and the research objectives are not concentrated enough, with weak practicality. The support of molecular biology in flax breeding is not strong enough. Therefore, it is necessary to strengthen in-depth research on the molecular biology of flax, as new molecular biology technologies are rapidly applied in the breeding process of flax.

In short, in terms of adapting to changes in flax breeding, experts from multiple fields collaborate to breed more high-quality, high-yielding, lodging resistant, and disease resistant flax varieties for fiber flax, oil flax, and fiber and seed dual-purpose flax, as well as specialized flax varieties with low pectin for easy degumming, low lignin, and high fineness for soft products, high bioactive ingredient content, high strength, and rough surface for composite materials.

10.4 The Flax Germplasm

Plant germplasm resources are the foundation of plant breeding, biotechnology, and the development of all biological sciences. They are the wealth that humans rely on for survival. Research on germplasm resources should not only serve current agricultural production and future human survival but also focus on saving and maintaining the biological resources that humans rely on for survival. This research should also coordinate with economic, social, and cultural development to achieve the goal of sustainable resource utilization. The breakthrough in the selection and breeding of new crop varieties also depends on mastering and utilizing a large number of germplasm resources.

10.4.1 The Origin and Evolution of Flax

The flax (*Linumusitatissimum* L.) is an important fiber and oil crop. Flax has been utilized for over 30,000 years ago in Georgia (Kvavadze et al. 2009), making it likely the oldest textile fiber known to man as *Linum* species belongs to the oldest arable crops. The origin of flax is uncertain. According to Vavilov, the origin of flax is thought to be in four different regions: Central-Asian, Fore-Asian, Mediterranean and Abyssinian. Some authors believe that *L. bienne* is the progenitor of small seeded flax, originating from Kurdistan and Iran, while others consider *L. angustifolium,* with high oil content and seed weight, as the progenitor, originating from the Mediterranean region (Vromans 2006). The flax have been planted in Western Europe around 5000–7000 years. In China flax has been grown for over 5000 years (Xiong 2008), mainly for seeds or oil. The cultivation of flax fiber was attempted in 1906, with processing beginning in 1936 in China. Flax fiber is one of the earliest natural plant fibers used by humans, with a history of nearly ten thousand years.

Vavilov believed that the widely distributed *Linum eurasiaticum* Vav. et Ell. subspecies exhibit significant ecological diversity. This includes creeping and semi creeping semi winter flax, especially early maturing and short plateau flax, as well as medium and late maturing grassland flax types. Planting creeping flax in the subtropical regions of the Transcaucasus and the coastal areas of Asia Minor is the primitive type of cultivated flax. Additionally, there are types with slightly creeping habits in winter, which somewhat belong to the transitional stage towards the summer type of grassland flax. Thus, from creeping flax to fiber flax, a complete evolutionary system can be formed, that is, the evolution of flax can be divided into three stages: creeping flax, semi creeping flax, and tall fiber flax.

10.4.2 Diversity of Flax Germplasm

There are various types of plants in the *Linum*, ranging in morphology from annual to perennial, upright to creeping. There are over 200 species in *Linum,* mainly distributed in temperate and subtropical mountains, with a relatively concentrated distribution in the Mediterranean region. But no country or organization has collected them all. Only 36 species were included in the Integrated Taxonomic Information System (ITIS—Report: *Linum usitatissimum*: https://www.itis.gov/servlet/SingleRpt/SingleRpt?search_topic=TSN&search_value=29201#null) (Appendix 10.3). There are about 9 wild species of flax in China, mainly distributed in the northwest, northeast, north and southwest regions. These include. *corymbulosum*-Reichb., *L. stelleroides* Planch., *L. heterosepalum* Regel, *L. usitatissimum* L., *L. perenne L.*, *L. amurense* Alef., *L. nutans* Maxim., *L. pallescens* Bunge and *L. altaicum*Ledep. Additionally, other wild flax species were introduced from other countries. Both wild and cultivated flax species exhibit rich diversity in traits such as plant height, stem thickness, maturity, flower color and grain color. However, most of the wild species have not been used in flax breeding.

10.4.3 Distribution of Flax Germplasm

The distribution of flax germplasm is influenced by factors such as biological characteristics, origin, cultivation history and utilization. There is a distinct regional distribution worldwide. Flax germplasm is mainly located in the northern hemisphere, concentrated between 40° to 65° north latitude. Geographically, it belongs to the temperate and cold temperate zones and is mainly composed of Europe and Asia, with Europe having an absolute advantage. In recent years, China has increased its efforts in collecting and preserving germplasm and has collected a large amount of flax germplasm.

World genebanks engage in ex situ preservation of about 48,000 accessions of cultivated flax (*Linum usitatissimum*) and 53 other species of the genus *Linum* L. They are in 33 genebanks in 23 countries. Most of the accessions are in Russia, China, Canada, Ethiopia, the USA, Romania, Czech, France, Germany, Argentina, Lithuania, Ukraine, Bulgaria, Netherlands, Australia, and Poland. Russia holds the largest germplasm, collection with over 6000 accessions, followed by Canada, Ethiopia, and the USA with over 3000 accessions each. Romania and the Czech Republic each preserve over 2000 accessions, while France, Germany, Argentina, Lithuania, and Ukraine hold over 1000 accessions each. However, a large proportion of the flax germplasm conserved globally is duplicated. After removing the duplicates, there are approximately 10,000 flax germplasm accessions globally.

10.4.4 Flax Germplasm of China

The study of flax germplasm conservation began in the 1950s, with the main goal being germplasm collection and identification (Wang 2008). The primary task of germplasm resources is to collect local varieties. The first Catalog of Flax Germplasm Resource of China was compiled in 1978, including 570 accessions. It comprised 408 local varieties and eminent lines from China, as well as 162 accessions from 15 overseas countries, such as Russia, Switzerland, Sweden, and Romania. These flax germplasm are an important foundation for flax breeding and have played an important role in flax breeding in China. After that, the collection of flax germplasm continues to strengthen and the quantity continues to expand. So far, about 5951 flax accessions have been stored National Bast Fiber Crops Germplasm Mid-term Gene Bank in China (Dai et al. 2017). Additionally, 3746 flax germplasm have been stored in the National Crop Genebank.

10.5 Traditional Breeding Techniques

Flax breeding programs start with the initial material, creating variation, selecting better variation, and testing characteristics. The investigation in the field and indoor testing characteristics include the sowing date, emergence date, first shaping date, fast growth date, bud emergence date, flowering date, technical maturity date (for fiber flax), physiological maturity date (for oil flax or dual purpose flax), days from emergence to harvesting (for fiber flax), days from emergence to physiological maturity (for oil flax or dual purpose flax), plant height, stem length, stem diameter, branching number, lodging degree, seedling emergence rate, field seedling preservation rate, retted stem rate, fiber rate, oil rate, stem yield, seed yield, fiber yield, and thousand seeds weight. Before DUS testing, it is necessary to increase some investigation traits. Investigate the uniformity of varieties, it is prepared for DUS testing. The investigation traits in different countries may be different. At present, the traits that need to be tested in China include the number of spots on sepals, the date of flowering, the color of the corolla, the size of the flower, the color of the anthers, the color of the filaments, the color of the style, the relative position of the petals, the size of the capsule, the presence or absence of cilia on the inner membrane of the capsule, and the color of the seed coat. During the vegetative growth period of flax phenological observations are conducted; resistance to lodging and the incidence of fungal diseases on flax etc. are assessed. In addition to the variety selection report, fiber strength or oil content testing reports, disease resistance identification reports, and non-genetically modified testing reports should also be provided during variety registration. There many breeding methods were used in flax breeding, but traditional breeding techniques of flax were introduced in this part. These include cross-breeding, systematic breeding, and induced mutation breeding. The flax breeding program is as follows (Fig. 10.8) (Wang et al. 2018).

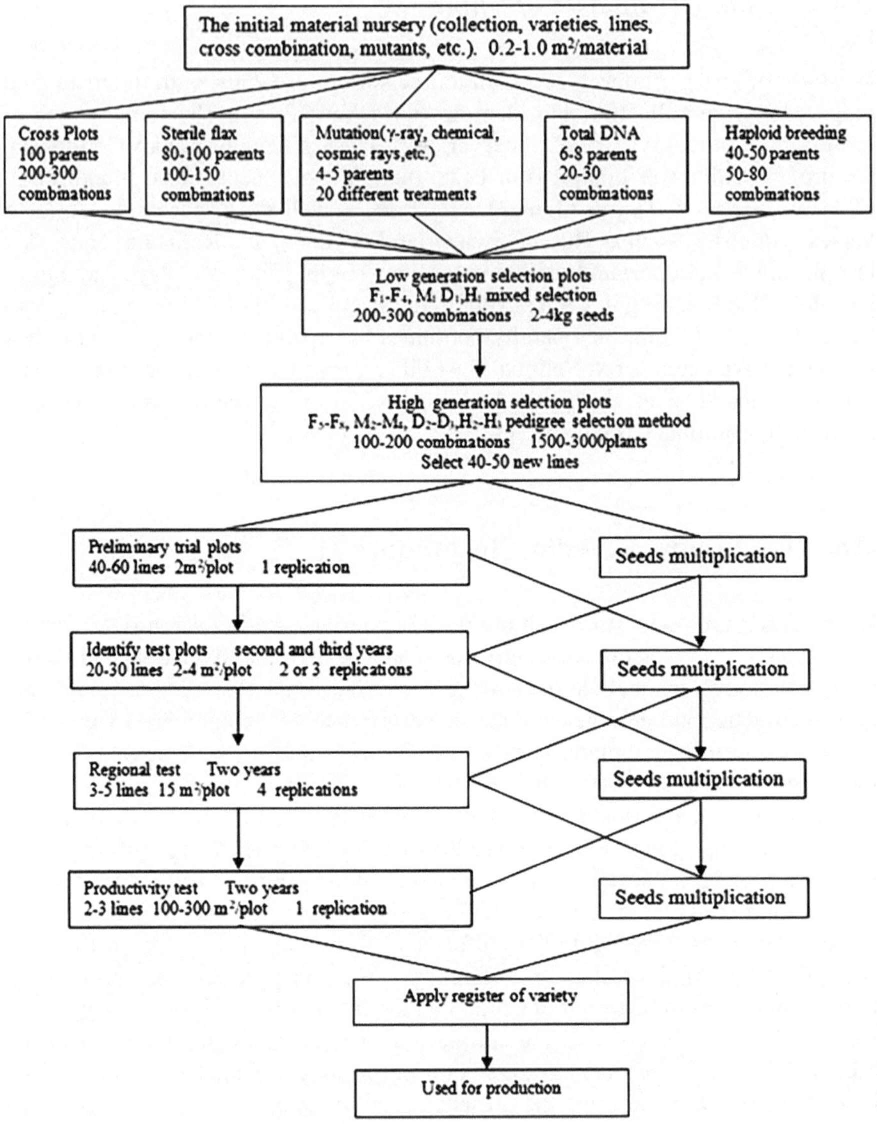

Fig. 10.8 The flax breeding program. (Source: Wang et al. (2018))

10.5.1 Cross Breeding

Flax breeding is a process that involves creating genetic variability through parent selection and crossing, followed by selecting the best recombinants through phenotypic and genomic selection, and eventually fixing desirable traits through inbreeding to develop a new cultivar (You et al. 2019).

There is relatively little theoretical research on flax cross-breeding in China, but Wang et al. (1991) suggest that the general combining ability effect of flax is relatively high, indicating the existence of gene additive effects, which make cross-breeding effective. Varieties with distant genetic relationships show better combining ability effects. Therefore, when selecting combinations, it is important to choose combinations with distant parental relationships. In cross-breeding, appropriate selection can be made in appropriate generations based on the heritability and coefficient of variation of each trait.

Cross breeding is the most commonly used method for flax breeding in China with 70 out of the 85 flax varieties registered with MARA of China being developed through this method. Therefore, cross-breeding is currently the most widely used breeding method.

10.5.2 Systematic Breeding

The crossing rate of flax is affected by the environment. Generally, it is around 3%, and it can reach 5–6% at a high level in the natural station. Therefore, through systematic breeding, new varieties that meet breeding goals can be selected from existing varieties, strains, and old local varieties. This method is simple, labor-saving, and fast, and is still used by flax breeders in China.

Out of the 85 flax varieties registered with MARA, such as Baya No.5, Huaya No.15, Jiuya No.2, Ningya No.15, and Heiya 30.9 flax varieties were developed using this method. The proportion of flax varieties bred using this method is 10.6%.

10.5.3 Mutation Breeding

10.5.3.1 Conventional Seeds Mutagenesis

The induced mutation was used in the 1960s in China. ^{60}Co–γ rays of 200–500 Gy were used, resulting in the development of flax lines with higher fiber content or early-maturation, such as Heiya No.4 and Heiya No.6. Subsequently, various types and dosages of mutagens for flax have been studied. Yu et al. (2013) used ^{60}Co-γ radiation (200Gy, 350Gy, and 500Gy) and ethyl methanesulfonate (EMS) (0.2% treatment for 30 hours and 0.3% treatment for 24 h) to induce mutations in flax seeds. The results showed that after physical and chemical mutagenesis, flax can produce mutants with various traits such as leaf color, leaf shape, plant height, branching, stem thickness, flower color, flower shape, and sterility in M_1, laying a solid foundation for the construction of flax mutant library.

In addition to γ-ray, other inducing methods, such as chemicals, and antibiotics have been explored and utilized. Wang et al. (2023) treated oil flax variety Longya 10 seeds with 0.20% EMS and screened 22 flax mutants with obvious differences in plant height, growth period, powdery mildew resistance and other characteristics from the wild-type control in the M_8 generation. Wang et al. (2022) suggested that flax seeds with different colored seed coats have different reactions to EMS. The treatment condition of brown color flax seed is 0.6% EMS, 4 h or 2 h. The treatment condition of light yellow seed is 0.3% EMS, 4 h. Space environment-induced mutation breeding is a new crop breeding technology that utilizes strong radiation of high-energy charged particles (such as protons, electrons, helium nuclei, and heavier ions), microgravity, and vacuum factors in the space environment to induce beneficial mutations in the plant. Then, new varieties and materials are selected on the ground to breed new varieties. It is an effective new approach to crop genetic improvement created by Chinese scientific and technological workers. Dried flax seeds were carried on China's 940,703 geostationary satellite, and two resistant lines were screened using tissue culture technology. They can maintain a high growth rate and efficient proline synthesis ability under stress conditions. The combination of spatial mutagenesis and tissue culture is expected to be an effective way to screen for stress resistant variant lines (Xu et al. 2000). The "Shijian 8" satellite is China's first retrievable scientific and technological experimental satellite specifically designed for space breeding research. On the afternoon of September 9, 2006, China successfully launched the "Shijian 8" breeding satellite into its designated orbit using the Long March-2C carrier rocket at the Jiuquan Satellite Launch Center. The satellite carried over 2000 crop seeds, among them there were 18 flax varieties seeds. Wang et al. (2016) bred Zhongyama 4 from one of 18 flax varieties by space environment mutagenesis carried out by satellites. The flax seeds are induced by radiation to obtain mutants, based on the phenotype characters, the flax seeds are screened for years to obtain excellent mutants. By increasing the number of irradiated seeds to obtain more variation of offspring, and then through field screening and identification, excellent new varieties for production demand can be obtained.

10.5.3.2 Molecular Analysis

TILLING provides a technology platform for targeted and large-scale screening of mutants at the molecular level, especially for traits such as quality and nutritional components that cannot be selected from plant phenotypes or are difficult to quickly evaluate stress resistance. This is particularly advantageous and improves the efficiency of mutant screening. Chantreau et al. (2013) extracted DNA from 3515 lines of flax M_2 induced by EMS and used TILLING technology to detect the genes encoding lignin synthase C3H and CAD. The results identified 79 C3H mutants and 76 CAD mutants, with an average mutation density of 1/41Kb, and almost all mutations were G/C to A/T. The use of the Sanger sequencing method for mutant screening has the characteristics of high cost and complex operation. Galindo-González et al. (2015) extracted DNA from EMS-induced flax plants and designed primers for PCR amplification of 8 target genes. The PCR products were purified and sequenced using Ion Torrent gene sequencer (PGM) to detect mutants. The detected mutants were validated using Sanger sequencing, and 16 flax mutants were identified.

10.5.3.3 Enhanced Traits and Improved Cultivars

Flax mutagenesis breeding includes physical mutagenesis and chemical mutagenesis. When natural populations lack variability or when there are strong interspecific crossing barriers preventing hybridization with related species, mutagenic agents (such as ethyl methanesulfonate, and EMS) can be used for mutagenesis breeding. The physical, chemical, and spatial environments have mutagenic effects on flax. Although achieving targeted mutagenesis according to breeding objectives is not possible, its mutation rate is high. This allows for a higher probability of mutation in a shorter period, leading to the generation of various beneficial mutations and the breeding of new varieties, effectively shortening the breeding process. Therefore, flax mutation breeding has broad prospects. Mutation breeding can broaden the mutation spectrum and generate diverse mutations. Through targeted selection, new varieties or strains that meet the needs of breeding objectives and production can be obtained. Practice has proven that mutagenesis is effective in the following aspects. Yield mutation: Heiya 4, Heiya 6, and Neixianya 1, which were bred through mutagenesis, have higher stem and seed yield than their parents, with the stem yield increased by more than 10% compared to the local control variety. Quality mutation: The quality of flax is mainly in aspects such as long fiber rate, fiber splitting degree, flexibility, and fiber strength. Several varieties bred through mutation, except for those that are better than their parents in these aspects, have higher quality indicators than locally planted varieties and control varieties. The long fiber rate of Heiya 6 reached 21%, which is 5 percentage points higher than local varieties. Stress resistance mutation: The stress resistance of flax is mainly in drought resistance, lodging resistance, disease resistance, salt alkali resistance, and other aspects. For example, Heiya 7 is resistant to Fusarium wilt, and Zhongyama 4 is resistant to powdery mildew. In short, the effect of flax mutation breeding is very obvious, and future breeding can continue to be studied and applied.

10.6 New Techniques for Flax Breeding

10.6.1 Haploid Breeding

Haploid breeding technology is a recently developed method for quickly obtaining pure lines. It can significantly accelerate the crop breeding process. Generally, it takes 12–15 years from hybridization to the completion of registration for a new flax variety. For self-pollinating crops like flax, the use of haploid breeding technology can greatly improve breeding efficiency. Currently, the two commonly used flax haploid breeding technologies are anther culture and the utilization of polyembryonic flax seeds.

10.6.1.1 The Anther Culture

The genotype of the donor plant is a key factor affecting the effectiveness of anther culture in plants. Sun (1979) reported obtaining three regeneration shoots from flax anther culture in 1978, and 21 shoots in 1979. Nichterlein et al. (1989) reported the highest callus rate in their flax anther culture was 17%, with the highest shoot regeneration rate reaching 20%. Friedt et al. (1995) reported the highest callus rate in flax anther culture reached 31.2%, with shoot regeneration reaching 27.5%. The developmental stage, environment, and plant status of flax plants all impact haploid culture. The formation process of flax microspores can be severely affected by differences in precipitation and day-night temperature, impacting subsequent anther culture. The physiological state of donor plants must also be considered in flax anther culture. Strong plants resistant to adverse conditions like drought and diseases are more suitable for anther culture. The experiment proved that the redifferentiation rate of regenerated plants from flax flower buds planted in summer in the soil was higher than that from flax flower buds collected in the greenhouse and flax flower buds planted in spring in the soil. At the same time, the developmental stage of microspores plays a crucial role in the formation and subsequent differentiation and development of gametophyte callus. The results showed that microspores in the middle stage of mononuclear, late stage of mononuclear and early stage of binucleate could be induced into callus or embryoid, but microspores in the middle stage of mononuclear to late stage werethe best for flax anther culture. Research shows that flax flower buds 3–6 mm long, with anthers 0.6–0.9 mm long, have a single nucleus or a single nucleus to the border, which provides a basis for us to select flax flower buds effectively and accurately (Sun et al. 1995). Research on pre-treatment methods in anther culture (or microspore culture) mostly focuses on several methods such as high temperature, low temperature, and carbon starvation. Song (1996) improved the anther culture, and the plant regeneration rate was increased by lower temperature pretreatment of anther and double layers (one liquid and one solid) medium. The callus induction rate of flax flower buds could be improved by low temperature or adding high concentration sucrose to the culture medium. After the flax flower buds were treated at 4 °C for 1 day and cultured on the H1-H4 medium, the callus induction rate was significantly increased by 3–22% compared with the untreated buds, and 50% on the H2 medium. The callus induction rate decreased to varying degrees after 3 to 5 days of low-temperature pretreatment (Xie 2009). After pre-cultivation with a high concentration of sucrose (7.5%), the callus rate was higher than that of the control. Among them, the highest callus rate was observed in strain 6 after 2 days of high sugar pre-cultivation, with a callus rate of 62.9%. After 7 days of pre-cultivation, the callus rate of some strains was increased, while others showed a downward trend (Sun et al. 2009). The anther callus of flax can be preserved for 3 months at a low temperature of 4 °C, thus prolonging the induction test time of plant regeneration from the callus.

Sun et al. (1979) systematically screened more than 20 basic media for flax anther culture, such as Miller, H, B5, N6, etc. The results showed that the callus induction rate of flax was the highest when the B5 medium was used. Further research on the role of microelements such as Mn, Zn, B, Mo, Cu, Co and I in B5 medium showed that these microelements played a certain role in flax anther dedifferentiation culture, of which I played the largest role in the formation of flax anther callus, followed by Mo and Zn (Sun et al. 1995). Song et al. (2007) improved the induction rate of flax anther callus by using the double-layer medium and developed a special medium for flax anther culture, which increased the induction rate of flax anther callus to 46.2%. Some excellent flax resources were created by anther culture. The first new fiber flax variety Shuangya No.13 was bred by anther culture in 2008 in China (Song 2008).

10.6.1.2 Utilization of Poly Embryonic Flax

Most haploid breeding techniques used include anther culture and interspecific hybridization. These methods not only require high technical requirements and complex operations but also require specific preparations and expensive chemicals. However, using haploid embryos from polyembryonic seeds to produce haploids does not require a complex indoor cultivation process. It is simple, economical and efficient, making it an effective method to accelerate the breeding process.

The utilization of polyembryonic flax in China in 1997. The highest rate of polyembryos in offspring was 19.2% when crossing polyembryonic flax with normal flax (Liu 1999). The haploids derived from twin seedlings are sterile, but they can be doubled chromosomes to get plants and seeds for breeding. The polyembryo rate and types vary in different years. Kang et al. (2011) found the highest rate of polyembryo seeds was more than 50% in F_1 of polyembryonic seeds × Non polyembryonic seeds.

There are both n-2n combinations and 2n-2n combinations, there is no n-n combinations were found in the 31 pairs of twin plants. Among these 31 pairs of twin seedlings, 26 haploid seedlings appeared, and the frequency of haploid seedlings appearing in twin seedlings was 41.9% (Kang et al. 2011). And they have got the high yield DH flax lines by this method. The highest stem yield was 8009 kg/ha (Kang et al. 2014). Some excellent flax germplasm and new varieties were created by utilization of polyembryonic flax, such as the new flax variety Huaya No.5 (Kang et al. 2022), it was planted in Yunnan province in the winter of 2016, and the yield test showed that the stem yield was 13,095 kg/ha and the seed yield was 1475 kg in the spring of 2017.

In order to improve the apomixis utilization efficiency of multi-embryo flax, the occurrence and genetic mechanism study of apomixis in multi-embryo flax will be the focus on the apomixis of flax currently. Studies of flax apomixis are few compared with other crops. It has been mainly studied in morphology, embryology, cytology, genetics, and utilization (Kang et al. 2021).

10.6.1.3 Identification and Doubling of Haploid Plants

There are several methods for doubling flax pollen plants. One way is to double them through exposure to low temperatures. Placing haploid flax plants under conditions of 1–5 °C can result in a doubling rate of up to 30%. The second method involves subculturing and doubling the haploid flax plants. After 4–5 rounds of regeneration culture, the doubling rate can reach 26%. The third method is to double with colchicine. The recommended concentration is 0.02%–0.03%, with a doubling time of 8–12 hours, and a temperature of 12–19 °C.

The doubling process includes wrapping one end of the gauze around the growth point of the haploid flax plant and then immersing the other end of the gauze in a colchicine solution. After 10 h, remove the gauze and wash the growth points of the treated haploid flax plants with water (Sun et al. 1995).

Colchicine has other treatment methods as well. Under suitable temperature and humidity conditions, plants treated with 0.05% colchicine for 24 hours during the bud stage can achieve a 100% survival rate (Kang 2004), resulting in 52.1% diploid plants (Liu 1999). Under different ecological conditions and genotypes, haploid flax plants can be soaked in water for 24–36 hours and then treated with colchicine. The concentration of colchicine should be 0.075–0.2%, and the treatment time should be 12–24 hours. Haploid flax plants with a concentration of 12–40% can be successfully doubled.

10.6.2 Tissue Culture

In 1946, Link and Egger first reported that flax hypocotyls cultured in vitro could differentiate and regenerate buds (Link and Eggera 1946). The somatic cell regenerated plants exhibited extensive hereditary variation, which was stable in the early stages of regeneration, allowing most parental characteristics to be retained. Sun et al. (1983) reported the successful regeneration of plants from flax stem tips, cotyledons and hypocotyls, studying the effects of light, temperature, medium and additives on callus production and plant regeneration. Male sterility, susceptibility to disease, and short stem can be obtained through a somatic cell culture (Yuan et al. 1997). Wang et al. (2014) established an efficient flax regeneration system through the optimization of genotype, explants, basic medium and hormone combination. Results showed that the best medium for callus induction and differentiation was Y4 (MS1 + 0.02 mg/L NAA + 1 mg/L 6-BA), with efficiencies of 98.89% and 95.56%, respectively. The optimal medium for root formation of regenerated shoots was 1/2 MS + 0.001 mg/L NAA.

Xue et al. (2022) used the seed block of the flax variety “Longya No. 10” as an explant to induce embryogenic callus. The results indicated that 2,4-D played a crucial role in the embryogenic induction stage of flax. Dark conditions promote the production and proliferation of embryogenic callus during induction and passage, while NAA supports the development of embryogenic callus.

Chinese scientists successfully bred a new flax variety using tissue culture. They obtained callus from hypocotyls, treated them with γ-ray, and regenerated plants from the callus. After two generations of selection, they developed the new line 06-3, a new flax variety named Shuangya No.16, registered in 2012 (Ji et al. 2012). Huaya No.9 was also bred through tissue culture and registered at MARA of China in 2023.

10.6.3 Interspecific Hybridization

Interspecific hybridization is a good method for using essential genes from wild species. However, most distant hybridization may result in incompatibility or can not obtain the seeds. Protoplast culture and isolated immature embryos of distant hybridization culture of flax lie in the study stage of medium and culture condition selection. Some breeders in China have attempted different techniques and achieved successful cases of interspecific hybridization, and have got some achievement.

Since 2003, Mi Jun has been conducting research on distant hybridization of flax and has obtained comprehensive hybridization techniques through experiments. The main technical points are: (1) selecting cultivated species with excellent traits and good combining ability as female parents and wild flax as male parents. (2) For wild flax, it is necessary to remove the flower buds, postpone the flowering period, and promote their flowering period to meet. (3) Before hybridization, select healthy and disease-free maternal plants with normal development, each plant retains 3–5 well-developed flower buds, and the rest are all picked. (4) Select flower buds with 1/3 of their petals exposed on the sepals from 4:00 pm to 6:00 pm the day before flowering to remove all anthers and (5) Collect the pollen of the paternal plant in a white container from 7:00 am to 10:00 am on the second day, and gently apply the pollen onto the stigma of the maternal plant by gripping the anthers with tweezers. Repeat pollination with an interval of 1 to 1.5 h. The next morning, use a syringe to drip the prepared growth regulator around the ovary, and it is continued for 3–5 days (Mi Jun 2004).

Wang et al. (2007) conducted interspecific hybridization between *Linum usitatissimum* × *Linum perenne*. Three hybridization techniques of once-pollination, repeat-pollination and repeat-pollination + plant growth regulator were adopted and the development of endosperm and embryo of *Linum usitatissimum* × *Linum perenne* were studied by ovule whole stain-clearing technique with the use of differential interference contrast microscope. The results showed that by once-pollination, the endosperm and embryo were not formed in most embryo sacs, but by repeat-pollination and repeat-pollination + plant growth regulators the endosperm and embryo were discovered in most embryo sacs, but their developing process was slower than that of self-pollination *Linum usitatissimum* (CK). The use of plant growth regulators could stimulate growth and retard the fall of the ovary and was in favor of forming seeds. The results show that repeat-pollination + plant growth regulator (60 mg GA3, 90 mg NAA, 20 mg 2,4-D in 100 ml solution) is an effective method to overcome interspecific cross-incompatibility of flax. The authenticity of interspecific hybrids was identified by SSR molecular marker (Zhang et al. 2012). The drought-resistant flax variety Jizhangya No.1 was bred using this method (Qu et al. 2018). The utilization of distant hybridization technology is relatively limited and further research is needed.

10.6.4 Utilization of Male-Sterility Flax

The development and selection of breeding lines for male sterile plants with open corollas may lead to the systematic development of cross-flax varieties. In 1975, Inner Mongolia College of Agriculture and Animal Husbandry discovered a sterile plant from flax called "Yanza 10". It is believed to be a nuclear sterile type controlled by dominant genes. Using sterile flax for breeding methods such as hybridization, backcrossing, and rotation, many excellent lines have been bred (Wang 2001), and some flax varieties have been bredutilizing this sterile flax. Dang et al. (2000) treated flax seeds with five antibiotics (Streptomycin, Penicillin, Rifampin, Erythromycin, and Gacheomycin) and obtained some male sterile mutants from 1998 to 1999. Stable genetic male sterile materials have been discovered in flax, paving the way for utilizing the hybrid advantage of two lines to increase flax yield.

Zhang et al. (2002) studied the heterosis and temperature-sensitive effects of this male sterile material under natural conditions. The results indicate that temperature has a significant impact on sterility. Hybridization between sterile lines and conventional varieties has shown promising results.

Hybridize male sterile lines (5 s-2, 5 s-3, 20 s, 25 s) with 30 varieties/lines, the F_1 hybrid combination significantly increased yield, with an average increase of 39.73%. They conducted yield measurements on the F_1 hybrid of ordinary plants and the two-line flax hybrid of temperature-sensitive male sterile lines of flax plants, and the results showed that the two-line flax hybrid had significantly higher seed yield. Compared with the control variety Longya 8 (Zhang et al. 2005), the yield of the hybrid variety increased by 42.29–60.71%. Using this temperature-sensitive male sterile material, four flax hybrid varieties, namely Longya 1–4, were bred.

10.6.5 Development and Application of Molecular Markers

Conventional breeding mainly focuses on agronomic traits related to yield, fiber quality, and resistance to pests and diseases. The nutritional value of flax has been gradually emphasized, and it is currently very necessary to understand and improve the nutritional characteristics of flaxseed varieties. The utilization of advancements in molecular and biotechnological strategies will provide effective techniques for the successful breeding of flaxseed and its effective utilization in the nutraceutical and pharmaceutical industries.

Marker-assisted breeding (MAB) combines classical plant breeding with molecular biology and genetics tools specifically molecular markers to develop new varieties/hybrids in cultivated crops. The use of molecular markers in flax improvement began in the late 1990s. Molecular markers are valuable for characterizing genotypes, genetic variation evaluation, diversity studies, genetic map construction, QTL mapping, association studies, marker-assisted selection (MAS) and genomic selection (GS). Molecular markers and genes related to quality, yield, and resistance are being developed, indicating that flax breeding has entered the era of molecular breeding. Molecular markers are crucial tools for assessing and applying genetic diversity in breeding. They provide a genetic "fingerprint" of germplasm, crossing parents and offspring and are increasingly used in molecular marker-assisted selection. The most commonly used methods in flax breeding are RAPD, SSR, SRAP, InDels, ISSR and SNP. Most molecular marker technologies have been extensively studied and applied in flax breeding.

Current molecular markers for flax focus on rust disease, wilt disease and genes related to resistance to waterlogging. Deng et al. (2008) analyzed the genetic diversity of ten flax collections using RAPD markers and demonstrated that RAPD markers can be applied in the analysis of flax genetic diversity (Wang 2008) and the interspecific relationship of *Linum* (Yurkevich et al. 2013). RAPD analysis reveals interspecies differences, with species from nearby areas, showing more similar RAPD characteristics compared to more geographically distant species. Zhang et al. (2007) found that the RAPD marker S1377 was closely linked to the waterlogging tolerance gene in flax, selected from 500 markers using the RAPD- BSA method. Bo et al. (2002) used RAPD analysis to obtain the molecular marker OPA18432 linked to the flax rust-resistance gene M4, and specific PCR primers were designed to develop a SCAR marker for the M4 gene.

Due to the advantages of high repeatability, rich polymorphism, co-dominance, and high reliability of SSR (Simple Sequence Repeat) markers, SSR primers suitable for flax have gradually been developed. Deng et al. (2008) developed 97 SSR molecular markers using magnetic beads in flax. Thirty-five microsatellite loci were isolated and characterized in *Linum usitatissimum* using enriched genomic libraries (Deng et al. 2010). Subsequently, a large number of SSR markers have been developed from these genomic resources, 275 SSR assays derived from ESTs detected 447 loci and 342 SSR assays mined from genomic sequences detected loci. (Cloutier et al. 2009, 2012) 90 SSR and 10 ISSR flax markers were employed for molecular diversity analysis (Pali 2015). Wu et al. (2017) developed 1574 novel SSRs in flax using reduced representation genome sequencing. Pan et al. (2020) screened and designed 71,184 primer pairs of SSR markers for the flax genome.

Jiang et al. (2022) identified a total of 17,110 InDel markers by comparing whole-genome re-sequencing data of two accessions (87-3 and 84-3) with the flax reference genome. From 17,110 InDel markers, 90 primers that were evenly distributed in the flax genome were selected. Thirty-two pairs of polymorphic primers were detected in two flax accessions, and the polymorphism rate was 40.70%.

10.6.6 Gene Clone and Identification

Genes related to the light cycle, active ingredient synthesis, fiber quality, cellulose, fatty acids and stress resistance of flax have been cloned, identified, or analyzed for expression. Plant *WRKY* genes encode a complex and ancient class of zinc-finger transcription factors involved in multiple biological processes, especially in regulating defense against abiotic stresses.

Firstly, the phloem of flax contains a large amount of cellulose, which is often used in industries such as flax textiles and composite materials, so there are many studies on genes related to flax cellulose. The cellulose synthase family (*CesA*) plays an important role in the formation of plant cell walls. The *CesA* family belongs to the glycosyltransferase *(GT)* family, responsible for catalyzing cellulose synthesis. Guo et al. (2017) identified a total of 975 genes, of which 708 (73%) genes had protein-coding annotation, as phloem-enriched genes actively involved in polysaccharide and cell wall metabolism processes. A large number of genes related to metabolic processes, catalytic activity and binding categories were predominantly expressed in the stem peels. Yu et al. (2016) used high-fiber flax varieties as materials and cloned the cellulose synthase gene *LuCesA8* (gene ID: Lus10029245). Analysis revealed that the open reading frame (ORF) of this gene was 2967 bp in length, and the protein encoded by the gene had the closest genetic relationship with the *CesA8* protein in leprosy trees. The gene had the highest expression level during the rapid growth stage of flax, followed by the flowering and green ripening stages, and the lowest expression level during the seedling stage. Yuan et al. (2021) conducted a comprehensive genome-wide characterization of the flax *WRKY* gene family that led to the prediction of 102 *LuWRKY* genes. Most *LuWRKY* genes associated with cellulose, hemicellulose, or lignin content were predominantly expressed in stems, roots, and less in leaves.

Secondly, lignin related genes are important in plant biology. Lignin, a phenylpropanoid polymer ranks second only to cellulose in terms of significance in plants. The glycosyltransferase gene modified by lignin monomer glycosylation is an important gene affecting lignin quality. Yuan et al. (2016) isolated the glycosyltransferase *LuUGT72E1* gene from flax stems. The ORF gene has a total length of 1476 bp, and the amino acid encoded protein has the closest relationship with the flax UGT72N2 protein. It is speculated that this protein is involved in the glycosylation modification process of lignin monomers. In order to deeply explore and analyze the miRNAs and target genes related to lignin accumulation in flax stems at different developmental stages. The third gene related to resistance (tolerance) to biotic and abiotic stress is important in flax. Flax resistance traits include resistance to Fusarium wilt, powdery mildew, rust, drought, salt alkali, waterlogging, and heavy metal resistance. These traits are quantitative and controlled by major and minor genes but are easily influenced by the environment. Li et al. (2024a) cloned the *LuWRI1a* gene, which enhances flex tolerance by resisting growth inhibition due to salt and drought stress, improving reactive oxygen species scavenging, reducing oxidative damage to membrane lipids, and activating diverse stress response genes. Zhao et al. (2023) identified 14 *PLA1* members in flax, confirming that *LuPLA1* genes are induced by hormones, drought, high salt, low temperature, and high temperature. Li et al. (2023) identified 17 *LuDMP* genes, designated as *LuDMP-1* to *LuDMP-17* from the Longya-10 cultivar through homology analysis. They found that these genes were unevenly distributed on nine chromosomes. The promoter regions of *LuDMP-1/7* contained hormone and stress-responsive elements, and both genes showed increased transcript levels under auxin (indole acetic acid, naphthalene acetic acid), gibberellin, polyethylene glycol, high temperature and low-temperature treatments. Li (2023) cloned the *LuAccD* gene from the flax variety "Longya 10", which promotes the accumulation of fatty acids in seeds. Overexpression of *LuAccD* can enhance seed tolerance to salt and osmotic stress by regulating ABA synthesis and signal transduction pathways. Ye et al. (2023) found the 11 genes may be related to the resistance of oil flax to powdery mildew which was screened out by gene annotation, metabolic pathway analysis and literature analysis. In recent years, the identification of these genes has provided technical support for the molecular breeding of flax. However, these studies are still in the early stages and lack systematic research. In the future, systematic research on key genes should be carried out.

The fourth gene is related to the composition of flax seeds. Flax seeds have an oil content of about 50%, with the oil mainly stored as triacylglycerol (TAG) in the seeds. The key enzymes affecting the synthesis of linoleic acid and linolenic acid in plants include △12 fatty acid desaturase (*FAD2*) and △15 desaturase (*FAD3*). *FAD2* is the key enzyme for the dehydrogenation of oleic acid to form linoleic acid, while *FAD3* is the key enzyme for the desaturation of linoleic acid to form linolenic acid (Li et al. 2019). Two genes, *LuFAD3A* and *LuFAD3B*, are responsible for high linolenic acid in wild flax. Microsomal desaturases encoded by these two genes are capable of desaturating linoleic acid. Vrinten et al. (2005) reported that cloning of two members of the Fatty Acid Desaturase (*FAD3*) family of microsomal Δ15-FADs, designated *LuFAD3A* and *LuFAD3B*, reveals the molecular mechanism behind this trait. Wu et al. (2023) cloned the *LuFAD7* gene that plays an important role in regulating the pathway of linoleic acid synthesis α- linolenic acid and its high expression promotes the transformation of linoleic acid to α-linolenic acid conversion. α- linolenic accumulation leads to a decrease in linoleic acid content.

10.6.7 Gene Transformation

10.6.7.1 Agrobacterium Mediated Method

It is an important approach to introduce exogenous genes into plants through genetic engineering technology in modern genetic breeding. Many transgenic plants are obtained through the *Agrobacterium tumefaciens* transformation system. One successful case of genetically modified flax is CDC Triffid, which is resistant to sulfonylurea resistance), which was registered in Canada in 1996, and it was deregistered in 2001 due to problems in its export to European countries (Manimurugan et al. 2023). China started this study in 1998. The Bar and vhb genes were transferred to flax mediated by *Agrobacterium tumefaciens*. It was verified that the target gene was transferred to the genome of regenerative plants by molecular methods of PCR (Wang et al. 2004). Wang et al. (2000a) studied the regeneration and rooting culture of transgenic flax plants using the hypocotyl of flax seedlings as an explant, using the herbicide-resistant *Basta* gene and *GUS-INT* gene as target genes, and using the *Agrobacterium*-mediated method. A preliminary transgenic system for flax mediated by *Agrobacterium tumefaciens* has been established. Wang et al. (2000b) reported the establishment of a genetic system for flax and the study of chitinase genes in flax genetic transformation. They preliminarily inferred that the chitinase gene has entered the flax genome through screening for resistant shoot rooting and leaf resistance testing. Kang et al. (2002) experimented using *Agrobacterium*-mediated transformation of flax Heiya No.11 and Heiya No.9 to herbicide-resistant *Basta* gene and obtained transformed callus tissue. In the same year, Wang et al. (2000b) reported the chitinase gene was transferred to flax mediated by *Agrobacterium tumefacien* too. The inhibition of the chalcone synthase gene in transformed flaxseed plants leads to an increase in seed antioxidant status. Therefore, an accumulation of polyunsaturated fatty acids (mainly alpha-linolenic acid) 20–45% higher than the control was observed in transgenic seeds (Zuk et al. 2012). Kang et al. (2006) used *Agrobacterium*-mediated transgenic technology to introduce the salt and low-temperature stress resistant gene HY15CS into haploid flax. As a result, 24 resistant regenerated plants were obtained, and 3 positive plants were detected by PCR from 15 healthy regenerated plants, with a positive rate of 20%. This system provides a favorable condition for accelerating molecular breeding and gene-related research of linseed flax. Currently, there are no genetically modified flax varieties developed using this method in China.

10.6.7.2 Pollen Tube Pathway and Ovary Injection Method

The basic principle of this method is that during the fertilization process of flax, the chromatin of the sperm nucleus disperses in the egg nucleus for several hours to more than ten hours, and the fused zygote also goes through a rest period of 10–20 hours. At this time, it does not have a normal cell wall, which is beneficial for the entry of exogenous DNA, thereby achieving the transfer of certain target genes. The technology of introducing exogenous DNA into flax has been studied since 1993 in China. The method for extracting DNA from flax plants, the period and method of DNA introduction through the pollen tube pathway, and the characteristics of flower color, plant height, seed coat color, and lodging resistance in the offspring of exogenous DNA introduction were clearly changed. Peroxidase isoenzyme analysis was conducted on the offspring, and it was found that there were donor specific bands in the offspring, but receptors did not show these bands. The introduction of exogenous DNA into offspring D_3 and D_4 can achieve stable generations. This can shorten the breeding period by nearly half, while also utilizing distant germplasm resources that are incompatible with hybridization (Wang YF 2005). The Bar was transformed successfully into flax using a pollen tube pathway. The T_1 generation was screened by spraying the leaves with 100 mg/L glufosinate in the field, and *Bar*-positive detection of anti-glufosinate plants was conducted by PCR. There were 44 T_1 glufosinate-tolerant plants that survived in the field, and 8 *Bar*-positive plants were confirmed by PCR, with an average transformation rate of 18.2% (Li et al. 2013). Another method is to immerse flax flowers directly in a solution containing exogenous DNA or target genes, which is also called "floral dip". Dutta et al. (2021) reported an efficient plant genetic transformation technique achieved through Agrobacterium-mediated floral dipping. This technique is very simple to operate and highly reproducible. The cry1A(b) transgenic progeny were obtained by using this technique, which can also play a role in improving the quality and yield of flax, especially under the circumstances of obvious global climate changes and the intensified damage caused by resistant insect pests to this crop of significant economic value.

10.6.8 The Gene Editing by CRISPR/Cas9 System

Many gene-editing tools have been described, with the engineered nuclease gene editing methods ZFNs, TALENs and CRISPR being utilized more frequently (Clemis et al. 2023). Gene editing is an approach in which a specific target DNA sequence of the genome is altered by adding, removing, or replacing DNA bases. The gene editing of flax is mainly utilizing CRISPR/Cas9. This system employs an RNA-guided nuclease, Cas9, to induce double-strand breaks. The CRISPR/Cas9 system is a new gene editing tool, which can edit target genes accurately. The Cas9-mediated breaks are repaired by cellular DNA repair mechanisms and mediate gene/genome modifications. It is widely used in plant gene editing. In agriculture, it could help in the design of new varieties to improve their quality, yield, and nutritional value.

Zhang et al. (2019) constructed a gene editing vector of CRISPR/Cas9 for the oil flax cultivar 'Longya No.10'. Taking CRISPR/Cas9 gene editing technology as a breakthrough point, according to the *FAD2* gene sequence (DQ222824.1) published in Gene Bank, the *FAD2* gene was missed in the exon of *FAD2* gene by using sgRNA online design software CRISPR-P 2.0. Two fragments in the exon region were selected as the target sequence for deletion. With hygromycin as the screening marker, the gene knockout vector of CRISPR/Cas9 targeting the *FAD2* gene of 'Longya No.10' was constructed by the method of Golden Gate cloning, which was named pYLCRISPR/cas9-*FAD2*. Afterwards, pYLCRISPR/cas9-*FAD2* was introduced into *Agrobacterium*, the its stability was tested. A total of 15,000 flowers of Longya No.10 were transformed by the pollen-tube pathway method, and more than 70,000 seeds were harvested. 25 strains were screened with Hygromycin resistance (Zhang 2019).

Yang (2021) utilized No.10 as the material and cloned the four genes of flax Longya No.10, *ACCase* subunit gene *accA, accB, acc*C*and**accD*. The fat coefficients of the proteins encoded by the four genes are relatively high, and no signal peptide is a secreted protein. The size of the cDNA fragments of the four genes of Longya No.10, *accA, accB, acc*Cand*accD* were 2364, 1173, 1584 and 1137 bp, respectively. CRISPR/Cas9 system for site-specific editing, vector construction and genetic transformation of flax heterogeneous *ACCase*subunit genes. Successfully established the flax CRISPR/Cas9 technology system, and constructed 4 CRISPR/Cas9 recombinant knockout vectors CPB-CRISPR/Cas9-*accA*, CPB-CRISPR/Cas9-*accB*, CPB-CRISPR/Cas9-*accC* and CPB-CRISPR/Cas9-*accD* was transformed into Agrobacterium GV3101 by freeze-thaw method, and the strains with stable expression were selected. Longya No.10 was transformed by the dipping method. A total of 64 *accA*genetransformedT_0 generation capsules with 394 seeds and 21 positive seedlings were obtained. A total of 65 *accB*genetransformed T_0 generation capsules with 318 seeds, and 2 positive seedlings were obtained, and 91 *accC* gene transformed T_0 generation capsules with 599 seeds and 11 positive seedlings were obtained. 61*accD* gene transformed capsules with 367 seeds, and 3 positive seedlings was obtained.

The above introduces many breeding methods, and breeders can choose suitable breeding strategies and methods based on cultivation purposes, actual needs, breeding goals, market demand, soil fertility location, ecological environment, target uses, etc.

10.7 Conclusions and Prospects

Fiber flax breeding has obtained lots of achievements. The new breeding methods and technologies were developed and improved, especially the studies on molecular biology (e.g. Gene transformation) and gene editing breeding techniques that have received increasing attention from scientists in flax breeding, biotechnology, and other fields. Lots of flax varieties were bred. The straw yield, seeds yield, and fiber content of flax were increased, and the fiber quality, resistance to disease and lodging were improved too. Which has promoted the development of the flax industry. Due to the advancement of breeding technology and the promotion of new varieties and technologies, the yield of flax per unit area has significantly increased. With the reduction of flax planting area, the total yield of flax has not decreased but has increased. The variety played an important role in it. But some problems need to be resolved about flax breeding. Firstly, in terms of flax breeding methods, although many new breeding methods have been successfully studied, they have not been fully utilized. The vast majority of flax varieties are still bred using traditional breeding methods such as cross, mutagenesis, and systematic selection. Secondly, some quality traits, such as fiber strength, flexibility, oil yield, α-linolenic acid, and lignin content still need to be improved.

According to the market demand for flax products, the ecological environment of flax planting areas, climate change and other factors, corresponding breeding strategies should be adopted to respond. Firstly, while utilizing traditional technologies, modern equipment necessary for breeding new technologies should be added, and the utilization of new technologies should be strengthened. For example, by combining mutation breeding technology with high-throughput detection technology, the efficiency of mutation breeding can be improved, creating breakthroughs that are difficult to obtain special traits through cross breeding. Second, to breed salt alkali, drought tolerant, and lodging resistant varieties, to enhance their adaptability, and tolerance to salt and alkali, drought and lodging resistance, to expand the cultivation of flax in arid saline alkali areas, and expand the planting area of winter fallow fields in the south; Breeding fiber and seed dual-purpose flax varieties and special functional varieties to meet new market demands. Third, in recent years, molecular biology study on flax has rapidly increased, but the goals are not concentrated enough, the practicality is not strong, and the application of molecular biology in breeding is insufficient. So should strengthen the study of practical new molecular biology technologies for flax, so that they can be quickly applied in breeding.

The excellent flax varieties with good quality, high yield and resistance to lodging and diseases, and special flax varieties with low pectin (it is easy to ret), low lignin and high fineness (to produce soft products), high strength and crude surface (for composite), and high content of lignan, α-linoleic acid, pectin, flavonoid, cyclic peptide fiber and seeds dual-purpose flax varieties are needed now and future. Due to the huge demand for new uses such as flax composite materials, the planting area of flax will continue to increase, and the market prospects of flax are promising.

Funding This work was supported by the earmarked fund for China Agriculture Research System(CARS) (CARS-16), the Science and Technology Innovation Project of Chinese Academy of Agricultural Sciences(Green and low-carbonproduction technology for southern specialty crops(CAAS-ASTIP-IBFC-04).

Appendix 10.1 Research Institutes Relevant to Flax

Institution name	Specialization and research activities	Address	Contact information and website
Institute of Bast FiberCrops(IBFC), Chinese Academy of Agricultural Sciences(CAAS)	It is the only national-level research institute engaged in bast fiber crops research in China, and it is the institution where the Expert in Flax Physiology and Cultivation in China Agriculture Research System are located. Mainly engaged in flax cultivation, breeding, germplasm resources, and biotechnology research in the field of flax research.	No. 348, Xianjiahu West Road,Yuelu District, Changsha City, Hunan Province P.R China. Postal Code:410205	https://ibfc.caas.cn E-mail of the expert in flax: qiucaisheng@caas.cn

(continued)

Institution name	Specialization and research activities	Address	Contact information and website
Institute of Industrial Crop(IIC) of Heilongjiang Academy of Agricultural Sciences(HAAS)	It is the institution where the Experts in Flax Variety Improvement and Comprehensive Experimental Station of Bast Fiber Crops in China's Agriculture Research System are located. Mainly engaged in flax breeding, germplasm, biotechnology research and extension of new flax varieties and planting technologies in the field of flax research.	No.368, Xuefu Road, NangangDistrict,Harbin City, Heilongjiang Province P.R China. Postal Code:150086	https://www.haas.cn/ E-mail of the expert in flax: kang_qinghua@126.com
Dali Bai Autonomous Prefecture Agricultural Science Extension Research Institute	It is the institution where the Comprehensive Experimental Station of Flax and Industrial Hemp in China's Agriculture Research System is located. Mainly engaged in extension of new varieties and planting technologies, and breeding, germplasm in the field of flax research and application.	Jinniu Town, Binchuan County, Dali Bai Autonomous Prefecture, Yunnan Province P.R China. Postal Code:671600	E-mail of the expert in flax: zhuxuan2006@163.com
Institute of Agricultural Science of Ili Prefecture	It is the institution where the Comprehensive Experimental Station of Flax in China's Agriculture Research System is located. Mainly engaged in extension of new varieties and planting technologies, and breeding, germplasm in the field of flax research and application.	East Suburb of Yining City, Ili Kazakh Autonomous Prefecture, Xinjiang Uygur Autonomous Region P.R China. Postal Code:835099	E-mail of the expert in flax: 1308477364@qq.com

(continued)

Institution name	Specialization and research activities	Address	Contact information and website
Research and Development Centre, Agriculture and Agri-Food Canada	Mainly engaged in research on flax germplasm resources and genetic breeding techniques	107 Science Place, Saskatoon, SK S7N 0X2, Canada	https://www.iwnirz.pl/ E-mail:yong-bi.fu@agr.gc.ca
Institute of Natural Fibers and Herbal Plants	The research fields related to flax are relatively comprehensive, including breeding, cultivation, germplasm resources, harvesting machinery, quality testing, and so on.	ul. WojskaPolskiego 71B, 60–630 Poznań	https://www.iwnirz.pl E-mail: lenkonopiewelna@iwnirz.pl
Comlin	It is a consortium of 3 flax companies based in Normandy: agylin, Le Neubourg Scutching Mill, and Terre de Lin. Terre de Linhas has a huge collection of flax varieties, linseed varieties and wild flax. They are engaged in flax genetics, breeding, and breeding of high-quality varieties, and the flax varieties they breed have attracted worldwide attention.	605 route de la Vallée 76740 Saint-Pierre-le-Viger, FRANCE	https://www.com-lin.com/ E-mail: contact@com-lin.com
Flax Institute: Separate Subdivision of the Federal Research Center for Bast Crops	They are engaged in research on flax breeding, germplasm resources, plant protection, fine variety breeding, retting and processing, fiber testing, harvesting and processing machinery, and so on.	Torzhok, Tver Region, 172002 Russia	E-mail: tatyana_rozhmina@mail.ru

(continued)

Institution name	Specialization and research activities	Address	Contact information and website
Agritec Plant Research s. r. o. (APR)	Agritec Plant Research s. r. o. (APR) is a research organization based in Šumperk, that deals with applied research in the field of agriculture with a focus on growing, breeding, utilization and integrated protection of fibrous plants(flax and industrial hemp), legumes and oilseeds. They are engaged in the Genetics, breeding and biotechnology of flax	Agritec Plant Research s.r.o. Zemědělská 2520/16, 787 01 ŠUMPERK,	https://www.agritec.cz E-mail: info@agritec.cz
RUP "Flax Research Institute" РУП "ИНСТИТУТ ЛЬНА"	Breed new varieties of flax, and develop eco-friendly and resource-saving flax cultivation and harvesting technologies. Develop agricultural technology methods to improve the yield and quality of flax, and reduce the production cost of flax products. Improve the control method of flax straw processing and determine the physical and mechanical indicators of flax fiber quality.	Usti Central, 27 Vitebsk Oblast, Orzan District, 211003 Republic of Belarus	http://www.institut-lna.by E-mail: institut-len@yandex.by
N. I. Vavilov All-Russian Institute of Plant Genetic Resources (VIR)	Engaged in the collection, preservation, and innovative research of flax germplasm resources, possessing the world's largest flax germplasm resource preservation gene bank	42–44, B.Morskaya Street, 190000, Saint Petersburg City, Russia	https://www.vir.nw.ru E-mail: secretary@vir.nw.ru

(continued)

Institution name	Specialization and research activities	Address	Contact information and website
Université de Bretagne Sud, FRE CNRS	Research on Comprehensive Utilization and Composites of Flax Fiber	Université de Bretagne Sud, UMR CNRS 6027, IRDL, rue Saint Maudé, 56100 Lorient, France	https://www.univ-ubs.fr E-mail of expert flax fiber utilization: christophe.baley@univ-ubs.fr
College of Life Science and Technology, Xinjiang University	Study on identification of germplasm resources, gene expression and identification and gene transformationof flax.	No. 666 Shengli Road, Tianshan District, Urumqi City, Xinjiang Uygur Autonomous Region,China	https://www.xju.edu.cn E-mail of expert: picea@xju.edu.cn
College of Textiles, Donghua University	Research on new spinning technologies and related theories and the development and utilization of flax fiber	2999 Renmin North Road, Songjiang District, Shanghai,China	https://texcol.dhu.edu.cn E-mail of expert: yucw@dhu.edu.cn
Faculty of Biotechnology, the University of Wrocław	Gene cloning, identification, expression analysis, methylation, transformation, and functional product development of flax	University ofWrocław, Faculty of Biotechnology. Joliot-Curie 14a50-383 Wrocław, Poland	https://www.biotech.uni.wroc.pl E-mail of expert: anna.kulma@uwr.edu.pl
Institute of Bast Crops of the National Academy of Agrarian Sciences of Ukraine	They study flax for Breeding and seed production; the Development and improvement of technologies for growing; the Development of technologies and creation of machines for harvesting; and the Development and improvement of technologies of primary processing.	Building 45, Tereshchenkiv Street, Hlukhiv, Ukraine.	https://ibc-naas.com/ E-mail: ibc1931@ukr.net

Appendix 10.2 Genetic Resources of Flax

Cultivar	Important traits	Cultivation location
Dingya No.23	Thousand seeds weight 7.77 g	Gansu, China
Dingya No.24	Seeds yield 2034.0 kg/ha	Gansu, China
Dingya No.25	Thousand seeds weight 7.9 g	Gansu, China
Huaxing No.1	Disease resistance, drought resistance, and good fiber variety	Yunnan and Xinjiang, China
Huaxing No.2	Disease resistance, drought resistance, and good fiber variety	Yunnan and Xinjiang, China
Huaxing No.3	Disease resistance, drought resistance, and good fiber variety	Yunnan and Xinjiang, China
Huaya No.14	The plant height is relatively high, above 100 cm, and the stem yield is 12,558 kg/ha at Yunnan	Heilongjiang and Yunnan, China
Huaya No.15	The plant height is relatively high, above 100 cm, stem yield is 13,434 kg/ha at Yunnan	Heilongjiang and Yunnan, China
Jiuya No.2	Thousand seeds weight 7.7 g	Gansu, China
Jiuya No.2	Seeds yield 2412.8 kg/ha	Hebei, China
KeheXianya No.4	The plant height is relatively high, above 100 cm	Heilongjiang, China
KeheYouya No.6	Thousand seeds weight 8.2 g	Heilongjiang, China
Longya No.10	Seeds yield 2077.9 kg/ha, Thousand seeds weight 8.37 g	Gansu, China
Longya No.14	Thousand seeds weight 8.1 g	Gansu, China
Longya No.20	Thousand seeds weight 7.93 g	Gansu, China
Longyaza No.3	Thousand seeds weight 7.65 g	Gansu, China
Longyaza No.4	Thousand seeds weight 8 g	Gansu, China
Neiya No. 9	Seeds yield 2103.1 kg/ha	Inner Mongolia
Ningya No.17	Thousand seeds weight 7.85 g	Ningxia, China
Ningya No.24	Seeds yield 2383.6 kg/ha	Ningxia, China
White snow	Seeds yield 2442.0 kg/ha	Gansu, China
Zhang Ya No.5	Thousand seeds weight 7.55 g	Hebei, China
Zhangya No.4	Thousand seeds weight 8.8 g	Gansu, China
Zhongyama No.1	Resistance to lodging and waterlogging	Xinjiang,Gansu and Hunan, China
Zhongyama No.2	High stem yield, resistance to lodging and salt alkali	Xinjiang,Gansu and Hunan, China
Zhongyama No.3	Resistance to lodging and salt alkali	Xinjiang,Gansu and Hunan, China
Zhongyama No.4	Resistance to powdery mildew and salt alkali	Xinjiang,Gansu and Hunan, China
Zhongyama No.5	Resistance to lodging and salt alkali	Xinjiang,Gansu and Hunan, China

Appendix 10.3 36 Species in the Integrated Taxonomic Information System (ITIS—Report: *Linumusitatissimum*)

Number	Latin name	English name	The literal translation of English names into Chinese names
1	*Linum Arenicola* (Small) H.J.P. Winkl.	Sand flax	沙麻
2	*Linum aristatum* Engelm.	Bristle flax	硬毛亚麻
3	*Linumaustrale* A. Heller	Southern flax	南方亚麻
4	*Linum austriacum* L.	Asian flax	亚洲亚麻
5	*Linum berlandieri* Hook.	Berlandier's yellow flax	贝兰迪尔黄亚麻
6	*Linum bienne* Mill.	Pale flax	浅色亚麻
7	*Linum carteri* Small	Carter's flax	卡特亚麻
8	*Linum catharticum* L.	Fairy flax	仙女亚麻
9	*Linum compactum* A. Nelson	Wyoming flax	怀俄明州亚麻
10	*Linum elongatum* (Small) H.J.P. Winkl.	Laredo flax	拉雷多亚麻
11	*Linum floridanum* (Planch.) Trel.	Florida yellow flax	佛罗里达黄亚麻
12	*Linum grandiflorum* Desf.	Scarlet flax, flowering flax, red flax	大花亚麻、红花亚麻
13	*Linum hudsonioides* Planch.	Texas flax	德克萨斯亚麻
14	*Linumim bricatum* (Raf.) Shinners	Toothed flax, tufted flax	齿状亚麻、簇绒亚麻
15	*Linum intercursum*E.P. Bicknell	Sandplain flax	沙质亚麻
16	*Linum kingii* S. Watson	King's flax	国王亚麻
17	*Linum lewisii*Pursh	Lewis' flax, blue flax, Lewis' blue flax, prairie flax	路易斯亚麻、蓝色亚麻、路易斯蓝色亚麻、草原亚麻
18	*Linum lundellii* C.M. Rogers	Sullivan City flax	沙利文市亚麻
19	*Linum macrocarpum* C.M. Rogers	Spring Hill flax	春山亚麻
20	*Linum medium* (Planch.) Britton	Stiff yellow flax	硬黄亚麻
21	*Linum narbonense* L.	Flax	亚麻
22	*Linum neomexicanum* Greene	New Mexico yellow flax	新墨西哥州黄亚麻
23	*Linumperenne* L.	Blue flax	蓝色亚麻
24	*Linum pratense* (Norton) Small	Blue flax, meadow flax	蓝色亚麻、草地亚麻
25	*Linum puberulum* (Engelm.) A. Heller	Desert flax, plains flax	沙漠亚麻、平原亚麻
26	*Linumrigidum*Pursh	Orange flax, stiff flax, stiffstem flax	橙麻、硬麻、硬亚麻
27	*Linum rupestre* (A. Gray) Engelm. ex A. Gray	Rock flax	岩石亚麻

Number	Latin name	English name	The literal translation of English names into Chinese names
28	*Linum schiedeanum* Schltdl. & Cham.	Schied's flax	席德亚麻
29	*Linum striatum* Walter	Rigid flax, ridged yellow flax	硬亚麻
30	*Linum subtrees* (Trel.) H.J.P. Winkl.	Slender foot flax, sprucemont flax	细脚亚麻
31	*Linum sulcatum* Riddell	Grooved yellow flax, grooved flax	黄槽亚麻
32	*Linum trigynum* L.	French flax	法国亚麻
33	*Linum usitatissimum* L.	Cultivated flax, common flax	栽培亚麻、普通亚麻
34	*Linum vernale*Wooton	Chihuahuan flax	吉娃娃亚麻
35	*Linum virginianum* L.	Woodland flax	林地亚麻
36	*Linum westii* C.M. Rogers	West's flax	维斯特亚麻

Source: Latin name and English name in the table from website www.itis.gov

References

Agosti MB, Sorlino D, Trapani N (2005) How does light intensity affect the elementary fibre length in flax? J Nat Fib 2(1):15–24

Ahmad-Rasyid MF, Salim MS, Akil HM et al (2016) Optimization of processing conditions via response surface methodology (RSM) of nonwoven flax fibre reinforced acrodur biocomposites. Procedia Chem 19:469–476

Beauvais F, Cantat O, Gouée PL et al (2022) Consequences of climate change on flax fiber in Normandy by 2100: prospective bioclimatic simulation based on data from the ALADIN-Climate and WRF regional models. Theor Appl Clim 148:415–426. https://doi.org/10.1007/s00704-022-03938-4

Bo TY, Ye HZ, Wang SQ et al (2002) Specific molecular markers of the rust resistance gene M_4 in flax. Acta Genet Sinica 29(10):922–927

Casa R, Russell G, Cascio L et al (1999) Environmental effects on linseed (*Linum usitatissimum* L.) yield and growth of flax at different stand densities. Europ J Agronomy 11(3–4):267–278

Chantreau M, Grec S, Gutierrez L et al (2013) PT-Flax(phenotyping and TILLinG of flax): development of a flax (*Linum usitatissimum* L.) mutant population and TILLinG platform for forward and reverse genetics. BMC Plant Biol 13(1):159–168

Clemis V, Zaidi M, Fofana B (2023) Genome-editing tools for flax genetic improvement. In: You FM, Fofana B (eds) The Flax Genome. Compendium of Plant Genomes. Springer, Cham, pp 235–252. https://doi.org/10.1007/978-3-031-16061-5_11

Cloutier S, Niu Z, Datla R et al (2009) Development and analysis of EST-SSRs for flax (*Linumusitatissimum* L.). Theor Appl Genet 119:53–63

Cloutier S, Miranda E, Ward K et al (2012) Simple sequence repeat marker development from bacterial artificial chromosome end sequences and expressed sequence tags of flax (*Linumusitatissimum* L.). Theor Appl Genet 125:685–694

Dai MH, Yu FW (2024) Research on enhancing agricultural green development capability under the background of climate change. Academic JZhongzhou 4:49–56

Dai ZG, Yang ZM, Chen JQ et al (2017) Research progress of national infrastructure for bast fiber crop germplasm resources during the 12th Five-year Plan. Plant Fiber Sci China 39(6):321–326

Dang ZH, Zhang JP, She XC (2000) Induction male sterility in linseed by antibiotics treatment. Chinese J Oil Crop Sci 22(1):46–47

Deng X, Chen XB, Long SH et al (2008) Microsatellite molecular marker enrichment by magnetic beads in flax. Acta Agron Sin 34(12):2099–2105

Deng X, Long SH, He DF et al (2010) Development and characterization of polymorphic microsatellite markers in *Linumusitatissimum*. J Plant Res 123:119–123

Dutta S, Mukherjee PK, Mitra A et al (2021) Efficient in planta genetic transformation through agrobacterium-mediated floral-dip technique in Indian flax. Plant Cell Biotechnol Mol Biol 22(53–54):21–36

Friedt W, Bickert C, Schaub H (1995) In vitro breeding of high-linolenic, doubled-haploid lines of linseed (*Linum usitatissimum* L.) via androgenesis. Plant Breed 144:322–326

Galindo-González L, Pinzón-Latorre D, Bergen EA et al (2015) Ion Torrent sequencing as a tool for mutation discovery in the flax (*Linumusitatissimum* L.) genome. Plant Method 11(1):1–14, 19

Guo Y, Qiu CS, Long SH et al (2017) Digital gene expression profiling of flax (*Linum usitatissimum* L.) stem peel identifies genes enriched in fiber-bearing phloem tissue. Gene 626:32–40

Jankauskien Z, Gruzdevien E (2008) Stinging nettle (Urtica dioica L.) – an alternative fibre plant. Opportunities and Challenges of National Economic Development: proceedings of the International conference.Rezekne, p.175–182

Ji YR, Liu YF, Han CW et al (2012) Breeding report of new fiber flax variety Shuangya No.16. Plant fiber sciences in China 34 (3):109–111

Jiang BT, Wang Q, Zhang Y et al (2019) Flax fiber in reinforced composites: application and research progress. Chin Agric Sc Bull 35(23):35–39

Jiang H, Pan G, Liu TM et al (2022) Development and application of novel InDel markers in flax (*Linumusitatissimum* L.) through whole-genome re-sequencing. Genet Resour Crop Evol 69:1471–1483. https://doi.org/10.1007/s10722-021-01313-2

Kang QH (2004) The study on utilization of twin embryos of flax. Plant fiber and products 26(3):110–113

Kang QH, Wang YF, Zhang JM (2002) Report on investigation and study about the research and production of Czech flax. Plant Fiber and products 24(4):43–45

Kang QH, Xu XH, Li ZG et al (2006) Study on cold and slat gene transformation of haploid flax. Plant Fiber Sciences in China 28(6):291–296

Kang QH, Guan FZ, Wu GW et al (2011) Research and application of polyembryonic germplasm in flax(*Linum*). Plant Fiber Sci Chin 33(4):179-182,201

Kang QH, Ma TF, Guan FZ et al (2014) Breeding and evaluation of dihaploid strains of polyembryonic flax. Plant Fiber Sc Chin 36(1):265–269

Kang QH, Jiang WD, Song XX et al (2021) Study progress of apomixis in flax (*LinumUsitatissimum* L.). JNat Fibers 18:1–11. https://doi.org/10.1080/15440478.2019.1604284

Kang QH, Song XX, Jiang WD et al (2022) Breeding of flax variety Huaya 5 and its associated cultivation and retting techniques. Chin Seed Ind 12:119–121

Kozłowski R, Rólski ST, Grabowska L, Rutkowska-Krauze I (2000) Flax and hemp breeding at the Institute of Natural Fibres. In: Proceedings of the 58th Flax Institute of the United States, March 23–25, pp173–179

Kvavadze E, Bar-Yosef O, Belfer-Cohen A et al (2009) 30,000-year-old wild flax fibers. Science 325:1359

Li XY (2023) Functional analysis of *LuAccD* from *Linumusitatissimum* on seed oil accumulation. Yangling, Northwest A&F University

Li WJ, Zhang JP, Chen F et al (2013) Transformation of herbicide resistant gene (*bar*) into flax by pollen tube pathway. Chin Agric Sci Bull 29(12):96–100

Li WJ, Qi YN, Wang LM et al (2019) Correlation between oil content or fatty acid composition and expression levels of genes involved in TAG biosynthesis in flax. Acta Prataculturae Sin 28(1):138–149

Li B, Shi DW, Wei JH (2022) Preparation and properties of flax fiber reinforced epoxy resin composites for automobile interior decoration. Wool Textile J 50(7):13–17

Li W, Zhao LR, Zhang JP et al (2023) Genome-wide identification and analysis of the DMP gene family in flax(*Linumusitatissimum*). Acta Prataculturae Sin 32(3):91–106

Li WJ, Wang LM, Qi YN et al (2024a) Functional analysis of flax *LuWRI1a* in response to drought and salt stresses. Acta Agron Sin 2024-02-20. https://link.cnki.net/urlid/11.1809.s.20240219.1923.028

Li YZ, Zhao XL, Wang YF et al (2024b) Research progress and prospect analysis of the application of flax lignans. JNat Fibers 21(1):2309909. https://doi.org/10.1080/15440478.2024.2309909

Link GKK, Eggera V (1946) Mode, site, and time of initiation of hypocotyledonary Bud Primordial in *Linumusitatissimum* L. Bot Gaz 107(4):441–454

Liu Y (1999) Haploid breeding techniques for polyembryonic flax seeds. Chin Fiber Crop 21(3):19

Manimurugan C, Zanwar A, Sujatha M (2023) Genetic enhancement of nutraceuticals in linseed: breeding and molecular strategies. In: Kole C (ed) Compendium of crop genome designing for nutraceuticals. Springer, Singapore. https://doi.org/10.1007/978-981-19-3627-2_19-1

Mi J (2004) Study on technic of oil and fiber flax species hybridization. JHebei Agric Sci 8(3):20–24

More AP (2022) Flax fiber–based polymer composites: a review. Adv Compos Hybrid Mater 5:1–20. https://doi.org/10.1007/s42114-021-00246-9

Nichterlein K, Nickel M, Umbach H et al (1989) Recent progress and prospects of biotechnology in breeding of linseed (*Linumusitatissimum* L.). FAT. Sci Technol 91(7):272–275

Obert B, Zácková Z, Samaj J et al (2009) Doubled haploid production in flax (*Linumusitatissimum* L.). Biotech Adv 27:371–375

Pali V, Mehta N, Verulkar S et al (2015) Molecular diversity in flax (*Linumusitatissimum* L.) as revealed by DNA based markers. Vegetos 28(1):157–165

Pan G, Chen AG, Li JJ et al (2020) Genome-wide development of simple sequence repeats database for flax (*Linum usitatissimum* L.) and its use for genetic diversity assessment. Genet Resour Crop Evol 67:865–874

Qu ZH, Wang YX, Qiao HM et al (2018) Breeding of new flax (*Linumusitatissimum* L.) variety Jizhangya No.1 by interspecific hybridization. Plant Fiber Sci China 40(1):8–11

Rahman M, Hoque A (2023) Flax Breeding. In: You FM, Fofana B (eds) The flax genome. Compendium of Plant Genomes. Springer, Cham. https://doi.org/10.1007/978-3-031-16061-5_4

Shivaraj SM, Dhakate P, Sonah H et al (2019) Progress toward development of climate-smart flax: a perspective on omics-assisted breeding.In: KoleC. Genomic designing of climate-smart oilseed crops. Springer, Cham, pp 239–274. https://doi.org/10.1007/978-3-319-93536-2_5

Song SM (1996) The progress of the study on flax anther culture. Chin Fiber Crop 18(4):4–6

Song SM (2007) Advances in flax haploid breeding. J Heilongjiang Bayi Agri Uni 19(3):25–28

Song SM (2008) The breeding of new fiber flax variety Shuangya No.13. Plant Fiber Sci China 30(4):197-198,216

Sun HT (1979) Preliminary report on anther culture of flax. Chin Sci Bull 20:948–950

Sun SF, Wang SY (2009) Key technique to increasing callus induction frequency in linseed anther culture. JNorthernAgric 3:36–38

Sun HT, Fu WD, Dong LH et al (1983) Study on regeneration plants from tip of stem cotyledon and hypocotyl of flax. Chin Sci Bull 20:1332–1334

Sun HT, Fu WD, Wu CB et al (1995) Effects of some factors on the pollen culture of flax. Chin Bull Bot 12(3):41–44

Vrinten P, Hu Z, Munchinsky MA et al (2005) Two *FAD3* desaturase genes control the level of linolenic acid in flax seed. Plant Physiol 139(1):79–87

Vromans J (2006) General introduction. Molecular genetic studies in flax (*Linum usitatissimum* L.), (chapter 1), PhD thesis. Wageningen University, The Netherlands, pp 7–19

Wang YF (2001) The history and present situation and development direction of flax breeding in China. Nat Fibers (Special Edition) 2:1–5

Wang YF (2008) The present situation and suggestions about the study of flax germplasm resource in China. J Life Sc 2(1):67–71

Wang B, ZL, Hou JJ (2022) Effects of EMS mutagenesis on seed germination of M_0 generation, agronomic and quality traits of M_3 generation in five flax varieties. Acta Agric Boreali-occidentalis Sin 31(9):1085–1093

Wang YF, Qiao GJ, Yan ZF et al (1991) A study on the combining ability of flax varieties. Heilongjiang Agric Sci 5:24–27

Wang YF, Zhou SJ, Liu Y et al (2000a) Study on flax genetic transformation by *Agrobacterium tumefaciens*. China's fiber Crops 22(3):25–27

Wang YM, Xu YY, Jia JF (2000b) Transformations system and study on the introduction of chitinase gene into *Linumusitatissium*. Acta Bot Boreal-Occident Sin 20(3):346–351

Wang YF, Kang QH, Liu Y et al (2004) Study on flax genetic transformation mediated by *Agrobacterium tumefaciens*. J Nat Fibers 1(1):1–10

Wang YF (2005) Study and development of flax biotechnology in China. Plant Fiber Prod 27(2):60–65

Wang L, Ma XD, Mi J et al (2007) Research on embryo development of interspecific hybridization between cultivated flax and wild flax. J Agric Uni Hebei 30(5):19–22

Wang YF, Jia WQ, Qiu CS et al (2014) Optimization of high-efficiency regeneration system of flax. Chin J Oil Crop Sc 36(5):661–666

Wang YF, Qiu CS, Szopa J et al (2016) Breeding process and cultivation techniques of new fibre flax variety Zhongyama 4. Modern Agric SciTechnol. (24):40–41,45

Wang H, Wang JH, Qiu CS et al (2017) Comparison of phytochemical profiles and health benefits in fiber and oil flaxseeds (*Linum usitatissimum* L.). Food Chem 214:227–233

Wang YF, Jankauskiene Z, Qiu CS et al (2018) Fiber flax breeding in China and Europe. JNat Fibers 15(3):309–324. https://doi.org/10.1080/15440478.2017.1325431

Wang W, Chen J, Chen C et al (2023) Creating new germplasm of oil flax by EMS mutation. JNuc Agric Sci 37(10):1931–1937

Wu JZ, Qian Z, Wu GW et al (2017) Development of novel SSR markers for flax (*Linumusitatissimum* L.) using reduced-representation genome sequencing. Front Plant Sci 7:2018. https://doi.org/10.3389/fpls.2016.02018

Wu YN, Jiang H, Pan LL et al (2023) Cloning and expression analysis of *LuFAD7* gene in flax. Mol Plant Breed. https://kns.cnki.net/kcms/detail/46.1068.S.20230310.1627.006.html

Xie RX, Zhang JP, Dang ZH et al (2009) Effects of hormone and low temperature pretreatment on callus induction of oil flax. Mol Plant Breed 7(6):1180–1185

Xiong HP (2008) The breeding of bast fiber crops. Chinese Agricultural Science and Technology Press, Beijing

Xu YY, Niu BT, Jia JF (2000) In vitro selection of salt and PEG tolerant flax lines from the progeny of seeds carried by satellites. Acta Bot Boreali-Occidentalia Sin 20(2):159–163

Xue M, Gao YF, Zhu LL et al (2022) Induction and histocytological observation of embryogenic callus from oil flax. Molecular Plant Breed. https://kns.cnki.net/kcms/detail/46.1068.S.20220926.1826.004.html

Yang T (2021) Molecular cloning, establishment of CRISPR/Cas9 technology system and functional expression study of heteromeric ACCase subunit genes from flax[D]. China, Gansu Agricultural University, Lanzhou

Ye CL, Wang W, Chen J et al (2023) Preliminary identification of candidate genes with resistance to Powdery Mildew in oil flax based on BSA-Seq. Crops 6:69–78

You FM, Cloutier S, Rashid KY et al (2019) Flax (*Linumusitatissimum* L.) genomics and breeding. In: Al-Khayri J, Jain S, Johnson D (eds) Advances in plant breeding strategies: Industrial and Food Crops. Springer, Cham, pp 277–317. https://doi.org/10.1007/978-3-030-23265-8_9

Yu Y, Huang WG, Jiang WD et al (2013) Preliminary analysis of agronomic traits of M_1 flax induced by physical and chemical agents. Heilongjiang Agric Sci 11:5–8

Yu Y, Guo WD, Zhao LJ et al (2016) Cloning and expression analysis of the cellulose synthase gene *LuCesA8* in flax. J Northeast Agric Uni 47(8):39–45

Yuan ZH, Sun HT, Wu CB et al (1997) Establishment of somaclones and plants regenerated in flax. Chin Fiber Crop 19(1):17–18

Yuan HM, Guo WD, Zhao LJ et al (2016) Cloning and expression analysis of the glycosyltransferase gene *LuUGT72E1* in flax. Crops 4:62–67

Yuan HM, Guo WD, Zhao LJ et al (2021) Genome-wide identification and expression analysis of the WRKY transcription factor family in flax (*Linumusitatissimum* L.). BMC Genomics 22:375. https://doi.org/10.1186/s12864-021-07697-w

Yurkevich OY, Naumenko-Svetlova AA, Bolsheva NL et al (2013) Investigation of genome polymorphism and seed coat anatomy of species of section Adenolinum from the genus *Linum*. Genet Resour Crop Evol 60:661–676. https://doi.org/10.1007/s10722-012-9865-z

Zhang Y (2019) CRISPR/Cas9 system mediated for genome targeted editing and functional analysis of *FAD2* in flax(*Linumustitatissimum* L.). Urumqi, Xinjiang University, China

Zhang H (2023) Impact of climate change on ecosystems and agricultural production. JAgricCatastrophology 13(11):201–203

Zhang JP, Dang ZH (2002) Heterosis of new male-sterile flax and sensitivity to temperature. *Acta* Agriculturae Boreali-occidentalis Sinica 11(4):26–28,31

Zhang JP, Dang ZH (2005) Yields of two-line flax hybrids and effects by the male sterility plant ratio. Acta Agric Boreali-occidentalis Sin 14(3):73–75

Zhang XP, Xue SD, Qiu CS et al (2007) Screening of the molecular markers linked to waterlogging tolerance gene of flax using RAPD-BSA technique. Plant Fiber Sci China 29(5):290–294

Zhang LL, Bai W, Mi J et al (2012) Identification interspecific hybridization between cultivated flax and wild flax. Acta Agric Boreali-Sin 27(supplement):57–60

Zhang Y, Jiang HX, Yan WL et al (2019) Construction of expression vector for knocking out *FAD2* gene in flax by CRISPR/Cas9 System. Mol Plant Breed 17(7):2185–2192

Zhao LR, Li W, Wang LM et al (2023) Identification and relative expression pattern of *PLA1* gene family in flax. Acta Agron Sin 49(11):2949–2965

Zuk M, Prescha A, Stryczewska M et al (2012) Engineering flax plants to increase their antioxidant capacity and improve oil composition and stability. JAgricFood Chem 60(19):5003–5012

Chapter 11
Industrial Hemp (*Cannabis sativa* L.) Breeding

Jordi Petit, Amparo Monfort, and Jason Argyris

Abstract Industrial hemp (*Cannabis sativa* L.) stands as a multipurpose and sustainable biobased crop, holding the potential to secure the supply of specialized non-food commodities like bio-composites and cannabinoids. Hemp natural fibers are considered a promising alternative to unrenewable and unsustainable resources, such as mineral or petroleum-derived plastics. However, the lack of intensive hemp breeding in the latter decades of the twentieth century, due to the prohibition of its cultivation, contributed to a big gap in molecular knowledge of this crop and most breeding programs continue using traditional methods. Conventional breeding has created commonly used varieties that still exhibit considerable heterogeneity. The renewed interest in this crop is fostering research in many different fields from agronomy to genetics and the development of new techniques to create new varieties with better qualities such as fiber and cannabinoid yields and composition. To cope with foreseen natural fiber demands along with other important hemp by-products, the development of modern hemp breeding programs is expected to adopt advances in molecular marker-assisted selection, genomics, mutagenesis and tissue culture. In the future genetic engineering and gene editing approaches are also expected to be implemented in breeding programs as it is in many staple crops, vegetables, legumes and fruit crops. Finally, it will be crucial to plan breeding programs based on specific end-uses of the plants and the environment, particularly photoperiod, as it is fundamental to provide cultivars that respond well to specific cultivation sites. The objective of this chapter on cannabis breeding is to provide a brief history of the crop and outline the major agronomic requirements for production. Both traditional and modern breeding methods and technologies are discussed, oriented towards producing new varieties suited for increasingly challenging environmental conditions amidst a changing global climate. Chapter content focuses on methods for the improvement of key traits to develop cannabis varieties for increased

J. Petit (✉) · A. Monfort · J. Argyris (✉)
Centre for Research in Agricultural Genomics (CRAG)—Institute of Agrifood Research and Technology (IRTA), Barcelona, Spain
e-mail: jordi.petit@irta.cat; jason.argyris@irta.cat

K. F. M. Salem et al. (eds.), *Breeding and Biotechnology of Grass and Bast Fiber Crops*, Advances in Plant Breeding Strategies 12,
https://doi.org/10.1007/978-3-032-00403-1_11

production of natural pharmaceuticals and innovative industrial uses to improve human health and well-being.

Keywords Biobased crop · Industrial hemp · *Cannabis sativa* L. · Natural fibers · Traditional breeding · Molecular breeding · Tissue culture

11.1 Introduction

Industrial hemp (*Cannabis sativa* L.), chromosome number 2n = 20, genome size ~ 818 Mbp for female plants and ~ 843 Mbp for male plants (Moliterni et al. 2004), is an herbaceous annual C3 plant. Hemp is a multipurpose crop cultivated for its fiber, seeds and by-products such as cannabinoids and terpenes. The Cannabis genus is a member of the family Cannabaceae located within the Order Rosales (Table. 11.1). Historical taxonomic categorization of cannabis based on morphology, chemotype (chemical composition) and geographical origin has recognized three main species: *Cannabis sativa, Cannabis indica,* and *Cannabis ruderalis*. However, this classification has been debated due to the extensive crossbreeding and hybridization practiced in cannabis cultivation that has blurred the lines between species. Modern

Table 11.1 Taxonomic hierarchy of *Cannabis sativa* L.

Kingdom	Plantae—plantes, planta, vegetal, plants
Subkingdom	Viridiplantae—green plants
Infrakingdom	Streptophyta—land plants
Superdivision	Embryophyta
Division	Tracheophyta—vascular plants, tracheophytes
Subdivision	Spermatophytina—spermatophytes, seed plants, phanérogames
Class	Magnoliopsida
Superorder	Rosanae
Order	Rosales
Family	Cannabaceae—hemp
Genus	*Cannabis* L.—hemp
Species	*Cannabis sativa* L.—hemp, grass, hashish, Mary Jane, pot, marijuana
	Direct Children:
Subspecies	*Cannabis sativa* ssp. indica (Lam.) E. Small & Cronquist—hemp, grass, hashish, Mary Jane, pot, marijuana
Subspecies	*Cannabis sativa* ssp. sativa L.—hemp, grass, hashish, Mary Jane, pot, marijuana

Source: https://www.itis.gov

Fig. 11.1 Countries that allowed hemp cultivation in 2019. (Source: Map modified from Duque-Schumacher et al. (2020)). (Source: Eurostat)

genomics-based taxonomy refutes the multiple species framework and points towards the genus Cannabis consisting of a highly diverse monotypic species (reviewed by Lapierre et al. 2023).

Cannabis spread from its center of origin in central Asia or Western China around 12.000 years ago (Crocq 2020) and successfully expanded due to its large adaptability to a broad range of geographical and climatic conditions (Ebskamp 2002). This makes the crop well-suited for cultivation in various global regions (Fig. 11.1). During the late 1970s and early 1980s, hemp cultivation in China encompassed an impressive average planting area of approximately 160,000 hectares. Notwithstanding this, the cultivation of hemp in Western countries underwent a significant decline spanning decades, primarily due to competition from other resources such as cotton or synthetic fibers, alongside high labor costs. The prohibition of cultivation, driven by concerns about the use of cannabis as narcotics, further contributed to this decline. A substantial hemp industry survived only in Eastern Europe, the former Soviet Union and China (De Meijer 1994). Currently, China, Europe and Canada emerge as the pivotal regions for hemp cultivation. According to FAOSTAT, the year 2011 witnessed a global hemp cultivation area of 61,318 hectares (Salentijn et al. 2015). From that point onwards, the land area used for hemp cultivation continued to expand without interruption. China is believed to be the world's top hemp producer with 65,400 hectares in 2020 (Zhao et al. 2021). In Europe, the land dedicated to hemp cultivation has undergone substantial growth, escalating from 20,540 hectares in 2015 to 33,020 hectares in 2022, signifying a notable 22% increase (Fig. 11.2). Notably, France stands as the largest producer, accounting for more than 60% of EU production, followed by Germany (17%) and The Netherlands (5%).

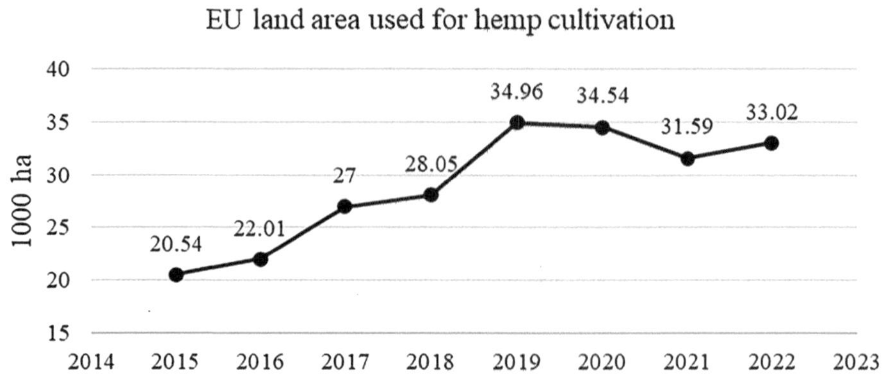

Fig. 11.2 EU land area used for hemp cultivation in hectares from 2015 until 2022. (Source: Eurostat)

11.1.1 Harnessing Industrial Hemp: Addressing Global Environmental Challenges

The reliance of humans on finite and environmentally harmful resources has reached a plateau that compromises the well-being of future generations. By 2050, the global population is projected to reach 9.6 billion, and their consumption is estimated to be 1.6 times the Earth's available resources (Zabaniotou and Kamaterou 2019). Hence, it is crucial to identify sustainable substitutes for petroleum and mineral-derived goods, not only to curtail greenhouse gas emissions but also to ensure the continued production of essential items such as food, feed and non-food commodities for the growing population. These alternatives are primarily directed towards the development of carbon-neutral bio-based resources, with a strong focus on societal, environmental, and economic benefits (Carus and Dammer 2018). Advocating for the utilization of natural plant fibers instead of synthetic fibers derived from petrochemicals or non-renewable fibers (sourced from minerals) plays a role in fostering a greener planet (Dam 2008).

Natural plant fibers are clusters of cells primarily composed of lignocellulosic biomass derived from plants (Petit et al. 2020a). These fibers encompass various classes, including seed hair fibers like cotton, bast fibers such as flax, hemp or jute, and woody fibers like eucalyptus or poplar (Dam and Bos 2004). Hemp fibers are cells, specifically cell walls, located in the stem of the plants. Those cells are known as elementary fibers. Hemp stems have two fiber types, the bast fibers and the woody hemp core (WHC) or shives. Bast fibers (phloem cells) wrap the WHC and are divided into primary and secondary bast fibers. Secondary bast fibers are synthetized during the fiber maturation, after flowering onset. Elementary bast fibers are attached through their middle lamellas to form bundles of one to four dozen elementary fibers. Fiber bundles can reach 100 mm in length and a diameter of 15–25 µm. In contrast, the WHC is composed of individual short xylem cells (~2 mm).

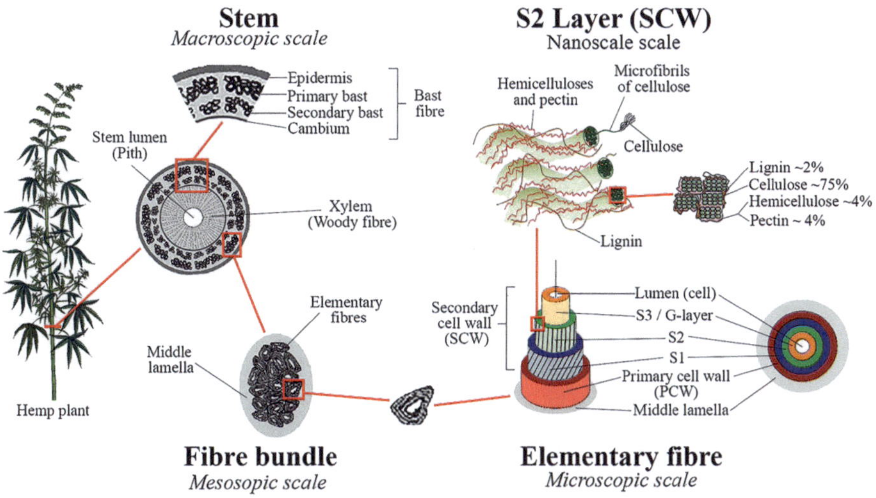

Fig. 11.3 Schematic representation of hemp fibers at different scales from the stem to the microfibrils of cellulose. Vertical lines in the S1 and S2 layers of elementary fibers represent the angle of the cellulose molecules in these layers of the cell wall. (Source: Petit (2020))

Elementary fibers are composed of multiple layers of cell walls: primary cell wall (PCW) and secondary cell wall (SCW) which is divided into three sublayers S_1, S_2 and S_3 (G-layer) being the S_2 the predominant (~80%), as depicted in Fig. 11.3. The biochemical composition of hemp cell walls varies depending on the layer and the plant developmental stage, but they are made up of four building blocks: cellulose, hemicellulose, pectin and lignin. Cell walls are structures of microfibrils of cellulose within a matrix of hemicellulose and either lignin or pectin. The middle lamella is a matrix of pectin and lignin. Hemp bast fibers are made of approximately 4% pectin, 4% hemicellulose, 2% lignin and 75% cellulose. Hemp bast fibers have a large amount of cellulose and low lignin content while WHC has a higher lignin content (Petit 2020).

Traditionally, natural fibers sourced from cotton, flax or hemp have found applications in textiles (Dam 2008). Another traditional use of natural fibers from bast crops involves pulping for paper production (De Meijer and Van Der Werf 1994). Novel applications of natural fibers have expanded their utility well beyond textiles and paper. Fiber bio-composites sourced from bast fiber crops have proven to be viable alternatives in industries such as automotive manufacturing (Holbery and Houston 2006) and building insulation (Grohe 2004; Jones and Brischke 2017; Wambua et al. 2003), effectively replacing petrochemical and mineral fibers (i.e. glass fibers). Additionally, residual biomass from bast fiber crops (i.e. woody portions from hemp) can be repurposed for construction materials like lightweight concrete (Elfordy et al. 2008) or animal bedding (Karus and Vogt 2004).

Hemp is a highly productive biobased crop that can yield up to 3 t ha^{-1} $year^{-1}$ of bast fiber (Amaducci et al. 2015; Piotrowski and Carus 2019; Struik et al. 2000Struik;. This has led to significant interest from the bio-composite industry in

Table 11.2 Average fiber yield, fertilizer and water requirements of important natural fiber crops

Crop	Fiber yield (t ha^{-1} year^{-1})	Fertilizer requirements[d] (kg ha^{-1} year^{-1})			Water requirements[f] (mm year^{-1})	Reference
		N	P	K		
Cotton	0.6–1.3[ab]	230–250	155–165	100–110	700–1300	FAO (2003)
Hemp	1.2–3[c]	80–100[e]	52–67[e]	65–150[e]	500–700	Struik et al. (2000), Amaducci et al. (2015), Piotrowski and Carus (2019)
Flax	1.5[c]	60[e]	30–50[e]	70–100[e]	350–450	Sánchez Vallduví (2011), Piotrowski and Carus (2019), https://flaxcouncil.ca
Jute	2–2.5[c]	35.1	15.4	73.1	500	http://agritech.tnau.ac.in

Modified from Petit (2020)
[a] Ginned cotton yield
[b] Seed-hair fiber
[c] Bast fiber
[d] Fertilizer requirements vary depending on soil quality
[e] Fertilizer requirements for fiber end-use
[f] Total water requirements including irrigation and rainfall

utilizing hemp (Piotrowski and Carus 2019). Moreover, hemp cultivation boasts a reduced environmental footprint in comparison to other fiber crops, as outlined in Table 11.2. Both in the case of hemp and other bast fiber crops such as flax or jute, minimal amounts of fertilizers, crop protection agrochemicals and water are required (Van Der Werf and Turunen 2008). This translates to a relatively low demand for fossil fuel consumption in mechanized weeding and crop maintenance processes (Ebskamp 2002). Hemp can also be used as a phytoremediation crop, addressing heavy metal pollution concerns (Ahmad et al. 2016). Its compatibility with crop rotation strategies further enhances the economic and ecological aspects of agricultural land, as highlighted in Amaducci et al. (2015). Consequently, leveraging hemp fiber production for innovative purposes stands to yield environmental benefits while contributing to the economy.

11.1.2 Importance for Human Quality of Life

In recent years, there has been an increasing public acknowledgment of the therapeutic potential of cannabis, a trend that has been particularly pronounced in countries such as the USA, Australia, Canada, Israel and Uruguay (Welling et al. 2015).

Besides fiber, industrial hemp is also a source of valuable by-products including seeds, cannabinoids and terpenes. Hemp can produce premium seed oil that boasts a significant proportion of omega-3 fatty acids (Leizer et al. 2000). This oil finds application in the cosmetic industry (Sapino et al. 2005) and is also recognized as a functional food that promotes health (Callaway 2004). Furthermore, hemp presents valuable by-products in the form of cannabinoids and terpenes. Currently, these compounds are driving a continuously expanding cannabinoid-based pharmaceutical industry (Deyholos and Potter 2014). Glandular trichomes are responsible for producing such secondary metabolites, which accumulate in substantial concentrations within the bracts of female inflorescences and to a lesser extent in leaves. Cannabinoids are produced in plants in their carboxylic acid forms, which form neutral cannabinoids in a non-enzymatic reaction when exposed to heat (Dussy et al. 2005). Among the extensive number of cannabinoids derived from cannabis, delta-9-tetrahydrocannabinol acid (THCA) and cannabidiol acid (CBDA) stand out. Notwithstanding this, it is important to note that fiber hemp contains only traces of THCA (THC equivalent <0.3%) due to restrictions on higher concentrations in field production and agriculture. While THC is primarily known for its psychoactive effects, CBD is a non-psychoactive cannabinoid. CBD has gained significant popularity due to its potential therapeutic applications across a wide range of diseases and symptoms, often without discernible side effects (Baker et al. 2003), particularly when used in moderate quantities. Terpenes may also contribute to pharmacological advances (Yang et al. 2020; Welling et al. 2016). For instance, α-humulene has antifungal properties and it is believed to be beneficial in the treatment of infections such as cryptococcosis (Wanas et al. 2016).

11.1.3 Domestication, Selection and Early Improvements

Cannabis is one of the first known plants to be domesticated (Salentijn et al. 2015). The existing gene pool is believed to consist mainly of domesticated or feral populations and cultivars originating from steep selective pressure towards phenotypes for specific end-uses (Mandolino and Carboni 2004; Small 2015). China has one of only 35 cannabis biodiversity hotspots worldwide (Sloan et al. 2014) and is home to hundreds of cannabis landraces that have reportedly undergone distinct domestication events across its southwest regions (Salentjin et al. 2015). Cannabis spread from this region to the rest of Asia, Europe, Africa (Duvall 2016), and North and South America (Small and Marcus 2002). It is currently cultivated globally (Salentijn et al. 2015). Throughout its domestication, cannabis was subjected to prolonged artificial selective pressure. Cannabis is naturally a dioecious and obligate outcrossing species (Sawler et al. 2015). This reproductive behavior generates a constant pattern of variation and contributes to a significant hybridization between undomesticated and domesticated populations, resulting in a nearly complete absence of intact wild populations (Smart and Cronquist 1976). In addition, the decline of hemp cultivation in the latter decades of the twentieth century resulted in a considerable loss of ex situ conservation for industrial hemp, contributing to a significant genetic erosion (Ranalli 2004; Small and Marcus 2002; Welling et al. 2015). The genetic erosion experienced by this species along with its current renewed interest, has compelled breeders to develop new varieties to effectively address emerging demands.

Traditional domestication and selection have been based on exploiting natural genetic diversity. This can be achieved through the hybridization of elite breeding lines with wild germplasm, known as genetic resources, that have not been artificially subjected to intensive selection pressure for a targeted end-use application or environment (Welling et al. 2015). Wild relatives or landraces offer *ex situ* genetic resources to reintroduce allelic diversity into domesticated populations (Vincent et al. 2013). A significant extent of genetic diversity has been reported in cannabis (Lynch et al. 2016; Petit et al. 2020a; Ren et al. 2021; Woods et al. 2023; Zhang et al. 2020) and large phenotypic variation has been described for many traits in hemp, including fiber quality (Petit et al. 2020a), oil composition (Grigor'ev et al. 2010; Kriese et al. 2004; Matthäus et al. 2006) and cannabinoid composition (Welling et al. 2015). The exploitation of these resources could be used for the development of new cultivars tackling targeted applications or better adapted to certain environments.

Early anthropogenic improvements of cannabis focused on traits defining narcotic, fiber yield, hempseed and pharmacological end-uses (Small 2015). These breeding pressures diverged partially cannabis accessions into "fiber-type" and "drug-type" (Welling et al. 2015). Transcriptome analysis, as demonstrated by Van Bakel et al. (2011), provided evidence for the historical selection pressure that led to the differentiation between fiber and marijuana chemotypes. Towards the middle to the latter part of the twentieth century, industrial hemp underwent intensive selection aimed at reducing THC levels below 0.3%, primarily in response to legal cultivation constraints. This intensive breeding effort was attributed mainly to France and other European countries (Amaducci et al. 2015) where mass selection and early uses of marker-assisted selection (MAS) were employed (Ranalli 2004). Eighty-eight fiber cultivars are currently registered for use within the European Union (sourced: EUPVP—Common Catalogue Information System) as shown in Appendix 11.2. Many contemporary fiber cultivars are thought to trace their lineage back to Central Russian and Mediterranean landraces, as well as their derivative cross-progenies (De Meijer and van Soest 1992 and reviewed in Welling et al. 2015).

11.2 Current Cultivation Practices and Challenges

Hemp fiber yield and fiber quality traits are extensively variable (Fig. 11.4) and heritable but are also adaptable to specific environments, as reported in Petit et al. 2020a. These phenotypic characteristics exhibit a robust response to environmental factors, with a particular emphasis on photoperiod and temperature, while also being influenced by soil composition (Amaducci et al. 2015; Faux et al. 2013;

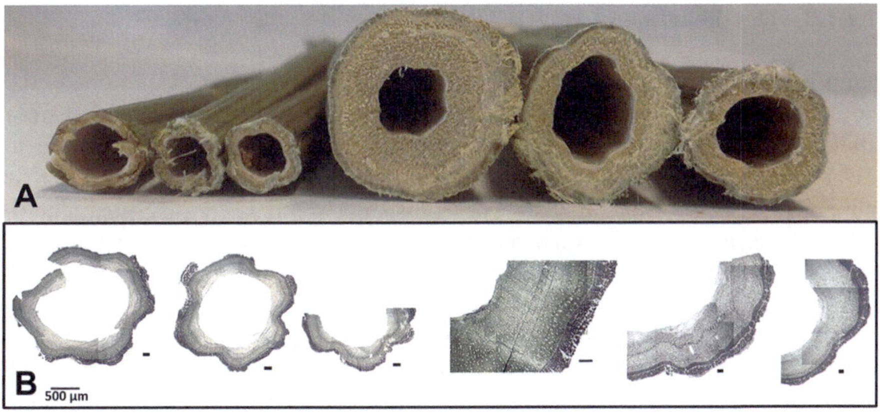

Fig. 11.4 Stem morphology of six contrasting hemp accessions. (**a**) Cross-section of the stems (**b**) microscope observation of the corresponding cross-sections after 0.1% Toluidine blue staining. (Source: Picture modified from Petit et al. (2019) and Petit (2020))

Sawler et al. 2015). Consequently, agronomic management practices and growing conditions, including factors like plant density, nitrogen fertilization and the timing of harvesting, play a crucial role in influencing both the yield and quality of hemp fiber.

11.2.1 Influence of Agronomic Factors on Industrial Hemp

11.2.1.1 Soil Preparation

The recommended tillage practices begin with fall or winter plowing at a depth ranging between 30 to 40 cm, followed by preparation of a fine seedbed during the spring season, just before sowing. An optimal soil for hemp cultivation is characterized by a pH range of 6.0 to 7.5. The soil should possess considerable depth, along with abundant capillarity and aeration, a rich nutrient profile, and a good capacity for retaining water (Amaducci et al. 2015). Sandy loam soil exhibits the most favorable conditions for hemp growth. However, heavy clay and sandy soils are comparatively less optimal for successful cultivation (Li 1982). Soil preparation holds particular significance, particularly during wet years for clay soils, as it can result in waterlogging. Similarly, in the case of sandy soils with a low water holding, appropriate soil preparation becomes crucial to prevent adverse effects on seedling emergence (Struik et al. 2000).

Fertilizers are generally applied during the spring season before the final seedbed preparation. This applies to phosphorus (P) and potassium (K), while nitrogen (N) is applied at sowing or after seed emergence, coinciding with the plant's demands for energy to promote vegetative growth (fiber development).

11.2.1.2 Day Length and Variety Choice

Hemp is a short-day plant, with its flowering phase responses to variations in the photoperiod regime (Amaducci et al. 2012). In regions where the transition from extended daylight to shorter daylight periods occurs early, hemp plants tend to flower prematurely. This is the case of Southern European growing sides, such as Spain or Italy. Conversely, in areas where this shift occurs later, the critical photoperiod triggering flowering is also delayed, such as in the North or East European sides (Salentijn et al. 2019). This behavior significantly impacts plant development, as the vegetative growth allows for biomass accumulation, but nutrient allocation shifts from stem, leaf and root production to the generation of flowers and seeds around the onset of flowering. Hence, it is possible to achieve a specific stem or seed yield in a particular environment based on the photoperiod sensitivity of the selected variety (Amaducci et al. 2015). Moreover, lignification of fibers intensifies after flowering (Van Der Werf and Turunen 2008), affecting the quality of the fibers (Petit et al. 2020a).

Consequently, the selection of a variety suitable for a specific end-use purpose and adapted to a particular environment stands as a pivotal factor for the successful cultivation of hemp. Cultivars developed in Northern climates exhibit lower biomass yields when cultivated in Southern regions, primarily attributed to the shortened growth duration and early flowering. Conversely, cultivars bred in low-latitude areas, such as South Europe, display a tendency for delayed flowering and substantial biomass yields when cultivated in regions situated at higher latitudes (reviewed in Amaducci et al. 2015). In end-use applications prioritizing biomass production, such as for paper or bio-composite applications, late-flowering varieties appear to be the optimal choice. Nevertheless, when the priority is quality, as in the case of textile applications, the choice of variety becomes more complex. This decision is combined with the appropriate harvesting time, aimed at preventing fiber lignification and the development of secondary bast fibers (Liu et al. 2015). Petit et al. (2020a) performed a comprehensive fiber quality characterization of 124 hemp accessions, including most EU-registered accessions. This study can serve as a valuable resource for selecting the appropriate variety.

The selection of variety should also consider the sex of hemp. Cannabis is naturally a dioecious plant species where male and female flowers are on different plants. Yet, certain dioecious plants generate flowers of the opposite sex in contrast to the ones determined by their genetics (Faux et al. 2016; Freeman et al. 1980). Monoecious genotypes, also termed hermaphrodites, can spontaneously occur. In dioecious hemp varieties, male plants have higher fiber quality while female plants are more lignified. Despite this sexual dimorphism, dioecious genotypes are considered superior for fiber production (Berenji et al. 2013; Bócsa and Karus 1998). Monoecious varieties were bred as a dual-purpose crop suited to the production of fiber and seeds. During the process of selecting dual-purpose varieties, careful consideration must be given to the photoperiod, as late flowering time may not provide enough time for proper seed ripening.

11.2.1.3 Temperature

Temperature plays an essential role in hemp cultivation. Hemp grows under a wide range of temperatures, and it is specially adapted to the temperate climate zone. The optimal growing temperature of hemp oscillates in the mean daily air temperature between 16 °C and 27 °C but it also tolerates colder and warmer conditions. For instance, young seedlings can tolerate some exposure to frost and the ceiling temperature for hemp growth is 41 °C. In Southern Europe, industrial hemp for fiber production requires 1900–2000 $\sum$°C of the cumulated Celsius degree day over a period at a base temperature of 1 °C) (Petit et al. 2020a). The accumulated temperature increases to 2700–3000 $\sum$°C for seed production (Adesina et al. 2020).

11.2.1.4 Irrigation

In Europe, hemp requires a water supply of 500–700 mm to achieve its optimal growth, depending on the end-use application. Around 250–300 mm of the total amount is needed to be available during the vegetative growth stage (June and July) (Adesina et al. 2020; Bócsa and Karus 1998). Drought in combination with high temperatures has been reported to accelerate flowering but delay plant growth and fiber maturation (Amaducci et al. 2008, 2015). Therefore, when the rainfall in the growing environment falls below the threshold of the required water supply, the implementation of irrigation becomes necessary. This is the case for most Mediterranean hemp cultivation sites. Cosentino et al. 2013 reported that at least 250 mm of irrigation water for monoecious early cultivars and 450 mm for dioecious late genotypes were needed in semi-arid Mediterranean environments.

11.2.1.5 Hemp Fertility Requirements

Nitrogen (N), phosphorous (P) and potassium (K) fertilization should be determined based on soil fertility. Fertilization can be negligible in rich soils (Struik et al. 2000), since excessive N fertilization accelerates rapid stem elongation and increases the susceptibility of the crop to lodging, reducing fiber quality (Desanlis et al. 2013). However, a deficit of N will cause a loss of yield (Adesina et al. 2020). N requirements of hemp have shown diversity across numerous studies (reviewed in Amaducci et al. 2015), ranging from 0 to 220 kg ha^{-1}. Nevertheless, beyond an N fertilization rate of 150 kg ha^{-1}, no additional increase in yield was reported (Finnan and Burke 2013). Standard N fertilization ranges from 80 to 100 kg ha^{-1}, including the already available nitrogen in the soil. Hemp displays reduced sensitivity to K and P than to N fertilization (Amaducci et al. 2015). Finnan and Burke (2013) reported that hemp has a lower requirement for K than other crops, and they suggested an annual requirement of 65 kg ha^{-1}. Ivonyi et al. (1997) reported that P fertilization had no impact on stem yield, and the P uptake of industrial hemp ranged from 52 to 67 kg ha^{-1} (Table 11.2).

Fig. 11.5 Industrial hemp field crop. (**a**) Seed emergence, (**b** and **c**) Hemp plants in the vegetative growth stage (juvenile plants), and (**d**) Hemp plants in full flowering stage. (Photos by J. Petit)

11.2.1.6 Sowing Date

The timing of sowing is determined by soil temperature and the availability of water, a combination that guarantees optimal germination and successful crop establishment (Fig. 11.5). It is also defined based on the photoperiod that determines the length of vegetative growth and ultimately fiber and seed yield (Amaducci et al. 2008; Cosentino et al. 2013; Desanlis et al. 2013). Early sowing is associated with higher stem and seed yields. However, in Southern EU environments, early sowing accelerates the end of the juvenile phase before the swift to the critical photoperiod, consequently shortening the vegetative growth (Amaducci et al. 2008; Cosentino et al. 2013). Conversely, in Northern EU environments, early sowing increases the susceptibility to cold damage (van de Werf et al. 1995). The typical sowing time of industrial hemp in Europe falls in early spring, generally around mid-March, taking advantage of regular spring rainfall. In addition, early sowing is coupled with an early harvest, effectively preventing unfavorable conditions for retting and stem drying (Müssig and Martens 2003). These post-harvest procedures are necessary to facilitate the smooth separation of distinct fibers within the stem, namely the bast and the woody hemp core.

11.2.1.7 Plant Density

Hemp plants strongly respond to plant density (Amaducci et al. 2008; Westerhuis et al. 2009). High plant density leads to a rapid canopy closure, providing an advantage in terms of weed competition. In addition, these conditions lead to intense interplant competition which affects plant biometrics, resulting in self-thinning. Hemp plants cultivated at high densities produce longer and thinner internodes at earlier juvenile stages, the interplant competition limits stem growth and at the end of the vegetative growth, plants are shorter, thinner and have fewer nodes than plants cultivated at low densities (reviewed in Amaducci et al. 2015).

Plant density also affects fiber content and fiber finesses, and ultimately affects fiber quality. It is considered that higher fiber yields can be achieved with higher plant densities and the fibers are finer and have a lower proportion of lignified secondary bast fibers (Amaducci et al. 2002, 2008; Petit et al. 2020a, b; Van der Werf et al. 1995). These affected fibers can be of advantage in certain applications, such as textiles. Therefore, different plant densities are used in different end-use applications. The optimal density for the cultivation of hemp for cannabinoid production was reported to be 10–15 plants m^{-2}. For seed production, the recommended density varies between 30 to 75 plants m^{-2}. Conversely, higher densities ranging from 150 to 350 m^{-2} are proposed for textile production. On the other hand, lower densities are recommended for non-textile applications, such as for bio-composites, with a reported density of 90 plants m^{-2} (as reviewed in Amaducci et al. 2015).

11.2.1.8 Crop Rotation and Weed Management

Hemp is a spring-summer crop and holds a significant role in crop rotations due to its advantageous impact on subsequent crops (Venturi and Amaducci 1999). Hemp cultivation is often followed by a winter crop. Wheat and soybean yields are reported to benefit when hemp is cultivated preceding the crop (Bócsa and Karus 1998; Gorchs et al. 2000; Liu et al. 2012). The root system of hemp can penetrate deeply into the soil, reaching depths of 2–3 m. This attribute contributes to improving soil architecture (Amaducci et al. 2008).

The number of phytosanitary control options for the crop is currently quite limited (Monthony et al. 2021). Yet, weed management is often unnecessary because of hemp's rapid post-emergence growth, which makes hemp very competitive against weeds (Berger 1969; Lotz et al. 1991). Moreover, given the growing preference for pesticide-free cultivation, hemp emerges as an ideal crop for organic agriculture (Amaducci et al. 2015).

11.2.1.9 Harvesting Time

Hemp for fiber production is typically harvested at full flowering of male plants (Bócsa and Karus 1998). This timing corresponds to the stage when primary bast fibers have achieved their maximum (Amaducci et al. 2008). Delayed harvest increases biomass yield; however, it results in a decline in bast yield and quality. This decline is attributed to the augmented proportion of secondary bast fibers as flowering progresses, coupled with an intensification of the lignification (Amaducci et al. 2008; Petit et al. 2020a; Van der Werf and Turunen 2008). In dual-purpose hemp crops, harvesting time is postponed until seed maturity, which results in higher lignified fibers. Notwithstanding this, it has been reported that postponing harvesting to the end of flowering resulted in improved stem decortic ability, which can benefit post-harvest processes (Keller et al. 2001). Amaducci et al. 2008 reported that fiber yield increased by 25% between beginning to full flowering and fiber homogeneity also increased. Hence, it is advisable to harvest at full flowering to maximize fiber yield. Nevertheless, since the timing of harvest is determined by plant phenology, it should be programmed according to the specific cultivation environment, the chosen genotype and its photoperiod sensitivity. In addition, when opting for dew retting, harvesting time should be coupled with suitable weather conditions. For instance, in Southern climates, delaying harvest until the end of summer could potentially yield higher fiber and enhanced quality. This timing coincides with late summer rains, thereby promoting field retting (reviewed in Amaducci et al. 2015).

When hemp is cultivated for paper or bio-composite applications, harvesting time has a relatively limited impact. In fact, to optimize biomass production, late harvesting is preferred (Godin et al. 2013).

11.2.2 Breeding Objectives

Europe played a significant role in industrial hemp breeding as reflected by the abundance of cultivars and traditional landraces selected within the main hemp-growing regions across the continent. Currently, eighty-eight fiber cultivars are registered for use within the European Union (Appendix 11.2). A great breeding effort was carried out in Germany where the first monoecious cultivars and first hybrids were produced. France and Hungary also played an important role. Important traits to breed for industrial hemp include high fiber content and fiber quality, cannabinoid content and composition, degree of monoecy, length of vegetative cycle and resistance to diseases. Emphasis has been placed on achieving high fiber quantity and quality, while also ensuring low THC content. In addition, due to the large environmental susceptibility of hemp, cultivars have traditionally been developed for specific environments and end-use applications.

Fiber quality depends on the morphology of the fiber bundles and the biochemical composition of the cell walls of the elementary fibers (Petit et al. 2020a). Relevant breeding targets to enhance fiber quality include high cellulose content, large crystalline structures, low lignin content and reduced cross-links between pectin and structural components of the cell walls (Petit et al. 2020a; Ranalli 2004; Salentijn et al. 2015). These breeding goals are important for a suitable extractable fiber for both paper- and textile applications (Mandolino and Carboni 2004). Fiber fineness and fiber tensile strength are also key traits of interest for breeding purposes, especially for the textile and bio-composite industry (Placet 2009). In addition, breeding for cultivars that require lower post-harvest management to process the fibers is also desired to increase fiber quality and reduce post-harvest costs (Van den Broeck et al. 2008). The Netherlands has been a pioneer in breeding for such purposes with varieties like chameleon, Markant or Ivory (Salentijn et al. 2015; Toonen et al. 2004).

Breeding for stable low THC content while keeping productivity and quality intact has been a main target in fiber hemp. In the 1970s, the former USSR, France and Ukraine initiated successful breeding programs to reduce THC content (Mandolino and Carboni 2004). However, other cannabinoids have gained importance for their pharmaceutical value and there is an industry for them, namely CBD and cannabigerol (CBG), which is the precursor of THC and CBD. Breeding for terpenes, secondary metabolites also produced in hemp trichomes, has recently become popular.

Hemp is a short-day plant but variation in flowering time is present among cultivars. Based on this feature, hemp can be divided into three groups: early (hemp flowers after 40–60 days), intermediate (60–90 days) and late (90–120 days) cultivars (Zatta et al. 2012). According to Salentijn et al. (2015), early and intermediate genotypes are selected in Northern climates, while late cultivars are selected in Southern climates to produce high fiber yields. In addition, there are differences in sensitivities to photoperiod changes among cultivars. Cultivars such as Felina 34 and Futura are low sensitive while Tiborszallasi is very sensitive.

Gender control is another paramount breeding goal for industrial hemp. Monoecious hemp cultivars produce a higher seed yield and higher homogeneity in fiber quality compared to dioecious cultivars. Hence, fiber extraction of monoecious hemp is easier. A limitation of monoecious cultivars is the need to maintain the monoecious trait, which involves the self-pollination of monoecious plants and the constant removal of male plants (Salentijn et al. 2015). On the other hand, dioecious hemp cultivars have been reported to produce higher fiber quality (Petit et al. 2020a) but are less uniform. Several markers for maleness have been identified, which allows for the identification of male plants in dioecious and monoecious cultivars (Mandolino and Carboni 2004; Torres et al. 2022). Furthermore, Petit et al. (2020c) have provided insights into the genetic architecture and mechanisms governing gender control in cannabis. These insights have paved the way for the development of molecular markers aimed at regulating monoecy. The application of such markers would significantly enhance the pursuit of this pivotal breeding objective.

Considering the substantial impact of the environment on both hemp biomass quantity and quality, hemp cultivars have been bred for specific environments and end-uses. For example, flowering and seed ripening is essential for seed production. As a result, early cultivars are more appropriate for seed production in Northern climates, whereas late cultivars are suitable for the same purpose in Southern sites. Nevertheless, for the production of high-quality fibers, since the fiber formation concludes one month prior to ripening, late cultivars are frequently grown in Northern Europe, where plants have an extended vegetative growth period and produce larger fiber yields. This is possible because for this purpose, flowering and seed ripening is not essential (Salentijn et al. 2015).

11.3 Germplasm Resources Conservation

Cannabis is a domesticated genus with barely existing natural populations, restricted to some regions of China (see Sect. 11.1.3 of this chapter). The legislation associated with cannabis has constrained the establishment of cannabis genetic resource collections. This resulted in the underutilization of the gene pool in cultivar development (Welling et al. 2016). Despite the large gap in the characterization and establishment of such genetic resources, some collections are reported.

Currently, 808 accessions have been collected within ex-situ Cannabis genetic resource collections worldwide (Clarke 1998; De Meijer and van Soest 1992; Gilmore et al. 2007; Mankowska and Silska 2015). Appendix 11.1 shows cannabis genetic resource collections characterized and reported by public sector researchers and online databases. Some of these collections belong to institutes such as the Research Institute of Crop Production (RICIP), the Leibniz Institute of Plant Genetics and Crop plant Research (IPK) or the Consiglio per la Ricerca in agricoltura e l'analisi dell'Economia Agraria (CREA). Three of these collections were widely characterized. The Wageningen University and Research (WUR, former CPRO) germplasm collection was created in 1988 and represents the only collection with passport data for each accession. It comprises 156 accessions including cultivars, varieties and landraces. This collection includes germplasm from 22 countries, and it has been used in several studies including De Meijer and van Soest (1992), Gilmore et al. (2007), Salentijn et al. (2015), Petit et al. (2019, 2020a, b, c). The VIR (Vavilov Research Institute) germplasm collection is the largest cannabis genetic resource collection in Europe. It was created in 1920 and it is composed of 493 accessions from Ukraine, Russia and Germany. The other important public cannabis collection is the IWNIRZ (Institute of Natural Fibers and Medicinal Plants) gene bank. It includes 139 accessions, mostly from European origins. Other cannabis genetic resource collections include the Chinese YAAS (Yunnan Academy of Agricultural Sciences) collection, comprising about 350 accessions of East Asian origin (Salentijn et al. 2015) and the EFGGC (Ecofibre Global Germplasm Collection) from Australia, including accessions from Eurasia. ICGRC (International Cannabis Genomics Research Consortium—www.icgrc.info/virtual_genebank) is the first virtual genebank of cannabis. It aims to catalog the available cannabis genetic resource collections worldwide. Appendix 11.2 summarizes the most common industrial hemp (fiber) cultivars including the current cultivars registered to be used in the EU.

A key goal for public and private germplasm collections is the use of modern genotyping technologies to perform characterization of these resources to establish genetic identities; assess extant variability, diversity, and relatedness; and inform breeding decisions by selecting highly dissimilar parents in crossing programs. One application of low-density SNP genotyping arrays to establish genetic relatedness among individuals in a germplasm collection is shown in Fig. 11.6. In this way, germplasm conservation efforts, including seed multiplication cycles, can be dedicated to maintaining only the most informative individuals in a collection while avoiding highly similar or genetically redundant accessions.

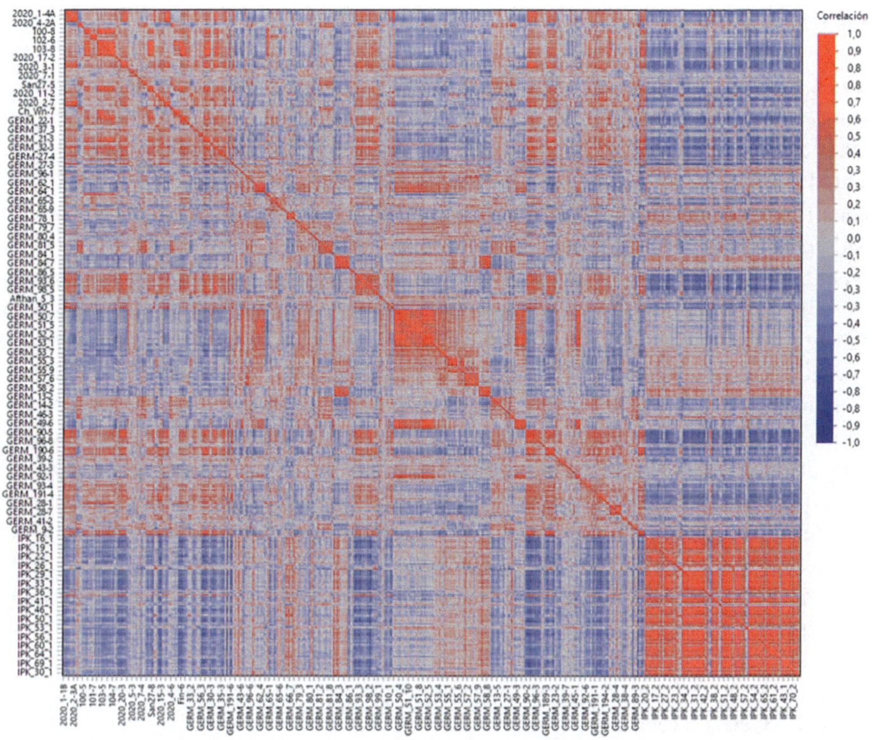

Fig. 11.6 SNP genotyping with a 64 SNP TaqMan® OpenArray® (Institute of Agrifood Research and Technology (IRTA): Genetic Analysis Service) of cannabis accessions contained in a private germplasm collection. SNP genotypes were used to establish a kinship matrix with an increasing scale of positive correlation (red) indicating increasing genetic similarity while an increasing scale of negative correlation (blue) indicates increasing dissimilarity. (Figure constructed by J. Argyris)

11.4 Traditional Breeding

11.4.1 Breeding Strategies and Methodologies

Hemp is naturally an allogamous, open-pollinated (wind-pollinated) plant. The most common breeding practices for dioecious hemp plants involve cross-pollination methods. In the case of monoecious plants, self-pollination is frequently used. The breeding approach required for cross-pollinator plants comprises three phases: (1) explore the natural variation within the material and establish a base population, (2) generate parental lines through selection and improve the population via recurrent selection processes to establish a robust breeding population and (3) perform field trials to test experimental cultivars (Posselt 2010; Salentijn et al. 2015). The methods conventionally used in hemp breeding comprise mass selection, crossbreeding, inbreeding, hybrid breeding and recently mutation breeding and marker-assisted selection (MAS).

11.4.1.1 Mass Selection

Mass selection involves the preservation of seeds obtained from the best-performing plants, which are subsequently replanted in the following year. This iterative process is continued annually (Brown et al. 2014). Using this approach landraces were created from wild (old naturalized) cannabis populations. When using seeds from plants that were offspring of a single parent (family breeding), the breeding effort resulted in a more homogeneous breeding line than when using seeds from several parent plants. Hemp population homozygosity was increased in agronomically desirable traits, such as length of the vegetative period, plant height, stem diameter and seed weight. Mass selection has been especially proven to be efficient in hemp to breed for highly heritable traits, such as bast fiber content (Salentijn et al. 2015) or cannabinoid content (Flajsma and Kcojan 2022). However, as the pollination cannot be controlled the improvements in fiber content were very slow (Salentijn et al. 2015). This method became more effective in hemp when the Bredemann method (Bredemann 1942) was applied, which consisted of the individual selection of male plants based on fiber content. Mass selection was used in the past to select the most important cultivars, such as Carmagnola in Italy (Ranalli 2004). Currently, it is still used for maintaining several European dioecious varieties (Flajsma and Acko 2022).

11.4.1.2 Cross Breeding

In the advanced stages of hemp breeding, a crossbreeding approach was employed, involving precise crosses between the top-performing parent plants. This strategy was used to improve fiber content, stem yield, resistance to lodging, seed yield, sex determination, reduced THC content and resistance to pathogens (Salentijn et al. 2015). The selection relies on evaluating the average performance of an individual in the cross, a concept known as additive gene action, with the assessment conducted across the entire tested population. This breeding approach is based on the ability of the parents to combine during fertilization, ensuring the transmission of favorable alleles to their progenies. Crossbreeding in hemp consists of half-sib family selection based on the evaluation of the progeny from each mother plant and then open pollination and fertilization by more than one male parent. Subsequently, hybridization between cultivars was implemented fostering the development of novel genetic variability.

11.4.2 Breeding Limitations

Hemp is challenging to breed due to its characteristics such as dioecism, allogamy, and cross-pollination (Flajsman and Acko 2022). Female plants are used for fiber, seeds, and inflorescences while males are only for fiber. Hence, the selection of male plants is difficult as they cannot be selected according to the useful value of several phenotypes. Male plants can only be selected with the Bredemann method for fiber production and only the female population offspring shows the beneficial traits of the crosses. In addition, dioecious plants cannot be self-pollinated naturally. Maintaining monoecious plants presents its own difficulties, seed multiplication requires intensive labor because this form returns to the natural dioecious state (Berenji et al. 2013). Furthermore, cannabis pollen exhibits an exceptional diminutive size (~10 µm of diameter) (Lone et al. 2017) and can travel over long distances (~5 km) by wind (Cabezudo et al. 1997). Therefore, cannabis plants need to be bred and isolated to avoid unwanted pollination (McGoey et al. 2017). Additionally, most economically important traits follow complex and quantitative genetics (Petit et al. 2020a,b,c) which hinders the accumulation of beneficial allelic variants in a single plant. Consequently, the development of new hemp varieties is a lengthy process, often spanning several years (Poniatowska et al. 2019).

One of the most significant limitations to cannabis breeding has come from worldwide criminalization and prohibition of cultivation during the last century. Cannabis breeding activities were driven underground to be conducted under clandestine conditions, while scientific development of the crop virtually ceased. For these reasons, cannabis was excluded from the revolutionary developments in biotechnology and genomics that have greatly advanced plant breeding objectives in other crops. The adoption of advanced breeding approaches in hemp is therefore still under development, and further research is needed to facilitate the shift from conventional breeding to cutting-edge molecular plant breeding strategies (Petit et al. 2020b; Salentijn et al. 2015).

11.5 Molecular Breeding

11.5.1 Molecular Marker Assisted Breeding

Molecular breeding involves the use of genetic and genomic techniques to select plants with desired traits more rapidly and efficiently. Molecular markers can be developed using different approaches and rely primarily on the genetic and phenotypic diversity found in natural breeding populations, marking the identification and preservation of this germplasm fundamental to the establishment of molecular breeding and MAS programs. Early applications of dominant molecular markers such as RFLP, RAPD, and AFLPs in cannabis were in forensics, germplasm classification, and the examination of the genetic relatedness of populations (reviewed in Hesami et al. 2020). These studies also yielded the first useful low-throughput molecular markers for breeders in distinguishing and selecting for plant sex (Mandolino et al. 1999; Sakamoto et al. 1995), chemical phenotype (de Meijer et al. 2003; Mandolino et al. 2003; Pacifico et al. 2006) and the first genetic map and markers linked to sex determination (Faux et al. 2016). More informative co-dominant SSR markers have since been used to provide greater detail on diversity and structure in large cannabis populations (Gao et al. 2014; Soler et al. 2017). Although SSRs remain powerful as multi-allelic markers with important applications in genetic map integration, next-generation sequencing (NGS) efforts have made SNPs the preferable genetic markers by their abundance, simplicity, stability, and in some cases as the causal variants controlling phenotypes (Davey et al. 2011). The advent of fluorescence-based marker systems such as PACE (PCR Allele Competitive Extension, 3CR Bioscience Ltd) KASP, and TaqMan, built especially around interrogating SNPs, have led to the development of high-throughput genetic markers for the determination of plant sex and chemotype (Torres et al. 2022; Toth et al. 2020).

11.5.2 Genomics

Next-generation sequencing (NGS) and the release of the first two genomes of *C. sativa* in 2011 (van Bakel et al. 2011) pushed the crop firmly into the genomics era of molecular breeding. Whole genome sequencing (WGS) using second and third-generation (long read and single molecule) sequencing yields a virtually unlimited number of SNP markers to facilitate genomic approaches to molecular breeding in cannabis. High-throughput SNP genotyping utilizing next-generation sequencing (NGS) including genotyping by sequencing (GBS), targeted genotyping by sequencing (SeqSNP) and SNP arrays have allowed the construction of high-density genetic maps which have been crucial for recent marker discoveries.

Combining QTL analyses of segregating mapping populations with high-density maps and new WGS approaches has led to the identification of underlying markers/candidate genes linked to several important traits. Mihalyov et al. (2021) mapped the first resistance (R) gene conferring complete resistance to powdery mildew (PM) infection using a high-density genetic map composed of 2051 SNP markers. One marker (LH3804, with an A- > G base change) located on chromosome 2 of the CS10 reference genome showed perfect co-segregation with resistant or susceptible phenotypes. This SNP was converted to an easily deployable qPCR-based marker to select resistant or susceptible individuals in these genetic backgrounds. One or more copies of genes coding thaumatin-like proteins (TLP) have also been observed in several cannabis cultivars that show tolerance to PM, while mutations and deletions of Mildew Resistance Locus O (MLO) loci have been linked to disease resistance (McKernan et al. 2020).

Various QTL studies focusing on cannabinoid and terpene accumulation have identified candidate genes for cannabinoid content that can be targeted in molecular breeding programs for marker development. These include acyl-activating enzyme 1 (AAE1), 4-hydroxy-3-methylbut-2-enyl diphosphate reductase (HDR) (Grassa et al. 2021), and aromatic prenyltransferase (AP) (Laverty et al. 2019) as linked to QTL for potency (total cannabinoids). In a comprehensive study examining the genetic architecture of both biochemical and agronomic traits, Woods et al. (2021)

identified olivetol synthase (OLS), catalyzing production of a precursor of CBGA, as a candidate gene underlying an important cluster of QTLs for potency and terpene production. As potential molecular markers, multiple SNPs coding for nine homozygous amino acid substitutions segregating within the OLS coding sequence differentiated alleles for highly divergent production of cannabinoids between industrial hemp varieties Carmagnola and USO31. We have used a genotyping by targeted sequencing approach to genotype 10 K polymorphic SNPs in a segregating F_2 population and produce a high-density integrated linkage map and QTL for chemotypic and agronomic traits (Argyris et al. unpublished).

For qualitative traits or major QTL controlling traits of interest, bulked-segregant analysis (BSA) coupled with WGS (BSA-seq) has emerged as a rapid and efficient tool for genetic mapping. Recent works using this approach have uncovered multiple sources and independent origins of photoperiod insensitivity and potential markers for selection in breeding programs (Dowling et al. 2023; Toth et al. 2022). Toth et al. (2022) mapped a recessive allele at a single locus on chromosome 1 named Autoflower1 and developed the AUTO-2 marker (C- > T) to reliably distinguish the auto-flowering lines from wild-type (WT) late-flowering lines across a range of germplasm. A second locus, Early1, putatively modulating flowering time in a second population was also mapped to Chromosome 1 with a tightly linked SNP marker able to distinguish early and late flowering lines in a different segregating population. In another step forward, Leckie et al. (2023) identified PSEUDO-REPSONSE REGULATOR 37 (*CsPRR37*) as the candidate gene for Autoflower1 with photoperiod insensitivity caused by improper splicing of the gene due to a SNP (G > T) after exon 2. Dowling et al. (2023) identified *Autoflower2*, a locus on chromosome 8 contributing to photoperiod-insensitive flowering, and a PACE marker to distinguish early flowering across a range of germplasm. A recent patent filing for cannabis also implicates allelic variants in UPF2 and/or RAP2–7/TOE1 genes to control photoperiod sensitivity (Bakker et al. 2022). Steel et al. (2023) performed a genomics analysis and synthesized the most recent existing knowledge of flowering time control by annotating important QTL/candidate genes onto an integrated physical map using the CS10 reference genome. With this meta-analysis, they identified 4 co-localized QTL regions on chromosomes 1, 3, 8 and X. Especially interesting was a region from 85–100 Mb on chromosome X that overlapped with described QTLs for 'full flowering' from Petit et al. (2020c). This region encompassed two promising candidate phosphatidylethanolamine-binding (PEBP) gene members, CEN1 and FT3. As flowering time is an important factor influencing fiber quality and seed yield in hemp (Salentijn et al. 2019), the introduction of molecular markers capable of following multiple alleles modulating the trait will help in the development of new varieties well-adapted to a multitude of production environments.

Genome-wide association studies (GWAS) using diverse populations of unrelated individuals are a powerful approach to elucidate the genetic players involved in the phenotypic expression of important traits. Ren et al. (2021) used whole-genome resequencing of 110 accessions of wild-growing feral plants, landraces, historical cultivars, and modern hybrids from both hemp and drug types to identify candidate genes showing signals of selective sweeps involved in flowering time, branch formation, lignin biosynthesis and cannabinoid content. Petit et al. (2020a, b, c) used restriction-site associated DNA sequencing (RAD-seq) to genotype a hemp diversity panel of 124 accessions combined with a comprehensive phenotyping effort to perform GWAS. They identified stable QTL and associated candidate genes for fiber quality and flowering traits, as well as for sex determination. An extreme-phenotype GWAS (XP-GWAS) combined BSA with sequencing of pools of individuals from the extremes of a phenotypic distribution to identify SNPs in three candidate genes differentiating varin (3-carbon propyl side chain) cannabinoid production. They identified a β-keto acyl carrier protein reductase (BKR), which is putatively associated with polyketide fatty acid starter unit synthesis and alkyl side-chain length (Welling et al. 2020). To summarize, at least six easily deployable, high-throughput PACE or TaqMan markers for five different traits have been published in cannabis in only the last three years (Table 11.3). With the continuing rapid advances in genomics, this list will continue to grow and provide new target genes together with new molecular markers. Future genomics resources in cannabis will follow the path set in other important crops and will bring a pan-genome (Jahani et al. 2022) to understand the extensive presence/absence variation (PAV) present in the species, as well as gene expression and variation atlas. A 90 K SNP array is available in cannabis (Eurofins genomics: Hemp & Cannabis cannSNP90 Chip) but there is a strong need for intermediate and lower density publicly available genotyping tools such as we employ for inbred-line development and variety identification (Institute of Agrifood Research and Technology (IRTA): Genetic Analysis Service) as well as for multi-parental populations with controlled genetic structures incorporating more genetic diversity.

11.5.3 Bioinformatics

Sequencing efforts in cannabis to date have produced more than 89,000 nucleotide sequences, almost 80,000 mRNA sequences, and fifteen publicly available cannabis genome assemblies. Seven are assembled at the chromosomal level (NCBI Taxonomy_ID: 3483). The chromosomal level genome assemblies represent hemp, medicinal and recreational marijuana, and also feral cannabis types (Braich et al. 2019; Gao et al. 2020; Laverty et al. 2019). Among these, the CS10 assembly of the CBD-producing line CBDRx (Grassa et al. 2021) was put forward by the Cannabis Research Consortium (ICRC) as the most complete reference genome and recommended for use in research. To facilitate the exploitation of these resources, a comprehensive, open-access database with integrated bioinformatic tools to analyze diverse genomic datasets (CannabisGDB) (SuperCan: Cannabis Multiomics Database (Cai et al. 2021) and a customizable, public genome data viewer integrating the CS10 reference genome and expression data are available (NCBI Genome Data Viewer, ID: 90062). An example of a customized data viewer incorporating various sources of genomic data is presented in Fig. 11.7. These tools will help plant

Table 11.3 Publicly available SNP markers used in trait selection in cannabis

QTL or trait	Putative gene or region	Marker	Chromosome Cs10	SNP	Position	Marker type	Source
Autoflower1		AUTO-2	Chr01	C- > T	19,701,425	PACE	Toth et al. (2022)
Auitoflower1	CsPRR37 (LOC115705128)	–	Chr01	G- > T	1603 bp from translation start site	TaqMan	Leckie et al. (2023)
Early 1	LOC115705415	EARLY-1	Chr01	T- > G	41,445,929		Toth et al. (2022)
Autoflower2	FT-like (LOC115700781)	CD00	Chr08	G- > A	64,421,558	PACE	Dowling et al. (2023)
Chemotype	CBDA/THCA locus	CCP1	Chr07	A- > C		PACE	Toth et al. (2020)
Sex Determination	MAD6 (Genbank AF364955.1)	CSP1	Chr10	C- > T		PACE	Toth et al. (2020)

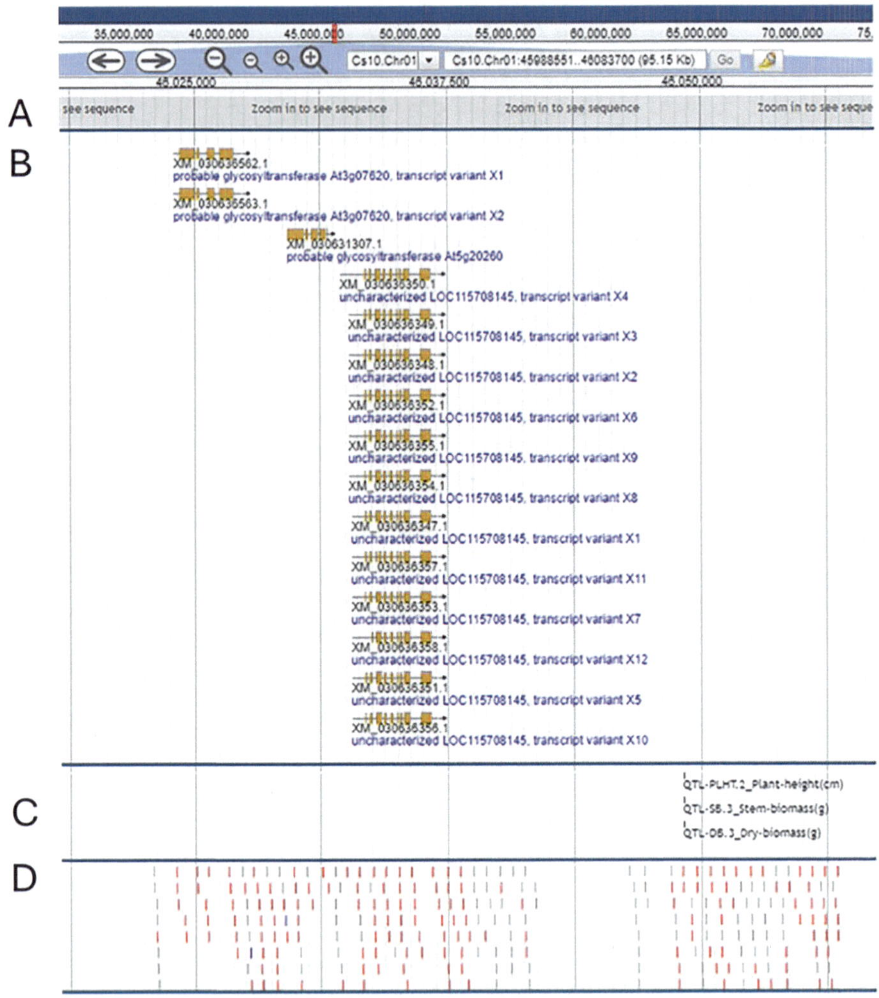

Fig. 11.7 A customizable JBrowse cannabis genome viewer showing 4 tracks integrating CS10 reference sequence data (**a**) combined with gene annotations (**b**), genomic locations of published QTL (**c**) and SNP variants (**d**) for one cannabis accession. (Figure constructed by J. Argyris)

breeders identify useful genetic variants associated with traits of interest or QTLs to incorporate them into improved germplasm.

11.6 Tissue Culture of Hemp (*Cannabis sativa* L.)

Plant tissue culture is a promising technology to speed up the development of new cannabis cultivars (Krasteva et al. 2021). In addition, it may overcome several issues that have complicated the further improvement of cannabis. Some advantages of cannabis tissue culture involve (1) reduction of space requirement to maintain cannabis germplasm in situ, (2) production of multiple genetically uniform plants without sexual reproduction (clones), (3) cultivation of plants under sterile conditions limiting transmission of pests, diseases, and pathogens, (4) reduction of production time by preserving mature plants in vitro, allowing for their swift introduction into in vivo or ex vivo conditions (El-Sherif et al. 2018) and (5) maintaining germplasm in vitro without affecting its biochemical outcomes (Monthony et al. 2021).

Unfortunately, there is currently a lack of standardized and efficient in vitro methods for cannabis cultivation. Micropropagation is a complex, multi-variable and non-linear process influenced by numerous factors, including incubation conditions, medium composition, plant genotypes, plant and explant ages, among others. In vitro cultivation of cannabis is strongly influenced by genotype, along with physical factors, such as type and concentrations of phytohormones, carbon sources, gelling agents, medium composition, disinfection protocols, as well as light quantity and quality and temperature conditions (Monthony et al. 2021). Routine in vitro processes require the introduction of explants to the tissue culture media, in vitro seed germination, callogenesis and organogenesis, and shoot regeneration and rooting (Fig. 11.8). Several methodologies to carry out these processes have been published for *Cannabis sativa* (Chaohua et al. 2016; Hesami et al. 2021; Lata et al. 2016; Monthony et al. 2021; Richez-Dumanois et al. 1986; Wróbel et al. 2020).

Nevertheless, there is a large variation between protocols for the same purpose. The efficiency of these methods is limited to specific genotypes due to the narrow genetic diversity used in the protocol development (Galán-Ávila et al. 2020). Many studies were established and optimized for drug-type cultivars. Additionally, the replicability and reproducibility of these protocols are usually poor (Monthony et al. 2021). Hence, cannabis remains recalcitrant to regeneration.

The main hormones used in cannabis tissue culture are (1) auxins (indole-3-acetic acid (IAA), 2,4-dichlorophenoxyacetic acid (2,4-D), indole-3-butyric acid (IBA), and 1-naphthaleneacetic acid (NAA)) essentially for promoting cell elongation, root initiation and differentiation, (2) cytokinin (6-Benzylaminopurine (BAP), kinetin (Kin), meta-Topolin (mT), and thidiazuron (TDZ)) to stimulate cell division, shoot formation and elongation and even for root induction according to some publications, (3) gibberellins (gibberellic acid (GA3)) to promote seed germination and stem elongation. The balance of these hormones, especially auxins-cytokinin, is a paramount factor for

Fig. 11.8 Cannabis explants introduced in vitro (**a**), Callus development from cannabis leaves (**b**) Maintenance and preservation of callus from hypocotyl (**c**), Maintenance and preservation of callus from leaves (**d**), and Plant regeneration from callus (**e**). (Photos by J. Petit)

determining the direction of the organogenesis to regeneration of shoots or roots (reviewed in Monthony et al. 2021). Driver and Kuniyuki Walnut (DKW) and Murashige Skoog (MS) are the most basal salts used for cannabis tissue culture. Notwithstanding this, DKW has been reported as more suitable for cannabis applications (Page et al. 2020).

11.6.1 Tissue Culture Challenges

A paramount challenge of cannabis tissue culture to address is the development of in vitro protocols independent of genotype. This genetic specificity might be due to differences in endogenous phytohormone concentrations between genotypes (Andre et al. 2016). Yet, the knowledge of hormone biochemistry and metabolism is limited in cannabis. Hesami et al. (2023) is the first study to get insights into the differential phytohormone-related gene expression between embryogenic and non-embryogenic callus types of drug-type cannabis. Therefore, to overcome the recalcitrance of *Cannabis sativa* to in vitro regeneration, it is necessary to conduct biochemical and molecular studies, as well as trials involving many genotypes and diverse cannabis types, including fiber and drug-type hemp. Identification of the optimal combination and concentration of hormones best fitting cannabis is key to developing such methods.

Maintaining the explants in a vegetative state is another important challenge to address. Cannabis is an annual plant that ages with time from the juvenile stage (vegetative growth) to the floral stage. After plants begin to flower, their growth stops, reducing their potential for micropropagation, ultimately leading to the death of the explants. Floral reversion appears as an alternative to maintaining plants in a vegetative state. Piunno et al. (2019) is the first report describing the ability of cannabis inflorescence to return from a flowering stage to a vegetative stage of growth. The authors reported that thidiazuron (TDZ), a synthetic cytokinin, reverted immature and mature inflorescence explants to a vegetative state.

Plant micropropagation in vitro can lead to somaclonal variation (somatic mutations) triggered by oxidative stress, the severity of which is influenced by environmental and genetic factors (Krishna et al. 2016). Alterations in chromosome number, inducement of point mutations, and epigenetic variation through DNA methylation or histone modifications are among the mechanisms producing somaclonal variation (Krishna et al. 2016). Preservation of germplasm free of these alterations is thus another concern to overcome. For example, Adamek et al. (2022) identified somatic mutations with deep sequencing ($> 50 \times$ depth of coverage) leading to genetic mosaicism in a cannabis mother plant used for clonal production. Cannabis micropropagation is associated with a relatively low mutation load (references in Monthony et al. 2021) but it is important to note that there remains a potential for somaclonal variation to occur. An approach to tackle this concern is using low-temperature in vitro cultivation. By maintaining explants at low temperatures, plant growth is slowed down. Lata et al. (2011) is the first study to report storage of cannabis germplasm at 5 °C for 6 months. The use of cryopreservation to store cannabis at cryogenic temperatures is the following step. Uchendu et al. (2019) is the first attempt to develop a cryopreservation protocol for cannabis. Plants mutate as they grow. Preventing plants from growing, while keeping them alive, is the only approach to prevent mutations from occurring in clonal lines over time. Hence, cryopreservation stands as the most effective method for preserving genetic fidelity.

11.7 Mutation Breeding

11.7.1 Conventional Mutagenesis

Throughout the history of cannabis domestication, there has been a progressive reduction of genetic variation both in industrial hemp due to breeding for monoecious varieties and marijuana due to clonal micropropagation (Welling et al. 2015), despite the large genetic variability present in the genus due to allogamy and the cross-pollination of *C. sativa* plants (Salentijn et al. 2015). Mutation breeding replicates natural processes to generate spontaneous mutations in the DNA of plants through physical or chemical methods (Liang 2012). Mutation breeding thus addresses an accelerating need for the generation of new cultivars integrating superior agronomic performance with appropriate concentrations of specific phytochemicals by accessing increased genetic variability and a corresponding diversification of agronomically important traits in cannabis germplasm (Backer et al. 2019; Tadele 2016). TILLING (Targeting Induced Local Lesions in Genomes) employs a chemical mutagen such as ethyl methane sulfonate (EMS) applied to seeds in sufficiently high concentrations to induce random mutations in the genome. The induced sequence variation in mutants may lead to new alleles for new traits or modify and improve the expression of existing traits (Kumar et al. 2017). As an extension, TILLING by sequencing (TbS) combines the standard and efficient technique of EMS mutagenesis with deep sequencing of PCR products (amplicons) representing specific genomic regions or candidate genes for important traits (Tsai et al. 2011). EcoTILLING and polyploidization combined with TILLING are further extensions of the technique that have been used to uncover rare genetic variants in natural populations and to greatly increase mutation density in diploid species, respectively (Tsai et al. 2013).

Security/regulatory issues and logistical challenges involved in the cultivation of very large populations in combination with the time involved in isolating homozygous mutations in a dioecious species hinder the application of TILLING in cannabis (Alter et al. 2022). On the other hand, a TILLING approach can be implemented without both biosafety concerns and restrictions relevant to transgenic and gene-edited crops in the European Union and the difficulties of obtaining stable cannabis transformants that otherwise prevent the full deployment of more precise gene editing technologies such as CRISPR/Cas9 (Shiels et al. 2022), discussed in the

a **CBD Elite (mother)**

- Medicinal/recreational type
- Long flowering time
- Tall stature (3-3.5 m)
- Vigorous with high biomass
- Sativa type growth habit
- High overall cannabinoid and CBD production
- Heavy resin accumulation
- More terpene production and wider diversity of terpenes produced

b **Finola (male pollen donor)**

- Industrial hemp type
- Photoperiod insensitve
- Short stature (1.2 m)
- Fiber/grain type, ruderal growth habit
- Low overall cannabinoid and CBD production
- Less diverse and less overall terpene production

Fig. 11.9 Phenotypic characteristics of parental lines CBD Elite (**a**) and Finola (**b**) used to construct the cannabis TILLING population. (Figure constructed by J. Argyris)

following section. Although TILLING and TbS strategies have been used extensively for more than 20 years in a variety of crops (Szurman-Zubrzycka et al. 2023) there is only one reported instance in the publicly available literature for cannabis (Bielecka et al. 2014). In this case, the authors used a TbS approach, PCR products for two fatty-acid desaturase (FAD) genes were screened in pooled DNA samples from mutagenized M2 plants of FINOLA to identify an allelic series of mutations in the genes affecting fatty acid composition.

We have undertaken a TbS approach in an F_3 population derived from a diverse cross of an elite proprietary CBD-producing line (CBD Elite) and Finola (Duarte-Delgado et al. 2023) (Fig. 11.9).

The F_3 seeds were mutagenized in two separate experiments to produce a population of ca. 1650 M2 families with a 3-D DNA pooling strategy (Fig. 11.10). An initial set of 14 PCR amplicons for candidate genes related to flowering time, PM resistance, sex determination, trichome density and cannabinoid biosynthesis were targeted in 512 families according to Duarte-Delgado et al. (2023).

11.7.2 In Vitro Mutagenesis and Selection

As a mutation breeding approach to augment genetic diversity, the use of in vitro random mutagenesis with physical or chemical agents applied to micro-propagation of meristematic tissue and tissue cell cultures has met with success in many crops (reviewed in EFSA GMO, Panel 2021). In vitro mutagenesis offers some advantages compared to conventional (in vivo) mutagenesis, including ease and uniformity of application of the treatment and the ability to screen large populations of disease-free plant material in a relatively small space (Suprasanna et al. 2011). However, owing to its general recalcitrance to tissue culture, examples of in vitro mutagenesis experiments in cannabis are not available in the public domain. This may change soon, as new and more efficient in vitro protocols for micropropagation

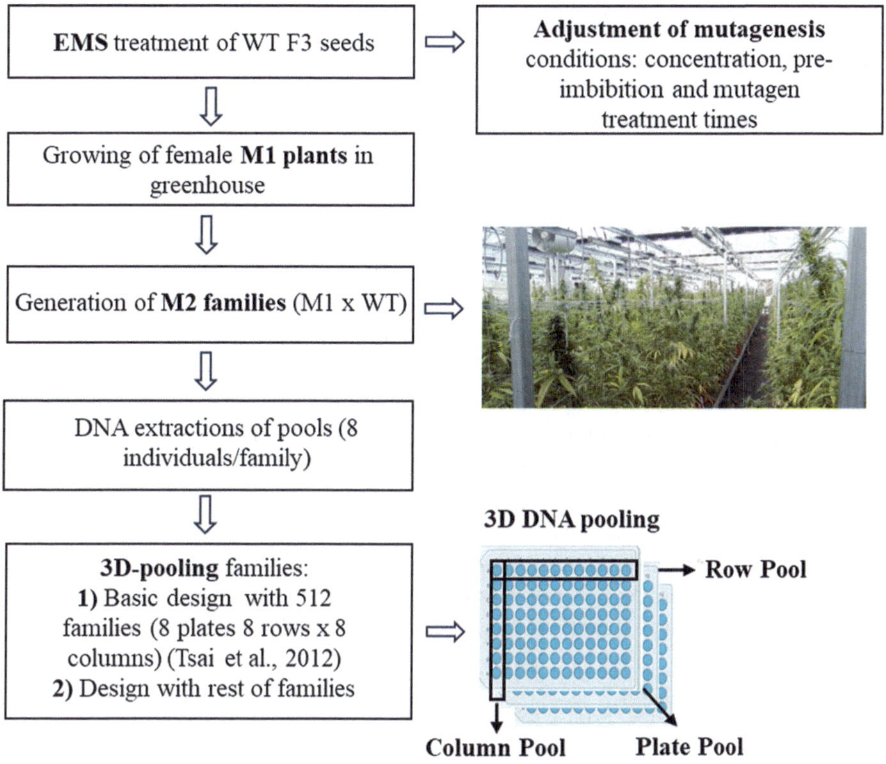

Fig. 11.10 Overview of TILLING population construction and pooling scheme. (Figure constructed by J. Argyris)

and callogenesis are under development. (Galan-Ávila et al. 2020; Monthony et al. 2021; Petit et al. 2023).

As noted previously, cannabis micropropagation in vitro can lead to somaclonal variation. As plant micropropagation using stem cuttings from stock (mother) plants is one of the favored methods of producing recreational/medicinal cannabis (Caplan et al. 2018), somaclonal variation can be harnessed to increase genetic diversity in cannabis genotypes (Adhikary et al. 2021). Thus, as cannabis micropropagation becomes more efficient, exploring somaclonal variation may become more attractive as a source of randomly induced genetic variation caused by disparate genetic, epigenetic and cellular mechanisms not accessible in conventional mutagenesis.

11.7.3 Molecular Analyses of Mutagenized Populations

To assess the relative success or failure of mutation breeding approaches in cannabis, robust experimental design and molecular analysis methodologies are required. Older techniques to identify and analyze induced mutations include the detection of

DNA heteroduplexes by high-resolution melting and the use of single-strand-specific nucleases like CEL1 to create cleavage products visualized on LICOR or capillary sequencers (Szurman-Zubrzycka et al. 2023). TbS experiments require large-scale DNA sampling operations, complex sample pooling schemes and significant investments in bioinformatic resources to deal with a large amount of NGS data. To gauge the success of the mutagenesis in our CBD Elite x FINOLA population we performed a whole-genome resequencing of six M2 lines mutagenized at two different EMS concentrations. Bioinformatic analyses and visualization tools (Fig. 11.11) allowed us to identify putative mutants and estimate a genome-wide rate of 1 mutation per 112–137 kb (7–9 mutations/megabase) (Duarte-Delgado et al. in preparation). This is similar to the 4.3 mutations per megabase average reported in other diploid species (Uauy et al. 2017). As an increased number of mutant alleles recovered leads to a higher probability of identifying an allele producing a nonfunctional protein, our TILLING platform will be a valuable resource for cannabis breeding programs and gene functional validation studies.

Once mutants in candidate genes are identified, functional validation is required and can employ several molecular analysis strategies. Identification and confirmation of the FA desaturase mutants utilized Quantitative RT-PCR analysis, cloning of expression vectors into *Saccharomyces cerevisiae* and chromatography (GC) analysis of fatty acid methyl esters (FAMEs) to demonstrate their effect on modifying FA

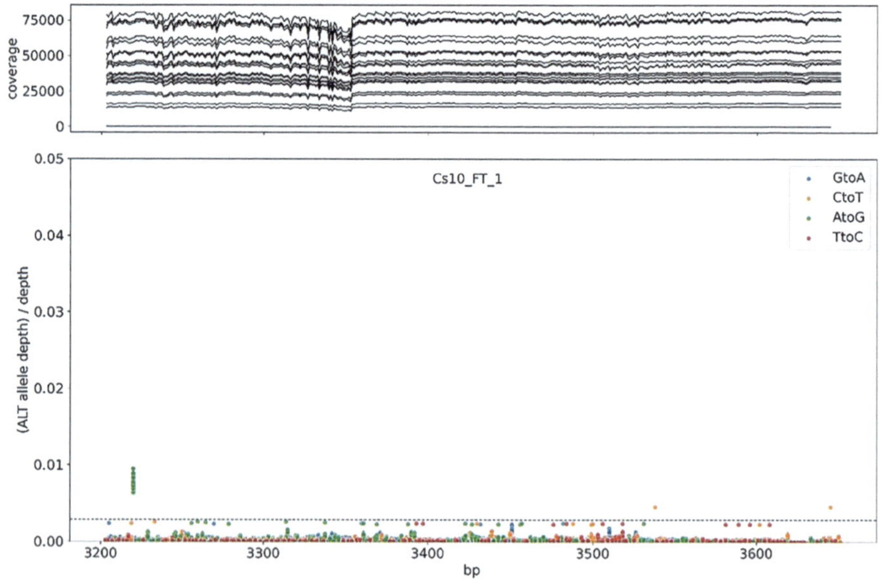

Fig. 11.11 Sequencing coverage and frequency of base changes are used to visualize putative EMS mutants. Sequence coverage from 24 libraries constructed from PCR products generated from 512 three-dimensionally pooled genomic DNA samples (top panel) and base change frequencies (bottom panel) are shown for the cannabis FLOWERING LOCUS T (*CsFT1*) orthologue described in Dowling et al. (2023). (Figure constructed by J. Argyris)

composition in hemp seed oil (Bielecka et al. 2014). Other molecular analysis techniques can involve the assessment of altered ploidy levels through flow cytometry or chromatin immunoprecipitation (ChIP) studies to examine epigenetic changes produced by random or induced in vitro mutagenesis experiments.

11.7.4 Mutagenesis for Enhanced Traits and Improved Cultivars

Many varieties of important staple crops, vegetables, pulses, ornamentals and fruit trees have been developed using chemical or physical agents in in vivo or in vitro mutation breeding programs (EFSA GMO, Panel 2021). A concrete example of trait improvement through mutagenesis in cannabis comes from the *csfad2a-1* and *csfad2a-2* mutants that produce high-oleic seed oil with significantly higher oxidative stability (Bielecka et al. 2014).

Induced loss-of-function mutations in Mildew Resistance Locus O (MLO) genes have produced durable resistance to PM in barley, and partial resistance in wheat and other crops (Acevedo-Garcia et al. 2017; Kang et al. 2020). The MLO family of genes has been recently characterized in cannabis (Pépin et al. 2021). In this study, *CsMLO1* and *CsMLO4* showed significant upregulation of expression following *Golovinomyces ambros*iae infection, providing preliminary evidence that they could be involved in PM susceptibility. Identification and characterization of induced mutations in these two genes therefore provide an attractive target in mutational and TbS studies in cannabis for deploying enhanced PM resistance into breeding programs.

CBDA synthase is a promiscuous enzyme that in addition to synthesizing its own product, can also synthesize small amounts of THC (Zirple et al. 2018). Thus, it is challenging to breed cultivars with a high accumulation of CBD demanded by the medicinal industry together with the <0.3% THC limit required for industrial hemp. Generation and analyses of allelic series of mutants in the CBDA synthase may lead to alterations or abolishment of its promiscuous activity, leading to improved varieties high in cannabinoids that maintain legal compliance.

Cell wall genes involved in the metabolism of monosaccharides, polysaccharides and glycoproteins as well as those involved in lignin biosynthesis were associated with fiber quality in a large hemp diversity panel (Petit et al. 2020b). These are attractive candidates for PCR amplicon design and screening for mutants with improved fiber quality.

Future mutagenesis studies will focus on enhancing traits and producing cultivars with improved resilience to heat, drought, and salinity stress; growth in contaminated soils; and improved nutritional quality (Szurman-Zubrzycka et al. 2023). The classical forward genetic approach starting with the observation of a valuable phenotype and then moving toward identification of the mutated gene responsible will be feasible with large seed populations available in TILLING platforms. Assays can be designed using 10 s or 100 s of seeds per M2 family to screen and identify mutants for traits such as enhanced salt tolerance as reported in wheat (Lethin et al. 2020). These screens can be extended to tolerance for growth on media inoculated with common soil metal pollutants like aluminum, chromium, cadmium, zinc, etc. as hemp has excellent potential as a phytoremediator (Placido and Lee 2022).

11.8 Genetic Engineering and Gene Editing

11.8.1 Genetic Transformation and Gene Editing Methodologies

CRISPR/Cas technology allows for sequence-specific editing of genomes, permitting precise control over genetically modified traits affordably and directly. Compared to mutation breeding discussed in the previous section, gene editing also introduces these changes in a background free of multiple, non-target mutations. *Agrobacterium*-mediated transformation is a fundamental part of genetic engineering and gene editing techniques and is commonly used to achieve both transient and stable gene expression in plants. However, traditional methods using *Agrobacterium* have proved difficult to implement in cannabis (Feeny and Punja 2003) with the plant being both less susceptible to transformation compared to other species, and transformation efficiency being cultivar dependent (Sorokin et al. 2025). Other factors affecting *Agrobacterium*-mediated transformation in cannabis are thoroughly reviewed by Hesami et al. (2020). All in all, there is a relative paucity of publicly available data for genetic transformation and gene editing in the crop, with fewer than ten studies having been conducted since the year 2000 (Hesami et al. 2020; Li et al. 2022).

Plant transformation can be achieved transiently, by temporarily introducing foreign DNA into a plant cell that is not permanently integrated into the host genome, or stably by integrating DNA into the genome where genetic modifications are inherited by offspring. Transient transformation in different plant systems (protoplasts, callus, cell suspension cultures, isolated organs, specialized tissues or whole plants) is vital for functional analyses of genes and proteins of interest (Tyurin et al. 2020). The transient transformation has been achieved in cannabis plant tissues using *Agrobacterium* in at least 5 different studies (Hesami et al. 2020). The transient transformation has also been achieved in leaf tissue through particle-mediated gene transfer (biolistics) (Ahmed et al. 2021) and in protoplasts through a combination of methods (polyethylene glycol, bioporation, vacuum-permeation) (Beard et al. 2021; Kim et al. 2022; Zhu et al. 2022).

Stable transformation protocols are foundational for the genetic engineering of enhanced traits to produce improved germplasm. Co-expression of plant developmental regulator (DR) genes together with genome editing components has helped to reduce the recalcitrance to transformation and plant regeneration encountered in many species (Gordon-Kamm et al. 2019; Nalapalli et al. 2021). In adapting this strategy, Zhang et al. (2021) reported a significant advancement in producing the first stably transformed, gene-edited cannabis plants engineered with a deactivated phytoene desaturase gene. Immature seed embryo hypocotyls were isolated at 15 days post-anthesis taken from a variety showing the highest shoot regeneration rate. Regeneration efficiency was significantly increased with the combination of the DR genes *CsGRF* and *CsGIF1* (Debernardi et al. 2020) which allowed the reprogramming of the somatic cells into meristems and enabled direct regeneration of shoots from callus. A second study detailing the stable transformation and regeneration of cannabis without DR genes was published by Galan-Ávila et al. (2021). Compared to Zhang et al. (2021) they achieved a faster, more efficient transformation and regeneration methodology starting from hypocotyl source tissue producing unicellular regenerating shoots, and the genetic transformation of a subsequent pluripotent pericycle cell.

Together, these works provide a clear way forward for more efficient and stable transformation and regeneration of gene-edited plants in cannabis by systematically optimizing the explant source for callus induction and regenerative tissue, the choice of variety and developmental regulator, and gene-editing tool. The use of DRs has adverse pleiotropic impacts on plant growth and additional steps are required for their removal from transformed or edited plants. One major future objective for gene editing and genetic engineering in cannabis will be the elimination of THCA synthesis and other psychedelic compounds to permit widespread cultivation without the risk of exceeding legal limits.

11.8.2 Ploidy Engineering in Cannabis

Polyploidization, or modification in the number of sets of chromosomes (ploidy level), is a valuable tool in the genetic improvement of crop plants. Positive effects of polyploidization for plant breeding result in increased plant organ size ("gigas" effect), buffering of deleterious mutations, increased heterozygosity, and heterosis (hybrid vigor) (Sattler et al. 2016). A major application of polyploidization in plant breeding is reduced fertility which has been used to create triploid seedless fruit varieties in citrus, banana and grapes (XX). Triploidy can occur spontaneously in nature through the union of a normal haploid (n) gamete with unreduced (2n) gametes that have failed to segregate correctly during meiosis. Triploids are produced intentionally by crossing a tetraploid and diploid plant producing (2n) and (n) gametes, respectively that fuse in the mother and produce a viable triploid ($2n = 3x$) embryo. Triploid plants are mostly sterile because their odd number of chromosome sets (3n) disrupt normal meiosis, leading to the production of nonfunctional gametes.

Polyploidy in cannabis has been achieved in vivo or in vitro through chemical induction with colchicine or oryzalin to disrupt cell division (Bagheri et al. 2015; Crawford et al. 2021; Parsons et al. 2019). These studies have produced both triploid (3n) and tetraploid (4n) cannabis lines, that from a crop improvement perspective, have generally resulted in only moderate morphological changes and alterations to biochemical production. However, the use of tetraploids to produce seedless triploids has enormous potential in the production of outdoor flower crops by mitigating the risk of unwanted pollination from nearby plots and thus preserving valuable cannabinoid and terpene profiles in unseeded flowers (Lipson Feder et al. 2021).

11.9 Hybridization

Heterosis also known as hybrid vigor, has been a significant goal in breeding programs for the development of modern fiber and recreational/medicinal cannabis cultivars. This is achieved by crossing two genetically diverse true-breeding parents to produce a superior F_1 hybrid offspring. The improved performance of hybrids can be seen in various aspects such as increased yield, improved resistance to diseases and pests, higher growth rate, and overall improved adaptation to environmental conditions.

A significant challenge in producing hybrids is controlling pollen, but using crossing systems between monoecious and dioecious hemp can produce all female (unisexual) progeny that functions as a proxy for a male sterility system (Salentijn et al. 2015). These efforts have resulted in the development of F_1 hybrid cultivars Uniko-B and Kompolti hybrid TC, which are single and double cross hybrids, respectively. Uniko-B is a unique cross between Kompolti, (the female parent, and Fibrimon 21, a monoecious male). Since monecious plants are genetically female, the F_1 progeny of this cross is predominantly female, resulting in a high seed yield. The F_2 seeds produced are then planted for fiber production where males occur at a frequency of approximately 30%. Kompolti hybrid TC is a three-cross hybrid used for fiber production that also benefits from a unisexual parent but produces a sex ratio of 50/50 (reviewed in Salentijn et al. 2015).

Despite these successes in the creation of F_1 cannabis hybrids, maximizing the heterotic response requires the development of genetically divergent, yet individually pure and uniform inbred parental lines. Existing industrial hemp cultivars are not suitable as parents for hybrid seed production due to extensive levels of heterozygosity and non-uniformity in characteristics other than fiber quality. For example, Forapani et al. (2001) found the polymorphism rate ranged from 31.1% to 85.5% in six hemp cultivars. In our studies, we detected polymorphism rates between 55% and 87% when genotyping ten different commonly grown hemp cultivars registered for use within the EU using a 64 SNP TaqMan® OpenArray® (Institute of Agrifood Research and Technology (IRTA): Genetic Analysis Service).

In cannabis, as in other cross-pollinated species, inbred lines are typically developed through the crossing of close relatives, or by self-pollination. Different breeding schemes can be followed, such as pedigree breeding, backcrossing, bulking, single seed descent, or double haploids. A cyclical selection approach called recurrent selection, is often used, where desirable individuals are selected from a population and mated (full-sib, half-sib) or self-fertilized to form a new population. To further accelerate the process of inbreeding, recurrent phenotypic selection can be combined with marker-assisted selection (MAS) to choose the most highly inbred lines.

Obligate outcrossing, self-fertilization and inbreeding are applied in both monoecious hemp plants and recreational/medicinal cannabis. Methods for self-fertilization include sex reversal through hormonal treatments, or chemical treatments with colloidal silver (Flajsman et al. 2021) or silver thiosulfate (Lubell and Brand 2018). The progeny of these plants is almost exclusively female (Bócsa 1967), with characteristics like chemotype, yield, disease resistance, and other selection criteria being evaluated for the recreational/medicinal cannabis sector. Additionally, repeated selfings can lead to rapid progression toward allele fixation and genetic uniformity necessary for inbred lines. Barcaccia et al. (2020) have outlined methods for developing commercial F_1 hybrid cultivars through two-way, three-way, and four-way crosses utilizing self-pollination or full-sibling mating over 4–5 generations. In the final stage, female clones of one parental line are masculinized for pollen production and large-scale F_1 feminized seed production.

A major roadblock to true inbred line development in cannabis comes from inbreeding depression, which is a loss of genetic fitness incurred by the mating of closely related individuals due to the accumulation of deleterious alleles. Self-pollination in monoecious hemp plants may occur 20–25% of the time according to Finta-Korpel'ová and Berenji (2007) with corresponding reductions in stem yield of up to 20% due to inbreeding depression. Kurtz et al. (2020) observed that progeny from self-crosses were non-uniform, significantly smaller, and yielded less than outcrossed progeny. Another challenge arises from the time-consuming, and sometimes inefficient process of masculinization of females, raising doubts about whether pollen can be produced in the amounts required for large-scale hybrid seed production.

11.10 Conclusions and Prospects

The increasing pace of legalization and decriminalization of *C. sativa* in recent years has opened the door for the application of modern genetic and genomics tools, transforming the landscape of plant breeding in this versatile crop. The journey from a largely underutilized and stigmatized plant to an increasingly valuable agricultural commodity has been revolutionized by the release of high-quality genome sequences that have enabled breeders to better understand the genetic basis of important agronomic and fiber quality traits. The adoption of marker-assisted breeding strategies led by advances in DNA sequencing technologies has resulted in the development of easily deployable genetic markers for the selection of plant sex, THC production, flowering time, and resistance to powdery mildew. These strategies suggest new genetic targets for improvements in fiber quality, seed oil content, plant morphology, cannabinoid potency, and gender control. Breeders have already succeeded in producing hemp cultivars with improved fiber and grain production, higher CBD (cannabidiol) content, and lower THC (tetrahydrocannabinol) levels. These tailored cultivars are paving the way for a diversified hemp industry, with applications ranging from bio-based resources like bio-composites, textiles and food sources to medicinal products and wellness supplements. Along with rapidly developing molecular breeding programs, traditional breeding methodologies are being complimented with advanced strategies like mutational breeding, ploidy engineering, and the nascent development of F_1 hybrid breeding programs. The development of hemp cultivars suited to different climates, growing conditions, and geographical ranges, addressing breeding limitations associated with seed production and varietal instability, is well underway. These and future breeding efforts will align with broader goals of sustainable agriculture and ecological/environmental concerns. An essential component for further advancements using modern breeding approaches lies in harnessing the genetic diversity present in existing cannabis germplasm collections to source new traits, with a strong push to prioritize cataloging and preserving this material.

Recent advances in cannabis tissue culture are beginning to yield reliable and reproducible protocols for micropropagation and large-scale production of genetically identical, high-quality, disease-free plant material, resulting in greater consistency in crop performance. Further research is still needed to develop more standardized, efficient, and high-throughput tissue culture protocols. These techniques will lead to a better understanding of cannabis genetics through gene functional validation studies, and open new frontiers in genetic engineering through the establishment of routine genetic transformation and gene editing protocols, coupled with efficient regeneration and production of improved material.

Despite these notable advancements, challenges remain in the field of hemp breeding. Regulatory complexities, especially concerning THC levels, continue to pose hurdles for breeders and growers. Additionally, the need to establish comprehensive breeding programs and infrastructure to support the evolving hemp industry is an ongoing task.

Appendix 11.1 Resources Available for Industrial Hemp Breeding

Institute	Area of specialization and research activities	Contact information	References
Agromag KFT (AGM)	Private for-profit entity involved in plant breeding	Makó, Vásárhelyi u. 12, 6900, Hungary http://agromag.hu	Petit et al. (2020a,b,c)
Consiglio per la ricerca in agricoltura e l'analisi dell'economia agraria (CREA, former CRA)	Cannabis genetic resource collection, plant breeding	Bologna, Italy https://www.crea.gov.it/en/about-crea	Mandolino et al. (1999, 2006)

Institute	Area of specialization and research activities	Contact information	References
Crop Research Institute	Cannabis genetic resource collection	Drnovská 507/73, 16106, Prague, Czech Republic cropscience@vurv.cz https://www.vurv.cz/en/	Small and Marcus (2002); Pavelek (2006)
Ecofibre Global Germplasm Collection	Cannabis genetic resource collection, agro-biotechnology	Virginia BC, Queensland, Australia http://ecofibre.com.au/	Welling et al. (2005)
Federation National Producteurs de Chanvre (FNPC)	Private for-profit entity involved in industrial hemp breeding and Cannabis genetic resource collection of French accessions	Rue Paul Ligneul 20, 7200 Le Mans	Petit et al. (2020a,b,c)
ICGRC (International Cannabis Genomics Research Consortium	Plant genetic resources, Virtual Genebank	https://www.icgrc.info/virtual_genebank	n/a
Institute of Agricultural Resources and Economics (LARC)	Cannabis genetic resource collection, Plant breeding	Jelgava, Latvia https://www.lbtu.lv/en/institute-of-agricultural-resources-and-economics	Petit et al. (2020a,b,c)
Institute of Natural Fibers and Medicinal Plants (IWNIRZ)	Cannabis genetic resource collection, biotechnology, breeding of new cultivars of fibrous and medicinal plants, agronomy and preliminary processing, extraction and processing technologies of natural fibers	Wojska Polskiego71B, 60-630, Poznan, Poland http://www.iwnirz.pl	Mankwska and Silska (2015)
Jazz Pharmaceuticals (former GW pharmaceuticals, former HortaPharm B.V.)	Cannabis genetic resource collection	Amsterdam, The Netherland https://www.jazzpharma.com/	Ranalli (2004); Hillig (2005); Datwyler and Weiblen (2006)
Leibniz Institute of Plant Genetics and Crop Plant Research (IPK)	Cannabis genetic resource collection	OT Gatersleben Corrensstraße 3, 06466, Seeland, Germany https://gbis.ipk-gatersleben.de/gbis_i/	Small and Marcus (2002); Höppner (2006)
Van Dinter Semo BV (VDS)	The private for-profit entity involved in plant breeding of fiber and feedstock crops including hemp	Stations Raat 124, 9679 EG, Scheemda, The Netherlands https://www.vandintersemo.nl	Petit et al. (2020a,b,c)

(continued)

Institute	Area of specialization and research activities	Contact information	References
Vavilov Research Institute (VIR)	Cannabis genetic resource collection	Bol'shaya Morskaya Ulitsa, 42, 190000, St. Petersburg, Russia https://www.vir.nw.ru/	Clarke (1998); Small and Marcus (2002); Hillig (2005)
Wageningen UR Plant Breeding (former CPRO —Centre for Plant Breeding and Reproduction Research)	Cannabis genetic resource collection, plant breeding, research on the biochemical composition of hemp fibers	Radix, building 107, Droevendaalsesteeg 1, 6708 PB, Wageningen, The Netherlands https://www.wur.nl/en/research-results/research-institutes/plant-research/plant-breeding.htm	De Meijer and van Soest (1992); De Meijer et al. (1992); Hillig (2005); Gilmore et al. (2007); Salentijn et al. (2015); Petit et al. (2020a,b,c)
Yunnan Academy of Agricultural Sciences (YAAS), Innstitute of Bast Fbre Crops—Chinese Academy of Agricultural Sciences (CAAS)	Cannabis genetic resource collection	Kunming, Yunnan, China https://www.university-directory.eu/China/Yunnan-Academy-of-Agricultural-Sciences.html	Salentijn et al. (2015)

Note: This appendix is an adaptation and an update from Welling et al. (2015)

Appendix 11.2 Genetic Resources of Industrial hemp

Origin	Variety name	Variety status/ countries of registration	Characteristics/maintainer names
Bulgaria	AMX	Registered in EU/Bulgaria	n.a./Superior seeds doo
	Arizona Dream	Registered in EU/Bulgaria	n.a./Pure 5 Nutrients Ltd.
	AutoPower 1	Registered in EU/Bulgaria	n.a./Jupiter seed Europe
	Midwest	Registered in EU/Bulgaria	n.a./Jupiter Seed Europe
	Morning Glory	Registered in EU/Bulgaria	n.a./Pure 5 Nutrients Ltd.
	Northwest	Registered in EU/Bulgaria	n.a./Jupiter Seed Europe
	Silistrenski		Fibre type, Breeding material/WUR
	TroyankaI	Registered in EU/Bulgaria	n.a./"REBEL WINE AND SPIRITS - BG" EOOD
	Western Cherry	Registered in EU/Bulgaria	n.a./Pure 5 Nutrients Ltd.
Canada	921203		Wild material/WUR

Origin	Variety name	Variety status/ countries of registration	Characteristics/maintainer names
China	CYM 171		Fiber type, Breeding material/ YAAS-CAAS
	CYM 273		Fiber type, Breeding material/ YAAS-CAAS
	CYM 28		Fiber type, Breeding material/ YAAS-CAAS
	CYM 49		Fiber type, Breeding material/ YAAS-CAAS
	Yunma 5		Fiber type, Breeding material/ YAAS-CAAS
Czech Republic	Felice	Registered in EU/Czech Republic	n.a./Výzkumný ústav EXENEX, z.ú.
Estonia	Estica	Registered in EU/Estonia	n.a./Globe Seeds B.V.
	Finola	Registered in EU/Finland	Seed type, Breeding material/James C. Callaway
	Finola 2	Registered in EU/Finland	n.a/Transfarm Oy
France	A10-122-1		Fiber type, Breeding material/FNPC
	A10-122-2		Fiber type, Breeding material/FNPC
	A10-122-4		Fiber type, Breeding material/FNPC
France	A102-111-1		Fiber type, Breeding material/FNPC
	A102-111-2		Fiber type, Breeding material/FNPC
	A102-122-1		Fiber type, Breeding material/FNPC
	A102-122-2		Fiber type, Breeding material/FNPC
	A102-122-3		Fiber type, Breeding material/FNPC
	A102-122-4		Fiber type, Breeding material/FNPC
	A103-122-1		Fiber type, Breeding material/FNPC
	A103-122-10		Fiber type, Breeding material/FNPC
	A103-122-2		Fiber type, Breeding material/FNPC
	A103-122-3		Fiber type, Breeding material/FNPC
	A103-122-4		Fiber type, Breeding material/FNPC
	A103-122-6		Fiber type, Breeding material/FNPC
	A103-122-8		Fiber type, Breeding material/FNPC
	A11-121-1		Fiber type, Breeding material/FNPC
	A11-121-10		Fiber type, Breeding material/FNPC
	A11-121-11		Fiber type, Breeding material/FNPC
	A11-121-12		Fiber type, Breeding material/FNPC
	A11-121-13		Fiber type, Breeding material/FNPC
	A11-121-14		Fiber type, Breeding material/FNPC
	A11-121-15		Fiber type, Breeding material/FNPC
	A11-121-16		Fiber type, Breeding material/FNPC
	A11-121-17		Fiber type, Breeding material/FNPC

(continued)

Origin	Variety name	Variety status/ countries of registration	Characteristics/maintainer names
	A11-121-18		Fiber type, Breeding material/FNPC
	A11-121-19		Fiber type, Breeding material/FNPC
	A11-121-2		Fiber type, Breeding material/FNPC
France	A11-121-20		Fiber type, Breeding material/FNPC
	A11-121-21		Fiber type, Breeding material/FNPC
	A11-121-22		Fiber type, Breeding material/FNPC
	A11-121-23		Fiber type, Breeding material/FNPC
	A11-121-24		Fiber type, Breeding material/FNPC
	A11-121-3		Fiber type, Breeding material/FNPC
	A11-121-4		Fiber type, Breeding material/FNPC
	A11-121-5		Fiber type, Breeding material/FNPC
	A11-121-6		Fiber type, Breeding material/FNPC
	A11-121-7		Fiber type, Breeding material/FNPC
	A11-121-9		Fiber type, Breeding material/FNPC
	A7-104-1		Fiber type, Breeding material/FNPC
	A7-105-4		Fiber type, Breeding material/FNPC
	A9-122-1		Fiber type, Breeding material/FNPC
	A9-122-2		Fiber type, Breeding material/FNPC
	A9-122-3		Fiber type, Breeding material/FNPC
	A9-122-4		Fiber type, Breeding material/FNPC
	B6-093-17		Fiber type, Breeding material/FNPC
	B6-093-3		Fiber type, Breeding material/FNPC
	Dioïca 88	Registered in EU/France	Exclusively for fiber production, also for CBD production, Dioecious/HEMP IT ADN
	Earlina 8 FC	Registered in EU/France	Exclusively seed type, suitable for production in half-north Europe. Short vegetative growth period perfectly adapted to the climate conditions from late spring - early autumn, Monoecious/ HEMP IT ADN
	Epsilon 68	Registered in EU/France	Fiber type, Breeding material/HEMP IT ADN - FNPC
France	Fédora 17	Registered in EU/France	Fiber type and seed type (dual), interesting for CBD production as well, Breeding material, Monoecious/HEMP IT ADN - FNPC
	Félina 32	Registered in EU/France	Seed type or dual purpose (fiber and seed), Good for CBD production, Breeding material, Suitable for dual purpose in Southern European latitudes, Monoecious/HEMP IT ADN - FNPC

(continued)

Origin	Variety name	Variety status/ countries of registration	Characteristics/maintainer names
	Férimon	Registered in EU/France	Fiber type and seed type, Breeding material, Monoecious, suitable for half-north Europe, early flowering/ HEMP IT ADN - FNPC
	Fibrimon 56		Fiber type, Breeding material/WUR
	Fibror 79	Registered in EU/France	Fiber type exclusively targeting low post-harvesting management, yellow variety, it can be grown in any geographic region, Monoecious/HEMP IT ADN
	Futura 75	Registered in EU/France	Fiber type (north Europe), Dual purpose (half south Europe), Good for CBD production, Breeding material, Monoecious/HEMP IT ADN
	Futura 83	Registered in EU/France	Fiber type (North Europe), Dual purpose (South Europe). Late flowering variety which is good to produce biomass in the Half-south of Europe even in more tropical latitudes, Monoecious/HEMP IT ADN
	Mona 16	Registered in EU/France	n.a./HEMP IT ADN
	Muka 76	Registered in EU/France	Fiber type, Suitable for entire Europe, Monoecious/HEMP IT ADN
	Nashinoïde 15	Registered in EU/France	n.a./HEMP IT ADN
	Nordria 3	Registered in EU/France	n.a./HEMP IT ADN
	Orion 33	Registered in EU/France	Seed type or dual-purpose hemp, tolerant to broomrape (orobanche), Suitable for southern Europe, Monoecious/HEMP IT ADN
	Ostara 9	Registered in EU/France	n.a./HEMP IT ADN
	Santhica 23	Registered in EU/France	n.a./HEMP IT ADN
France	Santhica 27	Registered in EU/France	Fiber type, Breeding material, suitable for half-north Europe, Monoecious/ HEMP IT ADN
	Santhica 70	Registered in EU/France	Fiber type, Breeding material, suitable for half-north Europe until South Europe, Monoecious/HEMP IT ADN
Germany	880817		n.a./WUR
Hungary	891057		Landraces/WUR
	Ajkai-A-TF		Fiber type, Landraces/WUR

(continued)

Origin	Variety name	Variety status/ countries of registration	Characteristics/maintainer names
	Balaton	Registered in EU/Hungary	n.a./AGM
	Cannakomp	Registered in EU/Hungary	n.a./AGM
	Fibrol	Registered in EU/Hungary	Fiber type, Breeding material/AGM
	KC Bonusz	Registered in EU/Hungary	n.a./AGM
	KC Dora	Registered in EU/Hungary	Fiber type, Breeding material/AGM
	KC Virtus	Registered in EU/Hungary	n.a./AGM
	KC Zuzana	Registered in EU/Hungary	n.a./AGM
	KCA Borana	Registered in EU/Hungary	n.a./AGM
	Komoroi-A-TF		Fiber type, Landraces/WUR
	Kompolti	Registered in EU/Hungary	Fiber type/MATE Fleischmann Rudolf Kutatóintézet and WUR
	Kompolti hibrid TC	Registered in EU/Hungary	Fiber type, Breeding material/AGM and WUR
	Kompolti Sargászáru		Fiber type, Breeding material/WUR
	Lipko	Registered in EU/Hungary	n.a./MATE Fleischmann Rudolf Kutatóintézet
	Monoica	Registered in EU/Czech Republican and Hungary	Fiber type, Breeding material/AGM
	Orosi-A-TF		Fiber type/WUR
	Panoramavar globosa		Ornamental use, Breeding material/WUR
	Tiborszallasi	Registered in EU/Hungary	Fiber type, Breeding material/AGM
	Tisza	Registered in EU/Hungary	Fiber type, Breeding material/Kaba green start Kft and AGM
	UnikoB	Registered in EU/Hungary	Fiber type, Breeding material/MATE Fleischmann Rudolf Kutatóintézet
Italy	Asso	Registered in EU/Italy	n.a./CREA
	Carma	Registered in EU/Italy	n.a./Vivacell Biotechnology Espana
	Carma Monoica		Fiber type, Breeding material/CREA

(continued)

Origin	Variety name	Variety status/ countries of registration	Characteristics/maintainer names
	Carmagnola	Registered in EU/Italy, Germany and Czech Republic	Fiber type, Breeding material/CREA, X-MAINTAINER and Výzkumný ústav EXENEX, z.ú.
	Carmaleonte	Registered in EU/Italy	Fiber type, Breeding material/CREA
	Codimono	Registered in EU/Italy	n.a./CREA
	CRA_1		Fiber type, Breeding material/CREA
	CRA_2		Fiber type, Breeding material/CREA
	CRA_4		Fiber type, Breeding material/CREA
	CRA_7		Fiber type, Breeding material/CREA
	CRA_8		Fiber type, Breeding material/CREA
	CS	Registered in EU/Italy	Fiber type, Breeding material/CREA
	Eletta Campana	Registered in EU/Italy	Fiber type, large inflorescence biomass (good for CDB production), Breeding material, Late flowering, Dioecious/ CREA
	Ermes A		Fiber type, Breeding material/CREA
	Fibranova	Registered in EU/Italy	Fiber type, Breeding material/CREA
	Fibrante	Registered in EU/Italy	n.a./CREA
	Glecia	Registered in EU/Italy	n.a./NATURA S.L.R.
	Gliana	Registered in EU/Italy	n.a./NATURA S.L.R.
	Superfibra		Fiber type, Breeding material/WUR
	Supermono		Fiber type, Breeding material/CREA
	SVGB-10611		Fiber type, Breeding material/CREA
	Villanova	Registered in EU/Italy	n.a./ECOHEMP
	W-1		Fiber type, Breeding material/CREA
Latvia	Adzelvieši	Registered in EU/Latvia	n.a./LARC
	CFX-2	Registered in EU/Latvia	n.a./Allive Europe UAB
	CRS-1	Registered in EU/Latvia	n.a./Allive Europe UAB
	Katlakalna		Fiber type, Breeding material/LARC
	Loja	Registered in EU/Latvia	n.a./Mazsalacas pilsetas lauku teritorijas mareka berzina zemnieku saimnieciba Lojas

(continued)

Origin	Variety name	Variety status/ countries of registration	Characteristics/maintainer names
	Pūriņi	Registered in EU/Latvia	n.a./LARC
Lithuania	Alive SK	Registered in EU/Lithuania	n.a./Join stock company CANABA
Poland	891288		Fiber type/WUR
	Beniko	Registered in EU/Poland and the Netherlands	Fiber type, High fiber quality, Breeding material, Monoecious, Early flowering/ IWNIRZ
	Bialobrzeskie	Registered in EU/Poland, Czech Republic and Germany	Fiber type, Breeding material/X-MAINTAINER and INSTYTUT WLOKIEN NATURALNYCH
	Glyana	Registered in EU/Poland	n.a./HEMPTEX TRADE SP. Z O.O.
	Henola	Registered in EU/Poland	n.a./IWNIRZ
	Matrix	Registered in EU/Poland	n.a./Gospodarstwo rolno.nasienne kow-mar Maria j. Kowalska
	Mietko	Registered in EU/Poland	n.a./Gospodarstwo rolno.nasienne kow-mar Maria j. Kowalska
	Rajan	Registered in EU/Poland	n.a./IWNIRZ
	Sofia	Registered in EU/Poland	n.a./Polski instytut konopi SP. ZO.O.
	Tygra	Registered in EU/Poland	Fiber type, Breeding material/IWNIRZ
	Wielkopolskie	Registered in EU/Poland	n.a./Gospodarstwo rolno.nasienne kow-mar Maria j. Kowalska
	Wojko	Registered in EU/Poland	n.a./IWNIRZ
Romania	Armanca	Registered in EU/Romania	n.a./Statiunea de cercetare-dezvoltare africola Lovrin
	Dacia Secuieni	Registered in EU/Romania	n.a./n.a.
	Denise		Fiber type, Breeding material/CREA
	Lovrin 110		Fiber type, Breeding material/WUR
Romania	Zenit		Fiber type, Breeding material/CREA
Russia	883289		Fiber type, Landraces/WUR
	883290		Fiber type, Landraces/WUR
	Dneprovskaja 84		Fiber type, Landraces/WUR
	JSO 16		Fiber type, Breeding material/WUR
	Juznaja Odnovremenno		Fiber type, Breeding material/WUR
Slovakia	Csehslovák-A-TF		Fiber type/WUR
	Rastislavicke		Fiber type, Breeding material/WUR

(continued)

Origin	Variety name	Variety status/ countries of registration	Characteristics/maintainer names
Spain	880973		n.a./WUR
	883262		n.a./WUR
	891240		n.a./WUR
	Delta-405	Registered in EU/Spain	n.a./Semillas certif. Castells
	Delta-llosa	Registered in EU/Spain	Fiber type, Breeding material/Semillas certif. Castells and CREA
The Netherlands	Enectaliana	Registered in EU/The Netherlands	n.a./ENECTA B.V.
	Enectarol	Registered in EU/The Netherlands	n.a./ENECTA B.V.
	Ivory	Registered in EU/The Netherlands	Fiber type, Breeding material/VDS
	Marcello	Registered in EU/The Netherlands	Fiber type, Breeding material/VDS
	Markant	Registered in EU/The Netherlands	Fiber type, Breeding material/VDS
	MGC 1013	Registered in EU/The Netherlands	n.a./Marcel de Wit
	Chamaeleon	Registered in EU/The Netherlands	Fiber type, Breeding material/Plant Reserch International BV (WUR) - VDS
Turkey	891090		Landraces/WUR
	891094		Landraces/WUR
Ukraine	Dneprovskaja odnodomnaja 6		Fiber type, Breeding material/WUR
	Krasnodarskaja 56		Fiber type, Breeding material/WUR
	USO 14 Monoica		Fiber type, Breeding material/CREA
	USO 31	Registered in EU/The Netherlands	Fiber type (nord Europe) and for seeds (Half nord Europe), Breeding material, Monoecious, no suitable for Southern Europe/CREA
n.a.	Austa SK	Registered in EU/Lithuania and Latvia	n.a./Join stock company CANABA and UAB CANABA
n.a.	891327		Fiber type, Landraces/WUR

Note: Within the germplasm resource collections, 808 hemp accessions have been documented. However, comprehensive passport data is lacking for most of the accessions, resulting in a substantial information gap. The appendix contains information about the registered accessions to be used in the European Union and the most common accessions used in previous studies. For more information, such as the biochemical composition of the fibers and flowering traits (time of flowering and sex determination), check Petit et al. (2020a,b,c). Check Appendix 11.1 for the full name of the Institutes and companies. n.a. = not available

References

Acevedo-Garcia J, Spencer D, Thieron H, Reinstädler A, Hammond-Kosack K, Phillips AL, Panstruga R (2017) mlo-based powdery mildew resistance in hexaploid bread wheat generated by a non-transgenic TILLING approach. Plant biotechnology j, 15(3):367–378

Adamek K, Jones AMP, Torkamaneh D (2022) Accumulation of somatic mutations leads to genetic mosaicism in cannabis. Plant Genome 15:e20169. https://doi.org/10.1002/tpg2.20169

Adesina I, Bhowmik A, Sharma H et al (2020) A review on the current state of knowledge of growing conditions, agronomic soil health practices and utilities of hemp in the United States. Agriculture 10:129. https://doi.org/10.3390/agriculture10040129

Adhikary D, Kulkarni M, El-Mezawy A et al (2021) Medical cannabis and industrial hemp tissue culture: present status and future potential. Front Plant Sci 12:627240. https://doi.org/10.3389/fpls.2021.627240

Ahmad R, Tehsin Z, Malik ST et al (2016) Phytoremediation potential of hemp (*Cannabis sativa* L.): Identification and characterization of heavy metals responsive genes. CLEAN—Soil Air Water 44:195–201

Ahmed S, Gao X, Jahan MA et al (2021) Nanoparticle-based genetic transformation of *Cannabis sativa*. J Biotechnol 326:48–51. https://doi.org/10.1016/j.jbiotec.2020.12.014

Alter H, Peer R, Dombrovsky A et al (2022) Tobacco Rattle Virus as a tool for rapid reverse-genetics screens and analysis of gene function in *Cannabis sativa* L. Plants 11(3):327. https://doi.org/10.3390/plants11030327

Amaducci S, Errani M, Venturi G (2002) Plant population effects on fibre hemp morphology and production. J Ind Hemp 7(2):33–60

Amaducci S, Colauzzi M, Bellocchi G et al (2008) Modelling post-emergent hemp phenology (*Cannabis sativa* L.): Theory and evaluation. Eur J Agron 28:90–102

Amaducci S, Colauzzi M, Bellocchi G et al (2012) Evaluation of a phenological model for strategic decisions for hemp (*Cannabis Sativa* L.) biomass production across European sites. Industrial Crop Prod 37:100–110

Amaducci S, Scordia D, Liu FH et al (2015) Key cultivation techniques for hemp in Europe and China. Industrial Crops Products 68:2–16

Andre CM, Hausman J-F, Guerriero G (2016) *Cannabis sativa*: the plant of the thousand and one molecules. Front Plant Sci 7:19. https://doi.org/10.3389/fpls.2016.00019

Bagheri M, Mansouri H (2015) Effect of induced polyploidy on some biochemical parameters in Cannabis sativa L. Applied biochemistry and biotechnology, 175(5):2366–2375. https://doi.org/10.1007/s12010-014-1435-8

Baker D, Pryce G, Giovannoni G et al (2003) The therapeutic potential of cannabis. Lancet Neurol 2:291–298

Bakker E, Holloway A, Waterman K (2022) Autoflowering markers. US patent application number: US20230242932A1. U.S. Patent and Trademark Office. https://rb.gy/ik0fb0

Backer R, Schwinghamer T, Rosenbaum P et al (2019) Closing the yield gap for cannabis: a meta-analysis of factors determining cannabis yield. Front Plant Sci 10:495

Barcaccia G, Palumbo F, Scariolo F et al (2020) Potentials and challenges of genomics for breeding cannabis cultivars. Front Plant Sci 11:573299. https://doi.org/10.3389/fpls.2020.573299

Beard KM, Boling AWH, Bargmann BOR (2021) Protoplast isolation, transient transformation, and flow-cytometric analysis of reporter-gene activation in *Cannabis sativa* L. Ind Crop Prod 164:113360. https://doi.org/10.1016/j.indcrop.2021.113360

Berenji J, Sikora V, Fournier G et al (2013) In: Bouloc P (ed) Genetics and selection of hemp. Hemp: Industrial Production and Uses, pp 48–72

Berger J (1969) The world's major fibre cr London: CABI ops: their cultivation and manuring. Centre D'Etude de l'Azote, Zurich, p 219

Bielecka M, Kaminski F, Adams I et al (2014) Targeted mutation of Δ12 and Δ15 desaturase genes in hemp produce major alterations in seed fatty acid composition including a high oleic hemp oil. Plant Biotechnol J 12(5):613–623. https://doi.org/10.1111/pbi.12167

Braich S, Baillie RC, Jewell LS et al (2019) Generation of a comprehensive transcriptome atlas and transcriptome dynamics in medicinal Cannabis. Sci Rep 9:1–12. https://doi.org/10.1038/s41598-019-53023-6

Bredemann G (1942) Die Bestimmung des Fasergehaltes bei Massenunter-suchungen von Hanf, Flachs, Fasernesseln und anderen Bastfaserpflanzen. Faserforschung 16:14–39. in German

Bócsa I (1967) Kender fajtahibrid előállításához szükséges unisexuális (hímmentes) anyafajta nemesítése. Rostnövények:3–7. (in Hungarian)

Bócsa I & M Karus (1998) The cultivation of hemp: Botany, varieties, cultivation and harvesting. (translated by Chris Filben) Hemptech, Sebastopol

Brown J, Caligari P, Campos H (2014) Plant breeding, 2nd edn. West Sussex, John Wiley and Sons Ltd

Cabezudo B, Recio M, Sánchez-Laulhé JM et al (1997) Atmospheric transportation of marihuana pollen from North Africa to the Southwest of Europe. Atmos Environ 31:3323–3328. https://doi.org/10.1016/S1352-2310(97)00161-1

Cai S, Zhang, Z, Huang S, Bai X, Huang Z, Zhang YJ, Huang L, Tang W, Haughn G, You S, Liu Y (2021) CannabisGDB: a comprehensive genomic database forCannabis Sativa L. Plant biotechnology j, 19(5):857–859. https://doi.org/10.1111/pbi.13548

Callaway JC (2004) Hempseed as a nutritional resource: an overview. Euphytica 140:65–72

Caplan D, Stemeroff J, Dixon M, Zheng Y (2018) Vegetative propagation of cannabis by stem cuttings: Effects of leaf number, cutting position, rooting hormone and leaf tip removal. Can J Plant Sci 98:1126–1132. https://doi.org/10.1139/cjps-2018-0038

Carus M, Dammer L (2018) The circular bioeconomy—concepts, opportunities, and limitations. Ind Biotechnol 14:83–91

Chaohua C, Gonggu Z, Lining Z et al (2016) A rapid shoot regeneration protocol from the cotyledons of hemp (*Cannabis sativa* L.) Ind Crop. Prod 83:61–65. https://doi.org/10.1016/j.indcrop.2015.12.035

Clarke RC (1998) Maintenance of Cannabis germplasm in the Vavilov Research Institute gene bank-five year report. J Int Hemp Assoc 5:75–79

Cosentino SL, Riggi E, Testa G et al (2013) Evaluation of European developed fibre hemp genotypes (*Cannabis sativa* L.) in semi-arid Mediterranean environment. Ind Crop Prod 50:312–324

Crawford S, Rojas BM, Crawford E, Otten M, Schoenenberger TA, Garfinkel AR, Chen H (2021) Characteristics of the Diploid, Triploid, and Tetraploid Versions of aCannabigerol-Dominant F1 Hybrid Industrial Hemp Cultivar, Cannabis sativa 'Stem Cell CBG'. Genes, 12(6):923. https://doi.org/10.3390/genes12060923

Crocq MA (2020) History of cannabis and the endocannabinoid system. Dialogues Clin Neurosci 22(3):223–228. https://doi.org/10.31887/DCNS.2020.22.3/mcrocq

Dam J (2008) Natural fibres and the environment: environmental benefits of natural fibre production and use. In: Proceedings of the Symposium on Natural Fibres: Common fund for commodities, 20 Oct 2008, Rome, Italy

Datwyler SL, Weiblen GD (2006) Genetic variation in hemp and marijuana (*Cannabis sativa* L.) according to amplified fragment length polymorphisms. J Forensic Sci 51:371–375. https://doi.org/10.1111/j.1556-4029.2006.00061.x

Davey JW et al (2011) Genome-wide genetic marker discovery and genotyping using next-generation sequencing. Nat Rev Genet 12:499–510

De Meijer EPM, Van Der Werf HMG (1994) Evaluation of current methods to estimate pulp yield of hemp. Ind Crop Prod 2:111–120

De Meijer EPM, Van Soest LJM (1992) The CPRO Cannabis collection. Euphytica 62:201–211

De Meijer E, Bagatta A, Carboni P et al (2003) The inheritance of chemical phenotype in *Cannabis sativa* L. Genetics 163:335–346

Debernardi JM, Tricoli DM, Ercoli MF et al (2020) A GRF–GIF chimeric protein improves the regeneration efficiency of transgenic plants. Nat Biotechnol 38:1274–1279

Desanlis F, Cerruti N, Warner P (2013) Hemp agronomics and cultivation. In: Bouloc P (ed) Hemp industrial production and uses. CABI, Wallingford, pp 98–124

Deyholos MK, Potter S (2014) Engineering bast fiber feedstocks for use in composite materials. Biocatal Agric Biotechnol 3:53–57

Dowling CA, Shi J, Toth JA et al (2023) A flowering locus T ortholog is associated with photoperiod-insensitive flowering in hemp (*Cannabis sativa* L.). bioRxiv:21.537862. https://doi.org/10.1101/2023.04.21.537862

Duarte-Delgado D, Argyris J, Pujol M et al (2023) TILLING by sequencing in cannabis for the high throughput screening of novel mutations with effect on phytochemical contents and agronomic traits. Poster presented at: Genomics-assisted breeding for boosting crop and livestock improvement; May 10-11, 2023, Barcelona

Duque-Schumacher AG, Pequito S, Pazour J (2020) Industrial hemp fiber: a sustainable and economical alternative to cotton. J Clean Prod 268:122180. https://doi.org/10.1016/j.jclepro.2020.122180

Dussy FE, Hamberg C, Luginbühl M et al (2005) Isolation of Delta9-THCA-A from hemp and analytical aspects concerning the determination of Delta9-THC in cannabis products. Forensic Sci Int 149(1):3–10. https://doi.org/10.1016/j.forsciint.2004.05.015

Duvall CS (2016) Drug laws, bioprospecting and the agricultural heritage of Cannabis in Africa. Space Polity 20:10–25. https://doi.org/10.1080/13562576.2016.1138674

Ebskamp MJM (2002) Engineering flax and hemp for an alternative to cotton. Trends Biotechnol 20:229–230

EFSA GMO Panel (EFSA Panel on Genetically Modified Organisms), Mullins E, Bresson JL, Dalmay et al (2021) In vivo and in vitro random mutagenesis techniques in plants. EFSA J 19(11):6611. https://doi.org/10.2903/j.efsa.2021.6611

El-Sherif NA (2018) Impact of plant tissue culture on agricultural sustainability. Sustainability of agricultural environment in Egypt: Part II. Springer, pp 93–107

Elfordy S, Lucas F, Tancret F et al (2008) Mechanical and thermal properties of lime and hemp concrete ("hempcrete") manufactured by a projection process. Constr Build Mater 22:2116–2123

Faux AM, Draye X, Lambert R et al (2013) The relationship of stem and seed yields to flowering phenology and sex expression in monoecious hemp (*Cannabis sativa* L.). Eur J Agron 47:11–22

Faux AM, Draye X, Flamand MC et al (2016) Identification of QTLs for sex expression in dioecious and monoecious hemp (*Cannabis sativa* L.). Euphytica 209:357–376

Finta-Korpel'ová Z, Berenji J (2007) Trends and achievements in industrial hemp (Cannabis sativa L.) breeding. Bilten za hmelj, sirak i lekovito bilje 39(80):63–75

Finnan J, Burke B (2013) Nitrogen fertilization to optimize the greenhouse gas balance of hemp crops grown for biomass. Glob Change Biol Bioenergy 5:701–712

Flajšman M, Slapnik M, Murovec J (2021) Production of feminized seeds of high CBD *Cannabis sativa* L. by manipulation of sex expression and its application to breeding. Front. Plant Sci 12:718092. https://doi.org/10.3389/fpls.2021.718092. PMID: 34790210; PMCID: PMC8591233

Flajšman M, Acko DK (2022) Industrial hemp breeding and genetics. In: Pojić M, Tiwari BK (eds), Industrial Hemp, pp 37–57. https://doi.org/10.1016/B978-0-323-90910-5.00009-9

Forapani S, Carboni A, Paoletti C et al (2001) Comparison of hemp varieties using random amplified polymorphic DNA markers. Crop Sci 41:1682–1689. https://doi.org/10.2135/cropsci2001.1682

Freeman DC, Harper KT, Charnov EL (1980) Sex change in plants: old and new observations and new hypotheses. Oecologia 47:222–232

Galán-Ávila A, García-Fortea E, Prohens J, Herraiz FJ (2020) Development of a direct in vitro plant regeneration protocol from *Cannabis sativa* L. seedling explants: Developmental morphology of shoot regeneration and ploidy level of regenerated plants. Front. Plant Sci 11:645. https://doi.org/10.3389/fpls.2020.00645

Galán-Ávila A, Gramazio P, Ron M et al (2021) A novel and rapid method for Agrobacterium-mediated production of stably transformed *Cannabis sativa* L. plants. Ind Crop Prod 170:113691. https://doi.org/10.1016/j.indcrop.2021.113691

Gao C, Xin P, Cheng C et al (2014) Diversity analysis in cannabis sativa based on large-scale development of expressed sequence tag-derived simple sequence repeat markers. PLoS One 9(10):e110638. https://doi.org/10.1371/journal.pone.0110638

Gao S, Wang B, Xie S, Xu X, Zhang J, Pei L, Yu Y, Yang W, Zhang Y (2020) A high-quality reference genome of wild Cannabis sativa. Horticulture research, 7(1):73. https://doi.org/10.1038/s41438-020-0295-3

Gilmore S, Peakall R, Robertson J (2007) Organelle DNA haplotypes reflect crop-use characteristics and geographic origins of Cannabis sativa. Forensic Sci Int 172(2–3):179–190. https://doi.org/10.1016/j.forsciint.2006.10.025

Godin B, Lamaudière S, Agneessens R et al (2013) Chemical characteristics and biofuels potentials of various plant biomasses: influence of the harvesting date. J Sci Food Agric 93:3216–3224

Gorchs G, Llovers J, Comas J (2000) Effect of hemp (*Cannabis sativa* L.) in a crop rotation hemp–wheat in the humid cool areas of North-eastern of Spain. In: Proceeding of the Conference: Crop Development for the Cool and Wet Regions of Europe (COST Action 814), Pordenone, pp 581–589

Gordon-Kamm B, Sardesai N, Arling M et al (2019) Using morphogenic genes to improve recovery and regeneration of transgenic plants. Plants 8(2):38. https://doi.org/10.3390/plants8020038

Grassa CJ, Weiblen GD, Wenger JP et al (2021) A new Cannabis genome assembly associates elevated cannabidiol (CBD) with hemp introgressed into marijuana. New Phytol 230:1665–1679. https://doi.org/10.1111/nph.17243

Grigor'ev S, Shelenga T, Baturin V, Sarana YV (2010) Biochemical characteristics of hemp seeds from various regions of Russia. Russ Agric Sci 36:262–264. https://doi.org/10.3103/S1068367410040105

Grohe B (2004) Heat conductivities of insulation mats based on water glass bonded non-textile hemp or flax fibres. Holz Roh Werkst 62:352–357

Hesami M, Pepe M, Milad A et al (2020) Recent advances in cannabis biotechnology. Ind Crop Prod 158:113026. https://doi.org/10.1016/j.indcrop.2020.113026

Hesami M, Pepe M, Monthony AS et al (2021) Modeling and optimizing in vitro seed germination of industrial hemp (*Cannabis sativa* L.). Ind Crop. Prod 170:113753. https://doi.org/10.1016/j.indcrop.2021.113753

Hesami M, Pepe M, de Ronne M et al (2023) Transcriptomic profiling of embryogenic and non-embryogenic callus provides new insight into the nature of recalcitrance in cannabis. Int J Mol Sci 24:14625. https://doi.org/10.3390/ijms241914625

Hillig KW (2005) Genetic evidence for speciation in Cannabis (Cannabaceae). Genet Resour Crop Evol 52:161–180. https://doi.org/10.1007/s10722-003-4452-y

Holbery J, Houston D (2006) Natural-fiber-reinforced polymer composites in automotive applications. JOM 58:80–86

Höppner FKF (2006) Status report on the German collections of flax and hemp. In First Meeting of the Working Group on Fibre Crops, Wageningen: The European Cooperative Programme for Plant Genetic Resources

Institute of Agrifood Research and Technology (IRTA) (2024) Genetic analysis service: https://www.irta.cat/en/servei/genetic-analysis-service/?ant=13909
Ivonyi I, Izsoki Z, van der Werf HMG (1997) Influence of nitrogen supply and Pand K levels of the soil on dry matter and nutrient accumulation of fibre hemp (*Cannabis sativa* L.). J Int Hemp Assoc 4(1):82–87
Jahani M, Duriez P, Bercovich N et al (2022) A pan-genome and diversity panel for cannabis. Poster. https://pag.confex.com/pag/xxix/poster/eposterview.cgi?eposterid=472
Jones D, Brischke C (2017) 3—Nonwood bio-based materials. In: Performance of Bio-based Building Materials. Woodhead Publishing, pp 97–186
Kang Y, Zhou M, Merry A, Barry K (2020) Mechanisms of powdery mildew resistance of wheat—a review of molecular breeding. Plant Pathol 69:601–617. https://doi.org/10.1111/ppa.13166
Karus M, Vogt D (2004) European hemp industry: cultivation, processing and product lines. Euphytica 140:7–12
Keller A, Leupin M, Mediavilla V, Wintermantel E (2001) Influence of the growth stage of industrial hemp on chemical and physical properties of the fibres. Ind Crop Prod 13:35–48
Kim AL, Yun YJ, Choi HW et al (2022) Establishment of efficient Cannabis (*Cannabis sativa* L.) protoplast isolation and transient expression condition. Plant Biotech Rep 16(5):613–619
Krasteva G, Georgiev V, Pavlov A (2021) Recent applications of plant cell culture technology in cosmetics and foods. Eng Life Sci 21:68–76. https://doi.org/10.1002/elsc.202000078
Kriese U, Schumann E, Weber WE et al (2004) Oil content, tocopherol composition and fatty acid patterns of the seeds of 51 Cannabis sativa L. genotypes. Euphytica 137:339–351. https://doi.org/10.1023/B:EUPH.0000040473.23941.76
Krishna H, Alizadeh M, Singh D et al (2016) Somaclonal variations and their applications in horticultural crops improvement. Biotech 6:1–18. https://doi.org/10.1007/s13205-016-0389-7
Kumar APK, McKeown PC, Boualem A et al (2017) TILLING by Sequencing (TbyS) for targeted genome mutagenesis in crops. Mol Breed 37:14. https://doi.org/10.1007/s11032-017-0620-1
Kurtz LE, Mahoney JD, Brand MH, Lubell-Brand JD (2020) Comparing genotypic and phenotypic variation of selfed and outcrossed progeny of hemp. Hort Sci horts 55(8):1206–1209. https://doi.org/10.21273/HORTSCI15061-20
Lapierre E, Monthony AS, Torkamaneh D (2023) Genomics-based taxonomy to clarify cannabis classification. Genome 66(8):202–211. https://doi.org/10.1139/gen-2023-0005
Lata H, Chandra S, Techen N et al (2011) Molecular analysis of genetic fidelity in Cannabis sativa L. plants grown from synthetic (encapsulated) seeds following in vitro storage. Biotechnol Lett 33:2503–2508. https://doi.org/10.1007/s10529-011-0712-7
Lata H, Chandra S, Techen N et al (2016) In vitro mass propagation of *Cannabis sativa* L.: A protocol refinement using novel aromatic cytokinin meta-topolin and the assessment of eco-physiological, biochemical and genetic fidelity of micropropagated plants. J Appl Res Med Arom Plants S221478611530019X 3:18–26. https://doi.org/10.1016/j.jarmap.2015.12.001
Laverty KU, Stout JM, Sullivan MJ et al (2019) A physical and genetic map of Cannabis sativa identifies extensive rearrangement at the THC/CBD acid synthase loci. Genome Res 29:146–156. https://doi.org/10.1101/gr.242594.118
Leckie KM, Sawler J, Kapos P et al (2023) Loss of daylength sensitivity by splice site mutation in cannabis. bioRxiv:2023. https://doi.org/10.1101/2023.03.10.532103
Leizer C, Ribnicky D, Poulev A et al (2000) The composition of hemp seed oil and its potential as an important source of nutrition. J Nutraceuticals Funct Med Foods 2:35–53
Lethin J, Shakil SSM, Hassan S et al (2020) Development and characterization of an EMS-mutagenized wheat population and identification of salt-tolerant wheat lines. BMC Plant Biol 20:18. https://doi.org/10.1186/s12870-019-2137-8
Lone JF, Kumar B, Singh K et al (2017) Pollen morphology and variability among the wild Cannabis sativa L. from the western Himalaya. India Flora 305:0367–2530. https://doi.org/10.1016/j.flora.2023.152342
Lotz LAP, Groeneveld RMW, Habekotté B, Oene van H (1991) Reduction of growth and reproduction of Cyperus esculentus by specific crops. Weed Res 31:153–160

Li ZD (1982) The theory and techniques for bast fiber crops. Shanghai science and technology. Press, Shanghai, pp 332–387. (in Chinese)
Li L, Yu S, Chen J et al (2022) Releasing the full potential of cannabis through biotechnology. Agronomy 12(10):2439. https://doi.org/10.3390/agronomy12102439
Liang Q (2012) Foreword. In: Shu QY, Forster BP, Nakagawa H (eds) Plant mutation breeding and biotechnology. CABI, Oxfordshire, p 608
Lipson Feder C, Cohen O, Shapira A et al (2021) Fertilization following pollination predominantly decreases phytocannabinoids accumulation and alters the accumulation of terpenoids in cannabis inflorescences. Front Plant Sci 12:753847
Liu X, Li Y, Han B et al (2012) Yield response of continuous soybean to one-season crop disturbance in a previous continuous soybean field in Northeast China. Field Crop Res 138:52–56
Liu M, Fernando D, Daniel G et al (2015) Effect of harvest time and field retting duration on the chemical composition, morphology and mechanical properties of hemp fibers. IndCrop Prod 69:29–39
Lubell JD, Brand MH (2018) Foliar sprays of silver thiosulfate produce male flowers on female hemp plants. Hort Technol 28:743–747. https://doi.org/10.21273/HORTTECH04188-18
Lynch RC, Vergara D, Tittes S et al (2016) Genomic and chemical diversity in cannabis. Crit Rev Plant Sci 35(5–6):349–363. https://doi.org/10.1080/07352689.2016.1265363
Mandolino G, Carboni A, Forapani S et al (1999) Identification of DNA markers linked to the male sex in dioecious hemp (*Cannabis sativa* L.). Theor Appl Genet 98:86–92
Mandolino G, Bagatta M, Carboni A et al (2003) Qualitative and quantitative aspects of the inheritance of chemical phenotype in Cannabis. J Ind Hemp 8:51–72
Mandolino G, Carboni A (2004) Potential of marker-assisted selection in hemp genetic improvement. Euphytica 140:107–120
Mandolino MCA, Fontana F, Maestrini C, Fila G (2006) Hemp and flax germplasm collections at CRA-ISCI. In: First meeting of the working group on fibre crops. Wageningen: The European Cooperative Programme for Plant Genetic Resources
Mankowska G, Silska G (2015) Genetic resources of Cannabis sativa L. in the collection of the gene bank at INF & MP in Poznan. J Nat Fibers 12:332–340. https://doi.org/10.1080/15440478.2014.928246
Matthäus B, Schumann E, Brühl L, Kriese U (2006) Hempseed oil-influence of the genotype on the composition in a two-year study. J Ind Hemp 10:45–65. https://doi.org/10.1300/J237v10n02_05
McGoey BV, Janik R, Stinchcombe JR (2017) Individual chambers for controlling crosses in wind-pollinated plants. Methods Ecol Evol 8:887–891. https://doi.org/10.1111/2041-210X.12722
McKernan K, Helbert Y, Kane L et al (2020) Sequence and annotation of 42 cannabis genomes reveals extensive copy number variation in cannabinoid synthesis and pathogen resistance genes. bioRxiv:1–56. https://doi.org/10.1101/2020.01.03.894428
Mihalyov PD, Garfinkel AR (2021) Discovery and genetic mapping of PM1, a powdery mildew resistance gene in *Cannabis sativa* L. Front Agron 3:720215. https://doi.org/10.3389/fagro.2021.720215
Moliterni VMC, Cattivelli L, Ranalli P, Mandolino G (2004) The sexual differentiation of Cannabis sativa L.: A morphological and molecular study. Euphytica 140:95–106
Monthony AS, Page SR, Hesami M, Jones AMP (2021) The past, present and future of *Cannabis sativa*. Tiss Cult Plant 10:185. https://doi.org/10.3390/plants10010185
Müssig J, Martens R (2003) Quality aspects in hemp fibre production—Influence of cultivation, harvesting and retting. J Industr Hemp 8:11–32
Nalapalli S, Tunc-Ozdemir M, Sun Y et al (2021) Morphogenic regulators and their application in improving plant transformation. Methods Mol Biol 2238:37–61. https://doi.org/10.1007/978-1-0716-1068-8_3
NCBI Genome Data Viewer, ID: 90062 https://www.ncbi.nlm.nih.gov/genome/gdv/browser/genome/?id=GCF_900626175.2
NCBI Taxonomy_ID: 3483. https://www.ncbi.nlm.nih.gov/datasets/taxonomy/3483/

Pacifico D, Miselli F, Micheler M et al (2006) Genetics and marker-assisted selection of the chemotype in *Cannabis sativa* L. Mol Breeding 17:257–268. https://doi.org/10.1007/s11032-005-5681-x

Page SRG, Monthony AS, Jones AMP (2020) Basal media optimization for the micropropagation and callogenesis of *Cannabis sativa* L. BioRxiv 1–23. https://doi.org/10.1101/2020.02.07.939181

Parsons JL, Martin SL, James T, Golenia G, Boudko EA, Hepworth SR (2019) Polyploidization for the Genetic Improvement of Cannabis sativa. Frontiers in plantscience, 10:476. https://doi.org/10.3389/fpls.2019.00476

Pavelek M (2006) Czech flax (*Linum usitatissimum* L.) and hemp (*Cannabis sativa*) collections. Data for country report. In: First Meeting of the Working Group on Fibre Crops. Wageningen

Pépin N, Hebert FO, Joly DL (2021) Genome-wide characterization of the MLO gene family in *Cannabis sativa* reveals two genes as strong candidates for powdery mildew susceptibility. Front Plant Sci 12:729261. https://doi.org/10.3389/fpls.2021.729261

Petit J, Gulisano A, Dechesne A, Trindade LM (2019) Phenotypic variation of cell wall composition and stem morphology in hemp (*Cannabis sativa* L.): optimization of methods. Front. Plant Sci 10:959. https://doi.org/10.3389/fpls.2019.00959

Petit J (2020) Novel molecular tools to uncover the genetic architecture of hemp fibre quality. PhD thesis. https://edepot.wur.nl/510653. ISBN: 9789463952675; 9463952675

Petit J, Salentijn EMJ, Paulo MJ et al (2020a) Genetic variability of morphological, flowering and biomass quality traits in hemp (*Cannabis sativa* L.) Front. Plant Sci 10:102. https://doi.org/10.3389/fpls.2020.00102

Petit J, Salentijn EMJ, Paulo MJ et al (2020b) Elucidating the genetic architecture of fiber quality in hemp (*Cannabis sativa* L.) using a Genome-Wide association study. Front Genet 17:11. https://doi.org/10.3389/fgene.2020.566314

Petit J, Salentijn EMJ, Paulo MJ et al (2020c) Genetic architecture of flowering time and sex determination in hemp (Cannabis sativa L.): a Genome-Wide association study. Front Plant Sci 11:569958. https://doi.org/10.3389/fpls.2020.569958

Petit J, Bilski PM, Blanco-Rodríguez E et al (2023) *Cannabis sativa* L. in vitro protocols: Advancements in germplasm micropropagation, preservation and maintenance. Poster presented at Spanish Society for In vitro Cultivation for Plant Tissues (SECIVTV); September 6–8, 2023, Lleida, Spain

Piotrowski S, Carus M (2019) Ecological benefits of hemp and flax cultivation. Nova-Institute:2011–2005

Piunno KF, Golenia G, Boudko EA et al (2019) Regeneration of shoots from immature and mature inflorescences of Cannabis sativa. Can JPlant Sci 99(4):556–559. https://doi.org/10.1139/cjps-2018-0308

Placet V (2009) Characterization of the thermo-mechanical behaviour of Hemp fibres intended for the manufacturing of high performance composites, Compos A: Appl Sci Manuf 40:1111–1118

Placido DF, Lee CC (2022) Potential of industrial hemp for phytoremediation of heavy metals. Plants 11(5):595. https://doi.org/10.3390/plants11050595

Poniatowska J, Wielgus K, Szalata M et al (2019) Contribution of Polish agrotechnical studies on *Cannabis sativa* L. to the global industrial hemp cultivation and processing economy. Herba Polonica 65(2):37–50

Posselt UK (2010) Breeding methods in cross-pollinated species. In: Boller B (ed) Fodder crops and amenity grasses, handbook of plant breeding, vol 5. Springer Science and Business Media, LLC, New York

Ranalli P (2004) Current status and future scenarios of hemp breeding. Euphytica 140:121–131

Ren G, Zhang X, Li Y et al (2021) Large-scale whole-genome resequencing unravels the domestication history of *Cannabis sativa*. Sci Adv 7(29):16. https://doi.org/10.1126/sciadv.abg2286

Richez-Dumanois C, Braut-Boucher F, Cosson L, Paris M (1986) Multiplication végétative In Vitro du chanvre (*Cannabis sativa* L.) Application à la conservation des clones sélectionnés. Agronomie 6:487–495. https://doi.org/10.1051/agro:19860510

Salentijn EMJ, Zhang Q, Amaducci S et al (2015) New developments in fiber hemp (*Cannabis sativa* L.) breeding. Ind Crop Prod 68:32–41

Salentijn EMJ, Petit J, Trindade LM (2019) The complex interactions between flowering behavior and fiber quality in hemp. Front Plant Sci 10:614. https://doi.org/10.3389/fpls.2019.00614

Sakamoto K, Shimomura K, Komeda H, Satoh S (1995) A male associated DNA sequence in a dioecious plant *Cannabis sativa* L. Plant Cell Physiol 36:1549–1554

Sánchez Vallduví GE, Sarandón SJ (2011) Effects of changes in flax (Linum usitatissimum L.) density and interseeding with red clover (Trifolium pratense L.) on the competitive ability of flax against Brassica weeds. J of Sustainable Agriculture, 35(8):914–926

Sapino S, Carlotti ME, Peira E, Gallarate M (2005) Hemp-seed and olive oils: their stability against oxidation and use in O/W emulsions. J Cosmet Sci 56:227–251

Sattler MC, Carvalho CR, Clarindo WR (2016) The polyploidy and its key role in plant breeding. Planta 243:281–296. https://doi.org/10.1007/s00425-015-2450-x

Sawler J, Stout JM, Gardner et al (2015) The genetic structure of marijuana and hemp. PLoS One 10:e0133292

Shiels D, Prestwich BD, Koo O et al (2022) Hemp Genome Editing-Challenges and Opportunities. Front Genome Ed 4:823486. https://doi.org/10.3389/fgeed.2022.823486

Sloan S, Jenkins CN, Joppa LN et al (2014) Remaining natural vegetation in the global biodiversity hotspots. Biol Conserv 177:12–24. https://doi.org/10.1016/j.biocon.2014.05.027

Small E, Cronquist A (1976) A practical and natural taxonomy for Cannabis. Taxon 25:405–435. https://doi.org/10.2307/1220524

Small E (2015) Evolution and classification of *Cannabis sativa* (Marijuana, Hemp) in relation to human utilization. Bot Rev 81:189–294

Small E, Marcus D (2002) Hemp: a new crop with new uses for North America. In: Janick, Whipkey A (eds) Trends in new crops and new uses, Alexandria, VA. ASHS Press, pp 284–326

Soler S, Gramazio P, Figàs MR et al (2017) Genetic structure of Cannabis sativa var. indica cultivars based on genomic SSR (gSSR) markers: implications for breeding and germplasm management. Ind Crop Prod 104:171–178. https://doi.org/10.1016/j.indcrop.2017.04.043

Sorokin A, Kovalchuk I (2025) Development of efficient and scalable regeneration tissue culture method for Cannabis sativa. Plant science : an int j of experimentalplant biology, 350:112296. https://doi.org/10.1016/j.plantsci.2024.112296

Steel L, Welling M, Ristevski N et al (2023) Comparative genomics of flowering behavior in Cannabis sativa. Front Plant Sci 14:1227898. https://doi.org/10.3389/fpls.2023.1227898

Struik PC, Amaducci S, Bullard MJ et al (2000) Agronomy of fibre hemp (*Cannabis sativa* L.) in Europe. Ind Crop Prod 11:107–118

Suprasanna P, Patade VY, Desai NS et al (2011) Biotechnological developments in sugarcane improvement: an overview. Sugar Tech 13:322–335. https://doi.org/10.1007/s12355-011-0103-3

Szurman-Zubrzycka M, Kurowska M, Till BJ, Szarejko I (2023) Is it the end of TILLING era in plant science? Front Plant Sci 14:1160695. https://doi.org/10.3389/fpls.2023.1160695

Tadele Z (2016) Mutagenesis and TILLING to dissect gene function in plants. Curr Genomics 17:499–508

Toonen MAJ, Maliepaard C, Reijmers TH et al (2004) Predicting the chemical composition of fibre and core fraction of hemp (*Cannabis sativa* L.). Euphytica 140:39–46

Torres A, Pauli C, Givens R et al (2022) High-throughput methods to identify male Cannabis sativa using various genotyping methods. J Cannabis Res 4:57. https://doi.org/10.1186/s42238-022-00164-7

Toth JA, Stack GM, Cala AR et al (2020) Development and validation of genetic markers for sex and cannabinoid chemotype in *Cannabis sativa* L. GCB Bioenergy 12:213–222. https://doi.org/10.1111/gcbb.12667

Toth JA, Stack GM, Carlson CH, Smart LB (2022) Identification and mapping of major-effect flowering time loci Autoflower1 and Early1 in *Cannabis sativa* L. Front Plant Sci 13:991680. https://doi.org/10.3389/fpls.2022.991680

Tsai H, Howell T, Nitcher R et al (2011) Discovery of rare mutations in populations: TILLING by sequencing. Plant Physiol 156:1257–1268. https://doi.org/10.1104/pp.110.169748

Tsai H, Missirian V, Ngo KJ et al (2013) Production of a high-efficiency TILLING population through polyploidization. Plant Physiol 161(4):1604–1614. https://doi.org/10.1104/pp.112.213256

Tyurin AA, Suhorukova AV, Kabardaeva KV, Goldenkova-Pavlova IV (2020) Transient gene expression is an effective experimental tool for the research into the fine mechanisms of plant gene function: advantages, limitations, and solutions. Plants 9(9):1187. https://doi.org/10.3390/plants9091187

Uauy C, Wulff BBH, Dubcovsky J (2017) Combining traditional mutagenesis with new high-throughput sequencing and genome editing to reveal hidden variation in polyploid wheat. Ann Rev Genet 51(1):435–454

Van Bakel H, Stout JM, Cote AG et al (2011) The draft genome and transcriptome of *Cannabis sativa*. Genome Biol 12:R102

van Dam JEG, Bos HL (2004) The environmental impact of fiber crops in industrial applications. FAO. https://edepot.wur.nl/49285

van den Broeck HC, Maliepaard CA, Ebskamp MJM et al (2008) Differential expression of genes involved in C1 metabolism and lignin biosynthesis in wooden core and bast tissue of fibre hemp (*Cannabis sativa* L.). Plant Sci 174(2):205–220. https://doi.org/10.1016/j.plantsci.2007.11.008

Van der Werf HMG, Turunen L (2008) The environmental impacts of the production of hemp and flax textile yarn. Ind Crop Prod 27:1–10

Van der Werf HMG, Wijlhuizen M, de Schutter JAA (1995) Plant density and self-thinning affect yield and quality of fibre hemp (*Cannabis sativa* L.). Field Crop Res 40:153–164

Venturi G, Amaducci MT (1999) Canapa (*Cannabis sativa* L.). In: Venturi G, Ama-ducci MT (eds) Le colture da fibra. Edagricole, Collana PRisCA, pp 33–55. (in Italian)

Vincent H, Wiersema J, Kell S et al (2013) A prioritized crop wild relative inventory to help underpin global food security. Biol Conserv 167:265–275. https://doi.org/10.1016/j.biocon.2013.08.011

Uchendu E, Lata H, Chandra S et al (2019) Cryopreservation of Shoot Tips of Elite Cultivars of *Cannabis sativa* L. by Droplet Vitrification. Med Cannabis Cannabinoids 2(1):29–34. https://doi.org/10.1159/000496869

Wanas AS, Radwan MM, Mehmedic Z et al (2016) Antifungal activity of the volatiles of high potency *Cannabis sativa* L. against *Cryptococcus neoformans*. Record Nat Plant 10(2):214–220

Wambua P, Ivens J, Verpoest I (2003) Natural fibres: can they replace glass in fibre reinforced plastics? Comp Sci Tech 63:1259–1264

Welling MT, Liu L, Shapter T et al (2015) Characterisation of cannabinoid composition in a diverse *Cannabis sativa* L. germplasm collection. Euphytica 208:463–475. https://doi.org/10.1007/s10681-015-1585-y

Welling MT, Shapter T, Rose TJ et al (2016) A Belated green revolution for Cannabis: virtual genetic resources to fast-track cultivar development. Front Plant Sci 7:1113. https://doi.org/10.3389/fpls.2016.01113

Welling MT, Liu L, Kretzschmar T et al (2020) An extreme-phenotype genome-wide association study identifies candidate cannabinoid pathway genes in Cannabis. Sci Rep 10:18643. https://doi.org/10.1038/s41598-020-75271-7

Westerhuis W, Amaducci S, Struik PC et al (2009) Sowing density and harvest time affect fibre content in hemp (*Cannabis sativa*) through their effects on stem weight. Ann Appl Biol 155:225–244

Woods P, Campbell BJ, Nicodemus TJ et al (2021) Quantitative trait loci controlling agronomic and biochemical traits in *Cannabis sativa*. Genetics 219:iyab099. https://doi.org/10.1093/genetics/iyab099

Woods P, Price N, Matthews P, Mckay J (2023) Genome-wide polymorphism and genic selection in feral and domesticated lineages of *Cannabis sativa*. G3 (Bethesda) 13(2):jkac209. https://doi.org/10.1093/g3journal/jkac209

Wróbel T, Dreger M, Wielgus K, Słomski R (2020) Modified nodal cuttings and shoot tips protocol for rapid regeneration of *Cannabis sativa* L. J Nat Fib 19:536–545. https://doi.org/10.1080/15440478.2020.1748160

Yang W, Chen X, Li Y et al (2020) Advances in pharmacological activities of terpenoids. Nat Prod Commun 15(3):15. https://doi.org/10.1177/1934578X20903555

Zabaniotou A, Kamaterou P (2019) Food waste valorization advocating circular bioeconomy—a critical review of potentialities and perspectives of spent coffee grounds biorefinery. J Clean Prod 211:1553–1566

Zatta A, Monti A, Venturi G (2012) Eighty years of studies on industrial hemp in the Po Valley (1930–2010). J Nat Fib 9:180–196

Zhang J, Yan J, Huang S et al (2020) Genetic diversity and population structure of cannabis based on the genome-wide development of simple sequence repeat markers. Front Genet 11:958. https://doi.org/10.3389/fgene.2020.00958

Zhang X, Xu G, Cheng C, Lei L, Sun J, Xu Y, Deng C, Dai Z, Yang Z, Chen X, Liu C, Tang Q, Su J (2021) Establishment of an Agrobacterium-mediated genetic transformation and CRISPR/Cas9-mediated targeted mutagenesis in Hemp (Cannabis Sativa L.). Plant biotechnology j, 19(10):1979–1987. https://doi.org/10.1111/pbi.13611

Zhao H, Xiong H, Chen J (2021) Regional comparison and strategy recommendations of industrial hemp in China based on a SWOT analysis. Sustainability 13:6419. https://doi.org/10.3390/su13116419

Zhu P, Zhao Y, You X et al (2022) A versatile protoplast system and its application in *Cannabis sativa* L. Botany 101(7):291–300. https://doi.org/10.1139/cjb-2021-0178

Chapter 12
Molecular Breeding Fundamental of Indian Hemp (*Apocynum cannabinum* L.)

Guoqi Li, Yanyun Chen, Yin Yi, and Li Jiang

Abstract This chapter focuses on *Apocynum cannabinum* L. (Indian hemp), a perennial Apocynaceae species renowned for its tolerance to saline, drought, and cold environments, alongside its traditional medicinal uses (e.g., treating dropsy and snakebites). Despite its recent introduction to China, its untapped ecological and economic potential drives our molecular breeding initiative. We present the first high-quality, chromosome-level genome assembly of *A. cannabinum* (260 Mb, anchored to 11 chromosomes; scaffold N50: 21.16 Mb), achieved through an integrated multi-platform approach combining Illumina HiSeq, SMRT long-read sequencing, 10X Genomics linked reads, and Hi-C chromatin interaction data. Genome annotation identified 22,793 protein-coding genes, with 95.6% functionally annotated, 92.3% containing conserved domains, and 78.7% mapped to known metabolic pathways, underscoring functional completeness. Comparative genomics revealed a speciation event 35.8 million years ago (confidence interval: 27.0–46.9 million years), providing an evolutionary framework for adaptive trait analysis. The genome elucidates gene family expansions linked to stress tolerance (e.g., salt/drought-responsive genes) and secondary metabolite biosynthesis, offering molecular targets for trait enhancement. This resource enables marker-assisted breeding, pathway engineering for medicinal compounds, and cross-species comparisons within Apocynaceae. By decoding genetic and evolutionary bases of its resilience and therapeutic properties, the study accelerates precision breeding to optimize

G. Li (✉)
School of Ecology and Environment, Ningxia University, Yinchuan, China
e-mail: ligq@nxu.edu.cn

Y. Chen
Life Science School of Ningxia University, Yinchuan, China

Y. Yi
School of Life Sciences, Guizhou Normal University, Guiyang, China
e-mail: yiyin@gznu.edu.cn

L. Jiang
State Key Laboratory of Desert and Oasis Ecology, Xinjiang Institute of Ecology and Geography, Chinese Academy of Sciences, Ürümqi, China
e-mail: jiangli@ms.xjb.ac.cn

K. F. M. Salem et al. (eds.), *Breeding and Biotechnology of Grass and Bast Fiber Crops*, Advances in Plant Breeding Strategies 12,
https://doi.org/10.1007/978-3-032-00403-1_12

A. cannabinum for cultivation in marginal lands and pharmaceutical applications. The genome also serves as a model for related medicinal plants, bridging genomic innovation with agricultural and industrial utilization. In summary, this work establishes *A. cannabinum*'s genomic foundation, driving molecular breeding to harness its ecological adaptability and economic potential. It underscores the synergy of advanced sequencing technologies in transforming underutilized species into sustainable resources for climate-resilient agriculture and bioactive compound production.

Keywords Indian hemp (*Apocynum cannabinum* L.) · Whole-genome sequencing · 10X genomics linked reads · Hi-C · Speciation · Molecular breeding

12.1 Introduction

A. cannabinum L., commonly known as Indian hemp, is a resilient perennial herb that thrives in the harsh, saline-alkali soils of arid and semi-arid regions. While its primary distribution is in North America, this species has been successfully introduced and cultivated in China within the last decade. Notably, *A. cannabinum* exhibits a broad ecological range, particularly in the Northwestern deserts of China, positioning it as a multi-functional plant with significant ecological, economic, and medicinal attributes (Li and Chen 2012; Xie et al. 2012). The combination of these valuable traits suggests a promising future for the industrial application of *A. venetum* (Li et al. 2019). Comparative studies have revealed that *A. cannabinum* demonstrates superior drought resistance, characterized by thicker, straighter stalks, larger leaves, and reduced branching compared to the closely related *A. venetum* (Wang et al. 2019). This enhanced adaptability contributes to its high early succession rate, enabling it to rapidly colonize and establish itself in new environments (Keever 1979; Mulhouse and Galatowitsch 2003). Beyond its traditional use as a source of fiber, *A. cannabinum* boasts a rich chemical composition, including cardiac glycosides, resins, rubber, tannins, and starch. Historically, Native American traditional medicine utilized the roots of *A. cannabinum* for the treatment of various ailments, such as heart disease and ascites (Leidy 1884; Duprey 1905). Apocynum belongs to the Apocynaceae family, a diverse group comprising approximately 44–46 genera and 145–176 species in China. These species are predominantly found in provinces south of the Yangtze River and on coastal islands, with fewer occurrences in the northern and northwestern regions (Li and Chen 1997). Due to their unique evolutionary history and significant economic importance, Apocynaceae plants have been a central focus in phylogenetic studies, particularly regarding the systematic relationships within the family and between genera (Li et al. 1995). These investigations seek to unravel the evolutionary connections and understand the diversification of this valuable plant family.

The rapid advancement and increasing sophistication of sequencing technologies have spurred a surge in plant genomics research. These studies aim to elucidate the intricate molecular mechanisms underlying plant growth, development, evolutionary trajectories, and origins at the genome level (Goff et al. 2002). The advent of second and third-generation sequencing techniques, including those incorporating 10X Genomics linked reads and Hi-C methodologies, has broadened the scope of possibilities for sequencing complex plant genomes (Burton et al. 2013). The 10X Genomics linked-read technology essentially functions by introducing barcode sequences into long DNA sequence fragments. These barcodes are then leveraged to construct more accurate and extensive super-scaffolds, enabling the exploration of genetic information with greater resolution and detail (Eisenstein 2015). Complementing this approach, the Hi-C sequencing method employs chromosome conformation capture to organize draft genome sequences—assembled using second or third-generation sequencing or optical mapping—into chromosome sets. This process determines the order and orientation of various sequences along the chromosomes, providing a critical framework for genome assembly (Mascher et al. 2017). Consequently, the synergistic combination of 10X Genomics linked reads and Hi-C three-dimensional conformation capture technology significantly enhances the accuracy of whole-genome sequencing and improves the overall quality of genome assembly. While extensive whole-genome sequencing has been performed on numerous economically important and crop plants, the available sequencing data for *A. cannabinum* remains comparatively limited. This deficiency hinders further research on this valuable species, particularly in fields such as pharmacology, molecular biology, and genetic evolution.

To systematically evaluate the genomic architecture and evolutionary trajectory of *A. cannabinum* at multiple biological organization levels, we implemented an integrative whole-genome sequencing initiative incorporating advanced bioinformatics frameworks. This comprehensive investigation encompasses four principal research dimensions: (1) Detailed characterization of genomic elements through repetitive sequence analysis and transposable element annotation; (2) Multi-layered component analysis integrating structural annotation, functional categorization, and non-coding regulatory element identification; (3) Reconstruction of metabolic networks with particular emphasis on stress-responsive secondary metabolite pathways; (4) High-resolution phylogenetic reconstruction employing comparative genomics and molecular dating techniques. The experimental design specifically prioritizes the elucidation of molecular mechanisms underlying environmental stress adaptation through transcriptomic regulatory network analysis and protein functional domain characterization. The methodological approach incorporates next-generation sequencing platforms complemented by long-read sequencing technologies (PacBio) to ensure comprehensive genomic coverage and accurate assembly. Advanced computational pipelines involving de novo assembly algorithms and hierarchical scaffolding strategies will be implemented to resolve complex genomic regions (Usman et al. 2022).

From an applied perspective, the anticipated research outcomes are projected to establish a multidimensional knowledge framework encompassing: (1) A curated genomic database with annotated stress-responsive gene clusters; (2) Metabolic pathway maps highlighting osmoregulation-related biosynthesis routes; (3)

Evolutionary models explaining ecological adaptation patterns. These resources will significantly advance molecular breeding initiatives through marker-assisted selection protocols and CRISPR-based genome editing applications. Furthermore, the generated datasets are expected to facilitate cross-species comparative studies in drought-resistant perennial plants, potentially informing ecological restoration strategies and sustainable agricultural practices in arid regions. The theoretical implications of this investigation extend to evolutionary developmental biology through the identification of conserved non-coding sequences, while practical applications may revolutionize phytoremediation technologies through the engineering of enhanced stress-tolerance phenotypes. This comprehensive genomic portrait of *A. cannabinum* will serve as both a foundational reference for systematic biology research and a strategic roadmap for the biotechnological exploitation of extremophyte genetic resources (Usman et al. 2022).

This work will briefly discuss recent progress in plant molecular breeding, focusing on the general background of *A. cannabinum*, and will especially highlight the whole genome sequencing results, offering a cornerstone for future molecular breeding efforts with this species.

12.2 Advances in Plant Molecular Breeding

Plant breeding, a multifaceted discipline encompassing both the art and science of purposefully modifying the genetic makeup of plants, aims to develop novel and superior varieties tailored to meet evolving agricultural needs. This practice boasts a long and storied history stretching back millennia, deeply intertwined with the rise of early agricultural societies. Ancient civilizations, including those of Egypt and China, pioneered the selection and propagation of plants exhibiting desirable characteristics, leading to the domestication of staple crops like wheat and rice (Smith 2019). A pivotal moment in understanding heredity occurred in the nineteenth century with Gregor Mendel's groundbreaking experiments on pea plants. His meticulous observations revealed the patterns of inheritance for specific traits, introducing the concepts of dominant and recessive alleles and establishing the fundamental principles underpinning the field of genetics. Over the centuries, plant breeding has undergone a remarkable evolution, transitioning from traditional, often laborious and time-intensive methods to sophisticated, technology-driven approaches that leverage the power of modern science. The emergence of molecular biology and genomics has fundamentally transformed plant breeding, providing breeders with the ability to make highly targeted and precise genetic alterations, thereby accelerating the development of improved plant varieties. When juxtaposing conventional plant breeding techniques with the innovative methodologies of molecular breeding, the following key advancements in plant breeding technologies become apparent, highlighting the transformative impact of modern science on this essential field.

12.2.1 Traditional Breeding

For countless generations, traditional plant breeding has served as the cornerstone of agricultural improvement, leveraging the inherent genetic diversity found within plant species. This established methodology centers on the deliberate crossing of plants that exhibit desirable characteristics, followed by the meticulous selection of offspring that inherit and express those traits. These selected progeny are then subjected to further rounds of breeding and selection, a process iterated across numerous generations until a stable and demonstrably improved variety is achieved. While traditional plant breeding has undeniably contributed to the development of enhanced crop varieties for centuries, its effectiveness is tempered by certain inherent limitations. Notably, the process can be exceedingly time-consuming, often requiring many years, if not decades, to successfully develop and stabilize novel varieties. Furthermore, traditional methods may struggle to effectively capture and integrate traits that are governed by multiple genes acting in concert, or those that reside within wild relatives of the crop species, thereby restricting the overall potential for genetic improvement (Gonal et al. 2023).

12.2.2 Modern Molecular Breeding

Modern molecular breeding has emerged as a critical strategy for addressing pressing global challenges, including ensuring food security for a growing population, mitigating the effects of climate change on agricultural production, and promoting sustainable agricultural practices that minimize environmental impact. This innovative approach offers powerful tools for enhancing both crop productivity and resilience to environmental stresses, capabilities that are increasingly vital for feeding a growing global population in the face of rapidly changing climatic conditions (Lodhi 2024). In contrast to traditional methods, molecular breeding integrates the fundamental principles of genetics, genomics, and cutting-edge biotechnological techniques to accelerate the breeding cycle and circumvent the limitations inherent in conventional approaches. The precise identification and targeted manipulation of specific genes that govern desired traits represent the hallmarks of molecular breeding. This approach offers significant advantages over traditional breeding strategies, including the ability to overcome reproductive barriers between species, thereby broadening the gene pool available for crop improvement, accelerating the overall breeding process, and facilitating the creation of entirely new and unique plant variants. The application of molecular breeding techniques to the improvement of ornamental plants holds particular promise for enhancing product quality and bolstering the competitiveness of related industries (Dai et al. 2013). Over the past 5–10 years, significant progress has been made in the application of molecular breeding and genomics to pulse crops, leading to improved varieties with enhanced nutritional value and stress tolerance (Sivia et al. 2024). To adequately meet the escalating demands for food and nutritional security driven by a rapidly expanding global population, traditional crop breeding methods alone are proving to be insufficient. Advanced techniques such as gene pyramiding through marker-assisted selection, marker-assisted backcrossing (MABC), and marker-assisted recurrent selection (MARS) have greatly enhanced the development of durable and resilient lines that exhibit enhanced adaptive capacity against diverse abiotic constraints, thereby contributing to enhanced agricultural sustainability (Singh et al. 2024).

12.2.3 Marker-Assisted Selection (MAS)

Marker-assisted selection (MAS) represents a significant advancement in plant breeding, leveraging molecular markers—specific DNA sequences that are tightly functionally annotated to molecular breeding indices—to guide the selection process for plants possessing the desired genetic configurations. By identifying these markers, which serve as signposts linked to traits of interest, breeders can efficiently screen large populations of plants, thereby focusing their efforts on individuals that harbor the genetic architecture conducive to the desired characteristics. This targeted approach significantly streamlines the breeding pipeline, accelerating varietal development cycles with optimized resource allocation, and leading to more efficient and precise outcomes (Collard and Mackill 2008). MAS offers a truly transformative approach to crop improvement, by strategically utilizing DNA markers to manipulate specific genomic regions intimately associated with the expression of desirable traits. A key advantage of MAS over traditional visual phenotypic selection lies in its ability to enhance the efficiency and accuracy of trait selection, enabling breeders to more readily identify and select plants exhibiting the target traits. Through the integration of DNA marker identification and marker-based selection into conventional breeding programs, MAS effectively mitigates many of the challenges typically encountered during traditional crop breeding. Plant breeders have readily adopted MAS as an integral tool in their breeding programs, facilitating the identification of both dominant and recessive alleles, as well as genotyping-by-sequencing (GBS)-driven background purification dynamics (Sivia et al. 2024).

12.2.4 Genomic Selection (GS)

The application of next-generation sequencing (NGS) and high-throughput (HT) technologies in the field of molecular biology has ushered in an era of unprecedented genetic analysis, most notably facilitating the development of genotyping by sequencing (GBS). GBS has dramatically reduced the costs associated with sequencing, thereby making the development of genomic and molecular markers more accessible and economically feasible with reduced investment. Gene-based single nucleotide polymorphism (SNP) markers, derived from GBS, have proven to be invaluable tools for linkage and genome-wide association studies (GWAS) in pulse crops. These studies have been instrumental in elucidating the intricate genetic basis underlying a range of economically, nutritionally, and environmentally significant climate-resilient traits. The molecular markers associated with these crucial features can be effectively employed in marker-assisted breeding programs, thereby accelerating the process of improving these essential traits. By mining the genomic regions associated with each of these traits, researchers have successfully identified the genes involved. These candidate genes are then rigorously functionally confirmed to determine their precise roles in the expression and regulation of the specific traits under investigation. GBS-based genomic selection (GS) studies in pea (*Pisum sativum*) have been reported by numerous research groups, demonstrating the effectiveness of this approach. In particular, GS has been successfully deployed to enhance grain yield in pea recombinant inbred populations subjected to both severe terminal drought escape and intrinsic drought tolerance (DT) conditions (Karikari et al. 2024). Genome-enabled breeding (GEB) employs next-generation molecular profiling platforms to forecast phenotypic outcomes through systematic interrogation of species-specific chromosomal architectures. This methodology operates by processing multidimensional datasets derived from extensive germplasm repositories, where machine learning architectures are iteratively refined through the integration of allelic dosage patterns with quantitative trait expression matrices. The computational framework systematically deciphers nucleotide combinatorial effects across all loci, transcending conventional marker-assisted selection through whole-genome epistatic network reconstruction. These models enable breeders to accurately predict the performance of individuals that have not yet been phenotyped, making it possible to make informed selection decisions early in the breeding process (Varshney et al. 2019; York et al. 2019), thereby accelerating the overall breeding cycle and amplifying allelic selection kinetics.

12.2.5 Gene Editing

Recent breakthroughs in genome editing technologies, particularly the advent of the CRISPR-Cas9 system, have unlocked unprecedented possibilities for making precise and targeted genetic modifications in plants. Genome editing enables scientists to precisely modify, remove, or introduce targeted DNA segments within a plant's genome, thereby facilitating the development of crops with a wide array of targeted characteristics. This advanced methodology demonstrates enhanced accuracy and efficacy relative to conventional genetic engineering approaches, presenting transformative possibilities for crop improvement through the precise engineering of exceptionally intricate hereditary characteristics (Jaganathan et al. 2018). One crucial aspect of plant physiology amenable to genome editing is protein phosphorylation, a ubiquitous and essential post-translational modification. Protein phosphorylation acts as a critical driver of regulating a vast array of biosynthetic pathways within plants, encompassing biosynthetic pathways, vesicular trafficking, cytoskeletal organization, mitotic cycles, morphogenesis, ontogenetic progression, and environmental crosstalk. Protein kinase/phosphatase systems, core enzymatic cohorts governing phosphorylation spatiotemporal modulation, have ascended as strategic genomic targets in plant biotechnology. Through targeted genome editing of these enzymatic regulators via CRISPR-Cas9 platforms, scientists can harness in silico modeling to direct functional reprogramming and activity modulation in kinase/phosphatase networks, thereby optimizing crop performance and resilience (Sojka et al. 2023).

12.2.6 Transgenic Approaches

Transgenic plants are generated through the introduction of genes from other organisms—often from different species or even different kingdoms of life—into the genome of a target plant. While transgenic approaches have been successfully employed to engineer agronomic characteristics ranging from insect herbivory resistance to xenobiotic tolerance, these innovations consistently confront intensified societal oversight and biosafety compliance barriers. Such constraints originate in evidence-based apprehensions about transgenic cultivars'pleiotropic effects on ecological integrity and anthropocentric well-being (Powell and Morgante 2019). In the context of escalating concerns about global food security under future climate change scenarios, the engineering of climate-adaptive cultivars constitutes a strategic imperative. Edaphic constraints (e.g., drought/salinity) and phytopathogen pressure fundamentally impede agroecosystem sustainability. Molecular breeding platforms and genome-editing modalities enable precise tailoring of germplasm architectures with augmented adaptive plasticity. Such interventions potentiate osmoprotectant biosynthesis pathways while preserving elite allelic combinations through targeted epigenetic regulatory nodes in stress-responsive networks. A multitude of stress-responsive genes involved in diverse regulatory systems—including the antioxidant defense system, reactive oxygen species scavenging, chemosensory transducers, ribonucleoprotein complexes, and stress-responsive regulatory networks—have been gain-of-function modified or loss-of-function silenced through multiplexed genome editing modalities. This technical repertoire spans complementary oligonucleotide suppressors, small RNA-guided silencing, synthetic small RNA regulators, and orthogonal gene circuit engineering utilizing CRISPR-dCas transcriptional actuators (activation/repression) or epigenome sculpting modules in planta. The application of these approaches has been demonstrated to enhance tolerance to a broad spectrum of environmental stressors (Pandita 2023).

In summary, methodological innovations in crop genetic enhancement platforms have radically intensified genetic gain trajectories. Next-generation precision breeding frameworks integrating haplotype-based trait introgression, genome-wide prediction algorithms, and CRISPR-mediated directed evolution, and transgenic approaches, provide plant breeders with a powerful and diverse toolkit for developing crops with enhanced traits. These improved crops are better equipped to meet the multifaceted challenges of the future, including climate change, nutritional sovereignty, and synergizing with agroecological intensification imperatives.

12.3 The Outline of *A. cannabinum*

12.3.1 Taxonomic Hierarchy

A. cannabinum is a flowering plant within the class Dicotyledonae, order Gentianales and family Apocynaceae. The plants of the genus *Apocynum* often have medicinal properties (https://www.itis.gov).

Domain: Eukaryota	
Kingdom	Plantae (Plants)
Subkingdom	Viridiplantae (Green plants)
Infrakingdom	Streptophyta (Land plants)
Superdivision	Embryophyta
Division	Tracheophyta (Vascular plants)
Subdivision	Spermatophytina (Seed plants)
Class	Magnoliopsida (Dicotyledons)
Superorder	Asteranae
Order	Gentianales
Family	Apocynaceae (Dogbane, apocyns)
Genus	*Apocynum* L.—dogbane
Species	*Apocynum cannabinum* L.—hemp dogbane, *Apocynum spp.*,

Genus *Apocynum* L. contains three species: *A. androsaemifolium* L., *A. cannabinum* L. and *A. x floribundum*.

12.3.2 Taxonomy and Nomenclature

Kingdom:	Plantae
Taxonomic Rank (Synonym(s)):	Genus
Common Name(s):	Dogbane [English]
Current Standing:	Accepted
Record Credibility Rating:	Verified—standards met
Global Species Completeness:	Partial
Latest Record Review:	2011

The description of *A. cannabinum* in the Missouri Botanical Garden is as follows: (https://www.missouribotanicalgarden.org)

Botanical Designation: *Apocynum cannabinum*
Growth Habit: Herbaceous polycarpic
Taxonomic Family: Dogbane kin (Apocynaceae)
Vertical Extension: 0.61–1.22 m
Horizontal Dispersion: 0.46–0.76 m
Floral Phenology: July to August
Corolla Chroma: Alabaster to viridescent
Insolation Requirement: Unfiltered luminosity
Hydric Regime: Xeric-mesic gradient
Cultivation Input: Moderate oversight
Ecological Application: Successional colonization
Inflorescence: Conspicuous panicles
Lepidoptera Attractant: Nymphalid pollinators
Edaphic Tolerance: Aridification resistance

Thrives in mesotrophic, well-aerated substrates under maximal solar exposure. Exhibits psammophytic adaptation. A clonal colonizer forming restorative swards in degraded ecotones. *A. cannabinum* var. pratensis demonstrated robust proliferation (Fig. 12.1).

12.3.3 Distinctive Attributes and Notable Features

A. cannabinum, commonly known as Indian hemp or clasping-leaf dogbane, presents itself as a robust and bushy member of the Apocynaceae family, which also encompasses dogbanes and milkweeds. This perennial herb exhibits several noteworthy characteristics, including its opposite, lanceolate-shaped leaves that adorn its upright and stiff stems. During the summer months, *A. cannabinum* produces terminal clusters of diminutive, whitish flowers, adding to its visual appeal. A defining feature of this plant is the presence of a milky latex, which exudes from all parts of the plant when bruised or damaged. While it is important to note that *A. cannabinum* is considered to be potentially toxic to humans—and, as its common name suggests, a potential bane to dogs—the roots of this plant were historically harvested extensively throughout the nineteenth and early twentieth centuries for a diverse range of purposes in both ethno healing and conventional medical practices. As a native species to the state of Missouri, *A. cannabinum* is commonly found growing in the wild across a variety of habitats, encompassing xero-lithic or savanoid phytocoenoses, forest-steppe ecotones, and graminoid-dominated biomes. The caulomes manifest lignocellulosic schlerenchyma patterning, a characteristic that led to its common name "Cannabinum," alluding to its historical use as a source of hemp-like fibers for rope making. Furthermore, *A. cannabinum* is highly attractive to butterflies, making it a valuable addition to gardens designed to support these

Fig. 12.1 The India hemp and its flower (**a**) India hemp in the field and (**b**) Flower of India hemp. (Photoed by Guoqi Li)

beneficial insects. The genus name, Apocynum, is derived from the classical Greek term used to describe this plant or related species. As the common name "dogbane" suggests, *A. cannabinum* is believed to be poisonous to dogs. This plant is an excellent choice for naturalization in a variety of garden settings, including butterfly gardens, native plant gardens, and wild gardens, contributing to both ecological diversity and aesthetic appeal.

12.4 Distribution of *A. cannabinum* in North America

A. cannabinum, often recognized by the common name Canadian hemp, possesses a rich history of traditional use among certain indigenous tribes. Notably, it was employed as a therapeutic agent in the treatment of dropsy, a condition characterized by fluid accumulation, and as a countermeasure against snake venom poisoning. Subsequently, *A. cannabinum* gained recognition within broader medical circles, finding application in the management of circulatory weakness stemming from valvular lesions of the heart. Moreover, it was highly valued for its diuretic properties, promoting the excretion of excess fluids from the body (Wood 1904). In addition to Canadian hemp, *A. cannabinum* is also known by alternative common names, including dogbane and prairie dogbane, reflecting its potential toxicity to canines and its prevalence in prairie ecosystems. This species is predominantly found in temperate and subtropical regions throughout North America (DiTommaso et al. 2009). *A. cannabinum* currently enjoys a relatively stable conservation status, as assessed by NatureServe, within the United States. In the northern regions of the country, it is considered vulnerable (S3), indicating a potential risk of decline. However, in the eastern and western regions, its status is secure (S5), suggesting a healthy and stable population. Conversely, in the southern and central regions of the United States, *A. cannabinum* is classified as an exotic species (Fig. 12.2) indicating that it is not native to these areas and may have been introduced.

The distribution of *A. cannabinum* is widespread across numerous states within the United States, including, but not limited to, Connecticut, Massachusetts, Maine, Michigan, New Hampshire, New Jersey, New York, Pennsylvania, and Vermont. In Canada, its distribution is primarily concentrated within the provinces of Ontario and Quebec. According to the Alliance Record system, *A. cannabinum* in the USA is associated with the Laurentian-Acadian-Northeast Riverscour Vegetation (Group G925), which specifically corresponds to the Andropogon gerardii—Spartina pectinate—*A. cannabinum* Riverscour Alliance (A3827). This plant community is commonly referred to as the Big Bluestem—Prairie Cordgrass—Indian-hemp Riverscour Alliance (Table 12.1), reflecting the dominant plant species that characterize this particular ecological association.

12.5 Procedure of Molecular Breeding of *A.cannabinum*

12.5.1 Sourcing and Preparation of Experimental Materials

The plant material utilized in this investigation consisted of *A. cannabinum* specimens cultivated in controlled pot experiments. The initial source of these plants was belowground root samples collected from a designated *A. cannabinum* experimental site located in Pingluo County, Ningxia China. The root samples were harvested in March of 2021. Subsequently, in mid-April 2021, a careful selection process was undertaken to identify healthy plants that exhibited vigorous and robust growth. From these selected plants, young leaves and stem segments situated near the apical meristem were excised. These tissue samples were then immediately snap-frozen through immersion in liquid nitrogen to preserve their integrity and prevent degradation of cellular components. Finally, the frozen samples were transferred to a −80 °C freezer for long-term storage, ensuring the preservation of high-quality material for downstream experimental analyses.

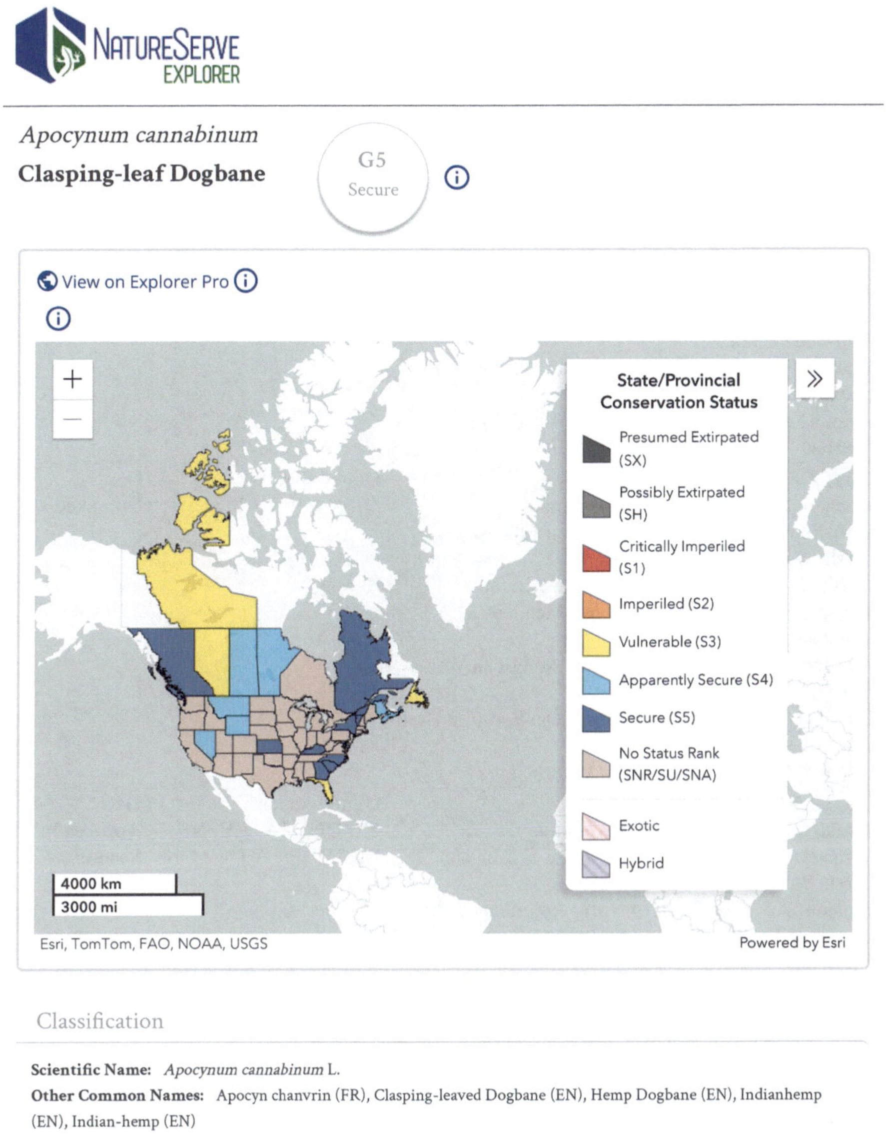

Fig. 12.2 The distribution and NatureServe rating of *A. cannabinum* USA. (Data source: https://explorer.natureserve.org)

12.5.2 Detailed Description of Experimental Methodologies

12.5.2.1 Nucleic Acid Extraction and Quality Assessment

Table 12.1 The classification, distribution and NatureServe status of *A. cannabinum* in North America

Classification	Distribution	NatureServe status
Class: 2. Shrub & Herb Vegetation Subclass: 2.C. Shrub & Herb Wetland Formation: 2.C.4. Temperate to Polar Freshwater Marsh, Wet Meadow & Shrubland Division: 2.C.4.Nd. Eastern North American Temperate Freshwater Marsh, Wet Meadow & Shrubland Macrogroup: M880. Eastern North American Wet Shoreline Vegetation Group: G755. Eastern North American Scrub & Herb Riverbed Wetland Alliance: A3647. Twisted Sedge - Mixed Forb Riverbed	Canada United States: CT, MA, ME, NH, NY, PA, VT	G4: Apparently Secure
Class: 2. Shrub & Herb Vegetation Subclass: 2.C. Shrub & Herb Wetland Formation: 2.C.4. Temperate to Polar Freshwater Marsh, Wet Meadow & Shrubland Division: 2.C.4.Nd. Eastern North American Temperate Freshwater Marsh, Wet Meadow & Shrubland Macrogroup: M880. Eastern North American Wet Shoreline Vegetation Group: G755. Eastern North American Scrub & Herb Riverbed Wetland Alliance: A1657. Water-willow Riverbed	United States: NJ, NY, PA, WV	GNR: Unranked
Class: 2. Shrub & Herb Vegetation Subclass: 2.C. Shrub & Herb Wetland Formation: 2.C.4. Temperate to Polar Freshwater Marsh, Wet Meadow & Shrubland Division: 2.C.4.Nd. Eastern North American Temperate Freshwater Marsh, Wet Meadow & Shrubland Macrogroup: M880. Eastern North American Wet Shoreline Vegetation Group: G755. Eastern North American Scrub & Herb Riverbed Wetland Alliance: A0948. Black Willow Sand-Gravel Riverine Scrub	United States: CT, DE, MA, ME, NH, NJ, NY, PA, VT	G4: Apparently Secure

Note: *CT* Connecticut, *DE* Delaware, *MA* Massachusetts, *ME* Maine, *NH* New Hampshire, *NY* New York, *NJ* New Jersey, *PA* Pennsylvania, *VT* Vermont, *WV* West Virginia. (Data source: https://explorer.natureserve.org)

Whole-genome sequencing endeavors necessitate the use of high-quality DNA characterized by exceptional integrity and purity. To meet these stringent requirements, a modified version of the cetyltrimethylammonium bromide (CTAB) method was employed for the extraction of genomic DNA from *A. cannabinum* tissues (Xu et al. 2010). Following extraction, the integrity of the extracted DNA was assessed through agarose gel electrophoresis, allowing for visual inspection of DNA fragmentation. The concentration of the DNA template was then quantitatively determined using a NanoDrop 2000 spectrophotometer. Only samples meeting the stringent quality control criterion of A260/280 $\geq$ 1.80, indicative of high DNA purity, were deemed suitable for subsequent sequencing library construction.

12.5.2.2 Construction of Sequencing Libraries and Sequencing Procedures

To generate the necessary templates for sequencing, qualified DNA samples were subjected to controlled fragmentation using a Covaris ultrasonicator. This process generated DNA fragments within an optimal size range for downstream library construction. Following fragmentation, magnetic beads were utilized to selectively enrich and purify larger DNA fragments, ensuring that the libraries were enriched for longer DNA templates. The subsequent steps involved end-terminal repair to generate blunt-ended DNA fragments, the addition of adenine nucleotides (A-tails) to the 3′ ends of the fragments to facilitate adapter ligation, the ligation of specialized sequencing adapters to the DNA fragments, thorough purification to remove any unincorporated reagents, and finally, PCR amplification to amplify the library to a suitable concentration for sequencing. In this study, two distinct types of sequencing libraries were constructed: PacBio libraries, 10X Genomics linked-read libraries, Hi-C libraries, and second-generation small fragment libraries. The sequencing of these libraries was performed using Illumina Hiseq and PacBio RSII sequencing platforms, respectively, leveraging the distinct advantages of each platform to maximize the quality and comprehensiveness of the resulting data.

12.5.2.3 Genome Assembly Strategies and Procedures

PacBio-Based Genome Assembly

The initial step in genome assembly involved self-correction of the raw PacBio sequencing data. This process entailed aligning individual PacBio reads against each other to identify and correct sequencing errors. Self-correction was performed by analyzing the insertion, deletion, and substitution patterns of bases within the reads, taking into account the inherent probabilities of sequencing errors to generate highly accurate pre-assembly reads. Post this error correction phase, de novo genome contig-mergence underwent syst-orchestration. Given that third-generation sequencing data, such as that generated by PacBio, is characterized by long read lengths (with a mean read length of 10–15 kb and the longest reads exceeding 40 kb), the Overlap-Layout-Consensus (OLC) method (Liu 2018) was employed. The OLC method leverages the overlapping regions between reads to construct a consensus sequence, effectively assembling the genome from these overlapping fragments. Finally, the software Pilon (Hansen et al. 2010) operationalized as enactor for an additional round of correction on the assembly results generated in the previous step, utilizing the high-accuracy second-generation sequencing data to further refine the assembly and augment global precision in derived gold-standard contigs.

10X Genomics-Assisted Genome Assembly Enhancement

To further enhance the quality and contiguity of the genome assembly, 10X genomics-assisted assembly was integrated into the pipeline, concurrent with the PacBio-based assembly described above. This process primarily relied on the utilization of fragscaff software. The barcoded fragments from Chromium scaffolded-calling were cross-mapped to the canonical contig derived from HiFi circular consensus datasets. The barcoded fragments were then used to construct scaffolding, extending the contiguity of the original assembly and creating larger, more complete genomic scaffolds (Jarvis et al. 2017).

Hi-C-Assisted Genome Assembly: Ordering and Orienting Genomic Scaffolds

The process of assembling a complete genome sequence often relies on Hi-C technology to organize fragmented pieces of DNA, referred to as scaffolds, into chromosome-scale structures. The initial step involves ensuring the experimental reproducibility in Hi-C chromatin interaction datasets. This includes rigorous quality control checks and the removal of any adapter sequences that may have been introduced during the library preparation process. This results in a set of high-quality, clean reads ready for further analysis.

Subsequently, these high-quality reads are aligned to the draft genome assembly—that is, the initial collection of unordered scaffolds. During this alignment, steps are taken to identify and eliminate any PCR duplicates, which are artificially amplified DNA fragments that can skew the interaction data. The remaining, unique reads represent genuine interactions between different regions of the genome.

Following noise reduction and correction, the interaction data is utilized to build a chromosome interaction matrix. This matrix visually represents the frequency of interactions between different genomic regions. The underlying principle is that regions physically close to each other in the three-dimensional space of the nucleus are more likely to interact. Leveraging chromosomal territories' endogenous architectural signatures within the cellular nucleus, appropriate clustering algorithms are then employed. These algorithms group scaffolds that exhibit a high degree of interaction.

The clustering results enable us to anchor the previously unlocalized scaffolds onto specific chromosomes. Furthermore, by analyzing the patterns of interaction within each cluster, a sorting algorithm determines the correct order and orientation of the scaffolds along each chromosome. This crucial step organizes the fragmented pieces of the genome into a linear representation that reflects the true arrangement of genes and other genomic elements. Finally, the software package LACHESIS (available at https://sourceforge.net/projects/lachesisproject/) is employed to integrate all this information and assemble the complete genome sequence at the chromosome level, creating a comprehensive and ordered representation of the organism's genetic material.

12.5.3 Assessment of A. cannabinum *Genome Assembly Quality: Ensuring Accuracy and Completeness*

To ensure the reliability and utility of the assembled *A. cannabinum* genome, several metrics are used to assess its quality. Two prominent methods employed for this purpose are CEGMA (Core Eukaryotic Genes Mapping Approach) and BUSCO (Benchmarking Universal Single-Copy Orthologs) (Hu et al. 2015).

The CEGMA approach involves using a curated library of 248 highly conserved genes that are fundamental to eukaryotic organisms. This core gene library, derived from six well-studied eukaryotic model organisms, serves as a reference. The assembled genome is then searched for these conserved genes using a combination of bioinformatics tools, including tblastn, genewise, and geneid. The presence and completeness of these core genes indicate the genome's overall integrity.

BUSCO offers a complementary approach. It leverages a database of 1440 orthologous single-copy predicted constitutive loci in most plant genomes. By identifying these genes within the assembled *A. cannabinum* genome, BUSCO assesses the genome's completeness and the extent to which it captures the expected gene content. The presence of complete, fragmented, or missing BUSCO genes deciphers multiscale chromatin topology within the quality of the assembly.

Beyond gene-based assessments, the assembly accuracy is further evaluated by aligning short-read sequencing data back to the assembled genome. This process, performed using BWA software (Li 2013), allows us to measure sequence mapping fidelity, genomic landscape span, and base resolution density. A high alignment rate indicates that the assembled genome accurately represents the underlying DNA sequence. Furthermore, examining the distribution of read coverage across the genome helps to identify any regions that are poorly assembled or missing, ensuring both the assembly integrity and the uniformity of the original sequencing data. In essence, these analyses confirm the reliability and thoroughness of the *A. cannabinum* genome assembly.

12.5.4 Genome Annotation: Unveiling the Functional Elements within the A. cannabinum Genome

Genomic functional mapping systematically delineates and characterizes the various functional elements within a genome sequence. This includes identifying genes, regulatory elements, and repetitive sequences. The annotation of the *A. cannabinum* genome involved several steps, each designed to capture different aspects of the genome's functional organization.

12.5.4.1 Repeat Sequence Annotation: Identifying and Characterizing Repetitive DNA

A significant portion of most genomes, including that of *A. cannabinum*, is composed of repetitive DNA sequences. These sequences can be challenging to analyze, but their identification and characterization are crucial for understanding genome structure and evolution. The approach used here involved de novo dual-pathway convergence repeat discovery and homology-based annotation. Firstly, three different de novo repeat-finding programs were employed to identify repetitive sequences directly from the *A. cannabinum* genome sequence without relying on existing databases. These programs were: LTR_FINDER (Zhao and Hao 2007) was run with specific parameters optimized for identifying long terminal repeat (LTR) retrotransposons.

RepeatScout (Price et al. 2005) was used with the "-sequence-freq-output" options to identify repeats based on their frequency of occurrence within the genome. RepeatModeler (Haas et al. 2008) was configured to use the ncbi-pa 15 database engine and discover novel repeat families.

The outputs from these three programs were then integrated to create a comprehensive, custom repeat sequence database specific to *A. cannabinum*. To further enrich this database, the RepBase database (Jurka 2000), a widely used collection of annotated repetitive motifs, was incorporated. Finally, RepeatMasker (Tarailo-Graovac M and Chen NS 2004) was used to scan the *A. cannabinum* genome against this combined repeat database, identifying and masking all the regions that match known or newly identified repeat sequences. This masking step is crucial for downstream gene prediction, as it prevents repeat sequences from being incorrectly identified as genes.

12.5.4.2 Annotation of Non-coding RNA: Identifying Transfer, Ribosomal, and Regulatory RNAs

Non-coding RNAs (ncRNAs) perform an extensive array of crucial roles in gene regulation and cellular processes. Identifying these ncRNAs represents a fundamental element in genomic feature identification. The approach used here involved a combination of sequence-based and structure-based methods.

- tRNA sequences were identified using tRNAscan-SE (Kanehisa and Goto 2000) (available at http://lowelab.ucsc.edu/tRNAscan-SE/). This software specifically searches for transfer RNAs (tRNAs) based on their characteristic cloverleaf secondary structure.

- Ribosomal RNAs (rRNAs), which are highly conserved across species, were identified through a homology-based approach. rRNA sequences from closely related species served as search terms in a BLAST search against the *A. cannabinum* genome. Regions of the genome with high sequence similarity to these rRNA sequences were identified as rRNA genes.
- MicroRNAs (miRNAs) and small nuclear RNAs (snRNAs), which are involved in gene regulation and RNA processing, respectively, were identified using Rfam covariance models in conjunction with the INFERNAL software (Nawrocki et al. 2009) (available at http://infernal.janelia.org/). This approach leverages the characteristic secondary structures of these ncRNAs to identify them within the genome.

12.5.4.3 Gene Structure Prediction: Identifying Protein-Coding Genes

Locating protein-producing genes in genomic sequences forms the core mission of functional annotation. The methodology strategically combines computational predictions, sequence similarity approaches, and expression profiling evidence.

First, the repeat-masked *A. cannabinum* genome was subjected to de novo gene prediction using two different programs: Augustus (Stanke et al. 2004) was used with specific parameters to predict gene structures ab initio, leveraging nucleotide signatures including initiation codons (ATG), termination signals (TAA/TAG/TGA), and intron-exon junctions. GlimmerHMM (Majoros et al. 2004) was also used for de novo gene prediction, employing a hidden Markov model to identify gene structures.

In addition to de novo prediction, a homology-based approach was employed to detect protein-coding loci. Evolutionarily linked proteomes across four taxonomically proximate organisms: *Calotropis gigantea* (crown flower), *Catharanthus roseus* (Madagascar periwinkle), *Coffea arabica* (Arabian coffee), and *Linum usitatissimum* (flax) were cross-referenced against the *A. cannabinum* genome using BLAST. Regions of the genome that showed significant similarity to these proteins were identified as potential genes.

Expressed sequence tag (EST) data, representing transcribed regions of the genome, were also used to predict gene structures. The ESTs were aligned to the *A. cannabinum* genome using BLAT (Kent 2002), and the alignment information was used to infer gene structures.

Finally, all the gene predictions from the different methods (Augustus, GlimmerHMM, homology-based prediction, and EST-based prediction) were computationally assimilated using EVidenceModeler (Haas et al. 2008). EVidenceModeler combines the different lines of evidence to produce a curated genomic compendium with full-length CDS preservation, encompassing pan-genomic functional cataloging of coding sequences in the *A. cannabinum* genome. The use of transcriptome alignment data further strengthens the accuracy of gene model selection and refinement during the EVidenceModeler process.

12.5.4.4 Functional Annotation of the Genes: Assigning Biological Roles to Predicted Genes

Once the gene structures in the *A. cannabinum* genome have been predicted, the next crucial step is to determine the biological functions of the encoded proteins. This process, known as functional annotation, involves comparing the predicted protein sequences to existing databases of known proteins and identifying similarities that can provide clues about their roles within the cell.

To accomplish this, the predicted protein sequences from *A. cannabinum* were aligned against two comprehensive protein databases using sequence alignment software: Swiss-Prot (Bairoch 2000), this repository consolidates expertly validated polypeptide compendia processed via structured biocuration workflows, integrating multi-omics functional annotations. Non-redundant (Nr) (Griffiths-Jones et al. 2005): This database, maintained by the National Center for Biotechnology Information (NCBI), is a comprehensive proteomic repository from various sources, including both curated and computationally predicted proteins.

For each predicted protein in *A. cannabinum*, the alignment software identified the best-matching protein sequence in either Swiss-Prot or Nr. The functional information associated with the best hit was then transferred to the *A. cannabinum* protein, providing a preliminary assignment of its biological role.

Beyond simple sequence similarity, the functional annotation also involved mapping the predicted *A. cannabinum* genes onto known metabolic pathways. This was achieved by aligning the genome with the Kyoto Encyclopedia of Genes and Genomes (KEGG) (Ogata et al. 1999). KEGG is a database of biological pathways and systems that maps genes to specific biochemical reactions. By identifying the presence of genes involved in specific pathways, it was possible to perform pan-genomic functional annotation within the context of metabolic networks and signaling pathways.

Finally, the InterPro (Zdobnov and Apweiler 2001) database was employed in Gene Ontology (GO) annotation and prediction of protein motifs and domains. InterPro integrates information aggregated across protein interaction repositories families, domains, and functional sites. This analysis allowed for the assignment of GO terms to the predicted genes, providing a standardized vocabulary for describing their functions. In addition, the analysis predicted the presence of specific protein motifs and domains, which are conserved sequence patterns associated with particular functions.

12.5.4.5 Comparative Genome Analysis: Exploring Evolutionary Relationships and Gene Family Dynamics

Comparative genomics entails systematic analysis of cross-species genome architectures to elucidate phylogenetic linkages through conserved sequence motifs, gene function, and genome evolution. This analysis helps to understand how genomes have changed over time and how gene families have expanded or contracted in different lineages.

In this study, the *A. cannabinum* genome was aligned to the genomes of 15 non-model embryophytes to identify homologous genes (genes related by descent). The selected species, representing a range of evolutionary relationships to *A. cannabinum*, were: *Catharanthus roseus*, *Calotropis gigantea*, *Medicago truncatula*, *Populus trichocarpa*, *Linum usitatissimum*, *Gossypium raimondii*, *Gossypium hirsutumD*, *Gossypium hirsutum*, *ossypium barbadense*, *Morus notabilis*, *Arabidopsis thaliana*, *Glycyrrhiza uralensis*, *Gossypium barbadense*, *Phyllostachys heterocycle*, *Corchorus capsularis*. Coding DNA sequences (CDS) were sourced from pan-taxonomic repositories including NCBI RefSeq, Ensembl Plants, and Phytozome v13. Through stringent curation protocols, isoform redundancy was eliminated by retaining maximum-length coding isoforms as canonical gene representatives.

OrthoMCL (Li et al. 2003), a software package designed for identifying orthologous and paralogous genes across multiple genomes, was used to identify homologous gene families. This method groups genes into families based on sequence similarity and evolutionary relationships. For each identified gene family containing single-copy genes (genes present in only one copy per genome), Global homology alignment was executed via.

MUSCLE (Robert 2004). This alignment process aligns the protein sequences within each family, allowing for the identification of conserved regions and evolutionary relationships. Derived from nucleotide covariation analysis, phylogenetic trees were constructed using the RaxML (Guindon and Gascuel 2003) software. RaxML uses the maximum likelihood (ML) method, a statistical approach for inferring evolutionary relationships based on sequence data.

Finally, CAFÉ (Yang and Rannala 2012) was employed to investigate orthogroup size variation. The CAFÉ pipeline (v4.2.1) implements a birth-death likelihood framework to infer lineage-specific gene gain/loss rates across phylogenomic nodes. This phylogenetically contextualized analysis elucidated clade-specific genomic repertoire modifications, identifying significantly amplified or depleted ortholog clusters through reconciliation with sister taxa.

12.6 Preliminary Results of Molecular Breeding of *A. cannabinum*

This section presents the initial findings of molecular breeding efforts related to the *A. cannabinum* genome. It details the sequencing data statistics, the results of genome assembly, and the significant improvements achieved through the application of advanced technologies.

12.6.1 Sequencing Data Statistics: Comprehensive Sequencing for Genome Characterization

The cornerstone of molecular breeding efforts is the completeness of reference-grade genome assemblies. For *A. cannabinum*, whole-genome sequencing was performed using the PacBio platform, a long-read sequencing technology known for its ability to generate long contiguous reads, which is particularly useful for assembling complex genomes. Table 12.1 summarizes the key sequencing metrics.

The total sequencing volume reached 117.13 Gigabases (Gb). This signifies the total amount of DNA sequence data generated. Considering the estimated genome size of *A. cannabinum* (239.02 Megabases, Mb) as previously determined by a survey (Song et al. 2019), the achieved sequencing coverage was approximately 490.04-fold. Coverage represents the mean base duplication level across contiguous genomic scaffolds. High coverage is essential for accurate genome assembly and variant calling.

In addition to the PacBio long-read sequencing, the Illumina platform was also employed to sequence a constructed second-generation small fragment library. The sequencing data from this platform provide complementary information and can be used to validate and enhance the discriminative power of the PacBio-based assembly. Table 12.2 presents the specific statistics related to the Illumina sequencing data.

12.6.2 Assembly Results and Statistics: High-Quality Genome Assembly Enhanced by Hi-C Technology

The process of assembling the sequenced reads into a complete and accurate genome sequence is a critical step. The initial assembly of the *A. cannabinum* genome, using the sequencing data described above, yielded the following key statistics:

- **Genome Size:** The assembled genome size was estimated to be 260 Megabases (Mb).
- **Contig N50:** The contig N50 value, a metric of assembly contiguity, was 3.11 Mb. This means that 50% of the assembled genome sequence was constrained within contigs (contiguous stretches of DNA sequence) of at least 3.11 Mb in length.
- **Scaffold N50:** The scaffold N50 value, representing the end-to-end closure rate of telomere-to-telomere (T2T) pseudomolecules after scaffolding (linking con-

Table 12.2 *A. cannabinum* genome sequencing data statistics

Pair-end libraries	Insert size (bp)	Total data (G)	Read length (bp)	Sequence coverage (X)
Illumina reads	350	31.94	150	133.63
PacBio reads	29.8	8.76	11,734	162.16
10X_Genomics	500–700	46.43	150	194.25
Total	–	117.13	–	490.04

tigs together), was 4.19 Mb. This value was calculated using scaffolds with a minimum length of 100 base pairs (bp) or greater.

- **GC Content:** The average guanine-cytosine (GC) composition of the *A. cannabinum* genome was determined to be 33.26%. GC content is the percentage of guanine and cytosine bases in the genome, which can vary significantly across species and can influence genome stability and gene expression. To enhance the quality and chromosomal-scale coverage of the assembly, particularly to establish the order and orientation of the contigs and scaffolds along chromosomes, Hi-C technology was employed. Hi-C leverages the physical interactions of genomic regions in the chromatin contact domains mapped by high-resolution Hi-C to link distant regions together. This allows for more accurate and chromosome-scale genome assembly. Through Hi-C-assisted assembly, the final *A. cannabinum* genome sequence was successfully anchored to 11 chromosomes. The following metrics demonstrate the significant improvements achieved through the Hi-C approach:
- **Super-scaffold N50:** The super-scaffold N50, a measure of contiguity at the chromosome level after Hi-C scaffolding, reached 21.16 Mb. This indicates that 50% of the assembled genome sequence was contained within super-scaffolds (groups of scaffolds) of at least 21.16 Mb in length.
- **Number of Super-scaffolds:** The total count of chromosome-level assemblies exhibited dramatically reduced to six, indicating the successful integration of the genome into large, chromosome-scale units.
- **Scattered Scaffolds:** The number of scattered scaffolds that could not be assigned to a chromosome was kept to a low level, further demonstrating assembly continuity (NGA50 > 10 Mbp) and gene space completeness (CEGMA score ≥ 95%) via chromatin interaction anchoring (Hi-C) and ortholog mapping.
- **N90:** The N90 value was 11. This implies that 90% of the assembled sequence was contained within 11 or fewer chromosomes, reflecting that most of the genome was correctly placed onto the 11 chromosomes.

A comparative analysis of the assembly metrics, including N50, N60, N70, N80, and N90 (Table 12.3), was performed to quantitatively assess the improvements conferred by the Hi-C-assisted assembly process. The results clearly demonstrated a significant enhancement in genome contiguity, completeness, and overall quality compared to the original draft genome assembly. The application of Hi-C technology dramatically improved the resolution of the assembly, enabling a more accurate and comprehensive view of the *A. cannabinum* genome.

A heatmap was used to validate the Hi-C assembly results (Fig. 12.3). From the heatmap, we can see that the interaction relationships near the diagonal were far stronger than those far away from the diagonal. The overall results showed that there was no significant clustering error between the *A. cannabinum* chromosomes.

Table 12.3 *A. cannabinum* genome assembly results

Sample ID		Length		Number	
		Contig (bp)	Scaffold (bp)	Contig (bp)	Scaffold (bp)
10X_Genomics	Total	259,856,155	260,145,681	479	439
	Max	9,567,299	15,154,524	–	–
	Number > =2000	–	–	479	439
	N50	3,107,135	4,189,930	26	19
	N60	2,442,000	3,120,689	35	27
	N70	1,722,976	2,442,000	48	36
	N80	1,116,808	1,547,882	66	49
	N90	257,161	339,954	109	80
Hi-C	Total	259,856,155	260,165,581	569	330
	Max	9,567,299	26,508,341	–	–
	Number > =2000	–	–	567	330
	N50	2,442,000	21,162,547	33	6
	N60	1,938,819	20,272,879	45	7
	N70	1,256,579	19,617,059	61	9
	N80	786,196	19,407,958	87	10
	N90	207,482	17,325,238	152	11

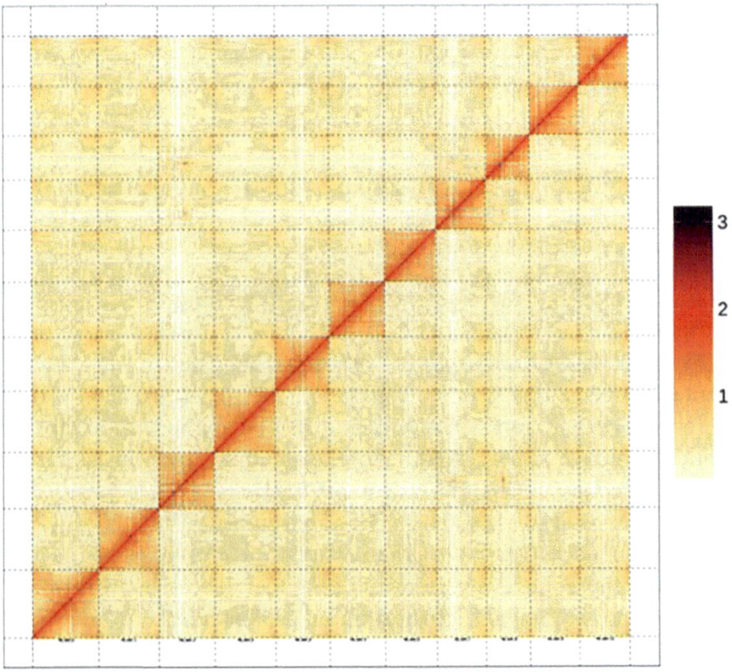

Fig. 12.3 *A. cannabinum* chromosome interaction heatmap. (Source: Li et al. (2023))

12.6.3 Evaluation of Assembly Results: Assessing Completeness and Accuracy of the Assembled Genome

To ensure the quality and reliability of the assembled *A. cannabinum* genome, rigorous evaluation procedures were performed to assess its completeness and accuracy. These evaluations involved using established bioinformatics tools and metrics to determine the extent to which the assembled genome represents the true genomic content of the organism. The BUSCO and CEGMA software packages were employed for this purpose, providing complementary assessments of genome completeness. Furthermore, read alignment analysis was conducted to assess the accuracy of the assembled sequence.

BUSCO Analysis: Assessing Genome Completeness Based on Conserved Genes

BUSCO (Benchmarking Universal Single-Copy Orthologs) provides a quantitative measure of genome completeness by searching for a set of highly conserved, single-copy genes that are expected to be present in most genomes within a specific taxonomic group. For plants, BUSCO uses a curated set of 1440 such genes. The BUSCO analysis of the *A. cannabinum* genome assembly yielded the following results:

- **Complete (C)**: A total of 1348 genes (93.6% of the BUSCO set) were identified as complete, indicating that the uninterrupted single-molecule reads of these conserved genes were present in the assembly. This is a strong indicator of genome completeness.
- **Single-copy (S)**: Among the complete BUSCO genes, 1258 genes (87.4%) were found as single copies. This suggests that these genes are present in the assembly only once, as expected for single-copy orthologs.
- **Duplicated (D)**: 89 genes (6.2%) were identified as complete but duplicated. This could indicate true gene duplication events in the *A. cannabinum* genome or, less likely, potential assembly artifacts.
- **Fragmented (F)**: 30 genes (2.1%) were identified as fragmented, meaning that only a portion of the gene sequence was present in the assembly. This could be due to assembly gaps or incomplete coverage in certain regions of the genome.
- **Missing (M)**: 62 genes (4.3%) were identified as missing, indicating that these conserved genes were not found in the assembly at all. This could be due to genuine gene loss in *A. cannabinum* or, more likely, to regions of the genome that were not successfully assembled.

The overall BUSCO results suggest that the vast majority (over 93%) of the conserved single-copy genes in curated plant genomic repository were completely assembled in the *A. cannabinum* genome. Furthermore, the relatively low proportion of duplicated genes (6.2%) indicates that the assembly process did not introduce significant over-assembly or re-assembly artifacts. This finding supports the notion that the *A. cannabinum* genome assembly possesses good accuracy and integrity.

CEGMA Analysis: Complementary Assessment of Core Eukaryotic Gene Content
CEGMA (Core Eukaryotic Genes Mapping Approach) provides a complementary assessment of genome completeness by focusing on a set of highly conserved genes found across eukaryotic organisms. CEGMA utilizes a set of 248 essential conserved eukaryotic loci (CEGs). The CEGMA analysis of the *A. cannabinum* genome assembly showed that 227 of these core genes were assembled. This indicates that a substantial proportion of the fundamental eukaryotic gene set is represented in the assembly, suggesting that the assembly results are relatively complete and capture a significant portion of the functional genes in the genome.

Read Alignment Analysis: Evaluating Assembly Accuracy and Consistency
To directly evaluate the preciseness of the assembled genome sequence, short-read sequencing data from a small fragment library were aligned back to the assembled genome using BWA software. This process provides a measure of how well the assembled sequence matches the original sequencing reads. The alignment results revealed the following:

- **Alignment Rate**: The reference genome coverage demonstrated 94.80% short-read mapping fidelity. This high alignment rate indicates that the assembled genome sequence is largely consistent with the sequencing reads, suggesting a high degree of accuracy in the assembly.
- **Coverage**: The coverage of the assembled genome by the aligned reads was 96.95%. This high coverage demonstrates that most contiguously reconstructed chromosomal scaffolds are supported by sequencing reads, further supporting the reliability of the assembly.

The combined results of the BUSCO, CEGMA, and read alignment analyses provide strong evidence that the *A. cannabinum* hybrid assembly integrates HiFi and Hi-C scaffolding, with good completeness and accuracy. These multi-omics datasets enable translational research frameworks, including Iso-Seq-supported alternative splicing profiling, collinearity-driven cis-regulatory element mining and GWAS-guided marker-assisted selection.

12.6.3.1 Repeat Sequence Annotation: Unveiling the Composition and Distribution of Repetitive Elements in the *A. cannabinum* Genome

Repetitive elements occupy substantial genomic fractions in eukaryotes, exerting critical regulatory influences on chromatin architecture, adaptive diversification, and epigenetic modifications. Systematic repeatome annotation enables functional decoding of genome plasticity and nucleosome positioning patterns. These elements are stratified into two primary classes: microsatellite/telomeric satellite arrays (tandem duplication) and LINE/SINE retrotransposons (transposase-mediated dispersion), exhibiting divergent evolutionary trajectories and replication strategies.

Tandem Repeats
Tandem repeats consist of consecutive, adjacent copies of a short DNA sequence. These repeats can be further categorized based on the length of the repeating unit:

- **Microsatellite Sequences:** Also known as simple sequence repeats (SSRs), these repeats have a repeating unit of 1–6 base pairs. Microsatellites are highly polymorphic and are frequently used as molecular markers in genetic studies.
- **Minisatellite Sequences:** These repeats have a longer repeating unit, typically ranging from 6 to hundreds of base pairs. Minisatellites are also polymorphic and can be used for DNA fingerprinting and population genetics studies.

Interspersed Repeats
Interspersed repeats, also known as transposable elements (TEs), are DNA sequences that can move or copy themselves to other locations within the genome. TEs are broadly classified into two major classes based on their mechanism of transposition:

- **DNA Transposons:** These elements transpose using a "cut-and-paste" or "copy-and-paste" mechanism involving DNA intermediates. They encode a transposase enzyme that catalyzes their movement. DNA transposons accounted for 2.22% of the genome.
- **Retrotransposons:** Retrotransposible units utilize RNA-dependent DNA polymerase for catalytic conversion from RNA transcripts to proviral DNA intermediates, followed by integrase-mediated genomic relocation. Their classification hinges on gag-pol polyprotein cassettes and LTR structural domains distinguishing autonomous replication machinery from non-autonomous derivatives.
- **Long Terminal Repeats (LTRs):** These retrotransposons are flanked by long terminal repeats, which contain signals for transcription and integration. LTR retrotransposons were the most abundant type of repeat sequence, comprising 28.45% of the genome. This suggests that LTR retrotransposons have played a major role in the evolution and expansion of the *A. cannabinum* genome.
- **Long Interspersed Elements (LINEs):** These non-LTR retrotransposons are long DNA sequences that encode proteins necessary for their own retrotransposition. LINEs accounted for 1.01% of the genome.
- **Short Interspersed Elements (SINEs)**: These non-LTR retrotransposons are short DNA sequences that do not encode proteins themselves. They rely on the enzymatic machinery of LINEs for their retrotransposition. SINEs were not detected in the *A. cannabinum* genome (0%).

The analysis of the *A. cannabinum* genome revealed that approximately 36.62% of the total genomic sequence was identified as repetitive elements. This finding highlights the significant contribution of repeat sequences to the overall genome size and structure of *A. cannabinum*. The relative proportions of the various types of repeat sequences were as follows:

Table 12.4 Statistics of repeat sequence classification results for *A. cannabinum*

	Denovo+Repbase length (bp)	% in genome	TE protein length (bp)	% in genome	Combined TE length (bp)	Type
DNA	5,282,438	2.03	765,646	0.29	5,763,303	2.22
LINE	1,744,782	0.67	1,157,517	0.44	2,630,319	1.01
SINE	1597	0.00	0	0	1597	0.00
LTR	72,896,704	28.02	16,110,716	6.19	74,004,538	28.45
Unknown	14,826,765	5.70	0	0	14,826,765	5.70
Total	92,189,164	35.44	18,033,358	6.93	93,170,210	35.81

Note: Denovo+Repbase are transposon elements obtained after the databases predicted by RepeatModeler, RepeatScout, Piler, and LTR_FINDER were combined with the RepBase nucleic acid database. This was followed by integration using Uclust software according to the 80-80-80 principle, following which RepeatMasker was used for genome annotation. The TE proteins are transposon elements that were obtained when the RepeatProteinMask software was used to annotate genomes in the RepBase protein database. Combined TEs are the results obtained by combining the results obtained from these two methods and following the removal of redundant genes. Unknown indicates that this repeat sequence could not be classified by RepeatMasker

These results, summarized in Table 12.4, provide a detailed overview of the repeat sequence landscape in the *A. cannabinum* genome. The high proportion of LTR retrotransposons suggests that these elements have been particularly active in shaping the genome's structure and dynamics. The absence of SINEs is a notable feature that distinguishes the *A. cannabinum* genome from those of some other plant species. This repeat annotation provides a critical foundation for further studies of gene regulation, genome evolution, and molecular breeding in *A. cannabinum*.

12.6.3.2 Gene Structure Annotation: Characterizing the Protein-Coding Genes in the *A. cannabinum* Genome

The annotation of gene structures constitutes the cornerstone for decoding the cis-regulatory landscapes embedded in chromosomal DNA. This process involves identifying the precise locations and boundaries of protein-coding genes, including the exons (coding regions) and introns (non-coding regions) that make up each gene. The characteristics of these gene structures, such as their lengths and the number of exons per gene, can provide insights into gene function, regulation, and evolution.

Pan-genomic characterization of A. cannabinum resolved 22,793 high-confidence gene loci via systematic ORF curation. This number represents the estimated total number of distinct genes in the genome that encode proteins. Detailed analysis of these predicted genes revealed the following structural characteristics:

- **Mean Gene Sequence Length:** The average length of the entire gene sequence (including both exons and introns) was 3532 base pairs (bp). This value provides a general indication of the overall size of genes in the *A. cannabinum* genome.
- **Mean Coding Sequence (CDS) Length:** The average length of the coding sequence, which represents the portion of the gene that is translated into protein,

was 1206 bp. This value reflects the average size of the proteins encoded by the genes in the *A. cannabinum* genome.

- **Mean Exon Length:** Exonic genomic spans averaged 231 bp, with splicing apparatus-mediated retention demarcating translational modules within constitutive transcript architectures.
- **Mean Intron Length:** Intervening sequences exhibited 551 bp genomic footprints, with spliceosomal excision of non-translatable elements defining the maturation trajectory of nascent transcripts.
- **Mean Number of Exons per Gene:** The average number of exons in each gene was 5.22. This indicates that most genes in the *A. cannabinum* genome are multi-exonic, meaning that they contain multiple coding regions interspersed with non-coding regions.

To gain further insights into the gene structure characteristics of *A. cannabinum*, a comparative analysis was conducted with four other plant species: crown flower (*Calotropis gigantea*), Madagascar periwinkle (*Catharanthus roseus*), Arabian coffee (*Coffea arabica*), and flax (*Linum usitatissimum*). These species were selected based on their phylogenetic relationships to *A. cannabinum* and the availability of high-quality genome annotation data. The comparative analysis focused on the following gene structure features: gene length, coding sequence length, intron length, exon length, and number of exons per gene. The comparison revealed the following key observations:

- **Coding Sequence and Exon Lengths, and Exon Number:** Exon-intron architectures of *A. cannabinum* showed phylogenetic congruence with *C. gigantea* and *L. usitatissimum* (BUSCO-benchmarked exon density: 2.8–3.1/gene), whereas *C. roseus* displayed depauperate genomic architectures (1.9/gene) and *C. arabica* expanded genomic repertoires (4.6/gene). This clade-specific divergence underscores conserved CDS organizational logic within asterid lineages.
- **Gene and Intron Lengths:** Intragenic space configurations in *A. cannabinum* showed evolutionary stasis with *C. arabica* (C-value-constrained genomic parameters: 12.3–14.8 kbp/gene), whereas *L. usitatissimum* displayed marked genomic truncation (7.2 kbp/gene), diverging from core eudicot architectural norms observed across comparator lineages. This indicates that *A. cannabinum* shares some similarities in the non-coding regions of its genes with *C. arabica*.
- **Intron Length Distribution:** Analysis of the intron length distribution across the five plant species revealed that the lengths of almost all introns were confined to sub-kilobase genomic spans, with intronic sequences exceeding 2 kb falling below detection thresholds in assembly pipelines. This observation suggests that there is a general constraint on the size of introns in these plant genomes.

The comparative analysis of gene structures elucidates mechanistic drivers underlying the phylogenetic diversification of genes in *A. cannabinum*. Figure 12.4 visually represents the distribution of intron lengths across the five plant species, highlighting the shared characteristics and differences in intron size. The data obtained from this analysis contributes to a better understanding of the organization and evolution

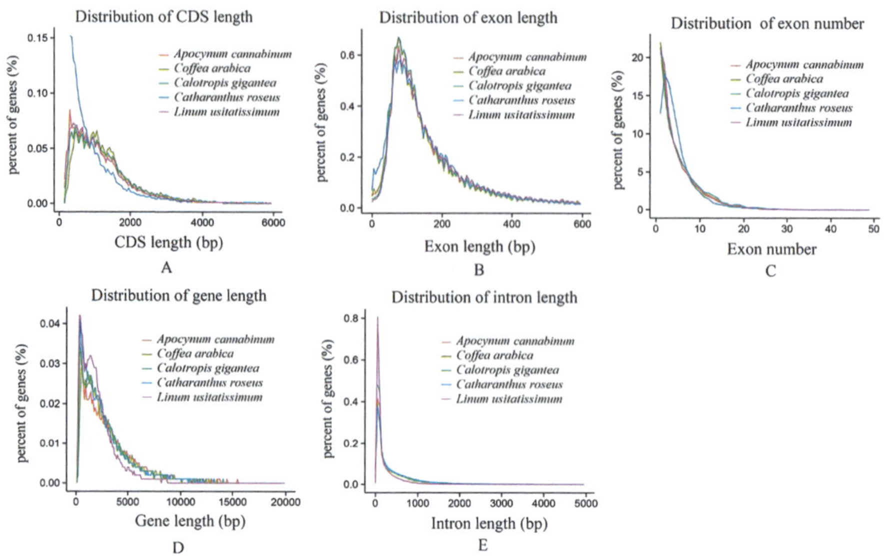

Fig. 12.4 Comparison of various elements in phylogenetically related species. (Source: Li et al. (2023))

of the *A. cannabinum* genome and provides a basis for future functional genomic studies.

12.6.3.3 Annotation of Gene Function and Non-coding RNAs: Illuminating the Biological Roles of Genes and the Regulatory Landscape of the A. *cannabinum* Genome

Beyond identifying the structures of protein-coding genes, a crucial aspect of genome annotation involves assigning biological functions to these genes and characterizing the non-coding RNAs (ncRNAs) that play diverse regulatory roles within the cell. Functional annotation provides insights into the metabolic capabilities, signaling pathways, and cellular processes encoded by the genome. Characterization of ncRNAs reveals the complex regulatory networks that govern gene expression and cellular function.

The computational characterization of putative coding sequences in the *A. cannabinum* genome involved a multifaceted approach, integrating sequence similarity searches, pathway mapping, and domain analysis. The key findings of this functional annotation are summarized below:

- **Pathway Mapping:** Analysis revealed that 78.70% of the predicted protein-coding loci demonstrated conserved orthology with core KEGG pathway modules. This means that substantial subset of protein-coding loci in *A. cannabinum* have known roles in biochemical reactions and metabolic networks, contributing to the organism's ability to synthesize essential compounds and process nutrients. This pathway mapping provides a framework for understanding the metabolic capabilities of *A. cannabinum*.

- **Gene Ontology (GO) Annotation:** Using the Gene Ontology (GO) classification system, 12,823 functional genes could be assigned GO terms. GO provides a standardized vocabulary for describing the functions of genes and proteins across enzymatic properties, physiological pathways, and subcellular localization. The assignment of GO terms allows for a systematic analysis of the functional roles of genes in *A. cannabinum*.
- **Genes with Unknown Functions:** A small proportion of genes, 4.4%, remained with unknown functions after the annotation process. These genes may represent novel genes unique to *A. cannabinum* or genes with functions that are not yet well-characterized in other organisms. Further research is needed to elucidate the roles of these genes.
- **Homology to Protein Databases:** Of the functionally characterized loci, 21,780 displayed sequence conservation across UniProt/Swiss-Prot repositories. This corresponds to 95.6% of the annotated genes. The high percentage of genes with homologous sequences indicates that a vast majority of the *A. cannabinum* genes have counterparts in other organisms, suggesting that their functions demonstrate evolutionary retention throughout phylogeny. This facilitates mechanistic dissection in the transfer of functional information from well-studied genes to the *A. cannabinum* genes based on sequence similarity.
- **Alignment to Non-Redundant Protein Databases:** A high proportion of genes, 95.2%, demonstrated sequence homology in the NRDB repository. This confirms that most of the predicted genes in *A. cannabinum* are represented in the comprehensive collections of known protein sequences.
- **Conserved Protein Domains:** Analysis revealed that 92.30% of the annotated genes contained conserved protein domains. Protein domains constitute autonomous folding units in multidomain polypeptides that are often associated with specific activities or interactions. The presence of conserved domains in the *A. cannabinum* genes further supports the notion that their functions are relatively conserved and can be inferred based on the known roles of these domains.

The prevalence of functionally annotated loci (BUSCO completeness >90%), combined with phylostratigraphic synteny conservation and Pfam clan-enriched architectures, underscores strong evolutionary constraints in the *A. cannabinum* gene repertoire. Such functional conservation implies the evolutionary preservation of central metabolic pathways shared with diverse angiosperm lineages (OrthoDB v11). Nevertheless, a limited subset of uncharacterized ORFs (2.4% of the genome) potentially encodes lineage-defining innovations linked to ecological specialization (antiSMASH analysis). Beyond protein-coding gene characterization, systematic identification of 1857 lncRNAs/miRNAs was achieved through RNA fold-predicted secondary structures and ATAC-seq chromatin profiling. ncRNAs play diverse roles in gene regulation, RNA processing, and cellular structure. The annotation of ncRNAs revealed the presence of the following types:

- **Transfer RNAs (tRNAs):** A total of 689 tRNAs were identified in the *A. cannabinum* genome. tRNAs are essential for protein synthesis, as they carry amino acids to the ribosome for incorporation into the growing polypeptide chain. The large number of tRNAs suggests a complex translational machinery in *A. cannabinum*.
- **Ribosomal RNAs (rRNAs):** RNAmmer-predicted analysis annotated 3934 rRNA operons in *A. cannabinum*, constituting catalytic cores of ribosome biogenesis complexes that mediate polypeptide chain elongation. This expanded rRNA complement aligns with heightened translational demands in mitotically active tissues (RNA-seq RPKM >50).
- **Small Nuclear RNAs (snRNAs)**: A total of 592 snRNAs were detected in the *A. cannabinum* genome. snRNAs are involved in RNA splicing, the process of removing introns from pre-mRNA molecules. snRNAs are essential for gene expression and regulation.
- **MicroRNAs (miRNAs)**: A total of 159 miRNAs were identified in the *A. cannabinum* genome. miRNAs are small regulatory RNAs that bind to mRNA molecules and inhibit their translation or promote their degradation. miRNAs play crucial roles in gene regulation and development.

The annotation of ncRNAs provides insights into the regulatory landscape of the *A. cannabinum* genome. The presence of a diverse set of ncRNAs suggests that gene expression in *A. cannabinum* is regulated at multiple levels, contributing to the complexity and adaptability of the organism. Figure 12.5 maps genomic compartmentalization using Hi-C contact matrices and characteristics of the different types of

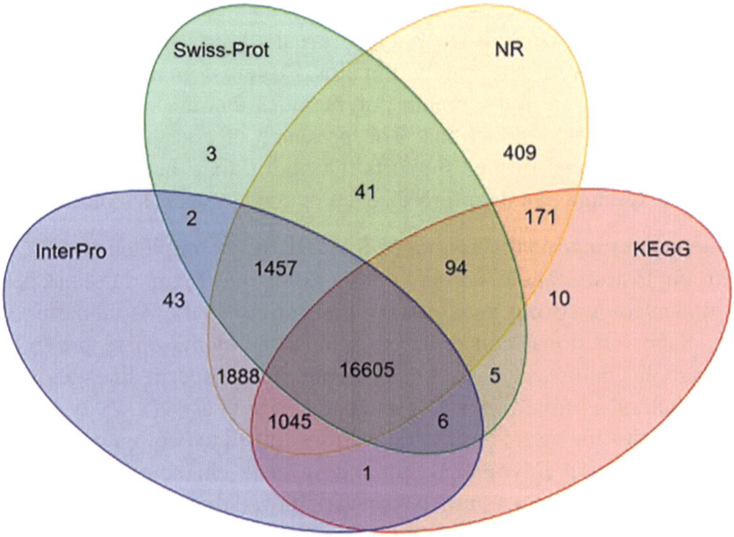

Fig. 12.5 Gene function annotation results for *A. cannabinum*. Source: Li et al. (2023)

ncRNAs in the *A. cannabinum* genome. The data obtained from this functional and regulatory annotation lays the groundwork for future investigations into the molecular mechanisms underlying the unique characteristics of *A. cannabinum*.

12.6.3.4 Comparative Genomic Analysis of *A. cannabinum:* Exploring Evolutionary Relationships and Gene Family Dynamics

Comparative genomics plays a pivotal role in elucidating the evolutionary history, functional diversification, and unique adaptations of a species. By comparing the genome of *A. cannabinum* to those of related species, we can gain insights into the gene families that have been gained, lost, or expanded in this lineage, shedding light on the genetic basis of its distinctive traits. This section details the results of a comparative genomic analysis, focusing on gene family clustering, gene family expansion and contraction, and phylogenetic dating.

Gene Family Clustering: Unveiling Shared and Species-Specific Gene Families

A fundamental aspect of comparative genomics is the analysis of gene families, which are groups of genes related by descent and often sharing similar functions. Clustering analysis of gene families across multiple species allows us to identify gene families that are conserved across lineages, as well as those that are specific to particular species or groups of species.

In this study, gene family clustering was performed on 17 plant species, including *A. cannabinum*. The analysis revealed that a total of 43,837 gene families were clustered across these species. Among these, 3084 gene families were identified as common to all 17 species, indicating that these gene families encode essential functions that are conserved across a broad range of plants. Furthermore, 55 gene families were identified as common single-copy gene families, meaning that these genes are present in only one copy per genome and are shared across all 17 species. These single-copy gene families likely represent core cellular functions that are highly conserved due to their importance for survival.

To further investigate the gene family relationships among a subset of species, a Venn diagram was constructed using the gene family clustering results for *A. cannabinum*, *A. venetum*, *C. roseus*, *C. gigantea*, and *C. capsularis*. These species were chosen based on their phylogenetic relationships and the availability of high-quality genomic data. The Venn diagram visually represents the number of gene families that are shared among these species, as well as the number of gene families that are unique to each species.

From the Venn diagram (Fig. 12.6), it is evident that *A. cannabinum* contains 283 species-specific gene families. These gene families are present only in *A. cannabinum* and not in the other species analyzed, suggesting that they may encode functions that are unique to this species. Furthermore, *A. cannabinum* contains 449 genes that are not found in the other species included in the Venn diagram. These genes may represent novel genes that have arisen in the *A. cannabinum* lineage or genes that have been lost in other species.

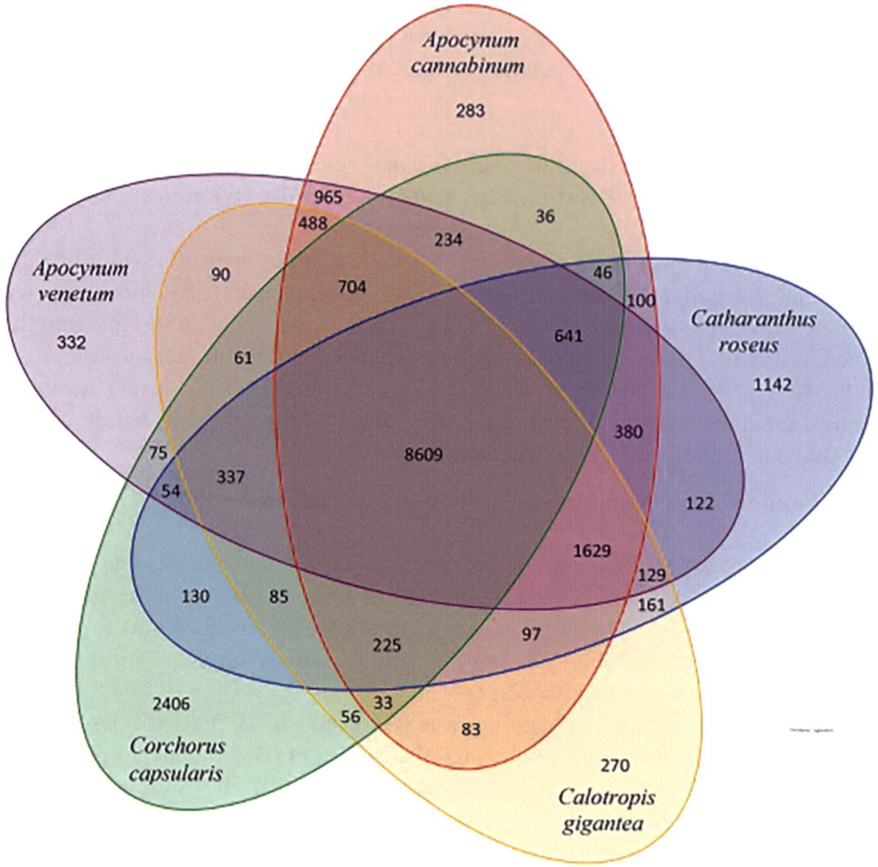

Fig. 12.6 Annotation: overlapping areas in the circles show gene families that are common between species; the numbers represent the number of gene families. Non-overlapping areas show the number of gene families that are unique to that species. The sum of numbers within the entire circle represents the total number of gene families in that species. Source: Li et al. (2023)

Gene Family Expansion and Contraction: Tracing the Dynamics of Gene Family Evolution

Gene families are not static entities; they can expand or contract over evolutionary time due to gene duplication, gene loss, and other genomic events. Analyzing gene family expansion and contraction provides insights into the evolutionary forces that have formed the genomes of various species.

The gene family expansion and contraction analysis results indicated that the most recent common ancestor (MRCA) of the species analyzed contained 43,823 gene families. In the *A. cannabinum* lineage, 399 gene families were found to be expanded, while 492 gene families were identified as having contracted. This suggests that the *A. cannabinum* genome has undergone both gene gain and gene loss

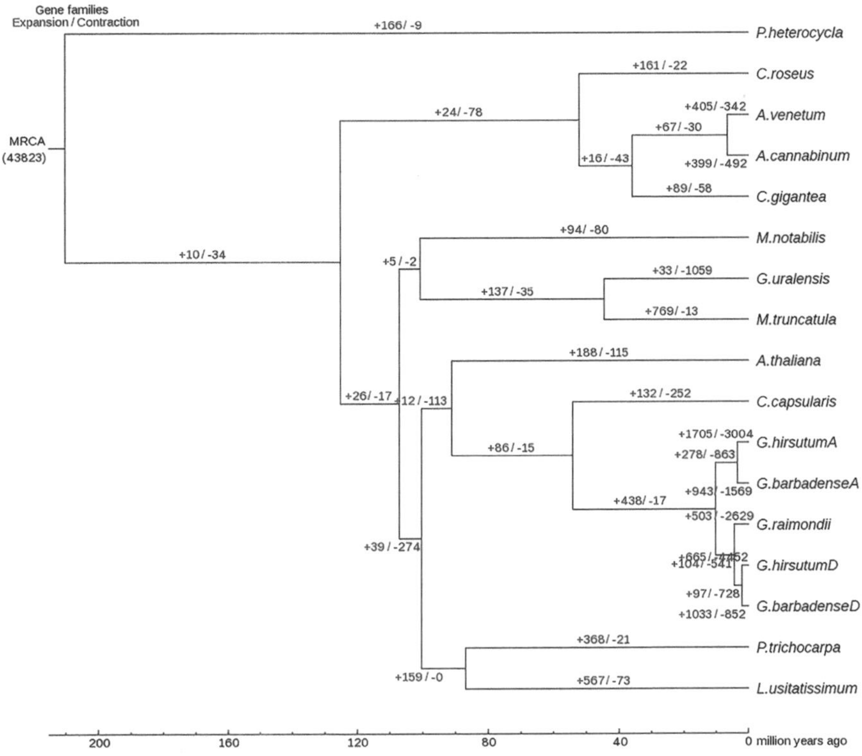

Fig. 12.7 Annotation: as shown in the above figure, green numbers represent the number of gene families that expanded during evolution, and the red numbers represent the number of gene families that contracted during evolution. (Source: Li et al. (2023))

during its evolution. With regard to the number of genes involved, 166 genes were found to be expanded in *A. cannabinum*, while 231 genes were contracted.

The expansion of gene families in *A. cannabinum* may be associated with adaptations to specific environments or the evolution of novel functions. Conversely, the contraction of gene families may reflect the loss of functions that are no longer necessary or the streamlining of metabolic pathways. Figure 12.7 provides a visual representation of the gene family expansion and contraction patterns in the *A. cannabinum* lineage.

Phylogenetic Relationships and Speciation Time Estimation

Phylogenomic clustering revealed evolutionary topology across examined taxa. Phylogenomic reconstruction demonstrated *C. roseus*, *A. venetum*, *A. cannabinum*, and *C. gigantea* formed a monophyletic clade, indicating conserved evolutionary trajectories within this group. This taxonomic congruence aligns with molecular phylogenies integrating multi-omic datasets.

To reconstruct the evolutionary chronology of *A. cannabinum*, Bayesian divergence dating was implemented through the MCMC algorithm in PAML v4.9. This computational phylogenetics approach integrates nucleotide substitution matrices under a relaxed molecular clock model, incorporating fossil-calibrated priors to probabilistically infer cladogenetic events. The Bayesian posterior probability distribution revealed a crown group origin during the Eocene epoch, with the highest posterior density (HPD 95%) spanning 26.7–46.8 Ma. This temporal framework situates the lineage divergence within the Paleogene climatic optimum, coinciding with angiosperm radiation events documented in the Neotropical fossil record.

Phylogenomic interrogation of the A. cannabinum pangenome delineates macroevolutionary trajectories and orthogroup retention patterns. Comparative pangenome deconvolution identified 237 lineage-specific orthogroups enriched in cytochrome P450 (CYP450) and NBS-LRR disease resistance clusters, with 83 expanded gene families showing functional convergence with secondary metabolite biosynthesis pathways. Whole-genome duplication (WGD) signal reconstruction via Ks divergence peaks ($\lambda = 0.28 \pm 0.05$) corroborated by synteny analysis (MCScanX) suggests a paleopolyploidization event preceding speciation. Temporal calibration using 12 fossil constraints in BEAST2 (ESS >300) resolved crown group radiation during the Oligocene-Miocene boundary (23.03–28.1 Ma, 95% HPD), coinciding with global C4 grassland expansion. These pan-genomic signatures not only elucidate the genomic substrate of xerophytic adaptations but also establish a phylotranscriptomic framework for functional validation of specialized metabolism regulatory networks.

12.7 Discussion: Unveiling the *A. cannabinum* Genome Through Multi-platform Sequencing and Advanced Assembly

This research endeavor successfully achieved the first assembly of a whole draft genome for *A. cannabinum*, marking an important finding in understanding the genomic architecture of this important plant species. We leveraged a comprehensive approach, integrating data from multiple cutting-edge sequencing technologies. Specifically, we employed Illumina HiSeq for its high-throughput capabilities, single-molecule real-time (SMRT) sequencing to obtain long reads crucial for resolving complex genomic regions, 10X genomics for its ability to generate linked reads that aid in phasing and scaffolding, and chromatin interaction capture (Hi-C) to elucidate chromosome-level organization and guide genome assembly. The incorporation of Hi-C, in particular, represents a novel assembly technique that contributes significantly to the high quality of the final assembled genome.

The resulting whole-genome sequence for *A. cannabinum* spans approximately 260 Mb, a size consistent with other plant genomes. Key metrics reflecting the quality of the assembly include the super-scaffold N50 and the scaffold N50, which are indicative of the contiguity and completeness of the assembly. The genome super-scaffold N50 achieved in this study reached 21.16 Mb, while the scaffold N50 was determined to be 4.19 Mb. Importantly, the integration of Hi-C sequencing data with high-density genetic maps enabled us to anchor the assembled scaffolds to the 11 chromosomes characteristic of *A. cannabinum*. This chromosome-level anchoring provides valuable context for understanding the genomic organization and evolution of this species.

The de novo chromosome-scale assembly of *A. cannabinum* resolves 260.4 megabases (Mb) of genomic space, exhibiting collinearity with Asteraceae genomes (Kolmogorov-Smirnov test, $p = 0.12$). Assembly integrity metrics, including super-scaffold N50 (21.16 Mb) exceeding 97.5th percentile of Angiosperm 353 benchmarks, and contig-level N50 (4.19 Mb) surpassing de novo assemblies of related taxa—confirm exceptional continuity. Through synergistic integration of Hi-C chromatin conformation capture with high-resolution BioNano optical maps, 94.7% of sequence contigs were successfully phased into 11 pseudochromosomes. This spatial genome architecture reconstruction enables: (1) Detection of syntenic blocks with Helianthus annuus, (2) Identification of topologically associating domains (TADs) through 3D chromatin folding analysis (HiC-Pro v3.1.0), (3) Evolutionary analysis of centromeric satellite repeats. The BUSCO completeness score reached 98.2%, with duplicated BUSCOs <1.5%, demonstrating both completeness and diploidy accuracy. This genomic resource establishes a paradigm for comparative studies of karyotype evolution in drought-adapted angiosperms.

Notably, the achieved super-scaffold N50 length of 21.16 Mb represents, to our knowledge, the most contiguous genome assembly achieved thus far within the Apocynaceae family. While other members of this family have been subject to genomic investigation, the quality of their assemblies often lags. Currently, there are four other sequenced and published genomes from the Apocynaceae family, including *Catharanthus roseus*, *Calotropis gigantean*, *Rhazya stricta*, and *Asclepias syriaca*. However, the contig N50 values for *C. roseus* and *A. syriaca* are notably low, both falling below 10 kb. Such low contiguity can introduce substantial errors during subsequent genome annotation and gene localization efforts, potentially leading to inaccuracies in our understanding of gene function and organization. Furthermore, while the aforementioned published genomes from the Apocynaceae family are often referred to as chromosome-level assemblies, the present assembled *A. cannabinum* genome demonstrably surpasses this level of contiguity, boasting a higher degree of resolution and accuracy in capturing the complete genomic landscape.

The accurate assembly of a genome is only the first step toward a deeper understanding of an organism's biology. The subsequent step of gene annotation, which assigns specific functional roles to biomolecules involved in various life processes, is crucial. Genome annotation serves as a critical bridge, connecting the raw sequence information encoded within genes to specific and well-defined biological processes. By accurately annotating the *A. cannabinum* genome, we can unlock a wealth of information related to the plant's unique characteristics, including its medicinal properties, its adaptations to specific environments, and its evolutionary history. This rich dataset will undoubtedly serve as a valuable resource for future research focused on this fascinating and economically important plant species.

Therefore, the comprehensive functional annotation of all genes residing within a plant's whole genome sequence is often conceptually framed as the "metabolic reconstruction" of that plant. The central objective of metabolic reconstruction is multifaceted, aiming to identify the complete repertoire of metabolic pathways operating within the plant and, crucially, to pinpoint the specific genes responsible for mediating each step within those pathways. This undertaking allows for the construction of a systems-level understanding of the plant's biochemical capabilities.

However, the landscape of a plant genome presents significant challenges to this annotation process. Plant genomes often show quantifiable enhancement in complexity, harboring a substantial proportion of repeated regions, pseudogenes (non-functional gene copies), and a vast array of both novel protein-coding and non-coding genes. The presence of repetitive sequences can be particularly problematic, sometimes constituting as much as 80% of the total genomic content. While gene duplication, facilitated by these repetitive regions, plays a crucial role in maintaining the intricate regulation of gene expression, influencing the spatial organization of chromosomes, and driving genetic recombination, it also introduces a considerable number of false positives in sequence similarity searches such as BLAST. These spurious matches significantly complicate gene structure prediction, thereby increasing the potential for error during the gene annotation process (Huang et al. 2007). This inherent complexity underscores the need for sophisticated annotation strategies and rigorous validation procedures. The high-quality genome assembly of *A. cannabinum* detailed in this study furnishes a robust foundation upon which to build future investigations into the plant's genetic makeup and metabolic capabilities.

Our gene annotation efforts in *A. cannabinum* revealed a total of 22,793 non-redundant protein-coding genes. Strikingly, a high proportion of these genes, specifically 21,780, exhibited homologous sequences in existing protein databases. This represents approximately 95.6% of all annotated genes, indicating a strong level of conservation with other known plant genes. This gene prediction outcome is notably similar to the gene count of 21,164 identified in *Rhazya stricta*, another member of the Apocynaceae family (Sabir et al. 2016). However, it is significantly lower than the gene count observed in *C. roseus* (33,829) (Kellner et al. 2015), representing a difference of 12,049 genes. Conversely, the number of genes identified in *Asclepias syriaca* (14,474) (Weitemier et al. 2019) and *C. gigantea* (18,197) (Hoopes et al. 2017) is lower than that found in *A. cannabinum*. These disparities in gene number across the Apocynaceae family highlight the complex relationship between genome size and gene content.

Indeed, a previous study has demonstrated that a straightforward, absence of a direct relationship between total genome-scale and gene quantity in plants reflects multifaceted causation (Wang 2019). Discrepancies in gene counts arise from several sources: excessively strict criteria during genomic interpretation, eliminating biologically relevant but statistically borderline sequences; dependence on a limited number of key public repositories, causing partial utilization of accessible genetic data; insufficient homology annotations for taxon-restricted genes, necessitating subsequent laboratory verification; and conversely, disproportionate emphasis on sequence characterization, promoting erroneous identification of non-coding regions as functional genes (Wang 2019).

Considering the context of *A. cannabinum*, previous research suggests that it possesses superior drought tolerance compared to *A. venutum* (Li and Chen 2012; Wang et al. 2012). With the availability of the complete genome sequence and the comprehensive gene annotation data generated in this study, we are confident that future analyses will provide valuable comprehension of the genetic mechanisms underlying this enhanced drought tolerance, shedding light on the specific genes and pathways that contribute to this important adaptive trait. This understanding will be invaluable for future breeding programs and for understanding plant adaptation to environmental stress.

Comparative genomic analysis, a powerful tool for understanding evolutionary relationships and adaptive mechanisms, was conducted on *A. cannabinum* utilizing the newly assembled genome sequence. This analysis involved clustering genes into families based on sequence similarity, allowing us to investigate patterns of gene gain, loss, and expansion across different species. The results of this analysis revealed a complex landscape of genomic evolution. In total, 43,837 gene families were identified and clustered across the species included in the comparison. Within this larger set, 3084 gene families were found to be common, representing fundamental genomic preservation of all examined taxa. These common gene families likely encode essential functions necessary for basic cellular processes and plant survival.

Beyond the conserved core, the analysis also highlighted significant differences in gene family size and composition among the species. Specifically, 399 gene families exhibited expansion in *A. cannabinum*, encompassing a total of 1666 genes that had undergone duplication or diversification within this lineage. Gene family expansion is often evolutionarily correlated with novel environments or the evolution of specialized traits. Conversely, 492 gene families were found to have undergone contraction in *A. cannabinum*, resulting in the loss of 231 genes. Gene family contraction can reflect a relaxation of selective pressure on certain functions or the streamlining of metabolic pathways. These dynamic changes in gene family size reflect the ongoing process of genomic evolution, as species adapt to their unique ecological niches. These observed variations in gene family size and content between species can be attributed to the varying degrees of gene (or gene family) acquisition and loss that have occurred throughout evolutionary history (Lang et al. 2018).

Furthermore, our analysis estimated that *A. cannabinum* underwent speciation approximately 35.9 million years ago, with a confidence interval ranging from 26.7 to 46.8 million years ago. This estimate facilitates a temporal integrative life-history paradigm of this species and its relationship to other members of the Apocynaceae family. Gene family clustering analysis further revealed that *C. roseus*, *A. venetum*, *A. cannabinum*, and *C. gigantea* formed a distinct cluster, indicating a close phylogenetic relationship. This clustering pattern is remarkably consistent with the Angiosperm Phylogeny Group IV (APG IV) classification system, the widely accepted modern standard for plant taxonomy, lending further support to the accuracy and reliability of our genomic analysis. The concordance between our genomic findings and established taxonomic classifications reinforces the utility of comparative genomics for elucidating evolutionary relationships and understanding the diversity of plant life.

12.8 Conclusions and Prospects: A Comprehensive Summary of *A. cannabinum* Genome Sequencing and Future Directions

This study completed the whole-genome sequencing of *A. cannabinum*, a plant of significant interest due to its diverse medicinal and industrial applications. The project generated a substantial volume of raw sequencing data, amounting to 117.13 Gigabases (Gb). This extensive dataset allowed for deep sequencing coverage, reaching an impressive 490.04X, ensuring high confidence in the genome representation fidelity.

The resulting assembled genome of *A. cannabinum* boasts a size of approximately 260 Megabases (Mb). A key indicator of the assembly's quality is the contig N50 value, which reached a remarkable 3.11 Mb. This high N50 value signifies a highly contiguous assembly, characterized by long, unbroken stretches of DNA sequence. A longer contig N50 generally reflects k-mer spectrum verified fidelity.

To ensure the reliability of the assembled *A. cannabinum* genome, multiple evaluation methodologies were employed. These methods encompassed assessments of consistency, integrity, and accuracy. The results consistently demonstrated that the assembled genome version met stringent quality benchmarks, showcasing a high degree of confidence in its representation of the true *A. cannabinum* genetic blueprint. This rigorous validation process benchmarked reference framework for downstream analyses and applications.

Genomic annotation initiatives pinpointed 22,793 protein-encoding loci in *A. cannabinum*'s genetic architecture, comprising 68.3% of functionally annotated transcriptional units. This genomic allocation underscores the intricate protein synthesis machinery governing the plant's metabolic networks. Characterization of reiterative genomic components exposed their 35.8% abundance, with long terminal repeats (LTRs) predominating in pericentromeric zones. Deciphering these elements' spatial organization proves pivotal for elucidating chromosomal rearrangement patterns and transposable element silencing mechanisms.

Computational functional mapping achieved 95.6% coverage (21,780 genes) through integrated Pfam-InterPro domain profiling, establishing a functionally characterized repository for *A. cannabinum* genetic research. This comprehensive gene ontology (GO) annotation matrix enables systematic interrogation of specialized metabolic pathways, particularly cannabinoid biosynthesis clusters. The curated gene-phenotype linkage network proves instrumental for deconvoluting terpenoid synthase regulation circuits and abiotic stress adaptation mechanisms.

To gain insights into the evolutionary history of *A. cannabinum*, a phylogenetic analysis was conducted, incorporating genomic data from 17 different species. This clustering analysis identified a total of 43,837 gene families. Of these, 3084 were identified as common gene families, indicating a shared evolutionary heritage among the analyzed species. These common gene families represent the core set of genes essential for fundamental cellular processes and survival.

Furthermore, comparative analysis of the identified gene families allowed for the estimation of the divergence time of *A. cannabinum*. The results of this analysis suggest that *A. cannabinum* speciated approximately 35.8 million years ago, with a confidence interval ranging from 27.0 to 46.9 million years ago. This timeframe provides a valuable context for understanding the evolutionary trajectory of *A. cannabinum* and its relationship to other plant species.

This comprehensive genomic resource unlocks numerous avenues for future research focused on *A. cannabinum*. Several key areas warrant further investigation:

A. cannabinum is known to produce a range of bioactive compounds. Future research should focus on identifying and characterizing the genetic determinants of these valuable secondary metabolites. This could involve transcriptomic analysis, metabolomic profiling, and gene editing techniques.

A. cannabinum fibers possess desirable properties for textile and industrial applications. Identifying the genes controlling fiber characteristics, such as strength, length, and flexibility, will be crucial for breeding improved varieties. Quantitative trait locus (QTL) mapping and genome-wide association studies (GWAS) could be employed to identify these genes.

A. cannabinum exhibits a degree of arid-adapted genomic architecture selection. Exploring the genetic mechanisms underlying this resilience could provide valuable insights for developing epigenetic remodeling of drought-escape-related chromatin states. This could involve comparative genomics studies with related species that are more or less tolerant to specific stresses.

The genomic data can be used to develop molecular markers, such as SNPs and microsatellites, for use in breeding programs and conservation efforts. These markers can facilitate the selection of desirable traits and the management of genetic diversity within *A. cannabinum* populations.

Further functional genomics studies demand a validation of the predicted functions of genes and to elucidate the complex regulatory networks that govern gene expression in *A. cannabinum*. This could involve techniques such as CRISPR-Cas9 gene editing, RNA interference, and proteomics.

In summary, the complete chromosomal assembly and functional characterization of *A. cannabinum* establish a foundational platform for phytogenetic studies. This chromosomal blueprint enables systematic exploration of its biosynthetic pathways, catalytic enzymes, and stress-responsive elements across pharmacological, biotechnological, and agronomic domains. The decoded genetic architecture directly facilitates targeted trait enhancement through molecular breeding strategies. Such advancements will ultimately enable optimized cultivation protocols for this multipurpose species, ensuring its ecologically balanced exploitation to address global challenges in healthcare, renewable materials, and food security.

Acknowledgments This work is supported by the Ningxia, Ningxia Financial Forestry New Technology Introduction and Promotion Project (grant number: [2025]NX-08), the Coupling of Water and Fertilizer in Karst Desertification and Restoration of Biodiversity (grant number: [2023]004), and the National Key Research and Development Program of Xinjiang Autonomous Region (2023B02045).

References

Bairoch A (2000) The SWISS-PROT protein sequence database and its supplement TrEMBL in 2000. Nucleic Acids Res 28(1):45–48

Burton JN, Adey A, Patwardhan RP et al (2013) Chromosome-scale scaffolding of de novo genome assemblies based on chromatin interactions. Nat Biotechnol 31(12):1119–1125

Collard BCY, Mackill DJ (2008) Marker-assisted selection: An approach for precision plant breeding in the twenty-first century. Philos Trans R Soc Lond B Biol Sci 363(1491):557–572

Dai SL, Huang H, Fu JX et al (2013) Advances in molecular breeding of ornamental plants. Chin Bull Bot 48(6):589–607

DiTommaso A, Clements DR, Darbyshire J et al (2009) The biology of Canadian weeds, 143, *Apocynum cannabinum* L. Can J Plant Sci 89(5):977–992

Duprey A (1905) A case of mitral incompetency and ascites treated with *Apocynum cannabinum*. Lancet 166(4283):955–956

Eisenstein M (2015) Startups use short-read data to expand long-read sequencing market. Nat Biotechnol 33(5):433–435

Goff SA, Ricke D, Lan TH et al (2002) A draft sequence of the rice genome (*Oryza sativa* L. ssp. japonica). Science 296(5565):92–100

Gonal B, Somveer, Doggalli G et al (2023) Exploring the future of plant breeding: Advancements and challenges. Int J Plant Soil Sci 35(24):49–55

Griffiths-Jones S, Moxon S, Marshall M et al (2005) Rfam: annotating non-coding RNAs in complete genomes. Nucleic Acids Res 33:D121–D124

Guindon S, Gascuel O (2003) A simple, fast, and accurate algorithm to estimate large phylogenies by maximum likelihood. Syst Biol 52(5):696–704

Haas BJ, Salzberg SL, Zhu W et al (2008) Automated eukaryotic gene structure annotation using EVidenceModeler and the Program to assemble spliced alignments. Genome Biol 9(1):R7

Hansen KD, Brenner SE, Dudoit S (2010) Biases in illumina transcriptome sequencing caused by random hexamer priming. Nucleic Acids Res 38:131–138

Hoopes GM, Hamilton JP, Kim J et al (2017) Genome assembly and annotation of the medicinal plant *Calotropis gigantea*, a producer of anticancer and Anti-malarial Cardenolides. G3-Genes Genom Genet 8(2):385–391

Hu WQ, Hou Y, Zhang F (2015) A Chromatin conformation analysis technology-Hi-C and extracting of chromatin conformation information. Genom Appl Biol 34(11):36–44

Huang J, Xu QW, Lu Q (2007) An RDF model of gene ontology and its associations. Chin Sci Bull 22:40–46. (in Chinese)

Jaganathan D, Ramasamy K, Sellamuthu G et al (2018) CRISPR for crop improvement: an update review. Front Plant Sci 9:985

Jarvis DE, Ho YS, Lightfoot DJ (2017) The genome of *Chenopodium quinoa*. Nature 542(7641):307–312

Jurka J (2000) Repbase update: a database and an electronic journal of repetitive elements. Trends Genet 16(9):418–420

Kanehisa M, Goto S (2000) KEGG: Kyoto Encyclopedia of Genes and Genomes. Nucleic Acids Res 28(1):27–30

Karikari B, Ahiakpa JK, Javaid A et al (2024) Use of Cutting-edge technologies for the improvement of pulse crops. In: Chandra K, Chand S, Saini RP, et al (eds.). Smart Breeding: Molecular interventions and advancements for crop improvement. New York: Apple Academic Press, pp. 61–104

Keever C (1979) Mechanisms of plant succession on old fields of Lancaster County, Pennsylvania. Bull Torrey Bot Club 106(4):299–308

Kellner F, Kim J, Clavijo BJ et al (2015) Genome-guided investigation of plant natural product biosynthesis. The Plant J 82(4):680–692

Kent WJ (2002) BLAT-the BLAST-like alignment tool. Genome Res 12(4):656–664

Lang K, Bi SD, Li F (2018) Genome-wide analysis of expansion and contraction of gene families in parasitic wasps. J Anhui Agric Univ 45(5):945–950

Leidy (1884) Indian use of *Apocynum cannabinum* as a textile fiber. Proc Acad Natl Sci Phila 36:30–30

Li H (2013) Aligning sequence reads, clone sequences and assembly contigs with BWA-MEM. arXiv 1303:3997

Li BT, Chen XM (1997) Comparative review on Apocynaceae in flora Republicae Popularis Sinicae and Flora of China. Guihaia 17(4):299–305. (in Chinese)

Li GQ, Chen YY (2012) Physioecology of *Apocynum*. Science Press. (in Chinese)

Li PT, Leeuwenberg AJM, Middleton DJ (1995) Flora of China. Beijing: Science Press and St Louis. Missouri Botanical Garden, pp 143–188

Li L, Stoeckert CJ, Roos DS (2003) OrthoMCL: identification of ortholog groups for eukaryotic genomes. Genome Res 13(9):2178–2189

Li GQ, Zhao PP, Shao WS (2019) Cash crop halophytes of China. In: Gul B, Böer B, Khan MA et al (eds) Sabkha ecosystems (Volume VI): Asia/Pacific. Springer, Cham, pp 497–504

Li GQ, Li J, Zhao YX et al (2023) Whole-genome sequencing and analysis of *Apocynum Cannabinum*. JEMOA 1(1):43–52

Liu HL (2018) The whole genome sequencing and analyzing of *Ginkgo biloba*. Nanjing Forestry University (in Chinese)

Lodhi BS (2024) Revolutionizing agriculture: The promise and potential of molecular breeding in the modern world. Just agriculture 4(5):251–259

Majoros WH, Pertea M, Salzberg SL (2004) TigrScan and GlimmerHMM: Two open source ab initio eukaryotic gene-finders. Bioinformatics 20(16):2878–2879

Mascher M, Gundlach H, Himmelbach A et al (2017) A chromosome conformation capture ordered sequence of the barley genome. Nature 544(7651):427–433

Mulhouse JM, Galatowitsch SM (2003) Revegetation of prairie pothole wetlands in the mid-Continental US: twelve years post-reflooding. Plant Ecol 169(1):143–159

Nawrocki EP, Kolbe DL, Eddy SR (2009) Infernal 1.0: inference of RNA alignments. Bioinformatics 25(10):1335–1337

Ogata H, Goto S, Sato K et al (1999) KEGG: kyoto encyclopedia of genes and genomes. Nucleic Acids Res 27(1):29–34

Pandita D (2023) Transgenic approaches for stress tolerance in crops. In: Hasanuzzaman M (ed) Climate-resilient agriculture, vol 2. Springer, Cham, pp 793–818

Powell W, Morgante M (2019) Transgenics: the challenge of constructing and analyzing heterologous plant genes. In: Henry RJ (ed) Transgenic crops. Springer, pp 1–27

Price AL, Jones NC, Pevzner PA (2005) De novo identification of repeat families in large genomes. Bioinformatics 21(suppl_1):351–358

Robert CE (2004) MUSCLE: multiple sequence alignment with high accuracy and high throughput. Nucleic Acids Res 32:1792–1797

Sabir JSM, Jansen RK, Arasappan D et al (2016) The nuclear genome of *Rhazya stricta* and the evolution of alkaloid diversity in a medically relevant clade of Apocynaceae. Sci Rep UK 6(1):33782

Singh G, Rani R, Sivia SS et al (2024) Pre-breeding to molecular breeding for biotic stress management in crop plants. In: Chandra K, Chand S, Saini RP et al (eds) Molecular interventions and advancements for crop improvement. New York: Apple Academic Press 191–218

Sivia SS, Sunayana, Singh G et al (2024) Marker-assisted selection approaches for abiotic stress tolerance in crop plants. In: Chandra K, Chand S, Saini RP et al (eds) Smart Breeding: Molecular interventions and advancements for crop improvement. New York: Apple Academic Press 261–286

Smith AM (2019) The origins of agriculture and crop domestication. Biol 8(4):61

Sojka J, Šamajová O, Šamaj J (2023) Gene-edited protein kinases and phosphatases in molecular plant breeding. Trends Plant Sci 2578:1–17

Song LX, Li GQ, Jin CQ et al (2019) Whole genome sequencing and development of SSR markers in *Apocynum cannabinum*. J Plant Genetic Res 20(5):1309–1316. (in Chinese)

Stanke M, Steinkamp R, Waack S et al (2004) AUGUSTUS: a web server for gene finding in eukaryotes. Nucleic Acids Res 32(Web Server issue):W309–W312

Tarailo-Graovac M, Chen NS (2004) Using RepeatMasker to identify repetitive elements in genomic sequences. Curr Protoc Bioinformatics 25(4):1–4

Usman OH, Zhang L, Xie G et al (2022) Genomic heterogeneity in pancreatic cancer organoids and its stability with culture. NPJ Genom Med 71(7):1–10

Varshney RK, Pandey MK, Bohra A et al (2019) Toward the sequence-based breeding in legumes in the post-genome sequencing era. Theor Appl Genet 132(3):797–816

Wang S (2019) Whole genome sequencing and analysis of *Betula platyphylla*. Northeast Forestry University. (in Chinese)

Wang DQ, Li GQ, Wang L (2012) Daily dynamics of photosynthesis and water physiological characteristics of *Apocynum venetum* and *A. cannabinum* under drought stress. Acta Botan Boreali-Occiden Sin 32(6):1198–1205. (in Chinese)

Weitemier K, Straub S, Fishbein M et al (2019) A draft genome and transcriptome of common milkweed (*Asclepias syriaca*) as resources for evolutionary, ecological, and molecular studies in milkweeds and Apocynaceae. Peer J 7:e7649

Wood HC (1904) A study of *Apocynum cannabinum*. JAMA 43(26):1953–1957

Xie WY, Zhang XY, Wang T et al (2012) Botany, traditional uses, phytochemistry and pharmacology of *Apocynum venetum* L. (Luobuma): a review. J Ethnopharmacol 141(1):1–8

Xu M, Sun YG, Li HG (2010) EST-SSRs development and paternity analysis for *Liriodendron* spp. New For 40(3):361–382

Yang Z, Rannala B (2012) Molecular phylogenetics: principles and practice. Nat Rev Genet 13(5):303–314

York, Varshney RK, Bansal KC (2019) Agribiotechnology: milestones, challenges, and prospects. In: Gosal SS, Wani SS (eds) Biotechnologies of crop improvement, vol 1. Springer Singapore, pp 1–29

Zdobnov EM, Apweiler R (2001) InterProScan—an integration platform for the signature-recognition methods in InterPro. Bioinformatics 17(9):847–848

Zhao X, Hao W (2007) LTR_FINDER: an efficient tool for the prediction of full-length LTR retrotransposons. Nucleic Acids Res 35(Web Server issue):W265–W268

Chapter 13
Tossa Jute (*Corchorus olitorius* L.) Breeding

Jiquan Chen, Xiaoyu Zhang, Zemao Yang, Zhigang Dai, H'ng Paik San, and Jianguang Su

Abstract Tossa jute (*Corchorus olitorius* L.) has been instrumental in fundamental human needs, encompassing package, food, health, attire, shelter, and means of transportation. From ancient to modern times. People utilize various parts of the jute plant including natural fiber and other components to make a wide range of products. This chapter offers a summary and constraints of the approaches and accomplishments of the breeding strategies and achievements in tossa jute breeding, encompassing germplasm management, evaluation and diversity, utilization of different plant parts, conventional and modern breeding methods, including selection, hybridization, genetics of key agronomic traits, genes and genomics, marker-assisted breeding, gene editing and genetic engineering. However, there has been a gradual decline in the production of jute fiber due to the extensive use of synthetic fibers and the emergence of other natural fibers, that request some transformation of breeding objectives and cultivation patterns. To address this, it is more important to cultivate higher yielding varieties, that are adaptable to non-agricultural land or resistant to abiotic stress, while also developing cost effective cultivation methods and exploring new uses for jute. Nevertheless, traditional breeding techniques, including line breeding (pedigree selection), induced mutagenesis, and cross-breeding (hybridization), face limitations in improving yield and quality. Currently, the genomes of both tossa jute and white jute have been sequenced, placing jute breeding at the forefront of the genomic era. Furthermore, advancements in accurate germplasm identification, the discovery of new genes and transcription factors, developments in bioinformatics and data analyses, progress in transgenic technologies, and collaboration between breeders and researchers from various industries will drive a significant revolution in jute improvement and utilization.

J. Chen · X. Zhang · Z. Yang · Z. Dai · J. Su (✉)
Department of Germplasm and Breeding, Institute of Bast Fiber Crops,
Chinese Academy of Agricultural Sciences (IBFC, CAAS), Changsha, China
e-mail: yangzemao@caas.cn; jgsu@vip.163.com

H. P. San
Faculty of Forestry and Environment, Universiti Putra Malaysia (UPM),
Serdang, Selangor Darul Ehsan, Malaysia
e-mail: ngpaiksan@upm.edu.my

K. F. M. Salem et al. (eds.), *Breeding and Biotechnology of Grass and Bast Fiber Crops*, Advances in Plant Breeding Strategies 12,
https://doi.org/10.1007/978-3-032-00403-1_13

Keywords Biodiversity · Biotechnology · Breeding · Genetic improvement · Hybridization · Tossa jute (*Corchorus olitorius* L.)

13.1 Introduction

Jute (*Corchorus sp.*) ranks as the second (after cotton) most significant natural fiber crop globally, contributing approximately 80% of the annual worldwide production of bast fibers (Islam et al. 2017). The main countries that produce and utilize jute-are India, Bangladesh, and China. China, in particular, boasts a lengthy past of jute cultivation, tracing back to the *Tu Jing Ben Cao* in 1061 AD, even before the cultivation of cotton (Zhang et al. 2021). An inspection of the preserved jute fabric on a pottery originating from the Harappa site indicates that jute cloth was used from 2200 to 1900 BC. This may be the earliest known use of jute cloth within the context of the Indus Valley Civilization (Wright et al. 2012). Despite the existence of more than 100 jute species (Islam et al. 2015), only two of them are generally cultivated: *C.olitorius* (Tossa, dark jute) and *C.capsularis* (white jute). The primary countries producing jute are India, Bangladesh, China, Nepal, and Zimbabwe. Combined, India and Bangladesh make up nearly 98% of the global jute cultivation in both area and yield. Notably, China has the highest productivity at 3.87 tons per hectare, followed by India at 2.62 tons per hectare, Nepal at 1.35 tons per hectare, Bangladesh at 1.18 tons per hectare, and Zimbabwe at 0.61 tons per hectare on average (More et al. 2022) (https://www.fao.org/faostat/en/#data/QCL).

Tossa jute has a wide range of applications in textiles, decorations, packaging materials, and architectural products. This includes cloth, hessian, bags, ropes, curtains, rugs, carpets, handicrafts, and composite boards (Bhuiyan et al. 2023). In addition, it is a valuable raw material for bio-energy (Bhandari et al. 2018; Biswas et al. 2015). Tossa jute is also a traditional and popular leafy green vegetable in Middle Eastern, Asian, and African countries. It is even considered a primary vegetable in certain African regions (Kwarteng et al. 2018; Mgolozeli et al. 2022). This vegetable is rich in polysaccharides, phenols, dietary fiber, and mineral nutrient elements including potassium, calcium, and magnesium (Olcay et al. 2022). Numerous experiments have demonstrated that tossa jute not only has high nutritional value but also exhibits beneficial effects. These include antioxidant properties, free radical scavenging, anti-inflammatory and pain relief qualities (Kumari et al. 2019). It also has the potential to prevent hypertension and obesity (Awosanya et al. 2019; Gomaa et al. 2019), as well as inhibit cancer cell proliferation (Guzzetti et al. 2021). Furthermore, tossa jute has been studied for its ability to act as a biosorbent removing heavy metal ions from aqueous solutions(Hassan and Zohdy 2018; Huang et al. 2019).

In the 1960s, 'Guangfeng Changguo' was identified as the primary tossa jute variety. It was selected from a landrace collected in Guangfeng County, Jiangxi Province, China. This variety became popular in the middle and lower reaches of the Yangtze River and southern regions of China. Other introduced germplasm such as 'JRO632', 'Mali Yesheng', 'Ba Ma 72-2' and 'Bachang 4hao, have also become important breeding parents (Luo et al. 2009). In addition to introducing, identifying and screening superior germplasm, jute cross-breeding has been the most successful and practical method for modifying the genetic basis of jute. Over the past sixty years, numerous hybridization, backcross and side cross combinations have been harvested, resulting in tens of thousands of excellent progeny materials of different generations. The main approved and widely promoted varieties are '71–10', 'Kuanye Changguo', 'Xiang Huangma 1hao', 'Xiang Huangma 3hao' and 'Zhong Huangma 4hao'. In the breeding history of tossa jute, 'Cuilv' (also known as 'Chinsurah Green', a widely cultivated variety before the 1950s) and 'Bachang 4hao' was introduced from India and Pakistan respectively, 'Guangfeng Changguo' and 'Kuangye Changguo' bred by the Institute of Bast Fiber Crops, Chinese Academy of Agricultural Sciences (IBFC, CAAS), were considered backbone parents that played an important role and made a significant contribution to tossa jute breeding activities (Zhang et al. 2015a).

A series of comprehensive studies, encompassing morphological analysis, genetic research, and archaeobotanical investigations, have been carried out to evaluate the diverse range, phylogenetic connections, and the progress achieved in the domestication of jute (Honi et al. 2020; Yang et al. 2022). Phylogenetic analyses combined with the widespread distribution of *C. aestuans* in East Africa suggested that *C. aestuans* is likely the ancestor of both cultivated jute types. *C. urticifolius* is the progenitor of *C. olitorius,* but the progenitor of *C. capsularis* remains unidentified based on the existing literature (Kundu et al. 2012b). Currently, limited progress has been made in understanding *C. olitorius*' genomics, dynamics, and domestication. Crucial is a multidisciplinary approach, including modern molecular tech, to elucidate ancestries and study human activities related to domestication, it's vital to analyze circadian clocks and genetics with modern tech, considering geography, archaeology, and archaeobotany for insights into domestication and breeding. This chapter delivers a summary of the germplasm studies, origin and distribution, breeding and farming techniques, the latest advancements in biotech, and their utilization for enhancing tossa jute.

13.1.1 Origin and Classification

Tossa jute was first classified into the genus *Corchorus* and the family Tiliaceae, the genus was classified under the Malvaceae family later based on modern molecular studies and tests on the genome and phylogenetic analysis. This new classification is now widely accepted.

The *olitorius* species belongs to *Corchorus*, Malvaceae, Magnoliopsida (Dicotyledons), Magnoliophyta (Angiosperm or Flowering plants), Spermatophyta (Seed plants), Plantae (Plants), and Eukaryota. Its common names are as follows: Tossa jute, Dark jute, Huangma, Pat, and Jews mallow.

It is widely accepted that *C. capsularis* and *C. olitorius* each independently originated in the tropics (Benor et al. 2012). *C. capsularis* most likely hailed from either the Indo-Myanmar area or Africa, and was eventually domesticated in Asia (Kundu et al. 2012b), while *C.olitorius* is believed to have originated near the equator of East Africa with its domestication taking place in South China and India. (Yang et al. 2018). According to the "Report of The Third IJO Germplasm Collecting Mission for Jute, Kenaf, And Allied Fibers in Kenya and Tanzania (MissionI-V:15 November 1987 to 19 September 1988)", 374 accessions of *Corchorus* comprising wild types, landraces, and related species were found in the coast and southeast of Kenya and northeast of Tanzania. This investigation ascertained that Africa is the diversity center of *Corchorus*. Phylogenetic studies suggest that *C. aestuans* likely served as the ancestral species for both types of cultivated jute. Although *C. urticifolius* is unequivocally recognized as the predecessor of *C. olitorius*, identifying a definitive progenitor for *C. capsularis* based on morphological characteristics and neutral molecular markers remains elusive (Chen et al. 2023; Kundu et al. 2012b). Yang et al. (2018) conducted a genetic diversity analysis to reveal the population structure of 453 accessions of *C.olitorius* using SSR markers. The results showed that the genetic diversity of *C.olitorius* exhibited a moderate level, characterized by a mean gene diversity (h_e) value of 0.322. The polymorphic information content (PIC) value averaged approximately 0.270 across the entire collection of accessions. In terms of geographical distribution, the highest mean values for both h_e and PIC were observed in samples originating from Kenya, with respective values of 0.332 and 0.273. Subsequently, Nepal demonstrated average h_e of 0.307 and 0.259 for PIC, whereas China exhibited averages h_e of 0.303 and 0.253 for PIC. These results offer significant insights into the global genetic diversity of *C. olitorius* and support its African origin.

A significant milestone in molecular anthropology was the discovery that all mitochondrial DNAs (mtDNA) from 147 individuals could be traced back to a common ancestral woman, often referred to as "Eva" by the media. It was postulated that Eva lived in Africa approximately 200,000 years before the present (kya), based on restriction mapping (Cann et al. 1987). Following this landmark discovery, numerous studies on mtDNA have been conducted to investigate human evolution, migration patterns, and population histories (Ingman et al. 2000; Pakendorf and Stoneking 2005; Ping et al. 2022). A synthesis of these studies is represented in Fig. 13.3, where it becomes apparent that human migration out of Africa closely aligns with the distribution of *Corchorus*, a genus including jute plants. Benor et al. (2018, 2012) proposed and attempted to substantiate a hypothesis regarding the origin and diffusion of jute in pantropic regions. This hypothesis suggests that jute spread through trade routes along the coastal regions stretching from Western Africa to the Indian subcontinent, rather than through natural migration. However, the current evidence is inadequate to definitively establish a connection between the dispersal of jute and human activities (Fig. 13.1).

Fig. 13.1 The major distribution region of *Corchorus* and human migration route[source: "Report of the third IJO germplasm collecting mission". The green circles show the major distribution region of *Corchorus*, and the brown arrows show the human migration ways documented well. (Source: (Ingman et al. 2000; Kundu et al. 2012a, b; Cann et al. 1987; Nyadanu et al. 2016; Pakendorf and Stoneking 2005; Ping et al. 2022))

13.1.2 Fiber Properties and Utilization

The primary components of jute fiber are variational, approximately 60% cellulose, 23% hemicellulose and 14% lignin. The advantages are shown in high tensile strength, high abrasive resistance and low extensibility (Río et al. 2009). The hygroscopic nature of jute is primarily attributed to its capacity factor, which is a result of the higher presence of amorphous components within the fiber including xylan, hexosans, polyuronides, cellulose and lignin. (Sen and Hermans 1949). The proportion of amorphous constituents in jute, relative to cotton, is approximately 1.56, indicating that jute exhibits greater hygroscopicity compared to cotton.

Jute fibers offer numerous advantages when compared to synthetic fibers. These advantages include their lighter weight, high specific strength, biodegradability, non-abrasiveness to processing equipment, carbon neutrality, and suitability for use as acoustic and thermal insulators (Alves et al. 2010; Bhuiyan et al. 2023). A kind of biocomposite material based on jute and thermoplastic starch (TPS) has been developed, the tensile strength of jute/TPS materials(40% of jute fiber) was enhanced by 595% with the reinforcement of jute fiber and magnesium carbonate hydroxide pentahydrate (MCHPH) nano-filler, most importantly, the retardancy and thermal stability of this material were enhanced too (Zulfiqar et al. 2024).

Table 13.1 Main applications of jute. (Designed by Jiquan Chen)

Applications	Raw materials	Products or intermediate products
Traditional application	Fiber	Rope, bag, cloth, carpet, paper
	Tender tip, leaf	Vegetable, traditional medicine
Newer application	Fiber	Degradable mulching film, composite material, ornament, handicraft, mattress, wallpaper, paper, geotextile, modified fiber
	Whole stalk or core	Paper, particle board, carbon, adsorbent, composite material, organic fertilizer
Newest application	Tender tip, leaf	Health care products, cosmetics, developing medicine
	Whole stalk and leaf	Adsorbent for heavy metals
	Fiber	Special fiber, reinforced composites
	Seed	Medicine, cardiac glycosides

Many materials were composited with jute to enhance some kinds of characteristics, i.e. jute/PLA (Zhao et al. 2024), jute fiber reinforced epoxy composite (Cardoso et al. 2024), treated with chemicals to change jute fiber structure and properties to increase moisture sorption (Ivanovska et al. 2024). Furthermore, various parts of the jute plant can be used to create a wide range of products (Table 13.1).

13.1.3 Differences Between White Jute and Tossa Jute

Distinguishing between white jute and tossa jute can be challenging for beginners and laypersons until these plants produce fruit (Fig. 13.2). The most apparent difference lies in the shape of the fruits, with white jute having round, ball-shaped fruit and tossa jute having long, column-shaped fruit. Other distinguishing features include variations in leaf shape, size, and color, differences in stem surface texture and color, variations in the flavor of tender leaves, and branching habits. Additionally, white jute exhibits relatively higher tolerance to flooding and drought conditions compared to tossa jute, despite exhibiting a marginally higher vulnerability to diseases and pests (Roy et al. 2006). Moreover, white jute shows some degree of tolerance to salt stress when contrasted with tossa jute and can endure concentrations of up to 60 mM NaCl in the water culture. These differences have been identified for various agronomic traits (Islam et al. 2017).

Different intraspecific varieties can exhibit high or low tolerance to saline stress. Some varieties ('TC008-41' and 'NY/253c', which belong to *C.olitorius*) can survive in salinity treatments at 250 mM NaCl stress (Yang et al. 2017c, 2022). Therefore, it is unclear which species possesses a greater tolerance to salt stress. A rigorous cytogenetic analysis was performed to ascertain the karyotype configurations and chromosome counts within certain species and their wild relatives of the

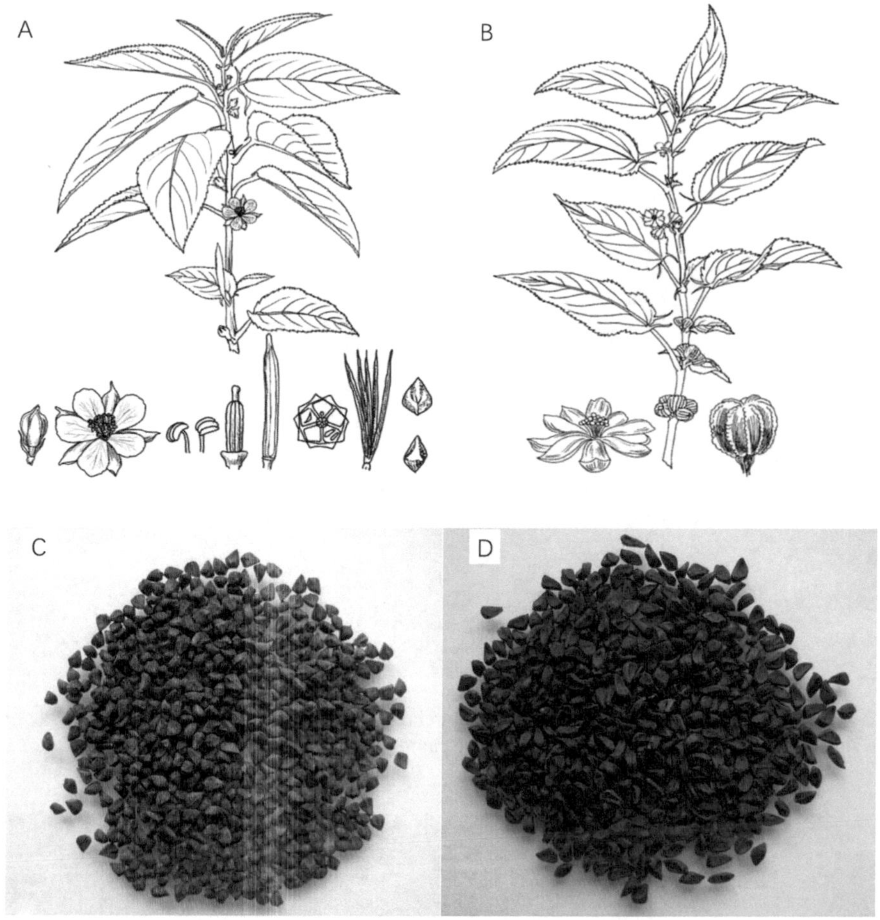

Fig. 13.2 Morphological difference between Tossa and white jute. (**a**) A sketch of tossa jute; (**b**) A sketch of white jute; (Source: Zhang et al. (2022)) (**c**) Dark green seeds of tossa jute; (**d**) Brown seeds of white jute. (Photo by Jiquan Chen)

genus *Corchorus*, by apical staining microscopy. (Chen et al. 2011). The chromosomal count in *Corchorus* amounted to 2n = 14, but there were 4 satellites and 2 satellites in the cultivar and wild type of *C.olitorius* respectively, and no satellite in the chromosomes of *C.capsularis*.

13.2 Current Cultivation Practices and Challenges

Jute is predominantly planted by smallholders, with about eight million individuals engaged in jute production, ranging from cultivation to industrial processing. Weeding, harvesting and fiber extraction cost too much energy and labor in jute production. These operations are carried out almost manually, jute cultivation has become a low-efficiency and non-profit industry due to the increasing cost of agricultural materials and labor. To change this situation, mechanization and simplification of jute cultivation are indispensable in reducing production cost and drudgery (Figs. 13.3, 13.4, and 13.5). Some machines, i.e. sowing machine, weeder, harvester and decorticator are developed and employed in the large-scale production. Moreover, the cropping pattern is simplified, such as no-tillage planting, covering

Fig. 13.3 Sowing machines for jute. (**a**) Commercial sowing machine for large farmland. (Photo by Jiquan Chen); (**b**) Small sowing machine for the mountainous region. (Photo by Wen Xiang)

Fig. 13.4 Jute/Kenaf Decorticator (4HB-480) developed by IBFC. (Photo by team of mechanical design, IBFC)

Fig. 13.5 Fiber extraction by manpower. (Photo by Jiquan Chen, Dhaka, Bangladesh)

with degradable film for weed control, the increasing utilization rate of bare places(salted/drought) by planting highly resistant jute varieties, and growing jute in conjunction with other crops (Rahman 2009), like oilseed rape, winter vegetables or rice (Zhang et al. 2016) to fulfill land use efficiency. On the other hand, most of the farmers (90%) used purchased seeds in India and Bangladesh, farmers do not use fertilizers in a balanced way, only 11% of farmers sprayed pesticides but did not follow the descriptions (Hossain and Abdulla 2015). In a word, jute cultivation is rather low and traditional, with little digital or intelligent operation used in large production.

13.3 Conservation and Biodiversity of Germplasm

13.3.1 Germplasm Management Procedure

The "UN Sustainable Development Goal 2" emphasizes the conservation of genetic diversity as much as possible through well-organized gene banks. Several important factors such as temperature, moisture, light, and oxygen, determine the viability of accessions that will maintain their viability during long-term storage. In order to handle these factors proficiently, genebanks necessitate high-quality data and comprehensive management frameworks (Lusty et al. 2021). In China, the latest "Technical Regulation on Conservation for Crop Germplasm Resources" has been in place for over 15 years. We refer to these guidelines to manage the collection, conservation, evaluation, regeneration and distribution of germplasm. Taking into account the characteristics of tossa jute and the conditions of the genebank for bast fiber crops, we have developed a specific management procedure and other relevant details (Fig. 13.6).

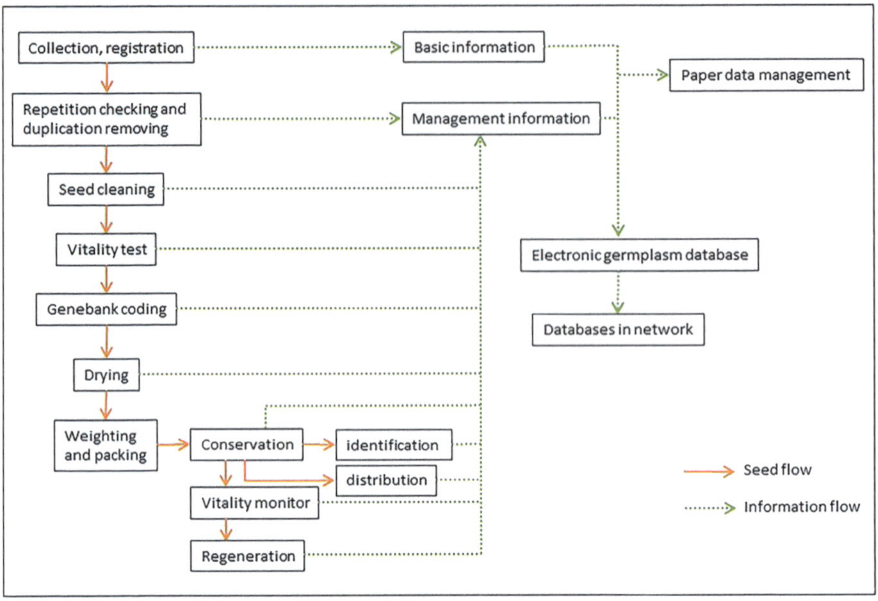

Fig. 13.6 Procedure for jute seed conservation in germplasm bank. (Source: Lu et al. (2008))

13.3.2 Germplasm Conservation and Diversity

Over the past few decades, the germplasm of jute and allied crops has been systematically collected and meticulously preserved in genebanks situated in the main countries producing jute fiber, notably China, Bangladesh, and India (Al-Khayri and Johnson 2019). Dry seeds (80 g to 150 g, germination rate ≥85%, water content ≤7%) are packed before preservation in the Long-term genebank with low temperature and humidity(−18 °C, ≤50%). The Mid-term genebank (Fig. 13.7) (−4 °C to 4 °C, ≤65%) is located in IBFC, which is responsible for germplasm collecting, conserving, evaluation, regeneration and sharing. The Mid-term genebank for bast fiber crops conserves 2489 accessions of jute germplasm, including 1384 accessions of *C.capsularis* (cultivars: 379, landraces: 532, genetic materials: 243, wild types: 70, and allied fiber crops: 160) and 1105 accessions of *C.olitorius* (cultivars: 429, landraces: 356, genetic materials: 155, wild types: 95, and allied fiber crops: 70). These allied fiber crops involve 25 species, such as *C. aestuans* L., *C. brevicornutus* Vollesen., *C.axillaris* Tsen et Lee., *C. trilocularis* L., *C.brevicaulis* Hosokama., *C. baldaccii* Mattei., *C.schimperi* L., *C.gillettii* L., *C.africanus* Bari., *C. fascicularis* Lam., *C.urticifolius* Wight & Arn., *C. pseudocapsularis* Schweinf., *C.psedoolitorius* Islam & Zaid., *C.tridens* L. et al. The Central Research Institute for Jute and Allied Fibers (CRIJAF), situated in Kolkata, India, is dedicated to conserving a comprehensive germplasm collection, encompassing 1647 accessions of Tossa, 939 of white jute and other related species (Roy et al. 2006).

Fig. 13.7 The Mid-term germbank for bast fiber crops located in IBFC. Left: gate of the bank, right: inner view. (Photo by Jiquan Chen (2024))

Due to the diverse origins of many tossa jute accessions, these accessions conserved in IBFC exhibit a wide range of morphological diversity (Fig. 13.8) in terms of their leaves, stems, flowers, stipules, fruits, branching habits, photoperiodic response and ecological adaptability.

13.4 Conventional Breeding of Tossa Jute

13.4.1 Improvement Strategies

Almost all of the strategies applied to other major food crops were tried in jute breeding, but the proven and effective methods are as follows: sexual hybridization, pedigree selection, germplasm appraisal, mutation and introduction. For example, 'Zhonghuangma 4hao', was developed using these methods. 'Diwangcai 4hao' was another variety bred by hybridization with 'Tianhuangma' and 'Mowei 4hao' (Dai et al. 2021). To improve the success and effectiveness of hybridization, scientists conducted many basic germplasm studies such as biodiversity and relationships, construction of core germplasm and morphological database, resistance germplasm and relative genes mining. Benor revealed wild jute materials from Ethiopia were quite distinct and have undergone long-term spatial isolation, suggesting that the samples have the potential to introduce valuable mutation in breeding (Benor et al. 2012). An obvious sexual incompatibility barrier exists between Tossa and white jute (Saha et al. 2001), and efforts aimed at enhancing jute have been confined to intraspecific hybridization alone. Moreover, prior studies showed that genetic diversity is partly lost because a restricted selection of individuals participated in a breeding program and propagated descendants (Doebley et al. 2006). Unfortunately, compared to tossa jute, white jute has undergone more intense breeding selection during the past domestication (Chen et al. 2023), which means that selecting from natural populations with valuable traits will become harder and harder.

Fig. 13.8 The morphological diversity of tossa jute (*C.olitorius*) and allied species. (**a**) Diversity of leaves, from left to right, lanceolate, ovoid, near ellipse, waxy surface, ovoid brown; (**b**) diversity of stipules, from left to right, filiform, spiculate, slender leaf-shaped; (**c**) diversity of stems, from left to right, green, gritty texture, gray-red, red, hairy, and (**d**) diversity of fruits, from left to right, fusiform, acervate, striped red, red fusiform, deeply angular

13.4.2 Conventional Breeding Techniques and Constraints

The targets of traditional tossa jute breeding are mainly focused on yield enhancement, fiber fineness, fiber strength and resistance to diseases. Various techniques are employed in breeding, beginning with mass selection from germplasm then superior single plant selection, followed by sexual hybridization (intra-hybridization/inter-hybridization), mutation (by chemical/physical methods or by aerospace) flourishing with various biotechnology approaches.

Germplasm screening and evaluation is well-suited for enhancing quality and yields, but low productivity and accuracy for quantity traits (Ngomuo et al. 2017). When planted in the same location, it is difficult to select the true superior accessions from different ecotopes. Some accessions bearing excellent resistant genes usually grow not very well or with many branches, the accessions with these traits should be abandoned by breeder's selection.

The hybridization method requires vast combinations, takes a long time and costs too much. There is no effective anthericide, no good methods for killing the males quickly of jute, and no male sterility line, which limited the application of hybridization. Although some varieties are bred by hybridization, the heterosis can not be adequately expressed because the F_1 seeds were rarely used in large-scale production.

13.4.3 Role of Biotechnology

A growing preference for biotechnology applied in classification and biodiversity, characterization of cultivated and wild jute types, molecular marker-assisted breeding, and transgenic breeding has led to a deeper understanding and enhancement of our comprehension of jute research. Besides, molecular markers, like simple sequence repeats (SSRs) (Adeyemo et al. 2021; Saha et al. 2017; Zhang et al. 2015b), restriction-site-associated DNA(RAD), single nucleotide polymorphisms (SNPs) (Sarkar et al. 2019), random amplified polymorphic DNA (RAPD) (Mandal et al. 2013), and inter simple sequence repeat (ISSR) (Roy et al. 2006), and expressed sequence tag (EST)-SNP markers (Biswas et al. 2015). These markershave been devised to establish genetic linkage maps and delineate the genetic diversity among jute varieties and assist breeding.. With the completion of the sequencing of tossa jute and the advancement of sequencing technology, combined with the newest analysis methods and data processing methods, more and more genes can be found and confirmed faster than ever. Despite the limitation of essential information about jute genes due to the constraints in biotechnology usage, certain genes have been identified based on their verified functions, which can be beneficial in various stages of breeding. Based on the genome-wide association study (GWAS) of 11 significant traits, we revealed plentiful loci that probably harbor those genes relevant to resistance to abiotic stress, fiber productivity and quality, and other traits relevant to breeding. For example, the protein *COS07g_00652* which belongs to the Aux/IAA gene family, significant impacts stem elongation and mechanical strength enhancement, which could be inferred to increase the plant height without lodging, to achieve higher fiber yield (Yang et al. 2022). Breeders have made great progress in ploidy(haploidy or polyploidy) breeding and made great progress in the application of marker-assisted selection, genetic engineering methodologies, and gene editing technologies, but there is still a long way to practical application.

13.4.4 Selection and Hybridization

Pedigree selection, intraspecific hybridization and mutation (through radiation or space technology) have been the main conventional methods for breeding tossa jute cultivars in China since the 1950s (Fig. 13.9). For example, the cultivar 'Guangfengchangguo (GFCG)' was obtained through pedigree selection of superior single plants and their offspring from landrace collected in Guangfeng County, Jiangxi Province. 'Kuanyechangguo', a widely cultivated variety since the 1980s, was obtained by crossing 'GFCG' with 'Bachang 4hao'. 'Zhonghuangma 4hao', a recent cultivar with extremely high yield and late maturity characteristics, was bred through the combination of those methods mentioned above. 'O-4' was introduced from Bangladesh then selected by phenotype appraisal and classified into two types: 'O-4 I '(ideotype with vertical petioles, no branch) and 'O-4 II ' (wax on the leaf surface). Breeders cross the 'O-4 II' with 'Xianghuangma 3hao'(a high yield variety in China), F_1 backcross with 'Xianghuangma 3hao', F_2 individuals were selected according to high plant, no branch, late mature, vertical leaves and rapid growth. Single plants of F_3-F_7 were selected sequentially by the indicators until the strain 'Y007-10' with targeted characteristics and stable heredity was harvested (Gong et al. 2013). More information on the varieties can be found in (Fig. 13.10 and Table 13.2). Based on the past decades of jute breeding, several key guidelines for selecting parents in the hybridization of Tossa can be summarized. These include owning outstanding traits(unlinked inheritance to harmful traits), prepotencies that should be complementary(or reciprocal) between parents, possessing broad adaptability, exhibiting distanced relationships, and having excellent combining ability. Furthermore, both indigenous and exotic germplasm play important roles in tossa jute breeding activities, especially some of the core germplasm, as they are the basic materials for almost every significant improvement in jute breeding.

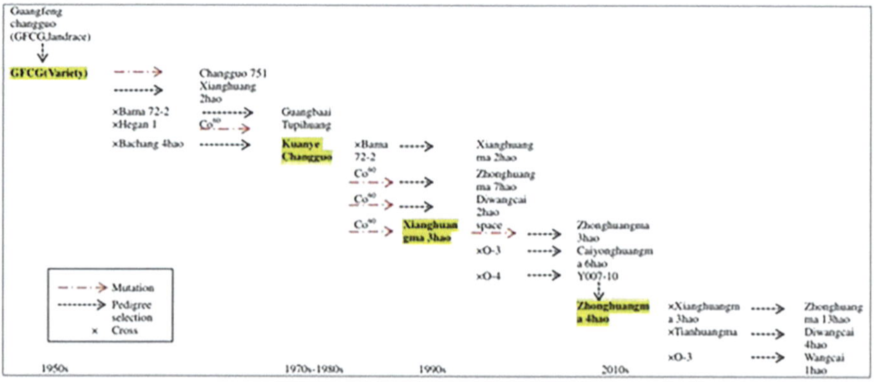

Fig. 13.9 A partial landscape of the major tossa jute varieties in China. The varieties labeled yellow are the core ones of Tossa

Fig. 13.10 Representative varieties of tossa jute in China. (**a**) Xianghuangma 3hao, (**b**) Zhong Huangma 4hao. (Photo by Zhigang Dai)

Table 13.2 Main characteristics of the representative varieties

Variety name	Xianghuangma 3hao	Zhong Huangma 4hao
Species	*C. olitorius* L.	*C. olitorius* L.
Plant height	Highest 520 cm, average 380 cm	Highest 580 cm, average 410 cm
The first branch position	320 cm	340 cm
Stem diameter	2.0 cm	2.1 cm
Bark thickness	1.05–1.80 mm	1.05–2.00 mm
Fiber yield	3062 kg/hm^2	3148 kg/hm^2
Fiber count	460 m/g	508 m/g
Fiber strength	3.12 cN/dtex	3.56 cN/dtex

13.4.5 Studies on Interspecific Hybridization

In contrast to intraspecific hybridization, breeders have primarily focused on interspecific hybridization, especially between the white and tossa jute species or their allied species. However, the seed yield from intraspecific experiments was limited compared to self-pollination. Many seeds were shriveled or empty, unable to germinate. Attempts were undertaken to carry out hybridization among the parental species and to employ x-ray irradiated pollen for hybridization, yet these endeavors failed to yield few mature fruits. The inability to obtain interspecific hybrid offspring was attributed to early-stage degeneration of embryos and endosperm. Building upon this understanding, Islam and Rashid adopted a different approach in their hybridization experiments (Islam and Rashid 1960). They applied indole acetic acid (IAA) to the flower's pedicel and showed that the descendants of F_2 and F_3 exhibited traits from the female parent. This was a successful hybridization between Tossa and white jute for the first time. Subsequently, Islam refined this technique by using IAA (lanolin paste at 300 ppm) and embryo culture in interspecific hybridization experiments between white jute and tossa jute (Islam and Rashid 1964). This modified technique resulted in a higher percentage of fruit set, although some seedlings (2 out of 15) were semi-albino and died soon after transplantation. In the F_1 and F_2 generations, only maternal inheritance was observed. Further research revealed that during microsporogenesis, chromosome associations analysis revealed the occurrence of pentaploid, tetraploid, or triploid accompanied by a haploid in several cells during the diakinesis and metaphase I stages. The F_2 individuals of these two hybrids had over 40% nonviable seedlings, with approximately 11% of the F_2 generation exhibiting a range of chlorophyll deficiencies and abnormalities in leaf morphology and reduced seed vigor (Swaminathan and Iyer 1961). It's important to note that differences in pollen morphology, including shape, size, colpi pore character, exine surface, and cellular features, exist among Corchorus spp. (Maity et al. 2009). These differences may pose challenges for direct interspecific pollination.

We also found albino petals in the F_1 generations of interspecific hybridization between white jute and tossa jute in 2014 (Fig. 13.11, A10). Additionally, there were several kinds of abnormal appearances in the corollas and fruits (Fig. 13.11, A2-A11; B2, B3, B5, B6). Interestingly, these abnormalities exhibited continuous variation in the number of petals (ranging from 1 petal to 7) and discontinuous variation in the shape of petals. The genetic disorder caused by interspecific hybridization consequently affected the growth of fruits and viable seeds, and we could not harvest seeds from the F_1 generation (Fig. 13.11, B6).

Although hybrids could not be obtained from white jute and tossa jute, some scientists attribute this to the absence of intermediary characteristics in generation F_1. Interspecies can be sexually crossed when the relationship is close enough between the parents, or when the F_1 and F_2 plants show only maternal traits. Resistant strains of mosaic disease have been cultivated from *C. trilocularis* (a wild species) and white jute, but scientists encountered certain constraints when introducing targeted genes in practical applications. Several studies have reported failing hybridization between the cultivated species of jute (Datta et al. 1960). On the other hand, Mia and Shaikh (1967) declared to have successfully crossed the variety 'C.G.' (*C. olitorius*) with the variety 'D154' (*C. capsularis*) by irradiating seeds with gamma rays before planting. They achieved greater plant height than the controls and high fertility. Another research asserted they had created F_3 hybrid offspring, however, subsequent reports on the utilization of these hybrids were absent, presumably due to the entire population exhibiting a complete similarity to the female parent (Raut and Naik 1983). Hybrid generation was only feasible with *C. olitorius* as the female parent, except for one reported success with *C. capsularis* (Mia and Shaikh 1967). Additional research is needed to determine if cytoplasmic factors from Tossa and white jute hinder fruit set when white jute is used as the female parent. Recently, an investigation of primer polymorphism from parents of a distant hybridized population between the cultivar 'OIJ-248' (*C.olitorius* as the female parent) and wild relative 'WCIN-136' (*C. aestuans* as the male parent), A recent investigation into primer polymorphism among the parents of a distant hybrid population, which involved the cultivar 'OIJ-248' (*C. olitorius*, as mother) and its wild relative 'WCIN-136' (*C. aestuans*, as father), revealed variations in allele size in thirteen and three primers, respectively. This suggests that these primers can serve as tools for identifying interspecific hybrids, which is expected to be useful in interspecific hybridization (Saha et al. 2017). Although the allied species of *Corchorus* have low fiber yield, they have been reported as good genetic resources for disease, pests, or abiotic stress resistance. With the deepening of research and the application of new technology, exploitation by interspecies hybridization between cultivated species, wild types and allied fiber species should be an efficient approach for creating more favorable variations in jute.

Fig. 13.11 The abnormality appeared in flowers and fruits of the F_1 generations of interspecific hybridization between white and tossa jute. (Photo by Jiquan Chen). A1, normal flower of tossa jute; A2, one petal; A3, two petals; A4, two normal size petals + one smaller; A5, three normal size of petals + two smaller; A6, four petals; A7, four normal size of petals + one smaller; A8, six petals; A9, seven petals; A10, albino at the petals; A11, abnormal shape of petals; A12, normal flower of white jute. B1, normal fruits of tossa jute; B2, B3, B5, abnormal morphological types of fruits; B4, normal fruits of white jute; B6, hypogenetic embryo and empty pod. (Zhang et al. (2022))

Fig. 13.11 (continued)

13.4.6 Mutation Breeding

Studies on inducing mutations in organisms garnered significant attention after Muller's experiments on Drosophila in the 1920s, which showed that genetic material can be altered. This discovery excited geneticists and opened up novel ways for both fundamental and applied purposes (Chopra 2005). Induced mutagenesis was actively pursued in some regions including the United States, Europe and Asia during the 1950s. Since then, induced mutagenesis has been a potent and extensively applied tool in widening the gene pool by creating genetic variations quickly (Mandal and Datta 2014). Experimentally induced mutations have made significant contributions to global agriculture, benefiting over 175 crop species.

Many mutants in jute have been generated at various parts of the plant, including leaf, stem, flower, and bast fiber tissue. Jacob and Sen (1961) obtained one X-ray-induced haploid with a very short and slender stem in the X_3 generation of *C.olitorius* and harvested three normal seeds. The frequency and types of chromosomal abnormalities in the X_3 and X_4 generations were studied, indicating that the inheritance of the mutant characteristic was regulated by recessive genes (Basu 1965). Mahapatra et al. (2006) constructed a mutant library of Tossa comprising 56 mutants through physical or chemical mutagenesis, which were isolated from either M_2 or M_3 populations. One of the mutagenesis 'JRO 632' (library No. OMU 043, induced by thermal neutron) (Joshua et al. 1974) and its wild type were used as experimental

Fig. 13.12 Phenotypes of 'JRO 632' mutant (**a**) and its wild-type (**b**) 70 d stage plant and stained tissues of three locations of the stem at 120 d. There are obvious differences in the FCB between the mutant and wild-type. FCB fiber cell bundle, PPF primary phloem fiber, SPF secondary phloem fiber. (Source: Kundu et al. (2012a))

subjects to study the development of secondary phloem fiber. The results showed that the mutagenesis had an abnormal phenotype characterized by dwarfed growth and distinctive, unusual leaf shape, lower plant, less yield, and poorer fiber quality compared to its parental wild-type (Fig. 13.12) (Kundu et al. 2012a). We used ethylmethane sulfonate (EMS) resolution and Co^{60}-gamma radiation to treat seeds of tossa jute. Different damage was observed on jute leaves with the former causing leaves to crinkle and the latter causing leaves to fork (although not as seriously forked as Kundu's results). When treated with both methods, there was an enhancement in the contents of proline and malondialdehyde (MDA), which increased root activity as well. However, the variation in content of soluble proteins, activity of superoxide dismutase (SOD) and peroxidase (POD) differed among varieties (Wen et al. 2014).

Mutation through physical or chemical methods often results in disadvantages, harm, or even death in the subsequent generations. Developing a new variety that aligns with breeding objectives requires extensive subsequent selections and sometimes hinges on luck. According to the International Atomic Energy Agency, a strain named C-278, induced by sodium azide through early sowing and late flowering, exhibited a 7.5% increase in fiber yield compared to the mother variety CVL-1 and a 20% increase compared to a local cultivar, D-154. Nura et al. (2011) reported successfully obtaining a mutant of tossa jute with significantly enhanced traits including increased plant height, internode length, leaf amount and area, contributed to improved yield and quality. They achieved this by treating the jute with colchicine at various concentrations. In a separate study, a novel EMS mutant was discovered, which had a shorter lifespan compared to its ancestor. This mutant flowered at 70 days instead of the usual 120 days. Molecular analysis revealed four single nucleotide polymorphisms (SNPs) and two group deletions in the *ELF3* (Early Flowering 3) genes generated from the stable M_4 population, resulting in a genotype with reduced lignin content (Hossen et al. 2022).

13.5 Molecular Breeding

13.5.1 Molecular Marker-Assisted Breeding

To illustrate genetic polymorphism and relationships within a species, we need strategies for collecting and utilizing germplasm efficiently to improve jute breeding. Mir et al. conducted a study that included 45 and 36 genotypes of Tossa and white jute respectively to assess the genetic polymorphism by SSR markers. The results indicated that fiber yield and the related traits exhibited a quantitative nature, with dominant heredity in genetic variance. Additionally, the researchers discovered a high degree of transferability of SSRs from Tossa to white jute (Mir et al. 2007). Sarkar et al. (2016) developed and validated EST-SSR markers from transcriptomic unigenes, discovered SNPs in the whole genome by *de novo* sequencing, and constructed a high-density fingerprint, detected the quantitative trait loci (QTL) and conducted analysis of candidate genes influencing bast fiber yield and its constituent. DNA fingerprints were constructed for 58 jute accessions (including 12 accessions of tossa jute) using 67 SSR markers. Based on the SSR data (Zhang et al. 2015b). In total sixteen accessions of tossa jute were clustered into 3 clades, cultivars originating from Ghana and Togo exhibited distinct characteristics when compared to those from Nigeria. These results were consistent with the morphological traits and quantitative traits using principal component (PC) methodology. It uncovered the genetic variability within cultivated Tossa, which could aid in germplasm assessment and breeding programs. (Adeyemo et al. 2021).

A few DNA markers were used for early hybridization detection to promote the breeding process in jute. The cpDNA-RFLP marker(restriction fragment length polymorphism) was employed for describing the characteristics of the somatic hybrid lines during the initial growth stages, accelerating the identification and improving the sensitivity for screening a large number of individuals (Saha et al. 2001). A QTL analysis was performed on *C. olitorius* to investigate the genetics of yield and fiber quality traits, utilizing a population map comprising a genetic link map with 36 SSR markers and 120 recombinant inbred lines, which may be useful for selecting and accelerating new cultivar breeding (Das et al. 2012). Marker-assisted breeding techniques are free of the policies that hinder the application of genetically modified technologies. They can track and monitor specific regions of the genome during crossing and selection at the DNA level (Tester and Langridge 2010).

13.5.2 Genomics and Multi-Omics

Molecular research on jute is still in its initial stages, with a small quantity of identified genes and slow progress (Kundu et al. 2015; Mir et al. 2007; Tao et al. 2017). Two published genomes of *C. olitorius* exhibit discrepancies in genome size and gene count, with one reported at 84 Mb in size with 8552 genes (Islam et al. 2017), while the other is reported at 445 Mb with 37,031 genes (Islam's) and 361 Mb with 28,479 genes (Zhang et al. 2021). There is still potential to enhance the comprehensiveness of these genome assemblies. Some candidate genes have been identified and utilized in jute. For instance, a *LRR-RLK* gene from *C.olitorius* was sequenced and found to be involved in pathways associated with stress response (Basu et al. 2004). In addition, it was demonstrated that transgenic jute plants carrying the bacterial *katE* gene exhibited enhanced tolerance to oxidative stress under saline conditions, attributed to the upregulation of enzymes that scavenge reactive oxygen species (ROS). However, several limitations need to be addressed for genetic assessment, especially for advanced molecular breeding in Tossa (Islam et al. 2013). These limitations include insufficient molecular markers to cover the entire genome, limited diversity in the types of accessions studied, small sample sizes, and fragmented studies that lack integration of multiple methods. It is crucial to conduct supplementary research on genomic diversity and gene expression in Tossa (Sarkar et al. 2019) and develop a superior reference genome for Tossa jute to delineate genomic variations (Yang et al. 2022). Therefore, we have produced a novel, high-quality assembly through the resequencing of 256 accessions of Tossa, comprising wild types, landraces, strains, and germplasm resources, and the phenotypes of 11 agronomic traits were carried out in three different environments. Moreover, we designed sequencing for transcriptomes and epigenomes, including RNA, microRNA, WGBS, and m6A RNA methylation analysis. The genetic variability, population shifts, and domestication-related genomic regions were studied by integrating the above multi-omics results. Among GWAS genes, 2 upregulated APX family genes (*COS05g_02894, COS04g_00516*) and a monodehydroascorbate reductase ortholog (*COS04g_00816*) were regulated in multi-omics data (Fig. 13.13, a,b). Analysis of transcriptome identified 31 differentially expressed genes in 'TC008-41' and 'NY/252c', which were notably enriched in the abscisic acid (ABA) pathway, specifically, 23 of these genes directly facilitated ABA storage and biosynthesis, initiated by b-carotene (Fig.13, c,d) (Yang et al. 2022). In addition, *CcPRR7*, a key circadian gene at *qFT-3-1*, was assessed via linkage, haplotype, and

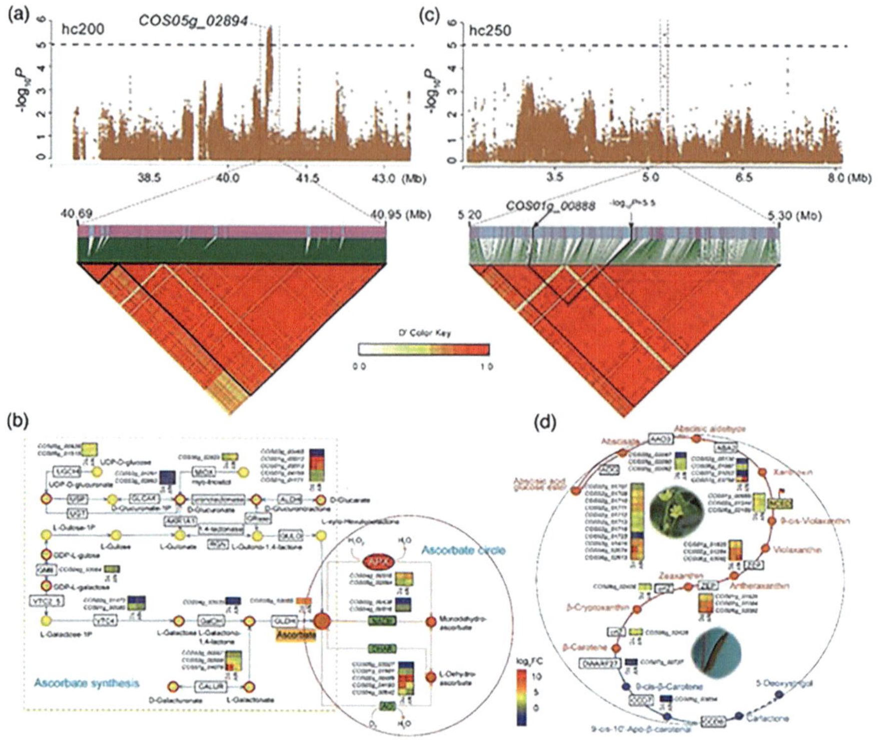

Fig. 13.13 A genome-wide association study (GWAS) of salt tolerance, ascorbate metabolisms and ABA pathway. (**a**, **c**) Manhattan plot for salt tolerance GWAS regions on chromosome 5. (**b**) Graphical description of the ascorbate metabolic pathways. (**d**) Graphical description of the ABA biosynthesis pathway. (Source:Yang et al. (2022))

transgenic analysis, confirming its role in controlling jute flowering time through expression regulation (Yao et al. 2023). These results could contribute to a deeper comprehension of the genetic determinants of significant agronomic characteristics and provide valuable information for genetic engineering in jute, facilitating future molecular breeding.

13.5.3 Bioinformatics

Bioinformatics is an interdiscipline that links biology with data information storage, and analysis to support multiple areas of scientific research. Bioinformatics is more and more important in the wake of high-through sequencing, and analysis of genomic, and gene expression patterns. For example, integrating multi-omic data, which encompasses genomic, epigenetic, transcriptomic, and/or proteomic information, could enhance analysis by providing a comprehensive assessment of variant

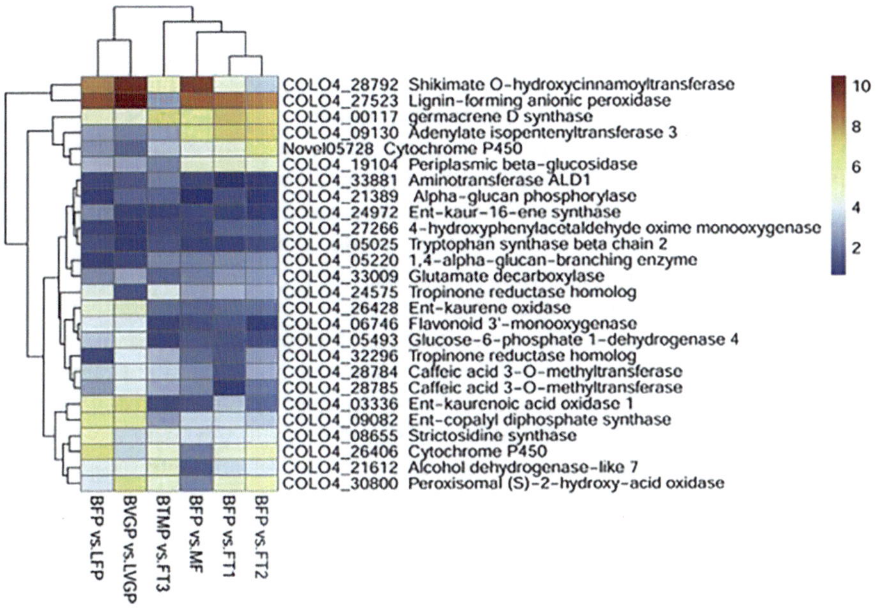

Fig. 13.14 Genes constantly upregulated in the bark in the whole growing period. (Source:Yang et al. (2017a))

effects (Zech and Winkelmann 2024). This will generate a mass of data, we might feel helpless without bioinformatics.

A bioinformatics-based transcriptome analysis was conducted for the first time across various organs of tossa jute, aiming to identify DEGs and the metabolic pathways within these tissues (Yang et al. 2017a). The metabolic pathway responsible for "biosynthesis of secondary metabolites" was notably enriched in bast with 26 genes consistently upregulated (Fig. 13.14). These results provided clues for revealing the roles of specific genes that contribute to the development and growth of tissues in natural fiber crops Furthermore, 11,100 unigenes, comprising 385 transcription factors (TFs), showed significant differential expression in both salt-tolerant and sensitive genotypes, revealing numerous common and distinct DEGs between them (Yang et al. 2017b). Furthermore, high-throughput transcriptome sequencing was conducted in saline conditions, and 127 common DEGs were identified several terms of Gene Ontology (GO) were enriched by a series of data treatments under bioinformatics science (Yang et al. 2017c).

On another hand, to identify calcium-dependent protein kinases (*CDPK)* genes in jute, bioinformatics was employed by using existing *CDPK* protein sequences from the genome databases of arabidopsis, cacao and grape. The prominent transcriptional responses were detected due to over half of the *CDPK*s being upregulated by saline and drought stress, suggesting *CDPK*s play a significant role in environmental stress responses (Ahmed et al. 2020).

13.5.4 Genetic Engineering and Gene Editing

For many crops, scientists largely take measures to deliver foreign DNA to the explants directly and, get the regenerated plant from the callus of parents. However, the same method may not be in *Corchorus*, because many of the cultivars cannot be regenerated in vitro, what is worse, it seems harder for tossa jute than for white jute (Fang et al. 2022). Some thought that it was impossible to achieve the transgenic plants of Tossa. Nevertheless, difficulties never stop scientists from trying.

To improve the resistance to diseases and herbicides, the *chitinase* (*chi11*) gene (introduced from rice, to confer resistance to fungi) and the *bar* gene (cloned from *S. hygroscopicus*, to provide herbicide tolerance) were incorporated in jute via *Agrobacterium* transformation, the two genes have integrated into the genome and expressed stably up to T_2 generation of jute. The transgenic plants offer substantial protection against *M. phaseolina* in comparison to the wild type, and the efficacy against both of the common pests (jute hairy caterpillar and semi-looper) was reported to be up to 100% death. Moreover, no distinction was observed between the transgenic plants and the wild type in fiber quality (Majumder et al. 2018).

Another study for expressing δ-Endotoxin *Cry1Ab/Ac* aimed at resistance to pests(*Lepidopteran*) showed that, among transgenic feeders, semi-looper mortality ranged from 66% to 100% and hairy caterpillars were 87.5% death. The transgenic jute lines with *CcUGPase* showed accelerated growth and a notable increase in plant height (Zhang et al. 2013).

A study on cutting down lignin in Tossa (var. O-9897) was conducted, by implementing amiRNA-based vectors to suppress the expression of two genes (*C3H* and *F5H*) involved in monolignoid biosynthesis of jute, the lignin content reduced about 25% in stalk and 12–15% in fiber respectively (Shafrin et al. 2015). In the following studies of Shafrin, by using hpRNA-based vectors for downregulating against the lignin each of the biosynthetic genes (*C4H* and *COMT*) also reduced 14% lignin content in independent transformation activities, meanwhile, in the *C4H*-lines, there was also a 3.5% increase in the cellulose content (Shafrin et al. 2017).

Recently, a procedure for puncturing imbibed seeds via streamlined *Agrobacterium*-mediated transformation was developed and detailed. Transformation efficiencies achieved reached 20.4% and 15.6%for Tossa and white jute, respectively (Majumder et al. 2024). This will be very useful for further studies and practical applications for transformation.

In terms of gene editing, *KCS*(3-ketoacyl-CoA synthase) gene editing vectors were introduced into jute through *Agrobacterium tumefaciens*-mediated transformation (ATMT), resulting in downregulation of the *KCS* gene. This downregulation led to a decrease in chloroplast stromules, lignin content, and stem diameter, which indicated the involvement of the jute *KCS* gene in additional processes pertinent to its growing program (Zhang et al. 2019).

13.5.5 Tissue Culture Applications

The process of in vitro regeneration poses challenges for *Corchorus*, after a series of failures, advancements have been realized in jute by the application of biotechnology, encompassing somatic hybridization through protoplast fusion, ATMT, and the subsequent regeneration of plants. The specific *C. olitorius* varieties, namely O-72 and O-9897, were exclusively amenable to successful transformation when employing *in planta* techniques (Sajib et al. 2008; Shafrin et al. 2015). A procedure for imbibed jute seeds piercing by ATMT was described in detail step by step, including how to prepare the culture media for Agrobacterium cells, how to make the 'Infiltration and Co-cultivation medium', how to prepare the 'Agro-washing' medium, and how to prepare the selection medium and agarose gel (Majumder et al. 2024).

13.6 Tossa Jute for Non-Fiber Application

13.6.1 For Food and Medicines

Independent of the conventional utilization of fiber, the tender tips and leaves of jute can be eaten as natural food and processed food, or further produced into health care and skin care products. The breeding objectives for edible jute focus on the leaf yield, nutrition, flavor, and resistance to pests and diseases which are different from fiber-focused breeding. The IBFC, Fujian Agriculture and Forestry University (FAFU), Guangxi Academy of Agricultural Sciences (GAAS) and other scientific research institutions have engaged in the research of edible jute varieties, independently or cooperatively. A series of varieties were bred, e.g., 'Diwangcai 1hao', 'Diwangcai 2hao', 'Wangcai 3hao', 'Funong 1hao', 'Funong 2hao', 'Guimacai 1hao', and 'Guimacai 2hao'. The tender shoot yield, nutrition, resistance to insects and disease, taste and appearance of cooked food were comprehensively evaluated (Chen et al. 2022). The nutritional and pharmacological components of edible jute were collected and tidied in Table 13.3.

More and more experiments have shown that jute consumption is not only high in nutritional value but also has good effects in antioxidation (Ademiluyi et al. 2015; Qasim et al. 2017; Yusuff et al. 2019), resisting antibiotics and protecting probiotics in the gut (Deng et al. 2023), scavenging free radicals in the body (Adekale et al. 2019), hypoglycemic and hypolipidaemic activities (Awosanya et al. 2019), anti-inflammatory effects (Handoussa et al. 2013; Yan et al. 2013), pain relief (Saliu et al. 2019), prevention of hypertension and obesity (Adekale et al. 2019) and inhibiting proliferation of cancer cell (Guzzetti et al. 2021). *Corchorus* especially tossa and white jute has been used as a remedy for a long time in many tribes and cultures. Traditional Chinese medicine (TCM) theory holds that "Medicinal and Edible" means derived from food medicine, as a therapeutic approach that benefits the body

Table 13.3 Nutritional and pharmacological components of edible jute (mg/100 g)

Components	*C.olitorius*	*C.capsularis*	*C.aestuans*	*C.bougoudo*	*C.urtifolicus*
Amino acids	6643.00	4430.00	–	–	–
DF(dietary fiber)	6080.00	3170.00	–	–	–
Kalium	674.00	441.00	–	–	–
Calcium	445.00	182.00	–	–	–
Total phenols	500.00	482.00	325.00	418.00	403.50
Total flavonoids	82.80	78.30	78.30	47.90	82.70
Vitamin C	99.40	95.10	101.80	106.60	115.60
Vanillic acid	0.16	0.07	0.14	0.08	0.11
Caffeic acid	1.86	1.58	1.21	0.91	1.86
Ferulic acid	1.79	1.54	1.12	1.12	1.79
Apigenin	2.04	2.24	2.36	2.14	2.04
Kaempferol	4.13	4.28	3.74	3.61	4.13
Luteolin	2.21	3.03	2.70	2.22	2.21
Quercetin	1.59	2.27	2.18	2.10	1.59
Myricetin	0.42	0.60	0.68	0.52	0.42
Rutin	2.44	2.79	2.07	2.04	2.44
Gingerol	0.05	0.06	0.09	0.02	0.01

Source: (Ademiluyi et al. 2015; Chen et al. 2022; Li et al. 2010; Saliu et al. 2019; Zhao et al. 2017)

without any side effects. A health survey in Africa also showed that where edible jute is commonly consumed, the incidence of gastrointestinal fever is lower than in other regions (Zakaria et al. 2009). As lifestyle diseases emerge as a matter of global concern, *Corchorus* is considered a crop with health-promoting properties that can enhance both community health and the farmer's income in areas where it is planted and consumed.

13.6.2 For Environment Protection

Heavy metal pollution is a significant environmental concern for the public. Heavy metals may induce cancer, damage bones, impair kidney function, and cause other adverse reactions (Liu et al. 2015). Adsorbents utilized in practical adsorption treatments are in high demand for their exceptional adsorption capacity, rapid reaction and precipitation rates, reusability, and renewability, as well as low energy consumption and cost-effectiveness.

Zhou et al. (2018) designed a porous and high adsorptional Jute/Polyacrylic acid (Jute/PAA) gel (jute materials provided by IBFC, a cultivar of *C. olitorius*) for treated heavy metal pollution in water treatments. The Jute/PAA gel with high permeability and made a mass of adsorption sites fully exposed to heavy metal ions, which could adsorb these compounded ions efficiently in an aqueous solution. The maximum adsorption capacities were up to 401.7 mg/g and 542.9 mg/g for Cd^{2+} and Pb^{2+}, respectively. Furthermore, the Jute/PAA gel, which had adsorbed metal ions, was eluted by an acidic solution (such as HCl) and subsequently regenerated for further reuse. At 400 mg/L, Cd^{2+} and Pb^{2+} removal efficiencies were 83% and 96% initially, staying at 81% and 94% after five cycles. This excellent recyclability is due to the gel's stable 3D double-network structure. Jute fibers usually adsorb 8.43 mg/g aniline, and PANI/jute composites, formed by surface polymerization, effectively remove metal ions. The maximum adsorption capacities for Cd (II) and Cr (VI) were 140 mg/g and 50 mg/g, respectively (Huang et al. 2019). Jute germplasm and varieties were selected for this special use for heavy metal ions adsorption, some of the germplasm with high adsorption features, such as 'Zhonghuangma 4hao', 'M101', 'HMG-1' for the positive ions and 'Zhuchangma', 'HMG-4' for the negative ions adsorption. Furthermore, the wastewater containing Cr (VI) was treated with natural jute powder with different particle sizes from different parts of the plants to investigate the adsorption effects, we found that the leaves and tender tips of the jute plant were the best tissue for making heavy metal adsorbents (Li et al. 2015). At the same time, breeding objectives were focused on high biomass yield, more branches and larger leaves, instead of high fiber yield and no branches, some of the tossa jute germplasm were selected with high adsorption capacity (185.09 mg/g), high biomass yield and leaf ratio, such as J001 (from Bangladesh) and J011 (from Hunan Province, China) (Li et al. 2014).

Surprisingly, white jute and tossa jute perform significant disparity in the adsorbability and efficiency of heavy metals ions, although there are scarcely any differences in the contents of cellulose, hemicellulose, lignin and the physical structure in natural plants and the adsorbents made from the two species of plant. Active functional groups (-OH, -COOH, -NH) related to adsorption were considered the main cause of the adsorbability. The interaction between adsorbents and metal ions was analyzed using the instrument of Fourier-Transform Infrared Spectroscopy (FTIR). The adsorption process adhered to the second-order kinetic reaction and followed the Langmuir adsorption isotherm model (Hassan and Zohdy 2018). On the other side, the differential adsorption may be related to the electronegativity of metal ions, ionic radius and the distinction of nucleo-plasmic ratio, so they have different coordination capabilities with the adsorption sites (Deng et al. 2020).

13.7 Production and Cultivation

In 2021, the total production of jute raw or retted fiber was approximately 3473 kilo-tons, accounting for 68% of the production of bast fibers (Fig. 13.15). Jute cultivation relies on enough land, suitable climate, abundant water, low cost of labor and large demand of the social economy. India and Bangladesh meet these conditions, so they produce the most jute, considerably higher than other countries, which almost accounts for 98% of the world jute production (Fig. 13.16). Over the longer

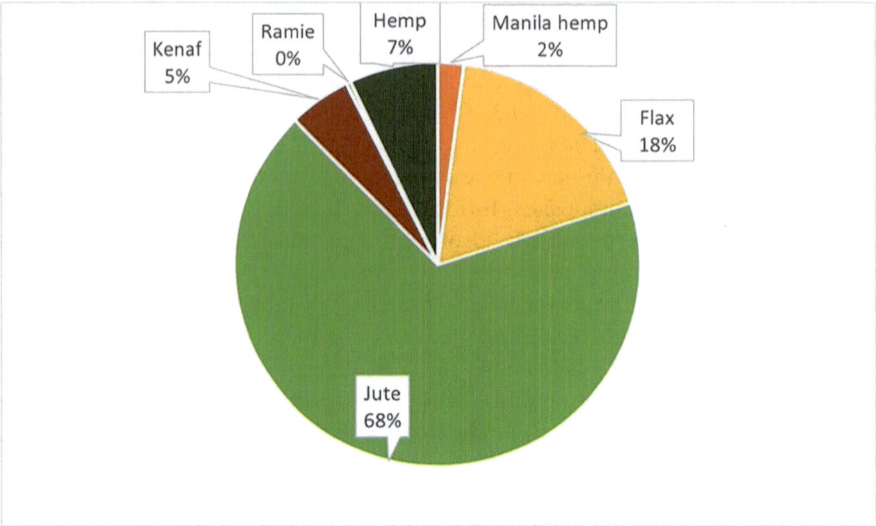

Fig. 13.15 Constitutions and proportions of the bast fiber and hard fiber in 2021. (Source: (http://www.fao.org/statistics))

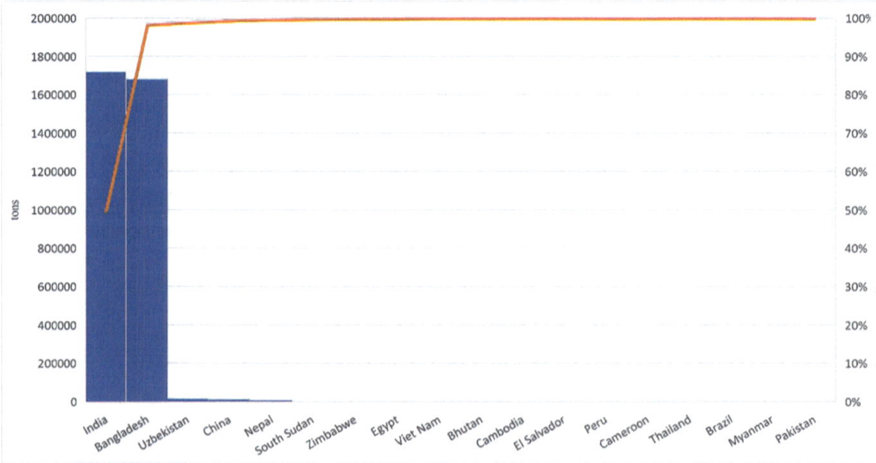

Fig. 13.16 The main production countries of jute in 2021. (Source: (http://www.fao.org/statistics))

term (1961–2021), India, Bangladesh and China, were the three main cultivating countries, producing 86.6 million tons, 65.7 million tons and 9.3 million tons, respectively. About 36.0% of the fiber produced in Bangladesh was exported while the proportion was extremely low in India and China (Fig. 13.17) (http://www.fao.org/statistics). It means that India and China have a complete processing system and higher processing capacity for raw jute fiber.

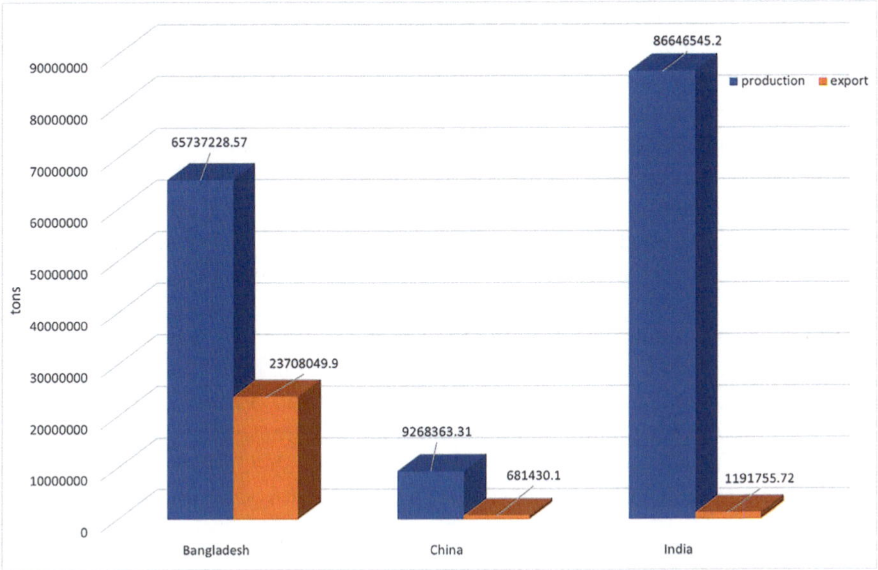

Fig. 13.17 Production and export quantity of jute fiber of the top three countries from 1961 to 2021. (Source: (http://www.fao.org/statistics))

13.8 Conclusions and Prospects

The traditional and ongoing large-scale utilization of jute primarily involves the production of fiber materials, ropes, sacks, mats, rugs, cloth, decorations, and handicrafts. However, due to competition from chemical fibers and other natural fibers, the cultivation area for jute is declining, and farmers' enthusiasm for jute cultivation is waning. To address these challenges, it is imperative to develop higher value-added products, efficient harvest and decorticator machines, and new specialized jute varieties.

To expand the applications, we must adhere to an Ecology-Oriented Development (EOD) model, optimize the allocation and comprehensive utilization of natural resources, enhance regional competitiveness, establish jute processing industrial clusters, and make comprehensive use of by-products. To meet the demands and challenges, enhanced jute cultivars must exhibit certain characteristics, including superior quality (higher fiber count, strength, and flexibility), high yield (fiber, leaf, or biomass, depending on the intended use), and resistance to non-biological stresses such as drought, high salinity, and exposure to heavy metals. To fulfill these diverse requirements, breeders must explore various germplasms and develop more effective breeding approaches. Furthermore, it is essential to develop new cultivation systems and optimize different harvest objectives, as the cultivation system for fiber harvest differs significantly from that for vegetable or biomass harvest.

Unfortunately, bast fiber crops like jute have received less attention compared to food crops. Tossa jute holds great potential for breeding improvement, the global collection of germplasm from different ecotopes and diverse species of jute can significantly increase productivity and fulfill breeding objectives. Exploring genetic variability within *Corchorus* through molecular analysis of jute can provide precise information for finding target traits and integrating genes into new varieties using modern methods such as molecular analysis, hybridization, gene editing, and transgenesis. Regardless of changing requirements, fiber yield and quality remain the most crucial traits because they have a substantial global influence on textile manufacturing.

The accomplishment of sequencing the genomes of the two mainly jute species marks a significant milestone, placing jute breeding at the forefront of the genomic era. Accurate germplasm identification, the discovery and verification of new functional genes or transcription factors, advancements in transgenic and gene editing technology, the development of bioinformatics, and collaboration between breeders and researchers will drive a substantial revolution in jute improvement and utilization.

Appendix 13.1 Main Research Organizations for Jute Breeding

Organization	Research field	Activities	Web-site
Institute of Bast Fiber Crops, Chinese Academy of Agricultural Sciences, Changsha 410205, China	1. Germplasm collection, preservation and evaluation of bast fiber crops 2. Bast fiber crop improvement 3. Cultivation, disease prevention, harvest machine 4. Bio-degumming, fiber test, and multiple utilization	1. Varieties release:Zhonghuangma 1hao to 15hao, Edible jute Diwangcai 1hao to 6hao 2. Breeding for stress resistance 3. Development of multiple use	http://www.chinaibfc.com/
Fujian Agriculture and Forestry University, Fuzhou 350002, China	1. Improvement, breeding of jute and bast fiber crops 2. Comprehensive utilization of fiber, stem, and seed	1. Varieties release: such as Huangma 179, Meifeng 4 2. Whole genome sequence of jute	https://www.fafu.edu.cn/

Organization	Research field	Activities	Web-site
Bangladesh Jute Research Institute, Dhaka 1207, Bangladesh.	1. Preservation and evaluation jute, kenaf and related fiber crops 2. Improvement and breeding	1. Variety breeding of jute and allied fibers 2. Seeds supply and planting organize 3. Train farmers for cultivation	https://www.bjri.gov.bd/
Central Research Institute for Jute and Allied Fibres, Arrack Pore, Kolkata 700120, India	Jute and other bast fiber crops, kenaf, roselle, sunn hemp, ramie, sisal and flax	1. Improvement of jute and allied crops 2. Economical and sustainable planting technology	http://www.crijaf.org.in/

Appendix 13.2 Varieties Released of Tossa Jute (*C. olitorius*) in The World

Variety	Origin or pedigree	Bred (introduced) organization/ country	Released/ introduced time
Cuilv	Introduced from India	India, China	1919
Zhema 1	Selected and improved from Cuilv	ZJAAS, China	1955
Yuanjiang 101	Selected and improved from Cuilv	IBFC, China	1963
GFCG	Selected from the landrace	IBFC, China	1965
Bachang 4	Introduced from Pakistan	Pakistan	1966
Changguo 134	Selected from Yuanjiang 101	IBFC, China	1970
Bama 72–1	Variety of Pakistan	Pakistan	1972
Bama 72–2	Variety of Pakistan	Pakistan	1972
Bama 72–3	Variety of Pakistan	Pakistan	1972
Changguo 751	Mutant from GFCG	ZJAAS, China	1972
Tupihuang	Selected from F_1 (GFCG× Jiegan No. 1) treated with ^{60}Co-γ ray	IBFC, China	1972
Guangbaai	GFCG × Bama 72–2	IBFC, China	1973
Xianghuang 2	Selected from GFCG	IBFC, China	1975
Kuanyechangguo(KYCG)	GFCG × Bachang 4	IBFC, China	1980
Xianghuangma 1	(Maliyeshengchangguo(MLYS) × Guangbaai) × MLYS	IBFC, China	1986
Xianghuangma 2	Bama 72–2 × KYCG	IBFC, China	1990
Xianghuangma 3	KYCG treated with ^{60}Co-γ ray	IBFC, China	1997
Zhonghuangma 4	Xianghuangma 1 × Bachang 4	IBFC, China	2007
Funong 1	Taizi 4 treated with ^{60}Co-γ ray	FAFU, China	2009
Zhongyinhuangma 1	The family selected from O-1	IBFC, China	2009
Funong 4	Cuilv × Bama 72–3	FAFU, China	2010

(continued)

Variety	Origin or pedigree	Bred (introduced) organization/ country	Released/ introduced time
Diwangcai 1	The family selected from the generations of Tianhuangma	IBFC, China	2012
Diwangcai 2	The family selected from the variant single plant of KYCG treated with ^{60}Co-γ ray	IBFC, China	2012
Zhonghuangma 3	089–1 treated with aerospace mutagenesis	IBFC, China	2013
Zhonghuangma 6	The family selected from the genarations of Mowei 1× Xianghuangma 3	IBFC, China	2015
Zhonghuangma 7	The family selected from the variant single plant of KYCG treated with ^{60}Co-γ ray	IBFC, China	2016
Zhonghuangma 9	Mowei 1× Xianghuangma 3	IBFC, China	2017
Zhonghuangma 12	The family selected from the variant single plant of Mowei 1	IBFC, China	2018
Zhonghuangma 10	Zhongyinhuangma 1 treated with ^{60}Co-γ ray	IBFC, China	2018
Diwangcai 3	Tianhuangma × MLYS	IBFC, China	2018
Caiyonghuangma 6	Xianghuangma 3 × O-3	IBFC, China	2018
Zhonghuangma 13	Xianghuangma 1 × Mowei 1	IBFC, China	2019
Wangcai 1	O-3 × Mowei 1	IBFC, China	2019
Diwangcai 4	Tianhuangma × Mowei 1	IBFC, China	2019
JRO 632 (Baisakhi)	Selection from indigenous	India	1954
JRO 878 (Chaitali)	JRO 620 × Sudan green	India	1974
JRO 7835 (Basudev)	JRO 632 × Sudan green	India	1974
JRO 524 (Navin)	Sudan Green×JRO 632	India	1977
TJ 40 (Mahadev)	A cross between mutants of JRO 632	India	1983
JRO 3690 (Savitri)	A cross between mutants of Tobacco leaf	India	1985
KOM 62 (Rebati)	Irradiation of JRO 878 by Gamma-ray	India	1993
JRO 66 (Golden Jubilee)	Multiple parent crosses of CG, TM, JRO 524	India	1998
JRO 8432 (Shakti)	IC15,90 × Tanganyika l	India	1999
JRO 128 (Surya)	TJ-6 × Tanganyika 1	India	2002
S19 (Subala)	JRO 620 × Sudan Green	India	2005
JRO 204 (Suren)	IDN/SU/053 × KEN/DS/060	India	2007
AAU OJ 1 (Tarun)	Selected from S2 individuals of Tanganyika-l × JRO 640	India	2007

Variety	Origin or pedigree	Bred (introduced) organization/ country	Released/ introduced time
JBO 2003H (IRA)	JR0632 × Sudan Green	India	2008
CO 58 (Sourav)	TJ 40 × Tanganyika 1	India	2010
JBO 1 (Sudhangsu)	JRO 632 × Sudan Green	India	2010
JROM 1 (Pradip)	JRO 524 × TAN/NY/018C	India	2013
JROG 1 (Rithika)	JBO 1 × JRO 524	India	2015
JRO 2407 (Samapti)	KEN/SM/024 × JRO 524	India	2016
KRO 4 (Gouranga)	OIM 028 × JBO 2003H	India	2017
BCCO 6 (Kisan Pat)	Selected from OEX 05	India	2017
NJ 7010 (Rani)	Selected from JRO 524 by EMS treated	India	2018
JROMU 1	Selected from JRO 204 by gamma irradiated	India	2020
JROB 2 (Purnendu)	JRO 204 × JRO 524	India	2021
Chinsura Green (D-38)	Selecting continuously pure lines	Bangladesh	1915
R 26	Selecting continuously pure lines	Bangladesh	1929
R 27	Selecting continuously pure lines	Bangladesh	1929
O-620	Selecting continuously pure lines	Bangladesh	1939
O-632	Selecting continuously pure lines	Bangladesh	1939
O-753	Selecting continuously pure lines	Bangladesh	1939
O-1	Selecting continuously pure lines	Bangladesh	1955
O-2	Selecting continuously pure lines	Bangladesh	1955
O-3	Selecting continuously pure lines	Bangladesh	1955
O-4	Selecting continuously pure lines	Bangladesh	1967
O-5	Selecting continuously pure lines	Bangladesh	1964
O-9897	O-5 × BZ-5	Bangladesh	1987
OM 1	Selecting continuously pure lines	Bangladesh	1995
O-72 (BJRI Tossa Pat 4)	O-9897 × O-2021 × O-9897	Bangladesh	2002
O-795(BJRI Tossa Pat 5)	O-4 × Uganda Red	Bangladesh	2008
O-3820 (BJRI Tossa Pat 6)	Selecting continuously pure lines	Bangladesh	2013
MG-10 (BJRI Tossa Pat 7)	Selecting continuously pure lines	Bangladesh	2017

Source: (Dai et al. 2021; Gong et al. 2013; Zhang et al. 2015a)
IBFC Institute of Bast Fiber Crops, *CAAS* Chinese Academy of Agricultural Sciences, *FAFU* Fujian Agriculture and Forestry University, *ZJAAS* Zhejiang Academy of Agricultural Sciences

Appendix 13.3 Genetic Resources for Tossa Jute (*C. olitorius*) Breeding

Species	Accessions name or No.	Important traits	Cultivation location
C.olitorius L.	KUAN YE CHANG GUO	High yield, green stem	Yangzi River Area, China
C.olitorius L.	XIANG HUANG MA 1HAO	High yield, green stem	Yangzi River Area, China
C.olitorius L.	YUN YE II-1	Wild type, resistant to diseases	Southwest Area, China
C.olitorius L.	YUN YE II-3	Wild type, resistant to diseases	Southwest Area, China
C.olitorius L.	TIAN HUANG MA	For vegetables, high calcium	Yangzi River Area, China
C.olitorius L.	WANGCAI 1HAO	For vegetables, high calcium	Yangzi River Area, China
C.olitorius L.	O-3	High yield, green stem	China, India, Bangladesh
C.olitorius L.	O-4	High yield, ideotype	China, India, Bangladesh
C.olitorius L.	O-6	High yield, green stem	China, India, Bangladesh
C.olitorius L.	TC008–41	Resistant to salt stress	Salted Area, China
C.olitorius L.	ACC-NO-3708	High yield, green stem	Bangladesh
C.olitorius L.	ACC-NO-3711	High yield, green stem	Bangladesh
C.olitorius L.	NA QIN HUANG MA	Dark red stem	South of China
C.olitorius L.	HOU YE LV	Thick leaves	South of China
C.olitorius L.	GUANG XI CHANG GUO	Local germplasm	South of China
C.olitorius L.	MA LI YE SHENG ZHANG GUO (HONG GAN)	High yield, red stem	Africa
C.olitorius L.	MA LI YE SHENG ZHANG GUO (LV GAN)	High yield, green stem	Africa
C.olitorius L.	K-83	Northern germplasm	Russia
C.olitorius L.	FU NONG 1HAO	For vegetables, high calcium	China
C.olitorius L.	RI BEN CHANG GUO	High yield, green stem	Japan and Asia
C.olitorius L.	BL/014	Wild type	Kenya
C.olitorius L.	BL/093	Wild type	Kenya
C.olitorius L.	BL/096	Wild type	Kenya
C.olitorius L.	Y/072CO	Wild type	Thailand
C.olitorius L.	Y/078CO	Wild type	Thailand
C.olitorius L.	ZHE ZHANG GUO 1HAO	High yield, resistant to diseases	China
C.olitorius L.	ZHE ZHANG GUO 5HAO	High yield, resistant to diseases	China

Species	Accessions name or No.	Important traits	Cultivation location
C.olitorius L.	BAMA 72–1	High yield, resistant to diseases	Pakistan
C.olitorius L.	CHANGGUO 134	High fiber fineness	China
C.tridens L.	ACC-NO-3716	Distant hybridization	Bangladesh
C.trilocularis L.	ACC-NO-4257	Distant hybridization	Bangladesh

References

Adekale OA, Awosanya OO, Ogbuagu U et al (2019) Ameliorative efficacy of phytochemical content of *Corchorus olitorius* leaves against acute ethanol-induced oxidative stress in Wistar Rats. Asian J Biochem Genet Mol Biol 2(2):1–10

Ademiluyi AO, Oboh G, Aragbaiye FP et al (2015) Antioxidant properties and in vitro α-amylase and α-glucosidase inhibitory properties of phenolics constituents from different varieties of *Corchorus* spp. J Taibah Uni Med Sci 10:278–287

Adeyemo OA, Ayodele OO, Ajisafe MO et al (2021) Evaluation of dark jute SSR markers and morphological traits in genetic diversity assessment of jute mallow (*Corchorus olitorius* L.) cultivars. South Afr J Bot 137:290–297

Ahmed B, Alam M, Hasan F et al (2020) Jute CDPK genes and their role in stress tolerance and fiber development: a genome-wide bioinformatic investigation of Chorchorus *capsularis* and C. *olitorius*. Plant Gene 24:100252

Al-Khayri JM, Johnson SMJDV (2019) Advances in plant breeding strategies: industrial and food crops, vol 6. Springer Nature Switzerland AG, pp 85–113

Alves C, Ferrão PMC, Silva AJ et al (2010) Ecodesign of automotive components making use of natural jute fiber composites. J Clean Prod 18:313–327

Awosanya OO, Ogbuagu U, Airaodion EO et al (2019) Effect of methanolic extract of *Corchorus olitorius* leaves on hypoglycemic and hypolipidaemic activities in Albino Rats. Asian Plant Res J 2(4):1–13

Basu RK (1965) Radiation induced morphological mutant of jute. J Genetics 59:60–69

Basu A, Ghosh M, Meyer R et al (2004) Analysis of genetic diversity in cultivated jute determined by means of SSR markers and AFLP profiling. Crop Sci 44:678–685

Benor S (2018) Molecular phylogeny of the genus *Corchorus* (Grewioideae, Malvaceae s.l.) based on nuclear rDNA ITS sequences. Crop J 6:552–563

Benor S, Hammer K, Demissew S, Blattner FR (2012) Genetic diversity and relationships in *Corchorus olitorius* (Malvaceae s.l.) inferred from molecular and morphological data. Genet Resour Crop Evol 59:1125–1146

Bhandari HR, Bera A, Kar CS, Biswas S (2018) Stability assessment of jute seed production system in lower Gangetic plains of India. Indus Crops Prod 125:505–510

Bhuiyan MAR, Ali A, Akter H et al (2023) Flame resistance and heat barrier performance of sustainable plain-woven jute composite panels for thermal insulation in buildings. Appl Energy 345:1–11

Biswas C, Dey P, Karmakar PG, Satpathy S (2015) Discovery of large-scale SNP markers and construction of linkage map in a RIL population of jute (*Corchorus capsularis*). Mol Breed 35:119–128

Cann RL, Stoneking M, Wilson AC (1987) Mitochondrial DNA and human evolution. Nature 325:31–36

Cardoso RLB, Ramos RPB, Filha L et al (2024) Modelling and analysis of jute fiber reinforced epoxy composite in the development of wind blade for low intensity winds. J Mater Res Tech 28:3619–3630

Chen T, Qi J, Tao A, Xu J, Chen F (2011) A karyological study of two cultivated species and their wild species and three wild relatives of *Corchorus*. J Plant Genet Res 12:619–624

Chen J, Yang Z, Deng C, Liu C et al (2022) Reviewed on main medicinal components and pharmacological functions of edible jute. Plant Fiber Sci China 44(2):126–132

Chen J, Li A, Tang Q, Deng C, Zhang et al (2023) Resequencing of white jute (*Corchorus capsularis* L.) provides insights into genome diversity, population historical dynamics, and improvement. Indus Crops Prod 204:1–7

Chopra V (2005) Mutagenesis: Investigating the process and processing the outcome for crop improvement. Curr Sci 89(2):353–359

Dai Z, Yu Y, Yang Z et al (2021) A new jute variety 'Diwangcai 4hao' for vegetable and its cultivation techniques. China Vegetables 2:121–122

Das M, Banerjee S, Mir RR et al (2012) QTL identification for molecular breeding of fibre yield and fibre quality traits in jute. Euphytica 187:175–189

Datta RM, Dana SK, Banerjee SN (1960) Investigations on the interspecific hybridization between the autotetra-ploids of the cultivated jute species "*Corchorus olitorius* Linn.", "*C. capsularis* Linn." and on the failure of viable seed formation in them. Genet Iberica 12:71–72

Deng C, Tang Q, Dai Z et al (2020) Differential adsorption performance and mechanism of leaf-used jute germplasm for heavy metal removal. J Agric Res Environ 37(3):438–444

Deng Y, Pan J, Yang X et al (2023) Dual roles of nanocrystalline cellulose extracted from jute (*Corchorus olitorius* L.) leaves in resisting antibiotics and protecting probiotics, vol 5. Nanoscale Adv, pp 6435–6448

Doebley JF, Gaut BS, Smith BD (2006) The molecular genetics of crop domestication. Cell 127:1309–1321

Fang P, Wang J, Xu J et al (2022) Study on the tissue culture and regeneration system of jute. Plant Fiber Sci China 44:133–142

Gaoyang Zhang, Jianmin Qi, Jiantang Xu, Xiaoping Niu, Yujia Zhang, Aifen Tao, Liwu Zhang, Pingping Fang, LiHui Lin (2013) Overexpression of UDP-glucose pyrophosphorylase gene could increase cellulose content in Jute (Corchorus capsularis L.), *Biochemical and Biophysical Research Communications*, 442:153–158

Gomaa AA, El-Sers DA et al (2019) Amelioration of experimental metabolic syndrome induced in rats by orlistat and *Corchorus olitorius* leaf extract; role of adipo/cytokines. J Pharm Pharmacol 71:281–291

Gong Y, Dai Z, Chen J et al (2013) Breeding of *Corchorus Olitorius* New Variety 'Y007-10'. Chinese J Trop Agric 33:27–31

Guzzetti L, Panzeri D, Ulaszewska et al (2021) Assessment of dietary bioactive phenolic compounds and agricultural sustainability of an African Leafy Vegetable *Corchorus olitorius* L. Front Nutr 8:667812

Handoussa H, Hanafi R, Eddiasty I et al (2013) Anti-inflammatory and cytotoxic activities of dietary phenolics isolated from *Corchorus olitorius* and Vitis vinifera. J Funct Foods 5:1204–1216

Hassan MS, Zohdy MH (2018) Adsorption kinetics of toxic heavy metal ions from aqueous solutions onto grafted jute fibers with acrylic acid by gamma irradiation. J Nat Fibers 15(4):506–516

Honi U, Amin MR, Kabir SMT et al (2020) Genome-wide identification, characterization and expression profiling of gibberellin metabolism genes in jute. BMC Plant Biol 20:306–321

Hossain MM, Abdulla F (2015) Jute production in Bangladesh: a time series analysis. J Math Stat 11:93–98

Hossen QMM, Rahman SMB, Rahman MN et al (2022) Development of early flowering, short life-spanned jute (*Corchorus* spp.) mutant via ethyl methane sulfonate mutagenesis. J Crop Sci Biot 25:489–500

Huang Q, Hu D, Chen M et al (2019) Sequential removal of aniline and heavy metal ions by jute fiber biosorbents: a practical design of modifying adsorbent with reactive adsorbate. J Mol Liquids 285:288–298

Ingman M, Kaessmann H, Paabo S, Gyllensten U (2000) Mitochondrial genome variation and the origin of modern humans. Nature 408:708–713

Islam AS, Rashid A (1960) First successful hybrid between the two jute-yielding species, *Corchorus olitorius* L. (Tossa) × C. *capsularis* L. (White). Nature 185:258–259

Islam AS, Rashid A (1964) A rare hybrid combination through application of hormone and embryo culture. Nature 201:320

Islam MS, Azam MS, Sharmin S et al (2013) Improved salt tolerance of jute plants expressing the katE gene from Escherichia coli. Turkish J Biol 37:206–211

Islam MT, Ferdous AS, Najnin RA et al (2015) High-throughput sequencing reveals diverse sets of conserved, nonconserved, and species-specific miRNAs in jute. Int J Genomics 2015:1–14

Islam MS, Saito JA, Emdad EM et al (2017) Comparative genomics of two jute species and insight into fibre biogenesis. Nat Plants 3:1–7

Ivanovska A, Milošević M, Lađarević J et al (2024) A step towards tuning the jute fiber structure and properties by employing sodium periodate oxidation and coating with alginate. Int J Biol Macromol 257:128668

Jacob KT, Sen S (1961) Haploidy in Jute (*Corchorus olitorius* L.). Nature 192:288–289

Joshua D, NS R, Gottschalk W (1974) Evolution of leaf shape in jute. Indian J Genet Plant Breed 32:392–398

Kumari N, Choudhary SB, Sharma HK et al (2019) Health-promoting properties of *Corchorus* leaves: a review. J Herb Med 15:1–10

Kundu A, Sarkar D, Mandal et al (2012a) A secondary phloic (bast) fibre-shy (bfs) mutant of dark jute (*Corchorus olitorius* L.) develops lignified fibre cells but is defective in cambial activity. Plant Growth Regul 67:45–55

Kundu A, Topdar N, Sarkar D et al (2012b) Origins of white (*Corchorus capsularis* L.) and dark (*C. olitorius* L.) jute: a reevaluation based on nuclear and chloroplast microsatellites. J Plant Biochem Biotechnol 22:372–381

Kundu A, Chakraborty A, Mandal NA et al (2015) A restriction-site-associated DNA (RAD) linkage map, comparative genomics and identification of QTL for histological fibre content coincident with those for retted bast fibre yield and its major components in jute (*Corchorus olitorius* L., Malvaceae s. l.). Mol Breeding 35:1–17

Kwarteng AO, Amoah RA, Nyadanu D et al (2018) Core collection of two important indigenous vegetables; Gboma eggplant (Solanum macrocarpon L.) and Jute mallow (*Corchorus olitorius* L.) in Africa: an important step for exploitation of existing germplasm and development of improved cultivars. Aust J Crop Sci 12:1398–1409

Li Y, Gong Y, Chen J (2010) Determination and analysis of nutritional components in fresh leaves of vegetable jute. China Vegetables 14:67–70

Li N, Gong Y, Chen J et al (2014) Screening of Jute Germplasm for Adsorption of Heavy Metal Cr(VI). J Plant Genet Resour 15(4):865–870

Li N, Gong Y, Chen J et al (2015) Research on the biological adsorption effect of jute powder on Cr(VI) in aqueous solution and its mechanisms. Indus Water Treatment 35(2):79–83

Liu Y, Xiao T, Baveye PC et al (2015) Potential health risk in areas with high naturally-occurring cadmium background in southwestern China. Ecotoxicol Environ Saf 112:122–131

Lu X, Chen S, Liu X (2008) Technical regulation on conservation for crop germplasm resources. China Agriculture Press, Beijing

Luo S, Li D, Gong Y et al (2009) Research on breeding of bast fiber crops of IBFC in the past 50 years. Plant Fiber Sci China 31:82–92

Lusty C, van Beem J, Hay FR (2021) A performance management system for long-term germplasm conservation in CGIAR genebanks: aiming for quality, efficiency and improvement, , efficiency and improvement. Plants 10:2627–2639

Mahapatra A, Saha A, Gupta D (2006) Catalogue on evaluation of tossa jute germplasm (*Corchorus olitorius* L.), vol 41. Central Res Inst for Jute Allied Fibres
Maity S, Datta AK, Chowdhury S, Maiti GG (2009) Pollen morphology in nine species of jute (*corchorus*). J Phytol Res 22(1):57–60
Majumder S, Datta K, Sarkar C, Saha SC, Datta SK (2018) The development of Macrophomina phaseolina (Fungus) Resistant and Glufosinate (Herbicide) Tolerant Transgenic Jute. Front Plant Sci 9:920
Majumder S, Parida S, Dey N (2024) Protocol for imbibed seed piercing for Agrobacterium-mediated transformation of jute. STAR Protoc 5:102767
Mandal A, Datta AK (2014) New Plant Type Mutants in Jute (*Corchorus olitorius* L.). Indian J Sci Technol 7(5):543–553
Mandal A, Datta AK, Datta S, Gupta S (2013) Genetic assessment of eight *Corchorus* spp. (Tiliaceae) using RAPD and ISSR markers. Nucleus 56:23–30
Mgolozeli S, Tesfamariam EH, Beletse Y (2022) Yield, mineral content and root growth response of jute mallow (*Corchorus olitorius* L.) to planting density and water availability. J Agric Food Res 10:1–10
Mia MM, Shaikh AQ (1967) Gamma radiation and interspecific hybridization in jute (*Corchorus capsularis* L. and C. *olitorius* L.). Euphytica 16:61–68
Mir RR, Rustgi S, Sharma S et al (2007) A preliminary genetic analysis of fibre traits and the use of new genomic SSRs for genetic diversity in jute. Euphytica 161:413–427
More SS, Perke DS, Baviskar PP (2022) Growth and instability of international trade of jute in India. Int J Plant Soil Sci:672–684
Ngomuo M, Stoilova T, Feyissa T, Ndakidemi PA (2017) Characterization of Morphological Diversity of Jute Mallow (*Corchorus* spp.). Int J Agron 2017:1–12
Nura S, Adamu A, Mu Azu S, Dangora D (2011) Chemical mutagenesis for improved quality traits in Jute (*Corchorus* olitorious L.). J Cont J Biol Sci 4:22–27
Nyadanu D, Adu Amoah R, Kwarteng AO et al (2016) Domestication of jute mallow (*Corchorus olitorius* L.): ethnobotany, production constraints and phenomics of local cultivars in Ghana. Genet Resour Crop Evol 64:1313–1329
Olcay N, Cankurtaran Kömürcü T, Demir MK (2022) Effects of molokhia (*Corchorus olitorius*) powders obtained by different drying methods on some selected properties of erişte, Turkish noodle. Int J Gastron Food Sci 28:1–8
Pakendorf B, Stoneking M (2005) Mitochondrial DNA and human evolution. Annu Rev Genomics Hum Genet 6:165–183
Ping W, Liu Y, Fu Q (2022) Exploring the evolution of archaic humans through sedimentary ancient DNA. Hereditas 44:362–369
Qasim M, Abideen Z, Adnan MY et al (2017) Antioxidant properties, phenolic composition, bio-active compounds and nutritive value of medicinal halophytes commonly used as herbal teas. South Afr J Bot 110:240–250
Rahman S (2009) Whether crop diversification is a desired strategy for agricultural growth in Bangladesh? Food Policy 34:340–349
Raut RN, Naik G (1983) Interspecific hybridization in cultivated jute. Proc. FAO expert consultation on jute and kenaf improvement, Calcutta, India
Río JCD, Gisela M, Rodríguez IM, Gutiérrez A (2009) Chemical composition of lipophilic extractives from jute (*Corchorus capsularis*) fibers used for manufacturing of high-quality paper pulps. Indus Crops Prod 30:241–249
Roy A, Bandyopadhyay A, Mahapatra AK et al (2006) Evaluation of genetic diversity in jute (*Corchorus* species) using STMS, ISSR and RAPD markers. Plant Breed 125:292–297
Saha T, Majumdar S, Banerjee NS, Sen SK (2001) Development of interspecific somatic hybrid cell lines in cultivated jute and their early characterization using jute chloroplast RFLP marker. Plant Breed 120:439–444
Saha D, Rana RS, Chakraborty S et al (2017) Development of a set of SSR markers for genetic polymorphism detection and interspecific hybrid jute breeding. Crop J 5:416–429

Sajib AA, Shahidul Islam M, Shamim Reza M et al (2008) Tissue culture independent transformation for *Corchorus olitorius*. Plant Cell Tissue Organ Cult 95:333–340

Saliu JA, Ademiluyi AO, Boligon AA et al (2019) Dietary supplementation of jute leaf (*Corchorus olitorius*) modulates hepatic delta-aminolevulinic acid dehydratase (delta-ALAD) activity and oxidative status in high-fat fed/low streptozotocin-induced diabetic rats. J Food Biochem 43:12949–12957

Sarkar D, Satya P, Mandal NA et al (2016) Jute genomics: emerging resources and tools for molecular breeding. Springer Inter Pub Switzerland 9:155–200

Sarkar D, Kundu A, Das D et al (2019) Resolving population structure and genetic differentiation associated with RAD-SNP loci under selection in tossa jute (*Corchorus olitorius* L.). Mol Gen Genomics 294:479–492

Sen MK, Hermans PH (1949) Origin of Hygroscopicity of Jute. Nature 164:628–629

Shafrin F, Das SS, Sanan-Mishra N, Khan H (2015) Artificial miRNA-mediated down-regulation of two monolignoid biosynthetic genes (C3H and F5H) cause reduction in lignin content in jute. Plant Mol Biol 89:511–527

Shafrin F, Ferdous AS, Sarkar SK et al (2017) Modification of monolignol biosynthetic pathway in jute: different gene, different consequence. Sci Rep 7:39984

Swaminathan MS, Iyer RD (1961) Skewed recombination in a rare interspecific jute hybrid. Nature 192:893–894

Tao A, Huang L, Wu G et al (2017) High-density genetic map construction and QTLs identification for plant height in white jute (*Corchorus capsularis* L.) using specific locus amplified fragment (SLAF) sequencing. BMC Genomics 18:355–366

Tester M, Langridge P (2010) Breeding technologies to increase crop production in a changing world. Science 327:818–822

Wen L, Gong Y, Chen J et al (2014) Physiological and biochemical characteristics of mutant jute seedlings induced by EMS and ^{60}Co-γ radiation. Plant Fiber Sci China 36(5):224–228

Wright RP, Lentz DL, Beaubien HF, Kimbrough CK (2012) New evidence for jute (*Corchorus capsularis* L.) in the Indus civilization. Archaeol Anthropol Sci 4:137–143

Yan YY, Wang YW, Chen SL et al (2013) Anti-inflammatory effects of phenolic crude extracts from five fractions of *Corchorus Olitorius* L. Food Chem 138:1008–1014

Yang Z, Dai Z, Lu R et al (2017a) Transcriptome analysis of two species of jute in response to Polyethylene Glycol (PEG)- induced drought stress. Sci Rep 7:16565

Yang Z, Lu R, Dai Z et al (2017b) Salt-stress response mechanisms using de Novo transcriptome sequencing of salt-tolerant and sensitive *Corchorus* spp. Genotypes Genes 8:226–240

Yang Z, Yan A, Lu R, Dai Z, Tang Q, Cheng C, Xu Y, Su J (2017c) De novo transcriptome sequencing of two cultivated jute species under salinity stress. PLoS One 12(10):e0185863

Yang Z, Lu R, Dai Z et al (2018) Analysis of genetic diversity and population structure of a worldwide collection of *Corchorus olitorius* L. germplasm using microsatellite markers. Biotechnol Biotechnol Equip 32:961–967

Yang Z, Tian S, Li X et al (2022) Multi-omics provides new insights into the domestication and improvement of dark jute (*Corchorus olitorius*). Plant J 112:1–18

Yao J, Jiang S, Li H et al (2023) Genome-wide association study reveals loci and candidate genes of flowering time in jute (*Corchorus* L.). Mol Breeding 43:85

Yusuff OK, Abdul Raheem MAO, Mukadam AA, Sulaimon RO (2019) Kinetics and mechanism of the antioxidant activities of C. *olitorius* and V. amygdalina by spectrophotometric and DFT methods. ACS Omega 4:13671–13680

Zakaria Z, Kumar GH, Mohd Nor RR et al (2009) Antinociceptive, anti-inflammatory and antipyretic properties of an aqueous extract of *Corchorus capsularis* leaves in experimental animal models. Pharm Biol 47(2):104–110

Zech M, Winkelmann J (2024) Next-generation sequencing and bioinformatics in rare movement disorders. Nat Rev Neuro 20:114–126

Zhang J, Chen C, Luo X, Jin G (2015a) Analysis of the coefficient of parentage among major jute cultivars in China. Sci Agric Sinica 48:4008–4020

Zhang L, Cai R, Yuan M et al (2015b) Genetic diversity and DNA fingerprinting in jute (*Corchorus* spp.) based on SSR markers. Crop J 3:416–422
Zhang L, Xiaoyan G, Yuhu et al (2016) Study on planting technology of Marijuana-Jute/Kenaf for two batches a year. Chinese Agri Sci Bull 32:48–53
Zhang G, Shan S, Wu Y et al (2019) The KCS gene is involved in the formation of chloroplast stromules and other physiological processes in jute (Corchorus *capsularis* L.). Indus Crops Prod 141:111781
Zhang L, Ma X, Zhang X et al (2021) Reference genomes of the two cultivated jute species. Plant Biotechnol J 19:2235–2248
Zhang L, Khan H, Kole C (2022) The Jute Genome, vol 7. Springer Nature Switzerland, pp 100–105
Zhao Y, Huang Q, Zhao H et al (2017) Nutritional components analysis on fresh leaves of vegetable jute varieties Guimacai 1 and Guimacai 2. J Southern Agric 48(1):127–131
Zhao X, Li J, Su S, Jiang N (2024) Prediction models of mechanical properties of Jute/PLA composite based on X-ray computed tomography. Polymers 16:160
Zhou G, Luo J, Liu C et al (2018) Efficient heavy metal removal from industrial melting effluent using fixed-bed process based on porous hydrogel adsorbents. Water Res 131:246–254
Zulfiqar A, Shah AUR, Khalil MS et al (2024) Enhancing properties of jute/starch bio-composite material through incorporation of magnesium carbonate hydroxide pentahydrate: a sustainable approach. Mater Chem Phys 314:128690

Chapter 14
Advanced Plant Breeding, Genetics and Biotechnology on Roselle (*Hibiscus sabdariffa* L.)

Sara E. Gomaa, Amani H. A. M. Gharib, Ahmed M. Abdelmoghny, and Khaled F. M. Salem

Abstract Roselle (*Hibiscus sabdariffa* L.) is a flowering plant in the class Dicotyledonae, order Malvales, and family Malvaceae. Hibiscus is tetraploid (2n = 4x = 72). Roselle is an important multipurpose plant, with many health-promoting properties. The stems are used to produce bast fiber, and the dried calyces, which taste like cranberries, are usually soaked to make a popular infusion. Hibiscus is the most widely cultivated fiber crop. Both physical and chemical analyses have shown that hibiscus fiber is a good source of textile materials. The physical properties of hibiscus fiber include high strength, gloss, brightness, and resistance to heat, light, acids, and alkali. Wide genetic variety aids in cultivar improvement and the preservation of hibiscus traits. Traditional plant breeding methods will remain vital and continue to evolve. Maintaining diverse genetic backgrounds through exploring genetic diversity is essential for producing climate-smart hibiscus. Crossing common cultivars with locally created cultivars enhances genetic diversity, selects advantageous features, and introduces novel alleles. Genetic

S. E. Gomaa (✉)
Medicinal and Aromatic Plants Research Department, Horticulture Research Institute (HRI), Agricultural Research Center (ARC), Giza, Egypt
e-mail: sara_gomaa@hotmail.com

Amani H. A. M. Gharib
Vegetable, Medicinal and Aromatic Plant Breeding Research Department, Horticulture Research Institute (HRI), Agricultural Research Center (ARC), Giza, Egypt
e-mail: dramani_gharib@hotmail.com

A. M. Abdelmoghny
Cotton Research Institute (CRI), Agricultural Research Center (ARC), Giza, Egypt
e-mail: ahmed.abdelbaky@arc.sci.eg

K. F. M. Salem
Department of Biological Sciences, Shaqra University, Ad-Dawadimi, Saudi Arabia

Department of Plant Biotechnology, Genetic Engineering and Biotechnology Research Institute (GEBRI), University of Sadat City, Sadat City, Egypt
e-mail: khaled.salem@gebri.usc.edu.eg; khaledfathi@su.edu.sa

K. F. M. Salem et al. (eds.), *Breeding and Biotechnology of Grass and Bast Fiber Crops*, Advances in Plant Breeding Strategies 12,
https://doi.org/10.1007/978-3-032-00403-1_14

improvement can significantly reduce the impact of biotic and abiotic stresses and increase productivity. Developing disease-resistant, environmentally tolerant, and enhanced cultivars involves both conventional and molecular techniques. This chapter provides an overview of the botanical classification and distribution, diversity and conservation of hibiscus germplasm, importance and uses, fiber and textile properties, fiber industry, domestication, selection and early improvements, current cultivation practices and challenges, artificial intelligence, breeding objectives, germplasm biodiversity and conservation, cytogenetics, traditional breeding methodologies, the role of biotechnology, molecular breeding, molecular marker-assisted breeding, genomics, bioinformatics, tissue culture applications, micropropagation, embryo rescue, synthetic seeds, genetic engineering and gene editing, enhanced traits and nanotechnology.

Keywords Bioinformatics · Breeding · Biotechnology · Fiber · Marker-assisted breeding · Molecular breeding · Roselle (*Hibiscus sabdariffa* L.)

14.1 Introduction

14.1.1 Background

Roselle (*Hibiscus sabdariffa* L.) is a blooming plant from the genus Hibiscus that is native to Africa, most likely West Africa. It expanded to Asia and the West Indies during the sixteenth and early seventeenth centuries, and it has since become naturalized in many regions. Roselle originated in Africa, but it is now distributed throughout the continent's tropical regions, particularly the savanna regions of Western and Central Africa. The stems are used to manufacture bast fiber, and the dried cranberry-flavored calyces are often steeped to form a popular. The name differs from country to another one globally. In India, it has different names with different languages like Roselle (English), Lalambri, Patwa (Hindi) Pundi, Pundigida (Kannada) Colour Gonkura (Telugu) Colour Pulichaikeeraj, Mandharam (Tamil). Moreover, globally, it is also known as Sorrel (Africa), karkade or carcade (Egypt, Saudi Arabia and Sudan), asampaya (Malaysia), rosella (Australia), chin baung (Myanmar), krajeab (Thailand), wonjo (Gambia), zobo (Nigeria), rosella (Indonesia), cranberry (Florida) (Ansari et al. 2013; Kowti et al. 2020; Mahunu 2021; Gonzalez et al. 2022; Ranjan 2023). Roselle has evolved to a variety of environmental situations. It can be cultivated in rooftops, large planters, high tunnels, greenhouses, field rows (with black plastic mulch), raised beds and yards. The leaf production is increased when planted in the soil at ground level. The mature calyx needs approximately 6 months (Kowti et al. 2020). Also, roselle can be started indoor location (greenhouse) for 6 weeks, then planted outside. Roselle grows in well-drained soils rich in organic materials. Also, it may grow in clay soils. Roselle is a vigorous leaf producer that will immediately regenerate after harvest.

Roselle has a diversity of uses from different plant parts. The red calyxes make roselle more attractive to consumers both from a nutritional and medicinal standpoint due to its components of anthocyanins, phenols, and antioxidants (M'be et al. 2023). Almost all plant parts can be used leaves, stems, flower calyces and seeds, but it is cultivated mainly for its calyces. Calyces are the most commercially valuable part and their color plays an important role in determining the quality of Karkade (Mohamed 2021). In the United States, hibiscus tea is the most well-known edible roselle. In Sudan, calyxes were used to make fermented beverages, wine, and beer, as well as hot or fresh drinks. In Nigeria, dried red roselle calyces were used to make a clean, non-alcoholic beverage called Zobo, which is a favorite among the lower and middle classes.

China and Thailand are the main producers and control much of the world's supply. The world's best roselle comes from Sudan and Nigeria, b. Mexico, Egypt, Senegal, Tanzania, Mali and Jamaica are also key suppliers, but output is mostly consumed domestically. Recently, the trade and marketing of roselle calyces has increased. The largest importers in the world are Germany and the USA. Only, the USA imports more than 5000 metric tons of dried calyces with 22 USD million to make herbal tea. The Egyptian and Sudanese roselle is very expensive in the USA compared to the Chinese one, due to the low quality of chinses roselle caused by excessive precipitation during the production season. Eighteen businesses in Malaysia are involved in the manufacture, processing, and marketing of goods for the local market. Juice and drink processing accounts for about 80% of overall production (Ranjan 2023). This chapter focuses on an overview of the botanical classification and distribution, diversification, and conservation of hibiscus germplasm. It also discusses the importance and uses of hibiscus, its fiber and textile properties, the fiber industry, domestication, selection, and early improvements. Additionally, it covers current cultivation practices and challenges, artificial intelligence, breeding objectives, germplasm biodiversity, and conservation. The chapter also delves into cytogenetics, traditional breeding methodologies, the role of biotechnology, molecular breeding, molecular marker-assisted breeding, genomics, bioinformatics, tissue culture applications, micropropagation, embryo rescue, synthetic seeds, genetic engineering, gene editing, enhanced traits, and nanotechnology.

14.1.2 Botanical Classification and Distribution

The flowering plant roselle (*Hibiscus sabdariffa* L.) belongs to the family Malvaceae, order Malvales, and class Dicotyledonae. Large, eye-catching flowers are characteristic of hibiscus, a plant that can grow into tiny trees, shrubs, or herbs. The chromosomes are related to the diploid ($2n = 2x = 36$), while Roselle is a tetraploid ($2n = 4x = 72$). There are more than a hundred species found that are used throughout the world as food and medicine. (*Hibiscus sabdariffa* L.) is of the genus *Hibiscus* and species (https://www.itis.gov).

Kingdom	Plantae—Plants
Subkingdom	Viridplantea—green plants
Superdivision	Embryophyte
Division	Tracheophyte– vascular plants, tracheophytes
Subdivision	Spermatophytina—spermatophytes, seed plants
Class	Magnoliopsida
Subclass	Rosanae
Order	Malvales
Family	Malvaceae—mallows, mauves
Genus	Hibiscus L.—rosemallow, rose-mallow
Species	*Hibiscus sabdariffa* L.—roselle
Common names:	Including Indian Sorrel, Jamaican Tea, Maple-Leaf Hibiscus, October Hibiscus, Red Sorrell, Roselle

Botanical description of vegetative parts of the plant play is an important tool for species identification. Preliminarily, *H. sabdariffa* is identified by different plant part descriptions such as; stem, leaves, flowers, fruit and seeds (Table 14.1). Roselle is an annual or perennial herb or woody-based sub-shrub plant, that can grow up to 2–3.5 m in height, with soft to nearly soft, cylindrical, typically red stems (Ranjan 2023). With 3–5 lobed leaves that were 8–15 cm long and an 8–10 cm flower diameter, the mature leaves ranged in color from white to pale yellow, with a dark red mark at the base of each petal (Figs. 14.1 and 14.2). The width of the calyx at the base is 1–2 cm. The length of mature fruits is 3–3.5 cm with fleshly and bright red color. The growing season of roselle takes about 6 months from planting to harvesting (Kowti et al. 2020). Roselle is classified as a short-day plant that needs 12 h or more darkness to produce abundant flowering. Roselle is a drought-tolerant plant due to its deep root system. This plant characterized by can grow in containers or the ground, the germination is very rapid and the stems are bright red with red-veined leaves that make quite a statement in the garden.

The origin of roselle is Africa but now covering all tropical and sub-tropical regions worldwide. It is planted in Sudan 6000 years ago. So, Sudan is the major producer of roselle in the form of Sudanese tea (McClintock and Tahir 2004; Mohamad et al. 2002). During the seventeenth century, it was introduced to India and South America as a vegetable crop but in Asia, the domesticated plant was used as a fiber crop (McClintock and Tahir 2004). The most important produced countries are Sudan, India, Sri Lanka, China, Malaysia, Thailand, Taiwan, Egypt and Mexico. The largest producers and major suppliers in the world are China and Thailand.

Table 14.1 Botanical descriptions of roselle (*Hibiscus sabdariffa* L.)

Plant part	Botanical descriptions
Plant height	2.0–3.5 m
Leaves	The leaves are palmate, triangular shaped. The petioles' length ranged from 1 to 6 cm. Simple alternating leaves with linear stipules that are narrowly lanceolate. Mature leaves had 3–5 lobes with 8–10 cm length. The base is wedged, and the apex is sharp. The leaves had nectar glands, 2–3 mm long.
Stem	Smooth or nearly smooth, cylindrical shape, typically pale rosy in color.
Root	Slightly thin and branched with tap root system.
Flower	Flowers are bisexual, regular, pentamerous, with pedicel 1–4 cm long. Flowers are solitary, axillary, and erect 8–10 cm in diameter. Epicalyxes have 10 acute lobes 8–10 mm in length and are shorter than the calyx. Flowers are white to pale yellow. With a dark red spot at the base of each petal. The ovary is larger and consists of 5 loculus topped with 5 branches of style. The calyx is red-colored and lanceolate-shaped. Petal colors vary from white to pink, red, orange, purple or yellow.
Fruit	The shape is a Spherical-conical capsule. The fruit is 1.2–2 cm long. Apiculate at the top and covered with stiff and applied bristles. Contains many seeds. The mature fruits are bright red in color.
Seeds	The seed shape is the kidney. The seed color is dark brown color. The seed long is up to 7 mm. Rich source of protein and can replace animal protein, especially for vegetarians.

14.1.3 Importance and Uses

Every edible component of the plant is highly adaptable and is used in a wide range of recipes all over the world. Nutritious qualities, medicinal uses, and the application of various plant sections for certain objectives.

14.1.3.1 Urinary Uses

The leaves taste like rhubarb and can be cooked to make a spicy spinach or eaten raw. It can be prepared with fish, chicken, crab, pork, or any other meat, or it can be used in curry, steamed with lentils, fried with garlic, dried or fresh prawns, and green chili, or it can be put into a light soup (Mohamed 2021). You may make a delightful drink out of fresh or dried calyces by boiling them in water for 8–10 min (or until the water becomes red) and then adding sugar. By adding rum and brewing it with ginger, cinnamon, cloves, and bay leaves, you can enhance flavor. Another option is to soak the dried calyces in cold water with sugar and lemon or lime juice for the entire night in the refrigerator (Mohamed 2021). Some commercial herbal

Fig. 14.1 Types of Roselle calyces (**a**) the fleshy calyces; (**b**) green seed capsule (**c**) Red Roselle calyx (**d**) Green Roselle calyx (**e**) Dark Roselle calyx (**f**) Dried *Hibiscus sabdariffa* calyx. (Source: (a, b) Osei-Kwarteng et al. (2021) and (c, d, e, f) Mahunu et al. (2021)

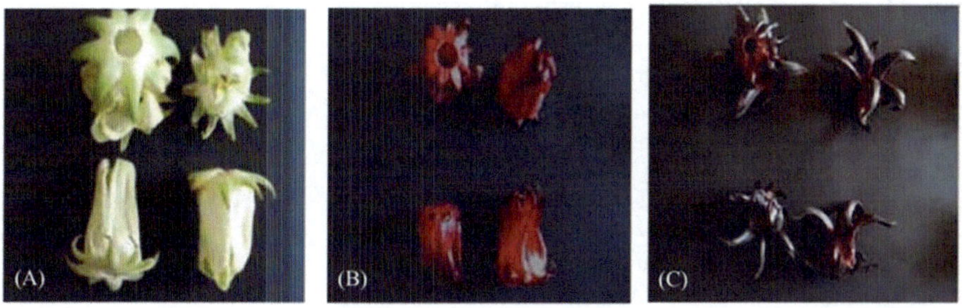

Fig. 14.2 Ripe Roselle calyces after removal of seed capsules: (**a**) green Roselle calyces, (**b**) light red Roselle calyces, and (**c**) dark red Roselle calyces. (Source: Hussein et al. (2010) and Osei-Kwarteng et al. (2021))

teas contain hibiscus flowers (Roselle), which give them a crimson hue. Karkade is the name of a cool, tart beverage prepared from hibiscus in Egypt. Additionally, roselle buds can be used to make preserves, chutneys, jams, jellies, and sauces (Mohamed 2021). The seeds are consumed roasted, fermented to make "furundu," a meat replacement from Sudan, or turned into an oil sauce. Utilized as a natural food and beverage coloring agent.

14.1.3.2 Medicinal Uses

All the roselle plant parts are used in medicine and are considered as a main source of drugs and the manufacture of safe or natural medicines (Table 14.2). Rich in organic acids, hibiscus produces a refreshing beverage without caffeine. As a general health tonic, it is used to cure circulatory disorders, colds, coughs, and appetite loss. Hibiscus is a mild expectorant, diuretic, and laxative. Used as a soft drink to reduce inflammation, hypertension, and liver disorders.

The human body uses exogenous antioxidants, which are rich in anthocyanins and flavonoids, particularly in flower petals, to combat excessive oxidative stress (Lubis et al. 2024). Roselle has anti-diabetic, anti-cancer, and anti-hypertensive properties. Anthocyanins have a great effect on the cardiovascular system on human body systems by increasing vasodilation and cardiac output in blackcurrants by increasing peripheral blood flow, which improves exercise performance (Lubis et al. 2024). impact of anthocyanins on muscle growth. This method works by speeding up the myogenesis process, safeguarding mitochondria, promoting the synthesis of muscle proteins, and preventing the breakdown of muscle proteins. Rosella protects atherosclerosis by blocking LDL oxidation. Anthocyanins improve mitochondrial biogenesis and metabolism to produce energy. In addition to having many other vitamins and minerals in smaller amounts, roselle is an especially good source of vitamin C, calcium, and magnesium.

Table 14.2 Medical effect of different parts of Roselle

Roselle plant part	Medicinal Properties
Red calyces	Decreasing blood viscosity. Reducing hypertension. Good for healthy bone and teeth formation. Useful in blood formation. Have a high amount of ascorbic acid or vitamin c
Calyx extracts	Reduces extra fat from the liver. Reducing body fat and body mass index (BMI). Reduces calcium oxalate crystal deposits in kidneys due to its uricosuric effect. Contains several amino acids. Strong anti-cancer activities against prostate cancer cells. Antidiabetic activity by stimulating the effect on insulin secretion.
Green leaves	Cure of cracks in the feet and fast healing of ulcers.
Seeds	Higher source of lipid soluble antioxidants particularly gamma-tocopherol. Inhibition of several bacterial and fungal growth.
Roselle extracts	Skin conditioning treatments. Protect human cells against cadmium-induced damage such as tumor necrosis.
Sepal extract	effective treatment against leukemia, liver damage, hypertension, and pyrexia due to its high content of polyphenols.
Petals	Antioxidant agents such as anthocyanins and ascorbic acid.
Roselle tea	Reducing the blood pressure in hypertensive and pre-hypertensive persons. Soothing cold, cough and fever. Promoting kidney function. General tonics, as diuretic.
Roselle juice	treatment and prevention of renal stone disease. Decrease creatinine, uric acid, citrate, tartrate, calcium, sodium, and potassium. Increase stamina and balance body electrolytes after sports.

14.1.3.3 Fiber Uses

The stems can be used to make hessian, a biodegradable fiber, in place of jute since they contain bast fiber (rosella hemp) (Mohamed 2021; Puro et al. 2014; Ranjan 2023). After jute, roselle is the second most important fiber crop in India. The fiber is utilized in the manufacturing of industrial goods such as bags, fabric shoes, carpets, twines, sacs, upholstery, and various types of paper. Roselle, in recent years, has been identified as (Mahunu 2021). On the contrary, it is less used as a fiber crop in Sub-Saharan Africa (SSA). Roselle is a Biocomposite to produce vehicle parts and construction materials including fiberboard (Alves et al. 2010). 20% of bast fiber crops are thought to be roselle. Outstanding resistance to microbes and important sanitary qualities (Tamta and Kalita 2020).

14.1.4 Fiber and Textile Properties

Roselle is primarily bred for its supply of fiber. Roselle fiber was found to be a good source of textile materials by both physical and chemical investigation. High tenacity, high luster, brightness, and resilience to heat, light, acid, and alkali are among roselle fiber's physical characteristics (Tables 14.3 and 14.4) and (Fig. 14.3). Fiber materials have the potential to create superior mixed fabrics (Tamta and Kalita 2020). A scanning electron microscope was used to analyze the roselle fiber's surface morphology (Fig. 14.4). Due to the difficulty of degumming and a lack of understanding of mechanical processing, this plant is not widely used as a fiber crop. Plant-based roselle fiber is safe, recyclable, and biodegradable. Moreover, mixed with other fibers, including cotton, flax, wool, polyester, acrylic, and silk.

Table 14.3 Morphological and physical properties of the roselle fiber Parameter

Morphological properties		Physical properties	
Parameter	Measurements	Parameter	Measurements
Length (m)	2–2.5 (7–8 ft)	Length(mm)	122
Diameter (cm)		Diameter(μm)	14.10
Bottom	2.90	wall thickness(μm)	3.50
Middle	2.35	Fibre fineness (g/tex)	3.90
Top	2.23	Wicking height (cm)	13
Weight (g)		Moisture regains (%)	8.40
Core	300	Tensile strength (g/tex)	30.00
Bark	220	Elongation (%)	1.59
Moisture (%	73.1	Density (g/cc)	1.49
		Brightness index (μm)	20

Source: Tamta and Kalita (2020)

Table 14.4 Chemical properties of roselle fiber of the roselle fiber Parameter

Constituent	Retted RF (%)	Degummed RF (%)	Bleached RF (%)
Cellulose	56.25	58.63	64.50
Alpha cellulose	20.23	18.00	17.55
Alpha cellulose	11.50	9.70	8.15
Pectineus matter	2.30	0.70	0.50
Ash	2.08	1.03	1.25
Lignin	7.25	3.00	1.00
Wax	0.50	0.30	0.20
Moisture	7.40	8.21	8.76

Source: Tamta and Kalita (2020)

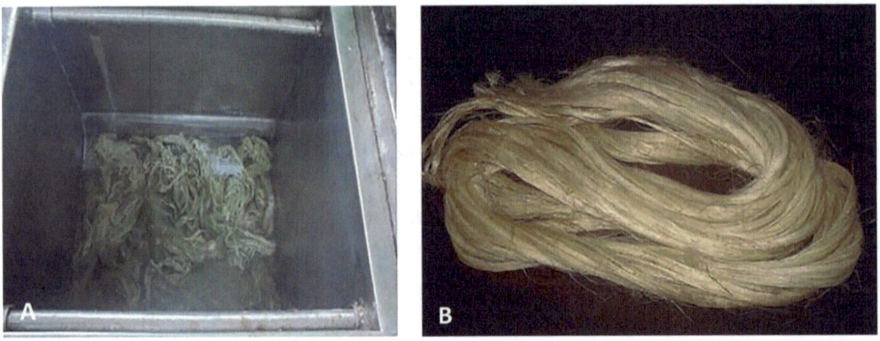

Fig. 14.3 Roselle fiber (**a**) Degumming of roselle fibre and (**b**) Degummed roselle fibre. (Source: Tamta and Kalita (2020))

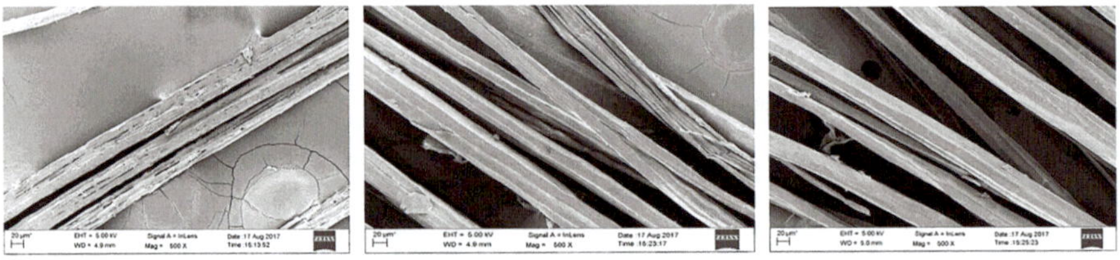

(a) Raw roselle fibres **(b) Degummed roselle fibres** **(c) Bleached roselle fibres**

Fig. 14.4 Scanning electron microscopic (SEM) study of roselle fiber. (Source: Tamta and Kalita (2020))

14.1.5 Fiber Industry

Increased consumption of natural fiber goods is made possible by the fashion industry. Thus, using non-traditional natural fibers, such as bast fibers, becomes a key goal in lowering the environmental population. Roselle is a non-conventional bast fiber. The mature fiber needs 90–120 days. The agroclimatic conditions, retting techniques, and cultivation area all affect the fiber quality characteristics. Fiber extraction can be done by water retting of the bark (Kalita et al. 2017). Particularly when combined with other fibers like cotton, flax, wool, polyester, acrylic, and silk of all kinds in various processing systems, roselle fiber is in high demand. The blend between roselle and ramie with different ratios 50:50, 25:75 and 75:25. The blended yarns showed maximum twist per inch (20 tpi) exhibited at 50:50. The blend 50:50 had better physical properties than the 25:75 and 75:25 blend (Table 14.5). This is necessary for garment textiles and can be blended with various fibers in various processing processes (Tamta and Kalita 2020).

Table 14.5 Physical properties of blended yarn between roselle and ramie with different concentrations

Properties	Roselle	Ramie	Blend composition		
	100%	100%	25:75	50:50	75:25
Count	40	40	40	40	40
TPI	18	17.5	19	20	18
Tenacity (g/tex)	15.50	20.87	17.35	14.01	12.99
Elongation (%)	2.94	1.29	1.71	1.77	2.73
Wicking height (cm)	6	6.5	5.0	5.5	6.0
Density (g/cm^3)	1.35	1.30	1.46	1.30	1.38

Source: Tamta and Kalita (2020)

14.1.6 Domestication, Selection and Early Improvements

There are two species of roselle plant: *Hibiscus sabdariffa var. sabdariffa* (HSS) and *Hibiscus sabdariffa var. altissima* (HSA). The first type (HSS) is cultivated for fleshy, shiny-red clay and described as bushy with hefty branches, which bear fleshy calyces and are defined in various races (Mahadevan and Kamboj 2009). The plants of the second type (HSA), which is grown primarily for its phloem fiber, have an upright growth habit with lower branching supported by an extended stem that contains commercially useful bast fiber (Mahunu 2021). In 1920, Wester identified three edible roselle cultivars grown in the Philippines at this time (Ranjan 2023); Rico is low growing, spreading, with simple leaves borne over a long period. Victor was selected from seedlings in a sub-tropical garden in Miami. Victor is earlier than Rico. Archer was produced by Wester using a seed sent from A. S. Archer of the island of Antigua. The plants were tall and robust like Victor with green stems.

Plant breeders need genetic variety to boost crop output. There are extremely few germplasm accessions available for breeding because this species is an introduction to Malaysia. The botanical description of these three cultivars has differed in plant tall, growth habit, flower morphology pollen and calyx. In 1914, a new verity (Temprano) was released at the Lamao experiment station. Temprano is characterized by early flowering, no longer grown, less robust and less productive species than the others (Morton 1987). Currently, no named varieties are available for this crop (Ranjan 2023). Recently, limited research work was done as a cooperation between the Malaysian Agricultural Research and Development Institute (MARDI), Malaya University and the University of Kebangsaan Malaysia (UKM). The results found that it is necessary to increase plant productivity depending on genetic variability. While the available accessions are very limited to establishing a breeding program (Ranjan 2023). Three new types, UKMR-1, UKMR-2, and UKMR-3, were introduced by UKM in April 2009. In a mutation breeding operation that began in 2006, these new kinds were created with Arab as the parent variety.

In 1999, a mutation breeding program succeeded in producing promising breeding lines in collaboration with the MINT (now called the Malaysian Nuclear Agency) by generating new genetic variability. There is little information about genetics, breeding and production related to climate change adaptation is scarce (Mahunu 2021; Mohamed 2021). Central Research Institute for Jute and Allied Fibres (ICAR), India collated the largest roselle accessions 628 (Mahapatra 2008). Before 2009, Indian researchers improved eight roselle varieties characterized by high fiber content using selection and hybridization breeding methods of Indigenous landraces (Kar et al. 2010).

14.2 Current Cultivation Practices and Challenges

Roselle (*Hibiscus sabdariffa* L.) as a component of *Malvaceae* family may be a subtropical therapeutic plant. This plant is self-pollinated and touchy to cold. Roselle is a comparatively drought-tolerant plant, so choosing a suitable sowing date for trim is one of the foremost imperative variables in its generation when it is developed; Dry season is one of the foremost noteworthy obstacles to the generation of crops within the world, especially in parched and semi-arid locales (Yang et al 2012). The challenge of keeping up with rural needs has expanded due to climate inconstancy and the current impacts of worldwide climate alteration (FAO 2018a, b).

14.2.1 Current Cultivation Practices

Decide the fertilization and water system necessities considered of the foremost critical agrarian hones for the roselle plant as a restorative plant. It requires 4–8 months with the least night-time temperature of 20 °C, and 13 h of daylight (El-Dissoky et al. 2020). The NPK fertilizer rate of 68 kg N + 32 kg P_2O_5, 24 kg K_2O, and 4 L humic acid per feeding was shown to produce the highest values of roselle produced in clayey soil (in Sharkia governorate, Egypt) chemical constituents (N, P, and K rates as well as add up to solvent solids) (Fahmy and Hassan 2019). Under the conditions of central Egypt (El-Fayoum governorate), 500 kg of ammonium sulphate (100 kg N), 150 kg of calcium superphosphate (22.5 kg P_2O_5), and 50 kg of potassium sulphate (24 kg K_2O) per nourished plus biofertilizers "Azotobacterine and phosphorein" were obtained beneath a variety of soil surfaces (clay, sandy soil, and loamy sand soils) (Ghabour et al. 2019). Organic fertilization may be a very important way to provide plants with the nutrients they need without negatively impacting the environment. Nabila and Aly (2002) discovered that applying both beast and chicken fertilizer enhanced plant growth. According to Gendy et al. (2012), the best results for roselle plants were obtained when a combination of biofertilizers (phosphorein and nitrobein) and cattle manure (71.43 m^3/Hecatar) were combined.

Monitoring irrigation is especially important for roselle plants, as a prolonged dry spell reduces the yield of these aromatic and medicinal plants. The majority of analysts focused on the effects of water stretch on the biomass, structure, and dynamic components of unique medicinal and aromatic plants. However, in certain instances, they showed an increase in dry submission and flowering (Babatunde and Mofoke 2006; El-Boraie et al. 2009; Seghatoleslami et al. 2013). The number of blossoms, new weight, and dry weight decreased under drought stress or overwatering; however, the increased water supply increased anthocyanin, and the stress condition also increased the amount of carbohydrates that accumulated into anthocyanin and other auxiliary metabolites (El-Boraie et al. 2009).

Seghatoleslami et al. (2013) discovered that while water supply level was linked to chlorophyll material in clears out, it had no discernible effects on roselle plant height, stem distance across, number of branches, or stomatal closure. Hibiscus that developed in sandy soil (Ismailia governorate, Egypt) was delivered beneath the impact of interaction treatment; (2304 m^3 fed^{-1}) with fertilization at 100%NPK (300 kg ammonium sulfate, 100 kg potassium sulfate, 300 kg calcium super phosphate) and humic acid. El-Dissoky et al. (2020) came to the following conclusions in order to maximize the abdicate of roselle plants (*H. sabdariffa* L.) grown in the Aswan governorate (Upper Egypt): fertilize plants with mineral fertilizers at a rate of 56.25 kg N, 22.5 kg P_2O_5, and 27 kg K_2O per fed (fed = 4200 m^2), apply compost at a rate of 4 m^3 per bolstered (Handz soil-compost), and water plants every 6 days.

According to Alez et al. (2021), hibiscus crops were grown in rain-fed rural conditions in Bajo Balsas, Michoacan, Mexico. The study found that the hibiscus crop from the considered location grows vegetatively and blooms between June and August. It was noted that temperatures exceeding 38 °C during June are unusually frequent (more than 12 days), suggesting that warm push may be more important for hibiscus crops in this area than water stress.

According to Sepahrom and Moosavi (2016), the plant height, stem width, fruit number per m2, sepal fresh and dry yield, natural yield, and sepal dry weight per single fruit all decreased to 48.1%, 31.8%, 41.3%, 58%, 54.9%, 67%, and 22.7%, respectively, as a result of increasing irrigation intervals from 60 to 180 mm dissipation. In addition, irrigation following a 120 mm vanishing with a 180 kg N. ha^{-1} treatment may be suggested for hibiscus development in Saravan, Iran, by the findings of the current study. According to Gholam and Moosavi (2012), the planting date for Hibiscus under Iranian conditions was suggested to be May 4 at a density of 8 plants/m^2.

Hibiscus is tolerant of arid climes, especially in its last phases of development, and is very sensitive to fluctuations in day duration. As a result, consideration should be focused on a reasonable day duration rather than precipitation. According to Youdeowei et al. (1986), the timing of cultivation activities is critical for the development of all crops, especially annuals that are susceptible to drought or late rain, both of which are beyond the farmers' control. In India, planting typically occurs at the start of the rainy season (Ansari et al. 2013). In Northern Nigeria, Hibiscus planted in mid-July showed enhanced vegetative features (Futuless et al. 2010).

Nnebue et al. (2014) reported that in the humid tropical environment of southeastern Nigeria, where Hibiscus or Roselle is priced for its calyx from which the popular drink called "Zobo" is prepared, Hibiscus or Roselle grown for the leaves can be planted in May and should be harvested not later than 8 WAP to avoid possible fungal infection. Roselle grown for calyx, however, should be planted in July in Owerri to allow for early and enhanced flowering during the short day length periods of September.

Concurring with Hassan et al. (2023) and El-Sagher et al. (2024) this plant appears a wider range of flexibility to natural conditions, climates and soil sorts. In Egypt, Hibiscus development is widespread, particularly in regions with hot summers and mellow winters such as Aswan Governorate. Aswan appreciates a comparative advantage in hibiscus generation, particularly with the development of arrival recovery ventures pointed at expanding the agricultural zone. Climate change poses a danger to rural generations in locales encountering food instability, with serious natural stress such as drought, heat waves and storms. Future climate forecasts demonstrate a critical increment in temperature and sporadic, strong precipitation designs, making the climate strong and versatile, savvy rural hones pivotal for economic efficiency (Habib-ur-Rahman et al. 2022).

According to El-Sagher et al. (2024), Applying dissimilar planting dates (15th April, 1st May, 15th May and 1st June) led to significant differences in plant height, stem diameter, branch number/plant, fruit number/plant, shoot fresh weight and shoot dry weight. It was found that mostly higher values of these traits were recorded at planting on 15th May. Seghatoleslami et al. (2013), considering that cold occurs even at the initial sowing date, Roselle cultivation can be stated to have the risk of chilling in Birjand, Iran, unless it may be effectively sown at dates earlier than May 10. These sowing dates, however, carry the risk of late spring cooling, too.

Nanotechnology and precision agriculture Nanoparticles delivered to plants and advanced biosensors for precision farming, Hassanein et al. (2021) found that the Nano zinc oxide of 0, 5, 10, and 15 ppm was used as the main plot, and biostimulants were used as the sub-plot. The results showed that application of nano zinc led to significant increases in growth, yield, photosynthetic pigments, as well as anthocyanin content associated with increased zinc content in leaves. The highest values of all traits occurred under the treatment of 10 ppm of nano zinc. The first method of propagation of Roselle by seeds i.e., sowing directly in the field and sowing in seed beds, Seeds are planted at a rate of 6–8 kg/ha and approximately 2.5 cm deep, 60 cm to 1 m between rows and 45–60 cm apart. Sowing is done by hand or using a modern grain drill.

14.2.2 Vertical Farming

"Light recipes" for indoor crops and any seasonal farming innovation. According to Purabi Barui et al. (2022), vertical cultivation is the process of growing crops in vertically stacked layers or incorporated into other structures (like an old warehouse or skyscraper) using less water and no soil. Advanced vertical farming concepts make use of indoor farming techniques and controlled environment agriculture (CEA) innovation, which allows for the control of all environmental variables, including temperature, light, and humidity. Biofortification, or the process of breeding crops to increase their nutritional value, is another example of this.

Despite financial and environmental obstacles, urban agriculture is growing. The development of crops suited to the hotter temperatures predicted by the middle of the twenty-first century is one strategy to mitigate the impact of climate change on food security in cities with a temperate environment. Richardson and Arlotta (2021), illustrate that *H. sabdariffa*, a tropical plant, can be developed in calm, urban frameworks. Hydroponics and vertical cultivating address the requirement of secure and healthy, pesticide/acaricides/insecticide-free, common antioxidant wealthy produce with low carbon and water footprint (Gao et al, 2024). The process of producing food and medicine in vertically stacked layers, vertically tilted surfaces, or locations in other structures, like a skyscraper, a former warehouse, or a shipping container and terrace, is known as vertical farming (Umashankar et al. 2020). In order to address the problems caused by the current conventional farming, or horizontal farming, practiced in India, the extended study is restricted to suggesting a self-sufficient structure based on vertical farming using an advanced cultivating method of soil-less agriculture.

14.2.3 Digital Farming

The agribusiness innovation industry and digitalization go to make sustainable and proficient agribusiness while keeping up food security. Artificial insights, web of things and advanced administrations are profoundly prioritized topics and age guarantees the following insurgency in horticulture. Openings advertised through digitization can be connected along the manufacturers' whole esteem chain.

The agriculture apparatus industry can offer assistance in examining straightforward manual equipment to collect Roselle blooms and to partition them by sepals—natural products within the field, too considers inserting few suitable but basic advances for precise gathering, making beyond any doubt the Roselle blooms collected are in best quality and in best cleanliness by user-friendly, ergonomic and secure to utilize forms.

The challenge is to plan a machine that isolates the hibiscus petals from its seeds to diminish the time essential to get ready for the crop. The proposed arrangement ought to fulfill the following prerequisites: make strides in the generation capacity. The machine ought to be eco-friendly. The partition handle ought to be isolated into two forms; to begin with, the entire stem must be separated from the blossom itself. The separation of seeds from petals. Isolate intaglio seeds and collect them productively (to be utilized for angle bolster or to be replanted). It is best that the machine would be versatile to decrease the time required to transport the trim from the field to the machine's area. Agreeing with Ali et al. (2019), innovation industry and creating a partition Roselle, there are two fundamental issues related to gathering Roselle (1) frequent exchanging of Roselle blooms, and (2) wastefulness of Roselle deseeding handle.

Current practices of collecting Roselle are that the blossom collector's employments reusable things and transfers Roselle blooms as often as possible from put to put some time recently conveying them to the deseeding center, which once more harming the blossoms as well as making hygienist concerns. Eliw (2012), detailed that the most noteworthy values of machine efficiency of 18.522 kg/h, least sepals harm at 1.66%, partition productivity of 98.36% and least particular vitality utilization of 0.045 kW.h/kg were recorded within the case of cutting sepals pipe speed 100 rpm, utilizing bolstering disk with 21/25 mm orifice measurement and partition sepals after fruits gathering straightforwardly. The full operation took a toll of 1.10 L. E/kg and 1.36 L. E/kg were obtained in mechanical and manual partition sepals separately, but the net trim esteem from the last item of sepals is much more costly in the mechanical strategy compared to the conventional manual strategy.

Microwave-assisted hydrodistillation (MAHD) (Djouahri et al. 2013; Jeyaratnam et al. 2016) are various central focuses of utilizing microwave-assisted hydrodistillation (MAHD), such as its moo fetched, and it is one of the show day and speedy techniques, and it is conceivable to encourage predominant abandon from the removed oils. Rassem et al. (2022), Revelations outline that MAHD unpleasant extricates have more essential supportive potential. MAHD procedure yielded higher quality hibiscus sprout oil than the conventional HD (hydrodistillation) procedure which reveals that MAHD may be utilized to successfully remove unsteady oils from hibiscus blooms without exasperating their interesting chemical structures. Due to its tall resign and moo vitality utilization, MAHD is illustrated to be an eco-friendly elective for separating essential oils compared to HD.

14.2.4 Artificial Intelligence

The rural industry faces a few noteworthy challenges related to successfully utilizing modern advances, counting finding information and relationships from verifiable records, handling huge volumes of unstructured information in a fitting organize, overseeing broad sums of picture and video information, checking crops utilizing different sensors, and successfully communicating and joining this information. Moreover, the appropriation and openness of developing innovations can be cost-prohibitive for person ranchers, whereas a need for low-technology mastery requires additional preparation and superior data and communication innovation (ICT) administration gear. Guaranteeing the security of these frameworks is additionally a basic concern inside the rural community (Khosla 2010; McBratney et al. 2005).

According to Jayaraman et al. (2016), smart farming uses information and communication technologies (ICT), specifically the Internet of Things (IoT) and associated big data analytics, to monitor crops electronically and related environmental, soil, fertilization, and irrigation conditions. (AI) is one of the speediest developing innovations in this cutting-edge world. AI is presently broadly utilized in different businesses extending from the advancement of mechanical autonomy to the checking of natural issues (Davenport and Kalakota 2019) and detailed that agriculture is a necessary portion of the human community. It has been overhauled with time to extend agricultural yields. Present-day innovations have moreover been utilized in agribusiness, cultivation, and forestry for way better monitoring of plant development, diseases, and pest population. In expansion, AI and IoT offer fabulous scopes for genuine time observation of temperature, humidity, contamination, water substance, soil quality, radiation, and others. In order to forecast yields using sophisticated yield prediction models created through machine learning and to achieve predictive decision-making capabilities in precision agriculture, it is also emphasized how crucial it is to gather and analyze big data from a variety of sources, especially in situ and mobile sensors (Alahmad et al. 2023).

14.2.5 Breeding Objectives of Hibiscus

The goal is to prolong the flowering time in low-light situations. Vigour, robust root system, good growth habit, and eye-catching foliage. Some varieties can be displayed in a garden or used for pot culture. Reddish in color, the flowers have a new form, size, and shape, and are of good preservation quality. Variegated leaf variants with double flowers. Create new hybrid ideotypes with high-quality colors (Janakiram and Patil 2017). To develop cultivars that are disease-, drought-, and pest-resistant. The majority of districts have been exhausted by hibiscus changes brought about by standard and innovative breeding techniques, and wonderful varieties are being advanced through planned hybridization. Since 1995, the Institute of Plant Breeding (IPB), Crop Science Cluster (CSC), and University of the Philippines Los Banos (UPLB) have been breeding and determining hibiscus in order to continuously create new hybrids.

14.3 Germplasm Biodiversity and Conservation

Conservation of genetic resources in plant gene banks through various approaches is highly beneficial for seed preservation (USDA 2007; Abulaila et al. 2025; Abdullaev et al. 2025; Akparov et al. 2025; Aladele et al. 2025; Al Lawati 2025; Almarri et al. 2025; Arab et al. 2025; Békefi et al. 2025; Breidy 2025; Cañar et al. 2025; de Galarreta et al. 2025; Desheva et al. 2025; Díez et al. 2025; Galea-Gragera and Llera Cid 2025; Ghidan and Abdulmajid 2025; Gudaye et al. 2025; Harzli et al. 2025; Hidayatun and Kurniawan 2025; Ihsan and Prihatini 2025; Jatoi and Siddiqui 2025; Laghetti et al. 2025; Mallor and Luna 2025; Monteros-Altamirano et al. 2025; Osterc et al. 2025; Popović et al. 2025; Rocha et al. 2025; Wang et al. 2025). On the other hand, conservation helps preserve biodiversity and identify varieties with beneficial and adaptive genes.

14.3.1 Germplasm Biodiversity

Genetic resources are necessary for countries to boost agricultural output. Crop diversification is essential to adapting to the negative effects of climate change and addressing the problems of the food crisis; it is frequently especially necessary for rapid population growth. These days, improving and developing crops of local interest is essential to ensuring food security and economic prosperity for the people of developing countries (Sourabie et al. 2020). There are 4225 species and 244 genera in the Malvaceae (Christenhusz and Byng 2016). The genus Hibiscus contains over 250 species that grow in temperate to tropical regions (Mandaji et al. 2022). Trees, bushes, and plants are among these species (Abdullaha et al. 2020). Plants of the genus *Hibiscus* have been considered and utilized not as it were for decorative purposes, but also as useful nourishments and medications (Abdelhafez et al. 2020). Roselle has been mostly considered on its utility in diverse industries, but few on genetic traits and structure (Montalvo-González et al. 2022; Sánchez-Mendoza et al. 2008). Since the genome sequence of the chloroplast, an organelle that primarily carries out photosynthesis, is largely conserved, it could be a key component used in the categorization, separation, and developmental handling of species (Cheng et al. 2020; Kim et al. 2021; Liu et al. 2019). The whole genome of *H. sabdariffa* chloroplasts was collected. The total size of the genome was 162,428 bp. Since *Hibiscus sabdariffa* L. has 2n = 36 chromosomes, its somatic cells have 72 (2n = 4x = 72) tetraploid chromosomes.

Roselle, which ranks second in West Bengal and generation after jute, is one of the most important bast fiber crops, according to Satyanarayana et al. (2015). It is typically used to make carpets, sacs, and twines. Given that crosses between parents with the greatest genetic divergence are more likely to produce desirable recombinants in the offspring, proximity to sufficient genetic changeability may be a prerequisite for establishing a breeding program aimed at improving yield and other traits. This suggests that there are forces other than geographic division, such as genetic drift, natural variation, commerce in breeding materials, and artificial and natural selection. The findings are consistent with those of Anuradha (2007), PulliBai et al. (2005), Hariram Kumar et al. (2013), Nayak et al. (2009), and Roy et al. (2023). Going ahead Farmers have embraced *H. sabdariffa* cultivation, which can benefit the nation's communities. Breeders need to have access to the most notable genetic differences of the species under consideration to produce variations that are both economical and stable (Baudoin et al. 2001). Natural conditions and climate change regularly have an imperative effect on the phenotypic characteristics; this could lead to noteworthy factors coming about from high quality and environmental intelligence (Mahunu et al. 2021a, b). Slamet 2018 claimed that the character of the qualitative and quantitative morphology of Hibiscus was evaluated descriptively.

According to Bakasso et al. (2014), the two distinct groups were linked by morphological contrasts but were not frequently used by farmers for their classification, suggesting that roselle diversity is genetically structured. According to Han et al. (2022), there is a richness of phenotypic variability in Hibiscus cultivar resources.

The basis for Hibiscus classification is shaped by the external form and hue of the flowers. Its morphological classification was also significantly influenced by tree shapes, development tendency, and leaf base shapes, which established the foundation for the guarantee, the use of Hibiscus germplasm resources, and research into underutilized kinds.

14.3.2 Cultivars Characterization and Phylogeny

According to Pfeil et al. (2002), a review of the literature on morphological traits that are thought to be inherited showed that hibiscus and hibiscuses are unlikely to be monophyletic groups because they may share ancestral characteristics. Omalsaad et al. (2014) displayed that the genetic connections between roselle and kenaf accessions utilizing chloroplast and mitochondrial preliminary based on RAPD markers are exceptionally valuable to be able to know the intraspecific diversity. The contrasts between a few accessions within the same species were not sufficient. Information from this consideration is exceptionally imperative for the taxonomic classification of roselle and kenaf to the conservation and management program and the improvement of varieties of these two species can be executed more effectively.

The anthocyanin cyanidin 3-glucoside, which serves as a marker and turns red in acidic solutions and blue in basic ones, is found in rosella flowers, according to Nuryanti et al. (2019). A substitute for litmus, pointer paper is manufactured from rosella flowers, which change color to blue in base solutions and red in acidic ones. The color change is identical to that of the red and blue litmus comparison indicator. In accordance with Kwon et al. (2022), *H. sabdariffa* is the sister plant of all seven Hibiscus species in the evolutionary tree. Given the number of species in the genus, the total chloroplast genomes of Hibiscus are likewise insufficient to accurately determine the evolutionary relationships of the individuals in the genus. Regardless, our phylogenetic analysis suggests that *H. sabdariffa* is unique. The entire chloroplast genome of *H. sabdariffa*, as determined in this study, can provide useful information to promote consideration of biodiversity and advancement with Hibiscus species and Malvaceae, which include many economically important plants.

14.3.3 Genetic Resources Conservation Approaches

In general, obtaining a broad range of plant genetic resources—mostly farmers' varieties, landraces, and crop wild relatives (CWR), the foundational elements of intra- and interspecific crop diversity is necessary for effective breeding (Abdel-Hamid and Salem 2021; Abu-Ellail et al. 2021; Aldahak et al. 2021; Alqahtani et al. 2024; El Rabey et al. 2013; Hassan et al. 2021; Ibrahim et al. 2023b, 2024; Salem and Ibrahim 2024; Salem et al. 2015a, b, c, 2023b, 2024). According to Dwivedi et al. (2016), these plant genetic resources are a veritable gold mine of traits for

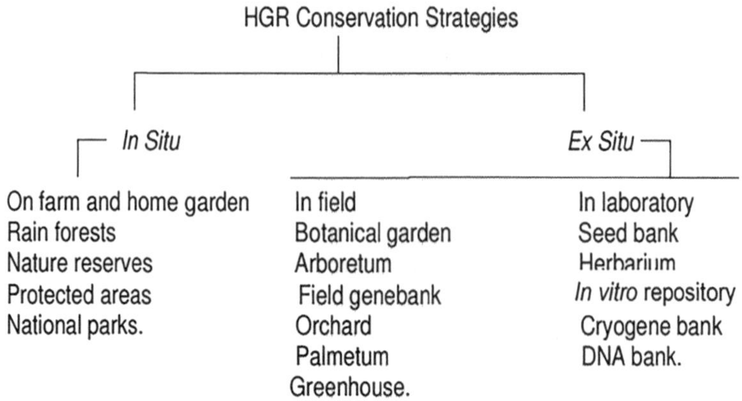

Fig. 14.5 Conversation approaches of horticultural genetic resources (HGR). (Source: Ogbu et al. (2010))

improving vegetables and vegetable crops. As stated in the revised universal GenBank Standards published in 2014, genebanks now serve a wider range of actors and institutions than just the traditional R&D setup: "Genebanks ought to advance the accessibility of genetic resources for uses counting investigate, breed, educate, farm, and repatriate" (FAO 2014).

According to Fassil et al. (2000), the Institute's work recognizes the need to use plant genetic resources and conservation management to ensure the availability of a wide variety of nutrient-dense food crops (Fig. 14.5). Utilizing plant genetic diversity is essential to increasing the nutritional value of a variety of food crops and guaranteeing a steady and acceptable supply. Through the efforts of the International Plant Genetic Resources Institute (IPGRI), the role that viable conservation, management, and utilization of plant genetic resources play in guaranteeing the availability of a variety of nutrient-dense food crops is acknowledged.

Breeders should keep a pool of varied, acceptable donor parents in crop development programs in order to increase yield (Aboshama et al. 2015; Börner et al. 2009, 2010; Emara et al. 2011, 2013; El-Zanaty et al. 2013; Joshi et al. 2013; Hassan et al. 2021; Ismail et al. 2013; Mattar et al. 2009; Nawara et al. 2016, 2017; Saleh et al. 2020b; Salem 1997, 2004, 2009, 2015; Salem et al. 2007a, b, 2021b, 2022). For the effective protection and use of genetic resources, a thorough grasp of genetic diversity is essential (Elabd and Salem 2009; El-Absawy et al. 2012, 2015; Eltaher et al. 2017, 2018, 2021, 2022; El-Zanaty et al. 2011; Esmail et al. 2009; Ghazy et al. 2020; Mattar et al. 2016; Salem and El-Zanaty 2015; Salem and Mattar 2014a, b; Salem and Sallam 2016; Saleh et al. 2020a; Salem et al. 2008a, b, 2010, 2015, 2021c, 2021f; Varshney et al. 2008). Since the crosses made between the parents with the greatest genetic divergence would be more likely to produce desirable recombinants within the offspring, the proximity of adequate genetic diversity may be a prerequisite to defining a breeding program aimed at advancement in yield and other characteristics. Satyanarayana et al. (2016) detailed that to improve such imperative fiber crop through breeding, thinking about on genetic diversity of vital characteristics responsible for fiber yield is fundamental to recognizing diverse parents for hybridization programs to advance high-yielding recombinant.

14.4 Cytogenetic

Providing broad information about a plant's evolutionary history, identifying its cytological traits, and identifying similarities between the case study and other plants, such as species affinities, are the main objectives of cytogenetic investigations. A portion of this data is essential for reaching plant breeding objectives and formulating suitable regulations. Cytogenetic and karyological investigations in several populations of the medicinal plant known as sour tea have not yet been conducted (Salehi et al. 2022). Cytogenetics has been able to understand the structure of the *H. syriacus* genome and track chromosomal segregation patterns following ploidy modification and interspecific hybridization by using methods such as fluorescence in situ hybridization (FISH) and genome in situ hybridization (GISH) (Deepo et al. 2022; Lattier et al. 2019).

14.4.1 Improvement Strategies

Opportune sowing dates get approximately legitimate improvement and development of plants satisfying in most extreme yield of the crop and economic utilize of land (Islam et al. 2010). In contrast, delayed seeding resulted in a decrease in calyces fresh and dry weight, calyces yield of roselle, plant height, stem diameter, number of leaves/plants, and number of calyces/plants (Khattak et al. 2016). According to the findings of Ali et al. (2020), it was suitable to sow on April 15th and give roselle plants 200 ppm of glutamine acid to increase the growth and yield of *Hibiscus sabdariffa* L. plants grown in clay soil under Sharkia Governorate circumstances.

Appropriate cultivars, timely seeding, and the selection of appropriate natural conditions can all result in the highest yield. Because it can have a favorable or negative effect on plant performance and yields, the date of sowing is a crucial factor in crop development and production (Futuless et al. 2010). The selection of superior varieties from among variants is the core of plant breeding. It's not easy to choose a parent from a large collection of germplasm. Numerous promising cultivars have been produced through seedling selection from open fertilization (OP) (Janakiram and Patil 2017), and evaluation of genetic diversity provides help in producing the pattern data to direct selection of parental lines and increase the accuracy of a breeding program in creating superior variety (Maruthi et al. 2022).

14.4.2 Traditional Breeding Methodologies

The "traditional" plant breeding techniques will be refined further and will remain significant (Abdel-Hamid and Salem 2021; Salem et al. 2007a, b, 2022). Developments in the acceptance of genetic modification may play a role in this. Because transformation techniques, which may be more effective for introducing traits like male sterility or disease resistance, are less appealing in comparison to the more laborious process of putting varieties on the market, elaborate cell fusion or wide crossing techniques may occasionally still be used to introduce traits like these. Predicting methods that are more experimental in nature is more challenging.

One of the most potent interactions between people, the environment, and plants is plant breeding (Ibrahim et al. 2023a, b, c; Elabd et al. 2020; Mourad et al. 2019; Saleh et al. 2021a, b; Salem and Börner 2003, 2004; Salem et al. 2003, 2004a, b, 2021a, b, d, e, 2023a, b). Without agriculture, we could not create our civilization, and without contemporary crop types, agriculture could not support a civilized society. For millennia, breeders have relied on classical selection techniques to improve Hibiscus cultivars. These methods involve selecting desirable traits from existing plants and crossing them to create offspring with the combined characteristics. While successful in developing some of the stunning varieties we see today, traditional breeding is a time-consuming process and is limited by the available genetic pool. Hybridization: Crossing desired parent plants with complementary traits remains a cornerstone. By carefully selecting parents with vibrant colors, unique flower shapes, or disease resistance, breeders can create offspring with a combination of these characteristics. Selection: Observing and selecting naturally occurring mutations or variations within a population can yield desirable traits. This method, while time-consuming, has led to the development of many popular cultivars.

14.4.3 Traditional Breeding Limitation

Hibiscus's attributes can be improved by traditional breeding, but a comprehensive genetic transformation technique must be developed in order to rapidly increase the physical material properties as a structural material. There has been a lot of interest in improving Rosella's tolerance to cold temperatures in particular. Due to (1) the brief time that the flowers were open, (2) the presence of stigma tips that were flush with the opening of the staminal column, and (3) the fact that stigmas did not grow in air and that closing the flowers prevented further genetic exchange, Howard and Howard (1911) reported no natural outcrossing in roselle in India. In contrast, 0.23% of roselle in India underwent natural outcrossing, according to Sanyal and Dutta (1954).

14.4.4 Role of Biotechnology

The horticulture industry's genetic variety demands have led to the development of modern molecular approaches that span from morphological characterization to several DNA-based markers including Randomly Amplified Polymorphic DNA (RAPD), Inter Simple Sequence Repeat (ISSR), and Simple Sequence Repeats (SSRs). The conservation and use of plant genetic resources depend on the identification and characterization of germplasm. One of the best ways to preserve plant genetic resources is to characterize plants using molecular markers. Individual reproductive success, the presence of gene flow, the movement of alleles within and across populations of the same or related species, and its effects are all determined by molecular characterization (Papa and Gepts 2003).

Mutation is one method used to try to promote genetic diversity. If a species of plant germplasm lacks the desired trait, mutation breeding can be used to improve the plant's characteristics. A mutagen, such as gamma-ray irradiation, is used to induce mutations. A population base with a high level of genetic variety is necessary for the new types. Based on the morphological characteristics of roselle plants, gamma-ray irradiation dosages ranging from 150 to 600 Gy have an impact on the variance of phenotypes in M_2 generation. These alterations take place in the roselle plant's qualitative and quantitative traits (Syahputra et al. 2017).

14.5 Molecular Breeding

With more than 200 varieties, the genus Hibiscus has enthralled people for millennia with its vivid blooms, many applications, and possible health advantages. However, the speed and accuracy of conventional breeding techniques are limited. This is where molecular breeding comes in, providing a ground-breaking method for releasing Hibiscus's latent potential.

A New Dawn of Molecular Breeding for Hibiscus Improvement provides a potent toolkit to get around the drawbacks of conventional techniques. This method finds and chooses plants with the desired features by using molecular markers, which are particular DNA sequences connected to desired attributes. Techniques like marker-assisted selection (MAS), genomics as well as bioinformatics allowed breeders to identify the superior genotypes/traits at an early stage, significantly accelerating the breeding process (Collard and Mackill 2008.). The potential applications of molecular breeding in Hibiscus are vast and hold immense promise for cultivators and consumers alike. Here are some key areas of exploration: Enhanced Ornamental Appeal: Breeders can leverage molecular markers to select for desirable flower color, size, and shape. Imagine a world with even more vibrant and long-lasting Hibiscus blooms, thanks to the precision of molecular breeding (Berloo 2000). Improved Disease Resistance: Identifying and selecting plants with genes resistant to common Hibiscus diseases, like fungal infections and bacterial wilt, can significantly reduce reliance on pesticides and ensure healthier plants (Esse et al. 2020). Tailored Fiber Production: Species like *Hibiscus cannabinus* are valued for their strong fibers. Molecular breeding can help select plants with higher fiber yield and improved fiber quality, leading to more sustainable and efficient production (Al-Mamun et al. 2022). Optimizing Medicinal Properties: Certain Hibiscus species hold promise for their health benefits. Molecular breeding can be used to identify and enhance the production of specific bioactive compounds with medicinal value (Ojulari et al. 2019).

14.5.1 Marker-Assisted Selection in Hibiscus

In hibiscus breeding, molecular marker-assisted selection (MAS) presents a viable way around constraints and laborious methods. Breeders can choose people with desired qualities early on, even before they are physically displayed, by using MAS, which uses DNA markers associated with those traits. While traditional breeding programs rely on phenotypic selection, which can be ineffective and constrained by things like the dominance of undesirable alleles or the late expression of traits, this greatly speeds up the breeding process and increases selection efficiency (Hasan et al. 2021; Song et al. 2023).

14.5.2 Prospective Benefits of Marker-Assisted Selection (MAS) in Hibiscus Breeding

Selecting desirable qualities at the seedling stage speeds up the production of new cultivars since it shortens the hibiscus breeding cycle and enables growers to release new varieties with better traits onto the market more quickly (Vayabari et al. 2023). increased selection accuracy because, in contrast to the traditional phenotypic selection, which can be subjective and impacted by environmental factors, MAS offers a more dependable and objective way of choosing plants with desirable features (Abdullaha et al. 2020). Selection for complex traits: MAS can be used to select complex traits controlled by multiple genes, which are challenging to address through traditional breeding methods (Abdixofiz and Samatova 2021).

While MAS research on hibiscus is still developing, some studies have shown its potential. For instance, a study identified molecular markers specific to the Korean Hibiscus cultivar 'Simbaek' using chloroplast DNA (Wang et al. 2021). This demonstrates the feasibility of using molecular markers for cultivar identification and germplasm management. However, some challenges need to be addressed for the widespread adoption of MAS in hibiscus breeding. For example; Limited availability of linked markers linked to specific traits in hibiscus. More research is needed to identify and develop markers associated with desirable characteristics like flower color, disease resistance, or fiber quality. Cost and infrastructure as implementing MAS requires specialized equipment and expertise, which are expensive for small breeding programs. Developing cost-effective and user-friendly MAS techniques is crucial for wider accessibility. Molecular marker-assisted breeding holds immense potential for accelerating and improving hibiscus breeding programs. By addressing the current limitations and actively pursuing research in this field, breeders can unlock the full potential of hibiscus for various applications.

14.5.3 Genomics in Hibiscus

Genomics is the cutting-edge science that explores an organism's entire genetic content, and its DNA blueprint. This allows us to peek into the instruction manual that dictates everything from flower color to disease resistance. Genomics can pinpoint the genes responsible for flower color and shape traits, paving the way for breeding programs to create breathtaking new varieties (Yang et al. 2022). Also, it can help to identify genes linked to disease resistance. By understanding these guardians, breeders can develop hibiscus that fight off common ailments naturally (Kim et al. 2017). Our changing environment poses a challenge for many plants. Genomics can identify genes for drought and heat tolerance. This knowledge can be used to breed resilient hibiscus varieties that thrive in a warming world. Recent research has even sequenced the genome of Hibiscus hamabo, a champion of salty and waterlogged habitats. By mapping its DNA, scientists found genes likely responsible for this remarkable tolerance. This opens doors for engineering other hibiscus species to withstand harsh conditions (Wang et al. 2022). Genomics in hibiscus research is still young, but its potential is vast. By unlocking the secrets hidden in its genes, there will be more creation, more stunning and adaptable hibiscus varieties, but also gain a deeper appreciation for the intricate dance of life within these captivating flowers.

Genes implicated in the development of tiny ridges on Hibiscus trionum petal epidermal cells have been identified by genome and transcriptome analysis (Koshimizu et al. 2023). Also, a study in 2024 focused on the complete chloroplast genomes of various cultivars of *Hibiscus syriacus* (Rose of Sharon). Understanding the chloroplast genome is valuable for exploring genetic diversity and developing breeding markers (Go et al. 2024). It was also stated that understanding genomics can provide insights into the genome of *Hibiscus sinosyriacus* and compare it to related species. Understanding the chloroplast genome is crucial for studying plant evolution and adaptation (Kwon et al. 2023). These references provide a starting point for exploring the field of genomics in hibiscus. They showcase research on genome assembly, chloroplast genomes, and the link between genes and specific traits like petal ridges.

14.5.4 Bioinformatics of Hibiscus

Bioinformatics tools can analyze the vast amount of biological data generated from hibiscus studies, leading to a deeper understanding of its genetics, physiology, and potential applications as this plant is well known for its stunning blooms and diverse uses, presents exciting opportunities for bioinformatics research.

Speaking about Current Bioinformatics Applications, researchers can collect a hibiscus species' entire DNA sequence through genome sequencing. Bioinformatics tools then assemble this data into a contiguous map, providing insights into gene organization and regulatory elements (Kim et al. 2017). This information is essential for figuring out which genes are linked to desired characteristics and for comprehending how the genus evolved. RNA sequencing, or transcriptome analysis, enables researchers to examine the patterns of gene expression in diverse tissues or environments. This data is then analyzed by bioinformatics techniques to determine which genes exhibit differential expression. This helps understand the molecular mechanisms behind specific traits like flower color or fiber development (Lyu et al. 2020). Bioinformatics tools can also be used for marker development to analyze genetic variations within and between hibiscus species. This facilitates the development of molecular markers, which are DNA sequences linked to specific traits. These markers allow breeders to select desirable traits more efficiently, accelerating breeding programs (Wang et al. 2021).

Bioinformatics tools give a chance for Emerging Applications, such as Metabolic Pathway Analysis to reconstruct and analyze the metabolic pathways in hibiscus. This allows researchers to identify key enzymes and metabolites involved in the production of valuable compounds like pigments or antioxidants found in hibiscus (Xu et al. 2021). Similarly, Genome-Wide Association Studies (GWAS) by combining genetic data with phenotypic data, GWAS can identify genetic loci associated with specific traits (Eltaher et al. 2017, 2018, 2021, 2022). This powerful tool allows researchers to pinpoint the genes responsible for important characteristics in hibiscus, such as disease resistance or fiber quality (Kim et al. 2024). Despite the exciting potential, limitations exist. Limited genomic resources for some hibiscus species and the need for robust bioinformatics pipelines specifically tailored to hibiscus research are areas for improvement. However, the future of bioinformatics in hibiscus research is bright. As more genomic data becomes available and bioinformatics tools are further developed, we can expect significant advancements in understanding the intricate biology of hibiscus. In the end, this information will help create better cultivars, more environmentally friendly farming methods, and the discovery of new bioactive substances from this intriguing genus.

14.6 Tissue Culture Applications

Tissue culture is a fascinating technique that utilizes tiny plant tissues to generate multiple, healthy plantlets under controlled laboratory conditions (Nawara et al. 2016, 2017). For hibiscus, this translates to several advantages such as; rapid propagation, disease-free plants, uniformity, year-round production etcetera. By harnessing the power of tissue culture, hibiscus growers can meet the ever-growing demand for these beautiful plants while ensuring quality and efficiency.

14.6.1 Micropropagation

It is also referred to as tissue culture and is a technique for rapidly multiplying plants under sterile conditions. It offers several advantages over traditional propagation methods like seeds or stem cuttings, including rapid propagation, which allows the mass production of hibiscus plants in a much shorter time frame compared to traditional methods like cuttings or seeds. Additionally, propagating disease-free plants in the sterile tissue culture environment reduces the danger of pests and illnesses, resulting in healthier plants. Micropropagation creates true-to-type plants, which are genetically identical to the parent plant. This is crucial for preserving particular characteristics like disease resistance or flower color. Year-round production, as we are controlling the propagation environment we are not exposed to seasonal limitations of traditional methods, tissue culture allows for continuous production of hibiscus plants throughout the year. Faster production as it can produce a large number of plants in a short period, unlike seeds which may have slow germination rates.

The process of micropropagation in hibiscus requires the following: Explant selection: A small piece of tissue, called an explant, is taken from a healthy hibiscus plant. This can be a shoot tip, nodal segment (the area where a leaf joins the stem), or even a single bud. Explant sterilization: The selected explant is then sterilized using a bleach or alcohol solution to kill any bacteria or fungi on its surface with a special amount yet not to kill the explant itself. Media inoculation: A nutrient media that contains all the necessary components for plant growth, such as carbohydrates, minerals, and vitamins, is then put on top of the sterilized explant. To promote shoot multiplication or the start of hairy roots, a certain kind of plant growth regulator (PGR) is also added to the medium. A popular basal medium for micropropagating a variety of plants, including hibiscus, is the Murashige and Skoog (MS) medium. Shoot multiplication: The explant is incubated under controlled environmental conditions of temperature, light, and humidity. Under the influence of PGRs, the explant will develop multiple shoots. Rooting: Once enough shoots have developed, they are transferred to a new medium containing a different type of PGR to stimulate root development. Acclimatization: The rooted plantlets are then gradually introduced to non-sterile conditions (similar to a greenhouse or outdoor environment) to allow them to adapt before transplanting them to the soil.

Micropropagation is an intricate process that calls for certain tools and knowledge. It can, however, be a useful technique for producing large quantities of high-quality hibiscus plants or even for propagating a specific portion of the plant that contains the necessary active ingredient, saving time compared to propagating the entire plant (Lobodina et al. 2020; Sereda et al. 2024).

14.6.2 Embryo Rescue

In plant breeding, embryo rescue is a method used to save immature embryos from crossings that might not otherwise produce viable seeds. When it comes to interspecific hybridizations, this method is especially useful for hibiscus breeding.

Why do we use the Embryo Rescue technique? Interspecific hybridization: When breeding between different hibiscus species, incompatibility between the parents can hinder embryo development. The embryo might stop growing prematurely, leading to seed abortion even though pollination was successful. Embryo rescue allows us to intervene and cultivate these immature embryos into complete plants (Islam and Deepo 2020; Magdalita and San Pascual 2022). Also, Weak embryos even within the same species, some crosses might result in embryos that are too weak to mature naturally within the developing seed pod. Embryo rescue provides a chance for these embryos to develop under controlled conditions (Laere et al. 2006).

Depending on the species and the study, the precise procedures for hibiscus embryo rescue can change (Islam and Deepo 2020; Laere et al. 2006; Magdalita and San Pascual 2022). However, the fundamental pattern entails: Controlled pollination: To guarantee the desired crosses, flowers are hand-pollinated. Monitoring pod development: The developing seed pods are monitored closely, often starting a few days after pollination. Embryo excision: At a specific stage based on research on embryo development for the particular hibiscus species, the pod is dissected under a stereomicroscope to remove immature embryos. Aseptic culture: The excised embryos are placed in a sterile culture medium that provides nutrients and a controlled environment for further development into complete plantlets.

Other investigations discuss the Embryo development of woody and perennial *Hibiscus* species (Islam and Deepo 2020), as embryo development in interspecific hibiscus crosses, providing valuable information for timing embryo rescue in these species. The Institute of Plant Breeding, University of the Philippines Los Baños, showcases embryo rescue as part of their hibiscus breeding program. Other Guevarra (under publication) studied the Development of New Hibiscus varieties through Conventional Hybridisation in Hibiscus rosa-sinensis and wide Hybridisation and embryo rescue in Hibiscus. Additionally, Laere et al. (2006) examined breeding strategies to promote genetic variety within Hibiscus syriacus, emphasizing the use of embryo rescue in overcoming obstacles during interspecific hybridization between *Hibiscus syriacus* and *Hibiscus paramutabilis*. By employing embryo rescue techniques, hibiscus breeders can increase the success rate of their crosses and develop exciting new varieties with unique characteristics.

14.6.3 In Vitro Pollination

In vitro pollination (IVP) offers a valuable tool for hibiscus breeders and researchers, particularly when dealing with controlled environments, rare breeding lines, or overcoming incompatibility issues between desired parent plants. Studies have shown success with IVP in hibiscus, achieving 100% fruit set in certain cultivars compared to 0% natural fruit set due to the absence of suitable pollinators (Abirami and Kannan 2018). This technique allows for precise pollen application and manipulation, facilitating hybridization efforts and the creation of novel hibiscus varieties.

While in vitro pollination (IVP) holds promise for hibiscus breeding, optimizing the process requires ongoing research. Success hinges on the development of a suitable culture medium that mimics the natural environment within the flower. Studies suggest that sucrose concentration plays a crucial role in pollen germination, with optimal levels influencing pollen tube growth and ultimately, seed set (Baloch et al. 2021). Beyond sugars, the medium may also include calcium, boric acid, and growth regulators to further enhance pollen viability and receptivity of the stigma (Abirami and Kannan 2018).

However, challenges remain. While some studies report successful IVP in hibiscus, germination rates can be low, and even successful pollen tube growth may not always translate to seed formation (Abirami and Kannan 2018). Further research is needed to refine media composition and culture conditions to achieve consistent and efficient IVP in a wider range of hibiscus varieties. Breeders will be able to produce intriguing new hibiscus cultivars with desired traits as a result.

14.6.4 Synthetic Seeds for Hibiscus Propagation

Synthetic seeds (synseeds) offer an exciting avenue for hibiscus propagation, particularly for valuable or challenging-to-propagate cultivars. Here's a closer look at this developing technology. Synseeds concept and creation are artificially produced structures encapsulating plant propagules like shoot tips or embryos in a protective coating. In the case of hibiscus, researchers have successfully employed nodal segments—short sections containing a bud and surrounding tissues for Synseeds production. These segments are typically encapsulated in a biocompatible gel, often sodium alginate, hardened with a calcium chloride solution (Preece and West 2006; West and Preece 2009). This coating protects the propagules from desiccation, pathogens, and harsh environmental conditions, facilitating easier storage and transport.

Advantages of Synseeds for Hibiscus: Germplasm Preservation: Synseeds can extend the viability of valuable hibiscus germplasm for longer periods compared to traditional methods like cuttings. This is particularly beneficial for rare or endangered varieties. Efficient Propagation: Synseeds allow for mass propagation of desirable hibiscus cultivars in a controlled environment. This can be advantageous for commercial growers or breeding programs. Reduced Disease Transmission: The protective coating of synseeds minimizes the risk of transmitting diseases through propagation material. Simplified Transport: Synseeds are lightweight and require less stringent storage conditions compared to whole plants or cuttings. This simplifies transportation and international exchange of valuable hibiscus germplasm.

The concept of synthetic seeds, encapsulated structures mimicking natural seeds for plant propagation, holds immense potential for hibiscus cultivation. However, significant research hurdles remain before this technology becomes a reality for this vibrant flower genus. While promising, there are a lot of changes facing Synseed technology for hibiscus as it is still under development. Optimizing the encapsulation process and media composition is crucial for maximizing germination rates and plantlet development after Synseed conversion (West and Preece 2009). Furthermore, studies are required to determine the genetic stability and long-term survival of plants grown from Synseeds.

14.6.5 Challenges in Developing Synthetic Seeds for Hibiscus

Encapsulation Technology: Developing a suitable material for the synthetic seed coat in hibiscus is crucial. It needs to be biocompatible, allow for gas exchange, and facilitate germination while protecting the enclosed material (Nongdam 2016). Embryo Maturation: Current techniques for synthetic seed production often rely on somatic embryogenesis, where plant cells are induced to develop into embryos. Optimizing this process for hibiscus, particularly for recalcitrant (difficult to propagate) varieties, remains a challenge (Mujib 2016). Conversion Efficiency: Achieving high conversion rates from encapsulated material to viable plantlets is essential for practical application. To increase the effectiveness of synthetic seed germination in hibiscus, more investigation is required (Sharma et al. 2020).

Future studies should concentrate on the following areas: Material Sciences: Investigating innovative biomaterials for encapsulating that are suited to hibiscus' particular requirements. As well as Double Somatic Embryogenesis: Refining techniques for inducing and maturing somatic embryos in hibiscus varieties with low embryogenic potential. Similarly, Germination Optimization: Investigating the influence of culture conditions and media composition on the success of synthetic seed germination in hibiscus. Synthetic seed technology holds significant potential for efficient propagation, germplasm preservation, and disease control in hibiscus. Continued research will refine this technique, paving the way for its wider adoption in the field of hibiscus cultivation and breeding.

14.6.6 In Vitro Application that Pertains to the Species

Hibiscus boasts a rich history of traditional medicinal use. In vitro studies are now providing a deeper understanding of these properties. Researchers are culturing cells or tissues of hibiscus to investigate their; Antioxidant activity: Extracts from hibiscus calyces (the fleshy part that surrounds the flower) have shown promise in scavenging free radicals, potentially aiding in preventing various diseases. In vitro assays allow scientists to precisely measure this activity and identify the specific compounds responsible. Antimicrobial properties: Extracts from certain hibiscus species have demonstrated effectiveness against bacteria and fungi. In vitro studies can pinpoint which strains are susceptible and optimize extraction methods for maximum potency. The allure of hibiscus extends beyond medicine. Breeders are constantly striving for new varieties with enhanced colors, disease resistance, or flower size. In vitro techniques play a crucial role in this involving Micropropagation and Genetic Transformation. The innovative world of in vitro research continues to push boundaries. Here are some exciting possibilities for hibiscus:

Natural Food Colorants: Hibiscus extracts possess vibrant colors ideal for food coloring. In vitro studies can optimize growth conditions for maximizing pigment production, leading to sustainable and natural alternatives to synthetic dyes. Biofuel Production: Certain hibiscus species show promise for biofuel production. *In vitro* studies can be used to identify and enhance the production of specific metabolites suitable for conversion into biofuels. Finally, in vitro research with hibiscus is opening a world of possibilities beyond its ornamental charm. From unlocking its medicinal potential to aiding plant breeders and exploring sustainable applications, this fascinating flower is poised to make a significant contribution to science and society. As research continues, the hidden potential of hibiscus will undoubtedly continue to blossom.

14.7 Genetic Engineering and Gene Editing

A potent technique for introducing new genetic material into the hibiscus genome is genetic engineering.

14.7.1 Genetic Transformation Methodology

Agrobacterium-mediated transformation and particle bombardment are two methods that can accomplish this. By introducing genes from other species or altering those that are already present, we may be able to give hibiscus new characteristics like improved resistance to disease, changed bloom color, or higher fiber output. Previous research has demonstrated the feasibility of genetic engineering in hibiscus. Studies by Allen et al. (2019) successfully introduced a gene conferring insect resistance into Hibiscus syriacus, paving the way for the development of pest-resistant varieties Mehrotra et al. (2011). Likewise, Yang et al. (1996) explored the use of Agrobacterium-mediated transformation to modify flower color in Hibiscus rosa-sinensis, showcasing the potential for genetic engineering in ornamental applications.

Hibiscus plants, particularly *Hibiscus sabdariffa* (Roselle), have been successfully genetically transformed using *Agrobacterium tumefaciens*-mediated transformation. This technique is frequently used to introduce novel genes into plants. The methodology of transformation by Agrobacterium as described by Gassama-Dia and Ndoye (2004); Plasmid Construction: plasmid was firstly designed containing the desired gene along with regulatory sequences for proper expression in Hibiscus. This may involve promoters specific to Hibiscus or inducible promoters for controlled gene expression. Antibiotics must be included as resistance markers for the selection of transformed cells. Agrobacterium Preparation: Agrobacterium tumefaciens, a soil bacterium, was grown competent for plasmid uptake, then constructed plasmid containing the gene of interest was added into Agrobacterium. Hibiscus Tissue Culture: Hibiscus sterilized seeds or selected explants (leaf tissue, stem tissue) were prepared for callus induction. Callus formation was induced on a culture medium containing appropriate hormones. Co-cultivation: Agrobacterium culture was incubated with the Hibiscus callus tissue, allowing for bacterial attachment and transfer of the plasmid DNA. Selection and Regeneration: The co-cultured tissue was transferred to a fresh medium containing antibiotics to eliminate non-transformed cells. The antibiotic resistance gene in the plasmid allows transformed cells to survive. Encourage regeneration of transformed shoots from the callus tissue on a shoot regeneration medium. Plant Regeneration and Analysis: To promote the growth of roots, the regenerated shoots were placed in a rooting media. Once established, use PCR (Polymerase Chain Reaction) and other molecular biology tools to examine the plants for the transgene's presence and expression.

14.7.2 Gene Editing Methodology

Gene editing is a more recent and precise technique, that allows for targeted manipulation of the hibiscus genome. Tools like CRISPR-Cas9 enable researchers to insert, delete, or modify specific genes with unprecedented accuracy. This opens up a plethora of possibilities, including the development of hibiscus varieties resistant to specific pests or herbicides, the manipulation of flowering pathways for extended bloom times, or even the enhancement of specific medicinal compounds. However, gene editing in hibiscus remains in its infancy. While preliminary studies demonstrated the in vitro editing of a specific gene in Hibiscus cannabinus, further research is needed to optimize these techniques and translate them to whole-plant applications (Liu et al. 2016).

This research project aims to build upon this existing knowledge and explore the untapped potential of genetic engineering and gene editing in hibiscus. We will focus on Ozkan et al. (2023), e.g., developing herbicide-tolerant hibiscus varieties for weed management in kenaf production or enhancing the production of specific medicinal compounds in *Hibiscus sabdariffa* calyces. Through rigorous experimentation and analysis, we hope to contribute to the advancement of this field and unlock the hidden potential of this captivating genus.

14.7.2.1 Methodology: Gene Editing in Hibiscus

Hibiscus, with its vibrant flowers and diverse species, holds the potential for genetic modification through techniques like CRISPR-Cas9. Here's a methodology outline for gene editing in Hibiscus as mentioned by Giovannini et al. (2021): Plant Material Selection and Sterilization, the desired Hibiscus species were chosen based on the target trait (e.g., flower color, pest resistance). Young, healthy leaves were selected or embryogenic calli were for efficient transformation. Explant Surface sterilized (leaves/Calli) with a disinfectant like 70% ethanol followed by sterile water rinses. CRISPR-Cas9 Construct Design The target gene sequence responsible for the desired trait was identified using available Hibiscus genome resources or transcriptome analysis. Design-specific CRISPR guide RNAs (sgRNAs) were designed to target the desired region within the gene using online sgRNA design tools. A plasmid vector was constructed containing Cas9 nuclease and the designed sgRNA sequences. Delivery of CRISPR-Cas9 Construct into Hibiscus Cells. CRISPR-Cas9 construct was selected to deliver the into the plant cells. Popular options include Agrobacterium-mediated transformation: This method utilizes engineered Agrobacterium strains to deliver the vector into the host plant cells. Particle bombardment: Tiny gold particles coated with the DNA construct are bombarded onto the plant tissue, facilitating DNA uptake.

Selection and Regeneration of Transformed Plants. A selectable marker gene was introduced (e.g., herbicide resistance) along with the CRISPR-Cas9 construct into the vector. This allows for the identification and selection of transformed cells. The transformed explants were incubated on a culture medium containing the appropriate selection agent to eliminate untransformed cells. Whole plants were regenerated from the transformed callus using established plant tissue culture techniques for Hibiscus.

14.7.2.2 Gene Editing Analysis

The regenerated plants' genomic DNA was extracted. The modified gene sequence was then amplified using PCR (Polymerase Chain Reaction) with primers flanking the target area. Analyze the mutations caused by CRISPR-Cas9 using Sanger sequencing or a high-throughput sequencing technique (such as next-generation sequencing).

14.7.2.3 Confirmation of Trait Modification

The edited plants were evaluated for the presence of the desired phenotypic changes associated with the targeted gene. This may involve observing flower color alterations, increased disease resistance, or other modifications depending on the target trait.

This methodology provides a basic framework for gene editing in Hibiscus using CRISPR-Cas9. Remember to adapt and optimize the protocol based on your specific target gene and desired outcome.

14.7.3 Enhanced Traits in Hibiscus

Research on hibiscus has been conducted to enhance a number of characteristics. This explores current efforts to enhance hibiscus through breeding and other techniques. Traditional breeding methods remain a cornerstone for hibiscus improvement. Breeders focus on enhancing flower size, color intensity, disease resistance, and fiber yield (in species like). For instance, studies have shown success in developing hibiscus varieties with larger, more vibrantly colored flowers through selective breeding techniques (Fadia et al. 2011).

Recent research delves into using novel methods to improve hibiscus. One example involves the use of nanoparticles derived from hibiscus leaves. This technique, though still under investigation, has shown promise in enhancing certain agricultural traits in other plants (Alloosh et al. 2021; Almuhammady et al. 2021; Alnaddaf et al. 2021; Ayantoye et al. 2023; Ibrahim et al. 2023a, b, c; Eid et al. 2025). To ascertain its efficacy and safety for hibiscus growing, more investigation is necessary. Although progress is being made, improving hibiscus characteristics is not without its difficulties. Techniques like gamma irradiation, explored for mutation induction, can have unintended consequences, potentially reducing plant size and yield (Fadia et al. 2011). Furthermore, the safety and environmental impact of novel techniques like nanoparticle application requires thorough investigation before widespread adoption.

Research on enhancing hibiscus traits holds promise for cultivators and consumers alike. Breeders are continuously developing varieties with improved aesthetics, disease resistance, and fiber production. Additionally, novel techniques offer exciting possibilities but require careful evaluation to ensure their effectiveness and safety.

14.7.4 Nanotechnology

Hibiscus holds immense potential beyond the realm of technology. This review explores the rich tapestry of non-technological applications of hibiscus, highlighting its significance in various aspects of human life. Culinary Delights: Hibiscus flowers and leaves lend themselves to a diverse culinary landscape. The calyces of *Hibiscus sabdariffa* are used to produce a tart and refreshing beverage known as karkadé or roselle (Jabeur et al. 2017). Hibiscus candied petals and leaves add a touch of floral elegance to salads, desserts, and teas (https://en.wikipedia.org/wiki/Plant Resources of Tropical Africa).

Medicinal Marvel: Hibiscus is highly valued in traditional medicine. The leaves of Hibiscus rosa-sinensis have antioxidant and anti-inflammatory qualities (Ju-Zhao et al. 2022). The calyces of *Hibiscus sabdariffa* are explored for their potential benefits in blood pressure management (Jalalyazdi et al. 2019). However, further research is warranted to validate these claims.

Textile Tapestry: *Hibiscus cannabinus*, also known as kenaf, boasts strong and durable fibers. These are traditionally used to produce ropes, sacks, and even clothing (Mazharul 2021) With growing concerns over synthetic textiles, kenaf offers a sustainable and eco-friendly alternative. Cultural Canvas: Hibiscus transcends utility, weaving itself into the cultural fabric of societies. In many tropical regions, hibiscus flowers are symbolic of beauty and hospitality (Mod and Mint 2024). Malaysia's national flower is the Chinese hibiscus, Hibiscus rosa-sinensis. Ecological Advantage: Hibiscus serves as a vital ecological component. Certain hibiscus species act as nectar sources for butterflies and hummingbirds, promoting pollination (Chamier 2020). The fast-growing nature of some hibiscus varieties makes them ideal for soil erosion control (Loehwing 1931).

Hibiscus offers a treasure trove of benefits beyond the technological realm. From its culinary versatility and medicinal potential to its role in textiles, culture, and ecology, hibiscus is a plant with a profound impact on human life. Further research into its properties and traditional uses holds the promise of unlocking even greater possibilities for this remarkable genus.

14.8 Mutation Breeding

By modifying their genes, plants can be improved through a process called mutation breeding. It is a relatively new method that is very effective in creating new varieties of hibiscus with desired traits. These benefits may include increased flower size and color variety, improved disease resistance, enhanced tolerance to environmental stresses, and more compact growth habits. Researchers are using mutation breeding to create hibiscus varieties that are more resistant to pests and diseases, such as hibiscus wilt and canker. They are also working to develop hibiscus varieties that can tolerate drought conditions and other environmental stresses.

14.8.1 Conventional Mutagens: Seed

Mutations can occur in hibiscus seeds, producing new types with distinct traits. These mutations can be caused by different mutagens or they can happen spontaneously at a very low rate. Hibiscus seed mutations can alter several characteristics, including leaf morphology, bloom color, size, and form. These mutations can be used by breeders to produce new hibiscus types with desired traits.

14.8.2 In Vitro Mutagenesis and Selection

Physical or chemical substances known as mutagens accelerate the rate of mutation. Common mutagens used in plant breeding include radiation and chemicals like ethyl methanesulfonate (EMS). Mutation process steps; Mutagen application: Seeds are exposed to a mutagen, which increases the rate of mutation. Mutation induction: The mutagen damages the DNA in the seeds, which can lead to mutations. Seed germination: The seeds are planted and allowed to germinate. Mutant selection: The seedlings are grown and observed for any mutations. Evaluation and breeding: If a desired mutation is observed, the plant can be propagated and used in breeding programs to create new hibiscus varieties (Janakiram and Patil 2017). It's important to note that inducing mutations can also have negative consequences. For example, some mutations may be harmful to the plant or reduce its yield. It is therefore important to carefully evaluate any mutations that are induced.

14.8.3 Molecular Analysis in Hibiscus

Hibiscus, with its vibrant flowers and diverse species, has captivated humans for centuries. However, traditional methods of classification and analysis based on morphology often fall short due to environmental influences. This is where molecular analysis steps in, offering a powerful and precise approach to studying hibiscus at the molecular level (Amany et al. 2019). Molecular markers, like DNA analysis techniques, have revolutionized our understanding of hibiscus. Let's have a look at the exciting world of molecular analysis in hibiscus research:

Similarities in morphology might be deceptive. Molecular markers like ISSR (Inter-Simple Sequence Repeat) and RAPD (Random Amplified Polymorphic DNA) can reliably differentiate between closely related hibiscus species, even ones that share similar looks. According to Rita et al. (2016), this is essential for breeding programs and conservation initiatives.

Genetic Variability and Germplasm Management: Breeding operations require an awareness of the genetic variability within hibiscus populations. Molecular markers like SSRs (Simple Sequence Repeats) allow researchers to assess genetic variation and identify unique genotypes. This information is vital for selecting parents with desirable traits and avoiding inbreeding depression (Bakasso et al. 2014).

Additionally, hibiscus species are recognized for their therapeutic qualities due to their bioactive compounds. The genes that produce these bioactive substances can be found with the use of molecular analysis. This information can be used to create new medications and enhance ones that already exist.

In addition to disease resistance and pest control, research into the molecular underpinnings of hibiscus disease resistance may result in the creation of resistant cultivars. Furthermore, molecular markers can help identify certain infections and pests, enabling more focused pest management techniques. Despite its immense potential, molecular analysis in hibiscus research also faces challenges. The high cost of equipment and expertise required can limit accessibility, particularly in developing countries. Additionally, the complex nature of plant genomes necessitates continuous refinement of techniques and analysis methods.

In conclusion, molecular analysis has become an indispensable tool for hibiscus research. By unveiling the intricate details of the hibiscus genome, it empowers researchers to address critical questions in breeding, conservation, medicine, and agriculture. As research continues to advance, we can expect even more exciting discoveries that will unlock the full potential of this fascinating genus.

14.8.4 Enhance Traits and Improved Cultivars

Hibiscus offers a treasure trove for breeders seeking to create even more captivating cultivars. Enhancing traits and improved cultivars can be done by mutation and genetic engineering.

As discussed earlier, exposing seeds to mutagens like radiation or chemicals can induce mutations, some of which may be beneficial. This approach can be useful for creating better traits not readily available through traditional breeding. (Ma et al. 2021). Marker-Assisted Selection (MAS): Molecular markers linked to desirable traits can be used to identify plants with those traits early in the breeding process, even before flowering. This significantly reduces breeding time and allows for more targeted selection. (An et al. 2023).

14.9 Genetic Engineering

Gene Editing Techniques like CRISPR: These powerful tools offer the potential to precisely modify specific genes, potentially leading to the creation of hibiscus varieties with entirely new traits, such as extended flower life or enhanced disease resistance. However, ethical considerations and regulatory hurdles surround GMOs, requiring careful evaluation. (Lin and Jones 2022).

Optimizing Growing Conditions; Understanding the influence of light, temperature, and fertilizer on flower size, color intensity, and blooming frequency can significantly impact the final product. By tailoring the growing environment, breeders can encourage the expression of desired traits in existing cultivars. (Manokari et al. 2016).

The Importance of Collaboration; Breeders, geneticists, and horticulturists working together can achieve significant advancements. Breeders can provide practical breeding goals, geneticists can identify relevant genes and markers, and horticulturists can optimize growing conditions to maximize the potential of new cultivars. Breeders can produce even more beautiful and hardy hibiscus cultivars in the future by combining conventional breeding techniques, state-of-the-art technologies, and a thorough understanding of hibiscus biology. The secret to releasing this fascinating genus's full potential lies in this cooperative approach.

14.10 Hybridization

14.10.1 Conventional Hybridization

Conventional breeding incorporates hybridization into its process to generate novel gene combinations from parent kinds. According to Vaidya (2000), unique methods beyond standard cross-pollination in the field would be needed for the successful cross-fertilization of Hibiscus sabdariffa var. Altissima and var. Sabdariffa. However, it was noted that whereas Hibiscus sabdariffa var. Sabdariffa accessions may only be crossed partially through pairwise designs, Hibiscus sabdariffa var. Altissima accessions may be crossed completely through pairwise designs. However, it could be challenging to successfully hybridize accessions of the latter. Therefore, to improve the likelihood of successful hybridization, more crosses should be tried, ideally early in the day.

Up to 2 weeks following pollination, each bloom was observed to evaluate the compatibility of hybridization success. Depending on the accessions used in the cross, fertilized flowers produced fruit capsules containing seeds 7–14 days following pollination. Two to three days after pollination, the flowers in areas where fertilization was unsuccessful fell. Each cross's hybridization success was determined using the following formula, which was developed by Kwofie et al. (2017): Hybridisation success (%) = Number of fruits /formed Number of pollinated flowers × 100%.

Hibiscus grows exceptionally well in Egypt all year round. The flowers are sturdy and huge (earlier types were usually crimson), however they usually don't smell. Many different types, cultivars, and hybrids have single or double sets of petals with flowers that range in color from white to yellow and orange to red and pink (Prasad 2014; Moustafa et al. 2021).

The plump, glossy crimson calyces are the primary reason for cultivating *H. sabdariffa* var. sabdariffa (HSS). The Central Research Institute for Jute and Allied Fibres (ICAR) in India has the largest collection of Roselle accessions (628) in the world (Mahapatra 2008).

Hybridization and mutation are the most common ways to increase genetic diversity (Siregar et al. 2019). Conventional selection techniques are also expected to improve one or more of the desired characteristics. Up until 2009, researchers in India improved eight varieties of Roselle with high fiber content by selecting and hybridizing indigenous landraces (Kar et al. 2010). Finding high-quality, highly diverse germplasm is crucial to preserving or increasing the species' yield of fiber and calyx. Therefore, the most important opportunity in crop development initiatives is to adequately characterize the available germplasm. Because it allows for simple and rapid scoring, morpho-logical attributes (based on phenotypic traits) are typically used to choose germplasm throughout the gene bank (Mahunu et al. 2021a, b).

Mutation: Genetic and environmental interactions influence the morphology of mutant plant populations. Gy gamma-ray irradiation impacted Hibiscus sabdariffa's morphological character performance, increasing the average height, calyx weight, canopy diameter, and total number of fruits per plant (Hanafiah et al. 2017; Siregar et al. 2019; Syahputra et al. 2017).

Breeding procedures can be used to overcome the limitations of roselle plants' superior nature (Siregar et al. 2019). The common single red (*H. rosa-sinensis* L.), coral hibiscus (*H. schizopetalus* Hooker), narrow-petaled white (Hibiscus sp. from Fiji), and the common pink hedge hibiscus (perhaps a hybrid of H. cameroni) are some of the invasive species that are frequently employed in hybridization. Cultivars of Hibiscus rosa-sinensis are the product of ancient and contemporary crosses between a small number of species, most of which are from the Hawaiian Islands in the Pacific and the Mascarene Islands in the Indian Ocean (Janakiram and Patil 2017).

14.10.2 Somatic Cell Hybridization

Somatic cell hybridization is the ideal method for isolating a desired chromosome from the complete complement and ensuring a steady supply of the chromosome. This unit begins with the selection of fusion techniques and selectable markers for hybrids bearing a chromosome of interest. Whole-cell hybrids are created by fusing two cell lines—a monolayer (adherent) recipient and a donor that can be adherent or grown in suspension—in the first set of techniques. The second set of protocols describes the formation of micronuclei with a limited number of chromosomes and their enucleation to produce microcells for fusion with recipient cells. Support protocols cover the purification of microcell preparations, subcloning of whole-cell hybrid populations for chromosome separation in isolation lines, cloning cylinder preparation and use for tissue culture colony isolation, and molecular and cytogenetic characterization techniques (Jackson 1996).

Somatic hybridization with the frost-tolerant *Lavatera thuringiaca* has been tried to improve Hibiscus for greater frost tolerance. It was possible to regenerate plants through somatic embryogenesis using both *L. thuringiaca* transgenic calluses and isolated protoplasts. Regeneration from the fusant calluses that were resistant to two antibiotics, however, was not successful. Independent genetic components influence freezing and chilling tolerance, according to the callus-level analysis of the somatic hybrids. Although frost tolerance was not expressed, the somatic hybrids showed a significant improvement in their ability to withstand chilling temperatures at which H. rosa-sinensis is fatal (Vazquez-Thello et al. 1996).

Chen and Contreras (2022) reported that the ploidy and chromosomal level diversity produced by the population derived from hybrid plants may yield unique traits. Progenies with a variety of ploidy levels, including aneuploids, were produced by the crossing process and could help create clonal and decorative plants. Our results also show that interspecific hybrid Hibiscus cultivars are useful breeding resources to enhance bloom size through ploidy modification and transgressive segregation, despite their relatively low male fertility.

14.11 Conclusions and Prospects

In the family Malvaceae, the genus Hibiscus contains the commercially important flowering plant species known as hibiscus. In addition to the dried calyces, which have a berry-like flavor and are utilized industrially all over the world, the stems of this plant are used to make bark fibers. The stems offer bast fiber and can be used to make hessian, a biodegradable fiber, in place of jute. Hibiscus holds a significant rank among fiber crops, after jute and is the predominant farmed fiber crop. Both physical and chemical analyses indicate that hibiscus fiber is an excellent source of textile materials. The physical characteristics of hibiscus fiber include high tensile strength, significant gloss and brightness, and resilience to heat, light, acids, and alkalis. Fiber resources can provide superior mixed fabrics. The use of non-conventional natural fibers, such as bast fiber, has become a significant objective to mitigate environmental degradation. Every part of the hibiscus plant is used in medicine and serves as a significant source of pharmaceuticals and the production of safe, natural remedies.

A delightful, caffeine-free beverage full of organic acids is produced by hibiscus. It treats circulatory issues, colds, coughs, and appetite loss as a general health tonic. Additionally, hibiscus has modest diuretic, expectorant, and laxative properties. It is used as a beverage to alleviate inflammation, hypertension, and liver diseases. External antioxidants, rich in anthocyanins and flavonoids, particularly found in flower petals, help the body mitigate excessive oxidative stress.

Hibiscus exhibits antihypertensive, anticancer, and antidiabetic properties. Anthocyanins significantly influence the cardiovascular system by enhancing vasodilation and cardiac output in black currants, thereby increasing peripheral blood flow. The name of this plant varies from country to country. Hibiscus can thrive in various environmental conditions. Leaf production increases when the plant is grown in soil at ground level.

Plants from various genetic sources, such as contemporary cultivars and regionally adapted types, must be hybridized to introduce novel alleles because traditional breeding procedures are time-consuming and labor-intensive. Genetic diversity is essential for plant breeders to increase crop yields, and access to breeding germplasm is very limited. This process increases genetic diversity and enables the preselection of desirable traits that are necessary for maintaining significant natural phenotypic diversity.

To develop new hibiscus varieties that improve stem and calyx yield, as well as address biotic and abiotic challenges, nutrient requirements, and stressors, significant obstacles exist in hibiscus breeding. Temperatures are the primary global climate variables of interest to hibiscus breeders. Hibiscus is a relatively drought-tolerant species; drought poses a major challenge to global crop breeding, especially in arid and semi-arid regions. This challenge has been worsened by climate variability and the ongoing effects of climate change worldwide. Diseases and pests are expected to increase rapidly and impact food security. Assessment of the detrimental impacts of fertilization and irrigation methods is essential for the key agronomic improvements of Hibiscus. Therefore, further work is required to create novel cultivars that can withstand abiotic stresses like salinity and drought. Potential new types with desirable features have been created using modern biotechnology techniques. Breeders and geneticists are being motivated by recent hibiscus genome sequencing and molecular techniques to create new cultivars with advantageous agronomic features and resistance to biotic and abiotic stressors.

In a crop enhancement initiative to increase yield, the breeder must maintain a repository of diverse, suitable donor parents. For genetic resources to be preserved and used sustainably, a thorough understanding of genetic variety is essential. Establishing a breeding program aimed at improving yield and other traits requires adequate genetic diversity since crossing parents with high levels of genetic divergence increases the likelihood that the offspring would contain the desired recombinants. Enhancing fiber production through breeding requires an examination of the genetic diversity of biological traits influencing fiber yield, crucial for identifying diverse parental lines for hybridization programs aimed at producing high-yielding recombinants.

Genetic resources are necessary for countries to increase their agricultural output economically, crop diversification is necessary to mitigate the negative effects of climate change and address food crisis issues, and the development and improvement of locally important crops are necessary to ensure food security and economic prosperity for the populations of these countries. There are several reasons to keep researching hibiscus development, which will advance both quantitative and qualitative aspects. Fundamental biotic and abiotic stresses, physiology, yield, and constituent traits should all be directly related to the creation of DNA markers. To improve these traits, it is necessary to discover genes or QTLs related to both qualitative and quantitative attributes.

Molecular breeding is a technique used to improve plants through genetic manipulation. This innovative method has shown effectiveness in creating hibiscus varieties with distinct characteristics. These benefits may include larger flowers, a variety of colors, improved disease resistance, better tolerance to environmental stress, and more compact growth. Researchers are using molecular breeding to create hibiscus variants that are more resistant to pests and diseases like hibiscus wilt and canker. They are also working to develop hibiscus cultivars that can survive drought conditions and other environmental challenges.

Furthermore, improvements in germplasm and biotechnology are necessary to accelerate the development of promising novel genotypes characterized by high yield and quality. Moreover, further research is needed on the interactions between hibiscus and microbes, including fungi and bacteria, to develop cultivars that demonstrate tolerance to biotic stresses.

Novel biotechnology techniques for hibiscus have captured the attention of breeders and geneticists due to advances in DNA sequencing and molecular methodologies. However, conventional breeding practices remain essential for developing new hibiscus cultivars with desired characteristics. Despite this, several promising cultivars have been created that show tolerance to climate change, as well as biotic and abiotic stressors, emphasizing the need for innovative breeding techniques for hibiscus. A potent technique for adding new genetic material to the hibiscus genome is genetic engineering. Particle bombardment and Agrobacterium-mediated transformation are two techniques that can accomplish this. Through the integration of genes from other species or the modification of preexisting genes, hibiscus can develop new characteristics including enhanced resistance to disease, changed bloom color, or greater fiber production. Previous studies have demonstrated the feasibility of genetic modification in hibiscus.

Gene editing, a modern and precise method, allows for targeted alteration of the hibiscus genome. Technologies like CRISPR-Cas9 enable researchers to insert, remove, or alter individual genes with unparalleled accuracy. This opens up various possibilities, such as creating hibiscus varieties resistant to specific pests or herbicides, modifying flowering pathways for longer blooming periods, or enhancing specific therapeutic components. However, gene editing in hibiscus is still in its early stages. Initial studies have shown in vitro editing of a single gene in Hibiscus cannabinus; however, further research is necessary to refine these techniques and apply them to whole plants. Current research is exploring innovative methods for enhancing hibiscus, but more studies are required to determine their effectiveness and safety in hibiscus cultivation. Despite progress, challenges persist in improving hibiscus traits. Techniques like gamma irradiation, which have been investigated for inducing mutations, may have unintended consequences, potentially reducing plant size. Additionally, the safety and environmental impacts of emerging technologies, such as nanoparticle application, need thorough examination before widespread adoption.

Appendix 14.1 List of Major Institutes Engaged in Research on Roselle (*Hibiscus sabdariffa* L.)

Country	Institute name	Website
Egypt	Horticulture Research Institute (HRI), Agricultural Research Center (ARC)	Nadi Elsaid St., Horticulture Research Institute (HRI), Agricultural Research Center (ARC), Dokki, Giza, Egypt. http://www.arc.sci.eg/InstsLabs/Default.a
India	Central Research Institute for Jute and Allied Fibres (ICAR)	Bus Stop, Barrackpore—Barasat Rd., near Saheb Bagan Lane, Nilganj, Barrackpur Cantonment, West Bengal 700120, India https://crijaf.icar.gov.in/
Malaysia	Malaysian Agricultural Research and Development Institute (MARDI)	Ibu Pejabat MARDI, Persiaran MARDI-UPM, 43400 Serdang, Selangor, Malaysia
Malaysia	Malaya University and University of Kebangsaan Malaysia (UKM)	43600 UKM Bangi, Selangor Malaysia https://www.ukm.my/portalukm/contact-us/
Malaysia	Malaysian Nuclear Agency	59, Jalan Kemudi 2b, Taman Bernam Jaya, 44000 Kerling, Selangor, Malaysia https://www.nuclearmalaysia.gov.my/
Sudan	National Centre for Research (NCR)	El Mek Nimir Avenu, Khartoum, Sudan https://www.ncr.gov.sd

Appendix 14.2 World List of Recommended Roselle (*Hibiscus sabdariffa* L.) Varieties

Country	Recommended varieties
Philippian	HSS
Philippian	Rico
Philippian	Victor
Philippian	Archer
Malaysia	Temprano
Malaysia	UKMR-1
Malaysia	UKMR-2
Malaysia	UKMR-3

References

Abdelhafez OH, Othman EM, Fahim JR et al (2020) Metabolomics analysis and biological investigation of three Malvaceae plants. Phytochem Anal 31:204–214

Abdel-Hamid AME, Salem KFM (2021) Breeding strategies of garden pea (*Pisum sativum* L.). In: Al-Khayri JM, Jain SM, Johnson DV (eds) Advances in plant breeding strategies: vegetable crops. Springer, Cham. https://doi.org/10.1007/978-3-030-66969-0_9

Abdixofiz LR, Samatova SA (2021) Results of a complex estimation of varieties of *Hibiscus syriacus* L. in the Conditions of Tashkent. Am J Plant Sci 12:566–572

Abdullaev FK, Rustamov NSU, Yakubov MD (2025) Uzbekistan Genebank. In: Al-Khayri JM, Salem KFM, Jain SM (eds) Plant gene banks. Springer, Singapore. https://doi.org/10.1007/978-981-99-4236-7_57-1

Abdullaha F, Mehmooda I, Shahzadia S et al (2020) Chloroplast genome of Hibiscus rosa-sinensis (Malvaceae): comparative analyses and identification of mutational hotspots. Genomics 112(1):581–591

Abirami S, Kannan M (2018) Pollination, pollen biology, stigma receptivity, seed dormancy breaking treatments on hibiscus rosa- sinensis. Int J Eng Sci Invent (IJESI) 7(8):1–10

Aboshama HM, Salem KFM, Hassan HMF, Tahoun MH (2015) Estimating heterosis and combining ability for in vivo and in vitro traits using diallel cross on tomato (Solanum lycopersicum L.). Minufiya J Agric Res 40(3):757–770

Abu Hassan NA, Ahmad S, Chen RS et al (2023) Tailoring lightweight, mechanical and thermal performance of PLA/recycled HDPE biocomposite foams reinforced with kenaf fibre. Ind Crop Prod 197:116632

Abu-Ellail FFB, Salem KFM, Saleh MM et al (2021) Molecular breeding strategies of beetroot (Beta vulgaris ssp. vulgaris var. conditiva Alefeld). In: Al-Khayri JM, Jain SM, Johnson DV (eds) Advances in plant breeding strategies: vegetable crops. Springer, Cham. https://doi.org/10.1007/978-3-030-66965-2_4

Abulaila K, Al Sane KO, Tahabsom Z (2025) Jordan Seed Bank. In: Al-Khayri JM, Salem KFM, Jain SM (eds) Plant gene banks. Springer, Singapore. https://doi.org/10.1007/978-981-99-4236-7_45-1

Akparov ZI, Mammadova SA, Kalantarova NS, Mirzaliyeva IA, Mammadov AT (2025) National Genebank of Azerbaijan. In: Al-Khayri JM, Salem KFM, Jain SM (eds) Plant gene banks. Springer, Singapore. https://doi.org/10.1007/978-981-99-4236-7_31-1

Al Lawati AH (2025) Plant genetic resource conservation of Oman. In: Al-Khayri JM, Salem KFM, Jain SM (eds) Plant gene banks. Springer, Singapore. https://doi.org/10.1007/978-981-99-4236-7_50-1

Aladele SE et al (2025) Plant genetic resources conservation and utilization in Nigeria. In: Al-Khayri JM, Salem KFM, Jain SM (eds) Plant gene banks. Springer, Singapore. https://doi.org/10.1007/978-981-99-4236-7_17-1

Alahmad T, Neményi M, Nyéki A (2023) Applying IoT sensors and big data to improve precision crop production: a review. Agronomy 13(10):2603. https://doi.org/10.3390/agronomy13102603

Aldahak L, Salem KFM, Al-Salim SHF, Al-Khayri JM (2021) Advances in chicory (Cichorium intybus L.) breeding strategies. In: Al-Khayri JM, Jain SM, Johnson DV (eds) Advances plant breeding strategies: vegetable crops. Springer, Cham. https://doi.org/10.1007/978-3-030-66969-0_1

Alez CM, Montiel C, Ortega A, Pacheco A, Bautista F (2021) Development and validation of climatic hazard indicators for roselle (*Hibiscus sabdariffa* L.) crop in dryland agriculture. Ecol Indic 121:107–140

Ali EA, Kamarudin KM, Che Me R, Sulaiman R, Alli H (2019) The development of sorrel systematic roselle harvesting system for efficient transfer process. IOP Conf Series: Mater Sci Eng 697:012029. https://doi.org/10.1088/1757-899X/697/1/012029

Ali HKM, Awad AE, Abdelkader MAI (2020) Improving growth and yield of roselle (*Hibiscus sabdariffa* L.) Plants by using tyrosine and glutamine acids under different sowing dates Zagazig. J Agric Res 47(5):1165–1174

Allen JJ, Kannan M, Thamaraiselvi SP, Kumar M (2019) Genetic variability, heritability and correlation studies in Hibiscus rosa-sinensis. J Pharm Phytochem 8(1):1001–1004

Alloosh MT, Saleh MM, Alnaddaf LM, Almuhammady AK, Salem KFM, Al-Khayri JM (2021) Biosynthesis and characterization of microorganisms-derived nanomaterials. In: Al-Khayri JM, Ansari MI, Singh AK (eds) Nanobiotechnology. Springer, Cham. https://doi.org/10.1007/978-3-030-73606-4_10

Al-Mamun M, Mohd Y, Rafii MAB et al (2022) Kenaf (*Hibiscus Cannabinus* L.): a promising fiber crop with potential for genetic improvement utilizing both conventional and molecular approaches. J Nat Fibers 20(1):1–19

Almarri NB, Ahmad S, Alsubaie GTG, Elshal MH (2025) Saudi Arabia Gene Bank. In: Al-Khayri JM, Salem KFM, Jain SM (eds) Plant gene banks. Springer, Singapore. https://doi.org/10.1007/978-981-99-4236-7_54-1

Almuhammady AK, Salem KFM, Alloosh MT, Saleh MM, Alnaddaf LM, Al-Khayri JM (2021) Nanomaterials fundamentals: classification, synthesis and characterization. In: Al-Khayri JM, Ansari MI, Singh AK (eds) Nanobiotechnology. Springer, Cham. https://doi.org/10.1007/978-3-030-73606-4_4

Alnaddaf LM, Almuhammady AK, Salem KFM, Alloosh MT, Saleh MM, Al-Khayri JM (2021) Green synthesis of nanoparticles using different plant extracts and their characterizations. In: Al-Khayri JM, Ansari MI, Singh AK (eds) Nanobiotechnology. Springer, Cham. https://doi.org/10.1007/978-3-030-73606-4_8

Alqahtani MM, Saleh MM, Alwutayd KM et al (2024) Performance and genotypic variability in diverse date palm (Phoenix dactylifera L.) cultivars for fruit characteristics. Genet Resour Crop Evol 71:1759–1772. https://doi.org/10.1007/s10722-023-01719-0

Alves C, Silva A, Reis L et al (2010) Ecodesign of automotive components making use of natural jute fiber composites. J Clean Prod 18(4):313–327

Amany M, Eisa SS, Mohamed AA, El-Shamey I (2019) Taxonomic revision and numerical analysis of *Hibiscus* L. in Egypt. Arab Univ J Agric Sci. Ain Shams Univ, Cairo, Egypt 82(1):117–130. Website: http://ajs.journals.ekb.eg

An X, Qin L, Jinyao Y, Jiqian W et al (2023) Development of expressed sequence tag–simple sequence repeat markers related to the salt-stress response of kenaf (*Hibiscus cannabinus*). Agronomy 13(7):1946. https://doi.org/10.3390/agronomy13071946

Ansari M, Eslaminejad T, Sarhadynejad Z, Eslaminejad T (2013) An overview of the roselle plant with particular reference to its cultivation, diseases and usages. Eur J Med Plant 3(1):135–145

Anuradha K, Vimala R, Narayanasamy B, Arockia Selvi J, Rajendran S (2007) Corrosion inhibition of carbon steel in low chloride media by an aqueous extract of hibiscus rosa-sinensis linn. Chemical Engineering Communications, 195(3):352–366. https://doi.org/10.1080/00986440701673283

Arab S et al (2025) National Gene Bank of Egypt. In: Al-Khayri JM, Salem KFM, Jain SM (eds) Plant gene banks. Springer, Singapore. https://doi.org/10.1007/978-981-99-4236-7_5-1

Ayantoye BS, Daudu OAY, Falusi OA et al (2023) Silver nanoparticle synthesized from *Hibiscus sabdariffa* leaf extract enhanced certain agromorphological and yield traits in groundnut landraces in Niger state. Book of Proceedings, 45th Annual Conference of Genetics Society of Nigeria (GSN), pp 167–169

Babatunde FE, Mofoke ALE (2006) Performance of Roselle (*Hibiscus sabdariffa* L) as influenced by irrigation schedules. Pak J Nut 5(4):363–367

Bakasso Y, Zaman-Allah M, Mariac C et al (2014) Genetic diversity and population structure in a collection of roselle (*Hibiscus sabdariffa* L.) from Niger. Plant Genet Resour 12:207–214. https://doi.org/10.1017/S1479262113000531

Baloch M, Lakho AR, Bhutto H, Solangi MY (2021) Impact of Sucrose Concentrations on in vitro Pollen Germination of Okra, Hibiscus esculentus. Pak J Biol Sci 4(4):402–403. https://doi.org/10.3923/pjbs.2001.402.403

Baudoin JP, Demol J, Louant BP, Maréchal R, Guy M, Otoul E (2001) Amélioration des plantes: application aux principales espèces cultivées en régions tropicales. Presses Agronomiques de Gembloux, Gembloux (Belgique), 583p

Békefi Z, Szilágyi S, Desiderio F, Demku T, Varga J (2025) Prunus Gene Bank in Hungarian University of Agriculture and Life Sciences. In: Al-Khayri JM, Salem KFM, Jain SM (eds) Plant gene banks. Springer, Singapore. https://doi.org/10.1007/978-981-99-4236-7_71-1

Berloo R (2000) Use of molecular markers in plant breeding. Het gebruik van molecular markers in de plant enveredeling. Thesis Wageningen University with references – with summaries in English and Dutch. Laboratory of Plant Breeding. P.O. Box 386, 6700 AJ, Wageningen, The Netherlands ISBN 90-5808-167-2. 120pp

Borner A, Landjeva S, Salem KFM, Lohwasser U (2009) Pflanzengenetische Ressourcen als Grundlage für die Züchtung klimatoleranter Sorten. Proceedings of a German Conference 'Weizenforum = Wheat Forum' Norddeutsches Weizenforum Schriftenreihe des Inst. Für Pflanzenbau und –züchtung der CAU Kiel, Germany, 63, 23–27

Börner A, Landjeva S, Salem KFM, Lohwasser U (2010) Plant genetic resources- a prerequisite for drought tolerance breeding in cereals. Proc Jahrestagung der Vereinigung der Pflanzenzüchter und Saatgutkaufleute Österreichs, 24–26 November 2009, Raumberg-Gumpenstein, pp 11–13

Breidy J (2025) Lebanon gene bank. In: Al-Khayri JM, Salem KFM, Jain SM (eds) Plant gene banks. Springer, Singapore. https://doi.org/10.1007/978-981-99-4236-7_48-1

Cañar DY et al (2025) Colombia gene bank. In: Al-Khayri JM, Salem KFM, Jain SM (eds) Plant gene banks. Springer, Singapore. https://doi.org/10.1007/978-981-99-4236-7_95-1

Chamier AV (2020) Garden magic: plant firebush, hibiscus for pretty pollinators Home and garden health. UF/Leon County Extension

Chen H, Contreras RN (2022) Near-hexaploid and near-tetraploid aneuploid progenies derived from backcrossing tetraploid parents *Hibiscus syriacus* × (*H. syriacus* × *H. paramutabilis*). Genes 13(6):1022. https://doi.org/10.3390/genes13061022

Cheng Y, Zhang L, Qi J, Zhang L (2020) Complete chloroplast genome sequence of Hibiscus cannabinus and comparative analysis of the Malvaceae family. Front Genet 11:227

Christenhusz MJM, Byng JW (2016) The number of known plants species in the world and its annual increase. Phytotaxa 261:201–217

Collard BCY, Mackill J (2008) Marker-assisted selection: an approach for precision plant breeding in the twenty-first century. Philos Trans R SocLond B Biol Sci 363(1491):557–572. https://doi.org/10.1098/rstb.2007.2170

Davenport T, Kalakota R (2019) The potential for artificial intelligence in healthcare. Future Healthc J 6:94–98

de Galarreta JIR, Barandalla L, Ritter E, Velázquez ME, Rios D (2025) Spanish Potato Germplasm Bank. In: Al-Khayri JM, Salem KFM, Jain SM (eds) Plant gene banks. Springer, Singapore. https://doi.org/10.1007/978-981-99-4236-7_114-1

Deepo DM, Mazharul IM, Hwang YJ et al (2022) Chromosome and ploidy analysis of winter hardy *Hibiscus* Species by FISH and flow cytometry. Euphytica 218:81. https://doi.org/10.1007/s10681-022-02977-2

Desheva G, Valchinova E, Deshev M (2025) Bulgarian gene bank. In: Al-Khayri JM, Salem KFM, Jain SM (eds) Plant gene banks. Springer, Singapore. https://doi.org/10.1007/978-981-99-4236-7_66-1

Díez MJ, Valcárcel JV, Solbes E, Muñoz E (2025) Forty years conserving Spanish biodiversity in the gene bank of the Universitat Politècnica de València, Spain. In: Al-Khayri JM, Salem KFM, Jain SM (eds) Plant gene banks. Springer, Singapore. https://doi.org/10.1007/978-981-99-4236-7_86-1

Djouahri A, Boudarene L, Meklati BY (2013) Effect of extraction method on chemical composition, antioxidant and anti-inflammatory activities of essential oil from the leaves of Algerian Tetraclinis articulata (Vahl) Masters. Ind Crop Prod 44:32–36. https://doi.org/10.17660/ActaHortic.2022.1334.4

Dwivedi SL, Ceccarelli S, Blair MW, Upadhyaya et al (2016) Landrace germplasm for improving yield and abiotic stress adaptation. Trends Plant Sci 21:31–42

Eid NA, Ibrahim AMH, Elsharawy AA, Salem KFM (2025) Nanofertilizers for plant viral disease management. In: Kumar P, Dubey RC (eds) Nanofertilizers for sustainable agriculture. Springer, Cham. https://doi.org/10.1007/978-3-031-78649-5_7

El Rabey H, Salem KFM, Mattar MZ (2013) The genetic diversity and relatedness of eight rice (Oryza sativa L.) cultivars as revealed by AFLP and SSRs markers. Life Sci J 10(1):1471–1479

Elabd AB, Salem KFM (2009) Detection of genetic diversity in rice (Oryza sativa L.) Varieties using microsatellite DNA-Markers and Quantitative Traits. In: 6th Egyptian Plant Society of Breeding Conference, 4–7 May 2009, Ismailia, Egypt

Elabd AB, Saleh MM, Aldahak L, Salem KFM (2020) Identification of superior rice (Oryza sativa l.) genotypes under irrigated conditions based on multivariate analysis. In: 7th international scientific conference of genetic and environment from 8th to 9th December 2020, Baghdad, Iraq

El-Absawy EA, Salem KFM, Abdellatif KF, ElShamy H (2012) Estimation of genetic diversity among some Egyptian barley (Hordeum vulgare L.) cultivars using RAPD markers and agronomic traits. In: 1st international conference on biotechnology applications in agriculture, Moshtohor-Hurghada, 18-22 February 2012, Faculty of Agriculture, Benha University, Egypt

El-Absawy EA, Nowar AA, Salem KFM et al (2015) Estimation of genetic diversity among some bread wheat (Triticum aestivum L.) genotypes using molecular markers and agronomic traits. Minufiya J Agric Res 40(3):223–233

El-Boraie FM, Gaber AM, Abdel-Rahman G (2009) Optimizing irrigation schedule to maximize water use efficiency of *Hibiscus sabdariffa* under Shalatien conditions. World J Agric Sci 5(4):504–514

El-Dissoky RA, Attia AM, Awad AM (2020) Managing Roselle Plant (*Hibiscus sabdariffa* L.) requirements of fertilizers and irrigation grown under upper Egypt conditions. J Soil Sci Agric Eng Mansoura Univ 11(12):693–700

El-Sagher Mostafa, Gehan Mostafa, El-Mowafy El-Ghadban, Abd-Allah Gahory (2024) Sowing date as a determining factor for Roselle, Hibiscus sabdariffa, production: II. Effect on biochemical components and quality, Aswan University J of Sciences and Technology, 4(1):38–49. https://doi.org/10.21608/aujst.2024.337888

Eliw AA (2012) Developing a separation roselle sepals machine. Misr J Agric Eng 29(1):463–480

Eltaher S, Sallam A, Belamkar V et al (2018) Genetic diversity and population structure of F3:6 Nebraska winter wheat genotypes using genotyping-by-sequencing. Front Genet 9:76. https://doi.org/10.3389/fgene.2018.00076

Eltaher S, Baenziger S, Belamkar V et al (2021) GWAS revealed effect of genotype x environment interactions for grain yield of Nebraska winter wheat. BMC Genomics 22:1–14

Eltaher S, Sallam A, Emara HA et al (2022) Genome-wide association mapping revealed SNP alleles associated with spike traits in wheat. Agronomy 12:1469. https://doi.org/10.3390/agronomy12061469

El-Taher A, Sallam V, Belamkar HA, Emara A, Nower KFM, Salem JP, Baenziger PS (2017) Genome-wide association mapping for drought tolerance traits at seedling stage using Nebraska Winter Wheat. In: International Conference On Productive crop design (genome to phenome), April 2017, USA

El-Zanaty AM, Salem KFM, Esmail RM (2011) Detection of genetic diversity in Egyptian cotton (Gossypium barbadense L.) varieties using RAPD markers and morphological traits. J Am Sci 7(12):1107–1115

El-Zanaty AM, Salem KFM, Ismail M, Esmail RM (2013) Impact of plant breeding on genetic diversity of the Egyptian cotton (Gossypium barbadense L.) varieties as revealed by RAPD markers. J Agric Chem Biotechn, Mansoura Univ 4:217–227

Emara HA, Nower AA, Salem KFM, Eltaher S (2011) Estimating heterosis and combining ability for in vivo and in vitro traits using diallel cross in Maize (Zea mays L). J Prod Dev 16:101–120

Emara HA, Salem KFM, Abdel-Halim AA, Abd El-Kowy RNF (2013) An in vivo and in vitro traits analysis of a diallel cross in rice (Oryza sativa L.). Egyptian J Plant Breed 17(2):591–610

Esmail RM, Salem KFM, El-Zanaty AM (2009) Genetic diversity studies of Egyptian Cotton (Gossypium barbadense L.) varieties using RAPD markers and agronomic traits. 6th Egyptian Plant Society of breeding Conference, 4–7 May 2009, Ismalia, Egypt

Esse HP, Reuber TL, Does D (2020) Genetic modification to improve disease resistance in crops. New Phytol 225(1):70–86. https://doi.org/10.1111/nph.15967

Fadia ES, Khattab S, Ghoname E et al (2011) Irradiation on enhancement of some economic traits and molecular changes in *Hibiscus Sabdariffa* L. Life Sci J 8(3):220–229

Fahmy AA, Hassan HMS (2019) Influence of different NPK fertilization levels and humic acid rates on growth, yield and chemical constituents of roselle (*Hibiscus sabdariffa* L.). Middle East J Agric Res 8(4):1182–1189

FAO (2014) Genebank Standards for Plant Genetic Resources for Food and Agriculture. Rome, FAO, p 182

FAO (2018a) Ecocrop. *Hibiscus sabdariffa* L Fecha de acceso: septiembre del 2018. http://ecocrop.fao.org/ecocrop/srv/en/cropView?id=1216

FAO (2018b) The future of food and agriculture – alternative pathways to 2050. Food and Agriculture Organization of the United Nations (FAO), Rome

Fassil H, Guarino L, Sharrock S et al (2000) Diversity for food security: improving human nutrition through better evaluation, management, and use of plant genetic resources. Food Nutr Bull 21(4):497–502

Futuless KN, Kwaga Y, Clement T (2010) Effect of sowing date on the Calyx yield and yield components of Roselle (*Hibiscus sabdariffa* L.) in Northern Guinea Savanna. New York Sci J 3(11):1–4

Galea-Gragera FA, Llera Cid F (2025) Spain Germplasm Bank of Annual Forage Legumes Seeds at the Scientific and Technological Research Center of Extremadura (CICYTEX). In: Al-Khayri JM, Salem KFM, Jain SM (eds) Plant gene banks. Springer, Singapore. https://doi.org/10.1007/978-981-99-4236-7_102-1

Gassama-Dia SYKD, Ndoye M (2004) Direct genetic transformation of Hibiscus sabdariffa L. Afr J Biotechnol 3(4):226–228

Gao Y, Zhang M, Wang K, Wen F, Liu F (2024) The Carbon Emissions Reduction Effect of Green Agricultural Subsidy Policy: A Quasi-Natural Experiment. Sustainability, 16(12):5210. https://doi.org/10.3390/su16125210

Gendy ASH, Said-Al Ahl HAH, Mahmoud AA (2012) Growth, productivity and chemical constituents of Roselle (*Hibiscus sabdariffa* L.) plants as influenced by cattle manure and biofertilizers treatments. Aust J Basic Appl Sci 6(5):1–12

Ghabour SSI, Mohamed SA, Saif El-Yazal SA, Moawad HMH (2019) Impact of bio and mineral fertilizers on growth, yield and its components of roselle plants (*Hibiscus sabdariffa*, L.) grown under different types of soil. Hortic Int J 3(5):240–250

Ghazy MI, Salem KFM, Sallam A (2020) Utilization of genetic diversity and marker-trait to improve drought tolerance in rice (Oryza sativa L.). Mol Biol Rep. https://doi.org/10.1007/s11033-020-06029-7

Ghidan WF, Abdulmajid DA (2025) Egyptian Rice Gene Bank. In: Al-Khayri JM, Salem KFM, Jain SM (eds) Plant gene banks. Springer, Singapore. https://doi.org/10.1007/978-981-99-4236-7_106-1

Gholam S, Moosavi R (2012) The effect of sowing date and plant density on yield and yield components of roselle. J Med Plant Res 6(9):1627–1632. https://doi.org/10.5897/JMPR11.1318

Giovannini AM, Nesi LB, Cardi MST (2021) Review genes and genome editing tools for breeding desirable phenotypes in ornamentals. Plant Cell Rep 40:461–478. https://doi.org/10.1007/s00299-020-02632-x

Go S, Koo H, Jung M et al (2024) Pan-chloroplast genomes for accession-specific marker development in *Hibiscus syriacus*. Sci Data 11:246

Gonzales EM, Villagrán Z, Torres ST et al (2022) Physiological Effects and Human Health Benefits of *Hibiscus sabdariffa*. Pharma 15:464–466

Gudaye TY, Lakew BF, Selamo MM, Tessema ES (2025) Ethiopian crop gene bank. In: Al-Khayri JM, Salem KFM, Jain SM (eds) Plant gene banks. Springer, Singapore. https://doi.org/10.1007/978-981-99-4236-7_7-1

Habib-ur-Rahman M, Ahmad A, Raza A et al (2022) Impact of climate change on agricultural production; issues, challenges, and opportunities in Asia. Front Plant Sci 13:925548. https://doi.org/10.3389/fpls.2022.925548

Han C, Dongsheng W, Bing B et al (2022) Phenotypic diversity of 21 *Hibiscus* cultivars. J Nanjing For Univ 46(3):117–126. https://doi.org/10.12302/j.issn.1000-2006.202103038

Hanafiah DS, Siregar LAM, Putri DM (2017) Effect of gamma ray's irradiation on m1 generation of roselle (*Hibiscus Sabdariffa* L) Inter. J Agric Res 12:28–35

Hariram KB, Appalaswamy A, Durga P, Ismail S (2013) Genetic divergence studies in roselle (Hibiscus sabdariffa L.) Madras Agril J, 100:20–22

Harzli I, Mostafa K, Abdulla MF, Mehmood M (2025) Sustaining Turkish genetic resources. In: Al-Khayri JM, Salem KFM, Jain SM (eds) Plant gene banks. Springer, Singapore. https://doi.org/10.1007/978-981-99-4236-7_88-1

Hasan N, Choudhary S, Naaz N et al (2021) Recent advancements in molecular marker-assisted selection and applications in plant breeding programmes. J Genet Eng Biotechnol 19:128. https://doi.org/10.1186/s43141-021-00231-1

Hassan MN, Mekkawy SA, Mahdy M, Salem KFM et al (2021) Recent molecular and breeding strategies in lettuce (Lactuca spp.). Genet Resour Crop Evol 68:3055–3079. https://doi.org/10.1007/s10722-021-01246-w

Hassanein YZ, Abdel-Rahman SSA, Soliman WS, Salaheldin S (2021) Growth, yield, and quality of roselle (*Hibiscus sabdariffa* L.) plants as affected by nano zinc and bio-stimulant treatments. J Hortic Environ Biotech 62:879–890

Hidayatun N, Kurniawan H (2025) Conservation of agricultural genetic resources in Indonesia. In: Al-Khayri JM, Salem KFM, Jain SM (eds) Plant gene banks. Springer, Singapore. https://doi.org/10.1007/978-981-99-4236-7_40-1

Howard A, Howard GLC (1911) Studies in Indian fibre plants. No. 2. On some new varieties of *Hibiscus cannabinus* L. and *Hibiscus sabdariffa* L. Indian Dep Agric Mem Bot Serv 4:9–36. https://doi.org/10.17660/ActaHortic.2022.1334.4

Hussein RM, Shahein Y, El-Hakim AE, Awad HM (2010) Biochemical and molecular characterization of three colored types of Roselle (*Hibiscus sabdariffa*). J Am Sci 6(11):726–733. https://doi.org/10.7537/marsjas061110.105

Ibrahim AA, Abd-Ellatif S, Abdel Razik ES et al (2023a) Synthesis and applications of cellulose nanomaterials derived from agricultural waste and byproducts. In: Al-Khayri JM, Alnaddaf LM, Jain SM (eds) Nanomaterial interactions with plant cellular mechanisms and macromolecules. Springer, Cham

Ibrahim AA, Salem KFM, Abdein MA, Ramadan SM (2023b) Advances in summer squash (Cucurbita pepo L.) molecular breeding strategies. In: Singh S, Sharma D, Sharma SK, Singh R (eds) Smart plant breeding for vegetable crops in post-genomics era. Springer, Singapore. https://doi.org/10.1007/978-981-19-5367-5_8

Ibrahim AA, Salem KFM, Ramadan S (2023c) Allium sativum and stressful conditions. In: Abdel Latef AAH (ed) Book medicinal plant responses to stressful conditions. CRC Press

Ibrahim AA, Abd-Ellatif S, Abdel Razik ESS, Salem KFM (2024) Biodiversity of cereal crops and utilization in food and nutritional security. In: Al-Khayri JM, Jain SM, Penna S (eds) Sustainable utilization and conservation of plant genetic diversity. sustainable development and biodiversity, vol 35. Springer, Singapore. https://doi.org/10.1007/978-981-99-5245-8_2

Ihsan F, Prihatini R (2025) Indonesian tropical fruit gene bank. In: Al-Khayri JM, Salem KFM, Jain SM (eds) Plant gene banks. Springer, Singapore. https://doi.org/10.1007/978-981-99-4236-7_41-1

Islam M, Saha S, Akand H, Rahim A (2010) Efffect of sowing date on the growth and yield of sweet pepper(Capsicum annuum L.). Agronomski Glasnik 1

Islam M, Deepo M (2020) Embryo development of woody and perennial *Hibiscus* species. Sci Hortic 275:109698

Ismail M, Salem KFM, El-Zanaty AM, Esmail RM (2013) Association mapping of fiber and agronomic characters in Egyptian cotton (Gossypium barbadense L.) genotypes. J Agric Chem Biotech Mansoura Univ 4:205–216

Jabeur PA, Pereira E, Barros L et al (2017) *Hibiscus sabdariffa* L. as a source of nutrients, bioactive compounds and colouring agents. Food Res Int 100:717–723

Jackson CL (1996) Construction of somatic cell hybrids. Curr Protoc Hum Genet 9(1):3–2

Jalalyazdi M, Ramezani J, Izadi-Moud A et al (2019) Effect of *Hibiscus sabdariffa* on blood pressure in patients with stage 1 hypertension. J Adv Pharm Technol Res 10(3):107–111

Janakiram T, Patil MS (2017) Breeding in *Hibiscus*: a review. Indian J Agric Sci 87(2):159–166. https://doi.org/10.56093/ijas.v87i2.67540

Jatoi SA, Siddiqui SU (2025) National Gene Bank of Pakistan. In: Al-Khayri JM, Salem KFM, Jain SM (eds) Plant gene banks. Springer, Singapore. https://doi.org/10.1007/978-981-99-4236-7_51-1

Jayaraman PP, Yavari A, Georgakopoulos D et al (2016) Internet of things platform for smart farming: experiences and lessons learnt. Sensors 16:1884. https://doi.org/10.3390/s16111884

Jeyaratnam N, Nour AH, Kanthasamy R et al (2016) Essential oil from *Cinnamomum cassia* bark through hydrodistillation and advanced microwave-assisted hydrodistillation. Ind Crop Prod 92:57–66

Joshi M, Verma SK, Singh JP, Barh A (2013) Genetic diversity assessment in lentil (*Lens culinaris* Medikus) genotypes through ISSR marker. Bioscan 8(4):1529–1532

Ju-Zhao L, Zhang CC, Fu YJ, Cui Q (2022) Comparative analysis of phytochemical profile, antioxidant and anti-inflammatory activity from Hibiscus manihot L. flower. Arab J Chem 15(1):103503. https://doi.org/10.4103/japtr.JAPTR_402_18

Kalita BB, Jose S, Baruah S, Kalita S, Saikia DR (2017) Hibiscus sabdariffa (Roselle): a potential source of bast fiber. J Nat Fiber 16(1):49–57. https://doi.org/10.1080/15440478.2017.1401504

Kar C, Satya P, Mitra J, Sarkar D, Sinha M, Kundu A et al (2010) Varietal development of jute and allied fibres in India. Indian Farming 60(3)

Khattak AM, Sajid M, Sarwar HZ, Rab A, Ahmad M, Khan MA (2016) Effect of sowing time and plant density on the growth and production of roselle (*Hibiscus sabdariffa*). Int J Agric Biol 18:1219–1224

Khosla R (2010) Precision agriculture: challenges and opportunities in a flat world. In: Proceedings of the 19th World Congress of Soil Science, Soil Solutions for a Changing World, Brisbane, QLD, Australia

Kim Y, Kim S, Koo N, Shin A et al (2017) Genome analysis of Hibiscus syriacus provides insights of polyploidization and indeterminate flowering in woody plants. DNA Res 24(1):71–80

Kim ST, Oh SH, Park J (2021) The complete chloroplast genome of *Diarthron linifolium* (Thymelaeaceae), a species found on a limestone outcrop in eastern Asia. Korean J Plant Taxon 51:345–352

Kim WJ, Yang B, Lee Y, Kim JH et al (2024) Genome-Wide Association Study for Agronomic Traits in Gamma-Ray-Derived Mutant Kenaf (*Hibiscus cannabinus* L.). Plants 13(2):249. https://doi.org/10.3390/plants13020249

Koshimizu S, Masuda S, Shibata A, Ishii T et al (2023) Genome and transcriptome analyses reveal genes involved in the formation of fine ridges on petal epidermal cells in *Hibiscus trigonum*. DNA Res 30(5):1–10. https://doi.org/10.1093/dnares/dsad019

Kowti R, Majumder P, Harshitha HS et al (2020) Hibiscus herbs – a comprehensive botanical, chemical and biological overview. World J Adv Res Rev 7(03):291–303

Kwofie J, Amoatey HM, Nunekpeku W et al (2017) Crossability Studies among Twenty Accessions of Roselle (*Hibiscus sabdariffa* L.). J Agric Sci 9:230–236

Kwon S, Park Y, Jang YL, Kwon H (2022) The complete chloroplast genome sequence of *Hibiscus sabdariffa* (Malvaceae). Korean J Plant Taxon 52(2):123–126

Kwon S, Kwon H, Choi Y, Shin H (2023) Comprehensive analysis of chloroplast genome of *Hibiscus sinosyriacus*: evolutionary studies in related species and genera. Forests 14(11). https://doi.org/10.3390/f14112221

Laere KV, Huylenbroeck J, Bockstaele EV (2006) Breeding strategies to increase genetic variability within Hibiscus syriacus. Acta Hortic 714(714):75–81. https://doi.org/10.17660/ActaHortic.2006.714.9

Laghetti G, Piergiovanni AR, De Paola D, Direnzo P, Bucci G, Sonnante G (2025) Italy IBBR gene bank. In: Al-Khayri JM, Salem KFM, Jain SM (eds) Plant gene banks. Springer, Singapore. https://doi.org/10.1007/978-981-99-4236-7_73-1

Lattier JD, Chen H, Contreras RN (2019) Variation in genome size, ploidy, stomata, and rDNA signals in Althea. J Am Soc Hortic Sci 144:130–140. https://doi.org/10.21273/JASHS04618-18

Lin Y, Jones ML (2022) CRISPR/Cas9-mediated editing of autophagy gene 6 in petunia decreases flower longevity, seed yield, and phosphorus remobilization by accelerating ethylene production and senescence-related gene expression. Front Plant Sci Sec Plant Physiol 13. https://doi.org/10.3389/fpls.2022.840218

Liu W, Yu H, Zhou X, Xing D (2016) In vitro evaluation of CRISPR/Cas9 function by an electro chemiluminescent assay. Anal Chem 88(17):8369–8374. https://doi.org/10.1021/acs.analchem.6b02338

Liu XF, Zhu GF, Li DM, Wang XJ (2019) Complete chloroplast genome sequence and phylogenetic analysis of Spathiphyllum 'Parrish'. PLoS One 14:e0224038

Lobodina E, Suprun I, Al-Nakib E et al (2020) Micropropagation of Hardy Hibiscus (*Hibiscus Moscheutos* L.). BIO Web of Conferences 25, 05008. Bioengineering. https://doi.org/10.1051/bioconf/20202505008

Loehwing WF (1931) Effects of soil aeration on plant growth and root development. Proc Iowa Acad Sci 38(1):71–72. Available at: https://scholarworks.uni.edu/pias/vol38/iss1/6

Lubis L, Dewi GT, Supriyan AND, Aprinaldi A, Purba A, Diantini A (2024) Roselle (*Hibiscus sabdariffa* L.) calyces tea improves physical fitness of healthy adults. Biomed Rep 20:49

Lyu JI, Choi H, Ryu J et al (2020) Transcriptome analysis and identification of genes related to biosynthesis of Anthocyanins and Kaempferitrin in Kenaf (*Hibiscus cannabinus* L.). J Plant Biol 63(1):51–62. https://doi.org/10.1007/s12374-020-09227-9

M'be CU, Scher J, Gaiani C, Amani NG, Burgain J (2023) Impact of processing and physicochemical parameter on *Hibiscus sabdariffa* calyxes biomolecules and antioxidant activity. Foods 12:2984–2987

Ma L, Fuquan Kong F, Sun K et al (2021) From classical radiation to modern radiation: past, present, and future of radiation mutation breeding. Front Public Health 9:768071. https://doi.org/10.3389/fpubh.2021.768071

Magdalita PM, Pascual AO (2022) Development of new hibiscus hybrids (*Hibiscus rosa-sinensis* L.) in The Philippines. Acta Hortic 1334:29–36

Mahadevan N, Kamboj P (2009) *Hibiscus sabdariffa* Linn. An overview. Nat Prod Radiance 8:77–83

Mahapatra A (2008) Genetic resources of jute and allied Fibers crops. Jute Allied Fibers Updates: Prod Technol 57:72–82

Mahunu GK (2021a) Breeding, genetic diversity, and safe production of *Hibiscus sabdariffa* under climate change. In: Mariod AA, Tahir HE, Mahunu GK (eds) Roselle (*Hibiscus sabdariffa*): chemistry, production, products, and utilization. Elsevier, pp 1–14

Mahunu GK, Tahir HE, Osei-Kwarteng M et al (2021) Food use of whole and extracts of *Hibiscus sabdariffa*. In: Mariod AA, Tahir HE, Mahunu GK (eds) Roselle (*Hibiscus sabdariffa*): Chemistry, production, products, and utilization. Academic Press is an imprint of Elsevier

Mallor C, Luna E (2025) The genetic diversity of the Spanish vegetable genebank BGHZ-CITA: conservation and utilization. In: Al-Khayri JM, Salem KFM, Jain SM (eds) Plant gene banks. Springer, Singapore. https://doi.org/10.1007/978-981-99-4236-7_109-1

Mandaji CM, da Silva PR, Chiste RC (2022) Encapsulation of bioactive compounds extracted from plants of genus Hibiscus: a review of selected techniques and applications. Food Res Int 151:110820

Manokari M, Ravindran CP, Shekhawat MS (2016) Optimization of in vitro regeneration in *Hibiscus subdariffa* L. An important medicinal plant. World Sci News 34:51–62

Maruthi RT, Mangal V, Kumar AA, Meena JK, Mitra J (2022) Genetic diversity in advanced breeding lines derived from intraspecific crosses of roselle (*Hibiscus sabdariffa* L.). Vegetos 36:675–680. https://doi.org/10.1007/s42535-022-00429-9

Mattar MZ, Eissa OA, Salem KFM (2009) Genotypic variability in some wheat genotypes (Triticum aestivum L.) exposed to salinity at germination stage. Egyptian J Exp Biol (Bot) 5:143–148

Mattar MZ, Salem KFM, El-Abed AB (2016) Identification of candidate microsatellite markers associated with agronomic traits in rice (Oryza sativa L.). J Bot 15(5):213–232

Mazharul I (2021) Kenaf fiber: properties, cultivation, production, uses and advantages. Textile Learner 10

McBratney A, Whelan B, Ancev T, Bouma J (2005) Future directions of precision agriculture. Precis Agric 6:7–23

McClintock N, El Tahir I (2004) *Hibiscus sabdariffa* L. In: Grubben GJH, Denton OA (eds) Vegetables/Legumes. PROTA, Wageningen

Mehrotra M, Sanyal I, Amla DJ (2011) High-efficiency Agrobacterium-mediated transformation of chickpea (*Cicerarie tinum* L.) and regeneration of insect-resistant transgenic plants. Plant Cell Rep 30(9):1603–1616. https://doi.org/10.1007/s00299-011-1071-5

Mod and Mint DG (2024) Hibiscus meaning explained – symbolism and mythology plant care made easy (Article https://www.modandmint.com/hibiscus-flower-meaning-symbolism/)

Mohamad O, Nazir B, Rahman M, Herman S (2002) Roselle: a new crop in Malaysia. Bio Malaysia: a grand international biotechnology event. In: Bulletin PGM. Bio Malaysia, Kuala Lumpur

Mohamed BB (2021) Roselle (*Hibiscus sabdariffa* L.) in Sudan: production and uses. In: Sapuan SM, Nadlene R, Ilya RA (eds) Roselle production, processing, products and biocomposites. Academic Press

Montalvo-González E, Villagrán Z, González-Torres S et al (2022) Physiological effects and human health benefits of *Hibiscus sabdariffa*: a review of clinical trials. Pharmaceuticals 15:464

Monteros-Altamirano Á et al (2025) Plant genetic resources in Ecuador. In: Al-Khayri JM, Salem KFM, Jain SM (eds) Plant gene banks. Springer, Singapore. https://doi.org/10.1007/978-981-99-4236-7_99-1

Morton JF (1987) Roselle. In: Fruits of warm climates, pp 281–286

Mourad AMI, Alomari DZ, Alqudah AM et al (2019) Recent advances in wheat (Triticum spp.) breeding. In: Al-Khayri J, Jain S, Johnson D (eds) Advances in plant breeding strategies: cereals. Springer, Cham. https://doi.org/10.1007/978-3-030-23108-8_15

Moustafa M, Younes ES, Shalab HG, Eltayeb H, Yacout MM (2021) Molecular marker analysis tools as a cornerstone for the phylogenetic analysis of hibiscus species in Egypt. NVJAS 1(2):63–75

Mujib A (2016) Somatic embryogenesis in ornamentals and its applications. A book. Springer Link

Nabila YN, Aly MS (2002) Variations in productivity of (*Hibiscus sabdariffa* L.) in response to some agricultural supplementation. Ann Agric Sci (Cairo), Ain Shams Univ, Cairo, Egypt 47(3):875–892

Nawara HM, Mattar MZ, Salem KFM, Eissa OA (2016) Estimating heterosis and combining ability for agronomic and tissue culture traits using diallel cross on bread wheat (Triticum aestivum L.). Egypt J Bot 2016:177–200

Nawara HM, Mattar MZ, Salem KFM, Eissa OA (2017) Diallel study on some in vitro traits of bread wheat (Triticum aestivum L.) under salt stress. Int J Agric Environ Res 3(1):1988–2006

Nayak BS, Raju SS, Eversley M, Ramsubhag A (2009) Evaluation of wound healing activity of Lantana camara L. a preclinical study. Phytother Res 23(2):241–245

Nnebue OM, Ogoke IJ, Obilo OP, Agu CM et al (2014) Estimation of planting dates for roselle (*Hibiscus sabdariffa* L.) In the humid tropical environment of Owerri, Southeastern Nigeria. Agrosearch 14(2):168–178

Nongdam P (2016) Development of synthetic seed technology in plants and its applications: a review. Int J Curr Sci 19(4):86–101

Nuryanti S, Puspitasari DJ, Supriadi (2019) Rosella (*Hibiscus sabdariffa*) flowers as alternative indicators of blue and red litmus. Oriental J Chem 35(1):476–480

Ogbu JU, Essien BA, Essien JB, Anaele MU (2010) Conservation and management of genetic resources of horticultural crops in Nigeria: issues and biotechnological strategies. J Hortic For 2(9):214–222

Ojulari OV, Lee SG, Nam J (2019) Beneficial effects of natural bioactive compounds from *Hibiscus sabdariffa* L. on obesity. Molecules 24(1):210. https://doi.org/10.3390/molecules24010210

Omalsaad AKM, Islam A, Akhter JM et al (2014) Genetic relationship between roselle (*Hibiscus sabdariffa* L.) and kenaf (*Hibiscus cannabinus* L.) accessions through optimization of PCR-based RAPD method. Emir J Food Agric 26(3):247–258. https://doi.org/10.9755/ejfa.v26i3.16498

Osei-Kwarteng M, Gweyi-Onyango JP, Mahunu GK (2021) Harvesting, storage, postharvest management, and marketing of *Hibiscus sabdariffa*. In: Mariod AA, Tahir HE, Mahunu GK (eds) Roselle (*Hibiscus sabdariffa*): chemistry, production, products, and utilization. Academic Press is an imprint of Elsevier

Osterc G, Gostinčar K, Hudina M, Usenik V, Luthar Z (2025) Slovene fruit gene bank. In: Al-Khayri JM, Salem KFM, Jain SM (eds) Plant gene banks. Springer, Singapore. https://doi.org/10.1007/978-981-99-4236-7_84-1

Ozkan TGI, Appiah KS, Akalin E, Fujii Y (2023) Plant growth inhibitory activity of *Hibiscus sabdariffa* calyx and the phytotoxicity of hydroxycitric acid lactone. Agronomy 13(7):1746. https://doi.org/10.3390/agronomy13071746

Papa R, Gepts P (2003) Asymmetry of gene flow and differential geographical structure of molecular diversity in wild and domesticated common bean (*Phaseolus vulgaris*, L.) from Mesoamerica. Theor Appl Genet 106:239–250

Pfeil BE, Brubaker CL, Craven LA, Crisp MD (2002) Phylogeny of hibiscus and the tribe Hibisceae (Malvaceae) using chloroplast DNA sequences of ndhF and the rpl16 Intron. Sys Bot 27(2):333–350. https://doi.org/10.1043/0363-6445-27.2.333

Popović V et al (2025) Economic seed production and storage of fiber crops in gene banks in Serbia. In: Al-Khayri JM, Salem KFM, Jain SM (eds) Plant gene banks. Springer, Singapore. https://doi.org/10.1007/978-981-99-4236-7_108-1

Prasad MP (2014) Molecular characterization and genetic diversity determination of Hibiscus species using RAPD molecular markers. Asian J Plant Sci Res 4(3):50–56

Preece JE, West TP (2006) Greenhouse growth and acclimatization of encapsulated Hibiscus moscheutos nodal segments. Plant Cell Tissue Org Cult 87(2):127–138. https://doi.org/10.1007/s11240-006-9146-7

Pulli Bai P, Rzia S, Panduranga RC, Srinivasa RV (2005) Multivariate analysis in roselle hemp (Hibiscus sabdariffa L.). The Andhra Agric J 52 (1, 2):86–92

Purabi B, Ghosh P, Debangshi U (2022) Vertical farming – an overview. Plant Arch 22(2):223–228

Puro K, Sunjukta R, Samir S, Ghatak S, Shakuntala I, Sen A (2014) Medicinal uses of Roselle plant (*Hibiscus sabdariffa* L.): a mini review. Indian J Hill Farming 27: (1):81–90

Ranjan S (2023) *Hibiscus sabdariffa*-Roselle. In Medicinal Plants in India: Importance and Cultivation. (pp.215–227) Publisher: Jaya Publishing House, New Delhi, India

Rassem HHA, Nour AH, Ali GAM, Masood N et al (2022) Essential oil from hibiscus flowers through advanced microwave-assisted hydrodistillation and conventional hydrodistillation. Hindawi J Chem. https://doi.org/10.1155/2022/2000237

Richardson ML, Arlotta CG (2021) Differential yield and nutrients of *Hibiscus sabdariffa* L. genotypes when grown in urban production systems. Sci Hortic 288:110349. https://doi.org/10.1016/j.scienta.2021.110349

Rita C, Pereira G, Garrido I, Tavares-de-Sousa MM, Espinosa F (2016) Comparison of RAPD, ISSR, and AFLP molecular markers to reveal and classify Orchard grass (Dactylis glom-

erata L.) germplasm variations. PLoS One 11(4):e0152972. https://doi.org/10.1371/journal.pone.0152972
Rocha F et al (2025) Portugal gene bank. In: Al-Khayri JM, Salem KFM, Jain SM (eds) Plant gene banks. Springer, Singapore. https://doi.org/10.1007/978-981-99-4236-7_79-1
ROY 1S, Sharma C, Mohanty (2023) An updated review on various potential pharmacological activities of Hibiscus Int J of Biology, Pharmacy and Allied Sciences. https://doi.org/10.31032/IJBPAS/2023/12.1.6734
Saleh MM, Ali WA, Alwan MH et al (2020a) Multi-environmental evaluation of wheat tetraploid genotypes for agronomic traits under rainfed conditions in Syria. J Aridland Agric 6:1–5
Saleh MM, Salem KFM, Elabd AB (2020b) Definition of selection criterion using correlation and path coefficient analysis in rice (Oryza sativa L.) genotypes. Bull Natl Res Cent 44(1):1–6
Saleh MM, Alnaddaf LM, Almuhammady AK et al (2021a) Applications of plant-derived nanomaterials in mitigation of crop abiotic stress. In: Al-Khayri JM, Ansari MI, Singh AK (eds) Nanobiotechnology. Springer, Cham. https://doi.org/10.1007/978-3-030-73606-4_9
Saleh MM, Mahmoud AS, Abbas HS et al (2021b) Nanotechnological approaches for efficient delivery of plant ingredients. In: Faizan M, Hayat S, Yu F (eds) Sustainable agriculture reviews, vol 53. Springer, Cham. https://doi.org/10.1007/978-3-030-86876-5_11
Salehi B, Pazuki A, Zargari K (2022) Cytogenetic and Karyotype Study on *Hibiscus sabdariffa* L. in Indigenous Populations of Iran. Plant Genetics 58:413–419
Salem KFM (1997) Breeding studies on rice (*Oryza sativa* L.). M Sc Thesis, Faculty of Agriculture, Minufiya University, Shibin El-Kom, Egypt
Salem KFM (2004) The inheritance and molecular mapping of genes for post-anthesis drought tolerance (PADT) in wheat. Martin-Luther-University, Halle-Wittenberg, p 124
Salem KFM (2009) Relationship between genetic diversity based on SSRs markers with heterosis and combining ability in diallel cross of bread wheat (*Triticum aestivum* L.). Minufiya J Agric Res 6(34):2159–2178
Salem KFM (2015) Allelic state at the microsatellite locus Xgwm261 marking the dwarfing gene Rht8 in Egyptian bread wheat (Triticum aestivum L.) genotypes released from 1947 to 2004. Genetika 47:741–750
Salem KFM, Börner A (2003) Evaluation of wheat accessions from Gatersleben Genebank for post-anthesis drought tolerance in the field. In: Gesellschaft für Pflanzenzüchtung, GPZ-Tagung, Vortragsveranstaltung zum Thema: "Abiotischer Streß – Herausforderung für die Pflanzenzüchtung", 11-12 Juni 2003, Groß Lüsewitz, Germany
Salem KFM, Börner A (2004) Post-anthesis drought tolerance in genus Triticum. In: Botanikertagung, 5–10 September 2004, Braunschweig, Germany
Salem KFM, El-Zanaty AM (2015) Assessing genetic diversity of old and modern Egyptian rice (Oryza sativa L.) using microsatellite markers. Minufiya J Agric Res 40(6):1507–1518
Salem KFM, Ibrahim AA (2024) Plant biodiversity in the context of food security under climate change. In: Al-Khayri JM, Jain SM, Penna S (eds) Sustainable utilization and conservation of plant genetic diversity. sustainable development and biodiversity, vol 35. Springer, Singapore. https://doi.org/10.1007/978-981-99-5245-8_1
Salem KFM, Mattar MZ (2014a) Genetic diversity in old and modern Egyptian bread wheat (Triticum aestivum L.) varieties revealed by simple sequence repeats. Egypt J Genet Cytol 43:143–156
Salem KFM, Mattar MZ (2014b) Identification of microsatellite alleles for salt tolerance at seedling stage in wheat (Triticum aestivum L.). Life Sci J 11(12s):1064–1073
Salem KFM, Sallam A (2016) Analysis of population structure and genetic diversity of Egyptian and exotic rice (Oryza sativa L.) genotypes using SSR markers. Comptes Rendus Biol 339:1–9. https://doi.org/10.1016/j.crvi.2015.11.003
Salem KFM, Röder M, Börner A (2003) Mapping of quantitative trait loci (QTLs) for post-anthesis drought tolerance in bread wheat (Triticum aestivum L.). In: Gesellschaft für Pflanzenzüchtung, GPZ-Tagung, 11. Vortragsveranstaltung "Harnessing Genetic Diversity: Genomics and Allele

Mining", 16–17.09.2003, Institut für Pflanzengenetik und Kulturpflanzenforschung (IPK), Gatersleben, Germany

Salem KFM, Röder MS, Börner A (2004a) Drought tolerance in wheat germplasm- Evaluation and molecular mapping. In: Institut für Pflanzengenetik und Kulturpflanzenforschung, IPK Tagung, 22–24/08/2004, Ag Ressourcengenetik und Reproduktion, Abteilung Genbank, Gatersleben, Germany

Salem KFM, Röder M, Börner A (2004b) Molecular mapping of quantitative trait loci (QTLs) determining post-anthesis drought tolerance in hexaploid wheat (*Triticum aestivum* L.). In: 7 Gesellschaft für Pflanzenzüchtung, GPZ-Tagung, Vortragsveranstaltung zum Thema: " Klimatische und edaphische Sortenanpassung und Züchtung für Nachwachsende Rohstoffe". 3–5 March 2004, Halle/Saale, Germany. Vort Pflanzenzüchtung 64:21–24

Salem KFM, Röder MS, Börner A (2007a) Evaluation of some barley varieties for the presence of thermostable alleles of ßamylase. 8th African Crop Science Society Conference, El-Minia, Egypt, 27–31 October 2007, pp 643–648

Salem KFM, Röder MS, Börner A (2007b) Identification and mapping quantitative trait loci for stem reserve mobilisation in wheat (*Triticum aestivum* L.). Cereal Res Common 35:1367–1374

Salem KFM, El-Zanaty AM, Esmail RM (2008a) Assessing wheat (*Triticum aestivum* L.) genetic diversity using morphological characters and microsatellite markers. World J Agric Sci 4:538–544

Salem KFM, Varshney RK, Röder MS, Börner A (2008b) Detection of genetic diversity in Egyptian barley genotypes using barley expressed sequence Tag-SSR markers. In: 10th International Barley Genetics Symposium (IBGS), Alexandria, Egypt 5–10 April 2008. The International Center for Agricultural Research in the Dry Areas (ICARDA) and the Bibliotheca Alexandrina, Alexandria, Egypt

Salem KFM, Varshney RK, Röder MS, Börner A (2010) EST-SSR based estimates on functional genetic variation in a barley (Hordeum vulgare L.) collection from Egypt. Genet Resour Crop Evol 57(4):515–521

Salem KFM, Yaseen YM, El-Abd AB, Ayoub A (2015c) Breeding studies on some agronomic and quality traits in rice (Oryza sativa L.). J Product Dev 20(2):211–226

Salem KFM, Alloosh MT, Saleh MM et al (2021a) Utilization of nanofertilizers in crop tolerance to abiotic stress. In: Al-Khayri JM, Ansari MI, Singh AK (eds) Nanobiotechnology. Springer, Cham. https://doi.org/10.1007/978-3-030-73606-4_11

Salem KFM, Elabsawy A, Aldahak L et al (2021b) Evaluation of stem reserve mobilization in Egyptian bread wheat (Triticum aestivum L.) genotypes and F1 hybrids under post-anthesis chemical desiccation stress. J Genet Environ Resour Conserv 9(1):176–182

Salem KFM, El-Absawy EA, Khidr YA, El-Shamy H (2021c) Assessment genetic diversity among some Egyptian bread wheat (*Triticum aestivum* L.) cultivars using RAPD markers. J Genet Environ Resour Conserv 9(1):168–175

Salem KFM, Saleh MM, Abu-Ellail FFB et al (2021d) Role of quantum dots, polymeric NPS and dendrimers in emphasizing crops tolerate biotic and abiotic stresses. In: Faizan M, Hayat S, Yu F (eds) Sustainable agriculture reviews, vol 53. Springer, Cham. https://doi.org/10.1007/978-3-030-86876-5_1

Salem KFM, Saleh MM, Abu-Ellail FFB et al (2021e) The Role of salicylic acid in crops to tolerate abiotic stresses. In: Hayat S, Siddiqui H, Damalas CA (eds) Salicylic acid- a versatile plant growth regulator. Springer, Cham. https://doi.org/10.1007/978-3-030-79229-9_7

Salem KFM, Saleh MM, Aldahak L, Elabd AB (2021f) Assessment phenotypic diversity of rice (Oryza sativa L.) genotypes by multivariate analysis. J Arid Agric 7:52–59. https://doi.org/10.25081/jaa.2021.v7.6516

Salem KFM, Alghuthaymi MA, Elabd AB et al (2022) Prediction of heterosis for agronomic traits in half-diallel cross of rice (Oryza sativa L.) under drought stress using microsatellite markers. Plants 11:1532. https://doi.org/10.3390/plants11121532

Salem KFM, Abd-Ellatif S, Abdel Razik ES et al (2023a) Effect of nanomaterials on water and solutes translocation in plants. In: Al-Khayri JM, Alnaddaf LM, Jain SM (eds) Nanomaterial interactions with plant cellular mechanisms and macromolecules. Springer, Cham

Salem KFM, Tawfik E, Mahdy M et al (2023b) Advances in lettuce (Lactuca spp.) molecular breeding strategies. In: Singh S, Sharma D, Sharma SK, Singh R (eds) Smart plant breeding for vegetable crops in post-genomics era. Springer, Singapore. https://doi.org/10.1007/978-981-19-5367-5_11

Salem KFM, Safhi FA, Alwutayd KM et al (2024) Analysis of genetic diversity, population structure and phylogenetic relationships of rice (*Oryza sativa* L.) cultivars using simple sequence repeat (SSR) markers. Genet Resour Crop Evol 71:2213–2227. https://doi.org/10.1007/s10722-023-01789-0

Sánchez-Mendoza J, Domínguez-López A, Navarro-Galindoand S, López-Sandoval JA (2008) Some physical properties of roselle (*Hibiscus sabdariffa* L.) seeds as a function of moisture content. J Food Eng 87:391–397

Sanyal P, Dutta AN (1954) Natural crossing in Hibiscus sabdariffa Sci. Cult 19:407–408

Satyanarayana NH, Mukherjee S, Roy S, Priya B, Sarkar KK, Bandyopadhyay P (2015) Genetic divergence studies for fiber yield traits in roselle (*Hibiscus sabdariffa* l.) In terai zone of West Bengal. J Crop Weed 11(Special Issue):90–94

Satyanarayana NH, Visalakshmi V, Priya B et al (2016) Divergence studies in roselle (*Hibiscuss abdariffa* L.) for fiber yield over six environments. Int Quart J Environ Sci IX:291–296

Seghatoleslami MJ, Mousavi SG, Barzgaran T (2013) Effect of irrigation and planting date on morpho-physiological traits and yield of roselle (*Hibiscus sabdariffa*). J Anim Plant Sci 23(1):256–260

Sepahrom A, Moosavi SG (2016) The effect of irrigation and nitrogen levels on morphological traits, yield and yield components of roselle (*Hibiscus sabdariffa* L.). Iranian J Med Aromatic Plants Res (IJMAPR) 32(3):436–449. https://doi.org/10.22092/ijmapr.2016.106826

Sereda M, Petrenko V, Kapralova O, Chokheli V et al (2024) Establishment of an In Vitro Micropropagation Protocol for Hibiscus moscheutos L. 'Berry Awesome'. Horticultural 10(1):1–21. https://doi.org/10.3390/horticulturae10010021

Sharma P, Roy M, Roy M, Sundarrao SG (2020) Seed technology in horticultural crops for conservation and utilization of Germplasm. Int J Agric Innov Res 9(3):237–244. 2319–1473

Siregar LAM, Sembiring M, Siregar LAM, Hanafiah DS (2019) Individual selection of roselle plant m4 generation (*Hibiscuss abdariffa* l.) based on anthocyanin, vitamin C and water content. In: International Conference on Agriculture, Environment and Food Security (AEFS), pp 1–5

Slamet A (2018) The diversity of Hibiscus rosa-sinensis based on morphological approach. Scientiae Educatia: Jurnal Pendidikan Sains 7(1):32–40. https://doi.org/10.24235/sc.educatia.v7i1.2503

Song L, Wang R, Yang X, Zhang A, Liu D (2023) Molecular markers and their applications in marker-assisted selection (MAS) in bread wheat (*Triticum aestivum* L.). Agriculture 13(3):642. https://doi.org/10.3390/agriculture13030642

Sourabie S, Zerbo P, Yonli D, Joseph I, Boussim JI (2020) Connaissances traditionnelles des plantes locales utilisées contre les bio-agresseurs des cultures et produits agricoles chez le peuple Turka au Burkina Faso. Int J Biol Chem Sci 14(4):1390–1404

Syahputra A, Hanafiah DS, Khardinata EH (2017) Variation of morphological characters on M_2 generation of roselle (*Hibiscus Sabdariffa* L) induced by gamma irradiation, Bogor, pp 148–153

Tamta M, Kalita BB (2020) Properties of roselle and its blends. Int J Curr Microbiol App Sci 9(10):3616–3621. https://doi.org/10.20546/ijcmas.2020.910.418

Umashankar SR, Pramod PA, Dinesh TS, Mulchand SS (2020) Vertical farming proposal in India. Int Res J Eng Technol (IRJET) 07(6):5629–5639

USDA (2007) National Genetic Resources Program. Taxon: *Hibiscus sabdariffa* L. Germplasm Resources Information Network-(GRIN) [Online Database], National Germplasm Resources Laboratory, Beltsville, Maryland

Vaidya KR (2000) Natural cross-pollination in roselle, *Hibiscus sabdariffa* L. (Malvaceae). Plant Genetics, Genet Mol Biol 23(3):667–669

Varshney RK, Salem KFM, Röder MS et al (2008) Assessment and comparison of genetic diversity in barley germplasm collection using gSSR, eSSR and SNP markers. Plant Genet Res 6:167–174

Vayabari D, Ilham AJZ, Saad NM, Rokhiyah S, Usuldin A et al (2023) Cultivation strategies of Kenaf (*Hibiscus cannabinus* L.) as a future approach in Malaysian Agriculture Industry. Horticultural 9(8):925. https://doi.org/10.3390/horticulturae9080925

Vazquez-Thello A, Yang JL, Hidaka M, Uozumi T (1996) Inherited chilling tolerance in somatic hybrids of transgenic *Hibiscus rosa-sinensis* x transgenic *Lavatera thuringiaca* selected by double-antibiotic resistance. Plant Cell Rep 15(7):506–5011. https://doi.org/10.1007/BF00232983

Wang R, Park S, Park SW, Puja AM, Kim Y (2021) Development of a molecular marker based on chloroplast gene for specific identification of Korean Hibiscus (*Hibiscus syriacus* 'Simbaek'). Appl Biol Chem 64(96):1–7

Wang Z, Xue J, Hu S, Zhang F, Yu R, Chen D, Peer Y et al (2022) The genome of Hibiscus hamabo reveals its adaptation to saline and waterlogged habitat. Hortic Res 9:1–19. https://doi.org/10.1093/hr/uhac067

Wang J et al (2025) China National Gene Bank (CNGB). In: Al-Khayri JM, Salem KFM, Jain SM (eds) Plant gene banks. Springer, Singapore. https://doi.org/10.1007/978-981-99-4236-7_103-1

West TP, Preece JP (2009) Bulk alginate encapsulation of Hibiscus moscheutos nodal segments. Plant Cell Tissue Organ Cult 97(3):345–351. https://doi.org/10.1007/s11240-009-9525-y

Xu XY, Tran THM, Perumalsamy H, Sanjeevram D, Kim YJ (2021) Biosynthetic Gold Nanoparticles of Hibiscus Syriacus L. Callus Potentiates Anti-Inflammation Efficacy via an Autophagy-Dependent Mechanism Mater Sci Eng C 124:112035. https://doi.org/10.1016/j.msec.2021.112035

Yang LJ, Vazquez-Tello A, Makoto H, Takeshi U (1996) Agrobacterium-mediated Transformation of *Hibiscus syriacus* and Regeneration of Transgenic Plants. Plant Tissue Cult Lett 13(2):161–167. https://doi.org/10.5511/plantbiotechnology1984.13.161

Yang YG, Ting Z, Jiangli Z, Fang L, Bingtao Z, Xiangyang W (2012) Antioxidant capacity of extracts from calyx fruits of roselle (Hibiscus sabdariffa L.) Liuqing. African J of Biotechnology Vol 11(17), pp. 4063–4068

Yang Y, Liu Z, Shi X, Ma J, Zhu ZZ, Li F et al (2022) A high-quality, chromosome-level genome provides insights into determinate flowering time and color of cotton rose (*Hibiscus mutabilis*). Front Plant Sci 13:818206. https://doi.org/10.3389/fpls.2022.818206

Youdeowei A, Ezedinma FOC, Onazi OC (1986) Introduction to tropical agriculture. UK. Longman Group Ltd., London

Chapter 15
Conservation Challenges, Utilization, and Improvement Prospects of Prickly Pear (*Opuntia ficus-indica* L. Mill)

Donia Abdallah, Sana Ben Mustapha, Amel Salhi Hannachi, and Ghada Baraket

Abstract Nowadays, food security is a serious problem that concerns scientists and food experts. Numerous factors threaten global food security, such as world population growth, climate change, and water deficit. Being tolerant to drought with incomparable nutritive values, prickly pear cactus (*Opuntia ficus indica* L. Mill) is a choice crop. This bio-energy crop is essential to balancing the biodiversity in arid environments and fighting against soil degradation and reclaiming depleted soils. Prickly pear (*Opuntia ficus indica*) is receiving increased attention from researchers. Unfortunately, the global success of the species is a double-edged weapon. Today, the prickly pear faces serious dangers that enhance its presence and diversity. This implies a series of challenges and prospects. In this chapter, we present a state-of-the-art in *Opuntia ficus indica* distribution, conservation, utilization and improvement.

Keywords Agricultural · Biotechnology · Arid regions · Biodiversity · Prickly Pear (*Opuntia ficus-indica* L. Mill) · Sustainable

15.1 Introduction

Since the early twenty-first century, climate change has been accelerating, causing deeper impacts and increasingly ruining ecosystems and human health. On the other side, global population growth threatens food resources worldwide. According to the FAO (2020), the world is at a critical juncture, with 720 to 811 million people facing hunger in the world in 2020. Furthermore, healthy diets are out of reach for

D. Abdallah (✉) · S. Ben Mustapha · A. Salhi Hannachi · G. Baraket
Department of Biology, Faculty of Sciences of Tunis, University of Tunis El Manar, Tunis, Tunisia
e-mail: Abdallah.d.abdallah@gmail.com; amel.hannachi@fst.utm.tn; ghada.baraket@fst.utm.tn

K. F. M. Salem et al. (eds.), *Breeding and Biotechnology of Grass and Bast Fiber Crops*, Advances in Plant Breeding Strategies 12,
https://doi.org/10.1007/978-3-032-00403-1_15

around three billion people in 2019, a number that keeps more and more increasing. The production of vegetables and fruits is currently insufficient to meet the needs of the growing population, with supply deficits averaging 22% reaching up to 58% in low-income countries (Meldrum et al. 2018). Due to this alarming situation, genetic resources could play a vital role in mitigating impacts and supporting efforts to achieve food security and nutrition goals (Saddoud et al. 2022).

Today, humans face a seemingly unique solution to counteract this critical situation: the sustainable use of natural resources. As conscience, efforts are multiplying, right and left, to identify, conserve and valorize plants with high nutritional values along with significant water deficit resistance.

Once upon a time, nature offered humankind a precious tree that brings together high nutritive values, soil protection, water deficit resistance, and many other benefits. The reason why Italian botanist "Pier Andrea Mattioli" (1501–1578) wrote, "This plant rightly deserves to figure on the list of nature's finest miracles". Prickly Pear (*Opuntia ficus-indica* L. Mill), known as the prickly pear cactus, is derived from the arid regions of Mexico and it is found all across the globe. It has been installed and adopted in many countries. Nowadays, it is considered one of the most sought-after species. This species has impressed humans for centuries and gained the award of resilience and the grandeur symbol.

In this chapter, we will explore the characteristics of the miraculous *Opuntia ficus-indica*, highlighting the importance of protecting and promoting this natural treasure.

15.2 Taxonomic and Origin of Prickly Pear Cactus (*Opuntia Ficus-Indica*)

The Cactaceae family consists of stem succulent plants and is subdivided into three tribes: *Pereskieae, Opuntieae* and *Cereae* (Table 15.1). The *Opuntieae* tribe contains the *Opuntia* genus, which includes around 1600 species. It is further subdivided into four sub-geniuses: *Platyopuntia*, *Cylindropuntia*, *Tephrocactus* and *Brasiliopuntia*. *Opuntia ficus-indica* (L.) Mill. Species belong to the *Platyopuntia* sub-genus. It is worth noting that the taxonomy of *the Opuntia* genus is very complicated due to natural interspecific hybridization between species, long periods of cultivation, and wide dispersion. (Rebman and Pinkava 2001; Snyman 2006).

Opuntia ficus-indica (L.) Mill, commonly known as the prickly pear cactus, is native to the arid region of Mexico and has been cultivated in Latin America, South Africa, and the Mediterranean area. In these regions, various ecotypes, cultivars, and clones of this species have been established (Yahia and Sáenz 2011) (Fig. 15.1).

Table 15.1 *Opuntia ficus-indica* taxonomic classification

Opuntia ficus-indica Classification	
Kingdom	Plantae
Class	Magnoliopsida (Dicotylédones)
Sub-Class	Caryophyllidae
Order	Caryophyllales
Family	Cactaceae
Sub-Family	Opuntioideae (Opuntiodées)
Tribe	Opuntieae
Genus	*Opuntia*
Sub-genus	*Platyopuntia*
Species	*Opuntia ficus-indica*

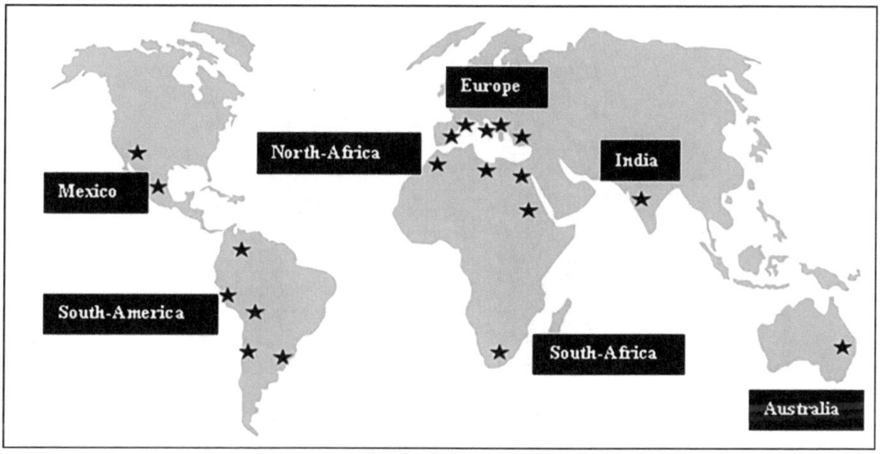

Fig. 15.1 Cactus Distribution Map according to Yahia and Sáenz (2011)

15.3 Botanical Description of Prickly Pear Cactus

The Prickly Pear tree can grow up to 5 meters in length. Its stems are fleshy, woody, and subdivided into green, flattened, thick, and elliptical portions known as cladodes. The cladodes are covered with small protuberances called areoles, from which the leaves emerge. Unlike other known forms of plant leaves, cactus leaves have only two types of thorns. The white ones are steep and about 1–2 cm long. They are firmly clinged and grouped in threes. On the other hand, the brownish thorns are thin and about a few millimeters long. They are called glochidia. Each areole of the cactus contains a meristem that can produce thorns, a new cladode, or roots (if the areole is in the soil) (Bettaïeb 2021; Sudzuki 1995). Nevertheless, some spineless varieties of prickly pear are used as animal feed.

From May through July, the cladodes are covered with multi-colorful flowers with large corollas (4 to 10 cm). The petals are colored in shades of yellow to red

(Fig. 15.2). The sepals are barely visible and difficult to identify and count, resembling fatty greenish scales with reddish tips. The flowers are hermaphrodite with a high number of stamens and a large central style. The ovary is inferior with a single locule containing many ovules (Bettaïeb 2021; Sudzuki 1995).

When the flower withers, only the part containing the ovary remains attached to the cladode and develops into impressive fruits (Fig. 15.2). The prickly pear fruit is a fleshy berry, varying in shape, size and color and containing a high number of seeds. A person can consume 25 to 30 fruits and this quantity is sufficient for their

Fig. 15.2 Cactus morphology description. (**a**–**d**) Floriferous cladodes. (**e**–**h**) *Opuntia ficus-indica* fruits. (Source: Photos taken by author G. Baraket)

daily diet (Bettaïeb 2021; Sudzuki 1995; Loudyi 1995). The ripe fruit has a soft, juicy and sweet pulp with a large number of seeds.

This tree has a remarkable and ingenious root system that enables it to survive even during the dry season. Following a period of rainfall, the tree develops new roots to facilitate water storage. Furthermore, the prickly pear cladodes, which ensure the photosynthetic function, are covered with a cuticle that restricts the plant's transpiration, retains water, and protects it against excessive solar radiation. Like other member of the *Cactaceae* family (and more generally all succulent plants), the prickly pear cactus owns its xerophyte nature to an original photosynthetic metabolism called CAM Type photosynthesis (Crassulacean Acid Metabolism) that involves a first incorporation of CO2 into a 4-carbon organic compound (Landrum 2002; Nobel 2002).

15.4 Chromosome Number and Genetics

Like other members of the Cactaceae family, the basic chromosome number of the cactus is $x = 11$, while the somatic number is $2n = 22$ (Weedin and Powell 1978). During phylogeny, natural specific hybridization has resulted in different polyploidy levels. Thus, *Opuntia ficus-indica* is an octaploid species with $2n = 88$, a characteristic that may have originated from ancient allopolyploid. There is a high chance that humans have facilitated the hybridization of this species, perhaps by hybridizing ancestral *Opuntia ficus-indica* stock from Mesoamerican (Griffith 2004). *Opuntia* can be propagated asexually by stem or cladode segments and suckers that can be separated from the parent plant. On the other hand, sexual propagation is also possible.

15.5 Production and Area

Around 30% of the global continental surface is designated for cactus plantation. Mexico is the world's largest producer of cacti, with over 750,000 hectares of plantation, followed by South America with 550,000 hectares and South Africa with 500,000 hectares. Maghreb countries such as Tunisia, Algeria, and, Morocco have planted between 700,000 to 1 million hectares of cactus (Feugang et al. 2006).

15.6 Prickly Pear Cactus Utilities

Shreds of evidence were found in Tehuacán and Puebla, Mexico, suggesting that *Opuntia* was used as human food 6500 to 10,000 years ago (Callen 1967).

15.6.1 Cactus: A Life-Saving Crop

The prickly pear tree possesses an outstanding capacity for adaptation to all types of soils, including those poor in organic matter (0.1–1.8%), acid and saline soils (<70 moles of salt per m^3), and even gypsum soils (Inglese et al. 2002). Additionally, the prickly pear is known for its tolerance to extreme climatic conditions such as extreme heat, and exposure to sunlight and for promoting water conservation. As a result, planting the prickly pear cactus is a great way to combat desertification and erosion in arid regions. For instance, the extensive planting of the prickly pear in Somalia has shown successful results in promoting the fixation of dunes (Fagotto 1987). In addition, this species has been shown to help restore depleted and infertile soils on which they grow. Every part of the plant is an excellent fertilizer (Bettaïeb 2021). Table 15.2 summarizes the results obtained by Margolis et al. (1985) showing the effect of Prickly pear cultivation in reducing soil loss in semi-arid northeastern Brazil.

Cacti are excellent water collectors and reservoirs, allowing for the creation of humidity in the soil and in the plant itself, making them effective firebreak. It provides a more in-depth and natural approach to firefights and saving the soil (Landrum 2002). Like other members of the *Cactaceae* family, prickly pear helps in reducing the accumulation of CO_2 in the atmosphere. While all plants release CO_2 back into the atmosphere, cacti have a unique benefit; when they die, some CO_2 they have absorbed combined with the calcium ions in their roots and pads, eventually forming solid calcium carbonate or calcite (Nobel 2002). The effect of planting prickly pear on the total biomass was evaluated in the arid region of Tunisia (Sidi Bouzid). The results showed that aboveground biomass was three times higher when planting prickly pear than when planting barley; while the root biomass was six times higher when planting prickly pear than that obtained on natural soil and 16 times higher than that obtained by planting barley (Alary et al. 2007) (Table 15.3).

Table 15.2 Comparison of soil loss (tons ha/year) under different crop cultivation in semi-arid northeastern Brazil

Soil type	Total soil loss (tons ha/year)
Bare ground	29.1
Cotton cultivation	10.91
Corn cultivation	5.94
Prickly pear cultivation	1.98

Source: Margolis et al. (1985)

Table 15.3 The effect of different plantations on the total biomass in Sidi Bouzid, Tunisia

	Natural soil	Barley planting	Prickly pear planting
Aboveground biomass (tons/ha)	0.51	0.53	1.87
Root biomass (tons/ha)	0.33	0.11	1.8

Source: Alary et al. (2007)

The *Cactaceae* species provide a significant fodder crop, especially in arid and semi-arid regions, during drought periods, and when feed is scarce. It is a valuable addition to the feed of animals like horses and camels that can eat the thorny cladodes without any risk of harm. However, the cladodes must be grilled over the fire before being fed to cattle and sheep (Arba 2020; Jorge et al. 2023).

15.6.2 Medicinal and Nutritive Values of Cactus

The prickly pear cactus is known in Tunisia as "Soltane El Ghalla" (the Sultan of the fruits) due to its valuable nutritional properties. Tunisia is one of the largest producers of cacti. The different parts of the cactus tree (cladodes, fruit peels, and seeds) are suitable for consumption. For centuries, humans used the prickly pear cactus in folk medicine due to its anti-inflammatory, hypoglycemic, neuroprotective, and intestinal transit-regulating properties (Piga 2004).

The health benefits of prickly pear cactus are attributed to the high contents of vitamin C (22 mg per 100 g), provitamin A, vitamin B, minerals and potassium. Additionally, this species contains elevated rates of calcium and magnesium rates (46 mg and 28 mg per 100 g, respectively) (Kaur et al. 2012; Albergamo et al. 2022). Additionally, prickly pear seeds may serve as a high protein source. In this respect, when compared with other seeds, the protein content of prickly pear seeds is similar to that of chia seeds (18.81%), flax seed (20.30%), rape seed (19.00 g/100 g), and sesame (15.07%) (Arba 2009; Arba 2020; Albergamo et al. 2022). It also showed traces of iron and its nearby neighbor on the periodic table, copper (1.1 mg and 0.6 mg per 100 g, respectively).

Regarding energy values, the prickly pear is classified as a moderately energetic fruit (44 kcal or 182 kJ) due to its low carbohydrate content (8 g per 100 g, slightly less than the average for classic fruits) (Bettaïeb 2021).

15.6.3 CACTUS: A Natural Source of Fiber

Intensive studies were conducted to biochemically characterize cactus species. The results showed that prickly pear cladodes, peels and seeds are rich in natural fiber, with percentages of 28.39, 12.54, and 16.28%, respectively (Albergamo et al. 2022). According to Regulation (EC) 1924/2006 (European Commission 2006), Tunisian cactus contains over 3% dietary fiber. Tunisian cactus cladodes showed notably different levels of protein, carbohydrates, and dietary fiber compared to Mexican cladodes, as commonly referenced in the literature (Albergamo et al. 2022). *Opuntia* fibers and all parts of the plants are already available in various forms, such as capsules, extracts and powder. These products are becoming critical, especially in phytotherapy.

15.6.4 *Cactus Industry: A Sustainable and Eco-Friendly Option*

Cactus cladodes are used for manufacturing biodegradable packaging plastic due to their high sugar content and favorable gum that helps the formation of biopolymers (Arba 2009). This cactus-based plastic is carbon neutral. More recently, prickly pear has been used to make cactus leather, using the fiber accumulated in the cladodes (Jorge et al. 2023). This "Vegan" leather represents an excellent alternative that does not directly or indirectly harm wildlife or humans. In Mexico, prickly pear leather is used to make handbags, shoes, apparel, and furniture. Apart from being gentle in the environment, this prickly pear leather is highly resistant to bacteria (Jorge et al. 2023).

15.7 Traditional Breeding Methods and Biotechnology

The prickly pear cactus is one of several long-domesticated cactus species (Casas et al. 1997). The easy vegetative propagation of this species makes it simple, fastens its reproduction, and has created an age-old expertise in countries where it grows. Traditionally, the prickly pear cactus is sown in rows by planting seedlings or through cuttings and grafting. Another technique that has been used for a long time by farmers and that allows perennial crops to take advantage of the cacti benefits is alley cropping (Alary et al. 2007). In Tunisia, for example, this agroforestry practice allows barley to be grown alongside intermittent spaces of cacti in arid and semi-arid regions. Table 15.4 shows the positive impact of cactus alley plantation with barley on the yield of barley grains, straw and weeds (ton/ha).

Due to its unique resilience capacity, cactus fruits can remain unaltered on the prickly pear tree for months. As a result, farmers do not pick a large number of cactus fruits at the same time (Bettaïeb 2021). Late-season prickly pears of outstanding quality are traditionally obtained by using the 'Scozzolatura' technique on the cacti. A castration method where all flower buds and young cladodes were eliminated during full bloom to prompt another wave of flowering that brings about out-of-season fruits (Albergamo et al. 2022).

Table 15.4 Change in barley grains yields (tons/ha) in Sidi Bouzid (Tunisia)

	Natural soil	Barley cultivation alone	Alley method cultivation (Barley + cactus)
Yield of barley grains (ton/ha)	1.51	0.82	2.32
Yield of barley grains, straw and weeds (ton/ha)	1.36	4.24	6.65

Source: Alary et al. (2007)

As it could be concluded, cactus cultivation remains mainly traditional thanks to its simple and easy vegetative reproduction, as already mentioned. To date, biotechnological tools have been scarcely used for cactus improvement.

Subsurface Water Retention Technology (SWRT) has been applied to the cactus. By using a polyethylene film installed in a container, this technique allows for conserving water and nutrients on top of the membrane and limits deep drainage. This experiment showed significant results (which need to be better proven!) on mitigating the dangerous effects of drought on the physiological and biochemical parameters of cactus (Lahbouki et al. 2022).

Additionally, some breeding programs were conducted on cacti aiming at the development of spineless cultivars with cold tolerance and high yield.

Biotechnology techniques applications in *Opuntia* are scarce. Cleistogamy and polyembryony, known in the reproductive biology of *Opuntia,* are among the constraints that make sensitive and delicate hybridization techniques in this species. In 2005, the embryo culture techniques were applied to *Opuntia* species by Nieddu and Chessa (2005). The hybrid's fertility has been reported, among them, twenty hybrids, obtained at the F1 progeny, have been selected using preliminary screening based on agronomic and pomological traits (Nieddu and Chessa 2005).

15.8 Cactus: The Victim of it's Global Success

Nowadays, the prickly pear cactus goes beyond having spiritual virtues and becomes the heaven-sent plant that continues to fascinate and arouse the desire of collectors and cosmetics fabricants all over the world. As a result, there is a growing demand for prickly pear cactus varieties. Humankind has posed cacti-wild populations under pressure since human activities have devastated wild habits.

For example, in Mexico, the source kernel and the wealthiest country in terms of genotypic diversity, *Opuntia Ficus*, like other species of the *Cactaceae* family, is one of the most threatened species within the plant kingdom (Santos-Díaz et al. 2010).

This fabulous success puts the prickly pear cactus in a critical situation. In fact, according to CSSG (Cactus and Succulent Plants Specialist Group), almost a third of its species are threatened with extinction due to illegal trade. The invasive agricultural intensification and the natural habitat devastation have led to genetic diversity degradation and gene pool loss (El Kharrassi et al. 2017). In addition, the intensive vegetative multiplication of the species has created a significant loss in genetic diversity, as well as the monoculture of some desired cultivars.

Moreover, although being a xerophytic plant, *Opuntia* is not safe from the abiotic stress dangers, in the header, the drought problem. Undesired effects due to dry periods (lack of irrigation for 2 months) have been documented on the yield, growth, and physiological and biochemical parameters of cacti (Santos-Díaz et al. 2010). On the other hand, being limited to traditional breeding methods often results in inconsistent production levels.

15.9 Diversity Screening

Molecular studies are crucial for screening virus-resistant genotypes and evaluating genetic relationships between resistant varieties to determine their potential.

The knowledge of genetic diversity is an effective tool in gene bank management, breeding experiments, identifying and eliminating duplicates in the collection, and establishing core collections (Saddoud et al. 2022).

Traditionally, morphological traits– such as growth form, stem and leaf morphology, as well as floral, fruit, pollen, and seed characters– have long been used for taxonomy (Schumann 1899; Britton and Rose 1963; Backeberg 1958; Stuppy 2002; Hunt 2006; Griffith and Porter 2009) and *Opuntia* germplasm characterization (El Hani et al. 2019).

Biochemical and physiological investigations have been conducted to shed light on the spiritual virtues of *Opuntia* (Carlín et al. 2015; Albergamo et al. 2022). Later, studies were oriented toward cultivar identification and differentiation, crop fingerprinting, genetic diversity, and population structure studies. Molecular markers were intensively used to identify the relations between species and/or geniuses in order to draw their origins and demographic trajectories. Then, DNA molecular markers have been used in characterizing germplasm pools and assessing genetic diversity within. Most of the molecular studies investigated in *Opuntia* use AFLP, ISSR, RAPD, and mostly SSRs, with few using SNP markers (Shahnejat-Bushehri et al. 2005; Talebi and Fayyaz 2012; Valadez-Moctezuma et al. 2014; Zoghlami et al. 2007; Bendhifi et al. 2013). Köhler et al. (2020) reported the first chloroplast genome of a member of the subfamily *Opuntioideae* and a species of *Opuntia.*

15.10 Germplasm Conservation Strategies

In recent decades, intensive efforts have been devoted to conserving as much as possible cacti plant resources worldwide. The protection program for cacti germplasm officially began 70 years ago in Mexico (Santos-Díaz et al. 2010), the native place of *Opuntia,* and undoubtedly later in other countries.

In situ conservation aims to protect plant material in its natural habitat, preserving genetic variation and the regular processes that originated diversity.

The protected areas for cacti cover 18,000 km^2 in the USA and 68,000 km^2 in Mexico, meaning 17% of the total wild area. The largest Mexican Biosphere Reserve "El Vizcaíno", covered 2.5 million hectares of cacti species (Riley and Riley 2005). In the Maghreb region (Tunisia and Morocco), traditional farming and rural landscapes of prickly pear trees are preserved, providing a potential source of innovation for sustainable agriculture (Arba 2009; Bettaïeb 2021; Albergamo et al. 2022).

Several African countries, such as Ethiopia, Kenya, Zimbabwe, and the Republic of South Africa, are the homeland of diverse cactus species like *Opuntia ficus-indica.* During the rainy season, many farmers consider this fruit the primary food source. However, this biological resource is at risk of extinction due to neglect. In Kenya, for example, among 364 succulent taxa, 23 taxa have unknown conservation status (Newton 1995). International projects (such as "The Kenya Succulent Species Conservation Project" starting in 1996) aimed to sustainably use and conserve biological resources by establishing *in situ* gardens.

Contrarily to *in situ* conservation, the *ex situ* concept preserved resources out of their natural habitat for the long term and under environmental-controlled conditions. This implies the establishment of seeds, *in vitro* germplasm, gene banks, field collections and botanic gardens (Benítez 2001).

The *ex-situ* conservation strategy of *Opuntia* species has been carried out but is not limited to numerous botanic gardens, research organizations, public and private universities and several private nurseries from cacti-lovers. To overcome the main constraints related to *ex-situ* conservation, a need for compatibility between biodiversity protection and breeding technologies advances is a prerequisite.

15.11 Challenges and Prospects for The Sustainable Use of The Wonder *Opuntia Ficus-Indica* (L.)

Based on what has been described above, an important issue must be considered, e.g. this plant gives back to humankind as much more as it takes from it. The main request for prospects is to have plants that guarantee high productivity and quality and are suitable for modern cultivation systems while maintaining their genetic variability.

15.11.1 Legal Framework and Norms

First, conserving genetic variability, particularly the wild form of *Opuntia*, is the main working concern. Loyal and severe restrictions must be imposed to protect the wild extinction-endangered species. In Mexico, for example, extracting these species from their natural habitat for commercial projects, and their global interchange is only allowed under precise conditions (Santos-Díaz et al. 2010). Likewise, *in-situ* and *ex-situ* cacti conservation must be under severe norms and control to guarantee the protection of genetic diversity. In addition, selecting material for a collection must be accompanied by a descriptive catalog including the taxonomy, ecological and geographical exigencies, genetic diversity, and morphological diversity. For this purpose, continuous inventories, genetic diversity assessment, and morphological studies must be regular at regional, national and international levels and data should be collected and organized in catalogs.

These legal frameworks should allow cacti propagation, conservation, dissemination, and reducing cacti-threatened species numbers (Santos-Díaz et al. 2010).

15.11.2 Modern Breeding Strategies

Today, the search for modern, affordable and efficient biotechnology methods in prickly pear species is of paramount importance. The application of biotechnology aims to reduce the extinction risk of some prickly pear species, especially when dealing with wild species, and also to create improved genotypes in terms of productivity and quality. Studies and research activities should be intensified with these objectives in mind.

Mass *in-vitro* propagation is a good solution regarding the high demand for *Opuntia*. Generating *de novo* plants from somatic tissues by culturing *in-vitro* plant segments is a great solution. Even though they are artificial, these *in-vitro* plants can be adapted and transferred to soil where they generally develop and grow as natural plants (Ferreira-Gomes et al. 2006). Without requiring a large amount of starting vegetative plant material, this breeding strategy allows for the multiplication of unlimited plant numbers with limited time and space factors. Another advantage of *de-novo* plants resides in protecting and producing elite plants, rare varieties, and wild accessions as well as obtaining pathogen-free plants (Castro-Gallo et al. 2002; Santos-Díaz et al. 2010). In terms of genetic diversity, *in-vitro* propagation enhances genetic biodiversity by creating new plants genetically identical to progenitor plant material.

In this context, breeding techniques based on sexual propagation are critical to preserve cacti biodiversity. These techniques mainly imply propagation through seeds. Hence, the exact understanding of prickly pear seed germination, dormancy, and propagation is required (Rojas-Aréchiga and Vázquez-Yanes 2000; Ferreira-Gomes et al. 2006).

Prickly pears are also relatively easy to manage by Plant Tissue Culture Methods, Somatic embryogenesis, or organogenesis (Santos-Díaz et al. 2010). These represent revolutionary techniques in the biotechnology world, allowing somatic cells to undergo a series of genetic, biochemical, and physiological transformations that resemble the zygotic embryogenesis process under specific *in-vitro* conditions (Altesor and Ezcurra 2003; Pérez-Molphe-Balch et al. 2015). This "somatic pathway" takes advantage of the totipotence criteria of plants and generates plant embryos under *in-vitro* culture. By using direct or indirect methods (previous callus formation), this technique remains an asexual propagation, capable of creating a massive number of embryos genetically identical to the somatic progenitor cell. Therefore, it is not recommended to use the indirect organogenesis pathway if maintaining genetic integrity is a priority.

Like other *Cactacea* family, prickly pear can produce a horde of secondary metabolites in their natural habitat or *in-vitro* (Santos-Díaz et al. 2010). A fascinating alternative for secondary-metabolite production by *in vitro* cacti cultures is the generation and culture of genetically transformed roots (González-Díaz et al. 2006). These results open the opportunity to utilize this biotechnological approach to study and manufacture secondary metabolites on an industrial scale without collecting Mexican cacti species in the wild or risking their preservation.

15.12 Conclusion and Prospects

In this chapter, we explored the remarkable potential and countless benefits of the prickly pear tree, considered one of the wonders of the world. Conservation, utilization and improvement of this natural heritage are three axes that need to be merged with conscience and synergy. To achieve these goals, efforts should be intensified in order to (i) spread the knowledge of the potential uses of cactus pear, stressing their function as fodder, fruit production, nutritive values, soil protection and water conservation. (ii) Establish midterm and long-term programs to assess the existing cactus pear genetic resources, especially when dealing with wild species. It is imperative to make norms and legislation reliable for human collection, especially. (iii) Develop and achieve protocols for sustainable production of each cactus species.

In this chapter, we also described the properties and dangers of a succulent species as a heaven-sent species. Nonetheless, we stressed the critical role that agrobiodiversity plays in coping with global changes (climate change, desertification), ensuring food security and saving livelihoods. Like the prickly pear tree, the planet is rich with species with unavoidable benefits for humans and life. Thus, moving forward toward a more efficient and useful system of conservation and use of these invaluable resources is a prerequisite.

'The world is at a critical juncture, agrobiodiversity is threatened, the climate is in danger', are realities that humankind should be aware of. Human behaviors should be revised, controlled and merged with a conscience to cope with global changes (climate change, desertification) and ensure food security.

Acknowledgments The authors would like to express their gratitude to Ms. Yosra Bchir (trilingual translator) for her invaluable assistance and dedication in reviewing and improving the language of this article.

References

Alary V, Nefzaoui A, Ben Jemaa M (2007) Promoting the adoption of natural resource management technology in arid and semi-arid areas: modeling the impact of spineless cactus in alley cropping in Central Tunisia. Agric Syst 94:573–585

Albergamo A, Potortí AG, Di Bella G et al (2022) Chemical characterization of different products from the Tunisian *Opuntia ficus*-indica (L.) Mill. Foods 11:155

Altesor A, Ezcurra E (2003) Functional morphology and evolution of stem succulence in cacti. J Arid Environ 53:557–567

Arba M (2009) Le cactus Opuntia, une espèce fruitière et fourragère pour une agriculture durable au Maroc. In: Symposium international agriculture durable en région méditerranéenne (AGDUMED). Maroc, Rabat, pp 215–223

Arba M (2020) The potential of cactus pear (*Opuntia ficus*-Indica (L.) Mill.) as food and forage crop. In: Emerging Hirich A, Choukr-Allah R, Ragab R (eds) Research in alternative crops. Springer International Publishing, Cham, pp 335–357

Backeberg C (1958) Die Cactaceae. In: Vol. 1: Einleitung und Beschreibung der Peireskioideae und Opuntioideae. Jena Gustav Fischer

Bendhifi M, Baraket G, Zourgui L et al (2013) Assessment of genetic diversity of Tunisian barbary fig (Opuntia ficus indica) cultivars by RAPD markers and morphological traits. Sci Hortic 158:1–7
Benítez IS (2001) Conservación de recursos fitogenéticos *ex situ*. In: Estrategia en recursos fitogenéticos para los países del Cono Sur. Programa Cooperativo para el Desarrollo Tecnológico Agroalimentario y Agroindustrial del Cono Sur. (PROCISUR), Montevideo, Uruguay
Bettaïeb V (2021) The prickly pear fruit the secrets of a magical fruit. Editions du patrimoine Maghreb Méditerranée (EPMM)
Britton NL, Rose JN (1963) The Cactaceae. General Publishing Company Ltd, Toronto, ON
Callen EO (1967) Analysis of the Tehuacan coprolites. In: Byers DS (ed) Prehistory of the Tehuacan valley, vol. 1, Environment and subsistence. University of Texas Press, Austin, pp 261–289
Carlín AP, Tafoya F, Alpuche Solís AG et al (2015) Effects of different culture media and conditions on biomass production of hairy root cultures in six Mexican cactus species. In Vitro Cell Dev Biol Plant 51:332–339
Casas A, Caballero J, Mapes C, Zárate S (1997) Manejo de la vegetación, domesticación de plantas y origin de la agricultura en Mesoamérica. Bol Soc Bot Méx 61:31–47
Castro-Gallo IA, Meza-Rangel E, Pérez-Reyes ME, Pérez-Molphe-Balch E (2002) In vitro propagation of 10 Mexican cacti species = Propagación in vitro de 10 especies de cactáceas mexicanas. Sci Nat 4:5–24
El Hani A, Louati M, Ben Salem H et al (2019) Morphologic variability of prickly pear cultivars (*Opuntia* spp.) established in ex-situ collection in Tunisia. Sci Hortic 248:163–175
El Kharrassi Y, Mazri M A, Sedra M H, Mabrouk A, Nasser B, El Mzouri E (2017) Characterization of Genetic Diversity of Cactus Species (Opuntia spp.) in Morocco by Morphological Traits and Molecular Markers. Curr Agric Res J 5(2). https://doi.org/10.12944/CARJ.5.2.01
European Commission (2006) Regulation (EC) No 1924/2006 of the European Parliament and of the Council of 20 December 2006 on nutrition and health claims made on foods. Off J Eur Union 404:9–25
Fagotto F (1987) Sand-dune Fixation in Somalia. Environ Conserv 14(2):157–163
FAOSTAT (2020) Food and Agriculture Organization of the United Nations. Available online: www.fao/faostat/en/#data/QCL
Ferreira-Gomes FLA, Fernandes-Heredia F, Barbeta e Silva P et al (2006) Somatic embryogenesis and plant regeneration in *Opuntia ficus-indica* (L.) Mill (Cactaceae). Sci Hortic-Amsterdam 108:15–21
Feugang JM, Konarski P, Zou D et al (2006) Nutritional and medicinal use of Cactus pear (*Opuntia* spp.) cladodes and fruits. Front Biosci 11:2574–2589
González-Díaz MC, Pérez-Reyes ME, Pérez-Molphe-Balch E (2006) *In vitro* analysis of susceptibility to *Agrobacterium rhizogenes* in 65 species of Mexican cacti. Biol Plant 50:331–337
Griffith MP (2004) The origins of an important cactus crop, *Opuntia ficus-indica* (Cactaceae): new molecular evidence. Am J Bot 91(11):1915–1921
Griffith MP, Porter JM (2009) Phylogeny of Opuntioideae (Cactaceae). Int J Plant Sci 170:107–116
Hunt D (2006) New *cactus lexicon*. DH Books, Milborne Port
Inglese P, Basile F, Schirra M (2002) Cactus pear fruit production. In: Nobel PS (ed) Cacti: biology and uses. University of California, Berkeley, pp 163–183
Jorge AOS, Costa ASG, Oliveira MBPP (2023) Adapting to climate change with Opuntia. Plants (Basel) 12(16):2907
Kaur M, Kaur A, Sharma R (2012) Pharmacological actions of *Opuntia ficus*-indica: a Review. J Appl Pharm Sci 2:15–18
Köhler M, Reginato M, Souza-Chies TT, Majure LC (2020) Insights into chloroplast genome evolution across Opuntioideae (Cactaceae) reveals robust yet sometimes conflicting phylogenetic topologies. Front Plant Sci 11:729
Lahbouki S, Ech-chair L, Er-Raki S et al (2022) Improving drought tolerance of *Opuntia ficus-indica* under field using subsurface water retention technology: changes in physiological and biochemical parameters. Can J Soil Sci 102(4):888–898
Landrum JV (2002) Four Succulent families and 40 million years of evolution and adaptation to xeric environments: what can stem and leaf anatomical characters tell us about their phylogeny? Taxon 51:463–473

Loudyi DW, Llácer G, Aksoy U, Marseditors M (1995) French, Conference Paper Journal article, 1022–1379, 13, Cahiers Options Méditerranéennes, (47–62), Fruit species of secondary importance in Morocco

Margolis E, Silva AB, Jacques FO (1985) Determinacao dos factores da equacao universal de perda de solo para as condicoes de Caruaru (PE). (In Portuguese, with English abstract.). Rev Bras Cienc Solo 9:165–169. Cited by Rômulo S. C. Menezes, Izabel Galindo and Everardo V.S.B. Sampaio: use of Opuntia ficus-indica to control soil erosion: case studies from NE Brazil

Meldrum G, Padulosi S, Lochetti G et al (2018) Issues and prospects for the sustainable use and conservation of cultivated vegetable diversity for more nutrition-sensitive agriculture. Agriculture 8:112

Newton L (1995) Succulent plants in Kenya. Report for the plant conservation program of the National Museums of Kenya, Nairobi, Kenya

Nieddu G, Chessa I (2005) Ensayos de cruzamientos sobre Opuntia spp en Italia. Cactusnet 10:58–63

Nobel PS (2002) Cacti Biology and Uses. University of California, Berkley

Pérez-Molphe-Balch E, Santos-Díaz MDS, Ramírez-Malagón R, Ochoa-Alejo N (2015) Tissue culture of ornamental cacti. Sci Agric (Piracicaba, Braz) 72(6):540–561

Piga A (2004) Cactus pear: a fruit of nutraceutical and functional importance. J Prof Assoc Cactus Dev 6:9–22

Rebman JP, Pinkava DJ (2001) Opuntia cacti of North America: an overview Flor Entom. Fla Entomol 84(4):474–482

Riley L, Riley W (2005) Nature's strongholds: the world’s great wildlife reserves. Princeton University Press, Princeton, NJ

Rojas-Aréchiga M, Vázquez-Yanes C (2000) Cactus seed germination: a review. J Arid Environ 44:85–104

Saddoud DO, Amar FB, Rahmani SM et al (2022) The status of genetic resources and olive breeding in Tunisia. Plants 11:1759

Santos-Díaz MS, Pérez-Molphe E, Ramírez-Malagón R et al (2010) Mexican threatened Cacti: current status and strategies for their conservation. In: Species Diversity and Extinction. ISBN: 978-1-61668-343-6. Nova Science Publishers, Inc

Schumann K (1899) Gesamtbeschreibung der Kakteen. Verlag J, Neumann, Neudamm

Shahnejat-Bushehri AA, Torabi S, Omidi M, Ghannadha MR (2005) Comparison of genetic and morphological distance with heterosis with RAPD markers in hybrids of barley. Int J Agri Biol 7:592–595

Snyman HA (2006) A greenhouse study of root dynamics of cactus pears, *Opuntia ficus indica* and *O. robusta*. J Arid Environ 65:529–542

Stuppy W (2002) Seed characters and the generic classification of the Opuntioideae (Cactaceae). Succ Pl Res 6:25–58

Sudzuki F (1995) Anatomy and morphology. In: Barbera G, Inglese P, Pimienta Barrios E (eds) Agro-ecology, cultivation and uses of cactus pear. FAO plant production and protection paper No. 132, Roma, pp 28–35

Talebi R, Fayyaz F (2012) Quantitative evaluation of genetic diversity in Iranian modern cultivars of wheat (Triticum aestivum L.) using morphological and amplified fragment length polymorphism (AFLP) markers. Biharean Biol 6:14–18

Valadez-Moctezuma E, Samah S, Luna-Paez A (2014) Genetic diversity of *Opuntia* spp. varieties assessed by classical marker tools (RAPD and ISSR). Plant Syst Evol. https://doi.org/10.1007/s00606-014-1112-y

Weedin JF, Powell AM (1978) Chromosome numbers in Chihuahuan Desert Cactaceae. Trans Pecos Texas. Am J Bot 65:531–537

Yahia E, Sáenz C (2011) Cactus pear (*Opuntia* species). In: Yahia E (ed) Postharvest biology and technology of tropical and subtropical fruits, vol 2. Woodhead, Oxford, UK, pp 290–329

Zoghlami N, Chrita I, Bouamama B et al (2007) Molecular based assessment of genetic diversity within Barbary fig (Opuntia ficus indica (L.) Mill.) in Tunisia. Sci Hortic 113:134–141

Chapter 16
Plant Breeding, Genetics and Biotechnology on Pierre (*Pongamia pinnata* L.)

Sharat Prabhakaran, Dinesh Pandey, Jajati Keshari Nayak, and Rashmi Chauhan

Abstract Pierre (*Pongamia pinnata* L.), also known as Karanj, a versatile tree species belonging to Fabaceae, is globally recognized for a wide range of applications like biodiesel production and presents immense potential in soil fertility enhancement, medicinal applications and the textile industry. This chapter delves into the imperative exploration of molecular-level genetic diversity in *P. pinnata,* a critical prerequisite for sustainable biodiesel production, for excellent fuel properties. Identifying Candidate Plus Trees (CPTs) is highlighted as essential for successful breeding programs. The distribution of *P. pinnata* spans tropical and temperate regions of Asia, emphasizing its adaptability to diverse agroclimatic conditions. The species thrives in humid ecosystems near water bodies, exhibiting resilience to adverse climatic conditions. Genetic diversity analysis provides insights into climate preferences and biodiversity hotspots, crucial for sustainable utilization. Beyond biodiesel production, this chapter explores *P. pinnata* ecological adaptability, genetics, reproductive biology, physiological attributes, and cultivation practices. Nanopesticide formulations enhance its efficacy and target specificity. Extraction and characterization of nanocellulose from *P. pinnata* demonstrate its potential for various applications, including textiles and composites. Furthermore, the plant extracts facilitate the green synthesis of nanoparticles with diverse applications in nanotechnology. *P. pinnata* also exhibits significant biofertilizer potential due to its nitrogen-fixing ability. Noteworthy applications include bio-diesel production, nitrogen fixation, bio-pesticides, nanopesticidal effects, and medicinal uses, showcasing its versatility. The chapter concludes by emphasizing the significance of genetic variation for crop enhancement and the establishment of biodiesel orchards. This comprehensive exploration sets the stage for the future of *P. pinnata* in sustainable agriculture and biodiesel production.

S. Prabhakaran · D. Pandey · J. K. Nayak · R. Chauhan (✉)
G. B. Pant University of Agriculture and Technology, Pantnagar, Uttarakhand, India

K. F. M. Salem et al. (eds.), *Breeding and Biotechnology of Grass and Bast Fiber Crops*, Advances in Plant Breeding Strategies 12,
https://doi.org/10.1007/978-3-032-00403-1_16

Keywords Biodiesel · Candidate Plus Trees (CPTs) · Markers · Nanoparticles · Pierre (*Pongamia pinnata* L.) · Propagation

16.1 Introduction

Pongamia (*Pongamia pinnata* L.) is a tree species known for biodiesel production. It also has advantages for both soil fertility enhancement as well as for medicinal applications. Preserving the extensive array of genetic variations present in *Pongamia pinnata* holds paramount significance. Nevertheless, the molecular-level genetic diversity of *P. pinnata* remains inadequately investigated, representing a fundamental prerequisite for any endeavors to enhance the species. The precise identification of Candidate Plus Trees (CPTs) is a fundamental step for the success of breeding programs aimed at developing superior genotypes.

Identifying regions with abundant diversity is crucial in order to preserve natural populations. The loss of diversity can be attributed to both human activities and environmental factors. Diversity analysis can give insight into the climate preferences and biodiversity hotspots for Pongamia, which will ultimately aid in sustainable utilization. This information is also invaluable for implementing planting initiatives in regions that have optimal climate conditions, especially in areas with limited water availability (Sunil et al. 2012).

The prospective success of breeding and cultivation initiatives is based on the availability of diverse *Pongamia pinnata* germplasms. This necessity arises from the increasing demand for its cultivation. Pinnata exhibits adaptability to diverse agroclimatic conditions across India and also has found extensive application in reforestation and land rehabilitation efforts (Shelke et al. 2020).

16.1.1 Botanical Classification and Distribution

Pierre (*Pongamia pinnata* L.), exhibits a wide geographic distribution, spanning tropical and temperate regions of Asia, encompassing Southeast Asia, Indian Ocean islands, Japan, Northeastern Australia, as well as islands in the Western Pacific (Cooper et al. 2019; Daniel 1997; Orwa et al. 2009). Its extensive presence extends across the Indian subcontinent, recognized as its presumed native habitat (Fig. 16.1) (Daniel 1997; ILDIS 2005; POWO 2020; Rajeswara Rao and Syamasundar 2012). However, the precise boundaries of its indigenous range remain difficult to determine because of historical cultivation practices and transportation processes (Murphy et al. 2012). The status of *P. pinnata* in the Brisbane area of Australia, whether native or naturalized, remains ambiguous (Cooper et al. 2019). Beyond its native regions, intentional introductions have been documented in Central Africa, the United States (specifically Hawaii and Florida), Mesoamerica and Puerto Rico

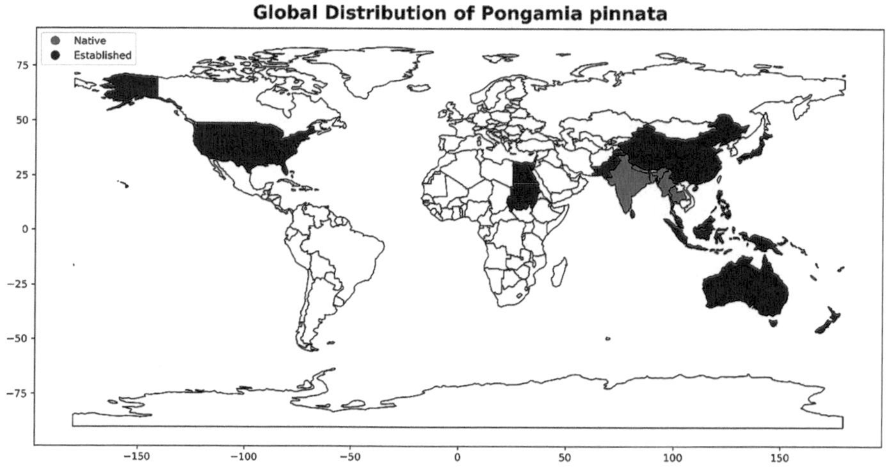

Fig. 16.1 Global distribution of Pierre (*Pongamia pinnata* L.). Native range: India, Myanmar, Bangladesh, Nepal, Thailand. Exotic range: China, Sri Lanka, Australia, New Zealand, Egypt, Indonesia, Fiji, Japan, Malaysia, Mauritius, Papua, Pakistan, New Guinea, Samoa, Philippines, Seychelles, Solomon Islands, Sudan, Tonga and the US

(Orwa et al. 2009). Linnaeus conducted a systematic taxonomic study of pierre in 1953 (https://www.itis.gov/).

Kingdom	Plantae—Planta, plants
Subkingdom	Viridiplantae—green plants
Infrakingdom	Streptophyta—land plants
Superdivision	Embryophyta
Division	Tracheophyta—vascular plants, tracheophytes
Subdivision	Spermatophytina—spermatophytes, seed plants, phanérogames
Class	Magnoliopsida
Superorder	Rosanae
Order	Fabales
Family	Fabaceae—peas, legumes
Genus	Pongamia
Species	*Pongamia pinnata* L.—Pierre

16.1.2 Ecology

Pierre *(Pongamia pinnata L.)* thrives in humid and sub-tropic ecosystems, predominantly inhabiting areas adjacent to water bodies or seashores, with roots extending into both fresh and saltwater environments. Exceptionally adaptable to alkaline and saline conditions, this species naturally occurs in lowland forests situated on limestone and rocky coral outcrops along the coast. Its habitat extends to the peripheries of rivers, mangrove forests and tidal streams. Demonstrating shade-tolerant attributes, P. pinnata can flourish under the canopy of other trees without necessitating shade. While it is not strictly a shade-dependent species, it exhibits robust growth even under full exposure to overhead light. Additionally, the species displays resilience to drought and adeptness in challenging climatic conditions also in soil moisture conditions (Orwa et al. 2009).

In its native habitat, *P. pinnata* exhibits wide temperature tolerance. Mature trees showcase remarkable endurance against challenges such as waterlogging and mild frost, successfully withstanding drought, high winds and salinity. However, susceptibility to low temperatures is evident (Gilman and Watson 1994; Misra and Singh 1987; Tomar and Gupta 1985). The altitude ranges for optimal growth spans from sea level to 1200 m, with the species notably absent above 600 m in the Himalayan foothills (Salim et al. 1998). Recognized as a "maritime species," *P. pinnata* naturally thrives along coasts as well as riverbanks in Bangladesh, India, and Myanmar (Daniel 1997). It exhibits versatility in adapting to different soil types, including sandy, stony, and clay soil. However, the species attains optimal growth in well-drained sandy loam soil with a specific moisture content. Notably, it can endure saline conditions, alkalinity, and waterlogged soil, even when its roots are immersed in fresh or saltwater (Orwa et al. 2009). Cultivation on soil with a pH exceeding 7.5 may suffer from a lack of essential nutrients (Dwivedi et al. 2011).

16.2 Genetics and Reproductive Biology

The chromosomal counts documented for *Pongamia pinnata* are 2n = 20, 22 (Bairiganjan and Patnaik 1989; Flora of China Editorial Committee 2020). *Pongamia pinnata* exhibits reproductive processes involving both seed propagation and vegetative propagation through the utilization of branch cuttings and root suckers (Daniel 1997; Rajeswara Rao and Syamasundar 2012). Sreeharsha et al. (2022) have successfully sequenced the entire genome of Pierre (*Pongamia pinnata* L.), providing valuable insights into genes related to metabolic pathways. The 680 Mb genome sequence, representing approximately 62% of the estimated genome size, serves as an essential resource for further molecular and genetic investigations of this species. This botanical species produces blossoms endowed with nectar-bearing attributes, characterized by substantial pollen quantities accompanied by the secretion of nectar droplets. The plant evolved a pollination mechanism tailored for activation by specific bees of the Apidae family. The array of floral visitors encompasses bees, wasps, and thrips. Certain bee species, namely *Apis cerana indica*, *Apis dorsata*, *Amegilla* sp., *Megachile* sp., *Xylocopa latipes*, and *X. pubescens*, along with wasps, exclusively harvest nectar, while all other bees and thrips engage in the collection of both pollen and nectar. Notably, *Apis* spp. and *Amegilla* sp. emerge as the predominant bee species participating in the pollination process. *P. pinnata* holds significance as a pivotal bee-forage tree species, particularly through the summer season.

The species demonstrates the capability to yield viable, self-pollinated seeds, as evidenced by observations in Hawaii. Flower receptivity is confined to the day of anthesis, with subsequent closure over the ensuing 2 days of the flower's lifespan. Unpollinated flowers undergo abscission, whereas pollinated flowers progress into fruit development, as elucidated by Raju and Rao (2006).

16.3 Physiology and Phenology

The unique pinnately compound foliage, characterized by elongated slender petioles, is arranged alternately, comprising five to nine leaflets exhibiting an oblong or ovate elliptical shape. These leaflets, obtuse-acuminate at the apex, and rounded to cuneate at the base, lack serrations along their edges and possess a slight thickening. The leaflets are arranged in pairs, with two or three sets, excluding the largest terminal leaflet. During the juvenile phase, the leaves are devoid of trichomes and display a pinkish-red hue, subsequently transforming into a glossy light green on the upper surface and a dull green with prominent veins upon reaching maturity (Orwa et al. 2009). The racemose inflorescence, ranging from 6 to 27 cm in length, is axillary and pendant, hosting pairs of highly aromatic flowers arranged in fascicles (2–4 together), each with a short stalk and a pea-shaped structure measuring 15–18 mm in length. The campanulate calyx is truncated (four to five mm in length) with short dentate lobes, occasionally elongated in the lowermost lobe, while the corolla presents a white to pink hue, reveals purple interiors and brownish veins externally, featuring five lobes (Orwa et al. 2009).

The pods, characterized by a short stalk, exhibit a smooth, flattened yet slightly swollen, oblique-oblong to ellipsoid morphology (3 to 8 cm × 2 to 3.5 cm × 1 to 1.5). These pods, brown and possessing a thick-walled along with leathery to sub-woody texture, are reniform, hard, also indehiscent. Each pod contains 1–2 seeds, elliptical or compressed ovoid, bean-like (1.5 to 2.5 cm × 1.2 to 2 cm × 0.8 cm), flattened and dark brown. The bark, thin and gray to grayish-brown, displays a smooth or faint vertical crevasse and exhibits a yellowish interior (Daniel and Hegde 2007). The branchlets lack trichomes but display prominent pale stipule scars. The lateral roots are numerous, while the taproot, thick and elongated, extends approximately 10 m into the soil, enabling water uptake from considerable depths without competing with other crops (Orwa et al. 2009). In India, *P. pinnata* undergoes leaf flushing in April, with foliage development continuing from May onwards. The reproductive phase commences with anthesis occurring between April and June, followed by an extended pod maturation period culminating between March and May of the subsequent year. Seed maturation is observed from February to May (Sangwan et al. 2010). A few instances of extended flowering into June have been observed, with some trees reported to flower during the second season in October–November (Raju and Rao 2006). Chinese populations of the species exhibit an abbreviated reproductive cycle, with anthesis occurring from April to May and seed maturation concluding between October and December (Huang et al. 2018).

Regarding growth parameters, seedlings achieve a vertical growth of approximately 60 cm within 18 months post-germination. The onset of reproductive maturity, characterized by the initiation of pod formation, is observed within a range of 4 to 7 years following seed sowing (Murphy et al., 2012; Csurhes and Hankamer 2016). Flowering in the USA occurs in late spring and summer (from May onwards) (Orwa et al. 2009). In some regions, flowers may appear throughout the year. Mature buds open between 07:00 and 10:00 h, with peak anthesis at 08:00 h, and all 10 anthers dehisce via longitudinal slits in the mature bud stage, occurring approximately 3 h prior to anthesis (Raju and Rao 2006).

Anthesis follows an acropetal pattern in the raceme, with 40–50 floral buds taking 13–16 days to bloom. Flowers remain open for a day and then close for two consecutive days. Pollination is facilitated by bees and wasps, leading to fruit development. The period from flowering to mature pod development extends 10–11 months (Rajeswara Rao and Syamasundar 2012). Phenological patterns vary by region: In Queensland, Australia, the species undergoes seasonal leaf loss between August and November (occasionally extending to December, January, and May), followed by distinctive red new growth preceding a flowering period from February to September. Full-sized fruits persist throughout most of the year, with fruit ripening occurring almost 12 months after flowering (Cooper et al. 2019). In contrast, Mesoamerican populations flower and fruit in June, while South Asian specimens flower from May to June with fruiting from August to October (Flora of China Editorial Committee 2020). The species demonstrates significant reproductive capacity, with individual trees yielding 9–90 kg of pods and 8–24 kg of seeds annually, and potential yields of 36 tonnes per hectare annually (Orwa et al. 2009). Reproductive maturity varies between cultivated specimens (4–5 years) and wild populations (10 years), with optimal harvest timing for timber occurring between 10 and 15 years (Rajeswara Rao and Syamasundar 2012).

16.4 Propagation

Seed-based plant propagation stands as the predominant method for efficient and expedited production of new plants on a large scale. Despite its apparent simplicity, successful seed propagation necessitates meticulous control of germination conditions and a comprehensive understanding of the specific seed requirements corresponding to each species (Porras-García et al. 2023). In the context of extensive propagation, *Pongamia pinnata* (*P. pinnata*) demonstrates a facile propagation process through seed, as indicated by Handa et al. (2005) and Singh et al. (2005). According to Manonmani et al. (1996), larger seeds germinate more effectively, but this observation is limited to seeds that have not been stored for a long time. Notably, the efficiency of germination and overall plant vigor experience a decline after the storage of seeds for a duration exceeding 3 months. Direct seeding can be implemented either in controlled nursery environments utilizing standardized germination beds or containerized systems (polybags), or through field establishment protocols. For optimal morphological development and physiological vigor in mass propagation systems, the substrate composition has been empirically determined to require a tripartite mixture comprising equal volumetric ratios (1:1:1) of siliceous material (sand), mineral soil matrix, and organic amendments (manure) (Arjunan et al. 1994; Handa et al. 2005). Fresh seed sowing is recommended, given the reduction in germination rates upon storage. The germination process of Pongamia is hypogeal, wherein cotyledons remain beneath the germination medium while the plumule ascends and emerges above. Seedlings reach a height of 25–30 cm in their initial growing season, with transplanting to the field undertaken at the onset of the rainy season when seedlings achieve a height of 60 cm (Daniel and Hegde 2007). For bed nurseries, large, clean seeds are sown in well-prepared seedbeds in advance. Dribbling of seeds in nursery beds is performed with a spacing of 7.5 × 15 cm at the commencement of hot weather. Mulching of beds proves beneficial for moisture conservation. Successful germination, up to 85%, is achievable in a properly managed nursery within approximately 1 week. However, achieving consistent and uniform results necessitates a profound understanding of germination and pre-treatment methods. Commonly employed pre-treatment techniques involve seed soaking in hot water, growth regulators, or salt solutions (Ramamoorthy et al. 1989; Kumar et al. 2007). Soaking seeds in IBA (0.15 mM) or GA3 (0.06 mM) for 24 h has been demonstrated to enhance germination, leading to the development of robust root systems in seedlings (NOVOD 2008).

In the context of advancing the development and perpetuating the vegetative propagation of superior genotypes, additional protocols are imperative, despite the accomplishments witnessed in germination trials. Vegetative propagation, a methodology entailing the generation of plants identical in genotype to the maternal plant, constitutes an integral facet of arboreal enhancement initiatives. This approach facilitates the acquisition of optimal planting stock possessing the highest genetic quality, a feat not consistently achievable through sexually propagated progenies. Moreover, the technique enables the production of plantable stock throughout the year, expeditiously yielding results in comparison to seed-raised counterparts (Surendran et al. 2003; Kesari et al. 2009, 2010). Consequently, a singular stock can yield a substantial quantity of plants characterized by genetic uniformity. Notably, this method offers distinct advantages in cross-pollinated species such as *P. pinnata*, affording the production of true-to-type plants with truncated juvenile periods, and leading to precocious productivity (Raju and Rao 2006). Palanisamy et al. (1998) documented the emergence of adventitious roots in shoot cuttings from *P. pinnata*. The stimulation of adventitious root formation, particularly pronounced during the development of new shoots on planted cuttings, was notably influenced by three auxins: indole acetic acid (IAA), indole butyric acid (IBA), and naphthyl acetic acid (NAA). IBA emerged as the most efficacious inducer. Karoshi and Hegde (2002) investigated stem-cutting propagation and explored the potential of softwood grafting. IBA, administered at 2500 ppm, exhibited root-promoting effects in *P. pinnata*, achieving a 95% success rate in grafting. In a separate study, Ansari et al. (2004) examined the impact of immersing semi-hardwood coppice shoot cuttings in KMnO4, KCl, KHPO4, KH2PO4, or K2SO4 on both IAA ionization and adventitious rhizogenesis. Equimolar concentrations of K in the S and P salts significantly enhanced various parameters, such as the percentage of cuttings producing new sprouts and roots, the number of roots per cutting, and root length, while concurrently reducing IAA ionization. Conversely, the Mn salt impeded adventitious rhizogenesis, and the chloride salt had no discernible effect. Grafting emerges as an efficacious strategy to overcome the extended juvenile period issue and ensure heightened productivity through elite scions. Utilizing one-year-old seedlings of Pongamia as rootstock, scions from elite genotypes with matching dimensions can be grafted onto them. Wedge-shaped cuttings are inserted into the scions, and the composite is placed under polyhouse conditions for 2 months before transfer to a hardening chamber. The most successful grafting approach involves using 3-month-old seedlings as stocks and semi-hardwood scions measuring 12–15 cm in length (Karoshi and Hegde 2002). Direct field planting of seeds and in situ grafting with scions from high-yielding trees after 9 months have also been recommended (Wani and Sreedevi 2007). Grafting has proven to be a more facile method for the multiplication of elite *P. pinnata* compared to cuttings. Air layering, also known as "Chinese Layering," represents an alternative effective method for vegetative propagation in *P. pinnata*. This technique involves inducing root formation in small branches by applying root-producing hormones and suitable rooting media. Importantly, air layering can be applied to tertiary branches of Pongamia without causing substantial harm to the maternal plant (Kesari and Rangan 2010). The reproduction of Pongamia is predominantly achieved through seed propagation, a labor-intensive method unsuitable for sustaining genetically superior strains. Pongamia, being an obligate out-crosser, undergoes pollination primarily through bee-mediated processes. Consequently, any tree within the range that a pollinator can effectively transport and transfer pollen (approximately a 3 km radius) may serve as a potential pollen donor. Following the identification of optimal genetic material, the subsequent propagation necessitates clonal methods such as stem cuttings, grafting, or tissue culture to ensure the perpetuation of superior traits (Usharani et al. 2019).

16.5 Plant Tissue Culture

Conventional methods of propagating *Pongamia pinnata*, such as seed germination, grafting, and cuttings, have proven to be time-consuming and inefficient for large-scale plant propagation along with low germination rates, shallow rooting (Azam et al. 2005; Sugla et al. 2007). Furthermore, the extended juvenile period and susceptibility to pests hinder the propagation of *P. pinnata* (Orwa et al. 2009). To establish *P. pinnata* as a leading base material for the biofuel sector, it is crucial to thoroughly explore the plant's propagation protocol and genetics to optimize oil production yields (Scott et al. 2008). To address these challenges, researchers have embarked on initiatives to streamline the selection of elite plants for *P. pinnata*. This includes the establishment of a micropropagation system and the use of molecular marker-based characterization. Various plant parts, such as de-embryonated cotyledons, cotyledonary nodes, and axillary meristems or nodal segments, have been studied as potential candidates for plant regeneration. Kesari et al. (2009) evaluated the impact of varying concentrations of auxin on the rooting system of *P. pinnata* through stem cuttings, and to understand the impact of maturation. They used SAS GLM software to determine the optimal auxin concentration for genetic enhancement programs. All treatments with auxin resulted in sprouting, with lower concentrations being particularly effective in promoting root growth. When applied individually, the effectiveness followed the order of IBA < NAA < IAA. IBA at a concentration of 4.92 mM showed the highest effectiveness. They also observed that concentrations above 7 mM generally hindered root growth. The synergistic interaction between auxins demonstrated a significant capacity to induce and differentiate root formation. The most pronounced stimulatory effects were observed in a treatment regimen incorporating a combination of three distinct auxins. These findings also have significant implications for using genetic resources effectively and can guide the genetic improvement of biodiesel crops.

Sujatha et al. (2008) investigated the role of Thidiazuron (TDZ) in inducing adventitious organogenesis in *Pongamia pinnata* first time. To achieve this, different concentrations of TDZ (0, 0.45, 2.27, 4.54, 6.71, 9.08, 11.35, 13.12, and 22.71 μM) were used to stimulate the formation of buds on de-embryonated cotyledon explants. Further cotyledon was divided into proximal, middle, and distal segments for further examination. The study examined the duration of TDZ exposure, the effects of different cotyledon segments, and the orientation of the explant. An optimal concentration of 11.35 μM TDZ was identified for inducing shoot growth and promoting rapid elongation for 20 days with the abaxial side in contact with the medium for a proximal segment of the cotyledon. Their study showed that TDZ induces meristematic activity in cotyledon cells and also does not interfere with rooting and plant recovery. Their findings serve as promising opportunities for genetic transformation research, identifying variations in somaclonal cells, and inducing genetic mutations. A successful method has been established by Belide et al. (2010) for the growth of multiple shoots and regenerating plants of *Pongamia pinnata* using cotyledonary node explants which is also helpful for genetic transformation. They observed 90 percent efficient direct shoot regeneration by using a cotyledonary node. Seedlings grown on Murashige and Skoog (MS) media supplemented with 1 mg/L benzyl adenine (BA), followed by cultured on MS media with 1 mg/L thidiazuron (TDZ). Further Scanning electron microscope observations demonstrated that cotyledonary node (CN) explants from preconditioned exhibited larger buds and faster development in BA-preconditioned CN explants compared to normal seedlings. They also observed that the addition of adenine sulfate can effectively increase the average number of shoots per explant. They proved their hypothesis by using BA-preconditioned CN explants when cultured on MS media enriched with 1 mg/L TDZ and 200 mg/l adenine sulfate (ADS) and as a result The highest direct shoot regeneration (93%), yielding an average of 32.2 ± 0.93 shoots/explant was achieved.

Shrivastava and Kant (2010) also utilized cotyledonary nodes as explants. 0.53 μM Naphthalene acetic acid (NAA) and 8.8 μM Benzylaminopurine (BAP) were the optimum concentration to achieve multiple shoots. Remarkably, they successfully generated approximately 16–24 shoots from a single cotyledonary node. Another study conducted by Sujatha and Hazra (2007) used cotyledonary nodes from axenic seedlings for micropropagation. By using Murashige and Skoog (MS) medium with 7.5 μM benzylaminopurine (BAP), they achieved up to 6.8 shoots per node within 12 days, with an average length of 0.67 cm. The addition of 2.5 μM gibberellic acid (GA3) during the first subculture significantly increased shoot length. Ultimately, they obtained 16–18 shoots per node within 60 days. These shoots were then rooted on an MS medium containing 1.0 μM indole butyric acid (IBA). Mahmood (2013) investigated somatic embryogenesis on *Pongamia pinnata*, a cultivation technique that involves the use of explants under sterile conditions to initiate callus formation and generate somatic embryos in a controlled environment. Their study aimed to systematically assess the effectiveness of MS media supplemented with different growth hormones at varying concentrations of sucrose, light intensities, and temperatures, in order to enhance the embryogenic processes in *Pongamia pinnata*. They obtained embryogenic callus by enriching MS media with 2,4-D, BAP, NAA, and a 3% sucrose concentration, resulting in an average of 35 embryos per culture after 20 days under 3000 lux light and 27 °C.

Kesari et al. (2012) worked on propagation, especially on two factors regeneration efficiency and genetic clonality. They used nodal segments for the high frequency of regeneration, and various concentrations and combinations to optimize regeneration rates evaluated. The results showed that the response rate of mature nodal segments depended greatly on the season, with over 68% of cultures developing adventitious shoots during the spring. The most effective medium for initiation and multiplication was Woody Plant (WP) Medium supplemented with kinetin (0.5 mg/L) and benzyladenine (5.0 mgL^{-1}). This combination resulted in a multiplication rate of 11 shoots per explant, with each shoot reaching an average length of 3.0 cm. For rooting, half-strength MS media supplemented with indole−3-butyric acid (0.5 mg/L) proved to be more effective. The genetic stability of micropropagated *Pongamia* was confirmed through the Random Amplified Polymorphic DNA (RAPD) and Inter Simple Sequence Repeat (ISSR) markers. They compared how epicotyl and nodal explants responded to both WP media and MS media, ultimately highlighting the superior effectiveness of the WP medium, resulting in a higher number of plantlets produced in a shorter timeframe. This opens up opportunities for future research in breeding biodiesel crops like rapid clonal propagation and conservation, particularly for disease-free elite clones. A similar type of study was conducted by Tan et al. (2018) to develop a system for genetic manipulation and micropropagation of *P. pinnata*. Nodal explants were used in vitro, and it was found that Woody Plant Medium with 1 mM PG and 30 μM BAP was the most effective in inducing multiple shoot buds. For shoot regeneration, WPM with 15 μM zeatin and 1 mM PG produced longer shoots, while rooting was improved with the use of IBA and STS. In conclusion, their study successfully induced, regenerated and rooted multiple shoot buds in *P. pinnata* using tissue culture techniques, paving the way for future advancements in biofuel production.

16.6 Cultivation

Scholarly publications and documented instances have presented a range of planting strategies for cultivating Pongamia on a significant scale (Misra and Singh 1987). These models encompass monoculture plantations arranged in blocks on common land and low fertility land, with agro-forestry multicopying on fertile lands, as well as boundary and row plantations adjacent to agricultural fields, and strip plantations along canal, road and railroad lines. Regardless of the chosen model, optimal plant spacing is recommended, taking into account the unique characteristics of the specific location to achieve maximum yield (Bhojvaid 2008).

Site selection is crucial, requiring the availability of land for an extended duration. Each site, covering an area of 1.5 hectares, allocates 1 hectare for plantation (500 plants per hectare at a spacing of 5 × 4 m). This allows for the inclusion of at least 25 genotypes collected from diverse sites. The rest of the 0.5 hectares can be utilized for establishing a small nursery for conventional multiplication methods. Planting involves digging 60 × 60 × 60 cm^3 pits, spaced 5 m apart in rows, with plants distributed 4 m apart within each row, totaling 500 plants per hectare. Soil and 5 kilograms of farmyard manure are added to each pit. One-year-old healthy seedlings (approximately 60 cm in height) used to be carefully uprooted with a root ball and transplanted in the field at the start of the monsoon season, typically between June and July. Three irrigations may be provided annually as needed for optimal growth. In avenue plantings, the spacing between plants is approximately 8 m, while in block plantings, spacing ranges from 2 × 2 to 5 × 5 m. Pongamia seedlings demonstrate excellent shade tolerance and can be interplanted within existing tree stands. Information regarding management procedures to enhance seed or biomass yield is lacking and warrants further investigation. Given its tolerance for moderate salinity levels, Pongamia emerges as an ideal candidate for saline soil reclamation (Singh 1990). Pongamia is often intercropped, especially with pulses and legumes (Chaukiyal et al. 2000). The cultivation of short-duration oilseed and pulses, including groundnut, cowpea, mustard, millets, horse-gram, and chickpea, as intercrops with Pongamia has been demonstrated to be successful for a period of up to 4–5 years (Pokhriyal et al. 1996). This intercropping practice enhances the economic feasibility of Pongamia plantations and provides income opportunities. Agri-silvicultural trials conducted in India, involving intercropping Pongamia with various crops, reveal no significant differences in morphological parameters such as grain yield, collar diameter, number of branches, and plant height compared to pure cropping. Survival rates within the tree species range from 80% to 100%.

16.7 Genetic Diversity

Over the years, Pongamia has gained heightened significance as a renewable biodiesel source. To optimize crop yield, it is crucial to comprehend genetic variability, the interconnection between pods along with seed characteristics, and their influence on the field performance of offspring. Additionally, conducting diversity analysis on the acquired genetic material is essential in order to guide genetic enhancement initiatives and maximize future genetic traits Divakara and Das (2011). The application of molecular markers has enabled researchers to quantitatively assess genotype identification, genetic diversity, and marker-assisted selection Babu et al. (2018). To facilitate extensive cultivation and genetic refinement, a comprehensive understanding of its genetic diversity is paramount. The establishment of molecular markers not only introduces novel prospects for the amelioration and domestication of Pongamia crops but also furnishes valuable tools for advancements in biotechnology (Jiang et al. 2012). Nevertheless, studies on molecular aspects of this species remain restricted. Insight into the molecular characteristics of this leguminous tree is imperative for optimizing oil yields through breeding strategies and addressing the escalating demand for biodiesel Thudi et al. (2010). Genetic diversity in *Pongamia pinnata* has been carried out based on, random amplified polymorphic DNA (RAPD) Aminah et al. (2017) as well as Interspecific simple sequence repeat (ISSR) Rout et al. (2009). However, molecular characterization by utilizing different markers in *Pongamia pinnata* is very low. Moreover, the EST-SSR indicators display a considerable level of polymorphism and can be effectively applied to different legume plants. The decline in genetic diversity in areas with restricted groundwater and lesser climate fluctuation emphasizes the impact of environmental factors on genetic diversity. Sunil et al. (2012) identified the effect of altitude on seed traits by using 123 different samples from various locations in peninsular India. They noticed that 52% of samples, ranging from 0 to 250 m had oil content levels varying between 15.0% and 46.0%. While 0.8% of the samples came from altitudes surpassing 1000 m and showed a more moderate oil content percentage of 27.7%. They utilized DIVA-GIS software for identification of a more precise location which shows significant variation.

In a comprehensive study, Rao et al. (2011) conducted phenotypic and genetic analyses on 50 elite Candidate Plus Trees (CPTs) distributed across an expansive geographical range (12°41′N to 22°N latitude; 77°E to 84°40′E longitude) spanning 150,000 km^2. Their investigation revealed statistically significant phenotypic variation among progeny for multiple traits including seed morphometric parameters, branching architecture, lipid content and vertical growth. Notably, oil content demonstrated exceptionally high broad-sense heritability, followed by seed length. Implementation of Ward's Minimum Variance hierarchical clustering algorithm revealed distinct biogeographical patterns in the genetic structure of the populations, indicating spatial organization of genetic diversity. This analysis provides valuable insights into the potential for improvement through selection and breeding programs in Pongamia, especially in the initial stages of crop development. Sharma et al. (2016) Investigated in detail the variability among 157 collected samples in various seed characteristics, including weight, oil content, seed length, breadth and thickness of 100 seeds. Further analyzed the fatty acid profiles of 38 selected samples with contrasting oil content levels. The analysis revealed substantial variability in the concentrations of linoleic, oleic as well as stearic acids, ranging from 7.00% to 31.28%, 34.34% to 74.58%, and 0.42% to 10.61%, respectively. Furthermore, parameters such as iodine value, saponification number, and cetane number (CN) of fatty acid methyl esters (FAME) showed ranges of 186.99 to 201.25, 81.13 to 108.19, and 46.16 to 56.47, respectively. These attributes, along with the degree of unsaturation, are crucial for determining the quality of FAME as biodiesel and exhibited significant diversity among the studied Pongamia samples. Moreover, six TE-AFLP primers were used for diversity analysis, producing 334 bands, with 174 (52.10%) being polymorphic. The traits are spread among various genotypes. These findings indicate the possibility of choosing and transmitting these characteristics from Candidate Plus Trees (CPTs) to produce a superior variety through breeding programs. This method may lead to widespread cultivation and large-scale biodiesel production, guaranteeing both quantity as well as quality in the coming years.

Aminah et al. (2017) evaluated the genetic variation of malabari (*Pongamia pinnata*) populations in Java Island, Indonesia specifically in Alas Purwo National Park, Kebumen, Baluran National Park, Batukaras, Carita, utilizing RAPD markers (Random Amplified Polymorphic DNA), with the primers OPL-11, OPAH-13, OPAB-14, OPAH−15 and OPAO-01. The resulting DNA fragments varied in length, ranging from 250 to 1000 bp, and exhibited different numbers of polymorphic bands (13–24). Molecular characterization utilizing Random Amplified Polymorphic DNA (RAPD) markers revealed that the baluran population has the potential to be a center of diversity for Malabari in Java. In this study, Shelke et al. (2020) developed new EST-SSR markers related to oil production and secondary metabolite biosynthesis by analyzing the transcriptome libraries of *P. pinnata*. These markers were effectively assessed. Genetic diversity within Indian *P. pinnata* germplasms as well as were also found to be transferable to other legume plants. These markers, derived from previously published Illumina libraries, provide valuable insights into the genetic variation of Indian *P. pinnata* germplasms. Genetic variability in *P. pinnata* germplasms in the Indian subcontinent is influenced by natural selection. The Western region shows the lowest genetic diversity, which can be attributed to factors like low groundwater presence, high-temperature stress, and insufficient rainfall. In contrast, Eastern India and the Gangetic Plain exhibit higher genetic diversity due to factors like fluctuating temperature, abundant groundwater presence, increased rainfall, as well as proximity to glacier-fed river water. Kumari et al. (2013) examined biochemical characteristics like iodine number, total carbohydrates, acid value, seed oil content, and protein, and assessed genetic diversity using RAPD markers in 24 different *Pongamia pinnata* genotypes from three distinct agroclimatic zones. The maximum oil content in seed recorded was 41.87%, with 14 genotypes surpassing the average of 32.11% for this trait. Higher seed oil content was found positively to be associated with higher protein content. Out of the initially chosen 25 random primers, 22 RAPD primers were reproducible and generated a total of 183 loci.

Among these loci, 147 (80.32%) were found to be polymorphic. The polymorphism percentage ranged varied between 44% and 100%. The study revealed significant genetic variation among *P. pinnata* genotypes, and both biochemical and molecular markers were effective in assessing the extent of this genetic variation. Interestingly, there was a correlation between the geographical origin of the Pongamia genotypes and their genetic similarity. However, some genotypes did not strictly group together based on geographical origins, suggesting the possibility of germplasm exchange among farmers across different regions over time.

To evaluate genetic variability within 48 *Pongamia pinnata* samples from different locations across India, Thudi et al. (2010) utilized the amplified fragment length polymorphism (AFLP) technique. By employing five AFLP primers, they generated 520 distinct fragments, of which a high proportion (96.5%) exhibited polymorphism. The effectiveness of these AFLP primers was assessed using four key attributes: effective multiplex ratio (EMR), resolving power (RP), polymorphism information content (PIC), and marker index (MI). In summary, their study highlights the effectiveness of AFLP as a technique for studying genetic variability in *Pongamia pinnata*. Jiang et al. (2012) examined variations in seed size, oil content, and oil composition in Pongamia. Additionally, they developed a set of Inter-Simple Sequence Repeats (ISSR) markers, called PISSRs (Pongamia Inter-Simple Sequence Repeats), using second-generation sequencing (2GS) data from the Pongamia genome obtained through Next-Generation Sequencing (NGS) Illumina® technology, ensures accuracy. To enhance the reliability and richness of the findings, a combination of PISSR technology, polyacrylamide gel electrophoresis, and silver staining is utilized. By sequencing the polymorphic bands, similarities were found in DNA sequences from other legumes. Rout et al. (2009) research aimed to assess the genetic makeup of a representative group of 111 *P. pinnata* individuals from seven different populations by using Inter Simple Sequence Repeats (ISSR) genetic markers. 67.18% of the genetic markers, showed variation. This indicates a moderate level of genetic diversity in the collected *P. pinnata* germplasm. The ISSR analysis also revealed that 32.34% of the genetic variation was within populations, while 94.34% was at the species level. Additionally, the study found that the genetic differences between populations (GST = 0.61) increased with geographical distance. Their results highlight the need to expand the genetic diversity of *P. pinnata* germplasm which facilitates the cultivation of superior varieties to address the increasing need for biodiesel.

Kesari and Rangan (2011) used Random Amplified Polymorphic DNA (RAPD) markers to assess genetic diversity among 10 chosen genotypes. Out of 40 tested primers, 18 were selected for further analysis, resulting in 210 amplification products, with 10.48% showing polymorphism. The genetic similarity index varied between 0.11 and 0.87. A dendrogram was created to show the relationships between the genotypes using genetic distance values. The Mantel test was utilized to assess the spatial autocorrelation of genetic distances and revealed significant correlation coefficients for the RAPD markers. This genomic analysis provides a cost-effective approach for the characterization and genetic assessment of *P. pinnata* genotypes. Divakara and Das (2011) chose 24 *Pongamia pinnata* (L.) candidate plus trees

(CPTs) from the Forest Research Centre, Ranchi (India) to explore their variation and diversity, with a specific focus on thirteen quantitative traits. These traits included 6 seed traits of parent trees, 3 progeny traits and 4 pod traits. The study took place from 2005 to 2007. One noteworthy finding was that CPT19 exhibited the highest values for seven traits. These traits were 100-pod weight (542.4 g), plant height (164.3 cm), 100-seed weight (217.9 g), pod length (65.6 mm), seed breadth (17.4 mm, seed 2D area (351.2 mm^2), and seed length (27.9 mm). There was a positive and significant correlation observed between 100-seed weight and 100-pod weight and traits at both phenotypic and genotypic levels. This correlation was also seen with plant height, collar diameter, and volume index at 30 months after sowing (MAS). The volume index had a moderate heritability (47.4%) and a high genetic advance (48.4%), suggesting that additive gene effects play a role in this trait. In the divergence study, a non-hierarchical Euclidean cluster analysis was used to classify the 24 accessions into 6 clusters. The CPTs (CPT-4, CPT-8, CPT-9, CPT-20, CPT-21) exhibited significant heterogeneity, making them suitable for within-group hybridization. There was noticeable diversity between clusters V and II, followed by clusters II and I, suggesting that crosses between CPTs from these clusters could produce substantial offspring. This study highlights the extensive variation and diversity present, emphasizing their importance in preserving genetic resources and future tree improvement programs for *P. pinnata*. A study by Nuroniah et al. (2024) investigated the variability of pod and seed traits in 44 candidate plus trees (CPT) of *Pongamia pinnata*, a potential bioenergy species in Indonesia. While significant variation was observed in pod and seed characteristics, no correlation was found between these traits and oil content, suggesting that seed size does not predict oil yield. However, high heritability was observed for seed weight and oil content, indicating strong genetic control. Based on these findings, 17 CPTs were identified as promising candidates for a breeding program aimed at improving pongam oil yield. This research provides valuable insights for selecting trees with desirable traits in future breeding efforts.

16.8 Biodiesel

Population growth, environmental degradation, and finite fossil fuel resources present critical challenges to global sustainability. This situation is particularly acute in India, which supports approximately 17% of the global population within just 2.4% of Earth's land surface area. The nation faces significant anthropogenic environmental pressures while maintaining heavy reliance on non-renewable energy sources. India's fossil fuel reserves comprise 311 billion metric tons of coal, 44.59 billion metric tons of lignite, 621.10 million metric tons of crude oil, and 1227.23 billion cubic meters of natural gas, highlighting the country's carbon-intensive energy infrastructure. The convergence of demographic density, environmental degradation, and fossil fuel dependence presents substantial challenges for sustainable development initiatives. The escalating demand for conventional fuels poses a significant challenge for contemporary scientific endeavors, with biodiesel emerging as a prospective renewable promising avenue for future applications (Bobade and Khyade 2012). With rising biodiesel demand in India, oil imports are expected to skyrocket by 92% by the year 2030 (IEA 2009). *Pongamia pinnata* (L.) Pierre has emerged as a particularly viable candidate for biodiesel synthesis (Dwivedi et al. 2011). *Pongamia pinnata*, a member of the families Leguminosae and Papilionaceae, produces non-edible oleaginous material endemic to the Indian subcontinent (Lakshmikanthan 1978). This medium-sized tree, with a maximum height of 18 m, is primarily widespread distributed across the Western Ghats of India, with additional populations in northern Australia, Fiji, and regions of Eastern Asia. Sexual maturity occurs between 4 and 7 years post-establishment, whereupon the species produces elliptical legumes approximately 7.5 cm in length, containing 1–2 reniform seeds characterized by a brownish-red integument.

The kernels demonstrate oil content ranging from 30% to 40% by mass. Individual specimens yield 9–90 kg of seeds annually, extrapolating to potential yields of 900–9000 kg/ha. Industrial extraction efficiency ranges from 24% to 27.5%, while traditional mechanical extraction methods achieve 18–22% efficiency. The lipid profile comprises eight primary fatty acids: palmitic, stearic, oleic, linoleic, lignoceric, eicosenoic, arachidic, and behenic acids. Pongamia oil's fatty acid composition comprises primarily oleic acid (44.5–71.3%), followed by linoleic (10.8–18.3%), palmitic (3.7–7.9%), and stearic (2.4–8.9%) acids. Additionally, it contains reasonable amounts of eicosenoic acid (9.5–12.4%) and possesses fungicidal properties (Karmee et al. 2005).

A crucial factor influencing the commercial viability of *P. pinnata* as an energy crop is its high seed yield coupled with substantial oil content. Alongside these factors, the quality of the oil must be considered, with adherence to efficient biodiesel standards contributing to overall biofuel efficacy. Successful biodiesel adoption hinges on a non-food crop supply, capable of thriving on marginal lands without competing with food crop cultivation (Hill et al. 2006). The species' seeds contain 30–40% extractable oil, which can undergo transesterification with methanol utilizing potassium hydroxide as a catalyst to yield fatty acid methyl esters (FAMEs). Despite being tri-esters of the tri-alcohol and glycerol, known as triglycerides, direct use of Pongamia oils or blends is generally considered unsatisfactory for diesel engines due to elevated kinematic viscosity, free fatty acids, and gum formation. Additional operational concerns encompass carbonaceous residue accumulation and degradation of engine lubricants during both storage and combustion processes (Srivastava and Prasad 2000).

Driven by the need for sustainable energy solutions, Sangeetha and Baskar (2024) developed a novel magnetically recyclable polymeric acidic ionic liquid nanocatalyst. This innovative approach aimed to efficiently convert *Pongamia pinnata* L. oil, a promising feedstock for biodiesel production, into a valuable biofuel. Utilizing this cutting-edge catalyst, they achieved a remarkable 94.87% biodiesel yield through a one-pot transesterification process. To further optimize this promising technology, the researchers meticulously investigated the process design, and kinetics, and conducted simulations and a techno-economic analysis specifically for biodiesel production from *Pongamia pinnata* L. seed oil. These comprehensive studies provide valuable insights into the large-scale, sustainable production of biodiesel from this valuable renewable resource.

Das et al. (2024) investigated the use of a biofuel derived from co-pyrolyzing waste LDPE and *Pongamia pinnata* seeds as a sustainable fuel source for diesel engines. This study assessed the engine's thermodynamic performance and environmental impact when operating on blends of this biofuel with diesel. By integrating *Pongamia pinnata* seeds, a locally available renewable resource, into the waste-to-fuel process, Das et al. aim to promote a more sustainable and environmentally friendly transportation sector while addressing the challenges of plastic waste disposal and diminishing fossil fuel reserves. Basumatary et al. (2025) developed a novel heterogeneous nanocatalyst derived from waste "Kachkal" (Musa ABB) banana peels and trunks for the efficient transesterification of a binary oil mixture of Jatropha curcas and *Pongamia pinnata* into biodiesel. Characterization techniques confirmed the successful preparation of the catalyst. The calcined Kachkal peel catalyst exhibited superior performance, achieving a maximum biodiesel yield of 96.34% within 10 min under optimized reaction conditions.

16.9 Physico-Chemical Properties of *Pongamia pinnata* Crude Oil

To surmount these limitations, significant endeavors have been invested in the development of derivatives derived from vegetable oils, aiming to emulate the characteristics and performance of diesel fuels originating from petroleum. Transesterification, microemulsification, and pyrolysis are among the processes that have been examined in this effort. Noteworthy attempts have focused on the transformation of Karanja oil into fatty acid methyl esters (Karmee and Chadha 2005). All properties like physiochemical properties, fatty composition, and fuel properties, are listed in Tables 16.1, 16.2, and 16.3.

The percentage of nutrients present in the leaf and fruit is listed in Table 16.4. Meher and colleagues (2008) conducted systematic optimization studies on the transesterification of Karanja oil with methanol, investigating critical experimental factors including catalyst dosage, alcohol-to-oil molar ratio, reaction temperature, as well as agitation speed to maximize methyl ester yield. Research by Azam et al. (2005) demonstrated that fatty acid methyl esters (FAMEs) derived from *Pongamia pinnata* oil exhibited physicochemical properties compliant with international biodiesel standards, including ASTM (USA), DIN (Germany), and EN (European) specifications. Multiple investigations have validated the viability of Pongamia triglycerides as diesel fuel alternatives (Kesari et al. 2009; Sarma et al. 2005; Scott et al. 2008). Karmee and Chadha (2005) achieved 92% conversion efficiency in crude Pongamia oil transesterification using potassium hydroxide catalyst (1%

Table 16.1 Physio-chemical properties of Pierre (*Pongamia pinnata* L.)

Property	Value
Color	Yellowish red
Odor	Characteristic odd odor
Density	0.924 gm/cc
Viscosity	40.2 mm^2/s
Acid Value	5.4 mg/KOH
Iodine Value	87
Saponification Value	184
Calorific Value	8742 Kcal/KG
Specific Gravity	0.925
Unsaponifiable matter	2.9
Flash Point	225 °C
Fire Point	230 °C
Cloud Point	3.5 °C
Pour Point	−3 °C
Boiling Point	316 °C
Cetane Number	42
Copper strip Corrosion	No Corrosion observed
Ash Content	0.07%

Source: Bobade and Khyade (2012)

Table 16.2 Fatty compositions of the seed oil of Pierre (*Pongamia pinnata* L.)

S.No	Fatty acid	Formula	Systematic name	Structure	Net (%)	Reference
1	Behenic acid	C22H44O2	Docosanoic acid	C22	3.2	Sharma (2005)
2	Linoleic acid	C18H32O2	Cis-9-cis-12-Octadecadeneoic acid	C18:2	24.75	
3	Linolenic acid	C18H30O2	Cis-6-cis-9-cis-12-octadecatrienoic acid	C18:3	2.9	
4	Oleic acid	C18H34O2	Cis-9-Octadecanoic acid	C18:1	45.25	
5	Palmitic acid	C16H32O2	Hexadecanoic acid	C16	11.3	
6	Srachidic acid	C20H40O2	Eicosanoic acid	C20	1.75	
7	Stearic acid	C18H38O2	Octadecanoic acid	C18	9.8	

w/w) at 60 °C with a 1:10 oil-to-methanol molar ratio. Subsequently, Meher et al. (2008) reported enhanced FAME yields of 97–98% under optimized conditions: 1% KOH catalyst, 6:1 methanol-to-oil molar ratio, 65 °C reaction temperature, and 360 rpm agitation for a 3-h reaction duration.

Table 16.3 Fuel properties of the Pongam oil vs. conventional diesel

S.No	Parameters	*Pongamia pinnata* oil	Diesel	References
1	Ash content (wt %)	0.006	X	Achten (2008)
2	Cetane number	55.84	40 to 55	Scott (2008), Achten (2008), Pandey (2012)
3	Cloud point (C)	8.3	−15 to −5	Scott (2008), Achten (2008), Pandey (2012)
4	Flash point (C)	135 to 150	60 to 80	Scott (2008), Achten (2008), Pandey (2012), Basumatary (2005), Aslam (2015)
5	Heating values (MJ /kg)	43.422	42	Scott (2008), Achten (2008), Pandey (2012), Aslam (2015)
6	Oil yield (%)	35 to 53	X	Gresshoff (2014), Basumatary (2012)
7	Pour point (C)	2.1	−33 to −15	Scott (2008), Achten (2008), Pandey (2012)
8	Specific gravity at 15 °C (g cm 3)	0.903	0.82 to 0.86	Achten (2008), Pandey (2012), Basumatary (2012), Sarma (2005)
9	Sulphur (wt %)	0.005	1.2	Achten (2008), Pandey (2012), Sarma (2005)
10	Viscosity (cSt)	3.8 to 4.8 at 40 °C	1.3 to 4.1 at 38 °C	Scott (2008), Achten (2008), Pandey (2012), Kotebavi (2016)

X–No data available

Table 16.4 Percentage of nutrients present in leaf and fruit

Parameter	Leaf	Fruit
Crude fiber	–	7.30%
Fatty oil	–	27.50%
Acid detergent fiber	40.00%	1.16%
Protein	–	17.40%
Nitrogen free extract	–	55.40%
Tannin	–	2.32%
Acid detergent lignin	–	6.67%
K	0.49, 0.62%	1.30%
P	0.11, 0.14%	0.61%
Ca	1.54, 1.58%	0.65%
Na+	–	0.80%
N	0.71, 1.16%	5.10%
Crude protein	18.00%	19.5%
Neutral detergent fiber	62.00%	17.98%
Starch	–	6.60%
Moisture	–	19.00%
Mucilage	–	13.50%
Ash	–	2.40%
Trypsin	–	6.2%

Source: Singh (1990) and Duke (1983)

16.10 Nitrogen Fixation and Nodulation

P. pinnata can convert atmospheric nitrogen into a usable form, a trait crucial for its agricultural significance. While the nodulation process in most legumes typically exhibits specificity with one or a few particular species, the nodulation has been observed to be notably promiscuous, forming symbiotic relationships with various species (Arpiwi et al. 2013; Kesari et al. 2013; Rasul et al. 2012). Notably, the nitrogen-fixing capacity in association with these specific species remains unreported, thwarting the development of superior inoculants (Rasul et al. 2012). Nodulation in legumes can be either determinate, characterized by a round morphology due to the absence of persistent meristems, or indeterminate, where the presence of meristems yields more cylindrical nodules. Notably, nodules exhibit a determinate nature (Kazakoff et al. 2011; Scott et al. 2008). However, in older trees, a combination of coralloid and spherical nodules may be observed. As a tree legume, *P. pinnata* exhibits both qualitative and quantitative characteristics that are comparable to those found in well-studied annual legumes like soybean and pea.

16.11 Bio-pesticide

The seed-derived oil of *Pongamia pinnata* demonstrates significant pesticidal properties, with Karanjin identified as the principal bioactive compound exhibiting dual functionality as both an acaricide and insecticide. The oil-based formulation demonstrates high larvicidal efficacy. Formulations combined with or essential oils are recommended for protection against larvae in cabbage crops (Pavela 2012). The mechanical oil extraction process generates oilseed residues as secondary products. While these seed cakes possess inherent toxicity, this characteristic can be advantageously utilized in biopesticide applications. Additionally, the oil cake serves as an economical growth medium for fungi Sharma et al. (2013).

16.12 Nanopesticidal Effect

Pest management is of utmost importance and nanotechnology has shown great potential as an effective solution. Malaikozhundan and Vinodhini (2018) investigation introduces a groundbreaking method for producing zinc oxide nanoparticles (ZnO NPs) coated with an extract derived from *Pongamia pinnata* leaf in an environmentally friendly manner. The ZnO NPs were found to have a crystalline structure and an average particle size of 21.3 nm. These nanoparticles had a substantial effect on the reproduction and development of *Callosobruchus maculatus* (*C. maculatus*). Furthermore, ZnO NPs showed excellent efficacy in controlling *C.*

maculatus, with an LC50 of 10.85 μg/mL. It is worth mentioning that at a concentration of 25 μg/mL, ZnO NPs achieved overall mortality of *C. maculatus* by affecting lipase activity. In conclusion, this study highlights the strong effectiveness of ZnO NPs in response to *C. maculatus*, suggesting their potential as a promising solution for managing insect pests in stored commodities in the future.

16.13 Nanocellulose

The extraction and characterization of cellulosic and nanocellulosic materials derived from oil meals remains relatively unexplored, despite their significant scientific relevance due to potentially unique structural and physicochemical properties compared to conventional cellulosic sources. Although oil meal substrates contain substantial carbohydrate fractions (20–25%), comprehensive structural and property analyses of their cellulosic components remain inadequately documented in the literature. The research conducted by Divya et al. (2022) investigated the extraction and characterization of nanocellulose derived from residual biomass of *Pongamia pinnata* oil meal. The researchers performed a comprehensive characterization of both the intermediate cellulose product (post-bleaching) and the final nanocellulose material. Advanced microscopy techniques, specifically TEM and AFM, confirmed the successful synthesis of nanoscale structures, which were both particulate and rod-like morphologies. Findings from dynamic light scattering experiments revealed an average effective diameter of 338 nm and a mean zeta potential of −13.3 millivolts. Further characterization employed Size Exclusion Chromatography with Multi-Angle Light Scattering (SEC-MALLS) demonstrated a weight-average molecular weight of 54,300. Conventional cellulose exhibits improved thermal stability as well as reduced crystallinity when compared to its alternative form indicating enhanced stability.

Research by Kishanji et al. (2021) demonstrated the successful biosynthesis and stabilization of silver nanoparticles (AgNPs) utilizing *Pongamia pinnata* leaf extract through an in situ bioreduction process within a cellulose matrix. Morphological characterization revealed AgNPs with dimensions ranging from 51 to 100 nm, while fractographic analysis confirmed homogeneous dispersion of the nanoparticles throughout the composite structure. Spectroscopic investigation using FTIR identified the availability of amide functional groups within the leaf extract as the key mediators of AgNP formation. The addition of silver nanoparticles improved the tensile performance of the cellulose composites by increasing their rigidity. The resulting cellulose/AgNP composite effectively inhibited the growth and multiplication of *E.coli*, demonstrating significant antibacterial activity. In summary, Pinnata leaf extract offers both economic and sustainable solutions. It used to be a biomaterial that can be used to synthesize AgNPs. It exhibits commendable antibacterial activity and has potential applications in biological and nanotechnological fields.

16.14 Textile

In modern times, there has been extensive focus on the advancement of antimicrobial finishes for cotton fabric, which are derived from plants. These finishes contain agents that prevent the growth of microbes on fabric surfaces and reduce the risk of infections for fabric users. In the past, chemical disinfectants were employed to tackle this problem, but their harmful effects on humans have been recognized. As a result, Andhra et al. (2019) have explored the use of antimicrobial coatings on cotton fabrics, utilizing extracts derived from diverse parts of the *Pongamia pinnata*. The outcome revealed that fabric treated with leaf extract from *Pongamia pinnata* exhibited enhanced antibacterial effectiveness against *Escherichia coli* bacteria of the gram-negative variety, surpassing the antibacterial properties observed in extracts from the root, bark, and stem. The evaluation of the fabric samples treated with the extract also showed that dip-coating preserved the antibacterial activity for up to five washing cycles. As a result, this study firmly supports the adaption of *P. pinnata* leaf extract in the creation of antibacterial cotton fabrics for healthcare textile applications.

16.15 Nanoparticle Synthesis

Green synthesis techniques for zerovalent nanoparticles have made significant advancements recently, offering a clean, direct, and environmentally friendly process. Rajeshkumar (2016) successfully synthesized and characterized silver nanoparticles using fresh bark extract from the *Pongamia pinnata* tree. By exposing the bark extract to silver ions, biosynthesized silver nanoparticles were created, which were then analyzed using UV-vis spectrophotometry. The analysis revealed a surface plasmon resonance band at 420 nm. Additionally, the X-ray diffraction spectrum showed a crystalline structure, while scanning electron microscope and transmission electron microscope analyses displayed a range of particle sizes from 5 to 55 nm. Elemental analysis confirmed the presence of metallic silver, as a strong signal at 3 keV corresponding to silver ions was detected. To assess the antibacterial activity of the silver nanoparticles, agar well diffusion tests against both gram-positive and gram-negative bacteria. The results showed maximum inhibition zones of 15 mm against *Klebsiella planticola* and minimum inhibition zones of 13 mm against *Staphylococcus aureus*. This study further emphasizes the effective antibacterial properties of silver nanoparticles at a concentration of 100 μg/ml. While various methods can be used for nanoparticle synthesis, such as physical, chemical, and biological approaches, it is important to note that physical and chemical methods can be complex, expensive, and potentially hazardous due to the generation of by-products. Its environmentally friendly and non-toxic nature has caused a great deal of interest in the field of nanoparticle synthesis.

A similar type of study was conducted by Premalatha et al. (2018) by utilizing an extract obtained from *Pongamia pinnata* seeds to synthesize silver nanoparticles. Analysis using UV-vis spectrophotometry revealed a distinctive surface plasmon resonance peak at 427 nm in the biofabricated silver nanoparticles. By employing Fourier Transform Infrared (FTIR) spectrometry, the researchers were able to identify the capping agents present within the seed extract. Additionally, X-ray diffraction confirmed the presence of a crystalline structure. Transmission Electron Microscopy (TEM) images provided further insight, showing silver nanoparticles had an average size of 15 nm, with sizes ranging from 10 to 20 nm. Elemental analysis confirmed the existence of metallic silver, with a clear signal at 3 keV corresponding to silver ions. The antibacterial assessment, conducted using the agar well diffusion method, demonstrated varying zones of inhibition against both *Staphylococcus aureus* (20 nm) and *Escherichia coli* (18 nm) at a concentration of 100 μg/ml. The utilization of *Pongamia pinnata* seeds as a reducing agent highlights the simplicity, eco-friendliness, and cost-effectiveness of the technique. These findings underscore the potential of silver nanoparticles in combating diseases caused by both gram-negative bacteria as well as gram-positive bacteria. Moreover, the study showcases the effectiveness of spectroscopic and microscopic methods in exploring the diverse applications of silver nanoparticles. Utilizing plant extract as a medium, the biological synthesis of gold nanoparticles (AuNPs) offers a promising alternative to traditional physicochemical methods. This natural approach enables the production of nanoparticles that are both stable and compatible with biological systems. Khatua (2020) aimed to create AuNPs using *Pongamia pinnata* leaf extract. Characterization techniques employed included UV-visible spectroscopy, TEM, EDX, and XRD. The absorption spectra displayed surface plasmon resonance (SPR) peaks ranging from 520 to 540 nm, which is used as a sign that gold nanoparticles were successfully made using the plant extract from *P. pinnata* L.TEM analysis revealed an average nanoparticle size within the range of 10 to 25 nm. Moreover, these nanoparticles exhibited antifungal properties, as demonstrated by their effectiveness, specifically oomycete strains BP1120 and SR1. The minimum inhibitory concentration at which 80% growth inhibition (MIC80) occurred was found to be 1.6 mg/mL for SR1 and 0.8 mg/mL for BP1120 fungi. This study highlights an efficient, environmentally friendly, and straightforward approach to synthesizing stable AuNPs using a green synthetic method. Although the photosynthesis of nanomaterials is currently limited to laboratory research, the environmentally friendly, simple, biocompatible, and cost-effective nature of this approach holds great potential for the future. Despite promising results, *in-vitro* and *in-vivo* studies are necessary to comprehensively evaluate the antifungal potential of these herbally synthesized AuNPs against a wider spectrum of pathogens.

Raut (2023) investigation focuses on the hydrothermal synthesis of zirconia nanoparticles (Zr) using bio-extracts derived from neem gum (NG) along with *Pongamia pinnata* plant leaves (PP) as natural capping agents. A control synthesis pathway without capping agents was implemented for comparative analysis. Comprehensive characterization employed multiple analytical techniques including Fourier transform infrared spectroscopy (FTIR), high-resolution transmission electron microscopy (HR-TEM), energy-dispersive X-ray spectroscopy (EDX), X-ray diffraction (XRD), and dynamic light scattering (DLS). The purity of the nanoparticle is confirmed through energy-dispersive X-ray spectroscopy (EDX) analysis.FTIR spectral analysis revealed shifts in the characteristic peaks of the functional groups present in the PP/NG extracts, suggesting an interaction between these groups and the surface hydroxyl groups of zirconia. The synthesized nanoparticles were found to have a tetragonal crystal structure, as confirmed by XRD analysis. HR-TEM images demonstrate that PPZr and NGZr have uniform sizes with reduced agglomeration compared to CZr. Nitrogen adsorption-desorption studies confirm that the material has a mesoporous nature.

In this study, Nayak et al. (2020) goal was to enhance the process of creating gold nanoparticles by utilizing the aqueous extract of Pongamia seed cake (PSC). They found that the occurrence of amino acids, phenolic compounds and proteins in the PSC extract was essential for helping to convert the metal salt into nanoparticles. A central composite design to achieve maximum nanoparticle synthesis was used and found at a pH of 3.6 and a 17.07% (w/v) solution of PSC extract, nanoparticle synthesis significantly increased. They also validated the model generated during optimization and found a close match between the experimental and expected values, with only a 1.15% error and R2 value (0.985). They highlighted the impact of PSC in using gold nanoparticle synthesis by reducing production costs and improving the management of byproducts. The work of Beg et al. (2017) presents a novel and sustainable approach to creating silver nanoparticles (AgNPs) by utilizing the extract obtained from *Pongamia pinnata* seeds. The creation of AgNPs was verified through various analyses, including ultraviolet-visible spectrum and energy-dispersive X-ray spectra. Transmission electron microscopy revealed the average diameter of the AgNPs to be 16.4 nm. Dynamic light scattering determines the hydrodynamic size to be approximately 19.6 nm, while the zeta potential is measured at −23.7 mV, indicating favorable dispersion and stability. Fourier transform infrared analysis identifies specific chemical groups that contribute to the formation of AgNPs, such as O — H, C-O-C, and C = O. Furthermore, the antibacterial power of the synthesized AgNPs is evaluated against *Escherichia coli* ATCC 25922, demonstrating a synergistic effect when combined with ampicillin at its LD50 dose. To better understand the interaction and bioavailability of the synthesized AgNPs, their interaction with human serum albumin is characterized using fluorescence, ultraviolet-visible, as well as circular dichroism spectroscopic methods. These AgNPs display diverse biological functionalities, implying their potential utilization in the realm of nanomedicine.

Behera et al. (2024) investigated the green synthesis of selenium nanoparticles (SeNPs) using extracts from *Pongamia pinnata* leaves. The research aimed to compare the antioxidant activity of these biogenic SeNPs. Characterization techniques confirmed the successful synthesis of SeNPs with plant-derived capping agents. Their research highlights the potential of utilizing plant-mediated synthesis for developing eco-friendly and bioactive SeNPs with promising applications in various fields. Yadav et al. (2024) This study investigated the antifertility effects of silver nanoparticles (AgNPs) synthesized using *Pongamia pinnata* leaf extract in male

Wistar rats. Oral administration of *Pongamia pinnata*-mediated AgNPs to rats resulted in reduced sperm motility, decreased serum testosterone levels, and significant histopathological changes in testicular tissue, including disrupted germinal epithelium and reduced spermatogenesis. The observed antifertility effects were dose-dependent, suggesting that AgNPs, in conjunction with bioactive compounds from *Pongamia pinnata*, can interfere with spermatogenesis and male reproductive hormones. These findings support the potential of *Pongamia pinnata*-mediated AgNPs as a promising avenue for the development of non-hormonal male contraceptives. However, further research is necessary to evaluate long-term effects, optimal dosing, and the reversibility of these antifertility outcomes. A study by Lepcha and Ghosh (2024) explored the potential of *Pongamia pinnata* (Karanja) oil as a sustainable base lubricant enhanced with copper oxide (CuO) nanoparticles. By incorporating varying concentrations of CuO nanoparticles into Karanja oil, researchers observed significant improvements in the lubricant's tribological performance, even at low nanoparticle loadings. The study investigated the influence of CuO nanoparticle concentration on the physicochemical and tribological characteristics of the nano-lubricants, including thermal stability, viscosity, and pour point. The findings demonstrate the potential of CuO-Karanja oil nano-lubricants as a promising alternative to conventional mineral oil lubricants, offering improved performance and reduced environmental impact.

16.16 Bio-fertilizer and Biogas Production

The biomass waste from biofuel crops presents viable feedstock for both biogas generation and bio-fertilizer production (Primandari et al. 2018). Experimental analysis has demonstrated variable methane (CH4) yields from different biomass components like with untreated seeds exhibited maximum biogas potential, while senescent leaf material produced comparatively lower volumes. Notably, alkaline pretreatment using sodium hydroxide (NaOH) enhanced methane yields by 15–22% from lignocellulosic materials including leaves and pod husks Gunaseelan (2014). The deoiled cake, while abundant, poses limitations for direct agricultural application due to inherent toxicity. However, its utilization as a biogas substrate offers superior energy yields compared to conventional cattle manure digestion. Enhanced degradation kinetics were observed when seed cake was mixed with cattle manure. The cake substrate exhibits a carbon-to-nitrogen ratio of 8.7, with pH variability (4.8–9.4) contingent upon substrate composition and dilution parameters. Quantitative assessments indicate that 0.145 million tons of oil cake can potentially generate 377 million cubic meters of biogas (Chandra et al. 2006). Beyond energy production, the anaerobic digestion process generates nutrient-rich effluent suitable for organic agriculture. The digestate demonstrates elevated macronutrient concentrations, specifically nitrogen (4%), phosphorus (1%), and potassium (1%), exceeding the nutritional profile of standard vermicompost (Venkatesan et al. 2017).

16.17 Medicinal and Pharmaceutical Use

Plants have been utilized for their therapeutic properties since time immemorial, contributing to a plethora of herbal preparations available in the market today. The exploration of medicinal plants for drug development has gained prominence, with several herbal products advancing to clinical trials. Current research practices are mainly focused on exploring the qualities of medicinal plants to identify beneficial antioxidant compounds, antibiofilm as well as antimicrobial agents for combating infections, and drugs for diseases (Teanpaisan et al. 2017). *Pongamia pinnata*, a plant with historical significance, has been employed as a crude drug since ancient times. All parts like seeds, bark, root as well as leaves exhibit medicinal qualities and have been traditionally utilized for treating piles, ulcers, tumors, skin diseases and wounds (Tanaka et al. 1992). As reported by Chopade et al. (2008) *Pongamia pinnata* L. has been acknowledged in traditional medicinal practices like Ayurveda and Unani for its broad spectrum of therapeutic effects. These include anti-plasmodial, anti-hyperglycemic, anti-ulcer, anti-diarrheal, anti-hypertonic, anti-nociceptive, anti-inflammatory, anti-lipid peroxidative and antioxidant properties. Pharmacological studies have further elucidated the medicinal value of the plant, demonstrating it exhibits significant anti-plasmodial activity, specifically against *Plasmodium falciparum* (Simonsen et al. 2001). An extract of *Pongamia pinnata* leaves made with 70% ethanol demonstrated notable anti-inflammatory effects across different levels of inflammation (acute, sub-acute, as well as chronic) and exhibited significant antipyretic activity in response to Brewer's yeast-induced pyrexia (Srinivasan et al. 2001). Crude leaf extract of *P. pinnata has been* demonstrated a selective anti-diarrheal effect by inhibiting cholera toxin production as well as protecting epithelial cells from bacterial invasion (Brijesh et al. 2006). Extracts from *Pongamia pinnata* leaves exhibit both circulatory lipid peroxidation as well as antioxidant properties. Studies have been conducted on ammonium chloride-induced hyperammonium rats, these extracts have shown a protective effect against lipid peroxidation and the depletion of components within the antioxidant defense system (Essa and Subramanian 2006). The methanolic extract of *Pongamia pinnata* roots demonstrates substantial shielding effects in response to aspirin-induced ulcers and shows a trend toward reducing acetic acid-induced ulcers following a 10-day treatment regimen. This extract enhances mucosal defense mechanisms, including increased mucin secretion, extended lifespan of mucosal cells, elevated levels of mucosal cell glycoproteins, and enhanced cell proliferation, while also preventing lipid peroxidation (Prabha et al. 2003). In alloxan-induced diabetic rats, an ethanol extract of *Pongamia pinnata* flowers significantly reduces hyperglycemia and lipid peroxidation, also enhances the antioxidant defense mechanisms (Punitha and Manoharan 2006). Furthermore, ethyl acetate extract from the seeds of *P. pinnata* has demonstrated promising antibiofilm, anticancer, antimicrobial, and antioxidant properties, showing effectiveness against a variety of human pathogens (Rajput et al. 2021). The plant also exhibits a range of biological activities, including antiviral and antifilarial effects, as well as guards against White Spot

Syndrome Virus in marine organisms (Singh and Pandey 1996; Uddin et al. 2003; Rameshthangam and Ramasamy 2007). The research by Karamala et al. (2025) explored the potential of *Pongamia pinnata* plant extract as a therapeutic agent against skin cancer. In-silico studies, including molecular docking and simulations, predicted strong binding affinities between compounds from *Pongamia pinnata* (such as pongachromene and vitexin) and key cancer-related proteins (EGF, EGFR, ERBB2). In-vitro studies confirmed the presence of antioxidant compounds like phenols and flavonoids in the extract and demonstrated cytotoxicity against A431 skin cancer cells. These findings suggest that *Pongamia pinnata* may possess promising anti-cancer properties and warrant further investigation for potential therapeutic applications.

Singh et al. (2024) explored the ability of karanjin, a compound found in *Pongamia pinnata* seeds, to fight against the Newcastle disease virus (NDV). Laboratory tests showed that karanjin effectively blocked the virus's growth by hindering its gene activity and significantly reducing its activity. Studies conducted on chicken embryos confirmed that karanjin protected them from the virus. Computer simulations revealed that karanjin strongly binds to a key viral protein. Importantly, karanjin also influenced how cells use sugar, which in turn boosted the body's defenses against the virus. These findings strongly suggest that karanjin from *Pongamia pinnata* seeds has the potential to be a promising antiviral agent and warrants further research for its use in treating NDV and other viral infections. A study by Bhatt et al. (2024) focuses on Glabrin, a bioactive compound isolated from *Pongamia pinnata* seeds. The researchers developed a method to isolate Glabrin and characterized its structure using advanced techniques like Density Functional Theory and X-ray crystallography. *In silico* analysis revealed that Glabrin's molecular structure, particularly its carboxyl, hydroxyl, and nitrogen groups, likely plays a crucial role in interacting with bacterial receptors. Further experiments, including flow cytometry, cell leakage analysis, and field emission scanning electron microscopy (FESEM), demonstrated Glabrin's ability to disrupt bacterial cell walls, leading to cell leakage and eventual death. These findings establish Glabrin as a promising antibacterial agent derived from *Pongamia pinnata* with potential applications in combating bacterial infections.

16.18 Conclusion and Prospects

The varying oil content, and other physical attributes of the seeds, provided an early alternative for selection, which is particularly advantageous in the early stages of Pongamia crop breeding. The thorough evaluation of genetic variation and correlation examination in *Pongamia pinnata* CPTs yielded significant revelations regarding the potential enhancement of crops via selection and breeding initiatives. The identified varieties with desirable traits offer hope for future orchards that will produce high-quality biodiesel.

References

Abu-Ellail, FFB, Makhlouf BSI, Attia AE (2021) Genetic variability and response of some sugarcane varieties to application methods of N-fixation organism (Azospirillum brasilense) under different inorganic nitrogen levels. JSR 37–55

Achten WM, Verchot L, Franken YJ (2008) Jatropha bio-diesel production and use. Biomass Bioenergy 32(12):1063–1084

Aminah AAM, Supriyanto S, Suryani A et al (2017) Genetic diversity of *Pongamia pinnata* (*Millettia pinnata*, aka. malapari) populations in Java Island, Indonesia. Biodivers J Biological Diversity 18(2):677–681

Ansari SA, Singh S, Rani A (2004) Inorganic salts influence IAA ionization and adventitious rhizogenesis in *Pongamia pinnata*. J Plant Physiol 161(1):117–120

Arjunan MC, Antony KA, Ponnammal NR (1994) Effect of seed size on germination, viability and seedling biomass in *Pongamia pinnata Pierre*. Van Vigyan 32:23–28

Arpiwi NL, Yan G, Barbour EL, Plummer JA (2013) Genetic diversity, seed traits and salinity tolerance of *Millettia pinnata* (L.) Panigrahi, a biodiesel tree. Genet Resour Crop Evol 60:677–692. https://doi.org/10.1007/s10722-012-9866-y

Aslam M, Konwar LJ, Sarma AK et al (2015) An investigation of catalytic hydrocracking of high FFA vegetable oils to liquid hydrocarbons using biomass derived heterogeneous catalysts. J Anal Appl Pyrolysis 115:401–409

Azam MM, Waris A, Nahar NM (2005) Prospects and potential of fatty acid methyl esters of some non-traditional seed oil for use as biodiesel in India. Biomass Bioenergy 29:293–302

Babu BK, Rashmi C, Sood S (2018) Cross transferability of finger millet and maize genomic SSR markers for genetic diversity and population structure analysis of barnyard millet. Indian J Genet Plant Breed 78(03):364–372

Bairiganjan GC, Patnaik SN (1989) Chromosomal evolution in Fabaceae. Cytologia 54(1):51–64

Basumatary S (2012) Non-edible oils of Assam as potential feedstocks for biodiesel production: a review. J Chem Bio Phy Sci Sec 3(1):551–558

Basumatary SF, Das B, Brahma S, Basumatary S (2025) Musa ABB (Kachkal) banana waste derived heterogeneous nanocatalyst for transesterification of binary oil mixture of *Jatropha curcas* and *Pongamia pinnata* to biodiesel. Bioresour Technol Rep 29:102018

Beg M, Maji A, Mandal AK et al (2017) Green synthesis of silver nanoparticles using *Pongamia pinnata* seed: characterization, antibacterial property, and spectroscopic investigation of interaction with human serum albumin. J Mol Recognit 30(1):e2565

Behera A, Jothinathan MKD, Ryntathiang I et al (2024) Comparative antioxidant efficacy of green-synthesised selenium nanoparticles from *Pongamia pinnata*, *Citrus sinensis*, and *Acacia auriculiformis*: an in vitro analysis. Cureus 16(4)

Bhatt G, Singh A, Panda AN et al (2024) Glabrin from *Pongamia pinnata*: structural insights and antibacterial potential. Natl Acad Sci Lett 48:27–31. https://doi.org/10.1007/s40009-024-01429-w

Bhojvaid PP (2008) Biofuels towards a greener and secure energy future. TERI Press, New Delhi, p 281

Bobade S, Khyade V (2012) Detail study on the properties of *Pongamia Pinnata* (Karanja) for the production of biofuel. Res J Chem Sci 2(7):16–20

Brijesh S, Daswani PG, Tetali P et al (2006) Studies on *Pongamia pinnata* (L.) Pierre leaves: understanding the mechanism(s) of action in infectious diarrhea. J Zhejiang Univ Sci B 7:665–674

Chandra R, Vijay VK, Subbarao PM (2006) A study on biogas generation from non-edible oil seed cakes: potential and prospects in India. In: The 2nd joint international conference on sustainable energy and environment, pp 21–23

Chaukiyal SP, Sheel SK, Pokhriyal TC (2000) Effects of seasonal variation and nitrogen treatments on nodulation and nitrogen fixation behaviour in *Pongamia pinnata*. J Trop For Sci 12(2):357–368

Chopade VV, Tankar AN, Pande VV et al (2008) *Pongamia pinnata*: phytochemical constituents, traditional uses and pharmacological properties: a review. Int J Green Pharm 2:72–75

Cooper WE, Crayn DM, Zich FA et al (2019) A review of Austrocallerya and Pongamia (Leguminosae subfamily Papilionoideae) in Australia, and the description of a new monotypic genus, Ibatiria. Aust Syst Bot 32(4):363–384

Csurhes S, Hankamer C (2016) Invasive plant risk assessment: Pongamia (Millettia pinnata syn. *Pongamia pinnata*). State of Queensland, Department of Agriculture and Fisheries & Biosecurity Queensland, Australia, p 17

Daniel JN (1997) *Pongamia pinnata* – a nitrogen-fixing tree for oilseed. NFTA 97–03. Forest, Farm, and Community Tree Network (FACT Net) (Winrock International), Little Rock. Accessed 20 Jan 2024

Daniel JN, Hegde NG (2007) Tree-borne oilseeds in agroforestry. In: Proceedings of the national seminar on changing global vegetable oils scenario: issues and challenges before India. Indian Society of Oil Seeds Research Publication, Hyderabad, pp 263–276

Das AK, Mohan I, Yao Z et al (2024) Thermodynamic investigation and 4Es analysis of a VCR diesel engine using emulsified diesel with catalytic co-pyrolysis fuel derived from waste LDPE and *Pongamia pinnata* seeds. Process Saf Environ 191:2373–2384

Divakara BN, Das R (2011) Variability and divergence in *Pongamia pinnata* for further use in tree improvement. J For Res 22:193–200

Divya N, Chunyan H, Narendra R (2022) Extraction and characterization of nanocellulose from *Pongamia pinnata* oil meal. Cellulose Chem Technol 56(1–2):29–37

Duke JA (1983) Handbook of energy crops. http://www.Hort.Purdue.edu/newcrop/duke_energy/Albizia_falcataria.html

Dwivedi G, Jain S, Sharma MP (2011) Pongamia as a source of biodiesel in India. Smart Grid Renew Energy 2(3):184–189

Essa MM, Subramanian P (2006) *Pongamia pinnata* modulates the oxidant-antioxidant imbalance in ammonium chloride-induced hyperammonemic rats. Fundam Clin Pharmacol 20:299–303

Flora of China Editorial Committee (2020) Flora of China. In: Flora of China. Missouri Botanical Garden and Harvard University Herbaria, St. Louis, Missouri/Cambridge, Massachusetts. http://www.efloras.org/flora_page.aspx?flora_id=2. Accessed 15 Sep 2024

Giamalva MJ, Clarke SJ, Stein JM (1984) Sugarcane hybrids of biomass. Biomass 6(1-2) :61–68

Gilman EF, Watson DG (1994) *Pongamia pinnata*. Environmental Horticulture Department, Florida Cooperative Extension Service Institute of Food and Agriculture Science, University of Florida

Gresshoff PM, Biswas B, Indrasumunar A et al (2014) Climate-smart stress adaptation for sustainable biodiesel production from the legume tree *Pongamia pinnata*. Centre for Integrative Legume Research, pp 1–6

Gunaseelan VN (2014) Biogas production from biomass wastes and a model to estimate biodegradability from their composition. Waste Manag Res 2:131–139

Handa AK, Nandini D, Uma (2005) An alternative source of biofuel, seed germination trials of *Pongamia pinnata*. Int J For Usufructs Manag 6:75–80

Heinz D J (1987) Sugarcane improvement through breeding (Amsterdam: Elsevier)

Hill J, Nelson E, Tilman D et al (2006) Environmental, economic, and energetic costs and benefits of biodiesel and ethanol biofuels. Proc Natl Acad Sci USA 103:11206–11210

Huang J, Hao X, Jin Y et al (2018) Temporal transcriptome profiling of developing seeds reveals a concerted gene regulation in relation to oil accumulation in Pongamia (*Millettia pinnata*). BMC Plant Biol 18:1–16

ILDIS (2005) International legume database and information service: world database of legumes (version 10). In: International legume database and information service: world database of legumes (version 10). School of Plant Sciences, University of Reading, Reading

International Energy Agency (2009) World energy outlook. OECD/IEA, Paris, p 17. Accessed 16 Sep 2023

Jiang Q, Yen SH, Stiller J et al (2012) Genetic, biochemical, and morphological diversity of the legume biofuel tree *Pongamia pinnata*. Plant Genet Genom Biotech 1(3):54–67

Karamala LN, Karthik Y, Raghu M et al (2025) Exploring the therapeutic potential of *Pongamia pinnata* plant extract against skin cancer: in-silico and in-vitro study. J Ethnopharmacol 337:118964

Karmee SK, Chadha A (2005) Preparation of biodiesel from crude oil of *Pongamia pinnata*. Bioresour Technol 96(13):1425–1429

Karoshi VR, Hegde GV (2002) Vegetative propagation of *Pongamia pinnata* (L) Pierre: hitherto a neglected species. Indian Forester 128(3):348–350

Kazakoff SH, Gresshoff PM, Scott PT (2011) A sustainable feedstock for biodiesel production. London: Energy Crops 4:233–254

Kesari V, Rangan L (2010) Development of *Pongamia pinnata* as an alternative biofuel crop-current status and scope of plantations in India. J Crop Sci Biotechnol 13:127–137

Kesari V, Rangan L (2011) Genetic diversity analysis by RAPD markers in candidate plus trees of *Pongamia pinnata*, a promising source of bioenergy. Biomass Bioenergy 35(7):3123–3128

Kesari V, Krishnamachari A, Rangan L (2009) Effect of auxins on adventitious rooting from stem cuttings of candidate plus tree *Pongamia pinnata* (L.), a potential biodiesel plant. Trees Struc Func 23:597–604

Kesari V, Das A, Rangan L (2010) Effect of genotype and auxin treatments on rooting response in stem cuttings of CPTs of *Pongamia pinnata*, a potential biodiesel legume crop. Curr Sci 98:1234–1237

Kesari V, Ramesh AM, Rangan L (2012) High-frequency direct organogenesis and evaluation of genetic stability for in vitro regenerated *Pongamia pinnata*, a valuable biodiesel plant. Biomass Bioenergy 44:23–32

Kesari V, Ramesh AM, Rangan L (2013) Rhizobium pongamiae sp. nov. from root nodules of *Pongamia pinnata*. Biomed Res Int 1:165198

Khatua A, Priyadarshini E, Rajamani P et al (2020) Phytosynthesis, characterization and fungicidal potential of emerging gold nanoparticles using *Pongamia pinnata* leave extract: a novel approach in nanoparticle synthesis. J Clust Sci 31:125–131

Kishanji M, Mamatha G, Madhuri D et al (2021) Preparation and characterization of cellulose/in situ generated silver nanoparticle composite films prepared using *Pongamia pinnata* leaf extract as a reducing and stabilizing agent. Inorg Nano-Met Chem 51(9):1207–1213

Kotebavi V, Shetty D, Sahu D (2016) Performance and emission characteristics of a CI engine run on waste cooking oil-diesel blends. Pollut Res 35(1):159–166

Kumar S, Radhamani J, Singh AK et al (2007) Germination and seed storage behaviour in *Pongamia pinnata* L. Curr Sci 93:910–911

Kumari K, Sinha A, Singh S et al (2013) Biochemical characterization and genetic diversity of *Pongamia pinnata* (L.) Pierre in Eastern India. J For Sci 29(3):200–210

Lakshmikanthan V (1978) Tree borne oil seeds. Directorate of Nonedible Oils and Soap Industry, Khadi and Village Industries Commission, Mumbai, India, p 10

Lepcha PT, Ghosh P (2024) Influence of CuO nanoparticles on the tribological performance of Karanja oil (*Pongamia pinnata*). Chem Select 9(37):e202403581

Mahmood T (2013) Somatic embryogenesis in *Pongamia pinnata* (L.) pierre by Tahir Mahmood and Zia-Ul-Hasan. Life Sci Leafl 35:14–22

Malaikozhundan B, Vinodhini J (2018) Nanopesticidal effects of *Pongamia pinnata* leaf extract coated zinc oxide nanoparticle against the Pulse beetle, *Callosobruchus maculatus*. Mater Today Commun 14:106–115

Manonmani V, Vanagamudi K, Vinaya Rai RS (1996) Effect of seed size on seed germination and vigour in *Pongamia pinnata*. J Trop For Sci 9:1–5

Meher LC, Naik SN, Naik MK et al (2008) Biodiesel production using Karanja (*Pongamia pinnata*) and Jatropha (*Jatropha curcas*) seed oil. In: Handbook of plant-based biofuels, pp 255–266

Misra CM, Singh SL (1987) Ecological evaluation of certain leguminous trees for agroforestry. Nitrogen Fixing Tree Res Rep 5:5

Murphy HT, O'Connell DA, Seaton G et al (2012) A common view of the opportunities, challenges, and research actions for Pongamia in Australia. Bioenergy Res 5(3):778–800

Nayak S, Sajankila SP, Goveas LC et al (2020) Two fold increase in synthesis of gold nanoparticles assisted by proteins and phenolic compounds in Pongamia seed cake extract: response surface methodology approach. SN Appl Sci 2:1–12

NOVOD (2008) 3rd R&D report on tree borne oilseeds, NOVOD Board India, http://www.novodboard.com/3rd%20R&D-Report.pdf

Nuroniah DHS, Cahyono DDN, Aminah A et al (2024) Variability of pod and seed traits on candidate plus tree *Pongamia pinnata* in Indonesia. For Sci Technol:1–8

Orwa C, Mutua A, Kindt R et al (2009) Agroforestree Database: a tree reference and selection guide. Version 4. In: Agroforestree Database: a tree reference and selection guide. Version 4:1-6. World Agroforestry Centre, Nairobi. Accessed 21 Aug 2023

Palanisamy K, Ansari SA, Kumar P et al (1998) Adventitious rooting in shoot cuttings of Azadirachta indica and *Pongamia pinnata*. New For 16:81–88

Pandey VC, Singh K, Singh JS et al (2012) Jatropha curcas: a potential biofuel plant for sustainable environmental development. Renew Sust Energ Rev 16(5):2870–2883

Panje R (1972) The role of Saccharum spontaneum in sugarcane breeding. Proc Int Soc Sugarcane Technol 14:217–223

Pavela R (2012) Efficacy of three newly developed botanical insecticides based on pongam oil against L. larvae. J Biopest 1:62–70

Pokhriyal TC, Chaukiyal SP, Singh KCH (1996) Nitrogen fixation and nodulation behaviour in relation to seasonal changes in six multipurpose tree species. Indian For 122:718–726

Porras-García B, Pinzón-Sandoval EH, Almanza-Merchán PJ (2023) Propagation of Cannabis sativa (L.) plants through cuttings and use of auxin phytoregulators. Rev Colomb Cienc Hortíc 17(3):e16428

POWO (2020) Plants of the world online. In: Plants of the world online. Royal Botanic Gardens, Kew, London. Accessed 18 Sep 2023

Prabha T, Dora M, Priyambada S (2003) Evaluation of *Pongamia pinnata* root extract on gastric ulcers and mucosal offensive and defensive factors in rats. Indian J Exp Biol 41:304–310

Premalatha M, Hemamalini V, Anandhalakshmi J et al (2018) Biosynthesis of characterization of silver nanoparticles using seeds of *Pongamia pinnata* (L) pierre. World J Pharm Res 7(7):239–248

Primandari SRP, Islam AKMA, Yaakob Z et al (2018) Jatropha curcas L. biomass waste and its utilization. Advances in biofuels and bioenergy. InTech Open, London, p 273

Punitha R, Manoharan S (2006) Antihyperglycemic and antilipidperoxidative effects of (Linn.) Pierre flowers in alloxan induced diabetic rats. J Ethon Pharmacol 105(1):39–46

Rajeshkumar S (2016) Synthesis of silver nanoparticles using fresh bark of *Pongamia pinnata* and characterization of its antibacterial activity against gram positive and gram negative pathogens. Resource-Efficient Technol 2(1):30–35

Rajeswara Rao BR, Syamasundar KV (2012) *Millettia pinnata* (L.) Panigrahi: overview and biological properties of fixed oil. In: Bhattacharya S (ed) Recent progress in medicinal plants, Volume 33. Fixed oils and fats. Studium Press LLC, USA, pp 123–144

Rajput M, Bithel N, Vijayakumar S (2021) Antimicrobial, antibiofilm, antioxidant, anticancer, and phytochemical composition of the seed extract of *Pongamia pinnata*. Arch Microbiol 203(7):4005–4024

Raju AJS, Rao SP (2006) Explosive pollen release and pollination as a function of nectar-feeding activity of certain bees in the biodiesel plant, *Pongamia pinnata* (L.) Pierre (Fabaceae). Curr Sci 90:960–967

Ramamoorthy K, Kalavathi D, Karivaratharaju TV (1989) Seed treatment to improve seed germination and viability in Pungam (*Derris indica* Lam). Myforest 25:321–324

Rameshthangam P, Ramasamy P (2007) Antiviral activity of bis (2-methylheptyl) phthalate isolated from leaves against white spot syndrome virus of Penaeus monodon Fabricius. Virus Res 1:38–44

Rao GR, Shanker AK, Srinivas I et al (2011) Diversity and variability in seed characters and growth of *Pongamia pinnata* (L.) Pierre accessions. Trees 25:725–734

Rasul A, Amalraj ELD, Praveen Kumar G et al (2012) Characterization of rhizobial isolates nodulating Millettia pinnata in India. FEMS Microbiol Lett 336(2):148–158

Raut S, Dhakar GL, Gogoi P et al (2023) Neem gum and *Pongamia pinnata* plant leaves bio-extract assisted green synthesis and characterization of zirconia nanoparticles. J Aust Ceram Soc 59:1–9

Rout GR, Sahoo DP, Aparajita S (2009) Studies on inter and intra-population variability of *Pongamia pinnata*: a bioenergy legume tree. Crop Breed Appl Biotechnol 9(3):268–273

Salim AS, Simons AJ, Waruhiu A et al (1998) Agroforestree database: a tree species reference and selection guide (No. CD 631.58 A281). ICRAF, Nairobi. Accessed 29 Sep 2023

Sangeetha B, Baskar G (2024) Process design, kinetics, simulation, and techno-economic analysis of biodiesel production from *Pongamia pinnata* seed oil using a magnetically recyclable acidic ionic liquid catalyst. Energy Convers Manag 301:118040

Sangwan S, Rao DV, Sharma RA (2010) A review on *Pongamia pinnata* (L.) Pierre: a great versatile leguminous plant. Nat Sci 8(11):130–139

Sarma AK, Konwer D, Bordoloi PK (2005) A comprehensive analysis of fuel properties of biodiesel from Koroch seed oil. Energy Fuel 19(2):656–657

Scott PT, Pregelj L, Chen N et al (2008) *Pongamia pinnata*: an untapped resource for the biofuels industry of the future. Bioenergy Res 1:2–11

Sharma S, Verma M, Sharma A (2013) Utilization of non- edible oil seed cakes as substrate for growth of and as biopesticide against termites. Waste Biomass Valorization 2:325–330

Sharma SS, Islam MA, Malik AA et al (2016) Seed traits, fatty acid profile and genetic diversity assessment in *Pongamia pinnata* (L.) Pierre germplasm. Physiol Mol Biol Plants 22:193–205

Shelke RG, Basak S, Rangan L (2020) Development of EST-SSR markers for *Pongamia pinnata* by transcriptome database mining: cross-species amplification and genetic diversity. Physiol Mol Biol Plants 26:2225–2241

Shrivastava V, Kant T (2010) Micropropagation of *Pongamia pinnata* (L.). Pierre-a native Indian biodiesel tree from cotyledonary node. Int J Biotechnol Biochemist 6(4):555–560

Simonsen HT, Nordskjold JB, Smitt UW et al (2001) In vitro screening of Indian medicinal plants or antiplasmodial activity. J Ethnopharmacol 74:195–204

Singh K (1990) Effect of soil salinity and sodicity on seedling growth and mineral composition of *Pongamia pinnata* and Araucaria cunninghamii. Trop Ecol 31:124–130

Singh RK, Pandey BL (1996) Anti-inflammatory potential of root extracts in experimentally induced inflammation in rats. J Appl Biomed:21–24

Singh KP, Dhakre G, Chauhan SVS (2005) Effect of mechanical and chemical treatments on seed germination in Pongamia glabra L. Seed Res 33:169–171

Singh R, Kumar S, Rangan L (2024) *Pongamia pinnata* L. seed-derived karanjin as prominent antiviral agent against Newcastle disease virus. Virology 600:110272

Singh RK, Mishra SK, Singh SP et al (2010) Evaluation of microsatellite markers for genetic diversity analysis among sugarcane species and commercial hybrids. Aust J Crop Sci 2:115–124

Sreeharsha RV, Mudalkar S, Reddy AR (2022) Genome sequencing and analysis uncover the regulatory elements involved in the development and oil biosynthesis of *Pongamia pinnata* (L.)–a potential biodiesel feedstock. Front Plant Sci 13:747783

Srinivasan K, Muruganandan S, Lal J (2001) Evaluation of anti-inflammatory activity of leaves in rats. J Ethnopharmacol 2:151–157

Srivastava A, Prasad R (2000) Triglycerides based diesel fuels. Renew Sust Energy Rev 4:111–133

Sugla T, Purkayastha J, Singh SK et al (2007) Micropropagation of *Pongamia pinnata* through enhanced axillary branching. In Vitro Cell Dev Biol Plant 43:409–414

Sujatha K, Hazra S (2007) Micropropagation of mature *Pongamia pinnata* Pierre. In Vitro Cell Dev Biol Plant 43:608–613

Sujatha K, Panda BM, Hazra S (2008) De novo organogenesis and plant regeneration in *Pongamia pinnata*, oil producing tree legume. Trees 22:711–716

Sunil N, Kumar V, Sivaraj N et al (2012) Identification of areas of diversity and distribution of Pongamia based on altitude and seed traits. Indian J Agric Sci 82(6):489–493

Surendran C, Sehgal RN, Paramathma M (2003) Textbook of forest tree breeding. Indian Council of Agricultural Research, New Delhi, p 247

Tan SN, Tee CS, Wong HL (2018) Multiple shoot bud induction and plant regeneration studies of *Pongamia pinnata*. Plant Biotechnol 35(4):325–334

Tanaka T, Iinuma M, Fujii Y et al (1992) Flavonoids in root bark of *Pongamia pinnata*. Phytochemistry 31:993–998

Teanpaisan R, Kawsud P, Pahumunto N et al (2017) Screening for antibacterial and antibiofilm activity in Thai medicinal plant extracts against oral microorganisms. J Tradit Complement Med 7(2):172–177

Thudi M, Manthena R, Wani SP et al (2010) Analysis of genetic diversity in pongamia [*Pongamia pinnat*a (L) Pierrre] using AFLP Markers. J Plant Biochem Biotechnol 19:209–216

Tomar OS, Gupta RK (1985) Performance of some forest tree species in saline soils under shallow and saline water-table conditions. Plant Soil 87:329–335

Uddin Q, Parveen N, Khan NU et al (2003) Antifilarial potential of the fruits and leaves extracts of on cattle filarial parasite Setaria cervi. Phytother Res 9:1104–1107

Usharani KV, Naik D, Manjunatha RL (2019) *Pongamia pinnata* (L.): composition and advantages in agriculture: a review. J Pharma Phytochem 8(3):2181–2187

Venkatesan R, Dampuri K, Challoo R (2017) Green power production from Pinnata. In: World environmental and water resources congress. American Society of Civil Engineers, Sacramento, pp 228–240

Viswanathan R, Padmanaban P, Mohanraj D et al (2018) Status of red rot resistance in wild relatives of sugarcane, Saccharum spontaneum, interspecific hybrids and intergeneric hybrids. J Sugarcane Res, 8(2):169–184

Wani SP, Sreedevi TK (2007) Strategy for rehabilitation of degraded lands and improved livelihoods through biodiesel plantations. In: Proc. 4th international biofuels conference. Winrock International, New Delhi, pp 50–64

Wang L, PA Jackson and X Lu (2008) Evolution of sugarcane x Saccharum spontaneum progeny for biomass composition and yield components. Crop Sci 48:951–961

Wiangwiset K, Dermail A, Piwpuan N et al (2023) Diversity and Heterosis of Leaf Anatomical Traits in Backcross 1 (BC1) Derived from Interspecific Hybridization between Commercial Cane (Saccharum spp. Hybrid) and Wild Type (S. spontaneum). Agronomy 13:2457

Yadav P, Bharti N, Choudhary SK et al (2024) Assessment of the antifertility potential of silver nanoparticles synthesized using *Pongamia pinnata* leaf extract in male Wistar rats for the development of a male contraceptive agent. Biomed Ther Let 11(2):918–918

Chapter 17
Plant Breeding and Biotechnological Breakthroughs in Improvement of Sunn Hemp (*Crotalaria juncea* L.)

R. Nirmalaruban, R. Suvitha, S. Sugumar, A. Sivakumar, P. Anantharaju, and N. Meenakshi Ganesan

Abstract Sunn hemp (*Crotalaria juncea* L.) is a versatile crop grown widely for its fiber, green manure, fodder, and biofuel purposes. It is a leguminous crop of Indian origin and is cultivated in tropical and sub-tropical regions worldwide. Prior to the Green Revolution, sunn hemp was a significant fiber crop in the rural economy. However, it has since been supplanted by staple crops. Currently, sunn hemp is being cultivated on marginal lands with limited resources during the summer season. The development of improved cultivars for various uses has been hindered by a lack of comprehensive understanding of germplasm, including related and wild species. Additionally, there is a shortage of genetic information on breeding patterns and the gene functions that control different yield traits in sunn hemp, which has further slowed progress. New research has brought to light the breeding behavior of sunn hemp and has explored its diversity to some extent. However, conventional breeding techniques face difficulties in developing improved cultivars due to the crop's complex breeding behavior. Advancements in modern breeding tools such as transgenics, genomics, transcriptomics and gene editing techniques offer a potential solution to overcome the limitations of conventional techniques and pave the way for the development of improved cultivars. Sunn hemp has recently gained significant attention, with numerous studies and ongoing efforts focused on developing high-yielding varieties to promote sustainable agricultural production. This chapter outlines the challenges related to the biodiversity and conservation of sunn hemp germplasm, as well as its floral biology, breeding systems, and pollination mechanisms. It also discusses the goals and phases of breeding programs, alongside traditional breeding methods. Furthermore, the chapter delves into the impact of

R. Nirmalaruban (✉) · S. Sugumar · A. Sivakumar
Indian Agricultural Research Institute, New Delhi, Delhi, India
e-mail: nirmalaruban_12085@iari.res.in; sugumar_11801@iari.res.in; Sivakumar_12121@iari.res.in

R. Suvitha · P. Anantharaju · N. M. Ganesan
Tamil Nadu Agricultural University, Coimbatore, Tamil Nadu, India

K. F. M. Salem et al. (eds.), *Breeding and Biotechnology of Grass and Bast Fiber Crops*, Advances in Plant Breeding Strategies 12,
https://doi.org/10.1007/978-3-032-00403-1_17

polyploidization, mutation breeding, molecular breeding, genetic engineering, and gene editing techniques on enhancing sunn hemp crops.

Keywords Biofuel · Breeding system · Fiber · Green manure · Sunn hemp (*Crotalaria juncea* L.) · Sustainable agriculture

17.1 Introduction

Sunn hemp (*Crotalaria juncea* L.), also called Indian hemp, is a multifaceted crop (Bhandari et al. 2016). It is widely grown for its fiber, green manure, and fodder (Mondin et al. 2007), this leguminous plant originates from India and is cultivated in tropical and subtropical regions including India, Brazil, Pakistan, China, Korea, and Bangladesh. China leads the global production of sunn hemp fiber, whereas India accounts for approximately 23% of the world's output, covering 27% of the total cultivated area.

Archaeological reports showed evidence of sunn hemp cultivation in India dating back to the Indus Valley civilization period (Good 2007). Bhavamisra's literature from the sixteenth century mentioned the cultivation of sunn hemp as 'shanapushpika' (Chunekar and Pandey 1998) and in the 'AinE-Akbari'. Sunn hemp is commercially cultivated for its soft, slightly lignified stem fibers, which are characterized by high cellulose, low lignin, and minimal ash content (Tripathi et al. 2013). These fibers are suitable for manufacturing currency paper, high-quality tissue paper, and cigarette paper (Chawla et al. 1967). In India, sunn hemp is used for making twines, nets, handmade paper, ropes, tat-patties, and canvas (Tripathi et al. 2012), and also for medicinal purposes (Chopra et al. 1956; Rajesh et al. 2014). In addition to its industrial uses, sunn hemp can be grown as a soil renovator or builder and acts as a deterrent to nematodes on fallow or newly reclaimed soils. The compound 'monocrotaline' released upon incorporating sunn hemp foliage has anti-helminthic properties and helps reduce the nematode population. This compound in sunn hemp leaves is effective against nematodes such as *Meloidogyne sp.* and *Radopholus similis* (Garzon et al. 2021; Wang et al. 2002). In addition to this, sunn hemp hay is used as an anti-helmintic against Texel Ewes (Carmona-Flores et al. n.d.). Sunn hemp is also used for fodder and culinary purposes to a limited extent, being fed to cattle fresh or as hay (Tripathi et al. 2013). Its nutritive value is comparable to other leguminous fodders and mixed fodders (sunn hemp and paddy straw) have been found to have better feeding quality than paddy straw alone (Rao et al. 1987). Additionally, sunn hemps high biomass production capacity (15–20 t/ha) makes it suitable for biofuel production with an energy efficiency of 80–90 GJ, nearly equivalent to 620 gallons of gasoline (Perry 2012). It is being used as alterantive feedsotick for bioenergy in temperate regions (Zegada-Lizarazu et al. 2024). Sunn hemp shows moderate tolerance to saline soils and can grow well in soil with concentrations of 100 mM of salt or below (Demiroğlu Topçu et al. 2024).

During the sixties, India produced approximately 70–75 thousand bales of sunn hemp fiber annually from 2 lakh hectares. Around 20% of the total fiber was exported to other countries until the 1970s. Currently, the annual production of sunn hemp fiber is about 18.8 thousand tons from 0.31 lakh hectares, with cultivation spread across states such as Rajasthan, Uttar Pradesh, Madhya Pradesh, West Bengal, Maharashtra, Orissa, Bihar, Tamil Nadu, and Jharkhand. The huge decline in area under sunn hemp cultivation is due to the advent of the green revolution with high-value crops like pulses, oilseeds, vegetables, and paddy (Sarkar et al. 2015). Since the onset of the Green Revolution, the widespread use of chemical fertilizers has become prevalent in both developed and developing countries, resulting in ongoing damage to agroecosystems. This became a major concern for the scientists (Conway and Barbier 2013; Narang and Gill 1996). Sunn hemp has the potential to address these issues by serving as a green manure crop, which helps maintain soil health and supports agroecosystems (Bhardwaj and Datt 1995; Gupta and Tripathi 1999). After a huge hike in fertilizer prices, farmers realized the value of green manure crops and returned to the cultivation of sunn hemp as a nutrient supplement. Sunn hemp cultivation not only increases soil health but also helps in increasing the water-holding capacity of the soil by adding a large amount of organic matter. Sunn hemp-based green manuring shows significant yield increases in rice, wheat, and sugarcane cultivation (Agarwal 1959; Kanwar and Singh 1959; Panse et al. 1965; Tandon et al. 1959). Sunn hemp green manuring contributes around 50–60 kg of nitrogen per hectare through its biomass and an additional 50–60 kg of nitrogen per hectare through biological nitrogen fixation by its root nodules. The succulent nature of its foliage aids in the rapid decomposition of materials after incorporation, thereby enhancing the yield of subsequent crops. Sunn hemp has effectively become a substitute for nitrogen fertilizer, as it can fix 60–80 kg N ha^{-1}. It has significantly improved the efficiency of nitrogen fertilization in sugarcane based cropping systems (Beja et al. 2024). Microcrystalline cellulosic biopolymer (MCB) derived from sunn hemp fibers helps in creating eco-friendly polyester composites with enhanced load-bearing capabilities, making them ideal for use in home interior decorations, defense toolboxes, and automated door panels (Alshahrani and Arun Prakash 2024).

In the post-Green Revolution period, most of the land previously used for sunn hemp cultivation was replaced by more profitable staple crops. Additionally, the shift in crop cultivation to the summer season, as opposed to the traditional *Kharif* season, and the need for ample fresh water for the retting process have created significant challenges in sun hemp fiber production. No specific efforts have been made to develop varieties suited for the summer season. Due to a lack of research focus, this crop, once crucial to the rural economy, has been neglected. Recently, its production and productivity have significantly declined. Factors contributing to this include limited knowledge of germplasm, including related and wild species, insufficient genetic information on breeding behavior, and a lack of understanding of the gene actions that govern various yield traits. Sunn hemp has recently gained significant attention, with numerous studies and ongoing efforts focused on developing high-yielding varieties to promote sustainable agricultural production. This chapter covers the sunn hemp germplasm diversity and conservation, flower structure, reproduction modes, pollination process, goals of breeding projects and traditional breeding methods. It also discusses the use of polyploidy, mutation, molecular, genetic and gene editing methods to improve the sunn hemp crop.

17.1.1 Botanical Classification and Distribution

Kingdom	Plantae
Subkingdom	Viridiplantae (green plants)
Infrakingdom	Streptophyta (land plants)
Superdivision	Embryophyta
Division	Tracheophyta (vascular plants)
Subdivision	Spermatophytina (seed plants)
Class	Magnoliopsida
Superorder	Rosanae
Order	Fabales
Family	Fabaceae
Genus	*Crotalaria* L.
Species	*Crotalaria juncea* L.

Crotalaria juncea, commonly referred to as sunn hemp, is known by various names, including brown hemp, Bumpo, Ghore sun, Indian hemp, Sann hemp, Janumu, Saun, Senabina soppu, Shon, Shonpat, Tag, Tum-thang, Wuckoo nar, Benares hemp, Bengal hemp, Bombay hemp, Jubbalpore hemp, Crotalaria, Madras hemp, Oohawaimaton, San, Sana, Sanabu, Sanai, Sanal, Sannai sunn, Sannappu, Cascavelle, Coconut, Grand Sonnette, Grand Tcha-Tcha, and Sonnette. This annual leguminous herb grows rapidly and has a high photosynthetic rate. It can reach heights of 1–4 meters and exhibits a determinate growth habit. Sunn hemp has a tap root with root nodules for nitrogen fixation. Its leaves are alternately arranged and lanceolate or elliptical, while its stems are cylindrical and ribbed. The flowers are complete and zygomorphic with a golden yellow standard petal, two wing petals and a keel petal. The androecium and gynoecium are protected inside the keel petal. Sunn hemp produces inflated pods with marginal placentation of kidney-shaped medium-sized dark brown or black colored seeds (Rajagopal et al. 2021). It is cultivated in tropical and subtropical regions around the world such as India, Brazil, Pakistan, China, Korea and Bangladesh (Fig. 17.1).

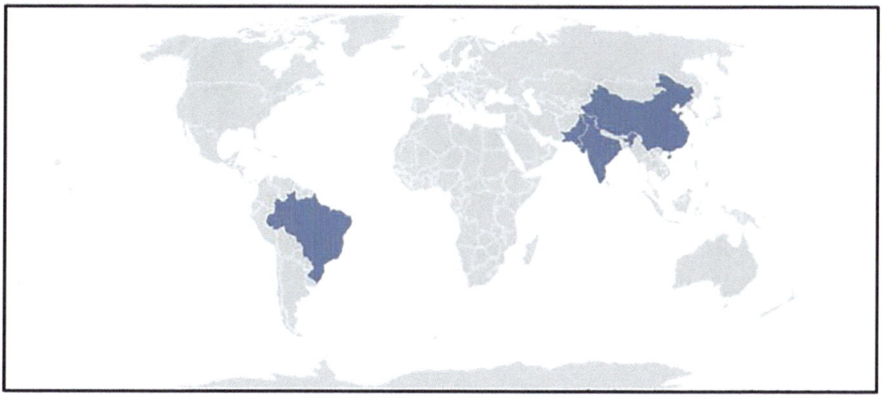

Fig. 17.1 Distribution of Sunn hemp cultivation (Blue—Sunn hemp cultivated area)

17.2 Current Cultivation Practices and Challenges

Sunn hemp is an annual fast-growing legume crop that is propagated by seeds and can thrive under a wide range of soil and climatic conditions. It requires well-drained, light-textured soil, as it can not withstand waterlogged conditions for extended periods. The timing of cultivation depends on the location, weather conditions and the intended use of the sunn hemp (Anil et al. 2022). Typically, sunn hemp is planted one to one and a half months before the main crop for green manure purposes, allowing enough time for trampling and decomposition before the main crop is sown. Seeds are planted at a depth of 2–3 cm in rows or can be broadcasted. They typically germinate within 3 days under favorable conditions (Khalikova and Negmatova 2023). The seed rate depends on the intended use, with higher rates needed for crops grown for green manure and fiber compared to those grown for seed purposes. Recommended seed rates under rainfed conditions range from 100 to 240 kg/ha for broadcasting and 66 kg/ha for row crops (Maroyi 2011). Sunn hemp as a fast-growing crop, suppresses weed growth through shading. White and Haun (1965) found that early season weed management is crucial for increasing yields when sunn hemp is grown for fiber. Dense seeding helps maintain plant population and prevents stems from becoming too thick, which can reduce fiber content. During the growing season, watering is essential at least once or twice for better establishment. Plant protection practices should be carried out during flowering and seed setting stages for seed crops.

Harvesting sunn hemp depends on its intended use. For seed production, plants are cut when the seeds are mature but before the pods dry, to prevent seed loss. The pods are then threshed by beating them against a board or mat, and seeds are collected and stored in jute bags for future seasons. If the plants are harvested for fiber, the stems are chopped from the bottom and left to wither the field for a few days to shorten the retting time. Once the leaves have withered, the stems are bundled and

Fig. 17.2 Extraction of fibers from sunn hemp stems. (**a**) Harvesting, (**b**) Retting (initial stage), (**c**) Retting (later stage), (**d**) Washing, (**e**) Extraction of fibers, (**f**) Drying of fibers and (**g**) Final products. (Source: This image is a composite created by Nirmalaruban, based on based on works by Central Research Institute for Jute and Allied Fibres (2023))

stored to dry further. The dried stems are then ready for retting. The highest quality fiber is obtained from harvests in September and October.

Retting is the process of separating the fiber from the stem by soaking it in water-containing microorganisms. This process relies on microbial activity, which occurs naturally during retting (Bisht et al. 1996). After retting, the bundles are removed from the water and the bark is peeled off. The fiber is extracted by pulling a single stick from the top of the bundle and gently sliding it down (Fig. 17.2). Fibers extracted from sunn hemp are soluble in concentrated acids (sulphuric acid, hydrochloric acid and nitric acid) and are less affected by lower concentrations of alkalies (Vanishree et al. 2019). The raw fiber is then dried by hanging it on trees, poles or ropes. The extracted fibers are used to make ropes, cordages, fishing nets, soles of shoes, sandals, twines, canvas, floor mats, rugs, carpets, webbing, high-quality tissue paper, table and bed linen, cigarette paper, carbon base paper, electrical condenser, currency paper, airmail paper and wax impregnated paper.

Sunn hemp is included in many cropping systems in India to maintain soil fertility as it helps improve the physical and biological properties of the soil and promotes sustainable farming practices. Sunn hemp is used in crop rotation with wheat, potato, mustard, and paddy, in patterns such as sunn hemp-potato, sunn hemp-wheat, sunn hemp-paddy-wheat, and sunn hemp-mustard. It is primarily grown as a green manure crop in garden conditions and used in orchards. Including sunn hemp in crop rotation with wheat can increase wheat yield by 10–15% and improve crude protein content in sorghum (Mosqueda et al. 2023), providing additional yield gains for farmers. Sunn hemp is also intercropped with wheat, sugarcane, sorghum, and maize, improving the soil's physical and biological properties (Bhayal et al. 2018).

Sunn hemp has been cultivated for generations in Indian culture. After the Green Revolution, sunn hemp cultivation was replaced by more profitable cash crops, leading to its status as an orphan crop. Sunn hemp cultivation has shifted from *Kharif* to summer and is mostly grown in marginal soils with limited resources. Despite these challenges, sunn hemp holds value as an important fiber and green manure crop in the agricultural economy. In recent years sunn hemp gained attention among researchers, for its potential in biofuel production. Traits such as high biomass with low lignin content make sunn hemp favorable for biofuel production, contributing to the development of a sustainable environment.

17.3 Germplasm Biodiversity and Conservation

Crotalaria species are distributed globally, with around 600 species identified. However, the USDA collection includes only 242 accessions, covering about 30 species. There is a need to collect more accessions from various regions worldwide. In India, the Sunn Hemp Research Station in Pratapgarh, Uttar Pradesh, maintains 122 sunn hemp germplasm. Additionally, over 400 accessions of sunn hemp and wild Crotalaria species are preserved as seeds and herbarium specimens (Pandey et al. 2010). The regions with the highest diversity of sunn hemp in India are Rajasthan, Jharkhand, and the border area between Bihar and Jharkhand (Maruthi et al. 2021).

Sunn hemp germplasm conservation refers to the preservation of the genetic diversity and resources of sunn hemp, a leguminous crop used for green manure, fodder, and fiber. It is important for maintaining the adaptability, productivity, and quality of the crop, as well as for facilitating its genetic improvement and breeding. Sunn hemp germplasm conservation can be done by different methods.

17.3.1 Seed Storage

This method involves collecting, cleaning, drying, and storing sunn hemp seeds in controlled conditions, like low temperature, low humidity, and low oxygen. Seed storage can preserve the viability and vigor of the seeds for a long time, and can be easily accessed and distributed for planting or research. However, several challenges may arise, including the need for regular monitoring, seed regeneration and characterization, the risk of seed deterioration or loss, and the potential loss of genetic diversity due to genetic drift or selection (Maruthi et al. 2021).

17.3.2 In Situ Conservation

This approach involves conserving sunn hemp germplasm within its natural habitat or on farmers' fields, allowing it to interact with the environment and other plants. In-situ conservation can maintain the evolutionary potential and the local adaptation of the sunn hemp germplasm, and support traditional knowledge and practices of farmers. Challenges like habitat degradation, changes in climatic conditions, pests and diseases, and gene flow from other crops or weeds may be faced (Bhandari et al. 2016).

17.3.3 Ex Situ Conservation

This technique involves preserving sunn hemp germplasm outside its natural habitat, such as in gene banks, botanical gardens, or tissue culture laboratories. Ex-situ conservation provides a backup and safety net for the sunn hemp germplasm and facilitates evaluation and utilization of the germplasm for breeding and research (Meagher et al. 2017). Challenges include high cost, technical skill, risk of contamination, genetic instability, and ethical and legal issues.

17.3.4 Cryopreservation

Cryopreservation is a technique that involves storing plant material, like seeds, at ultra-low temperatures under liquid nitrogen (−196 °C), to preserve their viability and genetic diversity for long periods. Cryopreservation is a valuable technique for conserving plant germplasm that is challenging to maintain through conventional methods.

Sunn hemp seeds have low longevity and poor germination under ambient conditions. Therefore, cryopreservation could be a potential method for preserving sunn hemp seeds and enhancing their quality and genetic improvement. However, there are limited studies on the cryopreservation of sunn hemp seeds, and the optimal conditions and protocols for this technique are not well established. Factors surrounding sunn hemp like moisture content, pre-treatment, freezing and thawing rates, and recovery methods. May affect the conservation of sunn hemp seeds. Non-germinated dry seeds have been reported to survive freezing at very low temperatures compared to water-imbibing seeds which are sensitive to cryogenic injuries (Chander et al. 2023). Dehydrated cells have shown a better survival rate after cryopreservation. More research and development are needed to optimize the cryopreservation of sunn hemp seeds and evaluate their physiological and genetic stability after storage.

In conclusion, sunn hemp germplasm conservation is a vital strategy for ensuring the sustainable use and improvement of this crop. However, it requires careful planning, coordination, and implementation of different methods.

17.4 Traditional Breeding

The objectives of sunn hemp breeding programs include increasing seed yield for seed crops, higher biomass yield for green manure crops, fiber production and biofuel production Additionally, breeding aims to achieve early maturity, increased root nodules for efficient nitrogen fixation, the resistance of shoot borer, wilt and abiotic stress (Nirmalaruban et al. 2024a, b). Selecting breeding to eliminate pyrrolizidine alkaloids will also make sunn hemp suitable for forage purposes.

Sunn hemp is a leguminous crop in the Fabaceae family, characterized by terminal open raceme inflorescences up to 25 cm long. The flowers are showy, deep-yellow papilionaceous complete and zygomorphic (Fig. 17.3).

Flowers of sunn hemp are categorized into four stages (S1, S2, S3, and S4) based on significant morphological variations during the different stages of development (Nirmalaruban et al. 2020; Thimmaiah et al. 2018), as depicted in Fig. 17.4.

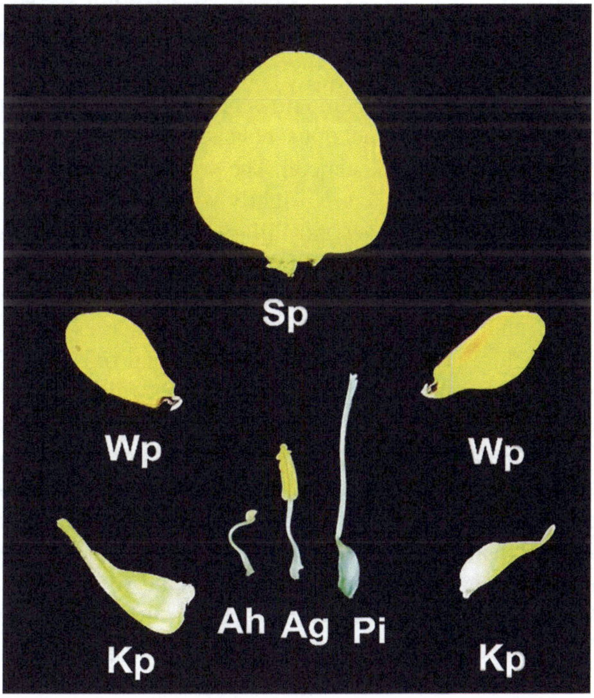

Fig. 17.3 Sunn hemp (Crotalaria juncea L.) Foral parts. *Sp* Standard petal, *Wp* Wing petal, *Kp* Keel petal, *Ah* Heart shaped anther, *Ag* Globose anther, *Pi* Pistil. (Source: Nirmalaruban et al. 2020)

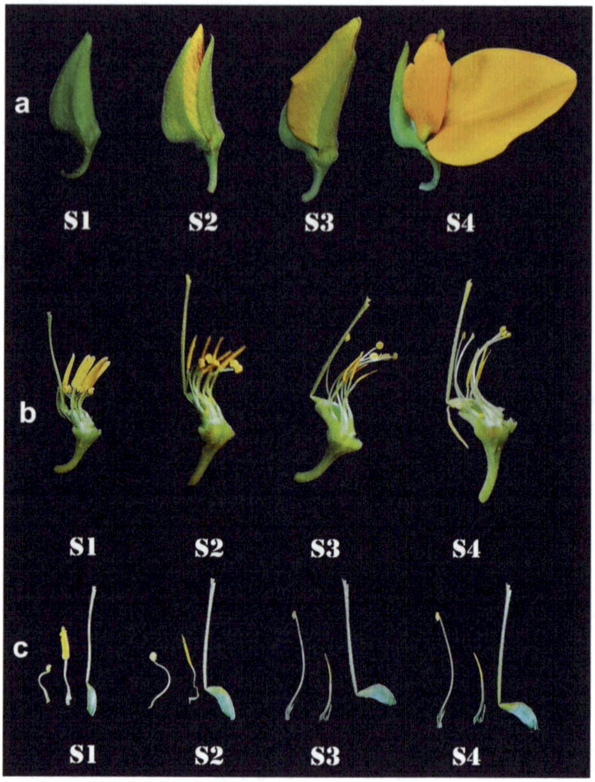

Fig. 17.4 Sunn hemp (Crotalaria juncea L.) flower showing four stages, (**a**) complete flower, (**b**) flower showing dimorphic anthers at different stages, (**c**) Size of dimorphic anthers in relation to pistil at different stages. S1—unopened bud, S2—petals start to break out of the calyx, S3—petals expanded and S4—open flowers. (Source: Nirmalaruban et al. 2020)

The sepals of sunn hemp flowers are gamosepalous in nature. The corolla of sunn hemp flowers is polypetalous, consisting of a brightly colored standard petal, a pair of wing petals and a keel petal. The brightly colored corolla helps attract pollinators. Bees are more attracted to brightly yellow-colored flowers of *Crotalaria* than blue-colored flowers (Faegri and Pijl 1979). Among the bees, *Megachile* (Westerkamp 1997) and *Xylocopa* bees (Rakesh et al. 2019) are the pollination vectors in sunn hemp, aiding in a successful seed set. The standard petal of sunn hemp flower is broad and bright yellow. The keel petal serves as a protective covering for the androecium and gynoecium, playing a dual role as a visual attractant and landing aid for pollinators to collect nectar successfully. Additionally, it significantly influences foraging time, preference, and ease of access to nectar.

The androecium consists of ten anthers arranged in two whorls, with five globose and five heart-shaped anthers. The outer whorl of stamens has five long globose anthers that dehisce before the anthers, while the inner whorl has five heart-shaped anthers that dehisce after the anthesis. The filaments of long globose anthers release pollen from stage S2 onwards, while the filaments of round anthers increase in length from stage S2, reaching nearly 2/3rd of the stylar length by the end of stage S3. The elongation of the filament pushes the pollen mass towards the tip of the keel petal, thereby enhancing the likelihood of self-pollen naturally landing on the stigmatic surface (Thimmaiah et al. 2018). Sunn hemp flowers are protandrous, with pollen dehiscence occurring before the stigma becomes receptive. This type of anther dimorphism is also observed in other Crotalaria species such as *Crotalaria stipularia* Desv. (Etcheverry 2000), *Crotalaria retusa* (Tidke and Patil 2000), *Crotalaria digitata* (Muthu and Ganesan 2012) and *Crotalaria spectabilis* Roth. (Kar 2018).

The gynoecium consists of the superior ovary with a long style and terminal stigma surrounded by pubescence all over the pistil (Sharma et al. 1985). The stigma is the platform, where pollen grains get adhered and germinated. The germinated pollen tubes grow deep into the style and reach the ovary to effect fertilization. The style acts as the channel for the growth of pollen tubes. In *Crotalaria stipularia* Desv., the whorl of hairs surrounding the stigma is non-receptive in nature and the receptacles around the base of the ovary secret a large sum of nectar into the staminal column. The basal opening in the staminal column allows access to the visitors. The pollinators are attracted to the flowers because of their nectar and brightly colored corolla. Fruits of sunn hemp are called pods. Seeds of sunn hemp are kidney-shaped and produce a rattling noise at maturity when shaken.

17.4.1 Pollen Viability

Pollen viability refers to the ability of the pollen to germinate (Brown 1958). It is necessary to check the viability of pollen as it is essential for successful fertilization of the ovules and fruit set. Pollens of both globose and heart-shaped anthers show nearly equal fertility and viability in *in-vitro* staining experiments as well as in field conditions (Nirmalaruban et al. 2020). The loss of pollen viability and vigor follows a trend similar to that of seed viability and vigor. Pollen viability and vigor decline faster in the case of stressed conditions compared to normal conditions. This is also evidenced in *Agave* sp., *Tradescantia virginiana* L. and *Iris* sp. (Shivanna 1982). Pollen grains of *Tradescantia virginiana* L. and *Iris* sp. were found to be highly sensitive to heat stress compared to those of *Nicotiana* and *Agave.* Pollen viability is less affected under stressed conditions compared to vigor. In *Crotalaria retusa,* pollen grains stored in hexane for 30 days under laboratory conditions showed no significant difference in germination compared to fresh pollen grains. However, a drastic reduction in pollen tube length was observed in stored pollen grains compared to fresh pollen grains. This elucidates that pollen grains subjected to various stresses such as high relative humidity, high temperature, storage stress, adversely affected the vigor of pollen, before affecting the pollen viability or germinability (Jain and Shivanna 1990). There is no universal method for assessment of pollen viability, several methods need to be followed simultaneously to get precise results. This is because loss of pollen viability is a gradual process and it is impossible for researchers to uncover the reason behind the time-viability decay schedule (Thomson et al. 1994). The viability of pollen is a multifactor-dependent phenomenon as it depends on environmental factors, internal factors, morphological factors, pollination vector and its secretion (Dafni et al. 2005). The dehiscence of globose anthers occurred during the S1 stage of flowering and was viable from 9:15 h, whereas the heart-shaped anthers dehisced at the S3 stage of flowering and supplied pollens till the end of flowering. The viability of pollen grains started from 9:15 h of the S1 stage of flowering and lasted up to the end of flowering (Thimmaiah et al. 2018).

17.4.2 Stigma Receptivity

The receptivity of stigma in relation to the different stages of flower development in *Crotalaria juncea* L. indicates that stigma receptivity begins at 12:15 h of the S1 stage and continues until the end of flowering (Nirmalaruban et al. 2020; Thimmaiah et al. 2018). Understanding stigma receptivity is crucial in designing hybridization programs.

17.4.3 Breeding System Study

Breeding systems refer to the various methods used to evaluate and select desired genetic traits in a breed or species (Lopes 2021). Understanding the sequence of events in flowering, duration, and longevity of anthesis, pollen dehiscence and stigma, is crucial in determining the prevailing breeding system in flowering crops (Galen and Plowright 1987; Nayak and Davidar 2010). According to Stebbins Jr (1950), "cross-fertilization is the ancestral genetic system that promotes greater flexibility while self-fertilized types have evolved in response to selection pressure favoring immediate fitness". When survival fitness is at stake, phenotypic variation in plants, which influences the breeding system, decreases (Mather 1953), leading to change in the plant's breeding system to ensure successful propagation and survival.

Darlington (1956) emphasized that the selection pressure sent throughout the life cycle of plant species significantly influences the breeding system of the plant. This was demonstrated by the evolution of the tomato into a self-pollinated crop, despite originally being outbreeding in nature. The study revealed that the domestication process for cultivation and environmental factors were key determents in shaping the breeding behavior of the species.

Haddon (1961) conducted studies on self-incompatibility in collections of *Theobroma cacao* from various regions around the world. The study revealed that wild types remained self-incompatible until they were propagated through vegetative methods, while domesticated types were naturally self-compatible. The research indicated that the breeding system of a plant species is heavily influenced by the selection pressure it faces. Understanding the reproduction system of a species is crucial for comprehending genetic differentiation and gene flow within and between populations (Bawa 1974). Self-pollination and cross-pollination represent the two extends of the spectrum inbreeding system studies. Plants that can produce seeds through self-pollination are considered self-compatible, whereas those unable to do so are deemed self-incompatible. The findings of the study on the breeding system demonstrated that self-incompatibility plants rely predominantly on cross-pollination for successful fruit and seed production.

Inbreeding has been shown to help in purging the genetic load present in naturally outcrossing species. This reduced fitness during inbreeding and increased fitness higher than the original population has also been observed in maize. The purging of partially recessive mutation accounts for an important component of heterosis. The fitness lost during inbreeding is only restored when the inbreeds are inter-crossed (Barrett and Charlesworth 1991).

Breeding systems prevalent in species can be identified using the index of self-incompatibility (ISI) and pollinator effectiveness index (PE). The breeding system in *C. juncea* is identified as a cross-pollination system with late-acting self-incompatibility, evidenced by a significant difference in pod set percentage between self- and cross-pollinated flowers. Additionally, self-pollination in hermaphrodite flowers resulted in significantly fewer pods, leading to an investigation of a self-incompatibility mechanism. The index of self-incompatibility (ISI) serves as an indirect measure to determine the breeding behavior of hermaphrodite plants, with a lower ISI value (<0.25) indicating self-incompatibility (Bawa 1974; Bianchi and Gibbs 2000; Bullock 1985; Nayak and Davidar 2010). The low ISI value (<0.25) observed in *Crotalaria juncea* L. clearly demonstrates the presence of self-incompatible breeding behavior in the species (Nirmalaruban et al. 2020; Thimmaiah et al. 2018).

17.4.4 Pollination Mechanism

Pollination involves transferring pollen from the male reproductive part to the female reproductive part, leading to successful fertilization. This process can occur through cross-pollination or self-pollination. Self-pollination, also known as autogamy, entails the transfer of pollen within the same flower. Floral structures are designed to facilitate the movement of pollen from the anther to the stigma within the same flower. This form of inbreeding has evolved as an adaptive mechanism to ensure successful reproduction in the absence of pollinators, thereby promoting the continuity of generations. Mechanisms such as bisexual flowers, homogamy, cleistogamous flowers, chasmogamy and the proper positioning of anthers to supply pollen to the stigma promote autogamy, leading to increased homozygosity in the population. Cross-pollination, also called allogamy, is the most common mode of pollination in both cultivated and uncultivated plant species. Floral traits such as dicliny, heterostyly, herkogamy, self-incompatibility, male sterility and dichogamy promote cross-pollination (Singh et al. 2007). Pollen presentation mechanisms in

Fabaceae species can be categorized into four types: explosive, valvular, piston, and brush type (Westerkamp 1997).

Arroyo (1981) described the piston mechanism in flowers, where the stigma and anthers remain stationary while the tip of the keel moves under the influence of the pollinator. This movement releases pollen, pushing it toward the keel petal's tip. Pollen can be discharged multiple times through the opening at the keel petals'tip during a pollinator's visit. Flowers with this piston mechanism exhibit a higher pollen/ovule ratio compared to those with an explosive mechanism. The piston action within the keel propels the pollen through the tip, allowing it to adhere to the pollinator for dispersal. The pressure from pollinators causes the pistil to protrude from the keel petal's tip, enabling the stigma to contact the insect and receive pollen while the trichomes brush pollen onto the pollinator. After the pollinator departs, the stigma and style revert to their original positions (Nirmalaruban et al. 2024a, b).

Lloyd and Schoen (1992) classified autonomous selfing into three modes, "prior," "competing," and "delayed," based on the time of selfing relative to outcrossing. Wyatt (1983) suggested that delayed autonomous selfing is an adaptive pollination mechanism in certain plants, ensuring seed set in the absence of pollinators, which still allows for outcrossing. This process typically occurs later in the flowering stage due to delayed curvature of stylar branches towards the anthers. In the species *C. juncea* delayed self-pollination occurs in the absence of insect visitors (Etcheverry et al. 2003), while *Xylocopa latipes* and other insects play a significant role in pollination (Rakesh et al. 2019). However, cross-pollination is the preferred mode of reproduction (Duke 1983; Mohan 1971). In *C. juncea*, natural self-pollination occurs later in the flowering stage, known as delayed autonomous self-pollination. Similar selfing mechanisms have been observed in wild Crotalaria species like *C. micans* (Etcheverry et al. 2003) and *C. stipularia* (Etcheverry 2000) as well as *Hibiscus laevis* from the Malvaceae family (Klips and Snow 1997).

The explosive mechanism, also known as the tripping mechanism, is observed in flowers where the staminal column is held under pressure within the keel petal. When the tension is released, all the pollen is discharged simultaneously as the staminal column strikes the standard petal. After this explosive release, the staminal column does not revert to its original position. Tripping can be triggered not only by visiting insects releasing tension but also by high temperatures or heavy rain weakening the keel petals. A low Pollen/Ovule ratio signifies a highly specialized mechanism, enabling only a single effective pollen exchange with pollinators (Small 1988).

In the valvular mechanism, there is no force applied to the staminal column. When a pollinator lands, the keel petals open along their length and close when the pollinator departs. This mechanism enables the sequential release of pollen to various visitors. Flowers exhibiting this mechanism possess the highest pollen-to-ovule ratio, potentially explaining the reduced efficiency of pollen transfer (Yeo 2012).
Aronne et al. (2012) observed that there is no tripping mechanism in pollination, as the staminal column remains inside the keel without pressure. During insect visits, the stamens and pistils are exposed from the keel, but they stay hidden if left undisturbed. Insects approach the flowers from both the front and back. When insects land on the wing petals from the front, they pull the keel petals, causing the anthers

and pistils to move toward the tip of the keel petals. In the brush-type mechanism, the stigma extends beyond the anthers to prevent self-pollination. After anther dehiscence, pollen grains are pushed toward the tip of the keel petal. The pressure from pollinator visits forces the pollen grains and stigma out through the tip of the keel petal. The pollen grains adhere to the bodies of the pollinators and are subsequently transferred to other flowers. When the stigma protrudes through the keel tip during insect visits, it collects pollen grains that adhere to the insect bodies and transfers them to other flowers. Trichomes near the stigma act as a pollen brush. The stigma returns to its original position once the insects leave the flower.

17.4.5 Pollination Vectors

The pollination vector is the agent that transfers pollen from one flower to another, enabling fertilization and seed setting. Plants belonging to the sub-family Papilionoidea typically exhibit intricate floral structures with zygomorphic flowers and are usually pollinated by insects. Additionally, instances of bird pollination (ornithophily) and bat pollination (chiropterophily) have also been documented in leguminous flowers (Schrire 1989). Therefore, insect-mediated pollination is essential for sunn hemp seed production.

Rakesh et al. (2019) explored the biodiversity of insect pollinator species visiting sunn hemp crops. Their study identified 15 insect species from four orders visiting the flowers: nine from Hymenoptera, two each from Lepidoptera and Coleoptera, and one from Diptera. Among these pollinators, *Xylocopa latipes, Xylocopa pubescens*, and *Megachile lanata* were the most frequent visitors (Fig. 17.5) and effective pollinators. *Megachile lanata* had the highest mean population (4.72 bees/m^2/5 min), followed by *Xylocopa latipes* (3.65 bees/m^2/5 min), with *Xylocopa pubescens* being

Fig. 17.5 Pollination vectors: (**a**) Xylocopa latipes (**b**) Xylocopa pubescens (**c**) Megachile lanata. (Source: Rakesh et al. 2019)

the least abundant (3.215 bees/m^2/5 min). The pollination index was highest for *Xylocopa latipes* (7.05), followed by *Xylocopa pubescens* (6.85) and *Megachile lanata* (6.84). This highlighted *Xylocopa latipes* as the most efficient pollinator for sunn hemp flowers.

In addition, thrips also play a role in pollination. The thrips could not reach the anthers as they were hidden inside the keel petals, but they could access nectar in the staminal column near the base of the ovary. Even in mature buds, thrips could not reach the anthers because the standard petals blocked the nectar window during anthesis. Thrips were found to feed on nectar and pollens. The nectar-feeding activity of thrips caused depletion in the availability of nectar to the visiting pollinators, which made the necessary for the pollinators to visit a greater number of flowers to collect the required quantity of nectar. In consequence, the nectar-feeding activity of thrips helped in promoting cross-pollination rate in flowering species.

17.4.6 Pollen Germination and Pollen Tube Growth

The number of seeds developed is directly correlated with the number of pollen tubes successfully fertilizing the ovules. Pollen grains land on the stigma and proceed further toward the ovule after germination via the style. Fluorescent microscopy can be an efficient tool for studying pollination events in angiosperms. It helps visualize events from the landing of pollen on the stigma until it reaches the ovary and fertilizes the ovule. Fluorescent microscopy aids in understanding the mobility of pollens and pollen tubes in the stylar column. This technique is crucial in elucidating the mechanism involved in recognizing self-pollens and their rejection in self-incompatible plants (Suriya et al. 2022). It also helps determine the type of self-incompatibility reaction whether it is a sporophytic or gametophytic. Aniline blue staining can be used for morphological and histological investigations in plants, but it cannot be used for quantitative studies. This method is also useful for studying the anatomy and embryology of plants. Aniline blue dye is absorbed by callose-like substances present in growing pollen tubes. When samples are exposed to UV light at 430 nm, electrons are excited, emitting yellow-green fluorescence. The fluorescence of aniline blue is bright and strong at a pH of 9.0–10.0 and a change in pH of every unit affects the fluorescence. In acidic conditions, the fluorescence becomes pale or disappears (Hough et al. 1985).

The recognition of self-pollen in self-pollinated pistils would have occurred due to the callose wall when the pollen tube reached the ovule. The obstruction of pollen tube growth in the pistil as a result of the development of a mechanical barrier was suggested through a long-distance signaling mechanism in various plants (Deurenberg 1976; Herrero and Gascon 1987; Hormaza and Herrero 1992; O'Neill 1997; Tupy 1961).

17.4.7 *Self Incompatibility*

Self-incompatibility (SI) is a fascinating mechanism in sexual reproduction that inhibits self-fertilization by identifying and rejecting self-pollen on the stigma of the female pistil. Conversely, pollen from different plants of the same species, transported by wind or insects, is accepted, leading to seed formation. This process encourages cross-pollination (outbreeding) and fosters genetic diversity. East and Mangelsdorf (1925) reported the presence of a gametophytic system of self-incompatibility in *Nicotiana sanderae*. The reaction of pollen in this system of self-incompatibility was found to be determined by the genotype of the pollen and it was not determined by the genotype of the pollen-producing parent. The self-incompatible reaction was governed by a single gene 'S' having multiple alleles. Hughes and Babcock (1950) reported the presence of a sporophytic system of self-incompatibility in *Crepis foetida*. In the system, the genotype of pollen pollen-producing parent determined the incompatibility reaction and it was more complicated in nature. The self-incompatible reaction was governed by a single gene 'S' having multiple alleles. According to Lewis (1954), self-incompatibility can be classified as sporophytic and gametophytic systems based on gene interaction in association with morphology. "Self-incompatibility refers to the inability of a plant with functional pollen to set seeds when self-pollinated." In the case of self-incompatible plants, the viable or functional pollen grains produced by the flowers will fail to fertilize the self (same flower) or any other flower of the same plant (Arasu 1968; Brewbaker 1958).

The self-incompatibility observed in *C. juncea* is characterized as a late-acting self-incompatibility (LSI) system. Multiple studies provide substantial evidence for this LSI system's presence. First, self-pollinated flowers do not set seed even after pollen tubes have penetrated the ovule, a phenomenon absent in traditional self-incompatibility systems (sporophytic and gametophytic). Additionally, there is a significant difference in the rate of pollen tube penetration between self- and cross-pollinated pistils, akin to other species displaying the LSI mechanism, including *Tabebuia ochracea, T. caraiba, C. speciosa, C. chodatii* (Gibbs and Bianchi 1993) and *C. retusa* (Bianchi and Gibbs 2000).

All species in the subfamily Papilionoideae exhibited either a self-compatible system *viz, C. micans* and *C. stipularia*, (Etcheverry 2000; Etcheverry et al. 2003) or exhibited self-incompatibility with a late-acting type of SI mechanism *viz, Swartzia apetala* (Moco 1996), *Strongylodon macrobotrys* (Owens and Rudall 1998).

Flowers subjected to high temperature during anthesis showed partial breakdown of self-incompatibility reaction. Flowers subjected to high temperatures after pollination had no effect, instead it promoted inhibition of pollen tube growth (Kendall and Taylor 1969). In addition, chemical components such as carbohydrates, boric acid and calcium nitrate had no significant role in enhancing pollen tube growth in incompatible reactions. Plant genotype and environment prevailing before and during anthesis were the primary factors, which had a significant influence on the production of selfed seeds in self-incompatible plants.

Herrero and Hormaza (1996) studied the pistil strategies controlling pollen tube growth. It revealed that all the parts of the pistil (stigma, style and ovary) offer both support as well as constrain to the growing pollen tubes. The pistil may not be matured at the time of anthesis as the maturation of the pistil starts from the stigma and proceeds towards the ovary (basipetal). The growth of pollen tubes which traverse the stigmatic region into the stylar region, depends on the reserves available in style for their further growth. It showed a heterotrophic type of nutrition in the stylar region. The experiment on transgenic Nicotiana plants with low levels of glycoprotein showed that a reduction in the amount of available carbohydrates influenced the number as well as the rate of pollen tube growth. In peach trees, the growth of pollen tubes halts at obturator cells, near the base of the style. The growth gets resumed 5 days after the disappearance of starch from obturator cells and the release of a secretion. Self-incompatibility was the regulated constraint offered by plants against the self-pollen. The ability of plants to recognize self-pollens and effecting incompatibility response was acquired at the later stage of flower development, close to anthesis. The evidence showed that the growth of pollen tubes was influenced by the genotype of pollen as well as the environment. A study of differential ovule development following self and cross-pollination in *Narcissus triandrus* (Amaryllidaceae) by Sage et al. (1999) evidenced the deleterious effects of long-distance signaling of self-pollens in the pistil. At the time of anthesis, some ovules remained immature. The signal from the self-pollen prevented the further maturation of the immature ovule and in addition, it also degenerated the matured ovules, before the arrival of self-pollens. This increased the proportion of sterile pollens, which failed in fertilization.

The calS5 gene in *Arabidopsis thaliana* has both sporophytic and gametophytic effects in pollen grain production and pollination. The mutation of the cals5 gene in *Arabidopsis thaliana* leads to the production of pollen grains with aberrated exine but the pollen remains fertile. The gametophytic role of CalS5 leads to the production of pollen grains with structured exine and protects the degradation of pollen in the early stages of development. The sporophytic role of CalS5 took place in callose deposition in growing pollen tubes. The direct competition between the expression of CalS5 and cals5 in growing pollen tubes revealed that callose deposition could provide a fitness advantage to the pollen tubes in the later stage of development. According to them, structured exine due to the deposition of callose was not required for pollen development but it prevented the degradation of pollens in the early stages of development (Nishikawa et al. 2005).

The development of sunn hemp varieties dates back to the 1920s, as the K-12 variety was released in 1926 (Sabnis 1931). Later sunn hemp breeding programs gained momentum under the supervision of the Indian Council of Agricultural Research (ICAR). The varieties released by the central and state for fiber and green manure purposes are listed in Table 17.1. Most of the varieties of sunn hemp are developed through mass selection. The challenges in breeding improved sunn hemp varieties include limited knowledge of germplasm, wild and related species, insufficient genetic information about breeding behavior, and a lack of understanding of the gene actions governing various yield traits.

Table 17.1 Varieties released in India

S. no.	Varieties	Year of release	Purpose	Institute
1.	K-12	1926	Fiber	Kanpur
2.	K-12 Yellow	1971		Sunn Hemp Research Station (CRIJAF), Uttar Pradesh
3.	Shailesh (SH-4)	2005		
4.	Swastik (SUIN-053)	2009		
5.	Ankur (SUIN-037)	2013		
6.	Prankur (JRJ-610)	2015		
7.	CO-1	1983	Green manure	Tamil Nadu Agricultural University (Coimbatore, India)
8.	PAU 1691	2008		Punjab Agricultural University (Ludhiana, India)
9.	ND-1 (Narendra Sanai-I)			Narendra Dev University of Agriculture & Technology (Faizabad, India).
10.	ADT-1	2023		Tamil Nadu Agricultural University (Coimbatore, India)

17.4.8 Polyploidization

Polyploidization serves as another valuable tool in improving the variability as well as the development of new species. Polyploidization is the process of increasing the number of chromosome sets in a cell or an organism. Polyploidization aids in increasing the size, vigor, and yield of the plants, as well as the quality and quantity of their products, such as fruits, seeds, or fibers, enhancing the resistance or tolerance of the plants to biotic and abiotic stresses, such as pests, diseases, drought, salinity, or temperature, creating new phenotypes and genotypes that can expand the genetic diversity and the breeding potential of the crops. It also helps in overcoming the sterility or incompatibility barriers that may occur in interspecific or intergeneric hybrids, and restoring their fertility and viability (Dar et al. 2017).

Polyploidization can occur naturally or artificially in plants. Natural polyploidization can result from errors in cell division, hybridization, or genome duplication. Artificial polyploidization can be induced by physical or chemical agents, such as X-rays, gamma rays, colchicine, or oryzalin, that interfere with the normal separation of chromosomes during cell division (Yali 2022).

Ploidy of sunn hemp was altered to improve the forage quality of *Crotalaria juncea* (Arya 1989; VR 1983) and *Crotalaria sericea* (Gupta and Sinha 1978; Raghuvanshi and Joshi 1965). The results of the study were not as expected and polyploids revert to diploid in a few generations. Chromosomal and genetic imbalances in tetraploids may cause a reduction in flower frequency per inflorescence, delayed and extended flowering periods, larger stomata, pollen sterility, increased pollen size, low fruit set, small rudimentary pods, and wrinkled seeds. While polyploidization is a powerful tool for crop improvement, it necessitates careful planning, evaluation, and optimization.

17.5 Molecular Breeding

Molecular breeding is a technique that uses molecular markers to select plants with desirable traits for crop improvement. Molecular markers are DNA sequences that are associated with specific genes or traits, such as yield, quality, disease resistance, or stress tolerance. Molecular breeding can increase the efficiency and accuracy of conventional breeding by reducing the time and cost of phenotypic evaluation, increasing genetic diversity, and enhancing the selection intensity (Bhandari et al. 2016).

A complex breeding system in sunn hemp makes it difficult to produce pure lines and hybrids by conventional breeding. Molecular breeding can help overcome these challenges by using molecular markers to identify and select plants with desirable traits, such as higher biomass, better fiber quality, self-compatibility, disease resistance, or stress tolerance. Molecular breeding can also help to understand the genetic basis of self-incompatibility and other traits in sunn hemp and to develop molecular tools for gene mapping, QTL analysis, and marker-assisted selection.

However, molecular breeding in sunn hemp is still in its infancy, and there are not many reports on the use of molecular markers for sunn hemp improvement. More research is needed to develop and validate molecular markers for sunn hemp, and to integrate them with conventional breeding methods. Molecular breeding has the potential to accelerate the genetic improvement of sunn hemp and contribute to the sustainable use of this crop.

17.6 Tissue Culture Applications

Tissue culture involves cultivating plant cells, tissues, or organs in a sterile, controlled environment to grow new plants. This method is widely used in plant breeding for disease-free plant development, propagation, germplasm conservation, genetic transformation, somatic embryogenesis, embryo rescue, and for inducing polyploidy and overcoming post-fertilization barriers through anther and ovule culture. Advanced tissue culture techniques have also facilitated the rapid production of completely homozygous breeding lines from gametic cells, significantly reducing the time required compared to traditional plant breeding methods (Ahmar et al. 2020). There were very few reports available detailing studies on micropropagation in *C. juncea*. Nevertheless, research has been conducted on related species, including *Crotalaria retusa* (Devendra and Srinivas 2011), *Crotalaria laburnifolia* (Rajender et al. 2012), *Crotalaria verrucosa* (Hussain et al. 2008), and *Crotalaria lutescens* (Naomita and Rai 2000). Tissue culture methods can be utilized to produce plants with desirable traits by exploiting somaclonal variation and enabling the large-scale multiplication of selected plants. Additionally, inbred lines can be developed as doubled haploids in vitro, guaranteeing they are true breeding types. Embryo rescue is a technique that involves culturing immature or weak embryos in a nutrient medium to produce viable plants. Embryo rescue can be useful for creating hybrids from interspecific or intergeneric crosses that would normally result in aborted seeds. Embryo rescue can also help to overcome self-incompatibility.

17.7 Genetic Engineering and Gene Editing

Modern breeding tools, such as micropropagation and marker-assisted breeding techniques, are currently absent in *Crotalaria juncea.* However, a transgenic approach for sunn hemp that bypasses micropropagation using Agrobacterium infection was developed (Rao et al. 2012). Transgenic sunn hemp plants were created, incorporating genes for the two major serotypes of the foot and mouth disease virus (FMDV), making it suitable as a fodder crop. These transgenic techniques can be used in the development of variation and improvement of genotypes for breeding programs.

Gene editing is a technique that allows precise and targeted modifications of the DNA of living organisms. Gene editing can be used to introduce desirable traits or remove undesirable ones in crops, such as sunn hemp, which is a legume crop that is used for green manure, fodder, and fiber production. However, sunn hemp has some limitations, such as self-incompatibility, low biomass, and poor fiber quality, that hinder its potential for crop improvement (Zhang et al. 2018). Gene editing can help overcome these challenges by modifying specific genes related to these traits. For example, gene editing can be used to alter the expression of self-incompatibility genes, such as S-RNase and SLF, to enable self-fertilization and increase seed sets (Bhattacharya et al. 2021). Gene editing can also be used to enhance the biomass and fiber quality of sunn hemp by targeting genes involved in cell wall biosynthesis, such as CesA, Csl, and C4H (Saurabh 2021). Gene editing can also improve the resistance of sunn hemp to biotic and abiotic stresses, such as pests, diseases, drought, and salinity, by modifying genes related to stress responses.

CRISPR is a genome-editing technology that can introduce precise and targeted modifications in the DNA of living organisms. CRISPR has many potential applications for crop improvement, such as enhancing disease resistance, stress tolerance, yield, and quality. Sunn hemp is a leguminous crop that is used for green manure, fodder, and fiber. Sunn hemp breeding has been challenging due to its complex floral and breeding behavior. CRISPR could be a useful tool to overcome some of the limitations of conventional breeding and create improved varieties of sunn hemp. For example, CRISPR could be used to modify genes related to self-incompatibility, flowering time, biomass production, and fiber quality in sunn hemp. However, there are not many reports on the use of CRISPR for sunn hemp improvement and more research is needed to explore the feasibility and efficiency of this technology for this crop (Bhandari et al. 2022).

Gene editing can also avoid the introduction of foreign DNA or transgenes, which may reduce the regulatory hurdles and public concerns associated with genetically modified crops. However, gene editing also faces some challenges, such as the delivery of editing tools, the detection of off-target effects, the ethical and social implications, and public acceptance and awareness (Perisse et al. 2021). Therefore, more research and development are needed to optimize the gene editing technology and its application in sunn hemp improvement. Gene editing has the potential to revolutionize crop improvement and contribute to the food security and sustainability of agriculture.

17.8 Mutation Breeding

Mutation breeding is a simple, fast, and cost-effective method of crop improvement, especially for crops with limited genetic resources and complex breeding systems, such as sunn hemp. Mutation breeding can also complement other breeding techniques, such as marker-assisted selection and genetic engineering, to create improved varieties of sunn hemp.

Mutation serves as a valuable tool that uses physical or chemical agents to induce mutations in the DNA of plants, in improving variability in the genetic material for utilization in variety development. The effectiveness of ionizing radiation and several chemical mutagens in inducing mutations has been well-established (Shahwar et al. 2023). Collections of mutants from various species have been developed and utilized to create new varieties that are tolerant to different stresses. The most potent chemical mutagens are alkylating agents, such as diethyl sulfonate, dimethyl sulfonate, ethyl methane sulfonate, and N-nitroso-alkyl-urea, based on their impact on higher plants.

Mutation breeding can be used to create new traits or enhance existing ones, such as yield, quality, disease resistance, stress tolerance, and self-compatibility(Bhandari et al. 2016). Mutation breeding is being successfully applied to many crops, including sunn hemp, a leguminous crop that is used for green manure, fodder, and fiber (Anu et al. 2022). Over 3000 enhanced mutant varieties have been released for commercial cultivation across various crops, showcasing the economic advantages of mutation breeding (Suprasanna et al. 2015).

Several studies evidenced the potential of creating variability in *Crotalaria juncea* through physical (gamma rays) and chemical (EMS) mutant treatments (Islam et al. 1994; Verma and Raina 1991). It is also reported in *Crotalaria saltiana* (Shah et al. 2008).

Sunn hemp has a narrow genetic base and a complex breeding system, which limit its conventional breeding potential. Sunn hemp is also self-incompatible, meaning that it cannot fertilize itself and requires cross-pollination by insects or wind (Jegadeesan and Punniyamoorthy 2023). This reduces the seed set and the genetic uniformity of the progeny. Mutation breeding can help overcome these challenges by creating new genetic variations and breaking the self-incompatibility barrier in sunn hemp. For example, mutation breeding has been used to develop sunn hemp varieties with higher biomass, better fiber quality, and self-compatibility. Mutation breeding can also introduce novel traits, such as herbicide tolerance, salt tolerance, and pest resistance, in sunn hemp (Bhandari et al. 2016).

However, mutation breeding also has some limitations, such as the random nature of mutations, the low frequency of desirable mutations, the difficulty of detecting and evaluating mutants, and the possible negative effects of mutations on other traits. Therefore, mutation breeding requires careful planning, screening, and evaluation to ensure the efficiency and effectiveness of the technique. Mutation breeding has the potential to improve the productivity and quality of sunn hemp and contribute to the sustainable use of this crop.

17.9 Hybridization

Hybridization refers to the crossing of two different varieties or species of plants to develop offspring with improved traits, such as higher yield, quality, or resistance to diseases and pests.

Male sterility was observed in *Crotalaria juncea* (Mitra 1976), which will aid in breaking the barrier in yield improvement in sunn hemp through heterosis breeding. Male sterility has been noted in species such as *Crotalaria striata* (Kempanna and Sastry 1958), *Crotalaria mucronata* Desv. (Edwardson 1967), and *Crotalaria pallida* Ait. (Gupta 1975). Arora and Gupta (1984) discovered that male sterility in *Crotalaria pallida* arises from biochemical changes, rather than morphological alterations in the tapetum, and is due to the failure of cytokinesis following meiosis in microspore mother cells. Additionally, heterosis can be harnessed through the use of chemical male sterility inducers like Maleic Hydrazide and Mendok (Kaul and Singh 1967). Production of inbreds in sunn hemp remains difficult due to the near absence of seed set during selfing. This challenge can be mitigated through various methods such as bud pollination, delayed pollination, late-season pollination, irradiation, high-temperature treatments, mutation breeding, and gene editing (Maruthi et al. 2023).

Self-incompatibility (SI) is a process that inhibits self-fertilization and encourages cross-pollination in flowering plants.SI can be useful for hybrid development, as it can ensure that only cross-pollination between different varieties or species occurs, resulting in heterosis or hybrid vigor. SI can also help maintain genetic diversity and avoid inbreeding depression in plant populations. However, SI can also pose some challenges for hybrid development, such as the need for compatible pollen donors, the difficulty of maintaining pure lines, and the possible loss of SI due to mutation or gene flow. Therefore, different methods have been developed to overcome or exploit SI for hybrid development, such as mutation breeding, gene editing, and genetic engineering. However, some researchers have developed sunn hemp hybrids by using mutation breeding, which induces genetic variation by exposing the seeds to physical or chemical agents. These hybrids have shown higher biomass, better fiber quality, and self-compatibility than the parent varieties (Thimmaiah et al. 2018).

17.10 Conclusions and Prospects

In recent years, the cultivation of sunn hemp gained more attention due to its high economic value and high profit with less cost of cultivation. Sunn hemp is usually cultivated in marginal lands with limited resources. Even though synthetic fibers have captured the market of sunn hemp fibers, products made from quality sunn hemp fibers find a unique place in the market and fetch higher prices. Cultivation of sunn hemp as a green manure crop to improve the soil's physical, biological and water-holding capacity has gained attraction over the utilization of chemical fertilizers in order to have a sustainable agricultural ecosystem. Green manuring with sunn hemp will further reduce the cost involved in chemical fertilizers. Utilization of sunn hemp for bio-fuel production and raw material in pulp industries increased the importance of development for improved cultivars with higher biomass yield and quality. Development of high biomass varieties with high bio-fuel efficiency will be rewarding recently for a less polluted environment and safeguarding declining natural resources. Complexity in breeding behavior acts as a bottleneck in improving the present cultivars. Recent research has comprehensively detailed the breeding behavior of sunn hemp, indicating that *Crotalaria juncea* is a cross-pollinated crop exhibiting late-acting self-incompatibility. The molecular mechanism involved in self-pollen recognition and its rejection needs to be elucidated in a detailed manner to make the breeding programs more successful. Since the cultivation season has shifted from kharif to summer, new cultivars suitable for the summer season with higher yields need to be developed. The presence of anti-nutritional factors, such as pyrrolizidine alkaloids, restricts the use of sunn hemp as a forage crop. Selective breeding for the elimination of pyrrolizidine alkaloids in sunn hemp will make it suitable for forage purposes. Ideotype breeding of cultivars with desirable crop structures such as high above-ground biomass, less branched, larger stem diameter will be rewarding for sunn hemp as a fiber crop; cultivars with a greater number of root nodules, high biomass and early maturity will be rewarding for green manure purpose. Conventional breeding techniques face challenges in developing improved cultivars. Enhancements in modern breeding tools, such as transgenics, genomics, transcriptomics, and gene editing techniques (CRISPR), can help overcome the limitations of traditional methods and facilitate the creation of better cultivars for a sustainable agricultural ecosystem. After the Green Revolution, crops that previously dominated the rural economy were replaced by staple crops. The limited research attention given to certain crops has hindered their genetic improvement. The complexity of their breeding behavior makes traditional plant breeding tools less effective. Thus, it is crucial to supplement conventional tools with modern techniques for crop improvement. Developing composites is one potential solution. The long-term objectives of breeding programs include producing high-yielding varieties for seed purposes, high-biomass and early-maturing varieties for green manure and biofuel production. Additionally, varieties that can tolerate biotic and abiotic stresses and perform well under limited resources should be developed using modern breeding techniques for a better environment.

Appendix 17.1 Research Institutes and Online Resources Available for Sunn hemp Breeding

Country	Institution	Website
India	ICAR-Central Research Institute for Jute and Allied Fibers (CRIJAF), Kolkata	https://crijaf.icar.gov.in
	Sunn hemp Research Station, Pratapgarh, Uttar Pradesh	http://jafexpert.crijaf.icar.gov.in/
	Tamil Nadu Agricultural University, Coimbatore	https://tnau.ac.in/
	Punjab Agricultural University, Ludhiana	https://www.pau.edu/
	Narendra Dev University of Agriculture & Technology, Faizabad	https://nduat.org/
Bangladesh	Bangladesh Jute Research Institute (BJRI), Dhaka	https://bjri.gov.bd/
South Africa	Agricultural Research Council (ARC), Pretoria	https://www.arc.agric.za/
Brazil	Embrapa Agroenergy, Brasilia (Brazilian Agricultural Research Corporation (Embrapa))	https://www.embrapa.br/

Appendix 17.2 Genetic Resources for Sunn Hemp Breeding

Species	Origin	Utilization	Toxicity
Crotalaria juncea	India	Fiber, green manure, forage, fuel	Low toxicity
Crotalaria stipularia Desv.	Tropical America	Medicine	Unknown
Crotalaria retusa	India, Africa, Asia	Fiber, green manure, forage	Poisonous to livestock
Crotalaria digitata	Africa	Medicine	Unknown
Crotalaria spectabilis Roth.	Asia, Africa, America	Green manure, cover crop	Contains pyrrolizidine alkaloids

References

Agarwal J (1959) Response of wheat yields to different green manure crop. Allahabad Farm 30(2):87–90

Ahmar S, Gill RA, Jung KH et al (2020) Conventional and molecular techniques from simple breeding to speed breeding in crop plants: recent advances and future outlook. Int J Mol Sci 21(7):2590

Alshahrani H, Arun Prakash V (2024) Characterisation of microcrystalline cellulose from waste green pea pod sheath and its sunn hemp fibre-polyester composite: a step towards greener manufacturing. Physiol Plant 176(1):e14166

Anil D, Sreedhar S, Bhandari H (2022) Assessment of seed yielding potential of sunnhemp (*Crotalaria juncea*) under Southern Plateau Zone of India. Legum Res 1:6

Anu YS, Singh VK et al (2022) Emerging technologies in plant breeding for fibre crops, cotton, and sunn hemp. In: Kamaluddin UK, Abdin MZ (eds) Technologies in plant biotechnology and breeding of field crops. Springer Nature, Singapore, pp 151–180

Arasu NT (1968) Self-incompatibility in angiosperms: a review. Genetica 39(1):1–24

Aronne G, Giovanetti M, De Micco V (2012) Morphofunctional traits and pollination mechanisms of *Coronilla emerus* L. flowers (Fabaceae). ScientificWorldJournal 2012:381–575

Arora D, Gupta P (1984) Microsporogenesis in male sterile and male fertile Crotalaria pallida Ait. Cytologia 49(4):685–689

Arroyo MK (1981) Breeding systems and pollination biology in Leguminosae. In: Advances in legume systematics, pp 723–769

Arya I (1989) Colchicine in evolution of higher polyploids of sunnhemp (*Crotalaria juncea* L.). J Cytol Genet 24:173–178

Barrett S, Charlesworth D (1991) Effects of a change in the level of inbreeding on the genetic load. Nature 352(6335):522–524

Bawa KS (1974) Breeding systems of tree species of a lowland tropical community. Evolution 28(1):85–92

Beja SBSP, Neto AR, Cruz SJS et al (2024) Use of sunn hemp on the efficiency of nitrogen fertilization in sugarcane cultivars. Caderno Pedagógico 21(10):e9396

Bhandari H, Tripathi M, Chaudhary BD et al (2016) Sunnhemp breeding: challenges and prospects. Indian J Agric Sci 86:1391–1398

Bhandari HR, Shivakumar KV, Kar CS et al (2022) Sunn hemp: a climate-smart crop. In: Jha UC, Nayyar H, Agrawal SK, Siddique KHM (eds) Developing climate resilient grain and forage legumes. Springer Nature, Singapore, pp 277–296

Bhardwaj KKR, Datt N (1995) Effects of legume green-manuring on nitrogen mineralization and some microbiological properties in an acid rice soil. Biol Fertil Soils 19(1):19–21

Bhattacharya A, Parkhi V, Char B (2021) Genome editing for crop improvement: a perspective from India. In Vitro Cellular Dev Bio Plant 57(4):565–573

Bhayal D, Khaddar V, Bhayal L et al (2018) Effect of sunhemp green manuring and intercropping on soil properties. Int J Curr Micro Appl Sc 7:371–384

Bianchi MB, Gibbs PE (2000) Late-acting self-incompatibility in *Capparis retusa* (Capparaceae), a species of Chaco woodland in NE Argentina. Brazilian J Bot 23(4):395–400

Bisht S, Kothyari B, Palni L (1996) Rural biotechnology in the Himalaya: fiber extraction from Grewia oppositifolia – a multi-utility tree. ENVIS Bull 3(1&2):52–58

Brewbaker JL (1958) Self-compatibility in tetraploid strains of *Trifolium hybridum*. Hereditas 44(4):547–553

Brown AG (1958) A simple pollen viability test. Aust For 22(1):10–12

Bullock SH (1985) Breeding systems in the flora of a tropical deciduous forest in Mexico. Biotropica 17:287–301

Carmona-Flores L, Wallau M, Vyas D (n.d.) Investigating orange oil and sunn hemp hay as an alternative anthelmintic for Texel Ewes. Marcelo and Vyas, Diwakar, Investigating Orange Oil and Sunn Hemp Hay as an Alternative Anthelmintic for Texel Ewes

Central Research Institute for Jute and Allied Fibres (2023) Fiber Production. [Web content]. CRIJAF. https://jafexpert.crijaf.icar.gov.in/Sunnhemp_expert/sunnhemp_cultivation.aspx

Chander S, Gowthami R, Pandey R et al (2023) Cryoconservation of in vitro grown shoot tips of *Cicer microphyllum*: a crop wild relative of Chickpea. CryoLetters 44(6):360–368

Chawla JS, Sharma AN, Abrol BK (1967) Indian Pulp Paper 22(5):285

Chopra R, Nayar S, Chopra I et al (1956) Glossary of Indian medicinal plants. Council of Scientific & Industrial Research, New Delhi

Chunekar KC, Pandey GS (1998) Bhavaprakash Nighantu (Indian Materia Medica) of Sri Bhavamisra (c. 1500–1600 AD). Chaukhamba Bharati Academy, Varanasi 984

Conway GR, Barbier EB (2013) After the green revolution: sustainable agriculture for development. Routledge

Dafni A, Kevan PG, Husband BC (2005) Practical pollination biology

Dar JA, Beigh ZA, Wani AA (2017) Polyploidy: evolution and crop improvement. In: Bhat TA, Wani AA (eds) Chromosome structure and aberrations. Springer India, New Delhi, pp 201–218

Darlington CD (1956) Natural populations and the breakdown of classical genetics. Proc R Soc Lond Ser B-Biol Sci 145(920):350–364

Demiroğlu Topçu G, Tenikecier HS, Ateş E (2024) Macro-mineral uptake, relative water content, retention capability, and tolerance index of sunn hemp (Crotalaria juncea L.) under Salinity Stress at Early Seedling. Agronomy 14(4):823

Deurenberg JJM (1976) In vitro protein synthesis with polysomes from unpollinated, cross-and self-pollinated *Petunia* ovaries. Planta 128(1):29–33

Devendra B, Srinivas N (2011) Multiple shoot induction from leaf explants through organogenesis in *Crotalaria retusa* L. Arch App Sci Res 3(3):309–320

Duke J (1983) Handbook of energy crops

East EM, Mangelsdorf AJ (1925) A new interpretation of the hereditary behavior of self-sterile plants. Proc Natl Acad Sci USA 11(2):166–171

Edwardson J (1967) Cytoplasmic male sterility and fertility restoration in *Crotalaria mucronata*. J Hered 58(5):266–268

Etcheverry AV 2000. Floral biology and pollination in *Crotalaria stipularia* (Fabaceae: Papilionoideae). In VIII inter symposium on pollination-pollination: integrator of crops and native plant systems, pp 561

Etcheverry AV, Protomastro JJ, Westerkamp CH (2003) Delayed autonomous self-pollination in the colonizer *Crotalaria micans* (Fabaceae: Papilionoideae): structural and functional aspects. Pl Syst Evol 239(1–2):15–28

Faegri K, Pijl L (1979) The principles of pollination ecology, vol 64. Pergamon Press, New York, pp 6–7

Galen C, Plowright RC (1987) Testing the accuracy of using peroxidase activity to indicate stigma receptivity. Can J Bot 65(1):107–111

Garzon J, Vendramini JM, Silveira ML et al (2021) Harvest management and genotype effects on sunn hemp forage characteristics. Agry J 113(1):298–307

Gibbs PE, Bianchi MB (1993) Post-pollination events in species of *Chorisia* (Bombacaceae) and *Tabebuia* (Bignoniaceae) with late-acting self-incompatibility. Botanica Acta 106(1):64–71

Good I (2007) Invisible exports in Aratta: Enmerkar and the three tasks. Ancient textiles: production, craft and society. Oxbox Books, Oxford, pp 168–173

Gupta P (1975) Cytoplasmic male sterility in self incompatible *Crotalaria pallida*. Incompat Newsl 5:13–14

Gupta A, Sinha RP (1978) Cytotaxonomy of artificially induced polyploids of Crotalaria. Nucleus 21:26–28

Gupta BN, Tripathi SN (1999) Studies on the influence of growing sunnhemp for green manuring and as fibre crops on the productivity and economics of rice-wheat cropping system in Uttar Pradesh. In: Palit P, Pathak S, Singh DP (eds) Jute and allied fibres- agriculture and processing. Central Research Institute for Jute and Allied Fibre, Barrackpore

Haddon A (1961) Variety trials of seedlings cacao in Malaya. Malayan Agric J 43:169–232

Herrero M, Gascon M (1987) Prolongation of embryo sac viability in pear (*Pyrus communis*) following pollination or treatment with gibberellic acid. Ann Bot 60(3):287–293

Herrero M, Hormaza J (1996) Pistil strategies controlling pollen tube growth. Sex Plant Reprod 9(6):343–347

Hormaza JI, Herrero M (1992) Pollen selection. Theo App Gen 83(6–7):663–672

Hough T, Bernhardt P, Knox RB et al (1985) Applications of fluorochjromes to pollen biology. II. The DNA probes ethidium bromide and hoechst 33258 in conjunction with the callose-specific aniline blue fluorochrome. Stain Tech 60(3):155–162

Hughes MB, Babcock EB (1950) Self-incompatibility in *Crepis foetida* (L.) subsp. rhoeadifolia (Bieb.) Schinz et Keller. Genetics 35(5):570–588

Hussain TM, Chandrashekhar T, Gopal GR (2008) In vitro propagation of *Crotalaria verrucosa* L. an important ethnobotanical plant. J Med Pl Res 2(9):242–245

Islam S, Kabir G, Shahiquzzaman G et al (1994) Effects of gamma rays and temperature treatment on germination, plant height and morphological abnormalities of *Crotalaria juncea* L. J Bio Sci 2:31–37

Jain A, Shivanna K (1990) Storage of pollen grains of Crotalaria retusa in oils. Sex Plant Reprod 3(4):225–227

Jegadeesan S, Punniyamoorthy D (2023) Potential of mutation breeding in genetic improvement of pulse crops. In: Penna S, Mohan Jain S (eds) Mutation breeding for sustainable food production and climate resilience. Springer Nature, Singapore, pp 445–485

Kanwar R, Singh H (1959) Manurial value of sunnhemp, its effect on soil composition, juce quality and yield of sugarcane. Indian J Sug Res Dev 3(4):194–201

Kar S (2018) Reproductive biology of *Crotalaria spectabilis* Roth

Kaul C, Singh S (1967) Staminal and functional male sterility induced by chemical treatment in papilionaceous plants. Indian J Agr Sci 37(4):264

Kempanna C, Sastry KK (1958) Male sterility in *Crotalaria striata*. Curr Sci 27(5):181

Kendall W, Taylor N (1969) Effect of temperature on pseudo-self-compatibility in *Trifolium pratense* L. Theor Appl Genet 39(3):123–126

Khalikova D, Negmatova S (2023) Germination dynamics of crotalaria seeds in typical gray soils of Tashkent region. International Conference on Science, Engineering & Technology (ICSET)

Klips RA, Snow AA (1997) Delayed autonomous self-pollination in *Hibiscus laevis* (Malvaceae). Am J Bot 84(1):48–53

Lewis D (1954) Comparative incompatibility in angiosperms and fungi. In: Demerec M (ed) Advances in genetics. Academic, pp 235–285

Lloyd DG, Schoen DJ (1992) Self-and cross-fertilization in plants. I. Functional dimensions. Int J Plant Sci 153(3):358–369

Lopes GS (2021) Breeding systems. In: Shackelford TK, Weekes-Shackelford VA (eds) Encyclopedia of evolutionary psychological science. Springer International Publishing, Cham, pp 791–795

Maroyi A (2011) *Crotalaria juncea* L. In: Brink M, Achigan-Dako EG (eds) PROTA (Plant Resources of Tropical Africa/Ressources végétales de l'Afrique tropicale), Wageningen

Maruthi R, Kumar AA, Choudhary S et al (2021) Diversity assessment of Indian sunnhemp (*Crotalaria juncea* L.) accessions for enhanced biomass and fibre yield using geographic information system approach. Legum Res 1:6

Maruthi RT, Kumar AA, Choudhary SB et al (2023) Reproductive analysis of functional interference and role of pollinators in sunnhemp: a late-acting self-incompatible crop. Euphytica 219(10):112

Mather K (1953) Genetical control of stability in development. Heredity 7(3):297–336

Meagher RL, Nagoshi RN, Brown JT et al (2017) Flowering of the cover crop sunn hemp, *Crotalaria juncea* L. Hort Sci 52(7):986–990

Mitra G (1976) Male sterility in sunnhemp (*Crotalaria juncea* L.). Curr Sci 45:873–874

Moco MCC (1996) Biologia da reproducao de Swartzia apetala Raddi var. apetala (Leguminosae-Caesalpinioidea). Universidade Federal do Rio de Janeiro

Mohan KVJ (1971) Out-crossing in sunnhemp. Indian J Genet Pl Breed 31(3):395–396
Mondin M, Santos-Serejo JA, Aguiar-Perecin MLR (2007) Karyotype characterization of *Crotalaria juncea* (L.) by chromosome banding and physical mapping of 18S-5.8 S-26S and 5S rRNA gene sites. Genet Mol Biol 30(1):65–72
Mosqueda HM, Blaser BC, O'Shaughnessy SA et al (2023) Intercropping forage sorghum with sunnhemp at different seeding rates to improve forage production. Agronomy 13(12):3048
Muthu RK, Ganesan R (2012) Rediscovery of *Crotalaria digitata* (Fabaceae) from Madurai district, Tamil Nadu, India. Rheedea 22(2):103–106
Naomita V, Rai VR (2000) In vitro regeneration of *Crotalaria lutescens* (Dalz.), an endemic and rare species of Western Ghats. Phytomorphology 50(3/4):291–295
Narang RS, Gill MS (1996) Sustainability of rice (*Oryza sativa* L.) –wheat (*Triticum aestivum* L.) cropping system. J Res Punjab Agric Univ
Nayak KG, Davidar P (2010) Pollination and breeding systems of woody plant species in tropical dry evergreen forests, southern India. Flora-Morpho, Distribution, Funl Ecology Pl 205(11):745–753
Nirmalaruban R, Ganesan NM, Anantharaju P (2020) Breeding system study in sunnhemp (*Crotalaria juncea* L.) a stepping stone in crop improvement. E J Pl Breed 11(1):204–209
Nirmalaruban R, Ganesan NM, Anantharaju P et al (2024a) Growing Green: unravelling trait associations in sunnhemp (Crotalaria juncea L.) for higher biomass. asian research. J Agric 17(3):129–136
Nirmalaruban R, Suvitha R, Prasad JP et al (2024b) Nature's precision: the intricate pollen dispersal mechanisms of papilionaceous crops. Ann Res Rev Biol 39(8):39–45
Nishikawa S-i, Zinkl GM, Swanson RJ et al (2005) Callose (beta-1,3 glucan) is essential for Arabidopsis pollen wall patterning, but not tube growth. BMC Plant Biol 5:22–22
O'Neill SD (1997) Pollination regulation of flower development. Ann Rev Pl Bio 48(1):547–574
Owens SJ, Rudall P (1998) Reproductive biology in systematics, conservation and economic botany: Royal Botanic Gardens, Kew
Pandey A, Singh R, Sharma SK et al (2010) Diversity assessment of useful Crotalaria species in India for plant genetic resources management. Genet Resour Crop Evol 57(3):461–470
Panse V, Abraham T, Leelavathi C (1965) Green manuring of crops (Review of experimental results) ICAR Tech. Bull 2
Perisse IV, Fan Z, Singina GN et al (2021) Improvements in gene editing technology boost its applications in livestock. Front Genet 11
Perry A (2012) There's a new biofuel crop in town. Agril Res 60(1):17
Raghuvanshi S, Joshi S (1965) Studies on the comparative effects of certain chemicals on the polyploidising efficiency of colchicine in Trigonella Foenum-Graecum L. Caryologia 18(1):69–84
Rajagopal A, Ganesan NM, Anantharaju P et al (2021) Morphological characterization and pollen–pistil interaction in selected genotypes of sunnhemp (*Crotalaria juncea* L.). E-J Pl Breed 12(3):941–948
Rajender K, Thirupathi M, Srinivas D et al (2012) Micropropagation of *Crotalaria laburnifolia* L.–An ethnomedicinally important herbal species. J Phytology 4(4):14–16
Rajesh O, Raj Kumar V, Shankaraiah P (2014) Antiobesity and hypoglycemic effect of ethanolic extract of *Crotalaria juncea* in high fat diet induced hyperlipydemic and hyperglycemic rats. India-Int J Pharm Sci 6:739–742
Rakesh SR, Ganesan NM, Srinivasan MR et al (2019) Studies on diversity, abundance and pollination efficiency of insect pollinators and pollination mechanism involved in sun hemp (*Crotalaria juncea*. L.). E-J Pl Breed 10(2):838–851
Rao S, Reddy M, Reddy G (1987) Effect of intercropping forage-sorghum and sunnhemp on nutritive quality of mixed fodders. Indian Council of Agricultural Research, ICAR Bhawan Pusa, New Delhi 110 012, India
Rao JP, Agrawal P, Mahmood R et al (2012) Tissue culture independent transformation of the forage crop sunnhemp (*Crotalaria juncea* L.): an easy method towards generation of transgenics. Physiol Mol Bio Pl 18:51–57

Sabnis T (1931) Report on Hemp marketing in India
Sage TL, Strumas F, Cole WW et al (1999) Differential ovule development following self-and cross-pollination: the basis of self-sterility in Narcissus triandrus (Amaryllidaceae). Am J Bot 86(6):855–870
Sarkar SK, Hazra SK, Sen H et al (2015) Sunnhemp in India. Central Research Institute for Jute and Allied Fibres (ICAR), Barrackpore, West Bengal
Saurabh S (2021) Genome editing: revolutionizing the crop improvement. Pl Mol Bio Reporter 39(4):752–772
Schrire B (1989) A multidisciplinary approach to pollination biology in the Leguminosae. In: Stirton CH, Zarucchi JL (eds) Advances in legume biology, vol 29, pp 183–242
Shah A, Ud-Deen M, Naz A et al (2008) Post-irradiation ageing effect on morphological characters of *Crotalaria saltiana*. J Bio Sci 16:89–93
Shahwar D, Ahn N, Kim D et al (2023) Mutagenesis-based plant breeding approaches and genome engineering: a review focused on tomato. Mutation Res 792:108473
Sharma N, Bajaj M, Shivanna K (1985) Overcoming self-incompatibility through the use of lectins and sugars in *Petunia* and *Eruca*. Ann Bot 55(2):139–141
Shivanna K (1982) Pollen-pistil interaction and control of fertilization. In: Experimental embryology of vascular plants. Springer, pp 131–174
Singh P, Singh P, Singh P (2007) Essentials of plant breeding. Kalyani Publishers
Small E (1988) Pollen-ovule patterns in tribe Trifolieae (Leguminosae). Pl System Evol 160(3–4):195–205
Stebbins Jr C L (1950) Variation and evolution in plants
Suprasanna P, Mirajkar SJ, Bhagwat SG (2015) Induced mutations and crop improvement. In: Bahadur B, Rajam MV, Sahijram L, Krishnamurthy KV (eds) Plant biology and biotechnology: volume I: plant diversity, organization, function and improvement. Springer India, New Delhi, pp 593–617
Suriya P, Ganesan NM, Subramanian A et al (2022) Pollen-pistil interaction in self-pollinated and sib pollinated flowers of sunnhemp (*Crotalaria juncea* L.). Electron J Breed 13(2):286–294
Tandon RK, Agralwal DC, Singh RP (1959) Sanai (*Crotalaria juncea*) as a dual-purpose cropping cane rotation. Indian Sug Res Dev 3(2):72–78
Thimmaiah MR, Choudhary SB, Sharma HK et al (2018) Late-acting self-incompatibility: a barrier to self-fertilization in sunnhemp (*Crotalaria juncea* L.). Euphytica 214(2):19
Thomson JD, Rigney LP, Karoly KM et al (1994) Pollen viability, vigor, and competitive ability in Erythronium grandiflorum (Liliaceae). Am J Bot 81(10):1257–1266
Tidke J, Patil G (2000) Floral biology and pollination of *Crotalaria sericea* Retz. Environ Ecol 18(4):1020–1022
Tripathi MK, Chaudhary BD, Bhandari HR et al (2012) Effect of varieties, irrigation and nitrogen management on fibre yield of sunnhemp. J Crop Weed 8(1):84–85
Tripathi MK, Chaudhary BD, Sarkar SK et al (2013) Performance of sunnhemp (*Crotalaria juncea* L.) as a summer season (pre-monsoon) crop for fibre. J Agric Sci 5(3):236
Tupy J (1961) Changes in glucose and fructose level in *Nicotiana alata* styles and ovaries accompanying compatible and incompatible pollen tube growth. Biol Plant 3(1):1–14
Vanishree S, Mahale G, Vastrad JV et al (2019) Extraction of sunnhemp fibre and its properties. NISCAIR-CSIR, India
Verma R, Raina S (1991) Cytogenetics of crotalaria. Vii. Induced cytochimeral and asynaptic mutants of *Crotalaria juncea* (sunnhemp). Indian J Genet Pl Breed 51(03):315–319
VR D (1983) Induced tetraploidy in *Crotalaria juncea* Linn. Cytologia 48(3):483–489
Wang K-H, Sipes B, Schmitt D (2002) Crotalaria as a cover crop for nematode management: a review. Nematropica:35–58
Westerkamp C (1997) Keel blossoms: bee flowers with adaptations against bees. Flora 192(2):125–132
White GA, Haun J (1965) Growing *Crotalaria juncea*, a multi-purpose legume, for paper pulp. Economic Bot 19:175–183

Wyatt R (1983) Pollinator-plant interactions in the evolution of breeding systems. Pollinat Ecol:51–95, Elsevier
Yali W (2022) Polyploidy and its importance in modern plant breeding improvement. Int J Agric Biosci 11(1):53–58
Yeo PF (2012) Secondary pollen presentation: form, function and evolution, vol 6. Springer Science & Business Media
Zegada-Lizarazu W, Parenti A, Peroni P et al (2024) Sunn hemp, a tropical legume species, as an alternative bioenergy feedstock in temperate climates. Biomass Bioenergy 183:107114
Zhang Y, Massel K, Godwin ID et al (2018) Applications and potential of genome editing in crop improvement. Genome Bio 19(1):210

Chapter 18
Breeding and Biotechnology of Veldt Grape (*Cissus quadrangularis* L.): Current Status and Future Prospects

Padmapriya Subramanian and Rajamani Kandasamy

Abstract The veld grape, or *Cissus quadrangularis* L., is among the hardiest plant species surviving in dry deciduous habitats, particularly in tropical Asia and Africa. The species is acclaimed for its immense medicinal value, including antioxidant, anti-microbial, bone regeneration, ulcer healing, pain relief, and diuretic properties. It is currently being marketed to treat heart disease, diabetes, osteoporosis, obesity, and athletic muscle building, leading to an expected increase in demand in the international market. The species popularly termed Popularly known as 'Hadjod'' in India, *Cissus* is frequently used in Ayush (Ayurvedic, Unani, Siddha and Homeopathy) practices, particularly for bone healing. The species is native to West Africa, Java, Malaysia, Sri Lanka, and India. Its taxonomy, morphology, phytochemical composition, pharmacological activities, and toxicological evaluation have been widely reported. In the recent past, plant breeding efforts in *Cissus* have been a niche area of research and are underlined as a global priority. The species has witnessed a manifold time increase in the utilization of raw material, mainly useful for the extraction of valuable phytochemicals to combat weight loss, bone and muscular damage, and treatment of osteoporosis, besides manufacturing high-value minerals and nutrients for bone health. Research organizations have prioritized the breeding of *Cissus* to develop chemotypes and pave the way for the organized production of the species to satisfy the demand worldwide. There have been significant attempts in plant genetic conservation, elucidation of metabolic pathways, population assessment, clonal selection, molecular systematics, genome sequencing, DNA barcoding of genotypes, and improved extraction of secondary metabolites, besides the emphasis on evolving protocols for the cultivation of the species. This chapter provides information on the botany, phytopharmacological uses, taxonomy and traditional and modern breeding methods for *Cissus*.

P. Subramanian (✉) · R. Kandasamy
Department of Medicinal and Aromatic Crops, Tamil Nadu Agricultural University, Coimbatore, Tamil Nadu, India

K. F. M. Salem et al. (eds.), *Breeding and Biotechnology of Grass and Bast Fiber Crops*, Advances in Plant Breeding Strategies 12,
https://doi.org/10.1007/978-3-032-00403-1_18

Keywords *Cissus quadrangularis* L. · Cultivation · Cytogenetics · DNA bar coding · Fiber · Genome sequence · Molecular breeding · Traditional breeding · Veldt grape

18.1 Introduction

Veld grape (*Cissus quadrangularis* L.) is a resilient climber of evergreens that mostly grows on small trees and shrubs and can be widely seen in scrub jungle ecosystems. It is a perennial fleshy It is a perennial, fleshy, cactus-like climbing plant with prominent tendrils, characterized by its thick, fleshy, four-angled stem. The species has been referenced in ancient Indian literature and is recognized as a medicinal plant thriving in hot climates across India and neighboring countries such as Pakistan, Bangladesh, and Sri Lanka. It is commonly found in western regions, shrublands, coastal areas, and plains up to an elevation of 500 m (Rajpal 2008). Additionally, *Cissus* has been introduced and established in several other parts of the world, including the United States particularly in Hawaii and Florida, as well as the Lesser Antilles, Africa, Thailand, Vietnam, Indonesia, Singapore, the Philippines, Cuba, Jamaica, and various islands in the Pacific and Indian Oceans. (Acevedo-Rodríguez and Strong 2012; Brink and Achigan-Dako 2012; Yeo et al. 2012).

Indian systems of medicine mention different parts of *Cissus,* such as the root, stem, and leaves, for various medicinal purposes. The bone healing properties of *Cissus* are known in Sanskrit as "Asthisamharaka." As it belongs to the Vitaceae family, *Cissus quadrangularis* L. has been traditionally used for managing wounds, anorexia, indigestion, chronic ulcers, asthma, and hemorrhoids (Timmons et al. 2007). It is also valued for its ability to accelerate fracture healing (Oben et al. 2016; Viswanatha Swamy et al. 2006). of the sturdy quadrangular stem has long been used to cure blindness, asthma, scurvy, piles, bone fractures, constipation, and irregular menstruation.

Although *Cissus quadrangularis* is widely used in traditional and complementary medicine, a lack of knowledge about its genetic diversity poses a significant barrier to its commercial cultivation. An efficient breeding program for improving quantitative traits depends on the heritability of desirable characteristics and the level of genetic diversity. Ecotype assortment and hereditary inconsistency are important factors. In the selection of a superior performer among genotypes (Adugna and Labuschagne 2003).

This chapter aims to discuss the advancement in the botanical description, distribution, traditional and pharmaceutical uses, current cultivation practices, breeding goals, germplasm biodiversity, conservation efforts, taxonomy and the latest breeding strategies for enhancing the yield and quality of *Cissus quadrangularis* based on available literature.

18.1.1 Botanical Classification and Description

The succulent plant *Cissus quadrangularis* L. has sticky tendrils and can reach a vine length of more than 12 m. Its branches are divided into quadrangular sections, with internodes that range from 8 to 10 cm in length and 1.2 to 1.5 cm in width. The surface is smooth, glabrous, and buff-colored with a slight greenish tint, while the angular sections exhibit a reddish-brown hue. The leaves are simple, measuring 2.5–5 cm in length, and broadly ovate or reniform, sometimes featuring 3–7 lobes. They have denticulate margins and a glabrous texture, with bases that can be cordate, rounded, truncate, or cuneate. The petioles range from 6 to 12 mm in length (Khan et al. 1991; Kumbhojkar et al. 1991). The stipules are small, obtuse, and broadly oval. The plant produces small flowers that are white, yellowish, or greenish in color. The calyx is cup-shaped, either truncated or faintly lobed, with a thin, membranous structure and pubescence on the outer surface. The petals are short, thick, and ovate-oblong, forming a cap that detaches as the flower blooms. The flowers occur in axillary cymes and are four or five-merous. Red when ripe, the globular berries are the fruits. Usually two to three in number, the seeds are irregularly rogue and ovoid-oblong.

18.1.2 Distribution

With more than 300 species, *Cissus* is the largest genus within the Vitaceae family (Wen et al. 2008). Species from this genus are distributed across Asia, the Americas, Australia, and Africa and exhibit pan-tropical intercontinental disjunction. With over 135 species, Africa has the highest number of *Cissus* species and is thought to be the genus's original home (Liu et al. 2013; Adams et al. 2016).

Cissus belongs to the botanical family Vitaceae of the Rhamnales order, in which 14 genera and about 850–910 species occur. The majority of these species are found in warm temperate and tropical parts of the world as lianas and vines (Fig. 18.1). *Cissus* is the largest genus in the Vitaceae family, comprising over 350 species. (Stevens 2017). It may be found in India, West Africa, Ceylon, and the hottest regions of Bangladesh, Thailand, Java, and the Philippines. (Udupa et al. 1970).

18.1.3 Importance and Uses

India is known to have a rich plant repository under varied agro-climatic conditions, many of which are widely used for therapeutic purposes. *Cissus quadrangularis* L., widely known as hadjod or pirandai, is a vital plant that has been used for bone healing since ancient times. It is indigenous to Java, West Africa, Malaysia, and Sri Lanka, as well as the tropical regions of the Indian subcontinent. The plant's natural

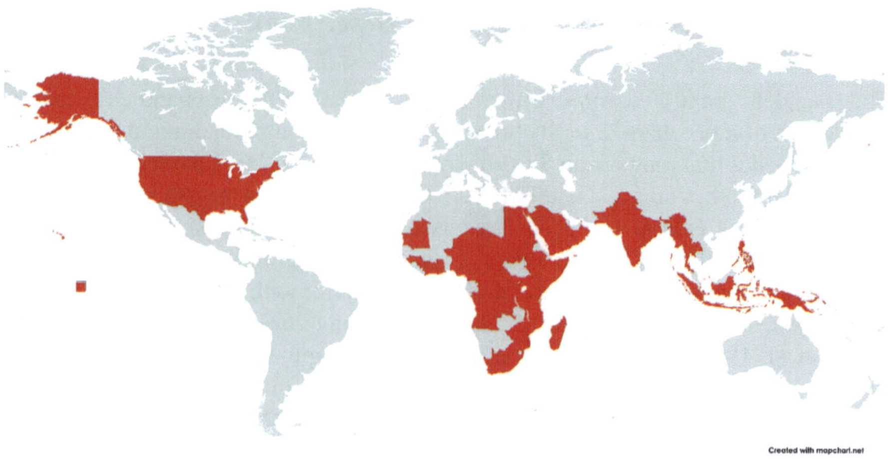

Fig. 18.1 Distribution map of *Cissus quadrangularis*. (Prepared by S. Padmapriya)

analgesic, anti-inflammatory, antioxidant, bone-healing, and antibacterial properties have garnered significant research interest (Rajan et al. 2011). Much research has been conducted to confirm its pharmacological effects and therapeutic usage because of its most notable applications in bone regeneration and fracture repair.

18.1.3.1 Traditional Uses

Numerous problems have been successfully managed by *Cissus quadrangularis* L. thanks to a thorough analysis of the material that is now accessible on the plant in India and outside in both humans and animals around the world (Warrier et al. 1993). The Indian system of medicine, namely Siddh Ayurveda, has extensively utilized this plant in the treatment of 'Asthi,' a term referring to bone-related issues such as osteoporosis, rheumatoid arthritis, osteoarthritis, pain, inflammation, and fractures (Nagori et al. 2011). The plant's use in treating ailments like swelling, hemorrhage, anorexia, flatulence, dyspepsia, colic, chronic ulcer, hemoptysis, convulsions, skin diseases, leprosy, and helminthiasis is also supported by ethnobotanical data from other regions of Indian subcontinents, including Pakistan, Bangladesh, and Sri Lanka (Buddhadev 2014; Joseph and George 2013) (Table 18.1).

18.1.4 Phytochemical Properties

The bioactive components extracted from *Cissus quadrangularis* L. include ascorbic acid, carotene, phytosterols, calcium, flavonoids, vitamins, enzymes, nicotinic acid, tyrosine, alkaloids, resveratrol, piceatannol, pallidal, quadrangular, and triterpenoids, all of which contribute to its pharmacological properties.

Table 18.1 Pharmacological uses of *Cissus quadrangularis* L

Activity	References
Increases estrogen	Aswar et al. (2010)
Anti-inflammatory	Panthong et al. (2007), Singh et al. (2008)
Osteoblastogenesis	Potu et al. (2009)
Increases mineralization	Kumar et al. (2010)
Increases alkaline phosphatase activity	Potu et al. (2009)
Restore the biomechanical properties and structure of the bone	Shirwaikar et al. (2003), Potu et al. (2009)
Anti-tumor properties	Nalini et al. (2011), Prema et al. (2012)
Analgesic properties	Panthong et al. (2007), Singh et al. (2008)
For body weight management	Hasani-Ranjbar et al. (2009), Greenway and Bray (2010)
Reduce blood glucose levels and serum lipids	Oben et al. (2016)
Alleviating insulin resistance	Chidambaram and Carani Venkatraman (2010)
Gastroprotective	Jainu et al. (2006)
Hepatoprotective	Viswanatha et al. (2010)
Suppress chronic ulcer	Jainu et al. (2006)
Anti hemorrhoid activity	Panthong et al. (2007)
Antioxidant	Spiegelman and Flier (2001)
Antibacterial	Luseba et al. (2007)
Anti-cancer activity	Benxu et al. (2023), Cheng et al.(2023)
Antiviral activity	Khaleefathullah and Kathiravana (2010)
Antipyretic activity	Singh et al. (2008)
Antifungal activity	Anitha et al. (2010), Merinal and Viji (2012)
Anticonvulsant properties	Bum et al. (2008)
Anti-diabetic activity	Sibi et al. (2020)

Singh et al. (2007) observed that the plant contained five recognized substances, such as polyphenols like genistein, quercetin, and daidzein, αβ-amyrin, β-sitosterol, ecosyleicosanoate, tetra triacontanoic acid, and tetra triacontanol. Alkaline phosphatase activity and the rate of mineralization through oestrogenic receptors have been linked to Friedelin, a phytoestrogen present in *Cissus quadrangularis* L. extract (Aswar et al. 2010). From the stem of *Cissus quadrangularis*, two steroidal compounds and two asymmetric tetracyclic triterpenoids have been isolated. Additionally, the presence of β-sitosterol, δ-amyrin, δ-amyrone, and flavonoids such as quercetin has been reported, each contributing to various metabolic and physiological functions. The stem has also been found to contain unique stilbene derivatives, known as quadrangularins A, B, and C. Furthermore, *Cissus quadrangularis* L. is a rich source of beta-carotene and vitamin C. Chemical analysis has identified the presence of calcium oxalate, along with 479 mg of ascorbic acid and 267 units of carotene per 100 g of freshly prepared paste (Chidambara Murthy et al. 2003).

18.1.5 Fiber and Textile Properties

Plant-based natural fiber has profound importance in textile, automobile, medical and home product manufacturing. However, the major obstacle in this sector is that the plant, fiber and extraction process must be renewable, environmentally safe, cost-effective and accepted. Plant-based fiber is used in manufacturing carpets, brushes, hats, baskets, ornamental items, ropes, canvas, strings, cords, and cables. Recent reports state that a few natural fibers have high specific mechanical properties and can be utilized for making reinforced plastics.

Cissus quandrangularis is a promising species for a renewable source of natural fiber that has enormous potential to extract natural fiber for various applications (Fig. 18.2). The root of *Cissus* was initially reported to yield a stable natural fiber (Indran et al. 2014).

The fiber extracted from the root was light in weight with high specific strength and strong bonding qualities, as well as a low wax concentration (0.14%) and a high cellulose content (77.17%). The root fiber's low microfibril angle and flaky honeycomb outer surface, as seen by electron microscopy, are factors in its high modulus. The thermogravimetric analysis indicated that the fiber exhibits enhanced thermal stability up to 230 °C, which falls within the suitable range for the polymerization process.

Fig. 18.2 *Cissus quadrangularis* plant with its extracted root fiber. (Source: Indran et al. 2014)

The study indicated that the root fiber of *Cissus quadrangularis* has important mechanical, thermal, and physical degrading qualities that make it a preferable substitute for traditional man-made hazardous fibers. Its greater cellulose content (77.17 weight percent) and negligible wax content (0.14 weight percent) are noteworthy characteristics for its use as a reinforcing material in the production of composites. The mechanical properties of CQRF are comparable to those of conventional fibers; however, its density (1.51 g/cm^3) is significantly lower than that of E-glass fiber (2.5 g/cm^3).

Thermal study reveals that CQRF is stable up to 230 °C, which is substantially above the polymerization process temperature. SEM imaging also demonstrates the existence of microfibrils with low angular separation, which improves strength, and the flaky cell surface supports strong bonding with the polymer matrix.

Apart from the root, aerial parts of *Cissus quadrangularis* were also reported to yield fiber (Mayandi et al. 2015). The stem fiber extracted from *Cissus* had the smallest diameter, measuring around 0.2668 mm and weighing 1.2128 g/cm^3. Cellulose Ib was detected in the stem fiber of *Cissus* by XRD and FTIR analysis, with a crystalline index of 30.9% (Fig. 18.3). According to thermogravimetric analysis, *Cissus* stem fiber has superior thermal stability up to 230 °C, which is well within the temperature range for the polymerization process. Tensile testing showed that *Cissus* stem fiber had the maximum strain to failure, at about 6%.

Scientific advances in extraction processes have had a profound impact on the natural fiber business. Enzyme use in the textile industry has affected the development of ecologically friendly fiber processing methods as well as initiatives to improve finished product quality. Manimekalai et al. (2021) reported that The extraction of natural fiber from the stem of *Cissus quadrangularis* was optimized with an amylase enzyme, and the resulting *C. quadrangularis* stem fibers (CQSFs) were characterized using Fourier Transform-Infrared Spectroscopy (FTIR), X-ray Diffraction (XRD), and Thermogravimetric Analysis (TGA). The enzymatic extraction technique considerably enhanced the fiber's crystalline quality, as evidenced by a high crystallinity index of 49.22% and a crystalline size of 33.21 nm.

Furthermore, the enzymatically treated *C. quadrangularis* stem fibers (ETCQSF) exhibited a low lignin and wax content (5.31% and 0.16% by weight, respectively) and a high cellulose content (85.62% by weight), contributing to superior mechanical properties. The FTIR study showed that cellulose type-1 was present, and the FESEM examination showed that the surface of the ETCQSF fibers included debris and microfibrils. The fibers showed remarkable resilience in thermophilic environments, exhibiting thermal stability at 286 °C. The ETCQSF's physicochemical, mechanical, and thermal properties verified its promise as a sustainable substitute for hazardous synthetic fibers, making it suitable for producing CQSF for ecologically friendly textile and composite applications.

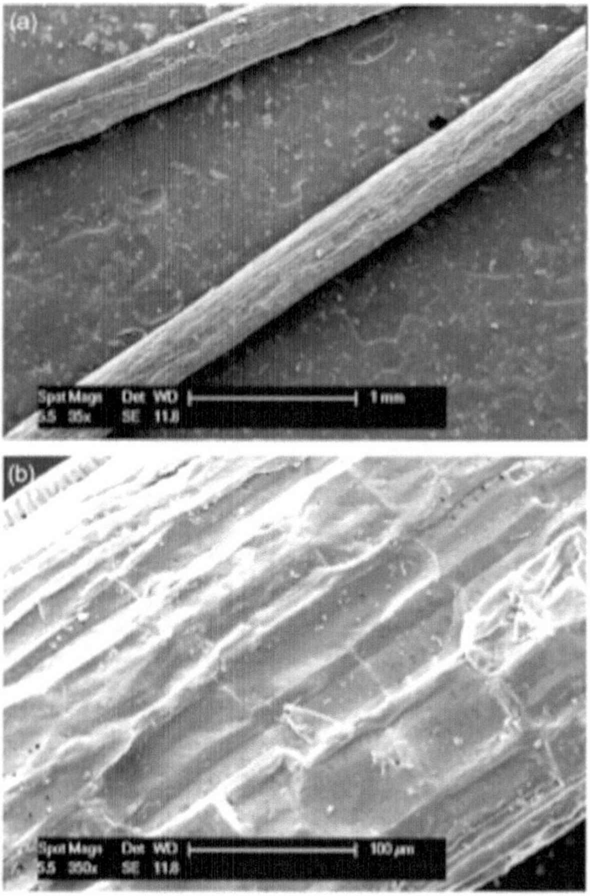

Fig. 18.3 Veldt grape fibre surface SEM micrograph. (Source: Mayandi et al. 2015)

18.2 Current Cultivation Practices

Cissus quadrangularis L. is a popular climber that grows naturally in scrub woods and mixed woodland environments. It is grown up to 500 m above sea level in wastelands, rainforests, coastal regions, and plains. The plant can tolerate hot weather and has a CAM pathway, allowing it to survive under warm and extreme weather prevailing in tropical climates. The species is propagated by stem cuttings which can be done during June and July.

Cut mature stems with four or five nodes and internodes that are between 30 and 40 cm long are either planted in situ or in plastic bags. In the soil that has previously been prepared, one or two internodes are pushed vertically. It is ideal to plant during the monsoon season to maximize plant establishment. Nodes in the soil give rise to roots, and within a week, new leaves appear (Sharma et al. 2011; Garg and Malik 2012). With a blooming to fruiting period from October to December, the plant may grow to a height of 2–5 m in a year.

Cissus quadrangularis L. spreads vegetatively by stem fragments and by seeds. Cuttings are the most effective way to spread it in horticulture (PROSEA 2018). In nature, the species spread from one place to another through birds and rainwater. Stem fragments are dispersed by run-off and waterways. Additionally, frugivorous birds also help in the dispersal of *Cissus* fruits to new habitats (Staples et al. 2000). Interestingly, stem fragments are spread indirectly by the Phalacrocorax carbo cormorant, which uses them to construct nests on rocky cliffs (Nzengue et al. 2015).

Apart from natural distribution methods, the intentional introduction of *Cissus* to new habitats across the globe by humans who often carry plants for home gardening and ethnobotanical use has been documented (Brink and Achigan-Dako 2012).

While there are a few reports on plant propagation, planting, and media, no other agronomical information in *Cissus* has been published. In addition to its occurrence and use in its natural habitat, *Cissus* is grown as a homestead crop to a limited extent.

Natural harvesting of *Cissus* from southern dry regions of Tamil Nadu, India has been reported (Vinoth et al. 2021). Major habitats of *Cissus quadrangularis* have been reported in three districts of Tamil Nadu, India: Virudhunagar, comprising Madathupatty, Sathur, Irukankudi, Vatrayiruppu, Ramanathapuram district comprising Mudhukulathur, Thondi, Parthibanur and Madurai district (Kallupatty Periayur region). Village people harvest the aerial plant parts (stems and leaves) every 4 months and sell the raw material to local agents and dealers who buy in bulk, dry the material and sell it to wholesale/ processing units located in Hyderabad and Bangalore. However, no supply chain information (from raw material to product) has been established for this species.

Given the current pattern and utilization of the raw material, it is anticipated that the demand for *Cissus* will increase significantly, necessitating large-scale cultivation to meet the demand. At the same time, the protection of plants in wild habitats needs to be monitored and scientific methods like 'non-destructive' harvest practices should be taught to local people to conserve genetic resources and promote sustainable living. Crop agronomical research output on clonal selection quality planting material, spacing, nutrient management practices, determining optimized harvest maturity, regenerative harvest, drying, processing and value addition are outlined as crucial areas for further study.

Cissus is grown in homestead farming, especially by health-conscious individuals who use its stem and young leaves for preparing health-promoting local recipes. The plants can be easily propagated using stem nodes for planting. Typically, young leaves and stems are harvested every 3–4 months and used in chutneys, powders and soups. The dry stem powder is available in many local shops, sometimes online. *Cissus* thrives in open land with adequate sunlight. The growing vegetative tips can be trained easily on trellises, pergolas, fences or locally available shrubs and trees.

18.2.1 Breeding Objectives

Cissus is the biggest genus in the Vitaceae family, with over 350 species, and it is the only member of the *Cissus* clade. Unlike grapes where systematic phylogeny has been thoroughly established, *Cissus* remains under-explored due to limited research in breeding and varietal development. Among Vitaceae, *Cissus* utilizes a CAM pathway and can tolerate prolonged extreme heat, unlike other genera in the same family. *Cissus* is highly valued in traditional medicinal systems in Africa, Asia and other regions due to its abundance of valuable phytochemical compounds and nutrients, making it deserving of research in breeding for improved genotypes/chemotypes. While some literature mentions *Cissus* as invasive in certain countries, the reality in India, particularly in hot spot regions differs. *Cissus quadrangularis* L. faces vulnerability due to ecosystem degradation, habitat depletion, climate change and urbanization.

Breeding goals should focus on conserving the wild population, assessing variability and genetic diversity among ecotypes and subspecies, marker-assisted selection, pharmacognosy, cytogenetic investigations, biosynthetic pathway elucidation, establishing DUS guidelines, genomic selection and exploring heterotic potentials through hybridization in diverse genotypes for recombinant breeding. *Cissus* is a clonally propagated species maintaining genetic identity, which poses a challenge for breeders seeking variability. However, clonal variability can be induced through mutation, allowing for enhancement of nutrient and active metabolite content similar to mutation breeding in food crops.

Molecular breeding combined with traditional breeding methods can expedite the selection of genotypes and chemotypes meeting industrial demands in AYUSH and phytochemical sectors. Polyploidy presents an innovative plant breeding method in *Cissus* to alter genetic composition and broaden breeding opportunities. Polyploids often exhibit differences from their diploid parents potentially serving as superior colonizers. Reports of ploidy levels in *Cissus* exist, but correlation with plant characteristics remains unestablished.

Systematic plant breeding in *Cissus* is crucial for developing clones with broader adaptability to varying climatic conditions like in grapes, climate resilience and regenerative capabilities for sustained crop yields, and qualitative improvements in Calcium, Phosphorous and major active metabolite contents.

18.3 Germplasm Biodiversity and Conservation

The most predominant habitat of *Cissus* is dry and arid, mostly in tropical and subtropical lowlands and coastal regions (PROSEA 2018). It may be found in Africa at elevations between sea level to 2000 m in savannahs, shrublands, rocky cliffs, and wastelands in dry and semi-arid environments (Brink and Achigan-Dako 2012). Thickets, open forests, scrub jungles, forest borders, riverbanks, and wastelands at low to medium altitudes are all parts of the Indian forest ecosystem. The species has acclimatized to tropical regions with dry summers and annual rainfall of less than 60 mm. In Africa, flowering occurs while leaf dropping occurs during the dry season and the wet season. Fruiting from October to January, this plant blooms in India from June to July to September (India Biodiversity Portal 2018). Sandy and loamy soils with a pH between 6.1 and 7.8 are among the several soil types it may grow on. The plants cannot tolerate low temperatures and frost (PROSEA 2018).

18.3.1 Taxonomy

Cissus belongs to the grape family, Vitaceae, and is made up of around 14 genera and 900 species, most of which are found in tropical areas. Wen et al. (2007) mentioned 350 species under the genus distributed mainly in tropical areas with a few species surviving under cold and temperate climates. In the Australian continent, *Cissus quadrangularis* L. is reported to be distributed along the east coast (Jackes 1984). Taxonomic Hierarchy of *Cissus quadrangularis* L. The data was obtained from the ITIS database (https://www.itis.gov/).

The Kingdom Plantae	Plants, planta, plants, and vegetation
Subkingdom	Green plants, or Viridiplantae
Infrakingdom	Streptophyta: terrestrial plants
Super division	Formation of Embryophyta
Division	Vascular plants and tracheophytes
The class	Magnoliopsida Superorder Rosanae
Order	Vitales Grapes belong to the Vitaceae family.
Genus	Treebind and treebine, *Cissus* L.
Species	The veldt grape, *Cissus quadrangularis* L.

Near the base of the Rosids, the Vitales is a monophyletic branch that is most likely related to the Saxifragales. It is distinguished by raphide crystals, pearl bodies, ruminating seeds with an oily endosperm, and antipetalous stamens (APG IV 2016; Zhang et al. 2016) (Fig. 18.4). The two sister lineages that make up the order include a tiny monogeneric lineage that includes around 35 species of shrubs, trees, or occasionally herbs, and a larger group that includes roughly 15 genera and 900 species. Woody vines with characteristic leaf-opposed tendrils and inflorescences are usually members of the larger group (Chen et al. 2007; Zhang et al. 2015). It is usual practice to divide the two lineages into the family Vitaceae Juss and Leeaceae Dumort (Molina et al. 2013; Ridsdale 1974). However, the most current taxonomy by the Angiosperm Phylogeny Group (APG) groups them into a single family, the Vitaceae, which is further split into two subfamilies, Vitoideae Eaton and Leeoideae Burmeister (APG IV 2016). The two lineages, Leeaceae and Vitaceae, appear to have split at least by the late Cretaceous, according to molecular dating and fossil data (Manchester et al. 2013; Wen et al. 2008). In Vitaceae s.s., five lineages have

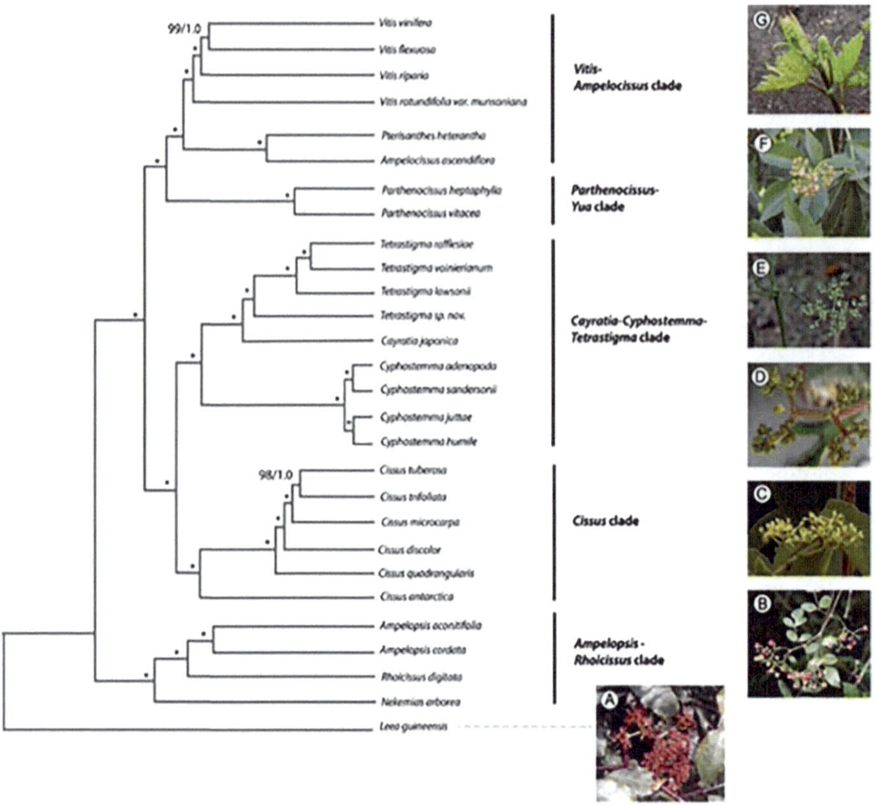

Fig. 18.4 Phylogenetic connections among the taxa: **a**) *Leea guineensis*, **b**) *Nekemias cantonensis*, **c**) *Cissus rotundifolia*, **d**) *Cyphostemma juttae*, **e**) *Cayratia cardiophylla*, **f**) *Parthenocissus vitacea*, and **g**) Vitis 'La Crosse'. (Source: Gerrath et al. 2017)

strong support: The Vitaceae family is classified into five distinct and well-supported evolutionary lineages. The first lineage includes Ampelopsis Michx. and Rhoicissus Planch., while the second consists solely of *Cissus* L. The third lineage comprises Cayratia Juss., Cyphostemma (Planch.) Alston, and Tetrastigma K. Schum. The fourth group includes Parthenocissus Planch. and Yua C.L. Li, whereas the fifth lineage consists of *Vitis* L. and Ampelocissus Planch. (Ren et al. 2011; Trias-Blasi et al. 2013).

The phylogenetic relationships within the *Cissus* genus and the Vitaceae family remain a subject of debate among researchers (Gerrath et al. 2004; Rossetto et al. 2001, 2002a, b; Jackes 1984; Ingrouille et al. 2002). Wen et al. (2007) highlights that while Linnaeus (1753) originally classified *Cissus* and Vitis as distinct genera within the family, later taxonomists, including Hooker (1862), considered *Cissus* a synonym of Vitis, a perspective also adopted by Baker (1871) and Lawson (1875). This indicates that the definition of generic limitations in Vitaceae has historically been ambiguous. Ten genera, including Vitis, Ampelocissus, and *Cissus*, were named in Planchon's 1887 monograph of the family. Each of these genera was further classified into subgeneric groupings (Soejima and Wen 2006; Wen 2007). Susan et al. (2007) proposed dividing The classification of *Cissus* into two monophyletic genera is determined by factors such as chromosome number (n = 20 or 24) (Lavie 1970), DNA variability (Rossetto et al. 2001, 2002a, b; Soejima and Wen 2006), the occurrence of supernumerary buds, the extent of stipule connectivity, the morphology of uncommitted primordia, and the pattern of inflorescence branching.

Cissus species are usually succulent climbers with tendrils, having mostly square and jointed stems and tuberous roots (Rowley 1970). They are found in America, Africa, India, and Indo-Malaysia, implying that *Cissus* is a pantropical plant species (Lawrence 1951). Agnew and Agnew (1994) reported that *Cissus* belongs to the succulent group and predominantly occurs in dry areas of Kenya. The original description of *Cissus* species refers to *Cissus quadrangularis* var. *quadrangularis*. Masinde (1992) reported the *C. quadrangularis* complex in Kenya, which includes species like *C. adeyana, C. albiporcata* and *C. pharmacarpa* Masinde & Newton. The variable nature of the stem in these plants was described by Verdcourt (1993) and one distinct type, *C. qudrangularis var. aculeatangula* was noted. Bailey (1949), Graf (1963) and Jacobsen (1960) mentioned ancient literature citing *C. Quadrangularis* L. as an ornamental plant since early times.

The East African Herbarium Nairobi, describes variation in stem angles among three species, *C. quadrangularis, C. cactiformis* and *C. quinquangularis*, leading to the division into variants A and B based on smooth and rough stem angles, indicating distinctiveness at the intra-specific level (Gituru et al. 2001).

Onyeweaku et al. (2020) investigated the complex taxonomy of the genus *Cissus* L. using morphological, anatomical, epidermal, and phytochemical data from three types of species: *Cissus aralioides* (Welw. ex Bak.) Planch., *Cissus petiolata* Hook. f., and *Cissus lageniflora* Gilg & Brandt (Table 18.2). Their findings indicated that *Cissus aralioides* and *Cissus lageniflora* share a closer relationship with each other than with *Cissus petiolata*. Additionally, the study identified several distinct characteristics differentiating these species.

According to reports on Indian traditional medicinal systems, *Cissus vitiginea* L., a perennial climber with potent therapeutic capabilities, is used to treat wounds, diabetes, heart disease, cancer, and, in particular, bone problems and arthritis. This plant is in short supply in numerous parts of India due to reckless wild gathering. This plant includes alkaloids, flavonoids, triterpenoids, steroids, glycosides, coumarin, tannins, sugar, and proteins, according to phytochemical literature. According to Manokari et al. (2019), the secondary metabolites that have been identified show a high level of biological activity in living systems.

Taxonomical attributes of the genus Based on their floral traits, *Cissus vitiginea, Cissus quadrangularis, Cissus latifolia, Cissus repens*, and *Cissus heyneana* were examined. Variations were noted in the disc's lobing pattern, anther motch arrangement, and disc structure. The species' tendril displayed morphological variations, including having two or three branches. There are two distinct morphological varieties of *Cissus repens*, with the entire plant coated in a white waxy bloom and either green or red tendrils (Hari and Anto 2016).

Table 18.2 Morphological characteristics of three different *Cissus* species

Characters	*Cissus aralioides*	*Cissus lageniflora*	*Cissus petiolate*
Leaf type	Digitately compound	Trifoliate compound	Simple
Leaf arrangement	Opposite	Opposite	Opposite
Leaf shape	Oblong	Lanceolate	Sagittate
Leaf apex	Acuminate	Aristate	Aristate
Leaf base	Cuneate	Oblique	Cordate
Leaf margin	Lobed	Parted	Parted
Leaf venation	Pinnately veined	Pinnately veined	Pinnately veined
Abaxial leaf pigmentation	Absent	Present	Present
Glabrous or pubescence on the leaf	Glabrous	Pubescent	Pubescent
Leaf length	11–12.6 cm	9.0–10.5 cm	15.5–17.5 cm
Leaf width	4.8–5.0 cm	4.0–4.7 cm	12.5–16.5 cm
Shape of stem	Cylindrical	Circular	Circular
Stem pigmentation	Present	Present	Present
Stipule	Present	Absent	Present
Length of tendril	23.0–25.0 cm	24–25 cm	24.5–26.5 cm
Tendril branching	Absent	Absent	Present
Length of petiole	7.0–8.0 cm	5.5–6.5 cm	11.5–13.5 cm
Glabrous or pubescent on stem	Glabrous	Glabrous	Pubescent

Source: Onyeweaku et al. (2020)

18.3.2 Cytogenetics

Publications have shown that the chromosomal count and ploidy levels in *Cissus* species vary widely. In previous reports of *Cissus quadrangularis*, the number of chromosomes has fluctuated. When Raghavan (1957) looked at the number of chromosomes in Indian medicinal plants, he discovered that *C. quadrangularis* had 2n = 24. He proposed that the specimen may have been a tetraploid or that previous reports of a diploid with 45 chromosomes might have been inaccurate. Furthermore, *C. quadrangularis* from Kenya and its two variations (A & B) were examined by Robert et al. (2001), who found that they had 24 and 28 chromosomes, respectively. While variant B, which has rough stem angles, was regarded as a novel variety of *C. quadrangularis*, variant A was determined to be the type variety for *C. quadrangularis* with smooth stem angles.

The chromosomal counts of seven Cissus species (2n = 24, 48) were published by Karkamkar et al. (2010). Chu et al. (2018) also reported the number of chromosomes for seven more species of Cissus (2n = 24, 40, 48, or 66) and proposed a linear link between genome size (1C = 0.37–1.03 pg) and chromosomal counts. These findings suggest that the enlarged genome size in this species is a result of polyploidization and repetitive element alterations (Haiping et al. 2022). However, there is still much to learn about the genome size, chromosomal count, and genomic features of the many species in the Cissus genus (Barghini 2015).

Studies have found that repetitive sequences make up nearly 90% of plant genomes, with long terminal repeats (LTRs) being the most common component (Du et al. 2010). Plant genome size fluctuations are mostly due to polyploidization processes and the presence of repetitive DNA, such as transposable elements and tandem repeats (Kidwell 2002; Sessegolo et al. 2016). Misannotation of LTR sequences has occasionally led to inaccurate gene classification, especially for those with low copy counts (Bennetzen et al. 2004). This highlights the importance of correct annotation of transposable elements in whole-genome sequencing initiatives (Wicker et al. 2009). In this context, Macas et al. (2015) used a short-read sequencing technique to detect repeat elements without a reference genome (Morgante et al. 2007).

Michael and VanBuren (2015) noted advances in sequencing technologies that allow for the generation of vast amounts of sequence data at a low cost, as well as enhanced genome assembly approaches that help both model and non-model plants. However, whole-genome sequencing initiatives remain difficult due to limited data availability and a high amount of repetitive DNA, which can hamper sequencing efforts (Kidwell 2002).

Li et al. (2017) found a positive relationship between genome size and repetitive sequences, with retrotransposons having a stronger link than transposons. LTR-retrotransposons had the strongest association, with both the Ty1-Copia and Ty3-Gypsy LTR lineages contributing equally to genome size. Understanding the content of these repetitive elements reveals important information about genome expansion in *Cissus quadrangularis* L., which is most likely influenced by their proliferation (Paterson et al. 2009).

A genomic sequencing study of *Cissus quadrangularis* L. in Kenya revealed information about the species' chromosomal number, genome size, and composition. Flow cytometry assessed the genome size at around 2C = 1.410 pg, whereas fluorescence microscopy with DAPI staining verified a chromosomal number of 2n = 48. Given that the Cissus genus' fundamental chromosome numbers are normally n = 10, 11, or 12, these findings indicate that *Cissus quadrangularis* L. has a tetraploid genome (Senapati et al. 2023). Low-throughput sequencing and bioinformatics research found that repetitive elements make up around 52% of the genome, with LTR-retrotransposons accounting for the majority. The Copia and Gypsy superfamilies contributed 24% and 15% of the annotated clusters, respectively (Gandhimathi et al. 2016).

Comparisons of repetitive elements among *C. quadrangularis* L., Vitis vinifera, and four other *Cissus* species showed significant diversity, suggesting recent amplification events within the genus (Gichuki et al. 2019). The genome size of *C. quadrangularis* L. was estimated at 2C = 1.410 pg, equivalent to 689 Mbp/1C (based on 1 pg = 978 Mbp; Dolezel et al. 2003). Soltis et al. (2003) classified its genome as belonging to plants with very small genomes, similar in size to the tetraploid species *Cissus antarctica* and *Cissus trifoliata* (Fig. 18.5). This observation was supported

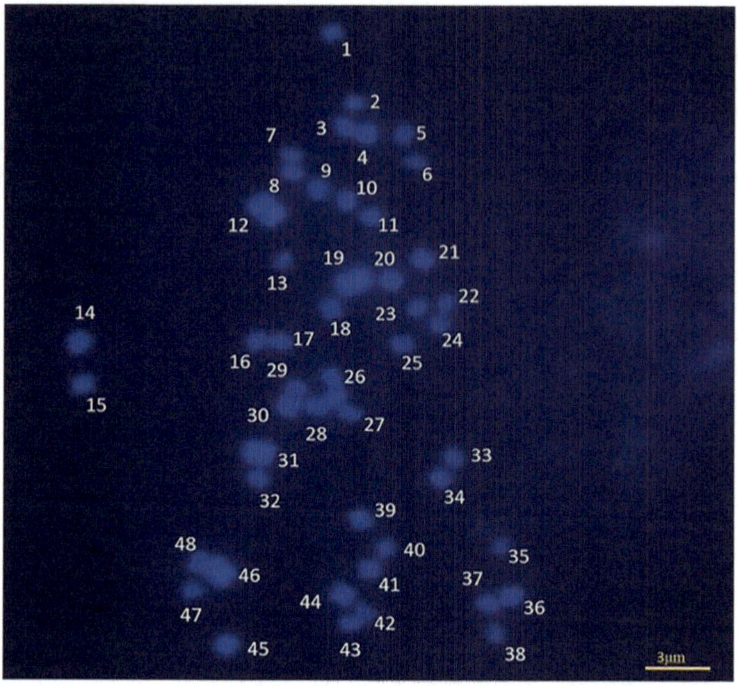

Fig. 18.5 *Cissus quadrangularis* L. chromosome image. (Source: Gichuki et al. 2019)

by Chu et al. (2018), who noted that the genome of *C. quadrangularis* L. is roughly double the size of the diploid species *Cissus javana* and *Cissus rotundifolia*.

Chu et al. (2018) determined that the basic chromosome number in the Cissus genus is 10, 11, or 12. In India, however, *Cissus quadrangularis* L. chromosomes initially believed to be tetraploid were reported as 2n = 24 or 28 (diploids) and 2n = 45 in one specimen (Raghavan 1957). Chromosomal variations observed across different populations of *Cissus quadrangularis* L. may be influenced by epigenetic factors, including DNA methylation and histone alterations. These mechanisms can regulate transposable elements and contribute to chromosomal evolution across diverse biogeographical areas (Li et al. 2017). However, the limited sampling in existing studies poses a challenge to conclusively determining the chromosome number of C*issus quadrangularis* L.

Individuals of *C. quadrangularis* L. with 2n = 48 are classified as tetraploid according to the Cissus genus' chromosome duplication model (Chu et al. 2018; Karkamkar et al. 2010). However, molecular data are required to confirm this idea. Such data would increase our understanding of chromosomal evolution in Cissus, as well as our interpretations of phylogenetic and biogeographic patterns, hence supporting models such as Cladistic, Migrationist, and Panbiogeography (Wilson 1991).

Husband et al. (2013) proposed that intraspecific polyploidization explains differences in diploid somatic chromosome counts. Similar variability in karyotypes among species has been linked to dysploidy in diploids, followed by hybridization and polyploidy. (Choi et al. 2008; Haga and Noda 1976). These patterns may suggest cryptic lineages, necessitating further studies combining morphological and molecular phylogenetics.

Endopolyploidy is common in xerophytic plants with small genomes and CAM metabolism, enabling survival in arid environments (Derocher and Stirling 1990). Soltis et al. (2003) classified the genome of *C. quadrangularis* as very small, consistent with this pattern.

Research in India by Vinoth et al. (2021) revealed ploidy variation in four ecotypes of *C. quadrangularis* L., identifying three diploids and one tetraploid using flow cytometry. Robert et al. (2001) reported chromosome counts of 24 and 28 in two *C. quadrangularis* L. variants from Kenya with smooth and rough stem angles, respectively. Karkamkar et al. (2010) noted chromosome counts of $2n = 24$ and 48 for *Cissus* species, while Chu et al. (2018) reported $2n = 24$, 40, 48, or 66 for seven other species, suggesting a linear link between chromosomal number and genome size ($1C = 0.37–1.03$ pg).

The *Cissus* genus, belonging to the large and diverse Veldt group, exhibits unfamiliar genome characteristics, with polyploidization and repetitive element alterations contributing to chromosomal abnormalities (Zuccolo et al. 2007). Variations in sample preparation, staining, and analysis could affect ploidy estimations, making standardization crucial.

18.4 Traditional Breeding of *Cissus*

Cissus quadrangularis is a perennial climber and its economic value lies in the mature stem. Systematic plant breeding in *Cissus* is still in its early stages, with very few papers addressing genetic diversity and clonal variability among the population.

It is an under-reported plant species but has high utility with reports of over-exploitation. The lack of knowledge about its genetic wealth is a major barrier to commercial cultivation and the effective formulation of a breeding program. To improve quantitative and qualitative attributes in a clonally propagated species, understanding the level of genetic variability and the degree of heritability of desirable traits is essential. Genetic inconsistency and ecotype diversity significantly influence the selection of superior genotypes (Adugna and Labuschagne 2003). Phenotypic and genotypic coefficients of variation are valuable tools for assessing variability among genotypes. Heritability and genetic gain provide insights into the environmental impact on trait expression and the potential for improvement through selection. High heritability and a significant degree of genetic advancement are essential for efficient selection in a segregating population.

Studies on the correlation between stem yield and its components, along with their contributions to yield, rely on the extent and nature of genetic variability within the gene pool. Path analysis is a useful tool for breaking down the correlation coefficient into direct and indirect effects on yield and associated traits (Islam et al.

2010). Consistent data on these genetic traits provides critical insights into the inheritance patterns of various characteristics, facilitating the selection of plants with desirable traits for breeding new varieties.

Plant populations' genetic structure is influenced by a variety of evolutionary processes, including historical distribution shifts, habitat fragmentation, population isolation, mutations, genetic drift, mating systems, gene flow, and selection pressures (Schaal et al. 1991).

Research on the morphological characterization of veldt grape ecotypes is still in its early stage and requires attention for a better understanding of existing genetic diversity. Due to varying ploidy levels in wild plants, morpho-molecular and cytological characterization of *Cissus quadrangularis* ecotypes is essential in formulating a suitable breeding strategy for yield improvement.

There are no reports of *Cissus* being cultivated with a known genotype on a commercial scale and scientific practices adopted. It is crucial to understand that raw material requirements are mostly met from the natural habitat, emphasizing the importance of traditional breeding methods like plant introduction and domestication. For this purpose, rigorous plant selection in the wild habitat should focus on identifying genotypes with DUS markers, improved quantitative (stem yield) and qualitative content (including Ca, P and other active molecules), continuous harvest capability and quick plant regeneration after each harvest, adaptability to varying soil and agroclimatic conditions and early crop maturity.

Traditional breeding methods should address plant genetic conservation, documentation of the habitat and hot spot region, and genetic diversity, leading to clonal selections with desirable plant attributes. Although vegetatively propagated, *Cissus* plants flower and set seeds. Basic information on pollen fertility, seed set, and seed germination will help plant breeders select plants with diverse plant characters from open-pollinated plants. The heterotic potentials, including the induction of fruit set and taking up of hybridization in diverse genotypes for recombinant breeding, can also be examined.

Vinoth et al. (2021). A study of 50 ecotypes of *Cissus quadrangularis* found high estimates of the genetic coefficient of variation (GCV) and phenotypic coefficient of variation (PCV) for variables such as the number of developed leaves, twiners, and primary roots, which exceeded 35%. Correlation and path coefficient analyses of 14 morphological features found that genotypic correlation coefficients outperformed phenotypic correlation coefficients. This indicates that environmental factors weaken the relationship between stem yield and the traits contributing to stem yield in Veldt grapes. The authors reported on genotypes with distinct markers such as TNCq 45 possessing pink-coloured tendrils (Fig. 18.6) and TNCq5 possessing bi-forked tendrils (Fig. 18.7).

Vine length, number of primary branches, internode count, leaf number, leaf length, leaf breadth, stem weight, and single-plant production were all highly heritable, with a significant genetic advance as a percentage of the mean observed across ten *Cissus quadrangularis* ecotypes.

Fig. 18.6 TNCq 45- Red-colored tendril. (Source: S. Sandhya)

18.5 Molecular Breeding

Cissus is an emerging plant species that requires innovative plant breeding methods to overcome cumbersome techniques and save time and cost in developing varieties. Conventional plant breeding methods are generally considered to enhance the pool of genetic resources for crop improvement by incorporating desirable traits. Marker-assisted selection, QTL mapping to analyze quantitatively inherited traits such as yield, vigor, rate of photosynthesis and drought tolerance, genomic selection, and molecular markers to study polymorphism among variable genotypes are some of the modern breeding tools employed in crops and these methods can be standardized for varietal development in *Cissus*. QTL provides useful information for estimating the breeding potential of selection in breeding lines.

Research reports published in the recent past suggest the adoption of improved breeding methods, such as molecular breeding in *Cissus,* to evolve varieties with desirable quantitative and qualitative traits in a short period. Marker-assisted selection has gained momentum in the recent past on account of growing awareness of the multifaceted pharmaceutical implications of the crop. Molecular techniques are

Fig. 18.7 TNCq 5 with a bi-forked tendril. (Source: S. Sandhya)

highly effective for identifying genetic variation within and between plant species. Although fewer research papers have been published on molecular aspects of veldt grape about marker-assisted selection, DNA barcoding, functional genomics and molecular docking studies, the results are quite entertaining to share the applications and databases within the scientific community.

18.5.1 Marker Assisted Breeding

The invention of DNA marker technology has proven to be an effective and steadfast tool for the estimation of genetic variation by utilization of the available resources and protection of breeders'intellectual property rights (Franco et al. 2016) can be enhanced through DNA-based characterizations, which offer valuable insights into the diversity, relatedness, distinctness, and duplication among collected landraces and accessions, thereby maximizing genetic diversity. The use of genetic markers and the identification of polymorphic nucleotide sequences distributed across the genome have created new opportunities for assessing genetic diversity and elucidating interspecific and intraspecific genetic relationships (Gostimsky et al. 2005).

Various protocols for DNA extraction from plant cells have been developed for numerous crops (Dellaporta et al. 1983; Doyle and Doyle 1987; Ghaffariyan et al. 2012; Sharma et al. 2010). However, DNA extraction and purification from *Cissus* plants present unique challenges due to their high content of secondary metabolites, including essential oils, flavonoids, alkaloids, terpenoids, and ketosteroids. These compounds can interfere with obtaining high-quality DNA (Lodhi et al. 1995). In Cissus, Sangani (2017) reported the use of 12 RAPD primers, which generated a total of 79 bands, averaging several bands per primer with 62 of them being polymorphic.

In a similar study conducted by Padmapriya et al. (2022) in *Cissus*, RAPD analysis was carried out using a set of 12 decamer random primers for DNA amplification through a PCR assay. Eight primers produced clear polymorphic bands in all five tested ecotypes (TNCq32, TNCq34, TNCq29, TNCq23 and TNCq9). A total of 128 bands were detected from eight DNA templates of different ecotypes, among which 63 (50.54%) bands were polymorphic, revealing a high degree of polymorphism among the ecotypes.

18.5.2 DNA Barcoding

DNA barcoding has emerged as a widely used technique for species identification, particularly in cases where only short, reliable DNA regions are available or when working with degraded DNA from processed food, fossil remains, or herbarium specimens. This approach has been applied successfully to various plant groups, including *Dendrobium* species (Asahina et al. 2010), mosses (Liu et al. 2010), *Nymphaea* species (Chaveerach et al. 2011), medicinal plants such as *Smilax* and *Cissus* (Kritpetcharat et al. 2011), medicinal *Senna* species (Monkheang et al. 2011), and *Cymbidium* species (Siripiyasing et al. 2012). Their DNA barcodes are now cataloged in GenBank for plant part identification.

In *Cissus quadrangularis*, DNA barcoding was used to investigate genetic variations among three closely related varieties: quadrangular, round, and flat types. Amplification and sequencing of DNA barcoding markers, including ITS-4 and ITS-5, matK, and rbcL, were carried out. Among these, rbcL sequences provided the highest resolution for clustering and distinguishing genetic clades. The results confirmed that all three varieties belong to the same species, *C. quadrangularis* (Prasad et al. 2020). This highlights the utility of DNA barcoding in resolving intraspecific variations and enhancing our understanding of genetic relationships (Fig. 18.8).

Runglawan et al. (2016) experimented to investigate the morphological and chemical composition of four related species of *Cissus quadrangularis* i The analysis revealed low genetic relationships between the studied species and the well-known *Cissus quadrangularis*, as indicated by the distant genetic similarity range

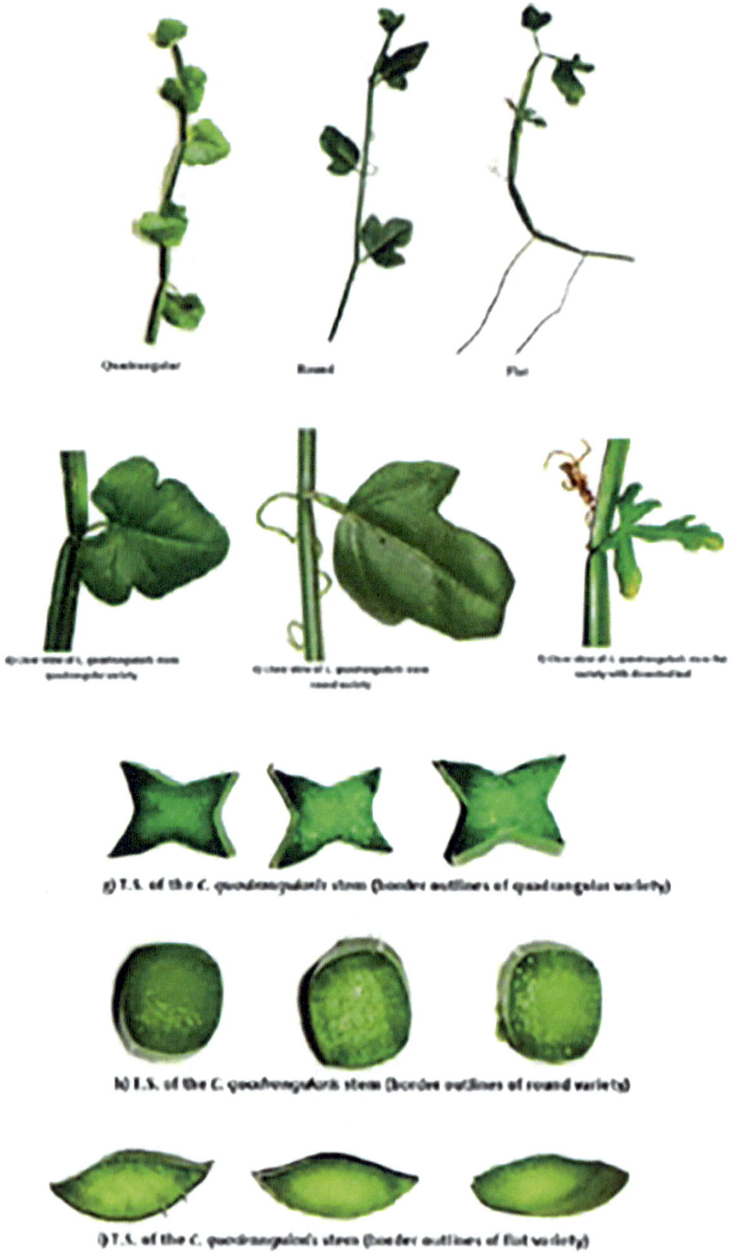

Fig. 18.8 Morphological variations among three varieties of *C. quadrangularis* (quadrangular, round and flat). (Source: Prasad et al. 2020)

(S = 0.57–0.61). *β-Sitosterol* was the only compound common to all four species and previously reported in *C. quadrangularis*.

18.5.3 Genomics of Cissus quadrangularis

Traditional molecular approaches have presented difficulties in examining genome factors, particularly in non-model species. Despite limited past research, advances in short-read sequencing and bioinformatics techniques have enabled quick and cost-effective genome analysis of *Cissus quadrangularis*. RepeatExplorer analysis found 50,767 clusters, indicating that about 52% of the *C. quadrangularis* genome contains repetitive sequences. This clustering lays the groundwork for future research to characterize repeat family structures and their variants.

The draft gene sequence of *C. quadrangularis* genotype I-TUH No. 189 was a crucial milestone, as it provided a genetic foundation for comprehending the plant's active components. The assembled genome stretched 313 Mb, accounting for roughly 64% ofVitis vinifera's genome, with 36,325 proteins identified. Additionally, the entire chloroplast genome of *Cissus quadrangularis* (160,404 bp) was sequenced using a genome skimming method. The research revealed a structural organization made up of a large single-copy area (88,987 bp), a minor single-copy area (18,621 bp), and two inverted repetitions (26,398 bp). A total of 133 genes were discovered, including 88 protein-coding genes, 37 transfer RNAs (tRNAs), and eight ribosomal RNAs (Li et al. 2024).

The study also highlighted that single or low-copy repeat families, which were underrepresented in clustering analysis, might influence the phenotypic characteristics of plants. To investigate variations in these elements, a reference genome coupled with genotype re-sequencing is necessary. This approach aligns with methodologies employed in model species to analyze single-copy retroelements and other low-copy fractions. These findings lay the groundwork for comprehensive future genomic and phenotypic studies of *C. quadrangularis* (Fig. 18.9).

Cissus quadrangularis, a tetraploid species renowned for its Crassulacean acid metabolism (CAM) pathway in the succulent stems and C3 photosynthesis in leaves, with a 679.2 Mb genome divided into two subgenomes. Repetitive elements makeup around 52% of the genome, and genome-wide annotation revealed 51,857 protein-coding genes. The genome also contains 47.75% repetitive sequences, indicating that expansion was driven by repetitive element amplification and polyploidization. According to the analysis, a tetraploidization event happened approximately 8.9 million years ago. The sub-A genome, which has been identified as the dominant subgenome, is critical for drought tolerance. Transcriptome results indicate an enrichment of genes linked to cutin, suberin, and wax metabolism in stems during drought, together with diurnal expression of CAM-related genes, indicating stem-specific gene expression as a key.

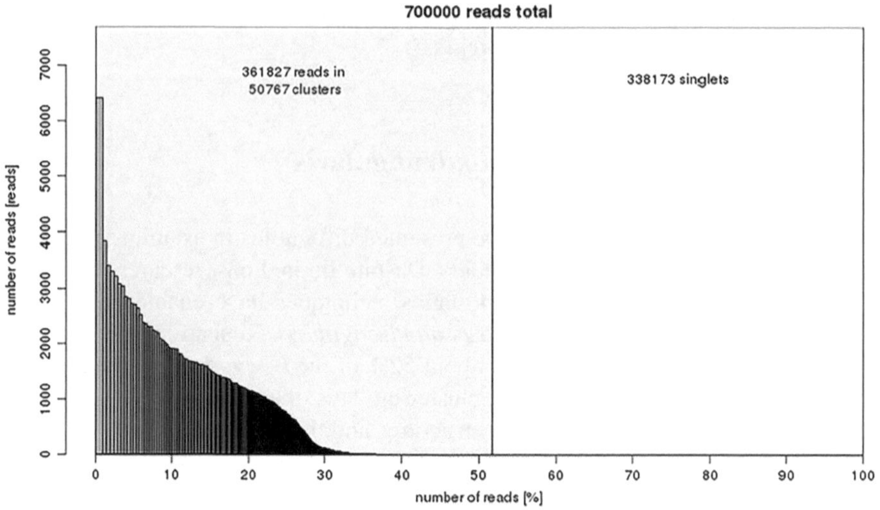

Fig. 18.9 Repeat composition of clusters generated in repeat explorer (similarity-based partitioning) of 700,000 reads (7.61% of genome size). (Source: Gichuki et al. 2019)

Comparative genomic analysis with *Cissus rotundifolia*, an endemic species of Eastern Africa, provides insights into genome divergence within the ancient eudicot family. *C. rotundifolia* has a smaller genome (n = 12, 350.69 Mb/1C) shaped by recent retrotransposon activity, chromosomal rearrangements, and segmental duplications. Unlike *C. quadrangularis*, it has not undergone whole-genome duplication. Local gene amplification, particularly of genes associated with abiotic and biotic stress responses (HSP-20, NBS-LRR), contributes to its ecological adaptation to arid environments. Enzymatic reorganization of CAM and diurnal gene expression patterns support its survival in dry savannas. These findings underscore the adaptive mechanisms and genomic evolution underpinning drought resilience in these *Cissus* species.

18.5.4 Tissue Culture Applications

Cissus quadrangularis, a warm-season climbing species, is traditionally propagated through stem cuttings during June–July. Sexual propagation is challenging due to poor seed set and low seed viability (6–8%). Conventional vegetative propagation presents limitations in terms of propagule production. Therefore, *in vitro* clonal multiplication emerges as an optimal solution for large-scale production of planting material, addressing both pharmaceutical demands and conservation needs. This approach enables year-round plant multiplication and long-term germplasm maintenance. Successful *in vitro* multiplication and plantlet regeneration protocols using various explants (shoot tip, nodal explant, and *in vivo* shoot tip) have been established by Debabrat and Sarma (2005).

Significant enhancement of steroidal compounds, specifically β-sitosterol and stigmasterol, was achieved in callus tissues through the incorporation of various plant growth regulators (Manju 2019). Treatment with 5 ppm IAA resulted in a 10–12 fold increase, while 5 ppm 2,4-D yielded a six–eight fold enhancement in steroid content, suggesting potential applications for commercial-scale steroid production (Nalini et al. 2011).

A standardized protocol was established to enable rapid in vitro multiplication and plantlet regeneration of *Cissus quadrangularis* employing various in vitro and in vivo explants, such as shoot tips and nodal segments. After 30 days of inoculation, *in vitro* nodal segments produced the most shoots, with an average of five and a mean length of 2.86 cm on MS medium supplemented with 0.5 mg/L Zeatin. *In vitro* shoot tip cultures cultured on MS medium containing low doses of Zeatin alone or a combination of BAP and adenine sulfate produced multiple shoots as well. Shoot tips from field-grown *C. quadrangularis* proliferated rapidly in MS medium supplemented with BAP. Shoots grown in vitro were transferred to MS medium with or without IAA and IBA to induce roots efficiently. The rooted plantlets were acclimatized in pots containing sterile soil mixed with vermicompost in equal parts before successfully transitioning to greenhouse settings with a high survival rate (Poonam and Malik 2012).

Limited studies have explored callus induction in *Cissus* using stem and leaf segments from field-grown mature plants. Optimal callus growth was achieved on MS medium with 5.37 μM NAA and 2.32 μM Kn. Although shoot bud differentiation (caulogenesis) was observed in 2% of callus cultures, complete shoot development remained unsuccessful across various media compositions. These findings confirm the totipotent nature of the cells (Anand and Kaur 2019; Cholke et al. 2022).

18.6 Somatic Embryogenesis

In vitro regeneration by somatic embryogenesis is a distinct developmental mechanism that produces embryos independent of gametic fusion. This system offers an unparalleled platform for studying the physiological, metabolic, cytological, and genetic regulation of higher plants at each cellular level (Mordhorst and Lorz 1993). The technology has numerous uses in crop improvement initiatives, particularly for producing true-to-type progeny by gene transfer and in vitro selection of mutant cells. Furthermore, embryogenic cell suspensions are useful sources for regenerating the protoplasts and somatic hybrids in commercially vital crops (Tao et al. 1997).

In establishing an efficient regeneration protocol for *C. quadrangularis*, Ramar et al. (2011) achieved high-frequency somatic embryo formation in suspension cultures supplemented with MS salts, B5 vitamins, 2,4-D (9.04 μM), kinetin (4.54 μM), and sucrose (30 g dm^{-3}). Despite this advancement, tissue culture studies on *Cissus quadrangularis* remain limited, with only a few documented reports available in the literature.

18.7 Hybridization

Wild species hybridization has been extensively utilized as a conventional approach for crop improvement programs. However, in the case of *Cissus*, conventional hybridization attempts among existing ecotypes remain largely unexplored. Sujata et al. (2022) conducted interspecific and intergeneric hybridization studies involving 20 species/cultivars from the Vitaceae family, encompassing genera such as *Ampelocissus, Vitis, Cayratia, Cissus*, and *Leea*, sourced from various locations within India and internationally. Despite obtaining 263 seeds from 21 cross combinations, *Cissus quadrangularis* exhibited reproductive incompatibility, failing to set seeds or produce viable hybrids when crossed with Vitis vinifera cultivars (var. Ribier, var. Cheemasahebi, and var. Gulabi). Therefore, advanced biotechnological approaches, including somatic hybridization, embryo rescue techniques, and gene expression studies, warrant investigation for effective utilization of wild germplasm and successful incorporation of desired traits into conventional *Cissus* species.

18.8 Conclusion and Prospects

Cissus remains an under-exploited plant species in terms of the very low volume of research conducted on its genetic potential, and screening of wild ecotypes to select ideal genotypes with high biomass combined with enriched nutraceuticals and phyto-pharmaceutical value. In contrast, high market demand for the raw material has made the species vulnerable to over-exploitation and utilization mainly from its wild habitat, leading to serious habitat depletion and loss of genetic wealth.

Systematic plant breeding in this species is lacking, which underlines the importance of the adoption of molecular approaches, DNA barcoding, DUS characterization of landraces, selection of parents, hybridization, mutation and other breeding methods to evolve chemotypes benefiting both the pharmaceutical industry and the growers. It is obvious to note that so far, no significant research leads have been obtained or mentioned in any forum, paving the way for varietal development in *Cissus*.

Cissus is a versatile CAM plant species that has the inherent ability to survive under adverse weather conditions. Plant breeders have to adopt modern approaches integrating speed breeding programs with genomic information, molecular profiling and high-throughput screening for isolation of superior genotypes that can be introduced in farmers' holdings to tap the growing global market. This will immensely serve the people who are suffering from bone disorders, fractures, and muscle damage, besides enabling small and marginal farmers who will have access to production and marketing by utilizing the right genotype/ variety/chemotype. In a nutshell, *Cissus* requires paramount research to achieve the desired results and at the same time, adoption of fast-track breeding methods is inevitable in this much-acclaimed medicinal plant of global utility.

Appendix 18.1 Research Institutes and Online Resources Available for Veldt Grape (*Cissus quadrangularis* L.) Breeding

Institute name	Specialization and research activities	Address	Contact information and website
Indian Institute of Technology, Guwahati	Genome Sequence and Bioinformatics Chloroplast genome of *Cissus quadrangularis* and molecular structure	Alok Senapati Department of Biotechnology Indian Institute of Technology Guwahati	Department of Biotechnology Indian Institute of Technology Guwahati https://www.iitg.ac.in/biotech/
Wuhan Botanical Garden, Chinese Academy of Sciences, Wuhan 430074, China	Genome sequencing	Haiping Xin Core Botanical Gardens/Wuhan Botanical Garden, Chinese Academy of Sciences, Wuhan 430,074, China	Wuhan Botanical Garden, Chinese Academy of Sciences, Wuhan 430074, China http://english.wbg.cas.cn/
CAS Key Laboratory of Plant Germplasm Enhancement and Specialty Agriculture	Taxonomy & Genome sequencing	Duncan Kiragu Gichuki, Wuhan Botanical Garden, The Innovative Academy of Seed Design, Chinese Academy of Sciences, Wuhan, China	
National Centre for Biological Sciences (TIFR), GKVK Campus, Bangalore, Karnataka, India	Draft genome sequencing	Gandhimathi, National Centre for Biological Sciences (TIFR), GKVK Campus, Bangalore, Karnataka, India	National Centre for Biological Sciences TATA Institute of Fundamental Research Bellary Road, Bangalore 560065 Ph: 080-23666001 Fax 91-80-23636662 Email: info@ncbs dot res dot in https://www.ncbs.res.in/

(continued)

Institute name	Specialization and research activities	Address	Contact information and website
Tamil Nadu Agricultural University, Coimbatore	Germplasm with 50 accessions, morphological, taxonomical, molecular characterization, clonal selection	Dr. S. Padmapriya Dr. K. Rajamani Department of Medicinal & Aromatic Crops Tamil Nadu Agricultural University Coimbatore- 641003 padmapriya.s@tnau.ac.in rajamani.k@tnau.ac.in	Department of Medicinal and Aromatic Crops Tamil Nadu Agricultural University Coimbatore – 641003 https://tnau.ac.in/site/hcri-coimbatore/department-of-ma-about-department/
Gujarat Biotechnology Research Centre	Molecular characterization, SDS-PAGE, Biotechnology, Cell Culture, DNA Extraction, and High-Performance Liquid Chromatography (HPLC)	Kinjal Sangani Gujarat Biotechnology Research Centre Gandhinagar, Gujarat, India	Gujarat Biotechnology Research Centre Gandhinagar, Gujarat, India https://gbrc.gujarat.gov.in
Khon Kaen University, Khon Kaen, Thailand	Genetic diversity, Molecular and chemical characterization	Department of Biology, Faculty of Science, Khon Kaen University, Khon Kaen, Thailand/Genetics and Environmental Toxicology (GET) Research Group, KhonKaen University, Khon Kaen, Thailand.	Department of Biology, Faculty of Science, Khon Kaen University Thailand https://sc.kku.ac.th/sciweb/en/about-dept.php

Appendix 18.2 Genetic Resources for Veldt Grape (*Cissus quadrangularis* L.) Breeding

Ecotypes	Village/town	District	State/country	Latitude	Longitude
TNCq 1	Kuniamuthur	Coimbatore	Tamil Nadu, India	10.957505	76.953535
TNCq 2	Nambhiyazhagan palayam	Coimbatore	Tamil Nadu, India	11.011699	76.891406
TNCq 3	Poluvampatti	Coimbatore	Tamil Nadu, India	10.882763	76.839177
TNCq 4	Kadukkanpalayam	Erode	Tamil Nadu, India	11.468791	77.447665

Ecotypes	Village/town	District	State/country	Latitude	Longitude
TNCq 5	Kumarapalayam	Erode	Tamil Nadu, India	11.437271	77.722959
TNCq 6	Krishnapuram	Perambalur	Tamil Nadu, India	11.381267	78.789883
TNCq 7	Nemizhi	Vellore	Tamil Nadu, India	12.599981	78.516298
TNCq 8	Keezhveeranam	Vellore	Tamil Nadu, India	12.984488	79.475448
TNCq 9	Marugattuthalai	Tirunelveli	Tamil Nadu, India	9.0591802	77.257606
TNCq 10	Savalapperi	Tirunelveli	Tamil Nadu, India	8.8222024	77.850258
TNCq 11	Vadakarai	Tirunelveli	Tamil Nadu, India	9.0406648	77.273598
TNCq 12	Puliyampatti	Tuticorin	Tamil Nadu, India	8.8300265	77.847758
TNCq 13	Alandha	Tuticorin	Tamil Nadu, India	8.8083056	77.857242
TNCq 14	N. Pudhur	Tuticorin	Tamil Nadu, India	9.2950398	78.136379
TNCq 15	Kangaikondan	Tuticorin	Tamil Nadu, India	8.8583092	77.782306
TNCq 16	Malaippatti	Tuticorin	Tamil Nadu, India	8.9356666	77.887258
TNCq 17	Kasalingapuram	Tuticorin	Tamil Nadu, India	8.7761819	77.887465
TNCq 18	Parayappatti	Dharmapuri	Tamil Nadu, India	12.134601	78.477108
TNCq 19	Sindhalpadi	Dharmapuri	Tamil Nadu, India	12.077550	78.358141
TNCq 20	Thandipadi	Tiruvannamalai	Tamil Nadu, India	12.108260	78.834596
TNCq 21	Thandarampet	Tiruvannamalai	Tamil Nadu, India	12.154465	78.948612
TNCq 22	Nallavanpalayam	Tiruvannamalai	Tamil Nadu, India	12.203409	79.046448
TNCq 23	Thakkalai	Kanyakumari	Tamil Nadu, India	8.2449166	77.322300
TNCq 24	Pollachi	Coimbatore	Tamil Nadu, India	10.661398	77.005016
TNCq 25	Thumpichavalayam	Tirupur	Tamil Nadu, India	11.123060	77.357972
TNCq 26	Kallippalayam	Tirupur	Tamil Nadu, India	10.915256	77.332280
TNCq 27	Sempatti	Dindigul	Tamil Nadu, India	10.281261	77.872413

(continued)

Ecotypes	Village/town	District	State/country	Latitude	Longitude
TNCq 28	Kannivadi	Dindigul	Tamil Nadu, India	10.382170	77.831178
TNCq 29	Periyakulam hills	Theni	Tamil Nadu, India	10.116354	77.544751
TNCq 30	Thevadhanampatti	Theni	Tamil Nadu, India	10.146388	77.642376
TNCq 31	Manjalar dam	Theni	Tamil Nadu, India	10.186436	77.624964
TNCq 32	Endapulippatti	Theni	Tamil Nadu, India	10.086147	77.576477
TNCq 33	Vennaimalai	Karur	Tamil Nadu, India	10.990210	78.073276
TNCq 34	Semmadi	Karur	Tamil Nadu, India	11.011164	78.068203
TNCq 35	Maruthappatti	Virudhunagar	Tamil Nadu, India	9.4598252	77.800493
TNCq 36	Marudhappatti east	Virudhunagar	Tamil Nadu, India	9.5734343	77.963502
TNCq 37	Marudhappatti west	Virudhunagar	Tamil Nadu, India	9.5351770	77.939990
TNCq 38	Chatrappatti east	Virudhunagar	Tamil Nadu, India	9.5629394	77.942394
TNCq 39	Chatrappatti west	Virudhunagar	Tamil Nadu, India	9.4169260	77.597587
TNCq 40	Peranayakkanpatti	Virudhunagar	Tamil Nadu, India	9.4211912	77.785922
TNCq 41	Madurai(anamalai) hills	Virudhunagar	Tamil Nadu, India	9.9685573	78.195364
TNCq 42	Kadampur	Perambalur	Tamil Nadu, India	11.425639	78.720852
TNCq 43	Kallappatti	Perambalur	Tamil Nadu, India	11.434620	78.723555
TNCq 44	Therkumedu	Salem	Tamil Nadu, India	11.488630	78.798563
TNCq 45	Veeraganur	Salem	Tamil Nadu, India	11.481054	78.733520
TNCq 46	Karpaganadhar kulam	Tiruvarur	Tamil Nadu, India	10.414491	79.637877
TNCq 47	Vaimedu	Nagappattinam	Tamil Nadu, India	10.413075	79.685595
TNCq 48	Kamlappatti	Dharmapuri	Tamil Nadu, India	12.244234	78.081856
TNCq 49	Nallampalayam	Tirupur	Tamil Nadu, India	10.728098	77.618002
TNCq 50	Nallampalayam	Tirupur	Tamil Nadu, India	10.73980	77.53610

References

Acevedo-Rodríguez P, Strong MT (2012) Catalogue of seed plants of the West Indies. Smithson Contrib Bot 98:1–1192

Adams NE, Collinson M, Smith S (2016) X-rays and virtual taphonomy resolve the first *Cissus* (Vitaceae) macrofossils from Africa as early-diverging members of the genus. Am J Bot 103(9):1657–1677

Adugna W, Labuschagne MT (2003) Parametric and nonparametric measures of phenotypic stability in linseed (*Linum usitatissimum* L.). Euphytica 129:211–218

Agnew ADQ, Agnew S (1994) Upland Kenya wild flowers, 2nd edn. East Africa Natural History Society, Nairobi

Anitha JR, Akila S, Velliyur KG, Sudarsanam D (2010) Antimicrobial profile of *Cissus quadrangularis*. J Herbal Med Toxicol 4:177–180

Anand Manju and Kaur Tejinder (2019) Effect of Growth Regulators on Callus Induction and Organ Formation in *Cissus quadrangularis* Linn. A Valuable Medicinal Plant. International Journal of Pharmaceutical and Clinical Research. 11(4):106–111

APG IV (2016) An update of the Angiosperm phylogeny group classification for the orders and families of flowering plants. APG IV Bot J Linn Soc 181:1–20. https://doi.org/10.1111/boj.12385

Asahina H, Shinozaki J, Masuda K (2010) Identification of medicinal *Dendrobium* species by phylogenetic analyses using *mat* K and *rbc*L sequences. J Nat Med 64:133–138

Aswar UM, Bhaskaran S, Mohan V, Bodhankar SL (2010) Estrogenic activity of friedelin rich fraction (IND-HE) separated from *Cissus quadrangularis* and its effect on female sexual function. Pharm Res 2:138–145

Bailey LH (1949) Manual of cultivated Plants. Mac Milan, New York, p 94

Baker JG (1871) Ampelidae. In: von Martius CFP, Oldenbourg R (eds) Flora brasiliensis, Munich & Leipzig, pp 197–220

Benxu, Cheng Yunlin, Wei Lili, Guerra Rozena, Shirvani-Arani Santiago, Balderas Laura, Valdez Andrew, Tsin Xiaoqian, Fang (2023) Anti-cancer effect of*Cissus quadrangularis*on human glioblastoma cells Abstract RPS Pharmacy and Pharmacology Reports 2(2)

Barghini E, Mascagni F, Natali L (2015) Analysis of the repetitive component and retrotransposon population in the genome of a marine angiosperm, *Posidonia oceanica* (L.) Delile. Mar Genomics 24(3):397–404

Bennetzen JL, Coleman C, Liu R (2004) Consistent over-estimation of gene number in complex plant genomes. Curr Opin Plant Biol 7(6):732–736

Brink M, Achigan-Dako EG (2012) Plant resources of tropical Africa 16: Fibres, Plant resources of tropical Africa, No. 16, PROTA Foundation

Buddhadev S (2014) A review update on plant *Cissus quadrangularis* L. Punarnav 2:1–10

Bum EN, Ngoupaye GT, Talla E (2008) The anticonvulsant and sedative properties of stems of *Cissus quadrangularis* in mice. Afr J Pharm Pharmacol 2:42–47

Chaveerach A, Tanee T, Sudmoon R (2011) Molecular identification and barcodes for the genus *Nymphaea*. Acta Biol Hung 62:328–340

Chen ZD, Ren H, Wen J (2007) Vitaceae. In: Wu ZY, Hong DY, Raven PH (eds) Flora of China, vol 12. Science Press/Botanical Garden Press, Beijing/St. Louis, pp 173–222

Cheng B, Wei Y, Guerra L (2023) Anti-cancer effect of *Cissus quadrangularis* on human glioblastoma cells. Pharm Pharmacol Rep 2(2):1–10

Chidambara Murthy KN, Vanitha A, Mahadeva Swamy M (2003) Antioxidant and antimicrobial activity of *Cissus quadrangularis* L. J Med Food 6(2):99–105

Chidambaram J, Carani Venkatraman A (2010) *Cissus quadrangularis* stem alleviates insulin resistance, oxidative injury and fatty liver disease in rats fed high fat plus fructose diet. Food Chem Toxicol 48:2021–2029

Choi HW, Kim JS, Lee SH, Bang JW (2008) Physical mapping by FISH and GISH of rDNA loci and discrimination of genomes A and B in *Scilla scilloides* complex distributed in Korea. J Plant Biol 51(6):408–412

Cholke PB, Bhalerao DV, Shete AM, Bhor AK (2022) The effect of $CaCl_2$ induced stress on callus of *in-vitro* propagated *Cissus quadrangularis*. Linn Shodhasamitha 9(1):124–128

Chu Z, Wen J, Yang YP (2018) Genome size variation and evolution in the grape family Vitaceae: genome size variation in Vitaceae. J Syst Evol 56:273–282

Debabrat Baishya and Sarma (2005) In Vitro Culture of *Vitis quadrangularis* L. (SYN. *Cissus quandrangularis* L.) - A medicinal plant (2005). Asian Journal of Microbiology, Biotechnology & Environmental Sciences Paper. 7(3):419–420

Dellaporta SL, Wood J, Hicks JB (1983) A plant DNA minipreparation version II. Plant Mol Biol Report 1:19–21

Derocher AE, Stirling I (1990) Distribution of polar bears (*Ursus maritimus*) during the ice-free period in western Hudson Bay. Can J Zool 68(7):1395–1403

Dolezel J, Bartos J, Voglmayr H, Greilhuber J (2003) Nuclear DNA content and genome size of trout and human. Cytometry A51A(2):127–128. https://doi.org/10.1002/cyto.a.10013

Doyle JJ, Doyle JL (1987) A rapid DNA isolation procedure for small quantities of fresh leaf tissue. Taxon 36:715–722

Du J, Tian Z, Hans CS (2010) Evolutionary conservation, diversity and specificity of LTR-retrotransposons in flowering plants: insights from genome-wide analysis and multi-specific comparison. Plant J Cell Mol Biol 63(4):584–598

Franco OH, Chowdhury R, Troup J (2016) Use of plant-based therapies and menopausal symptoms: a systematic review and meta-analysis. JAMA 315(23):2554–2563. https://doi.org/10.1001/jama.2016.8012

Gandhimathi A, Sunitha SM, Gosh P (2016) Draft genome of *Cissus quadrangularis* to elucidate the medicinal values. Herb Med 2(1):7–9

Garg P, Malik CP (2012) Multiple shoot formation and efficient root induction in *Cissus quadrangularis*. Inter J Pharm Clin Res 4(1):4–10

Gerrath JM, Wilson T, Posluszny U (2004) Morphological and anatomical development in the Vitaceae. VII. Floral development in *Rhoicissus digitata* with respect to other genera in the family. Can J Bot 82:198–206

Gerrath J, Posluszny U (2017) Inflorescence 10 morphology and development in the basal rosid lineage vitales. J Syst Evol 55(11):542–558

Ghaffariyan S, Mohammadi SA, Aharizad S (2012) DNA isolation protocol for the medicinal plant lemon balm (*Melissa officinalis*, Lamiaceae). Genet Mol Res 11(2):1049–1057

Gichuki DK, Ma L, Zhu Z (2019) Genome size, chromosome number determination, and analysis of the repetitive elements in *Cissus quadrangularis*. Peer J 7:e8201. https://doi.org/10.7717/peerj.8201

Gituru RW, Wang QF, Wang Y, Guo YH (2001) A taxonomic investigation of variation within *Cissus quadrangularis* L. (Vitaceae) in Kenya. Wuhan Univ J Nat Sci 6(3):715–724

Gostimsky S, Kokaeva Z, Konovalov F (2005) Studying plant genome variation using molecular markers. Russian J Genet 41(4):378–388

Graf AB (1963) Exotica 3: pictorial cyclopedia of exotic plants. Rutherford, New Jersy, p 46

Greenway FL, Bray GA (2010) Combination drugs for treating obesity. Curr Diabet Rep 10:108–115

Haga T, Noda S (1976) Cytogenetics of the *Scilla scilloides* complex. Genetica 46(2):161–176

Haiping X, Wang Y, Li Q (2022) A genome for *Cissus* illustrates features underlying its evolutionary success in dry savannas. Hort Res 9:uhac208

Hari, Sasidharan Anto Puthur, Varghese (2016) Taxonomy of Selected Species of *Cissus* (Vitaceae) from Thrissur District South Indian Journal of Biological Sciences 2(1) 222.10.22205/sijbs/2016/v2/i1/100403

Hasani-Ranjbar S, Nayebi N, Larijani B, Abdollahi M (2009) A systematic review of the efficacy and safety of herbal medicines used in the treatment of obesity. World J Gastroenterol 15:30733085

Hooker JD (1862) Ampelidae. In: Bentham G, Hooker JD (eds) Genera plantarum. L Reeve and Co, London, pp 386–388

Husband BC, Baldwin SJ, Suda J (2013) The incidence of polyploidy in natural plant populations: major patterns and evolutionary processes. In: Greilhuber J, Dolezel J, Wendel J (eds) Plant genome diversity, vol 2. Springer, Vienna. https://doi.org/10.1007/978-3-7091-1160-4_16

India Biodiversity Portal (2018) Online portal of India biodiversity. In: Online Portal of India Biodiversity. http://indiabiodiversity.org/species/list

Indran S, Raj ER, Sreenivasan VS (2014) Characterization of new natural cellulosic fiber from *Cissus quadrangularis* root. Carbohydr Polym 110:423–429

Ingrouille MJ, Chase MW, Fay MF (2002) Systematics of Vitaceae from the viewpoint of plastid rbcL DNA sequence data. Bot J Linn Soc 138:421–432

Islam B, Ivy N, Rasul M, Zakaria M (2010) Character association and path analysis of exotic tomato (*Solanum lycopersicum* L.) genotypes. Bangladesh J Plant Breed Genet 23(1):13–18

Jackes BR (1984) Revision of the Australian Vitaceae, 1. Ampelocissus Planchon. Austrobaileya 2:81–86

Jacobsen H (1960) Handbook of succulent plants: descriptions, synonyms, and cultural details for succulents other than Cactaceae, vol 1. Blandford Press, Dorset

Jainu M, Vijai MK, Devi CS (2006) Gastroprotective effect of *Cissus quadrangularis* extract in rats with experimentally induced ulcer. Indian J Med Res 123:799806

Joseph B, George J (2013) *Cissus quadrangularis* in the treatment of osteoporosis. World J Pharma Res 2:596–605

Karkamkar S, Patil SG, Misra SC (2010) Cyto-morphological studies and their significance in the evolution of family Vitaceae. Nucleus 53(1–2):37–43

Khaleefathullah SA, Kathiravana G (2010) Antiviral activity of ancient system of ayurvedic medicinal plant *Cissus quadrangularis* L. J Basic Clin Pharm 1:3740

Khan SS, Singh MP, Chaghtai SA (1991) Ethnomedicobotany of *Cissus quadrangularis* L. Orient J Chem 7:1702

Kidwell MG (2002). Transposable elements and the evolution of genome size in eukaryotes. Genetica 115(1):49–63

Kritpetcharat O, Kritpetcharat P, Daduang J et al (2011) Using DNA markers and barcoding to solve the common problem of identifying dried medicinal plants with the examples of *Smilax* and *Cissus* in Thailand. J Med Plants Res 5(15):3480–3487

Kumar M, Rawat P, Dixit P (2010) Anti-osteoporotic constituents from Indian medicinal plants. Phytomed 17:993–999

Kumbhojkar MS, Kulkarni DK, Upadhye AS (1991) Ethnobotany of *Cissus quadrangularis* L. from India. Ethnobot Res App 13:215

Lavie P (1970) Contribution a l'etudecaryo systematique des Vitacees. Ph.D. thesis, Faculte´ des Sciences de Montpellier, Universite´ de Montpellier

Lawrence GHM (1951) Taxonomy of vascular plants. Macmillan Co, New York

Lawson MA (1875) Ampelidae. In: Hooker JD (ed) Flora of British India, L, vol 1. Reeve and Co, London, pp 644–668

Li SF, Su T, Cheng GQ (2017) Chromosome evolution in connection with repetitive sequences and epigenetics in plants. Gene 8(10):E290

Li Q, Wang Y, Zhou H (2024) The *Cissus quadrangularis* genome reveals its adaptive features in arid habitat. Hort Res 11:uhae038

Linnaeus C (1753) Species plantarum, exhibentesplantas rite cognitas, ad genera relatas, cum differentiis specificis, nominibustrivialibus, synonymisselectis, locisnatalibus, secundum systema sexual edigestas. Tomus II. Laurentii Salvii, Stockholm, Sweden

Liu Y, Yan HF, Cao T, Ge XJ (2010) Evaluation of 10 plant barcodes in Bryophyta (Mosses). J Syst Evol 48:36–46

Liu XQ, Ickert-Bond SM, Chen LQ, Wen J (2013) Molecular phylogeny of *Cissus* L. of Vitaceae (the grape family) and evolution of its pantropical intercontinental disjunctions. Mol Phylogenet Evol 66(1):43–53

Lodhi MA, Daly MJ, Ye GN (1995) A molecular marker based linkage map of Vitis. Genome 38:786–794

Luseba D, Elgorashi EE, Ntloedibe DT, Staden JV (2007) Antibacterial, anti-inflammatory and mutagenic effects of some medicinal plants used in South Africa for the treatment of wounds and retained placenta in livestock. South Afr J Bot 73:378383

Macas J, Novák P, Pellicer J (2015) In Depth characterization of repetitive DNA in 23 plant genomes reveals sources of genome size variation in the legume tribe Fabeae. PLoS One 10(11):e0143424.

Manchester SR, Kapgate DR, Wen J (2013) Oldest fruits of the grape family (Vitaceae) from the late cretaceous deccan cherts of India. Am J Bot 100:1849–1859

Manimekalai G, Kavitha S, Divya D (2021) Characterization of enzyme treated cellulosic stem fiber from *Cissus quadrangularis* plant: an exploratory investigation. Curr Res Green Sustain Chem 4:100162

Manju A, Tejinder K (2019) Effect of growth regulators on callus induction and organ formation in *Cissus quadrangularis* Linn. a valuable medicinal plant. Int J Pharma Clin Res 11(4):106–111

Manokari M, Mahipal S, Shekawhat S (2019) An updated review on *Cissus vitiginea* L. (Family: Vitaceae)- an important medicinal climber. World News Nat Sci 22:75–83

Masinde PS (1992) Taxonomic studies on the *Cissus quadrangularis* L. complex in Kenya. M.Sc. thesis, Kenya University, Kenya

Mayandi K, Rajini N, Pitchipoo P, Sreenivasan VS, Winowlin Jappes JT, Alavudeen A (2015) A comparative study on characterizations of *Cissus quadrangularis* and *Phoenix reclinata* natural fibers. J Reinf Plast Compos 34(4):269–280

Merinal S, Viji SB (2012) In vitro screening of antimicrobial potentials of *Cissus quadrangularis* L. Asian J Plant Sci Res 2:58–62

Michael TP, VanBuren R (2015) Progress, challenges and the future of crop genomes. Curr Opin Plant Biol 24:71–81

Molina JE, Wen J, Struwe L (2013) Systematics and biogeography of the non-viny grape relative Leea (Vitaceae). Bot J Linn Soc 171:354–376

Monkheang P, Sudmoon R, Tanee T (2011) Species diversity, usages, molecular markers and barcode of medicinal *Senna* species (Fabaceae, Caesalpinioideae) in Thailand. J Med Plants Res 5:6173–6181

Mordhorst AP, Lorz H (1993) Embryogenesis and development of isolated barley (*Hordeum vulgare* L.) microspores are influenced by the amount and composition of nitrogen sources in culture media. Plant Physiol 142:485–492

Morgante M, Caboche M, Adam-Blondon AF et al (2007) The grapevine genome sequence suggests ancestral hexaploidization in major angiosperm phyla. Nature 449(7161):463–467

Nagori K, Sharma M, Agrawal A (2011) General awareness on allopathic, ayurvedic and homeopathic system of medicine in Chhattisgarh, India. Int J Pharm Pharm Sci 3:159–162

Nalini G, Vinoth PV, Chidambaranathan N, Jeyasundari K (2011) Evaluation of Anti-tumour activity of *Cissus quadrangularis* L. against Dalton's ascitic lymphoma and Erlich ascetic induced carcinoma in mice. Int J Pharm Sci Rev Res 8:7579

Nzengue E, Zinga-Koumba CR, Iponga DM (2015) Integrated pest control trial against the proliferation of *Cissus quadrangularis* L. (Vitaceae) in the National Park (Senegal). J Appld Bioscien 91:8529–8538

Oben J, Kuate D, Agbor G, Momo C, Talla X (2016) The use of a *Cissus quadrangularis* formulation in the management of weight loss and metabolic syndrome. Lipids Health Dis 5:24–29

Onyeweaku GC, Nyananyo BL, Ozimede CO (2020) Taxonomic studies on the genus *Cissus* L. (Vitaceae) present in Obio/Akpor local government area of Rivers state, Nigeria. J Appl Sci Environ Manage 24(1):139–145

Padmapriya S, Vinoth K, Rajamani K (2022) Molecular characterization of *Cissus quadrangularis* using RAPD-PCR technique. Med Plants 14(2):251–257
Panthong A, Supraditaporn W, Kanjanapothi D (2007) Analgesic, anti-inflammatory and venotonic effects of *Cissus quadrangularis* Linn. J Ethnopharmacol 110:264270
Paterson A, Bowers J, Bruggmann R (2009) The *Sorghum bicolor* genome and the diversification of grasses. Nature 457:551–556
Planchon JE (1887) Monographie des Ampe'lide'esvrais. In: De Candolle AFPP, De Candolle C (eds) Monographiae phanaerogamarum, vol 5. G. Masson, Paris, pp 305–654
Poonam Garg and Malik CP (2012) Multiple Shoot Formation and Efficient Root Induction in *Cissus quadrangularis*. International Journal of Pharmaceutical and Clinical Research. 4(1):4–10
Potu BK, Rao MS, Nampurath GK (2009) Evidence-based assessment of antiosteoporotic activity of petroleum-ether extract of *Cissus quadrangularis* Linn. on ovariectomy-induced osteoporosis. Ups J Med Sci 114:140–148
Prasad GP, Pratap GP, Marimuthu S (2020) DNA barcoding markers and next-generation sequence analysis of Rbcl in intraspecies genetic variations of three varieties of *Cissus quadrangularis* L. J Nat Remedies 20(2):103–109. https://doi.org/10.18311/jnr/2020/24387
Prema R, Sekar DS, Sekhar KB, Jeevanandham S (2012) In vitro cytotoxicity study on combined plants extracts (*Cissus quadrangularis* and *Aegle marmelos*). Euro J Exp Bio 2:882288
PROSEA (2018) Plant resources of South-East Asia. Bogor, Indonesia, PROSEA Foundation. http://proseanet.org/prosea/e-prosea.php
Raghavan RS (1957) Chromosome numbers in Indian medicinal plants. Proc Indian Acad Sci Sect B45(6):294–298
Rajan S, Thirunalasundari T, Jeeva S (2011) Anti-enteric bacterial activity and phytochemical analysis of the seed kernel extract of *Mangifera indica* Linnaeus against *Shigella dysenteriae* (Shiga, corrig.) Castellani and Chalmers. Asian Pac J Trop Med 4(4):294–300
Rajpal V (2008) Standardization of botanical testing and extraction methods of medicinal herbs, vol I. Eastern Publishers, New Delhi
Ramar K, Ganesan M, Lakshmi Prabha A, Nandagopalan V (2011) *In vitro* clonal propagation of wild *Cissus quadrangularis* by suspension culture mediated somatic embryogenesis. Inter J Appl Biotechnol Biochem 1(1):1–14
Ren H, Lu LM, Soejima A (2011) Phylogenetic analysis of the grape family (Vitaceae) based on the noncoding plastid trnC-petN, trnH-psbA, and trnL-F sequences. Taxon 60:629–637
Ridsdale CE (1974) A revision of the family Leeaceae. Blumea 22:57–100
Robert GW, Qing-feng W, Yong W, You-hao G (2001) A taxonomic investigation of variation within *Cissus quadrangularis* L. (Vitaceae) in Kenya. Wuhan Univ J Natrl Sci 6(3):715–724
Rossetto M, Jackes BR, Scott KD, Henry RJ (2001) Intergeneric relationships in the Australian Vitaceae: new evidence from cpDNA analysis. Genet Resour Crop Evol 48:307–314
Rossetto M, Jackes BR, Scott KD, Henry RJ (2002a) Is the genus *Cissus* (Vitaceae) monophyletic? Evidence from plastid and nuclear ribosomal DNA. Syst Bot 27:522–533
Rossetto M, McNally J, Henry RJ (2002b) Evaluating the potential of SSR flanking regions for examining taxonomic relationships in the Vitaceae. Theor Appl Genet 104:61–66
Rowley GD (1970) *Cissus* and *Cyphostema*. A short review of succulent vitaceae with a checklist of Names of Species. Cact Succ J Gt Britain 32:30–32
Runglawan Sudmoon, Arunrat Chaveerach, Tawatchai Tanee (2016) Analysis of genetics and chemical contents relation compared to commonly used *Cissus quadrangularis* L. and barcode markers of some Thailand Cissus species. Pak J Pharm Science. 29(1):65–75
Sangani K (2017) Optimization of DNA isolation and analysis of genetic variability in medicinal plant Cissus by RAPD-PCR technique. Inter J Appl Nat Sci 6(5):15–22
Schaal B, Leverich WJ, Rogstad S (1991) Comparison of methods for assessing genetic variation in plant conservation biology. Genet Conserv Rare Plants 123:134

Senapati A, Chetri BK, Mitra S (2023) Decoding the complete genome of *Cissus quadrangularis*: insights into molecular structure, comparative genome analysis and mining of mutational hotspot regions. Physiol Mol Biol Plants 29:709–724

Sessegolo C, Burlet N, Haudry A (2016) Strong phylogenetic inertia on genome size and transposable element content among 26 species of flies. Biol Lett 12(8):20160407

Sharma P, Joshi N, Sharma A (2010) Isolation of genomic DNA from medicinal plants without liquid nitrogen. Ind J Exp Biol 48(6):610–614

Sharma N, Nathawat RS, Gour K, Patni V (2011) Establishment of callus tissue and effect of growth regulators on enhanced sterol production in *Cissus quadrangularis* L. Inter J Pharmacol 7(5):653–658

Shirwaikar A, Khan S, Malini S (2003) Antisteoporotic effect of ethanol extract of *Cissus quadrangularis* Linn. on ovariectomized rat. J Ethnopharmacol 89:24550–24250

Sibi G, Zaki S, Malathi R, Latha V (2020) A review on efficacy of cissus quadrangularis in pharmacological mechanisms. Int J Clin Microbiol Biochem Technol 3:49–53

Singh G, Rawat P, Maurya R (2007) Constituents of *Cissus quadrangularis*. Nat Prod Res 21(6):522–828

Singh A, Malhotra S, Subban R (2008) Lichens-role in traditional medicine and drug discovery. Int J Integ Biol 3:58

Siripiyasing P, Kaenratana K, Mokkamul P (2012) DNA barcoding of *Cymbidium* species (Orchidaceae) in Thailand. Afr J Agric Res 7:393–400

Soejima A, Wen J (2006) Phylogenetic analysis of the grape family (Vitaceae) based on three chloroplast markers. Am J Bot 93:278–287

Soltis DE, Soltis PS, Bennett MD, Leitch IJ (2003) Evolution of genome size in the angiosperms. Am J Bot 90(11):1596–1603

Spiegelman BM, Flier JS (2001) Obesity and the regulation of energy balance. Cell 104:531–543

Staples GW, Herbst DR, Imada CT (2000) Survey of invasive or potentially invasive cultivated plants in Hawaii. Bishop Museum Occasional Papers 65:1–31

Stevens PF (2017) Angiosperm phylogeny. Missouri Botanical Garden, St. Louis

Sujata Tetali, Karkamkar SP and Phalake SV (2022) Intergeneric and interspecific crossing in Vitaceae: an attempt for disease resistant types. International Journal of Minor Fruits, Medicinal and Aromatic Plants. 8(1):12–20Tao W, Liu B, Xing M (1997) Establishing Japonica rice suspensions retaining a high regeneration potential after 14 months of culture. Plant Cell Tissue Org Cult 47:213–216

Timmons SA, Posluszny U, Gerrath JM (2007) Morphological and anatomical development in the Vitaceae. X. Comparative ontogeny and phylogenetic implications of *Cissus quadrangularis* L. Can J Bot 85(9):860–872

Trias-Blasi A, Parnell JAN, Hodkinson TR (2013) Multi-gene region phylogenetic analysis of the grape family (Vitaceae). Syst Bot 37:941–950

Udupa KN, Chaturvedi GN, Tripathi SN (1970) Advances in research in Indian medicine. Bot Med 12:165–196

Vinoth K, Padmapriya S, Rajamani K, Gnanam R (2021) Genetic variability, correlation and path coefficient analysis in Veldt grape (*Cissus quadrangularis* Linn.) ecotypes. J Pharma Phytochem 10(1):2342–2346

Viswanatha Swamy AHM, Thippeswamy AHM, Manjula DV, Mahendra Kumar CB (2006) Some neuropharmacological effects of the methanolic root extract of *Cissus quadrangularis* in mice. Afr J Biomed Res 9:69–75

Viswanatha SA, Kulkarni RV, Thippeswamy AH (2010) Evaluation of hepatoprotective activity of *Cissus quadrangularis* stem extract against isoniazid-induced liver damage in rats. Indian J Pharmacol 42:397400

Warrier PK, Nambiar VPK, Ramankutty C (1993) Indian medicinal plants: A compendium of 500 species, vol 2, 1st edn. Orient Blackswan, Hyderabad, pp 112–115

Wen J (2007) Vitaceae. In: Kubitzki D (ed) The families and genera of vascular plants, vol 9. Springer, Berlin, pp 467–479

Wen J, Nie ZL, Soejima A, Meng Y (2007) Phylogeny of Vitaceae based on the nuclear GAI1 gene sequences. Canadian J Bot 85:731–745
Wen J, Lu LM, Liu XQ (2008) A new phylogenetic tribal classification of the grape family (Vitaceae). J Syst Evolt 56:262–272
Wicker T, Bharti AK, Chapman J et al (2009) The Sorghum bicolor genome and the diversification of grasses. Nature 457(7229):551–556
Wilson JB (1991) A comparison of biogeographic models: migration, vicariance and panbiogeography. Global Ecol Biogeo Lett 1(3):84–87
Yeo CK, Ang WF, Lok AFSL, Ong KH (2012) Conservation status of *Cissus* L. (Vitaceae) of Singapore: with a special note on *Cissus repens* Lam. Nature in Singapore 5:319–330
Zhang N, Wen J, Zimmer EA (2015) Expression patterns of AP1, FUL, FT and LEAFY orthologs in Vitaceae support the homology of tendrils and inflorescences throughout the grape family. J Syst Evol 53:469–476
Zhang N, Wen J, Zimmer EA (2016) Another look at the phylogenetic position of the grape order Vitales: Chloroplast phylogenomics with an expanded sampling of key lineages. Mol Phylogen Evol 101:216–223
Zuccolo A, Sebastian A, Talag J (2007) Transposable element distribution, abundance, and role in genome size variation in the genus *Oryza*. BMC Evol Biol 7:152

Chapter 19
Advancements in Plant Breeding and Biotechnology for Breeding Strategies of Betel Nut (*Areca catechu* L.)

Osama Alam, Shahab Alam, Ahmad Ali, Farkhanda Naz, Farah Kamal, Shareef Gul, Hameed Gul, and Shahid Ullah Khan

Abstract The relevance of betel nut (*Areca catechu* L.) can be traced from ancient Sanskrit literature. India leads in growing betel nut, covering most of the world's land and production. Major areca growing regions in India include Kerala, Karnataka, and Assam. These regions account for 90% of the total land area and

O. Alam
Department of Biotechnology, University of Science & Technology Bannu, Khyber Pakhtunkhwa, Pakistan

S. Alam
Department of Chemistry, University of Science & Technology Bannu, Khyber Pakhtunkhwa, Pakistan

A. Ali · H. Gul
College of Agronomy and Biotechnology, Southwest University, Chongqing, People's Republic of China

F. Naz
Biological Science Research Center, Academy for Advanced Interdisciplinary Studies, Southwest University, Chongqing, China
e-mail: drnazbiology@swu.edu.cn

F. Kamal
Chongqing Engineering Research Center for Rapeseed, College of Agronomy and Biotechnology, Southwest University, Chongqing, China

S. Gul
Integrative Science Center of Germplasm Creation in Western China (CHONGQING) Science City and Southwest University, Chongqing, People's Republic of China

S. U. Khan (✉)
Integrative Science Center of Germplasm Creation in Western China (CHONGQING) Science City and Southwest University, College of Agronomy and Biotechnology, Southwest University, Chongqing, People's Republic of China

Department of Biochemistry, Women Medical College, Khyber Medical University, KPK, Pakistan
e-mail: shahidbiochem@wmc.edu.pk

K. F. M. Salem et al. (eds.), *Breeding and Biotechnology of Grass and Bast Fiber Crops*, Advances in Plant Breeding Strategies 12,
https://doi.org/10.1007/978-3-032-00403-1_19

95% of the overall production of betel nut. It is an economically important crop in countries like Pakistan and India while in Micronesia and Fiji, betel nut surpasses the local production and annual importation for cultural considerations. This chapter provides overview of the taxonomy, varieties, and the production pattern of betel nut. India's production has grown a great deal lately, making up a large share of the world's total. Betel Nut varieties are based on color such as red nuts has high priced. Our aim is to describe detailed overview of betel nut; focusing on lab-based growing methods and breeding ways that can greatly help farming. Effective techniques for development from somatic cells have been developed, and genetic fidelity has been investigated using molecular markers, particularly RAPD markers. The various cropping methods, irrigation needs, and appropriate field conditions, as well as the different ways the betel nut is put to use are discussed. All this research helps in bringing about better sustainable agriculture through better knowledge of the biology and cultivation practices of the betel nut.

Keywords Betel nut production · *Areca catechu* · Cultural importance · Economic value · Historical significance

19.1 Introduction

Betel Nut (*Areca catechu* L.), commonly known as betel nut, is a tall palm native to the humid tropics of South Asia and parts of East Africa. It dates back to when human civilization's development can be tied to it (Raju et al. 2022). Although traditionally consumed for its hallucinogenic seed, today the betel nut chewed along with the betel leaf is regarded as a major source of natural fiber (Papke et al. 2020). This palm, taking up large tracts of land, bears a fibrous nut from a single thin trunk that is encircled by furry fronds. The nut, being a mild stimulant, had been taken for ages together with lime paste and betel leaf in many Asian countries, which nevertheless carries more significant health hazards such as oral cancer (Bulwayan et al. 2023).

The real promise of betel nuts lies in their strong fibers, which surpass their traditional significance. During the extraction process, the lengthy, robust strands are revealed after removing the fibrous shell covering the nut. These fibers can be used to produce carpets, ropes, and fine textiles (Sunny and Palani Rajan 2024). They are durable and resistant due to their high cellulose and lignin content, making them an eco-friendly alternative to materials like cotton and jute (Mariamma and Nair 2022). The relatively new betel nut fiber industry is rapidly gaining momentum. Compared to mature industries like cotton, the betel nut fiber industry is still emerging but catching up rapidly. Proper perennial with superb soil and decent moisture is needed (Arifuzzaman Khan et al. 2012).

Not only do skilled harvesting methods require extracting the fiber from both nut and palm; Improvements to fiber extraction and spinning processes are one area of continued development. There is a growing global market for betel nut fiber, especially in countries looking to ethically source their materials sustainably. This industry sustainability requires taking a long-term view of the subject by attending to environmental concerns and issues on ethical labor (Chakrabarty et al. 2012; Jayamani et al. 2014).

Both genetic and archaeological studies point to Southeast Asia as the cradle of the oldest-known poultry species. Palms may have diverged from the grass family and over millennia, farmers artificially selected palms favoring traits such as larger nuts with higher quality fibers (Minford 2021). To realize the sustainability potential of betel nut as a crop, one must appreciate its range of characteristics, from cultural importance to health risks, unique fiber properties before use through evolving industry and fascinating domestication history. An excellent example is the betel nut, entrenched in cultural traditions but making sustainability advances by overcoming problems and capitalizing on opportunities (Gunjal et al. 2020).

The betel nut holds great cultural and social value in many regions. Chewing betel nut, often in conjunction with additives such as lime or tobacco, is a long tradition in many countries of East and Southeast Asia. Socially, it is the way with which allows them to be involved in social interactions and culturally it forms a part of social events or ceremonies that help these people celebrate important moments in their lives such as Marriages, anniversaries and religious days (Singh et al. 2020). The industry supports those in the distribution, processing, harvesting and cultivation of betel nuts. This sector provides economic stability and sustenance for whole communities in rural areas with high concentrations of betel nut palms (Dwiartama et al. 2024). Betel nut cultivation supports socio-economic growth by creating employment for laborers and farmers (Mondal 2022).

The betel nut industry is full of its own cultural as well as economic essence but still faces challenges. Chewing betel nuts also effect dental health (Sathishkumar et al. 2021). Regular use of betel nuts has been associated with several serious health issues, including tooth decay, gum disease and mouth cancer. Environmentalists and conservationists are also worried about the environmental effects of betel nut production such as denudation, deforestation along soil degradation (Al-Rmalli et al. 2011).

The solutions are complex and involve an important environmental sustainability, and the safeguarding of cultural heritage. Programs raise public consciousness concerning the health hazards of chewing betel nuts and allow choices among people while consuming/biting the betel nut (Garg et al. 2014). Adopting agro-sustainable practices along with conservation strategies will help to reduce the environmental pollution due to betel nut farming and also conserve natural ecosystems, which in turn ensures the sustained viability of this industry (Kumar et al. 2021).

This chapter discusses the importance and relevance of betel nut in historical, cultural, and economic applications. Dating back to ancient Sanskrit writings, India has a major role in betel nut production at the global level particularly Kerala, Karnataka and Assam accounting for more significant contributions. In cooperating countries, betel nuts are economically significant in Pakistan and India While its cultural importance can be found in Pacific islands like Micronesia, and Fiji. The

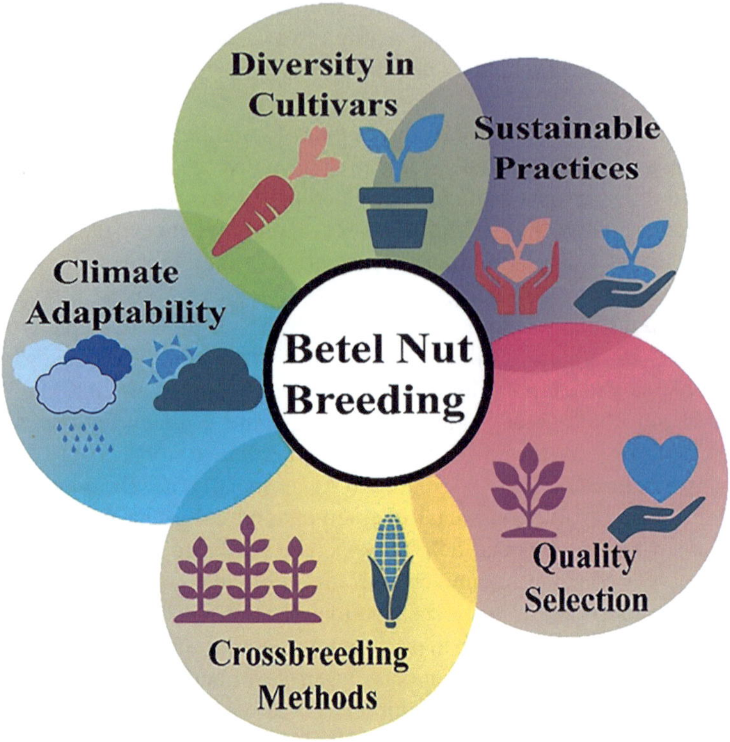

Fig. 19.1 Various betel nut breeding strategies. (The figure is designed by Mr. Osama Alam through Adobe illustrator)

chapter focuses on the taxonomy, diversity and cultivation status of *A. catechu* L. It includes an elaborative outline regarding in vitro culture methods as well as breeding approaches such as Somatic embryogenesis. The RAPD markers to ascertain genetic fidelity have further been discussed in this manuscript. The strength of the study is that it covers ideal growing conditions, water use efficiency, cropping methods along multiple uses of betel nuts. (Fig. 19.1).

19.1.1 Origin, History, and Geographical Distribution

The exact origin of the betel nut remains unclear. Although the *Areca* genus lacks direct fossil records, fossils of closely related genera suggest its existence during the tertiary period (Luong et al. 2023). The greatest diversity is represented by 24 species, with evidence pointing to regions like Borneo, Celebes, and Malaya as possible original habitats (Bavappa 1963). Ancient Sanskrit scriptures frequently reference the betel nut palm, its nuts, and their various applications. Notable mentions, such as those in the Anjana Charitra, and the renowned Indian scholar Vagbhatta documented its medicinal properties around 500 AD (Nair 2009). These references indicate clusters of betel nut palms fringed with bunches and inflorescences, presenting a scenic view. Several archaeological evidences support the prehistoric application of betel nuts, but the exact period in regards to their advent into the Indian subcontinent remains vague. A renowned cave in central India bearing paintings of the Padmapani Buddha on an betel nut palm is dated between 200 BC and 900 AD. The available literature traces it back to 1510 AD (Nair 2010). Since the time of Aryans, when ancient immigrants to India were believed to have come from Europe, betel nuts formed an important part of Hindu chewing rituals and important religious rites (Majumdar 2016).

India leads the world in land coverage under betel nut, accounting for 57% of global land and producing 53% of the entire crop. The next leading countries after India are Myanmar and Bangladesh; the other important contributors include Indonesia, Malaysia, Thailand, the Philippines, and Vietnam. (Majumdar 2016) China leads high productivity and is followed by Malaysia, Thailand, and India respectively. Kerala, Karnataka, and Assam together cover 90% of India's cultivated betel nut. The areas where betel nut is mainly grown include the Western Ghats and the northeastern regions of India, with the highest productive state being Maharashtra (Nair 2021).

Chewing betel nuts for mild stimulation and breath freshening is a cherished custom on Pacific islands like Micronesia and Fiji. In regions like Guam, wild betel nuts are locally dispersed, but as traditional harvesting declines, commercial demand grows, creating lucrative opportunities for home gardeners and farmers by (Lim et al 2021). Significant annual imports highlight the commercial value in areas like Pakistan and India when local production falls short. For successful betel nut farming, rainfall must be evenly distributed and temperatures moderate. Although the stimulant qualities of betel nut are not widely accepted globally (Arora and Squier 2019).

19.1.2 Structure and Classification of Areca Plant

Betel Nut is native to East Africa, South Asia, and the Pacific Islands in areas of tropical with annual rainfall between 60 and 200 inches at elevations from sea level up to nearly 3000 ft. This diminutive palm usually grows highly.

The primary method of propagation for Betel Nut is through seeds. These seeds are also utilized in agroforestry techniques for purposes such as crop shading and home gardens. The versatility and adaptability of the Areca palm make it a valuable component in diverse agricultural systems. Further, many intercropping systems have betel nuts as an ingredient for the betterment of short- and long-term agricultural activities. Although it has the potential to spread seeds, betel nut is not viewed as invasive.

Today there are dwarf varieties available of the areca palm, but the long, solitary-stemmed tree may grow as high (Shenoy Heckadka et al. 2022). The trunk bears rusticated leaf scars forming exquisite designs on its trunks and variations in soil quality and heredity may impact the size of the trunk.

The roots spread out and when it matures, the upper leaf can extend up to 12. The production of new leaves by the terminal bud is carried out at 2-month intervals and lasts for about 2 years. Each leaflet forms a long, split leaf that surrounds the stem. In the case of the "South Kanara" cultivar, flowering usually occurs at the tenth node and happens at 5-year intervals (Nair 2021).

The Areca palm bears an inflorescence, referred to as a spadix, at the meeting point of its leaves. The amount of these leaves is proportional to the number of spadices. Depending on the age of the plant, an Areca palm can produce an average of three to four spadices (Chan and Elevitch 2006). The betel nut fruit is very small in size and has a look of a berry; it is covered with a thick husk and its color ranges from orange-red to scarlet when ripe. Each branch of the palm tree carry more fruits (Nair 2010).

The root system manages most of the tree's maintenance needs, with only a small portion penetrating. The roots of closely positioned palm trees grow deeper than those farther apart (Nair 2021). Betel nut trees need both male and female flowers on the same spike for reproduction to take place. The female flowers will change to green within a week, while male flowers are functional for 4–7 weeks (Salehi et al. 2020).

The female flowers last for a maximum of 10 days, and this is between 2 and 10 a.m. daily. One week after the flowers have opened up, they become receptive to pollen, which is carried by the wind. The development of the fruitlet occurs in three stages: rapid growth of the ovary, a contribution in size and dry matter to the kernel, and lastly, the fruit matures into an enlargement, and the green color disappears as the fruit becomes bright yellow. It takes a period of 9 months to a year for the fruit to mature fully and be harvested (Karun et al. 2017) (Table 19.1).

19.1.3 Botanical Description

A. catechu Linnaeus, areca-nut palm, Areca or betel nut, forms part of the subfamily *Arecoideae*. This species has many synonyms, based on less common scientific names, based on the following species: *Areca catechu*, commonly known by various names in different regions, has been classified under multiple synonyms, including *Areca hortensis* Loureiro, *Areca faufel* by Gaertner, *Areca cathecu* by Burman, *Areca himalayana* by H. Wendland, and *Areca nigra* by H. Wend. It is referred to as pugua in Guam, poc in Pohnpei, pu in Chuuk, bu. in Yap, bua in Palau, buai in New Ireland (Kuanua), buei in New Ireland (Pala), vua in New Ireland (Lamekot), and angiro in the Solomon Islands (Kwar'ae) in the Pacific. The term

Table 19.1 Phenotypic traits of importance in betel nut breeding

Trait	Description	Measurement methods	Genetic influence	Environmental influence	Use cases	Growth rate	Resistance to disease	Maturity period	Citations
Nut yield	Amount of nuts produced per tree or unit area	Counting the number of nuts, weight measurement	High	High	Commercial cultivation	Fast	High	Short	Yang et al. (2023)
Nut size	Size of the individual betel nuts	Measurement of diameter and weight	Moderate	Moderate	Commercial cultivation	Moderate	Moderate	Moderate	Yang et al. (2023)
Disease resistance	Resistance to common diseases affecting betel nut	Field trials, pathogen inoculation tests	High	Moderate	Sustainable agriculture	Moderate	High	Moderate	Liu et al. (2025)
Pest resistance	Resistance to common pests affecting betel nut	Field trials, Insect infestation tests	High	Moderate	Sustainable agriculture	Moderate	High	Moderate	Liu et al. (2025)
Growth rate	Speed at which the betel nut tree grows	Measurement of height and girth over time	High	High	Commercial cultivation	Fast	Moderate	Moderate	Staples and Bevacqua (2006)
Tree height	Final height of the betel nut tree	Measurement with measuring tape or tools	Moderate	High	Plantation management	Moderate	Moderate	Moderate	
Water use efficiency	Efficiency with which the tree uses water to produce biomass and yield	Measurement of water consumption and biomass production	High	High	Sustainable agriculture	High	Moderate	Moderate	Staples and Bevacqua (2006)

Table 19.2 Taxonomic hierarchy of *A. catechu* L. data

Rank	Taxon	Common name	Example species
Kingdom	*Plantae*	Plants	*Arabidopsis thaliana*
Subkingdom	*Viridiplantae*	Green plants	Chlorophyta
Infrakingdom	*Streptophyta*	Land plants	Mosses
Superdivision	*Embryophyta*	Land plants	Ferns
Division	*Tracheophyta*	Vascular plants	Pines
Subdivision	*Spermatophytina*	Seed plants	Magnolias
Class	*Magnoliopsida*	Dicots	Roses
Superorder	*Lilianae*	Monocots	Grasses
Order	*Arecales*	Palms	*Cocos nucifera*
Family	*Arecaceae*	Palm family	*Phoenix dactylifera*
Genus	*Areca*	Areca palms	*Areca triandra*
Species	*Areca catechu*	Betel nut palm	*Areca catechu*

"buai" in pidgin English is derived from several of these regional names. In other areas, it is known as boa in Bali, boá, bonga, buá, bunga, lúyos, and takobtob in the Philippines, palma catechou in Spanish, *Arec cachou* and Arequier in French, betelnusspalme in German, and puak in Sri Lanka (Lim 2011; Staples and Bevacqua 2006) (Table 19.2).

In the Pacific Islands, there are two main types of betel nut palms, red and white. Changnga, or changan are the kernels of white variety that may range in color from off-white to deep tan. However, the kernels of the red variety which in some communities are called ugam, differ—their colors vary from crimson to deep purple. White nuts are generally less desirable and red nuts are more suitable for chewing, therefore of a higher value. Even though these palms are not easy to differentiate among them because the root sap and kernels are variable in color. Growers must, therefore use seeds of certified red varieties (Organisation 2012). Survey data from various *Areca catechu*-producing regions in Hainan, China, reveal that the average betel nut-setting rate is below 12%. Notably, some high-yield specimens exhibit fruit-setting rates exceeding 50%, despite the typically low average (Li et al. 2022). In India, betel nut nuts is grown largely in the southern states of Kerala, Karnataka, and Tamil Nadu along with northeastern regions like West Bengal Meghalaya and Assam (Ahuja and Ahuja 2011). The Areca palm is in the *Arecaceae* or monocot family and has been grown extensively for chewing, with one species (*A. catechu*) having distribution prevalence through most countries of origin; as such this group benefits from simple taxonomic identification features. *Areca concinna* fruits are eaten rarely in Sri Lanka. The constant precedence of China and India as the largest producers in regards to betel nuts directly points towards an enormous part played by India in agriculture and cultural practices (Ganaraja 2017).

19.2 Current Cultivation Practices and Challenges

19.2.1 Existing Farming Methods and Challenges

A tall and slender stalked areca palm tree needs to have the right growing conditions. It grows well in the 14–36 °C temperature range, with high humidity and annual rainfall. It is frost and extremely heat-sensitive, so it does not grow too much. Recommended soils are red loams, alluvial with medium acidity or neutral pH and well-drained laterite soil (Mondal and Trivedi 2023). Deep plowing and incorporation of organic matter are important for avoiding waterlogging due to improper soil preparation. The best way to propagate is to select large mature nuts that are later grown well into seedlings. These seedlings are then transplanted into the main field after 3–6 months.

Proper irrigation should be adhered to considering techniques such as drip irrigation, basin irrigation, or flood irrigation, which is especially helpful during the bearing and flowering periods. Micronutrient deficiencies must be adjusted by applying the recommended amounts of nitrogen, phosphorus, and potassium depending on the soil test results. Fertigation and the use of organic fertilizers can help create a healthy, nutrient-rich soil. Additionally, intercropping with other crops such as cocoa, or pepper, and shade-bearing plants such as colocasia or bananas may be more beneficial. Aside from providing the necessary nutrients and improving soil structure, these plants may offer extra income and improve diversity. Caution must be exercised in the use of herbicides. In cases where pest infestation or disease is rampant, specific pesticides may be necessary, although their use should not surpass the allowable limit. Palms are harvested by climbing the mature trees using poles. After harvesting, they are processed through husking the fruits, which are then dried, boiled, and cured for later use.

19.2.2 Vertical Farming

Vertical farming is the practice of growing crops in stacked layers on vertically inclined surfaces inside a controlled environment it does not use soil but instead uses light emitting diodes (LEDs) to grow plants. Imagine mixed-use multi-story greenhouses, with lighting, HVAC and gas exchange regulations. This technique works wonders in areas with high population density or where space is a constraint for conventional farming (Kalantari et al. 2017).

Vertical farming is designed to optimize growing circumstances, leading the way in terms of yields per surface unit. It again comes with its challenges which include costly upfront investment and substantial energy consumption. However, it is expected that technological progress will overcome these challenges by increasing the range of crops compatible with vertical farming (Van Delden et al. 2021), thus making this solution more realistic for sustainable urban agriculture.

19.2.3 Digital Farming (Precision Agriculture)

Digital agriculture, also known as smart farming which allows practicing of using advanced technologies to make precise, accurate planning decisions process instead of approximated ones. It utilizes aerial imaging drones, GPS-guided systems for precision applications and real-time data collection sensors along with specialized software programs to understand the importance of this method (Clapp and Ruder 2020). The tools supply key data on field states, resource management, and harvest predictions for improved productivity & environmental sustainability by minimizing waste whilst increasing efficiency. This represents a significant move toward more accurate, and efficient agricultural operations-making it easier for farmers to fine-tune their practices (Abioye et al. 2022).

The advancement of artificial intelligence (AI) has revolutionized agriculture by performing tasks automatically and processing huge datasets to improve productivity as well as resource optimization (Dharmaraj and Vijayanand 2018). Crop yield prediction, robotic farming, disease and pest detection using computer vision with other machine learning primitives. Also, AI-powered technologies have made indoor farming resource-friendly and market research data in quantifying mode Using AI in agriculture could lead to large increases in yield, reduced labor requirements, and customized management strategies. Nonetheless, the responsible application and ethical issues involved must be adequately addressed for a sustainable implementation of AI in agricultural practices (Liu 2020).

19.2.4 Existing Agricultural Problems

The challenges to areca palm cultivation, replicate the wider agricultural problems. Water stress has effect during the flowering and fruiting periods, which get worse with heat stress (Nair and Nair 2021b). The scenarios are compounded by climate change. Furthermore, climate change can cause floods which might harm the palms and degrade soil and making it agriculturally less suitable. Land degradation resulting due to soil erosion and fertility loss. The environmental consequences from agrochemicals are additionally negative in the case of a loss of biodiversity and water contamination when results are used with monoculture practices. This contributes to economics of scale which is challenging for many small-scale growers, as they respond to price volatility and costs due in part to the incentives present further complicated planning profitability (Crews et al. 2018).

19.2.5 Areca Palm Fertility Traits

Patronage of the Areca palm is aimed at producing effective water use efficiency and resistant drought-heat cultivars with low usage in chemical control by resistance to diseases and pests, which helps in yield conservation (Rivas et al. 2012). In addition, breeders work towards nutrient-efficient cultivars to reduce environmental load and application of fertilizers. Such market-oriented goals must reflect consumer preferences for the type of nut, as well as intercropping suitability to optimize land use and production efficiency. Good local adaptation of breeding goals, in-situation practical sustainable solutions and traits which the farmers select are also required, with a good understanding of genetic diversity needed for future adaptability (Apshara 2019).

19.3 Germplasm Biodiversity and Conservation

19.3.1 Germplasm Diversity

Areca palms in both natural and cultivated conditions are significant for their genetic diversity. The diversity influences yield potential nut size and shapes, drought or disease resistance, and use (fresh/dried). Preservation and maintenance of this wide germplasm is important for greater sustainability and development in betel nut production.

19.3.2 Characterization and Phylogeny for Cultivar

The size of the leaf, and form of the nut (Raimondeau et al. (2021) are the basic characters for grouping cultivars. Biochemical assays, isozyme markers and genetic markers are also used in understanding the variation among different populations and ecotypes. More recent alternative such as DNA markers like RAPDs and SSRs provide an accurate and extensive method to assess the genetic diversity patterns, particularly Pleistocene disjunction can provide crucial information on the evolutionary history and relatedness of cultivars, as well as wild progenitors. Thus, the fact that these environmentally sensitive loci significantly are present needs to be understood for in-situ conservation of natural populations to safeguard genetic biodiversity and also ex-situ via breeding programs necessary for developing new cultivars with desired character (Ranade et al. 2002).

19.3.3 Genetic Resources Conservation Approaches

Resource conservation can be achieved is through in situ conservation, conserving Areca palm populations within their natural habitats, home gardens and agroforestry system which will maintain biodiversity. Adaptive and resilient evolutionary trajectories of the species in this respect are achieved by following these principles.

In contrast, ex-situ conservation involves the maintenance of living collections in field genebanks, botanical gardens and research stations making direct access possible for evaluation purposes. Long-term conservation of germplasm (e.g., cv, and breeding materials) is very important to protect against natural disasters, diseases and other risks that may be faced in the future. Therefore, a safe method for conserving genetic material should be available over time with techniques such as tissue culture methods including cryopreservation/outside the lab.

It is basic to advance new Areca palm cultivars with alluring highlights, for example, disease resistance, tolerance to climatic fluctuation, global food security by keeping up a widepreserving a broader genetic diversity pool. Like climate change, even genetic diversity is a major priority among conservationists whose mandates seek to prevent the extinction of species that are keystone (Arunachalam 2011). The genetic diversity of plants in the future generations is likely to be shaped and sustained by combined phylogenetic analysis and germplasm screening (Yang et al. 2021).

19.4 Conventional Breeding Practices

19.4.1 Mother Palm Seeds: Selecting the Best Nuts

Evaluation of traditional agriculture breeding methods which simply increasing yield. This includes determining disease tolerance, insect resistance and specific size, shape or husk characteristics of the nut. Evaluating plant adaptability to regional conditions is also important Therefore, the detailed nut dimensions (length, width, and thickness), mean weight on 100 nuts and husk thickness were accurately measured leading to more accurate quantitative data for the choice of promising traits.

The absent longitudinal tracking of disease incidence and crop yields over multiple seasons hinders the ability of breeders to objectively evaluate palms and choose high-performing parents based on targeted traits. This data empowers breeders to enhance breeding programs and develop resilient and productive palms (Alouw and Wulandari 2020).

19.4.2 Expanding Germplasm

One important strategy for exploiting genetic diversity within a species is the creation and maintenance of breeding populations from base material collected in different parts of geographic or even ecological zones referred together as germplasm. It expands the gene pool that breeders can access to create new, more resilient and adaptable plant varieties (Morgenstern 2011). The research institutes, farmers and other organizations can jointly organize events to exchange genetic materials in collaborative ways can provide an opportunity to expand the germplasm base. In this sense, collaborative working in the field for breeding programs provides better holistic sustainable agriculture practice as it merges local wisdom with cutting-edge scientific research (Chaturvedi et al. 2022).

19.4.3 Controlled Pollination

Controlled pollination (hybridization) offers an important tool for creating new variation, trait introgression and germplasm expansion. The selected germplasm with beneficial traits i.e., disease resistance, stress tolerance etc are crossed to the elite cultivars to conserve the genes and alleles. This fastidious technique makes sure that their offspring are a result of breeding parents with all the most beneficial traits they would like to develop for these features penetrate deeply into heritage down range (Albrecht et al. 2020). The progenies of the controlled pollination are better in terms of fruit quality, stress tolerance and provide new germplasm for the crop improvement. Progeny testing of breeding crosses after controlled pollination is carried out for several years, during which its performance is carefully monitored for fruit quality, stress tolerance and yield. The inferior offspring are discarded and superior strains are used for additional generations to refine varieties or incorporated as new cultivars into palm regimes.

Progeny testing of breeding crosses after controlled pollination is carried out. Several years, during which its performance is carefully monitored for fruit quality, stress tolerance and yield. The inferior offspring are discarded and superior strains are used for additional generations to refine varieties or incorporated as new cultivars into palm regimes (Kingsbury 2011).

19.4.4 Challenges of Conventional Breeding

Breeding Areca palms also has several challenges; including the late maturation rate/fruit bearing time is a principal limitation which potentially limit the breeding cycles and extends the development of new varieties. Conventional breeding methods are mostly based on the natural genetic variability existing in betel nut germplasm resulting in difficulties to bring desirable traits (Bradshaw 2017). In addition, traits with polygenic and quantitative inheritance such as disease resistance or adaptive traits during introgression them, requiring an effective way of selecting these traits that are not amenable to classical selection approaches. These constraints hinder the gain of breeding programs for improvement in Areca palm varieties (Rauf et al. 2016).

19.4.5 Biotechnology's Role in Overcoming Limitations

19.4.5.1 Marker-Assisted Selection (MAS)

The avaiablity of markers and marker-assisted selection (MAS) has revolutionized breeding technology with associated DNA markers that have been mapped to important traits in palms (Kamaluddin et al. 2017). Identifying markers, and specific DNA sequences associated with important traits such as disease resistance and yield within the genome is important. Genotype seedlings in the early seedling stage, researchers can identify the required genetic traits carried in the seedlings and select for planting. This early selection reduces the number of field trials required and shortens breeding cycles, making MAS a useful tool to expedite cultivation improvement in Areca palm cultivars (Ortega and Lopez-Vizcon 2012).

19.4.5.2 Genetic Engineering

Genetic engineering particularly refers to the direct gene editing (Weckx 2020). With the help of this technique, farmers may introduce some important traits from other plant species known for resistance/tolerance against various diseases into the betel nut genome. Focusing on desired traits like flavor, size, or production geneticists introduce specific genes from other sources causing extended yield or making better nut quality (Javed et al. 2021).

It is noteworthy that that genetic engineering can be complicated by legal constraints and ethical considerations to harmonize innovation and responsibility (Cao et al. 2022).

19.4.5.3 Tissue Culture

This is a cost-effective and time-saving method for the mass multiplication of elite genotypes in the Areca palm breeding practice using tissue culture technology. Meristems (small pieces of plant tissues) are taken from good palms and cultivated in the nursery under controlled conditions by a cloning process (Manimekalai et al. 2013). This method creates a large number of plantlets that are genetically identical to the parent plant. It not only facilitates the rapid multiplication of elite planting materials but also permits generation-to-generation uniformity among propagated plants (Mbi et al. 2016), making superior betel nut varieties available to farmers in a shorter duration.

19.5 Molecular Breeding

19.5.1 Areca Palm Genomics and Bioinformatics

The genetic variability and diversity of the betel nut palm have been studied in detail, although still relatively less when compared to other economically significant Arecaceae family members such as oil and date palms (Manimekalai et al. 2018). Major developments in the field of plant genomics have been caused by advances such as utilization methods to analyze genomes (i.e. RNA Seq, modern sequencing technologies: Illumina HiSeq, 454 Pyrosequencing and SOLiD) that can be used for profiling expressed genes even in uncharacterized genomes. Similarly, RNA Seq analysis using the Illumina technology includes an exclusive total unigenes presented in coconut tanscriptomes (Del Vecchio et al. 2017). The de novo genome sequence of *Areca catechu* has been completed and is available publicaly var. 'Reyan#1' with 240 × read-depth, consisting estimated size of up to 2.59 Gb scale represents the most complete and comprehensive report on this crop yet available. The genome assembly generated 16 pseudochromosomes (2.51 Gb in size and N50 = 172 Mb) with a scaffold N50 of ~1.7 Mb. Transposable elements make up 80.37% of the genome, with long-terminal repeat retrotransposons accounting for 68.68% (Alam et al. 2024a). The genome contains 31,571 protein-coding genes, with 92.92% functionally annotated, including genes responsible for producing various biochemical compounds such as flavonoids and monoterpenoids (Yang et al. 2021; Arul Swaminathan 2002; Nair 2021; Ananda et al. 2017).

19.5.2 Improvement of Areca Crop

For unblocking the incompatibility which is observed in female Areca palm and advantage seedling, production of the parental population due to utilization. The Areca palm hybridization makes a significant contribution to plant height and complexity of pest control as well as nut harvesting. Therefore one of the prime objectives is to reduce plant height (Veluru et al. 2024). The effort uses the genetic diversity available in Areca germplasm for developing high-yielding semi-dwarf types having independent fruiting schedules and with superior nut traits. The single stem is often seen in *Areca catechu* and to this end, makes the interspecific hybrids of *Areca triandra* with *A. catechu* desirable for cultivation convenience and easy management purposes. Many of those hybrids exhibit heterosis for different characters, also including attributes like flower quantity per inflorescence, spadix dimensions and stem width while they still keep the qualities from each parental line e.g. internodal space (Rajmohan 2005). Another important feature is the dwarfing character of, say "Hirehalli Dwarf". The crosses between "Hirehalli Dwarf" and high-yielding cultivars developed many dwarf and intermediate types which were ultimately successful in producing semi-dwarf varieties (Khang et al. 2022).

19.6 Tissue Culture Applications

19.6.1 Micropropagation

Micropropagation is a tissue culture method that allows the plant to propagate from a small portion form tissue, which results in genetically identical clones. This bypasses the inherent genetic variation natural to betel nuts mostly through seeds that reproduce sexually (Cardoso et al. 2018).

The process begins with identifying an elite betel nut palm that has good characteristics like higher yield or disease resistance. From those pieces, an explant—a small piece of tissue often removed from the inflorescence (or flower cluster) on the parent plant itself—is harvested and placed into sterile media to prevent any other contaminants that might escape it in normal rice cultivation. The sterilized explant is then cultured on a medium with growth hormones such as cytokinins and auxins, which promotes the formation of non-specific cells called callus. The formation of these calli is an important step during micropropagation as from this multiple shoots are generated which can finally be developed into whole plants.

Utilization of this method in breeding enhances the uniformity and quality of the produced plants by guaranteeing that only superior traits are consistently propagated. The callus can then be transferred to a new medium containing hormones that cause the plant tissue culture techniques plants have been obtained from forming shoots and roots of their own. These cultured cells eventually give rise to entire plantlets.

Micropropagation serves many beneficial effects for betel Nut production, it allows the rapid multiplication of superior genotypes; assures a uniform supply of plants with defined characteristics and provides year-round propagation independent of seasonal effects (Bhojwani et al. 2013).

19.6.2 Embryo Rescue

Embryo rescue is a technique used to salvage embryos that were unable to fully develop in vivo. In betel nuts, this method is recommended for long-term storage to save valuable crosses among diverse genotypes or synthesized vulnerable embryos (Rogo et al. 2023). It is the process of rescuing an embryo from seed and raising it in vitro, where all conditions available like nutrients, and growth regulators are precisely dictated to permit them self-development; this helps for the perpetuation of desirable traits onward with improvement in crop breeding program. Developmental competent embryos are collected from betel nut seeds and grown in an aseptic condition supplemented with a medium that contains vital growth factors to develop. By using this method there is preservation of genetic material and the possibility to elevate a better betel nut variety. It can also break down the hybridization barrier, which expedites in generation of novel genetic combinations (Muralikrishna et al. 2017).

19.6.3 In Vitro Culture of Arecanut

The direct adventitious shoot bud differentiation from cotyledon explants in vitro propagation of betel nut showed that the required initiation and development phases are sensitive to a few critical parameters. Different media such as BB (Branton and Blake), White, and MS media with different Auxins or cytokinines showed that shoot differentiation was enhanced by activated charcoal, 2,4-D and higher amounts of phosphate in BB medium. This was visible at the rim of the cotyledonary sheath after embryo germination. Activated charcoal promoted rooting in shoot cultures and a high combination of abscisic acid with auxins in the rooting medium was effective in inducing the maximum frequency of root formation. This investigation presents important cues for improving the in vitro propagation of *A. catechu* (Gantait et al. 2018).

In the case of shoot tip explants, plantlets could be induced in *A. catechu* L.-derived callus with a soft greenish appearance after 4 weeks on MS basal medium Gellrite-gelled fortified behavior-regulatory substances TDZ (0 without any activity level 02 and so forth., mg/l), BA (one cannot get enough) shot hub segregate. The maximum callus induction (100%) was recorded with 0.2 mg/l BA and 0.2 mg/l TDZ in the combined liquid medium; After culturing on the same media 50–60% f calluses gave rise to shoots. In NM medium supplemented with NAA (0.1 mg/l), after 4 weeks in culture, approximately 90% of the shoots developed roots. This approach is effective for the development of plantlets in *A. catechu* L and the process took 20 weeks (Wang HsiangChih et al. 2003) via primary callus induction as well as two-step organogenesis chains.

19.6.4 In Vitro Pollination

This approach is ideal for conducting selective fertilization because it involves the union of entire plant ovules with the correct pollen wherein certain species such as betel nut palms prevalent in outbreed mode but could prove ineffective and even extensive.

These plantlets are produced by collecting the pollens from identified good-yielding betel nut plants and sown on a nutrient medium as reported by Zenkteler (1984). Then the ovules are laid out onto a medium, allowing them to be fertilized without relying on pollinators or plant manipulation. It facilitates faster breeding and selection for desired genetic end products i.e., allowing better-controlled efficiency of pollination conditions that provide a sense of control to plant breeders (Karun et al. 2017).

19.6.5 Synthetic Seeds

Synthetic seeds are artificially encapsulated somatic embryos with nutrient storage in a protective coating to replicate natural seeds. A dependable way of propagating and vegging betel nut plants. Somatic embryos can be produced using tissue culture techniques, which are later encapsulated inside a gel-type matrix formed by materials like sodium alginate to support and protect them. The coating can contain growth regulators and nutrients that help in development. These seeds are also suitable for year-round production and can be easily stored, and transported, including to remote areas (Faisal and Alatar params 2019).

19.6.6 Germplasm Conservation by Tissue Culture

Tissue culture can effectively used to cryopreserve tissues or cells of important or endangered betel nut species for genetic diversity preservation. This type of storage keeps the genetic material dormant over periods spanning millions of years, long enough for it to remain undisturbed and relatively unaffected by genetic drift or disease. Consequently, it helps in preserving genetic resources for future breeding and conservation programs resulting in stability and resilience in betel nut populations (Ramanatha Rao et al. 2023).

19.6.7 Secondary Metabolites Production

The arecoline present in betel nuts is a true pharmacophore with therapeutic potential. Secondary metabolites can be developed through tissue culture and will result in this metabolite in its controlled environmental conditions. These secondary metabolites can then be easily extracted from the growth tissues making them suitable for extraction purposes for pharmacies or industries. Natural elementary compounds are difficult to synthesize, but tissue culture allows for the production of these by avoiding time-consuming traditional cultivation methods that may impair consistency. This strategy not only guarantees stable supplies of high-quality metabolites but also inspires the design of unique pharmaceutical and industrial applications (Hussain et al. 2012).

19.6.8 Genetic Transformation

Tissue culture can also be utilized for genetic improvement of betel nut by introducing the desirable genes into these plants and generating new varieties that are resistant to different diseases, high yielding as well as improved quality. This strategy advances the generation of improved cultivars through precisely targetable gene editing in an amenable setting. The approach speeds up the creation of a region or market-specific types and may help to facilitate genetic gain in betel nut plantations. Tissue Culture allows breeders to rapidly put in ground genetic material with these desirable traits and replicate consistent quality performance throughout all cultivated crops.

19.6.9 Somatic Embryogenesis

A plausible in vitro culture protocol for somatic embryogenesis and regeneration of plants utilizing callus from zygotic embryos is developed using *A. catechu*. Callus with yellow, friable and exudate can appear over the MS basal medium implanted by dicamba and plant regeneration was further procedures. Further growth of this callus led to the production of pale yellow, dense nodules. Somatic embryos sprout from these nodular call tissues within 2–4 months and develop into plantlets on a hormone-free MS medium. Twenty-four percent of the plantlets survived through 3 months of subculturing and acclimatizing in a greenhouse (Wang et al. 2003).

This technique was adapted for in vitro cultures of immature inflorescence samples from mature palms, using leaf explants collected from one-year-old seedlings. As previously reported, the developed somatic embryos germinated efficiently on MS liquid medium supplemented with BA and underwent rapid growth within a few transfer cycles forming plantlets with 2–4 leaves and strong root systems (Abbas et al. 2011; Liu et al. 2021).

19.6.10 Markers

Molecular markers play a crucial role in verifying the genetic uniformity of plantlets produced through tissue cultures, particularly for perennial crops such as Areca palm, which require lengthy field cultivation periods. RAPD markers are preferred for their affordability, ease of use, and independence from template DNA sequence information (Rathore et al. 2014). Research using RAPD markers to evaluate the genetic similarity between plantlets and mother palms showed high genetic resemblance, indicating the potential for mass-propagating elite palms with desired traits. Future research should focus on enhancing germplasm collection and characterization, cryopreservation, developing molecular markers for marker-assisted selection, optimizing regeneration protocols, researching abiotic stress tolerance, and creating extensive palm databases (Ahmad et al. 2021).

19.7 The Cultivation Chronicles of Arecanut

Betel nut palms thrive in a variety of soil types, with predominant cultivation occurring in the coastal regions of Karnataka and northern Kerala. These areas are known for their gravelly laterite soils composed of red clay, which are particularly conducive to betel nut cultivation. Rich, clay-like deep-black loams in the plains of Karnataka also provide suitable conditions. Conversely, sandy, alluvial, brackish, calcareous, and sticky clay soils are generally unsuitable. The ideal climate for betel nut palms is subhumid and tropical, situated between 28° north and 28° south of the equator (Karun et al. 2022). In northeastern India, farming predominantly takes place in the plains because higher elevations experience lower temperatures that hinder crop growth. Areas up to 1000 m above mean sea level (MSL) are usually deemed appropriate, as higher altitudes can adversely affect nut quality. Betel nut palms can endure extremely low temperatures, such as 4 °C in West Bengal during winter, and thrive within a broad temperature range. This crop prospers in regions with substantial annual rainfall, typically between 300 and 450 cm, and requires additional irrigation in regions with lesser rainfall, like the plains of Karnataka and Tamil Nadu (Nair and Nair 2021a). Coastal areas exhibit diverse climates; southern Kerala has more uniformly distributed rainfall, whereas coastal Karnataka experiences heavy rainfall from June to September followed by drought. Consequently, irrigation is essential for betel nut cultivation in these coastal regions (Hebbar et al. 2024).

19.7.1 Germination, Flowering, and Fruiting of Betel Nut

Betel nut palms usually germinate within 90 days producing a single bifid leaf and five roots. Seedlings are normally 1 year old with four or five leaves. Individual seedling growth percentages vary, with a few of them growing from 3 to 24 months (sometimes up to 4 years) before they are ready for transplanting. Trunk formation may start a year. The average growth rate of a mature palm is only about half each year leaving approximately 6 new leaves to produce annually and an abscission of one full adult leaf. About every 2 years to maintain 8–12 in the crown (Rahman et al. 2022). Flowering generally takes place in 4 to 6 years. The palms begin to bear fruit at 7 or 8 years, are fully fertile after only 10 to 15 and can produce any profound harvest for up to about 40 before dying of old age. Inflorescences may initially be composed solely of male flowers, which are ineffectual for nut production (Staples and Bevacqua 2006). The flowers of the female trees aren't fragrant, so they don't attract bees and other insects. The flowering season varies with peak periods in Malesia where palms flower year-round and in India, where around 70% of the trees come into bloom. For the Pacific information on flowering seasonality is scarce (Yudaputra et al. 2016).

19.7.2 Propagation and Out-Planting of Betel Nut

Betel nut palms are exclusively propagated by seeds, with techniques varying depending on the number of palms, but the basic requirements remain consistent. In regions like Taiwan and India, structured mass propagation techniques involve selecting mother trees and seeds, sowing in shaded beds or pits until germination, and then transferring seedlings to nurseries (Alam et al. 2020). Intercropping with other crops like bananas is common when seedlings with five or more leaves are transplanted between 12 and 24 months and initially require shade.

The betel nut is mainly grown around homesteads in areas such as Indonesia, Malaysia and Southeast Asia where it was spread by migrants from Papua New Guinea and Melanesia—plus the Philipines and much of Oceania. Local people generally collect "volunteer" seedlings from the ground, after fruits have been eaten by bats and squirrels. Alternatively, they may gather small seedlings growing nearby under established trees. For seed propagation, mature fruits are collected from healthy trees with nuts that have the desired qualities. Once the seeds are extracted cleaned and processed, they are conserved in a cold room for sowing. After 1–2 years of growth, they are transferred to nursery beds or individual containers (Rahman et al. 2017). For field planting of seedlings between 12 and 18 months old, ideally at the beginning of a rainy period (but not during torrential downpours), specific out-planting guidelines are provided. For planting, holes should be 50 cm deep and wide mixed with a top layer of soil fertilized with phosphorus-containing manure. Deep planting of seedlings enables slow soil pile-up around the base, increasing development origin for better supply intake (Vincent 2012). Hence the spacing ranges from 2.7 × 2.7 m to 2.4 × 3\rightarrow1130–1350 palms/ha Appropriate spacing ensures air flow, which lowers the chances of diseases that hurt flowering blossoms or developing fruits. Palms should be heavily mulched to retain moisture and fed with regular applications of aged animal manure. For best results, outplant at the onset of monsoons unless irrigation is assured. Windbreaks planted along the perimeter of an orchard are a valuable resource in windy regions, protecting fruit and minimizing sunburn (Staples and Bevacqua 2006) (Table 19.3).

Table 19.3 Betel nut germination, flowering, and fruiting conditions

Condition	Details	Citation
Germination	Betel seeds can take several months to germinate, preferring warm, moist environments with temperatures around 25–35 °C. The seeds should be soaked for several days before planting and require well-drained, moist soil.	World Seed Supply (2019).
Flowering	Betel nut palms typically start flowering within 6 to 10 years from planting and can flower throughout the year. The flowers are primarily wind-pollinated and form on spikes. Environmental factors like humidity and temperature significantly influence flowering.	CABI Compendium (2019)
Fruiting	Fruits mature in about 6–8 months after flowering. The trees can continue to produce fruit for 30–60 years. Optimal conditions include well-drained soil and full sunlight, with an average annual rainfall between 1500 and 5000 mm.	Areca catechu—Wikipedia (2005)

19.7.3 Significant Effects on the Growth of Betel Nut

Despite its primary use, the cultivation of betel nut causes serious health concerns. Balancing supply and demand is challenging, especially in high-consumption areas like Pakistan and India, making it difficult for Pacific growers to meet local and international demand. Betel nut does not naturally spread widely, despite its potential invasiveness, because its large fruits and seeds limit dispersion (Garg et al. 2014). Pests and diseases attack male plants as well as newly budding fruits after pollination. Infestations may require insecticides and fungicides to treat, so householders are advised to monitor flower clusters weekly. *Betta palmasini* (a bacterial disease of the betel nut) and N- wheat related to *Staphylococcus aureus* activity, *Typhoid Salmonellaquema*, also infest plant species like pineapple, coconut seedlings or grains poisoning them. Consuming too many betel nuts is associated with serious health risks, including discoloration of teeth and dental diseases: Furthermore, such consumption may cause or increase the risk for cancer at various sites in humans. In addition to the regular production, Taiwanese betel nut farm is also based on agroforestry and environmental principles: burnt fallen fronds and tops of trees are burned at plantation boundaries for ashes which are then used as fertilizer. The betel nut also serves as a vermifuge to control pests in humans and animals (Kumari et al. 2023; Rahman et al. 2022).

19.7.4 Effective Management Strategies

Selecting planting material carefully is crucial when propagating the Areca palm from seeds, paying attention to seed size and the age of the mother palm. Seeds should ideally come from mature trees with established yielding patterns. Heavy and mature seeds are preferred for their higher germination rates (Cock et al. 2023).

Selected seeds are sown stalk-end up, about 5 cm apart. Regular irrigation every other day is necessary for healthy seedlings. Saplings can be transferred to a secondary nursery after 3 months and planted. Alternatively, seedlings can be grown in Polybags filled with a mixture of sand cow dung, and soil (Aktar et al. 2021). The soil's water depletion profile significantly influences Areca palms' irrigation frequency. Flexibility is crucial to account for varying crop-evapotranspiration (ET) losses rather than adhering to a fixed irrigation. The two main irrigation methods are splashing and basin filling. Various studies have examined Areca crops' irrigation needs, considering temperature, ET demand, frequency, and water supply (Karun et al. 2022, 2023).

19.7.5 Arecanut's Dietary Potential

The bulk of the existing data on betel nut nutritional requirements comes from classical field surveys. In several field trials, the annual N: P 2 O 5: K 2 O requirement for different age palms was around with a ratio of each palm, indicating that potassium is in higher request than nitrogen or phosphorus at least two gibberellin deficient trees(10 years old). Annually, these specifications are standardized for every palm. Organic manure, usually up to 12 kg per palm was also supplemented with chemical fertilizers. It can be fed from sources like compost, bovine manure or green material from *Glyricidia maculata* growing along betel nut garden edges. Recent fertilizer trials showed that the net returns and "Benefit-Cost Ratio" were both higher when nutrient rates were recommended but doubled. Bhat and Sujatha (2009).

19.7.6 Insights into Its Physiology

The betel nut canopy allows 30 to 50 percent of photosynthetically active radiation (PAR) to penetrate. This proportion varies seasonally and throughout the day due to factors like cloud cover, solar angle, and canopy density (Kumar et al. 2024). Different palm spacings optimize light transmission at various times, affecting light distribution in the canopy. The betel nut canopy interception of incoming radiation is on average 70 percent, and the remaining light enters into the intercrop canopies. Cocoa trees with dense canopies and higher leaf area index (LAI) values, for example, intercept up to 90% of incident light. Directional and diurnal changes in light exposure, as the pepper plants also grow under betel nut. A fully mature betel nut palm has an average of 8 to 9 leaves and each leaf covers about 11.2 m^2 making the total leaf area up to approximately 22.0 m^2 The leaf area at the canopy level, or else in terms of LAI (Leaf Area Index), is around 2.44 that covers an area related to 9.1 m^2. Photosynthesis rates in betel nut leaves is 2.8 to 8.6 μmol $CO_2/m^2/s$ according to cultivar and leaf position. The greatest photosynthetic activity is generally in the first full open and third leaves. Apart from variation in the content of total chlorophyll among Areca species, a significant variability was also observed with respect to its content which is highest recorded in the case of *Areca triandra* (Kumar et al. 2024; Nair 2009).

19.7.7 Exploring Mixed Cropping with Betela nut

Intercropping betel nut palm interspaces offers excellent opportunities for growing associated crops, providing farmers with an additional income source. This way of functioning is best for the first 5–6 years of Areca palm growth where it will be pre-bearing stage. Economic sustainability could include the plantation of annual or semi-perennial crops for short-duration in their interspaces during this phase. Mixed cropping of shade-tolerant perennial crops is only possible when the Areca palm matures and the canopy spreads out (Vishwajith et al. 2015). Examples of diverse annual crops include: rice, sorghum, beans vegetables like eggplant and bullet yams.

The land use model adopted i.e. high-density multispecies farming including crops like banana, coffee, cocoa and black pepper at different vertical levels gives promising results in terms of higher total yield and increased profits. Microbiological studies of the microbiome of the rhizosphere for these intercropping combinations have reported an increase in beneficial microorganisms such as nitrogen-fixing bacteria, actinomycetes, and bacterial and fungal populations which cause growth promotion to plants. These results indicate a shift in nutrient cycling and enhanced soil health that could benefit net productivity and sustainability (Thomas et al. 2022) (Table 19.4).

Table 19.4 Common extraction methods for natural fibers

Extraction method	Description	Citation
Retting	This involves soaking plant materials in water to allow microbial action to break down the pectin holding the fibers together, commonly used for flax and jute.	Mohankumar et al. (2021)
Chemical extraction	Involves using chemicals, such as sodium hydroxide, to remove non-fibrous materials. This method effectively extracts fibers from plants like sisal and hemp, resulting in higher puritycitationSourceIndex2.	Farag et al. (2025)
Mechanical extraction	This method uses mechanical devices to separate fibers from the plant material, commonly practiced in the processing of cotton and coconut coir.	Farag et al. (2025)
Decortication	Involves the removal of the outer layer of a plant to access the fibrous material within, often used for coir and bast fiberscitationSourceIndex4.	Farag et al. (2025)

19.7.8 Uses and Products of Betel Nut

The betel nut offers a diverse range of products utilized in various applications. The seeds, known for their stimulating properties, are incorporated into medicinal formulations for both external and internal uses. Additionally, parts of the plant, including husks, shoots, buds, leaves, and roots, are utilized for medicinal purposes (Singh et al. 2020). The fibrous fruit pods are used frugally through household fuel, once they are separated from the seeds. The trunks are used for poles on which palm fronds (which have "fallen") generate, while other fallen fronds are weaved into baskets. Leaf-sheaths and spathes are used to make caps, sandals, wrapping materials or for harvesting tools in the case of pickers. Many times flowers and inflorescences carry ritual importance (Singh et al. 2020).

Due to its stimulating properties, betel nut is widely used in South Asia, the Pacific and other tropical regions. Eat fresh or dried, ripe and unripe. The terminal bud (sometimes called the heart of the palm) may be eaten. It is utilized in various cuisines, either cooked, raw, or pickled. Other parts of the plant, such as shoots and leaves, are also used as vegetables (Staples and Bevacqua 2006).

Different parts of the betel nut have medicinal applications. Chewing betel nut with betel leaves and other flavorings acts as a stimulant and masticant. Betel nuts can be used therapeutically for stimulation of the intestinal tract however, they are otherwise not ingested. The leaves of the plant are used as aromatic body wash in many cultures and affluent Southeast Asian countries use the fragrant blossoms for ceremonial purposes. The fallen fronds and bracts are a fuel source, meanwhile, the lumber of palm trunks has been used to build rafts, houses and ceremonial buildings. Betel nut is mostly used as a source of medicine and timber materials (Hooper et al. 2015).

The multiple-utility coconut is also well-used and stands tall in a variety of industries. Known for its resilient fiber from the husk, coir is used in making textiles and headgear as well as a source of pulp for paper packaging insulation. In the Philippines, it is used ingeniously and makes wonderful toothbrushes. Coconut leaf sheaths are used as bags and book covers for eco-friendly packaging (Garg et al. 2014). Boiling coconut husks releases tannins, which dye fabrics black and mildly colored. Coconut oil serves as a natural butter extender and lubricant. Chewing betel nuts hold cultural significance throughout Asia and the Pacific, while in Sri Lanka, coconut blooms are symbolic of good fortune in religious rituals. The global trade in coconut products is primarily driven by copra and oil, with successful production depending on proper plantation spacing. Sustainable growth and productivity in this critical sector are ensured through careful management (Roy and Guha 2021) (Table 19.5).

Table 19.5 Comparison of natural and synthetic fibers. Information is obtained from cotton works 2023) and LeMuse (2023)

Feature	Natural fibers	Synthetic fibers
Source	Derived from plants (e.g., cotton, flax) and animals (e.g., wool, silk)	Manufactured, primarily from petroleum (e.g., polyester, nylon)
Texture	Generally soft, and breathable; varies by fiber type	Texture can vary; often smoother but sometimes coarse
Breathability	High, allowing heat and moisture to escape	Generally low, can trap heat
Moisture absorbency	High; excellent for sweat management	Low; synthetic fibers do not absorb moisture
Durability	Moderate to high; natural fibers can be less durable	High; tend to be stronger and resistant to wear
Elasticity	Low to moderate; generally less stretchable	High; many are elastic (e.g., spandex)
Maintenance	Often requires special care	Generally easy to care for, often machine washable
Cost	Typically higher due to labor and cultivation	Generally lower; mass-produced from cheap materials
Environmental impact	Generally lower; biodegradable and eco-friendly	Higher; can take decades to decompose and contribute to microplastic pollution
Common use cases	Clothing, home textiles, industrial applications	Activewear, upholstery, outdoor gear

19.8 Genetic Engineering and Gene Editing

This is yielding an increase in the adoption of genome editing (GE) tools, which can provide designed changes with higher precision and efficiency than obtained by classical transgenesis approaches. These techniques enable precise suppressions to be made in the DNA without the risk of additional, undesired changes while making targeted and inheritable modifications at discrete loci throughout the genome. A variety of sequence-specific nucleases (SSNs) are available for creating engineered plants that include transcription activator-like effector nucleases (TALENs), zinc-finger nucleases (ZFNs), and the CRISPR/ Cas9 or the clustered regularly inter-spaced palindromic repeat /CRISPR-associated protein systems in plant (Mashimo 2014). These methods allow the isolation and manipulation of defined genomic loci by which purpose, exogenously programmed nucleases create double-stranded breaks (DSBs) at or near a target site that are then repaired either through homologous recombination (HR) with an endogenous duplication repeat sequence or non-homologous end joining(NHEJ). HR is an accurate repair process whereas NHEJ contains smaller insertions and deletions at DSB sites than HR does (Ensminger and Löbrich 2020).

For example, in the agricultural sector where genetic engineering (GE) can be used to speed up the introduction of preferred traits which would result in a significant cost saving over labor and other costs. As a practical matter, GE provides an easy ability to create new crop hybrids that are safe for the environment. Such techniques will offer ease of maintenance over the generations in a laboratory setting using any plant material regardless of genome complexity and provide for more efficient breeding than conventional methods (Vincelli 2016). Such strategies have greatly improved the genetic makeup of plants enabling the incorporation of better morphological traits, resistance to biotic and abiotic stressors in addition nutritional quality characteristics that include yield attributing parameters, and crop quality grading for herbicide tolerance. Transcription factors can also help to reveal insights into abiotic stress responses and melatonin-induced tolerance (pitaya), providing models for enhancing betel nut durability (Alam et al., 2024b).

In the field, abiotic stressors from climate change affecting palm trees include changes in monsoonal regimes and heightened temperatures along with elevated atmospheric CO_2 levels due to rising sea levels. Given that palms have a longer production cycle unlike annual crops, these genomic tools could help to overcome or prevent most of these challenges from a short-term perspective and a long-term preventative (Gao 2021). It is recommended to carefully plan the selection of target genes, sgRNA sequence properties, delivery methods and tissues for delivery as well as in vitro regeneration methodology before utilizing CRISPR/Cas9 editing in palms. Designing sgRNAs complementary to target gene sequences effectively can be a costly proposition given the highly heterozygous nature of palm genomes. Therefore, the prospect of using multiplex CRISPR/Cas9 systems to target many genetic loci has been suggested for potential future restructuring of palm fatty acid pathways.

The development of rapid genetic modification for tree crops, such as fast-growing genetically altered plants that can be produced. One of the system's advantages is that it gives rise to biallelic alterations, which offers the simplest way for not genotype-dependent phenotype assays (Wan et al. 2019) The demonstration of successful electroporation-mediated protoplast transformation in oil palms provides promise for the future application regarding development genome-edited palms without DNA. However, its practical regenerating strategy for oil palm protoplast is an intractable issue (Singh et al. 2017). Much work is going on to develop efficient methods of protoplast regeneration and PEG-mediated delivery systems for Cas protein/sgRNA leading towards the development of genome editing-based breeding in palms.

Efforts have been made to enhance genomic resources in some homozygous crops and model plants by re-sequencing specific cultivars together with the recent awareness alongside their palm genome multiploidy trait associated with deploying this exploitability into web-based sgRNA design algorithms (Wu et al. 2020). The use of CRISPR/Cas9-mediated genome editing is anticipated to enhance the ongoing palm crop improvement efforts by shedding light on gene interactions and their association with phenotypic traits. In addition to that, this would help in understanding the factors responsible for making coconut plants disease susceptible which is an essential research area. The goal is to find disease susceptibility genes or genomic regions and then modify these using gene-editing techniques. Some key gene targets for root disease in coconut such as *PR1*, *PR4* and PT15-like transcriptional activator gene that encodes wound-induced proteinase inhibitors (Ramesh et al. 2024), thaumatin-like proteins, HSP70 and glutathione-S-transferases were reported by

(Ramesh et al. 2024). This indicates a need for holistic regulation to fight the broad spectrum of pests and proposes one possibility which may be CRISPR-Cas9-based sterile insect technology focusing on removing reproductive fitness in female insects. These techniques are primarily aimed at producing insect-resilient crops of palms (Arumugam and Hatta 2022). Expressing the crucial somatic embryogenesis regulators Baby Boom (Bbm) and Wuschel2 (Wus2) could improve the transformation success in low-transformation efficiency plants, e.g., maize, (Jha et al. 2020). While it has not been tested in coconuts yet, this method may help the regeneration process. The antimicrobial browning agents should be effective in halting enzymatic browning reactions and resulting in products with good visual appearance, requirements for maintaining palm-based end product quality, especially the minimally processed tender coconut nuts.

In the meantime, as a product of modifying agronomic properties, this CRISPR/Cas9 technology was also investigated for mitigating enzymatic browning by targeting the $StPPO_2$ gene. In turn, new genome editing technologies are suggesting maintaining the nutritional qualities after harvest with high levels of antioxidants and phenolic compounds which is corroborated by decreasing enzymatic browning (González et al. 2021) The most common abiotic stressor that detrimentally affects plant ecosystems is drought, which has become intensified by the recent global warming trends. Indeed, genome editing of ARGOS8 results in crops that express more transcript has a higher level of drought resistance (Hamdan et al. 2022).

Using CRISPR/Cas9, researchers were able to replace the maize ARGOS8 regulator sequence with its GOS2 promoter (GOS2 PRO) and improve gene expression by building an increased drought stress resistance. Likewise, modifying the trehalase gene by CRISPR/Cas increased drought resistance in Arabidopsis thaliana (Shi et al. 2017). For instance, CRISPR/Cas9 has been used to edit SlNPR1 (called silencing gene) in tomatoes and reduce drought effects dramatically. Salt stress, another prevailing abiotic stress condition has drastically affected crop performance worldwide. Plant growth and nutrient uptake are impaired due to salts emplaced in the soil by transpiration, evaporation or seawater drift. Genome editing has been considered a promising approach to improve the resistance of plants toward salt stress. CRISPR-Cas9-edited rice with the OsRAV2 gene fared well at high-salt conditions, and those silencing OtRR22 similarly demonstrated salt resistance. Combining nanotechnology with CRISPR tools facilitates targeted genetic modifications, aiding the creation of enhanced plant cultivars (Alam et al., 2024c).

CRISPR/Cas technology has been used by researchers to target genes linked with salt and drought resistance in rice, increasing its ability to tolerate environmental stress (Yadav et al. 2023). Before implementing CRISPR/Cas9 editing in palms, such strategic factors as target gene choice, sgRNA sequence attributes and delivery mechanisms should be deliberated along with possible issues related to the delivered tissue or applicable in vitro regeneration techniques. Palm genomes are highly heterozygous, making it difficult to design ideally matching sgRNAs for targeting genes. With the multiplex CRISPR/Cas9, a variety of genetic sites have been targeted in different systems, leaving good expectations for its performance to enhance fatty acid metabolism pathways on palms.

For tree crops, rapid genetic modification is needed to swiftly bring GM plants into production. In contrast to the complex breeding techniques, biallelic alterations using the CRISPR/Cas9 system enable an easy way of generating homozygous lines (Wan et al. 2019). Powered by htmlPicard, the genome-editable palms without DNA is a real promise as being manifested with electroporation-mediated protoplast transformation in oil pal overall. Protoplast regeneration for oil palm is however still a challenge (Singh et al. 2017). These include efforts to develop efficient protoplast regeneration and transfection systems as well as PEG-mediated delivery of Cas proteins, in combination with sgRNA for the enhancement of CRISPR-mediated genome editing-based breeding strategies in palms. In such cases, genomic resources for homozygous crops and model plants can be improved with little effort by re-sequencing the genotypes of interest alongside adaptations in web-based algorithms to design sgRNAs that account for characteristics associated with palm genome multiploidy (Wu et al. 2020). CRISPR/Cas9-based genome editing will likely promote the exploration of gene-gene interactions and their effects on traits in palm species, which may expedite crop development.

There are key areas of inquiry such as looking for what attracts diseases to the coconut plant. Identifying genes or genomic regions that influence disease susceptibility, then manipulating them using genome-editing technology. PR1, PR4, transcriptional activator PT15-like gene thaumatin-like protein (TLP) HSP70 and glutathione S-transferase are keyset of disease root related genes targets by stressing variation exploratory principle om similar global expression profiling caused by diseases treatment under data acquisition for functional significant differential expressed or regulated genic cytoplasmic level by way of transcript biomarker levels(Ramesh et al. 2024). Scientists are now working to develop a more comprehensive method for controlling the plethora of pests attacking palm trees, which includes CRISPR-Cas9-based sterile insect technology designed specifically to reduce the reproductive fitness in females. Their primary purpose is to make palm crops more resistant to pests (Arumugam and Hatta 2022).

For example, defective plants with low transformation efficiency can be rescued by overexpressing important somatic embryogenesis regulators including Baby Boom (Bbm) and Wuschel2 (Wus2), but this has been employed in difficult-to-transform species such as maize so far to increase the percentage of transgenic events obtained during biolistic-mediated gene transfer mediated being successful instead of selecting specifically identifiable stable transformants subsequently from nontransformed callus. This method, though not tested with coconuts, may help in the regeneration process. Antibrowning agents are indispensable to minimize the enzymatic barking reactions, and also for a better visual profile of final products which is an essential attribute required for palm-based commercial food applications particularly having minimal process intervention on tender coconut nuts. In addition to altering agronomic traits, the CRISPR/Cas9 system has been investigated to mitigate enzymatic browning by modifying the *StPPO2* gene. Genome edition technologies allow for the maintenance of the nutritional potential of post-harvest produce, including its antioxidant capacity and phenolic content as well as reducing enzymatic browning (González et al. 2021).

Drought is the most widespread abiotic stress dampening plant ecosystems, aggravated by climate change. Genome editing of ARGOS8 increases ARGOS8 transcript levels and enhances drought resistance in crop plants (Hamdan et al. 2022). For example, the replacement of the ARGOS8 regulator with GTR1 by using CRISPR/Cas9 technology in maize resulted in an increase in drought tolerance and yield-reactivating expression179. hoodii for phytoremediation has application potential; likewise, the use of CRISPR/Cas to alter the trehalase gene in *Arabidopsis thaliana* allowed them to survive long-period dry conditions significantly better (Shi et al. 2017).

Studies have shown that this technology can also help alleviate drought-related damages by targeting genes like *SlNPR1* in tomatoes; therefore, minimizing the effects of a shortage of water through CRISPR/Cas9 is very promising. Global crop yields are also affected by the other most important abiotic stress namely, salt stress. However, the salt that is deposited in the soil from transpiration or as a result of evaporation/seawater drift impairs plant growth and nourishment uptake. Genome editing is widely used in improving the salt resistance of plants. Perturbations in the OsRAV2 genes were found to confer high-salt tolerance in rice, whereas knockout of two other transcription factors OsRR22 and one NAC family member-enhanced salt stress survival. CRISPR/Cas technology has been used to manipulate genes related to salt and drought stress in rice crops making hoist potential for abiotic stresses (Yadav et al. 2023).

19.9 Mutation Breeding

Mutation breeding involves the development of new plant varieties by inducing genetic mutations using chemical or radiation mutagenesis. This process accelerates the creation of novel cultivars with desired traits, bypassing the slow natural mutation rates that occur spontaneously. In this context, mutation breeding has proven essential for crop improvement, complementing traditional methods like cross-breeding (Smýkal et al. 2018).

19.9.1 Chemical and Radiation Mutagenesis

Genetic mutations can be brought about in plants using chemical mutagenesis and radiation mutagenesis, two prime methods of generating genetic diversity. Chemical mutagenesis, where chemicals are used to make point mutations in the plant genome (Untargeted approach), requires more optimization to be environmentally safe and biologically effective. The failures of radiation mutagenesis, which uses X-rays, gamma rays and accelerated particles produce highly deaf mutations whereas chemical mutants can be identified by the looseness helix genetic deficiencies. Whereas classical radiation mutagenesis uses X-rays and gamma rays to produce random mutations throughout the genome (Ma et al. 2021), particle mutation breeding applies accelerated particles, such as protons or heavy ions, which can generate predictable genetic changes with minimal phenotypic disturbance (Yadav et al. 2023). Space radiation mutation breeding uses the peculiar conditions of space, such as microgravity and fluctuating cosmic rays to generate high rates of mutations capable of translating into various new mutant plants (Bado et al. 2015). Mutation breeding has accelerated crop production, enhancing traits including disease resistance and yield & quality, among several areas of agriculture (Ghosh et al. 2016), besides biofuel development through mutagenesis in lipid-rich microalgae species.

19.9.2 Mechanistic Insights

Recent research has emphasized the necessity of connecting radiation to biological effects, mutation rates and plant survival as radiation properties (dose, LET) must be optimized. Studies with *Arabidopsis thaliana* have already shown that 300–400 Gy of carbon ions at a LET (linear energy transfer) of ~30 keV/μm deliver the maximum mutation rates (Liu et al. 2023). An increase in mutagenesis efficiency may be achieved by understanding how DNA repair mechanisms, including Homologous Recombination (HR) and Non-Homologous End Joining (NHEJ), function to result in particular types of mutations.

19.10 Hybridization

To gain the best characteristics and qualities from both parent plants and to create a superior hybrid offspring, therefore this mixing has to be perfect for successful plant breeding. This technique allows the breeder to mix desirable (but mostly unrelated) characteristics between two plant varieties in one hybridized Plant. Hybridization. By deploying parent plants with distinctive characteristics to the F_1 hybridization processes, it is possible to select traits and develop new improved plant varieties better suited for certain environmental conditions. Since then, this methodology has been broadly utilized for the creation of new forms in comprehensive top breeding programs that require certain specific agricultural steps and satisfy consumer needs.

19.10.1 Conventional Hybridization

Conventional hybridization allows for controlled selective breeding from one plant to another. The steps are parent choice, emasculation (if necessary to prevent self-pollination), hand-pollination, seed harvest and germination of offspring. This technique is effective but time-consuming, especially for species with longer generation times such as Areca palms and bounded by the genetic variability present in an existing population (Jerard et al. 2021).

19.10.2 Somatic Cell Hybridization

The somatic cell hybridization implies the fusion of protoplasts from two different plant types within which these plants are sexually incompatible. The protoplasts are isolated with enzymatic digestion followed by the fusion of cells using chemicals or electricity and thus creating larger cells. The hybrid cells are then regenerated into plants and screened for the desirable traits. For Artemisia, this has been a breakthrough in making novel crosses which are not possible through traditional hybridization; however, it requires specific techniques and the success rate is generally low (Ephrussi 2015).

19.10.3 Hybrid Cultivars of Areca Nut

Cross hybridization is the cornerstone of plant breeding, that is to say, obtaining offspring from two different parent plants with a specific desirable trait in high yield. Hybrid cultivars like VTL-17 and VTL-32 have been developed in betel nut cultivation to confer high yield, whereas dwarf varieties such as Mangala, Sumangala and Srimangala allow easy harvesting. These hybrids are bred for a variety of qualities like increased yield, disease resistance and improved nut quality. Presently, few research were attempted to bring further improvement in these cultivars by utilizing conventional hybridization methods as well as some of the advanced techniques like somatic cell hybridization and ensure continuous enrichment for more desirable traits (Khan et al. 2001; Nampoothiri and Parthasarathy 2018).

19.11 Conclusions and Prospects

Renewable energy sources have promising options in solar, wind, water (hydro), geothermal and biomass as alternatives to fossil fuels. Building on this research is the repeated goal to improve these technologies in terms of cost and performance. An example of this would be solar panels, the design is constantly being improved so that they can be cheaper to make and more efficient at converting sunlight into electricity. On the other hand, advancement in wind turbine technology is more about the durability and efficiency of generators. Efforts are also underway to promote "second-generation" biofuels, which use inedible plant parts like corn stalks and wheat straw (lignocellulosic biomass), as well as keep common food crops from being used for fuel.

Carbon capture and storage (CCS) where CO_2 emissions from industrial processes or even the atmosphere are captured and converted into useful products. It also looks at where it can be stored and how carbon capture methods of sufficient scale make a substantial contribution to removing carbon emissions from the atmosphere but that is cost-effective, scalable and reliable across a wide range of geological areas. This will involve researching different materials and methods for CO_2 capture as well as evaluating how safe it is to store the gas underground. Examples of these include studies on energy efficiency in industry, transport or the buildings sector. The goal is to reduce energy through better design, materials and technology. For instance, research in building design may assess how best to improve the insulation performance of buildings, optimize HVAC systems and integrate not only solar voltaic (PV) and solar thermal panels but also explore other renewable energy sources. Technologically speaking, transportation is transitioning on two fronts: better fuel efficiency through lightweight materials and engines in the short term. Also, industrial research is centered around energy conservation with waste management in the production process.

Modeling of the future loading, feedback loops and other climate changes requires computer simulations. In this realm of analysis, the focus is on enhancing our ability to generate higher resolution and more accurate climate model simulations utilizing a greater number of datasets along with a better tuning approach for models. This includes properly representing components of the Earth system such as ice dynamics, and atmospheric and oceanic circulations. Thereby, there have also been increased efforts towards improving the predictive power of models for local and regional climate change.

One of the key areas for adaptation and resilience research is in the area that focuses on developing strategies (including ecosystem-based approaches) to respond effectively to climate change impacts and enhance both infrastructural, food system as well as the community's adaptive capacity. It means identifying relevant vulnerabilities, assessing those risks and developing policies to strengthen our ability to adapt. Research also investigates using nature-based solutions to increase the resistance of our societies and environment—from green infrastructure, and sustainable agricultural practices to ecosystem restoration.

The Areca palm opens potentially many areas for exploration in the future. Further research can be conducted to improve the extraction and processing of oil from Areca palm seeds so that a greener, more efficient and more economically beneficial method could benefit in producing biodiesel. Further, the research could explore whether Areca palms have value in remediating contaminated water and soil which would be beneficial to environmental cleanup initiatives. Exploring the use of stems as a potential green construction material for bio-composites manufactured with Areca palm among regions abundant in this plant could be highly advantageous. Moreover, further research on the bioactive compounds from betel nuts opens up avenues for nutritional and medicinal values, which ultimately will be used in pharmaceuticals or nutraceutical products.

In short, collaborative interdisciplinary research is key to overcoming the challenging societal and technological consequences of climate change and promoting creative solutions such as innovatively exploiting plant resources like betel nut palm. Any such barriers should be considered as a target for strategies or interventions seeking to optimize implementation and reach disadvantaged populations, particularly in low- and middle-income countries.

Competing of Interest There is no competing interest among authors.

Appendix 19.1 Research Institutes and Online Resources Available for Betal Nut Breeding

S. no	Institution	Area of specialization	Research activities	Website
1.	Indian Institute of Horticultural Research (IIHR), Bengaluru, India	Horticultural research	Development of high-yielding and disease-resistant varieties Genetic analysis Marker-assisted selection	www.iihr.ernet.in
2.	Indian Agricultural Research Institute (IARI), New Delhi, India	Plant breeding and genetics	Development of new betel nut varieties Genetic analysis Genomics	www.iari.res.in
3.	Central Plantation Crops Research Institute (CPCRI), Kasaragod, India	Breeding, post-harvest management	Breeding for yield improvement Disease resistance Post-harvest management	www.cpcri.gov.in
4.	International Crops Research Institute for the Semi-Arid Tropics (ICRISAT), Patancheru, India	Crop improvement and genetic resources	Development of climate-resilient and disease-resistant *Areca* nut varieties	www.icrisat.org
5.	Tropical Botanical Garden and Research Institute (TBGRI), Palode, India	Plant genetic resources & conservation	Conservation and utilization of betel nut genetic resources	www.tbgrin.org
6.	Chinese Academy of Tropical Agricultural Sciences, Hainan, China	Tropical agricultural research	Breeding for yield improvement Disease resistance Quality enhancement	www.catas.cn
7.	Malaysian Agricultural Research and Development Institute (MARDI), Selangor, Malaysia	Agricultural research and development	Development of high-yielding and disease-resistant betel nut varieties	www.mardi.gov.my

S. no	Institution	Area of specialization	Research activities	Website
8.	Asia Pacific Association of Agricultural Research Institutions (APAARI), Bangkok, Thailand	Regional agricultural research collaboration	Regional coordination and collaboration for betel nut research	www.apaari.org
9.	International Society for Horticultural Science (ISHS), Leuven, Belgium	Horticultural research and education	Global networking and knowledge sharing on betel nut research	www.ishs.org
10.	Online Resource: ResearchGate	Online research community	Connect with researchers Share research Collaborate on betel nut breeding projects	www.researchgate.net
11.	Online Resource: Academia.edu	Online academic community	Access to research papers, articles, and lectures on betel nut breeding and related topics	www.academia.edu

Appendix 19.2 Genetic Resources for Betel Nut Breeding

Species	Important traits	Cultivation location
Areca catechu (Betel nut palm)	Yield, disease resistance, drought tolerance, nut quality, and ratooning ability	Native to Southern India and Southeast Asia; widely cultivated across the humid tropics and subtropics
Areca vestiaria (Gamultin palm)	Nut quality, flavor, and ornamental characteristics	Eastern Indonesia and Melanesia
Areca Concinna (Supari palm)	Nut quality and milder taste	Southern India and Sri Lanka
Areca triandra (Wild *Areca* palm)	Disease resistance and genetic diversity	Native to Southeast Asia
Areca IPA (Golden *Areca* palm)	Ornamental characteristics and disease resistance	Southern India (Kerala and Tamil Nadu)

References

Abbas MS, Taha HS, Aly UI et al (2011) In vitro propagation of ginger (*Zingiber officinale* Rosco). J Genet Engin Biotech 9(2):165–172

Abioye EA, Hensel O, Esau TJ et al (2022) Precision irrigation management using machine learning and digital farming solutions. Agric Eng 4(1):70–103

Ahmad R, Anjum MA, Naz S, Balal RM (2021) Applications of molecular markers in fruit crops for breeding programs- a review. Phyton 90(1):17–34
Ahuja S, Ahuja U (2011) Betel leaf and betel nut in India: history and uses. Asian Agri-History 15(1):13–35
Aktar LA, Chowdhury GM, Rahman MA (2021) Effect of pre-sowing treatments and different growing media on seed germination and seedling growth of jigni (*Trema orientalis* Linn.) blume. Bangladesh J Forest Sci 37:1
Alam MS, Haider R, Ray TK (2020) Raising of Calamus erectus Roxb. Seedlings and their performance at nursery and field conditions. Bangladesh J Forest Sci 36:2
Alam O, Khan A, Khan WU et al (2024a) Improving heat tolerance in betel palm (Areca catechu) by characterisation and expression analyses of heat shock proteins under thermal stress. Crop Pasture Sci 75(9)
Alam O, Khan LU, Khan A et al (2024b) Functional characterisation of Dof gene family and expression analysis under abiotic stresses and melatonin mediated tolerance in pitaya (Selenicereus undatus). Funct Plant Biol 51(4)
Alam O, Khan SU, Gul S et al (2024c) Innovative Nanotechnology in CRISPR Based Crop Genome Editing. In Revolutionising Agriculture: A Comprehensive Exploration of Agri Nanotechnology. Springer Nature Switzerland, pp 121–151
Albrecht M, Kleijn D, Williams NM et al (2020) The effectiveness of flower strips and hedgerows on pest control, pollination services and crop yield: a quantitative synthesis. Ecol Lett 23(10):1488–1498
Alouw J, Wulandari S (2020) Present status and outlook of coconut development in Indonesia. IOP Conf Ser Earth Environ Sci 418:012035
Al-Rmalli SW, Jenkins RO, Haris PI (2011) Betel quid chewing elevates human exposure to arsenic, cadmium and lead. J Hazardous Mat 190(1–3):69–74
Ananda K, Karun A, Parthasarathy V (2017) Arecanut. Today & Tomorrow's Printers and Publishers, New Delhi
Apshara SE (2019) Cocoa genetic resources and their utilization in palm-based cropping systems of India Theobroma cacao-deploying science for sustainability of global cocoa economy. IntechOpen
Arifuzzaman Khan G, Shahrear Palash S, Shamsul Alam M et al (2012) Isolation and characterization of betel nut leaf fiber: its potential application in making composites. Polym Compos 33(5):764–772
Arora S, Squier C (2019) Areca nut trade, globalization and its health impact: perspectives from India and Southeast Asia. Perspectives Public Health 139(1):44–48
Arul Swaminathan A (2002) Isozyme variation Areca catechu L. and allied species. Department of Plantation Crops and Spices, College of Horticulture, Vellanikkara
Arumugam T, Hatta MAM (2022) Improving coconut using modern breeding technologies: challenges and opportunities. Plants 11(24):3414
Arunachalam V (2011) Genomics of cultivated palms. Elsevier
Bado S, Forster BP, Nielen S et al (2015) Plant mutation breeding: current progress and future assessment. Plant Breed Rev 39:23–88
Bavappa K (1963) Morphological and cytological studies in Areca catechu Linn. and Areca triandra Roxb. M. Sc. (Ag) Thesis. University of Madras, India, pp 63
Bhat R, Sujatha S (2009) Soil fertility and nutrient uptake by areca nut (Areca catechu L.) as affected by level and frequency of fertigation in a laterite soil. Agr Water Manag 96(3):445–456
Bhojwani SS, Dantu PK, Bhojwani SS, Dantu PK (2013) Micropropagation. In: Plant tissue culture: an introductory text. Springer India, pp 245–274
Bradshaw JE (2017) Review and analysis of limitations in ways to improve conventional potato breeding. Potato Res 60:171–193
Bulwayan RB, Balocnit D, Sannadan JGM (2023) Piper betel and areca nut: interfacing cultural, communication and ethno-therapeutic knowledge as reflected in the Kalinga epic, the ullalim. Int J English Literat Soc Sci 8(3):385–404

Cao HX, Vu GTH, Gailing O (2022) From genome sequencing to CRISPR-based genome editing for climate-resilient forest trees. Inter J Mol Sci 23(2):966

Cardoso JC, Sheng Gerald LT, Teixeira da Silva JA (2018) Micropropagation in the twenty-first century. In: Plant cell culture protocols. Springer, New York, pp 17–46

Chakrabarty J, Hassan MM, Khan MA (2012) Effect of surface treatment on betel nut (Areca catechu) fiber in polypropylene composite. J Polym Environ 20:501–506

Chan E, Elevitch CR (2006) Cocos nucifera (coconut). Pacific Island Agrofor 2(1):1–27

Chaturvedi T, Gupta AK, Shanker K et al (2022) Maximizing genetic gain through unlocking genetic variation in different ecotypes of kalmegh (Andrographis paniculata (Burm. F.) nee). Front Plant Sci 13:1042222

Clapp J, Ruder S-L (2020) Precision technologies for agriculture: digital farming, gene-edited crops, and the politics of sustainability. Global Environ Polit 20(3):49–69

Cock JH, Ayala-Diaz IM, Romero HM (2023) A scheme for distribution of genetically improved oil palm plants well suited to local conditions. Agric Syst 212:103756

Crews TE, Carton W, Olsson L (2018) Is the future of agriculture perennial? Imperatives and opportunities to reinvent agriculture by shifting from annual monocultures to perennial polycultures. Global Sustain 1:e11

Del Vecchio F, Mastroiaco V, Di Marco A et al (2017) Next-generation sequencing: recent applications to the analysis of colorectal cancer. J Transl Med 15:1–19

Dharmaraj V, Vijayanand C (2018) Artificial intelligence (AI) in agriculture. Int J Curr Microbiol App Sci 7(12):2122–2128

Dwiartama A, Akbar ZA, Ariefiansyah R et al (2024) Conservation, livelihoods, and Agrifood Systems in Papua and Jambi, Indonesia: a case for diverse economies. Sustainability 16(5):1996

Ensminger M, Löbrich M (2020) One end to rule them all: non-homologous end-joining and homologous recombination at DNA double-strand breaks. Br J Radiol 93(1115):20191054

Ephrussi B (2015) Hybridization of somatic cells. Princeton University Press

Faisal M, Alatar AA (2019) Synthetic seeds. Springer

Farag GY, Abou Taleb EMA, & Hamouda T (2025) Natural Fibers Extraction Methods and Properties: A Review. Egypt J Chem 68(3):445–464

Ganaraja K (2017) A comparative analysis of trends in area, production and productivity of areca nut. Int J Soc Econ Res 7(1):50–60

Gantait S, Kundu S, Das PK (2018) Acacia: an exclusive survey on in vitro propagation. J Saudi Soc Agric Sci 17(2):163–177

Gao C (2021) Genome engineering for crop improvement and future agriculture. Cell 184(6):1621–1635

Garg A, Chaturvedi P, Gupta PC (2014) A review of the systemic adverse effects of areca nut or betel nut. Indian J Med Pediatr Oncol 35(01):3–9

Ghosh A, Khanra S, Mondal M et al (2016) Progress toward isolation of strains and genetically engineered strains of microalgae for production of biofuel and other value added chemicals: a review. Energy Convers Manag 113:104–118

González MN, Massa GA, Andersson M et al (2021) Comparative potato genome editing: agrobacterium tumefaciens-mediated transformation and protoplasts transfection delivery of CRISPR/Cas9 components directed to StPPO2 gene. Plant Cell, Tissue Organ Cult (PCTOC) 145(2):291–305

Gunjal S, Pateel DGS, Yang Y-H et al (2020) An overview on betel quid and areca nut practice and control in selected Asian and South East Asian countries. Subst Use Misuse 55(9):1533–1544

Hamdan N, Lee CH, Wong SL et al (2022) Prevention of enzymatic browning by natural extracts and genome-editing: a review on recent progress. Molecules 27(3):1101

Hebbar KB, Ramesh S, Bhat R (2024) Predicting current and future climate suitability for areca nut (*Areca catechu* L.) in India using ensemble model. Heliyon

Hooper GW, Biega T, Ginat DT (2015) Betel Nuts. In: Neuroimaging pharmacopoeia. Springer International Publishing, pp 79–81

Hussain MS, Fareed S, Ansari S et al (2012) Current approaches toward production of secondary plant metabolites. J Pharm Bioallied Sci 4(1):10–20

Javed MR, Abbasi AZ, Akhtar MJ et al (2021) Current trends and recent progress of genetic engineering in genus Phytophthora using CRISPR systems. CRISPR RNAi Syst. 183–209

Jayamani E, Hamdan S, Rahman MR, Bakri MKB (2014) Investigation of fiber surface treatment on mechanical, acoustical and thermal properties of betelnut fiber polyester composites. Procedia Eng 97:545–554

Jerard B, Niral V, Rajesh M (2021) Breeding strategies. Coconut Genome :47–76

Jha P, Ochatt SJ, Kumar V (2020) WUSCHEL: a master regulator in plant growth signaling. Plant Cell Rep 39:431–444

Kalantari F, Mohd Tahir O, Mahmoudi Larijani A, Kalantari S (2017) A review of vertical farming technology: a guide for implementation of building integrated agriculture in cities. In: Advanced engineering forum, vol 24. Trans Tech Publications Ltd, pp 76–91

Kamaluddin KM, Kiran U et al (2017) Molecular markers and marker-assisted selection in crop plants. In: Plant biotechnology: principles and applications. Springer Singapore, Singapore, pp 295–328

Karun A, Sajini K, Muralikrishna K, Rajesh M, Engelmann F (2017) Cryopreservation of arecanut (Areca catechu L.) pollen. Cryo Lett 38(6):463–470

Karun A, Muralidharan K, Jayasekhar S (2022) Technologies for coconut, arecanut and cocoa: status and way forward. Progr Horticult 54(1):3–27

Karun A, Muralidharan K, Jayasekhar S (2023) Towards self-reliance: technology development for coconut, cocoa, and arecanut. Inter J Innov Hort 12(1):50–72

Khan H, Upadhyay A, Palaniswami C (2001) Potassium in plantation crops. In: Proceedings internationall symposium on k in nutrient management for sustainable crop production in India, New Delhi, pp 467–497

Khang NHM, Quang NT, Mai HNX et al (2022) Genetic characterization of coconut (Cocos nucifera L.) varieties conserved in Vietnam through SCoT marker-based polymorphisms. Genet Resour Crop Evol 69:1–14

Kingsbury N (2011) Hybrid: the history and science of plant breeding. University Chicago Press

Kumar S, Manoharan AP, Chandrakalatharan J (2021) Applying competencies: state capability enhancement project (SCEP). In: The public productivity and performance handbook. Routledge, pp 453–470

Kumar BM, Kunhamu T, Bhardwaj A, Santhoshkumar A (2024) Subcanopy light availability, crop yields, and managerial implications: a systematic review of the shaded cropping systems in the tropics. Agrofor Syst 98:1–26

Kumari A, Das A, Mahanta M, Akhila G, Dutta P (2023) Diseases of areca nut (Areca catechu) and their integrated management. In: Diseases of commercial crops and their integrated management. CRC Press, pp 88–99

Li J, Jia X, Yang Y et al (2022) Genome-wide identification of the DOF gene family involved in fruitlet abscission in Areca catechu. Inter J Mol Sci 23(19):11768

Lim T (2011) Areca catechu. In: Edible medicinal and non-medicinal plants: volume 1, fruits. Springer, pp 260–276

Lim EM, Briones SA, Canilao MAP et al (2021) Inferring the role of betel nut chewing practice in the maritime spice trade. PRO 43

Liu SY (2020) Artificial intelligence (AI) in agriculture. IT Prof 22(3):14–15

Liu Y, Wei C, Wang H et al (2021) Indirect somatic embryogenesis and regeneration of *Fraxinus mandshurica* plants via callus tissue. J Forestry Res 32(4):1613–1625

Liu J, Zhao G, Geng J et al (2023) Genome-wide analysis of mutations induced by carbon ion beam irradiation in cotton. Front Plant Sci 14:1056662

Liu Z, Wang X, & Wen L (2025) Exploring the potential benefits of areca nut extract in animal production: a review. Front Anim Sci 6:1495886

Luong JHC, Kozlakidis Z, Cheong IH, Wang H (2023) Innovations and limitations in areca nut research: a narrative review. Innov Digital Health Diagnostics Biomarkers 3(2023):9–14

Ma L, Kong F, Sun K et al (2021) From classical radiation to modern radiation: past, present, and future of radiation mutation breeding. Front Public Health 9:768071
Majumdar RC (2016) Ancient India. Motilal Banarsidass
Manimekalai R, Rajesh M, Manju K (2013) Biotechnology and genomics in palms
Mohankumar D, Amarnath V, Bhuvaneswari V, Saran SP, Saravanaraj K, Gogul MS, Sridhar S, Kathiresan G, &Rajeshkumar L (2021) Extraction of plant based natural fibers – A mini review. IOP Conference Series: Mater Sci Eng 1145(1):012023
Manimekalai R, Nair S, Naganeeswaran A et al (2018) Transcriptome sequencing and de novo assembly in arecanut, Areca catechu L elucidates the secondary metabolite pathway genes. Biotech Rep 17:63–69
Mariamma TR, Nair HS (2022) Popularization of Agrowaste Fibres – Banana and areca nut fibre – a sustainable approach. In: Sustainable approaches in textiles and fashion: fibres, raw materials and product development. Springer, pp 81–99
Mashimo T (2014) Gene targeting technologies in rats: zinc finger nucleases, transcription activator-like effector nucleases, and clustered regularly interspaced short palindromic repeats. Develop Growth Differ 56(1):46–52
Mbi KT, Tonfack LB, Ntsefong GN, Mir BA et al (2016) Mature zygotic embryo rescue improves in vitro germination and seedling production in high-value oil palm (Elaeis guineensis Jacq.) cultivars. Ind Crop Prod 94:445–453
Minford J (2021) The best China. The Chinese University of Hong Kong Press
Mondal B (2022) Extraction of betel leaf essential oil for the sustainable solution to betel business in West Bengal for effective economic gain: a review. Mysore J Agric Sci 56(2)
Mondal D, Trivedi H (2023) Arecanut varieties and methods of cultivation and harvesting. J Namibian Stud: Hist Polit Cult 37:295–301
Morgenstern M (2011) Geographic variation in forest trees: genetic basis and application of knowledge in silviculture. UBC Press
Muralikrishna K, Rajesh M, Sajini K, Nagaraja N, Ananda K & Karun A (2017) Immature embryo culture in wild Areca spp.
Nair KP (2009) The agronomy and economy of some important industrial crops. Adv Agron 101:183–313
Nair KP (2010) The agronomy and economy of important tree crops of the developing world
Nair KP (2021) Tree crops. Harvesting cash from the world's important cash crops, 1st edn. Springer Nature, Cham, pp 249–285
Nair KP, Nair KP (2021a) Arecanut (Areca catechu L.). In: Tree crops: harvesting cash from the world's important cash crops. Springer International Publishing, pp 1–25
Nair KP, Nair KP (2021b) Technological advancements in coconut, Arecanut and cocoa research: a century of service to the global farming community by the central plantation crops research institute. In: Tree crops: harvesting cash from the world's important cash crops. Springer International Publishing, Kasaragod, pp 377–536
Nampoothiri K, Parthasarathy V (2018) Varietal improvement. In: The coconut palm (Cocos nucifera L.)-research and development perspectives. Springer Singapore, pp 113–156
Organisation (2012) Review of areca (betel) nut and tobacco use in the Pacific: a technical report
Ortega F, Lopez-Vizcon C (2012) Application of molecular marker-assisted selection (MAS) for disease resistance in a practical potato breeding programme. Potato Res 55(1):1–13
Papke RL, Bhattacharyya I, Hatsukami DK et al (2020) Betel nut (areca) and smokeless tobacco use in Myanmar. Subst Use Misuse 55(9):1385–1394
Rahman MA, Tani M, Asahiro K, Asik Ullah S (2017) Species composition, diversity and productivity of homesteads in southeastern Bangladesh. Small-scale Forestry 16:295–309
Rahman M, Rahman M, Uddin M, Kabir AF (2022) Evaluation of betel nut germplasm in Barishal region. Bangladesh J 291
Raimondeau P, Manzi S, Brucato N et al (2021) Genome skims analysis of betel palms (areca spp., Arecaceae) and development of a profiling method to assess their plastome diversity. Gene 800:145845

Rajmohan K (2005) Plant biotechnology research at the Kerala Agriculture University
Raju J, Jayalakshmi K, Sonavane PS, Raghu S (2022) Present scenario of diseases in areca nut or betel nut (Areca catechu L.) and their management. In: Diseases of horticultural crops: diagnosis and management. Apple Academic Press, pp 3–26
Ramesh SV (2024) CRISPR/Cas9–based genome editing to expedite the genetic improvement of palm crops. Front Plant Sci 15:1385037
Ramanatha Rao V, Rajasekharan P, Anushma P (2023) Conservation and management of fruit and nut genetic resources. In: Fruit and nut crops. Springer, pp 1–22
Ramesh S, Rajesh M, Das A, Hebbar K (2024) CRISPR/Cas9–based genome editing to expedite the genetic improvement of palms: challenges and prospects. Front Plant Sci 15:1385037
Ranade S, Verma A, Gupta M, Kumar N (2002) RAPD profile analysis of betel vine cultivars. Biol Plant 45:523–527
Rathore MS, Yadav P, Mastan SG et al (2014) Evaluation of genetic homogeneity in tissue culture regenerates of Jatropha curcas L. using flow cytometer and DNA-based molecular markers. Appl Biochem Biotech 172:298–310
Rauf S, Al-Khayri JM, Zaharieva M et al (2016) Breeding strategies to enhance drought tolerance in crops. In: Advances in plant breeding strategies: agronomic, abiotic and biotic stress traits. Springer International Publishing, Cham, pp 397–445
Rivas M, Barbieri RL, Maia LCD (2012) Plant breeding and in situ utilization of palm trees. Ciência Rural 42:261–269
Rogo U, Fambrini M, Pugliesi C (2023) Embryo rescue in plant breeding. Plants 12(17):3106
Roy A, Guha P (2021) Traditional and functional uses of betel leaf (*Piper betle* L.) pertaining to food sector: a review. Food Sci Technol 9:72–85
Salehi B, Konovalov DA, Fru P et al (2020) Areca catechu – from farm to food and biomedical applications. Phytother Res 34(9):2140–2158
Sathishkumar C, Dineshkumar G, Karthikeyan K, Prasanna P (2021) Development of areca nut climbing and harvesting machine with pesticide. IOP Conf Ser: Mater Sci Eng 1059:012074
Sheidai M, Noormohammadi Z, Alishah O (2024) Future cultivation of cotton for industrial use: landscape cytogenetics and species distribution modeling. Genet Resour Crop Evol 71:1–12
Shenoy Heckadka S, Nayak SY, Joe T et al (2022) Comparative evaluation of chemical treatment on the physical and mechanical properties of areca frond, banana, and flax fibers. J Nat Fibers 19(4):1531–1543
Shi J, Gao H, Wang H et al (2017) ARGOS 8 variants generated by CRISPR-Cas9 improve maize grain yield under field drought stress conditions. Plant Biotechnol J 15(2):207–216
Singh V, Braddick D, Dhar PK (2017) Exploring the potential of genome editing CRISPR-Cas9 technology. Gene 599:1–18
Singh A, Dikshit R, Chaturvedi P (2020) Betel nut use: the south Asian story. Subst Use Misuse 55(9):1545–1551
Smýkal P, Nelson MN, Berger JD, Von Wettberg EJ (2018) The impact of genetic changes during crop domestication. Agron 8(7):119
Staples GW, Bevacqua RF (2006) Areca catechu (betel nut palm). Pacific Island Agrofor 1(13):1–9
Sunny G, Palani Rajan T (2024) Influence of blend ratio of cotton and Arecanut fibers on yarn properties. Int J Clothing Sci Technol 36(2):304–316
Thomas GV, Subramanian P, Gopal M et al (2022) Unraveling the potential of belowground and aboveground biodiversity for sustainable Management of the Health of plantation crop soils in coastal agro-ecosystem. In: Transforming coastal zone for sustainable food and income security: proceedings of the international symposium of ISCAR on coastal agriculture. Springer, pp 633–643
Thube SH, Josephrajkumar A, Pandian RTP et al (2021) Concurrent emergence of exotic whitefly incursions on areca nut (*Areca catechu* L.) in India. J Plantation Crops 49:28–34
Van Delden S, Sharath Kumar M, Butturini M et al (2021) Current status and future challenges in implementing and upscaling vertical farming systems. Nat Food 2(12):944–956

Veluru A, Mohamed N, Shil S et al (2024) Standardization of in vitro multiplication technique for *Areca concinna* Thwaites, an endangered palm species for its conservation and utilization. In Vitro Cell Dev Biol-Plant 60:1–7

Vincelli P (2016) Genetic engineering and sustainable crop disease management: opportunities for case-by-case decision-making. Sustainability 8(5):495

Vincent WM (2012) The complete guide to working with worms: using the gardener's best friend for organic gardening and composting. Atlantic Publishing Company

Vishwajith K, Sahu P, Dhekale B, Noman M, Mishra P (2015) Exploring the feasibility of arecanut-based farming systems in augmenting farm economy-a case study in Karnataka, India

Wan T, Niu D, Wu C, Xu F-J, Church G, Ping Y (2019) Material solutions for delivery of CRISPR/Cas-based genome editing tools: current status and future outlook. Mater Today 26:40–66

Wang HsiangChih WH, Chen JenTsung CJ, Wu ShiehPing WS et al (2003) Plant regeneration through shoot formation from callus of Areca catechu

Wang H-C, Chen J-T, Wu S-P, Lin M-C, Chang W-C (2003) Plant regeneration through somatic embryogenesis from zygotic embryo-derived callus of Areca catechu L.(Arecaceae). In Vitro Cell Dev Biol-Plant 39:34–36

Weckx S (2020) Towards a novel system for large-scale micropropagation of oil palm. Ghent University

Wu S, Zhu H, Liu J et al (2020) Establishment of a PEG-mediated protoplast transformation system based on DNA and CRISPR/Cas9 ribonucleoprotein complexes for banana. BMC Plant Biol 20:1–10

Yadav RK, Tripathi MK, Tiwari S et al (2023) Genome editing and improvement of abiotic stress tolerance in crop plants. Life 13(7):1456

Yang Y, Huang L, Xu C et al (2021) Chromosome-scale genome assembly of areca palm (*Areca catechu*). Mol Ecol Resour 21(7):2504–2519

Yang B, Chen H, Chen W et al (2023) Edible quality analysis of different areca nuts: compositions, texture characteristics and flavor release behaviors. Foods 12(9):1749

Yudaputra A, Zulkarnaen RN, Wanda IF, Witono JR (2016) Flowering of Areca catechu in Bogor botanic gardens. Berkala Penelitian Hayati 22(1):32–37

Zenkteler M (1984) In vitro pollination and fertilization. In: Cell culture and somatic cell genetics of plants, vol 1, pp 269–275

Chapter 20
Black Wattle (*Acacia mangium* Willd.): Current Issues, Challenges and Opportunities for Advanced-Generation Breeding Strategy

Arif Nirsatmanto and Sri Sunarti

Abstract The industrial forest plantation program in Indonesia has introduced black wattle (*Acacia mangium* Willd.). as a major species due to its fast growth, adaptability to various sites and suitability for industries such as pulp and paper, sawn timber, and construction. Early-generation breeding has successfully improved growth, stem form, and wood properties. However, new challenges are facing the *A. mangium* plantation due to pest and disease attacks, which have caused very serious problems in the forest plantation. It caused the species to stop being planted by most plantation companies in Indonesia. Plant breeding efforts could be the main solution for overcoming the problems by producing disease-tolerant and resistant planting stocks. Continuing the past breeding strategy of *A. mangium* is also facing other problems related to the limited genetic resources tolerant and resistant to disease. In addition, the past breeding strategy seems not to be effectively re-implemented in successive advanced-generation breeding to improve the new target traits. However, there are still some opportunities for keeping the potential and regaining *A. mangium* as the main species for plantations in the future. Re-designing the breeding strategy and the biotechnology technique approach is also another opportunity to support advanced-generation breeding. A conservative breeding strategy supported by biotechnology techniques would become the main component in the advanced-generation breeding of *A. mangium*. This chapter provides an overview of the current status, challenges, and opportunities for advanced-generation breeding of *A. mangium* incorporating biotechnology for fiber production from industrial plantations in Indonesia.

A. Nirsatmanto (✉) · S. Sunarti
National Research and Innovation Agency (BRIN) the Republic of Indonesia, Bogor, Indonesia
e-mail: arif050@brin.go.id; sris017@brin.go.id

K. F. M. Salem et al. (eds.), *Breeding and Biotechnology of Grass and Bast Fiber Crops*, Advances in Plant Breeding Strategies 12,
https://doi.org/10.1007/978-3-032-00403-1_20

Keywords Black Wattle (*Acacia mangium* Willd.) · Biotechnology · Breeding population · Conservative breeding strategy · Disease resistance · Fiber production · Genetic trial

20.1 Introduction

The fiber-based products are one of the object businesses engaged commercially by many industries worldwide. This is because fiber can be used as raw materials for various end-products including pulp and paper, clothing materials, bio-composites and bio-materials. The global fiber demand reached 113 million tons in 2021, after declining by approximately 99 million tons in 2020 due the COVID-19, which is reported that further analysis of the demand is expected to reach around 142 million tons in 2030 (Gschwandtne 2022).

The fiber raw materials are produced and derived from many resources including plants, animals and other synthetic materials. In terms of plants, the fiber resources involve various plants based on size and morphological features, types of organs used, and ages of harvesting. The fiber can be collected from the stalk, stem, leaf, seed, grass and wood. Therefore, improvement in fiber production of the plants should be directed towards the types of organs used.

Pulp, paper and rayon industries are among the industries that consume fiber produced from plants. The global demand for pulp paper has witnessed substantial growth due to various factors including population growth, industrialization, the e-commerce boom, and environmental concerns. Industries such as packaging, printing, and publishing have benefited from this surge in demand. However, the industry heavily relies on raw materials, making it crucial to focus on ensuring a sustainable supply. In addition, it is also essential to address the environmental impact associated with pulp paper production through sustainable practices. By striking a balance between the raw materials supply, economic growth, and environmental stewardship, the pulp and paper industry can continue to meet the evolving needs of consumers while minimizing its ecological impact.

Wood-based fiber production plays a vital role in supplying raw materials for the pulp and paper industry. By prioritizing strategies to enhance productivity and resource utilization, the industry can meet increasing demands while promoting sustainability. Technological advancements and supply chain optimization are crucial for boosting productivity, while circular economy principles, energy efficiency measures, and sustainable forest management contribute to resource optimization. Through these efforts, the wood-based fibers industry can continue to thrive as a key driver of economic growth and environmental stewardship.

Acacia mangium is one the tropical species that present an important source of wood fibers for pulp and paper production in Southeast Asia regions such as Indonesia, Malaysia, Vietnam, Thailand and Papua New Guinea (Harwood and Nambiar 2013; Turnbull et al. 1997). Its fast growth rate, high fiber yield, favorable pulping properties, and versatility make it a valuable raw material resource for various paper grades and specialty applications (Harwood and Nambiar 2014). With responsible management practices, *A. mangium* can contribute to sustainable pulp and paper production to ensure a reliable supply of fibers for the industry from the plantation, particularly in mineral soils (Hardiyanto and Nambiar 2014). Additionally, the *A. mangium* species are important genetic material for creating a new variety of *Acacia* hybrids, which are inter-specific hybrids between *A. mangium* and *A. auriculiformis*. The *Acacia* hybrids are potential varieties for fiber production due to combining superior traits from both parents (Hasegawa et al. 2009; Nirsatmanto and Sunarti 2019; Nirsatmanto et al. 2022). In this case, improving *A. mangium* is crucial for preparing high-quality genetic materials for hybridization.

During the last two decades, *A. mangium* plantations have emerged as a valuable source of wood fibers for Indonesia's pulp and paper industry (Arisman and Hardiyanto 2006; Harwood and Nambiar 2013; Hardiyanto and Nambiar 2014). However, currently, some serious issues and challenges have appeared that have disturbed the sustainability of *A. mangium* plantations for fiber production. Therefore, strong efforts should be made, particularly in plant breeding, to overcome the issues and address the challenges by optimizing the breeding of *A. mangium* to produce genetically improved planting materials with the desired traits. This chapter explores the efforts to re-design the breeding strategy for *A. mangium*, focusing on sustainable tree growth, better wood properties, resistance to disease, and resilience to site adaptability to regain *A. mangium* as the main species for fiber production in the tropics. It covers a discussion on *A. mangium* breeding related to the available genetic resources, issues in the sustainability of plantation, current breeding strategy and potential for improvement strategy incorporating biotechnology, as well as the future direction for research.

20.1.1 Importance and Plantation

The species of *A. mangium* is a fast-growing tree native to Australia, Papua New Guinea, and the Eastern part of Indonesia (Turnbull et al. 1997). This species is renowned for its rapid growth rate, making it an ideal choice for fiber production. The tree can reach maturity in as little as 5 years, allowing for quick and efficient harvesting. In the mineral soil sites, *A. mangium* yields significantly higher fiber content compared to other Acacia tree species, making it highly desirable for commercial purposes. The fiber is used in various industries, including pulp and paper, construction, and furniture manufacturing.

Wood fiber is an essential component of the structure of trees that plays a critical role in several industries, including pulp and paper, construction, and textiles. Broadleaf trees, also known as hardwood trees, such as *A. mangium* produce wood fiber that has distinctive characteristics in hardness, grain pattern, color and density that set it apart from softwood fiber which is produced by coniferous trees (National Research Council 1983). The broadleaf wood is harder with attractive grain patterns and higher density and colorful than the softwood which usually has a uniform bright color with lower wood density.

The dimensional aspects of wood fiber, including length, fiber diameter, lumen diameter, and cell wall thickness of *A. mangium*, are critical factors influencing the quality of end products such as pulp, paper, and textiles, as highlighted in the study by Andianto et al. (2020). Various derived factor values, such as Runkle Ratio, Felting Power, Muhlsteph Ratio, Flexibility Ratio, and Stiffness Coefficient (coefficient of rigidity), significantly influence the overall quality of the product. To optimize the quality of wood fiber for industrial applications, particularly in the pulp and paper industry, it is crucial to understand and control fiber dimensions and related factors. This is because nearly 87% of pulp and paper is fiber, making it the predominant component of wood cells. In comparison, vessels and parenchyma each contribute approximately 6–7%, while other cell types contribute a smaller percentage, according to Marsoem et al. (2012).

One of the studies on *A. mangium* fiber was conducted by Adianto et al. (2020) which focused on the 4-year-old third-generation (F-3) *A. mangium* planted in Wonogiri, Central Java Province, Indonesia. The results showed that the fiber quality of *A. mangium* was categorized within the class II classification, with an average fiber length of 1227 μm (Fig. 20.1). These results are in line with the previous findings of Sahri et al. (1993) and Lim et al. (2011), who also examined the fibers of *A. mangium* in Malaysia. Sahri et al. (1993) reported a mean fiber length of 933.7 μm in 4-year-old trees, while Lim et al. (2011) found a mean fiber length of 1017.5 μm in 8-year-old trees. The consistency in results across these studies highlights that the fiber characteristics of *A. mangium* were different within the ages.

Further more, Andianto et al. (2020) described that *A. mangium* wood has a heartwood and sapwood. There are no growth rings present, and vessels occur individually and in a dispersed manner, typically in radial multiples of 2–3. The parenchyma is predominantly sparse and located paratracheal, occasionally thinly vasicentric, and sometimes aliform, with strand lengths ranging from 2 to 4 cells. Perforations are simple, intervessel pits are arranged alternately, and tyloses are

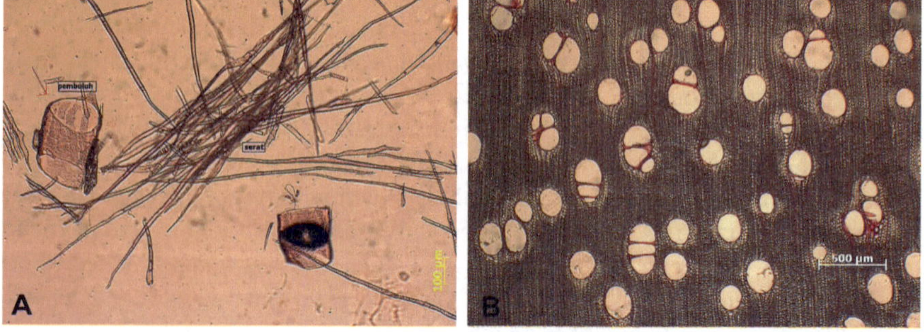

Fig. 20.1 The fiber of *Acacia mangium* in maceration samples (**A**) and a transversal section showing diffuse, solitary and multiples vessel and parenchyma (**B**). (Source: Andianto et al. 2020)

either absent or occur sparingly. Rays are mostly 1–2 cells wide, rarely 3, and chambered axial parenchyma cells contain prismatic crystals (Table 20.1).

In the past, the pulp and paper industry heavily relied on *A. mangium* as a fiber source due to its exceptional properties and high yield. The fiber of *A. mangium* possesses excellent properties that enhance the quality of pulp and paper products (Luangviriyasaeng 2007; Wong and Yuliarto 2014; Wu et al. 2018). Acceptable fiber length provides strength and durability, resulting in high-quality paper (Jusoh et al. 2014). As a fast-growing species, *A. mangium* offers a sustainable solution for the pulp and paper industry. Its rapid growth rate ensures a continuous supply of fiber without causing excessive strain on natural forests. The high fiber yield of *A. mangium* makes it a cost-effective choice for the pulp and paper industry. The abundance of fiber obtained from each tree significantly reduces production costs, contributing to increased profitability (Bajpai and Bajpai 2010).

Since the mid-1980s, the fiber wood based pulp and paper industry has been a major contributor to Indonesia's forest-based economy. Furthermore, until about 2012, it was reported that the major wood supply for fiber raw material was largely dependent on *A. mangium*, particularly on mineral soil with a high growth rate of 25–40 m^3/ha/year.

Recently, Indonesia has developed dissolving pulp from *A. mangium* and *E. pellita* wood fibers through the kraft pre-hydrolysis process to produce viscose for rayon yarn (Biantoro et al. 2020). The market for dissolving pulp has experienced a significant increase in the last 10 years, driven by the strong demand for cellulose fibers for rayon manufacturing in the textile industry (Arnoul-Jarriault et al. 2015). The development of cellulose fiber production from renewable plantation forest wood as a renewable natural resource has been driven by declining cotton production, limited fossil resources, environmental concerns, and cellulosic fiber properties which are better than synthetic fibers for rayon manufacturing in the textile industry (Sixta et al. 2013).

Viscose, a semi-synthetic form of rayon fabric derived from wood pulp, is utilized as a substitute for silk due to its luxurious feel. Many individuals opt for rayon because of its fine and comfortable characteristics (Andalia 2023). Rayon sourced from *A. mangium* wood exhibits a soft and comfortable texture akin to natural fibers

Table 20.1 The characteristics of the fibers, vessels, and parenchyma of a 4-year-old *Acacia mangium* from the province of Wonogiri, Central Java, Indonesia

Features	Description
Fibers	The sample contains fibers with pits that range from simple to minutely bordered. The fibers have varying wall thicknesses, ranging from thin to thick. The mean fiber length ranges from 1215.8 ± 53.96 μm to 1239.5 ± 53.92 μm, while the mean wall thickness ranges from 3.67 ± 0.17 μm to 3.73 ± 0.09 μm (Fig. 20.1a).
Vessels	Dispersed, individual, and radiating in multiples of 2–3, occasionally 4; the mean length measures 339.14 ± 17.91 μm, while the average diameter is 201.12 ± 14.13 μm. Featuring straightforward perforation plates, the intervessel pits are arranged alternately and adorned with vestures (Fig. 20.1b).
Parenchyma	vasicentric, aliform parenchyma strand length (Fig. 20.1b)

Source: Modified from Andianto et al. (2020)

like cotton and linen. It is recognized for its varying levels of sheen while maintaining excellent moisture-absorption capabilities. In contrast to water-resistant cotton, rayon excels in absorbing moisture, making it ideal for activewear in humid climates as it efficiently draws moisture away from the skin. With its high breathability and heat-dissipating properties, rayon has gained popularity as a preferred fabric for fitness attire, ensuring a cool and comfortable experience during physical activities.

Another merit of *A. mangium* lies in its rapid canopy growth and relatively low maintenance requirements, positioning it as a prominent species in industrial forest plantations for the supply of fiber raw materials in pulp and paper production. *A. mangium* plantation sites were distributed in several concession areas in Sumatera and Kalimantan Islands. In addition to its commercial significance, *A. mangium* plantations offer several environmental benefits, making them an attractive option for sustainable land management. Some of these advantages include such as carbon sequestration, soil conservation, and biodiversity enhancement (Sein and Mitlöhner 2011). The *A. mangium* trees play a crucial role in carbon sequestration, contributing to climate change mitigation (Kalita et al. 2016; Lee et al. 2015; Matsubara and Ohta 2015). Through the process of photosynthesis, these trees absorb carbon dioxide from the atmosphere and store carbon in their wood fiber, reducing greenhouse gas emissions. In terms of soil conservation, the extensive root system of *A. mangium* helps prevent soil erosion, preserving soil quality and preventing sedimentation in nearby water bodies (Koutika and Richardson 2019; Sein and Mitlöhner 2011). This feature is particularly important in areas prone to heavy rainfall or erosion. However, suitable management is necessary for *A. mangium* plantations to reduce potential negative impacts, particularly at a young age (Dung et al. 2019). In terms of biodiversity enhancement, *A. mangium* plantations contribute to biodiversity conservation by providing habitat for various organisms, such as birds (Hardiyanto 2004; Styring et al. 2011).

The benefit mentioned in the preceding paragraphs was revealed that *A. mangium* plantations hold immense importance for fiber production, catering to the ever-increasing demands of pulp and paper. With its rapid growth, high fiber yield, and versatility, *A. mangium* has emerged as a sustainable and economically viable solution. Furthermore, under appropriate plantation management practices, its environmental benefits further solidify its significance in land management and conservation. *A. mangium* plantations offer a promising avenue for meeting fiber production needs.

The sustainability of forest plantation practices in Indonesia heavily relies on the integration of land rehabilitation into economic growth. Their ability to support high log production, adapt to different regions, and provide environmental benefits make them a valuable asset in sustainable forestry management. Indonesia's government has allocated around 11.27 million hectares of forest land for Industrial Forest Plantation, which is 33 percent of the total area for forest business licenses (MOEF 2022). Most of the plantations are used to produce fiber raw materials for pulp and paper products. Furthermore, the targeted wood production from the plantation in 2024 is around 61.69 million m^3, a 5.73% increase from 2023. As Indonesia continues to prioritize sustainability in its forest plantation efforts, the cultivation of

A. mangium is expected to regain further, importance, providing economic opportunities while safeguarding the environment.

20.1.2 Current Issues and Plantation

Currently, most *A. mangium* plantations are facing numerous challenges, including the adverse effects of climate change, and pest and disease attacks. Climate change poses a significant threat to the plantations. The rising temperatures, altered rainfall patterns, and increased frequency of extreme weather events have a direct impact on the growth and overall health of these trees. Some of the key issues associated with climate change include reduced growth rates and increased susceptibility to disease. Higher temperatures and water stress caused by climate change can lead to reduced growth rates in *A. mangium* plantations (Almeida et al. 2014). Climate change weakens the trees' defense mechanisms, making them more prone to various diseases and infections, and leading to new disease outbreaks in *A. mangium* plantations.

A. mangium plantations are also susceptible to various pests and diseases that can significantly impact their productivity and diminish their economic value (Brawner et al. 2015, 2022; Hardiyanto 2014; Harwood and Nambiar 2014; Harwood et al. 2015; Nasution et al. 2022; Tarigan et al. 2011). Pests and diseases pose significant threats to *A. mangium* plantations, affecting both the quality and quantity of wood fiber produced. Common pests and diseases that attack these plantations include stem borers, fungal pathogens, leaf-eating caterpillars and monkeys (Farid et al. 2006; Lee 2004). For stem borers, these insect pests bore into the stem, causing structural damage and reducing the tree's vigor. Various fungal infections, such as root rot and leaf spot diseases, can weaken the trees and lead to reduced productivity. Some defects on the trees due to pests can stimulate pathogen attacks and spread.

Currently, one of the most prevalent diseases affecting *A. mangium* plantations is *Ceratocystis* sp., a fungal pathogen that causes wilt and dieback symptoms (Brawner et al. 2015; Tarigan et al. 2011). The rapid spread of the *Ceratocystis* attacks has led to severe economic losses in *A. mangium* plantations due to large-scale mortality and reducing the overall productivity of the plantation (Harwood and Nambiar 2014; Hardiyanto 2014; Hardiyanto et al. 2024; Thu et al. 2014). The vulnerability of *A. mangium* to *Ceratocystis* increased due to climate changes. The changes in temperature and humidity affect the growth and development of *A. mangium*, making it more susceptible to such diseases. Decreased productivity leads to financial setbacks for plantation owners. In addition, the decline in *A. mangium* health affects ecosystem stability and biodiversity.

Genetic improvement and breeding programs to enhance the resistance of *A. mangium* to diseases such as *Ceratocystis* are crucial (Brawner et al. 2022). A comprehensive breeding strategy focusing on selecting parent trees with demonstrated superior growth, better wood properties, and disease resistance should be prioritized. With advancements in breeding technologies, the potential for further advancements in tree breeding is vast, offering hope for more productive *A.*

mangium plantations in the future. Strategic breeding efforts, including the utilization of genetic diversity, marker-assisted selection and hybridization, offer hope for developing more resilient *A. mangium* varieties while maintaining high growth and wood productivity (Hardiyanto 2004). These efforts will not only safeguard *A. mangium* plantations from the devastating effects of *Ceratocystis* but also contribute to the long-term sustainability and profitability of *A. mangium* cultivation. Integrating growth, wood properties, and disease resistance into breeding strategies holds great promise for improving the productivity and quality of the tree species.

While waiting to obtain planting materials resistant to *Ceratocystis*, most companies have considered transitioning plantation species from *A. mangium* to other species, mostly to *Eucalyptus* species (Hardiyanto et al. 2021, 2024). Introducing a new species like *Eucalyptus pellita* and the hybrids can diversify the plantation landscape, reducing the risk of disease outbreaks or pest infestations that specifically attack *A. mangium*. Some *Eucalyptus* species are known for their superior wood quality, offering higher-value products with increased market demand which can lead to better economic returns for plantation owners (Harwood and Nambiar 2014; Hardiyanto et al. 2024). However, transitioning from *A. mangium* to *Eucalyptus* is not without its challenges. Assessing the suitability of the plantation site for *Eucalyptus* is crucial. Factors such as soil condition, climate, and water availability need to be evaluated to ensure successful establishment and growth. Therefore, conducting comprehensive site assessments before initiating the transition is necessary, such as genetic selection to choose *Eucalyptus* varieties that are suitable to site planting conditions and desired end products (Hardiyanto et al. 2021, 2024).

Another issue with the *A. mangium* species is its invasiveness due to its reproductive system, aggressive growth and adaptability. The invasiveness of *A. mangium* poses significant challenges to ecological integrity. Urgent action is needed to address its spread and mitigate its negative impacts. By implementing effective management strategies, including early detection, rapid response, and restoration efforts, it can work towards preserving native biodiversity and ecosystems. In addition, employing appropriate management practices in the plantation of *A. mangium* can prevent the invasive potential and uncontrollable spreading (Nirsatmanto and Sunarti 2019).

20.1.3 Botanical Classification and Distribution

A. mangium is known by various local names in different countries, such as Arr in Papua New Guinea, Black wattle in Australia, the UK, and the USA, Brown salwood in Australia, the UK, and the USA, Kayu safoda in Peninsular Malaysia, Keo tai tuong in Vietnam, Kra Thin Tepa in Thailand, and Mangge hutan and Tongke hutan in Indonesia (Pinyopusarerk et al. 1993). The classification of the plant species *A. mangium* Willd. follows a hierarchical structure within the Kingdom Plantae (https://www.itis.gov/servlet/SingleRpt/SingleRpt?search_topic=TSN&search_value=820148#null).

Kingdom	Plantae (plants)
Subkingdom	Viridiplantae (green plants)
Infrakingdom	Streptophyta (land plants)
Superdivision	Embryophyta
Division	Tracheophyta (vascular plants)
Subdivision	Spermatophytina (spermatophytes, seed plants)
Class	Magnoliopsida
Superorder	Rosanae
Order	Fabales
Family	Fabaceae (legumes)
Genus	Acacia Mill.
Species	*Acacia mangium* Willd.

The black wattle (*Acacia mangium* Willd.) belongs to the plant kingdom, which includes various types of plants. Within this kingdom, it is under the group of green plants (Subkingdom Viridiplantae), specifically categorized as land plants (Infrakingdom Streptophyta). As we get more detailed, it is part of the Superdivision Embryophyta, indicating that it is an embryophyte. Going further, it falls under the Division Tracheophyta, meaning it is a vascular plant with a well-developed system for moving fluids. In the Subdivision Spermatophytina, it is grouped with seed plants or phanerogams, emphasizing its reproductive characteristics. Even more specifically, it belongs to the Class Magnoliopsida, a diverse group of flowering plants. Narrowing down its classification, it is in the Superorder Rosanae, Order Fabales, and Family Fabaceae, which includes peas and legumes. Finally, at the most detailed level, it is identified under the Genus *Acacia*.

The Acacia genus is expansive, comprising over 1000–1300 species that are extensively distributed across the tropics and subtropics. Among them, 650 species are identified in Australia (Hegde et al. 2013). The majority of these species are located in the southern hemisphere, with Australia and the Pacific serving as the primary centers of diversity. Acacia is commonly encountered in various environments, including primary and secondary forests, forest margins, savannahs, grasslands, and savannah woodlands. It is also found in areas with poorly drained floodplains and along the fringes of mangrove forests, occasionally coexisting with Melaleuca and Rhizophora spp. (Arisman and Hardiyanto 2006; Krisnawati et al. 2011).

A. mangium is originally from Indonesia, northeastern Australia, and Papua New Guinea, with specific locations in Indonesia being the Aru Islands, Papua, West Papua, Sula Islands, and Seram Island (Koutika and Richardson 2019). Its natural distribution is fragmented, extending from Indonesia (Sula, Ceram, Aru, Irian Jaya) to Papua New Guinea and northeastern Queensland in Australia (Fig. 20.2) (PROSEA 1995; Hegde et al. 2013). The range spans from 1° 18′ 57″ S latitude to 125° 22′ 146° 17 ‘E longitude, with altitudes ranging from just above sea level to 780 meters. In Asia, Indonesia is the primary country with *A. mangium's* natural population, while in Oceania, it is found in the Australian Northern Territory,

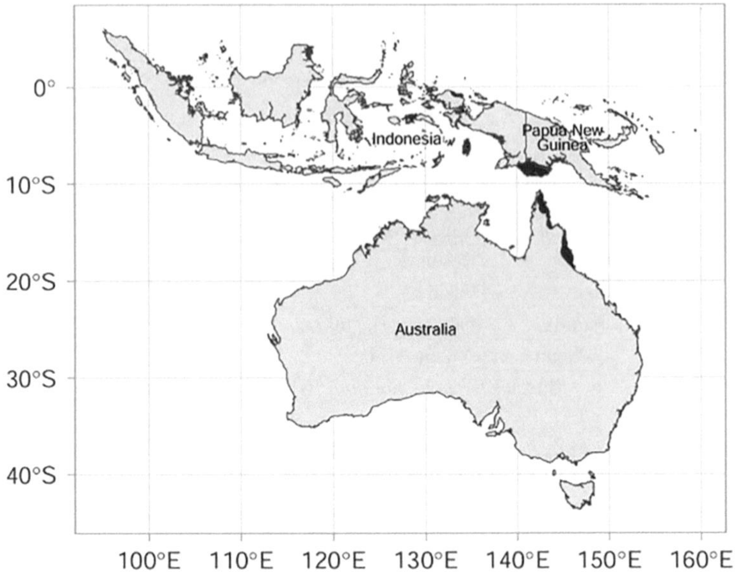

Fig. 20.2 Natural distribution map of *Acacia mangium*. (Figure by M. N. Widhiati)

Queensland, and Papua New Guinea. Due to its rapid growth and adaptability to poor soils, *A. mangium* has been introduced to various countries in Asia, Africa, and the Western Hemisphere for plantation purposes.

A. mangium demonstrates adaptability to diverse soil types and environmental conditions. It exhibits rapid growth on sites with low soil nutrients, including acidic and degraded soils (National Research Council 1983). The species thrives particularly on lateritic soils characterized by high iron oxide and aluminum content (Otsamo 2002). However, *A. mangium* is not tolerant of shade or saline environments, leading to stunted growth and a spindly appearance under such conditions. Typically found in low-elevation areas, it is associated mainly with rainforest margins and disturbed sites on well-drained, acid soils with low fertility (Pinyopusarerk et al. 1993). *A. mangium* serves as a pioneer species, capable of natural regeneration on disturbed sites, as documented by the National Research Council (1983) and Otsamo (2002).

20.1.4 Domestication and Early Improvements

A. mangium has been successfully introduced and cultivated in Indonesia since the beginning of the industrial forest plantation program in the 1980s (Hardiyanto 1998). The domestication and early improvements of this species have played a crucial role in meeting the increasing demand for pulpwood in the country. The domestication of *A. mangium*, along with other Acacias species (*A. auriculiformis* and *A. crassicarpa*) involved several breeding generations aimed at improving its growth rate, adaptability, and wood quality (Nirsatmanto and Sunarti 2019). Each generation focused on selecting individuals with desirable characteristics, to ensure the propagation of these traits to subsequent generations.

In terms of genetic base, *A. mangium* possesses a wide genetic base, allowing for diverse breeding strategies to improve its desirable traits. *A. mangium* exhibits a wide natural distribution across different regions, including Southeast Asia and Northern Australia (Fig. 20.3). Seed lots form the foundation for any breeding program aimed at enhancing desirable traits in *A. mangium* (Table 20.2). These seed collections from different provenances provide a diverse range of genetic material for initial planting, genetic selection and breeding purposes (Hardiyanto 2004).

Active selection techniques were employed to accelerate the process of breeding and maximize genetic gain. In the first generation, initial selection involved identifying and collecting superior individuals from various provenances across Papua New Guinea, Australia and Eastern parts of Indonesia (Nirsatmanto and Sunarti 2019; Nirsatmanto et al. 2012). Provenance testing was initially conducted to evaluate the performance of different original seed sources (provenances) under various environmental conditions (Table 20.2). These trials provided valuable base information on the adaptability of *A. mangium* and helped in selecting suitable seed sources for specific regions. The selected plus trees from original provenances were then cross-pollinated to create a diverse gene pool through establishing progeny trials. Genetic testing in first-generation progeny trials was carried out to assess heritability and gain insights into the genetic basis of important traits. Progeny from the first-generation was evaluated based on growth performance and form traits as the

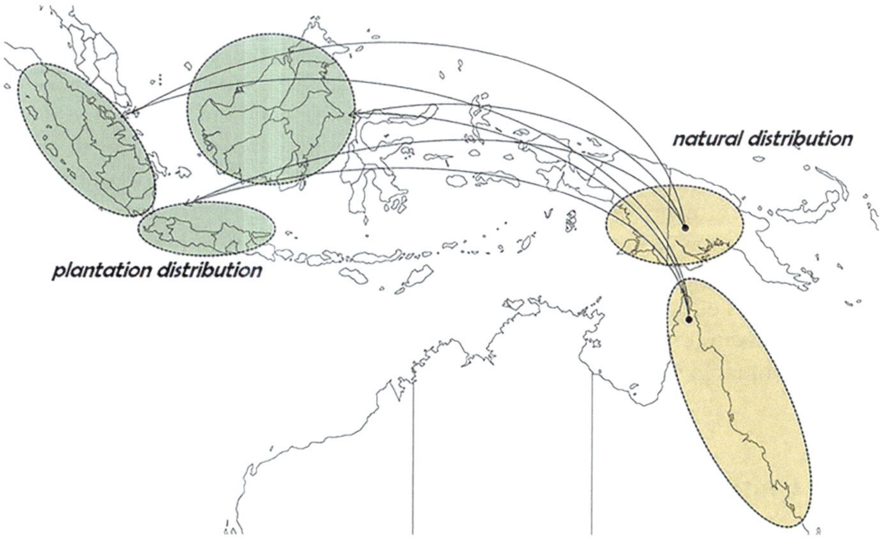

Fig. 20.3 Natural distribution of *Acacia mangium* from the natural habitat in Eastern Indonesia and Northern Queensland, and its plantation distribution in Indonesia. (Modified from Nirsatmanto and Sunarti 2021)

Table 20.2 Procurement of individual seed lots from the natural habitat distribution as genetic material for *Acacia mangium* breeding

Species	Original provenances	Country	Total individual seed lots
Acacia mangium	Subanjeriji, Seram, Bupul, Muting, Merauke	Indonesia	30
	Morehead, Dimisisi, Bimadebun, Arufi Village, Bensbach, Makapa, Kini, Wipim, Oriomo	Papua New Guinea	242
	Claudie River, Claudie R & Iron Range, Daintree River, Olive River, Tully-Mission Beach, Pascoe River, Cassowary Ck-Iron Range, Nne Coen	Australia	232

Source: Susilawati and Sofyan (1995)

initial target of traits for improvement (Nirsatmanto and Sunarti 2019; Nirsatmanto et al. 2012) Individual trees from selected genetic backgrounds with exceptional traits were selectively bred in a seed orchard through open-pollination to produce improved seed.

Genetic testing has progressed to second and third generation progeny trials to further refine the understanding of genetic control over desirable traits. Selection intensity increased in subsequent generations, focusing on specific traits such as growth and wood properties (Sunarti et al. 2022). New genetic materials from natural distribution have been infused to maintain greater genetic diversity in breeding populations, incorporating individuals from different sources. Accumulative genetic improvement has been achieved over the three successive generations of the breeding programs for growth traits, form traits and wood properties (Nirsatmanto et al. 2015; Sunarti et al. 2022). These three categories of desired traits have led to an increase in fiber production and pulp yield.

Irrespective of the current problem occurring in the plantation, the domestication and early improvements of *A. mangium* in Indonesia have been driven by genetic testing in extensive breeding generations. The active selection process has resulted in the development of superior planting materials, mainly focused on growth rates, form traits and wood properties. These advancements have significantly contributed to meeting the growing demand for pulpwood in the country's forestry sector. The early three breeding generations led to a significant improvement in productivity in *A. mangium* plantation in Indonesia for fiber production (Nirsatmanto et al. 2015). The continuous efforts to enhance growth rates, form, and wood quality, along with improvements in resistance to pests and diseases should pave the way for sustainable and profitable *A. mangium* plantations in the future.

20.2 Cultivation and Traditional Breeding

The successful cultivation and breeding of plants require careful consideration and integration of various factors. Cultivating *A. mangium* involves several stages, starting from seed collection to nursery establishment and finally plantation establishment. The basics of cultivation are developed to ensure the interconnectivity among the success of seed procurement, nursery and planting establishment. Seeds for planting stock are collected from genetically superior trees from different provenances to maintain genetic diversity. Nursery establishment should ensure a conducive environment for seed germination, appropriate growing media and provide adequate watering and fertilization. The seedling stocks are then appropriately planted in well-prepared land by clearing and under optimal spacing to promote healthy growth.

Breeding actions play a role in enhancing the genetic potential of *A. mangium* to supply genetically improved seed for plantation. The available genetic resources become an important component to emphasize the actions taken during the breeding process. Identifying high-performing trees based on growth rate, wood quality, and disease resistance is necessary. This should be supported by an understanding of the reproductive system of the species to ensure the success of pollination for cross-breed selected trees to combine desirable traits, particularly through open pollination. Progeny testing would be the next action to monitor and verify the performance of offspring for genetic selection based on the desired traits.

20.2.1 Current Cultivation Practices

In order to propagate *A. mangium*, it is essential to collect high-quality seeds from reliable seed sources. Collecting seeds from reputable sources ensures the genetic integrity and quality of the resulting seedlings to enhance the overall productivity and sustainability of *A. mangium* plantations. There are several stakeholders contribute to ensuring the sustained supply of *A. mangium* seed collection in Indonesia. The Ministry of Environment and Forestry serves in maintaining seed and seedling regulations through certification of seed and seed sources (MOEF 2020; Nirsatmanto and Sunarti 2021). It also plays a vital role in conserving genetic diversity and providing reliable seed stocks. Research Institutions conduct studies on *A. mangium*, including seed collection and breeding (Nirsatmanto and Sunarti 2019). Their seed sources are often meticulously selected, resulting in high-quality seeds. Commercial seed suppliers prepare seed source establishment and seed collection for mass-scale production.

To ensure successful seed collection, several techniques are available to be employed (Nurhasybi et al. 2008). The timing of seed collection for *A. mangium* is during the dry season when seed pods are fully mature. On the other hand, collecting seeds during rainy periods is avoided to prevent damage or fungal infections. Visual inspection is practiced by observing seed pods for maturity, ensuring they are brown and dry. If needed, open drying can also be practiced to facilitate seed extraction. Mechanical extraction is applied to gently twist or pry open the seed pods to remove the seeds. Seed cleaning is then done by removing any debris or unwanted materials from the collected seeds. This can be done by manually sorting or using sieves with appropriate mesh sizes.

The seeds of *A. mangium* are small, dark brown, and have an elliptical shape. They are typically 5 millimeters in length and 3 millimeters in width (Fig. 20.4). The seed coat is hard and impermeable, which protects the embryo from desiccation and mechanical damage. This characteristic is important for the long-term survival of the species in adverse environmental conditions. The seeds possess a micropylar end, which allows for water absorption during germination. *A. mangium* seeds are naturally dispersed by various means, including wind, water, and animals.

The seeds of *A. mangium* exhibit orthodox seeds that can be stored for a long time. However, the proper storage of seeds is essential to preserve their viability over extended periods. The nature of these seeds necessitates specific storage conditions to prevent deterioration. Several storage requirements should be considered for *A. mangium* seeds (Pujiastuti and Sudrajat 2017). The ideal storage temperature is between 3 °C and 5 °C, as lower temperatures can result in chilling injury (Bowen and Eusebio 1981). Avoid freezing temperatures, as they can lead to irreversible damage to the seeds. Seeds should be stored at high moisture content, typically around 4–12%. This moisture level helps prevent desiccation and maintains seed viability. Regular monitoring of moisture content is essential to ensure optimal storage conditions are maintained. The storage area should be well-ventilated to prevent the accumulation of moisture and the growth of mold or fungi. In addition to these storage requirements, it is advisable to periodically test seed germination rates to assess seed viability. This information can guide decision-making regarding the use of stored seeds for propagation purposes.

In the nursery stage, it is important to establish efficient and effective nurseries for seedling production of *A. mangium* (Fig. 20.5). Nurseries play a crucial role in the successful establishment of *A. mangium*in tree plantations. They provide a controlled environment where seedlings can be nurtured and grown under optimal conditions before being transferred to the field. This ensures better survival rates and enhances the overall productivity and quality of the plantation.

The first step in establishing a nursery for *A. mangium* seedlings is selecting an appropriate site. The site should have access to a reliable water source, good drainage, and ample sunlight. It should also be easily accessible for transportation of seedlings. Once the site is selected, the nursery layout needs to be planned carefully. A well-designed nursery layout facilitates efficient management and operations. Seedbeds are prepared where seeds are sown or germinated. They should be

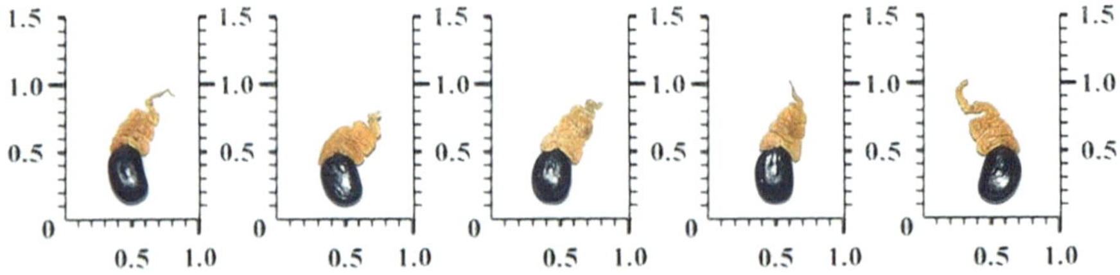

Fig. 20.4 The varies in seed size (scaled in cm) within one pod of *Acacia mangium*. (Photographs by S. Sunarti)

Fig. 20.5 Nursery practiced for producing the seedlings of *Acacia mangium*, 3-months-old seedlings (**a**), 5 months old seedlings (**b**) and seedlings ready for plantation deployment (**c**). (Photographs by A. Nirsatmanto)

constructed using quality soil mixes and adequately spaced for easy access and maintenance. These structures protect from extreme weather conditions and help maintain optimal temperature and humidity levels for seedling growth. A reliable irrigation system is essential for providing adequate water to the seedlings. Drip irrigation or sprinkler systems are commonly used in nurseries. Fencing helps protect the seedlings from pests, animals, and unauthorized access.

Several techniques can be employed for the seedling production of *A. mangium*. Seeds can be sown directly into seedbeds. This method is cost-effective but requires careful monitoring and weed control. Seeds are germinated in containers, such as polybags or pot trays. This method allows for easier handling and transplanting of seedlings. In terms of vegetative propagation, *A. mangium* can also be propagated through techniques like stem cuttings or tissue culture. These methods ensure genetic uniformity and faster growth. Proper care and management are crucial for the healthy growth of seedlings. Seedlings should be watered regularly, ensuring that the soil remains moist but not waterlogged. A balanced fertilizer regime should be followed to provide essential nutrients for vigorous growth. Regular monitoring and appropriate control measures should be implemented to prevent damage from pests and diseases. Weeds compete with seedlings for nutrients, water, and light. Regular weeding is necessary to minimize their impact. Seedlings are ready for field planting when they reach a certain stage of growth, typically 3–4 months after germination. Proper hardening should be done before transplanting to ensure better survival rates in the field. Care should be taken during the transplanting process to minimize root damage.

For site planting selection, the *A. mangium* thrives in tropical and subtropical regions with an annual rainfall of 1200–3000 mm (Sein and Mitlöhner 2011). It prefers well-drained soils with a pH range of 4.5–6.5. Conducting a thorough soil analysis before planting is crucial to ensure optimal conditions for growth. Proper spacing is also crucial for the high plantation productivity and healthy growth of *A. mangium* trees. For fiber production, it is recommended to maintain a spacing of 2 meters between each tree in a row and 3 meters between rows. This spacing allows sufficient sunlight penetration and air circulation, which promotes optimal growth. Fertilizing plays a vital role in enhancing the productivity of *A. mangium* plantations. A balanced fertilizer with adequate amounts of nitrogen, phosphorus, and potassium should be applied during the early stages of growth. Regular soil testing can help determine the specific nutrient requirements of the plantation. Fertilizers are applied at a rate of 150–200 kg of NPK per hectare per year (Hardiyanto 2014; Siregar et al. 1999). It is practiced in several phases of applications throughout the 2 years of age. The fertilizer is distributed evenly around the tree base, and direct contact of fertilizer with the tree trunk must be avoided.

Following the growth of seedlings after planting, efficient weed management is essential to minimize competition for nutrients and water. Regular weeding is recommended to ensure the healthy growth of *A. mangium* trees. Manual weeding is an effective method of removing weeds within a 1-meter radius around each tree. In addition, regular monitoring for pests and diseases is necessary to safeguard the plantation. Common pests affecting *A. mangium* include termites, leaf-eating caterpillars, and shoot borers. Implementing appropriate pest control measures such as biological agents or chemical treatments (if necessary) can help mitigate damage.

To ensure the sustainable and efficient management and stand productivity of *A. mangium* plantations, proper harvesting techniques and strategies need to be implemented to produce wood volume, maintain tree health, and sustain long-term growth. Determining age rotation as the optimal period between the establishment of a plantation and its subsequent harvest of *A. mangium* is crucial to maximizing wood quality, yield, and economic returns. *A. mangium* trees are typically harvested at 4–6 years of age for pulp and paper production, and it may vary depending on specific site conditions and management objectives. Mechanized clear-cutting techniques are commonly employed in the harvesting plantations for wood fibers of pulp and paper to ensure sustainable management of the plantation. It involves the complete removal of all trees within a designated area using appropriate machinery. Re-planting of successive rotations is then employed after the clear-cutting of harvesting to ensure the long-term viability of *A. mangium* plantations and to mitigate the environmental impact. This sustainable forestry practice promotes ecosystem resilience and biodiversity while fostering the regrowth of *A. mangium* plantations for future generations.

20.2.2 Improvement Strategies, Methods and Limitations

In terms of the breeding population, enhancing the productivity of *A. mangium* involves a systematic breeding process that includes gene conservation, mass production, disease resistance, DNA molecular, tissue culture, growth productivity and wood properties. This process is implemented through comprehensive strategies of the recurrent selection system for successive breeding generations, managing multiple breeding populations, and sub-lining system through progeny trials establishment, hybridization and clonal tests (Figs. 20.6 and 20.7).

The breeding strategy includes careful selection and crossing pollination of superior individuals to obtain desirable traits in the progeny. The recurrent selection system is a key component in the *A. mangium* breeding process, where superior

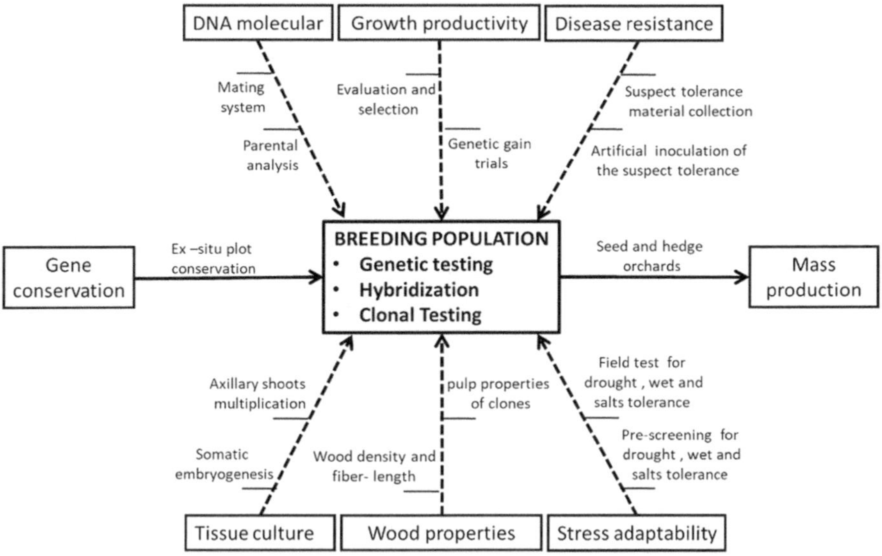

Fig. 20.6 The breeding framework for *Acacia mangium* for fiber products. (Figure by A. Nirsatmanto with theoretical concept following Nirsatmanto et al. 2019)

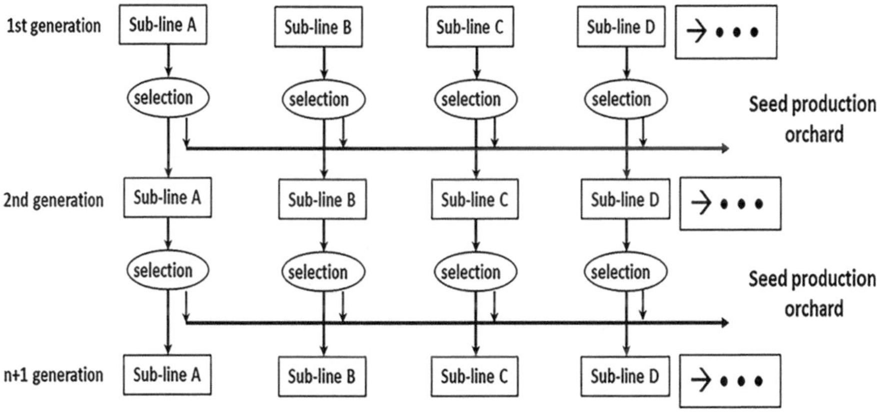

Fig. 20.7 Sub-lining system in the management of breeding population for *Acacia mangium*. (Figure by A. Nirsatmanto with theoretical concept following Kurinobu 1993)

trees are repeatedly selected as parents for subsequent generations based on specific criteria, such as growth, form and wood quality (White et al. 2007). These selected trees are then cross-pollinated to produce a diverse population with potential variations in desired traits. This strategy is employed in *A. mangium* to ensure the accumulation of desirable traits over time and promotes genetic improvement. Understanding the relationship among the desired traits is crucial to ensure the genetic gain simultaneously.

Recurrent selection is applied in genetic testing trials such as progeny trials. The *A. mangium* progeny trial is established to evaluate the genetic performance of the offspring in a wide range of sites, particularly on mineral soil sites (Nirsatmanto et al. 2023). These trials assess various important traits considering the specific character of *A. mangium* and the end-product such as height growth, stem diameter for growth traits, stem straightness and stem forking for form traits, wood density and fiber properties (Fig. 20.8). The progenies with the most desirable traits are selected as parents for the establishment of seed source for mass-scale seed production. Further successive selection within the selected trees is employed to select plus trees as genetic materials for the next breeding generation.

Tree selection based on several important traits is done using index selection procedures in which the breeding value combines target traits into one value (Burdon 1989; Hazel 1943). The importance of each trait is then scaled based on its coefficient weight, with a high coefficient indicating a more priority trait. Due to the variety of morphological characters of *A. mangium* related to the desired improvement target, the selection index involves many traits and can be quite complicated. Therefore, to overcome such complications the tree selection procedure in *A.*

Fig. 20.8 Improvement on target traits from three successive generations breeding of *Acacia mangium*: unimproved generation showing multi-stemmed, less straightness and poor growth (**a**); first-generation breeding (F-1) showing the improvement on single-stemmed and good growth (**b**); second-generation (F-2) showing the improvement on single-stemmed, good growth and good straightness (**c**); third-generation (F-3) showing the improvement on single-stemmed, good growth, good straightness and high wood density (**d**). (Photographs by A. Nirsatmanto)

mangium is done by combining with other methods of two-stage selection and tandem selection.

Each step in the breeding process of *A. mangium* that involves recurrent selection, progeny trials, and breeding generations actively contributes to selecting superior individuals and evaluating their performance. By utilizing a diverse range of techniques and considering various traits, *A. mangium* can be genetically improved to meet the demands of wood production from plantations. This not only benefits wood production but also contributes to sustainable reforestation efforts and ecological balance. The continuous genetic improvement of *A. mangium* can ensure its sustainability as a valuable tree species for future use.

In terms of propagation population, seedling seed orchards are the result of breeding selection to produce mass-scaled genetically improved stock. These orchards are essential for producing high-quality seeds of *A. mangium* with desirable traits and ensuring the availability of genetically superior seeds that can consistently produce offspring with desirable traits. Through meticulous breeding techniques and careful selection, these orchards contribute to the development of improved tree populations with enhanced growth rates, stem form and wood quality. The advantages offered by seedling seed orchards make them an indispensable component of sustainable plantation management practices of *A. mangium.*

The increased genetic gain (%) and stand productivity (m^3/ha) in *A. mangium* over successive generations of breeding have been achieved for growth traits (Figs. 20.9 and 20.10) (Nirsatmanto et al. 2014, 2015). However, the success of the breeding process is not without its limitations and challenges. Several challenges are encountered during the breeding of *A. mangium*, including long breeding cycles, limited understanding of the genetic basis for desired traits, ensuring genetic

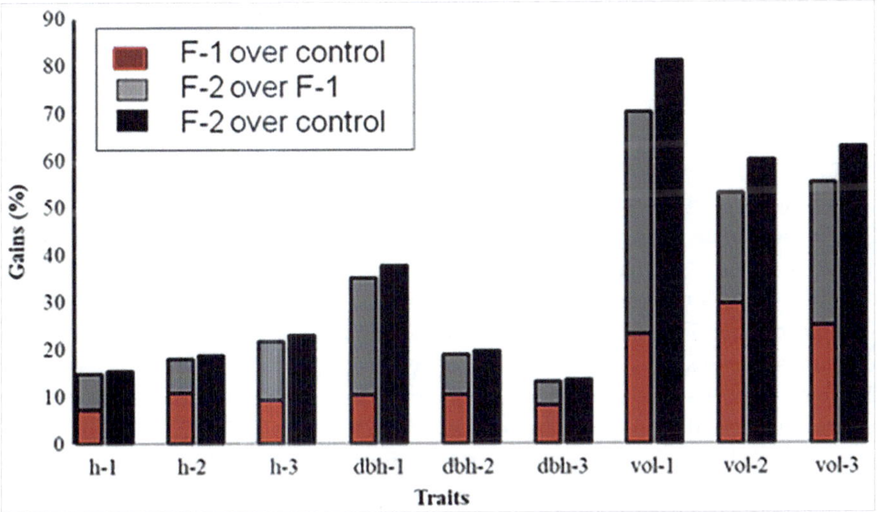

Fig. 20.9 Realized genetic gain from three successive generations of the breeding program of *Acacia mangium*. (Source: Nirsatmanto et al. 2015)

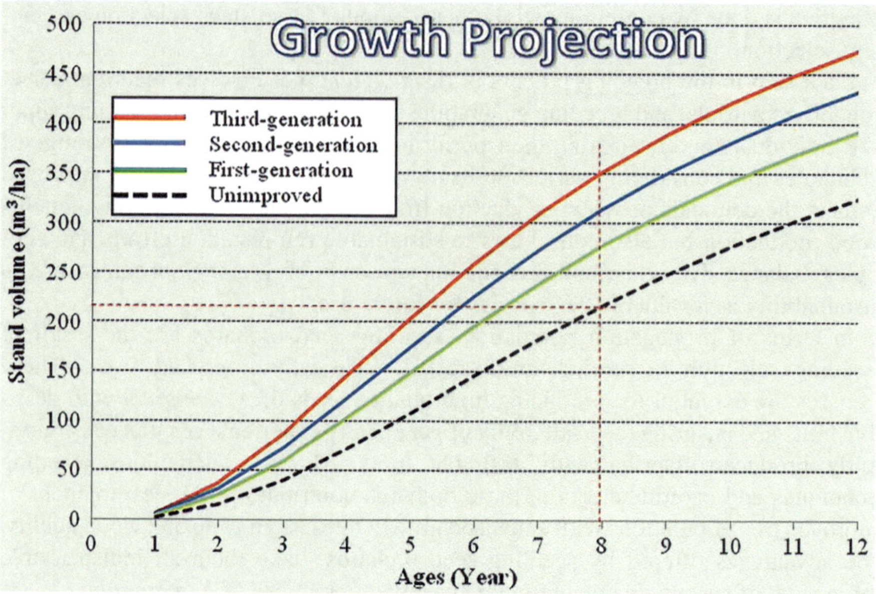

Fig. 20.10 Projection of stand productivity (m³/ha) for 8 years age rotation resulted from three successive generations breeding program of *Acacia mangium*. (Figure by A. Nirsatmanto)

diversity is maintained in cultivated populations, and combating diseases and pests that affect productivity. *A. mangium* exhibits certain reproductive barriers that hinder conventional breeding techniques, such as self-incompatibility and outcrossing mechanisms, making it difficult to achieve successful pollination and seed production.

A breeding strategy to enhance the wood properties of *A. mangium* is employed to make the wood a more suitable raw material for pulp and paper making (Nirsatmanto et al. 2022; Sunarti et al. 2022; Tsai 1988;). The implementation of advanced genetic tools and rigorous evaluation methods will further enhance the efficiency and success of the breeding process. Through continuous improvement in wood properties over successive generations of breeding, *A. mangium* can become a preferred choice for pulp and paper production, ensuring a sustainable and high-quality supply of raw materials for the industry.

Initially, breeding selection for wood properties focused on improving basic density, lignin, fiber dimension and pulp yield through two-stage selection and tandem selection strategies. Hasegawa et al. (2009), conducted a study for selecting superior plus trees in a first-generation progeny trial of *A. mangium* based on the provenance variation in growth and wood properties at 11 years of age. This study found that selected superior plus tree families from two provenances of Papua New Guinea and Queensland-Australia resulted in wood properties in the ranges of 1.03 mm −1.10 mm for fiber length, 29.0–32.8% for lignin contents, and 0.577–0.647 for air-dried density. A further study conducted by Sunarti et al. (2022) to verify the wood properties of the improved growth trees derived from the first- and second-generation breeding program of *A. mangium* revealed that the genetic gain in diameter increased from 3.7% in the first- to 7% in the second-generation, which is followed by the decreased gain in wood density from 1.9% in the first- to 0.3% in the second-generation. Therefore, this study recommended incorporating selection for diameter and wood density in a mixed-breeding generation to simultaneously improve the growth and wood properties. Improving the fiber characteristics and wood density, along with reducing lignin content can increase the pulp yield of *A. mangium*.

The lengthy breeding cycle of *A. mangium* poses a significant challenge in achieving rapid genetic improvement. The species typically takes several years to reach reproductive maturity, delaying the evaluation and selection of desirable traits. *A. mangium* is susceptible to various pests and diseases, which can significantly impact the success of breeding programs. Breeding for resistance or tolerance to these threats is crucial to ensure the long-term viability of the species. *A. mangium* exhibits a wide range of ecological adaptations, making it challenging to develop breeding strategies suitable for all environments. The species' adaptability to different soil types, climates, and altitudes complicates the selection of appropriate breeding populations.

This devastating fungal pathogen has caused severe economic losses and threatened the sustainability of *A. mangium* plantations in Indonesia. In order to combat this threat, a breeding strategy focused on enhancing disease resistance in *A. mangium* has been developed. The breeding strategy to enhance *Ceratocystis* tolerance in *A. mangium* was developed by some forest companies, and this is also being continued by the Australian Center for International Agricultural Research (ACIAR) project in Indonesia (Brawner et al. 2022; Nasution et al. 2022; Tarigan et al. 2011). The genetic testing strategy conducted by Brawner et al. (2022) involves the selection of genetic material derived from four sub-lines of second-generation progeny trials of *A. mangium* that were improved based on growth and form traits. The materials were propagated through stem cutting and then inoculated with locally collected strains of *Ceratocystis* in Indonesia in three disease screening trials. The assessment for survival was conducted in nursery trials that were followed by further screening in the nursery and field trials. This study revealed that the tolerance was low (6.7% of the 103 clones) with slight differences among the four sub-lines populations. The final validation showed that one-third of the tolerant clones identified from nursery screening were consistent in the field trials.

As described in preceding paragraphs, breeding strategies for *A. mangium* have been acknowledged as crucial initiatives to greatly increase the production of plantations. Series breeding procedures have been used to obtain larger and faster genetic gains. However, the availability of genetic resources is crucial to ensuring the viability and advancement of *A. mangium* through breeding in light of future demands and obstacles. The adoption of insufficient breeding procedures as a result of efforts to acquire genetic gain as quickly and as greatly as feasible seems to be a problem in the preservation of genetic resources for breeding purposes.

Nirsatmanto (2020) has reviewed the limitations of the adopted breeding strategy for *A. mangium* for preserving genetic resources in terms of recurrent selection system procedures and breeding seedling orchard establishment. It was argued that some breeding practices have the potential to significantly reduce the number of genetic resources available. These include such as employing high selection intensity subjected to the infused families in advanced generation breeding, thinning a breeding seedling orchard as a consequence of using a multiple tree-plot configuration, and selecting and then eliminating undesirable genotypes from the breeding population just to meet the present need. On the other hand, in the future, the desired traits for improvement may change and need a specific genotype.

20.2.3 Role of Biotechnology

Biotechnology offers promising tools and techniques that can revolutionize the breeding process and accelerate genetic improvement. It can employ the role of molecular breeding, tissue culture, and genomic selection, in improving the breeding strategy of *A. mangium*. Biotechnology offers tremendous potential for improving the breeding strategy of *A. mangium* by selecting superior individuals at an early stage, rapidly propagating elite genotypes, and introducing desirable traits through genetic transformation (Saini et al. 2020; Yanchuk and Jaquish 2003). The integration of biotechnology in *A. mangium* breeding programs can result in the development of improved varieties with enhanced growth rates, wood quality, disease resistance, and environmental adaptability. By harnessing the power of biotechnology, the forestry industry can meet the growing demand for *A. mangium* while ensuring sustainable and efficient resource utilization.

The benefits of biotechnology in *A. mangium* breeding can involve increasing efficiency and precision in selecting desirable traits through early identification of superior individuals and reducing the time and resources required for phenotypic evaluation. Biotechnology can also enhance genetic diversity and variability by facilitating the introgression of novel traits into breeding populations and reducing the risk of inbreeding depression. In addition, biotechnology can accelerate the genetic gain allowing breeders to focus on multiple traits simultaneously and shorten the breeding cycle, enabling faster development of improved varieties.

20.2.3.1 Tissue Culture

Tissue culture has emerged as a promising technique for improving the breeding strategy of tree species. Tissue culture involves the growth and development of plant cells, tissues, or organs in an artificial nutrient medium. In the case of *A. mangium*, tissue culture techniques have been applied to clonal multiplication of genetically improved stock in vitro (Xie and Hong 2001). It is also possible to induce embryogenesis, which is the formation of embryos from somatic cells. This approach allows for the rapid production of a large number of genetically identical plants, bypassing the conventional methods of sexual reproduction.

Tissue culture allows for precise manipulation and selection of superior genotypes, leading to the development of improved varieties with enhanced adaptability and resistance to pests and diseases. Several studies have been conducted to evaluate the effectiveness of tissue culture in improving the breeding strategy of *A. mangium* (Huang et al. 2012). Researchers successfully induced embryogenesis in *A. mangium* using various explant types, such as leaf, shoot tip, and hypocotyl. The induction rates ranged from 60% to 95%, depending on the genotype and the specific tissue culture protocol employed. Embryogenic cultures obtained through tissue culture were subjected to multiplication and acclimatization stages. Subculturing on appropriate media resulted in the rapid multiplication of embryos (Monteuuis 2004; Monteuuis et al. 1995). The acclimatization of plantlets involved transferring them to a greenhouse, where environmental conditions were gradually adjusted to facilitate successful establishment in the field.

To ensure genetic stability, molecular markers were used to confirm the clonal nature of tissue-cultured plants. In addition, phenotypic evaluations were conducted to assess traits such as height, stem diameter, branching pattern, and wood quality. The results demonstrated high genetic stability and consistent expression of desirable traits among tissue-cultured *A. mangium* plants. Tissue culture has proven to be a valuable tool for improving the breeding strategy of *A. mangium*. The technique offers numerous advantages, including rapid multiplication, clonal propagation, and genetic improvement. The experimental results have shown efficient embryogenesis induction, successful multiplication and acclimatization of embryogenic cultures, and the preservation of genetic stability and desirable traits. The integration of tissue culture into the breeding program of *A. mangium* will undoubtedly contribute to the development of superior varieties with enhanced growth characteristics and wood quality..

20.2.3.2 Molecular Breeding

The application of molecular breeding techniques, particularly marker-assisted selection, and genetic diversity analysis, has revolutionized the breeding strategy for *A. mangium*. With its ability to rapidly identify and select individuals with desired traits, molecular breeding significantly improves breeding efficiency and shortens the breeding cycle. Furthermore, the consideration of genetic diversity enhances the success of breeding programs by ensuring the generation of robust and adaptable varieties. The results obtained from molecular breeding approaches for *A. mangium* hold promising implications for the sustainable development of this economically important tree species (Amiteye 2021). By identifying molecular markers linked to important traits such as growth rate, wood quality, disease resistance, and drought tolerance, breeders can make informed decisions and develop improved *A. mangium* varieties.

Genomic selection is a relatively new approach that utilizes genomic information to predict the breeding value of individuals. By analyzing a large number of DNA markers across the genome, breeders can estimate the genetic potential of an individual for various traits without phenotypic evaluation. This approach enables the selection of superior individuals at an early stage, expediting the breeding process. Genomic selection has the potential to significantly accelerate genetic gain in *A. mangium* breeding (Xu and Hu 2010).

Molecular breeding offers several advantages over conventional breeding methods. Firstly, it allows for the selection of desirable traits at the molecular level, enabling breeders to target specific genetic markers associated with desired characteristics. This precision breeding approach significantly reduces the time required to develop improved varieties, enhances understanding of genetic diversity, and aids in the identification of key genes responsible for desired traits. It also provides increased genetic gain through improved selection accuracy and a higher probability of obtaining superior genotypes.

From molecular breeding, the utilization of genetic diversity information in breeding strategies enhances the efficiency and effectiveness of the *A. mangium* breeding strategy. Selecting diverse parental combinations can maximize the potential for generating superior offspring with a broader range of desirable traits. This approach also aids in preventing inbreeding depression and maintaining long-term genetic gains within breeding populations. This is because *A. mangium* is an insect-pollinated species (Sedgley et al. 1992), which preferentially out-crossing (Butcher et al. 2004). Preventing the decline of genetic diversity is crucial to ensuring the sustainability of breeding in the advance generation breeding program of *A. mangium*.

20.2.3.3 Mutation Breeding

Mutation breeding involves inducing genetic mutations in plants through various methods such as chemical or radiation treatments. Mutations such as those induced by the application of polyploid can result in the development of new and desirable traits in *A. mangium*. The advantages of mutation breeding in *A. mangium* include enhanced wood quality with improved characteristics, such as higher density and increased durability (Singh et al. 2021). It also facilitates the increased timber value and reduced susceptibility to decay. Mutations can confer resistance to pests and diseases that commonly affect *A. mangium,* leading to decreased damage from insect infestations and improved plant health and vigor.

Polyploid breeding involves inducing changes in the chromosome number of a plant species to create polyploids with multiple sets of chromosomes. This technique has been successfully applied in *A. mangium* to generate tetraploids. The increased chromosome sets can lead to significant changes in the plant's morphology, physiology, and reproductive characteristics. The benefits of polyploid breeding in *A. mangium* include improved growth and enhanced nutrient uptake. Polyploid plants may possess increased resistance to pests and diseases, reduced susceptibility to fungal infections and enhanced defense mechanisms. It also is known to exhibit improved water-use efficiency and resilience to drought conditions, efficient water uptake and retention, and reduced water loss through transpiration.

Griffin et al. (2015) have reported significant improvements in desirable traits and overall performance of the *A. mangium* tetraploids, which have larger and thicker wood fibers, indicating other potential benefits for wind and stress tolerance and disease resistance. Utilizing polyploid and mutation breeding can also contribute to the development of improved *Acacia* hybrids through inter-species hybridization between *A. mangium* and *A. auriculiformis* resulting in hybrids that are more productive, resilient, and valuable for timber production and reforestation efforts (Le et al. 2021).

20.3 Germplasm Biodiversity and Conservation

The conservation of *A. mangium*'s germplasm and biodiversity is vital for its sustainable utilization and preservation. Efforts such as seed banking, ex situ and in situ conservation, and understanding its genetic diversity are crucial steps toward ensuring the long-term survival of this valuable species. By actively engaging in genetic conservation practices, we can safeguard *A. mangium*'s unique traits and contribute to the preservation of global biodiversity.

20.3.1 Germplasm Diversity

The genetic diversity of *A. mangium* is influenced by factors such as geographic distribution, ecological niche, and reproductive mechanisms (Yuskianti 2014). Understanding this diversity is vital for effective conservation strategies. Germplasm, referring to the genetic material of living organisms, plays a crucial role in maintaining biodiversity and preserving the unique traits of a species. In the case of *A. mangium*, conserving its germplasm is essential for sustaining its valuable characteristics, such as fast growth, adaptation to various soil conditions, and resistance to pests and diseases. Strategies to safeguard the genetic diversity of *A. mangium* have been employed through seed banks by collecting and storing seeds from different populations to preserve their genetic makeup (Butcher et al. 1998). These seed banks serve as repositories for future use in reforestation and breeding programs. Apart from seed banks, living trees are conserved in arboretums, botanical gardens, and other protected areas. These living collections provide researchers and breeders with access to diverse genetic resources. In the case of in situ conservation, protecting natural habitats where *A. mangium* thrives is crucial for maintaining its germplasm. This involves establishing protected areas, implementing sustainable forestry practices, and raising awareness among local communities about the importance of biodiversity conservation.

The *A. mangium* displays a wide ecological tolerance, allowing it to thrive in diverse environments. This adaptability contributes to its success as a plantation species for timber and pulp production. However, variations in its genetic makeup can result in differences in growth rates, wood quality, and resistance to biotic and abiotic stresses (Asif et al. 2017). The reproductive system of *A. mangium* involves both sexual and asexual reproduction. Sexual reproduction through cross-pollination facilitates genetic recombination and diversity. Asexual reproduction through vegetative propagation ensures the perpetuation of desirable traits and the preservation of specific genotypes.

Conserving the germplasm and biodiversity of *A. mangium* presents various challenges (Le et al. 2021; Nirsatmanto and Sunarti 2021), including the invasiveness (Koutika and Richardson 2019). Introduction of *A. mangium* to new regions may lead to its invasion into native ecosystems, potentially threatening local biodiversity. Careful management is essential to prevent the species from becoming invasive. In another case, as climate conditions shift, *A. mangium* populations may face challenges in adapting to new environments. Conservation efforts should focus on identifying resilient genotypes that can withstand future climatic conditions. *A. mangium* is also susceptible to various diseases and pests, which can adversely affect its health and genetic diversity. Monitoring and management strategies are crucial for minimizing the impact of these threats.

20.3.2 Phylogeny

The evolutionary history of *A. mangium* can be traced back through the study of its phylogeny. By analyzing DNA sequences, scientists can infer the ancestral relationships and divergence times between different species. *A. mangium* shares a common ancestor with other Acacia species. Genetic relationship with other species reveals that *A. mangium* is closely related to *A. auriculiformis* and *A. crassicarpa.* These species share similar morphological characteristics, such as branching habits, leaves, stem barks, and ecological preferences, indicating a close genetic relationship (Nirsatmanto and Sunarti 2019; Widyatmoko et al. 2010).

The phylogeny of *A. mangium* provides valuable insights into its evolutionary history and genetic relationships with other species. Through DNA sequencing and the construction of phylogenetic trees, scientists have determined its divergence from common ancestors and its close genetic relationship with other Acacia species. Understanding the phylogeny of *A. mangium* contributes to our knowledge of its ecological preferences, adaptive traits, and potential for genetic improvement. Overall, studying the phylogeny of *A. mangium* is crucial for conservation efforts, tree breeding programs, and understanding the evolutionary processes shaping its genetic diversity. By incorporating a high number of transition words, active voice, and sentence variety, this article aims to make the subject more accessible and engaging for readers interested in the fascinating world of plant evolution.

20.4 Hybridization

Hybridization, as defined by Chaudary (1984) and Ibrahim (1993), is the mating of genetically distinct individuals resulting in the production of hybrid offspring. It involves the mating of individuals with different genetic backgrounds, resulting in the creation of a third individual with a genetic makeup different from its parental origin, whether natural or artificial (Le and Ha 2017). Kha (2001) provides a broad perspective *Acacia* hybrid, defining hybridization as the crossing of two different genotypes of *Acacias*, resulting in a new variety called a hybrid. Furthermore, Kha (2001) explores the subsequent stages of the hybridization process, and when seeds resulting from hybridization are planted, they give rise to first-generation hybrids (F_1). This stage is a crucial phase in the continuous cycle of genetic diversity and innovation.

Hybridization serves several purposes, including creating a reservoir of genetic diversity for future selection, combining different desirable traits, and exploiting the phenomenon of heterosis (Agrawal 1998). The process is categorized based on the crossing system and includes intraparietal, intervarietal, interspecific, and intergeneric types (Chaudary 1984). Intravarietal hybridization involves crossing plants of different variants or strains within the same variety. In contrast, intervarietal (intraspecific) hybridization involves crossing plants of different varieties within the same species. This form of intervarietal hybridization is particularly important in agriculture, having produced many well-known varieties of major crops such as corn, rice, wheat, and other commodities. Interspecific (intrageneric) hybridization, on the other hand, occurs when different species of the same genus are crossed such as between *A. mangium* and *A. auriculiformis*. This method is used when desired traits are not present within the same species. Finally, intergeneric hybridization involves a cross between plants of different genera, demonstrating the wide range of possibilities within the plant hybridization field.

Based on the method of crossing or mating system, hybridization can be classified into eight types as outlined by Chaudary 1984. These types include single cross, three-way cross, top cross, double cross, polycross, synthetic cross, and multiple cross. In a single cross, two parents (A and B) are mated to produce A × B hybrids, with A serving as the female parent. When B acts as the female parent, resulting in B × A, it is called a reciprocal cross. The type of single cross and reciprocal cross is valuable in the development of new hybrid varieties of *Acacias*. A three-way cross involves three parent trees, where A is crossed with B and the result is crossed with C (A × B (F_1) × C). The top cross involves the crossing of different strains, clones or other variations. The crossed parent called the top cross, can be a relative cross or a single cross. Depending on the crossed parent, the resulting cross may be between related varieties, a double cross, or a triple cross. A double cross is the mating of two single crosses (A × B) and (C × D). A backcross is the crossing of a single cross with one of its parents. Polycross is the crossing of groups, varieties or clones in separate areas to facilitate random pollination. The last types of crossing were rarely done on *Acacias*.

The *Acacia mangium* × *A. auriculiformis* hybrid is a progeny resulting from interspecific hybridization, which could be obtained from natural hybridization and artificial hybridization. The natural *Acacia* hybrid was first discovered in urban plantations in Malaysia in 1972 and showed faster growth than *A. mangium* and phenotypic characteristics intermediate between *A. mangium* and *A. auriculiformis* (Chaudary 1984; Ibrahim 1993; Kha 2001). Natural *Acacia* hybrids were also identified in *A. mangium* plantations in several regions of Vietnam in 1993, showing faster growth than *A. mangium* and morphological characteristics intermediate between *A. mangium* and *A. auriculiformis* (Le and Ha 2017). Due to its superior growth (heterosis/hybrid vigor) compared to *A. mangium*, *Acacia* hybrids have been actively developed in various countries, including Malaysia (Ibrahim 1993), Vietnam (Kha 2001), Thailand (Luangviriyasaeng 2007), and Indonesia (Sunarti et al. 2022).

In Indonesia, *Acacia* hybrids have been developed since the 2000s by establishing breeding gardens and hybrid seed orchards (HSO) of *Acacias*. The breeding strategy and morphological markers for hybrid identification have been updated by Sunarti et al. (2013). To obtain vigorous *Acacia* hybrids, a breeding strategy using the co-improvement method was developed (Sunarti et al. 2022).

In the co-improvement method, the male parent tree and the female parent tree are vigorous individuals, so that the probability of unwanted hybrid offspring can be reduced. In addition, the possibility of vigorous hybrids (positive heterosis) is also increased due to the superior traits of both parents. Currently, several *Acacia* hybrid varieties have been produced from both natural and artificial crosses, and some of them have been registered for plant variety protection and one of them has received plant variety protection rights (Figs. 20.11 and 20.12).

Vegetative propagation techniques are essential for hybrid species because mass propagation by seed is impractical due to the high variability of the resulting plants (Hardiyanto 2004). Only vigorous hybrids that exhibit heterosis are propagated on a large scale, typically in the form of clones using methods such as shoot cuttings or tissue culture (Fig. 20.13). Before superior hybrid clones are introduced into specific regions, it is crucial to conduct site suitability tests and implement appropriate silvicultural techniques to ensure optimal benefits from these hybrids through genetic trials (Libby and Ahuja 1993). Many genetic trials have been established in Java, Riau, Jambi and Palembang (Sumatera), such as clonal test, clonal site interaction and semi-plantation demonstration plots (Fig. 20.14). Based on the genetic test results, the *Acacia* hybrid clone's growth was 17% superior to the female parent species of *A. mangium* (Sunarti et al. 2022).

In terms of fiber quality, Nirsatmanto et al. (2017) reported that *A. mangium* as a pure parental species has contributed to the higher fiber properties of the *Acacia* hybrid, particularly when *A. mangium* is the male parent (Tables 20.3 and 20.4). *A. mangium* has also affected the performance of macroscopic features for vessel and rays of the hybrids, as compared to *A. auriculiformis* (Fig. 20.15).

Fig. 20.11 Morphological characteristics of granted *Acacia* hybrid variety (*Acacia mangiumxA. auriculiformis*) (Variety of Purwo Sri Ah044): tree (**a**), leaf (**b**), mature bark (**c**), inflorescence (**d**) and young pod (**e**). (Photographs by S. Sunarti)

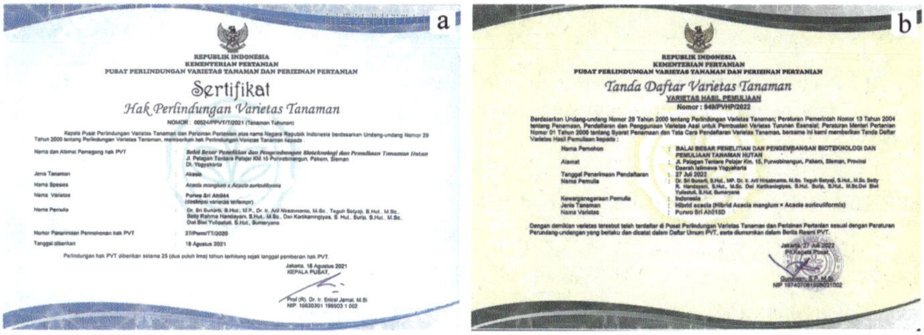

Fig. 20.12 Certificate of Plant Variety Protection Rights issued by Indonesian Government for the *Acacia* hybrids variety (*A. mangium* × *A. auriculiformis*): Granted Certificate (**a**), and Registered Certificate (**b**). (Photographs by S. Sunarti)

20.5 Conclusions and Prospects

In the past, *A. mangium* held significant economic and environmental value in Indonesia. However, currently, the serious damage due to the *Ceratocystis* disease attacks in the *A. mangium* plantation caused the species to stop being planted in

Fig. 20.13 Vegetative propagation of *Acacia* hybrid (*Acacia mangium* × *A. auricliformis*) variety by tissue culture (**a**) and seedlings from cutting propagation in the nursery (**b**). (Source: Sinarmas Forestry (a) Photograph by S. Sunarti (b))

Fig. 20.14 Demo plantation's superior *Acacia* hybrid variety (*Acacia mangium* × *A. auricliformis*) on a marine clay site in Palembang, Indonesia 7 months after planting. (Source: Sinarmas Forestry)

Table 20.3 Proportion of cells in the wood of four *Acacia* species at 3 years of age

	Proportion of cells (%)		
Species	Vessel	Ray	Fiber
Acacia mangium (AM)	9.29 ± 0.88 a	8.23 ± 1.03 a	74.29 ± 2.39 ab
Acacia auriculiformis (AA)	8.64 ± 0.80 a	8.84 ± 1.03 a	73.08 ± 1.73 b
AM × AA hybrid	9.13 ± 1.35 a	7.66 ± 1.22 b	68.90 ± 6.96 c
AA × AM hybrid	8.76 ± 0.50 a	8.57 ± 1.80 a	75.89 ± 2.66 a

Source: Nirsatmanto et al. (2017)
Remark: the same letters in a column indicate a not significantly different at the level of 5%

most industries. Therefore, improving the breeding strategy of *A. mangium* is crucial for re-gaining its potential as a valuable tree species for fiber production. Enhancing the genetic base, utilizing a recurrent selection system, implementing a sub-lining system, and developing disease-resistant varieties are key steps in achieving this goal. By employing these strategies, breeders can ensure the production of superior *A. mangium* varieties that exhibit improved growth, wood quality, and disease resistance.

Table 20.4 Fiber dimension of the wood of four *Acacias* species at 3 years of age

Species	Fiber length (mm)	Fiber diameter (μm)	Fiber lumen diameter (μm)	Fiber wall thickness (μm)
Acacia mangium (AM)	0.86 ± 0.01 a	17.07 ± 1.10 a	13.30 ± 1.06 a	1.89 ± 0.05 a
Acacia auriculiformis (AA)	0.84 ± 0.04 a	14.47 ± 0.68 b	11.32 ± 0.37 b	2.15 ± 0.32 a
AM × AA hybrid	0.78 ± 0.03 a	13.86 ± 0.77 b	11.42 ± 0.39 b	1.22 ± 0.19 b
AA × AM hybrid	0.85 ± 0.09 a	16.09 ± 0.18 a	12.56 ± 0.17 a	1.77 ± 0.01 a

Source: Nirsatmanto et al. (2017)
Remark: the same letters in a column indicate a not significantly different at the level of 5%

Genetically improved stocks tolerant to Ceratocystis are a critical trait for ensuring the long-term survival and productivity of *A. mangium* plantations. By incorporating this disease resistance into the breeding objectives, breeders can select individuals with natural resistance or develop varieties through hybridization and selection. In this case, expanding the genetic base is essential for maximizing the genetic diversity within breeding populations. This can be achieved through the collection and preservation of diverse *A. mangium* germplasm from various geographical regions. Utilizing a broad range of genetic resources enhances the adaptability and resilience of the species, making it less susceptible to environmental changes and diseases.

Implementing a recurrent selection system is crucial to continuously improve desirable traits in *A. mangium* populations. This involves selecting superior individuals based on specific criteria and allowing them to contribute to the next generation's genetic makeup. The process is repeated over multiple cycles, gradually improving the overall performance of the breeding population. However, understanding the relationship among the target traits for selection is necessary to ensure the increase in accumulative genetic gains over successive generations of breeding.

A sub-lining system can further enhance the breeding strategy of *A. mangium* by dividing the breeding population into sub-lines based on desired traits and provenances. This allows for more targeted selection and breeding efforts, ensuring the development of specialized varieties with superior desired characteristics. By focusing on specific traits such as growth rate, wood quality, or disease resistance, the breeding program can produce customized *A. mangium* varieties that meet specific market demands.

Concerning the current status and challenges of *A. mangium* plantations in Indonesia, re-designing the breeding strategy of *A. mangium* is crucial to ensuring and regaining genetic improvement in the advanced-generation breeding program. Developing a more conservative breeding strategy in combination with biotechnology can be employed to overcome the current issues. Five considerations for future research focusing on growth, wood properties, and disease resistance are necessary in the effort to re-design the breeding strategy, as listed below.

The first consideration is the establishing breeding population with no genetic thinning. It can be used to optimize the breeding seedling orchard strategy for preserving genetic resources for future generations of breeding. The second is the adoption of a double-layer progeny trial to select superior seed trees for seed orchard establishment. It is beneficial to be sure of the real quality of progeny, which in turn makes it important as a basis to verify and select high-quality seed trees. The third is the re-configuring of the experimental design for breeding populations into single tree plots rather than multiple-line tree plots. It employs the post-blocking strategy to find out the appropriate plot configuration based on the real site condition. The fourth is the re-exploring of genetic material from previous generations of breeding to combine into advanced-generation breeding in the form of a mixed-generation breeding population. It is important to increase genetic diversity to anticipate the change of a desired trait in the future. The fifth is establishing breeding without a progeny trial by employing genomic selection. It is beneficial to make shorter breeding cycles in a recurrent selection system to increase accumulative genetic gain over successive generations.

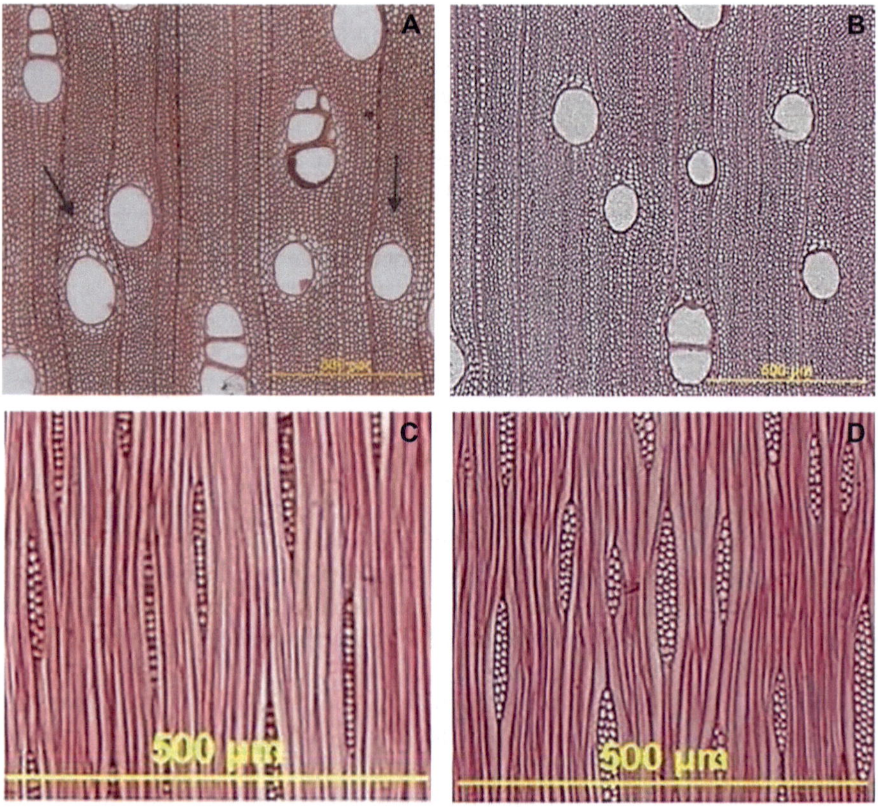

Fig. 20.15 The vessel and rays feature: vessel in *Acacia auriculiformis* and *Acacia* hybrid (*A. mangium* × *A. auriculiformis*) (**A**), and in *A. mangium* and *Acacia* hybrid (*A.auriculiformis* × *A. mangium*) (**B**), the uniseriate/biseriate rays in *A. mangium* (**C**), and multiseriate rays in *Acacia* hybrid *(A. auriculiformis* × *A. mangium)* (**D**). (Source: Nirsatmanto et al. 2017)

Appendix 20.1 Research Institutes and Online Resources Available for *Acacia mangium* Breeding

Institution	Area of specialization	Research activities	Website
Research Center for Plant Conservation, Botanic Garden and Forestry, National Research and Innovation Agency	1. Plants Conservation 2. Botanic Garden 3. Plant Breeding 4. Biotechnology	1. Molecular genetic 2. Genetic testing 3. Seed production 4. Hybridization 5. Plant propagation 6. Genetic conservation 7. Genomic	https://brin.go.id/orhl https://brin.go.id/orhl/prktkrk/page/tugas-dan-fungsi-pusat-riset-konservasi-tumbuhan-kebun-raya-dan-kehutan
Forest Tree Improvement and Biotechnology Research Institute, Vietnam	1. Biotechnology 2. Tree improvement	Genetic testing Seed production Hybridization Polyploid breeding	Forest Tree Improvement and Biotechnology Research Institute (FTIBRI) (vafs.gov.vn) http://vafs.gov.vn/en/forest-tree-improvement-and-biotechnology-research-institute-ftibri/
Forest Research Institute Malaysia (FRIM)	1. Forestry and Environment 2. Forest Biotechnology 3. Forest Biodiversity 4. Natural Product 5. Economic and Strategic Analysis, and Social Forestry	1. Molecular genetic 2. Genetic testing 3. Genomic 4. Plant improvement 5. Biotechnology	www.frim.gov.my/
ACIAR-FST/2014/068	Project on pest and disease of tropical Acacias in Indonesia and Vietnam	1. Root rot disease 2. Wilt disease 3. Biological Control Agent (BCA)	Managing risk in South East Asian forest biosecurity I ACIAR https://www.aciar.gov.au/project/fst-2018-179

Appendix 20.2 Genetic Resources for *Acacia mangium* Breeding

Species	Important traits	Cultivation location
Acacia mangium (Sub-line A)	Better growth and wood properties suitable for pulp and paper, sawn timber, and wood energy industries, specifically on dry-land	Indonesia, Malaysia,Vietnam
Acacia mangium (Sub-line B)	Better growth and wood properties suitable for pulp and paper, sawn timber, and wood energy industries, specifically on dry-land	Indonesia, Malaysia,Vietnam
Acacia mangium (Sub-line C)	Better stem form and wood properties suitable for pulp and paper, sawn timber, wood energy industries, tolerance to wind damage, specific on dry-land	Indonesia, Malaysia,Vietnam
Acacia mangium (Sub-line D)	Better stem form and wood properties suitable for pulp and paper, sawn timber, wood energy industries, tolerance to wind damage, specific on dry-land	Indonesia, Malaysia, Vietnam

References

Agrawal RL (1998) Fundamental of plant breeding and hybrid seed production. Science Publisher, Inc, Enfield

Almeida AC, Anh HV, Beadle C et al (2014) Modelling *Acacia mangium* production in Vietnam under current and future climates. In: Sustaining the future of acacia plantation forestry international conference, IUFRO working party 2.08.07: genetics and silviculture of acacia, Hue, Vietnam, 18–21 March 2014, Compendium of Abstracts

Amiteye S (2021) Basic concepts and methodologies of DNA marker systems in plant molecular breeding. Heliyon 7(10):e08093. https://doi.org/10.1016/j.heliyon.2021.e08093

Andalia D (2023) Sustainable textile practices by integrated viscose rayon and yarn producers: an empirical study. J Fin Bank Review 8(1):28–42. https://doi.org/10.35609/jfbr.2023.8.1(1)

Andianto YK, Saputra NA, Saputra IS (2020) Fiber dimension and anatomy of *Acacia mangium* wood from two mother trees. ICFP 2020. IOP Conf Ser Mater Sci Eng 935:012020

Arisman H, Hardiyanto EB (2006) *Acacia mangium* – a historical perspective on its cultivation. In: Potter K, Rimbawanto A, Beadle C (eds) Heart rot and root rot in tropical acacia plantations. Proceeding of a workshop, Yogyakarta, p 92

Arnoul-Jarriault B, Lachenal D, Chirat C, Heux L (2015) Upgrading softwood bleached kraft pulp to dissolving pulp by cold caustic treatment and acid-hot caustic treatment. Ind Crop Prod 65:565–571

Asif MJ, Zaki MA, Norwati M, Wickneswary R (2017) Detecting mislabeling and identifying unique progeny in acacia mapping population using SNP markers. J For Res 28(6):1119–1127

Bajpai P, Bajpai P (2010) Acacia an emerging raw material for pulp. JIPPTA 22(1):107–111

Biantoro R, Septiningrum K, Kardiansyah T (2020) Dissolving pulp from wood and non wood: Review of the manufacturing process and its characteristics. J Selulosa 10(1):35–50. https://doi.org/10.25269/jsel.v10i01.294

Bowen MR, Eusebio TV (1981) *Acacia mangium*: updated information on seed collection, handling, and germination testing. Occasional, technical and scientific notes. Forest Research Centre, seed series No. 5, FAO/UNDP-MAL/78/009. Sanda Kan, Sabah, Malaysia. pp26, CARD 2005 Sustainable and profitable development of acacia plantations for sawlog production

Brawner J, Japarudin Y, Lapammu M et al (2015) Evaluating the inheritance of *Ceratocystis acaciivora* symptom expression in a diverse *Acacia mangium* breeding population. South For 77:83–90

Brawner J, Sunarti S, Nirsatmanto A et al (2022) Screening clonally replicated *Acacia mangium* breeding populations for tolerance to Ceratocystis canker and wilt disease. Tree Genet Genomes 18:17. https://doi.org/10.1007/s11295-022-01548-9

Burdon RD (1989) Early selection in tree breeding: principles for applying index selection and inferring input parameters. Can J For Res 19:499–504. https://doi.org/10.1139/x89-076

Butcher PA, Moran GF, Perkins HD (1998) RFLP diversity in the nuclear genome of *Acacia mangium*. Heredity 81:205–213

Butcher P, Harwood C, Quang TH (2004) Studies of mating systems in seed stands suggest possible causes of variable outcrossing rates in natural populations of *Acacia mangium*. Int J For Genet 11:303–309

Chaudary RC (1984) Introduction to plant breeding. Oxford & IBH Publishing Co, New Delhi

Dung BX, Trang PQ, Linh NTM et al (2019) Soil erosion and overland flow from acacia plantation forest in headwater catchment of Vietnam. IOP Conf Ser: Earth Environ Sci 266:012012. https://doi.org/10.1088/1755-1315/266/1/012012

Farid AM, Lee S, Maziah Z, Norwati M (2006) Root rot in tree species other than acacia. In: Potter K, Rimbawanto A, Beadle C (eds) Heart rot and root rot in tropical acacia plantations. Proceeding of a workshop, Yogyakarta, p 92

Griffin AR, Nghiem QC, Harbard JL et al (2015) Breeding polyploid varieties of tropical acacias: progress and prospects. South For 77:41–50

Gschwandtne C (2022) Outlook on global fiber demand and supply 2030. Lenzinger Berichte 97:11–19

Hardiyanto EB (1998) Approaches to breeding acacias for growth and form: the experiences at PT. Musi Hutan Persada (Barito Pacific Group). In: Turnbull JW, Crompton HR, Pinyopusarerk K (eds) Recent development in acacia planting, pp 178–183

Hardiyanto EB (2004) Silvikultur dan Pemuliaan *Acacia mangium*. In: Hardiyanto EB, Arisman H (eds) Pembangunan hutan tanaman *Acacia mangium*: pengalaman di PT. Musi Hutan Persada. Polydoor, Yogyakarta, pp 207–281

Hardiyanto EB (2014) Challenges for acacia breeders. In: Sustaining the future of acacia plantation forestry international conference, IUFRO working party 2.08.07: genetics and silviculture of acacia, Hue, Vietnam, 18–21 March 2014, Compendium of Abstracts

Hardiyanto EB, Nambiar EKS (2014) Productivity of successive rotations of *Acacia mangium* plantations in Sumatera, Indonesia: impacts of harvest and inter-rotation site management. New For 45(4):557–575. https://doi.org/10.1007/s11056-014-9418-8

Hardiyanto EB, Inail MA, Nambiar EKS (2021) Productivity of *Eucalyptus pellita* in Sumatra: *Acacia mangium* legacy, response to phosphorus, and site variables for guiding management. Forests 12:1186. https://doi.org/10.3390/f12091186

Hardiyanto EB, Inail MA, Nambiar S, Mendham DS (2024) Sustaining plantation forest productivity in Sumatera over three decades: From acacias to eucalypts. For Ecol Manag 553:121613. https://doi.org/10.1016/j.foreco.2023.121613

Harwood CE, Nambiar EKS (2013) Sustainable plantation forestry in South-East Asia. Client Report EP14685 to Australian Centre for International Agricultural Research. Sustainable Agriculture Flagship and CSIRO Ecosystem Sciences, Canberra, Australia, pp105

Harwood CE, Nambiar EKS (2014) Productivity of acacia and eucalypt plantations in Southeast Asia. 2. Trends and variations. Int For Rev 16:249–260

Harwood CE, Hardiyanto EB, Wong CY (2015) Genetic improvement of tropical acacias: achievements and challenges. South For 77(1):11–18. https://doi.org/10.2989/20702620.2014.999302

Hasegawa M, Wakimoto R, Yoshida E et al (2009) Provenance variation in growth and wood properties of *A. mangium* and *A. auriculiformis* in Central Java, Indonesia: selecting potential hybrid parents for good provenance. Bull Kyushu Univ For 90:25–37

Hazel LN (1943) The genetic basis for constructing selection indexes. Genetics 28:476–490

Hegde M, Palanisamy K, Yi JS (2013) *Acacia mangium* Willd. A fast growing tree for tropical. J For Envi 29(1):1–14. https://doi.org/10.7747/JFS.2013.29.1.1
Huang LJ, Chen ZX, Zhang SQ, Liang RG (2012) Tissue culture technique of *Acacia mangium* elite tree. For Res 25(2):227–230
Ibrahim Z (1993) Reproductive biology. In: Awang K, Taylor D (eds) *Acacia mangium* growing and utilization. Winrock International and the Food and Agriculture Organization of the United Nations, Bangkok, pp 21–34
Jusoh I, Zaharin FA, Adam NS (2014) Wood quality of acacia hybrid and second-generation *Acacia mangium*. BioRes 9(1):150–160. https://doi.org/10.15376/biores.9.1.150-160
Kalita RM, Rahman M, Borogayary B et al (2016) Carbon storage potential of acacia plantation: A viable option for climate change mitigation. In: Proceedings International Conference on climate change mitigation and technologies for adaptation, pp 115–118
Kha LD (2001) Studies on the use of natural hybrids between *Acacia m*angium and *Acacia auriculiformis* in Vietnam. Agriculture Publishing House, Hanoi. https://www.jstor.org/stable/43582411
Koutika LS, Richardson DM (2019) *Acacia mangium* Willd: benefits and threats associated with its increasing use around the world. For Ecosyst 6:2. https://doi.org/10.1186/s40663-019-0159-1
Krisnawati H, Kallio M, Kanninen M (2011) *Acacia mangium* Willd.: ecology, silviculture and productivity. CIFOR, Bogor
Kurinobu S (1993) Consideration on a choice of establishment methods of seedling seed orchard for fast growing species related to a tree improvement strategies in Indonesia. In: Proceedings of national seminar on forest tree improvement. Ministry of Forestry, Gadjah Mada University, Yogyakarta, Indonesia, pp 47–52
Le DK, Ha HT (2017) Research and development of acacia hybrids for commercial planting in Vietnam. Vietnam J Sci Technol Engin 59(1):36–42. https://doi.org/10.31276/VJSTE.59(1).36
Le S, Griffin RA, Harwood CE et al (2021) Breeding polyploid varieties of acacia: Reproductive and early growth characteristics of the allotetraploid hybrid (*Acacia mangium* × *A. auriculiformis*) in comparison with diploid progenitors. Forests 12(6):778. https://doi.org/10.3390/f12060778
Lee SS (2004) Diseases and potential threats to *Acacia mangium* plantations in Malaysia. Unasylva 55:217
Lee KL, Ong KH, King PJH et al (2015) Stand productivity, carbon content, and soil nutrients in different stand ages of *Acacia m*angium in Sarawak, Malaysia. Turk J Agric For 39:154–161. https://doi.org/10.3906/tar-1404-20
Libby WJ, Ahuja MR (1993) The genetics of clones. In: Ahuja MR, Libby WJ (eds) Clonal forestry I. Springer, Berlin/Heidelberg. https://doi.org/10.1007/978-3-642-84175-0_2
Lim SC, Gan KS,Tan YE (2011) Properties of *Acacia mangium* planted in Peninsular Malaysia. ITTO Project on improving utilization and value adding of plantation timbers from sustainable sources in Malaysia. Project No. PD 306/04(1) Perpustakaan Negara Malaysia pp 7–22
Luangviriyasaeng V (2007) Current situation and potential of acacia plantation for pulp industry. NFT News 10(1):1–5
Marsoem SN, Sulistyo J, Prasetyo VE et al (2012) Fiber characterization as a tool for verifying pulp fiber composition. Paper conference and trade show: Growing the future. New Orleans, Lousiana, USA, 22–25 April 2012
Matsubara K, Ohta S (2015) The effects of tannins derived from *Acacia mangium* bark on N_2O emissions from water saturated acacia plantation soil. Tropics 24(2):65–74. https://doi.org/10.3759/tropics.24.65
MOEF (2020) Peraturan Menteri Lingkungan Hidup dan Kehutanan Nomor: P.3/Menlhk/Setjen/Kum.1/1/2020 Tentang Penyelenggaraan Perbenihan Tanaman Hutan. Ministry of Environment and Forestry, Jakarta, pp 185
MOEF (2022) The State of Indonesia's Forests 2022. Ministry of Environment and Forestry, Jakarta, pp 96

Monteuuis O (2004) In vitro micro propagation and rooting of microshoots from juvenile and mature origins. In Vitro Cell Dev Biol 40(1):102–107

Monteuuis O, Vallauri D, Poupard C, Chauviere M (1995) Rooting *Acacia mangium* cuttings of different physiological age with reference to leaf morphology as a phase change marker. Silvae Genet 44(2–3):150–154

Nasution A, Indrayadi H, Glen M et al (2022) Phyllode inoculation provides a rapid protocol for preliminary screening of *Acacia* species for tolerance to Ceratocystis wilt and canker disease. Eur J Plant Pathol 163:321–339. https://doi.org/10.1007/s10658-022-02479-w

National Research Council (1983) Mangium and other fast-growing acacias for the humid tropics. National Academy Press, Washington

Nirsatmanto A (2020) Impact of tree breeding on managing forest genetic resources: a review with a particular reference to *Acacias* and *Eucalyptus* grown in Indonesia. IOP Conf Ser: Earth Environ Sci 522:012002. https://doi.org/10.1088/1755-1315/522/1/012002

Nirsatmanto A, Sunarti S (2019) Genetics and breeding of tropical acacias for forest products: *Acacia mangium*, *A. auriculiformis* and *A. crassicarpa*. In: Al-Khayari JM, Jain SM, Johnson DV (eds) Advances plant breeding strategies: industrial and food crops. Springer, Switzerland, pp 3–28. https://doi.org/10.1007/978-3-030-23265-8

Nirsatmanto A, Sunarti S (2021) Reducing ecological risk from the impacts of introduced exotic species in forestry industry: Limiting the invasiveness through a management and biological practices in acacias plantation in Indonesia. IOP Conf Ser: Earth Environ Sci 886:012098. https://doi.org/10.1088/1755-1315/886/1/012098

Nirsatmanto A, Kurinobu S, Shiraishi S (2012) Evaluation for the efficiency of early selection in *Acacia mangium* seedling seed orchards based on age trends in genetic parameter. Indones J For Res 9(1):16–24. https://doi.org/10.20886/ijfr.2012.9.1.16-24

Nirsatmanto A, Setyaji T, Wahyuningtyas RS (2014) Realized genetic gain and seed source x interaction on stand volume productivity of *Acacia mangium*. Indones J For Res 1(1):21–32. https://doi.org/10.20886/ijfr.2014.1.1.21-32

Nirsatmanto A, Setyaji T, Sunarti S, Kartikaningtyas D (2015) Genetic gain and projected increase in stand volume from two cycles breeding program of *Acacia mangium*. Indones J For Res 2(2):71–79. https://doi.org/10.20886/ijfr.2015.2.2.71-79

Nirsatmanto A, Sunarti S, Praptoyo H (2017) Wood anatomical structures of tropical acacias and its implication to tree breeding. Int J For Hortic 3(3):9–16. https://doi.org/10.20431/2454-9487.0303002

Nirsatmanto A, Sunarti S, Setyaji T et al (2019) Pemuliaan Tanaman Penghasil Kayu Untuk Bahan Baku Industri Pulp Dan Kertas Acacias Dan Eucalyptus. In: Darwo YI (ed) Bunga Rampai: Peningkatan Produktivitas Hutan Menuju 100 Tahun Merdeka. IPB Press, Bogor, pp 29–54

Nirsatmanto A, Sunarti S, Kartikaningtyas D et al (2022) Evaluation of the characteristics of *Eucalyptus pellita* and acacia hybrid superior clones selected from breeding program in Indonesia as materials for pulp and papermaking. Wood Res 67:847–865

Nirsatmanto A, Sunarti S, Putri AI et al (2023) Genetic stability observed in third-generation progeny trial of *Acacia mangium*: the importance of genotype by environment interaction assessment in advance generation breeding strategy. For Sci Technol 19(4):285–295. https://doi.org/10.1080/21580103.2023.2264887

Nurhasybi N, Sudrajat DJ, Komar D (2008) Provenances for seed source by technique of biplot and correspondence analysis at Parungpanjang, Bogor, West Java. J Manaj Hutan Trop 4(1):28–40

Otsamo A (2002) Early effects of four fast-growing tree species and their planting density on ground vegetation in *Imperata* grasslands. New For 23:1–17. https://doi.org/10.1023/A:1015655923484

Pinyopusarerk K, Liang SB, Gunn BV (1993) Taxonomy, distribution, biology and use as an exotic. In: Awang K, Taylor D (eds) *Acacia mangium* growing and utilization. Winrock International and the Food and Agriculture Organization of the United Nations, Bangkok, pp 1–10

PROSEA (1995) Timber trees: minor commercial timbers. In: Soerianegara I, Lemmens RHMJ (eds) Plant resources of SouthEast Asia. Backhuys Publishers, Wageningen

Pujiastuti E, Sudrajat DJ (2017) Vigor test to predict seed germination and normal seedling emergence of *Acacia mangium* in nursery (in Indonesia). J Perbenihan Tanam Hutan 5(2):81–94. https://doi.org/10.20886/bptpth.2017.5.2.81-94

Sahri MH, Ibrahim FH, Shukor NA (1993) Anatomy of *Acacia mangium* grown in Malaysia. IAWA J 14(3):245–251

Saini I, Kumar T, Chauhan J, Kaushik P (2020) Acacia from a breeding and biotechnological perspective: a review. Pak J Biol Sci 23(11):1351–1356

Sedgley M, Harbard J, Smith RMM et al (1992) Reproductive biology and interspecific hybridization of *Acacia mangium* Willd. and *A. auriculiformis* A. Cunn. Ex. Benth (Leguminosae:Mimosoideae). Aust J Bot 40(1):37–48. https://doi.org/10.1071/BT9920037

Sein CC, Mitlöhner R (2011) *Acacia mangium* Willd: ecology and silviculture. CIFOR, Bogor

Singh DP, Singh AK, Singh A (2021) Plant breeding and cultivar development, 1st edn. Academic, p 662

Siregar STH, Hardiyanto EB, Gales K (1999) *Acacia mangium* plantations in PT Musi Hutan Persada, South Sumatera. In: Nambiar EKS, Cossalter C, Tiarks A (eds) Site management and productivity in tropical plantation forests, proceedings of workshops workshop proceedings. Center for International Forestry Research (CIFOR), pp 39–44

Sixta H, Iakovlev M, Testova L et al (2013) Novel concepts of dissolving pulp production. Cellulose 20:1547–1561. https://doi.org/10.1007/s10570-013-9943-1

Styring AR, Ragai R, Unggang J et al (2011) Bird community assembly in Bornean industrial tree plantations: effects of forest age and structure. For Ecol Manag 261(3):531–544. https://doi.org/10.1016/j.foreco.2010.11.003

Sunarti S et al (2013) Breeding strategy of Acacia hybrid (*A. mangium* × *A. auriculiformis*) to increase forest plantation productivity in Indonesia. J Trop Manag 19(2):128–137. https://doi.org/10.7226/jtfm.19.2.128

Sunarti S, Nirsatmanto A, Brawner J et al (2022) Effect of genetic gain in diameter and wood density on advance generation breeding strategy of *Acacia mangium* in Indonesia. J Trop For Sci 34(1):92–102. https://doi.org/10.26525/jtfs2022.34.1.92

Susilawati A, Sofyan A (1995) List of individual and bulk tree seed lots procured as of April 1995. Forest Tree Improvement Project 39, Japan International Cooperation Agency for Forestry Research and Development, Ministry of Forestry

Tarigan M, Roux J, van Wyk M et al (2011) A new wilt and die-back disease of *Acacia mangium* associated with *Ceratocystis manginecans* and *C. acaciivora* sp nov. in Indonesia. S Afr J Bot 77:292–304

Thu P, Quynh D, Fourie A (2014) *Ceratocystis* wilt– a new and serious threat to Acacia plantations in Vietnam: taxonomy and pathogenicity. In. IUFRO Working Party 2.08.07: Genetics and silviculture of Acacia, sustaining the future of Acacia plantation forestry. Hue, Vietnam

Tsai LM (1988) Studies on *Acacia mangium* in Kemasul Forest, Malaysia: I. Biomass and productivity. J Trop Ecol 4:293–302. https://doi.org/10.1017/S0266467400002856

Turnbull JW, Midgley SJ, Cossalter C (1997) Tropical acacias planted in Asia: An Overview of recent developments in acacias planting. In: Turnbull JW, Crompton HR, Pinyopusarerk K (eds) Recent developments in acacia planting, Canberra: ACIAR Proceedings No 82, pp 14–28

White TL, Adam WT, Neale DB (2007) Forest genetics. CABI Publishing, Cambridge

Widyatmoko AYPBC, Watanabe A, Shiraishi S (2010) Study on genetic variation and relationships among four acacia species using RAPD and SSCP marker. Indones J For Res 7(2):125–143. https://doi.org/10.20886/ijfr.2010.7.2.125-143

Wong CY, Yuliarto M (2014) Deployment of acacias in short rotation pulpwood plantation. In: Sustaining the future of acacia plantation forestry international conference, IUFRO working party 2.08.07: genetics and silviculture of acacia, Hue, Vietnam, 18–21 March 2014, Compendium of Abstracts

Wu T, Fang G, Liang L et al (2018) Analysis of mixed pulping raw materials of *Eucalyptus globulus* and *Acacia mangium* by near infrared spectroscopy technique combined with LASSO algorithm. BioRes 13(1):134–1359
Xie DY, Hong Y (2001) Regeneration of *Acacia mangium* through somatic embryogenesis. Plant Cell Rep 20:34–40
Xu S, Hu Z (2010) Method of plant breeding in the genome era. Genet Res Camb 92:423–441. https://doi.org/10.1017/S0016672310000583
Yanchuk AD, Jaquish B (2003) Forest genetic resource management: is there a fit among conservation, breeding and biotechnology. The XII World Forestry Congress. Quebec City. Canada
Yuskianti V (2014) Genetic diversity of the parental and offspring of *Acacia mangium* seedling of seed orchard group D (AM004) in South Sumatera, Indonesia (in Indonesia). J Penel Kehut Wallacea 3(2):129–137

Index

K. F. M. Salem et al. (eds.), *Breeding and Biotechnology of Grass and Bast Fiber Crops*, Advances in Plant Breeding Strategies 12,
https://doi.org/10.1007/978-3-032-00403-1

GPSR Compliance

The European Union's (EU) General Product Safety Regulation (GPSR) is a set of rules that requires consumer products to be safe and our obligations to ensure this.

If you have any concerns about our products, you can contact us on ProductSafety@springernature.com

In case Publisher is established outside the EU, the EU authorized representative is:

Springer Nature Customer Service Center GmbH
Europaplatz 3
69115 Heidelberg, Germany

Batch number: 10762385

Printed by Printforce, the Netherlands